3.2 POINT-SLOPE FORM

If a straight line has slope m and passes through the point (x_1, y_1), then an equation of the line is $y - y_1 = m(x - x_1)$.

3.2 SLOPE-INTERCEPT FORM

If a linear equation is written as $y = mx + b$, then m is the slope of the line and b is the y-intercept.

3.3–3.6 CONIC SECTIONS

All conic sections have an equation of the form $Ax^2 + Bx + Cy^2 + Dy + E = 0$ where either A or C is nonzero. The particular conic sections are listed below. See graphs in back endpapers.

Conic Section	Characteristic
Parabola	Either A or C is 0, but not both.
Circle	$A = C \neq 0$
Ellipse	$A \neq C, AC > 0$
Hyperbola	$AC < 0$

4.1 FUNCTION

A function is a relation in which each element in the domain corresponds to exactly one element in the range.

4.3 COMPOSITION OF FUNCTIONS

$(g \circ f)(x) = g[f(x)]$

4.5 ONE-TO-ONE FUNCTION

f is a one-to-one function if $a \neq b$ implies $f(a) \neq f(b)$.

4.5 INVERSE FUNCTION

Let f be a one-to-one function. The inverse function, f^{-1}, can be found by exchanging x and y in the equation for f, then solving for y if possible.

5.1 VALUE OF e

$e \approx 2.718281828$

5.2 DEFINITION OF LOGARITHM

$y = \log_a x$ if and only if $x = a^y$.

5.2 PROPERTIES OF LOGARITHMS

(a) $\log_a xy = \log_a x + \log_a y$ (b) $\log_a \dfrac{x}{y} = \log_a x - \log_a y$

(c) $\log_a x^r = r \log_a x$ (d) $\log_a a = 1$

(e) $\log_a 1 = 0$

9.1 ARITHMETIC SEQUENCES

nth term: $a_n = a_1 + (n - 1)d$

9.2 GEOMETRIC SEQUENCES

nth term: $a_n = a_1 r^{n-1}$

9.3, 9.4 SERIES

Arithmetic   Sum of n terms: $S_n = \dfrac{n}{2}(a_1 + a_n)$ or $S_n = \dfrac{n}{2}[2a_1 + (n - 1)d]$

Geometric   Sum of n terms: $S_n = \dfrac{a_1(1 - r^n)}{1 - r}$, where $r \neq 1$

Sum of an infinite geometric sequence: $S_\infty = \dfrac{a_1}{1 - r}$, where $-1 < r < 1$

9.6 PERMUTATIONS

The number of arrangements of n things taken r at a time is $P(n, r) = \dfrac{n!}{(n - r)!}$.

9.7 COMBINATIONS

The number of ways to choose r elements from a group of n elements is $\dbinom{n}{r} = \dfrac{n!}{(n - r)!r!}$.

9.8 PROPERTIES OF PROBABILITY

For any events E and F:

1. $0 \leq P(E) \leq 1$

2. $P(\text{a certain event}) = 1$

3. $P(\text{an impossible event}) = 0$

4. $P(E') = 1 - P(E)$

5. $P(E \text{ or } F) = P(E \cup F) = P(E) + P(F) - P(E \cap F)$.

COLLEGE ALGEBRA

SIXTH EDITION

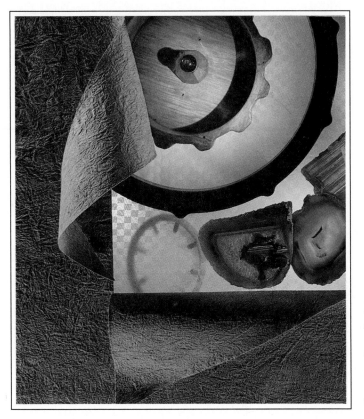

MARGARET L. LIAL
AMERICAN RIVER COLLEGE

CHARLES D. MILLER

E. JOHN HORNSBY, JR.
UNIVERSITY OF NEW ORLEANS

HarperCollinsCollegePublishers

TO THE INSTRUCTOR

This is the Annotated Instructor's Edition of *College Algebra,* Sixth Edition. It includes all material found in the student's edition, plus an Instructor's Answer Section at the end of the book. All of the exercises from the student's edition are reprinted in this section *with the answers in place.*

Sponsoring Editor: Anne Kelly
Developmental Editor: Louise Howe
Project Editor: Cathy Wacaser
Art Director: Julie Anderson
Text and Cover Design: Lesiak/Crampton Design Inc: Cynthia Crampton
Cover and Chapter Opener Photos: The Chicago Photographic Company
Production: Steve Emry, Linda Murray
Compositor: The Clarinda Company
Printer and Binder: R.R. Donnelley & Sons Company
Cover Printer: The Lehigh Press, Inc.

College Algebra, Sixth Edition
Copyright © 1993 by HarperCollins College Publishers

Library of Congress Cataloging-in-Publication Data

Lial, Margaret L.
   College algebra/Margaret L. Lial, Charles D. Miller, E. John
Hornsby, Jr.—6th ed., Annotated instructor's ed.
      p.   cm.
   Includes index.
   ISBN 0-673-46752-X (hard cover)
   1. Algebra.   I. Miller, Charles David.   II. Hornsby, E.
John.   III. Title.
QA154.2.L528   1992
512′.9—dc20   92-22248
                                                        CIP

92 93 94 95 9 8 7 6 5 4 3 2 1

PREFACE

The Sixth Edition of *College Algebra* is designed for a one-semester or one-quarter course. Some students may be taking this course to fulfill a mathematics requirement, while others may be going on to take courses in trigonometry, statistics, finite mathematics, calculus, discrete mathematics, or computer science.

This text has been written with the assumption that students have had an earlier course in algebra. For those whose background is not as solid as it might be, a chapter of review topics has been included. In fact, Chapter 2 on equations will be considered a review by some students. A pretest that can be used to determine the necessary topics for review is provided in the Instructor's Manual.

In this edition of *College Algebra,* we have attempted to maintain the strengths of past editions while enhancing the pedagogy, readability, usefulness, and attractiveness of the text. Many new features have been added to make the text easier and more enjoyable for both students and teachers to use, including new exercises, enhanced chapter summaries, and the use of full color. We continue to provide an extensive supplemental package. For students, we offer a solution manual, a study guide, interactive tutorial software, and videotapes. For instructors, we present an instructor's edition with all answers provided next to the exercises in a special section, overhead transparencies, a computerized test generator, a test manual, and complete solutions to all exercises.

All of the successful features of the previous edition are carried over in the new edition: careful exposition, fully developed examples with comments printed at the side (about 365 examples), and carefully graded section and chapter review exercises (more than 4300 in all). Screened boxes set off important definitions, formulas, rules, and procedures to further aid students in learning and reviewing the course material.

NEW FEATURES

Several new features, designed to assist students in the learning process, have been integrated into this edition. The use of full color and changes in format create a fresh look for the book. The design has been crafted to enhance the book's pedagogical features and increase its accessibility. The next three pages illustrate these features.

NEW FEATURES

CHAPTER SUMMARIES
Key terms and ideas are presented in a new "grid" format that makes it easier to review the chapter section by section.

PROBLEM-SOLVING STRATEGIES
Special paragraphs labeled "Problem Solving" expand our discussion of strategies to include connections to techniques learned earlier.

EXAMPLE TITLES
Each example now has a title to help students see the purpose of the example. The titles also facilitate working the exercises and studying for examinations.

"IN SIMPLEST TERMS"
Application boxes based on the PBS video series *"College Algebra: In Simplest Terms,"* produced by the Consortium for Mathematics and its Applications (COMAP), appear in relevant sections throughout the book. In each of these, a real-life situation is explained and two or three problems based on that situation are presented.

Content shown in the sample pages

CHAPTER 3 SUMMARY — 203

SECTION	TERMS	KEY IDEAS
3.1 Relations and the Rectangular Coordinate System	ordered pair, relation, domain, range, origin, x-axis, y-axis, rectangular coordinate system (Cartesian coordinate system), coordinate plane (xy-plane), quadrants, coordinates, collinear	**Distance Formula** Suppose $P(x_1, y_1)$ and $R(x_2, y_2)$ are two points in a coordinate plane. Then the distance between P and R, written $d(P, R)$, is $$d(P, R) = \sqrt{(x_2 - x_1)^2 + (y_2 - y_1)^2}.$$ **Midpoint Formula** The midpoint of the line segment with endpoints (x_1, y_1) and (x_2, y_2) is $$\left(\frac{x_1 + x_2}{2}, \frac{y_1 + y_2}{2}\right).$$
3.2 Linear Relations	linear relation, x-intercept, y-intercept, slope, change in x, change in y, point-slope form, slope-intercept form	**Definition of Slope** The slope m of the line through the points (x_1, y_1) and (x_2, y_2) is $$m = \frac{\Delta y}{\Delta x} = \frac{y_2 - y_1}{x_2 - x_1},$$ where $\Delta x \neq 0$. **Linear Equations**

General Equation	Type of Equation
$Ax + By = C$	Standard ... $B \neq 0$ ... y-inter
$x = k$	Vertical ... y-inter
$y = k$	Horizont ... no x-i
$y = mx + b$	Slope-in ... b, slop
$y - y_1 = m(x - x_1)$	Point-slo ... throug

3.2 LINEAR RELATIONS — 159

PROBLEM SOLVING
A straight line is often the best approximation of a set of data points that result from a real situation. If the equation is known, it can be used to predict the value of one variable, given a value of the other. For this reason, the equation is written as a linear relation in slope-intercept form. One way to find the equation of such a straight line is to use two typical data points and the point-slope form of the equation of a line.

Example 10
FINDING AN EQUATION FROM DATA POINTS

Scientists have found that the number of chirps made by a cricket of a particular species per minute is almost linearly related to the temperature. Suppose that for a particular species, at 68° F a cricket chirps 124 times per minute, while at 80° F the cricket chirps 172 times per minute. Find the linear equation that relates the number of chirps to the temperature.

Think of the ordered pairs in the relation as (chirps, temperature), or (c, t). Then c takes on the role of x and t takes on the role of y. Since we are using a linear relationship, find the slope of the line by using the slope formula with the points (124, 68) and (172, 80).

$$m = \frac{68 - 80}{124 - 172} = \frac{-12}{-48} = \frac{1}{4}$$

Choose one of the points, say (124, 68), and substitute into the point-slope form, with $m = 1/4$.

$$t - 68 = \frac{1}{4}(c - 124)$$

$$t - 68 = \frac{1}{4}c - 31$$

$$t = \frac{1}{4}c + 37$$

The equation is $t = (1/4)c + 37$. By substituting the number of chirps per minute into this equation, the temperature t can be approximated. ■

IN SIMPLEST TERMS

10

SOLVE EACH PROBLEM

In Massachusetts, speeding fines are determined by the linear relationship $y = 50 + 10(x - 65)$, where y is the fine in dollars and x is the speed of the vehicle. A motorist driving 72 miles per hour would have his fine calculated as $50 + $10(72 - 65) = $50 + $70 = $120.

A. José had to make an 8:00 a.m. final exam, but overslept after a big weekend in Boston. Radar clocked his speed at 76 miles per hour. How much was his fine?

B. While balancing the checkbook, Harry ran across a cancelled check that his wife, Rachel, had written to the Department of Motor Vehicles for a speeding fine. The check was written for $100. How fast was Rachel driving?

ANSWERS A. José had to pay $160. B. Rachel was driving at 70 miles per hour.

GRAPHING CALCULATOR
Selected chapters (primarily those involving functions and relations) include a brief section at the end giving detailed information on the use of graphing calculators.

CHALLENGING EXERCISES
Most sections include a few challenging exercises that extend the ideas presented in the section. These are identified in the instructor's edition, although not in the student's edition.

APPLICATIONS
These exercises have been extensively rewritten to make them interesting and grounded in real-world, familiar situations. About 170 of these are new to this edition or revised.

■ **THE GRAPHING CALCULATOR** ■

Graphing calculators can easily be used to graph polynomial and rational functions. In Example 4 in Section 6.1, we graphed the polynomial function defined by

$$P(x) = (2x + 3)(x - 1)(x + 2)$$

by using intercepts and test points in each interval formed by the x-intercepts. The graph is shown in Figure 6.11. Use a graphing calculator to graph this function, with the minimum and maximum x-values of -4 and 2, and minimum and maximum y-values of -6 and 6. Compare the result to the figure in the text.

The rational function defined by

$$f(x) = \frac{3(x + 1)(x - 2)}{(x + 4)^2}$$

was analyzed in Example 6 in Section 6.6. Program your graphing calculator to graph this function, using minimum and maximum x-values of -15 and 12, and minimum and maximum y-values of -1 and 15. Compare the display to Figure 6.23.

Graphing calculators have the capability of "zooming in" on a particular portion of the graph of a function. Zoom in on the portion of the graph of this rational function in the vicinity of the origin to see how the function changes from decreasing to increasing. These calculators also have the capability to "trace" the graph while displaying the coordinates of the points on the graph.

In calculus we are often required to find the highest and lowest points on a graph in a particular region of the domain. By experimenting with your graphing calculator, you can begin to appreciate just how useful it can be in helping us to find excellent approximations of the coordinates of these points.

You may want to use the functions defined below to experiment with your calculator. These functions have been selected from examples in the text, so you can compare your calculator graphs with those given in the example figures. The figure number of the graph in the text is given with each equation.

1. $P(x) = x^3$ (Figure 6.1)
2. $P(x) = x^6$ (Figure 6.2)
3. $P(x) = \frac{1}{2}x^3$ (Figure 6.3)
4. $P(x) = -\frac{3}{2}x^4$ (Figure 6.4)
5. $P(x) = x^5 - 2$ (Figure 6.5)
6. $P(x) = (x + 1)^6$ (Figure 6.6)
7. $P(x) = -(x - 1)^3 + 3$ (Figure 6.7)
8. $P(x) = 3x^4 + x^3 - 2x^2$ (Figure 6.12)
9. $P(x) = 8x^3 - 12x^2 + 2x + 1$ (Figure 6.16)
10. $P(x) = 3x^4 - 14x^3 + 54x - 3$ (Figure 6.17)
11. $f(x) = \frac{1}{x}$ (Figure 6.18)
12. $f(x) = -\frac{2}{x}$ (Figure 6.19)
13. $f(x) = \frac{2}{1 + x}$ (Figure 6.20)
14. $f(x) = \frac{x + 1}{(2x - 1)(x + 3)}$ (Figure 6.21)
15. $f(x) = \frac{2x + 1}{x - 3}$ (Figure 6.22)
16. $f(x) = \frac{x^2 + 1}{x - 2}$ (Figure 6.24)
17. Graph $f(x) = \frac{x^2 - 4}{x - 2}$ on your graphing calculator and compare your result with Figure 6.25.
18. On the same screen, graph $y = \frac{1}{x^2}$ and $y = -\frac{1}{x^2}$.

Describe in words how they are alike and how they are different.

For each pair of functions defined as follows, show that $(f \circ g)(x) = x$ and $(g \circ f)(x) = x$.

33. $f(x) = 8x$, $g(x) = \frac{1}{8}x$
34. $f(x) = \frac{3}{4}x$, $g(x) = \frac{4}{3}x$
35. $f(x$

36. $f(x) = \frac{x - 3}{4}$, $g(x) = 4x + 3$
37. $f(x) = x^3 + 6$, $g(x) = \sqrt[3]{x - 6}$
38. $f(x$

39. Explain how to find the domain of $(f \circ g)(x) = \frac{1}{x^2 - 1}$, where $f(x) = \frac{1}{x}$ and $g(x) = x^2$

40. For each of the following functions f, decide what operation was used with $g(x)$ and m get $f(x)$.
(a) $f(x) = ax^2$; $g(x) = a, m(x) = x^2$
(b) $f(x) = x^2 + k$; $m(x) = x^2, g(x) = k$
(c) $f(x) = (x - h)^2$; $m(x) = x^2, g(x) = x - h$
(d) $f(x) = a(x - h)^2 + k$; $g(x) = ax^2 + k, m(x) = x - h$

In each of the following exercises, a function h is defined. Find $f(x)$ and $g(x)$ such that $h(x) = (f \circ g)(x)$. Many such pairs of functions exist. See Example 6.

41. $h(x) = (6x - 2)^2$
42. $h(x) = (11x^2 + 12x)^2$
43. $h(x) = \sqrt{x^2 - 1}$

44. $h(x) = \frac{1}{x^2 + 2}$
45. $h(x) = \frac{(x - 2)^2 + 1}{5 - (x - 2)^2}$
46. $h(x) = (x + 2)^3 - 3(x + 2)^2$

47. A couple planning their wedding has found that the cost to hire a caterer for the reception depends on the number of guests attending. If 100 people attend, the cost per person will be $2. For each person less than 100, the cost will increase by $.20. Assume that no more than 100 people will attend. Let x represent the number less than 100 that attend. For example, if 95 attend, $x = 5$.
(a) Write a function $N(x)$ for the possible number of guests.
(b) Write a function $G(x)$ for the possible cost per guest.
(c) Write the function $N(x) \cdot G(x)$ for the total cost, $C(x)$.

48. The manager of a music store has found that the price $p(x)$, in dollars, of a certain compact disc depends on the demand for the disc, x. For one particular disc,

$$p(x) = -.001x + 20.$$

Express the revenue as $R(x)$, where $R(x)$ is the product of price and demand, that is, $p(x) \cdot x$.

49. Suppose the population P of a certain species of fish depends on the number x (in hundreds) of a smaller kind of fish which serves as its food supply, so that

$$P(x) = 2x^2 + 1.$$

Suppose, also, that the number x (in hundreds) of the smaller species of fish depends upon the amount a (in

appropriate units) of its food supply, a kind of plankton. Suppose

$$x = f(a) = 3a + 2.$$

A biologist wants to find the relationship between the population P of the large fish and the amount a of plankton available, that is, $(P \circ f)(a)$. What is the relationship? If the amount of plankton decreases, what will happen to the fish population?

50. Suppose the demand for a certain brand of graphing calculator is given by

$$D(p) = \frac{-p^2}{100} + 500,$$

where p is the price in dollars. If the price, in terms of the cost, c, is expressed as

$$p(c) = 2c - 10,$$

find $D(c)$, the demand in terms of the cost. What happens to the demand for the calculator as the cost goes down?

51. When a thermal inversion layer is over a city, pollutants cannot rise vertically but are trapped below the layer and must disperse horizontally. Assume that a factory smokestack begins emitting a pollutant at 8 A.M. Assume that the pollutant disperses horizontally over a circular area. If t represents the time, in hours, since the factory began emitting pollutants ($t = 0$ represents

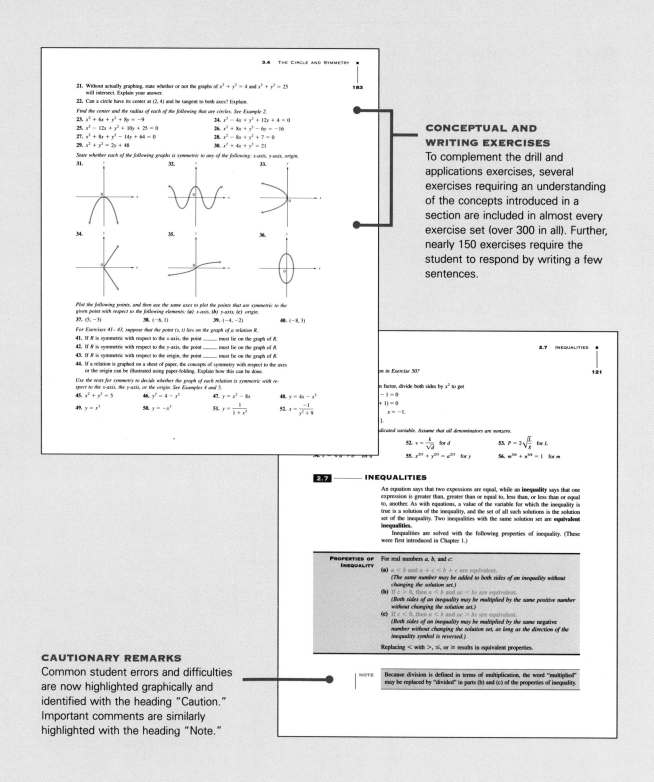

21. Without actually graphing, state whether or not the graphs of $x^2 + y^2 = 4$ and $x^2 + y^2 = 25$ will intersect. Explain your answer.

22. Can a circle have its center at $(2, 4)$ and be tangent to both axes? Explain.

Find the center and the radius of each of the following that are circles. See Example 2.

23. $x^2 + 6x + y^2 + 8y = -9$ **24.** $x^2 - 4x + y^2 + 12y + 4 = 0$

25. $x^2 - 12x + y^2 + 10y + 25 = 0$ **26.** $x^2 + 8x + y^2 - 6y = -16$

27. $x^2 + 8x + y^2 - 14y + 64 = 0$ **28.** $x^2 - 8x + y^2 + 7 = 0$

29. $x^2 + y^2 = 2y + 48$ **30.** $x^2 + 4x + y^2 = 21$

State whether each of the following graphs is symmetric to any of the following: x-axis, y-axis, origin.

31.   **32.**   **33.**

34.   **35.**   **36.**

Plot the following points, and then use the same axes to plot the points that are symmetric to the given point with respect to the following elements: (a) x-axis, (b) y-axis, (c) origin.

37. $(5, -3)$ **38.** $(-6, 1)$ **39.** $(-4, -2)$ **40.** $(-8, 3)$

For Exercises 41–43, suppose that the point (s, t) lies on the graph of a relation R.

41. If R is symmetric with respect to the x-axis, the point _____ must lie on the graph of R.

42. If R is symmetric with respect to the y-axis, the point _____ must lie on the graph of R.

43. If R is symmetric with respect to the origin, the point _____ must lie on the graph of R.

44. If a relation is graphed on a sheet of paper, the concepts of symmetry with respect to the axes or the origin can be illustrated using paper-folding. Explain how this can be done.

Use the tests for symmetry to decide whether the graph of each relation is symmetric with respect to the x-axis, the y-axis, or the origin. See Examples 4 and 5.

45. $x^2 + y^2 = 5$ **46.** $y^2 = 4 - x^2$ **47.** $y = x^2 - 8x$ **48.** $y = 4x - x^2$

49. $y = x^3$ **50.** $y = -x^3$ **51.** $y = \dfrac{1}{1 + x^2}$ **52.** $x = \dfrac{-1}{y^2 + 9}$

CONCEPTUAL AND WRITING EXERCISES
To complement the drill and applications exercises, several exercises requiring an understanding of the concepts introduced in a section are included in almost every exercise set (over 300 in all). Further, nearly 150 exercises require the student to respond by writing a few sentences.

...on in Exercise 50?

...n factor, divide both sides by x^2 to get

...$- 1 = 0$

...$+ 1) = 0$

...$x = -1$.

...}.

...dicated variable. Assume that all denominators are nonzero.

52. $v = \dfrac{k}{\sqrt{d}}$ for d **53.** $P = 2\sqrt{\dfrac{L}{g}}$ for L

55. $x^{2/3} + y^{2/3} = a^{2/3}$ for y **56.** $m^{3/4} + n^{3/4} = 1$ for m

2.7 — **INEQUALITIES**

An equation says that two expressions are equal, while an **inequality** says that one expression is greater than, greater than or equal to, less than, or less than or equal to, another. As with equations, a value of the variable for which the inequality is true is a solution of the inequality, and the set of all such solutions is the solution set of the inequality. Two inequalities with the same solution set are **equivalent inequalities.**

Inequalities are solved with the following properties of inequality. (These were first introduced in Chapter 1.)

PROPERTIES OF INEQUALITY	For real numbers a, b, and c:
	(a) $a < b$ and $a + c < b + c$ are equivalent. *(The same number may be added to both sides of an inequality without changing the solution set.)*
	(b) If $c > 0$, then $a < b$ and $ac < bc$ are equivalent. *(Both sides of an inequality may be multiplied by the same positive number without changing the solution set.)*
	(c) If $c < 0$, then $a < b$ and $ac > bc$ are equivalent. *(Both sides of an inequality may be multiplied by the same negative number without changing the solution set, as long as the direction of the inequality symbol is reversed.)*
	Replacing $<$ with $>$, $\leq$, or $\geq$ results in equivalent properties.

NOTE Because division is defined in terms of multiplication, the word "multiplied" may be replaced by "divided" in parts (b) and (c) of the properties of inequality.

CAUTIONARY REMARKS
Common student errors and difficulties are now highlighted graphically and identified with the heading "Caution." Important comments are similarly highlighted with the heading "Note."

NEW CONTENT HIGHLIGHTS

- In order to keep Chapter 1, a review chapter, as short as possible, the section on integer exponents has been absorbed into the sections on polynomials and rational exponents. The binomial theorem now follows the discussion of polynomials, extending the ideas on multiplication presented there. This section is optional and could be used at any point in the text.

- The presentation of radicals has been rewritten and now includes more examples.

- New problem-solving techniques have been added in Chapter 2, enhancing the presentation of applications.

- Although Chapter 3 now has seven sections, it has not been substantially lengthened; the material has been rearranged into more sections to reduce the subject matter covered in one class meeting and to improve the flow. We have increased the emphasis on domain and range of relations in this chapter in preparation for the work in functions in Chapter 4. There is also more emphasis on symmetry and translation.

- Chapter 4, on functions, has been rearranged to present the easier topics earlier in the chapter. Variation is introduced in Section 4.2, giving students an opportunity to absorb the function concept before getting into the more theoretical sections in the chapter. (This section may be covered later in the chapter if the instructor wishes.) Inverse functions are discussed last so as to lead naturally into Chapter 5 on exponential and logarithmic functions.

- Exponential and logarithmic functions are now presented before polynomial functions, since the section on inverses (in Chapter 4) is a natural lead-in to the topic. However, Chapters 5 and 6 are independent and can be used by an instructor in either order. We have assumed that students will use calculators in the work with logarithms, so no tables are provided. We discuss the use of calculators as necessary. There is more emphasis on the graphs of logarithmic functions in this edition. A new section on applications of exponential and logarithmic functions includes an expanded discussion of several important applications, and many new exercises.

- The section on rational functions has been thoroughly reorganized with more emphasis on finding asymptotes. In Chapter 9, arithmetic sequences are now introduced in Section 9.1, and all work with series is combined in a new Section 9.3.

- To aid those students who have not worked with geometry for some time, a brief review of geometry is included in Appendix B.

SUPPLEMENTS

Our extensive supplemental package includes an annotated instructor's edition that contains answers to all exercises in a special exercise section at the back of the book, testing materials, solution manuals, software, and videotapes.

FOR THE INSTRUCTOR

ANNOTATED INSTRUCTOR'S EDITION   With this volume, instructors have immediate access to the answers to every exercise in the text, excluding proofs and writing exercises. In a special section at the end of the book, each answer is printed in color next to the corresponding text exercise. In addition, challenging exercises, which will require most students to stretch beyond the concepts discussed in the text, are marked with the symbol ▲. The conceptual (◉) and writing (✐) exercises are also marked in this edition so instructors may assign these problems at their discretion. (Calculator exercises will be marked by ▦ in both the student's and instructor's editions.) Each section of the instructor's edition also includes a list of "Resources," containing cross-references to relevant sections in each of the supplements for *College Algebra*.

INSTRUCTOR'S TEST MANUAL   Included here are two forms of a pretest; six versions of a chapter test for each chapter—four open-response and two multiple-choice; additional test items for each chapter, and two forms of a final examination. Answers to all tests and additional exercises also are provided.

INSTRUCTOR'S SOLUTION MANUAL   This manual includes complete, worked-out solutions to every exercise in the textbook (excluding most writing exercises).

HARPERCOLLINS TEST GENERATOR FOR MATHEMATICS   The HarperCollins Test Generator is one of the top testing programs on the market for IBM and Macintosh computers. It enables instructors to select questions for any section in the text or to use a ready-made test for each chapter. Instructors may generate tests in multiple-choice or open-response formats, scramble the order of questions while printing, and produce up to twenty-five versions of each test. The system features printed graphics and accurate mathematical symbols. The program also allows instructors to choose problems randomly from a section or problem type or to choose questions manually while viewing them on the screen, with the option to regenerate variables if desired. The editing feature allows instructors to customize the chapter data disks by adding their own problems.

QUIZMASTER ON-LINE TESTING SYSTEM   The QuizMaster program, available in both IBM and Macintosh formats, coordinates with the HarperCollins Test Generator and allows instructors to create tests for students to take at the computer. The test results are stored on disk so the instructor can view or print test results for a student, a class section, or an entire course.

TRANSPARENCIES   Approximately fifty color overhead transparencies of figures, examples, definitions, procedures, properties, and problem-solving methods are available to assist instructors in presenting important points during their lectures.

FOR THE STUDENT

STUDENT'S SOLUTION MANUAL Complete, worked-out solutions are given for odd-numbered exercises and chapter review exercises and all chapter test exercises in a volume available for purchase by students. In addition, a practice chapter test is provided for each chapter.

STUDENT'S STUDY GUIDE Written in a semiprogrammed format, the study guide includes a pretest and posttest for each chapter, plus exercises that give additional practice and reinforcement for students.

COLLEGE ALGEBRA WITH TRIGONOMETRY: GRAPHING CALCULATOR INVESTIGATIONS This new supplemental text, written by Dennis Ebersole of Northampton County Area Community College, provides investigations that help students visualize and explore key concepts, generalize and apply concepts, and identify patterns.

VIDEOTAPES A new videotape series has been developed to accompany *College Algebra,* Sixth Edition. In a separate lesson for each section of the book, the series covers all objectives, topics, and problem-solving techniques within the text.

COMPUTER-ASSISTED TUTORIALS These tutorials offer self-paced, interactive review in IBM, Apple, and Macintosh formats. Solutions are given for all examples and exercises, as needed.

GRAPHEXPLORER With this sophisticated software, available in IBM and Macintosh versions, students can graph rectangular, conic, polar, and parametric equations; zoom; transform functions; and experiment with families of equations quickly and easily.

ACKNOWLEDGMENTS

We wish to thank the many users of the Fifth Edition for their insightful comments and suggestions for improvements to this book. We also wish to thank our reviewers for their contributions:

Judy Barclay, *Cuesta College*
Edward L. Bloxom, *Coastal Carolina Community College*
Shoko A. Brant, *Essex Community College*
Don Buckholtz, *University of Kentucky*
Barbara C. Carter, *University of North Carolina at Greensboro*
Allan C. Cochran, *University of Arkansas*
David Cope, *University of North Alabama*
Sally Copeland, *Johnson County Community College*
Thad A. Crosnoe, *Western Oklahoma State College*

John S. Cross, *University of Northern Iowa*
Elaine Deutschman, *Oregon Institute of Technology*
William Drezdzon, *Oakton Community College*
Odene Forsythe, *Westark Community College*
Linda D. Green, *Santa Fe Community College*
Jean Harvey, *Jones County Junior College*
Walter James Helmers, *San Antonio College*
Norma F. James, *New Mexico State University*
Diane Jamison, *Cumberland College*
Jimmy P. Kan, *City College of San Francisco*
Dix J. Kelly, *Central Connecticut State University*
Jean Krichbaum, *Broome Community College*
Sue Little, *North Harris County College*
Wanda J. Long, *St. Charles County Community College*
Lewis Lum, *University of Portland*
Sheila McNicholas, *University of Illinois at Chicago*
Carol Jean Martin, *Dodge City Community College*
James E. Moran, *Diablo Valley College*
Bart Nelson, *Snow College*
E. James Peake, *Iowa State University*
Michael Perkowski, *University of Missouri—Columbia*
David Price, *Tarrant County Junior College*
Robyn E. Reid, *University of South Carolina*
Gloria P. Rivkin, *Lawrence Technological University*
Bruce Sisko, *Belleville Area College*
W. Arlene Starwalt-Jeskey, *Rose State College*
Katalin Szucs, *East Carolina University*
Raymond Tanner, *Mississippi Gulf Coast Community College*
Margaret M. Van Parys, *Lakeland Community College*
Richard C. Weimer, *Frostburg State University*
Evelyn Woodward, *St. Petersburg Junior College*

Paul Eldersveld, College of DuPage, has our gratitude for coordinating the print supplements, an enormous and time-consuming task. Special thanks go to Jane Brandsma, of Suffolk Community College, who was instrumental in preparing the material corresponding to the COMAP video series. Paul Van Erden, of American River College, has done his usual fine job in creating the index. Kitty Pellissier, of the University of New Orleans, did an outstanding job of checking all answers that appear in the Annotated Instructor's Edition and the answer section of the student text.

We especially want to thank the fine professional staff at HarperCollins for their assistance: Anne Kelly, Louise Howe, Linda Youngman, Julie Anderson, Cathy Wacaser, and Tammy McClenning have all made important contributions.

Margaret L. Lial
E. John Hornsby, Jr.

USING THE ANNOTATED INSTRUCTOR'S EDITION

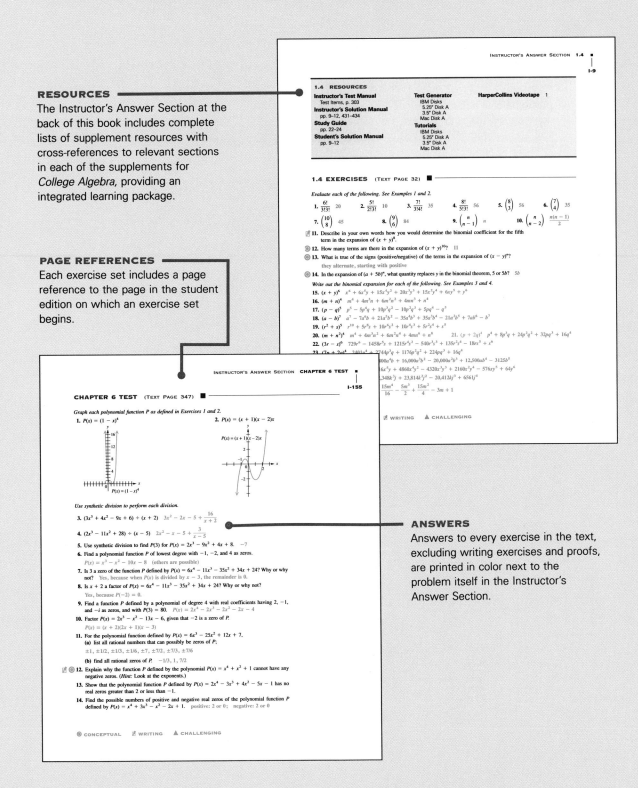

RESOURCES

The Instructor's Answer Section at the back of this book includes complete lists of supplement resources with cross-references to relevant sections in each of the supplements for *College Algebra,* providing an integrated learning package.

PAGE REFERENCES

Each exercise set includes a page reference to the page in the student edition on which an exercise set begins.

ANSWERS

Answers to every exercise in the text, excluding writing exercises and proofs, are printed in color next to the problem itself in the Instructor's Answer Section.

1.4 RESOURCES

Instructor's Test Manual	Test Generator	HarperCollins Videotape 1
Test Items, p. 303	IBM Disks	
Instructor's Solution Manual	5.25" Disk A	
pp. 9–12, 431–434	3.5" Disk A	
Study Guide	Mac Disk A	
pp. 22–24	**Tutorials**	
Student's Solution Manual	IBM Disks	
pp. 9–12	5.25" Disk A	
	3.5" Disk A	
	Mac Disk A	

1.4 EXERCISES (TEXT PAGE 32)

Evaluate each of the following. See Examples 1 and 2.

1. $\frac{6!}{3!3!}$ 20
2. $\frac{5!}{2!3!}$ 10
3. $\frac{7!}{3!4!}$ 35
4. $\frac{8!}{5!3!}$ 56
5. $\binom{8}{3}$ 56
6. $\binom{7}{4}$ 35
7. $\binom{10}{8}$ 45
8. $\binom{9}{6}$ 84
9. $\binom{n}{n-1}$ n
10. $\binom{n}{n-2}$ $\frac{n(n-1)}{2}$

11. Describe in your own words how you would determine the binomial coefficient for the fifth term in the expansion of $(x + y)^8$.

12. How many terms are there in the expansion of $(x + y)^{10}$? 11

13. What is true of the signs (positive/negative) of the terms in the expansion of $(x - y)^n$?
    they alternate, starting with positive

14. In the expansion of $(a + 5b)^n$, what quantity replaces y in the binomial theorem, 5 or 5b? 5b

Write out the binomial expansion for each of the following. See Examples 3 and 4.

15. $(x + y)^6$ $x^6 + 6x^5y + 15x^4y^2 + 20x^3y^3 + 15x^2y^4 + 6xy^5 + y^6$
16. $(m + n)^4$ $m^4 + 4m^3n + 6m^2n^2 + 4mn^3 + n^4$
17. $(p - q)^5$ $p^5 - 5p^4q + 10p^3q^2 - 10p^2q^3 + 5pq^4 - q^5$
18. $(a - b)^7$ $a^7 - 7a^6b + 21a^5b^2 - 35a^4b^3 + 35a^3b^4 - 21a^2b^5 + 7ab^6 - b^7$
19. $(r^2 + s)^5$ $r^{10} + 5r^8s + 10r^6s^2 + 10r^4s^3 + 5r^2s^4 + s^5$
20. $(m + n^2)^4$ $m^4 + 4m^3n^2 + 6m^2n^4 + 4mn^6 + n^8$
21. $(p + 2q)^4$ $p^4 + 8p^3q + 24p^2q^2 + 32pq^3 + 16q^4$
22. $(3r - s)^6$ $729r^6 - 1458r^5s + 1215r^4s^2 - 540r^3s^3 + 135r^2s^4 - 18rs^5 + s^6$
23. $(7p + 2q)^4$ $2401p^4 + 2744p^3q + 1176p^2q^2 + 224pq^3 + 16q^4$
    $800a^4b + 16,000a^3b^2 - 20,000a^2b^3 + 12,500ab^4 - 3125b^5$
    $6x^5y + 4860x^4y^2 - 4320x^3y^3 + 2160x^2y^4 - 576xy^5 + 64y^6$
    $348k^3j + 23,814k^2j^2 - 20,412kj^3 + 6561j^4$
    $\frac{15m^4}{16} - \frac{5m^3}{2} + \frac{15m^2}{4} - 3m + 1$

WRITING   CHALLENGING

CHAPTER 6 TEST (TEXT PAGE 347)

Graph each polynomial function P as defined in Exercises 1 and 2.

1. $P(x) = (1 - x)^4$

2. $P(x) = (x + 1)(x - 2)x$

Use synthetic division to perform each division.

3. $(3x^3 + 4x^2 - 9x + 6) \div (x + 2)$ $3x^2 - 2x - 5 + \frac{16}{x + 2}$

4. $(2x^3 - 11x^2 + 28) \div (x - 5)$ $2x^2 - x - 5 + \frac{3}{x - 5}$

5. Use synthetic division to find $P(3)$ for $P(x) = 2x^3 - 9x^2 + 4x + 8$. -7

6. Find a polynomial function P of lowest degree with -1, -2, and 4 as zeros.
   $P(x) = x^3 - x^2 - 10x - 8$ (others are possible)

7. Is 3 a zero of the function P defined by $P(x) = 6x^4 - 11x^3 - 35x^2 + 34x + 24$? Why or why not?   Yes, because when $P(x)$ is divided by $x - 3$, the remainder is 0.

8. Is $x + 2$ a factor of $P(x) = 6x^4 - 11x^3 - 35x^2 + 34x + 24$? Why or why not?
   Yes, because $P(-2) = 0$.

9. Find a function P defined by a polynomial of degree 4 with real coefficients having 2, -1, and $-i$ as zeros, and with $P(3) = 80$. $P(x) = 2x^4 - 2x^3 - 2x^2 - 2x - 4$

10. Factor $P(x) = 2x^3 - x^2 - 13x - 6$, given that -2 is a zero of P.
    $P(x) = (x + 2)(2x + 1)(x - 3)$

11. For the polynomial function defined by $P(x) = 6x^3 - 25x^2 + 12x + 7$.
    (a) list all rational numbers that can possibly be zeros of P;
    ± 1, $\pm 1/2$, $\pm 1/3$, $\pm 1/6$, ± 7, $\pm 7/2$, $\pm 7/3$, $\pm 7/6$
    (b) find all rational zeros of P. $-1/3$, 1, 7/2

12. Explain why the function P defined by the polynomial $P(x) = x^4 + x^2 + 1$ cannot have any negative zeros. (*Hint:* Look at the exponents.)

13. Show that the polynomial function P defined by $P(x) = 2x^4 - 3x^3 + 4x^2 - 5x - 1$ has no real zeros greater than 2 or less than -1.

14. Find the possible numbers of positive and negative real zeros of the polynomial function P defined by $P(x) = x^4 + 3x^3 - x^2 - 2x + 1$.   positive: 2 or 0;   negative: 2 or 0

CONCEPTUAL   WRITING   CHALLENGING

40. For the months of June through October, the percent of maximum possible chlorophyll production in a leaf is approximated by C, where

$$C = 10x + 50.$$

Here x is time in months with $x = 1$ representing June. From October through December, C is approximated by

$$C = -20(x - 5)^2 + 100,$$

with x as above. Find the percent of maximum possible chlorophyll production in each of the following months:

(a) June 60   **(b)** July 70
(c) September 90   **(d)** October 100
(e) November 80   **(f)** December. 20

41. Use your results from Exercise 40 to sketch a graph of $y = C$ from June through December. In what month is chlorophyll production at a maximum? October

42. The temperature T at which water boils varies with the elevation E, where T is in degrees Celsius and E is in meters above sea level, according to the equation

$$E = 580(100 - T)^2 + 1000(100 - T).$$

▲ 🖩 **(a)** Determine the boiling point at each of the following elevations. (*Hint:* Let $X = 100 - T$. Solve for X, and then solve for T.)
1. Mount McKinley (Alaska), 6194 meters 97.5° C
2. Mount Everest (Nepal-Tibet), 8848 meters 96.9° C
3. Mount Andrew Jackson (Antarctica), 4191 meters 98.0° C
4. Mount Mauna Loa (Hawaii), 4170 meters 98.0° C

🖉 ◎ **(b)** Discuss the interpretation of letting $T = 100$ in the formula.

▲ **(c)** What is the boiling point of water in Denver, elevation 1600 meters? 99.0° C

▲ *Use the geometric definition of parabola given in the text to find the equation of each parabola having the given point as focus.*

43. $(0, 3)$, directrix $y = -3$ $x^2 = 12y$
44. $(0, -5)$, directrix $y = 5$ $x^2 = -20y$
45. $(-2, 0)$, directrix $x = 2$ $y^2 = -8x$
46. $(8, 0)$, directrix $x = -8$ $y^2 = 32x$
47. $(3, 6)$, vertex $(3, 4)$ $(x - 3)^2 = 8(y - 4)$
48. $(-5, 2)$, vertex $(-5, 5)$ $(x + 5)^2 = -12(y - 5)$

▲ *Just as the discriminant of the quadratic equation $ax^2 + bx + c = 0$ ($a \neq 0$) determines the number of real solutions of the equation (see Section 2.4), the discriminant $b^2 - 4ac$ for the vertical parabola $y = ax^2 + bx + c$ determines the number of x-intercepts. If the discriminant is 0, there is one x-intercept; if it is positive, there are two x-intercepts; if it is negative, there are no x-intercepts.*

$b^2 - 4ac = 0$
One x-intercept

$b^2 - 4ac > 0$
Two x-intercepts

$b^2 - 4ac < 0$
No x-intercepts

49. Find a value of c so that $y = x^2 - 10x + c$ has exactly one x-intercept. 25
50. Find b so that $y = x^2 + bx + 9$ has exactly one x-intercept. ±6

◎ CONCEPTUAL   🖉 WRITING   ▲ CHALLENGING

CONTENTS

An Introduction to Scientific Calculators   xvii

■ **CHAPTER 1**   **ALGEBRAIC EXPRESSIONS**   **1**

    1.1   **The Real Numbers**   2

    1.2   **Order and Absolute Value**   13

    1.3   **Polynomials**   20

    1.4   **The Binomial Theorem**   27

    1.5   **Factoring**   33

    1.6   **Rational Expressions**   40

    1.7   **Rational Exponents**   48

    1.8   **Radicals**   57

    Chapter 1 Summary   65

    Chapter 1 Review Exercises   68

    Chapter 1 Test   70

■ **CHAPTER 2**   **EQUATIONS AND INEQUALITIES**   **72**

    2.1   **Linear Equations**   73

    2.2   **Applications of Linear Equations**   80

    2.3   **Complex Numbers**   91

    2.4   **Quadratic Equations**   98

    2.5   **Applications of Quadratic Equations**   108

    2.6   **Other Types of Equations**   115

    2.7   **Inequalities**   121

    2.8   **Absolute Value Equations and Inequalities**   132

    Chapter 2 Summary   137

    Chapter 2 Review Exercises   139

    Chapter 2 Test   142

■ **CHAPTER 3 RELATIONS AND THEIR GRAPHS 143**

3.1 **Relations and the Rectangular Coordinate System**   144

3.2 **Linear Relations**   150

3.3 **Parabolas: Translations and Applications**   162

3.4 **The Circle and Symmetry**   176

3.5 **The Ellipse and the Hyperbola**   184

3.6 **The Conic Sections**   193

3.7 **Inequalities**   199

Chapter 3 Summary   203

Chapter 3 Review Exercises   205

Chapter 3 Test   207

The Graphing Calculator   208

■ **CHAPTER 4 FUNCTIONS 209**

4.1 **Functions**   210

4.2 **Variation: Applications of Functions**   218

4.3 **Combining Functions: Algebra of Functions**   224

4.4 **Graphing Basic Functions and Their Variations**   231

4.5 **Inverse Functions**   237

Chapter 4 Summary   247

Chapter 4 Review Exercises   248

Chapter 4 Test   251

The Graphing Calculator   252

■ **CHAPTER 5 EXPONENTIAL AND LOGARITHMIC FUNCTIONS 253**

5.1 **Exponential Functions**   254

5.2 **Logarithmic Functions**   265

5.3 **Evaluating Logarithms; Change of Base**   273

5.4 **Exponential and Logarithmic Equations**   280

5.5 **Exponential Growth and Decay**   287

Chapter 5 Summary   292

Chapter 5 Review Exercises   293

Chapter 5 Test   295

The Graphing Calculator   296

■ **CHAPTER 6   POLYNOMIAL AND RATIONAL FUNCTIONS   297**

6.1   Graphing Polynomial Functions   298

6.2   Polynomial Division   306

6.3   Zeros of Polynomial Functions   313

6.4   Rational Zeros of Polynomial Functions   320

6.5   Real Zeros of Polynomial Functions   324

6.6   Graphing Rational Functions   333

Chapter 6 Summary   344

Chapter 6 Review Exercises   345

Chapter 6 Test   347

The Graphing Calculator   348

■ **CHAPTER 7   SYSTEMS OF EQUATIONS AND INEQUALITIES   349**

7.1   Linear Systems with Two Variables   350

7.2   Linear Systems with Three Variables   359

7.3   Solution of Linear Systems by Matrices   366

7.4   Nonlinear Systems of Equations   376

7.5   Systems of Inequalities; Linear Programming   383

Chapter 7 Summary   391

Chapter 7 Review Exercises   392

Chapter 7 Test   394

The Graphing Calculator   396

■ **CHAPTER 8   MATRICES   397**

8.1   Basic Properties of Matrices   398

8.2   Matrix Products   404

8.3   Multiplicative Inverses of Matrices   412

8.4   Determinants   423

8.5   Properties of Determinants   429

8.6   Solution of Linear Systems by Cramer's Rule   437

Chapter 8 Summary   443

Chapter 8 Review Exercises   446

Chapter 8 Test   448

The Graphing Calculator   449

■ **CHAPTER 9**   SEQUENCES AND SERIES; PROBABILITY   **451**

9.1   **Sequences; Arithmetic Sequences**   452

9.2   **Geometric Sequences**   458

9.3   **Series; Applications of Sequences and Series**   461

9.4   **Sums of Infinite Geometric Sequences**   471

9.5   **Mathematical Induction**   475

9.6   **Permutations**   481

9.7   **Combinations**   486

9.8   **An Introduction to Probability**   491

Chapter 9 Summary   498

Chapter 9 Review Exercises   500

Chapter 9 Test   502

Appendix A Sets   503

Appendix B Geometry Review and Formulas   508

Appendix C Symbols Used in This Book   513

Answers to Selected Exercises   A-1

Index   A-45

Instructor's Answer Section   I-1

An Introduction to Scientific Calculators

There is little doubt that the appearance of hand-held calculators twenty years ago and the later development of scientific and graphing calculators have changed the methods of learning and studying mathematics forever. Where the study of computations with tables of logarithms and slide rules made up an important part of mathematics courses prior to 1970, today the widespread availability of calculators make their study a topic only of historical significance.

In the past two decades, the hand-held calculator has become an integral part of our everyday existence. Today calculators come in a large array of different types, sizes, and prices. For the course for which this textbook is intended, the most appropriate type is the *scientific calculator,* which costs between ten and twenty dollars. While some scientific calculators have advanced features such as programmability and graphing capability, these two features are not essential for the study of the material in this text.

In this introduction, we explain some of the features of scientific calculators. However, remember that calculators vary among manufacturers and models, and that while the methods explained here apply to many of them, they may not apply to your specific calculator. For this reason, it is important to remember that *this is only a guide, and is not intended to take the place of your owner's manual.* Always refer to the manual when you need an explanation of how to perform a particular operation.

The explanations that follow apply to *basic* scientific calculators. Modern graphing calculators follow different sequences of keystrokes.

FEATURES AND FUNCTIONS OF MOST SCIENTIFIC CALCULATORS

Most scientific calculators use *algebraic logic.* (Models sold by Texas Instruments, Sharp, Casio, and Radio Shack, for example, use algebraic logic.) A notable exception is Hewlett Packard, a company whose calculators use *Reverse Polish Notation* (RPN). In this introduction, we discuss calculators that use algebraic logic.

ARITHMETIC OPERATIONS To perform an operation of arithmetic, simply enter the first number, touch the operation key ([+], [−], [×], or [÷]), enter the second number, and then touch the [=] key. For example, to add 4 and 3, use the following keystrokes.

(The final answer is displayed in color.)

CHANGE SIGN KEY The key marked $\boxed{+/-}$ allows you to change the sign of a display. This is particularly useful when you wish to enter a negative number. For example, to enter -3, use the following keystrokes.

MEMORY KEY Scientific calculators can hold a number in memory for later use. The label of the memory key varies among models; two of these are $\boxed{M}$ and $\boxed{STO}$. $\boxed{M+}$ and $\boxed{M-}$ allow you to add to or subtract from the value currently in memory. The memory recall key, labeled $\boxed{MR}$, $\boxed{RM}$, or $\boxed{RCL}$, allows you to retrieve the value stored in memory.

Suppose that you wish to store the number 5 in memory. Enter 5, then touch the key for memory. You can then perform other calculations. When you need to retrieve the 5, touch the key for memory recall.

If a calculator has a constant memory feature, the value in memory will be retained even after the power is turned off. Some advanced calculators have more than one memory. It is best to read the owner's manual for your model to see exactly how memory is activated.

CLEARING/CLEAR ENTRY KEYS These keys allow you to clear the display or clear the last entry entered into the display. They are usually marked $\boxed{C}$ and $\boxed{CE}$. In some models, touching the $\boxed{C}$ key once will clear the last entry, while touching it twice will clear the entire operation in progress.

SECOND FUNCTION KEY This key is used in conjunction with another key to activate a function that is printed *above* an operation key (and not on the key itself). It is usually marked $\boxed{2nd}$. For example, suppose you wish to find the square of a number, and the squaring function (explained in more detail later) is printed above another key. You would need to touch $\boxed{2nd}$ before the desired squaring function can be activated.

SQUARE ROOT KEY Touching the square root key, $\boxed{\sqrt{x}}$, will give the square root (or an approximation of the square root) of the number in the display. For example, to find the square root of 36, use the following keystrokes.

The square root of 2 is an example of an irrational number (Chapter 1). The calculator will give an approximation of its value, since the decimal for $\sqrt{2}$ never terminates and never repeats. The number of digits shown will vary among models. To find an approximation of $\sqrt{2}$, use the following keystrokes.

$\boxed{2}$ $\boxed{\sqrt{x}}$ $\boxed{\quad 1.4142136 \quad}$ An approximation

SQUARING KEY This key, $\boxed{x^2}$, allows you to square the entry in the display. For example, to square 35.7, use the following keystrokes.

The squaring key and the square root key are often found on the same key, with one of them being a second function (that is, activated by the second function key, described above).

RECIPROCAL KEY   The key marked $\boxed{1/x}$ (or $\boxed{x^{-1}}$) is the reciprocal key. (When two numbers have a product of 1, they are called *reciprocals.*) Suppose that you wish to find the reciprocal of 5. Use the following keystrokes.

$\boxed{5}$ $\boxed{1/x}$ $\boxed{\qquad\quad 0.2}$

INVERSE KEY   Some calculators have an inverse key, marked $\boxed{\text{INV}}$. Inverse operations are operations that "undo" each other. For example, the operations of squaring and taking the square root are inverse operations. The use of the $\boxed{\text{INV}}$ key varies among different models of calculators, so read your owner's manual carefully.

EXPONENTIAL KEY   This key, marked $\boxed{x^y}$ or $\boxed{y^x}$, allows you to raise a number to a power. For example, if you wish to raise 4 to the fifth power (that is, find 4^5), use the following keystrokes.

$\boxed{4}$ $\boxed{x^y}$ $\boxed{5}$ $\boxed{=}$ $\boxed{\qquad\quad 1024}$

ROOT KEY   Some calculators have this key specifically marked $\boxed{\sqrt[x]{x}}$ or $\boxed{\sqrt[x]{y}}$; with others, the operation of taking roots is accomplished by using the inverse key in conjunction with the exponential key. Suppose, for example, your calculator is of the latter type and you wish to find the fifth root of 1024. Use the following keystrokes.

$\boxed{1}$ $\boxed{0}$ $\boxed{2}$ $\boxed{4}$ $\boxed{\text{INV}}$ $\boxed{x^y}$ $\boxed{5}$ $\boxed{=}$ $\boxed{\qquad\quad 4}$

Notice how this "undoes" the operation explained in the exponential key discussion earlier.

PI KEY   The number π is an important number in mathematics. It occurs, for example, in the area and circumference formulas for a circle. By touching the $\boxed{\pi}$ key, you can get in the display the first few digits of π. (Because π is irrational, the display shows only an approximation.) One popular model gives the following display when the $\boxed{\pi}$ key is activated: $\boxed{\quad 3.1415927\quad}$.

log AND ln KEYS   Many students taking this course have never studied logarithms. Logarithms are covered in Chapter 5 in this book. In order to find the common logarithm (base ten logarithm) of a number, enter the number and touch the $\boxed{\log}$ key. To find the natural logarithm, enter the number and touch the $\boxed{\ln}$ key. For example, to find these logarithms of 10, use the following keystrokes.

Common logarithm: $\boxed{1}$ $\boxed{0}$ $\boxed{\log}$ $\boxed{\qquad\quad 1}$
Natural logarithm: $\boxed{1}$ $\boxed{0}$ $\boxed{\ln}$ $\boxed{\quad 2.3025851}$ An approximation

10^x AND e^x KEYS   These keys are special exponential keys, and are inverses of the log and ln keys. (On some calculators, they are second functions.) The number e is an irrational number and is the base of the natural logarithm function. Its value is approximately 2.71828. To use these keys, enter the number to which 10 or e is to be raised, and then touch the $\boxed{10^x}$ or $\boxed{e^x}$ key. For example, to raise 10 or e to the 2.5 power, use the following keystrokes.

Base is 10: $\boxed{2}$ $\boxed{\cdot}$ $\boxed{5}$ $\boxed{10^x}$ $\boxed{\mathbf{316.22777}}$ An approximation
Base is e: $\boxed{2}$ $\boxed{\cdot}$ $\boxed{5}$ $\boxed{e^x}$ $\boxed{\mathbf{12.182494}}$ An approximation

(Note: If no $\boxed{10^x}$ key is specifically shown, touching $\boxed{\text{INV}}$ followed by $\boxed{\text{log}}$ accomplishes raising 10 to the power x. Similarly, if no $\boxed{e^x}$ key is specifically shown, touching $\boxed{\text{INV}}$ followed by $\boxed{\text{ln}}$ accomplishes raising e to the power x.)

FACTORIAL KEY   The factorial key, $\boxed{x!}$, evaluates the factorial of any non-negative integer within the limits of the calculator. Factorials are defined in Section 1.4 (and Section 9.6) in this book. For example, $5! = 1 \cdot 2 \cdot 3 \cdot 4 \cdot 5$. To use the factorial key, just enter the number and touch $\boxed{x!}$. To evaluate $5!$ on a calculator, use the following keystrokes.

$\boxed{5}$ $\boxed{x!}$ $\boxed{\mathbf{120}}$

OTHER FEATURES OF SCIENTIFIC CALCULATORS

When decimal approximations are shown on scientific calculators, they are either *truncated* or *rounded*. To see which of these a particular model is programmed to do, evaluate 1/18 as an example. If the display shows .0555555 (last digit 5), it truncates the display. If it shows .0555556 (last digit 6), it rounds off the display.

When very large or very small numbers are obtained as answers, scientific calculators often express these numbers in scientific notation. For example, if you multiply 6,265,804 by 8,980,591, the display might look like this:

$\boxed{\mathbf{5.6270623 \quad 13}}$

The "13" at the far right means that the number on the left is multiplied by 10^{13}. This means that the decimal point must be moved 13 places to the right if the answer is to be expressed in its usual form. Even then, the value obtained will only be an approximation: 56,270,623,000,000.

ADVANCED FEATURES   Two features of advanced scientific calculators are programmability and graphing capability. A programmable calculator is capable of running small programs, much like a mini-computer. A graphing calculator can be used to plot graphs of functions on a small screen. One of the issues in mathematics education today deals with how graphing calculators should be incorporated into the curriculum. Their availability in the 1990s parallels the availability of scientific calculators in the 1980s, and they will no doubt play a prominent role in mathematics education as we move into the twenty-first century. Several chapters in this book conclude with features on graphing calculators.

ALGEBRAIC EXPRESSIONS

1.1 THE REAL NUMBERS

1.2 ORDER AND ABSOLUTE VALUE

1.3 POLYNOMIALS

1.4 THE BINOMIAL THEOREM

1.5 FACTORING

1.6 RATIONAL EXPRESSIONS

1.7 RATIONAL EXPONENTS

1.8 RADICALS

Today, people in fields ranging from accounting to zoology use algebra to solve a variety of problems. Equations, graphs, and functions are important tools in business, science, and other areas. As preparation for discussing these topics of algebra, in this chapter we review the properties of real numbers and the basic rules of algebra. We will be using set notation throughout this text. For a review of sets, see Appendix A.

1.1 ———— THE REAL NUMBERS

SETS OF NUMBERS   The idea of counting goes back into the mists of antiquity. When people first counted they used only the **natural numbers,** written in set notation as

$$\{1, 2, 3, 4, 5, \ldots\}.$$

Much more recent is the idea of counting *no* object—that is, the idea of the number 0. Including 0 with the set of natural numbers gives the set of **whole numbers.**

$$\{0, 1, 2, 3, 4, 5, \ldots\}$$

(These and other sets of numbers are summarized later in this section.)

About 500 years ago, people came up with the idea of counting backwards, from 4 to 3 to 2 to 1 to 0. There seemed no reason not to continue this process, calling the new numbers -1, -2, -3, and so on. Including these numbers with the set of whole numbers gives the very useful set of **integers,**

$$\{\ldots, -4, -3, -2, -1, 0, 1, 2, 3, \ldots\}.$$

Integers can be shown pictorially with a **number line.** (A number line is similar to a thermometer on its side.) As an example, the elements of the set $\{-3, -1, 0, 1, 3, 5\}$ are located on the number line in Figure 1.1.

■ **FIGURE 1.1**

The result of dividing two integers, with a nonzero divisor, is called a *rational number.* By definition, the **rational numbers** are the elements of the set

$$\left\{ \frac{p}{q} \,\middle|\, p, q \text{ are integers and } q \neq 0 \right\}.$$

This definition, which is given in *set-builder notation,* is read "the set of all elements p/q such that p and q are integers and $q \neq 0$." Examples of rational numbers include 3/4, $-5/8$, 7/2, and $-14/9$. All integers are rational numbers, since any integer can be written as the quotient of itself and 1.

Rational numbers can be located on a number line by a process of subdivision. For example, 5/8 can be located by dividing the interval from 0 to 1 into 8 equal parts, then labeling the fifth part 5/8. Several rational numbers are located on the number line in Figure 1.2.

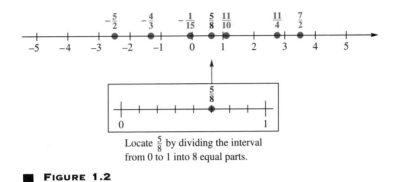

Locate $\frac{5}{8}$ by dividing the interval from 0 to 1 into 8 equal parts.

■ **FIGURE 1.2**

The set of all numbers that correspond to points on a number line is called the set of **real numbers.** The set of real numbers is shown in Figure 1.3.

■ **FIGURE 1.3**

A real number that is not rational is called an **irrational number.** The set of irrational numbers includes $\sqrt{3}$ and $\sqrt{5}$ but not $\sqrt{1}, \sqrt{4}, \sqrt{9}, \ldots$, which equal 1, 2, 3, . . . , and hence are rational numbers. Another irrational number is π, which is approximately equal to 3.14159. A calculator shows that $\sqrt{2} \approx 1.414$ (where $\approx$ is read "is approximately equal to") and $\sqrt{5} \approx 2.236$. Using these approximations, the numbers in the set $\{-2/3, 0, \sqrt{2}, \sqrt{5}, \pi, 4\}$ can be located on a number line as shown in Figure 1.4.

■ **FIGURE 1.4**

Real numbers can also be defined in another way, in terms of decimals. Using repeated subdivisions, any real number can be located (at least in theory) as a point on a number line. By this process, the set of real numbers can be defined as the set of all decimals. Further work would show that the set of rational numbers is the set of all decimals that repeat or terminate. For example,

$$.25 = 1/4,$$
$$.833333 \ldots = 5/6,$$
$$.076923076923 \ldots = 1/13,$$

and so on. Repeating decimals are often written with a bar to indicate the digits that repeat endlessly. With this notation, $5/6 = .8\overline{3}$, and $1/7 = .\overline{142857}$. The set of irrational numbers is the set of decimals that neither repeat nor terminate. For example,

$$\sqrt{2} = 1.414213562373 \ldots \quad \text{and} \quad \pi = 3.141592654359 \ldots$$

The sets of numbers discussed so far are summarized as follows.

SETS OF NUMBERS		
Real numbers	$\{x \mid x$ corresponds to a point on a number line$\}$	
Integers	$\{\ldots, -3, -2, -1, 0, 1, 2, 3, \ldots\}$	
Rational numbers	$\left\{\dfrac{p}{q} \middle\vert\, p \text{ and } q \text{ are integers and } q \neq 0\right\}$	
Irrational numbers	$\{x \mid x$ is real but not rational$\}$	
Whole numbers	$\{0, 1, 2, 3, 4, \ldots\}$	
Natural numbers	$\{1, 2, 3, 4, \ldots\}$	

■ ***Example 1***

IDENTIFYING ELEMENTS OF SUBSETS OF THE REAL NUMBERS

Let set $A = \{-8, -6, -3/4, 0, 3/8, 1/2, 1, \sqrt{2}, \sqrt{5}, 6, 9/0\}$. List the elements from set A that belong to each of the sets of numbers just discussed.

(a) All elements of A are real numbers except $9/0$. Division by 0 is not defined, so $9/0$ is not a number.

(b) The irrational numbers are $\sqrt{2}$ and $\sqrt{5}$.

(c) The rational numbers are $-8, -6, -\dfrac{3}{4}, 0, \dfrac{3}{8}, \dfrac{1}{2}, 1,$ and 6.

(d) The integers are $-8, -6, 0, 1,$ and 6.

(e) The whole numbers are 0, 1, and 6.

(f) The natural numbers in set A are 1 and 6.   ■

The relationships among the various subsets of the real numbers are shown in Figure 1.5.

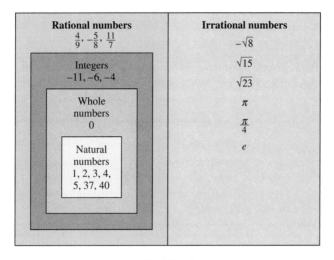

The Real Numbers

■ FIGURE 1.5

EXPONENTS Exponential notation is used to write the products of repeated factors. For example, the product $2 \cdot 2 \cdot 2$ can be written as 2^3, where the 3 shows that three factors of 2 appear in the product. The notation a^n is defined as follows.

DEFINITION OF a^n	If n is any positive integer and a is any real number, then

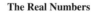
$$a^n = a \cdot a \cdot a \cdots a,$$
where a appears n times.

The integer n is the **exponent,** and a is the **base.** (Read a^n as "a to the nth power," or just "a to the nth.")

■ **Example 2**
EVALUATING
EXPONENTIAL
EXPRESSIONS

Evaluate each exponential expression, and identify the base and the exponent.

(a) $4^3 = 4 \cdot 4 \cdot 4 = 64$; the base is 4 and the exponent is 3

(b) $(-6)^2 = (-6)(-6) = 36$; the base is -6 and the exponent is 2

(c) $-6^2 = -(6 \cdot 6) = -36$; the base is 6 and the exponent is 2

(d) $4 \cdot 3^2 = 4 \cdot 3 \cdot 3 = 36$; the base is 3 and the exponent is 2

(e) $(4 \cdot 3)^2 = 12^2 = 144$; the base is $4 \cdot 3$ or 12 and the exponent is 2. ■

CAUTION | In Example 2, notice that $4 \cdot 3^2 \neq (4 \cdot 3)^2$.

ORDER OF OPERATIONS When a problem involves more than one operation symbol, use the following *order of operations.*

ORDER OF OPERATIONS	If grouping symbols such as parentheses, square brackets, or fraction bars are present:

1. Work separately above and below each fraction bar.
2. Use the rules below within each set of parentheses or square brackets. Start with the innermost and work outward.

If no grouping symbols are present:

1. Simplify all powers and roots, working from left to right.
2. Do any multiplications or divisions in the order in which they occur, working from left to right.
3. Do any additions or subtractions in the order in which they occur, working from left to right.

■ **Example 3**

USING ORDER OF OPERATIONS

Use the order of operations given above to evaluate each of the following.

(a) $6 \div 3 + 2^3 \cdot 5 = 6 \div 3 + 8 \cdot 5 = 2 + 8 \cdot 5 = 2 + 40 = 42$

(b) $(8 + 6) \div 7 \cdot 3 - 6 = 14 \div 7 \cdot 3 - 6$
$$= 2 \cdot 3 - 6$$
$$= 6 - 6 = 0$$

(c) $\dfrac{-(-3)^3 + (-5)}{2(-8) - 5(3)} = \dfrac{-(-27) + (-5)}{2(-8) - 5(3)}$ Evaluate the exponential.

$$= \dfrac{27 + (-5)}{-16 - 15}$$ Multiply.

$$= \dfrac{22}{-31} = -\dfrac{22}{31}$$ Add and subtract. ■

■ **Example 4**

USING ORDER OF OPERATIONS

Use the order of operations to evaluate each expression if $x = -2$, $y = 5$, and $z = -3$.

(a) $-4x^2 - 7y + 4z$
Replace x with -2, y with 5, and z with -3.

$$-4x^2 - 7y + 4z = -4(-2)^2 - 7(5) + 4(-3)$$
$$= -4(4) - 7(5) + 4(-3)$$
$$= -16 - 35 - 12 = -63$$

(b) $\dfrac{2(x - 5)^2 + 4y}{z + 4} = \dfrac{2(-2 - 5)^2 + 4(5)}{-3 + 4}$

$$= \dfrac{2(-7)^2 + 20}{1}$$ Work inside parentheses; multiply; add.

$$= 2(49) + 20$$ Evaluate the exponential.

$$= 118$$ ■

| CAUTION | Notice the use of parentheses when numbers are substituted for the variables. This is especially important when substituting negative numbers for the variables in a product. |

■ IN SIMPLEST TERMS

2

Near the end of a major league baseball season, fans are often interested in the current first-place team's "magic number." The magic number is the sum of the required number of wins of the first-place team and the number of losses of the second-place team (for the remaining games) necessary to clinch the pennant. (In a major league season, each team plays 162 games.)

To calculate the magic number M for a first-place team prior to the end of a season, we can use the formula

$$M = W_2 + N_2 - W_1 + 1,$$

where W_2 = the number of current wins of the second-place team;

N_2 = the number of remaining games of the second-place team;

W_1 = the number of current wins of the first-place team.

On September 17, 1991, the Pittsburgh Pirates led the second-place St. Louis Cardinals.

	Won	Lost	Pct.
Pittsburgh	86	58	.597
St. Louis	76	67	.531
Chicago	70	74	.486
New York	69	74	.483
Philadelphia	67	77	.465
Montreal	63	79	.444

To calculate Pittsburgh's magic number M, we have

$W_2 = 76$ Number of St. Louis wins to date

$N_2 = 162 - (76 + 67) = 19$ Number of games St. Louis has remaining

$W_1 = 86$ Number of Pittsburgh wins to date

Therefore,

$$M = 76 + 19 - 86 + 1$$
$$= 10.$$

Any total of Pirate wins and Cardinal losses that add up to 10 would assure that the Pirates would clinch the pennant. (Pittsburgh actually clinched the pennant on September 22, 1991.)

SOLVE EACH PROBLEM

A. On September 17, 1991, the Minnesota Twins led the American League West Division with a record of 87 wins and 58 losses. The Chicago White Sox were in second place with a record of 79 wins and 66 losses. What was the Twins' magic number on that day?

B. Suppose that the Cleveland Indians were in first place with a record of 81 wins and 63 losses, leading the second-place Red Sox who had a record of 68 wins and 74 losses. What would be the Indians' magic number?

ANSWERS A. $M = 10$ B. $M = 8$

PROPERTIES OF REAL NUMBERS   There are several properties that describe how real numbers behave. We will discuss each property and then summarize them at the end of the section. The first of these properties, the *commutative properties,* state that two numbers may be added or multiplied in any order:

$$4 + (-12) = -12 + 4, \qquad 8(-5) = -5(8),$$
$$-9 + (-1) = -1 + (-9), \qquad (-6)(-3) = (-3)(-6),$$

and so on. Generalizing, the **commutative properties** say that for all real numbers a and b,

$$a + b = b + a \qquad \text{and} \qquad ab = ba.$$

■ *Example 5*

ILLUSTRATING THE COMMUTATIVE PROPERTIES

The following statements illustrate the commutative properties. Notice that the *order* of the numbers changes from one side of the equals sign to the other.

(a) $(6 + x) + 9 = (x + 6) + 9$ **(b)** $(6 + x) + 9 = 9 + (6 + x)$

(c) $5 \cdot (9 \cdot 8) = (9 \cdot 8) \cdot 5$ **(d)** $5 \cdot (9 \cdot 8) = 5 \cdot (8 \cdot 9)$ ■

By the *associative properties,* if three numbers are to be added or multiplied, either the first two numbers or the last two may be "associated." For example, the sum of the three numbers -9, 8, and 7 may be found in either of two ways:

$$-9 + (8 + 7) = -9 + 15 = 6,$$
or
$$(-9 + 8) + 7 = -1 + 7 = 6.$$

Also,

$$5(-3 \cdot 2) = 5(-6) = -30$$
or
$$(5 \cdot -3)2 = (-15)2 = -30.$$

In summary, the **associative properties** say that for all real numbers a, b, and c,

$$(a + b) + c = a + (b + c) \qquad \text{and} \qquad (ab)c = a(bc).$$

CAUTION

It is a common error to confuse the associative and commutative properties. To avoid this error, check the order of the terms: with the commutative properties the order changes from one side of the equals sign to the other; with the associative properties the order does not change, but the grouping does.

■ *Example 6*

DISTINGUISHING BETWEEN THE COMMUTATIVE AND THE ASSOCIATIVE PROPERTIES

This example shows a list of statements using the same symbols. Notice the difference between the commutative and associative properties.

Commutative Properties	Associative Properties
$(x + 4) + 9 = (4 + x) + 9$	$(x + 4) + 9 = x + (4 + 9)$
$7 \cdot (5 \cdot 2) = (5 \cdot 2) \cdot 7$	$7 \cdot (5 \cdot 2) = (7 \cdot 5) \cdot 2$ ■

■ *Example 7*

USING THE
COMMUTATIVE AND
ASSOCIATIVE
PROPERTIES TO
SIMPLIFY
EXPRESSIONS

Simplify each expression using the commutative and associative properties as needed.

(a) $6 + (9 + x) = (6 + 9) + x = 15 + x$ Associative property

(b) $\dfrac{5}{8}(16y) = \left(\dfrac{5}{8} \cdot 16\right)y = 10y$ Associative property

(c) $(-10p)\left(\dfrac{6}{5}\right) = \dfrac{6}{5}(-10p)$ Commutative property

$\qquad\qquad = \left[\dfrac{6}{5}(-10)\right]p$ Associative property

$\qquad\qquad = -12p$ ■

The *identity properties* show special properties of the numbers 0 and 1. The sum of 0 and any real number a is a itself. For example,

$$0 + 4 = 4, \qquad -5 + 0 = -5.$$

The number 0 preserves the identity of a number under addition, making 0 the **identity element for addition** (or the **additive identity**).

The number 1, the **identity element for multiplication** (or **multiplicative identity**), preserves the identity of a number under multiplication, since the product of 1 and any number a is a. For example,

$$5 \cdot 1 = 5, \qquad 1\left(-\frac{2}{3}\right) = -\frac{2}{3}.$$

In summary, the **identity properties** say that for every real number a, there exists a unique real number 0 such that

$$a + 0 = a \qquad \text{and} \qquad 0 + a = a,$$

and there exists a unique real number 1 such that

$$a \cdot 1 = a \qquad \text{and} \qquad 1 \cdot a = a.$$

The sum of the numbers 5 and -5 is the identity element for addition, 0, just as the sum of $-2/3$ and $2/3$ is 0. In fact, for any real number a there is a real number, written $-a$, such that the sum of a and $-a$ is the identity element 0. The number $-a$ is the **additive inverse** or **negative** of a.

CAUTION

Do not confuse the *negative of a number with a negative number*. Since a is a variable, it can represent either a positive or a negative number. The negative of a, written $-a$, can also be either a negative or a positive number (or zero). Do not make the common mistake of thinking that $-a$ *must* be a negative number. For example, if a is -3, then $-a$ is $-(-3) = 3$.

For each real number a (except 0) there is a real number $1/a$ such that the product of a and $1/a$ is the identity element for multiplication, 1. The number $1/a$ is called the **multiplicative inverse** or **reciprocal** of the number a. Every real number except 0 has a reciprocal.

The existence of $-a$ and of $1/a$ comes from the **inverse properties,** which say that for every real number a, there exists a unique real number $-a$ such that

$$a + (-a) = 0 \qquad \text{and} \qquad -a + a = 0,$$

and for every nonzero real number a, there exists a unique real number $1/a$ such that

$$a \cdot \frac{1}{a} = 1 \qquad \text{and} \qquad \frac{1}{a} \cdot a = 1.$$

The area of the entire region shown in Figure 1.6 can be found in two ways. One way is to multiply the length of the base of the entire region, or $5 + 3 = 8$, by the width of the region:

$$4(5 + 3) = 4(8) = 32.$$

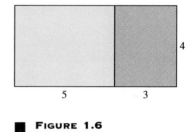

■ **FIGURE 1.6**

Another way to find the area of the region is to add the areas of the smaller rectangles on the left and right,

$$4(5) = 20 \qquad \text{and} \qquad 4(3) = 12,$$

to get a total area of

$$4(5) + 4(3) = 20 + 12 = 32,$$

the same result. The equal results from finding the area in two different ways,

$$4(5 + 3) = 4(5) + 4(3),$$

illustrate the **distributive property of multiplication over addition,** which says that for all real numbers a, b, and c,

$$a(b + c) = ab + ac.$$

We will refer to this as simply "the distributive property" from now on.

Using a commutative property, the distributive property can be rewritten as $(b + c)a = ba + ca$. Another form of the distributive property, $a(b - c) = ab - ac$, comes from the definition of subtraction as "adding the negative." Also, the distributive property can be extended to include more than two numbers in the sum. For example,

$$9(5x + y + 4z) = 9(5x) + 9y + 9(4z)$$
$$= 45x + 9y + 36z.$$

NOTE	The distributive property is one of the key properties of the real numbers, because it is used to change products to sums and sums to products.

■ *Example 8*

ILLUSTRATING THE DISTRIBUTIVE PROPERTY

The following statements illustrate the distributive property.

(a) $3(x + y) = 3x + 3y$

(b) $-(m - 4n) = -1 \cdot (m - 4n) = -m + 4n$

(c) $7p + 21 = 7p + 7 \cdot 3 = 7(p + 3)$

(d) $\dfrac{1}{3}\left(\dfrac{4}{5}m - \dfrac{3}{2}n - 27\right) = \dfrac{1}{3}\left(\dfrac{4}{5}m\right) - \dfrac{1}{3}\left(\dfrac{3}{2}n\right) - \dfrac{1}{3}(27)$

$$= \dfrac{4}{15}m - \dfrac{1}{2}n - 9 \quad ■$$

A summary of the properties of the real numbers follows.

PROPERTIES OF THE REAL NUMBERS

For all real numbers a, b, and c:

Commutative properties

$a + b = b + a$

$ab = ba$

Associative properties

$(a + b) + c = a + (b + c)$

$(ab)c = a(bc)$

Identity properties

There exists a unique real number 0 such that

$a + 0 = a$ and $0 + a = a$.

There exists a unique real number 1 such that

$a \cdot 1 = a$ and $1 \cdot a = a$.

Inverse properties

There exists a unique real number $-a$ such that

$a + (-a) = 0$ and $(-a) + a = 0$.

If $a \neq 0$, there exists a unique real number $1/a$ such that

$a \cdot \dfrac{1}{a} = 1$ and $\dfrac{1}{a} \cdot a = 1$.

Distributive property

$a(b + c) = ab + ac$

1.1 EXERCISES ■ ─────────────────────────────

Let set B = {−6, −12/4, −5/8, −$\sqrt{3}$, 0, 1/4, 1, 2π, 3, $\sqrt{12}$ }. List all the elements of B that belong to each of the following sets. See Example 1.

1. Natural numbers   **2.** Whole numbers   **3.** Integers

4. Rational numbers   **5.** Irrational numbers   **6.** Real numbers

For Exercises 7–12, choose all words from the following list that apply: (a) natural number, (b) whole number, (c) integer, (d) rational number, (e) irrational number, (f) real number, (g) not defined. See Example 1.

7. 12   **8.** 0   **9.** $-\dfrac{3}{4}$ **10.** $\dfrac{5}{9}$ **11.** $\sqrt{8}$ **12.** $-\sqrt{2}$

13. Explain why division by zero is not defined.

Simplify each of the following. See Example 2.

14. 3^4 **15.** -3^5 **16.** -2^6 **17.** $(-3)^4$

18. $(-2)^5$ **19.** $(-3)^5$ **20.** $(-3)^6$

21. Based on your answers to Exercises 14–20, complete the following statements. A negative base raised to an odd exponent is _____. A negative base raised to an even
                                                        positive/negative
exponent is _____.
                positive/negative

Use the order of operations to evaluate each of the following. See Example 3.

22. $9 - 4^2 - (-12)$ **23.** $8^2 - (-4) + 11$ **24.** $16(-9) - 4$

25. $-15 - 3(-8)$ **26.** $-2 \cdot 5 + 12 \div 3$ **27.** $9 \cdot 3 - 16 \div 4$

28. $-4(9 - 8) + (-7)(2)^3$ **29.** $6(-5) - (-3)(2)^4$ **30.** $(4 - 2^3)(-2 + \sqrt{25})$

31. $[-3^2 - (-2)] [\sqrt{16} - 2^3]$ **32.** $\left(-\dfrac{2}{9} - \dfrac{1}{4}\right) - \left[-\dfrac{5}{18} - \left(-\dfrac{1}{2}\right)\right]$ **33.** $\left[-\dfrac{5}{8} - \left(-\dfrac{2}{5}\right)\right] - \left(\dfrac{3}{2} - \dfrac{11}{10}\right)$

34. $\dfrac{-8 + (-4)(-6) \div 12}{4 - (-3)}$ **35.** $\dfrac{15 \div 5 \cdot 4 \div 6 - 8}{-6 - (-5) - 8 \div 2}$

▦ *Evaluate each of the following if p = −2, q = 4, and r = −5. See Example 4.*

36. $-3(p + 5q)$ **37.** $2(q - r)$ **38.** $\dfrac{p}{q} + \dfrac{3}{r}$ **39.** $\dfrac{q + r}{q + p}$

40. $\dfrac{3q}{3p - 2r}$ **41.** $\dfrac{3q}{r} - \dfrac{5}{p}$ **42.** $\dfrac{\dfrac{q}{4} - \dfrac{r}{5}}{\dfrac{p}{2} + \dfrac{q}{2}}$ **43.** $\dfrac{\dfrac{3r}{10} - \dfrac{5p}{2}}{q + \dfrac{2r}{5}}$

Identify the properties that are illustrated in each of the following statements. Some will require more than one property. Assume that all variables represent real numbers. See Examples 5–8.

44. $6 \cdot 12 + 6 \cdot 15 = 6(12 + 15)$ **45.** $8(m + 4) = 8m + 8 \cdot 4$

46. $(x + 6) \cdot \left(\dfrac{1}{x + 6}\right) = 1, \quad \text{if } x + 6 \neq 0$ **47.** $\dfrac{2 + m}{2 - m} \cdot \dfrac{2 - m}{2 + m} = 1, \quad \text{if } m \neq 2 \text{ or } -2$

───────────────

*The symbol ▦ identifies problems that can be solved more quickly with the help of a calculator.

48. $(7 - y) + 0 = 7 - y$

49. $[9 + (-9)] \cdot 5 = 5 \cdot 0$

50. $(-5 + 7)(3 + 4) = (-5 + 7) \cdot 3 + (-5 + 7) \cdot 4$

51. $x \cdot \dfrac{1}{x} + x \cdot \dfrac{1}{x} = x\left(\dfrac{1}{x} + \dfrac{1}{x}\right)$, if $x \neq 0$

Use the distributive property to rewrite sums as products and products as sums. See Example 8.

52. $10k + 3k$ **53.** $8p - 14p$ **54.** $15 - 10x$ **55.** $18y + 6$

56. $6(x + y)$ **57.** $9(r - s)$ **58.** $-3(z - y)$ **59.** $-2(m + n)$

60. $-(8r - k)$ **61.** $-(-2 + 5m)$ **62.** $a(r + s - t)$ **63.** $p(q - w + x)$

Use the various properties of real numbers to simplify each of the following expressions. See Example 7.

64. $\dfrac{5}{3}(12y)$ **65.** $\dfrac{10}{11}(22z)$ **66.** $\left(\dfrac{3}{4}r\right)(-12)$ **67.** $\left(-\dfrac{5}{8}p\right)(-24)$

68. $\dfrac{2}{3}(12y - 6z + 18q)$ **69.** $-\dfrac{1}{4}(20m + 8y - 32z)$ **70.** $\dfrac{3}{8}\left(\dfrac{16}{9}y + \dfrac{32}{27}z - \dfrac{40}{9}\right)$ **71.** $\dfrac{2}{3}\left(\dfrac{9}{4}a + \dfrac{15}{14}b - \dfrac{27}{20}\right)$

72. Is there a commutative property for subtraction? That is, does $a - b = b - a$? Support your answer with examples.

73. Is there an associative property for subtraction? Does $(a - b) - c = a - (b - c)$? Support your answer with examples.

74. This exercise is designed to show that $\sqrt{2}$ is irrational. Give a reason for each of steps (a) through (h).
There are two possibilities:
(1) A rational number a/b exists such that $(a/b)^2 = 2$.
(2) There is no such rational number.
We work with assumption (1). If it leads to a contradiction, then we will know that (2) must be correct. Start by assuming that a rational number a/b exists with $(a/b)^2 = 2$. Assume also that a/b is written in lowest terms.
(a) Since $(a/b)^2 = 2$, we must have $a^2/b^2 = 2$, or $a^2 = 2b^2$.
(b) $2b^2$ is an even number.

(c) Therefore, a^2, and a itself, must be even numbers.
(d) Since a is an even number, it must be a multiple of 2. That is, we can find a natural number c such that $a = 2c$. This changes $a^2 = 2b^2$ into $(2c)^2 = 2b^2$.
(e) Therefore, $4c^2 = 2b^2$ or $2c^2 = b^2$.
(f) $2c^2$ is an even number.
(g) This makes b^2 an even number, so that b must be even.
(h) We have reached a contradiction. Show where the contradiction occurs.
(i) Since assumption (1) leads to a contradiction, we are forced to accept assumption (2), which says that $\sqrt{2}$ is irrational.

1.2 ORDER AND ABSOLUTE VALUE

ORDER Figure 1.7 shows a number line with the points corresponding to several different numbers marked on the line. A number that corresponds to a particular point on a line is called the **coordinate** of the point. For example, the leftmost marked point in Figure 1.7 has coordinate -4. The correspondence between points on a line and the real numbers is called a **coordinate system** for the line. (From now on, the phrase "the point on a number line with coordinate a" will be abbreviated as "the point with coordinate a," or simply "the point a.")

■ **FIGURE 1.7**

If the real number a is to the left of the real number b on a number line, then ***a* is less than *b*,** written $a < b$. If a is to the right of b, then ***a* is greater than *b*,** written $a > b$. For example, in Figure 1.7, $-\sqrt{5}$ is to the left of $-11/7$ on the number line, so $-\sqrt{5} < -11/7$, while $\sqrt{20}$ is to the right of π, indicating $\sqrt{20} > \pi$.

NOTE	Remember that the "point" of the symbol goes toward the smaller number.

As an alternative to this geometric definition of "is less than" or "is greater than," there is an algebraic definition: if a and b are two real numbers and if the difference $a - b$ is positive, then $a > b$. If $a - b$ is negative, then $a < b$. The geometric and algebraic statements of order are summarized as follows.

Statement	Geometric Form	Algebraic Form
$a > b$	a is to the right of b	$a - b$ is positive
$a < b$	a is to the left of b	$a - b$ is negative

■ *Example 1*

IDENTIFYING THE
SMALLER OF TWO
NUMBERS

Part (a) of this example shows how to identify the smaller of two numbers with the geometric approach, and part (b) uses the algebraic approach.

(a) In Figure 1.7, $-\sqrt{5}$ is to the left of 2/3, so

$$-\sqrt{5} < \frac{2}{3}.$$

Since 2/3 is to the right of $-\sqrt{5}$,

$$\frac{2}{3} > -\sqrt{5}.$$

(b) The difference $2/3 - (-11/7) = 2/3 + 11/7$ is positive, showing that

$$\frac{2}{3} > -\frac{11}{7}.$$

The difference $-11/7 - 2/3 = -(11/7 + 2/3)$ is negative, showing that

$$-\frac{11}{7} < \frac{2}{3}. \quad ■$$

The following variations on $<$ and $>$ are often used.

Symbol	Meaning (Reading Left to Right)
$\leq$	is less than or equal to
$\geq$	is greater than or equal to
$\not<$	is not less than
$\not>$	is not greater than

Statements involving these symbols, as well as $<$ and $>$, are called **inequalities.**

■ *Example 2*

SHOWING WHY
INEQUALITY
STATEMENTS ARE
TRUE

The list at the right shows several statements
and the reason why each is true.

Statement	Reason
$8 \leq 10$	$8 < 10$
$8 \leq 8$	$8 = 8$
$-9 \geq -14$	$-9 > -14$
$-8 \ngtr -2$	$-8 < -2$
$4 \nless 2$	$4 > 2$

The inequality $a < b < c$ says that b is *between* a and c, since

$$a < b < c$$

means $\qquad\qquad a < b \qquad$ and $\qquad b < c.$

In the same way, $\qquad\qquad a \leq b \leq c$

means $\qquad\qquad a \leq b \qquad$ and $\qquad b \leq c.$

CAUTION

When writing these "between" statements, make sure that both inequality symbols point in the same direction, toward the smallest number. For example,

both $2 < 7 < 11 \qquad$ and $\qquad 5 > 4 > -1$

are true statements, but $3 < 5 > 2$ is meaningless. Generally, it is best to rewrite statements such as $5 > 4 > -1$ as $-1 < 4 < 5$, which is the order of these numbers on the number line.

The following *properties of order* give the basic properties of $<$ and $>$.

PROPERTIES OF
ORDER

For all real numbers a, b, and c:

Transitive property $\qquad$ If $a < b$ and $b < c$, then $a < c$.

Addition property $\qquad$ If $a < b$, then $a + c < b + c$.

Multiplication property $\qquad$ If $a < b$, and if $c > 0$, then $ac < bc$.

$\qquad\qquad\qquad\qquad$ If $a < b$, and if $c < 0$, then $ac > bc$.

In these properties, replacing $<$ with $>$ and $>$ with $<$ results in equivalent properties.

■ *Example 3*

ILLUSTRATING THE
PROPERTIES OF
ORDER

(a) By the transitive property, if $3z < k$ and $k < p$, then $3z < p$.

(b) By the addition property, any real number can be added to both sides of an inequality. For example, adding -2 to both sides of

$$x + 2 < 5$$

gives $\qquad\qquad x + 2 + (-2) < 5 + (-2)$

$$x < 3.$$

This process is explained in more detail in Chapter 2.

(c) While any number may be *added* to both sides of an inequality, more care must be used when *multiplying* both sides by a number. For example, multiplying both sides of

$$\frac{1}{2}x < 5$$

by 2 gives

$$2 \cdot \frac{1}{2}x < 2 \cdot 5$$

$$x < 10,$$

by the first part of the multiplication property. On the other hand, multiplying both sides of

$$-\frac{3}{5}r \geq 9$$

by $-\frac{5}{3}$ gives

$$-\frac{5}{3}\left(-\frac{3}{5}r\right) \leq -\frac{5}{3} \cdot 9$$

$$r \leq -15.$$

Here the $\geq$ symbol has changed to $\leq$, using the second part of the multiplication property. ■

CAUTION | Don't forget to reverse the inequality symbol when multiplying (or dividing) an equation by a negative quantity.

ABSOLUTE VALUE The distance on the number line from a number to 0 is called the **absolute value** of that number. The absolute value of the number a is written $|a|$. For example, the distance on the number line from 9 to 0 is 9, as is the distance from -9 to 0. (See Figure 1.8.) Therefore,

$$|9| = 9 \quad \text{and} \quad |-9| = 9.$$

■ **FIGURE 1.8**

NOTE | Since distance cannot be negative, the absolute value of a number is always nonnegative.

■ ***Example 4***
EVALUATING
ABSOLUTE VALUE

(a) $|2\pi| = 2\pi$

(b) $\left|-\frac{5}{8}\right| = \frac{5}{8}$

(c) $-|8| = -(8) = -8$

(d) $-|-2| = -(2) = -2$ ■

The algebraic definition of the absolute value of the real number a can be stated as follows.

ABSOLUTE VALUE

$$|a| = \begin{cases} a \text{ if } a \geq 0 \\ -a \text{ if } a < 0 \end{cases}$$

The second part of this definition requires some thought. If a is a negative number, that is, if $a < 0$, then $-a$ is positive. Thus, for a *negative* number a,

$$|a| = -a,$$

or the negative of a. For example, if $a = -5$, then $|a| = |-5| = -(-5) = 5$.

∎ *Example 5*
FINDING ABSOLUTE VALUE

Write each of the following without absolute value bars.

(a) $|-8 + 2|$

Work inside the absolute value bars first. Since $-8 + 2 = -6$, $|-8 + 2| = |-6| = 6$.

(b) $|\sqrt{5} - 2|$

Since $\sqrt{5} > 2$ ($\sqrt{5} \approx 2.24$), $\sqrt{5} - 2 > 0$, and $|\sqrt{5} - 2| = \sqrt{5} - 2$.

(c) $|\pi - 4|$

Here, $\pi < 4$, so that $\pi - 4 < 0$, and

$$|\pi - 4| = -(\pi - 4)$$
$$= -\pi + 4 \quad \text{or} \quad 4 - \pi.$$

(d) $|m - 2|$ if $m < 2$

If $m < 2$, then $m - 2 < 0$, so

$$|m - 2| = -(m - 2)$$
$$= -m + 2 \quad \text{or} \quad 2 - m. \quad \blacksquare$$

The definition of absolute value can be used to prove the following properties of absolute value.

PROPERTIES OF ABSOLUTE VALUE

For all real numbers a and b:

$|a| \geq 0$ $\qquad\qquad$ $|-a| = |a|$ $\qquad\qquad$ $|a| \cdot |b| = |ab|$

$\left|\dfrac{a}{b}\right| = \dfrac{|a|}{|b|}$ $\quad (b \neq 0)$ $\qquad$ $|a + b| \leq |a| + |b|$ $\quad$ (the triangle inequality).

■ **IN SIMPLEST TERMS**

9 ▭

Systolic blood pressure is the maximum pressure produced by each heartbeat. Both low blood pressure and high blood pressure are cause for medical concern. Therefore, health-care professionals are interested in a patient's "pressure difference from normal," or P_d. If 120 is considered a normal systolic pressure, $P_d = |P - 120|$ where P is the patient's recorded systolic pressure. For example, a patient with a systolic pressure, P, of 113 would have a pressure difference from normal of

$$P_d = |P - 120|$$
$$= |113 - 120|$$
$$= |-7|$$
$$= 7.$$

SOLVE EACH PROBLEM

A. Calculate the P_d value for a woman whose actual systolic pressure is 116 and whose normal value should be 125.

B. If a patient's P_d value is 17 and the normal pressure for his sex and age should be 130, what are the two possible values for his systolic blood pressure?

ANSWERS   A. $P_d = 9$ B. The patient's blood pressure is either 113 or 147.

■ *Example 6*

ILLUSTRATING THE PROPERTIES OF ABSOLUTE VALUE

(a) $|-15| = 15 \geq 0$

(b) $|-10| = 10$ and $|10| = 10$, so $|-10| = |10|$.

(c) $|5x| = |5| \cdot |x| = 5|x|$ since 5 is positive.

(d) $\left|\dfrac{2}{y}\right| = \dfrac{|2|}{|y|} = \dfrac{2}{|y|}$ for $y \neq 0$

(e) For $a = 3$ and $b = -7$:

$$|a + b| = |3 + (-7)| = |-4| = 4$$
$$|a| + |b| = |3| + |-7| = 3 + 7 = 10$$
$$|a + b| < |a| + |b|$$

(f) For $a = 2$ and $b = 12$:

$$|a + b| = |2 + 12| = |14| = 14$$
$$|a| + |b| = |2| + |12| = 2 + 12 = 14$$
$$|a + b| = |a| + |b| \quad ■$$

Sometimes it is necessary to evaluate expressions involving absolute value. The next example shows how this is done.

■ **Example 7**

EVALUATING
ABSOLUTE VALUE
EXPRESSIONS

Let $x = -6$ and $y = 10$. Evaluate each expression.

(a) $|2x - 3y| = |2(-6) - 3(10)|$ Substitute.

$\qquad\qquad = |-12 - 30|$ Work inside the bars; multiply.

$\qquad\qquad = |-42|$ Subtract.

$\qquad\qquad = 42$

(b) $\dfrac{2|x| - |3y|}{|xy|} = \dfrac{2|-6| - |3(10)|}{|-6(10)|}$ Substitute.

$\qquad\qquad = \dfrac{2 \cdot 6 - |30|}{|-60|}$ $|-6| = 6$; multiply.

$\qquad\qquad = \dfrac{12 - 30}{60}$ Multiply; $|30| = 30$.

$\qquad\qquad = \dfrac{-18}{60} = -\dfrac{3}{10}$ ■

1.2 EXERCISES ■ ─────────────────────────────────

Write the following numbers in numerical order, from smallest to largest. Use a calculator as necessary. See Example 1.

1. $|-8|, -|9|, -|-6|$

2. $-|-9|, -|7|, -|-2|$

3. $\sqrt{8}, -4, -\sqrt{3}, -2, -5, \sqrt{6}, 3$

4. $\sqrt{2}, -1, 4, 3, \sqrt{8}, -\sqrt{6}, \sqrt{7}$

5. $\dfrac{3}{4}, \sqrt{2}, \dfrac{7}{5}, \dfrac{8}{5}, \dfrac{22}{15}$

6. $-\dfrac{9}{8}, -3, -\sqrt{3}, -\sqrt{5}, -\dfrac{9}{5}, -\dfrac{8}{5}$

7. Explain why it is necessary to reverse the direction of the inequality symbol when multiplying an inequality by a negative number. Give examples to illustrate your explanation.

8. What is wrong with writing the statement "$x < 2$ or $x > 5$"   as $5 < x < 2$?

For each of the given inequalities, multiply both sides by the indicated number to get an equivalent inequality. See Example 3(c).

9. $-2 < 5, 3$

10. $-7 < 1, 2$

11. $-11 \geq -22, -\dfrac{1}{11}$

12. $-2 \geq -14, -\dfrac{1}{2}$

13. $3x < 9, \dfrac{1}{3}$

14. $5x > 45, \dfrac{1}{5}$

15. $-9k > 63, -\dfrac{1}{9}$

16. $-3p \leq 24, -\dfrac{1}{3}$

Let $x = -4$ and $y = 2$. Evaluate each of the following. See Examples 4, 5, and 7.

17. $|2x|$

18. $|-3y|$

19. $|x - y|$

20. $|2x + 5y|$

21. $|3x + 4y|$

22. $|-5y + x|$

23. $|-4x + y| - |y|$

24. $\dfrac{|-8y + x|}{-|x|}$

25. $\dfrac{|x| + 2|y|}{5 + x}$

26. Explain how to remove the absolute value bars from the expression $|1 - \sqrt{2}|$.

27. Explain the steps necessary to write $|-x^2 - 4|$ without absolute value bars.

Write each of the following without absolute value bars. See Examples 4 and 5.

28. $-|-8| + |-2|$

29. $3 - |-4|$

30. $|2 - \sqrt{3}|$

31. $|\sqrt{7} - 5|$

32. $|\sqrt{2} - 3|$

33. $|\pi - 3|$

34. $|\pi - 5|$

35. $|x - 4|$, if $x > 4$

36. $|y - 3|$, if $y < 3$

37. $|2k - 8|$, if $k < 4$

38. $|3r - 15|$, if $r > 5$

39. $|-8 - 4m|$, if $m > -2$

40. $|6 - 5r|$, if $r < -2$

41. $|x - y|$, if $x < y$

42. $|x - y|$, if $x > y$

43. $|3 + x^2|$

44. $|x^2 + 4|$

45. $|-1 - p^2|$

Justify each of the following statements by giving the correct property from this section. Assume that all variables represent real numbers. See Examples 3 and 6.

46. If $2k < 8$, then $k < 4$.

47. If $x + 8 < 15$, then $x < 7$.

48. If $-4x < 24$, then $x > -6$.

49. If $x < 5$ and $5 < m$, then $x < m$.

50. If $m > 0$, then $9m > 0$.

51. If $k > 0$, then $8 + k > 8$.

52. $|8 + m| \leq |8| + |m|$

53. $|k - m| \leq |k| + |-m|$

54. $|8| \cdot |-4| = |-32|$

55. $|12 + 11r| \geq 0$

56. $\left|\dfrac{-12}{5}\right| = \dfrac{|-12|}{|5|}$

57. $\left|\dfrac{6}{5}\right| = \dfrac{|6|}{|5|}$

Under what conditions are the following statements true?

58. $|x| = |y|$

59. $|x + y| = |x| + |y|$

60. $|x + y| = |x| - |y|$

61. $|x| \leq 0$

62. $|x - y| = |x| - |y|$

63. $\|x + y\| = |x + y|$

Evaluate each of the following if x is a nonzero real number.

64. $\dfrac{|x|}{x}$

65. $\left|\dfrac{|x|}{x}\right|$

1.3 —————— POLYNOMIALS

RULES FOR EXPONENTS Positive integer exponents were introduced in Section 1.1. Work with exponents can be simplified by using the rules for exponents. By definition, the notation a^m (where m is a positive integer and a is a real number) means that a appears as a factor m times. In the same way, a^n (where n is a positive integer) means that a appears as a factor n times. In the product $a^m \cdot a^n$, the number a would appear $m + n$ times.

PRODUCT RULE   For all positive integers m and n and every real number a:

$$a^m \cdot a^n = a^{m+n}.$$

■ *Example 1*

USING THE
PRODUCT RULE

Find the following products.

(a) $y^4 \cdot y^7 = y^{4+7} = y^{11}$

(b) $(6z^5)(9z^3)(2z^2)$

$$(6z^5)(9z^3)(2z^2) = (6 \cdot 9 \cdot 2) \cdot (z^5 z^3 z^2) \qquad \text{Commutative and associative properties}$$
$$= 108z^{5+3+2}$$
$$= 108z^{10} \qquad \text{Product rule}$$

(c) $(2k^m)\,(k^{1+m})$

$$(2k^m)\,(k^{1+m}) = 2k^{m+(1+m)} \qquad \text{Product rule (if } m \text{ is a positive integer)}$$
$$= 2k^{1+2m} \quad \blacksquare$$

An exponent of zero is defined as follows.

DEFINITION OF a^0   For any nonzero real number a:

$$a^0 = 1.$$

We will show why a^0 is defined this way in Section 1.7. The symbol 0^0 is not defined.

∎ **_Example 2_**
USING THE
DEFINITION OF a^0

(a) $3^0 = 1$

(b) $(-4)^0 = 1$
Replace a with -4 in the definition.

(c) $-4^0 = -1$
As shown in Section 1.1, $-4^0 = -(4^0) = -1$.

(d) $-(-4)^0 = -(1) = -1$

(e) $(7r)^0 = 1, \quad \text{if } r \neq 0 \quad \blacksquare$

The expression $(2^5)^3$ can be written as

$$(2^5)^3 = 2^5 \cdot 2^5 \cdot 2^5.$$

By a generalization of the product rule for exponents, this product is

$$(2^5)^3 = 2^{5+5+5} = 2^{15}.$$

The same exponent could have been obtained by multiplying 3 and 5. This example suggests the first of the **power rules** given below. The others are found in a similar way.

POWER RULES   For all nonnegative integers m and n and all real numbers a and b:

$$(a^m)^n = a^{mn} \qquad (ab)^m = a^m b^m \qquad \left(\frac{a}{b}\right)^m = \frac{a^m}{b^m} \quad (b \neq 0).$$

■ *Example 3*

USING THE POWER
RULES

(a) $(5^3)^2 = 5^{3(2)} = 5^6$

(b) $(3^4 x^2)^3 = (3^4)^3(x^2)^3 = 3^{4(3)}x^{2(3)} = 3^{12}x^6$

(c) $\left(\dfrac{2^5}{b^4}\right)^3 = \dfrac{(2^5)^3}{(b^4)^3} = \dfrac{2^{15}}{b^{12}}, \quad$ if $b \neq 0$ ■

CAUTION

Be careful not to confuse examples like mn^2 and $(mn)^2$. The two expressions are *not* equal. The second power rule given above can be used only with the second expression: $(mn)^2 = m^2 n^2$.

POLYNOMIALS An **algebraic expression** is any combination of variables or constants joined by the basic operations of addition, subtraction, multiplication, division (except by zero), or extraction of roots. Here are some examples of algebraic expressions.

$$2x^2 - 3x, \qquad \frac{5y}{2y - 3}, \qquad \sqrt{m^3 - 8}, \qquad (3a + b)^4$$

The simplest algebraic expressions, *polynomials,* are discussed in this section.

The product of a real number and one or more variables raised to powers is called a **term.** The real number is called the **numerical coefficient,** or just the **coefficient.** The coefficient in $-3m^4$ is -3, while the coefficient in $-p^2$ is -1. **Like terms** are terms with the same variable raised to the same power. For example, $-13x^3$, $4x^3$, and $-x^3$ are like terms, while $6y$ and $6y^2$ are not.

A **polynomial** is defined as a term or a finite sum of terms, with only nonnegative integer exponents permitted on the variables. If the terms of a polynomial contain only the variable x, then the polynomial is called a **polynomial in x.** (Polynomials in other variables are defined similarly.) Examples of polynomials include

$$5x^3 - 8x^2 + 7x - 4, \qquad 9p^5 - 3, \qquad 8r^2, \qquad \text{and} \qquad 6.$$

The expression $9x^2 - 4x - 6/x$ is not a polynomial because of $-6/x$. The terms of a polynomial cannot have variables in a denominator.

The highest exponent in a polynomial in one variable is the **degree** of the polynomial. A nonzero constant is said to have degree 0. (The polynomial 0 has no degree.) For example, $3x^6 - 5x^2 + 2x + 3$ is a polynomial of degree 6.

A polynomial can have more than one variable. A term containing more than one variable has degree equal to the sum of all the exponents appearing on the variables in the term. For example, $-3x^4 y^3 z^5$ is of degree $4 + 3 + 5 = 12$. The degree of a polynomial in more than one variable is equal to the highest degree of any term appearing in the polynomial. By this definition, the polynomial

$$2x^4 y^3 - 3x^5 y + x^6 y^2$$

is of degree 8 because of the $x^6 y^2$ term.

A polynomial containing exactly three terms is called a **trinomial;** one containing exactly two terms is a **binomial;** and a single-term polynomial is called a **monomial.** For example, $7x^9 - 8x^4 + 1$ is a trinomial of degree 9.

■ *Example 4*
DETERMINING THE DEGREE AND TYPE OF A POLYNOMIAL

The list below shows several polynomials, gives the degree of each, and identifies each as a monomial, binomial, trinomial, or none of these.

Polynomial	Degree	Type
$9p^7 - 4p^3 + 8p^2$	7	Trinomial
$29x^{11} + 8x^{15}$	15	Binomial
$-10r^6s^8$	14	Monomial
$5a^3b^7 - 3a^5b^5 + 4a^2b^9 - a^{10}$	11	None of these ■

ADDITION AND SUBTRACTION   Since the variables used in polynomials represent real numbers, a polynomial represents a real number. This means that all the properties of the real numbers mentioned in this chapter hold for polynomials. In particular, the distributive property holds, so

$$3m^5 - 7m^5 = (3 - 7)m^5 = -4m^5.$$

Thus, polynomials are added by adding coefficients of like terms; polynomials are subtracted by subtracting coefficients of like terms.

■ *Example 5*
ADDING AND SUBTRACTING POLYNOMIALS

Add or subtract, as indicated.

(a) $(2y^4 - 3y^2 + y) + (4y^4 + 7y^2 + 6y)$
$$= (2 + 4)y^4 + (-3 + 7)y^2 + (1 + 6)y$$
$$= 6y^4 + 4y^2 + 7y$$

(b) $(-3m^3 - 8m^2 + 4) - (m^3 + 7m^2 - 3)$
$$= (-3 - 1)m^3 + (-8 - 7)m^2 + [4 - (-3)]$$
$$= -4m^3 - 15m^2 + 7$$

(c) $8m^4p^5 - 9m^3p^5 + (11m^4p^5 + 15m^3p^5) = 19m^4p^5 + 6m^3p^5$

(d) $4(x^2 - 3x + 7) - 5(2x^2 - 8x - 4)$
$$= 4x^2 - 4(3x) + 4(7) - 5(2x^2)$$
$$\qquad - 5(-8x) - 5(-4) \qquad \text{Distributive property}$$
$$= 4x^2 - 12x + 28 - 10x^2 + 40x + 20 \qquad \text{Associative property}$$
$$= -6x^2 + 28x + 48 \qquad \text{Add like terms.} \ ■$$

As shown in parts (a), (b), and (d) of Example 5, polynomials in one variable are often written with their terms in *descending order,* so the term of highest degree is first, the one with the next highest degree is next, and so on.

MULTIPLICATION The associative and distributive properties, together with the properties of exponents, can also be used to find the product of two polynomials. For example, to find the product of $3x - 4$ and $2x^2 - 3x + 5$, treat $3x - 4$ as a single expression and use the distributive property as follows.

$$(3x - 4)(2x^2 - 3x + 5) = (3x - 4)(2x^2) - (3x - 4)(3x) + (3x - 4)(5)$$

Now use the distributive property three separate times on the right of the equals sign to get

$$
\begin{aligned}
(3x &- 4)(2x^2 - 3x + 5) \\
&= (3x)(2x^2) - 4(2x^2) - (3x)(3x) - (-4)(3x) + (3x)5 - 4(5) \\
&= 6x^3 - 8x^2 - 9x^2 + 12x + 15x - 20 \\
&= 6x^3 - 17x^2 + 27x - 20.
\end{aligned}
$$

It is sometimes more convenient to write such a product vertically, as follows.

$$
\begin{array}{r}
2x^2 - 3x + 5 \\
3x - 4 \\
\hline
- 8x^2 + 12x - 20 \\
6x^3 - 9x^2 + 15x \\
\hline
6x^3 - 17x^2 + 27x - 20
\end{array}
$$

Add in columns.

■ **Example 6**
MULTIPLYING
POLYNOMIALS

Multiply $(3p^2 - 4p + 1)(p^3 + 2p - 8)$.

Multiply each term of the second polynomial by each term of the first and add these products. It is most efficient to work vertically with polynomials of more than two terms, so that like terms can be placed in columns.

$$
\begin{array}{r}
3p^2 - 4p + 1 \\
p^3 + 2p - 8 \\
\hline
- 24p^2 + 32p - 8 \\
6p^3 - 8p^2 + 2p \\
3p^5 - 4p^4 + p^3 \\
\hline
3p^5 - 4p^4 + 7p^3 - 32p^2 + 34p - 8
\end{array}
$$

Multiply $3p^2 - 4p + 1$ by -8.
Multiply $3p^2 - 4p + 1$ by $2p$.
Multiply $3p^2 - 4p + 1$ by p^3.
Add in columns. ■

The FOIL method is a convenient way to find the product of two binomials. The memory aid FOIL (for First, Outside, Inside, Last) gives the pairs of terms to be multiplied to get the product, as shown in the next examples.

■ **Example 7**
USING FOIL TO
MULTIPLY TWO
BINOMIALS

Find each product.

(a) $(6m + 1)(4m - 3) = (6m)(4m) + (6m)(-3) + 1(4m) + 1(-3)$
$$= 24m^2 - 14m - 3$$

(b) $(2x + 7)(2x - 7) = 4x^2 - 14x + 14x - 49$
$$= 4x^2 - 49$$

(c) $(2k^n - 5)(k^n + 3) = 2k^{2n} + 6k^n - 5k^n - 15$
$$= 2k^{2n} + k^n - 15 \quad \text{(if } n \text{ is an integer)} \quad ■$$

In parts (a) and (c) of Example 7, the product of two binomials was a trinomial, while in part (b) the product of two binomials was a binomial. The product of two binomials of the forms $x + y$ and $x - y$ is always a binomial. Check by multiplying that the following is true.

PRODUCT OF THE SUM AND DIFFERENCE OF TWO TERMS	$(x + y)(x - y) = x^2 - y^2$

This product is called the **difference of two squares.** Since products of this type occur frequently, it is important to *memorize this formula.*

■ *Example 8*

MULTIPLYING THE SUM AND DIFFERENCE OF TWO TERMS

Find each product.

(a) $(3p + 11)(3p - 11)$
　　Using the pattern discussed above, replace x with $3p$ and y with 11.

$$(3p + 11)(3p - 11) = (3p)^2 - 11^2 = 9p^2 - 121$$

(b) $(5m^3 - 3)(5m^3 + 3) = (5m^3)^2 - 3^2 = 25m^6 - 9$

(c) $(9k - 11r^3)(9k + 11r^3) = (9k)^2 - (11r^3)^2 = 81k^2 - 121r^6$ ■

The **squares of binomials** are also special products. The products $(x + y)^2$ and $(x - y)^2$ are shown below.

SQUARES OF BINOMIALS	$(x + y)^2 = x^2 + 2xy + y^2$ $(x - y)^2 = x^2 - 2xy + y^2$

This formula is also one that occurs frequently and *should be memorized.* In Section 1.5 on factoring polynomials, you will need to be able to recognize these patterns.

■ *Example 9*

USING THE FORMULAS FOR SQUARES OF BINOMIALS

Find each product.

(a) $(2m + 5)^2 = (2m)^2 + 2(2m)(5) + (5)^2$
$$= 4m^2 + 20m + 25$$

(b) $(3x - 7y^4)^2 = (3x)^2 - 2(3x)(7y^4) + (7y^4)^2$
$$= 9x^2 - 42xy^4 + 49y^8$$ ■

CAUTION   As shown in Example 9, the square of a binomial has three terms. Students often mistakenly give $a^2 + b^2$ as the product of $(a + b)^2$. Be careful to avoid that error.

Use the properties of exponents to simplify each exponential expression. Leave answers with exponents. See Examples 1–3.

1. $(-4)^3 \cdot (-4)^2$ **2.** $(-5)^2 \cdot (-5)^6$ **3.** 2^0 **4.** -2^0 **5.** $(5m)^0$ **6.** $(-4z)^0$

7. $(2^2)^5$ **8.** $(6^4)^3$ **9.** $(2x^5y^4)^3$ **10.** $(-4m^3n^9)^2$ **11.** $-\left(\dfrac{p^4}{q}\right)^2$ **12.** $\left(\dfrac{r^8}{s^2}\right)^3$

Identify each of the following as a polynomial *or* not a polynomial. *For each polynomial, give the degree and identify as* monomial, binomial, trinomial, *or* none of these. *See Example 4.*

13. $5x^{11}$ **14.** $-9y^{12} + y^2$ **15.** $8p^5q + 6pq$

16. $2a^6 - 5a^2 + 4a$ **17.** $\sqrt{2}x + \sqrt{3}x^6$ **18.** $-\sqrt{7}m^5n^2 + 2\sqrt{3}m^4n^2$

19. $\dfrac{1}{3}r^2s - \dfrac{3}{5}r^4s^2 + rs^3$ **20.** $\dfrac{3}{10}p^7 - \dfrac{2}{7}p^5$ **21.** $\dfrac{5}{p} + \dfrac{2}{p^2} - \dfrac{5}{p^3}$

22. $\dfrac{q^2}{6} - \dfrac{q^5}{9} + \dfrac{q^8}{10}$ **23.** $5\sqrt{z} + 2\sqrt{z^3} - 5\sqrt{z^5}$ **24.** $\dfrac{8}{\sqrt{y}} - \dfrac{6}{\sqrt{y^3}} + \dfrac{3}{\sqrt{y^5}}$

Find each of the following sums and differences. See Example 5.

25. $(3x^2 - 4x + 5) + (-2x^2 + 3x - 2)$ **26.** $(4m^3 - 3m^2 + 5) + (-3m^3 - m^2 + 5)$

27. $(12y^2 - 8y + 6) - (3y^2 - 4y + 2)$ **28.** $(8p^2 - 5p) - (3p^2 - 2p + 4)$

29. $(6m^4 - 3m^2 + m) - (2m^3 + 5m^2 + 4m) + (m^2 - m)$

30. $-(8x^3 + x - 3) + (2x^3 + x^2) - (4x^2 + 3x - 1)$

Find each of the following products. See Examples 6–7.

31. $(4r - 1)(7r + 2)$ **32.** $(5m - 6)(3m + 4)$ **33.** $\left(3x - \dfrac{2}{3}\right)\left(5x + \dfrac{1}{3}\right)$

34. $\left(2m - \dfrac{1}{4}\right)\left(3m + \dfrac{1}{2}\right)$ **35.** $4x^2(3x^3 + 2x^2 - 5x + 1)$ **36.** $2b^3(b^2 - 4b + 3)$

37. $(2z - 1)(-z^2 + 3z - 4)$ **38.** $(k + 2)(12k^3 - 3k^2 + k + 1)$ **39.** $(m - n + k)(m + 2n - 3k)$

40. $(r - 3s + t)(2r - s + t)$ **41.** $(a - b + 2c)^2$ **42.** $(k - y + 3m)^2$

43. State the formula for the square of a binomial in words.

44. State the formula for the product of the sum and difference of two terms in words.

Use the special products formulas for each of the following. See Examples 8 and 9.

45. $(2m + 3)(2m - 3)$ **46.** $(8s - 3t)(8s + 3t)$ **47.** $(4m + 2n)^2$

48. $(a - 6b)^2$ **49.** $(5r + 3t^2)^2$ **50.** $(2z^4 - 3y)^2$

51. $[(2p - 3) + q]^2$ **52.** $[(4y - 1) + z]^2$ **53.** $[(3q + 5) - p][(3q + 5) + p]$

54. $[(9r - s) + 2][(9r - s) - 2]$ **55.** $[a + (b + c)][a - (b + c)]$ **56.** $[(k + 2) + r][(k + 2) - r]$

57. $[(3a + b) - 1]^2$ **58.** $[(2m + 7) - n]^2$

Perform the indicated operations.

59. $(6p + 5q)(3p - 7q)$ **60.** $(2p - 1)(3p^2 - 4p + 5)$

61. $(3x - 4y)^3$ **62.** $(r^5 - r^3 + r) + (3r^5 - 4r^4 + r^3 + 2r)$

63. $(6k - 3)^2$ **64.** $(4x + 3y)(4x - 3y)$

65. $(p^3 - 4p^2 + p) - (3p^2 + 2p + 7)$

66. $(2z + y)(3z - 4y)$

67. $(7m + 2n)(7m - 2n)$

68. $(3p + 5)^2$

69. $-3(4q^2 - 3q + 2) + 2(-q^2 + q - 4)$

70. $2(3r^2 + 4r + 2) - 3(-r^2 + 4r - 5)$

71. $p(4p - 6) + 2(3p - 8)$

72. $m(5m - 2) + 9(5 - m)$

73. $-y(y^2 - 4) + 6y^2(2y - 3)$

74. $-z^3(9 - z) + 4z(2 + 3z)$

In each of the following, find the coefficient of x^3 without finding the entire product.

75. $(x^2 + 4x)(-3x^2 + 4x - 1)$

76. $(4x^3 - 2x + 5)(x^3 + 2)$

77. $(1 + x^2)(1 + x)$

78. $(3 - x)(2 - x^2)$

79. $x^2(4 - 3x)^2$

80. $-4x^2(2 - x)(2 + x)$

Find each of the following products. Assume that all variables used in exponents represent nonnegative integers.

81. $(k^m + 2)(k^m - 2)$

82. $(y^x - 4)(y^x + 4)$

83. $(b^r + 3)(b^r - 2)$

84. $(q^y + 4)(q^y + 3)$

85. $(3p^x + 1)(p^x - 2)$

86. $(2^a + 5)(2^a + 3)$

87. $(m^x - 2)^2$

88. $(z^r + 5)^2$

89. $(q^p - 5p^q)^2$

90. $(3y^x - 2x^y)^2$

91. $(3k^a - 2)^3$

92. $(r^x - 4)^3$

Suppose one polynomial has degree m and another has degree n, where m and n are natural numbers with $n < m$. Find the degree of the following for the polynomials.

93. Sum

94. Difference

95. Product

96. What would be the degree of the square of the polynomial of degree m?

1.4 — THE BINOMIAL THEOREM

In this section we introduce a method for writing out the terms of expressions of the form $(x + y)^n$, where n is a natural number. The formula for writing out the powers of a binomial $(x + y)^n$ as a polynomial is called the *binomial theorem*. This theorem is important when working with probability and statistics. Some expansions of $(x + y)^n$, for various nonnegative integer values of n, are given below.

$$(x + y)^0 = 1$$
$$(x + y)^1 = x + y$$
$$(x + y)^2 = x^2 + 2xy + y^2$$
$$(x + y)^3 = x^3 + 3x^2y + 3xy^2 + y^3$$
$$(x + y)^4 = x^4 + 4x^3y + 6x^2y^2 + 4xy^3 + y^4$$
$$(x + y)^5 = x^5 + 5x^4y + 10x^3y^2 + 10x^2y^3 + 5xy^4 + y^5$$

Studying these results reveals a pattern that can be used to write a general expression for $(x + y)^n$.

First, notice that after the special case $(x + y)^0 = 1$, each expression begins with x raised to the same power as the binomial itself. That is, the expansion of $(x + y)^1$ has a first term of x^1, $(x + y)^2$ has a first term of x^2, $(x + y)^3$ has a first term of x^3, and so on. Also, the last term in each expansion is y to the same power as the binomial. Thus the expansion of $(x + y)^n$ should begin with the term x^n and end with the term y^n.

Also, the exponents on x decrease by one in each term after the first, while the exponents on y, beginning with y in the second term, increase by one in each succeeding term. That is, the *variables* in the terms of the expansion of $(x + y)^n$ have the following pattern.

$$x^n, x^{n-1}y, x^{n-2}y^2, x^{n-3}y^3, \ldots, xy^{n-1}, y^n$$

This pattern suggests that the sum of the exponents on x and y in each term is n. For example, in the third term in the list above, the variable is $x^{n-2}y^2$, and the sum of the exponents is $n - 2 + 2 = n$.

Now examine the *coefficients* in the terms of the expansions shown above. Writing the coefficients alone gives the following pattern.

PASCAL'S TRIANGLE

$$
\begin{array}{ccccccccccc}
 & & & & & 1 & & & & & \\
 & & & & 1 & & 1 & & & & \\
 & & & 1 & & 2 & & 1 & & & \\
 & & 1 & & 3 & & 3 & & 1 & & \\
 & 1 & & 4 & & 6 & & 4 & & 1 & \\
1 & & 5 & & 10 & & 10 & & 5 & & 1 \\
\end{array}
$$

With the coefficients arranged in this way, it can be seen that each number in the triangle is the sum of the two numbers directly above it (one to the right and one to the left.) For example, if we number the rows starting with row 0, in row four, 1 is the sum of 1, the only number above it, 4 is the sum of 1 and 3, 6 is the sum of 3 and 3, and so on. This triangular array of numbers is called **Pascal's triangle,** in honor of the seventeenth-century mathematician Blaise Pascal (1623–1662), one of the first to use it extensively.

To get the coefficients for $(x + y)^6$, we add row six in the array of numbers given above. Adding adjacent numbers, we find that row six is

$$1 \quad 6 \quad 15 \quad 20 \quad 15 \quad 6 \quad 1.$$

Using these coefficients, the expansion of $(x + y)^6$ is

$$(x + y)^6 = x^6 + 6x^5y + 15x^4y^2 + 20x^3y^3 + 15x^2y^4 + 6xy^5 + y^6.$$

Although it is possible to use Pascal's triangle to find the coefficients of $(x + y)^n$ for any positive integer value of n, this becomes impractical for large values of n because of the need to write out all the preceding rows. A more efficient way of finding these coefficients uses factorial notation. The number $n!$ (read "n-factorial") is defined as follows.

***n*-FACTORIAL**   For any positive integer n:

$$n! = n(n - 1)(n - 2) \ldots (3)(2)(1),$$

and

$$0! = 1.$$

■ **Example 1**

EVALUATING
FACTORIALS

Evaluate each factorial.

(a) $5! = 5 \cdot 4 \cdot 3 \cdot 2 \cdot 1 = 120$ **(b)** $7! = 7 \cdot 6 \cdot 5 \cdot 4 \cdot 3 \cdot 2 \cdot 1 = 5040$

(c) $2! = 2 \cdot 1 = 2$ **(d)** $1! = 1$ ■

| NOTE | Many scientific calculators have a key to evaluate factorials. |

Now look at the coefficients of the expression

$$(x + y)^5 = x^5 + 5x^4y + 10x^3y^2 + 10x^2y^3 + 5xy^4 + y^5.$$

The coefficient on the second term, $5x^4y$, is 5, and the exponents on the variables are 4 and 1. Note that

$$5 = \frac{5!}{4!1!}.$$

The coefficient on the third term is 10, with exponents of 3 and 2, and

$$10 = \frac{5!}{3!2!}.$$

The last term (the sixth term) can be written as $y^5 = 1x^0y^5$, with coefficient 1, and exponents of 0 and 5. Since $0! = 1$, check that

$$1 = \frac{5!}{0!5!}.$$

Generalizing from these examples, we find that the coefficient for the term of the expansion of $(x + y)^n$ in which the variable part is x^ry^{n-r} (where $r \le n$) will be

$$\frac{n!}{r!(n - r)!}.$$

This number, called a **binomial coefficient,** is often written as $\binom{n}{r}$ (read "n above r").

BINOMIAL
COEFFICIENT

For nonnegative integers n and r, with $r \le n$, the symbol $\binom{n}{r}$ is defined as

$$\binom{n}{r} = \frac{n!}{r!(n - r)!}.$$

These binomial coefficients are just numbers from Pascal's triangle. For example, $\binom{3}{0}$ is the first number in row three, and $\binom{7}{4}$ is the fifth number in row seven. Notice that $\binom{7}{3} = \binom{7}{4}$, and in general,

$$\binom{n}{r} = \binom{n}{n - r},$$

for positive integers n and r, $n \ge r$.

30

■ ***Example 2***

EVALUATING
BINOMIAL
COEFFICIENTS

Evaluate each of the following.

(a) $\dfrac{8!}{6!2!} = \dfrac{8 \cdot 7 \cdot 6 \cdot 5 \cdot 4 \cdot 3 \cdot 2 \cdot 1}{6 \cdot 5 \cdot 4 \cdot 3 \cdot 2 \cdot 1 \cdot 2 \cdot 1} = \dfrac{8 \cdot 7}{2 \cdot 1} = 28$

(b) $\dfrac{4!}{3!1!} = \dfrac{4 \cdot 3 \cdot 2 \cdot 1}{3 \cdot 2 \cdot 1 \cdot 1} = 4$

(c) $\dbinom{6}{2} = \dfrac{6!}{2!4!} = \dfrac{6 \cdot 5 \cdot 4 \cdot 3 \cdot 2 \cdot 1}{2 \cdot 1 \cdot 4 \cdot 3 \cdot 2 \cdot 1} = 15$

(d) $\dbinom{7}{5} = \dfrac{7!}{5!2!} = \dfrac{7 \cdot 6 \cdot 5 \cdot 4 \cdot 3 \cdot 2 \cdot 1}{5 \cdot 4 \cdot 3 \cdot 2 \cdot 1 \cdot 2 \cdot 1} = 21$ ■

NOTE

The $_nC_r$ key on many calculators gives $\dbinom{n}{r}$.

Our observations about the expansion of $(x + y)^n$ are summarized as follows.

EXPANSION OF
$(x + y)^n$

1. There are $(n + 1)$ terms in the expansion.
2. The first term is x^n, and the last term is y^n.
3. The exponent on x decreases by 1 and the exponent on y increases by 1 in each succeeding term.
4. The sum of the exponents on x and y in any term is n.
5. The coefficient of $x^r y^{n-r}$ is $\dbinom{n}{r}$.

These observations about the expansion of $(x + y)^n$ for any positive integer value of n suggest the **binomial theorem.**

BINOMIAL
THEOREM

For any positive integer n:

$$(x + y)^n = x^n + \binom{n}{n-1}x^{n-1}y + \binom{n}{n-2}x^{n-2}y^2 + \binom{n}{n-3}x^{n-3}y^3$$

$$+ \cdots + \binom{n}{n-r}x^{n-r}y^r + \cdots + \binom{n}{1}xy^{n-1} + y^n.$$

The binomial theorem can be proved using mathematical induction (discussed in Section 9.5). Note that the expansion of $(x + y)^n$ has $n + 1$ terms as expected.

CAUTION

Avoid the common error of expanding $(x + y)^n$ as $x^n + y^n$. Assuming that x and y are nonzero, these expressions are equivalent only if $n = 1$.

■ **Example 3**

EXPANDING A
BINOMIAL

Write out the binomial expansion of $(x + y)^9$.

Using the binomial theorem,

$$(x + y)^9 = x^9 + \binom{9}{8} x^8 y + \binom{9}{7} x^7 y^2 + \binom{9}{6} x^6 y^3 + \binom{9}{5} x^5 y^4$$

$$+ \binom{9}{4} x^4 y^5 + \binom{9}{3} x^3 y^6 + \binom{9}{2} x^2 y^7 + \binom{9}{1} xy^8 + y^9.$$

Now evaluate each coefficient.

$$(x + y)^9 = x^9 + \frac{9!}{8!1!} x^8 y + \frac{9!}{7!2!} x^7 y^2 + \frac{9!}{6!3!} x^6 y^3 + \frac{9!}{5!4!} x^5 y^4$$

$$+ \frac{9!}{4!5!} x^4 y^5 + \frac{9!}{3!6!} x^3 y^6 + \frac{9!}{2!7!} x^2 y^7 + \frac{9!}{1!8!} xy^8 + y^9$$

$$= x^9 + 9x^8 y + 36x^7 y^2 + 84x^6 y^3 + 126x^5 y^4 + 126x^4 y^5$$

$$+ 84x^3 y^6 + 36x^2 y^7 + 9xy^8 + y^9 \quad ■$$

■ **Example 4**

EXPANDING A
BINOMIAL

Expand $\left(a - \dfrac{b}{2} \right)^5$.

Again, use the binomial theorem.

$$\left(a - \frac{b}{2} \right)^5 = a^5 + \binom{5}{4} a^4 \left(-\frac{b}{2} \right) + \binom{5}{3} a^3 \left(-\frac{b}{2} \right)^2 + \binom{5}{2} a^2 \left(-\frac{b}{2} \right)^3$$

$$+ \binom{5}{1} a \left(-\frac{b}{2} \right)^4 + \left(-\frac{b}{2} \right)^5$$

$$= a^5 + 5a^4 \left(-\frac{b}{2} \right) + 10a^3 \left(-\frac{b}{2} \right)^2 + 10a^2 \left(-\frac{b}{2} \right)^3$$

$$+ 5a \left(-\frac{b}{2} \right)^4 + \left(-\frac{b}{2} \right)^5$$

$$= a^5 - \frac{5}{2} a^4 b + \frac{5}{2} a^3 b^2 - \frac{5}{4} a^2 b^3 + \frac{5}{16} ab^4 - \frac{1}{32} b^5 \quad ■$$

*r*TH TERM   Any single term of a binomial expansion can be determined without writing out the whole expansion. For example, the seventh term of $(x + y)^9$ has y raised to the sixth power (since y has the power 1 in the second term, the power 2 in the third term, and so on). The exponents on x and y in each term must have a sum of 9, so the exponent on x in the seventh term is $9 - 6 = 3$. Thus, writing the coefficient as given in the binomial theorem, the seventh term should be

$$\frac{9!}{3!6!} x^3 y^6.$$

This is in fact the seventh term of $(x + y)^9$ found in Example 3 above. This discussion suggests the next theorem.

*r*TH TERM OF THE BINOMIAL EXPANSION	The *r*th term of the binomial expansion of $(x + y)^n$, where $n \geq r - 1$, is $$\binom{n}{n - (r - 1)} x^{n-(r-1)} y^{r-1}.$$

■ ***Example 5***

FINDING A SPECIFIC TERM

Find the fourth term of $(a + 2b)^{10}$.

In the fourth term $2b$ has an exponent of 3, while a has an exponent of $10 - 3$, or 7. Using $n = 10$, $r = 4$, $x = a$, and $y = 2b$ in the formula above, we find that the fourth term is

$$\binom{10}{7} a^7 (2b)^3 = 120a^7(8b^3) = 960a^7b^3. \quad ■$$

1.4 EXERCISES ■

Evaluate each of the following. See Examples 1 and 2.

1. $\dfrac{6!}{3!3!}$ **2.** $\dfrac{5!}{2!3!}$ **3.** $\dfrac{7!}{3!4!}$ **4.** $\dfrac{8!}{5!3!}$

5. $\dbinom{8}{3}$ **6.** $\dbinom{7}{4}$ **7.** $\dbinom{10}{8}$ **8.** $\dbinom{9}{6}$

9. $\dbinom{n}{n-1}$ **10.** $\dbinom{n}{n-2}$

11. Describe in your own words how you would determine the binomial coefficient for the fifth term in the expansion of $(x + y)^8$.

12. How many terms are there in the expansion of $(x + y)^{10}$?

13. What is true of the signs (positive/negative) of the terms in the expansion of $(x - y)^n$?

14. In the expansion of $(a + 5b)^n$, what quantity replaces y in the binomial theorem, 5 or $5b$?

Write out the binomial expansion for each of the following. See Examples 3 and 4.

15. $(x + y)^6$ **16.** $(m + n)^4$ **17.** $(p - q)^5$ **18.** $(a - b)^7$

19. $(r^2 + s)^5$ **20.** $(m + n^2)^4$ **21.** $(p + 2q)^4$ **22.** $(3r - s)^6$

23. $(7p + 2q)^4$ **24.** $(4a - 5b)^5$ **25.** $(3x - 2y)^6$ **26.** $(7k - 9j)^4$

27. $\left(\dfrac{m}{2} - 1\right)^6$ **28.** $\left(3 + \dfrac{y}{3}\right)^5$ **29.** $\left(\sqrt{2}r + \dfrac{1}{m}\right)^4$ **30.** $\left(\dfrac{1}{k} - \sqrt{3}p\right)^3$

For each of the following, write the indicated term of the binomial expansion. See Example 5.

31. Fifth term of $(m - 2p)^{12}$ **32.** Fourth term of $(3x + y)^6$

33. Sixth term of $(x + y)^9$ **34.** Twelfth term of $(a - b)^{15}$

35. Ninth term of $(2m + n)^{10}$ **36.** Seventh term of $(3r - 5s)^{12}$

37. Seventeenth term of $(p^2 + q)^{20}$ **38.** Tenth term of $(2x^2 + y)^{14}$

39. Eighth term of $(x^3 + 2y)^{14}$ **40.** Thirteenth term of $(a + 2b^3)^{12}$

📱 *In calculus it is shown that*

$$(1 + x)^n = 1 + nx + \frac{n(n - 1)}{2!}x^2 + \frac{n(n - 1)(n - 2)}{3!}x^3 + \cdots$$

for any real number n (not just positive integers) and any real number x, where $|x| < 1$. This result, a generalized binomial theorem, may be used to find approximate values of powers or roots. For example,

$$\sqrt[4]{630} = (625 + 5)^{1/4} = \left[625\left(1 + \frac{5}{625}\right)\right]^{1/4} = 625^{1/4}\left(1 + \frac{5}{625}\right)^{1/4}.$$

41. Use the expression given above for $(1 + x)^n$ to approximate $\left(1 + \frac{5}{625}\right)^{1/4}$ to the nearest thousandth. Then approximate $\sqrt[4]{630}$.

42. Approximate $\sqrt[3]{9.42}$, using this method.

43. Approximate $(1.02)^{-3}$.

44. Show that $\binom{n}{2} = \frac{n(n - 1)}{2!}$ and $\binom{n}{3} = \frac{n(n - 1)(n - 2)}{3!}$.

1.5 ——————— **FACTORING**

The process of finding polynomials whose product equals a given polynomial is called **factoring.** For example, since $4x + 12 = 4(x + 3)$, both 4 and $x + 3$ are called **factors** of $4x + 12$. Also, $4(x + 3)$ is called the **factored form** of $4x + 12$. A polynomial that cannot be written as a product of two polynomials with integer coefficients is a **prime** or **irreducible polynomial.** A polynomial is **factored completely** when it is written as a product of prime polynomials with integer coefficients.

FACTORING OUT THE GREATEST COMMON FACTOR   Polynomials are factored by using the distributive property. For example, to factor $6x^2y^3 + 9xy^4 + 18y^5$, we look for a monomial that is the greatest common factor of each term. The terms of this polynomial have $3y^3$ as the greatest common factor. By the distributive property,

$$6x^2y^3 + 9xy^4 + 18y^5 = (3y^3)(2x^2) + (3y^3)(3xy) + (3y^3)(6y^2)$$
$$= 3y^3(2x^2 + 3xy + 6y^2).$$

∎ *Example 1*

FACTORING OUT THE GREATEST COMMON FACTOR

Factor out the greatest common factor from each polynomial.

(a) $9y^5 + y^2$

The greatest common factor is y^2.

$$9y^5 + y^2 = y^2 \cdot 9y^3 + y^2 \cdot 1$$
$$= y^2(9y^3 + 1)$$

(b) $6x^2t + 8xt + 12t = 2t(3x^2 + 4x + 6)$

(c) $14m^4(m + 1) - 28m^3(m + 1) - 7m^2(m + 1)$

The greatest common factor is $7m^2(m + 1)$. Use the distributive property as follows.

$$14m^4(m + 1) - 28m^3(m + 1) - 7m^2(m + 1)$$
$$= [7m^2(m + 1)](2m^2 - 4m - 1)$$
$$= 7m^2(m + 1)(2m^2 - 4m - 1) \quad ■$$

CAUTION

In Example 1(a), avoid the common error of forgetting to include the 1. Since $y^2(9y^3) \neq 9y^5 + y^2$, the 1 is essential in the answer. Remember that factoring can always be checked by multiplication.

FACTORING BY GROUPING When a polynomial has more than three terms, it can sometimes be factored by a method called **factoring by grouping.** For example, to factor

$$ax + ay + 6x + 6y,$$

collect the terms into two groups so that each group has a common factor.

$$ax + ay + 6x + 6y = (ax + ay) + (6x + 6y)$$

Factor each group, getting

$$ax + ay + 6x + 6y = a(x + y) + 6(x + y).$$

The quantity $(x + y)$ is now a common factor, which can be factored out, producing

$$ax + ay + 6x + 6y = (x + y)(a + 6).$$

It is not always obvious which terms should be grouped. Experience and repeated trials are the most reliable tools for factoring by grouping.

■ ***Example 2***

FACTORING BY GROUPING

Factor by grouping.

(a) $mp^2 + 7m + 3p^2 + 21$

Group the terms as follows.

$$mp^2 + 7m + 3p^2 + 21 = (mp^2 + 7m) + (3p^2 + 21)$$

Factor out the greatest common factor from each group.

$$(mp^2 + 7m) + (3p^2 + 21) = m(p^2 + 7) + 3(p^2 + 7)$$
$$= (p^2 + 7)(m + 3) \qquad p^2 + 7 \text{ is a}$$
$$\text{common factor.}$$

(b) $2y^2 - 2z - ay^2 + az$

Grouping terms as above gives

$$2y^2 - 2z - ay^2 + az = (2y^2 - 2z) + (-ay^2 + az)$$
$$= 2(y^2 - z) + a(-y^2 + z).$$

The expression $-y^2 + z$ is the negative of $y^2 - z$, so the terms should be grouped as follows.

$$2y^2 - 2z - ay^2 + az = (2y^2 - 2z) - (ay^2 - az)$$
$$= 2(y^2 - z) - a(y^2 - z) \qquad \text{Factor each group.}$$
$$= (y^2 - z)(2 - a). \qquad \text{Factor out } y^2 - z. \; ■$$

Later in this section we show another way to factor by grouping three of the four terms.

FACTORING TRINOMIALS Factoring is the opposite of multiplication. Since the product of two binomials is usually a trinomial, we can expect factorable trinomials (that have terms with no common factor) to have two binomial factors. Thus, factoring trinomials requires using FOIL backwards.

■ **Example 3**

FACTORING TRINOMIALS

Factor each trinomial.

(a) $4y^2 - 11y + 6$

To factor this polynomial, we must find integers a, b, c, and d such that

$$4y^2 - 11y + 6 = (ay + b)(cy + d).$$

By using FOIL, we see that $ac = 4$ and $bd = 6$. The positive factors of 4 are 4 and 1 or 2 and 2. Since the middle term is negative, we consider only negative factors of 6. The possibilities are -2 and -3 or -1 and -6. Now we try various arrangements of these factors until we find one that gives the correct coefficient of y.

$$(2y - 1)(2y - 6) = 4y^2 - 14y + 6 \qquad \text{Incorrect}$$
$$(2y - 2)(2y - 3) = 4y^2 - 10y + 6 \qquad \text{Incorrect}$$
$$(y - 2)(4y - 3) = 4y^2 - 11y + 6 \qquad \text{Correct}$$

The last trial gives the correct factorization.

(b) $6p^2 - 7p - 5$

Again, we try various possibilities. The positive factors of 6 could be 2 and 3 or 1 and 6. As factors of -5 we have only -1 and 5 or -5 and 1. Try different combinations of these factors until the correct one is found.

$$(2p - 5)(3p + 1) = 6p^2 - 13p - 5 \qquad \text{Incorrect}$$
$$(3p - 5)(2p + 1) = 6p^2 - 7p - 5 \qquad \text{Correct}$$

Finally, $6p^2 - 7p - 5$ factors as $(3p - 5)(2p + 1)$. ■

NOTE In Example 3, we chose positive factors of the positive first term. Of course, we could have used two negative factors, but the work is easier if positive factors are used.

Each of the special patterns of multiplication given earlier can be used in reverse to get a pattern for factoring. Perfect square trinomials can be factored as follows.

| PERFECT SQUARE TRINOMIALS | $x^2 + 2xy + y^2 = (x + y)^2$ |
| | $x^2 - 2xy + y^2 = (x - y)^2$ |

These formulas should be memorized.

■ *Example 4*

FACTORING PERFECT SQUARE TRINOMIALS

Factor each polynomial.

(a) $16p^2 - 40pq + 25q^2$

Since $16p^2 = (4p)^2$ and $25q^2 = (5q)^2$, use the second pattern shown above with $4p$ replacing x and $5q$ replacing y to get

$$16p^2 - 40pq + 25q^2 = (4p)^2 - 2(4p)(5q) + (5q)^2$$
$$= (4p - 5q)^2.$$

Make sure that the middle term of the trinomial being factored, $-40pq$ here, is twice the product of the two terms in the binomial $4p - 5q$.

$$-40pq = 2(4p)(-5q)$$

(b) $169x^2 + 104xy^2 + 16y^4 = (13x + 4y^2)^2$, since $2(13x)(4y^2) = 104xy^2$. ■

FACTORING BINOMIALS The pattern for the product of the sum and difference of two terms gives the following factorization.

| DIFFERENCE OF TWO SQUARES | $x^2 - y^2 = (x + y)(x - y)$ |

■ *Example 5*

FACTORING A DIFFERENCE OF SQUARES

Factor each of the following polynomials.

(a) $4m^2 - 9$

First, recognize that $4m^2 - 9$ is the difference of two squares, since $4m^2 = (2m)^2$ and $9 = 3^2$. Use the pattern for the difference of two squares with $2m$ replacing x and 3 replacing y. Doing this gives

$$4m^2 - 9 = (2m)^2 - 3^2$$
$$= (2m + 3)(2m - 3).$$

(b) $256k^4 - 625m^4$

Use the difference of two squares pattern twice, as follows:

$$256k^4 - 625m^4 = (16k^2)^2 - (25m^2)^2$$
$$= (16k^2 + 25m^2)(16k^2 - 25m^2)$$
$$= (16k^2 + 25m^2)(4k + 5m)(4k - 5m).$$

(c) $(a + 2b)^2 - 4c^2 = (a + 2b)^2 - (2c)^2$

$= [(a + 2b) + 2c][(a + 2b) - 2c]$

$= (a + 2b + 2c)(a + 2b - 2c)$

(d) $y^{4q} - z^{2q} = (y^{2q} + z^q)(y^{2q} - z^q)$ (if q is a positive integer)

(e) $x^2 - 6x + 9 - y^4$

Group the first three terms to get a perfect square trinomial. Then use the difference of squares pattern.

$$x^2 - 6x + 9 - y^4 = (x^2 - 6x + 9) - y^4$$
$$= (x - 3)^2 - (y^2)^2$$
$$= [(x - 3) + y^2][(x - 3) - y^2]$$
$$= (x - 3 + y^2)(x - 3 - y^2)   ▪$$

Two other special results of factoring are listed below. Each can be verified by multiplying on the right side of the equation.

DIFFERENCE AND SUM OF TWO CUBES	Difference of two cubes	$x^3 - y^3 = (x - y)(x^2 + xy + y^2)$
	Sum of two cubes	$x^3 + y^3 = (x + y)(x^2 - xy + y^2)$

▪ **Example 6**

FACTORING THE SUM OR DIFFERENCE OF CUBES

Factor each polynomial.

(a) $x^3 + 27$

Notice that $27 = 3^3$, so the expression is a sum of two cubes. Use the second pattern given above.

$$x^3 + 27 = x^3 + 3^3 = (x + 3)(x^2 - 3x + 9)$$

(b) $m^3 - 64n^3$

Since $64n^3 = (4n)^3$, the given polynomial is a difference of two cubes. To factor, use the first pattern in the box above, replacing x with m and y with $4n$.

$$m^3 - 64n^3 = m^3 - (4n)^3$$
$$= (m - 4n)[m^2 + m(4n) + (4n)^2]$$
$$= (m - 4n)(m^2 + 4mn + 16n^2)$$

(c) $8q^6 + 125p^9$

Write $8q^6$ as $(2q^2)^3$ and $125p^9$ as $(5p^3)^3$, so that the given polynomial is a sum of two cubes.

$$8q^6 + 125p^9 = (2q^2)^3 + (5p^3)^3$$
$$= (2q^2 + 5p^3)[(2q^2)^2 - (2q^2)(5p^3) + (5p^3)^2]$$
$$= (2q^2 + 5p^3)(4q^4 - 10q^2p^3 + 25p^6)   ▪$$

METHOD OF SUBSTITUTION   Sometimes a polynomial can be factored by substituting one expression for another. The next example shows this **method of substitution.**

■ ***Example 7***
FACTORING BY
SUBSTITUTION

Factor each polynomial.

(a) $6z^4 - 13z^2 - 5$

Replace z^2 with y, so that $y^2 = (z^2)^2 = z^4$. This replacement gives

$$6z^4 - 13z^2 - 5 = 6y^2 - 13y - 5.$$

Factor $6y^2 - 13y - 5$ as

$$6y^2 - 13y - 5 = (2y - 5)(3y + 1).$$

Replacing y with z^2 gives

$$6z^4 - 13z^2 - 5 = (2z^2 - 5)(3z^2 + 1).$$

(Some students prefer to factor this type of trinomial directly using trial and error with FOIL.)

(b) $10(2a - 1)^2 - 19(2a - 1) - 15$

Replacing $2a - 1$ with m gives

$$10m^2 - 19m - 15 = (5m + 3)(2m - 5).$$

Now replace m with $2a - 1$ in the factored form and simplify.

$$
\begin{aligned}
10(2a - 1)^2 - 19(2a - 1) - 15 &= [5(2a - 1) + 3][2(2a - 1) - 5] &&\text{Let } m = 2a - 1. \\
&= (10a - 5 + 3)(4a - 2 - 5) &&\text{Multiply.} \\
&= (10a - 2)(4a - 7) &&\text{Add.} \\
&= 2(5a - 1)(4a - 7) &&\text{Factor out the common factor.}
\end{aligned}
$$

(c) $(2a - 1)^3 + 8$

Let $2a - 1 = K$ to get

$$
\begin{aligned}
(2a - 1)^3 + 8 &= K^3 + 8 \\
&= K^3 + 2^3 \\
&= (K + 2)(K^2 - 2K + 2^2).
\end{aligned}
$$

Replacing K with $2a - 1$ gives

$$
\begin{aligned}
(2a - 1)^3 + 8 &= (2a - 1 + 2)[(2a - 1)^2 - 2(2a - 1) + 4] &&\text{Let } K = 2a - 1. \\
&= (2a + 1)(4a^2 - 4a + 1 - 4a + 2 + 4) &&\text{Multiply.} \\
&= (2a + 1)(4a^2 - 8a + 7). &&\text{Combine terms.} ■
\end{aligned}
$$

Factor the greatest common factor from each polynomial. See Example 1.

1. $4k^2m^3 + 8k^4m^3 - 12k^2m^4$ **2.** $28r^4s^2 + 7r^3s - 35r^4s^3$

3. $2(a + b) + 4m(a + b)$ **4.** $4(y - 2)^2 + 3(y - 2)$

5. $(2y - 3)(y + 2) + (y + 5)(y + 2)$ **6.** $(6a - 1)(a + 2) + (6a - 1)(3a - 1)$

7. $(5r - 6)(r + 3) - (2r - 1)(r + 3)$ **8.** $(3z + 2)(z + 4) - (z + 6)(z + 4)$

9. $2(m - 1) - 3(m - 1)^2 + 2(m - 1)^3$ **10.** $5(a + 3)^3 - 2(a + 3) + (a + 3)^2$

Factor each of the following by grouping. See Example 2.

11. $6st + 9t - 10s - 15$ **12.** $10ab - 6b + 35a - 21$ **13.** $rt^3 + rs^2 - pt^3 - ps^2$

14. $2m^4 + 6 - am^4 - 3a$ **15.** $6p^2 - 14p + 15p - 35$ **16.** $8r^2 - 10r + 12r - 15$

17. $20z^2 - 8zx - 45zx + 18x^2$ **18.** $16a^2 + 10ab - 24ab - 15b^2$ **19.** $15 - 5m^2 - 3r^2 + m^2r^2$

Factor each trinomial. See Example 3.

20. $6a^2 - 48a - 120$ **21.** $8h^2 - 24h - 320$ **22.** $3m^3 + 12m^2 + 9m$

23. $9y^4 - 54y^3 + 45y^2$ **24.** $6k^2 + 5kp - 6p^2$ **25.** $14m^2 + 11mr - 15r^2$

26. $5a^2 - 7ab - 6b^2$ **27.** $12s^2 + 11st - 5t^2$ **28.** $9x^2 - 6x^3 + x^4$

29. $30a^2 + am - m^2$ **30.** $24a^4 + 10a^3b - 4a^2b^2$ **31.** $18x^5 + 15x^4z - 75x^3z^2$

32. When asked to factor $6x^4 - 3x^2 - 3$ completely, a student gave the following result: $6x^4 - 3x^2 - 3 = (2x^2 + 1)(3x^2 - 3)$. Is this answer correct? Explain why.

Factor each perfect square trinomial. See Example 4.

33. $9m^2 - 12m + 4$ **34.** $16p^2 - 40p + 25$ **35.** $32a^2 - 48ab + 18b^2$

36. $20p^2 - 100pq + 125q^2$ **37.** $4x^2y^2 + 28xy + 49$ **38.** $9m^2n^2 - 12mn + 4$

39. $(a - 3b)^2 - 6(a - 3b) + 9$ **40.** $(2p + q)^2 - 10(2p + q) + 25$

Factor each difference of two squares. See Example 5.

41. $9a^2 - 16$ **42.** $16q^2 - 25$ **43.** $25s^4 - 9t^2$ **44.** $36z^2 - 81y^4$

45. $(a + b)^2 - 16$ **46.** $(p - 2q)^2 - 100$ **47.** $p^4 - 625$ **48.** $m^4 - 81$

49. Which of the following is the correct complete factorization of $x^4 - 1$?
   (a) $(x^2 - 1)(x^2 + 1)$ **(b)** $(x^2 + 1)(x + 1)(x - 1)$
   (c) $(x^2 - 1)^2$ **(d)** $(x - 1)^2(x + 1)^2$

50. Which of the following is the correct factorization of $x^3 + 8$?
   (a) $(x + 2)^3$ **(b)** $(x + 2)(x^2 + 2x + 4)$
   (c) $(x + 2)(x^2 - 2x + 4)$ **(d)** $(x + 2)(x^2 - 4x + 4)$

Factor each sum or difference of cubes. See Example 6.

51. $8 - a^3$ **52.** $r^3 + 27$ **53.** $125x^3 - 27$ **54.** $8m^3 - 27n^3$

55. $27y^9 + 125z^6$ **56.** $27z^3 + 729y^3$ **57.** $(r + 6)^3 - 216$ **58.** $(b + 3)^3 - 27$

59. $27 - (m + 2n)^3$ **60.** $125 - (4a - b)^3$

61. Is the following factorization of $3a^4 + 14a^2 - 5$ correct? Explain. If it is incorrect, give the correct factors.
$$3a^4 + 14a^2 - 5 = 3u^2 + 14u - 5 \quad \text{Let } u = a^2.$$
$$= (3u - 1)(u + 5)$$

40

Factor each of the following, using the method of substitution. See Example 7.

62. $m^4 - 3m^2 - 10$ **63.** $a^4 - 2a^2 - 48$ **64.** $7(3k - 1)^2 + 26(3k - 1) - 8$

65. $6(4z - 3)^2 + 7(4z - 3) - 3$ **66.** $9(a - 4)^2 + 30(a - 4) + 25$ **67.** $20(4 - p)^2 - 3(4 - p) - 2$

Factor by any method.

68. $a^3(r + s) + b^2(r + s)$ **69.** $4b^2 + 4bc + c^2 - 16$ **70.** $(2y - 1)^2 - 4(2y - 1) + 4$

71. $x^2 + xy - 5x - 5y$ **72.** $8r^2 - 3rs + 10s^2$ **73.** $p^4(m - 2n) + q(m - 2n)$

74. $36a^2 + 60a + 25$ **75.** $4z^2 + 28z + 49$ **76.** $6p^4 + 7p^2 - 3$

77. $1000x^3 + 343y^3$ **78.** $b^2 + 8b + 16 - a^2$ **79.** $125m^6 - 216$

80. $q^2 + 6q + 9 - p^2$ **81.** $12m^2 + 16mn - 35n^2$ **82.** $216p^3 + 125q^3$

83. $4p^2 + 3p - 1$ **84.** $100r^2 - 169s^2$ **85.** $144z^2 + 121$

86. $(3a + 5)^2 - 18(3a + 5) + 81$ **87.** $(x + y)^2 - (x - y)^2$ **88.** $4z^4 - 7z^2 - 15$

Factor each of the following polynomials. Assume that all variables used in exponents represent positive integers. (See Example 5(d).)

89. $r^2 + rs^q - 6s^{2q}$ **90.** $6z^{2a} - z^a x^b - x^{2b}$ **91.** $9a^{4k} - b^{8k}$

92. $16y^{2c} - 25x^{4c}$ **93.** $4y^{2a} - 12y^a + 9$ **94.** $25x^{4c} - 20x^{2c} + 4$

95. $6(m + p)^{2k} + (m + p)^k - 15$ **96.** $8(2k + q)^{4z} - 2(2k + q)^{2z} - 3$

Find a value of b or c that will make the following expressions perfect square trinomials.

97. $4z^2 + bz + 81$ **98.** $9p^2 + bp + 25$ **99.** $100r^2 - 60r + c$ **100.** $49x^2 + 70x + c$

1.6 ———— RATIONAL EXPRESSIONS

An expression that is the quotient of two algebraic expressions (with denominator not 0) is called a **fractional expression.** The most common fractional expressions are those that are the quotients of two polynomials; these are called **rational expressions.** Since fractional expressions involve quotients, it is important to keep track of values of the variable that satisfy the requirement that no denominator be 0. For example, $x \neq -2$ in the rational expression

$$\frac{x + 6}{x + 2}$$

because replacing x with -2 makes the denominator equal 0. Similarly, in

$$\frac{(x + 6)(x + 4)}{(x + 2)(x + 4)}$$

$x \neq -2$ and $x \neq -4$.

The restrictions on the variable are found by determining the values that make the denominator equal to zero. In the second example above, finding the values of x that make $(x + 2)(x + 4) = 0$ requires using the property that $ab = 0$ if and only if $a = 0$ or $b = 0$, as follows.

$$(x + 2)(x + 4) = 0$$

$$x + 2 = 0 \quad \text{or} \quad x + 4 = 0$$

$$x = -2 \quad \text{or} \quad x = -4$$

Just as the fraction 6/8 is written in lowest terms as 3/4, rational expressions may also be written in lowest terms. This is done with the fundamental principle.

FUNDAMENTAL PRINCIPLE OF FRACTIONS	$\dfrac{ac}{bc} = \dfrac{a}{b}$ $(b \neq 0, c \neq 0)$

■ **Example 1**
WRITING IN LOWEST TERMS

Write each expression in lowest terms.

(a) $\dfrac{2p^2 + 7p - 4}{5p^2 + 20p}$

Factor the numerator and denominator to get

$$\frac{2p^2 + 7p - 4}{5p^2 + 20p} = \frac{(2p - 1)(p + 4)}{5p(p + 4)}.$$

By the fundamental principle,

$$\frac{2p^2 + 7p - 4}{5p^2 + 20p} = \frac{2p - 1}{5p}.$$

In the original expression p cannot be 0 or -4, because $5p^2 + 20p \neq 0$, so this result is valid only for values of p other than 0 and -4. From now on, we shall always assume such restrictions when reducing rational expressions.

(b) $\dfrac{6 - 3k}{k^2 - 4}$

Factor to get $\dfrac{6 - 3k}{k^2 - 4} = \dfrac{3(2 - k)}{(k + 2)(k - 2)}.$

The factors $2 - k$ and $k - 2$ have opposite signs. Because of this, multiply numerator and denominator by -1, as follows.

$$\frac{6 - 3k}{k^2 - 4} = \frac{3(2 - k)(-1)}{(k + 2)(k - 2)(-1)}$$

Since $(k - 2)(-1) = -k + 2$, or $2 - k$,

$$\frac{6 - 3k}{k^2 - 4} = \frac{3(2 - k)(-1)}{(k + 2)(2 - k)},$$

giving $\dfrac{6 - 3k}{k^2 - 4} = \dfrac{-3}{k + 2}.$

Working in an alternative way would lead to the equivalent result

$$\frac{3}{-k - 2}. \quad ■$$

Probably the most common error made in algebra is the incorrect use of the fundamental principle to write a fraction in lowest terms. Remember, the fundamental principle requires a pair of common *factors,* one in the numerator and one in the denominator. For example,

$$\frac{2x + 4}{6} = \frac{2(x + 2)}{6} = \frac{2(x + 2)}{2 \cdot 3} = \frac{x + 2}{3}.$$

On the other hand, $\dfrac{2x + 5}{6}$ cannot be simplified further by the fundamental principle, because the numerator cannot be factored.

MULTIPLICATION AND DIVISION Rational expressions are multiplied and divided using definitions from earlier work with fractions.

MULTIPLICATION AND DIVISION	For fractions $\dfrac{a}{b}$ and $\dfrac{c}{d}$ $(b \neq 0, d \neq 0)$, $$\frac{a}{b} \cdot \frac{c}{d} = \frac{ac}{bd}$$ $$\frac{a}{b} \div \frac{c}{d} = \frac{a}{b} \cdot \frac{d}{c} \quad \left(\text{if } \frac{c}{d} \neq 0 \right).$$

■ *Example 2*
MULTIPLYING OR DIVIDING RATIONAL EXPRESSIONS

Multiply or divide, as indicated.

(a) $\dfrac{2y^2}{9} \cdot \dfrac{27}{8y^5} = \dfrac{2y^2 \cdot 27}{9 \cdot 8y^5}$

$\qquad = \dfrac{2 \cdot 9 \cdot 3 \cdot y^2}{2 \cdot 9 \cdot 4 \cdot y^2 \cdot y^3}$ Factor.

$\qquad = \dfrac{3}{4y^3}$ Fundamental principle

The product was written in lowest terms in the last step.

(b) $\dfrac{3m^2 - 2m - 8}{3m^2 + 14m + 8} \cdot \dfrac{3m + 2}{3m + 4} = \dfrac{(m - 2)(3m + 4)}{(m + 4)(3m + 2)} \cdot \dfrac{3m + 2}{3m + 4}$ Factor.

$\qquad\qquad = \dfrac{(m - 2)(3m + 4)(3m + 2)}{(m + 4)(3m + 2)(3m + 4)}$ Multiply fractions.

$\qquad\qquad = \dfrac{m - 2}{m + 4}$ Fundamental principle

(c) $\dfrac{5}{8m + 16} \div \dfrac{7}{12m + 24} = \dfrac{5}{8(m + 2)} \div \dfrac{7}{12(m + 2)}$ Factor.

$= \dfrac{5}{8(m + 2)} \cdot \dfrac{12(m + 2)}{7}$ Definition of division

$= \dfrac{5 \cdot 12(m + 2)}{8 \cdot 7(m + 2)}$ Multiply.

$= \dfrac{15}{14}$ Fundamental principle

(d) $\dfrac{3p^2 + 11p - 4}{24p^3 - 8p^2} \div \dfrac{9p + 36}{24p^4 - 36p^3} = \dfrac{(p + 4)(3p - 1)}{8p^2(3p - 1)} \div \dfrac{9(p + 4)}{12p^3(2p - 3)}$

$= \dfrac{(p + 4)(3p - 1)(12p^3)(2p - 3)}{8p^2(3p - 1)(9)(p + 4)}$

$= \dfrac{12p^3(2p - 3)}{9 \cdot 8p^2} = \dfrac{p(2p - 3)}{6}$ ∎

ADDITION AND SUBTRACTION   Adding and subtracting rational expressions also depends on definitions from earlier work with fractions.

ADDITION AND SUBTRACTION   For fractions $\dfrac{a}{b}$ and $\dfrac{c}{d}$ $(b \neq 0, d \neq 0)$,

$$\frac{a}{b} + \frac{c}{d} = \frac{ad + bc}{bd}$$

$$\frac{a}{b} - \frac{c}{d} = \frac{ad - bc}{bd}.$$

In practice, rational expressions are normally added or subtracted after rewriting all the rational expressions with a common denominator found with the steps given below.

FINDING A COMMON DENOMINATOR   **1.** Write each denominator as a product of prime factors.
2. Form a product of all the different prime factors. Each factor should have as exponent the *highest* exponent that appears on that factor.

∎ *Example 3*

ADDING OR SUBTRACTING RATIONAL EXPRESSIONS

Add or subtract, as indicated.

(a) $\dfrac{5}{9x^2} + \dfrac{1}{6x}$

Write each denominator as a product of prime factors, as follows.

$$9x^2 = 3^2 \cdot x^2$$

$$6x = 2^1 \cdot 3^1 \cdot x^1$$

For the common denominator, form the product of all the prime factors, with each factor having the highest exponent that appears on it. Here the highest exponent on 2 is 1, while both 3 and x have a highest exponent of 2. The common denominator is

$$2^1 \cdot 3^2 \cdot x^2 = 18x^2.$$

Now use the fundamental principle to write both of the given expressions with this denominator, then add.

$$\frac{5}{9x^2} + \frac{1}{6x} = \frac{5 \cdot 2}{9x^2 \cdot 2} + \frac{1 \cdot 3x}{6x \cdot 3x}$$

$$= \frac{10}{18x^2} + \frac{3x}{18x^2}$$

$$= \frac{10 + 3x}{18x^2}$$

Always check at this point to see that the answer is in lowest terms.

(b) $\dfrac{y + 2}{y^2 - y} - \dfrac{3y}{2y^2 - 4y + 2}$

Factor each denominator, giving

$$\frac{y + 2}{y^2 - y} - \frac{3y}{2y^2 - 4y + 2} = \frac{y + 2}{y(y - 1)} - \frac{3y}{2(y - 1)^2}.$$

The common denominator, by the method above, is $2y(y - 1)^2$. Write each rational expression with this denominator and subtract, as follows.

$$\frac{y + 2}{y(y - 1)} - \frac{3y}{2(y - 1)^2} = \frac{(y + 2) \cdot 2(y - 1)}{y(y - 1) \cdot 2(y - 1)} - \frac{3y \cdot y}{2(y - 1)^2 \cdot y}$$

$$= \frac{2(y^2 + y - 2)}{2y(y - 1)^2} - \frac{3y^2}{2y(y - 1)^2}$$

$$= \frac{2y^2 + 2y - 4 - 3y^2}{2y(y - 1)^2} \qquad \text{Subtract.}$$

$$= \frac{-y^2 + 2y - 4}{2y(y - 1)^2} \qquad \text{Combine terms.}$$

(c) $\dfrac{3}{(x - 1)(x + 2)} - \dfrac{1}{(x + 3)(x - 4)}$

The common denominator here is $(x - 1)(x + 2)(x + 3)(x - 4)$. Write each fraction with this common denominator, then perform the subtraction.

$$\frac{3}{(x - 1)(x + 2)} - \frac{1}{(x + 3)(x - 4)}$$

$$= \frac{3(x + 3)(x - 4)}{(x - 1)(x + 2)(x + 3)(x - 4)} - \frac{(x - 1)(x + 2)}{(x + 3)(x - 4)(x - 1)(x + 2)}$$

$$= \frac{3(x^2 - x - 12) - (x^2 + x - 2)}{(x - 1)(x + 2)(x + 3)(x - 4)}$$

$$= \frac{3x^2 - 3x - 36 - x^2 - x + 2}{(x - 1)(x + 2)(x + 3)(x - 4)}$$

$$= \frac{2x^2 - 4x - 34}{(x - 1)(x + 2)(x + 3)(x - 4)} \quad ∎$$

CAUTION | When subtracting fractions where the second fraction has more than one term in the numerator, as in Example 3(c), be sure to distribute the negative sign to each term. Notice in Example 3(c) how parentheses were used in the second step to avoid an error in the subtraction step.

COMPLEX FRACTIONS Any quotient of two rational expressions is called a **complex fraction.** Complex fractions often can be simplified by the methods shown in the following example.

∎ *Example 4*
SIMPLIFYING
COMPLEX
FRACTIONS

Simplify each complex fraction.

(a) $\dfrac{6 - \dfrac{5}{k}}{1 + \dfrac{5}{k}}$

Multiply both numerator and denominator by the least common denominator of all the fractions, k.

$$\frac{k\left(6 - \dfrac{5}{k}\right)}{k\left(1 + \dfrac{5}{k}\right)} = \frac{6k - k\left(\dfrac{5}{k}\right)}{k + k\left(\dfrac{5}{k}\right)} = \frac{6k - 5}{k + 5}$$

(b) $\dfrac{\dfrac{a}{a + 1} + \dfrac{1}{a}}{\dfrac{1}{a} + \dfrac{1}{a + 1}}$

Multiply both numerator and denominator by the least common denominator of all the fractions, in this case $a(a + 1)$.

$$\frac{\dfrac{a}{a+1}+\dfrac{1}{a}}{\dfrac{1}{a}+\dfrac{1}{a+1}}=\frac{\left(\dfrac{a}{a+1}+\dfrac{1}{a}\right)a(a+1)}{\left(\dfrac{1}{a}+\dfrac{1}{a+1}\right)a(a+1)}$$

$$=\frac{\dfrac{a}{a+1}(a)(a+1)+\dfrac{1}{a}(a)(a+1)}{\dfrac{1}{a}(a)(a+1)+\dfrac{1}{a+1}(a)(a+1)} \qquad \text{Distributive property}$$

$$=\frac{a^2+(a+1)}{(a+1)+a}$$

$$=\frac{a^2+a+1}{2a+1}$$

As an alternative method of solution, first perform the indicated additions in the numerator and denominator, and then divide.

$$\frac{\dfrac{a}{a+1}+\dfrac{1}{a}}{\dfrac{1}{a}+\dfrac{1}{a+1}}=\frac{\dfrac{a^2+1(a+1)}{a(a+1)}}{\dfrac{1(a+1)+1(a)}{a(a+1)}} \qquad \begin{array}{l}\text{Get a common denominator;}\\ \text{add terms in numerator}\\ \text{and denominator.}\end{array}$$

$$=\frac{\dfrac{a^2+a+1}{a(a+1)}}{\dfrac{2a+1}{a(a+1)}} \qquad \begin{array}{l}\text{Combine terms in numerator}\\ \text{and denominator.}\end{array}$$

$$=\frac{a^2+a+1}{a(a+1)}\cdot\frac{a(a+1)}{2a+1} \qquad \text{Definition of division}$$

$$=\frac{a^2+a+1}{2a+1} \qquad \begin{array}{l}\text{Multiply fractions and}\\ \text{write in lowest terms.} \quad \blacksquare\end{array}$$

1.6 EXERCISES ■

For each of the following, give the restrictions on the variable.

1. $\dfrac{x-2}{x+6}$ 　　　　 **2.** $\dfrac{x+5}{x-3}$ 　　　　 **3.** $\dfrac{2x}{5x-3}$

4. $\dfrac{6x}{2x-1}$ 　　　　 **5.** $\dfrac{-8}{x^2+1}$ 　　　　 **6.** $\dfrac{3x}{3x^2+7}$

7. Which one of the following expressions is equivalent to $\dfrac{x^2+4x+3}{x+1}$?

　　(a) $x+3$ 　　**(b)** $x+7$ 　　**(c)** $5x+3$ 　　**(d)** x^2+7

8. Explain why $\dfrac{2x+3}{2x+5}=\dfrac{3}{5}$ is false for all real numbers x.

Write each of the following in lowest terms. See Example 1.

9. $\dfrac{25p^3}{10p^2}$

10. $\dfrac{14z^3}{6z^2}$

11. $\dfrac{8k + 16}{9k + 18}$

12. $\dfrac{20r + 10}{30r + 15}$

13. $\dfrac{3(t + 5)}{(t + 5)(t - 3)}$

14. $\dfrac{-8(y - 4)}{(y + 2)(y - 4)}$

15. $\dfrac{8x^2 + 16x}{4x^2}$

16. $\dfrac{36y^2 + 72y}{9y}$

17. $\dfrac{m^2 - 4m + 4}{m^2 + m - 6}$

18. $\dfrac{r^2 - r - 6}{r^2 + r - 12}$

19. $\dfrac{8m^2 + 6m - 9}{16m^2 - 9}$

20. $\dfrac{6y^2 + 11y + 4}{3y^2 + 7y + 4}$

Find each of the following products or quotients. See Example 2.

21. $\dfrac{15p^3}{9p^2} \div \dfrac{6p}{10p^2}$

22. $\dfrac{3r^2}{9r^3} \div \dfrac{8r^3}{6r}$

23. $\dfrac{2k + 8}{6} \div \dfrac{3k + 12}{2}$

24. $\dfrac{5m + 25}{10} \cdot \dfrac{12}{6m + 30}$

25. $\dfrac{x^2 + x}{5} \cdot \dfrac{25}{xy + y}$

26. $\dfrac{3m - 15}{4m - 20} \cdot \dfrac{m^2 - 10m + 25}{12m - 60}$

27. $\dfrac{4a + 12}{2a - 10} \div \dfrac{a^2 - 9}{a^2 - a - 20}$

28. $\dfrac{6r - 18}{9r^2 + 6r - 24} \cdot \dfrac{12r - 16}{4r - 12}$

29. $\dfrac{p^2 - p - 12}{p^2 - 2p - 15} \cdot \dfrac{p^2 - 9p + 20}{p^2 - 8p + 16}$

30. $\dfrac{x^2 + 2x - 15}{x^2 + 11x + 30} \cdot \dfrac{x^2 + 2x - 24}{x^2 - 8x + 15}$

31. $\dfrac{m^2 + 3m + 2}{m^2 + 5m + 4} \div \dfrac{m^2 + 5m + 6}{m^2 + 10m + 24}$

32. $\dfrac{y^2 + y - 2}{y^2 + 3y - 4} \div \dfrac{y^2 + 3y + 2}{y^2 + 4y + 3}$

33. $\dfrac{2m^2 - 5m - 12}{m^2 - 10m + 24} \div \dfrac{4m^2 - 9}{m^2 - 9m + 18}$

34. $\dfrac{6n^2 - 5n - 6}{6n^2 + 5n - 6} \cdot \dfrac{12n^2 - 17n + 6}{12n^2 - n - 6}$

35. $\left(1 + \dfrac{1}{x}\right)\left(1 - \dfrac{1}{x}\right)$

36. $\left(3 + \dfrac{2}{y}\right)\left(3 - \dfrac{2}{y}\right)$

37. $\dfrac{x^3 + y^3}{x^2 - y^2} \cdot \dfrac{x + y}{x^2 - xy + y^2}$

38. $\dfrac{8y^3 - 125}{4y^2 - 20y + 25} \cdot \dfrac{2y - 5}{y}$

39. $\dfrac{x^3 + y^3}{x^3 - y^3} \cdot \dfrac{x^2 - y^2}{x^2 + 2xy + y^2}$

40. $\dfrac{x^2 - y^2}{(x - y)^2} \cdot \dfrac{x^2 - xy + y^2}{x^2 - 2xy + y^2} \div \dfrac{x^3 + y^3}{(x - y)^4}$

41. Which of the following rational expressions equals -1?

(a) $\dfrac{x - 4}{x + 4}$ **(b)** $\dfrac{-x - 4}{x + 4}$ **(c)** $\dfrac{x - 4}{4 - x}$ **(d)** $\dfrac{x - 4}{-x - 4}$

42. In your own words, explain how to find the least common denominator for two fractions.

Perform each of the following additions or subtractions. See Example 3.

43. $\dfrac{3}{2k} + \dfrac{5}{3k}$

44. $\dfrac{8}{5p} + \dfrac{3}{4p}$

45. $\dfrac{a + 1}{2} - \dfrac{a - 1}{2}$

46. $\dfrac{y + 6}{5} - \dfrac{y - 6}{5}$

47. $\dfrac{3}{p} + \dfrac{1}{2}$

48. $\dfrac{9}{r} - \dfrac{2}{3}$

49. $\dfrac{1}{6m} + \dfrac{2}{5m} + \dfrac{4}{m}$

50. $\dfrac{8}{3p} + \dfrac{5}{4p} + \dfrac{9}{2p}$

51. $\dfrac{1}{a + 1} - \dfrac{1}{a - 1}$

52. $\dfrac{1}{x + z} + \dfrac{1}{x - z}$

53. $\dfrac{m + 1}{m - 1} + \dfrac{m - 1}{m + 1}$

54. $\dfrac{2}{x - 1} + \dfrac{1}{1 - x}$

55. $\dfrac{3}{a - 2} - \dfrac{1}{2 - a}$

56. $\dfrac{q}{p - q} - \dfrac{q}{q - p}$

57. $\dfrac{x + y}{2x - y} - \dfrac{2x}{y - 2x}$

58. $\dfrac{m - 4}{3m - 4} + \dfrac{3m + 2}{4 - 3m}$

59. $\dfrac{1}{a^2 - 5a + 6} - \dfrac{1}{a^2 - 4}$

60. $\dfrac{-3}{m^2 - m - 2} - \dfrac{1}{m^2 + 3m + 2}$

61. $\dfrac{1}{x^2 + x - 12} - \dfrac{1}{x^2 - 7x + 12} + \dfrac{1}{x^2 - 16}$

62. $\dfrac{2}{2p^2 - 9p - 5} + \dfrac{p}{3p^2 - 17p + 10} - \dfrac{2p}{6p^2 - p - 2}$

63. $\dfrac{3a}{a^2 + 5a - 6} - \dfrac{2a}{a^2 + 7a + 6}$

64. $\dfrac{2k}{k^2 + 4k + 3} + \dfrac{3k}{k^2 + 5k + 6}$

Perform the indicated operations. See Example 4.

65. $\dfrac{1 + \dfrac{1}{x}}{1 - \dfrac{1}{x}}$

66. $\dfrac{2 - \dfrac{2}{y}}{2 + \dfrac{2}{y}}$

67. $\dfrac{\dfrac{1}{x + 1} - \dfrac{1}{x}}{\dfrac{1}{x}}$

68. $\dfrac{\dfrac{1}{y + 3} - \dfrac{1}{y}}{\dfrac{1}{y}}$

69. $\dfrac{1 + \dfrac{1}{1 - b}}{1 - \dfrac{1}{1 + b}}$

70. $m - \dfrac{m}{m + \dfrac{1}{2}}$

71. $\dfrac{m - \dfrac{1}{m^2 - 4}}{\dfrac{1}{m + 2}}$

72. $\dfrac{\dfrac{3}{p^2 - 16} + p}{\dfrac{1}{p - 4}}$

73. $\left(\dfrac{3}{p - 1} - \dfrac{2}{p + 1}\right)\left(\dfrac{p - 1}{p}\right)$

74. $\left(\dfrac{y}{y^2 - 1} - \dfrac{y}{y^2 - 2y + 1}\right)\left(\dfrac{y - 1}{y + 1}\right)$

75. $\dfrac{\dfrac{1}{x + h} - \dfrac{1}{x}}{h}$

76. $\dfrac{1}{h}\left(\dfrac{1}{(x + h)^2 + 9} - \dfrac{1}{x^2 + 9}\right)$

1.7 ———— **RATIONAL EXPONENTS**

In Section 1.3 we introduced some rules for exponents: the product rule and the power rules. In this section we complete our review of exponential expressions, beginning with a rule for division.

NEGATIVE EXPONENTS AND THE QUOTIENT RULES In the product rule, $a^m \cdot a^n = a^{m+n}$, the exponents are *added.* By the definition of exponent in Section 1.3, if $a \neq 0$,

$$\frac{a^3}{a^7} = \frac{a \cdot a \cdot a}{a \cdot a \cdot a \cdot a \cdot a \cdot a \cdot a} = \frac{1}{a \cdot a \cdot a \cdot a} = \frac{1}{a^4}.$$

This suggests that we should *subtract* exponents when dividing. Subtracting exponents gives

$$\frac{a^3}{a^7} = a^{3-7} = a^{-4}.$$

The only way to keep these results consistent is to define a^{-4} as $1/a^4$. This example suggests the following definition.

NEGATIVE EXPONENTS

If a is a nonzero real number and n is any integer, then

$$a^{-n} = \frac{1}{a^n}.$$

■ *Example 1*

USING THE
DEFINITION OF A
NEGATIVE
EXPONENT

Evaluate each expression in (a)–(c). In (d) and (e), write the expression without negative exponents.

(a) $4^{-2} = \dfrac{1}{4^2} = \dfrac{1}{16}$

(b) $\left(\dfrac{2}{5}\right)^{-3} = \dfrac{1}{\left(\dfrac{2}{5}\right)^3} = \dfrac{1}{\dfrac{8}{125}} = \dfrac{125}{8}$

(c) $-4^{-2} = -\dfrac{1}{4^2} = -\dfrac{1}{16}$

(d) $x^{-4} = \dfrac{1}{x^4}$ $(x \neq 0)$

(e) $xy^{-3} = x \cdot \dfrac{1}{y^3} = \dfrac{x}{y^3}$ $(y \neq 0)$ ■

| CAUTION

A negative exponent indicates a reciprocal, *not* a negative expression.

Part (b) of Example 1 showed that

$$\left(\frac{2}{5}\right)^{-3} = \frac{125}{8} = \left(\frac{5}{2}\right)^3.$$

This result can be generalized. If $a \neq 0$ and $b \neq 0$, then

$$\left(\frac{a}{b}\right)^{-n} = \left(\frac{b}{a}\right)^n,$$

for any integer n.

The quotient rule for exponents follows from the definition of exponents, as shown above.

QUOTIENT RULE

For all integers m and n and all nonzero real numbers a,

$$\frac{a^m}{a^n} = a^{m-n}.$$

By the quotient rule, if $a \neq 0$,

$$\frac{a^m}{a^m} = a^{m-m} = a^0.$$

On the other hand, any nonzero quantity divided by itself equals 1. This is why we defined $a^0 = 1$ in Section 1.3.

■ *Example 2*

USING THE
QUOTIENT RULE

Use the quotient rule to simplify each expression. Assume that all variables represent nonzero real numbers.

(a) $\dfrac{12^5}{12^2} = 12^{5-2} = 12^3$

(b) $\dfrac{a^5}{a^{-8}} = a^{5-(-8)} = a^{13}$

(c) $\dfrac{16m^{-9}}{12m^{11}} = \dfrac{16}{12} \cdot m^{-9-11} = \dfrac{4}{3}m^{-20} = \dfrac{4}{3} \cdot \dfrac{1}{m^{20}} = \dfrac{4}{3m^{20}}$

(d) $\dfrac{25r^7z^5}{10r^9z} = \dfrac{25}{10} \cdot \dfrac{r^7}{r^9} \cdot \dfrac{z^5}{z^1} = \dfrac{5}{2}r^{-2}z^4 = \dfrac{5z^4}{2r^2}$

(e) $\dfrac{x^{5y}}{x^{3y}} = x^{5y-3y} = x^{2y}$,   if y is an integer   ■

The rules for exponents from Section 1.3 also apply to negative exponents.

■ ***Example 3***

USING THE RULES FOR EXPONENTS

Use the rules for exponents to simplify each expression. Write answers without negative exponents. Assume that all variables represent nonzero real numbers.

(a) $3x^{-2}(4^{-1}x^{-5})^2 = 3x^{-2}(4^{-2}x^{-10})$ Power rule

$\qquad\qquad\qquad = 3 \cdot 4^{-2} \cdot x^{-2+(-10)}$ Rearrange factors;   product rule

$\qquad\qquad\qquad = 3 \cdot 4^{-2} \cdot x^{-12}$

$\qquad\qquad\qquad = \dfrac{3}{16x^{12}}$ Write with positive exponents.

(b) $\dfrac{5m^{-3}}{10m^{-5}} = \dfrac{5}{10}m^{-3-(-5)}$ Quotient rule

$\qquad\qquad = \dfrac{1}{2}m^2$ or $\dfrac{m^2}{2}$

(c) $\dfrac{12p^3q^{-1}}{8p^{-2}q} = \dfrac{12}{8} \cdot \dfrac{p^3}{p^{-2}} \cdot \dfrac{q^{-1}}{q^1}$

$\qquad\qquad = \dfrac{3}{2} \cdot p^{3-(-2)}q^{-1-1}$ Quotient rule

$\qquad\qquad = \dfrac{3}{2}p^5q^{-2}$

$\qquad\qquad = \dfrac{3p^5}{2q^2}$ Write with positive exponents.

(d) $\dfrac{(3x^2)^{-1}(3x^5)^{-2}}{(3^{-1}x^{-2})^2} = \dfrac{3^{-1}x^{-2}3^{-2}x^{-10}}{3^{-2}x^{-4}}$ Power rule

$\qquad\qquad\qquad = \dfrac{3^{-1+(-2)}x^{-2+(-10)}}{3^{-2}x^{-4}} = \dfrac{3^{-3}x^{-12}}{3^{-2}x^{-4}}$ Product rule

$\qquad\qquad\qquad = 3^{-3-(-2)}x^{-12-(-4)} = 3^{-1}x^{-8}$ Quotient rule

$\qquad\qquad\qquad = \dfrac{1}{3x^8}$ Write with positive exponents.   ■

CAUTION	Notice the use of the power rule $(ab)^n = a^n b^n$ in Example 3(d): $(3x^2)^{-1} =$ $3^{-1}(x^2)^{-1} = 3^{-1}x^{-2}$. It is a common error to forget to apply the exponent to a numerical coefficient.

RATIONAL EXPONENTS The definition of a^n can be extended to rational values of n by defining $a^{1/n}$ to be the nth root of a. By one of the power rules of exponents (extended to a rational exponent)

$$(a^{1/n})^n = a^{(1/n)n} = a^1 = a,$$

suggesting that $a^{1/n}$ is a number whose nth power is a.

$a^{1/n}$, n EVEN	**(i)** If n is an *even* positive integer, and if $a > 0$, then $a^{1/n}$ is the positive real number whose nth power is a. That is, $(a^{1/n})^n = a$. In this case, $a^{1/n}$ is the principal nth root of a.
$a^{1/n}$, n ODD	**(ii)** If n is an *odd* positive integer, and *a is any real number*, then $a^{1/n}$ is the positive or negative real number whose nth power is a. That is, $(a^{1/n})^n = a$.

■ *Example 4*
USING THE
DEFINITION OF $a^{1/n}$

Evaluate each expression.

(a) $36^{1/2} = 6$ because $6^2 = 36$ **(b)** $-100^{1/2} = -10$

(c) $-(225)^{1/2} = -15$ **(d)** $625^{1/4} = 5$

(e) $(-1296)^{1/4}$ is not defined, but $-1296^{1/4} = -6$

(f) $(-27)^{1/3} = -3$

(g) $-32^{1/5} = -2$ ■

What about more general rational exponents? The notation $a^{m/n}$ should be defined so that all the past rules for exponents still hold. For the power rule to hold, $(a^{1/n})^m$ must equal $a^{m/n}$. Therefore, $a^{m/n}$ is defined as follows.

RATIONAL EXPONENTS	For all integers m, all positive integers n, and all real numbers a for which $a^{1/n}$ is defined: $$a^{m/n} = (a^{1/n})^m.$$

■ *Example 5*
USING THE
DEFINITION OF $a^{m/n}$

Evaluate each expression.

(a) $125^{2/3} = (125^{1/3})^2 = 5^2 = 25$

(b) $32^{7/5} = (32^{1/5})^7 = 2^7 = 128$

(c) $-81^{3/2} = -(81^{1/2})^3 = -9^3 = -729$

(d) $(-4)^{5/2}$ is not defined because $(-4)^{1/2}$ is not defined.

(e) $(-27)^{2/3} = [(-27)^{1/3}]^2 = (-3)^2 = 9$

(f) $16^{-3/4} = \dfrac{1}{16^{3/4}} = \dfrac{1}{(16^{1/4})^3} = \dfrac{1}{2^3} = \dfrac{1}{8}$ ■

NOTE

By starting with $(a^{1/n})^m$ and $(a^m)^{1/n}$, and raising each expression to the nth power, it can be shown that $(a^{1/n})^m$ is equal to $(a^m)^{1/n}$. This means that $a^{m/n}$ could be defined in either of the following ways.

For all real numbers a, integers m, and positive integers n for which $a^{1/n}$ is defined:

$$a^{m/n} = (a^{1/n})^m \qquad \text{or} \qquad a^{m/n} = (a^m)^{1/n}.$$

Now $a^{m/n}$ can be evaluated in either of two ways: as $(a^{1/n})^m$ or as $(a^m)^{1/n}$. It is usually easier to find $(a^{1/n})^m$. For example, $27^{4/3}$ can be evaluated in either of two ways:

$$27^{4/3} = (27^{1/3})^4 = 3^4 = 81$$
$$27^{4/3} = (27^4)^{1/3} = 531{,}441^{1/3} = 81.$$

The form $(27^{1/3})^4$ is easier to evaluate.

■ IN SIMPLEST TERMS

3

Meteorologists can determine the duration of a storm by using the formula $.07D^{3/2} = T$, where D is the diameter of the storm in miles and T is the time in hours. For example, if radar shows that the diameter of a storm is 16 miles, we can expect the storm to last

$$.07(16)^{3/2} = .07(16^{1/2})^3$$
$$= .07(4)^3$$
$$= .07(64)$$
$$= 4.48 \text{ hours.}$$

SOLVE EACH PROBLEM

A. The National Weather Service reports that a storm 4 miles in diameter is headed toward New Haven. How long can the residents expect the storm to last?

B. After weeks of dry weather, a thunderstorm is predicted for the farming community of Apple Valley. The crops need at least an hour and a half of rain. Local radar shows that the storm is 7 miles in diameter. Will it rain long enough to meet the farmers' need?

ANSWERS   A. They can expect the storm to last .56 hour.   B. The storm will last approximately 1.3 hours, not quite long enough to meet their need.

It can be shown that all the earlier results concerning integer exponents also apply to rational exponents. These definitions and rules are summarized here.

DEFINITIONS AND RULES FOR EXPONENTS

Let r and s be rational numbers. The results below are valid for all positive numbers a and b.

$$a^r \cdot a^s = a^{r+s} \qquad (ab)^r = a^r \cdot b^r \qquad (a^r)^s = a^{rs}$$

$$\frac{a^r}{a^s} = a^{r-s} \qquad \left(\frac{a}{b}\right)^r = \frac{a^r}{b^r} \qquad a^{-r} = \frac{1}{a^r}$$

■ **Example 6**

USING THE DEFINITIONS AND RULES FOR EXPONENTS

Use the definitions and rules for exponents to simplify each expression.

(a) $\dfrac{27^{1/3} \cdot 27^{5/3}}{27^3} = \dfrac{27^{1/3+5/3}}{27^3}$ Product rule

$= \dfrac{27^2}{27^3} = 27^{2-3}$ Quotient rule

$= 27^{-1} = \dfrac{1}{27}$

(b) $81^{5/4} \cdot 4^{-3/2} = (81^{1/4})^5 (4^{1/2})^{-3} = 3^5 \cdot 2^{-3} = \dfrac{3^5}{2^3}$ or $\dfrac{243}{8}$.

(c) $6y^{2/3} \cdot 2y^{1/2} = 12y^{2/3+1/2} = 12y^{7/6}$,   where $y \geq 0$

(d) $\left(\dfrac{3m^{5/6}}{y^{3/4}}\right)^2 \cdot \left(\dfrac{8y^3}{m^6}\right)^{2/3} = \dfrac{9m^{5/3}}{y^{3/2}} \cdot \dfrac{4y^2}{m^4} = 36m^{5/3-4}y^{2-3/2} = \dfrac{36y^{1/2}}{m^{7/3}}$ $(m > 0, y > 0)$

(e) $m^{2/3}(m^{7/3} + 2m^{1/3}) = (m^{2/3+7/3} + 2m^{2/3+1/3}) = m^3 + 2m$

(f) $\dfrac{(x^{2/p})^p(x^{p-1})}{x^{-1/4}} = \dfrac{x^{(2/p)p} \cdot x^{p-1}}{x^{-1/4}}$ Power rule

$= \dfrac{x^2 x^{p-1}}{x^{-1/4}}$

$= \dfrac{x^{2+p-1}}{x^{-1/4}}$ Product rule

$= \dfrac{x^{1+p}}{x^{-1/4}}$

$= x^{1+p-(-1/4)}$ Quotient rule

$= x^{(5/4)+p}$ or $x^{(5+4p)/4}$ ■

The next example shows how to factor with negative or rational exponents.

54

■ *Example 7*

FACTORING AN EXPRESSION WITH NEGATIVE OR RATIONAL EXPONENTS

Factor out the smallest power of the variable. Assume that all variables represent positive real numbers.

(a) $9x^{-2} - 6x^{-3}$

The smallest exponent here is -3. Since 3 is a common numerical factor, factor out $3x^{-3}$.

$$9x^{-2} - 6x^{-3} = 3x^{-3}(3x - 2)$$

Check by multiplying on the right. The factored form can now be written without negative exponents as $\dfrac{3(3x - 2)}{x^3}$.

(b) $4m^{1/2} + 3m^{3/2} = m^{1/2}(4 + 3m)$

To check this result, multiply $m^{1/2}$ by $4 + 3m$.

(c) $y^{-1/3} + y^{2/3} = y^{-1/3}(1 + y)$

The factored form can be written with only positive exponents as $\dfrac{1 + y}{y^{1/3}}$. ■

Negative exponents are sometimes used to write complex fractions. Recall, complex fractions are simplified either by first multiplying the numerator and denominator by the greatest common multiple of all the denominators, or by performing any indicated operations in the numerator and the denominator and then using the definition of division for fractions.

■ *Example 8*

SIMPLIFYING A FRACTION WITH NEGATIVE EXPONENTS

Simplify $\dfrac{(x + y)^{-1}}{x^{-1} + y^{-1}}$. Write the result with only positive exponents.

Begin by using the definition of a negative integer exponent. Then perform the indicated operations.

$$\frac{(x + y)^{-1}}{x^{-1} + y^{-1}} = \frac{\dfrac{1}{x + y}}{\dfrac{1}{x} + \dfrac{1}{y}}$$

$$= \frac{\dfrac{1}{x + y}}{\dfrac{y + x}{xy}}$$

$$= \frac{1}{x + y} \cdot \frac{xy}{x + y}$$

$$= \frac{xy}{(x + y)^2} \quad ■$$

——————————————————————————————————

Simplify each of the following. Assume that all variables represent positive real numbers. See Examples 1, 4, and 5.

1. $(-4)^{-3}$

2. $(-5)^{-2}$

3. $\left(\frac{1}{2}\right)^{-3}$

4. $\left(\frac{2}{3}\right)^{-2}$

5. $-4^{1/2}$

6. $25^{1/2}$

7. $8^{2/3}$

8. $-81^{3/4}$

9. $27^{-2/3}$

10. $(-32)^{-4/5}$

11. $\left(-\frac{4}{9}\right)^{-3/2}$

12. $\left(\frac{1}{8}\right)^{-5/3}$

13. $\left(\frac{27}{64}\right)^{-4/3}$

14. $\left(\frac{121}{100}\right)^{-3/2}$

▦ 15. $7^{-3.1}$

▦ 16. $2^{5.41}$

17. $(16p^4)^{1/2}$

18. $(36r^6)^{1/2}$

19. $(27x^6)^{2/3}$

20. $(64a^{12})^{5/6}$

21. Why is $a^{1/n}$ defined to be the nth root of a (with appropriate restrictions)?

22. Explain why a must be positive if n is even for $a^{1/n}$ to be defined.

23. Which of the following expressions is equivalent to $(2x^{-3/2})^2$?

    (a) $2x^{-3}$ **(b)** 2^{-3} **(c)** $2^2x^{-3/4}$ **(d)** $\dfrac{2^2}{x^3}$

24. Explain how you would evaluate $(-27)^{5/3}$.

Perform the indicated operations. Write answers using only positive exponents. Assume that all variables represent positive real numbers and that variables used as exponents represent rational numbers. See Examples 2, 3, and 6.

25. $2^{-3} \cdot 2^{-4}$

26. $5^{-2} \cdot 5^{-6}$

27. $27^{-2} \cdot 27^{-1}$

28. $9^{-4} \cdot 9^{-1}$

29. $\dfrac{4^{-2} \cdot 4^{-1}}{4^{-3}}$

30. $\dfrac{3^{-1} \cdot 3^{-4}}{3^2 \cdot 3^{-2}}$

31. $(m^{2/3})(m^{5/3})$

32. $(x^{4/5})(x^{2/5})$

33. $(1 + n)^{1/2}(1 + n)^{3/4}$

34. $(m + 7)^{-1/6}(m + 7)^{-2/3}$

35. $(2y^{3/4}z)(3y^{-2}z^{-1/3})$

36. $(4a^{-1}b^{2/3})(a^{3/2}b^{-3})$

37. $(4a^{-2}b^7)^{1/2} \cdot (2a^{1/4}b^3)^5$

38. $(x^{-2}y^{1/3})^5 \cdot (8x^2y^{-2})^{-1/3}$

39. $\left(\dfrac{r^{-2}}{s^{-5}}\right)^{-3}$

40. $\left(\dfrac{p^{-1}}{q^{-5}}\right)^{-2}$

41. $\left(\dfrac{-a}{b^{-3}}\right)^{-1}$

42. $\dfrac{7^{-1/3}7r^{-3}}{7^{2/3}r^{-2}}$

43. $\dfrac{12^{5/4}y^{-2}}{12^{-1}y^{-3}}$

44. $\dfrac{6k^{-4}(3k^{-1})^{-2}}{2^3k^{1/2}}$

45. $\dfrac{8p^{-3}(4p^2)^{-2}}{p^{-5}}$

46. $\dfrac{k^{-3/5}h^{-1/3}t^{2/5}}{k^{-1/5}h^{-2/3}t^{1/5}}$

47. $\dfrac{m^{7/3}n^{-2/5}p^{3/8}}{m^{-2/3}n^{3/5}p^{-5/8}}$

48. $\dfrac{m^{2/5}m^{3/5}m^{-4/5}}{m^{1/5}m^{-6/5}}$

49. $\dfrac{-4a^{-1}a^{2/3}}{a^{-2}}$

50. $\dfrac{8y^{2/3}y^{-1}}{2^{-1}y^{3/4}y^{-1/6}}$

51. $\dfrac{(k + 5)^{1/2}(k + 5)^{-1/4}}{(k + 5)^{3/4}}$

52. $\dfrac{(x + y)^{-5/8}(x + y)^{3/8}}{(x + y)^{1/8}(x + y)^{-1/8}}$

53. $\left(\dfrac{x^4y^3z}{16x^{-6}yz^5}\right)^{-1/2}$

54. $\left(\dfrac{p^3r^9}{27p^{-3}r^{-6}}\right)^{-1/3}$

Perform the indicated operations. Write answers without denominators. Assume that all variables used in denominators are not zero and all variables used as exponents represent rational numbers. See Examples 2(e) and 6(f).

55. $(r^{3/p})^{2p}(r^{1/p})^{p^2}$

56. $(m^{2/x})^{x/3}(m^{x/4})^{2/x}$

57. $\dfrac{m^{1-a}m^a}{m^{-1/2}}$

58. $\dfrac{(y^{3-b})(y^{2b-1})}{y^{1/2}}$

59. $\dfrac{(x^{n/2})(x^{3n})^{1/2}}{x^{1/n}}$

60. $\dfrac{(a^{2/3})(a^{1/x})}{(a^{x/3})^{-2}}$

61. $\dfrac{(p^{1/n})(p^{1/m})}{p^{-m/n}}$

62. $\dfrac{(q^{2r/3})(q^r)^{-1/3}}{(q^{4/3})^{1/r}}$

Find each of the following products. Assume that all variables represent positive real numbers. See Example 6(e). (Hint: Use the special binomial product formulas in Exercises 67, 69, and 70.)

63. $y^{5/8}(y^{3/8} - 10y^{11/8})$

64. $p^{11/5}(3p^{4/5} + 9p^{19/5})$

65. $-4k(k^{7/3} - 6k^{1/3})$

66. $-5y(3y^{9/10} + 4y^{3/10})$

67. $(x + x^{1/2})(x - x^{1/2})$

68. $(2z^{1/2} + z)(z^{1/2} - z)$

69. $(r^{1/2} - r^{-1/2})^2$

70. $(p^{1/2} - p^{-1/2})(p^{1/2} + p^{-1/2})$

Factor, using the given common factor. Assume all variables represent positive real numbers. See Example 7.

71. $4k^{-1} + k^{-2}$; k^{-2}

72. $y^{-5} - 3y^{-3}$; y^{-5}

73. $9z^{-1/2} + 2z^{1/2}$; $z^{-1/2}$

74. $3m^{2/3} - 4m^{-1/3}$; $m^{-1/3}$

75. $p^{-3/4} - 2p^{-7/4}$; $p^{-7/4}$

76. $6r^{-2/3} - 5r^{-5/3}$; $r^{-5/3}$

77. $(p + 4)^{-3/2} + (p + 4)^{-1/2} + (p + 4)^{1/2}$; $(p + 4)^{-3/2}$

78. $(3r + 1)^{-2/3} + (3r + 1)^{1/3} + (3r + 1)^{4/3}$; $(3r + 1)^{-2/3}$

One important application of mathematics to business and management concerns supply and demand. Usually, as the price of an item increases, the supply increases and the demand decreases. By studying past records of supply and demand at different prices, economists can construct an equation that describes (approximately) supply and demand for a given item. The next two exercises show examples of this.

79. The price (in dollars) of a certain type of solar heater is approximated by p, where

$$p = 2x^{1/2} + 3x^{2/3}$$

and x is the number of units supplied. Find the price when the supply is 64 units.

80. For a certain commodity the demand and the price (in dollars) are related by the equation

$$p = 1000 - 200x^{-2/3} \quad (x > 0),$$

where x is the number of units of the product demanded. Find the price when the demand is 27.

In our system of government, the president is elected by the electoral college, and not by individual voters. Because of this, smaller states have a greater voice in the selection of a president than they otherwise would have. Two political scientists have studied the problems of campaigning for president under the current system and have concluded that candidates should allot their money according to the formula

$$\text{Amount for large state} = \left(\frac{E_{\text{large}}}{E_{\text{small}}}\right)^{3/2} \times \text{amount for small state.}$$

Here E_{large} represents the electoral vote of the large state, and E_{small} represents the electoral vote of the small state. Find the amount that should be spent in each of the following larger states if $1,000,000 is spent in the small state and the following statements are true.

81. The large state has 48 electoral votes, and the small state has 3.

82. The large state has 36 electoral votes, and the small state has 4.

83. 6 votes in a small state;   28 in a large

84. 9 votes in a small state;   32 in a large

▦ *The Galapagos Islands are a chain of islands ranging in size from 2 to 2249 square miles. A biologist has shown that the number of different land-plant species on an island in this chain is related to the size of the island approximately by*

$$S = 28.6A^{.32},$$

where A is the area of the island in square miles and S is the number of different plant species on that island. Estimate S (rounding to the nearest whole number) for islands of the following areas.

85. 1 square mile   **86.** 25 square miles   **87.** 300 square miles   **88.** 2000 square miles

Perform all indicated operations and write all answers with positive integer exponents. See Example 8.

89. $\dfrac{a^{-1} + b^{-1}}{(ab)^{-1}}$

90. $\dfrac{p^{-1} - q^{-1}}{(pq)^{-1}}$

91. $\dfrac{r^{-1} + q^{-1}}{r^{-1} - q^{-1}} \cdot \dfrac{r - q}{r + q}$

92. $\dfrac{xy^{-1} + yx^{-1}}{x^2 + y^2}$

93. $\dfrac{x - 9y^{-1}}{(x - 3y^{-1})(x + 3y^{-1})}$

94. $\dfrac{(m + n)^{-1}}{m^{-2} - n^{-2}}$

1.8 ──────── RADICALS

RADICAL NOTATION   In the last section the notation $a^{1/n}$ was used for the nth root of a for appropriate values of a and n. An alternative (and more familiar) notation for $a^{1/n}$ uses *radical notation*.

RADICAL NOTATION FOR $a^{1/n}$	If a is a real number, n is a positive integer, and $a^{1/n}$ is defined, then $$\sqrt[n]{a} = a^{1/n}.$$

The symbol $\sqrt[n]{}$ is a **radical sign,** the number a is the **radicand,** and n is the **index** of the radical $\sqrt[n]{a}$. It is customary to use the familiar notation $\sqrt{a}$ instead of $\sqrt[2]{a}$ for the square root.

For even values of n (square roots, fourth roots, and so on) there are two nth roots, one positive and one negative. In such cases, the notation $\sqrt[n]{a}$ represents the positive root, the **principal nth root.** The negative root is written $-\sqrt[n]{a}$.

■ **Example 1**
| **EVALUATING ROOTS**

Evaluate each root.

(a) $\sqrt[4]{16} = 16^{1/4} = 2$ **(b)** $-\sqrt[4]{16} = -16^{1/4} = -2$

(c) $\sqrt[4]{-16}$ is not defined.   **(d)** $\sqrt[5]{-32} = -2$

(e) $\sqrt[3]{1000} = 10$ **(f)** $\sqrt[6]{\dfrac{64}{729}} = \dfrac{2}{3}$ ■

With $a^{1/n}$ written as $\sqrt[n]{a}$, $a^{m/n}$ also can be written using radicals.

RADICAL NOTATION FOR $a^{m/n}$	If a is a real number, m is an integer, n is a positive integer, and $\sqrt[n]{a}$ is defined, then $$a^{m/n} = (\sqrt[n]{a})^m = \sqrt[n]{a^m}.$$

■ *Example 2*

CONVERTING FROM
RATIONAL
EXPONENTS TO
RADICALS

Write in radical form and simplify.

(a) $8^{2/3} = (\sqrt[3]{8})^2 = 2^2 = 4$

(b) $(-32)^{4/5} = (\sqrt[5]{-32})^4 = (-2)^4 = 16$

(c) $-16^{3/4} = -(\sqrt[4]{16})^3 = -(2)^3 = -8$

(d) $x^{5/6} = \sqrt[6]{x^5}$ $(x \geq 0)$

(e) $3x^{2/3} = 3\sqrt[3]{x^2}$

(f) $2p^{-1/2} = \dfrac{2}{p^{1/2}} = \dfrac{2}{\sqrt{p}}$ $(p > 0)$

(g) $(3a + b)^{1/4} = \sqrt[4]{3a + b}$ $(3a + b \geq 0)$ ■

CAUTION

It is not possible to "distribute" exponents over a sum, so in Example 2(g), $(3a + b)^{1/4}$ *cannot be written as* $(3a)^{1/4} + b^{1/4}$. More generally,

$$\sqrt[n]{x^n + y^n} \text{ is } not \text{ equal to } x + y.$$

Be alert for this common error.

■ *Example 3*

CONVERTING FROM
RADICALS TO
RATIONAL
EXPONENTS

Write in exponential form.

(a) $\sqrt[4]{x^5} = x^{5/4}$ $(x \geq 0)$ **(b)** $\sqrt{3y} = (3y)^{1/2}$ $(y \geq 0)$

(c) $10(\sqrt[5]{z})^2 = 10z^{2/5}$ **(d)** $5\sqrt[3]{(2x^4)^7} = 5(2x^4)^{7/3} = 5 \cdot 2^{7/3}x^{28/3}$

(e) $\sqrt{p^2 + q} = (p^2 + q)^{1/2}$ $(p^2 + q \geq 0)$ ■

By the definition of $\sqrt[n]{a}$, for any positive integer n, if $\sqrt[n]{a}$ is defined, then

$$(\sqrt[n]{a})^n = a.$$

If a is positive, or if a is negative and n is an odd positive integer,

$$\sqrt[n]{a^n} = a.$$

Because of the conditions just given, we *cannot* simply write $\sqrt{x^2} = x$. For example, if $x = -5$,

$$\sqrt{x^2} = \sqrt{(-5)^2} = \sqrt{25} = 5 \neq x.$$

To take care of the fact that a negative value of x can produce a positive result, we use absolute value. For any real number a,

$$\sqrt{a^2} = |a|.$$

For example,

$$\sqrt{(-9)^2} = |-9| = 9, \quad \text{and} \quad \sqrt{13^2} = |13| = 13.$$

This result can be generalized to any even nth root.

$$\sqrt[n]{a^n} \quad \text{If } n \text{ is an even positive integer, } \sqrt[n]{a^n} = |a|, \text{ and if } n \text{ is an odd positive integer,}$$
$$\sqrt[n]{a^n} = a.$$

■ **Example 4**

USING ABSOLUTE VALUE TO SIMPLIFY ROOTS

Use absolute value as applicable to simplify the following expressions.

(a) $\sqrt{p^4} = |p^2| = p^2$ **(b)** $\sqrt[4]{p^4} = |p|$

(c) $\sqrt{16m^8r^6} = |4m^4r^3| = 4m^4|r^3|$ **(d)** $\sqrt[6]{(-2)^6} = |-2| = 2$

(e) $\sqrt[5]{m^5} = m$ **(f)** $\sqrt{(2k+3)^2} = |2k+3|$

(g) $\sqrt{x^2 - 4x + 4} = \sqrt{(x-2)^2} = |x-2|$ ■

NOTE   To avoid difficulties when working with variable radicands, we usually will assume that all variables in radicands represent only nonnegative real numbers.

Three key rules for working with radicals are given below. These rules are just the power rules for exponents written in radical notation.

RULES FOR RADICALS

For all real numbers a and b, and positive integers m and n for which the indicated roots are defined,

$$\sqrt[n]{a} \cdot \sqrt[n]{b} = \sqrt[n]{ab} \qquad \sqrt[n]{\frac{a}{b}} = \frac{\sqrt[n]{a}}{\sqrt[n]{b}} \quad (b \neq 0) \qquad \sqrt[m]{\sqrt[n]{a}} = \sqrt[m \cdot n]{a}.$$

■ **Example 5**

USING THE RULES FOR RADICALS TO SIMPLIFY RADICAL EXPRESSIONS

(a) $\sqrt{6} \cdot \sqrt{54} = \sqrt{6 \cdot 54} = \sqrt{324} = 18$

(b) $\sqrt[3]{m} \cdot \sqrt[3]{m^2} = \sqrt[3]{m^3} = m$

(c) $\sqrt{\frac{7}{64}} = \frac{\sqrt{7}}{\sqrt{64}} = \frac{\sqrt{7}}{8}$

(d) $\sqrt[4]{\frac{a}{b^4}} = \frac{\sqrt[4]{a}}{\sqrt[4]{b^4}} = \frac{\sqrt[4]{a}}{b}$ $(a \geq 0, b > 0)$

(e) $\sqrt[7]{\sqrt[3]{2}} = \sqrt[21]{2}$ Use the third rule given above.

(f) $\sqrt[4]{\sqrt{3}} = \sqrt[8]{3}$ ■

NOTE   In Example 5, converting to fractional exponents would show why these rules work. For example, in part (e)

$$\sqrt[7]{\sqrt[3]{2}} = (2^{1/3})^{1/7} = 2^{(1/3)(1/7)} = 2^{1/21} = \sqrt[21]{2}.$$

SIMPLIFYING RADICALS In working with numbers, it is customary to write a number in its simplest form. For example, 10/2 is written as 5, −9/6 is written as −3/2, and 4/16 is written as 1/4. Similarly, expressions with radicals should be written in their simplest forms.

SIMPLIFIED RADICALS	An expression with radicals is simplified when all of the following conditions are satisfied. 1. The radicand has no factor raised to a power greater than or equal to the index. 2. The radicand has no fractions. 3. No denominator contains a radical. 4. Exponents in the radicand and the index of the radical have no common factor. 5. All indicated operations have been performed (if possible).

■ ***Example 6***
SIMPLIFYING
RADICALS

Simplify each of the following. Assume that all variables represent nonnegative real numbers.

(a) $\sqrt{175} = \sqrt{25 \cdot 7} = \sqrt{25} \cdot \sqrt{7} = 5\sqrt{7}$

(b) $-3\sqrt[5]{32} = -3\sqrt[5]{2^5} = -3 \cdot 2 = -6$

(c) $\sqrt[3]{81x^5y^7z^6} = \sqrt[3]{27 \cdot 3 \cdot x^3 \cdot x^2 \cdot y^6 \cdot y \cdot z^6}$ Factor.

$\qquad = \sqrt[3]{(27x^3y^6z^6)(3x^2y)}$ Group all perfect cubes.

$\qquad = 3xy^2z^2\sqrt[3]{3x^2y}$ Remove all perfect cubes from the radical. ■

Radicals with the same radicand and the same index, such as $3\sqrt[4]{11pq}$ and $-7\sqrt[4]{11pq}$, are called **like radicals.** Like radicals are added or subtracted by using the distributive property. Only like radicals can be combined. As shown in parts (b) and (c) of the next example, it is sometimes necessary to simplify radicals before adding or subtracting.

■ ***Example 7***
ADDING AND
SUBTRACTING LIKE
RADICALS

Add or subtract as indicated. Assume all variables are positive real numbers.

(a) $3\sqrt[4]{11pq} + (-7\sqrt[4]{11pq}) = -4\sqrt[4]{11pq}$

(b) $\sqrt{98x^3y} + 3x\sqrt{32xy}$

First remove all perfect square factors from under the radical. Then use the distributive property, as follows.

$$\sqrt{98x^3y} + 3x\sqrt{32xy} = \sqrt{49 \cdot 2 \cdot x^2 \cdot x \cdot y} + 3x\sqrt{16 \cdot 2 \cdot x \cdot y}$$

$$= 7x\sqrt{2xy} + (3x)(4)\sqrt{2xy}$$

$$= 7x\sqrt{2xy} + 12x\sqrt{2xy}$$

$$= 19x\sqrt{2xy} \quad \text{Distributive property}$$

(c) $\sqrt[3]{64m^4n^5} - \sqrt[3]{-27m^{10}n^{14}} = \sqrt[3]{(64m^3n^3)(mn^2)} - \sqrt[3]{(-27m^9n^{12})(mn^2)}$

$= 4mn\sqrt[3]{mn^2} - (-3)m^3n^4\sqrt[3]{mn^2}$

$= 4mn\sqrt[3]{mn^2} + 3m^3n^4\sqrt[3]{mn^2}$

$= (4 + 3m^2n^3)mn\sqrt[3]{mn^2}$ Distributive property ∎

If the index of the radical and an exponent in the radicand have a common factor, the radical can be simplified by writing it in exponential form, simplifying the rational exponent, then writing the result as a radical again.

∎ *Example 8*

SIMPLIFYING RADICALS BY WRITING THEM WITH RATIONAL EXPONENTS

Simplify the following radicals by first rewriting with rational exponents.

(a) $\sqrt[6]{3^2} = 3^{2/6} = 3^{1/3} = \sqrt[3]{3}$

(b) $\sqrt[6]{x^{12}y^3} = (x^{12}y^3)^{1/6} = x^2y^{3/6} = x^2y^{1/2} = x^2\sqrt{y}$ $(y \geq 0)$

(c) $\sqrt[9]{\sqrt{6^3}} = \sqrt[9]{6^{3/2}} = (6^{3/2})^{1/9} = 6^{1/6} = \sqrt[6]{6}$ ∎

In Example 8(a), we simplified $\sqrt[6]{3^2}$ as $\sqrt[3]{3}$. However, to simplify $(\sqrt[6]{x})^2$, the variable x must be nonnegative. For example,

$$(-8)^{2/6} = [(-8)^{1/6}]^2.$$

This result is not a real number, since $(-8)^{1/6}$ is not defined. On the other hand,

$$(-8)^{1/3} = -2.$$

Here, even though $2/6 = 1/3$,

$$(\sqrt[6]{x})^2 \neq \sqrt[3]{x}.$$

If a is nonnegative, then it is always true that $a^{m/n} = a^{mp/(np)}$. Reducing rational exponents on negative bases must be considered case by case.

Multiplying radical expressions is much like multiplying polynomials.

∎ *Example 9*

MULTIPLYING RADICAL EXPRESSIONS

(a) $(\sqrt{2} + 3)(\sqrt{8} - 5) = \sqrt{2}(\sqrt{8}) - \sqrt{2}(5) + 3\sqrt{8} - 3(5)$ FOIL

$= \sqrt{16} - 5\sqrt{2} + 3(2\sqrt{2}) - 15$ Multiply.

$= 4 - 5\sqrt{2} + 6\sqrt{2} - 15$

$= -11 + \sqrt{2}$ Combine terms.

(b) $(\sqrt{7} - \sqrt{10})(\sqrt{7} + \sqrt{10}) = (\sqrt{7})^2 - (\sqrt{10})^2$ Product of the sum and difference of two terms

$= 7 - 10$

$= -3$ ∎

RATIONALIZING THE DENOMINATOR   Condition 3 of the rules for simplifying radicals described above requires that no denominator contain a radical. The process of achieving this is called **rationalizing the denominator.** It is accomplished by multiplying by a form of 1, as explained in Example 10.

■ *Example 10*
| RATIONALIZING
| DENOMINATORS

Rationalize each denominator.

(a) $\dfrac{4}{\sqrt{3}}$

To rationalize the denominator, multiply by $\sqrt{3}/\sqrt{3}$ (which equals 1) so that the denominator of the product is a rational number.

$$\frac{4}{\sqrt{3}} \cdot \frac{\sqrt{3}}{\sqrt{3}} = \frac{4\sqrt{3}}{3}$$

(b) $\sqrt[4]{\dfrac{3}{5}}$

Start by using the fact that the radical of a quotient can be written as the quotient of radicals.

$$\sqrt[4]{\frac{3}{5}} = \frac{\sqrt[4]{3}}{\sqrt[4]{5}}$$

The denominator will be a rational number if it equals $\sqrt[4]{5^4}$. That is, four factors of 5 are needed under the radical. Since $\sqrt[4]{5}$ has just one factor of 5, three additional factors are needed, so multiply by $\sqrt[4]{5^3}/\sqrt[4]{5^3}$.

$$\frac{\sqrt[4]{3}}{\sqrt[4]{5}} = \frac{\sqrt[4]{3} \cdot \sqrt[4]{5^3}}{\sqrt[4]{5} \cdot \sqrt[4]{5^3}} = \frac{\sqrt[4]{3 \cdot 5^3}}{\sqrt[4]{5^4}} = \frac{\sqrt[4]{375}}{5} \quad ■$$

■ *Example 11*
| SIMPLIFYING
| RATIONAL
| EXPRESSIONS

Simplify. Assume all variables represent positive real numbers.

(a) $\dfrac{\sqrt[4]{ab^3} \cdot \sqrt[4]{ab}}{\sqrt[4]{a^3b^3}}$

To begin, use the product and quotient rules to write all radicals under one radical sign.

$$\frac{\sqrt[4]{ab^3} \cdot \sqrt[4]{ab}}{\sqrt[4]{a^3b^3}} = \sqrt[4]{\frac{ab^3 \cdot ab}{a^3b^3}}$$

$$= \sqrt[4]{\frac{a^2b^4}{a^3b^3}} \qquad \text{Multiply.}$$

$$= \sqrt[4]{\frac{b}{a}} \qquad \text{Write in lowest terms.}$$

$$= \frac{\sqrt[4]{b}}{\sqrt[4]{a}} \cdot \frac{\sqrt[4]{a^3}}{\sqrt[4]{a^3}} \qquad \sqrt[4]{a} \cdot \sqrt[4]{a^3} = \sqrt[4]{a^4} = a.$$

$$= \frac{\sqrt[4]{a^3b}}{a}$$

(b) $\dfrac{-1}{\sqrt[3]{16}} - \dfrac{5}{\sqrt[3]{128}} + \dfrac{4}{\sqrt[3]{2}}$

Begin by simplifying all radicals:

$$\sqrt[3]{16} = \sqrt[3]{8 \cdot 2} = 2\sqrt[3]{2}$$

and

$$\sqrt[3]{128} = \sqrt[3]{64 \cdot 2} = 4\sqrt[3]{2}.$$

Now proceed as follows.

$$\dfrac{-1}{\sqrt[3]{16}} - \dfrac{5}{\sqrt[3]{128}} + \dfrac{4}{\sqrt[3]{2}} = \dfrac{-1}{2\sqrt[3]{2}} - \dfrac{5}{4\sqrt[3]{2}} + \dfrac{4}{\sqrt[3]{2}}$$

$$= \dfrac{-2}{4\sqrt[3]{2}} - \dfrac{5}{4\sqrt[3]{2}} + \dfrac{16}{4\sqrt[3]{2}} \qquad \text{Write with a common denominator.}$$

$$= \dfrac{9}{4\sqrt[3]{2}} \qquad \text{Combine terms.}$$

$$= \dfrac{9\sqrt[3]{2^2}}{4\sqrt[3]{2} \cdot \sqrt[3]{2^2}} \qquad \text{Rationalize the denominator.}$$

$$= \dfrac{9\sqrt[3]{4}}{4 \cdot \sqrt[3]{8}}$$

$$= \dfrac{9\sqrt[3]{4}}{4 \cdot 2}$$

$$= \dfrac{9\sqrt[3]{4}}{8} \quad ■$$

NOTE In Example 11(a), $\sqrt[4]{a^4} = a$ (not $|a|$) because of the assumption that a is positive.

In Example 9(b), we saw that

$$(\sqrt{7} - \sqrt{10})(\sqrt{7} + \sqrt{10}) = -3,$$

a rational number. This suggests a way to rationalize a denominator that is a binomial in which one or both terms is a radical. The expressions $a\sqrt{m} + b\sqrt{n}$ and $a\sqrt{m} - b\sqrt{n}$ are **conjugates.**

■ *Example 12*

RATIONALIZING A BINOMIAL DENOMINATOR

Rationalize the denominator of $\dfrac{1}{1 - \sqrt{2}}$.

As mentioned above, the best approach here is to multiply both numerator and denominator by the conjugate of the denominator, in this case $1 + \sqrt{2}$.

$$\dfrac{1}{1 - \sqrt{2}} = \dfrac{1(1 + \sqrt{2})}{(1 - \sqrt{2})(1 + \sqrt{2})} = \dfrac{1 + \sqrt{2}}{1 - 2} = -1 - \sqrt{2} \quad ■$$

1.8 EXERCISES ■

Write in radical form. Assume all variables are positive real numbers. See Example 2.

1. $(-m)^{2/3}$ **2.** $p^{5/4}$ **3.** $5m^{4/5}$ **4.** $-2k^{1/6}$ **5.** $-4z^{-1/3}$ **6.** $10a^{-3/4}$ **7.** $(2m + p)^{2/3}$ **8.** $(5r + 3t)^{4/7}$

Write in exponential form. Assume all variables are nonnegative real numbers. See Example 3.

9. $\sqrt[5]{k^2}$ **10.** $-\sqrt[4]{z^5}$ **11.** $-\sqrt[3]{a^2}$ **12.** $2\sqrt{m^9}$

13. $-3\sqrt{5p^3}$ **14.** $m\sqrt{2y^5}$ **15.** $18\sqrt{m^2n^3p}$ **16.** $-12\sqrt[5]{x^3y^2z^4}$

17. What is wrong with the statement $\sqrt[3]{4} \cdot \sqrt[3]{4} = 4$?

18. Which of the following expressions is *not* simplified? Give the simplified form.

   (a) $\sqrt[3]{2y}$ **(b)** $\dfrac{\sqrt{5}}{2}$ **(c)** $\sqrt[4]{m^3}$ **(d)** $\sqrt{\dfrac{3}{4}}$

19. Explain how to rationalize the denominator of $\sqrt[3]{\dfrac{3}{2}}$.

20. How can we multiply $\sqrt{2}$ and $\sqrt[3]{2}$?

Simplify each of the following. Assume that all variables represent positive real numbers. See Examples 1, 5, 6, 8, 10, and 11(a).

21. $\sqrt[3]{125}$ **22.** $\sqrt[4]{81}$ **23.** $\sqrt[5]{-3125}$ **24.** $\sqrt[3]{343}$ **25.** $\sqrt{50}$

26. $\sqrt{45}$ **27.** $\sqrt[3]{81}$ **28.** $\sqrt[3]{250}$ **29.** $-\sqrt[4]{32}$ **30.** $-\sqrt[4]{243}$

31. $-\sqrt{\dfrac{9}{5}}$ **32.** $-\sqrt[3]{\dfrac{3}{2}}$ **33.** $-\sqrt[3]{\dfrac{4}{5}}$ **34.** $\sqrt[4]{\dfrac{3}{2}}$ **35.** $\sqrt[3]{16(-2)^4(2)^8}$

36. $\sqrt[3]{25(3)^4(5)^3}$ **37.** $\sqrt{8x^5z^8}$ **38.** $\sqrt{24m^6n^5}$ **39.** $\sqrt[3]{16z^5x^8y^4}$ **40.** $-\sqrt[6]{64a^{12}b^8}$

41. $\sqrt[4]{m^2n^7p^8}$ **42.** $\sqrt[4]{x^8y^7z^9}$ **43.** $\sqrt[4]{x^4 + y^4}$ **44.** $\sqrt[3]{27 + a^3}$ **45.** $\sqrt{\dfrac{2}{3x}}$

46. $\sqrt{\dfrac{5}{3p}}$ **47.** $\sqrt{\dfrac{x^5y^3}{z^2}}$ **48.** $\sqrt{\dfrac{g^3h^5}{r^3}}$ **49.** $\sqrt[3]{\dfrac{8}{x^2}}$ **50.** $\sqrt[3]{\dfrac{9}{16p^4}}$

51. $\sqrt[4]{\dfrac{g^3h^5}{9r^6}}$ **52.** $\sqrt[4]{\dfrac{32x^5}{y^5}}$ **53.** $\dfrac{\sqrt[3]{mn} \cdot \sqrt[3]{m^2}}{\sqrt[3]{n^2}}$ **54.** $\dfrac{\sqrt[3]{8m^2n^3} \cdot \sqrt[3]{2m^2}}{\sqrt[3]{32m^4n^3}}$ **55.** $\dfrac{\sqrt[4]{32x^5y} \cdot \sqrt[4]{2xy^4}}{\sqrt[4]{4x^3y^2}}$

56. $\dfrac{\sqrt[4]{rs^2t^3} \cdot \sqrt[4]{r^3s^2t}}{\sqrt[4]{r^2t^3}}$ **57.** $\sqrt[3]{\sqrt{4}}$ **58.** $\sqrt[4]{\sqrt[3]{2}}$ **59.** $\sqrt[6]{\sqrt[3]{x}}$ **60.** $\sqrt[8]{\sqrt[4]{y}}$

Simplify each of the following, assuming that all variables represent nonnegative numbers. See Examples 7, 9, and 11(b).

61. $4\sqrt{3} - 5\sqrt{12} + 3\sqrt{75}$ **62.** $2\sqrt{5} - 3\sqrt{20} + 2\sqrt{45}$ **63.** $3\sqrt{28p} - 4\sqrt{63p} + \sqrt{112p}$

64. $9\sqrt{8k} + 3\sqrt{18k} - \sqrt{32k}$ **65.** $2\sqrt[3]{3} + 4\sqrt[3]{24} - \sqrt[3]{81}$ **66.** $\sqrt[3]{32} - 5\sqrt[3]{4} + 2\sqrt[3]{108}$

67. $\dfrac{1}{\sqrt{3}} - \dfrac{2}{\sqrt{12}} + 2\sqrt{3}$ **68.** $\dfrac{1}{\sqrt{2}} + \dfrac{3}{\sqrt{8}} + \dfrac{1}{\sqrt{32}}$ **69.** $\dfrac{5}{\sqrt[3]{2}} - \dfrac{2}{\sqrt[3]{16}} + \dfrac{1}{\sqrt[3]{54}}$

70. $\dfrac{-4}{\sqrt[3]{3}} + \dfrac{1}{\sqrt[3]{24}} - \dfrac{2}{\sqrt[3]{81}}$ **71.** $(\sqrt{2} + 3)(\sqrt{2} - 3)$ **72.** $(\sqrt{5} + \sqrt{2})(\sqrt{5} - \sqrt{2})$

73. $(\sqrt[3]{11} - 1)(\sqrt[3]{11^2} + \sqrt[3]{11} + 1)$ **74.** $(\sqrt[3]{7} + 3)(\sqrt[3]{7^2} - 3\sqrt[3]{7} + 9)$ **75.** $(\sqrt{3} + \sqrt{8})^2$

76. $(\sqrt{2} - 1)^2$ **77.** $(3\sqrt{2} + \sqrt{3})(2\sqrt{3} - \sqrt{2})$ **78.** $(4\sqrt{5} - 1)(3\sqrt{5} + 2)$

79. $(2\sqrt[3]{3} + 1)(\sqrt[3]{3} - 4)$ **80.** $(\sqrt[3]{4} + 3)(5\sqrt[3]{4} + 1)$

Rationalize the denominator of each of the following. Assume that all variables represent nonnegative numbers and that no denominators are zero. See Example 12.

81. $\dfrac{\sqrt{3}}{\sqrt{5}+\sqrt{3}}$

82. $\dfrac{\sqrt{7}}{\sqrt{3}-\sqrt{7}}$

83. $\dfrac{1+\sqrt{3}}{3\sqrt{5}+2\sqrt{3}}$

84. $\dfrac{\sqrt{7}-1}{2\sqrt{7}+4\sqrt{2}}$

85. $\dfrac{p}{\sqrt{p}+2}$

86. $\dfrac{\sqrt{r}}{3-\sqrt{r}}$

87. $\dfrac{a}{\sqrt{a+b}-1}$

88. $\dfrac{3m}{2+\sqrt{m+n}}$

In advanced mathematics it is sometimes useful to write a radical expression with a rational numerator. The procedure is similar to rationalizing the denominator. Rationalize the numerator of each of the following expressions.

89. $\dfrac{1+\sqrt{2}}{2}$

90. $\dfrac{1-\sqrt{3}}{3}$

91. $\dfrac{\sqrt{x}}{1+\sqrt{x}}$

92. $\dfrac{\sqrt{p}}{1-\sqrt{p}}$

93. $\dfrac{\sqrt{x}+\sqrt{x+1}}{\sqrt{x}-\sqrt{x+1}}$

94. $\dfrac{\sqrt{p}+\sqrt{p^2-1}}{\sqrt{p}-\sqrt{p^2-1}}$

Write the following without radicals. Use absolute value if necessary. See Example 4.

95. $\sqrt{(m+n)^2}$

96. $\sqrt[4]{(a+2b)^4}$

97. $\sqrt{z^2-6zx+9x^2}$

98. $\sqrt[3]{(r+2s)(r^2+4rs+4s^2)}$

CHAPTER 1 SUMMARY ■ ─────────────

SECTION	TERMS	KEY IDEAS
1.1 The Real Numbers	exponent base identity element for addition (additive identity) identity element for multiplication (multiplicative identity) additive inverse (negative) multiplicative inverse (reciprocal)	**Sets of Numbers** **Real Numbers** $\{x \mid x$ corresponds to a point on a number line$\}$ **Integers** $\{\ldots, -3, -2, -1, 0, 1, 2, 3, \ldots\}$ **Rational Numbers** $\left\{\dfrac{p}{q} \mid p \text{ and } q \text{ are integers and } q \neq 0\right\}$ **Irrational Numbers** $\{x \mid x$ is real but not rational$\}$ **Whole Numbers** $\{0, 1, 2, 3, 4, \ldots\}$ **Natural Numbers** $\{1, 2, 3, 4, \ldots\}$

SECTION	TERMS		KEY IDEAS
			Properties of the Real Numbers For all real numbers a, b, and c: **Commutative Properties** $$a + b = b + a$$ $$ab = ba$$ **Associative Properties** $$(a + b) + c = a + (b + c)$$ $$(ab)c = a(bc)$$ **Identity Properties** There exists a unique real number 0 such that $$a + 0 = a \quad \text{and} \quad 0 + a = a.$$ There exists a unique real number 1 such that $$a \cdot 1 = a \quad \text{and} \quad 1 \cdot a = a.$$ **Inverse Properties** There exists a unique real number $-a$ such that $$a + (-a) = 0 \quad \text{and} \quad (-a) + a = 0.$$ If $a \neq 0$, there exists a unique real number $1/a$ such that $$a \cdot \frac{1}{a} = 1 \quad \text{and} \quad \frac{1}{a} \cdot a = 1.$$ **Distributive Property** $$a(b + c) = ab + ac$$
1.2 Order and Absolute Value	coordinate inequalities absolute value		$a > b$ if a is to the right of b or if $a - b$ is positive. $a < b$ if a is to the left of b or if $a - b$ is negative.
1.3 Polynomials	algebraic expressions term coefficient like terms	polynomial degree trinomial binomial monomial	**Special Products** $$(x + y)(x - y) = x^2 - y^2$$ $$(x + y)^2 = x^2 + 2xy + y^2$$ $$(x - y)^2 = x^2 - 2xy + y^2$$

SECTION	TERMS	KEY IDEAS
1.4 The Binomial Theorem	Pascal's triangle *n*-factorial binomial coefficient	**Binomial Theorem** For any positive integer n: $$(x + y)^n = x^n + \binom{n}{n-1}x^{n-1}y + \binom{n}{n-2}x^{n-2}y^2$$ $$+ \binom{n}{n-3}x^{n-3}y^3$$ $$+ \cdots + \binom{n}{n-r}x^{n-r}y^r$$ $$+ \cdots + \binom{n}{1}xy^{n-1} + y^n.$$
1.5 Factoring	factoring factor factored form prime (irreducible) polynomial	**Factoring Patterns** $$x^2 - y^2 = (x + y)(x - y)$$ $$x^2 + 2xy + y^2 = (x + y)^2$$ $$x^2 - 2xy + y^2 = (x - y)^2$$ $$x^3 - y^3 = (x - y)(x^2 + xy + y^2)$$ $$x^3 + y^3 = (x + y)(x^2 - xy + y^2)$$
1.6 Rational Expressions	rational expression complex fraction	
1.7 Rational Exponents		**Rules for Exponents** Let r and s be rational numbers. The results below are valid for all positive numbers a and b. $$a^r \cdot a^s = a^{r+s} \quad (ab)^r = a^r \cdot b^r \quad (a^r)^s = a^{rs}$$ $$\frac{a^r}{a^s} = a^{r-s} \quad \left(\frac{a}{b}\right)^r = \frac{a^r}{b^r} \quad a^{-r} = \frac{1}{a^r}$$

1.8 Radicals	radical sign
	radicand
	index
	rationalizing the denominator
	conjugate
		Radical Notation
		If a is a real number, n is a positive integer, and $a^{1/n}$ is defined, then
		$$\sqrt[n]{a} = a^{1/n}.$$
		If m is an integer, n is a positive integer, and a is a real number for which $\sqrt[n]{a}$ is defined, then
		$$a^{m/n} = (\sqrt[n]{a})^m = \sqrt[n]{a^m}.$$

Let set $K = \{-12, -6, -.9, -\sqrt{7}, -\sqrt{4}, 0, 1/8, \pi/4, 6, \sqrt{11}\}$. *List the elements of K that are members of the following sets.*

1. Integers

2. Rational numbers

3. Irrational numbers

For Exercises 4–7 choose all words from the following list that apply.

(a) natural numbers   **(b)** whole numbers   **(c)** integers   **(d)** rational numbers
(e) irrational numbers   **(f)** real numbers

4. 0

5. $-\sqrt{25}$

6. $\dfrac{5}{8}$

7. $\dfrac{3\pi}{4}$

Use the order of operations to simplify each of the following.

8. $-4 - [2 - (-3^2)]$

9. $[2^3 - (-5)] - 2^2$

10. $(-4 - 1)(-3 - 5) - 2^3$

11. $(6 - 9)(-2 - 7) - (-4)$

12. $\left(-\dfrac{5}{9} - \dfrac{2}{3}\right) - \dfrac{5}{6}$

13. $\left(-\dfrac{2^3}{5} - \dfrac{3}{4}\right) - \left(-\dfrac{1}{2}\right)$

14. $\dfrac{6(-4) - 3^2(-2)^3}{-5[-2 - (-6)]}$

15. $\dfrac{(-7)(-3) - (-2^3)(-5)}{(-2^2 - 2)(-1 - 6)}$

Evaluate each of the following if $a = -1$, $b = -2$, *and* $c = 4$.

16. $9a - 5b + 4c$

17. $-4(2a - 5b)$

18. $\dfrac{9a + 2b}{a + b + c}$

19. Suppose your friend missed class the day order of operations was discussed. Explain to her how to work the following problem. Evaluate $\dfrac{4a - 5c}{2a + 3b}$ if $a = -1$, $b = -2$, and $c = 4$.

Identify the properties illustrated in each of the following.

20. $8(5 + 9) = (5 + 9)8$

21. $4 \cdot 6 + 4 \cdot 12 = 4(6 + 12)$

22. $3 \cdot (4 \cdot 2) = (3 \cdot 4) \cdot 2$

23. $-(r - 2) = -r + 2$

24. $(9 + p) + 0 = 9 + p$

25. $[6 \cdot 5 + 8 \cdot 5] \cdot 2 = [(6 + 8) \cdot 5] \cdot 2$

Use the distributive property to rewrite sums as products and products as sums.

26. $-11(m - n)$

27. $k(r + s - t)$

28. $mn - 9m$

Write the following numbers in numerical order from smallest to largest.

29. $|6 - 4|, -|-2|, |8 + 1|, -|3 - (-2)|$

30. $\sqrt{7}, -\sqrt{8}, -|\sqrt{16}|, |-\sqrt{12}|$

Write without absolute value bars.

31. $-|-6| + |3|$

32. $7 - |-8|$

33. $|\sqrt{8} - 3|$

34. $|\sqrt{52} - 8|$

35. $|m - 3|$ if $m > 3$

36. $|-6 - x^2|$

37. $|\pi - 4|$

38. $|3 + 5k|$ if $k < -3/5$

Perform each of the following operations. Assume that all variables appearing as exponents represent positive integers.

39. $(3q^3 - 9q^2 + 6) + (4q^3 - 8q + 3)$

40. $2(3y^6 - 9y^2 + 2y) - (5y^6 - 10y^2 - 4y)$

41. $(8y - 7)(2y + 7)$

42. $(2r + 11s)(4r - 9s)$

43. $(3k - 5m)^2$

44. $(4a - 3b)^2$

45. $(3w - 2)(5w^2 - 4w + 1)$

46. $(2k + 5)(3k^3 - 4k^2 + 8k - 2)$

47. $(p^q + 1)(p^q - 3)$

48. $(a^y + 6)^2$

Use the binomial theorem to expand each of the following.

49. $(x + 2y)^4$

50. $\left(\dfrac{k}{2} - g\right)^5$

Find the indicated term or terms for each of the following expansions.

51. Fifth term of $(3x - 2y)^6$

52. Eighth term of $(2m + n^2)^{12}$

53. First four terms of $(3 + x)^{16}$

54. Last three terms of $(2m - 3n)^{15}$

Factor as completely as possible.

55. $7z^2 - 9z^3 + z$

56. $3(z - 4)^2 + 9(z - 4)^3$

57. $r^2 + rp - 42p^2$

58. $z^2 - 6zk - 16k^2$

59. $6m^2 - 13m - 5$

60. $48a^8 - 12a^7b - 90a^6b^2$

61. $169y^4 - 1$

62. $49m^8 - 9n^2$

63. $8y^3 - 1000z^6$

64. $6(3r - 1)^2 + (3r - 1) - 35$

65. $ar - 3as + 5rb - 15sb$

66. $15mp + 9mq - 10np - 6nq$

67. $(16m^2 - 56m + 49) - 25a^2$

68. A student factored $64a^3 + 8b^3$ as $(4a + 2b)(16a^2 - 8ab + 4b^2)$. Is the polynomial factored completely? Explain.

69. Which one of the following is equal to $\dfrac{2a + b}{4a^2 - b^2}$?

(a) $\dfrac{1}{2a - b}$ **(b)** $\dfrac{2}{2a - b}$ **(c)** $\dfrac{1}{2ab}$ **(d)** $-\dfrac{1}{2ab}$

Perform each of the following operations.

70. $\dfrac{k^2 + k}{8k^3} \cdot \dfrac{4}{k^2 - 1}$

71. $\dfrac{3r^3 - 9r^2}{r^2 - 9} \div \dfrac{8r^3}{r + 3}$

72. $\dfrac{x^2 + x - 2}{x^2 + 5x + 6} \div \dfrac{x^2 + 3x - 4}{x^2 + 4x + 3}$

73. $\dfrac{27m^3 - n^3}{3m - n} \div \dfrac{9m^2 + 3mn + n^2}{9m^2 - n^2}$

74. $\dfrac{p^2 - 36q^2}{(p - 6q)^2} \cdot \dfrac{p^2 - 5pq - 6q^2}{p^2 - 6pq + 36q^2} \div \dfrac{5p}{p^3 + 216q^3}$

75. $\dfrac{1}{4y} + \dfrac{8}{5y}$

76. $\dfrac{m}{4 - m} + \dfrac{3m}{m - 4}$

77. $\dfrac{3}{x^2 - 4x + 3} - \dfrac{2}{x^2 - 1}$

78. $\left[\dfrac{1}{(x + h)^2 + 16} - \dfrac{1}{x^2 + 16}\right] \div h$

79. $\dfrac{\dfrac{1}{p} + \dfrac{1}{q}}{1 - \dfrac{1}{pq}}$

80. $\dfrac{3 + \dfrac{2m}{m^2 - 4}}{\dfrac{5}{m - 2}}$

Simplify each of the following. Write results with only positive exponents. Assume that all variables represent positive real numbers.

81. 2^{-6}

82. -3^{-2}

83. $\left(-\dfrac{5}{4}\right)^{-2}$

84. $3^{-1} - 4^{-1}$

85. $(5z^3)(-2z^5)$

86. $(8p^2q^3)(-2p^5q^{-4})$

87. $(-6p^5w^4m^{12})^0$

88. $(-6x^2y^{-3}z^2)^{-2}$

89. $\dfrac{-8y^7p^{-2}}{y^{-4}p^{-3}}$

90. $\dfrac{a^{-6}(a^{-8})}{a^{-2}(a^{11})}$

91. $\dfrac{(p + q)^4(p + q)^{-3}}{(p + q)^6}$

92. $\dfrac{[p^2(m + n)^3]^{-2}}{p^{-2}(m + n)^{-5}}$

93. $\left(\dfrac{r^{-1}s^2}{t^{-2}}\right)^{-2}$

94. $\dfrac{(2x^{-3})^2(3x^2)^{-2}}{6(x^2y^3)}$

95. $\dfrac{p^4(p^{-2})}{p^{5/3}}$

96. $(2+p)^{5/4}(2+p)^{-9/4}$

97. $(7r^{1/2})(2r^{3/4})(-r^{1/6})$

98. $(a^{3/4}b^{2/3})(a^{5/8}b^{-5/6})$

99. $\dfrac{y^{5/3}\cdot y^{-2}}{y^{-5/6}}$

100. $\left(\dfrac{25m^3n^5}{m^{-2}n^6}\right)^{-1/2}$

Simplify. Assume that all variables represent positive real numbers. Write without denominators.

101. $\dfrac{k^{2+p}\cdot k^{-4p}}{k^{6p}}$

102. $\dfrac{(r^{3+2z})(r^{-5-z})}{(r^{-z/2})^{-4}}$

Find each product. Assume that all variables represent positive real numbers.

103. $2z^{1/3}(5z^2-2)$

104. $-m^{3/4}(8m^{1/2}+4m^{-3/2})$

105. $(p+p^{1/2})(3p-5)$

106. $(m^{1/2}-4m^{-1/2})^2$

Simplify. Assume that all variables represent positive numbers.

107. $\sqrt{200}$

108. $\sqrt[3]{16}$

109. $\sqrt[4]{1250}$

110. $-\sqrt{\dfrac{16}{3}}$

111. $-\sqrt[3]{\dfrac{2}{5p^2}}$

112. $\sqrt{\dfrac{2^7y^8}{m^3}}$

113. $\sqrt[4]{\sqrt[3]{m}}$

114. $\dfrac{\sqrt[4]{8p^2q^5}\cdot\sqrt[4]{2p^3q}}{\sqrt[4]{p^5q^2}}$

115. $(\sqrt[3]{2}+4)(\sqrt[3]{2^2}-4\sqrt[3]{2}+16)$

116. $\dfrac{3}{\sqrt{5}}-\dfrac{2}{\sqrt{45}}+\dfrac{6}{\sqrt{80}}$

117. $\sqrt{18m^3}-3m\sqrt{32m}+5\sqrt{m^3}$

118. $\dfrac{2}{7-\sqrt{3}}$

119. $\dfrac{6}{3-\sqrt{2}}$

120. $\dfrac{z}{\sqrt{z+1}}$

121. $\dfrac{k}{\sqrt{k}-3}$

122. $\dfrac{\sqrt{x}-\sqrt{x-2}}{\sqrt{x}+\sqrt{x-2}}$

CHAPTER 1   TEST ■

Let set $A=\left\{-10,-\dfrac{9}{3},0,\dfrac{2}{3},\dfrac{4}{\pi},\sqrt{25}\right\}$. List the elements of set A that are members of the following sets.

1. Irrational numbers

2. Integers

3. Evaluate the following when $x=-2$, $y=-3$, and $z=-4$.

$$\dfrac{2x-|3y|-|z|}{-|x+y|+z}$$

Identify the properties illustrated in each of the following.

4. $2(x+3)=2(3+x)$

5. $3(x+4)=3x+12$

In Exercises 6–8, perform each of the following operations. Assume that all variables appearing as exponents represent positive integers.

6. $(a^2 - 3a + 1) - (a - 2a^2) + (-4 + 3a - a^2)$ **7.** $(2x^n - 3)^2$

8. $(3y^2 - y + 4)(y + 2)$

9. Use the binomial theorem to expand $(2x - 3y)^4$.

10. Find the third term in the expansion of $(w - 2y)^6$.

Factor as completely as possible.

11. $x^4 - 16$ **12.** $24m^3 - 14m^2 - 24m$ **13.** $x^3y^2 - 9x^3 - 8y^2 + 72$

14. What is wrong with the following simplification?

$$\frac{4(x - 1) - (2x + 3)}{x(x - 1)} = \frac{4 - (2x + 3)}{x}$$
$$= \frac{4 - 2x - 3}{x}$$
$$= \frac{1 - 2x}{x}$$

Perform each of the following operations.

15. $\dfrac{5x^2 - 9x - 2}{30x^3 + 6x^2} \cdot \dfrac{2x^8 + 6x^7 + 4x^6}{x^4 - 3x^2 - 4}$

16. $\dfrac{x}{x^2 + 3x + 2} + \dfrac{2x}{2x^2 - x - 3}$

17. $\dfrac{a + b}{2a - 3} - \dfrac{a - b}{3 - 2a}$

18. $\dfrac{y - 2}{y - \dfrac{4}{y}}$

Simplify each of the following. Write results without negative exponents. Assume that all variables represent positive numbers.

19. $\left(\dfrac{x^{-1}x^3}{2x^0x^{-5}}\right)^{-2}$

20. $\left(\dfrac{x^{-2}y^{-1/3}z}{x^{-5/3}y^{-2/3}z^{2/3}}\right)^3$

Simplify. Assume that all variables represent positive numbers.

21. $\sqrt{18x^5y^8}$ **22.** $\sqrt[5]{\dfrac{x}{4y^2}}$ **23.** $\sqrt{32x} + \sqrt{2x} - \sqrt{18x}$

24. $\dfrac{3}{\sqrt{2} - 1}$ **25.** $(\sqrt{x} - \sqrt{y})(\sqrt{x} + \sqrt{y})$

EQUATIONS AND INEQUALITIES

2.1 LINEAR EQUATIONS

2.2 APPLICATIONS OF LINEAR EQUATIONS

2.3 COMPLEX NUMBERS

2.4 QUADRATIC EQUATIONS

2.5 APPLICATIONS OF QUADRATIC EQUATIONS

2.6 OTHER TYPES OF EQUATIONS

2.7 INEQUALITIES

2.8 ABSOLUTE VALUE EQUATIONS AND INEQUALITIES

$\mathbf{A}$pplications of algebra in everyday life frequently require solving one or more *equations*. In fields such as business, many applications of algebra also require the use of *inequalities*. Methods of solving several different kinds of equations and inequalities are discussed in this chapter.

2.1 ———— # LINEAR EQUATIONS

An **equation** is a statement that two expressions are equal. Examples of equations include

$$x + 2 = 9, \qquad 11y = 5y + 6y, \qquad \text{and} \qquad x^2 - 2x - 1 = 0.$$

To **solve** an equation means to find all numbers that make the equation a true statement. Such numbers are called **solutions** of the equation. A number that is a solution of an equation is said to **satisfy** the equation, and the solutions of an equation make up its **solution set.**

An equation satisfied by every number that is a meaningful replacement for the variable is called an **identity.** Examples of identities are

$$3x + 4x = 7x \qquad \text{and} \qquad x^2 - 3x + 2 = (x - 2)(x - 1).$$

Equations that are satisfied by some numbers but not by others are called **conditional equations.** Examples of conditional equations are

$$2m + 3 = 7 \qquad \text{and} \qquad \frac{5r}{r - 1} = 7.$$

The equation $3(x + 1) = 5 + 3x$

is neither an identity nor a conditional equation. Multiplying on the left side gives

$$3x + 3 = 5 + 3x,$$

which is false for every value of x. Such an equation is called a **contradiction.**

■ *Example 1*

IDENTIFYING AN EQUATION AS CONDITIONAL, AN IDENTITY, OR A CONTRADICTION

Decide whether each of the following equations is an identity, a conditional equation, or a contradiction.

(a) $9p^2 - 25 = (3p + 5)(3p - 5)$

Since the product of $3p + 5$ and $3p - 5$ is $9p^2 - 25$, the given equation is true for *every* value of p and is an identity. Its solution set is {all real numbers}.

(b) $5y - 4 = 11$

Replacing y with 3 gives

$$5 \cdot 3 - 4 = 11$$
$$11 = 11,$$

a true statement. On the other hand, $y = 4$ leads to

$$5 \cdot 4 - 4 = 11$$
$$16 = 11,$$

a false statement. The equation $5y - 4 = 11$ is true for some values of y, but not all, and thus is a conditional equation. (The word *some* in mathematics means "at least one." We can therefore say that the statement $5y - 4 = 11$ is true for *some* replacements of y, even though it turns out to be true only for $y = 3$.) Since 3 is the only number that is a solution (as can be shown using methods discussed later), the solution set is $\{3\}$.

(c) $(a - 2)^2 + 3 = a^2 - 4a + 2$

The left side can be rewritten as follows.

$$a^2 - 4a + 4 + 3 = a^2 - 4a + 2$$
$$a^2 - 4a + 7 = a^2 - 4a + 2$$

The final equation is false for every value of a, so this equation is a contradiction. The solution set contains no elements. It is called the *empty* or *null* set, and is symbolized $\emptyset$. ■

Equations with the same solution set are called **equivalent equations.** For example, $x + 1 = 5$ and $6x + 3 = 27$ are equivalent equations because they have the same solution set, $\{4\}$.

■ *Example 2*

DETERMINING
WHETHER TWO
EQUATIONS ARE
EQUIVALENT

Decide which of the following pairs of equations are equivalent.

(a) $2x - 1 = 3$ and $12x + 7 = 31$

Each of these equations has solution set $\{2\}$. Since the solution sets are equal, the equations are equivalent.

(b) $x = 3$ and $x^2 = 9$

The solution set for $x = 3$ is $\{3\}$, while the solution set for the equation $x^2 = 9$ is $\{3, -3\}$. Since the solution sets are not equal, the equations are not equivalent. ■

One way to solve an equation is to rewrite it as a series of simpler equivalent equations. These simpler equations often can be obtained with the *addition and multiplication properties of equality.*

ADDITION AND
MULTIPLICATION
PROPERTIES OF
EQUALITY

For real numbers a, b, and c:

$a = b$ and $a + c = b + c$ are equivalent. *(The same number may be added to both sides of an equation without changing the solution set.)*

If $c \neq 0$, then $a = b$ and $ac = bc$ are equivalent. *(Both sides of an equation may be multiplied by the same nonzero number without changing the solution set.)*

■ *Example 3*

SOLVING A LINEAR
EQUATION

Solve $3(2x - 4) = 7 - (x + 5)$.

Use the distributive property and then collect like terms to get the following sequence of simpler equivalent equations.

$$3(2x - 4) = 7 - (x + 5)$$

$$6x - 12 = 7 - x - 5 \qquad \text{Distributive property}$$

$$6x - 12 = 2 - x$$

$$x + 6x - 12 = x + 2 - x \qquad \text{Add } x \text{ to each side.}$$

$$7x - 12 = 2 \qquad \text{Combine terms.}$$

$$12 + 7x - 12 = 12 + 2 \qquad \text{Add 12 to each side.}$$

$$7x = 14 \qquad \text{Combine terms.}$$

$$\frac{1}{7} \cdot 7x = \frac{1}{7} \cdot 14 \qquad \text{Multiply both sides by } \frac{1}{7}.$$

$$x = 2$$

To check, replace x with 2 in the original equation, getting

$$3(2x - 4) = 7 - (x + 5) \qquad ? \qquad \text{Original equation}$$

$$3(2 \cdot 2 - 4) = 7 - (2 + 5) \qquad ? \qquad \text{Let } x = 2.$$

$$3(4 - 4) = 7 - (7)$$

$$0 = 0. \qquad ? \qquad \text{True}$$

Since replacing x with 2 results in a true statement, 2 is the solution of the given equation. The solution set is therefore $\{2\}$. ■

The equation in Example 3 is a *linear equation,* as are most of the equations in this section.

LINEAR EQUATION IN ONE VARIABLE	A **linear equation** in one variable is an equation that can be written in the form $$ax + b = 0,$$ where $a \neq 0$.

The next examples will show how some equations can be simplified before solving.

■ *Example 4*

SIMPLIFYING AN
EQUATION BEFORE
SOLVING

Solve each equation.

(a) $\dfrac{3p - 1}{3} - \dfrac{2p}{p - 1} = p$

This equation does not satisfy the definition of a linear equation. However, the equation can be written as a linear equation and solved as one if the restrictions on the denominator are noted. Multiply both sides of the equation by the common denominator, $3(p - 1)$, assuming $p \neq 1$. Doing this gives

$$3(p - 1)\left(\frac{3p - 1}{3}\right) - 3(p - 1)\left(\frac{2p}{p - 1}\right) = 3(p - 1)p$$

$$(p - 1)(3p - 1) - 3(2p) = 3p(p - 1)$$

$$3p^2 - 4p + 1 - 6p = 3p^2 - 3p.$$

A simpler equivalent equation comes from combining terms and adding $-3p^2$ to both sides, producing

$$-10p + 1 = -3p.$$

Now add $10p$ to both sides to get

$$1 = 7p.$$

Finally, multiplying both sides by 1/7 gives

$$\frac{1}{7} = p.$$

Check 1/7 in the given equation to verify that the solution set is $\{1/7\}$. The restriction $p \neq 1$ does not affect the solution set here, since $1/7 \neq 1$. Since the original equation and the linear equation found by multiplying both sides by $3(p - 1)$ have the same solution set, $\{1/7\}$, they are equivalent equations.

(b) $\dfrac{x}{x - 2} = \dfrac{2}{x - 2} + 2$

Multiply both sides of the equation by $x - 2$, assuming that $x \neq 2$.

$$x = 2 + 2(x - 2)$$

$$x = 2 + 2x - 4$$

$$x = 2$$

It is necessary to assume $x \neq 2$ in order to be able to multiply both sides of the equation by $x - 2$. Since $x = 2$, however, the multiplication property of equations does not apply. The solution is $\emptyset$. (Substituting 2 for x in the original equation would result in a denominator of 0.) The original equation and the linear equation $x = 2 + 2(x - 2)$ are *not* equivalent. ■

CAUTION | It is essential to check proposed solutions (such as 2 in Example 4(b)) whenever each side of an equation is multiplied by a variable expression. Do not forget this important step.

■ *Example 5*

SIMPLIFYING AN
EQUATION BEFORE
SOLVING

Solve $\dfrac{2}{x-1} - \dfrac{4}{3x} = \dfrac{1}{x^2 - x}$.

First find a common denominator. Since $x^2 - x$ can be factored as $x(x - 1)$, the least common denominator is $3x(x - 1)$. Multiply both sides of the equation by $3x(x - 1)$.

$$3x(x-1)\frac{2}{x-1} - 3x(x-1)\frac{4}{3x} = 3x(x-1)\frac{1}{x(x-1)} \quad (x \neq 0, 1)$$

$$6x - 4(x - 1) = 3$$

$$6x - 4x + 4 = 3$$

$$2x = -1$$

$$x = -\frac{1}{2}$$

Check the answer in the original equation to verify that the solution set is $\{-1/2\}$. ■

Sometimes an equation with more than one letter must be *solved for a specified variable.* Such an equation is called a **literal equation.** This process is shown in the next example. (As a general rule letters from the beginning of the alphabet, such as *a, b, c,* and so on, are used to represent constants, while letters such as *x, y,* and *z* represent variables.)

■ *Example 6*

SOLVING A LITERAL
EQUATION FOR A
SPECIFIED VARIABLE

Solve the equation $3(2x - 5a) + 4b = 4x - 2$ for x.

Using the distributive property gives

$$6x - 15a + 4b = 4x - 2.$$

Treat x as the variable and the other letters as constants. Get all terms with x on one side and all terms without x on the other side.

$$6x - 4x = 15a - 4b - 2$$

$$2x = 15a - 4b - 2$$

$$x = \frac{15a - 4b - 2}{2} \quad ■$$

2.1 EXERCISES ■

Decide whether each of the following equations is an identity, a conditional equation, or a contradiction. Give the solution set. See Example 1.

1. $x^2 + 5x = x(x + 5)$

2. $3y + 4 = 5(y - 2)$

3. $2(x - 7) = 5x + 3 - x$

4. $2x - 4 = 2(x + 2)$

5. $\dfrac{m + 3}{m} = 1 + \dfrac{3}{m}$

6. $\dfrac{p}{2 - p} = \dfrac{2}{p} - 1$

7. $4q + 20q - 25 = 24q + 5$

8. $3(k + 2) - 5(k + 2) = -2k - 4$

Decide which of the following pairs of equations are equivalent. See Example 2.

9. $\dfrac{3x}{x-1} = \dfrac{2}{x-1}$

$3x = 2$

10. $\dfrac{x+1}{12} = \dfrac{5}{12}$

$x + 1 = 5$

11. $\dfrac{x}{x-2} = \dfrac{2}{x-2}$

$x = 2$

12. $\dfrac{x+3}{x+1} = \dfrac{2}{x+1}$

$x = -1$

13. $x = 4$

$x^2 = 16$

14. $z^2 = 9$

$z = 3$

15. Which one of the following is not a linear equation?
   (a) $5x + 7(x - 1) = -3x$ **(b)** $8x^2 - 4x + 3 = 0$
   (c) $7y + 8y = 13y$ **(d)** $.04t - .08t = .40$

16. True or false: The solution set of the equation in Exercise 15(c) is $\emptyset$.

17. In solving the equation $3(2t - 4) = 6t - 12$, a student obtains the result $0 = 0$, and gives the solution set $\{0\}$. Is this correct? Explain.

18. Write the steps you would use to solve the equation in Exercise 15(a).

Solve each of the following equations. See Examples 3–5.

19. $2m - 5 = m + 7$

20. $.01p + 3.1 = 2.03p - 2.96$

21. $\dfrac{5}{6}k - 2k + \dfrac{1}{3} = \dfrac{2}{3}$

22. $\dfrac{3}{4} + \dfrac{1}{5}r - \dfrac{1}{2} = \dfrac{4}{5}r$

23. $3r + 2 - 5(r + 1) = 6r + 4$

24. $5(a + 3) + 4a - 5 = -(2a - 4)$

25. $2[m - (4 + 2m) + 3] = 2m + 2$

26. $4[2p - (3 - p) + 5] = -7p - 2$

27. $\dfrac{3x-2}{7} = \dfrac{x+2}{5}$

28. $\dfrac{2p+5}{5} = \dfrac{p+2}{3}$

29. $\dfrac{1}{4p} + \dfrac{2}{p} = 3$

30. $\dfrac{2}{t} + 6 = \dfrac{5}{2t}$

31. $\dfrac{m}{2} - \dfrac{1}{m} = \dfrac{6m+5}{12}$

32. $\dfrac{-3k}{2} + \dfrac{9k-5}{6} = \dfrac{11k+8}{k}$

33. $\dfrac{2r}{r-1} = 5 + \dfrac{2}{r-1}$

34. $\dfrac{3x}{x+2} = \dfrac{1}{x+2} - 4$

35. $\dfrac{5}{2a+3} + \dfrac{1}{a-6} = 0$

36. $\dfrac{2}{x+1} = \dfrac{3}{5x+5}$

37. $\dfrac{4}{x-3} - \dfrac{8}{2x+5} + \dfrac{3}{x-3} = 0$

38. $\dfrac{5}{2p+3} - \dfrac{3}{p-2} = \dfrac{4}{2p+3}$

39. $\dfrac{2p}{p-2} = 3 + \dfrac{4}{p-2}$

40. $\dfrac{5k}{k+4} = 3 - \dfrac{20}{k+4}$

41. $\dfrac{3}{y-2} + \dfrac{1}{y+1} = \dfrac{1}{y^2-y-2}$

42. $\dfrac{2}{p+3} - \dfrac{5}{p-1} = \dfrac{1}{3-2p-p^2}$

43. $.08w + .06(w + 12) = 7.72$

44. $.04(x - 12) + .06x = 1.52$

45. $(3x - 4)^2 - 5 = 3(x + 5)(3x + 2)$

46. $(2x + 5)^2 = 3x^2 + (x + 3)^2$

Solve each of the following equations for x. See Example 6.

47. $2(x - a) + b = 3x + a$

48. $5x - (2a + c) = a(x + 1)$

49. $ax + b = 3(x - a)$

50. $4a - ax = 3b + bx$

51. $\dfrac{x}{a-1} = ax + 3$

52. $\dfrac{2a}{x-1} = a - b$

53. $a^2x + 3x = 2a^2$

54. $ax + b^2 = bx - a^2$

55. $y = \dfrac{ax+b}{cx+d}$

56. $y = \dfrac{px-q}{rx-s}$

A baseball pitcher's earned run average (E.R.A.) *gives the average number of earned runs given up per 9 innings. An equation that gives this average is*

$$\text{E.R.A.} = \frac{9(\text{number of earned runs allowed})}{\text{number of innings pitched}}.$$

In each of the following exercises, two of the three values in this equation are given for a particular major league pitcher during the 1990 season. Find the remaining value.

57. Roger Clemens;   E.R.A.: 3.13;   innings pitched: $253\frac{1}{3}$

58. Dwight Gooden;   E.R.A.: 2.89;   innings pitched: $118\frac{1}{3}$

59. Greg Swindell;   innings pitched: $184\frac{1}{3}$;   earned runs allowed: 69

60. Frank Wills;   innings pitched: $71\frac{1}{3}$;   earned runs allowed: 29

61. Bert Blyleven;   E.R.A.: 2.73;   earned runs allowed: 73

62. Tim Belcher;   E.R.A.: 2.82;   earned runs allowed: 72

In the metric system of weights and measures, temperature is measured in degrees Celsius (° C) *instead of degrees Fahrenheit* (° F). *To convert back and forth between the two systems, use the equations*

$$C = \frac{5(F - 32)}{9} \qquad and \qquad F = \frac{9}{5}C + 32.$$

In each of the following exercises, convert to the other system. Round answers to the nearest tenth of a degree if necessary.

63. 20° C   **64.** 100° C   **65.** 59° F

66. 86° F   **67.** 100° F   **68.** 350° F

When a consumer borrows money, the lender must tell the consumer the true annual interest rate of the loan. The method of finding the exact true annual interest rate requires special tables available from the government, but a quick approximate rate can be found by using the equation

$$A = \frac{2pf}{b(q + 1)},$$

where p is the number of payments made in one year, f is the finance charge, b is the balance owed on the loan, and q is the total number of payments. Find the value of the variables not given in Exercises 69–74. Round A to the nearest percent and round other variables to the nearest whole number. (This formula is not *accurate enough for the requirements of federal law.)*

69. $p = 12, f = \$800, b = \$4000, q = 36$;   find A

70. $p = 12, f = \$60, b = \$740, q = 12$;   find A

71. $A = 14\%$ (or .14), $p = 12, b = \$2000, q = 36$;   find f

72. $A = 11\%, p = 12, b = \$1500, q = 24$;   find f

73. $A = 16\%, p = 12, f = \$370, q = 36$;   find b

74. $A = 10\%, p = 12, f = \$490, q = 48$;   find b

80 *When a loan is paid off early, a portion of the finance charge must be returned to the borrower. By one method of calculating finance charge (called the* rule of 78), *the amount of unearned interest (finance charge to be returned) is given by*

$$u = f \cdot \frac{n(n+1)}{q(q+1)},$$

where u represents unearned interest, f is the original finance charge, n is the number of payments remaining when the loan is paid off, and q is the original number of payments. Find the amount of the unearned interest in each of the following.

75. Original finance charge = $800, loan scheduled to run 36 months, paid off with 18 payments remaining

76. Original finance charge = $1400, loan scheduled to run 48 months, paid off with 12 payments remaining

77. Original finance charge = $950, loan scheduled to run 24 months, paid off with 6 payments remaining

78. Original finance charge = $175, loan scheduled to run 12 months, paid off with 3 payments remaining

79. Find the error in the following.

$$x^2 + 2x - 15 = x^2 - 3x$$
$$(x + 5)(x - 3) = x(x - 3)$$
$$x + 5 = x$$
$$5 = 0$$

Solution set: $\emptyset$

Find the value of k that will make each equation equivalent to x = 2.

80. $9x - 7 = k$

81. $-5x + 11x - 2 = k + 4$

82. $\dfrac{8}{k + x} = 4$

83. $\sqrt{x + k} = 0$

84. $\sqrt{3x - 2k} = 4$

2.2 ——— **APPLICATIONS OF LINEAR EQUATIONS**

In Section 2.1 we solved literal equations. Now we will extend this idea to **solving a formula for a specified variable.** For example, the formula

$$A = \frac{1}{2}(b_1 + b_2)h$$

gives the area of a trapezoid having bases of lengths b_1 and b_2 and height h. To solve the equation for b_1, get b_1 alone, with all other variables or numbers on the other side of the equals sign. First multiply both sides of the equation by 2, and then proceed as shown.

$$2A = (b_1 + b_2)h$$

$$\frac{2A}{h} = b_1 + b_2 \qquad \text{Multiply by } \frac{1}{h}.$$

$$\frac{2A}{h} - b_2 = b_1 \qquad \text{Subtract } b_2.$$

$$b_1 = \frac{2A}{h} - b_2$$

Using the distributive property on the right side of the equation after the first step above would give

$$2A = (b_1 + b_2)h$$

$$2A = b_1 h + b_2 h \qquad \text{Distributive property}$$

$$2A - b_2 h = b_1 h \qquad \text{Subtract } b_2 h.$$

$$\frac{2A - b_2 h}{h} = b_1, \qquad \text{Multiply by } \frac{1}{h}.$$

which is equivalent to the result found by the first method.

∎ *Example 1*

SOLVING A
FORMULA FOR A
SPECIFIED VARIABLE

Solve $J\left(\dfrac{x}{k} + a\right) = x$ for x.

To get all terms with x on one side of the equation and all terms without x on the other, first use the distributive property.

$$J\left(\frac{x}{k}\right) + Ja = x$$

Eliminate the denominator, k, by assuming $k \neq 0$ and multiplying both sides by k.

$$kJ\left(\frac{x}{k}\right) + kJa = kx$$

$$Jx + kJa = kx$$

Then add $-Jx$ to both sides to get the two terms with x together.

$$kJa = \quad x - Jx$$

$$kJa = x(k - J) \qquad \text{Factor the right side.}$$

Assuming $k \neq J$, multiply both sides by $1/(k - J)$ to find the solution.

$$x = \frac{kJa}{k - J} \quad ∎$$

CAUTION   Errors often occur in problems like Example 1 because students forget to factor out the variable for which they are solving. It is necessary to have that variable as a factor so that in the next step we can multiply by the reciprocal of its co-efficient.

PROBLEM SOLVING

One of the main reasons for learning mathematics is to be able to use it in solving practical problems. For most students, however, learning how to apply mathematical skills to real situations is the most difficult task they face. In the rest of this section a few hints are given that may help with applications.

A common difficulty with applied problems is trying to do everything at once. It is usually best to attack the problem in stages.

SOLVING APPLIED PROBLEMS

1. Decide on an unknown, and name it with some variable that you *write down*. Most students try to skip this step. They are eager to get on with writing an equation. But this is an important step. If you don't know what "*x*" represents, how can you write a meaningful equation or interpret a result?
2. Draw a sketch or make a chart, if appropriate, showing the information given in the problem.
3. Decide on a variable expression to represent any other unknowns in the problem. For example, if *W* represents the width of a rectangle, *L* represents the length, and you know that the length is one more than twice the width, *write down L = 1 + 2W*.
4. Use the results of Steps 1 and 3 to write an equation with one variable.
5. Solve the equation.
6. Check the solution in the words of the original problem. Be sure that the answer makes sense. ■

Notice how each of the steps listed above is carried out in the following examples.

■ *Example 2*

SOLVING A GEOMETRY PROBLEM

If the length of a side of a square is increased by 3 centimeters, the perimeter of the new square is 40 centimeters more than twice the length of a side of the original square. Find the dimensions of the original square.

First, decide what the variable should represent (Step 1). Since the length of a side of the original square is to be found, let the variable represent it.

$$x = \text{length of side of the original square in centimeters}$$

Now draw a figure using the given information, as in Figure 2.1 (Step 2).

The length of a side of the new square is 3 centimeters more than the length of a side of the old square. Write a variable expression for that relationship (Step 3).

$$x + 3 = \text{length of side of the new square}$$

Now write a variable expression for the perimeter of the new square. Since the perimeter of a square is 4 times the length of a side,

$$4(x + 3) = \text{perimeter of the new square.}$$

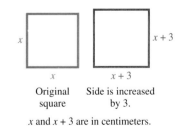

x and *x* + 3 are in centimeters.

■ **FIGURE 2.1**

Now you can use the information given in the problem to write an equation (Step 4). The perimeter of the new square is 40 centimeters more than twice the length of a side of the original square, so the equation is written as follows.

The new perimeter	is	40	more than	twice the side of the original square.
$4(x + 3)$	$=$	40	$+$	$2x$

Solve the equation (Step 5).

$$4(x + 3) = 40 + 2x$$
$$4x + 12 = 40 + 2x$$
$$2x = 28$$
$$x = 14$$

This solution should be checked using the words of the original problem (Step 6). The length of a side of the new square would be $14 + 3 = 17$ centimeters; its perimeter would be $4(17) = 68$ centimeters. Twice the length of a side of the original square is $2(14) = 28$ centimeters. Since $40 + 28 = 68$, the solution satisfies the problem. Each side of the original square measures 14 centimeters. ∎

PROBLEM SOLVING

If a problem asks for one unknown quantity, then only one condition is needed—a condition that tells you how to write the equation. If there is more than one unknown, then there must be at least as many conditions given in the problem as there are unknown quantities. In such cases, some conditions tell how the variables are related, while others tell how to write the necessary equations. In Example 2, one condition gave the length of the side of the new square, another gave the perimeter of the new square, and a third gave the wording that led to an equation. ∎

In the remaining examples, the steps are not numbered. See if you can identify them. (Some steps may not apply in some problems.)

∎ *Example 3*

SOLVING A
GRADE-AVERAGING
PROBLEM

Gerry Vidrine has grades of 88, 86, and 92 on her first three algebra tests. What grade must she make on her fourth test to raise her average to 90?

Let $t =$ the grade on her fourth test. Since we want the average of her four tests to equal 90, we must add the four test grades, divide by 4, and set the result equal to 90.

$$\frac{88 + 86 + 92 + t}{4} = 90$$

Now solve the equation.

$$\frac{266 + t}{4} = 90$$

$$266 + t = 360$$

$$t = 94$$

If she makes a 94 on her fourth test, the average of the four tests will be $(88 + 86 + 92 + 94)/4 = 360/4 = 90$, so the answer checks; the required grade is 94. ■

■ **IN SIMPLEST TERMS** ─────────────────────────────── ■

5 ▭

Biologists can use algebra to estimate the number of fish in a lake. They first catch a sample of fish and mark each specimen with a harmless tag. Some weeks later, they catch a similar sample of fish from the same areas of the lake and determine the proportion of previously tagged fish in the new sample. The total fish population is estimated by assuming that the proportion of tagged fish in the new sample is the same as the proportion of tagged fish in the entire lake.

For example, suppose the biologists tag 300 fish on May 1. When they return and take a new sample of 400 fish on June 1, 5 of the 400 were previously tagged. Let x be the total fish population in the lake. Then we can set up the following proportion:

$$\frac{300}{x} = \frac{5}{400}$$

$$400x\left(\frac{300}{x}\right) = 400x\left(\frac{5}{400}\right)$$

$$120{,}000 = 5x$$

$$24{,}000 = x.$$

There are approximately 24,000 fish in the lake.

SOLVE EACH PROBLEM

A. Biologists tagged 250 fish in Willow Lake on October 5. On a later date they found 7 tagged fish in a sample of 350. Estimate the total number of fish in Willow Lake.

B. On May 13, researchers at Argyle Lake tagged 420 fish. When they returned a few weeks later, their sample of 500 fish contained 9 that were tagged. Give a good approximation of the fish population in Argyle Lake.

ANSWERS   A. The total fish population in Willow Lake would be approximately 12,500. B. The total fish population in Argyle Lake would be approximately 23,333.

■ *Example 4*

SOLVING A CONSTANT VELOCITY PROBLEM

Chuck travels 80 kilometers in the same time that Mary travels 180 kilometers. Mary travels 50 kilometers per hour faster than Chuck. Find the rate at which each person travels.

In a problem such as this, it is important to distinguish among units; here distance is in kilometers, time is in hours, and rate is in kilometers per hour. Let

x represent Chuck's rate of travel. Since Mary travels 50 kilometers per hour faster,

$$x + 50 = \text{rate for Mary.}$$

Constant velocity problems of this kind are solved with the formula $d = rt,$ where d is distance traveled in time t at a constant rate $r.$ A chart is helpful for organizing the information given in this problem and determining the necessary variable expressions.

For Chuck, $d = 80$ and $r = x.$ The formula $d = rt$ gives $t = d/r,$ so that for Chuck, $t = 80/x.$ For Mary, $d = 180,$ $r = x + 50,$ and $t = 180/(x + 50).$ This information is shown in the following chart.

	d	r	t
Chuck	80	x	$\dfrac{80}{x}$
Mary	180	$x + 50$	$\dfrac{180}{x + 50}$

Times are the same.

Since they both travel for the same amount of time,

$$\frac{80}{x} = \frac{180}{x + 50}.$$

Multiply both sides of this equation by $x(x + 50),$ getting

$$x(x + 50) \cdot \frac{80}{x} = x(x + 50) \cdot \frac{180}{x + 50}$$

$$80(x + 50) = 180x$$

$$80x + 4000 = 180x$$

$$4000 = 100x$$

$$40 = x.$$

Chuck's rate, which is represented by $x,$ is 40 kilometers per hour. Mary's rate is $x + 50,$ or $40 + 50 = 90$ kilometers per hour. Check these results. ■

PROBLEM SOLVING

In Example 4 (a constant velocity problem), we used the formula relating rate, time, and distance. In problems involving rate of work (as in Example 5, which follows), we use a similar idea. If a person or a machine can do a job in t units of time, then the rate of work is $1/t$ job per time unit. Therefore,

rate × time = portion of the job completed. ■

■ *Example 5*
SOLVING A PROBLEM
ABOUT WORK

One computer can do a job twice as fast as another. Working together, both computers can do the job in 2 hours. How long would it take each computer, working alone, to do the job?

Let x represent the number of hours it would take the faster computer, working alone, to do the job. The time for the slower computer to do the job alone is then $2x$ hours. Therefore, the rates for the two computers are as follows:

$$\frac{1}{x} = \text{rate of faster computer (job per hour)}$$

$$\frac{1}{2x} = \text{rate of slower computer (job per hour).}$$

The time for the computers to do the job together is 2 hours. Multiplying each rate by the time will give the fractional part of the job accomplished by each. This is summarized in the chart that follows.

	Rate	Time	Part of the Job Accomplished
Faster computer	$\frac{1}{x}$	2	$2\left(\frac{1}{x}\right) = \frac{2}{x}$
Slower computer	$\frac{1}{2x}$	2	$2\left(\frac{1}{2x}\right) = \frac{1}{x}$

The sum of the two parts of the job accomplished is 1, since one whole job is done. The equation can now be written and solved.

$$\frac{2}{x} + \frac{1}{x} = 1 \qquad \text{The sum of the two parts is 1.}$$

$$2 + 1 = x \qquad \text{Multiply by } x.$$

$$x = 3$$

The faster computer could do the entire job, working alone, in 3 hours. The slower computer would need $2(3) = 6$ hours. ■

■ *Example 6*
SOLVING A MIXTURE PROBLEM

Constance Morganstern is a chemist. She needs a 20% solution of alcohol. She has a 15% solution on hand, as well as a 30% solution. How many liters of the 15% solution should she add to 3 liters of the 30% solution to get her 20% solution?

Let x be the number of liters of the 15% solution to be added. See Figure 2.2. Arrange the information of the problem in a chart.

x liters   3 liters $3 + x$ liters

■ **FIGURE 2.2**

Strength	Liters of Solution	Liters of Pure Alcohol
15%	x	$.15x$
30%	3	$.30(3)$
20%	$3 + x$	$.20(3 + x)$

Since the number of liters of pure alcohol in the 15% solution plus the number of liters in the 30% solution must equal the number of liters in the final 20% solution,

Liters in 15%		Liters in 30%		Liters in 20%
$.15x$	$+$	$.30(3)$	$=$	$.20(3 + x).$

Solve this equation as follows.

$$.15x + .90 = .60 + .20x \qquad \text{Distributive property}$$
$$.30 = .05x$$
$$6 = x$$

By this result, 6 liters of the 15% solution should be mixed with 3 liters of the 30% solution, giving $6 + 3 = 9$ liters of 20% solution. ■

NOTE In Example 6 (a mixture problem), we multiplied rate of concentration by the number of liters to get the amount of pure chemical present. Similarly, in Example 7 (an investment problem) which follows, we will multiply interest rate by principal to find the amount of interest earned.

■ **Example 7**
SOLVING AN INVESTMENT PROBLEM

A financial manager has $14,000 to invest for her company. She plans to invest part of the money in tax-free bonds at 6% interest and the remainder at 9%. She wants to earn $1005 per year in interest from the investments. Find the amount she should invest at each rate.

Let x represent the dollar amount to be invested at 6%, so that $14,000 - x$ is the amount to be invested at 9%. Interest is given by the product of principal, rate, and time in years ($i = prt$). Summarize this information in a chart.

Amount Invested	Interest Rate	Interest Earned in 1 yr
x	$6\% = .06$	$.06x$
$14,000 - x$	$9\% = .09$	$.09(14,000 - x)$

Since the total interest is to be $1005,

$$.06x + .09(14,000 - x) = 1005.$$

To clear decimal points, we first multiply both sides of the equation by 100.

$$6x + 9(14,000 - x) = 100,500$$
$$6x + 126,000 - 9x = 100,500$$
$$126,000 - 3x = 100,500$$
$$-3x = -25,500$$
$$x = 8500$$

The manager should invest $8500 at 6%, and $14,000 - \$8500 = \$5500$ at 9%. ■

> **NOTE** The interest formula used in Example 7 ($i = prt$) applies to simple interest. In most real-life applications, compound interest is used. We will study the formula for compound interest and its applications in Chapter 5.

2.2 EXERCISES ■

Solve each formula for the indicated variable. Assume that all denominators are nonzero. See Example 1.

1. $i = prt$ for p (simple interest)

2. $V = lwh$ for l (volume of a rectangular box)

3. $P = 2l + 2w$ for w (perimeter of a rectangle)

4. $P = a + b + c$ for c (perimeter of a triangle)

5. $A = \dfrac{1}{2}(B + b)h$ for h (area of a trapezoid)

6. $A = \dfrac{1}{2}(B + b)h$ for B (area of a trapezoid)

7. $S = 2lw + 2wh + 2hl$ for h (surface area of a rectangular box)

8. $S = 2\pi rh + 2\pi r^2$ for h (surface area of a right circular cylinder)

9. $s = \dfrac{1}{2}gt^2$ for g (distance traveled by a falling object)

10. $C = \dfrac{5}{9}(F - 32)$ for F (Fahrenheit to Celsius)

11. $\dfrac{1}{R} = \dfrac{1}{r_1} + \dfrac{1}{r_2}$ for R (electricity)

12. $u = f \cdot \dfrac{k(k + 1)}{n(n + 1)}$ for f (unearned interest)

13. $A = \dfrac{24f}{B(p + 1)}$ for f (approximate annual interest rate)

14. $A = \dfrac{24f}{B(p + 1)}$ for B (approximate annual interest rate)

15. $A = P\left(1 + \dfrac{i}{m}\right)$ for m (compound interest)

16. $V = \dfrac{1}{3}\pi r^2 h$ for h (volume of a right circular cone)

17. Refer to Exercise 16. Why is it not possible, using the methods of this section, to solve this formula for r?

18. Refer to Example 1. Suppose that someone tells you that there is no reason to solve for x, since the right side of the formula is already equal to x. How would you respond?

19. Suppose two acid solutions are mixed. One is 26% acid and the other is 32% acid. Which one of the following concentrations cannot possibly be the concentration of the mixture? Explain.
  (a) 36%   **(b)** 28%   **(c)** 30%   **(d)** 31%

20. Suppose that a computer that originally sells for x dollars has been discounted 30%. Which one of the following expressions does not represent its sale price?
  (a) $x - .30x$ **(b)** $.70x$
  (c) $\dfrac{7}{10}x$ **(d)** $x - .30$

Solve each of the following problems. See Examples 2–7.

21. A puzzle piece in the shape of a triangle has a perimeter of 30 cm. Two sides of the triangle are each twice as long as the shortest side. Find the length of the shortest side. (In the figure, let x = length of the shortest side. Then $2x$ = length of the other sides.)

Side lengths are in centimeters.

22. The length of a rectangular label is 3 cm less than twice the width. The perimeter is 54 cm. Find the width. (In the figure, let w = width. Then $2w - 3$ = length.)

Side lengths are in centimeters.

23. Hien has grades of 84, 88, and 92 on his first three calculus tests. What grade on his fourth test will give him an average of 90?

24. Doug scored 78, 94, and 60 on his three trigonometry tests. If his final exam score is to be counted as two test grades in determining his course average, what grade must he make on his final exam to give him an average of 80?

25. A pharmacist wishes to strengthen a mixture that is 10% alcohol to one that is 30% alcohol. How much pure alcohol should be added to 7 liters of the 10% mixture?

26. A student needs 10% hydrochloric acid for a chemistry experiment. How much 5% acid should be mixed with 60 ml of 20% acid to get a 10% solution?

27. A recycling bin is in the shape of a rectangular box. Find the height of the box if its length is 18 ft, its width is 8 ft, and its surface area is 496 sq ft. (In the figure, let h = height.)

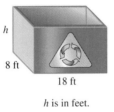

h is in feet.

28. A right circular cylinder has radius 6 in and volume 144π cu in. What is its height? (In the figure, let h = height.)

h is in inches.

Exercises 29 and 30 depend on the idea of the octane rating of gasoline, a measure of its antiknock qualities. In one measure of octane, a standard fuel is made with only two ingredients: heptane and isooctane. For this fuel, the octane rating is the percent of isooctane. An actual gasoline blend is then compared to a standard fuel. For example, a gasoline with an octane rating of 98 has the same antiknock properties as a standard fuel that is 98% isooctane.

29. How many liters of 94-octane gasoline should be mixed with 200 liters of 99-octane gasoline to get a mixture that is 97-octane?

30. A service station has 92-octane and 98-octane gasoline. How many liters of each should be mixed to provide 12 liters of 96-octane gasoline needed for chemical research?

31. On a vacation trip, Jose averaged 50 mph traveling from Denver to Minneapolis. Returning by a different route that covered the same number of miles, he averaged 55 mph. What is the distance between the two cities if his total traveling time was 32 hr?

32. Cindy left by plane to visit her mother in Hartford, 420 km away. Fifteen minutes later, her mother left to meet her at the airport. She drove the 20 km to the airport at 40 km/hr, arriving just as the plane taxied in. What was the speed of the plane?

33. Russ and Janet are running in the Apple Hill Fun Run. Russ runs at 7 mph, Janet at 5 mph. If they start at the same time, how long will it be before they are 1/2 mi apart?

34. If the run in Exercise 33 has a staggered start, and Janet starts first, with Russ starting 10 min later, how long will it be before he catches up with her?

35. Joann took 20 min to drive her boat upstream to water-ski at her favorite spot. Coming back later in the day, at the same boat speed, took her 15 min. If the current in that part of the river is 5 km/hr, what was her boat speed?

36. Joe traveled against the wind in a small plane for 3 hr. The return trip with the wind took 2.8 hr. Find the speed of the wind if the speed of the plane in still air is 180 mph.

37. Johnny gets to work in 20 min when he drives his car. Riding his bike (by the same route) takes him 45 min. His average driving speed is 4.5 mph greater than his average speed on his bike. How far does he travel to work?

38. In the morning, Marge drove to a business appointment at 50 mph. Her average speed on the return trip in the afternoon was 40 mph. The return trip took 1/4 hr longer because of heavy traffic. How far did she travel to the appointment?

39. Le can clean the house in 9 hr, while Tran needs 6 hr. How long will it take them to clean the house if they work together?

40. Helen can paint a room in 5 hr. Jay can paint the same room in 4 hr. How long will it take them to paint the room together?

41. Two chemical plants are polluting a river. If plant A produces a predetermined maximum amount of pollutant twice as fast as plant B, and together they produce the maximum pollutant in 26 hr, how long will it take plant B alone?

42. A sewage treatment plant has two inlet pipes to its settling pond. One can fill the pond in 10 hours, the other in 12 hours. If the first pipe is open for 5 hours and then the second pipe is opened, how long will it take to fill the pond?

43. An inlet pipe can fill Dominic's pool in 5 hr, while an outlet pipe can empty it in 8 hr. In his haste to watch television, Dominic left both pipes open. How long did it take to fill the pool?

44. Suppose Dominic discovered his error (see Exercise 43) after an hour-long program. If he then closed the outlet pipe, how much more time would be needed to fill the pool?

45. A VCR is on sale for $245. If the sale price is 30% less than the regular price, what was the regular price?

46. A jeweler prices his items 60% over their wholesale price. If a watch sells for $152, what is its wholesale price?

A person's intelligence quotient (IQ) *is found by multiplying the mental age by* 100 *and dividing by the chronological age. Use this to solve Exercises 47 and 48.*

47. Jack is 7 years old. His IQ is 130. Find his mental age.

48. If a person is 16 years old with a mental age of 20, what is the person's IQ?

49. Bill Cornett won $200,000 in a state lottery. He first paid income tax of 30% on the winnings. Of the rest, he invested some at 8.5% and some at 7%, making $10,700 interest per year. How much is invested at each rate?

50. Marjorie Williams earned $48,000 from royalties on her cookbook. She paid a 28% income tax on these royalties. The balance was invested in two ways, some of it at 6.5% and some at 6.25%. The investments produced $2210 interest per year. Find the amount invested at each rate.

51. Janet Tilden bought two plots of land for a total of $120,000. When she sold the first plot, she made a profit of 15%. When she sold the second, she lost 10%. Her total profit was $5500. How much did she pay for each piece of land?

52. Suppose $10,000 is invested at 6%. How much additional money must be invested at 8% to produce a yield of 7.2% on the entire amount invested?

53. Cathy Wacaser earns take-home pay of $592 a week. If her deductions for taxes, retirement, union dues, and medical plan amount to 26% of her wages, what is her weekly pay before deductions?

54. George Duda gives 10% of his net income to charity. This amounts to $167.20 per month. In addition, his paycheck deductions are 24% of his gross monthly income. What is his gross monthly income?

55. A bank pays 5% interest on passbook accounts and 7% interest on long-term deposits. Suppose a depositor divides $20,000 among the two types of deposits. Find the amount deposited at each rate if the total annual income from interest is $1340.

56. Adam Bryer wishes to sell a piece of property for $125,000. He wishes the money to be paid off in two ways—a short-term note at 12% and a long-term note at 10%. Find the amount of each note if the total annual interest paid is $13,700.

57. In planning her retirement, Louise Howe deposits some money at 4.5% with twice as much deposited at 5%. Find the amount deposited at each rate if the total annual interest income is $2900.

58. A church building fund has invested $75,000 in two ways: part of the money at 7% and four times as much at 11%. Find the amount invested at each rate if the total annual income from interest is $7650.

59. Which of the following cannot be a correct equation to solve a geometry problem, if x represents the length of a rectangle? (*Hint:* Solve each equation and consider the solutions.)

(a) $2x + 2(x - 1) = 14$
(b) $-2x + 7(5 - x) = 62$
(c) $4(x + 2) + 4x = 8$
(d) $2x + 2(x - 3) = 22$

60. If x represents the number of pennies in a jar in an applied problem, which of the following equations cannot be a correct equation for finding x? (*Hint:* Solve each equation and consider the solutions.)

(a) $5x + 3 = 9$ (b) $12x + 3 = -4$
(c) $100x = 50(x + 3)$ (d) $6(x + 4) = x + 24$

2.3 ———— COMPLEX NUMBERS

In the next section we will solve *quadratic equations,* which have a term raised to the second power (for example, $x^2 - 4x + 3 = 0$). Solutions of quadratic equations may not be real numbers. For example, there are no real number solutions to the quadratic equation

$$x^2 + 1 = 0.$$

A set of numbers is needed that permits the solution of *all* quadratic equations. To get such a set of numbers, the number i is defined as follows.

DEFINITION OF i $\qquad\qquad i^2 = -1 \qquad$ or $\qquad i = \sqrt{-1}.$

Numbers of the form $a + bi,$ where a and b are real numbers, are called **complex numbers.** Each real number is a complex number, since a real number a may be thought of as the complex number $a + 0i.$ A complex number of the form $a + bi,$ where b is nonzero, is called an **imaginary number.** Both the set of real numbers and the set of imaginary numbers are subsets of the set of complex numbers. (See Figure 2.3, which is an extension of Figure 1.5 in Section 1.1.) A complex number that is written in the form $a + bi$ or $a + ib$ is in **standard form.** (The form $a + ib$ is used to simplify certain symbols such as $i\sqrt{5},$ since $\sqrt{5}i$ could be too easily mistaken for $\sqrt{5i}.$)

Imaginary numbers	Rational numbers $\frac{4}{9}, -\frac{5}{8}, \frac{11}{7}$	Irrational numbers
		$-\sqrt{8}$
		$\sqrt{15}$
$4i, -11i, 3 + 2i$	Integers $-11, -6, -4$	$\sqrt{23}$
		π
	Whole numbers 0	$\frac{\pi}{4}$
		e
	Natural numbers 1, 2, 3, 4, 5, 37, 40	

Complex numbers (Real numbers are shaded.)

■ **FIGURE 2.3**

■ *Example 1*

IDENTIFYING KINDS OF COMPLEX NUMBERS

The following statements identify different kinds of complex numbers.

(a) -8, $\sqrt{7}$, and π are real numbers and complex numbers.

(b) $3i$, $-11i$, $i\sqrt{14}$, and $5 + i$ are imaginary numbers and complex numbers. ■

■ *Example 2*

WRITING COMPLEX NUMBERS IN STANDARD FORM

The list below shows several numbers, along with the standard form of each number.

Number	Standard Form
$6i$	$0 + 6i$
9	$-9 + 0i$
0	$0 + 0i$
$-i + 2$	$2 - i$
$8 + i\sqrt{3}$	$8 + i\sqrt{3}$

■

Many of the solutions to quadratic equations in the next section will involve expressions such as $\sqrt{-a}$, for a positive real number a, defined as follows.

DEFINITION OF $\sqrt{-a}$

If $a > 0$, then

$$\sqrt{-a} = i\sqrt{a}.$$

■ *Example 3*

WRITING $\sqrt{-a}$ AS $i\sqrt{a}$

Write each expression as the product of i and a real number.

(a) $\sqrt{-16} = i\sqrt{16} = 4i$

(b) $\sqrt{-70} = i\sqrt{70}$ ■

Products or quotients with negative radicands are simplified by first rewriting $\sqrt{-a}$ as $i\sqrt{a}$ for positive numbers a. Then the properties of real numbers can be applied, together with the fact that $i^2 = -1$.

The rule $\sqrt{c} \cdot \sqrt{d} = \sqrt{cd}$ is valid only when c and d are *not* both negative. For example,

$$\sqrt{(-4)(-9)} = \sqrt{36} = 6,$$

while

$$\sqrt{-4} \cdot \sqrt{-9} = 2i(3i) = 6i^2 = -6,$$

so that

$$\sqrt{(-4)(-9)} \text{ is not equal to } \sqrt{-4} \cdot \sqrt{-9}.$$

CAUTION When working with negative radicands, be sure to use the definition $\sqrt{-a} = i\sqrt{a}$ *before* using any of the other rules for radicals.

■ *Example 4*

FINDING PRODUCTS AND QUOTIENTS INVOLVING NEGATIVE RADICANDS

Multiply or divide as indicated.

(a) $\sqrt{-7} \cdot \sqrt{-7} = i\sqrt{7} \cdot i\sqrt{7}$

$= i^2 \cdot (\sqrt{7})^2$

$= (-1) \cdot 7 \qquad i^2 = -1$

$= -7$

(b) $\sqrt{-6} \cdot \sqrt{-10} = i\sqrt{6} \cdot i\sqrt{10}$

$= i^2 \cdot \sqrt{60}$

$= -1 \cdot 2\sqrt{15}$

$= -2\sqrt{15}$

(c) $\dfrac{\sqrt{-20}}{\sqrt{-2}} = \dfrac{i\sqrt{20}}{i\sqrt{2}} = \sqrt{\dfrac{20}{2}} = \sqrt{10}$

(d) $\dfrac{\sqrt{-48}}{\sqrt{24}} = \dfrac{i\sqrt{48}}{\sqrt{24}} = i\sqrt{2}$ ■

OPERATIONS ON COMPLEX NUMBERS   Complex numbers may be added, subtracted, multiplied, and divided using the properties of real numbers, as shown by the following definitions and examples.

The *sum* of two complex numbers $a + bi$ and $c + di$ is defined as follows.

ADDITION OF COMPLEX NUMBERS

$$(a + bi) + (c + di) = (a + c) + (b + d)i$$

■ *Example 5*

ADDING COMPLEX NUMBERS

Find each sum.

(a) $(3 - 4i) + (-2 + 6i) = [3 + (-2)] + [-4 + 6]i$

$= 1 + 2i$

(b) $(-9 + 7i) + (3 - 15i) = -6 - 8i$ ■

Since $(a + bi) + (0 + 0i) = a + bi$ for all complex numbers $a + bi$, the number $0 + 0i$ is called the **additive identity** for complex numbers. The sum of $a + bi$ and $-a - bi$ is $0 + 0i$, so the number $-a - bi$ is called the **negative** or **additive inverse** of $a + bi$.

Using this definition of additive inverse, *subtraction* of complex numbers $a + bi$ and $c + di$ is defined as

$$(a + bi) - (c + di) = (a + bi) + (-c - di) = (a - c) + (b - d)i.$$

SUBTRACTION OF COMPLEX NUMBERS	$(a + bi) - (c + di) = (a - c) + (b - d)i$

■ *Example 6*

SUBTRACTING COMPLEX NUMBERS

Subtract as indicated.

(a) $(-4 + 3i) - (6 - 7i) = (-4 - 6) + [3 - (-7)]i$

$$= -10 + 10i$$

(b) $(12 - 5i) - (8 - 3i) = (12 - 8) + (-5 + 3)i$

$$= 4 - 2i \quad ■$$

The *product* of two complex numbers can be found by multiplying as if the numbers were binomials and using the fact that $i^2 = -1$, as follows.

$$(a + bi)(c + di) = ac + adi + bic + bidi$$
$$= ac + adi + bci + bdi^2$$
$$= ac + (ad + bc)i + bd(-1)$$
$$(a + bi)(c + di) = (ac - bd) + (ad + bc)i$$

Based on this result, the product of the complex numbers $a + bi$ and $c + di$ is defined in the following way.

MULTIPLICATION OF COMPLEX NUMBERS	$(a + bi)(c + di) = (ac - bd) + (ad + bc)i$

This definition is not practical to use. To find a given product, it is easier just to multiply as with binomials.

■ *Example 7*

MULTIPLYING COMPLEX NUMBERS

Find each of the following products.

(a) $(2 - 3i)(3 + 4i) = 2(3) + 2(4i) - 3i(3) - 3i(4i)$

$$= 6 + 8i - 9i - 12i^2$$

$$= 6 - i - 12(-1) \qquad\qquad i^2 = -1$$

$$= 18 - i$$

(b) $(5 - 4i)(7 - 2i) = 5(7) + 5(-2i) - 4i(7) - 4i(-2i)$

$$= 35 - 10i - 28i + 8i^2$$

$$= 35 - 38i + 8(-1)$$

$$= 27 - 38i$$

(c) $(6 + 5i)(6 - 5i) = 6^2 - 25i^2$ Product of the sum and
                                            difference of two terms

$= 36 - 25(-1)$ $i^2 = -1$

$= 36 + 25$

$= 61$ or $61 + 0i$ Standard form

(d) $(4 + 3i)^2 = 4^2 + 2(4)(3i) + (3i)^2$ Square of a binomial

$= 16 + 24i + (-9)$

$= 7 + 24i$ ■

Powers of i can be simplified using the facts that $i^2 = -1$ and $i^4 = 1$. The next example shows how this is done.

■ *Example 8*
SIMPLIFYING
POWERS OF *i*

Simplify each power of i.

(a) i^{15}

Since $i^2 = -1$, the value of a power of i is found by writing the given power as a product involving i^2 or i^4. For example, $i^3 = i^2 \cdot i = (-1) \cdot i = -i$. Also, $i^4 = i^2 \cdot i^2 = (-1)(-1) = 1$. Using i^4 and i^3 to rewrite i^{15} gives

$$i^{15} = i^{12} \cdot i^3 = (i^4)^3 \cdot i^3 = (1)^3(-i) = -i.$$

(b) $i^{-3} = i^{-4} \cdot i = (i^4)^{-1} \cdot i = (1)^{-1} \cdot i = i$ ■

We can use the method of Example 8 to construct the following table of powers of i.

POWERS OF *i*			
$i^1 = i$	$i^5 = i$	$i^9 = i$	
$i^2 = -1$	$i^6 = -1$	$i^{10} = -1$	
$i^3 = -i$	$i^7 = -i$	$i^{11} = -i$	
$i^4 = 1$	$i^8 = 1$	$i^{12} = 1,$	and so on.

Example 7(c) showed that $(6 + 5i)(6 - 5i) = 61$. The numbers $6 + 5i$ and $6 - 5i$ differ only in their middle signs; for this reason these numbers are called **conjugates** of each other. The product of a complex number and its conjugate is always a real number.

PROPERTY OF COMPLEX CONJUGATES	For real numbers a and b: $$(a + bi)(a - bi) = a^2 + b^2.$$

■ *Example 9*

EXAMINING
CONJUGATES AND
THEIR PRODUCTS

The following list shows several pairs of conjugates, together with their products.

Number	Conjugate	Product
$3 - i$	$3 + i$	$(3 - i)(3 + i) = 9 + 1 = 10$
$2 + 7i$	$2 - 7i$	$(2 + 7i)(2 - 7i) = 53$
$-6i$	$6i$	$(-6i)(6i) = 36$

The conjugate of the divisor is used to find the *quotient* of two complex numbers. The quotient is found by multiplying both the numerator and the denominator by the conjugate of the denominator. The result should be written in standard form.

■ *Example 10*

DIVIDING COMPLEX
NUMBERS

(a) Find $\dfrac{3 + 2i}{5 - i}$.

Multiply numerator and denominator by the conjugate of $5 - i$.

$$\frac{3 + 2i}{5 - i} = \frac{(3 + 2i)(5 + i)}{(5 - i)(5 + i)}$$

$$= \frac{15 + 3i + 10i + 2i^2}{25 - i^2} \qquad \text{Multiply.}$$

$$= \frac{13 + 13i}{26} \qquad i^2 = -1$$

$$= \frac{13}{26} + \frac{13i}{26} \qquad \frac{a + bi}{c} = \frac{a}{c} + \frac{bi}{c}$$

$$= \frac{1}{2} + \frac{1}{2}i \qquad \text{Lowest terms}$$

To check this answer, show that

$$(5 - i)\left(\frac{1}{2} + \frac{1}{2}i\right) = 3 + 2i.$$

(b) $\dfrac{3}{i} = \dfrac{3(-i)}{i(-i)}$ $-i$ is the conjugate of i.

$$= \frac{-3i}{-i^2}$$

$$= \frac{-3i}{1} \qquad -i^2 = -(-1) = 1$$

$$= -3i \quad \text{or} \quad 0 - 3i \qquad \text{Standard form} \quad ■$$

2.3 EXERCISES ■

Identify each complex number as real or imaginary. See Example 1.

1. $-9i$ **2.** 6   **3.** π **4.** $-\sqrt{7}$

5. $i\sqrt{6}$ **6.** $-3i$ **7.** $2 + 5i$ **8.** $-7 - 6i$

Write each of the following without negative radicands. See Examples 3 and 4.

9. $\sqrt{-100}$ **10.** $\sqrt{-169}$ **11.** $-\sqrt{-400}$ **12.** $-\sqrt{-225}$

13. $-\sqrt{-39}$ **14.** $-\sqrt{-95}$ **15.** $5 + \sqrt{-4}$ **16.** $-7 + \sqrt{-100}$

17. $9 - \sqrt{-50}$ **18.** $-11 - \sqrt{-24}$ **19.** $\sqrt{-5} \cdot \sqrt{-5}$ **20.** $\sqrt{-20} \cdot \sqrt{-20}$

21. $\dfrac{\sqrt{-40}}{\sqrt{-10}}$ **22.** $\dfrac{\sqrt{-190}}{\sqrt{-19}}$ **23.** $\dfrac{\sqrt{-6} \cdot \sqrt{-2}}{\sqrt{3}}$ **24.** $\dfrac{\sqrt{-12} \cdot \sqrt{-6}}{\sqrt{8}}$

Add or subtract. Write each result in standard form. See Examples 5 and 6.

25. $(3 + 2i) + (4 - 3i)$ **26.** $(4 - i) + (2 + 5i)$

27. $(-2 + 3i) - (-4 + 3i)$ **28.** $(-3 + 5i) - (-4 + 3i)$

29. $(2 - 5i) - (3 + 4i) - (-2 + i)$ **30.** $(-4 - i) - (2 + 3i) + (-4 + 5i)$

Multiply. Write each result in standard form. See Example 7.

31. $(2 + i)(3 - 2i)$ **32.** $(-2 + 3i)(4 - 2i)$ **33.** $(2 + 4i)(-1 + 3i)$ **34.** $(1 + 3i)(2 - 5i)$

35. $(-3 + 2i)^2$ **36.** $(2 + i)^2$ **37.** $(2 + 3i)(2 - 3i)$ **38.** $(6 - 4i)(6 + 4i)$

39. $(\sqrt{6} + i)(\sqrt{6} - i)$ **40.** $(\sqrt{2} - 4i)(\sqrt{2} + 4i)$ **41.** $i(3 - 4i)(3 + 4i)$ **42.** $i(2 + 7i)(2 - 7i)$

43. $3i(2 - i)^2$ **44.** $-5i(4 - 3i)^2$

Find each of the following powers of i. See Example 8.

45. i^5 **46.** i^8 **47.** i^9 **48.** i^{11} **49.** i^{12} **50.** i^{25}

51. i^{43} **52.** $\dfrac{1}{i^9}$ **53.** $\dfrac{1}{i^{12}}$ **54.** i^{-6} **55.** i^{-15} **56.** i^{-49}

57. Suppose that your friend, Susan Katz, tells you that she has discovered a method of simplifying a positive power of *i*. "Just divide the exponent by 4," she says, "and then look at the remainder. Then refer to the table of powers of *i* in this section. The large power of *i* is equal to *i* to the power indicated by the remainder. And if the remainder is 0, the result is $i^0 = 1$." Explain why Susan's method works.

58. Explain why the method of dividing complex numbers (that is, multiplying both the numerator and the denominator by the conjugate of the denominator) works. That is, what property justifies this process?

Divide, Write each result in standard form. See Example 10.

59. $\dfrac{1 + i}{1 - i}$ **60.** $\dfrac{2 - i}{2 + i}$ **61.** $\dfrac{4 - 3i}{4 + 3i}$ **62.** $\dfrac{5 - 2i}{6 - i}$ **63.** $\dfrac{3 - 4i}{2 - 5i}$ **64.** $\dfrac{1 - 3i}{1 + i}$

65. $\dfrac{-3 + 4i}{2 - i}$ **66.** $\dfrac{5 + 6i}{5 - 6i}$ **67.** $\dfrac{2}{i}$ **68.** $\dfrac{-7}{3i}$ **69.** $\dfrac{1 - \sqrt{-5}}{3 + \sqrt{-4}}$ **70.** $\dfrac{2 + \sqrt{-3}}{1 - \sqrt{-9}}$

Perform the indicated operations. Write answers in standard form.

71. $\dfrac{2 + i}{3 - i} \cdot \dfrac{5 + 2i}{1 + i}$ **72.** $\dfrac{1 - i}{2 + i} \cdot \dfrac{4 + 3i}{1 + i}$ **73.** $\dfrac{6 + 2i}{5 - i} \cdot \dfrac{1 - 3i}{2 + 6i}$

74. $\dfrac{5 - 3i}{1 + 2i} \cdot \dfrac{2 - 4i}{1 + i}$ **75.** $\dfrac{5 - i}{3 + i} + \dfrac{2 + 7i}{3 + i}$ **76.** $\dfrac{4 - 3i}{2 + 5i} + \dfrac{8 - i}{2 + 5i}$

Give the necessary conditions for a and b so that the square $(a + bi)^2$ is

77. Real;

78. Imaginary.

79. Show that $\dfrac{\sqrt{2}}{2} + \dfrac{\sqrt{2}}{2} i$ is a square root of i.

80. Show that $\dfrac{\sqrt{3}}{2} + \dfrac{1}{2} i$ is a cube root of i.

81. Evaluate $3z - z^2$ if $z = 3 - 2i$.

82. Evaluate $-2z + z^3$ if $z = -6i$.

2.4 ——— QUADRATIC EQUATIONS

As mentioned earlier, an equation of the form $ax + b = 0$ is a linear equation. A *quadratic equation* is defined as follows.

QUADRATIC EQUATION IN ONE VARIABLE	An equation that can be written in the form
	$$ax^2 + bx + c = 0,$$
	where a, b, and c are real numbers with $a \neq 0$, is a **quadratic equation.**

(Why is the restriction $a \neq 0$ necessary?) A quadratic equation written in the form $ax^2 + bx + c = 0$ is in *standard form.*

The simplest method of solving a quadratic equation, but one that is not always easily applied, is by factoring. This method depends on the following property.

ZERO-FACTOR PROPERTY	If a and b are complex numbers, with $ab = 0$, then $a = 0$ or $b = 0$ or both.

The next example shows how the zero-factor property is used to solve a quadratic equation.

■ *Example 1*

USING THE
ZERO-FACTOR
PROPERTY

Solve $6r^2 + 7r = 3$.

First write the equation in standard form as

$$6r^2 + 7r - 3 = 0.$$

Now factor $6r^2 + 7r - 3$ to get

$$(3r - 1)(2r + 3) = 0.$$

By the zero-factor property, the product $(3r - 1)(2r + 3)$ can equal 0 only if

$$3r - 1 = 0 \quad \text{or} \quad 2r + 3 = 0.$$

Solve each of these linear equations separately to find that the solutions of the original equation are 1/3 and $-3/2$. Check these solutions by substituting in the original equation. The solution set is $\{1/3, -3/2\}$. ■

A quadratic equation of the form $x^2 = k$ can be solved by factoring with the following sequence of equivalent equations.

$$x^2 = k$$
$$x^2 - k = 0$$
$$(x - \sqrt{k})(x + \sqrt{k}) = 0$$
$$x - \sqrt{k} = 0 \qquad \text{or} \qquad x + \sqrt{k} = 0$$
$$x = \sqrt{k} \qquad \text{or} \qquad x = -\sqrt{k}$$

This proves the following statement, which we call the **square root property.**

SQUARE ROOT PROPERTY The solution set of $x^2 = k$ is $\{\sqrt{k}, -\sqrt{k}\}$.

This solution set is often abbreviated as $\{\pm\sqrt{k}\}$. Both solutions are real if $k > 0$ and imaginary if $k < 0$. (If $k = 0$, there is only one distinct solution, sometimes called a *double* solution.)

■ *Example 2*

USING THE SQUARE ROOT PROPERTY

Solve each quadratic equation.

(a) $z^2 = 17$

The solution set is $\{\pm\sqrt{17}\}$.

(b) $m^2 = -25$

Since $\sqrt{-25} = 5i$, the solution set of $m^2 = -25$ is $\{\pm 5i\}$.

(c) $(y - 4)^2 = 12$

Use a generalization of the square root property, working as follows.

$$(y - 4)^2 = 12$$
$$y - 4 = \pm\sqrt{12}$$
$$y = 4 \pm \sqrt{12}$$
$$y = 4 \pm 2\sqrt{3}$$

The solution set is $\{4 \pm 2\sqrt{3}\}$. ■

COMPLETING THE SQUARE As suggested by Example 2(c), any quadratic equation can be solved using the square root property if it is first written in the form $(x + n)^2 = k$ for suitable numbers n and k. The next example shows how to write a quadratic equation in this form.

■ *Example 3*

USING THE METHOD OF COMPLETING THE SQUARE

Solve $x^2 - 4x = 8$.

To write $x^2 - 4x = 8$ in the form $(x + n)^2 = k$, we must find a number that can be added to the left side of the equation to get a perfect square. The equation $(x + n)^2 = k$ can be written as $x^2 + 2xn + n^2 = k$. Comparing this equation with $x^2 - 4x = 8$ shows that

$$2xn = -4x$$
$$n = -2.$$

If $n = -2$, then $n^2 = 4$. Adding 4 to both sides of $x^2 - 4x = 8$ and factoring on the left gives

$$x^2 - 4x + 4 = 8 + 4$$
$$(x - 2)^2 = 12.$$

Now the square root property can be used as follows.

$$x - 2 = \pm\sqrt{12}$$
$$x = 2 \pm 2\sqrt{3}$$

The solution set is $\{2 \pm 2\sqrt{3}\}$. ■

The steps used in solving a quadratic equation by completing the square follow.

SOLVING BY COMPLETING THE SQUARE

To solve $ax^2 + bx + c = 0$, $a \neq 0$, by completing the square:

1. If $a \neq 1$, multiply both sides of the equation by $1/a$.
2. Rewrite the equation so that the constant term is alone on one side of the equals sign.
3. Square half the coefficient of x, and add this square to both sides of the equation.
4. Factor the resulting trinomial as a perfect square and combine terms on the other side.
5. Use the square root property to complete the solution.

■ *Example 4*

USING THE METHOD OF COMPLETING THE SQUARE

Solve $9z^2 - 12z - 1 = 0$.

The coefficient of z^2 must be 1. Multiply both sides by $1/9$.

$$z^2 - \frac{4}{3}z - \frac{1}{9} = 0$$

Now add $1/9$ to both sides of the equation.

$$z^2 - \frac{4}{3}z = \frac{1}{9}$$

Half the coefficient of z is $-2/3$, and $(-2/3)^2 = 4/9$. Add $4/9$ to both sides, getting

$$z^2 - \frac{4}{3}z + \frac{4}{9} = \frac{1}{9} + \frac{4}{9}.$$

Factoring on the left and combining terms on the right gives

$$\left(z - \frac{2}{3}\right)^2 = \frac{5}{9}.$$

Now use the square root property and the quotient property for radicals to get

$$z - \frac{2}{3} = \pm\sqrt{\frac{5}{9}}$$

$$z - \frac{2}{3} = \pm\frac{\sqrt{5}}{3}$$

$$z = \frac{2}{3} \pm \frac{\sqrt{5}}{3}.$$

These two solutions can be written as

$$\frac{2 \pm \sqrt{5}}{3},$$

with the solution set abbreviated as $\left\{\dfrac{2 \pm \sqrt{5}}{3}\right\}$. ■

QUADRATIC FORMULA   The method of completing the square can be used to solve any quadratic equation. However, in the long run it is better to start with the general quadratic equation,

$$ax^2 + bx + c = 0, \quad a \neq 0,$$

and use the method of completing the square to solve this equation for x in terms of the constants a, b, and c. The result will be a general formula for solving any quadratic equation. For now, assume that $a > 0$ and multiply both sides by $1/a$ to get

$$x^2 + \frac{b}{a}x + \frac{c}{a} = 0.$$

Add $-c/a$ to both sides.

$$x^2 + \frac{b}{a}x = -\frac{c}{a}$$

Now take half of b/a, and square the result:

$$\frac{1}{2} \cdot \frac{b}{a} = \frac{b}{2a} \quad \text{and} \quad \left(\frac{b}{2a}\right)^2 = \frac{b^2}{4a^2}.$$

Add the square to both sides, producing

$$x^2 + \frac{b}{a}x + \frac{b^2}{4a^2} = \frac{b^2}{4a^2} - \frac{c}{a}.$$

The expression on the left side of the equals sign can be written as the square of a binomial, while the expression on the right can be simplified.

$$\left(x + \frac{b}{2a}\right)^2 = \frac{b^2 - 4ac}{4a^2}$$

By the square root property, this last statement leads to

$$x + \frac{b}{2a} = \sqrt{\frac{b^2 - 4ac}{4a^2}} \quad \text{or} \quad x + \frac{b}{2a} = -\sqrt{\frac{b^2 - 4ac}{4a^2}}.$$

Since $4a^2 = (2a)^2$, or $4a^2 = (-2a)^2$,

$$x + \frac{b}{2a} = \frac{\sqrt{b^2 - 4ac}}{2a} \quad \text{or} \quad x + \frac{b}{2a} = \frac{-\sqrt{b^2 - 4ac}}{2a}.$$

Adding $-b/(2a)$ to both sides of each result gives

$$x = \frac{-b + \sqrt{b^2 - 4ac}}{2a} \quad \text{or} \quad x = \frac{-b - \sqrt{b^2 - 4ac}}{2a}.$$

It can be shown that these two results are also valid if $a < 0$. A compact form of these two equations, called the *quadratic formula*, follows.

QUADRATIC FORMULA

The solutions of the quadratic equation $ax^2 + bx + c = 0$, where $a \neq 0$, are

$$\frac{-b \pm \sqrt{b^2 - 4ac}}{2a}.$$

CAUTION   Notice that the fraction bar in the quadratic formula extends under the $-b$ term in the numerator.

■ *Example 5*

USING THE
QUADRATIC
FORMULA (REAL
SOLUTIONS)

Solve $x^2 - 4x + 2 = 0$.

Here $a = 1$, $b = -4$, and $c = 2$. Substitute these values into the quadratic formula to get

$$x = \frac{-b \pm \sqrt{b^2 - 4ac}}{2a}$$

$$= \frac{-(-4) \pm \sqrt{(-4)^2 - 4(1)2}}{2(1)} \qquad a = 1, b = -4, c = 2$$

$$= \frac{4 \pm \sqrt{16 - 8}}{2}$$

$$= \frac{4 \pm 2\sqrt{2}}{2} \qquad \sqrt{16 - 8} = \sqrt{8} = 2\sqrt{2}$$

$$= \frac{2(2 \pm \sqrt{2})}{2} \qquad \text{Factor out a 2 in the numerator.}$$

$$= 2 \pm \sqrt{2} \qquad \text{Lowest terms}$$

The solution set is $\{2 + \sqrt{2}, 2 - \sqrt{2}\}$, abbreviated as $\{2 \pm \sqrt{2}\}$. ■

■ Example 6

USING THE
QUADRATIC
FORMULA (COMPLEX
SOLUTIONS)

Solve $2y^2 = y - 4$.

To find the values of a, b, and c, first rewrite the equation in standard form as $2y^2 - y + 4 = 0$. Then $a = 2$, $b = -1$, and $c = 4$. By the quadratic formula,

$$y = \frac{-(-1) \pm \sqrt{(-1)^2 - 4(2)(4)}}{2(2)}$$

$$= \frac{1 \pm \sqrt{1 - 32}}{4}$$

$$= \frac{1 \pm \sqrt{-31}}{4}$$

$$= \frac{1 \pm i\sqrt{31}}{4}.$$

The solution set is $\left\{\dfrac{1}{4} \pm \dfrac{i\sqrt{31}}{4}\right\}$. ■

The equation in Example 7 is called a *cubic* equation, because of the term of degree 3. In Chapter 6 we will discuss such higher degree equations in more detail. However, the equation $x^3 + 8 = 0$, for example, can be solved using factoring and the quadratic formula.

■ Example 7

USING THE
QUADRATIC
FORMULA IN
SOLVING A
PARTICULAR CUBIC
EQUATION

Solve $x^3 + 8 = 0$.

Factor on the left side, and then set each factor equal to zero.

$$x^3 + 8 = 0$$

$$(x + 2)(x^2 - 2x + 4) = 0$$

$$x + 2 = 0 \quad \text{or} \quad x^2 - 2x + 4 = 0$$

The solution of $x + 2 = 0$ is $x = -2$. Now use the quadratic formula to solve $x^2 - 2x + 4 = 0$.

$$x^2 - 2x + 4 = 0$$

$$x = \frac{2 \pm \sqrt{4 - 16}}{2} \qquad a = 1, b = -2, c = 4$$

$$x = \frac{2 \pm \sqrt{-12}}{2}$$

$$x = \frac{2 \pm 2i\sqrt{3}}{2} \qquad$$

$$x = 1 \pm i\sqrt{3} \qquad \text{Factor out a 2 in the numerator and reduce to lowest terms.}$$

The solution set is $\{-2, 1 \pm i\sqrt{3}\}$. ■

Sometimes it is necessary to solve a literal equation for a variable that is squared. In such cases, we usually apply the square root property of equations or the quadratic formula.

■ *Example 8*

SOLVING FOR A
VARIABLE THAT IS
SQUARED

(a) Solve for d: $A = \dfrac{\pi d^2}{4}$.

Start by multiplying both sides by 4 to get

$$4A = \pi d^2.$$

Now divide by π.

$$d^2 = \frac{4A}{\pi}$$

Use the square root property and rationalize the denominator on the right.

$$d = \pm\sqrt{\frac{4A}{\pi}}$$

$$d = \frac{\pm 2\sqrt{A}}{\sqrt{\pi}}$$

$$d = \frac{\pm 2\sqrt{A\pi}}{\pi}$$

(b) Solve for t: $rt^2 - st = k \ (r \neq 0)$.

Because this equation has a term with t as well as t^2, we use the quadratic formula. Subtract k from both sides to get

$$rt^2 - st - k = 0.$$

Now use the quadratic formula to find t, with $a = r$, $b = -s$, and $c = -k$.

$$t = \frac{-b \pm \sqrt{b^2 - 4ac}}{2a}$$

$$t = \frac{-(-s) \pm \sqrt{(-s)^2 - 4(r)(-k)}}{2(r)}$$

$$t = \frac{s \pm \sqrt{s^2 + 4rk}}{2r} \quad ■$$

NOTE

In practical applications of formulas solved for a squared variable, it is often necessary to reject one of the solutions because it does not satisfy the physical conditions of the problem.

THE DISCRIMINANT   The quantity under the radical in the quadratic formula, $b^2 - 4ac$, is called the **discriminant.** When the numbers a, b, and c are *integers* (but not necessarily otherwise), the value of the discriminant can be used to determine whether the solutions will be rational, irrational, or imaginary numbers. If the discriminant is 0, there will be only one distinct solution. (Why?)

The discriminant of a quadratic equation gives the following information about the solutions of the equation.

DISCRIMINANT	Discriminant	Number of Solutions	Kind of Solutions
	Positive, perfect square	Two	Rational
	Positive, but not a perfect square	Two	Irrational
	Zero	One (a double solution)	Rational
	Negative	Two	Imaginary

■ *Example 9*
USING THE
DISCRIMINANT

Use the discriminant to determine whether the solutions of $5x^2 + 2x - 4 = 0$ are rational, irrational, or imaginary.

The discriminant is

$$b^2 - 4ac = (2)^2 - (4)(5)(-4) = 84.$$

Because the discriminant is positive and *a, b,* and *c* are integers, there are two real-number solutions. Since 84 is not a perfect square, the solutions will be irrational numbers. ■

■ *Example 10*
USING THE
DISCRIMINANT

Find all values of *k* so that the equation

$$16p^2 + kp + 25 = 0$$

has exactly one solution.

A quadratic equation with real coefficients will have exactly one solution if the discriminant is zero. Here, $a = 16$, $b = k$, and $c = 25$, giving the discriminant

$$b^2 - 4ac = k^2 - 4(16)(25) = k^2 - 1600.$$

The discriminant is 0 if

$$k^2 - 1600 = 0$$
$$k^2 = 1600,$$

from which $k = \pm 40$. ■

Recall from Section 1.6 that a rational expression is not defined when its denominator is 0. Restrictions on the variable are found by determining the value or values that cause the expression in the denominator to equal 0.

■ *Example 11*

DETERMINING
RESTRICTIONS ON
THE VARIABLE

For each of the following, give the real number restrictions on the variable.

(a) $\dfrac{2x - 5}{2x^2 - 9x - 5}$

Set the denominator equal to 0 and solve.

$$2x^2 - 9x - 5 = 0$$
$$(2x + 1)(x - 5) = 0$$

$$2x + 1 = 0 \qquad \text{or} \qquad x - 5 = 0$$

$$x = -\frac{1}{2} \qquad \text{or} \qquad x = 5$$

The restrictions on the variable are $x \neq -1/2$ and $x \neq 5$.

(b) $\dfrac{1}{3x^2 - x + 4}$

Solve $3x^2 - x + 4 = 0$. Since the polynomial does not factor, use the quadratic formula.

$$x = \frac{-(-1) \pm \sqrt{(-1)^2 - 4(3)(4)}}{2(3)} = \frac{1 \pm \sqrt{-47}}{6}$$

Both solutions are imaginary numbers, so there are no real numbers that make the denominator equal to zero. Thus there are no real number restrictions on x. ■

2.4 EXERCISES ■

Solve the following equations by factoring or by using the square root property. See Examples 1 and 2.

1. $p^2 = 16$ **2.** $k^2 = 25$ **3.** $x^2 = 27$ **4.** $r^2 = 48$

5. $t^2 = -16$ **6.** $y^2 = -100$ **7.** $x^2 = -18$ **8.** $(p + 2)^2 = 7$

9. $(3k - 1)^2 = 12$ **10.** $(4t + 1)^2 = 20$ **11.** $p^2 - 5p + 6 = 0$ **12.** $q^2 + 2q - 8 = 0$

13. $6z^2 - 5z - 50 = 0$ **14.** $21p^2 = 10 - 29p$ **15.** $(5r - 3)^2 = -3$ **16.** $(-2w + 5)^2 = -8$

Solve the following equations by completing the square. See Examples 3 and 4.

17. $p^2 - 8p + 15 = 0$ **18.** $m^2 + 5m = 6$ **19.** $x^2 - 2x - 4 = 0$

20. $r^2 + 8r + 13 = 0$ **21.** $2p^2 + 2p + 1 = 0$ **22.** $9z^2 - 12z + 8 = 0$

23. Erin solved Exercise 16 correctly and wrote the solution as $\dfrac{5}{2} \pm i\sqrt{2}$. Paul wrote his solution as $\dfrac{-5}{-2} \pm i\sqrt{2}$. Was Paul correct? Explain.

Solve the following equations by using the quadratic formula. See Examples 5 and 6.

24. $m^2 - m - 1 = 0$ **25.** $y^2 - 3y - 2 = 0$ **26.** $x^2 - 6x + 7 = 0$

27. $11p^2 - 7p + 1 = 0$ **28.** $4z^2 - 12z + 11 = 0$ **29.** $x^2 = 2x - 5$

30. $\frac{1}{2}t^2 + \frac{1}{4}t - 3 = 0$

31. $\frac{2}{3}x^2 + \frac{1}{4}x = 3$

32. $.2x^2 + .4x - .3 = 0$

33. $4 + \frac{3}{x} - \frac{2}{x^2} = 0$

34. $4 - \frac{11}{x} - \frac{3}{x^2} = 0$

35. $3 - \frac{4}{p} = \frac{2}{p^2}$

36. Why is the restriction $a \neq 0$ necessary in the statement of the quadratic formula?

37. Which one of the following equations has two real, distinct solutions? Do not actually solve.
   (a) $(3x - 4)^2 = -4$ **(b)** $(4 + 7x)^2 = 0$ **(c)** $(5x + 9)(5x + 9) = 0$ **(d)** $(7x + 4)^2 = 11$

38. Which equations in Exercise 37 have only one distinct, real solution?

39. Which one of the equations in Exercise 37 has two imaginary solutions?

40. Discuss the advantages and disadvantages of the three methods of solving quadratic equations discussed in this section: using the square root property, completing the square, and using the quadratic formula.

Solve each of the following equations by factoring first. See Example 7.

41. $x^3 - 1 = 0$

42. $x^3 + 64 = 0$

43. $x^3 + 27 = 0$

Solve the following equations by any method.

44. $8p^3 + 125 = 0$

45. $2 - \frac{5}{k} + \frac{2}{k^2} = 0$

46. $(m - 3)^2 = 5$

47. $t^2 - t = 3$

48. $x^2 + x = -1$

49. $64r^3 - 343 = 0$

50. $2s^2 + 2s = 3$

51. $(3y + 1)^2 = -7$

52. $\frac{1}{3}x^2 + \frac{1}{6}x + \frac{1}{9} = 0$

Each of the following quadratic equations has at least one coefficient that is an irrational number. To solve for example,

$$\sqrt{2}m^2 + 5m - 3\sqrt{2} = 0,$$

we can use the quadratic formula with $a = \sqrt{2}$, $b = 5$, and $c = -3\sqrt{2}$. Solve each equation using the quadratic formula.

53. $m^2 - \sqrt{2}m - 1 = 0$

54. $z^2 - \sqrt{3}z - 2 = 0$

55. $\sqrt{2}p^2 - 3p + \sqrt{2} = 0$

56. $-\sqrt{6}k^2 - 2k + \sqrt{6} = 0$

57. $x^2 + \sqrt{5}x + 1 = 0$

58. $3\sqrt{5}m^2 - 2m = \sqrt{5}$

Solve each equation for the indicated variable. Assume that no denominators are zero. See Example 8.

59. $s = \frac{1}{2}gt^2$ for t

60. $A = \pi r^2$ for r

61. $F = \frac{kMv^4}{r}$ for v

62. $s = s_0 + gt^2 + k$ for t

63. $P = \frac{E^2R}{(r + R)^2}$ for R

64. $S = 2\pi rh + 2\pi r^2$ for r

For each of the following equations, (a) solve for x in terms of y, and (b) solve for y in terms of x. See Example 8.

65. $4x^2 - 2xy + 3y^2 = 2$

66. $3y^2 + 4xy - 9x^2 = -1$

Identify the values of a, b, and c for each of the following: then evaluate the discriminant $b^2 - 4ac$ and use it to predict the type of solutions. Do not solve the equations. See Example 9.

67. $x^2 + 8x + 16 = 0$

68. $x^2 - 5x + 4 = 0$

69. $3m^2 - 5m + 2 = 0$

70. $8y^2 = 14y - 3$

71. $4p^2 = 6p + 3$

72. $2r^2 - 4r + 1 = 0$

73. $9k^2 + 11k + 4 = 0$

74. $3z^2 = 4z - 5$

75. $8x^2 - 72 = 0$

Find all values of k for which each of the following equations has exactly one solution. See Example 10.

76. $9x^2 + kx + 4 = 0$ **77.** $25m^2 - 10m + k = 0$ **78.** $y^2 + 11y + k = 0$

79. Show that the discriminant for the equation $\sqrt{2}m^2 + 5m - 3\sqrt{2} = 0$ is 49. If this equation is completely solved, it can be shown that the solution set is $\{-3\sqrt{2}, \sqrt{2}/2\}$. Here we have a discriminant that is positive and a perfect square, yet the two solutions are irrational. Does this contradict the discussion in this section? Explain.

80. Solve the equation in Exercise 79.

Give the real number restrictions on the variable for each of the following rational expressions. See Example 11.

81. $\dfrac{3 - 2x}{3x^2 - 19x - 14}$ **82.** $\dfrac{7 - 3x}{2x^2 + 7x - 15}$ **83.** $\dfrac{-3}{16y^2 + 8y + 1}$

84. $\dfrac{18}{25y^2 - 30y + 9}$ **85.** $\dfrac{8 + 7x}{x^2 + x + 1}$ **86.** $\dfrac{-9x + 4}{7x^2 + 2x + 2}$

87. Let r_1 and r_2 be the solutions of the quadratic equation $ax^2 + bx + c = 0$. Show that the following statements are correct.

(a) $r_1 + r_2 = -\dfrac{b}{a}$ **(b)** $r_1 \cdot r_2 = \dfrac{c}{a}$

88. State in words the result of Exercise 87(a).

89. State in words the result of Exercise 87(b).

90. Is it possible for the solution set of a quadratic equation with real coefficients to consist of one real and one imaginary solution?

2.5 ———————— **APPLICATIONS OF QUADRATIC EQUATIONS**

Many applied problems lead to quadratic equations. In this section we give examples of several kinds of such problems.

CAUTION | When solving problems that lead to quadratic equations, we may get a solution that does not satisfy the physical constraints of the problem. For example, if x represents a width and the two solutions of the quadratic equation are -9 and 1, the value -9 must be rejected, since a width must be a positive number.

■ *Example 1*
SOLVING A
GEOMETRY PROBLEM

A landscape contractor wants to make an exposed gravel border of uniform width around a rectangular pool in a garden. The pool is 10 feet long and 6 feet wide. There is enough material to cover 36 square feet. How wide should the border be?

A diagram of the pool with the border is shown in Figure 2.4. Since we are asked to find the width of the border, let

$$x = \text{the width of the border in feet.}$$

Then $6 + 2x = $ the width of the larger rectangle in feet,

and $10 + 2x = $ the length of the larger rectangle in feet.

10 + 2x ft

10 ft

6 + 2x ft   6 ft

x is in feet.

■ **FIGURE 2.4**

The area of the larger rectangle is $(6 + 2x)(10 + 2x)$ square feet, and the area of the pool is $6 \cdot 10 = 60$ square feet. The area of the border is found by subtracting the area of the pool from the area of the larger rectangle. This difference should be 36 square feet.

Area of the larger rectangle	minus	area of the pool	is	36 square feet.
$(6 + 2x)(10 + 2x)$	$-$	60	$=$	36

Now solve this equation.

$$60 + 32x + 4x^2 - 60 = 36$$
$$4x^2 + 32x - 36 = 0$$
$$x^2 + 8x - 9 = 0$$
$$(x + 9)(x - 1) = 0$$

The solutions are -9 and 1. The width of the border cannot be negative, so the border should be 1 foot wide. ■

Problems involving rate of work were first introduced in Section 2.2. Recall that if a job can be done in x units of time, the rate of work is $1/x$ job per unit of time.

■ *Example 2*

SOLVING A WORK PROBLEM

Pat and Mike clean the offices in a downtown office building each night. Working alone, Pat takes 1 hour less time than Mike to complete the job. Working together, they can finish the job in 6 hours. One night Pat calls in sick. How long should it take Mike to do the job alone?

Let

$$x = \text{the time for Mike to do the job alone}$$

and

$$x - 1 = \text{the time for Pat to do the job alone.}$$

The rates for Mike and Pat are, respectively, $1/x$ and $1/(x - 1)$ job per hour. If we multiply the time worked together, 6 hours, by each rate, we get the fractional part of the job done by each person. This is summarized in the following chart.

	Rate	Time	Part of the Job Accomplished
Mike	$\dfrac{1}{x}$	6	$6\left(\dfrac{1}{x}\right) = \dfrac{6}{x}$
Pat	$\dfrac{1}{x - 1}$	6	$6\left(\dfrac{1}{x - 1}\right) = \dfrac{6}{x - 1}$

Since one whole job can be done by the two people, the sum of the parts must equal 1, as indicated by the equation

$$\frac{6}{x} + \frac{6}{x-1} = 1.$$

To clear fractions, multiply both sides of the equation by the least common denominator, $x(x-1)$.

$$x(x-1)\frac{6}{x} + x(x-1)\frac{6}{x-1} = 1x(x-1)$$

$$6(x-1) + 6x = x(x-1)$$

$$6x - 6 + 6x = x^2 - x$$

$$0 = x^2 - 13x + 6$$

$$x = \frac{13 \pm \sqrt{(-13)^2 - 4(1)(6)}}{2(1)} \qquad a = 1, b = -13, \; c = 6$$

$$x = \frac{13 \pm \sqrt{169 - 24}}{2}$$

$$x = \frac{13 \pm \sqrt{145}}{2}$$

$$x \approx \frac{13 \pm 12.04}{2}$$

Use a calculator to find that to the nearest tenth, $x = 12.5$ or $x = .5$. The solution $x = .5$ does not satisfy the conditions of the problem, since then Pat takes $x - 1 = -.5$ hour to complete the work. It will take Mike 12.5 hours to do the job alone. ■

■ *Example 3*

SOLVING A MOTION PROBLEM

A river excursion boat traveled upstream from Galt to Isleton, a distance of 12 miles. On the return trip downstream, the boat traveled 3 miles per hour faster. If the return trip took 8 minutes less time, how fast did the boat travel upstream?

The chart below summarizes the information in the problem, where x represents the rate upstream.

	d	r	t
Upstream	12	x	$\dfrac{12}{x}$
Downstream	12	$x + 3$	$\dfrac{12}{x+3}$

$$t = \frac{d}{r}$$

The entries in the column for time are found from solving the distance formula, $d = rt$, for t in each case. Since rates are given in miles per hour, convert 8 minutes to hours as follows, letting H represent the equivalent number of hours.

$$\frac{H \text{ hr}}{1 \text{ hr}} = \frac{8 \text{ min}}{60 \text{ min}}$$

$$H = \frac{8}{60} = \frac{2}{15}$$

Now write an equation using the fact that the time for the return trip (downstream) was 8 minutes or 2/15 hour less than the time upstream.

Time downstream   is   time upstream   less $\frac{2}{15}$ hour.

$$\frac{12}{x + 3} \qquad = \qquad \frac{12}{x} \qquad - \qquad \frac{2}{15}$$

Solve the equation, first multiplying on both sides by the common denominator, $15x(x + 3)$, to get

$$12(15x) = 12(15)(x + 3) - 2x(x + 3)$$
$$180x = 180x + 540 - 2x^2 - 6x$$
$$2x^2 + 6x - 540 = 0 \qquad\qquad \text{Standard form}$$
$$x^2 + 3x - 270 = 0 \qquad\qquad \text{Divide by 2.}$$
$$(x + 18)(x - 15) = 0$$
$$x = -18 \qquad \text{or} \qquad x = 15.$$

Reject the negative solution. The boat traveled 15 miles per hour upstream. ■

CAUTION   When problems involve different units of time (as in Example 3, where rate was given in miles per hour and time was given in minutes), it is necessary to convert to the same unit before setting up the equation.

Example 4 requires the use of the **Pythagorean theorem** from geometry.

PYTHAGOREAN THEOREM   In a right triangle, the sum of the squares of the lengths of the legs is equal to the square of the length of the hypotenuse.

$$a^2 + b^2 = c^2$$

■ *Example 4*

SOLVING A PROBLEM INVOLVING THE PYTHAGOREAN THEOREM

A lot is in the shape of a right triangle. The longer leg of the triangle is 20 meters longer than twice the length of the shorter leg. The hypotenuse is 10 meters longer than the longer leg. Find the lengths of the three sides of the lot.

Let s = length of the shorter leg in meters. Then $2s + 20$ meters represents the length of the longer leg, and $(2s + 20) + 10 = 2s + 30$ meters represents the length of the hypotenuse. See Figure 2.5.

■ FIGURE 2.5

Application of the Pythagorean theorem gives the equation

$$s^2 + (2s + 20)^2 = (2s + 30)^2$$
$$s^2 + 4s^2 + 80s + 400 = 4s^2 + 120s + 900$$
$$s^2 - 40s - 500 = 0$$
$$(s - 50)(s + 10) = 0$$
$$s = 50 \quad \text{or} \quad s = -10.$$

Since s represents a length, the value -10 is not reasonable. The shorter leg is 50 meters long, the longer leg 120 meters long, and the hypotenuse 130 meters long. ■

■ **IN SIMPLEST TERMS** ━━━━━━━━━━━━━━━━━━━━━━━━━━━ ■

7 ▭

To determine the appropriate landing speed of an airplane, the formula $.1s^2 - 3s + 22 = D$ is used, where s is the initial landing speed in feet per second and D is the distance needed in feet. If the landing speed is too fast, the pilot may run out of runway; if the speed is too slow, the plane may stall.

Suppose the runway is 800 feet long. The appropriate landing speed may be calculated by completing the square. In the first step, multiply the equation by 10 to eliminate the decimal.

$$.1s^2 - 3s + 22 = 800$$
$$s^2 - 30s + 220 = 8000$$
$$s^2 - 30s = 7780$$
$$s^2 - 30s + 225 = 8005$$
$$(s - 15)^2 = 8005$$
$$s - 15 = \pm\sqrt{8005}$$
$$s = 15 \pm \sqrt{8005}$$
$$s \approx 15 \pm 89.5$$

The only realistic solution for the landing speed is approximately 104.5 feet per second.

SOLVE EACH PROBLEM

A. Use the formula given above to calculate the appropriate landing speed of an airplane if 650 feet of runway are available.

B. If an airplane stalls at speeds less than 70 feet per second, find the minimum safe landing field length.

ANSWERS   A. The landing speed should be about 96 feet per second.   B. The landing field should be at least 302 feet long.

∎ *Example 5*
**SOLVING A
PROBLEM INVOLVING
MOTION OF A
PROJECTILE**

If a projectile is shot vertically upward with an initial velocity of 100 feet per second, neglecting air resistance, its height s (in feet) above the ground t seconds after projection is given by

$$s = -16t^2 + 100t.$$

(a) After how many seconds will it be 50 feet above the ground?

We must find the value of t so that $s = 50$. Let $s = 50$ in the equation, and use the quadratic formula.

$$50 = -16t^2 + 100t$$
$$16t^2 - 100t + 50 = 0 \qquad \text{Standard form}$$
$$8t^2 - 50t + 25 = 0 \qquad \text{Divide by 2.}$$
$$t = \frac{-(-50) \pm \sqrt{(-50)^2 - 4(8)(25)}}{2(8)}$$
$$t = \frac{50 \pm \sqrt{1700}}{16}$$

$$t \approx .55 \qquad \text{or} \qquad t \approx 5.70 \qquad \text{Use a calculator.}$$

Here, both solutions are acceptable, since the projectile reaches 50 feet twice: once on its way up (after .55 second) and once on its way down (after 5.70 seconds).

(b) How long will it take for the projectile to return to the ground?

When it returns to the ground, its height s will be 0 feet, so let $s = 0$ in the equation.

$$0 = -16t^2 + 100t$$

This can be solved by factoring.

$$0 = -4t(4t - 25)$$

$$-4t = 0 \qquad \text{or} \qquad 4t - 25 = 0$$
$$t = 0 \qquad\qquad\qquad 4t = 25$$
$$t = 6.25$$

The first solution, 0, represents the time at which the projectile was on the ground prior to being launched, so it does not answer the question. The projectile will return to the ground 6.25 seconds after it is launched. ∎

2.5 EXERCISES ∎

Solve each of the following problems. See Examples 1–4.

1. A shopping center has a rectangular area of 40,000 sq yd enclosed on three sides for a parking lot. The length is 200 yd more than twice the width. What are the dimensions of the lot?

$2x + 200$

x is in yards.

2. An ecology center wants to set up an experimental garden using 300 m of fencing to enclose a rectangular area of 5000 sq m. Find the dimensions of the rectangle.

$150 - x$

x

x is in meters.

3. Pat Kelley went into a frame-it-yourself shop. He wanted a frame 3 in longer than it was wide. The frame he chose extended 1.5 in beyond the picture on each side. Find the outside dimensions of the frame if the area of the unframed picture is 70 sq in.

4. Kathy Monaghan wants to buy a rug for a room that is 12 ft wide and 15 ft long. She wants to leave a uniform strip of floor around the rug. She can afford to buy 108 sq ft of carpeting. What dimensions should the rug have?

5. A rectangular page in a book is to have an 18 cm by 23 cm illustration in the center with equal margins on all four sides. How wide should the margins be if the page has an area of 594 sq cm?

6. A landscape architect has included a rectangular flower bed measuring 9 ft by 5 ft in her plans for a new building. She wants to use two colors of flowers in the bed, one in the center and the other for a border of the same width on all four sides. If she has enough plants to cover 24 sq ft for the border, how wide can the border be?

7. Felipe Santiago can clean a garage in 9 hr less time than his brother Felix. Working together, they can do the job in 20 hr. How long would it take each one to do the job alone?

8. An experienced roofer can do a complete roof in a housing development in half the time compared to an inexperienced roofer. If the two work together on a roof, they complete the job in 2 2/3 hr. How long would it take the experienced roofer to do a roof by himself?

9. Two typists are working on a special project. The experienced typist could complete the project in 2 hr less time than the new typist. Together they complete the project in 2.4 hr. How long would it have taken the experienced typist working alone?

10. It takes two copy machines 6/5 hr to make the copies for a company newsletter. One copy machine would take 1 hr longer than the other to do the job alone. How long would it take the faster machine to complete the job alone?

11. Paula drives 10 mph faster than Steve. Both start at the same time for Atlanta from Chattanooga, a distance of about 100 mi. It takes Steve 1/3 hr longer than Paula to make the trip. What is Steve's average speed?

12. Marjorie Seachrist walks 1 mph faster than her daughter Sibyl. In a walk for charity, both walked the full distance of 24 mi. Sibyl took 2 hr longer than Marjorie. What was Sibyl's average speed?

13. A plane flew 1000 mi. It later took off and flew 2025 mi at an average speed of 50 mph faster than the first trip. The second trip took 2 hr more time than the first. Find the time for the first trip.

14. The Branson family traveled 100 mi to a lake for their vacation. On the return trip their average speed was 50 mph faster. The total time for the round trip was 11/3 hr. What was the family's average speed on their trip to the lake?

15. To solve for the lengths of the sides of the right triangle shown, which equation is correct?

(a) $x^2 = (2x - 2)^2 + (x + 4)^2$
(b) $x^2 + (x + 4)^2 = (2x - 2)^2$
(c) $x^2 = (2x - 2)^2 - (x + 4)^2$
(d) $x^2 + (2x - 2)^2 = (x + 4)^2$

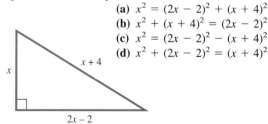

$x + 4$

x

$2x - 2$

16. If a rectangle is r feet long and s feet wide, which one of the following is the length of its diagonal in terms of r and s?

(a) $\sqrt{rs}$ 　　 (b) $r + s$
(c) $\sqrt{r^2 + s^2}$ 　 (d) $r^2 + s^2$

17. A kite is flying on 50 ft of string. How high is it above the ground if its height is 10 ft more than the horizontal distance from the person flying it? Assume the string is being released at ground level.

18. A boat is being pulled into a dock with a rope attached at water level. When the boat is 12 ft from the dock, the length of the rope from the boat to the dock is 3 ft longer than twice the height of the dock above the water. Find the height of the dock.

19. Chris and Josh have received walkie-talkies for Christmas. If they leave from the same point at the same time, Chris walking north at 2.5 mph and Josh walking east at 3 mph, how long will they be able to talk to each other if the range of the walkie-talkies is 4 miles? Round your answer to the nearest minute.

20. There is a bamboo 10 ft high, the upper end of which, being broken, reaches the ground 3 ft from the stem. Find the height of the break. (Adapted from an ancient Chinese work, *Arithmetic in Nine Sections*.)

A projectile is launched from ground level with an initial velocity of v_0 feet per second. Neglecting air resistance, its height t seconds after launch in feet is given by

$$s = -16t^2 + v_0 t.$$

In Exercises 21–24, find the time(s) that the projectile will (a) reach a height of 80 ft and (b) return to the ground for the given value of v_0. Round answers to the nearest hundredth if necessary. See Example 5.

21. $v_0 = 96$ **22.** $v_0 = 128$

23. $v_0 = 32$ **24.** $v_0 = 16$

25. Exercise 21(a) has two answers. A student was once asked why such a problem can have two answers, and his response was "What goes up must come down." Explain in more detail than the student why such a problem may have two answers, one answer, or no answers.

26. Consider the following problem.

The manager of an 80-unit apartment complex knows from experience that at a rent of $300, all the units will be full. On the average, one additional unit will remain vacant for each $20 increase in rent over $300. Currently, the revenue from the complex is $35,000. How many apartments are rented?

If x represents the number of $20 increases over $300, which one of the following equations indicates that the revenue generated will be $35,000?
(a) $(80 - x)(300 + 20x) = 35{,}000$
(b) $(80 + x)(300 + 20x) = 35{,}000$
(c) $(80 - x)(300 - 20x) = 35{,}000$
(d) $(80 + x)(300 - 20x) = 35{,}000$

27. The manager of a cherry orchard is trying to decide when to schedule the annual harvest. If the cherries are picked now, the average yield per tree will be 100 lb, and the cherries can be sold for 40 cents per pound. Past experience shows that the yield per tree will increase about 5 lb per week, while the price will decrease about 2 cents per pound per week. How many weeks should the manager wait to get an average revenue of $38.40 per tree?

28. A local group of scouts has been collecting old aluminum cans for recycling. The group has already collected 12,000 lb of cans, for which they could currently receive $4 per hundred pounds. The group can continue to collect cans at the rate of 400 lb per day. However, a glut in the old-can market has caused the recycling company to announce that it will lower its price, starting immediately, by $.10 per hundred pounds per day. The scouts can make only one trip to the recycling center. How many days should they wait in order to get $490 for their cans?

29. A local club is arranging a charter flight to Miami. The cost of the trip is $225 each for 75 passengers, with a refund of $5 per passenger for each passenger in excess of 75. How many passengers must take the flight to produce revenue of $16,000?

30. Suppose one solution of the equation $km^2 + 10m = 8$ is -4. Find the value of k, and the other solution.

2.6   ——— OTHER TYPES OF EQUATIONS

Many equations that are not actually quadratic equations can be solved by the methods discussed earlier in this chapter.

EQUATIONS QUADRATIC IN FORM   The equation $12m^4 - 11m^2 + 2 = 0$ is not a quadratic equation because of the m^4 term. However, with the substitutions

$$x = m^2 \quad \text{and} \quad x^2 = m^4$$

the given equation becomes

$$12x^2 - 11x + 2 = 0,$$

which is a quadratic equation. This quadratic equation can be solved to find x, and then $x = m^2$ can be used to find the values of m, the solutions to the original equation.

QUADRATIC IN FORM

An equation is said to be **quadratic in form** if it can be written as

$$au^2 + bu + c = 0$$

where $a \neq 0$ and u is some algebraic expression.

■ ***Example 1***

SOLVING AN EQUATION QUADRATIC IN FORM

Solve $12m^4 - 11m^2 + 2 = 0$.

As mentioned above, this equation is quadratic in form. By making the substitution $x = m^2$, the equation becomes

$$12x^2 - 11x + 2 = 0,$$

which can be solved by factoring in the following way.

$$12x^2 - 11x + 2 = 0$$
$$(3x - 2)(4x - 1) = 0$$
$$x = \frac{2}{3} \quad \text{or} \quad x = \frac{1}{4}$$

The original equation contains the variable m. To find m, use the fact that $x = m^2$ and replace x with m^2, getting

$$m^2 = \frac{2}{3} \qquad \text{or} \qquad m^2 = \frac{1}{4}$$

$$m = \pm\sqrt{\frac{2}{3}} \qquad\qquad m = \pm\sqrt{\frac{1}{4}}$$

$$m = \pm\frac{\sqrt{2}}{\sqrt{3}} \cdot \frac{\sqrt{3}}{\sqrt{3}}$$

$$m = \frac{\pm\sqrt{6}}{3} \qquad \text{or} \qquad m = \pm\frac{1}{2}.$$

These four solutions of the given equation $12m^4 - 11m^2 + 2 = 0$ make up the solution set $\{\sqrt{6}/3, -\sqrt{6}/3, 1/2, -1/2\}$, abbreviated as $\{\pm\sqrt{6}/3, \pm 1/2\}$. ■

NOTE

Some equations that are quadratic in form, such as the one in Example 1, can be solved quite easily by direct factorization. The polynomial there can be factored as $(3m^2 - 2)(4m^2 - 1)$, and by setting each factor equal to zero, the same solution set is obtained.

■ *Example 2*

SOLVING AN
EQUATION
QUADRATIC IN
FORM

Solve $6p^{-2} + p^{-1} = 2$.

Let $u = p^{-1}$ so that $u^2 = p^{-2}$. Then substitute and rearrange terms to get

$$6u^2 + u - 2 = 0.$$

Factor on the left, and then place each factor equal to 0, giving

$$(3u + 2)(2u - 1) = 0$$

$$3u + 2 = 0 \qquad \text{or} \qquad 2u - 1 = 0$$

$$u = -\frac{2}{3} \qquad \text{or} \qquad u = \frac{1}{2}.$$

Since $u = p^{-1}$,

$$p^{-1} = -\frac{2}{3} \qquad \text{or} \qquad p^{-1} = \frac{1}{2},$$

from which

$$p = -\frac{3}{2} \qquad \text{or} \qquad p = 2.$$

The solution set of $6p^{-2} + p^{-1} = 2$ is $\{-3/2, 2\}$.   ■

CAUTION

When solving an equation that is quadratic in form, if a substitution variable is used, do not forget the step that gives the solution in terms of the original variable that appears in the equation.

EQUATIONS WITH RADICALS OR RATIONAL EXPONENTS   To solve equations containing radicals or rational exponents, such as $x = \sqrt{15 - 2x}$, or $(x + 1)^{1/2} = x$, use the following property.

If P and Q are algebraic expressions, then every solution of the equation $P = Q$ is also a solution of the equation $(P)^n = (Q)^n$, for any positive integer n.

CAUTION

Be very careful when using this result. It does *not* say that the equations $P = Q$ and $(P)^n = (Q)^n$ are equivalent; it says only that each solution of the original equation $P = Q$ is also a solution of the new equation $(P)^n = (Q)^n$.

When using this property to solve equations, we must be aware that the new equation may have *more* solutions than the original equation. For example, the solution set of the equation $x = -2$ is $\{-2\}$. If we square both sides of the equation $x = -2$, we get the new equation $x^2 = 4$, which has solution set $\{-2, 2\}$. Since the solution sets are not equal, the equations are not equivalent. Because of this, when an equation contains radicals or rational exponents, it is *essential* to check all proposed solutions in the original equation.

■ *Example 3*

SOLVING AN
EQUATION
CONTAINING A
RADICAL

Solve $x = \sqrt{15 - 2x}$.

The equation $x = \sqrt{15 - 2x}$ can be solved by squaring both sides as follows.

$$x^2 = (\sqrt{15 - 2x})^2$$
$$x^2 = 15 - 2x$$
$$x^2 + 2x - 15 = 0$$
$$(x + 5)(x - 3) = 0$$
$$x = -5 \quad \text{or} \quad x = 3$$

Now the proposed solutions *must* be checked in the original equation, $x = \sqrt{15 - 2x}$.

If $x = -5$,

$$x = \sqrt{15 - 2x}$$
$$-5 = \sqrt{15 - 2(-5)} \quad ?$$
$$-5 = \sqrt{15 + 10} \quad ?$$
$$-5 = \sqrt{25} \quad ?$$
$$-5 = 5. \qquad \text{False}$$

If $x = 3$,

$$x = \sqrt{15 - 2x}$$
$$3 = \sqrt{15 - 2(3)} \quad ?$$
$$3 = \sqrt{15 - 6} \quad ?$$
$$3 = \sqrt{9} \quad ?$$
$$3 = 3. \qquad \text{True}$$

As this check shows, only 3 is a solution, giving the solution set {3}. ■

To solve an equation containing radicals, follow these steps.

SOLVING AN
EQUATION
INVOLVING
RADICALS

1. Isolate the radical on one side of the equation.
2. Raise each side of the equation to a power that is the same as the index of the radical so that the radical is eliminated.
3. Solve the resulting equation. If it still contains a radical, repeat Steps 1 and 2.
4. Check each proposed solution in the *original* equation.

■ *Example 4*

SOLVING AN
EQUATION
CONTAINING TWO
RADICALS

Solve $\sqrt{2x + 3} - \sqrt{x + 1} = 1$.

When an equation contains two radicals, begin by isolating one of the radicals on one side of the equation. For this one, let us isolate $\sqrt{2x + 3}$ (Step 1).

$$\sqrt{2x + 3} = 1 + \sqrt{x + 1}$$

Now square both sides (Step 2). Be very careful when squaring on the right side of this equation. Recall that $(a + b)^2 = a^2 + 2ab + b^2$; replace a with 1 and b with $\sqrt{x + 1}$ to get the next equation, the result of squaring both sides of $\sqrt{2x + 3} = 1 + \sqrt{x + 1}$.

$$2x + 3 = 1 + 2\sqrt{x + 1} + x + 1$$
$$x + 1 = 2\sqrt{x + 1}$$

One side of the equation still contains a radical; to eliminate it, square both sides again (Step 3).

$$x^2 + 2x + 1 = 4(x + 1)$$
$$x^2 - 2x - 3 = 0$$
$$(x - 3)(x + 1) = 0$$
$$x = 3 \quad \text{or} \quad x = -1$$

Check these proposed solutions in the original equation (Step 4).

Let $x = 3$.

$$\sqrt{2x + 3} - \sqrt{x + 1} = 1$$
$$\sqrt{2(3) + 3} - \sqrt{3 + 1} = 1 \quad ?$$
$$\sqrt{9} - \sqrt{4} = 1 \quad ?$$
$$3 - 2 = 1 \quad ?$$
$$1 = 1 \qquad \text{True}$$

Let $x = -1$.

$$\sqrt{2x + 3} - \sqrt{x + 1} = 1$$
$$\sqrt{2(-1) + 3} - \sqrt{-1 + 1} = 1 \quad ?$$
$$\sqrt{1} - \sqrt{0} = 1 \quad ?$$
$$1 - 0 = 1 \quad ?$$
$$1 = 1 \qquad \text{True}$$

Both proposed solutions 3 and -1 are solutions of the original equation, giving $\{3, -1\}$ as the solution set. ∎

∎ **Example 5**

SOLVING AN EQUATION CONTAINING A RATIONAL EXPONENT

Solve $(5x^2 - 6)^{1/4} = x$.

Since the equation involves a fourth root, begin by raising both sides to the fourth power.

$$[(5x^2 - 6)^{1/4}]^4 = x^4$$
$$5x^2 - 6 = x^4$$
$$x^4 - 5x^2 + 6 = 0$$

Now substitute y for x^2.

$$y^2 - 5y + 6 = 0$$
$$(y - 3)(y - 2) = 0$$
$$y = 3 \quad \text{or} \quad y = 2$$

Since $y = x^2$,

$$x^2 = 3 \quad \text{or} \quad x^2 = 2$$
$$x = \pm\sqrt{3} \quad \text{or} \quad x = \pm\sqrt{2}.$$

Checking the four proposed solutions, $\sqrt{3}, -\sqrt{3}, \sqrt{2}$, and $-\sqrt{2}$ in the original equation shows that only $\sqrt{3}$ and $\sqrt{2}$ are solutions, so the solution set is $\{\sqrt{3}, \sqrt{2}\}$. ∎

NOTE In the equation of Example 5, we can use the fact that $b^{1/4} = \sqrt[4]{b}$ is a principal fourth root, and thus the right side, x, cannot be negative. Therefore, the two negative proposed solutions must be rejected.

Solve each of the following equations. Use substitution to first write each as a quadratic equation. See Examples 1–2.

1. $m^4 + 2m^2 - 15 = 0$

2. $3k^4 + 10k^2 - 25 = 0$

3. $2r^4 - 7r^2 + 5 = 0$

4. $4x^4 - 8x^2 + 3 = 0$

5. $(g - 2)^2 - 6(g - 2) + 8 = 0$

6. $(p + 2)^2 - 2(p + 2) - 15 = 0$

7. $-(r + 1)^2 - 3(r + 1) + 3 = 0$

8. $-2(z - 4)^2 + 2(z - 4) + 3 = 0$

9. $6(k + 2)^4 - 11(k + 2)^2 + 4 = 0$

10. $8(m - 4)^4 - 10(m - 4)^2 + 3 = 0$

11. $7p^{-2} + 19p^{-1} = 6$

12. $5k^{-2} - 43k^{-1} = 18$

13. $(r - 1)^{2/3} + (r - 1)^{1/3} = 12$

14. $(y + 3)^{2/3} - 2(y + 3)^{1/3} - 3 = 0$

15. $3 + \dfrac{5}{p^2 + 1} = \dfrac{2}{(p^2 + 1)^2}$

16. $6 = \dfrac{7}{2y - 3} + \dfrac{3}{(2y - 3)^2}$

What is wrong with each solution in Exercises 17 and 18?

17. Solve $4x^4 - 11x^2 - 3 = 0$.

Let $t = x^2$.

$$4t^2 - 11t - 3 = 0$$
$$(4t + 1)(t - 3) = 0$$
$$4t + 1 = 0 \quad \text{or} \quad t - 3 = 0$$
$$t = -\frac{1}{4} \quad \text{or} \quad t = 3$$

The solution set is $\{-1/4, 3\}$.

18. Solve $x = \sqrt{3x + 4}$.

Square both sides to get

$$x^2 = 3x + 4$$
$$x^2 - 3x - 4 = 0$$
$$(x - 4)(x + 1) = 0$$
$$x - 4 = 0 \quad \text{or} \quad x + 1 = 0$$
$$x = 4 \qquad\qquad x = -1.$$

The solution set is $\{4, -1\}$.

Solve the following equations. See Examples 3–5.

19. $\sqrt{3z + 7} = 3z + 5$

20. $\sqrt{4r + 13} = 2r - 1$

21. $\sqrt{4k + 5} - 2 = 2k - 7$

22. $\sqrt{6m + 7} - 1 = m + 1$

23. $\sqrt{4x} - x + 3 = 0$

24. $\sqrt{2t} - t + 4 = 0$

25. $\sqrt{y} = \sqrt{y - 5} + 1$

26. $\sqrt{2m} = \sqrt{m + 7} - 1$

27. $\sqrt{m + 7} + 3 = \sqrt{m - 4}$

28. $\sqrt{r + 5} - 2 = \sqrt{r - 1}$

29. $\sqrt{2z} = \sqrt{3z + 12} - 2$

30. $\sqrt{5k + 1} - \sqrt{3k} = 1$

31. $\sqrt{r + 2} = 1 - \sqrt{3r + 7}$

32. $\sqrt{2p - 5} - 2 = \sqrt{p - 2}$

33. $\sqrt{2\sqrt{7x + 2}} = \sqrt{3x + 2}$

34. $\sqrt{3\sqrt{2m + 3}} = \sqrt{5m - 6}$

35. $\sqrt[3]{4n + 3} = \sqrt[3]{2n - 1}$

36. $\sqrt[3]{2z} = \sqrt[3]{5z + 2}$

37. $\sqrt[3]{t^2 + 2t - 1} = \sqrt[3]{t^2 + 3}$

38. $\sqrt[3]{2x^2 - 5x + 4} = \sqrt[3]{2x^2}$

39. $(2r + 5)^{1/3} = (6r - 1)^{1/3}$

40. $(3m + 7)^{1/3} = (4m + 2)^{1/3}$

41. $\sqrt[4]{q - 15} = 2$

42. $\sqrt[4]{3x + 1} = 1$

43. $\sqrt[4]{y^2 + 2y} = \sqrt[4]{3}$

44. $\sqrt[4]{k^2 + 6k} = 2$

45. $(z^2 + 24z)^{1/4} = 3$

46. $(3t^2 + 52t)^{1/4} = 4$

47. $(2r - 1)^{2/3} = r^{1/3}$

48. $(z - 3)^{2/5} = (4z)^{1/5}$

49. How can we tell that the equation $x^{1/4} = -2$ has no real solution without actually going through a solution process?

What is wrong with the solution in Exercise 50?

50. Solve $x^4 - x^2 = 0$.

    Since x^2 is a common factor, divide both sides by x^2 to get

$$x^2 - 1 = 0$$
$$(x - 1)(x + 1) = 0$$
$$x = 1 \quad \text{or} \quad x = -1.$$

The solution set is $\{-1, 1\}$.

Solve each equation for the indicated variable. Assume that all denominators are nonzero.

51. $d = k\sqrt{h}$ for h

52. $v = \dfrac{k}{\sqrt{d}}$ for d

53. $P = 2\sqrt{\dfrac{L}{g}}$ for L

54. $c = \sqrt{a^2 + b^2}$ for a

55. $x^{2/3} + y^{2/3} = a^{2/3}$ for y

56. $m^{3/4} + n^{3/4} = 1$ for m

2.7 —————— INEQUALITIES

An equation says that two expressions are equal, while an **inequality** says that one expression is greater than, greater than or equal to, less than, or less than or equal to, another. As with equations, a value of the variable for which the inequality is true is a solution of the inequality, and the set of all such solutions is the solution set of the inequality. Two inequalities with the same solution set are **equivalent inequalities.**

    Inequalities are solved with the following properties of inequality. (These were first introduced in Chapter 1.)

PROPERTIES OF INEQUALITY

For real numbers a, b, and c:

(a) $a < b$ and $a + c < b + c$ are equivalent.
(The same number may be added to both sides of an inequality without changing the solution set.)

(b) If $c > 0$, then $a < b$ and $ac < bc$ are equivalent.
(Both sides of an inequality may be multiplied by the same positive number without changing the solution set.)

(c) If $c < 0$, then $a < b$ and $ac > bc$ are equivalent.
(Both sides of an inequality may be multiplied by the same negative number without changing the solution set, as long as the direction of the inequality symbol is reversed.)

Replacing $<$ with $>$, $\leq$, or $\geq$ results in equivalent properties.

NOTE Because division is defined in terms of multiplication, the word "multiplied" may be replaced by "divided" in parts (b) and (c) of the properties of inequality.

Pay careful attention to part (c): if both sides of an inequality are multiplied by a negative number, the direction of the inequality symbol must be reversed. For example, starting with the true statement $-3 < 5$ and multiplying both sides by the positive number 2 gives

$$-3 \cdot 2 < 5 \cdot 2$$
$$-6 < 10,$$

still a true statement. On the other hand, starting with $-3 < 5$ and multiplying both sides by the *negative* number -2 gives a true result only if the direction of the inequality symbol is reversed.

$$-3(-2) > 5(-2)$$
$$6 > -10$$

A similar situation exists when dividing both sides by a negative number. In summary, the following statement can be made.

> When multiplying or dividing both sides of an inequality by a negative number, we must reverse the direction of the inequality symbol to obtain an equivalent inequality.

LINEAR INEQUALITIES A linear inequality is defined in a way similar to a linear equation.

LINEAR INEQUALITY

A **linear inequality** in one variable is an inequality that can be written in the form
$$ax + b > 0,$$
where $a \neq 0$. (Any of the symbols $\geq$, $<$, or $\leq$ may also be used.)

■ **Example 1**
SOLVING A LINEAR INEQUALITY

Solve the inequality $-3x + 5 > -7$.

Use the properties of inequality. Adding -5 on both sides gives

$$-3x + 5 + (-5) > -7 + (-5)$$
$$-3x > -12.$$

Now multiply both sides by $-1/3$. (We could also divide by -3.) Since $-1/3 < 0$, reverse the direction of the inequality symbol.

$$-\frac{1}{3}(-3x) < -\frac{1}{3}(-12)$$
$$x < 4$$

■ **FIGURE 2.6**

The original inequality is satisfied by any real number less than 4. The solution set can be written $\{x \mid x < 4\}$. A graph of the solution set is shown in Figure 2.6, where the parenthesis is used to show that 4 itself does not belong to the solution set. ■

The set $\{x|x < 4\}$, the solution set for the inequality in Example 1, is an example of an **interval.** A simplified notation, called **interval notation,** is used for writing intervals. With this notation, the interval in Example 1 can be written as $(-\infty, 4)$. The symbol $-\infty$ is not a real number; it is used to show that the interval includes all real numbers less than 4. The interval $(-\infty, 4)$ is an example of an **open interval,** since the endpoint, 4, is not part of the interval. Examples of other sets written in interval notation are shown below. A square bracket is used to show that a number *is* part of the graph, and a parenthesis is used to indicate that a number is *not* part of the graph. Whenever two real numbers a and b are used to write an interval in the chart that follows, it is assumed that $a < b$.

Type of Interval	Set	Interval Notation	Graph
Open interval	$\{x|x > a\}$	(a, ∞)	
	$\{x|a < x < b\}$	(a, b)	
	$\{x|x < b\}$	$(-\infty, b)$	
Half-open interval	$\{x|x \geq a\}$	$[a, \infty)$	
	$\{x|a < x \leq b\}$	$(a, b]$	
	$\{x|a \leq x < b\}$	$[a, b)$	
	$\{x|x \leq b\}$	$(-\infty, b]$	
Closed interval	$\{x|a \leq x \leq b\}$	$[a, b]$	
All real numbers	$\{x|x \text{ is real}\}$	$(-\infty, \infty)$	

■ **Example 2**

SOLVING A LINEAR INEQUALITY

Solve $4 - 3y \leq 7 + 2y$. Write the solution in interval notation and graph the solution on a number line.

Write the following series of equivalent inequalities.

$$4 - 3y \leq 7 + 2y$$
$$-4 - 2y + 4 - 3y \leq -4 - 2y + 7 + 2y \qquad \text{Subtract 4 and 2y.}$$
$$-5y \leq 3$$
$$\left(-\frac{1}{5}\right)(-5y) \geq \left(-\frac{1}{5}\right)(3); \qquad \text{Multiply by } -\frac{1}{5}; \text{ reverse}$$
$$\text{the inequality symbol.}$$
$$y \geq -\frac{3}{5}$$

124

■ **FIGURE 2.7**

In set-builder notation, the solution set is $\{y|y \geq -3/5\}$, while in interval notation the solution set is $[-3/5, \infty)$. See Figure 2.7 for the graph of the solution set. ■

From now on, the solutions of all inequalities will be written with interval notation.

THREE-PART INEQUALITIES The inequality $-2 < 5 + 3m < 20$ in the next example says that $5 + 3m$ is between -2 and 20. This inequality can be solved using an extension of the properties of inequality given above, working with all three expressions at the same time.

■ *Example 3*

SOLVING A
THREE-PART
INEQUALITY

Solve $-2 < 5 + 3m < 20$.

Write equivalent inequalities as follows.

$$-2 < 5 + 3m < 20$$
$$-7 < 3m < 15 \qquad \text{Add } -5.$$
$$-\frac{7}{3} < m < 5 \qquad \text{Multiply by } \frac{1}{3}.$$

■ **FIGURE 2.8**

The solution, graphed in Figure 2.8, is the interval $(-7/3, 5)$. ■

A product will break even, or begin to produce a profit, only if the revenue from selling the product at least equals the cost of producing it. If R represents revenue and C is cost, then the **break-even point** is the point where $R = C$.

■ *Example 4*

FINDING THE
BREAK-EVEN POINT

If the revenue and cost of a certain product are given by $R = 4x$ and $C = 2x + 1000$, where x is the number of units produced, where does R at least equal C?

Set $R \geq C$ and solve for x.

$$R \geq C$$
$$4x \geq 2x + 1000$$
$$2x \geq 1000$$
$$x \geq 500$$

The break-even point is at $x = 500$. This product will at least break even only if the number of units produced is in the interval $[500, \infty)$. ■

QUADRATIC INEQUALITIES The solution of *quadratic inequalities* depends on the solution of quadratic equations, which were introduced in Section 2.4.

QUADRATIC
INEQUALITY

A **quadratic inequality** is an inequality that can be written in the form

$$ax^2 + bx + c < 0$$

for real numbers $a \neq 0$, b, and c. (The symbol $<$ can be replaced with $>$, $\leq$, or $\geq$.)

▪ **IN SIMPLEST TERMS** ━━━━━━━━━━━━━━━━━━━━━━━━━━ ▪

8

In recent years, many states have increased their maximum speed limit on rural highways to 65 miles per hour. Inequalities can help determine whether driving at this faster speed is more cost effective.

Large shipping companies pay drivers to deliver goods. The cost of shipping can be simply expressed as the cost of gas, G, plus the cost of paying the driver an hourly wage, W. If D represents the distance of the trip and M represents the gas mileage, then the cost of driving X miles per hour can be expressed $C = G(D/M) + (D/X)W$. To compare the cost of driving 55 miles per hour to driving 65 miles per hour, set up the inequality $G(D/M) + (D/65)W < G(D/M) + (D/55)W$.

Suppose gas is $1.30 per gallon and the gas mileage at 55 miles per hour for a 400-mile trip is 25, while gas mileage for the same trip at 65 miles per hour is 18. At what hourly wage would it be more cost effective for the driver's speed to be 65 miles per hour?

$$\text{At 55 miles per hour,} \quad C = 1.30\left(\frac{400}{25}\right) + \left(\frac{400}{55}\right)W.$$

$$\text{At 65 miles per hour,} \quad C = 1.30\left(\frac{400}{18}\right) + \left(\frac{400}{65}\right)W.$$

The inequality is

$$1.30\left(\frac{400}{18}\right) + \left(\frac{400}{65}\right)W < 1.30\left(\frac{400}{25}\right) + \left(\frac{400}{55}\right)W$$

$$1.30(22.2) + 6.2W < 1.30(16) + 7.3W$$

$$28.86 + 6.2W < 20.8 + 7.3W$$

$$8.06 < 1.1W$$

$$7.33 < W.$$

Thus, if the driver is paid more than $7.33 per hour, it is more cost effective for the company if he travels at 65 miles per hour.

SOLVE EACH PROBLEM

A. If gas costs $1.25 per gallon and gas mileage at 55 mph for a 200-mile trip is 27, while gas mileage at 65 mph for the same trip is 20, find the hourly wage at which it is more cost effective for the company if the driver travels at 65 mph.

B. If the driver is paid $6.00 per hour in part (A) above and all values remain the same except the price of gasoline, determine the price at which it is more cost effective for the driver to travel at 65 mph.

ANSWERS   A. It is more cost effective to drive 65 mph if the driver is paid $6.11 per hour or more.   B. It is more cost effective to drive 65 mph if the price of gasoline is $1.27 per gallon or less.

126

■ *Example 5*

SOLVING A
QUADRATIC
INEQUALITY

Solve the quadratic inequality $x^2 - x - 12 < 0$.

Begin by finding the values of x that satisfy $x^2 - x - 12 = 0$.

$$x^2 - x - 12 = 0$$

$$(x + 3)(x - 4) = 0 \qquad \text{Factor.}$$

$$x = -3 \quad \text{or} \quad x = 4 \qquad \text{Use the zero-factor property.}$$

These two points, -3 and 4, divide a number line into the three regions shown in Figure 2.9. If a point in region A, for example, makes the polynomial $x^2 - x - 12$ negative, then all points in region A will make that polynomial negative.

To find the regions that make $x^2 - x - 12$ negative (< 0), draw a number line that shows where factors are positive or negative, as in Figure 2.9. First decide on the sign of the factor $x + 3$ in each of the three regions; then do the same thing for the factor $x - 4$. The results are shown in Figure 2.9.

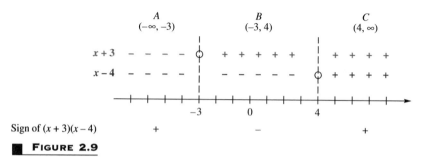

■ **FIGURE 2.9**

Now consider the sign of the product of the two factors in each region. As Figure 2.9 shows, both factors are negative in the interval $(-\infty, -3)$; therefore their product is positive in that interval. For the interval $(-3, 4)$, one factor is positive and the other is negative, giving a negative product. In the last region, $(4, \infty)$, both factors are positive, so their product is positive. The polynomial $x^2 - x - 12$ is negative (what the original inequality calls for) when the product of its factors is negative, that is, for the interval $(-3, 4)$. The graph of this solution set is shown in Figure 2.10. ■

■ **FIGURE 2.10**

Figure 2.9 is an example of a **sign graph**, a graph that shows the values of the variable that make the factors of a quadratic inequality positive or negative. The steps used in solving a quadratic inequality are summarized below.

SOLVING A QUADRATIC INEQUALITY	
1. Solve the corresponding quadratic equation.
2. Identify the intervals determined by the solutions of the equation.
3. Use a sign graph to determine which intervals are in the solution set.

∎ *Example 6*

SOLVING A
QUADRATIC
INEQUALITY

Solve the inequality $2x^2 + 5x - 12 \geq 0$.

Begin by finding the values of x that satisfy $2x^2 + 5x - 12 = 0$.

$$2x^2 + 5x - 12 = 0$$

$$(2x - 3)(x + 4) = 0$$

$$x = \frac{3}{2} \quad \text{or} \quad x = -4$$

These two points divide the number line into the three regions shown in the sign graph in Figure 2.11. Since both factors are negative in the first interval, their product, $2x^2 + 5x - 12$, is positive there. In the second interval, the factors have opposite signs, and therefore their product is negative. Both factors are positive in the third interval, and their product also is positive there. Thus, the polynomial $2x^2 + 5x - 12$ is positive or zero in the interval $(-\infty, -4]$ and also in the interval $[3/2, \infty)$. Since both of the intervals belong to the solution set, the result can be written as the *union** of the two intervals,

$$(-\infty, -4] \cup \left[\frac{3}{2}, \infty\right).$$

The graph of the solution set is shown in Figure 2.12. ∎

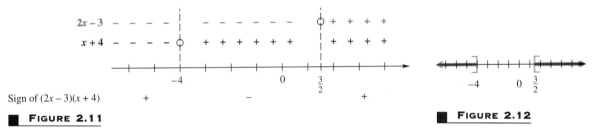

Sign of $(2x - 3)(x + 4)$

∎ **FIGURE 2.11**

∎ **FIGURE 2.12**

RATIONAL INEQUALITIES The inequalities discussed in the remainder of this section involve quotients of algebraic expressions, and for this reason they are called **rational inequalities.** These inequalities can be solved with a sign graph in much the same way as quadratic inequalities.

SOLVING A
RATIONAL
INEQUALITY

1. Rewrite the inequality, if necessary, so that 0 is on one side.
2. Make a sign graph with intervals determined by the numbers that cause either the numerator or the denominator of the rational expression to equal zero.
3. Determine the appropriate interval(s) of the solution set.

---

*The **union** of sets A and B, written $A \cup B$, is defined as $A \cup B = \{x | x$ is an element of A or x is an element of $B\}$.

CAUTION It would be incorrect to try to solve a rational inequality such as

$$\frac{5}{x + 4} \geq 1$$

by multiplying both sides by $x + 4$ to get $5 \geq x + 4$, since the sign of $x + 4$ depends on the value of x. If $x + 4$ were negative, then we would have to reverse the inequality sign. The procedure described in the box and used in the next two examples eliminates the need for considering separate cases.

■ **Example 7**

SOLVING A
RATIONAL
INEQUALITY

Solve the rational inequality $\frac{5}{x + 4} \geq 1$.

Start by rewriting the inequality so that 0 is on one side, getting

$$\frac{5}{x + 4} - 1 \geq 0.$$

Writing the left side as a single fraction gives

$$\frac{5 - (x + 4)}{x + 4} \geq 0$$

or

$$\frac{1 - x}{x + 4} \geq 0.$$

The quotient can change sign only when the denominator is 0 or when the numerator is 0. This occurs at

$$1 - x = 0 \quad \text{or} \quad x + 4 = 0$$
$$x = 1 \quad \text{or} \quad x = -4.$$

Make a sign graph as before. This time, consider the sign of the quotient of the two quantities rather than the sign of their product. See Figure 2.13.

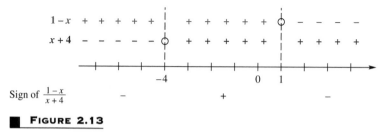

Sign of $\frac{1-x}{x+4}$

■ **FIGURE 2.13**

The quotient of two numbers is positive if both numbers are positive or if both numbers are negative. On the other hand, the quotient is negative if the two numbers have opposite signs. The sign graph in Figure 2.13 shows that values in the interval $(-4, 1)$ give a positive quotient and are part of the solution. With a quotient, the endpoints must be considered separately to make sure that no denominator is 0. Here, -4 gives a 0 denominator but 1 satisfies the given inequality. In interval notation, the solution set is $(-4, 1]$. ■

| CAUTION | As suggested by Example 7, be very careful with the endpoints of the intervals in the solution of rational inequalities. |

■ *Example 8*

SOLVING A
RATIONAL
INEQUALITY

Solve $\dfrac{2x - 1}{3x + 4} < 5$.

Begin by subtracting 5 on both sides and combining the terms on the left into a single fraction.

$$\frac{2x - 1}{3x + 4} < 5$$

$$\frac{2x - 1}{3x + 4} - 5 < 0 \qquad \text{Subtract 5.}$$

$$\frac{2x - 1 - 5(3x + 4)}{3x + 4} < 0 \qquad \text{Common denominator is } 3x + 4.$$

$$\frac{-13x - 21}{3x + 4} < 0 \qquad \text{Combine terms.}$$

To draw a sign graph, first solve the equations

$$-13x - 21 = 0 \qquad \text{and} \qquad 3x + 4 = 0,$$

getting the solutions

$$x = -\frac{21}{13} \qquad \text{and} \qquad x = -\frac{4}{3}.$$

Use the values $-21/13$ and $-4/3$ to divide the number line into three intervals. Now complete a sign graph and find the intervals where the quotient is negative. See Figure 2.14.

Sign of $\dfrac{-13x - 21}{3x + 4}$

■ **FIGURE 2.14**

From the sign graph, values of x in the two intervals $(-\infty, -21/13)$ and $(-4/3, \infty)$ make the quotient negative, as required. Neither endpoint satisfies the given inequality, so the solution set should be written $(-\infty, -21/13) \cup (-4/3, \infty)$. ■

2.7 EXERCISES ■ ————————————————————————

Write each of the following in interval notation. Graph each interval.

1. $-1 < x < 4$ **2.** $x \geq -3$ **3.** $x < 0$ **4.** $8 > x > 3$

5. $2 > x \geq 1$ **6.** $-4 \geq x > -5$ **7.** $-9 > x$ **8.** $6 \leq x$

Using the variable x, write each of the following intervals as an inequality with set-builder notation.

9. $(-4, 3)$ **10.** $[2, 7)$ **11.** $(-\infty, -1]$

12. $(3, \infty)$ **13.**   **14.**

17. Explain how to determine whether a parenthesis or a square bracket is used when graphing the solution set of a linear inequality.

18. The three-part inequality $a < x < b$ means "a is less than x and x is less than b." Which one of the following inequalities is not satisfied by some real number x?
     (a) $-3 < x < 5$ **(b)** $0 < x < 4$ **(c)** $-3 < x < -2$ **(d)** $-7 < x < -10$

Solve the following inequalities. Write the solutions in interval notation. Graph each solution. See Examples 1–3.

19. $-3p - 2 \leq 1$ **20.** $-5r + 3 \geq -2$

21. $2(m + 5) - 3m + 1 \geq 5$ **22.** $6m - (2m + 3) \geq 4m - 5$

23. $8k - 3k + 2 < 2(k + 7)$ **24.** $2 - 4x + 5(x - 1) < -6(x - 2)$

25. $\dfrac{4x + 7}{-3} \leq 2x + 5$ **26.** $\dfrac{2z - 5}{-8} \geq 1 - z$

27. $2 \leq y + 1 \leq 5$ **28.** $-3 \leq 2t \leq 6$

29. $-10 > 3r + 2 > -16$ **30.** $4 > 6a + 5 > -1$

31. $-3 \leq \dfrac{x - 4}{-5} < 4$ **32.** $1 < \dfrac{4m - 5}{-2} < 9$

Find all intervals where the following products will at least break even. See Example 4.

33. The cost to produce x units of wire is $C = 50x + 5000$, while the revenue is $R = 60x$.

34. The cost to produce x units of squash is $C = 100x + 6000$, while the revenue is $R = 500x$.

35. $C = 85x + 900$; $R = 105x$

36. $C = 70x + 500$; $R = 60x$

Solve the following quadratic inequalities. Write the solutions in interval notation. Graph each solution. (Hint: In Exercises 47–50, use the quadratic formula.) See Examples 5 and 6.

37. $x^2 \leq 9$ **38.** $p^2 > 16$ **39.** $r^2 + 4r + 6 \geq 3$

40. $z^2 + 6z + 16 < 8$ **41.** $x^2 - x \leq 6$ **42.** $r^2 + r < 12$

43. $2k^2 - 9k > -4$ **44.** $3n^2 \leq -10 - 13n$ **45.** $x^2 > 0$

46. $p^2 < -1$ **47.** $x^2 + 5x - 2 < 0$ **48.** $4x^2 + 3x + 1 \leq 0$

49. $m^2 - 2m \leq 1$ **50.** $p^2 + 4p > -1$

Solve the following inequalities. Give answers in interval notation. See Examples 7 and 8.

51. $\dfrac{m - 3}{m + 5} \le 0$

52. $\dfrac{r + 1}{r - 4} > 0$

53. $\dfrac{k - 1}{k + 2} > 1$

54. $\dfrac{a - 6}{a + 2} < -1$

55. $\dfrac{3}{x - 6} \le 2$

56. $\dfrac{1}{k - 2} < \dfrac{1}{3}$

57. $\dfrac{1}{m - 1} < \dfrac{5}{4}$

58. $\dfrac{6}{5 - 3x} \le 2$

59. $\dfrac{10}{3 + 2x} \le 5$

60. $\dfrac{1}{x + 2} \ge 3$

61. $\dfrac{7}{k + 2} \ge \dfrac{1}{k + 2}$

62. $\dfrac{5}{p + 1} > \dfrac{12}{p + 1}$

63. $\dfrac{3}{2r - 1} > -\dfrac{4}{r}$

64. $-\dfrac{5}{3h + 2} \ge \dfrac{5}{h}$

65. $\dfrac{4}{y - 2} \le \dfrac{3}{y - 1}$

66. $\dfrac{4}{n + 1} < \dfrac{2}{n + 3}$

67. $\dfrac{y + 3}{y - 5} \le 1$

68. $\dfrac{a + 2}{3 + 2a} \le 5$

In each of the following inequalities, the intervals of the sign graph are

$$(-\infty, 2), \quad (2, 5), \quad and \quad (5, \infty).$$

Without actually solving the inequality, state whether the numbers 2 and 5 will be included in or excluded from the solution set of the inequality.

69. $(x - 2)(x - 5) \ge 0$

70. $(x - 2)(x - 5) < 0$

71. $\dfrac{x - 2}{x - 5} > 0$

72. $\dfrac{x - 2}{x - 5} < 0$

73. $\dfrac{x - 2}{x - 5} \ge 0$

74. $\dfrac{x - 5}{x - 2} \le 0$

The following inequalities are not quadratic inequalities, but they still may be solved in much the same way (with four regions). Solve each inequality. Write solutions in interval notation.

75. $(2r - 3)(r + 2)(r - 3) \ge 0$

76. $(y + 5)(3y - 4)(y + 2) > 0$

77. $x^3 - 4x \le 0$

78. $r^3 - 9r \ge 0$

79. $4m^3 + 7m^2 - 2m > 0$

80. $6p^3 - 11p^2 + 3p > 0$

81. Which of the following inequalities have solution set $(-\infty, \infty)$?
   (a) $(x + 3)^2 \ge 0$ **(b)** $(5x - 6)^2 \le 0$ **(c)** $(6y + 4)^2 > 0$
   (d) $(8p - 7)^2 < 0$ **(e)** $\dfrac{x^2 + 7}{2x^2 + 4} \ge 0$ **(f)** $\dfrac{2x^2 + 8}{x^2 + 9} < 0$

82. Which of the inequalities in Exercise 81 have solution set $\emptyset$?

83. A student attempted to solve the inequality
$$\frac{2x - 1}{x + 2} \le 0$$
by multiplying both sides by $x + 2$ to get
$$2x - 1 \le 0$$
$$x \le \frac{1}{2}.$$
He wrote the solution set as $(-\infty, 1/2]$. Is his solution correct? Explain.

84. A student solved the inequality $p^2 \le 16$ by taking the square root of both sides to get $p \le 4$. She wrote the solution set as $(-\infty, 4]$. Is her solution correct? Explain.

85. The commodity market is very unstable; money can be made or lost quickly when investing in soybeans, wheat, pork bellies, and the like. Suppose that an investor kept track of her total profit, P, at time t, measured in months, after she began investing, and found that
$$P = 4t^2 - 29t + 30.$$
Find the time intervals when she has been ahead. (*Hint:* $t > 0$ in this case.)

86. Suppose the velocity of an object is given by
$$v = 2t^2 - 5t - 12,$$
where t is time in seconds. (Here t can be positive or negative.) Find the intervals where the velocity is negative.

87. A projectile is fired from ground level. After t sec its height above the ground is $220t - 16t^2$ ft. For what time period is the projectile at least 624 ft above the ground?

88. An analyst has found that his company's profits, in hundreds of thousands of dollars, are given by

$$P = 3x^2 - 35x + 50,$$

where x is the amount, in hundreds of dollars, spent on advertising. For what values of x does the company make a profit?

89. The manager of a large apartment complex has found that the profit is given by

$$P = -x^2 + 250x - 15{,}000,$$

where x is the number of units rented. For what values of x does the complex produce a profit?

90. If $a > b$, is it always true that $1/a < 1/b$? Explain.

91. If $b > 0$, when is it true that $b^2 > b$? Explain.

92. If $a > b$, is it always true that $a^2 > b^2$? Explain.

2.8 ——————— **ABSOLUTE VALUE EQUATIONS AND INEQUALITIES**

In this section we describe methods of solving equations and inequalities involving absolute value. Recall from Chapter 1 that the absolute value of a number a, written $|a|$, gives the distance from a to 0 on a number line. By this definition, the absolute value equation $|x| = 3$ can be solved by finding all real numbers at a distance of 3 units from 0. As shown in Figure 2.15, there are two numbers satisfying this condition, 3 and -3, so that the solution set of the equation $|x| = 3$ is the set $\{3, -3\}$.

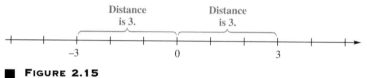

■ FIGURE 2.15

ABSOLUTE VALUE EQUATIONS If a and b represent two real numbers, then the absolute value of their difference, $|a - b|$ or $|b - a|$, represents the distance between the points on the number line whose coordinates are a and b. (Verify this for 3 and -3 in Figure 2.15.) This concept is used in simple equations involving absolute value.

■ *Example 1*

USING THE DISTANCE DEFINITION TO SOLVE AN ABSOLUTE VALUE EQUATION

Solve $|p - 4| = 5$.

The expression $|p - 4|$ represents the distance between p and 4. The equation $|p - 4| = 5$ can be solved by finding all real numbers that are 5 units from 4. As shown in Figure 2.16, these numbers are -1 and 9. The solution set is $\{-1, 9\}$. ■

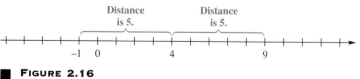

■ FIGURE 2.16

The definition of absolute value leads to the following properties of absolute value that can be used to solve absolute value equations algebraically.

SOLVING ABSOLUTE VALUE EQUATIONS	**1.** For $b > 0$, $|a| = b$ if and only if $a = b$ or $a = -b$ **2.** $|a| = |b|$ if and only if $a = b$ or $a = -b$.

■ **Example 2**
SOLVING ABSOLUTE
VALUE EQUATIONS

Solve each equation.

(a) $|5 - 3m| = 12$

Use property (1) above, with $a = 5 - 3m$, to write

$$5 - 3m = 12 \quad \text{or} \quad 5 - 3m = -12.$$

Solve each equation.

$$
\begin{array}{ccc}
5 - 3m = 12 & \text{or} & 5 - 3m = -12 \\
-3m = 7 & & -3m = -17 \\
m = -\dfrac{7}{3} & & m = \dfrac{17}{3}
\end{array}
$$

The solution set is $\{-7/3, 17/3\}$.

(b) $|4m - 3| = |m + 6|$

By property (2) above, this equation will be true if

$$4m - 3 = m + 6 \quad \text{or} \quad 4m - 3 = -(m + 6).$$

Solve each equation.

$$
\begin{array}{ccc}
4m - 3 = m + 6 & \text{or} & 4m - 3 = -(m + 6) \\
3m = 9 & & 4m - 3 = -m - 6 \\
m = 3 & & 5m = -3 \\
& & m = -\dfrac{3}{5}
\end{array}
$$

The solution set of $|4m - 3| = |m + 6|$ is thus $\{3, -3/5\}$. ■

ABSOLUTE VALUE INEQUALITIES The method used to solve absolute value equations can be extended to solve inequalities with absolute value.

■ **Example 3**
USING THE
DISTANCE
DEFINITION FOR
ABSOLUTE VALUE
INEQUALITIES

(a) Solve $|x| < 5$.

Since absolute value gives the distance between a number and 0, the inequality $|x| < 5$ is satisfied by all real numbers whose distance from 0 is less than 5. As shown in Figure 2.17, the solution includes all numbers from -5 to 5, or $\{x | -5 < x < 5\}$. In interval notation, the solution is written as the open interval $(-5, 5)$. A graph of the solution set is shown in Figure 2.17.

FIGURE 2.17

(b) Solve $|x| > 5$.

In a manner similar to part (a), we see that the solution of $|x| > 5$ consists of all real numbers whose distance from 0 is greater than 5. This includes those numbers greater than 5 or those less than -5: $x < -5$ or $x > 5$.

In interval notation, the solution is written $(-\infty, -5) \cup (5, \infty)$. The solution set is shown in Figure 2.18. ■

FIGURE 2.18

The following properties of absolute value, which can be obtained from the definition of absolute value, are used to solve absolute value inequalities.

SOLVING ABSOLUTE VALUE INEQUALITIES	For any positive number b: **1.** $|a| < b$ if and only if $-b < a < b$; **2.** $|a| > b$ if and only if $a < -b$ or $a > b$.

■ *Example 4*
SOLVING AN
ABSOLUTE VALUE
INEQUALITY

Solve $|x - 2| < 5$.

This inequality is satisfied by all real numbers whose distance from 2 is less than 5. As shown in Figure 2.19, the solution set is the interval $(-3, 7)$. Property (1) above can be used to solve the inequality as follows. Let $a = x - 2$ and $b = 5$, so that $|x - 2| < 5$ if and only if

$$-5 < x - 2 < 5.$$

Adding 2 to each part of this three-part inequality produces

$$-3 < x < 7,$$

giving the interval solution $(-3, 7)$. ■

FIGURE 2.19

■ *Example 5*

SOLVING AN
ABSOLUTE VALUE
INEQUALITY

Solve $|x - 8| \geq 1$.

All numbers whose distance from 8 is greater than or equal to 1 are solutions. To find the solution using property (2) above, let $a = x - 8$ and $b = 1$ so that $|x - 8| \geq 1$ if and only if

$$x - 8 \leq -1 \qquad \text{or} \qquad x - 8 \geq 1$$
$$x \leq 7 \qquad \text{or} \qquad x \geq 9.$$

The solution set, $(-\infty, 7] \cup [9, \infty)$, is shown in Figure 2.20. ■

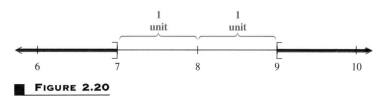

■ **FIGURE 2.20**

The properties given above for solving absolute value inequalities require that the absolute value expression be *alone* on one side of the inequality. Example 6 shows how to meet this requirement when this is not the case at first.

■ *Example 6*

SOLVING AN
ABSOLUTE VALUE
INEQUALITY
REQUIRING A
TRANSFORMATION

Solve $|2 - 7m| - 1 > 4$.

In order to use the properties of absolute value given above, first add 1 to both sides; this gives

$$|2 - 7m| > 5.$$

Now use property (2) above. By this property, $|2 - 7m| > 5$ if and only if

$$2 - 7m < -5 \qquad \text{or} \qquad 2 - 7m > 5.$$

Solve each of these inequalities separately to get the solution set $(-\infty, -3/7) \cup (1, \infty)$. ■

If an absolute value equation or inequality is written with 0 or a negative number on one side, such as $|2 - 5x| \geq -4$, we do not solve by applying the methods of the earlier examples. Use the fact that the absolute value of any expression must be a nonnegative number to solve the equation or inequality.

■ *Example 7*

SOLVING SPECIAL
CASES OF
ABSOLUTE VALUE
EQUATIONS AND
INEQUALITIES

Use the fact that absolute value is always nonnegative to solve each equation or inequality.

(a) $|2 - 5x| \geq -4$.

Since the absolute value of a number is always nonnegative, $|2 - 5x| \geq -4$ is always true. The solution set includes all real numbers, written $(-\infty, \infty)$.

(b) $|4x - 7| < -3$

The absolute value of any number will never be less than -3 (or less than *any* negative number). For this reason, the solution set of this inequality is $\emptyset$.

(c) $|5x + 15| = 0$

The absolute value of a number will be zero only if that number is 0. Therefore, this equation is equivalent to $5x + 15 = 0$, which has solution set $\{-3\}$. ■

We end this section with an example showing how certain statements involving distance can be described using absolute value inequalities.

■ *Example 8*

USING ABSOLUTE
VALUE
INEQUALITIES TO
DESCRIBE
DISTANCES

Write each statement using an absolute value inequality.

(a) k is not less than 5 units from 8.

Since the distance from k to 8, written $|k - 8|$ or $|8 - k|$, is not less than 5, the distance is greater than or equal to 5. Write this as

$$|k - 8| \geq 5.$$

(b) n is within .001 of 6.

This statement indicates that n may be .001 more than 6 or .001 less than 6. That is, the distance of n from 6 is no more than .001, written

$$|n - 6| \leq .001. \quad ■$$

2.8 EXERCISES ■

Solve each of the following equations. See Examples 1, 2, and 7(c).

1. $|a - 2| = 1$

2. $|x - 3| = 2$

3. $|3m - 1| = 2$

4. $|4p + 2| = 5$

5. $|5 - 3x| = 3$

6. $|-3a + 7| = 3$

7. $\left|\dfrac{z - 4}{2}\right| = 5$

8. $\left|\dfrac{m + 2}{2}\right| = 7$

9. $\left|\dfrac{5}{r - 3}\right| = 10$

10. $\left|\dfrac{3}{2h - 1}\right| = 4$

11. $|4w + 3| - 2 = 7$

12. $|8 - 3t| - 3 = -2$

13. $|6x + 9| = 0$

14. $|12t - 3| = 0$

15. $\left|\dfrac{6y + 1}{y - 1}\right| = 3$

16. $\left|\dfrac{3a - 4}{2a + 3}\right| = 1$

17. $|2k - 3| = |5k + 4|$

18. $|p + 1| = |3p - 1|$

19. $|4 - 3y| = |7 + 2y|$

20. $|5k + 16| = |3k|$

21. $|8b + 7| = |4b|$

22. $|2 + 5a| = |4 - 6a|$

23. $|x + 2| = |x - 1|$

24. $|y - 5| = -8$

25. Without actually going through the solution process, we can say that the equation $|5x - 6| = 6x$ cannot have a negative solution. Why is this true?

26. Determine by inspection the solution set of each of the following absolute value equations.
   (a) $-|x| = |x|$ **(b)** $|-x| = |x|$ **(c)** $|x^2| = |x|$ **(d)** $-|x| = 3$

Solve each of the following inequalities. Give the solution in interval notation. See Examples 3–7.

27. $|x| \leq 3$

28. $|y| \leq 10$

29. $|m| > 1$

30. $|z| > 5$

31. $|a| < -2$

32. $|b| > -5$

33. $|x| - 3 \leq 7$

34. $|r| + 3 \leq 10$

35. $|2x + 5| < 3$

36. $\left|x - \dfrac{1}{2}\right| < 2$

37. $|3m - 2| > 4$

38. $|4x - 6| > 10$

39. $|3z + 1| \geq 7$

40. $|8b + 5| \geq 7$

41. $\left|\dfrac{2}{3}t + \dfrac{1}{2}\right| \leq \dfrac{1}{6}$

42. $\left|\dfrac{5}{3} - \dfrac{1}{2}x\right| > \dfrac{2}{9}$

43. $\left|5x + \dfrac{1}{2}\right| - 2 < 5$

44. $\left|x + \dfrac{2}{3}\right| + 1 < 4$

45. $|6x + 3| \geq -2$

46. $|7 - 8x| < -4$

47. $\left|\dfrac{1}{2}x + 6\right| > 0$

48. $\left|\dfrac{2}{3}r - 4\right| \leq 0$

49. Write an equation involving absolute value that says that the distance between p and q is 5 units.

50. Write an inequality involving absolute value that says that the distance between r and s is less than 9 units.

51. Suppose that you hear someone say "The absolute value of a number is always positive." How might you politely correct this person's misconception?

52. The following triangle inequality holds for all real numbers x and y:

$$|x| + |y| \geq |x + y|.$$

Illustrate cases for this inequality by choosing values of x and y that satisfy the following.
(a) x and y are both positive   (b) x and y are both negative
(c) x is positive and y is negative   (d) x is negative and y is positive

Write each of the statements in Exercises 53–60 as an absolute value equation or inequality. See Example 8.

53. x is within 4 units of 2   54. m is no more than 8 units from 9   55. z is no less than 2 units from 12

56. p is at least 5 units from 9   57. k is 6 units from 1   58. r is 5 units from 3

59. If x is within .0004 unit of 2, then y is within .00001 unit of 7.

60. y is within 10^{-6} unit of 10 whenever x is within 2×10^{-4} unit of 5.

Solve each problem.

61. The temperatures on the surface of Mars in degrees Celsius approximately satisfy the inequality

$$|C + 84| \leq 56.$$

What range of temperatures corresponds to this inequality?

62. Dr. Tydings has found that, over the years, 95% of the babies that he has delivered have weighed y pounds, where $|y - 8.0| \leq 1.5$. What range of weights corresponds to this inequality?

63. The industrial process that is used to convert methanol to gasoline is carried out at a temperature range of 680° F to 780° F. Using F as the variable, write an absolute value inequality that corresponds to this range.

64. When a model kite was flown across the wind in tests to determine its limits of power extraction, it attained speeds of 98 to 148 feet per second in winds of 16 to 26 feet per second. Using x as the variable in each case, write absolute value inequalities that correspond to these ranges.

65. If $|x - 2| < 3$, find values of m and n such that $m < 3x + 5 < n$.

66. If $|x + 8| < 16$, find values of p and q such that $p < 2x - 1 < q$.

67. Explain why it is *incorrect* to write the absolute value inequality $|x| > 6$ as any of these:
$-6 > x > 6$, $-6 > x < 6$, or $-6 < x > 6$

68. Is $|a - b|^2$ always equal to $(b - a)^2$? Why?

CHAPTER 2   SUMMARY ■

SECTION	TERMS		KEY IDEAS
2.1 Linear Equations	equation solution solution set identity conditional equation	contradiction equivalent equations linear equation literal equation	**Addition and Multiplication Properties of Equality** For real numbers a, b, and c: $a = b$ and $a + c = b + c$ are equivalent. If $c \neq 0$, then $a = b$ and $ac = bc$ are equivalent.

SECTION	TERMS	KEY IDEAS
2.2 Applications of Linear Equations	solving for a specified variable	**Steps in Problem Solving** **1.** Decide what the variable is to represent. **2.** Make a sketch or chart, if appropriate. **3.** Decide on a variable expression for any other unknown quantity. **4.** Write an equation. **5.** Solve the equation. **6.** Check.
2.3 Complex Numbers	complex number imaginary number conjugates	**Definition of i** $$i^2 = -1 \quad \text{or} \quad i = \sqrt{-1}$$ **Definition of $\sqrt{-a}$** For $a > 0$, $\sqrt{-a} = i\sqrt{a}$.
2.4 Quadratic Equations	quadratic equation completing the square quadratic formula discriminant	**Zero-Factor Property** If a and b are complex numbers, with $ab = 0$, then $a = 0$ or $b = 0$ or both. **Square Root Property** The solution set of $x^2 = k$ is $\{\sqrt{k}, -\sqrt{k}\}$. **Quadratic Formula** The solutions of the quadratic equation $ax^2 + bx + c = 0$, where $a \neq 0$, are given by $$x = \frac{-b \pm \sqrt{b^2 - 4ac}}{2a}.$$
2.5 Applications of Quadratic Equations		**Pythagorean Theorem** In a right triangle, the sum of the squares of the legs a and b is equal to the square of the hypotenuse c: $$a^2 + b^2 = c^2.$$
2.6 Other Types of Equations	quadratic in form	

SECTION	TERMS	KEY IDEAS
2.7 Inequalities	inequality equivalent inequalities linear inequality interval interval notation open interval half-open interval closed interval three-part inequality quadratic inequality sign graph rational inequality	**Properties of Inequality** For real numbers a, b, and c: **(a)** $a < b$ and $a + c < b + c$ are equivalent. **(b)** If $c > 0$, then $a < b$ and $ac < bc$ are equivalent. **(c)** If $c < 0$, then $a < b$ and $ac > bc$ are equivalent. **For Quadratic or Rational Inequalities:** **1.** Be sure that 0 is on one side. **2.** Make a sign graph. **3.** Determine the intervals of the solution set.
2.8 Absolute Value Equations and Inequalities		**Properties of Absolute Value** If b is a positive number, then $\quad \lvert a \rvert = b \quad$ if and only if $\quad a = b \;$ or $\; a = -b$; $\quad \lvert a \rvert < b \quad$ if and only if $\quad -b < a < b$; $\quad \lvert a \rvert > b \quad$ if and only if $\quad a < -b \;$ or $\; a > b$. For any real numbers a and b, $\quad \lvert a \rvert = \lvert b \rvert \quad$ if and only if $\quad a = b$ or $a = -b$.

CHAPTER 2 REVIEW EXERCISES ■ ———————

Solve each of the following equations.

1. $2m + 7 = 3m + 1$

2. $4k - 2(k - 1) = 12$

3. $5y - 2(y + 4) = 3(2y + 1)$

4. $\dfrac{2}{x - 3} = \dfrac{3}{x + 3}$

5. $\dfrac{10}{4z - 4} = \dfrac{1}{1 - z}$

6. $\dfrac{2}{p} - \dfrac{4}{3p} = 8 + \dfrac{3}{p}$

7. $\dfrac{5}{3r} - 10 = \dfrac{3}{2r}$

8. $\dfrac{p}{p - 2} - \dfrac{3}{5} = \dfrac{2}{p - 2}$

Solve for x.

9. $3(x + 2b) + a = 2x - 6$

10. $9x - 11(k + p) = x(a - 1)$

11. $\dfrac{x}{m - 2} = kx - 3$

12. $r^2x - 5x = 3r^2$

Solve each of the following for the indicated variable.

13. $A = P + Pi$ for P

14. $A = I\left(1 - \dfrac{j}{n}\right)$ for j

15. $A = \dfrac{24f}{b(p + 1)}$ for f

16. $P(r + R)^2 = E^2R$ for P

17. $\dfrac{xy^2 - 5xy + 4}{3x} = 2p$ for x

18. $\dfrac{zx^2 - 5x + z}{z + 1} = 9$ for z

Solve each of the following problems.

19. A computer printer is on sale for 15% off. The sale price is $425. What was the original price?

20. To make a special candy mix for Valentine's Day, the owner of a candy store wants to combine chocolate hearts that sell for $5 per lb with candy kisses that sell for $3.50 per lb. How many pounds of each kind should be used to get 30 lb of a mix which can be sold for $4.50 per pound?

21. Alison can ride her bike to the university library in 20 min. The trip home, which is all uphill, takes her 30 min. If her rate is 8 mph slower on the return trip, how far does she live from the library?

22. Ed and Faye Clement are stuffing envelopes for a political campaign. Working together, they can stuff 5000 envelopes in 4 hr. When Faye worked at the job alone, it took her 6 hr to stuff the 5000 envelopes. How long would it take Ed working alone to stuff 5000 envelopes?

23. A realtor borrowed $90,000 to develop some property. He was able to borrow part of the money at 11.5% interest and the rest at 12%. The annual interest on the two loans amounts to $10,525. How much was borrowed at each rate?

24. An excursion boat travels upriver to a landing and then returns to its starting point. The trip upriver takes 1.2 hr, and the trip back takes .9 hr. If the average speed on the return trip is 5 mph faster than on the trip upriver, what is the boat's speed upriver?

Perform each operation. Write each result in standard form.

25. $(6 - i) + (4 - 2i)$

26. $(-11 + 2i) - (5 - 7i)$

27. $15i - (3 + 2i) - 5$

28. $-6 + 4i - (5i - 2)$

29. $(5 - i)(3 + 4i)$

30. $(8 - 2i)(1 - i)$

31. $(5 - 11i)(5 + 11i)$

32. $(7 + 6i)(7 - 6i)$

33. $(4 - 3i)^2$

34. $(1 + 7i)^2$

35. $\dfrac{6 + i}{1 - i}$

36. $\dfrac{5 + 2i}{5 - 2i}$

37. The product of a complex number and its conjugate is always a _____ number.

38. True or false: A real number is a complex number.

Find each of the following powers of i.

39. i^7

40. i^{150}

41. i^{-35}

42. i^{-20}

Solve each equation.

43. $(b + 7)^2 = 5$

44. $(3y - 2)^2 = 8$

45. $2a^2 + a - 15 = 0$

46. $12x^2 = 8x - 1$

47. $2q^2 - 11q = 21$

48. $3x^2 + 2x = 16$

49. $2 - \dfrac{5}{p} = \dfrac{3}{p^2}$

50. $\dfrac{4}{m^2} = 2 + \dfrac{7}{m}$

51. $\sqrt{2}x^2 - 4x + \sqrt{2} = 0$

52. Find all values of k for which $4x^2 + 3kx + 9 = 0$ will have one real solution (a double solution).

Evaluate the discriminant for each of the following, and then use it to predict the type of solutions for the equation.

53. $8y^2 = 2y - 6$

54. $6k^2 - 2k = 3$

55. $16r^2 + 3 = 26r$

56. $8p^2 + 10p = 7$

57. $25z^2 - 110z + 121 = 0$

58. $4y^2 - 8y + 17 = 0$

Work each problem.

59. Paula Story plans to replace the vinyl floor covering in her 10- by 12-ft kitchen. She wants to have a border of even width of a special material. She can afford only 21 sq ft of this material. How wide a border can she have?

60. Calvin wants to fence off a rectangular playground beside an apartment building. The building forms one boundary, so he needs to fence only the other three sides. The area of the playground is to be 11,250 sq m. He has enough material to build 325 m of fence. Find the length and width of the playground.

61. It takes two gardeners 3 hr (working together) to mow the lawns in a city park. One gardener could do the entire job in 1 hr less time than the other. How long would it take the slower gardener to complete the work alone? Give the answer to the nearest tenth.

62. Steve and Paula sell pies. It takes Paula 1 hr longer than Steve to bake a day's supply of pies. Working together, it takes them 1 1/5 hr to bake the pies. How long would it take Steve working alone?

63. In a marathon (a 26-mile run), the winner finished 2/5 hr before his friend. If the difference in their rates was 4/3 mph, what was the winner's average speed in the race?

64. Suppose that one solution of the equation $km^2 - 11m = 3$ is 3. Find the value of k and the other solution.

Solve each equation.

65. $4a^4 + 3a^2 - 1 = 0$

66. $2x^4 - x^2 = 0$

67. $(2z + 3)^{2/3} + (2z + 3)^{1/3} = 6$

68. $5\sqrt{m} = \sqrt{3m + 2}$

69. $\sqrt{4y - 2} = \sqrt{3y + 1}$

70. $\sqrt{2x + 3} = x + 2$

71. $\sqrt{p + 2} = 2 + p$

72. $\sqrt{k} = \sqrt{k + 3} - 1$

73. $\sqrt{x + 3} - \sqrt{3x + 10} = 1$

74. $\sqrt{5x - 15} - \sqrt{x + 1} = 2$

75. $\sqrt[3]{6y + 2} = \sqrt[3]{4y}$

76. $(x - 2)^{2/3} = x^{1/3}$

Solve the following inequalities. Write answers in interval notation.

77. $-9x < 4x + 7$

78. $11y \geq 2y - 8$

79. $-5z - 4 \geq 3(2z - 5)$

80. $-(4a + 6) < 3a - 2$

81. $3r - 4 + r > 2(r - 1)$

82. $7p - 2(p - 3) \leq 5(2 - p)$

83. $5 \leq 2x - 3 \leq 7$

84. $-8 > 3a - 5 > -12$

85. $x^2 + 3x - 4 \leq 0$

86. $p^2 + 4p > 21$

87. $6m^2 - 11m - 10 < 0$

88. $k^2 - 3k - 5 \geq 0$

89. $x^2 - 6x + 9 \leq 0$

90. $(x - 2)(x - 4)(x - 3) > 0$

91. $\dfrac{3a - 2}{a} > 4$

92. $\dfrac{5p + 2}{p} < -1$

93. $\dfrac{3}{r - 1} \leq \dfrac{5}{r + 3}$

94. $\dfrac{3}{x + 2} > \dfrac{2}{x - 4}$

95. If $0 < a < b$, on what interval is $(x - a)(x - b)$ positive? negative? zero?

96. Without actually solving the inequality, explain why 3 cannot be in the solution set of $\dfrac{2x + 5}{x - 3} < 0$.

Work the following problems.

97. A company produces videotapes. The revenue from the sale of x units of tapes is $R = 8x$. The cost to produce x units of tapes is $C = 3x + 1500$. In what interval will the company at least break even?

98. A projectile is launched upward. Its height in feet above the ground after t seconds is $320t - 16t^2$. **(a)** After how many seconds in the air will it hit the ground? **(b)** During what time interval is the projectile more than 576 ft above the ground?

Solve each equation.

99. $|a + 4| = 7$

100. $|-y + 2| = -4$

101. $\left|\dfrac{7}{2 - 3a}\right| = 9$

102. $|5 - 8x| + 1 = 3$

103. $|5r - 1| = |2r + 3|$

104. $|k + 7| = 2k$

Solve each inequality. Write solutions with interval notation.

105. $|m| \leq 7$

106. $|r| < 2$

107. $|p| > 3$

108. $|z| > -1$

109. $|2z + 9| \leq 3$

110. $|5m - 8| \leq 2$

111. $|7k - 3| < 5$

112. $|2p - 1| > 2$

113. $|3r + 7| - 5 > 0$

114. Write as an absolute value inequality: k is at least 4 units from 1.

Solve each of the following equations.

1. $x - (2x + 1) = 7 - 3(x + 1)$　　**2.** $\dfrac{x}{x - 3} = \dfrac{3}{x - 3} + 4$　　**3.** $\dfrac{1}{c} - \dfrac{1}{a} = \dfrac{1}{b}$　for c

Solve each of the following problems.

4. How many quarts of a 60% alcohol solution must be added to 40 qt of a 20% alcohol solution to obtain a mixture that is 30% alcohol?

5. Fred and Wilma start from the same point and travel on a straight road. Fred travels at 30 mph, while Wilma travels at 50 mph. If Wilma starts 3 hr after Fred, find the distance they travel before Wilma catches up with Fred.

Perform each operation. Write each result in standard form.

6. $(7 - i) - (6 - 10i)$

7. $(4 + 3i)(-5 + 2i)$

8. $\dfrac{5 - 5i}{1 - 3i}$

9. Is i^{297} equal to i, -1, $-i$, or 1?

Solve each equation.

10. $6x(2 - x) = 7$

11. $\dfrac{3x - 2}{3x + 2} = \dfrac{2x + 3}{4x - 1}$

12. Evaluate the discriminant, and use it to predict the type of solutions for the equation $3x^2 = 5x - 2$.

13. Tony Lally has a rectangular-shaped flower box that measures 4 ft by 6 ft. He wants to double the available area by increasing the length and width by the same amount. What should the new dimensions be?

Solve each equation.

14. $\sqrt{5 + 2x} = x + 1$　　**15.** $\sqrt{2x + 1} - \sqrt{x} = 1$　　**16.** $x^4 - 3x^2 - 10 = 0$　　**17.** $2 - \sqrt[3]{2x + x^2} = 0$

Solve each of the following inequalities. Write each answer in interval notation.

18. $-2(x - 1) - 10 \le 2(2 + x)$

19. $4 \ge 3 + \dfrac{x}{2} \ge -2$

20. $2x^2 - x - 3 \ge 0$

21. $\dfrac{6}{2x - 5} \le 2$

22. What is wrong with the following "solution" of $\dfrac{1}{x - 3} \ge 2$?

$$\dfrac{1}{x - 3} \ge 2$$

$$(x - 3)\left(\dfrac{1}{x - 3}\right) \ge (x - 3)(2) \qquad \text{Multiply by } x - 3.$$

$$1 \ge 2x - 6 \qquad \text{Distributive property}$$

$$7 \ge 2x \qquad \text{Add 6.}$$

$$\dfrac{7}{2} \ge x \qquad \text{Divide by 2.}$$

The solution set is $\left(-\infty, \dfrac{7}{2}\right]$.

23. Solve $|-5 - 3x| = 4$.

Solve each inequality. Write solutions in interval notation.

24. $|2x - 5| < 9$

25. $|2x + 1| - 11 \ge 0$

RELATIONS AND THEIR GRAPHS

3.1 RELATIONS AND THE RECTANGULAR COORDINATE SYSTEM

3.2 LINEAR RELATIONS

3.3 PARABOLAS: TRANSLATIONS AND APPLICATIONS

3.4 THE CIRCLE AND SYMMETRY

3.5 THE ELLIPSE AND THE HYPERBOLA

3.6 THE CONIC SECTIONS

3.7 INEQUALITIES

A major goal of this text is to make a careful study of *relations* and *functions* (a special kind of relation discussed in the next chapter). In this chapter we discuss several important first- and second-degree relations.

<table>
<tr><td>**3.1**</td><td>——————</td></tr>
</table>

RELATIONS AND THE RECTANGULAR COORDINATE SYSTEM

Many things in daily life are related. For example, a student's grade in a course usually is related to the amount of time spent studying, while the number of miles per gallon of gas used on a car trip depends on the speed of the car. Driving at 55 mph might give 31 miles per gallon, while driving at 65 mph might reduce the gas mileage to 28 miles per gallon.

Pairs of related numbers, such as 55 and 31 or 65 and 28 in the gas mileage illustration, can be written as *ordered pairs*. An **ordered pair** of numbers consists of two numbers, written inside parentheses, in which the sequence of the numbers is important. For example, (4, 2) and (2, 4) are different ordered pairs because the order of the numbers is different. Notation such as (3, 4) has already been used in this book to show an interval on the number line. Now the same notation is used to indicate an ordered pair of numbers. In virtually every case, the intended use will be clear from the context of the discussion.

RELATIONS A set of ordered pairs is called a **relation.** The **domain** of a relation is the set of first elements in the ordered pairs, and the **range** of the relation is the set of all possible second elements. In the driving example above, the domain is the set of all possible speeds and the range is the set of resulting miles per gallon. In this text, we confine domains and ranges to real number values.

Ordered pairs are used to express the solutions of equations in two variables. For example, we say that (1, 2) is a solution of $2x - y = 0$, since substituting 1 for x and 2 for y in the equation gives

$$2(1) - 2 = 0$$
$$0 = 0$$

a true statement. When an ordered pair represents the solution of an equation with the variables x and y, the x-value is written first.

Although any set of ordered pairs is a relation, in mathematics we are most interested in those relations that are solution sets of equations. We may say that an equation *defines a relation,* or that it is the *equation of the relation.* For simplicity, we often refer to equations such as

$$y = 3x + 5 \qquad \text{or} \qquad x^2 + y^2 = 16$$

as relations, although technically the solution set of the equation is the relation.

■ *Example 1*

FINDING ORDERED PAIRS, DOMAINS, AND RANGES

For each relation defined below, give three ordered pairs that belong to the relation, and state the domain and the range of the relation.

(a) $\{(2, 5), (7, -1), (10, 3), (-4, 0), (0, 5)\}$

Three ordered pairs from the relation are any three of the five ordered pairs in the set. The domain is the set of first elements,

$$\{2, 7, 10, -4, 0\},$$

and the range is the set of second elements,

$$\{5, -1, 3, 0\}.$$

(b) $y = 4x - 1$

To find an ordered pair of the relation, choose any number for x or y and substitute in the equation to get the corresponding value of the other variable. For example, let $x = -2$. Then

$$y = 4(-2) - 1 = -9,$$

giving the ordered pair $(-2, -9)$. If $y = 3$, then

$$3 = 4x - 1$$
$$4 = 4x$$
$$1 = x,$$

and the ordered pair is $(1, 3)$. Verify that $(0, -1)$ also belongs to the relation. Since x and y can take any real-number values, both the domain and range are $(-\infty, \infty)$.

(c) $x = \sqrt{y - 1}$

Verify that the ordered pairs $(1, 2)$, $(0, 1)$, and $(2, 5)$ belong to the relation. Since x equals the principal square root of $y - 1$, the domain is restricted to $[0, \infty)$. Also, only nonnegative numbers have a real square root, so the range is determined by the inequality

$$y - 1 \geq 0$$
$$y \geq 1,$$

giving $[1, \infty)$ as the range. ■

THE RECTANGULAR COORDINATE SYSTEM Since the study of relations often involves looking at their *graphs,* this section includes a brief review of the coordinate plane. As mentioned in Chapter 1, each real number corresponds to a point on a number line. This correspondence is set up by establishing a coordinate system for the line. This idea is extended to the two dimensions of a plane by drawing two perpendicular lines, one horizontal and one vertical. These lines intersect at a point O called the **origin.** The horizontal line is called the **x-axis,** and the vertical line is called the **y-axis.**

Starting at the origin, the x-axis can be made into a number line by placing positive numbers to the right and negative numbers to the left. The y-axis can be made into a number line with positive numbers going up and negative numbers going down.

146

■ **FIGURE 3.1**

■ **FIGURE 3.2**

The *x*-axis and *y*-axis together make up a **rectangular coordinate system,** or **Cartesian coordinate system** (named for one of its co-inventors, René Descartes; the other co-inventor was Pierre de Fermat). The plane into which the coordinate system is introduced is the **coordinate plane,** or **xy-plane.** The *x*-axis and *y*-axis divide the plane into four regions, or **quadrants,** labeled as shown in Figure 3.1. The points on the *x*-axis and *y*-axis belong to no quadrant.

Each point *P* in the *xy*-plane corresponds to a unique ordered pair (*a*, *b*) of real numbers. The numbers *a* and *b* are the **coordinates** of point *P*. To locate on the *xy*-plane the point corresponding to the ordered pair (3, 4), for example, draw a vertical line through 3 on the *x*-axis and a horizontal line through 4 on the *y*-axis. These two lines cross at point *A* in Figure 3.2. Point *A* corresponds to the ordered pair (3, 4). Also in Figure 3.2, *B* corresponds to the ordered pair (−5, 6), *C* to (−2, −4), *D* to (4, −3), and *E* to (−3, 0). The point *P* corresponding to the ordered pair (*a*, *b*) often is written as *P*(*a*, *b*) as in Figure 3.1 and referred to as "the point (*a*, *b*)."

As we shall see later in this chapter the **graph** of a relation is the set of points in the plane that corresponds to the ordered pairs of the relation.

Two formulas, the *distance* and *midpoint* formulas, will be useful in our study of relations in this chapter.

THE DISTANCE FORMULA By using the Pythagorean theorem, we can develop a formula to find the distance between any two points in a plane. For example, Figure 3.3 shows the points $P(-4, 3)$ and $R(8, -2)$.

To find the distance between these two points, complete a right triangle as shown in the figure. This right triangle has its 90° angle at (8, 3). The horizontal side of the triangle has length

$$|8 - (-4)| = 12,$$

where absolute value is used to make sure that the distance is not negative. The vertical side of the triangle has length

$$|3 - (-2)| = 5.$$

By the Pythagorean theorem, the length of the remaining side of the triangle is

$$\sqrt{12^2 + 5^2} = \sqrt{144 + 25} = \sqrt{169} = 13.$$

The distance between (−4, 3) and (8, −2) is 13.

To obtain a general formula for the distance between two points on a coordinate plane, let $P(x_1, y_1)$ and $R(x_2, y_2)$ be any two distinct points in a plane, as shown in Figure 3.4. Complete a triangle by locating point *Q* with coordinates (x_2, y_1). Using the Pythagorean theorem gives the distance between *P* and *R*, written $d(P, R)$, as

$$d(P, R) = \sqrt{(x_2 - x_1)^2 + (y_2 - y_1)^2}.$$

$$d(P, R) = \sqrt{(x_2 - x_1)^2 + (y_2 - y_1)^2}$$

■ **FIGURE 3.3**

■ **FIGURE 3.4**

NOTE	The use of absolute value bars is not necessary in this formula, since for all real numbers a and b, $|a - b|^2 = (a - b)^2$.

The distance formula can be summarized as follows.

DISTANCE FORMULA	Suppose that $P(x_1, y_1)$ and $R(x_2, y_2)$ are two points in a coordinate plane. Then the distance between P and R, written $d(P, R)$, is given by the **distance formula,** $$d(P, R) = \sqrt{(x_2 - x_1)^2 + (y_2 - y_1)^2}$$

Although the proof of the distance formula assumes that P and R are not on a horizontal or vertical line, the result is true for any two points.

▪ **Example 2**

USING THE
DISTANCE FORMULA

Find the distance between $P(-8, 4)$ and $Q(3, -2)$.

According to the distance formula,

$$d(P, Q) = \sqrt{[3 - (-8)]^2 + (-2 - 4)^2} \qquad x_1 = -8, y_1 = 4, x_2 = 3, y_2 = -2$$
$$= \sqrt{11^2 + (-6)^2}$$
$$= \sqrt{121 + 36} = \sqrt{157}. \quad ▪$$

NOTE	As shown in Example 2, it is customary to leave the distance between two points in radical form rather than approximating it with a calculator (unless, of course, it is otherwise specified).

A statement of the form "If p, then q" is called a *conditional* statement. The related statement "If q, then p" is called its converse. In Chapter 2 we studied the Pythagorean theorem. The converse of the Pythagorean theorem is also a true statement: If the sides a, b, and c of a triangle satisfy $a^2 + b^2 = c^2$, then the triangle is a right triangle with legs having lengths a and b and hypotenuse having length c. This can be used to determine whether three points are the vertices of a right triangle, as shown in the next example.

▪ **Example 3**

DETERMINING
WHETHER THREE
POINTS ARE THE
VERTICES OF A
RIGHT TRIANGLE

Are the three points $M(-2, 5)$, $N(12, 3)$, and $Q(10, -11)$ the vertices of a right triangle?

A triangle with the three given points as vertices is shown in Figure 3.5. This triangle is a right triangle if the square of the length of the longest side equals the sum of the squares of the lengths of the other two sides. Use the distance formula to find the length of each side of the triangle.

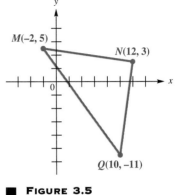

▪ **FIGURE 3.5**

$$d(M, N) = \sqrt{[12 - (-2)]^2 + (3 - 5)^2} = \sqrt{196 + 4} = \sqrt{200}$$

$$d(M, Q) = \sqrt{[10 - (-2)]^2 + (-11 - 5)^2} = \sqrt{144 + 256} = \sqrt{400} = 20$$

$$d(N, Q) = \sqrt{(10 - 12)^2 + (-11 - 3)^2} = \sqrt{4 + 196} = \sqrt{200}$$

By these results,

$$[d(M, Q)]^2 = [d(M, N)]^2 + [d(N, Q)]^2,$$

since $20^2 = \sqrt{200}^2 + \sqrt{200}^2$, or $400 = 400$, is a true statement. This proves that the triangle is a right triangle with hypotenuse connecting M and Q. ■

Using a procedure similar to that of Example 3, it can be determined whether three points lie on a straight line. Points that lie on a line are called collinear. Three points are **collinear** if the sum of the distances between two pairs of the points is equal to the distance between the remaining pair of points.

■ ***Example 4***
DETERMINING
WHETHER THREE
POINTS ARE
COLLINEAR

Are the points $(-1, 5)$, $(2, -4)$, and $(4, -10)$ collinear?

The distance between $(-1, 5)$ and $(2, -4)$ is

$$\sqrt{(-1 - 2)^2 + [5 - (-4)]^2} = \sqrt{9 + 81} = \sqrt{90} = 3\sqrt{10}.$$

The distance between $(2, -4)$ and $(4, -10)$ is

$$\sqrt{(2 - 4)^2 + [-4 - (-10)]^2} = \sqrt{4 + 36} = \sqrt{40} = 2\sqrt{10}.$$

Finally, the distance between the remaining pair of points, $(-1, 5)$ and $(4, -10)$ is

$$\sqrt{(-1 - 4)^2 + [5 - (-10)]^2} = \sqrt{25 + 225} = \sqrt{250} = 5\sqrt{10}.$$

Because $3\sqrt{10} + 2\sqrt{10} = 5\sqrt{10}$, the three points are collinear. ■

THE MIDPOINT FORMULA   The midpoint formula is used to find the coordinates of the midpoint of a line segment. (Recall that the midpoint of a line segment is equidistant from the endpoints of the segment.) To develop the midpoint formula, let (x_1, y_1) and (x_2, y_2) be any two distinct points in a plane. (Although Figure 3.6 shows $x_1 < x_2$, no particular order is required.) Assume that the two points are not on a horizontal or vertical line. Let (x, y) be the midpoint of the segment connecting (x_1, y_1) and (x_2, y_2). Draw vertical lines from each of the three points to the x-axis, as shown in Figure 3.6.

Since (x, y) is the midpoint of the line segment connecting (x_1, y_1) and (x_2, y_2), the distance between x and x_1 equals the distance between x and x_2, so that

$$x_2 - x = x - x_1$$

$$x_2 + x_1 = 2x$$

$$x = \frac{x_1 + x_2}{2}.$$

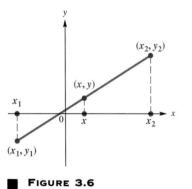

■ FIGURE 3.6

By this result, the x-coordinate of the midpoint is the average of the x-coordinates of the endpoints of the segment. In a similar manner, the y-coordinate of the midpoint is $(y_1 + y_2)/2$, proving the following statement.

MIDPOINT FORMULA	The midpoint of the line segment with endpoints (x_1, y_1) and (x_2, y_2) is $$\left(\frac{x_1 + x_2}{2}, \frac{y_1 + y_2}{2} \right).$$

In other words, the midpoint formula says that the coordinates of the midpoint of a segment are found by calculating the *average* of the x-coordinates and the average of the y-coordinates of the endpoints of the segment. In Exercise 43, you are asked to verify that the coordinates above satisfy the definition of midpoint.

■ **Example 5**
USING THE MIDPOINT FORMULA

Find the midpoint M of the segment with endpoints $(8, -4)$ and $(-9, 6)$.
Use the midpoint formula to find that the coordinates of M are

$$\left(\frac{8 + (-9)}{2}, \frac{-4 + 6}{2} \right) = \left(-\frac{1}{2}, 1 \right). \quad ■$$

■ **Example 6**
USING THE MIDPOINT FORMULA

A line segment has an endpoint at $(2, -8)$ and midpoint at $(-1, -3)$. Find the other endpoint of the segment.
The formula for the x-coordinate of the midpoint is $(x_1 + x_2)/2$. Here the x-coordinate of the midpoint is -1. Letting $x_1 = 2$ gives

$$-1 = \frac{2 + x_2}{2}$$
$$-2 = 2 + x_2$$
$$-4 = x_2.$$

In the same way, $y_2 = 2$ and the endpoint is $(-4, 2)$. ■

3.1 EXERCISES ■

For each relation, give three ordered pairs that belong to the relation and state the domain and the range of the relation. See Example 1.

1. $\{(-3, 5), (-2, 4), (-1, 6), (0, -8), (1, 2)\}$ **2.** $\{(5, 12), (4, 12), (3, 9), (-1, 6), (-5, 5)\}$

3. $y = 9x + 2$ **4.** $y = -2x + 7$ **5.** $x = y$ **6.** $x = -y$ **7.** $y = \sqrt{x}$

8. $x = \sqrt{y}$ **9.** $y = x^2$ **10.** $x = \sqrt{y + 3}$ **11.** $y = \sqrt{x + 3}$ **12.** $y = |x|$

13. Explain in your own words the terms *relation, domain,* and *range.*

14. Fill in the blank with the appropriate number: The relation in Exercise 2 has _____ elements. There are _____ elements in its domain and _____ elements in its range.

Find the distance d(P, Q) and the midpoint of segment PQ. See Examples 2 and 5.

15. $P(5, 7)$, $Q(13, -1)$ **16.** $P(-2, 5)$, $Q(4, -3)$ **17.** $P(-8, -2)$, $Q(-3, -5)$ **18.** $P(-6, -10)$, $Q(6, 5)$

19. $P(3, -7)$, $Q(-5, 19)$ **20.** $P(-4, 6)$, $Q(8, -5)$ **21.** $P(a, b)$, $Q(3a, -4b)$ **22.** $P(4x, 6y)$, $Q(-3x, 2y)$

23. $P(\sqrt{2}, -\sqrt{5})$, $Q(3\sqrt{2}, 4\sqrt{5})$ **24.** $P(5\sqrt{7}, -\sqrt{3})$, $Q(-\sqrt{7}, 8\sqrt{3})$

Decide whether or not the given points are the vertices of a right triangle. See Example 3.

25. $(2, 8)$, $(0, 4)$, $(4, 7)$ **26.** $(6, 4)$, $(0, 2)$, $(10, -8)$

27. $(-4, 0)$, $(1, 3)$, $(-6, -2)$ **28.** $(-8, 2)$, $(5, -7)$, $(3, -9)$

29. $(\sqrt{3}, 2\sqrt{3} + 3)$, $(\sqrt{3} + 4, -\sqrt{3} + 3)$, $(2\sqrt{3}, 2\sqrt{3} + 4)$

30. $(4 - \sqrt{3}, -2\sqrt{3})$, $(2 - \sqrt{3}, -\sqrt{3})$, $(3 - \sqrt{3}, -2\sqrt{3})$

Decide whether or not the given points are collinear. See Example 4.

31. $(0, 7)$, $(3, -5)$, $(-2, 15)$ **32.** $(1, -4)$, $(2, 1)$, $(-1, -14)$ **33.** $(0, -9)$, $(3, 7)$, $(-2, -19)$

34. $(1, 3)$, $(5, -12)$, $(-1, 11)$ **35.** $(2, 7)$, $(-4, -2)$, $(10, 19)$ **36.** $(-2, -13)$, $(0, 1)$, $(5, 36)$

Find the other endpoint of each segment having endpoint and midpoint as given. See Example 6.

37. Endpoint $(-3, 6)$, midpoint $(5, 8)$ **38.** Endpoint $(-5, 3)$, midpoint $(-7, 6)$

39. Endpoint $(5, -4)$, midpoint $(12, 6)$ **40.** Endpoint $(-2, 7)$, midpoint $(-9, 8)$

41. Endpoint (a, b), midpoint (c, d) **42.** Endpoint $(-r, -s)$, midpoint (r, s)

43. Show that if M is the midpoint of the segment with endpoints $P(x_1, y_1)$ and $Q(x_2, y_2)$, then $d(P, M) + d(M, Q) = d(P, Q)$, and $d(P, M) = d(M, Q)$.

44. The distance formula as given in the text involves a square root radical. Write the distance formula using rational exponents.

Give all quadrants in which the set of points satisfying the following conditions for ordered pairs (x, y) are located.

45. $x > 0$ **46.** $y < 0$ **47.** $xy < 0$

48. $\dfrac{x}{y} > 0$ **49.** $|x| < 3$, $y < -2$ **50.** $|y| < 2$, $x > 1$

51. Show that the points $(-2, 2)$, $(13, 10)$, $(21, -5)$, and $(6, -13)$ are the vertices of a rhombus (all sides equal in length).

52. Are the points $A(1, 1)$, $B(5, 2)$, $C(3, 4)$, $D(-1, 3)$ the vertices of a parallelogram? Of a rhombus (all sides equal in length)?

53. "If p then q and if q then p" is called a *biconditional,* and is often stated "p if and only if q." State the Pythagorean theorem and its converse using this language.

3.2 ———— **LINEAR RELATIONS**

In Section 3.1 we introduced relations. In this section we examine one of the simplest types of relations, the linear relation. Every *linear relation* has a graph that is a straight line, and so we need only find two points on the graph in order to sketch it. Examples of linear relations are $y = 2x + 3$, $y = x$, and $3x + 2y = 6$.

<table>
<tr><td>**LINEAR RELATION**</td><td>A **linear relation** in two variables is a relation that can be written in the form

$$y = ax + b,$$

where a and b are real numbers.</td></tr>
</table>

NOTE Linear relations are often written in the form $Ax + By = C$, where A, B, and C are real, and A and B are not both 0. This is called the **standard form** of a linear relation.

In the equation $Ax + By = C$, any number can be used for x or y, so both the domain and range of a linear relation in which neither A nor B is 0 are the set of real numbers $(-\infty, \infty)$.

GRAPHING LINEAR RELATIONS   The graph of a linear relation can be found by plotting at least two points. Two points that are especially useful for sketching the graph of a line are found with the *intercepts*. An **x-intercept** is an x-value at which a graph crosses the x-axis. A **y-intercept** is a y-value at which a graph crosses the y-axis. Since $y = 0$ on the x-axis, an x-intercept is found by setting y equal to 0 in the equation and solving for x. Similarly, a y-intercept is found by setting $x = 0$ in the equation and solving for y.

▪ *Example 1*

GRAPHING A LINEAR RELATION USING INTERCEPTS

Graph $3x + 2y = 6$.

   Use the intercepts. The y-intercept is found by letting $x = 0$.

$$3 \cdot 0 + 2y = 6$$
$$2y = 6$$
$$y = 3$$

For the x-intercept, let $y = 0$, getting

$$3x + 2 \cdot 0 = 6$$
$$3x = 6$$
$$x = 2.$$

Plotting $(0, 3)$ and $(2, 0)$ gives the graph in Figure 3.7. A third point could be found as a check if desired.   ▪

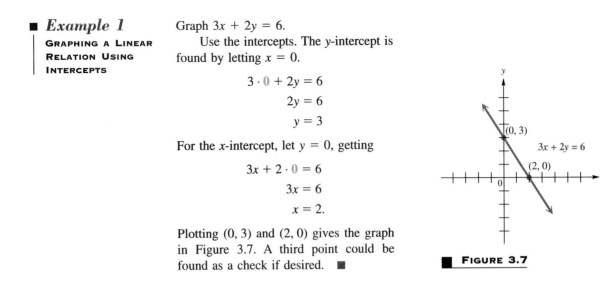

▪ **FIGURE 3.7**

152

■ *Example 2*

GRAPHING
HORIZONTAL AND
VERTICAL LINES

(a) Graph $y = -3$.

Since y always equals -3, the value of y can never be 0. This means that the graph has no x-intercept. The only way a straight line can have no x-intercept is for it to be parallel to the x-axis, as shown in Figure 3.8. Notice that the domain of this linear relation is $(-\infty, \infty)$, but the range is $\{-3\}$.

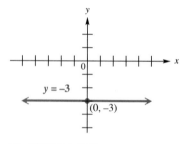

■ **FIGURE 3.8**

(b) Graph $x = -3$.

Here, since x always equals -3, the value of x can never be 0, and the graph has no y-intercept. Using reasoning similar to that of part (a), we find that this graph is parallel to the y-axis, as shown in Figure 3.9. The domain of this relation is $\{-3\}$, while the range is $(-\infty, \infty)$.

From this example we may conclude that a linear relation of the form $y = k$ has as its graph a horizontal line through $(0, k)$, and one of the form $x = k$ has as its graph a vertical line through $(k, 0)$. ■

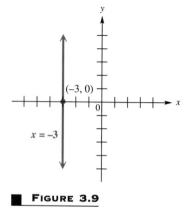

■ **FIGURE 3.9**

■ *Example 3*

GRAPHING A LINE
THROUGH THE
ORIGIN

Graph $4x - 5y = 0$.

Find the intercepts. If $x = 0$, then

$$4(0) - 5y = 0$$
$$-5y = 0$$
$$y = 0.$$

Letting $y = 0$ leads to the same ordered pair, $(0, 0)$. The graph of this relation has just one intercept—at the origin. Find another point by choosing a different value for x (or y). Choosing $x = 5$ gives

$$4(5) - 5y = 0$$
$$20 - 5y = 0$$
$$20 = 5y$$
$$4 = y,$$

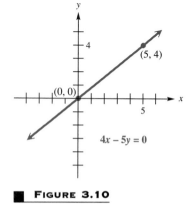

■ **FIGURE 3.10**

which leads to the ordered pair $(5, 4)$. Complete the graph using the two points $(0, 0)$ and $(5, 4)$, with a third point as a check. See Figure 3.10. ■

■ **FIGURE 3.11**

SLOPE An important characteristic of a straight line is its *slope,* a numerical measure of the steepness of the line. (Geometrically, this may be interpreted as the ratio of *rise* to *run.*) To find this measure, start with the line through the two distinct points (x_1, y_1) and (x_2, y_2), as shown in Figure 3.11, where $x_1 \neq x_2$. The difference

$$x_2 - x_1$$

is called the **change in x** and denoted by Δx (read "delta x"), where Δ is the Greek letter *delta.* In the same way, the **change in y** can be written

$$\Delta y = y_2 - y_1.$$

The *slope* of a nonvertical line is defined as the quotient of the change in y and the change in x, as follows.

SLOPE The **slope** m of the line through the points (x_1, y_1) and (x_2, y_2) is

$$m = \frac{\Delta y}{\Delta x} = \frac{y_2 - y_1}{x_2 - x_1},$$

where $\Delta x \neq 0$.

CAUTION When using the slope formula, be sure that it is applied correctly. It makes no difference which point is (x_1, y_1) or (x_2, y_2); however, it is important to be consistent. Start with the x- and y-value of *one* point (either one) and subtract the corresponding values of the *other* point.

The slope of a line can be found only if the line is nonvertical. This guarantees that $x_2 \neq x_1$ so that the denominator $x_2 - x_1 \neq 0$. It is not possible to define the slope of a vertical line.

The slope of a vertical line is undefined.

■ *Example 4*
FINDING SLOPES WITH THE SLOPE FORMULA

Find the slope of the line through each of the following pairs of points.

(a) $(-4, 8), (2, -3)$

Let $x_1 = -4$, $y_1 = 8$, and $x_2 = 2$, $y_2 = -3$. Then

$$\Delta y = -3 - 8 = -11$$

and

$$\Delta x = 2 - (-4) = 6.$$

The slope is

$$m = \frac{\Delta y}{\Delta x} = -\frac{11}{6}.$$

(b) $(2, 7)$, $(2, -4)$

A sketch would show that the line through $(2, 7)$ and $(2, -4)$ is vertical. As mentioned above, the slope of a vertical line is not defined. (An attempt to use the definition of slope here would produce a zero denominator.)

(c) $(5, -3)$ and $(-2, -3)$

By definition of slope,

$$m = \frac{-3 - (-3)}{-2 - 5} = \frac{0}{-7} = 0. \quad \blacksquare$$

Drawing a graph through the points in Example 4(c) would produce a line that is horizontal, which suggests the following generalization.

The slope of a horizontal line is 0.

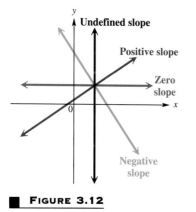

■ **FIGURE 3.12**

Figure 3.12 shows lines of various slopes. As the figure shows, a line with a positive slope goes up from left to right, but a line with a negative slope goes down from left to right.

It can be shown, using theorems for similar triangles, that the slope is independent of the choice of points on the line. That is, the slope of a line is the same no matter which pair of distinct points on the line are used to find it.

Since the slope of a line is the ratio of vertical change to horizontal change, if we know the slope of a line and the coordinates of a point on the line, the graph of the line can be drawn. The next example illustrates this.

■ *Example 5*

GRAPHING A LINE USING A POINT AND THE SLOPE

Graph the line passing through $(-1, 5)$ and having slope $-5/3$.

First locate the point $(-1, 5)$ as shown in Figure 3.13. Since the slope of this line is $-5/3$, a change of -5 units vertically (that is, 5 units down) produces a change of 3 units horizontally (3 units to the right).

This gives a second point, $(2, 0)$, which can then be used to complete the graph.

Because $-5/3 = 5/(-3)$, another point could be obtained by starting at $(-1, 5)$ and moving 5 units *up* and 3 units to the *left*. We would reach a different second point, but the line would be the same. ■

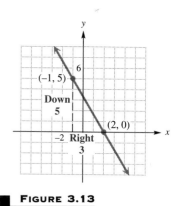

■ **FIGURE 3.13**

EQUATIONS OF A LINE   Since equations can define relations, we now consider methods of finding equations of linear relations. Figure 3.14 shows the line passing through the fixed point (x_1, y_1) and having slope m. (Assuming that the line has a slope guarantees that it is not vertical.) Let (x, y) be any other point on the line. By the definition of slope, the slope of the line is

$$\frac{y - y_1}{x - x_1}.$$

Since the slope of the line is m,

$$\frac{y - y_1}{x - x_1} = m.$$

Multiplying both sides by $x - x_1$ gives

$$y - y_1 = m(x - x_1).$$

This result, called the *point-slope form* of the equation of a line, identifies points on a given line: a point (x, y) lies on the line through (x_1, y_1) with slope m if and only if

$$y - y_1 = m(x - x_1).$$

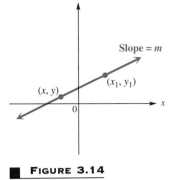

∎ **FIGURE 3.14**

POINT-SLOPE FORM	The line with slope m passing through the point (x_1, y_1) has an equation $$y - y_1 = m(x - x_1),$$ the **point-slope form** of the equation of a line.

∎ ***Example 6***

USING THE POINT-SLOPE FORM (GIVEN A POINT AND THE SLOPE)

Write an equation of the line through $(-4, 1)$ with slope -3.
   Here $x_1 = -4$, $y_1 = 1$, and $m = -3$. Use the point-slope form of the equation of a line to get

$$y - 1 = -3[x - (-4)] \qquad x_1 = -4, y_1 = 1, m = -3$$
$$y - 1 = -3(x + 4)$$
$$y - 1 = -3x - 12 \qquad \text{Distributive property}$$

or $\qquad 3x + y = -11,$

in standard form.   ∎

CAUTION | The definition of "standard form" is not standard from one text to another. Any linear equation can be written in many different (all equally correct) forms. For example, the equation $2x + 3y = 8$ can be written as $2x = 8 - 3y$, $3y = 8 - 2x$, $x + \frac{3}{2}y = 4$, $4x + 6y = 16$, and so on. In addition to writing it in the form $Ax + By = C$ (with $A \geq 0$), let us agree that the form $2x + 3y = 8$ is preferred over any multiples of both sides, such as $4x + 6y = 16$.

156 ■ *Example 7*

USING THE
POINT-SLOPE FORM
(GIVEN TWO
POINTS)

Find an equation of the line through $(-3, 2)$ and $(2, -4)$.

Find the slope first. By the definition of slope,

$$m = \frac{-4 - 2}{2 - (-3)} = -\frac{6}{5}.$$

Either $(-3, 2)$ or $(2, -4)$ can be used for (x_1, y_1). Choosing $x_1 = -3$ and $y_1 = 2$ in the point-slope form gives

$$y - 2 = -\frac{6}{5}[x - (-3)]$$

$5(y - 2) = -6(x + 3)$	Multiply by 5.
$5y - 10 = -6x - 18$	Distributive property
$6x + 5y = -8.$	Standard form

Verify that the same equation results if $(2, -4)$ is used instead of $(-3, 2)$ in the point-slope form. ■

As a special case of the point-slope form of the equation of a line, suppose that a line passes through the point $(0, b)$, so the line has y-intercept b. If the line has slope m, then using the point-slope form with $x_1 = 0$ and $y_1 = b$ gives

$$y - y_1 = m(x - x_1)$$
$$y - b = m(x - 0)$$
$$y = mx + b$$

as an equation of the line. Since this result shows the slope of the line and the y-intercept, it is called the *slope-intercept form* of the equation of the line.

SLOPE-INTERCEPT FORM	The line with slope m and y-intercept b has an equation $$y = mx + b,$$ the **slope-intercept** form of the equation of a line.

■ *Example 8*

USING THE
SLOPE-INTERCEPT
FORM TO GRAPH A
LINE

Find the slope and y-intercept of $3x - y = 2$. Graph the line using this information.

First write $3x - y = 2$ in the slope-intercept form, $y = mx + b$, by solving for y, getting $y = 3x - 2$. This result shows that the slope is $m = 3$ and the y-intercept is $b = -2$. To draw the graph, first locate the y-intercept. See Figure 3.15. Then, as in Example 5, use the slope of 3, or 3/1, to get a second point on the graph. The line through these two points is the graph of $3x - y = 2$. ■

■ **FIGURE 3.15**

In the preceding discussion, it was assumed that the given line had a slope. The only lines having undefined slope are vertical lines. The vertical line through the point (a, b) passes through all the points of the form (a, y), for any value of y. This fact determines the equation of a vertical line.

EQUATION OF A VERTICAL LINE	An equation of the vertical line through the point (a, b) is $x = a$.

For example, the vertical line through $(-4, 9)$ has equation $x = -4$, while the vertical line through $(0, 1/4)$ has equation $x = 0$. (This is the y-axis.)

The horizontal line through the point (a, b) passes through all points of the form (x, b), for any value of x. Therefore, the equation of a horizontal line involves only the variable y.

EQUATION OF A HORIZONTAL LINE	An equation of the horizontal line through the point (a, b) is $y = b$.

For example, the horizontal line through $(1, -3)$ has the equation $y = -3$. See Figure 3.8 for the graph of this equation. The equation of the x-axis is $y = 0$.

PARALLEL AND PERPENDICULAR LINES Slopes can be used to decide whether or not two lines are parallel. Since two parallel lines are equally "steep," they should have the same slope. Also, two distinct lines with the same "steepness" are parallel. The following result summarizes this discussion.

PARALLEL LINES	Two distinct nonvertical lines are parallel if and only if they have the same slope.

Slopes are also used to determine if two lines are perpendicular. Whenever two lines have slopes with a product of -1, the lines are perpendicular.

PERPENDICULAR LINES	Two lines, neither of which is vertical, are perpendicular if and only if their slopes have a product of -1.

For example, if the slope of a line is $-3/4$, the slope of any line perpendicular to it is $4/3$, since $(-3/4)(4/3) = -1$. We often refer to numbers like $-3/4$ and $4/3$ as "negative reciprocals." A proof of this result is outlined in Exercises 63–66.

158 ■ *Example 9*

USING THE SLOPE
RELATIONSHIPS FOR
PARALLEL AND
PERPENDICULAR
LINES

Find the equation of the line that passes through the point $(3, 5)$ and satisfies the given condition.

(a) parallel to the line $2x + 5y = 4$

Since it is given that the point $(3, 5)$ is on the line, we need only find the slope to use the point-slope form. Find the slope by writing the equation of the given line in slope-intercept form. (That is, solve for y.)

$$2x + 5y = 4$$
$$y = -\frac{2}{5}x + \frac{4}{5}$$

The slope is $-2/5$. Since the lines are parallel, $-2/5$ is also the slope of the line whose equation is to be found. Substituting $m = -2/5$, $x_1 = 3$, and $y_1 = 5$ into the point-slope form gives

$$y - y_1 = m(x - x_1)$$
$$y - 5 = -\frac{2}{5}(x - 3)$$
$$5(y - 5) = -2(x - 3)$$
$$5y - 25 = -2x + 6$$
$$2x + 5y = 31.$$

(b) perpendicular to the line $2x + 5y = 4$

In part (a) it was found that the slope of this line is $-2/5$, so the slope of any line perpendicular to it is $5/2$. Therefore, use $m = 5/2$, $x_1 = 3$, and $y_1 = 5$ in the point-slope form.

$$y - 5 = \frac{5}{2}(x - 3)$$
$$2(y - 5) = 5(x - 3)$$
$$2y - 10 = 5x - 15$$
$$-5x + 2y = -5$$

or $$5x - 2y = 5 \quad ■$$

All the lines discussed above have equations that could be written in the form $Ax + By = C$ for real numbers A, B, and C. As mentioned earlier, the equation $Ax + By = C$ is the standard form of the equation of a line. The various forms of linear equations are listed below.

LINEAR EQUATIONS	General Equation	Type of Equation
	$Ax + By = C$	*Standard form* (if $A \neq 0$ and $B \neq 0$), x-intercept C/A, y-intercept C/B, slope $-A/B$
	$x = k$	*Vertical line*, x-intercept k, no y-intercept, undefined slope
	$y = k$	*Horizontal line*, y-intercept k, no x-intercept, slope 0
	$y = mx + b$	*Slope-intercept form*, y-intercept b, slope m
	$y - y_1 = m(x - x_1)$	*Point-slope form*, slope m, through (x_1, y_1)

PROBLEM SOLVING

A straight line is often the best approximation of a set of data points that result from a real situation. If the equation is known, it can be used to predict the value of one variable, given a value of the other. For this reason, the equation is written as a linear relation in slope-intercept form. One way to find the equation of such a straight line is to use two typical data points and the point-slope form of the equation of a line. ∎

∎ *Example 10*

FINDING AN
EQUATION FROM
DATA POINTS

Scientists have found that the number of chirps made by a cricket of a particular species per minute is almost linearly related to the temperature. Suppose that for a particular species, at 68° F a cricket chirps 124 times per minute, while at 80° F the cricket chirps 172 times per minute. Find the linear equation that relates the number of chirps to the temperature.

Think of the ordered pairs in the relation as (chirps, temperature), or (c, t). Then c takes on the role of x and t takes on the role of y. Since we are using a linear relationship, find the slope of the line by using the slope formula with the points $(124, 68)$ and $(172, 80)$.

$$m = \frac{68 - 80}{124 - 172} = \frac{-12}{-48} = \frac{1}{4}$$

Choose one of the points, say $(124, 68)$, and substitute into the point-slope form, with $m = 1/4$.

$$t - 68 = \frac{1}{4}(c - 124)$$

$$t - 68 = \frac{1}{4}c - 31$$

$$t = \frac{1}{4}c + 37$$

The equation is $t = (1/4)c + 37$. By substituting the number of chirps per minute into this equation, the temperature t can be approximated. ∎

◼ **IN SIMPLEST TERMS** ─────────────────────────────── ▪

10

In Massachusetts, speeding fines are determined by the linear relationship $y = 50 + 10(x - 65)$, where y is the fine in dollars and x is the speed of the vehicle. A motorist driving 72 miles per hour would have his fine calculated as $50 + 10(72 - 65) = 50 + 70 = 120$.

SOLVE EACH PROBLEM

A. José had to make an 8:00 a.m. final exam, but overslept after a big weekend in Boston. Radar clocked his speed at 76 miles per hour. How much was his fine?

B. While balancing the checkbook, Harry ran across a cancelled check that his wife, Rachel, had written to the Department of Motor Vehicles for a speeding fine. The check was written for $100. How fast was Rachel driving?

ANSWERS   A. José had to pay $160.   B. Rachel was driving at 70 miles per hour.

160 **3.2 EXERCISES** ■ ──────────────────────────────

Graph each of the following. Give the domain and range. See Examples 1–3.

1. $x - y = 4$ **2.** $x + y = 4$ **3.** $3x - y = 6$ **4.** $2x - 3y = 6$

5. $2x + 5y = 10$ **6.** $4x - 3y = 9$ **7.** $x = 2$ **8.** $y = -3$

9. $y = 3x$ **10.** $x = -2y$

Find the slope of each of the following lines. See Examples 4 and 8.

11. Through $(-2, 1)$ and $(3, 2)$ **12.** Through $(-2, 3)$ and $(-1, 2)$

13. Through $(8, 4)$ and $(-1, -3)$ **14.** Through $(-4, -3)$ and $(5, 0)$

15. $3x + 4y = 6$ **16.** $2x + y = 8$ **17.** $y = 4$ **18.** $x = -6$

19. Match each equation with the line that would most closely resemble its graph. (*Hint:* Consider the signs of m and b in the slope-intercept form.)

  (a) $y = 3x + 2$ **(b)** $y = -3x + 2$ **(c)** $y = 3x - 2$ **(d)** $y = -3x - 2$

20. Match each equation with the line that would most closely resemble its graph.

  (a) $y = 2$ **(b)** $y = -2$ **(c)** $x = 2$ **(d)** $x = -2$

21. Explain in your own words what is meant by the slope of a line.

22. Explain how to graph a line using a point on the line and the slope of the line.

Graph the line passing through the given point and having the indicated slope. Indicate two points on the line. See Example 5.

23. Through $(-1, 3)$, $m = 3/2$ **24.** Through $(-2, 8)$, $m = -1$ **25.** Through $(3, -4)$, $m = -1/3$

26. Through $(-2, -3)$, $m = -3/4$ **27.** Through $(-1, 4)$, $m = 0$ **28.** Through $(9/4, 2)$, undefined slope

Write an equation in standard form for each of the following lines. See Examples 6–7.

29. Through $(1, 3)$, $m = -2$

30. Through $(2, 4)$, $m = -1$

31. Through $(-5, 4)$, $m = -3/2$

32. Through $(-4, 3)$, $m = 3/4$

33. Through $(-8, 1)$, undefined slope

34. Through $(6, 1)$, $m = 0$

35. Through $(-1, 3)$ and $(3, 4)$

36. Through $(8, -1)$ and $(4, 3)$

37. x-intercept 3, y-intercept -2

38. x-intercept -2, y-intercept 4

39. Vertical, through $(-6, 5)$

40. Horizontal, through $(8, 7)$

41. Fill in each blank with the appropriate response: The line $x + 2 = 0$ has x-intercept
_____. It _____ have a y-intercept. The slope of this line is
(does/does not)
_____. The line $4y = 2$ has y-intercept _____. It _____ have an
(zero/undefined)   (does/does not)
x-intercept. The slope of this line is _____.
(zero/undefined)

42. What is the equation of the x-axis?

43. What is the equation of the y-axis?

44. What can be said about the sign of the slope of a line perpendicular to a line whose slope is a positive number?

Give the slope and y-intercept of each of the following lines. See Example 8.

45. $y = 3x - 1$

46. $y = -2x + 7$

47. $4x - y = 7$

48. $2x + 3y = 16$

49. $4y = -3x$

50. $2y - x = 0$

Write an equation in standard form for each of the following lines. See Example 9.

51. Through $(-1, 4)$, parallel to $x + 3y = 5$

52. Through $(3, -2)$, parallel to $2x - y = 5$

53. Through $(1, 6)$, perpendicular to $3x + 5y = 1$

54. Through $(-2, 0)$, perpendicular to $8x - 3y = 7$

55. Through $(-5, 7)$, perpendicular to $y = -2$

56. Through $(1, -4)$, perpendicular to $x = 4$

Use slopes to decide whether or not the given points lie on a straight line. (Hint: In each exercise, first find the slope of the line through the first and second points and then the slope of the line through the second and third points. If these slopes are the same, the points lie on a straight line.)

57. $M(1, -2)$, $N(3, -18)$, $P(-2, 22)$

58. $A(0, -2)$, $B(3, 7)$, $C(-4, -14)$

59. $X(1, 3)$, $Y(-4, 73)$, $Z(5, -50)$

60. $R(0, -7)$, $S(8, -11)$, $T(-9, -25)$

61. Find k so that the line through $(4, -1)$ and $(k, 2)$ is
(a) parallel to $3y + 2x = 6$;
(b) perpendicular to $2y - 5x = 1$.

62. Find r so that the line through $(2, 6)$ and $(-4, r)$ is
(a) parallel to $2x - 3y = 4$;
(b) perpendicular to $x + 2y = 1$.

To prove that two perpendicular lines, neither of which is vertical, have slopes with a product of -1, go through the following steps. Let line L_1 have equation $y = m_1x + b_1$, and let line L_2 have equation $y = m_2x + b_2$. Assume that L_1 and L_2 are perpendicular, and complete right triangle MPN as shown in the figure. $\overline{PQ}$ is horizontal and $\overline{MN}$ is vertical.

63. Show that MQ has length m_1.

64. Show that QN has length $-m_2$.

65. Show that triangles MPQ and PNQ are similar.

66. Show that $m_1/1 = 1/-m_2$ and that $m_1m_2 = -1$.

In each of the following problems, assume that the data can be approximated fairly closely by a straight line. Find the equation of the line. See Example 10.

67. A company finds that it can make a total of 20 solar heaters for $13,900, while 10 heaters cost $7500. Let y be the total cost to produce x solar heaters.

68. Consumer prices in the United States at the beginning of each year, measured as a percent of the 1967 average, have produced a graph that is approximately linear. In 1984 the consumer price index was 300%, and in 1987 it was 333%. Let y represent the consumer price index in year x, where $x = 0$ corresponds to 1980.

69. When a certain industrial pollutant is introduced into a river, the reproduction of catfish declines. In a given period of time, dumping three tons of the pollutant results in a fish population of 37,000. Also, 12 tons of pollutant produce a fish population of 28,000. Let y be the fish population when x tons of pollutant are introduced into the river.

70. In the snake *Lampropelbis polyzona*, total length y is related to tail length x in the domain 30 mm $\leq x \leq$ 200 mm by a linear relation. Find such a linear relation, if a snake 455 mm long has a 60-mm tail, and a 1050-mm snake has a 140-mm tail.

71. According to research done by the political scientist James March, if the Democrats win 45% of the two-party vote for the House of Representatives, they win 42.5% of the seats. If the Democrats win 55% of the vote, they win 67.5% of the seats. Let y be the percent of seats won, and x the percent of the two-party vote.

72. If the Republicans win 45% of the two-party vote, they win 32.5% of the seats (see Exercise 71). If they win 60% of the vote, they get 70% of the seats. Let y represent the percent of the seats, and x the percent of the vote.

73. When the Celsius temperature is 0°, the corresponding Fahrenheit temperature is 32°. When the Celsius temperature is 100°, the corresponding Fahrenheit temperature is 212°. Let C represent the Celsius temperature and F the Fahrenheit temperature. (Solve for F in terms of C; this is an exact linear relationship.)

74. Solve the equation found in Exercise 73 for C in terms of F.

75. For what temperature does Celsius equal Fahrenheit? See the results of Exercises 73–74.

76. Show that the line $y = x$ is the perpendicular bisector of the segment with endpoints (a, b) and (b, a).

3.3 ——————— **PARABOLAS: TRANSLATIONS AND APPLICATIONS**

In the previous section we studied linear relations, defined by equations of the form $y = mx + b$. Now we consider quadratic relations.

QUADRATIC RELATION	A **quadratic relation** in two variables is a relation that can be written in the form $$y = ax^2 + bx + c \quad \text{or} \quad x = ay^2 + by + c,$$ where a, b, and c are real numbers, and $a \neq 0$.

The graphs of quadratic relations are called **parabolas**. The simplest quadratic relation of the form $y = ax^2 + bx + c$ is $y = x^2$, with $a = 1$, $b = 0$, and $c = 0$, so this relation is graphed first.

■ **Example 1**

GRAPHING THE
SIMPLEST
QUADRATIC
RELATION

Graph $y = x^2$.

Set x equal to 0 in $y = x^2$ to get $y = 0$, which shows that the only intercept is at the origin. By plotting other selected points, as shown in the table accompanying Figure 3.16, we obtain the graph. The domain is $(-\infty, \infty)$ and the range is $[0, \infty)$. ■

x	y
0	0
1	1
2	4
3	9
-1	1
-2	4
-3	9

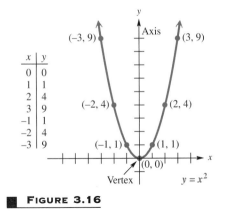

■ **FIGURE 3.16**

NOTE

The domain and the range of a parabola with a vertical axis, such as the one in Figure 3.16, can be determined by looking at the graph. Since the graph extends indefinitely to the right and to the left, we see that the domain is $(-\infty, \infty)$. Since the lowest point on the graph is $(0, 0)$, the minimum range value (y-value) is 0. The graph extends upward indefinitely, indicating that there is no maximum y-value, and so the range is $[0, \infty)$. (Domains and ranges of other types of relations can also be determined by observing their graphs.)

Notice in Figure 3.16 that the part of the graph in quadrant II is a "mirror image" of the part in quadrant I. We say that this graph is *symmetric with respect to the y-axis*. (More will be said about symmetry in the next section.) The line of symmetry for a parabola is the **axis** of the parabola. The lowest point on this parabola, the point $(0, 0)$, is called the **vertex** of the parabola.

Starting with $y = x^2$, there are several possible ways to get a more general expression:

$y = ax^2$ Multiply by a positive or negative coefficient.

$y = x^2 + k$ Add a positive or negative constant.

$y = (x - h)^2$ Replace x with $x - h$, where h is a constant.

$y = a(x - h)^2 + k$ Do all of the above.

The graph of each of these relations is still a parabola, but it is modified from that of $y = x^2$. The next few examples show how these changes modify the parabola. The first example shows the result of changing $y = x^2$ to $y = ax^2$.

164

■ *Example 2*

GRAPHING
RELATIONS OF THE
FORM y = ax²

Graph each relation.

(a) $y = 2x^2$

A table of selected ordered pairs is given with the graph in Figure 3.17. The y-values of the ordered pairs of this relation are twice as large as the corresponding y-values for the graph of $y = x^2$. This makes the graph rise more rapidly, so the parabola is narrower than the parabola for $y = x^2$, as can be seen in the figure.

x	y
2	8
1	2
0	0
−1	2
−2	8

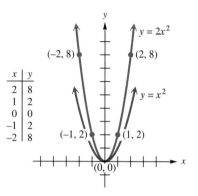

■ **FIGURE 3.17**

(b) $y = -\dfrac{1}{2}x^2$

The coefficient $-1/2$ causes the y-values to be closer to the x-axis than for $y = x^2$, making the graph broader than that of $y = x^2$. Because the y-values are negative for each nonzero x-value, this graph opens downward. Again the axis is the line $x = 0$, and the vertex, the *highest* point on the graph, is $(0, 0)$. The domain is $(-\infty, \infty)$ and the range is $(-\infty, 0]$. See Figure 3.18. ■

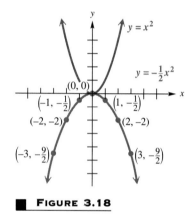

■ **FIGURE 3.18**

As Example 2 suggests, $|a|$ in $y = ax^2$ determines the width of a parabola, so that it is narrower than the graph of $y = x^2$ when $|a| > 1$ and broader than the graph of $y = x^2$ when $|a| < 1$.

The next two examples show how changing $y = x^2$ to $y = x^2 + k$, or to $y = (x - h)^2$, respectively, affects the graph of a parabola.

■ *Example 3*

GRAPHING A
RELATION OF THE
FORM y = x² + k

Graph $y = x^2 - 4$.

Each value of y will be 4 less than the corresponding value of $y = x^2$. This means that $y = x^2 - 4$ has the same shape as $y = x^2$ but is shifted 4 units down. See Figure 3.19. The vertex of the parabola (on this parabola, the lowest point) is at $(0, -4)$. The axis of the parabola is the vertical line $x = 0$. When the vertex of a parabola is shifted vertically, the intercepts are usually good choices for additional points to plot. Here,

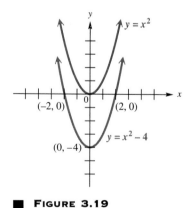

■ **FIGURE 3.19**

the y-intercept is -4, which is the y-value of the vertex. The x-intercepts are found by setting $y = 0$:

$$y = x^2 - 4$$
$$0 = x^2 - 4$$
$$4 = x^2$$

$$x = 2 \quad \text{or} \quad x = -2.$$

The x-intercepts are 2 and -2. ■

The vertical shift of the graph in Example 3 is called a **translation.** Example 4 below shows a horizontal translation, which is a shift to the right or left.

■ *Example 4*
GRAPHING A
RELATION OF THE
FORM $y = (x - h)^2$

Graph $y = (x - 4)^2$.

Comparing the two tables of ordered pairs shown in Figure 3.20, for $y = x^2$ and $y = (x - 4)^2$, indicates that this parabola is translated 4 units to the right as compared to the graph of $y = x^2$. For example, the vertex is $(4, 0)$ instead of $(0, 0)$, and $x = -2$ in $y = x^2$ corresponds to the same y-value as $x = 2$ in $y = (x - 4)^2$, a difference of $2 - (-2) = 4$ units. The axis of $y = (x - 4)^2$ is the vertical line $x = 4$. See Figure 3.20. ■

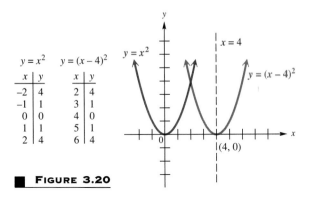

■ FIGURE 3.20

CAUTION

Errors frequently occur when horizontal translations are involved. In order to determine the direction and magnitude of horizontal translations, find the value that would cause the expression $x - h$ to equal 0. For example, the graph of $y = (x - 5)^2$ would be shifted 5 units to the *right,* because $+5$ would cause $x - 5$ to equal 0. On the other hand, the graph of $y = (x + 4)^2$ would be shifted 4 units to the *left,* because -4 would cause $x + 4$ to equal 0.

A combination of all the transformations illustrated in Examples 2, 3, and 4 is shown in the next example.

166

■ *Example 5*

GRAPHING A
RELATION OF THE
FORM
$y = a(x - h)^2 + k$

Graph $y = -(x + 3)^2 + 1$.

This parabola is translated 3 units to the left and 1 unit up. Because of the negative sign, it opens downward, so that the vertex, the point $(-3, 1)$, is the *highest* point on the graph. The axis is the line $x = -3$. The y-intercept is $y = -(0 + 3)^2 + 1 = -8$. By symmetry about the axis $x = -3$, the point $(-6, -8)$ also is on the graph. The x-intercepts are found by solving the equation

$$0 = -(x + 3)^2 + 1$$
$$0 = -(x^2 + 6x + 9) + 1 \qquad \text{Square the binomial.}$$
$$0 = -x^2 - 6x - 9 + 1 \qquad \text{Distributive property}$$
$$0 = -x^2 - 6x - 8$$
$$0 = x^2 + 6x + 8 \qquad \text{Multiply by } -1.$$
$$0 = (x + 2)(x + 4). \qquad \text{Factor.}$$

from which $x = -2$ or $x = -4$. The graph is shown in Figure 3.21. Notice from the graph that the domain is $(-\infty, \infty)$ and the range is $(-\infty, 1]$. The y-value of the vertex determines the range. ■

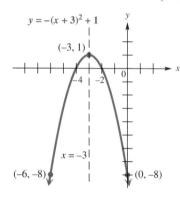

■ **FIGURE 3.21**

Examples 2–5 suggest the following generalizations.

GRAPH OF A
PARABOLA

The graph of

$$y = a(x - h)^2 + k, \quad \text{where } a \neq 0,$$

(a) is a parabola with vertex (h, k), and the vertical line $x = h$ as axis;
(b) opens upward if $a > 0$ and downward if $a < 0$;
(c) is broader than $y = x^2$ if $0 < |a| < 1$ and narrower than $y = x^2$ if $|a| > 1$.

Given the relation $y = ax^2 + bx + c$, where a, b, and c are real numbers and $a \neq 0$, the process of *completing the square* (discussed in Chapter 2) can be used to change $ax^2 + bx + c$ to the form $a(x - h)^2 + k$, so that the vertex and axis may be identified. Follow the steps given in the next example.

■ *Example 6*

GRAPHING A
PARABOLA BY
COMPLETING THE
SQUARE

Graph $y = -3x^2 - 2x + 1$.

Our goal is to write the equation in the form $y = a(x - h)^2 + k$. We may start by dividing both sides by -3 to get

$$\frac{y}{-3} = x^2 + \frac{2}{3}x - \frac{1}{3}.$$

Now complete the procedure, as explained in Section 2.4.

$$\frac{y}{-3} + \frac{1}{3} = x^2 + \frac{2}{3}x \qquad \text{Add } \frac{1}{3} \text{ to both sides.}$$

$$\frac{y}{-3} + \frac{1}{3} + \frac{1}{9} = x^2 + \frac{2}{3}x + \frac{1}{9} \qquad \left[\frac{1}{2}\left(\frac{2}{3}\right)\right]^2 = \frac{1}{9}, \text{ so add } \frac{1}{9} \text{ to both sides.}$$

$$\frac{y}{-3} + \frac{4}{9} = \left(x + \frac{1}{3}\right)^2 \qquad \begin{array}{l}\text{Combine terms on the left,}\\ \text{factor on the right.}\end{array}$$

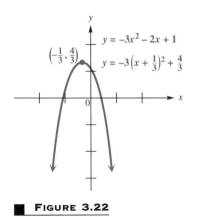

FIGURE 3.22

$$\frac{y}{-3} = \left(x + \frac{1}{3}\right)^2 - \frac{4}{9} \qquad \text{Subtract } \frac{4}{9}.$$

$$y = -3\left(x + \frac{1}{3}\right)^2 + \frac{4}{3} \qquad \text{Multiply by } -3.$$

or

$$y = -3\left[x - \left(-\frac{1}{3}\right)\right]^2 + \frac{4}{3}$$

Now the equation of the parabola is written in the form $y = a(x - h)^2 + k$, and this rewritten equation shows that the axis of the parabola is the vertical line $x = -1/3$ and that the vertex is $(-1/3, 4/3)$. Use these results, together with the intercepts and additional ordered pairs as needed, to get the graph in Figure 3.22. From the graph, the domain of the relation is $(-\infty, \infty)$ and the range is $(-\infty, 4/3]$. ∎

A formula for the vertex of the graph of the quadratic relation $y = ax^2 + bx + c$ can be found by completing the square for the general form of the equation.

$$y = ax^2 + bx + c \quad (a \neq 0)$$

$$\frac{y}{a} = x^2 + \frac{b}{a}x + \frac{c}{a} \qquad \text{Divide by } a.$$

$$\frac{y}{a} - \frac{c}{a} = x^2 + \frac{b}{a}x \qquad \text{Subtract } \frac{c}{a}.$$

$$\frac{y}{a} - \frac{c}{a} + \frac{b^2}{4a^2} = x^2 + \frac{b}{a}x + \frac{b^2}{4a^2} \qquad \text{Add } \frac{b^2}{4a^2}.$$

$$\frac{y}{a} + \frac{b^2 - 4ac}{4a^2} = \left(x + \frac{b}{2a}\right)^2 \qquad \begin{array}{l}\text{Combine terms on left} \\ \text{and factor on right.}\end{array}$$

$$\frac{y}{a} = \left(x + \frac{b}{2a}\right)^2 - \frac{b^2 - 4ac}{4a^2} \qquad \text{Get } y \text{ term alone on the left.}$$

$$y = a\left(x + \frac{b}{2a}\right)^2 + \frac{4ac - b^2}{4a} \qquad \text{Multiply by } a.$$

$$y = a\underbrace{\left[x - \left(-\frac{b}{2a}\right)\right]}_{h}^2 + \underbrace{\frac{4ac - b^2}{4a}}_{k}$$

The final equation shows that the vertex (h, k) can be expressed in terms of a, b, and c. However, it is not necessary to memorize the expression for k, since it can be obtained by replacing x with $-b/(2a)$.

VERTEX OF A PARABOLA
$(y = ax^2 + bx + c)$
The x-value of the vertex of the parabola $y = ax^2 + bx + c$, where $a \neq 0$, is $-\dfrac{b}{2a}$.

■ *Example 7*

USING THE VERTEX
FORMULA

Use the formula above to find the vertex of the parabola $y = 2x^2 - 4x + 3$.

In this equation, $a = 2$, $b = -4$, and $c = 3$. By the formula given above, the x-value of the vertex of the parabola is

$$x = -\frac{b}{2a} = -\frac{-4}{2(2)} = 1.$$

The y-value is found by substituting 1 for x into the equation $y = 2x^2 - 4x + 3$ to get $y = 2(1)^2 - 4(1) + 3 = 1$, so the vertex is $(1, 1)$. ■

APPLICATION OF QUADRATIC RELATIONS Quadratic relations can be applied in situations as illustrated in the next example.

PROBLEM SOLVING

The fact that the vertex of a vertical parabola is the highest or lowest point on the graph makes equations of the form $y = ax^2 + bx + c$ important in problems where the maximum or minimum value of some quantity is to be found. When $a < 0$, the y-value of the vertex gives the maximum value of y and the x-value tells where it occurs. Similarly, when $a > 0$, the y-value of the vertex gives the minimum y-value. ■

■ *Example 8*

FINDING THE
VERTEX IN AN
APPLICATION

Ms. Whitney owns and operates Aunt Emma's Pie Shop. She has hired a consultant to analyze her business operations. The consultant tells her that her profit P in dollars is given by

$$P = 120x - x^2,$$

where x is the number of units of pies that she makes. How many units of pies should be made in order to maximize the profit? What is the maximum possible profit?

The profit relation P can be rewritten as

$$P = -x^2 + 120x + 0,$$

with $a = -1$, $b = 120$, and $c = 0$. The graph of this relation will be a parabola opening downward, so that the vertex, of the form (x, P), will be the highest point on the graph. To find the vertex, use the fact that $x = -b/(2a)$:

$$x = \frac{-120}{2(-1)} = 60.$$

Let $x = 60$ in the equation to find the value of P at the vertex.

$$P = 120x - x^2$$
$$P = 120(60) - 60^2$$
$$P = 3600$$

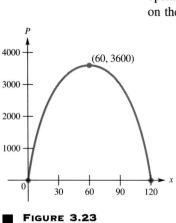

■ FIGURE 3.23

The vertex is $(60, 3600)$. Figure 3.23 shows the portion of the profit graph located in quadrant I. (Why is quadrant I the only one of interest here?) The maximum profit of $3600 occurs when 60 units of pies are made. In this case, profit increases as more and more pies are made up to 60 units and then decreases as more pies are made past this point. ■

In this section, we started by defining a quadratic relation $y = ax^2 + bx + c$ and, by point-plotting, found the graph, which we called a parabola. It is possible to start with a parabola, a set of points in the plane, and find the corresponding relation, using the formal geometric definition of a parabola.

GEOMETRIC DEFINITION OF A PARABOLA Geometrically, a parabola is defined as the set of all points in a plane that are equally distant from a fixed point and a fixed line not containing the point. The point is called the **focus** and the line is the **directrix.** The line through the focus and perpendicular to the directrix is the axis of the parabola. The point on the axis that is equally distant from the focus and the directrix is the vertex of the parabola.

The parabola in Figure 3.24 has the point $(0, p)$ as focus and the line $y = -p$ as directrix. The vertex is $(0, 0)$. Let (x, y) be any point on the parabola. The distance from (x, y) to the directrix is $|y - (-p)|$, while the distance from (x, y) to $(0, p)$ is $\sqrt{(x - 0)^2 + (y - p)^2}$. Since (x, y) is equally distant from the directrix and the focus,

$$|y - (-p)| = \sqrt{(x - 0)^2 + (y - p)^2}.$$

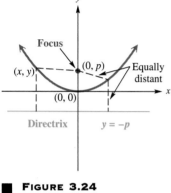

Square both sides, getting

$$(y + p)^2 = x^2 + (y - p)^2$$
$$y^2 + 2py + p^2 = x^2 + y^2 - 2py + p^2,$$
$$4py = x^2,$$

FIGURE 3.24

the equation of the parabola with focus $(0, p)$ and directrix $y = -p$. Solving $4py = x^2$ for y gives

$$y = \frac{1}{4p} x^2,$$

so that $1/(4p) = a$ when the equation is written in the form $y = ax^2 + bx + c$.

This result could be extended to a parabola with vertex at (h, k), focus p units above (h, k), and directrix p units below (h, k), or to a parabola with vertex at (h, k), focus p units below (h, k), and directrix p units above (h, k).

The geometric properties of parabolas lead to many practical applications. For example, if a light source is placed at the focus of a parabolic reflector, as in Figure 3.25, light rays reflect parallel to the axis, making a spotlight or flashlight. The process also works in reverse. Light rays from a distant source come in parallel to the axis and are reflected to a point at the focus. (If such a reflector is aimed at the sun, a temperature of several thousand degrees may be obtained.) This use of parabolic reflection is seen in the satellite dishes used to pick up signals from communications satellites.

Parabolic reflector

FIGURE 3.25

HORIZONTAL PARABOLAS The directrix of a parabola could be the *vertical* line $x = -p$, where $p > 0$, with focus on the x-axis at $(p, 0)$, producing a parabola opening to the right. This parabola is the graph of the relation $y^2 = 4px$ or $x = [1/(4p)]y^2$. The next examples show the graphs of horizontal parabolas with equations of the form $x = ay^2 + by + c$.

■ *Example 9*

GRAPHING A
HORIZONTAL
PARABOLA

Graph $x = y^2$.

The equation $x = y^2$ can be obtained from $y = x^2$ by exchanging x and y. Choosing values of y and finding the corresponding values of x gives the parabola in Figure 3.26. The graph of $x = y^2$, shown in red, is symmetric with respect to the line $y = 0$ and has vertex at $(0, 0)$. For comparison, the graph of $y = x^2$ is shown in blue. These graphs are mirror images of each other with respect to the line $y = x$. From the graph, the domain of $x = y^2$ is $[0, \infty)$, and the range is $(-\infty, \infty)$. ■

■ **FIGURE 3.26**

NOTE

The domain and the range of a horizontal parabola, such as $x = y^2$ in Figure 3.26, can be determined by looking at the graph. Since the vertex $(0, 0)$ has the smallest x-value of any point on the graph, and the graph extends indefinitely to the right, the domain is $[0, \infty)$. Because the graph extends upward and downward indefinitely, the range is $(-\infty, \infty)$.

■ *Example 10*

COMPLETING THE
SQUARE TO GRAPH
A HORIZONTAL
PARABOLA

Graph $x = 2y^2 + 6y + 5$.

To write this equation in the form $x = a(y - k)^2 + h$, complete the square on y as follows:

$$x = 2y^2 + 6y + 5$$

$$\frac{x}{2} = y^2 + 3y + \frac{5}{2} \qquad \text{Divide by 2.}$$

$$\frac{x}{2} - \frac{5}{2} = y^2 + 3y \qquad \text{Subtract } \frac{5}{2}.$$

$$\frac{x}{2} - \frac{5}{2} + \frac{9}{4} = y^2 + 3y + \frac{9}{4} \qquad \text{Add } \frac{9}{4}.$$

$$\frac{x}{2} - \frac{1}{4} = \left(y + \frac{3}{2}\right)^2 \qquad \text{Combine terms; factor.}$$

$$\frac{x}{2} = \left(y + \frac{3}{2}\right)^2 + \frac{1}{4} \qquad \text{Add } \frac{1}{4}.$$

$$x = 2\left(y + \frac{3}{2}\right)^2 + \frac{1}{2}. \qquad \text{Multiply by 2.}$$

As this result shows, the vertex of the parabola is the point (1/2, −3/2). The axis is the horizontal line

$$y + \frac{3}{2} = 0 \quad \text{or} \quad y = -\frac{3}{2}.$$

There is no y-intercept, since the vertex is on the right of the y-axis and the graph opens to the right. However, the x-intercept is

$$x = 2(0)^2 + 6(0) + 5 = 5.$$

Using the vertex, the axis of symmetry, and the x-intercept, and plotting a few additional points gives the graph in Figure 3.27. The domain is [1/2, ∞) and the range is (−∞, ∞). ■

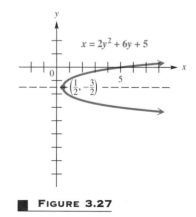

■ **FIGURE 3.27**

The vertex of a horizontal parabola can also be found by using the values of a and b in $x = ay^2 + by + c$.

VERTEX OF A PARABOLA $(x = ay^2 + by + c)$	The y-value of the vertex of the parabola $$x = ay^2 + by + c, \text{ where } a \neq 0,$$ is $-\dfrac{b}{2a}$. The x-value is found by substitution of $-\dfrac{b}{2a}$ for y.

CAUTION Be careful when using the two vertex formulas of this section. It is essential that you recognize whether the parabola is a vertical parabola or a horizontal one, so that you can decide whether $-b/(2a)$ represents the x- or y-coordinate of the vertex. (It always represents the coordinate of the variable that is squared.)

The types of parabolas discussed in this section are summarized on the following page.

GRAPHS OF PARABOLAS	Equation	Graph

GRAPHS OF PARABOLAS

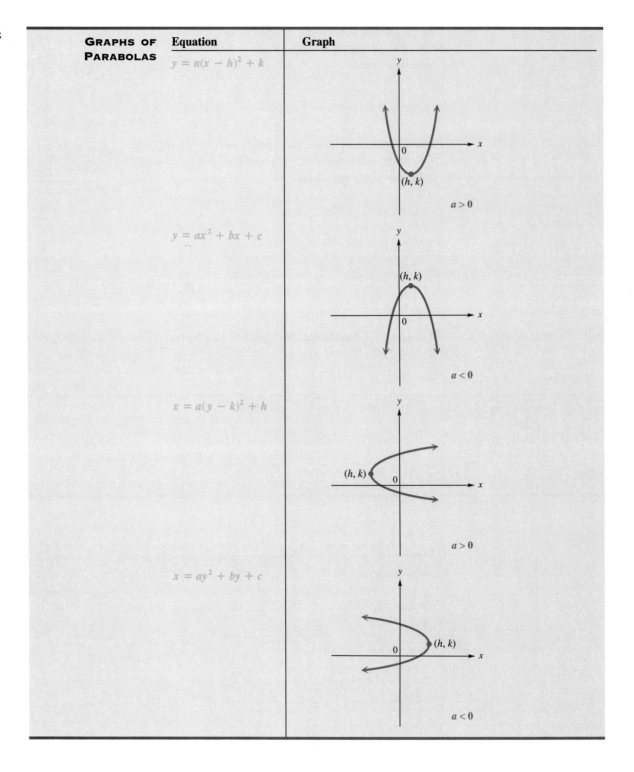

$y = a(x - h)^2 + k$

$a > 0$

$y = ax^2 + bx + c$

$a < 0$

$x = a(y - k)^2 + h$

$a > 0$

$x = ay^2 + by + c$

$a < 0$

1. Graph the following on the same coordinate system.

   (a) $y = 2x^2$ (b) $y = 3x^2$ (c) $y = \dfrac{1}{2}x^2$ (d) $y = \dfrac{1}{3}x^2$

   (e) How does the coefficient of x^2 affect the shape of the graph?

2. Graph the following on the same coordinate system.
   (a) $y = x^2 + 2$ (b) $y = x^2 - 1$ (c) $y = x^2 + 1$ (d) $y = x^2 - 2$
   (e) How do these graphs differ from the graph of $y = x^2$?

3. Graph the following on the same coordinate system.
   (a) $y = (x - 2)^2$ (b) $y = (x + 1)^2$ (c) $y = (x + 3)^2$ (d) $y = (x - 4)^2$
   (e) How do these graphs differ from the graph of $y = x^2$?

4. Match each equation with the description of the parabola that is its graph.
   (a) $y = (x - 4)^2 - 2$ **A.** vertex $(2, -4)$, opens down
   (b) $y = (x - 2)^2 - 4$ **B.** vertex $(2, -4)$, opens up
   (c) $y = -(x - 4)^2 - 2$ **C.** vertex $(4, -2)$, opens down
   (d) $y = -(x - 2)^2 - 4$ **D.** vertex $(4, -2)$, opens up
   (e) $x = (y - 4)^2 - 2$ **E.** vertex $(-2, 4)$, opens left
   (f) $x = (y - 2)^2 - 4$ **F.** vertex $(-2, 4)$, opens right
   (g) $x = -(y - 4)^2 - 2$ **G.** vertex $(-4, 2)$, opens left
   (h) $x = -(y - 2)^2 - 4$ **H.** vertex $(-4, 2)$, opens right

5. For the graph of $y = a(x - h)^2 + k$, in what quadrant is the vertex if:
   (a) $h < 0,\ \ k < 0$;   (b) $h < 0,\ \ k > 0$;   (c) $h > 0,\ \ k < 0$;   (d) $h > 0,\ \ k > 0$?

6. Repeat parts (a)–(d) of Exercise 5 for the graph of $x = a(y - k)^2 + h$.

Graph each of the following parabolas. Give the vertex, axis, domain, and range of each. See Examples 1–7.

7. $y = (x - 2)^2$ 8. $y = (x + 4)^2$ 9. $y = (x + 3)^2 - 4$

10. $y = (x - 5)^2 - 4$ 11. $y = -2(x + 3)^2 + 2$ 12. $y = -3(x - 2)^2 + 1$

13. $y = -\dfrac{1}{2}(x + 1)^2 - 3$ 14. $y = \dfrac{2}{3}(x - 2)^2 - 1$ 15. $y = x^2 - 2x + 3$

16. $y = x^2 + 6x + 5$ 17. $y = 2x^2 - 4x + 5$ 18. $y = -3x^2 + 24x - 46$

19. Explain what causes the graph of a parabola with equation of the form $y = ax^2$ to be wider or narrower than the graph of $y = x^2$.

20. Explain what causes the graph of a parabola with equation of the form $x = ay^2$ to open to the left or the right.

Graph each horizontal parabola. Give the vertex, axis, domain, and range of each. See Examples 9 and 10.

21. $x = y^2 + 2$ 22. $x = -y^2$ 23. $x = (y + 1)^2$ 24. $x = (y - 3)^2$

25. $x = (y + 2)^2 - 1$ 26. $x = (y - 4)^2 + 2$ 27. $x = -2(y + 3)^2$ 28. $x = \dfrac{2}{3}(y - 3)^2 + 2$

29. $x = y^2 + 2y - 8$ 30. $x = -4y^2 - 4y - 3$

Solve each application of quadratic relations.
See Example 8.

31. Glenview Community College wants to construct a rectangular parking lot on land bordered on one side by a highway. It has 320 ft of fencing with which to fence off the other three sides. What should be the dimensions of the lot if the enclosed area is to be a maximum? (*Hint:* Let x represent the width of the lot and let $320 - 2x$ represent the length. Graph the area parabola, $A = x(320 - 2x)$, and investigate the vertex.)

32. What would be the maximum area that could be enclosed by the college's 320 ft of fencing if it decided to close the entrance by enclosing all four sides of the lot? (See Exercise 31.)

33. Ed Doskey owns a sandwich company. By studying data concerning his past costs, he has found that the cost of operating his company is given by
$$C = 2x^2 - 1200x + 180,100,$$
where C is the daily cost in dollars to make x sandwiches. Find the number of sandwiches Ed must sell to minimize the cost. What is the minimum cost?

34. The revenue of a charter bus company depends on the number of unsold seats. If the revenue R, in dollars, is given by
$$R = 5000 + 50x - x^2,$$
where x is the number of unsold seats, find the maximum revenue and the number of unsold seats which produce maximum revenue.

35. The number of mosquitoes M, in millions, in a certain area of Alabama depends on the July rainfall x, in inches, approximately as follows.
$$M = 10x - x^2$$
Find the rainfall that will produce the maximum number of mosquitoes.

36. Refer to Example 5 in Section 2.5. After how many seconds will the projectile reach its maximum height? What is the maximum height?

37. If an object is thrown upward with an initial velocity of 32 ft per sec, then its height h in ft after t sec is given by
$$h = 32t - 16t^2.$$
After how many seconds does it reach its maximum height? Find the maximum height attained by the object. Find the number of seconds it takes the object to hit the ground.

38. A charter flight charges a fare of $200 per person plus $4 per person for each unsold seat on the plane. If the plane holds 100 passengers, and if x represents the number of unsold seats, find the following.
(a) an expression for the total revenue R, in dollars, received for the flight (*Hint:* Multiply the number of people flying, $100 - x$, by the price per ticket.)
(b) the graph for the expression of part (a)
(c) the number of unsold seats that will produce the maximum revenue
(d) the maximum revenue

39. The demand for a certain type of cosmetic is given by
$$p = 500 - x,$$
where p is the price when x units are demanded.
(a) Find the revenue R, in dollars, that would be obtained at a price of x. (*Hint:* Revenue = Demand × Price.)
(b) Graph the revenue relation R.
(c) From the graph of the revenue relation, estimate the price that will produce maximum revenue.
(d) What is the maximum revenue?

40. For the months of June through October, the percent of maximum possible chlorophyll production in a leaf is approximated by C, where
$$C = 10x + 50.$$
Here x is time in months with $x = 1$ representing June. From October through December, C is approximated by
$$C = -20(x - 5)^2 + 100,$$
with x as above. Find the percent of maximum possible chlorophyll production in each of the following months:
(a) June   **(b)** July   **(c)** September
(d) October   **(e)** November   **(f)** December.

41. Use your results from Exercise 40 to sketch a graph of $y = C$ from June through December. In what month is chlorophyll production at a maximum?

42. The temperature T at which water boils varies with the elevation E, where T is in degrees Celsius and E is in meters above sea level, according to the equation
$$E = 580(100 - T)^2 + 1000(100 - T).$$
(a) Determine the boiling point at each of the following elevations. (*Hint:* Let $X = 100 - T$. Solve for X, and then solve for T.)

1. Mount McKinley (Alaska), 6194 meters
2. Mount Everest (Nepal-Tibet), 8848 meters
3. Mount Andrew Jackson (Antarctica), 4191 meters
4. Mount Mauna Loa (Hawaii), 4170 meters

(b) Discuss the interpretation of letting $T = 100$ in the formula.

(c) What is the boiling point of water in Denver, elevation 1600 meters?

Use the geometric definition of parabola given in the text to find the equation of each parabola having the given point as focus.

43. $(0, 3)$, directrix $y = -3$

44. $(0, -5)$, directrix $y = 5$

45. $(-2, 0)$, directrix $x = 2$

46. $(8, 0)$, directrix $x = -8$

47. $(3, 6)$, vertex $(3, 4)$

48. $(-5, 2)$, vertex $(-5, 5)$

Just as the discriminant of the quadratic equation $ax^2 + bx + c = 0$ $(a \neq 0)$ determines the number of real solutions of the equation (see Section 2.4), the discriminant $b^2 - 4ac$ for the vertical parabola $y = ax^2 + bx + c$ determines the number of x-intercepts. If the discriminant is 0, there is one x-intercept; if it is positive, there are two x-intercepts; if it is negative, there are no x-intercepts.

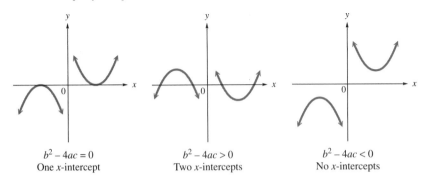

$b^2 - 4ac = 0$
One *x*-intercept

$b^2 - 4ac > 0$
Two *x*-intercepts

$b^2 - 4ac < 0$
No *x*-intercepts

49. Find a value of c so that $y = x^2 - 10x + c$ has exactly one *x*-intercept.

50. Find b so that $y = x^2 + bx + 9$ has exactly one *x*-intercept.

The figures below show several possible graphs of $y = ax^2 + bx + c$. For the restrictions on a, b, and c in Exercises 51–54, select the possible corresponding graph.

51. $a < 0$, $b^2 - 4ac = 0$

52. $a > 0$, $b^2 - 4ac < 0$

53. $a > 0$, $b^2 - 4ac > 0$

54. $a > 0$, $b^2 - 4ac = 0$

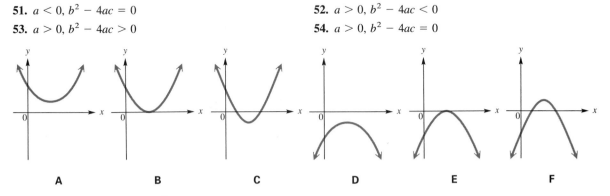

A   **B**   **C**   **D**   **E**   **F**

55. Graph the parabola $y = 2x^2 + 5x - 3$.

    (a) Use the graph to find the solution of the quadratic inequality $2x^2 + 5x - 3 < 0$.

    (b) Use the graph to find the solution of the quadratic inequality $2x^2 + 5x - 3 > 0$.

    (c) Use the graph to find the solutions of the quadratic equation $2x^2 + 5x - 3 = 0$.

56. Find the equation of the vertical parabola having *x*-intercepts 2 and 5, and *y*-intercept 5.

⬛ **3.4** ——————— **THE CIRCLE AND SYMMETRY**

In Section 3.3 we studied the graphs of quadratic relations in which one of the variables was first-degree and the other was second-degree. Three important kinds of relations involve equations in which both variables are second-degree: *circles, ellipses,* and *hyperbolas.* In this section we look at circles, and in the next we study the other two.

We will start with the geometric definition of a circle and use the distance formula to derive the equation of the corresponding relation. This is similar to the way we derived the equation of a quadratic relation from the geometric definition of a parabola in Section 3.3. From now on we will refer to both the relation and its graph as the circle.

CIRCLES By definition, a **circle** is the set of all points in a plane that lie a given distance from a given point. The given distance is the **radius** of the circle and the given point is the **center.** Since a circle is a set of points, it corresponds to a relation. The equation of a circle can be found from its definition by using the distance formula.

Figure 3.28 shows a circle of radius 3 with center at the origin. To find the equation of this circle, let (x, y) be any point on the circle. The distance between (x, y) and the center of the circle, $(0, 0)$, is given by

$$\sqrt{(x - 0)^2 + (y - 0)^2}.$$

Since this distance equals the radius, 3,

$$\sqrt{(x - 0)^2 + (y - 0)^2} = 3$$
$$\sqrt{x^2 + y^2} = 3$$
$$x^2 + y^2 = 9.$$

As suggested by Figure 3.28, the domain of the relation is $[-3, 3]$, and the range of the relation is $[-3, 3]$.

⬛ **FIGURE 3.28**

■ *Example 1*
FINDING THE
EQUATION OF A
CIRCLE

Find an equation for the circle having radius 6 and center at $(-3, 4)$. Graph the circle.

This circle is shown in Figure 3.29. Its equation can be found by using the distance formula. Let (x, y) be any point on the circle. The distance from (x, y) to $(-3, 4)$ is given by

$$\sqrt{[x - (-3)]^2 + (y - 4)^2} = \sqrt{(x + 3)^2 + (y - 4)^2}.$$

This same distance is given by the radius, 6. Therefore,

$$\sqrt{(x + 3)^2 + (y - 4)^2} = 6$$

or $\quad (x + 3)^2 + (y - 4)^2 = 36.$

The domain is $[-9, 3]$, and the range is $[-2, 10]$, as seen in Figure 3.29. ∎

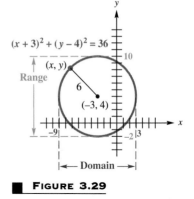

FIGURE 3.29

Generalizing from the work in Example 1 gives the following result.

CENTER-RADIUS FORM OF THE EQUATION OF A CIRCLE	The circle with center (h, k) and radius r has equation $$(x - h)^2 + (y - k)^2 = r^2,$$ the **center-radius form** of the equation of a circle.

In particular, many circles that we encounter have their centers at the origin, as shown in Figure 3.30.

EQUATION OF A CIRCLE WITH CENTER AT THE ORIGIN	A circle with center $(0, 0)$ and radius r has equation $$x^2 + y^2 = r^2.$$

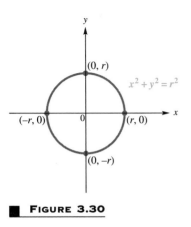

FIGURE 3.30

Starting with the center-radius form of the equation of a circle, $(x - h)^2 + (y - k)^2 = r^2$, and squaring $x - h$ and $y - k$ gives an equation of the form

$$x^2 + y^2 + cx + dy + e = 0, \qquad (*)$$

where c, d, and e are real numbers. This result is the **general form of the equation of a circle.** Also, starting with an equation in the form of (*), the process of completing the square can be used to get an equation of the form

$$(x - h)^2 + (y - k)^2 = m$$

for some number m. If $m > 0$, then $r^2 = m$, and the equation represents a circle with radius $\sqrt{m}$. If $m = 0$, then the equation represents the single point (h, k). If $m < 0$, no points satisfy the equation.

178

■ *Example 2*

FINDING THE
CENTER AND THE
RADIUS BY
COMPLETING THE
SQUARE

Find the center and radius for $x^2 - 6x + y^2 + 10y + 25 = 0$.

Since this equation has the form of equation (*) above, it represents either a circle, a single point, or no points at all. To decide which, complete the square on x and y separately. Start with

$$(x^2 - 6x \quad) + (y^2 + 10y \quad) = -25.$$

Half of -6 is -3, and $(-3)^2 = 9$. Also, half of 10 is 5, and $5^2 = 25$. Add 9 and 25 on the left, and to compensate, add 9 and 25 on the right:

$$(x^2 - 6x + 9) + (y^2 + 10y + 25) = -25 + 9 + 25$$
$$(x - 3)^2 + (y + 5)^2 = 9.$$

Since $9 > 0$, the equation represents a circle with center at $(3, -5)$ and radius $\sqrt{9} = 3$. ■

In Section 3.3 we saw how the graph of $y = x^2$ can be translated up, down, right, or left by adding or subtracting constants on the right side. This procedure of translation can be extended to other types of graphs. The next example shows how this is done with a circle.

■ *Example 3*

EXAMINING A
TRANSLATION OF A
CIRCLE

(a) Graph $x^2 + y^2 = 16$.

This is the equation of a circle with center at $(0, 0)$ and radius $\sqrt{16} = 4$. See Figure 3.31(a).

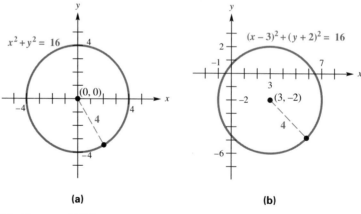

(a)   (b)

■ FIGURE 3.31

(b) Graph $(x - 3)^2 + (y + 2)^2 = 16$.

By the center-radius form of the equation of a circle, this relation is a circle with center at $(3, -2)$ and radius 4. The graph is shown in Figure 3.31(b).

(c) Compare the graphs in Figures 3.31 (a) and (b).

The graph of $(x - 3)^2 + (y + 2)^2 = 16$ is the same as that of $x^2 + y^2 = 16$, except that it has been translated 3 units to the right (due to $(x - 3)$) and 2 units down (due to $(y + 2) = y - (-2)$). ■

Generalizing from Example 3, we can make the following observations concerning the graph of $(x - h)^2 + (y - k)^2 = r^2 \ (r > 0)$.

TRANSLATIONS OF A CIRCLE	The graph of the circle $(x - h)^2 + (y - k)^2 = r^2 \ (r > 0)$ is the same as the graph of $x^2 + y^2 = r^2$, with the following translations: **1.** h units to the right if $h > 0$; $\lvert h \rvert$ units to the left if $h < 0$ **2.** k units up if $k > 0$; $\lvert k \rvert$ units down if $k < 0$

∎ IN SIMPLEST TERMS

11

Seismologists can locate the epicenter of an earthquake by determining the intersection of three circles. The radii of these circles represent the distances from the epicenter to each of three receiving stations. The centers of the circles represent the receiving stations.

Suppose receiving stations *A, B,* and *C* are located on a coordinate plane at the points (1, 4), (−3, −1), and (5, 2). Let the distance from the earthquake epicenter to each station be 2 units, 5 units, and 4 units, respectively. Where on the coordinate plane is the epicenter located?

Graphically, it appears that the epicenter is located at (1, 2). To check this algebraically, determine the equation for each circle and substitute $x = 1$ and $y = 2$.

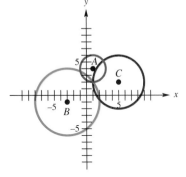

Station 1:

$(x - 1)^2 + (y - 4)^2 = 4$

$(1 - 1)^2 + (2 - 4)^2 = 4$

$0 + 4 = 4$

$4 = 4$

Station 2:

$(x + 3)^2 + (y + 1)^2 = 25$

$(1 + 3)^2 + (2 + 1)^2 = 25$

$16 + 9 = 25$

$25 = 25$

Station 3:

$(x - 5)^2 + (y - 2)^2 = 16$

$(1 - 5)^2 + (2 - 2)^2 = 16$

$16 + 0 = 16$

$16 = 16$

Thus, we can be sure that the epicenter lies at (1, 2).

SOLVE EACH PROBLEM

A. Show algebraically that if three receiving stations at (1, 4), (−6, 0), and (5, −2) record distances to an earthquake epicenter of 4 units, 5 units, and 10 units, respectively, that the epicenter would lie at (−3, 4).

B. Three receiving stations record the presence of an earthquake. The location of the receiving center and the distance to the epicenter are contained in the following three equations: $(x - 2)^2 + (y - 1)^2 = 25$, $(x + 2)^2 + (y - 2)^2 = 16$ and $(x - 1)^2 + (y + 2)^2 = 9$. Graph the circles and determine the location of the earthquake epicenter.

ANSWER   B. The epicenter is at (−2, −2).

SYMMETRY In Section 3.3 we discussed the axis of symmetry of a parabola. For example, the axis of symmetry of the graph of $y = -(x + 3)^2 + 1$ is the vertical line $x = -3$ (see Figure 3.21 in Section 3.3), because if the graph were folded along this line, the two halves would coincide. The idea of symmetry is helpful in drawing graphs, and can be extended to other graphs as well.

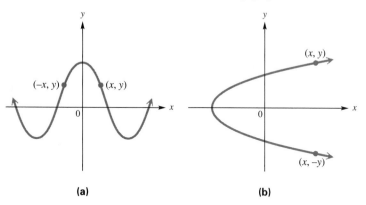

(a)   **(b)**

■ **FIGURE 3.32**

Figure 3.32(a) shows a graph that is *symmetric with respect to the y-axis.* As suggested by Figure 3.32(a), for a graph to be symmetric with respect to the y-axis, the point $(-x, y)$ must be on the graph whenever (x, y) is on the graph.

Similarly, if the graph in Figure 3.32(b) were folded in half along the x-axis, the portion from the top would exactly match the portion from the bottom. Such a graph is *symmetric with respect to the x-axis.* As the graph suggests, symmetry with respect to the x-axis means that the point $(x, -y)$ must be on the graph whenever the point (x, y) is on the graph.

The following tests tell when a graph is symmetric with respect to the x-axis or y-axis.

SYMMETRY WITH RESPECT TO AN AXIS	The graph of an equation is **symmetric with respect to the y-axis** if the replacement of x with $-x$ results in an equivalent equation.
	The graph of an equation is **symmetric with respect to the x-axis** if the replacement of y with $-y$ results in an equivalent equation.

■ *Example 4*

TESTING FOR SYMMETRY WITH RESPECT TO AN AXIS

Test for symmetry with respect to the x-axis or y-axis.

(a) $y = x^2 + 4$

Replace x with $-x$:

$$y = x^2 + 4$$

becomes $y = (-x)^2 + 4 = x^2 + 4.$

The result is the same as the original equation, so the graph (shown in Figure 3.33) is symmetric with respect to the y-axis. The graph is *not* symmetric with respect to the x-axis, since replacing y with $-y$ gives

$$-y = x^2 + 4,$$

which is not equivalent to the original equation.

$y = x^2 + 4$

Symmetric with respect to y-axis

■ **FIGURE 3.33**

(b) $x = y^2 - 3$

Replace y with $-y$ to get $x = (-y)^2 - 3 = y^2 - 3$, the same as the original equation. The graph is symmetric with respect to the x-axis, as shown in Figure 3.34. The graph is not symmetric with respect to the y-axis because $-x = y^2 - 3$ is not equivalent to $x = y^2 - 3$.

(c) $2x + y = 4$

Replace x with $-x$ and then replace y with $-y$; in neither case does an equivalent equation result. This graph is symmetric neither with respect to the x-axis nor the y-axis. See Figure 3.35. ■

Another kind of symmetry is found when it is possible to rotate a graph 180° about the origin and have the result coincide exactly with the original graph. Symmetry of this type is called *symmetry with respect to the origin.* It can be shown that rotating a graph 180° is equivalent to saying that the point $(-x, -y)$ is on the graph whenever (x, y) is on the graph. Figure 3.36 shows two graphs that are symmetric with respect to the origin. A test for this type of symmetry is given below.

SYMMETRY WITH RESPECT TO THE ORIGIN	The graph of an equation is **symmetric with respect to the origin** if the replacement of both x with $-x$ and y with $-y$ results in an equivalent equation.

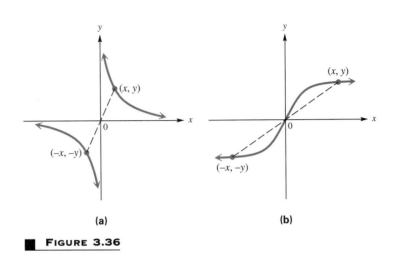

(a)   **(b)**

■ **FIGURE 3.36**

182

■ *Example 5*

TESTING FOR
SYMMETRY WITH
RESPECT TO THE
ORIGIN

Show algebraically that a circle with center at the origin is symmetric with respect to the origin. (See Figure 3.30.)

A circle with center at the origin and radius $r > 0$ has equation $x^2 + y^2 = r^2$. By substituting $-x$ for x and $-y$ for y, we get an equivalent equation.

$$x^2 + y^2 = r^2 \qquad \text{Original equation}$$
$$(-x)^2 + (-y)^2 = r^2 \qquad \text{Replace } x \text{ with } -x, \, y \text{ with } -y.$$
$$x^2 + y^2 = r^2 \qquad \text{Equivalent to original equation} \quad ■$$

The tests for symmetry are listed below.

TESTS FOR SYMMETRY		Symmetric with Respect to		
		x-Axis	**y-Axis**	**Origin**
	Test	Replace y with $-y$.	Replace x with $-x$.	Replace x with $-x$ and replace y with $-y$.
	Example			

3.4 EXERCISES ■

Find the center-radius form of the equation for each of the following circles. See Example 1.

1. Center $(1, 4)$, radius 3

2. Center $(-2, 5)$, radius 4

3. Center $(0, 0)$, radius 1

4. Center $(0, 0)$, radius 5

5. Center $(2/3, -4/5)$, radius $3/7$

6. Center $(-1/2, -1/4)$, radius $12/5$

7. Center $(-1, 2)$, passing through $(2, 6)$

8. Center $(2, -7)$, passing through $(-2, -4)$

9. Center $(-3, -2)$, tangent to the *x*-axis (*Hint: Tangent to* means touching at one point.)

10. Center $(5, -1)$, tangent to the *y*-axis

Graph each of the following circles. Give the domain and the range. See Examples 1 and 3.

11. $x^2 + y^2 = 36$

12. $x^2 + y^2 = 81$

13. $(x - 2)^2 + y^2 = 36$

14. $x^2 + (y + 3)^2 = 49$

15. $(x + 2)^2 + (y - 5)^2 = 16$

16. $(x - 4)^2 + (y - 3)^2 = 25$

17. $(x + 3)^2 + (y + 2)^2 = 36$

18. $(x - 5)^2 + (y + 4)^2 = 49$

19. Describe the graph of the equation $(x - 3)^2 + (y - 3)^2 = 0$.

20. Describe the graph of the equation $(x - 3)^2 + (y - 3)^2 = -1$.

21. Without actually graphing, state whether or not the graphs of $x^2 + y^2 = 4$ and $x^2 + y^2 = 25$ will intersect. Explain your answer.

22. Can a circle have its center at $(2, 4)$ and be tangent to both axes? Explain.

Find the center and the radius of each of the following that are circles. See Example 2.

23. $x^2 + 6x + y^2 + 8y = -9$

24. $x^2 - 4x + y^2 + 12y + 4 = 0$

25. $x^2 - 12x + y^2 + 10y + 25 = 0$

26. $x^2 + 8x + y^2 - 6y = -16$

27. $x^2 + 8x + y^2 - 14y + 64 = 0$

28. $x^2 - 8x + y^2 + 7 = 0$

29. $x^2 + y^2 = 2y + 48$

30. $x^2 + 4x + y^2 = 21$

State whether each of the following graphs is symmetric to any of the following: x-axis, y-axis, origin.

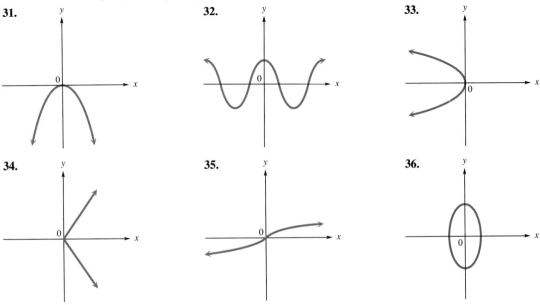

31.

32.

33.

34.

35.

36.

Plot the following points, and then use the same axes to plot the points that are symmetric to the given point with respect to the following elements: (a) x-axis, (b) y-axis, (c) origin.

37. $(5, -3)$

38. $(-6, 1)$

39. $(-4, -2)$

40. $(-8, 3)$

For Exercises 41– 43, suppose that the point (s, t) lies on the graph of a relation R.

41. If R is symmetric with respect to the x-axis, the point _____ must lie on the graph of R.

42. If R is symmetric with respect to the y-axis, the point _____ must lie on the graph of R.

43. If R is symmetric with respect to the origin, the point _____ must lie on the graph of R.

44. If a relation is graphed on a sheet of paper, the concepts of symmetry with respect to the axes or the origin can be illustrated using paper-folding. Explain how this can be done.

Use the tests for symmetry to decide whether the graph of each relation is symmetric with respect to the x-axis, the y-axis, or the origin. See Examples 4 and 5.

45. $x^2 + y^2 = 5$

46. $y^2 = 4 - x^2$

47. $y = x^2 - 8x$

48. $y = 4x - x^2$

49. $y = x^3$

50. $y = -x^3$

51. $y = \dfrac{1}{1 + x^2}$

52. $x = \dfrac{-1}{y^2 + 9}$

Sketch examples of graphs that satisfy the following conditions.

53. Symmetric with respect to the x-axis but not to the y-axis

54. Symmetric with respect to the y-axis but not to the x-axis

55. Symmetric with respect to the origin but to neither the x-axis nor the y-axis

In Exercises 56–58, find the equation of the circle having the given points as endpoints of a diameter. (Hint: Use the midpoint formula to find the center, use the distance formula to find the radius, and then use the center-radius form of the equation of a circle to find the equation.)

56. (4, 3), (4, −3)   **57.** (−1, 3), (5, −9)   **58.** (3, −5), (−7, 3)

59. Find all points (x, y) with $x = y$ that are 4 units from $(1, 3)$.

60. Find all points satisfying $x + y = 0$ that are 8 units from $(−2, 3)$.

61. Let F be some algebraic expression involving x as the only variable. Suppose the equation $y = F$ is changed to $y = −F$. What is the effect on the graph of $y = F$?

62. Suppose a circle is tangent to both axes, is in the third quadrant, and has a radius of $\sqrt{2}$. Find its equation.

63. One circle has center at (3, 4) and radius 5. A second circle has center at (−1, −3) and radius 4. Do the circles intersect?

64. Does a circle with radius 6 and center at (0, 5) intersect a circle with center at (−5, −4) and radius 4?

3.5 ——— THE ELLIPSE AND THE HYPERBOLA

We have studied two types of second-degree relations thus far: parabolas and circles. We now look at another type, the *ellipse.*

ELLIPSES   The definition of an ellipse is also based on distance.

ELLIPSE   An **ellipse** is the set of all points in a plane the sum of whose distances from two fixed points is constant. The two fixed points are called the **foci** of the ellipse.

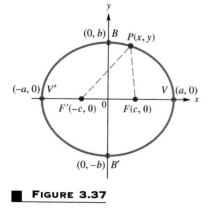

■ **FIGURE 3.37**

For example, the ellipse in Figure 3.37 has foci at points F and F'. By the definition, the ellipse is made up of all points P such that the sum $d(P, F) + d(P, F')$ is constant. The ellipse in Figure 3.37 has its **center** at the origin. Points V and V' are the **vertices** of the ellipse, and the line segment connecting V and V' is the **major axis.** The foci always lie on the major axis. The line segment from B to B' is the **minor axis.** The major axis has length $2a$, and the minor axis has length $2b$.

If the foci are chosen to be on the x-axis (or y-axis), with the center of the ellipse at the origin, then the distance formula and the definition of an ellipse can be used to obtain the following result. (See Exercise 43.)

EQUATION OF AN ELLIPSE	The ellipse centered at the origin with x-intercepts a and $-a$, and y-intercepts b and $-b$, has equation

$$\frac{x^2}{a^2} + \frac{y^2}{b^2} = 1,$$

where $a \neq b$.

In an equation of an ellipse, the coefficients of x^2 and y^2 must be different positive numbers. (What happens if the coefficients are equal?)

An ellipse is the graph of a relation. As suggested by the graph in Figure 3.37, if the ellipse has equation $(x^2/a^2) + (y^2/b^2) = 1$, the domain is $[-a, a]$ and the range is $[-b, b]$. Notice that the ellipse in Figure 3.37 is symmetric with respect to the x-axis, the y-axis, and the origin. More generally, every ellipse is symmetric with respect to its major axis, its minor axis, and its center.

Ellipses have many useful applications. As the earth makes its year-long journey around the sun, it traces an ellipse. Spacecraft travel around the earth in elliptical orbits, and planets make elliptical orbits around the sun. An interesting recent application is the use of an elliptical tub in the nonsurgical removal of kidney stones.

■ **Example 1**

GRAPHING AN ELLIPSE CENTERED AT THE ORIGIN

Graph $4x^2 + 9y^2 = 36$.

To get the form of the equation of an ellipse, divide both sides by 36.

$$\frac{x^2}{9} + \frac{y^2}{4} = 1$$

This ellipse is centered at the origin, with x-intercepts 3 and -3, and y-intercepts 2 and -2. Additional ordered pairs that satisfy the equation of the ellipse may be found and plotted as needed (a calculator with a square root key will be helpful). The domain of this relation is $[-3, 3]$, and the range is $[-2, 2]$. The graph is shown in Figure 3.38. ■

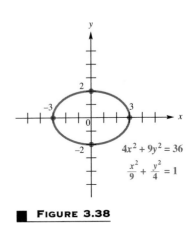

$4x^2 + 9y^2 = 36$

$\dfrac{x^2}{9} + \dfrac{y^2}{4} = 1$

■ FIGURE 3.38

■ **Example 2**

FINDING THE EQUATION OF AN ELLIPSE

Give the equation of the ellipse with center at the origin, a vertex at $(5, 0)$, and minor axis of length 6.

The equation will have the form $(x^2/a^2) + (y^2/b^2) = 1$. One vertex is at $(5, 0)$, so $a = 5$. The minor axis has length $2b$, so

$$2b = 6$$

$$b = 3.$$

The equation is

$$\frac{x^2}{25} + \frac{y^2}{9} = 1. \quad ■$$

Recall from Section 3.4 that the circle $x^2 + y^2 = r^2$, whose center is at the origin, can be translated away from the origin so that the circle $(x - h)^2 + (y - k)^2 = r^2$ has its center at (h, k). In a similar manner, an ellipse can be translated so that its center is away from the origin.

■ IN SIMPLEST TERMS

12

One of the most useful properties of an ellipse is its reflexive property. If a beam is projected from one focus onto the ellipse, it will reflect to the other focus. This feature has helped scientists develop the lithotripter, a machine that uses shock waves to crush kidney stones. The waves originate at one focus and are reflected to hit the kidney stone which is positioned at the second focus.

To determine the focus of an ellipse, use the formula $b^2 = a^2 - c^2$, where the distance from the center to one end of the major axis is represented by a, the distance from the center to one end of the minor axis is represented by b, and the distance from the center to each focus is represented by c. The foci will lie on the major axis. Solving this formula for c gives $c = \pm\sqrt{a^2 - b^2}$.

In finding the foci of the ellipse $\frac{x^2}{25} + \frac{y^2}{9} = 1$, we see that $a^2 = 25$, $b^2 = 9$, and $c = \pm\sqrt{16} = \pm4$. Since the center of the ellipse is at $(0, 0)$, the foci are located at $(-4, 0)$ and $(4, 0)$.

SOLVE EACH PROBLEM

A. If a lithotripter is based on the ellipse $\frac{x^2}{36} + \frac{y^2}{27} = 1$, determine how many units the kidney stone and the wave source must be placed from the center of the ellipse.

B. A patient is placed 12 units away from the source of shock waves. The lithotripter is based on an ellipse with a minor axis of 16 total units. Find an equation of an ellipse that would satisfy this situation.

ANSWERS   A. The kidney stone and the wave source must each be placed 3 units from the center of the ellipse.   B. One such ellipse would have an equation of $\frac{x^2}{100} + \frac{y^2}{64} = 1$.

■ *Example 3*

GRAPHING AN ELLIPSE TRANSLATED AWAY FROM THE ORIGIN

Graph $\dfrac{(x-2)^2}{9} + \dfrac{(y+1)^2}{16} = 1.$

If the equation were

$$\frac{x^2}{9} + \frac{y^2}{16} = 1,$$

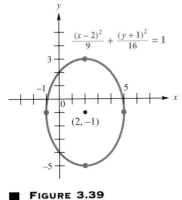

$$\frac{(x-2)^2}{9} + \frac{(y+1)^2}{16} = 1$$

we would have an ellipse with center at $(0, 0)$. The terms in the numerators of the fractions on the left side, however, indicate that this relation represents an ellipse centered at $(2, -1)$. Graph the ellipse using the fact that $a = 3$ and $b = 4$. Start at $(2, -1)$ and locate two points each 3 units away from $(2, -1)$ on a horizontal line, one to the right of $(2, -1)$ and one to the left. Locate two other points on a vertical line through $(2, -1)$, one 4 units up and one 4 units down. Since $b > a$, the vertices are on the vertical line through the center. The vertices are $(2, 3)$ and $(2, -5)$. Find additional points as necessary. The final graph is shown in Figure 3.39. As the graph suggests, the domain is $[-1, 5]$, and the range is $[-5, 3]$. ■

■ **FIGURE 3.39**

HYPERBOLAS The definition of an ellipse requires that the *sum* of the distances from two fixed points be constant. The definition of hyperbola involves the *difference* rather than the sum.

HYPERBOLA A **hyperbola** is the set of all points in a plane such that the absolute value of the difference of the distances from two fixed points (called **foci**) is constant.

Some applications of hyperbolas are given in the exercises.

As with ellipses, the equation of a hyperbola can be found from the distance formula and the definition of a hyperbola. (See Exercise 45.)

EQUATIONS OF HYPERBOLAS A hyperbola centered at the origin, with x-intercepts a and $-a$, has an equation of the form

$$\frac{x^2}{a^2} - \frac{y^2}{b^2} = 1,$$

while a hyperbola centered at the origin, with y-intercepts b and $-b$, has an equation of the form

$$\frac{y^2}{b^2} - \frac{x^2}{a^2} = 1.$$

Some texts use $y^2/a^2 - x^2/b^2 = 1$ for this last equation. For a brief introduction such as this, the form given is commonly used.

The *x*-intercepts are the **vertices** of a hyperbola with the equation $(x^2/a^2) - (y^2/b^2) = 1$, and the *y*-intercepts are the vertices of a hyperbola with the equation $(y^2/b^2) - (x^2/a^2) = 1$. The line segment between the vertices is the **transverse axis** of the hyperbola, and the midpoint of the transverse axis is the **center** of the hyperbola.

■ *Example 4*
GRAPHING A
HYPERBOLA
CENTERED AT THE
ORIGIN

Graph the hyperbola $\dfrac{x^2}{16} - \dfrac{y^2}{9} = 1$.

By the first equation of a hyperbola given earlier, the hyperbola is centered at the origin and has *x*-intercepts 4 and -4. However, if $x = 0$,

$$-\frac{y^2}{9} = 1 \qquad \text{or} \qquad y^2 = -9,$$

which has no real solutions. For this reason, the graph has no *y*-intercepts. To complete the graph, find some other ordered pairs that belong to it. For example, if $x = 6$,

$$\frac{6^2}{16} - \frac{y^2}{9} = 1 \qquad\qquad \text{Let } x = 6.$$

$$-\frac{y^2}{9} = 1 - \frac{36}{16} \qquad\qquad \text{Subtract } \frac{36}{16}.$$

$$\frac{y^2}{9} = \frac{20}{16} \qquad\qquad \text{Multiply by } -1 \text{ and combine terms.}$$

$$y^2 = \frac{180}{16} = \frac{45}{4} \qquad\qquad \text{Multiply by 9; lowest terms.}$$

$$y = \pm\frac{3\sqrt{5}}{2} \approx \pm 3.4. \qquad \text{Take square roots and use a calculator.}$$

The graph includes the points $(6, 3.4)$ and $(6, -3.4)$. If $x = -6$, $y \approx \pm 3.4$, so the points $(-6, 3.4)$ and $(-6, -3.4)$ also are on the graph. These points, along with other points on the graph, were used to help sketch the final graph shown in Figure 3.40. The graph suggests that the domain of this hyperbola is $(-\infty, -4] \cup [4, \infty)$, while the range is $(-\infty, \infty)$. Using the tests for symmetry would show that this hyperbola is symmetric with respect to the *x*-axis, the *y*-axis, and the origin. ■

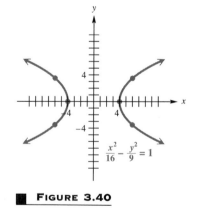

■ **FIGURE 3.40**

Starting with

$$\frac{x^2}{a^2} - \frac{y^2}{b^2} = 1$$

and solving for y gives

$$\frac{x^2}{a^2} - 1 = \frac{y^2}{b^2}$$

$$\frac{x^2 - a^2}{a^2} = \frac{y^2}{b^2}$$

$$y = \pm \frac{b}{a}\sqrt{x^2 - a^2}. \qquad (*)$$

If x^2 is very large in comparison to a^2, the difference $x^2 - a^2$ would be very close to x^2. If this happens, then the points satisfying equation (*) above would be very close to one of the lines

$$y = \pm \frac{b}{a}x.$$

Thus, as $|x|$ gets larger and larger, the points of the hyperbola $x^2/a^2 - y^2/b^2 = 1$ come closer and closer to the lines $y = (\pm b/a)x$. These lines, called the *asymptotes* of the hyperbola, are very helpful when graphing the hyperbola. An **asymptote** is a line that the graph of a relation approaches but never reaches as $|x|$ gets large. Asymptotes are discussed again in Sections 5.1 and 6.6.

∎ *Example 5*

USING ASYMPTOTES TO GRAPH A HYPERBOLA

Graph $\dfrac{x^2}{25} - \dfrac{y^2}{49} = 1$.

For this hyperbola, $a = 5$ and $b = 7$. With these values, $y = (\pm b/a)x$ becomes $y = (\pm 7/5)x$. Use the x-intercepts: if $x = 5$, then $y = (\pm 7/5)(5) = \pm 7$, and if $x = -5$, $y = \pm 7$. These four points, $(5, 7)$, $(5, -7)$, $(-5, 7)$, and $(-5, -7)$, lead to the rectangle shown in Figure 3.41. The extended diagonals of this rectangle are the asymptotes of the hyperbola. The hyperbola has x-intercepts 5 and -5. The domain is $(-\infty, -5] \cup [5, \infty)$, and the range is $(-\infty, \infty)$. The final graph is shown in Figure 3.41. ∎

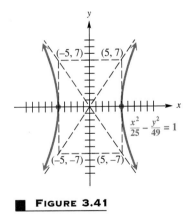

∎ **FIGURE 3.41**

The rectangle used to find the asymptotes of the hyperbola in Example 5 is called the *fundamental rectangle*.

ASYMPTOTES OF A HYPERBOLA

The **asymptotes** of the hyperbola with equation

$$\frac{x^2}{a^2} - \frac{y^2}{b^2} = 1 \qquad \text{or} \qquad \frac{y^2}{b^2} - \frac{x^2}{a^2} = 1$$

are the extended diagonals of the **fundamental rectangle** with vertices at (a, b), $(a, -b)$, $(-a, b)$, and $(-a, -b)$.

By using slopes of the diagonals of the fundamental rectangle, we can verify that the equations of the diagonals are as follows.

EQUATIONS OF ASYMPTOTES

The equations of the asymptotes of either of the hyperbolas $(x^2/a^2) - (y^2/b^2) = 1$ or $(y^2/b^2) - (x^2/a^2) = 1$ are

$$y = \pm \frac{b}{a} x.$$

Like the relations studied earlier in this chapter, hyperbolas may be translated. The type of translation can be determined from the equation, just as it was with parabolas, circles, and ellipses.

■ *Example 6*
GRAPHING A HYPERBOLA TRANSLATED AWAY FROM THE ORIGIN

Graph $\dfrac{(y + 2)^2}{9} - \dfrac{(x + 3)^2}{4} = 1$.

This hyperbola has the same graph as

$$\frac{y^2}{9} - \frac{x^2}{4} = 1,$$

except that it is centered at $(-3, -2)$. See Figure 3.42. ■

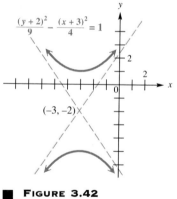

■ **FIGURE 3.42**

■ *Example 7*
FINDING THE EQUATION OF A HYPERBOLA

Write the equation of the hyperbola centered at $(-2, 1)$, with a vertex at $(-2, 3)$, and with a equal to half of b.

Since both the vertex and the center are on the transverse axis, it must be the vertical line $x = -2$. The equation will have the form

$$\frac{(y - 1)^2}{b^2} - \frac{(x + 2)^2}{a^2} = 1.$$

The distance from the center to the given vertex at $(-2, 3)$ gives b for this hyperbola. Using the distance formula,

$$b = \sqrt{(-2 + 2)^2 + (1 - 3)^2} = 2.$$

From the information given, $a = (1/2)b = (1/2)(2) = 1$, so the equation is

$$\frac{(y - 1)^2}{4} - \frac{(x + 2)^2}{1} = 1. \quad ▪$$

3.5 EXERCISES ▪

Graph each ellipse. Give the domain and range. See Examples 1 and 3.

1. $\dfrac{x^2}{9} + \dfrac{y^2}{4} = 1$

2. $\dfrac{x^2}{16} + \dfrac{y^2}{36} = 1$

3. $\dfrac{x^2}{6} + \dfrac{y^2}{9} = 1$

4. $\dfrac{x^2}{8} + \dfrac{y^2}{12} = 1$

5. $\dfrac{x^2}{1/9} + \dfrac{y^2}{1/16} = 1$

6. $\dfrac{x^2}{4/25} + \dfrac{y^2}{9/49} = 1$

7. $\dfrac{64x^2}{9} + \dfrac{25y^2}{36} = 1$

8. $\dfrac{121x^2}{25} + \dfrac{16y^2}{9} = 1$

9. $\dfrac{(x - 1)^2}{9} + \dfrac{(y + 3)^2}{25} = 1$

10. $\dfrac{(x + 3)^2}{16} + \dfrac{(y - 2)^2}{36} = 1$

11. $\dfrac{(x - 2)^2}{16} + \dfrac{(y - 1)^2}{9} = 1$

12. $\dfrac{(x + 3)^2}{25} + \dfrac{(y + 2)^2}{36} = 1$

Graph each hyperbola. Give the domain and range. See Examples 4– 6.

13. $\dfrac{x^2}{16} - \dfrac{y^2}{9} = 1$

14. $\dfrac{y^2}{9} - \dfrac{x^2}{9} = 1$

15. $\dfrac{y^2}{36} - \dfrac{x^2}{49} = 1$

16. $\dfrac{x^2}{49} - \dfrac{y^2}{144} = 1$

17. $\dfrac{4x^2}{9} - \dfrac{25y^2}{16} = 1$

18. $x^2 - y^2 = 1$

19. $\dfrac{(x - 1)^2}{9} - \dfrac{(y + 3)^2}{25} = 1$

20. $\dfrac{(x + 3)^2}{16} - \dfrac{(y - 2)^2}{36} = 1$

21. $\dfrac{(x - 3)^2}{16} - \dfrac{(y + 2)^2}{49} = 1$

22. $\dfrac{(y - 5)^2}{4} - \dfrac{(x + 1)^2}{9} = 1$

23. $\dfrac{(y + 1)^2}{25} - \dfrac{(x - 3)^2}{36} = 1$

24. $\dfrac{(x + 2)^2}{16} - \dfrac{(y + 2)^2}{25} = 1$

Write an equation for each of the following ellipses or hyperbolas. See Examples 2 and 7.

25. Ellipse, center at the origin, length of major axis = 12, endpoint of minor axis at $(0, 4)$

26. Ellipse, center at the origin, vertex $(6, 0)$, length of minor axis = 8

27. Hyperbola, center at the origin, vertex $(0, 2)$, $a = 2b$

28. Hyperbola, center at the origin, length of transverse axis = 8, $b = 3a$

29. Ellipse, center at $(2, -2)$, $a = 4$, $b = 3$, major axis parallel to the x-axis

30. Hyperbola, center at $(-3, 1)$, $a = 4$, $b = 2$, transverse axis parallel to the y-axis

31. Hyperbola, center at $(4, 3)$, vertex $(1, 3)$, $b = 2$

32. Ellipse, center at $(-1, -2)$, length of minor axis = 4, vertex $(-1, 1)$

33. Explain how it can be determined, before graphing, whether the branches of a hyperbola open up and down or right and left.

34. How can a circle be considered a special case of an ellipse?

35. Draftspeople often use the method shown in the sketch to draw an ellipse. Explain why the method works.

36. Explain how the method of Exercise 35 can be modified to draw a circle.

37. The Roman Coliseum is an ellipse with major axis 620 ft and minor axis 513 ft. Find the distance between the foci of this ellipse.

38. A formula for the approximate circumference of an ellipse is

$$C \approx 2\pi \sqrt{\frac{a^2 + b^2}{2}},$$

where a and b are the lengths as shown in the figure. Use this formula to find the approximate circumference of the Roman Coliseum (see Exercise 37).

39. Ships and planes often use a location-finding system called LORAN. With this system, a radio transmitter at M on the figure sends out a series of pulses. When each pulse is received at transmitter S, it then sends out a pulse. A ship at P receives pulses from both M and S. A receiver on the ship measures the difference in

the arrival times of the pulses. The navigator then consults a special map, showing certain curves according to the differences in arrival times. In this way, the ship can be located as lying on a portion of which curve?

40. Microphones are placed at points $(-c, 0)$ and $(c, 0)$. An explosion occurs at point $P(x, y)$ having positive x-coordinate. (See the figure.) The sound is detected at the closer microphone t sec before being detected at the farther microphone. Assume that sound travels at a speed of 330 m per sec, and show that P must be on the hyperbola

$$\frac{x^2}{330^2 t^2} - \frac{y^2}{4c^2 - 330^2 t^2} = \frac{1}{4}.$$

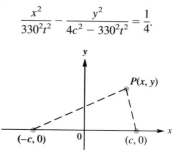

41. The orbit of Venus is an ellipse with the sun at one focus. An approximate equation for its orbit is

$$\frac{x^2}{5013} + \frac{y^2}{4970} = 1,$$

where x and y are measured in millions of miles. Find the widest possible distance across the ellipse.

42. A one-way road passes under an overpass in the form of half of an ellipse, 15 ft high at the center and 20 ft wide. Assuming a truck is 12 ft wide, what is the tallest truck that can pass under the overpass?

43. Suppose that $(c, 0)$ and $(-c, 0)$ are the foci of an ellipse. Suppose that the sum of the distances from any point (x, y) of the ellipse to the two foci is $2a$. See the accompanying figure.

(a) Use the distance formula to show that the equation of the resulting ellipse is

$$\frac{x^2}{a^2} + \frac{y^2}{a^2 - c^2} = 1.$$

(b) Show that a and $-a$ are the x-intercepts.

(c) Let $b^2 = a^2 - c^2$, and show that b and $-b$ are the y-intercepts.

44. Use the result of Exercise 43(a) to find an equation of an ellipse with foci $(3, 0)$ and $(-3, 0)$, where the sum of the distances from any point of the ellipse to the two foci is 10.

45. Suppose a hyperbola has center at the origin, foci at $F'(-c, 0)$ and $F(c, 0)$, and the value

$|d(P, F') - d(P, F)| = 2a$. Let $b^2 = c^2 - a^2$, and show that the equation of the hyperbola is

$$\frac{x^2}{a^2} - \frac{y^2}{b^2} = 1.$$

46. Use the result of Exercise 45 to find an equation of a hyperbola with center at the origin, foci at $(-2, 0)$ and $(2, 0)$, and the absolute value of the difference of the distances from any point of the hyperbola to the two foci equal to 2.

3.6 — THE CONIC SECTIONS

The graphs of the second-degree relations studied in this chapter, parabolas, hyperbolas, ellipses, and circles, are called **conic sections** since each can be obtained by cutting a cone with a plane, as shown in Figure 3.43.

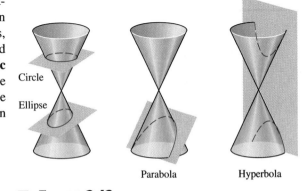

Circle

Ellipse

Parabola   Hyperbola

■ **FIGURE 3.43**

All conic sections presented in this chapter have equations of the form

$$Ax^2 + Bx + Cy^2 + Dy + E = 0,$$

where either A or C must be nonzero. The special characteristics of each of the conic sections are summarized below.

EQUATIONS OF CONIC SECTIONS	Conic Section	Characteristic	Example
	Parabola	Either $A = 0$ or $C = 0$, but not both.	$y = x^2$ $x = 3y^2 + 2y - 4$
	Circle	$A = C \neq 0$	$x^2 + y^2 = 16$
	Ellipse	$A \neq C, AC > 0$	$\dfrac{x^2}{16} + \dfrac{y^2}{25} = 1$
	Hyperbola	$AC < 0$	$x^2 - y^2 = 1$

The following chart summarizes our work with conic sections.

Summary of the Conic Sections

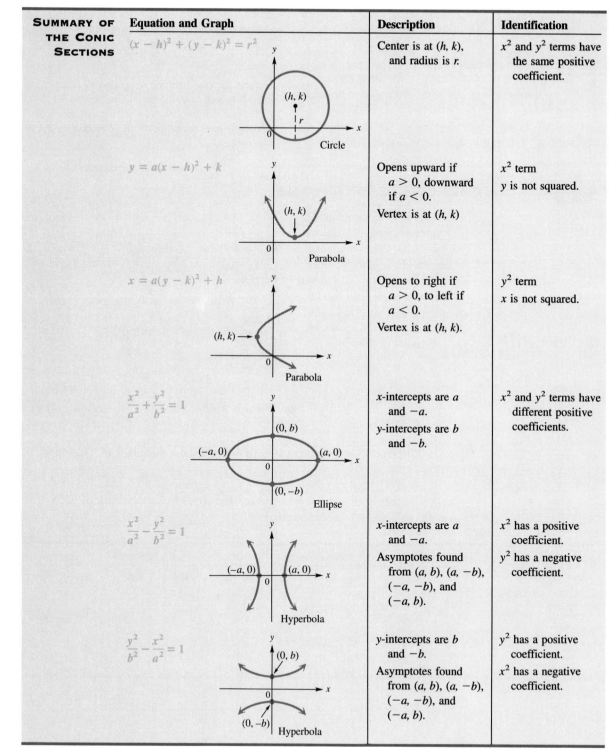

Equation and Graph	Description	Identification
$(x - h)^2 + (y - k)^2 = r^2$ — Circle	Center is at (h, k), and radius is r.	x^2 and y^2 terms have the same positive coefficient.
$y = a(x - h)^2 + k$ — Parabola	Opens upward if $a > 0$, downward if $a < 0$. Vertex is at (h, k)	x^2 term y is not squared.
$x = a(y - k)^2 + h$ — Parabola	Opens to right if $a > 0$, to left if $a < 0$. Vertex is at (h, k).	y^2 term x is not squared.
$\dfrac{x^2}{a^2} + \dfrac{y^2}{b^2} = 1$ — Ellipse	x-intercepts are a and $-a$. y-intercepts are b and $-b$.	x^2 and y^2 terms have different positive coefficients.
$\dfrac{x^2}{a^2} - \dfrac{y^2}{b^2} = 1$ — Hyperbola	x-intercepts are a and $-a$. Asymptotes found from (a, b), $(a, -b)$, $(-a, -b)$, and $(-a, b)$.	x^2 has a positive coefficient. y^2 has a negative coefficient.
$\dfrac{y^2}{b^2} - \dfrac{x^2}{a^2} = 1$ — Hyperbola	y-intercepts are b and $-b$. Asymptotes found from (a, b), $(a, -b)$, $(-a, -b)$, and $(-a, b)$.	y^2 has a positive coefficient. x^2 has a negative coefficient.

In order to recognize the type of graph that a given conic section has, it is sometimes necessary to transform the equation into a more familiar form, as shown in the next examples.

■ *Example 1*

DETERMINING THE
TYPE OF A CONIC
SECTION FROM ITS
EQUATION

Decide on the type of conic section represented by each of the following equations, and sketch each graph.

(a) $25y^2 - 4x^2 = 100.$

Divide each side by 100 to get

$$\frac{y^2}{4} - \frac{x^2}{25} = 1.$$

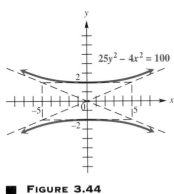

This is a hyperbola centered at the origin, with foci on the y-axis, and y-intercepts 2 and -2. The points $(5, 2)$, $(5, -2)$, $(-5, 2)$, $(-5, -2)$ determine the fundamental rectangle. The diagonals of the rectangle are the asymptotes, and their equations are

$$y = \pm \frac{2}{5}x.$$

The graph is shown in Figure 3.44.

■ FIGURE 3.44

(b) $x^2 = 25 + 5y^2$

Rewriting the equation as

$$x^2 - 5y^2 = 25$$

or

$$\frac{x^2}{25} - \frac{y^2}{5} = 1$$

shows that the equation represents a hyperbola centered at the origin, with asymptotes

$$y = \pm \frac{b}{a}x$$

or

$$y = \frac{\pm\sqrt{5}}{5}x.$$

The x-intercepts are ± 5; the graph is shown in Figure 3.45.

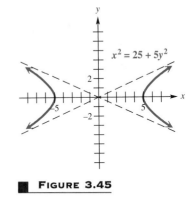

■ FIGURE 3.45

(c) $4x^2 - 16x + 9y^2 + 54y = -61$

Since the coefficients of the x^2 and y^2 terms are unequal and both positive, this equation might represent an ellipse. (It might also represent a single point or no points at all.) To find out, complete the square on x and y.

$$4(x^2 - 4x \quad) + 9(y^2 + 6y \quad) = -61$$

Factor out a 4; factor out a 9.

$$4(x^2 - 4x + 4 - 4) + 9(y^2 + 6y + 9 - 9) = -61$$

Add and subtract the same quantity.

$$4(x^2 - 4x + 4) - 16 + 9(y^2 + 6y + 9) - 81 = -61$$

Regroup and distribute.

$$4(x - 2)^2 + 9(y + 3)^2 = 36$$

Add 97 and factor.

$$\frac{(x - 2)^2}{9} + \frac{(y + 3)^2}{4} = 1$$

Divide by 36.

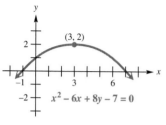

$$4x^2 - 16x + 9y^2 + 54y = -61$$

■ **FIGURE 3.46**

This equation represents an ellipse having center at $(2, -3)$ and graph as shown in Figure 3.46.

(d) $x^2 - 8x + y^2 + 10y = -41$

Complete the square on both x and y, as follows:

$$(x^2 - 8x + 16) + (y^2 + 10y + 25) = -41 + 16 + 25$$
$$(x - 4)^2 + (y + 5)^2 = 0.$$

This result shows that the equation is that of a circle of radius 0; that is, the point $(4, -5)$. Had a negative number been obtained on the right (instead of 0), the equation would have represented no points at all, and there would be no graph.

(e) $x^2 - 6x + 8y - 7 = 0$

Since only one variable is squared (x, and not y), the equation represents a parabola. Rearrange the terms to get the term with y (the variable that is not squared) alone on one side. Then complete the square on the other side of the equation.

$$8y = -x^2 + 6x + 7$$
$$8y = -(x^2 - 6x \quad) + 7 \qquad \text{Regroup and factor out } -1.$$
$$8y = -(x^2 - 6x + 9) + 7 + 9 \qquad \text{Add 0 in the form } -9 + 9.$$
$$8y = -(x - 3)^2 + 16 \qquad \text{Factor.}$$
$$y = -\frac{1}{8}(x - 3)^2 + 2 \qquad \text{Multiply both sides by } \frac{1}{8}.$$

The parabola has vertex at $(3, 2)$, and opens downward, as shown in Figure 3.47. ■

$x^2 - 6x + 8y - 7 = 0$

■ **FIGURE 3.47**

CAUTION   The next example is designed to serve as a warning about a very common error.

■ *Example 2*

DETERMINING THE
TYPE OF A CONIC
SECTION FROM ITS
EQUATION

Graph $4y^2 - 16y - 9x^2 + 18x = -43$.

Complete the square on x and on y.

$$4(y^2 - 4y \qquad) - 9(x^2 - 2x \qquad) = -43$$

$$4(y^2 - 4y + 4) - 9(x^2 - 2x + 1) = -43 + 16 - 9$$

$$4(y - 2)^2 - 9(x - 1)^2 = -36$$

Because of the -36, it is very tempting to say that this equation does not have a graph. However, the minus sign in the middle on the left shows that the graph is that of a hyperbola. Dividing through by -36 and rearranging terms gives

$$\frac{(x - 1)^2}{4} - \frac{(y - 2)^2}{9} = 1,$$

a hyperbola centered at $(1, 2)$, with graph as shown in Figure 3.48. ■

$4y^2 - 16y - 9x^2 + 18x = -43$

■ **FIGURE 3.48**

Relations are sometimes defined as square roots of expressions so that their graphs consist of only a portion of the graph of a complete conic section. The final example illustrates one such relation.

■ *Example 3*

GRAPHING A
RELATION DEFINED
BY A SQUARE ROOT

Graph $y = -\sqrt{1 + 4x^2}$.

Squaring both sides gives

$$y^2 = 1 + 4x^2$$

or

$$y^2 - 4x^2 = 1.$$

Use the fact that $4 = 1/(1/4)$ to write the equation as

$$\frac{y^2}{1} - \frac{x^2}{1/4} = 1,$$

the equation of a hyperbola with y-intercepts 1 and -1. Use the points $(1/2, 1)$, $(1/2, -1), (-1/2, 1)$, and $(-1/2, -1)$ to sketch the asymptotes. Since the given equation $y = -\sqrt{1 + 4x^2}$ restricts y to nonpositive values, the graph is the lower branch of the hyperbola, as shown in Figure 3.49. The domain of $y = -\sqrt{1 + 4x^2}$ is the set of all x such that $1 + 4x^2 \geq 0$. This condition is satisfied for all x in the interval $(-\infty, \infty)$. The range is $(-\infty, -1]$. ■

$y = -\sqrt{1 + 4x^2}$

■ **FIGURE 3.49**

In Exercises 1–12, the equation of a conic section is given in a familiar form. Identify the type of graph that the equation has, without actually graphing.

1. $x^2 + y^2 = 144$

2. $(x - 2)^2 + (y + 3)^2 = 25$

3. $y = 2x^2 + 3x - 4$

4. $x = 3y^2 + 5y - 6$

5. $x = -3(y - 4)^2 + 1$

6. $\dfrac{x^2}{25} + \dfrac{y^2}{36} = 1$

7. $\dfrac{x^2}{49} + \dfrac{y^2}{100} = 1$

8. $x^2 - y^2 = 1$

9. $\dfrac{x^2}{4} - \dfrac{y^2}{16} = 1$

10. $\dfrac{(x + 2)^2}{9} + \dfrac{(y - 4)^2}{16} = 1$

11. $\dfrac{x^2}{25} - \dfrac{y^2}{25} = 1$

12. $y = 4(x + 3)^2 - 7$

For each of the following equations that has a graph, identify the corresponding graph. It may be necessary to transform the equation. See Examples 1 and 2.

13. $\dfrac{x^2}{4} = 1 - \dfrac{y^2}{9}$

14. $\dfrac{x^2}{4} = 1 + \dfrac{y^2}{9}$

15. $\dfrac{x^2}{4} + \dfrac{y^2}{4} = 1$

16. $\dfrac{x^2}{4} + \dfrac{y^2}{4} = -1$

17. $x^2 + 2x = x^2 + y - 6$

18. $y^2 - 4y = y^2 + 3 - x$

19. $x^2 = 25 + y^2$

20. $x^2 = 25 - y^2$

21. $9x^2 + 36y^2 = 36$

22. $x^2 = 4y - 8$

23. $\dfrac{(x + 3)^2}{16} + \dfrac{(y - 2)^2}{16} = 1$

24. $\dfrac{(x - 4)^2}{8} + \dfrac{(y + 1)^2}{2} = 0$

25. $y^2 - 4y = x + 4$

26. $11 - 3x = 2y^2 - 8y$

27. $(x + 7)^2 + (y - 5)^2 + 4 = 0$

28. $4(x - 3)^2 + 3(y + 4)^2 = 0$

29. $3x^2 + 6x + 3y^2 - 12y = 12$

30. $2x^2 - 8x + 2y^2 + 20y = 12$

31. $x^2 - 6x + y = 0$

32. $x - 4y^2 - 8y = 0$

33. $4x^2 - 8x - y^2 - 6y = 6$

34. $x^2 + 2x = x^2 - 4y - 2$

35. $4x^2 - 8x + 9y^2 + 54y = -84$

36. $3x^2 + 12x + 3y^2 = -11$

37. $6x^2 - 12x + 6y^2 - 18y + 25 = 0$

38. $4x^2 - 24x + 5y^2 + 10y + 41 = 0$

39. Identify the type of conic section consisting of the set of all points in the plane for which the sum of the distances from the points $(5, 0)$ and $(-5, 0)$ is 14.

40. Identify the type of conic section consisting of the set of all points in the plane for which the absolute value of the difference of the distances from the points $(3, 0)$ and $(-3, 0)$ is 2.

41. **(a)** Show by a sketch how a cone can be cut by a plane in exactly one point. (For this reason, a point is sometimes called a *degenerate circle* or *degenerate ellipse.*)
(b) Show by a sketch how a cone can be cut by a plane to produce exactly one straight line. (For this reason, a straight line is sometimes called a *degenerate parabola.*)

42. Following the definitions in Exercise 41, what is a *degenerate hyperbola*?

Graph each of the following, and give the domain and range. See Example 3.

43. $y = \sqrt{4 - x}$

44. $y = \sqrt{16 - x}$

45. $y = \sqrt{x^2 - 9}$

46. $y = -\sqrt{16 - x^2}$

47. $\dfrac{y}{3} = -\sqrt{1 + \dfrac{x^2}{16}}$

48. $\dfrac{y}{2} = \sqrt{1 - \dfrac{x^2}{25}}$

49. $y = \sqrt{1 - \dfrac{x^2}{64}}$

50. $y = \sqrt{1 + \dfrac{x^2}{36}}$

51. If you are given the graph of the relation $y = \sqrt{1 - x^2}$, how would you describe a procedure you can use to graph $y = -\sqrt{1 - x^2}$?

52. The graph of $y = \sqrt{x}$ is shown here.

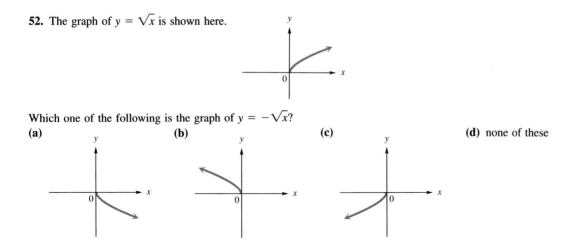

Which one of the following is the graph of $y = -\sqrt{x}$?

(a)

(b)

(c)

(d) none of these

3.7 INEQUALITIES

Many mathematical descriptions of real situations are best expressed as inequalities rather than equations. For example, a firm might be able to use a machine *no more* than 12 hours a day, while production of *at least* 500 cases of a certain product might be required to meet a contract. The simplest way to see the solution of an inequality in two variables is to draw its graph.

A line divides a plane into three sets of points: the points of the line itself and the points belonging to the two regions determined by the line. Each of these two regions is called a **half-plane.** In Figure 3.50 line *r* divides the plane into three different sets of points: line *r*, half-plane *P*, and half-plane *Q*. The points on *r* belong neither to *P* nor to *Q*. Line *r* is the **boundary** of each half-plane.

A **linear equality in two variables** is an inequality of the form

$$Ax + By \leq C,$$

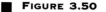

■ **FIGURE 3.50**

where *A*, *B*, and *C* are real numbers, with *A* and *B* not both equal to 0. (The symbol $\leq$ could be replaced with $\geq$, $<$, or $>$.) The graph of a linear inequality is a half-plane, perhaps with its boundary. For example, to graph the linear inequality $3x - 2y \leq 6$, first graph the boundary, $3x - 2y = 6$, as shown in Figure 3.51.

Since the points of the line $3x - 2y = 6$ satisfy $3x - 2y \leq 6$, this line is part of the solution. To decide which half-plane (the one above the line $3x - 2y = 6$ or the one below the line) is part of the solution, solve the original inequality for *y*.

$$3x - 2y \leq 6$$
$$-2y \leq -3x + 6$$
$$y \geq \frac{3}{2}x - 3 \qquad \text{Multiply by } -\frac{1}{2}; \text{ change } \leq \text{ to } \geq.$$

■ **FIGURE 3.51**

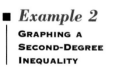

■ FIGURE 3.52

For a particular value of x, the inequality will be satisfied by all values of y that are *greater than* or equal to $(3/2)x - 3$. This means that the solution includes the half-plane *above* the line, as well as the line itself. The domain and range are both $(-\infty, \infty)$. See Figure 3.52.

There is an alternative method for deciding which side of the boundary line to shade. Choose as a test point any point not on the graph of the equation. The origin, $(0, 0)$, is often a good test point (as long as it does not lie on the boundary line). Substituting 0 for x and 0 for y in the inequality $3x - 2y \leq 6$ gives

$$3(0) - 2(0) \leq 6$$
$$0 \leq 6,$$

a true statement. Since $(0, 0)$ leads to a true result, it is part of the solution of the inequality. Shade the side of the graph containing $(0, 0)$, as shown in Figure 3.52.

■ **Example 1**
GRAPHING A LINEAR INEQUALITY

Graph $x + 4y > 4$.

The boundary here is the straight line $x + 4y = 4$. Since the points on this line do not satisfy $x + 4y > 4$, it is customary to graph the boundary as a dashed line, as in Figure 3.53. To decide which half-plane satisfies the inequality, use a test point. Choosing $(0, 0)$ as a test point gives $0 + 4 \cdot 0 > 4$, or $0 > 4$, a false statement. Since $(0, 0)$ leads to a false statement, shade the side of the graph *not* containing $(0, 0)$, as in Figure 3.53. ■

■ FIGURE 3.53

■ **Example 2**
GRAPHING A SECOND-DEGREE INEQUALITY

Graph $y \leq 2x^2 - 3$.

First graph the boundary, the parabola $y = 2x^2 - 3$, as shown in Figure 3.54. Then select any test point not on the parabola, such as $(0, 0)$. Since $(0, 0)$ does not satisfy the original inequality, shade the portion of the graph that does not include $(0, 0)$, as shown in Figure 3.54. Because of the $=$ portion of $\leq$, the points on the parabola itself also belong to the graph. ■

■ FIGURE 3.54

NOTE Another method of determining the region to shade for a second-degree inequality that is solved for y is observing the inequality symbol. For instance, as in Example 2, since y is *less than* or equal to $2x^2 - 3$, we shade *below* the boundary. The graph of $y \geq 2x^2 - 3$ (y is *greater than* or equal to $2x^2 - 3$) consists of the boundary and all points *above* it.

■ *Example 3*

**GRAPHING
SECOND-DEGREE
INEQUALITIES**

Graph each inequality

(a) $y^2 \leq 1 + 4x^2$

Write the boundary $y^2 = 1 + 4x^2$ as $y^2 - 4x^2 = 1$, a hyperbola with y-intercepts 1 and -1, as shown in Figure 3.55(a). Select any point not on the hyperbola and test it in the original inequality. Since $(0, 0)$ satisfies this inequality, shade the area between the two branches of the hyperbola, as shown in Figure 3.55(a). The points on the hyperbola are part of the solution.

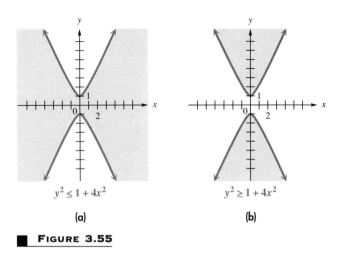

$y^2 \leq 1 + 4x^2$

(a)

$y^2 \geq 1 + 4x^2$

(b)

■ **FIGURE 3.55**

(b) $y^2 \geq 1 + 4x^2$

The boundary is the same as in part (a). Since $(0, 0)$ makes this inequality false, the *two* regions above and below the branches of the hyperbola are shaded, as shown in Figure 3.55(b). Again, the points on the hyperbola itself are included in the solution. ■

■ *Example 4*

**GRAPHING A LINEAR
INEQUALITY WITH A
RESTRICTION**

Graph $2x - y \leq 4$, where $x < 3$.

Figure 3.56 shows the graphs of both inequalities. The shaded area to the left of the line $x = 3$ and above the line $2x - y = 4$ represents the region that satisfies both inequalities. The point where the boundary lines intersect, $(3, 2)$, belongs to the first inequality but not to the second. For this reason, $(3, 2)$ is not part of the solution. This is shown with an open dot on the graph at $(3, 2)$. ■

$x < 3$

$(3, 2)$

$2x - y \leq 4$

■ **FIGURE 3.56**

■ **Example 5**

GRAPHING A
SECOND-DEGREE
INEQUALITY WITH A
RESTRICTION

Graph $16x^2 + y^2 \leq 16$, where $y \geq 0$.

The inequality $16x^2 + y^2 \leq 16$ has as its boundary the ellipse $16x^2 + y^2 = 16$, partially graphed in Figure 3.57. Since the test point $(0, 0)$ satisfies the inequality, we shade *inside* the portion of the ellipse that is graphed. The inequality $y \geq 0$ includes all points on or above the x-axis. The final graph consists of the region inside the ellipse and above the x-axis, together with its boundary, as shown in Figure 3.57. ■

■ **FIGURE 3.57**

3.7 EXERCISES ■

Graph each of the following inequalities. See Examples 1–3.

1. $x \leq 3$
2. $y \leq -2$
3. $x + 2y \leq 6$
4. $x - y \geq 2$

5. $2x + 3y \geq 4$
6. $4y - 3x < 5$
7. $3x - 5y > 6$
8. $x < 3 + 2y$

9. $5x \leq 4y - 2$
10. $2x > 3 - 4y$
11. $y < 3x^2 + 2$
12. $y \leq x^2 - 4$

13. $y > (x - 1)^2 + 2$
14. $y > 2(x + 3)^2 - 1$
15. $x^2 + (y + 3)^2 \leq 16$
16. $(x - 4)^2 + (y + 3)^2 \leq 9$

17. $4x^2 \leq 4 - y^2$
18. $x^2 + 9y^2 > 9$
19. $9x^2 - 16y^2 > 144$
20. $4x^2 \leq 36 + 9y^2$

21. Which one of the following is a description of the graph of the inequality

$$(x - 5)^2 + (y - 2)^2 < 4?$$

  (a) the region inside a circle with center $(-5, -2)$ and radius 2
  (b) the region inside a circle with center $(5, 2)$ and radius 2
  (c) the region inside a circle with center $(-5, -2)$ and radius 4
  (d) the region outside a circle with center $(5, 2)$ and radius 4

22. Without graphing, write a description of the graph of the inequality $y > 2(x - 3)^2 + 2$.

23. Which one of the following inequalities satisfies the following description: the region outside an ellipse centered at the origin, with x-intercepts 4 and -4, and y-intercepts 9 and -9?

  (a) $\dfrac{x^2}{4} + \dfrac{y^2}{9} > 1$
  (b) $\dfrac{x^2}{16} - \dfrac{y^2}{81} > 1$
  (c) $\dfrac{x^2}{16} + \dfrac{y^2}{81} > 0$
  (d) $\dfrac{x^2}{16} + \dfrac{y^2}{81} > 1$

24. Explain how it is determined whether the boundary of an inequality is a solid line or a dashed line.

Graph each of the following inequalities. See Examples 4 and 5.

25. $x - 3y < 4$, where $x \leq 0$
26. $2x + y > 5$, where $y \geq 0$

27. $3x + 2y \geq 6$, where $y \leq 2$
28. $4x + 3y \leq 6$, where $x \leq -2$

29. $y^2 \leq x + 3$, where $y \geq 0$
30. $x^2 \leq 1 - y$, where $x \geq 0$

31. $y^2 \leq 49 - x^2$, where $y \leq 0$
32. $x^2 > 16 - y^2$, where $x \leq 0$

33. $\dfrac{x^2}{36} < 1 - \dfrac{y^2}{121}$, where $x \geq 0$
34. $\dfrac{y^2}{16} \leq 1 + \dfrac{x^2}{25}$, where $y \geq 0$

35. $x + 2y < 4$, where $3x - y > 5$
36. $4x + 1 < 2y$, where $-x > y - 3$

37. $2x + 3y < 6$, where $x - 5y \geq 10$
38. $3x - 4y > 6$, where $2x + 3y > 4$

CHAPTER 3 SUMMARY ■

SECTION	TERMS	KEY IDEAS
3.1 Relations and the Rectangular Coordinate System	ordered pair relation domain range origin x-axis y-axis rectangular coordinate system (Cartesian coordinate system) coordinate plane (xy-plane) quadrants coordinates collinear	**Distance Formula** Suppose $P(x_1, y_1)$ and $R(x_2, y_2)$ are two points in a coordinate plane. Then the distance between P and R, written $d(P, R)$, is $$d(P, R) = \sqrt{(x_2 - x_1)^2 + (y_2 - y_1)^2}.$$ **Midpoint Formula** The midpoint of the line segment with endpoints (x_1, y_1) and (x_2, y_2) is $$\left(\frac{x_1 + x_2}{2}, \frac{y_1 + y_2}{2} \right).$$
3.2 Linear Relations	linear relation x-intercept y-intercept slope change in x change in y point-slope form slope-intercept form	**Definition of Slope** The slope m of the line through the points (x_1, y_1) and (x_2, y_2) is $$m = \frac{\Delta y}{\Delta x} = \frac{y_2 - y_1}{x_2 - x_1},$$ where $\Delta x \neq 0$. **Linear Equations**

General Equation	Type of Equation
$Ax + By = C$	*Standard form* (if $A \neq 0$ and $B \neq 0$), x-intercept C/A, y-intercept C/B, slope $-A/B$
$x = k$	*Vertical line*, x-intercept k, no y-intercept, undefined slope
$y = k$	*Horizontal line*, y-intercept k, no x-intercept, slope 0
$y = mx + b$	*Slope-intercept form*, y-intercept b, slope m
$y - y_1 = m(x - x_1)$	*Point-slope form*, slope m, through (x_1, y_1)

SECTION	TERMS	KEY IDEAS
3.3 Parabolas: Translations and Applications	quadratic relation parabola axis vertex, vertices translation focus, foci directrix	**Equations of Parabolas** $$y = a(x - h)^2 + k$$ $$y = ax^2 + bx + c$$ $$x = a(y - k)^2 + h$$ $$x = ay^2 + by + c$$
3.4 The Circle and Symmetry	circle radius center center-radius form	**Equation of a Circle** $$(x - h)^2 + (y - k)^2 = r^2$$ **Symmetry** The graph of an equation is **symmetric with respect to the y-axis** if the replacement of x with $-x$ results in an equivalent equation. The graph of an equation is **symmetric with respect to the x-axis** if the replacement of y with $-y$ results in an equivalent equation. The graph of an equation is **symmetric with respect to the origin** if the replacement of both x with $-x$ and y with $-y$ results in an equivalent equation.
3.5 The Ellipse and the Hyperbola	ellipse asymptotes major axis fundamental minor axis rectangle hyperbola transverse axis	**Equation of an Ellipse** $$\frac{x^2}{a^2} + \frac{y^2}{b^2} = 1$$ **Equation of a Hyperbola** $$\frac{x^2}{a^2} - \frac{y^2}{b^2} = 1 \quad \text{or} \quad \frac{y^2}{b^2} - \frac{x^2}{a^2} = 1$$
3.6 The Conic Sections	conic sections	For $Ax^2 + Bx + Cy^2 + Dy + E = 0$, where A or C is nonzero:
3.7 Inequalities	half-plane boundary	

Conic Section	Characteristic	Example
Parabola	Either $A = 0$ or $C = 0$, but not both.	$y = x^2$ $x = 3y^2 + 2y - 4$
Circle	$A = C \neq 0$	$x^2 + y^2 = 16$
Ellipse	$A \neq C, AC > 0$	$\dfrac{x^2}{16} + \dfrac{y^2}{25} = 1$
Hyperbola	$AC < 0$	$x^2 - y^2 = 1$

Give the domain and the range of each of the following relations.

1. $\{(-3, 6), (-1, 4), (8, 5)\}$ **2.** $y = \sqrt{-x}$

Find the distance between each of the following pairs of points, and state the midpoint of each pair.

3. $P(3, -1)$, $Q(-4, 5)$ **4.** $M(-8, 2)$, $N(3, -7)$ **5.** $A(-6, 3)$, $B(-6, 8)$

6. Are the points $(5, 7)$, $(3, 9)$, $(6, 8)$ the vertices of a right triangle?

7. Find all possible values of k so that $(-1, 2)$, $(-10, 5)$, and $(-4, k)$ are the vertices of a right triangle.

8. Use the distance formula to determine whether the points $(-2, -5)$, $(1, 7)$, and $(3, 15)$ are collinear.

Find the slope for each of the following lines that has a slope.

9. Through $(8, 7)$ and $(1/2, -2)$ **10.** Through $(2, -2)$ and $(3, -4)$

11. Through $(5, 6)$ and $(5, -2)$ **12.** Through $(0, -7)$ and $(3, -7)$

13. $9x - 4y = 2$ **14.** $11x + 2y = 3$ **15.** $x - 5y = 0$

16. $x - 2 = 0$ **17.** $y + 6 = 0$ **18.** $y = x$

Graph each of the following. Give the domain and range.

19. $3x + 7y = 14$ **20.** $2x - 5y = 5$ **21.** $3y = x$

22. $y = 3$ **23.** $x = -5$ **24.** $y = x$

For each of the following lines, write the equation in standard form.

25. Through $(-2, 4)$ and $(1, 3)$ **26.** Through $(-2/3, -1)$ and $(0, 4)$

27. Through $(3, -5)$ with slope -2 **28.** Through $(-4, 4)$ with slope $3/2$

29. Through $(1/5, 1/3)$ with slope $-1/2$ **30.** x-intercept -3, y-intercept 5

31. No x-intercept, y-intercept $3/4$ **32.** Through $(2, -1)$, parallel to $3x - y = 1$

33. Through $(0, 5)$, perpendicular to $8x + 5y = 3$

34. Through $(2, -10)$, perpendicular to a line with undefined slope

35. Through $(3, -5)$, parallel to $y = 4$

36. Through $(-7, 4)$, perpendicular to $y = 8$

Graph each line satisfying the given conditions.

37. Through $(2, -4)$, $m = 3/4$ **38.** Through $(0, 5)$, $m = -2/3$

39. Through $(-4, 1)$, $m = 3$ **40.** Through $(-3, -2)$, $m = -1$

Graph each of the following quadratic relations. Give the domain and range.

41. $y = x^2 - 4$ **42.** $y = 6 - x^2$ **43.** $y = 3(x + 1)^2 - 5$

44. $y = -\dfrac{1}{4}(x - 2)^2 + 3$ **45.** $y = x^2 - 4x + 2$ **46.** $y = -3x^2 - 12x - 1$

47. $x = y^2 - 2$ **48.** $x = (y + 3)^2$ **49.** $x = -(y + 1)^2 + 2$

50. $x = y^2 - 4y$ **51.** $x = -y^2 + 5y + 1$ **52.** $x = -3y^2 + 6y - 2$

Solve each problem.

53. Find the dimensions of the rectangular region of maximum area that can be enclosed with 180 m of fencing.

54. Find the dimensions of the rectangular region of maximum area that can be enclosed with 180 m of fencing, if no fencing is needed along one side of the region.

55. Which one of the following has a graph that is a parabola with a vertical axis and opens downward?

    (a) $y = 3x^2 - 4x + 2$ **(b)** $y = -3x^2 - 4x + 2$
    (c) $x = 3y^2 - 4y + 2$ **(d)** $x = -3y^2 - 4y + 2$

56. If the graph of $y = ax^2 + bx + c$ (for some particular values of a, b, and c) has its vertex in the third quadrant and $a < 0$, how many real solutions does the equation $ax^2 + bx + c = 0$ have?

Find equations for each circle satisfying the given conditions.

57. Center $(-2, 3)$, radius 5

58. Center $(\sqrt{5}, -\sqrt{7})$, radius $\sqrt{3}$

59. Center $(-8, 1)$, passing through $(0, 16)$

60. Center $(3, -6)$, tangent to the x-axis

Find the center and radius of each of the following circles.

61. $x^2 - 4x + y^2 + 6y + 12 = 0$

62. $x^2 - 6x + y^2 - 10y + 30 = 0$

63. $2x^2 + 14x + 2y^2 + 6y + 2 = 0$

64. $3x^2 + 33x + 3y^2 - 15y = 0$

65. Find all possible values of x so that the distance between $(x, -9)$ and $(3, -5)$ is 6.

66. Find all points (x, y) with $x = 6$ so that (x, y) is 4 units from $(1, 3)$.

67. Find all points (x, y) with $x + y = 0$ so that (x, y) is 6 units from $(-2, 3)$.

68. Describe the graph of $(x - 4)^2 + (y + 5)^2 = 0$.

Decide whether the relations defined below have graphs that are symmetric with respect to the x-axis, the y-axis, or the origin.

69. $3y^2 - 5x^2 = 15$ **70.** $x + y^2 = 8$ **71.** $y^3 = x + 1$ **72.** $x^2 = y^3$

73. $|y| = -x$ **74.** $|x + 2| = |y - 3|$ **75.** $|x| = |y|$ **76.** $xy = 8$

Graph each of the following relations. Give the domain and range.

77. $\dfrac{x^2}{25} + \dfrac{y^2}{4} = 1$ **78.** $\dfrac{x^2}{3} + \dfrac{y^2}{16} = 1$ **79.** $\dfrac{x^2}{4} - \dfrac{y^2}{9} = 1$

80. $\dfrac{y^2}{100} - \dfrac{x^2}{25} = 1$ **81.** $\dfrac{(x-2)^2}{9} + \dfrac{(y+3)^2}{4} = 1$ **82.** $\dfrac{(x+1)^2}{16} - \dfrac{(y-2)^2}{4} = 1$

Identify the type of graph, if any, of each of the following.

83. $x^2 = 64 - y^2$ **84.** $x^2 = 4y - 15$ **85.** $\dfrac{(x+3)^2}{9} + \dfrac{(y-2)^2}{9} = 1$

86. $\dfrac{(x-4)^2}{9} - \dfrac{(y+1)^2}{16} = 1$ **87.** $y^2 - 3y = x + 4$ **88.** $(x + 7)^2 + (y - 5)^2 + 16 = 0$

Graph each of the following. Give the domain and range.

89. $y = \sqrt{100 - x^2}$ **90.** $y = -\sqrt{16 + x^2}$ **91.** $y = -\sqrt{4 - x}$ **92.** $y = \sqrt{1 + x}$

Graph each inequality.

93. $5x - y \leq 20$ **94.** $3x - 5y > 10$ **95.** $y \leq (x - 4)^2$

96. $y < (x - 1)^2 + 3$ **97.** $25y^2 - 36x^2 > 900$ **98.** $(x - 2)^2 + (y + 3)^2 \geq 9$

99. $2x - y \geq 4$, where $y \geq -2$ **100.** $y \leq x^2$, where $x \geq 0$

1. For the points $(-2, 1)$ and $(3, 4)$: **(a)** Find the distance between them. **(b)** Find the midpoint of the line segment joining them.

Find the slope of each of the following lines.

2. Through $(-3, 4)$ and $(5, -6)$

3. $5x - 4y = 6$

4. The line shown in the figure has a positive slope and a negative y-intercept. Explain why this is true.

For each of the following lines, write the equation in standard form.

5. Through $(-2, 3)$ and $(6, -1)$

6. Through $(-6, 2)$, perpendicular to $x = 4$

7. Through $(-6, 2)$, parallel to $2x - y = 8$

8. Graph the line with the equation $4x - 5y = 10$. Give the domain and range.

9. Graph the line through $(2, -3)$ with slope $-2/3$.

10. Graph $y = -(x + 3)^2 + 4$. Give the axis, vertex, domain, and range.

11. Give the vertex and axis of the graph of $y = 3x^2 - 12x + 9$.

12. Graph $x = -(y + 2)^2 - 4$. Give the axis, vertex, domain, and range.

13. A developer has 80 ft of fencing. He wants to put a fence around a rectangular plot of land adjacent to a river. Use a quadratic relation to find the maximum area he can enclose. (*Note:* the side next to the river will not be fenced.)

14. Find the equation of a circle centered at $(5, -1)$ with radius 4. Graph the circle, and give the domain and range.

15. Find the center and radius of the circle with equation
$$x^2 + y^2 + 4x - 6y = -4.$$

16. Decide whether the equation $y = 2x^3 - x$ has a graph that is symmetric with respect to the x-axis, y-axis, or the origin.

Graph each of the following relations. Give the domain and range.

17. $\dfrac{(x + 3)^2}{4} + \dfrac{(y - 2)^2}{9} = 1$

18. $9y^2 = 25x^2 + 225$

19. $y = \sqrt{25 - x^2}$

20. Graph the inequality $(x - 4)^2 + (y + 3)^2 > 9$.

■ THE GRAPHING CALCULATOR ■

The emergence of scientific calculators with graphing capabilities over the past few years has opened new avenues for investigating and understanding the concepts of graphing. This is the first in a series of features on the graphing calculator in this text. Since graphing calculators vary between makes and models with respect to methods of input of information, we will, in general, discuss features and capabilities in general terms and not attempt to address particular features of individual makes and models. You should read your owner's manual to learn about your particular model.

Suppose that we wish to graph the line with equation $3x + 2y = 6$. In order to input this information we must first solve the equation for y to get $y = -1.5x + 2$, since graphing calculators are designed to accept equations that are solved for y. (The reason for this will become more apparent after you have studied Chapter 4, *Functions*.) We must also set the domain and range using the range function on the calculator. When the relation is graphed we get the line seen in Figure 3.7 (in Section 3.2).

Vertical parabolas, such as $y = -3x^2 - 2x + 1$, can also be graphed using the same techniques as those for graphing lines. For example, if we graph this parabola, using x-values from -2 to 2, and y-values from -3 to 2, we get the parabola found in Figure 3.22 (in Section 3.3). By using the zooming and tracing features of the calculator, we can verify that the vertex of this parabola is approximately $(-.33, 1.33)$.

Horizontal parabolas, ellipses, and hyperbolas require special consideration if we wish to graph them on a graphing calculator. You should understand, after studying functions in Chapter 4, why it is necessary to rewrite each equation as two separate equations to graph them.

The types of equations in one variable, studied in Chapter 2, can be solved graphically as well. In Example 3 of Section 2.1 we algebraically solved

$$3(2x - 4) = 7 - (x + 5).$$

In order to solve this equation graphically, simplify the left side to get $6x - 12$, and graph $y_1 = 6x - 12$. Then simplify the right side to get $2 - x$, and graph on the same axes $y_2 = 2 - x$. Now use the tracing feature of the calculator to find the coordinates of the point of intersection of these two lines. The x-value of this point, 2, is the value for which the left and right sides are equal, and thus is the solution of the equation.

You might want to use the following relations and linear equations to experiment with your calculator. These examples have been selected from examples in the text, so you can compare your calculator graphs with those given in the example figures. The figure number of the graph in the text is given with each problem.

---

1. $y = -3$ (Figure 3.8)
2. $4x - 5y = 0$ (Figure 3.10)
3. $5x + 3y = 10$ (Figure 3.13)
4. $3x - y = 2$ (Figure 3.15)
5. $y = x^2$ (Figure 3.16)
6. $y = 2x^2$ (Figure 3.17)
7. $y = -\frac{1}{2}x^2$ (Figure 3.18)
8. $y = x^2 - 4$ (Figure 3.19)
9. $y = (x - 4)^2$ (Figure 3.20)
10. $y = -(x + 3)^2 + 1$ (Figure 3.21)
11. $y = -\sqrt{1 + 4x^2}$ (Figure 3.49)
12. $3x - 2y = 6$ (Figure 3.51)

In the remaining items, solve the equation for x algebraically, and then verify your solution graphically as explained above.

13. $2x + 3(x - 4) = 2(x - 3)$
14. $6x - 4(3 - 2x) = 5(x - 4) - 10$
15. $x^2 - 8 = 2x$
16. $|x + 2| = 4$

FUNCTIONS

4.1 FUNCTIONS

4.2 VARIATION: APPLICATIONS OF FUNCTIONS

4.3 COMBINING FUNCTIONS; ALGEBRA OF FUNCTIONS

4.4 GRAPHING BASIC FUNCTIONS AND THEIR VARIATIONS

4.5 INVERSE FUNCTIONS

In Chapter 3 we defined a relation as a set of ordered pairs. A *function* is a special type of relation that is very important in applications of mathematics because each first element in an ordered pair of a function corresponds to exactly one second element. This describes many situations in our everyday lives. For example, each amount of pressure on the accelerator of an automobile corresponds to exactly one speed; if you are paid by the hour, then each number of hours worked corresponds to exactly one amount on your paycheck.

4.1 ———— FUNCTIONS

In this section, we will concentrate on identifying those relations that are functions, using the following definition.

FUNCTION	A **function** is a relation in which each element in the domain corresponds to exactly one element in the range.*

Suppose a group of students get together each Monday evening to study algebra (and perhaps watch football). A number giving the student's weight to the nearest kilogram can be associated with each member of this set of students. Since each student has only one weight at a given time, the relationship between the students and their weights is a function. The domain is the set of all students in the group, while the range is the set of all the weights of the students.

If x represents any element in the domain, the set of students, x is called the **independent variable.** If y represents any element in the range, the weights, then y is called the **dependent variable,** because the value of y *depends on* the value of x. That is, each weight depends on the student associated with it.

In most mathematical applications of functions, the correspondence between the domain and range elements is defined with an equation, like $y = 5x - 11$. The equation is usually solved for y, as it is here, because y is the dependent variable. As we choose values from the domain for x, we can easily determine the corresponding y values of the ordered pairs of the function. (These equations need not use only x and y as variables; any appropriate letters may be used. In physics, for example, t is often used to represent the independent variable *time*.)

■ ***Example 1***
DECIDING WHETHER A RELATION IS A FUNCTION

Decide whether the following sets are functions. Give the domain and range of each relation.

(a) $\{(1, 2), (3, 4), (5, 6), (7, 8), (9, 10)\}$

The domain is the set $\{1, 3, 5, 7, 9\}$, and the range is $\{2, 4, 6, 8, 10\}$. Since each element in the domain corresponds to just one element in the range, this set is

---

*A function may also be defined as a *mapping* from one set, the domain, into another set, the range.

a function. The correspondence is shown below using D for the domain and R for the range.

$$D = \{1, 3, 5, 7, 9\}$$
$$\downarrow \downarrow \downarrow \downarrow \downarrow$$
$$R = \{2, 4, 6, 8, 10\}$$

(b) $\{(1, 1), \ (1, 2), (1, 3) \ (2, 4)\}$

The domain here is $\{1, 2\}$, and the range is $\{1, 2, 3, 4\}$. As shown in the correspondence below, one element in the domain, 1, has been assigned three different elements from the range, so this relation is not a function.

$$D = \{1, 2\}$$

$$R = \{1, 2, 3, 4\}$$

(c) $\{(-5, 2), (-4, 2), (-3, 2), (-2, 2), (-1, 2)\}$

Here, the domain is $\{-5, -4, -3, -2, -1\}$, and the range is $\{2\}$. Although every element in the domain corresponds to the same range element, this is a function because each element in the domain has exactly one range element assigned to it.

(d) $\{(x, y)|y = x - 2, x \text{ any real number}\}$

Since y is always found by subtracting 2 from x, each x corresponds to just one y, so this relation is a function. Any number can be used for x, and each x will give a number 2 smaller for y; thus, both the domain and the range are the set of real numbers, or in interval notation, $(-\infty, \infty)$. ■

FUNCTION NOTATION It is common to use the letters f, g, and h to name functions. If f is a function and x is an element in the domain of f, then $f(x)$, read "f of x," is the corresponding element in the range. The notation $f(x)$ is an abbreviation for "(the function) f evaluated at x." For example, if f is the function in which a value of x is squared to give the corresponding value in the range, f could be defined as

$$f(x) = x^2.$$

If $x = -5$ is an element from the domain of f, the corresponding element from the range is found by replacing x with -5. This is written

$$f(-5) = (-5)^2 = 25.$$

The number -5 in the domain corresponds to 25 in the range, and the ordered pair $(-5, 25)$ belongs to the function f.

This *function notation* can be summarized as follows.

> **NOTE** Keep in mind that $f(x)$ is just another (more meaningful) notation for y. You can always replace the notation $f(x)$ with y or vice versa.

■ ***Example 2***

USING FUNCTION NOTATION

Let $g(x) = 3\sqrt{x}$, $h(x) = 1 + 4x$, and $k(x) = x^2 + 3$. Find each of the following.

(a) $g(16)$

To find $g(16)$, replace x in $g(x) = 3\sqrt{x}$ with 16, getting

$$g(16) = 3\sqrt{16} = 3 \cdot 4 = 12.$$

(b) $h(-3) = 1 + 4(-3) = -11$

(c) $k(-2) = (-2)^2 + 3 = 4 + 3 = 7$

(d) $g(-4)$ is undefined; -4 is not in the domain of g since $\sqrt{-4}$ is undefined.[*]

(e) $h(\pi) = 1 + 4\pi$

(f) $g(m) = 3\sqrt{m}$, if m represents a nonnegative real number.

(g) $k(a + b)$

Let $x = a + b$ in $k(x) = x^2 + 3$.

$$k(a + b) = (a + b)^2 + 3 = a^2 + 2ab + b^2 + 3. \quad ■$$

■ ***Example 3***

FINDING THE DIFFERENCE-QUOTIENT

Let $f(x) = 2x^2 - 3x$, and find the *difference-quotient*

$$\frac{f(x + h) - f(x)}{h}, \quad \text{where } h \neq 0,$$

which is important in calculus.

To find $f(x + h)$, replace x in $f(x)$ with $x + h$, to get

$$f(x + h) = 2(x + h)^2 - 3(x + h).$$

Then

$$\frac{f(x + h) - f(x)}{h} = \frac{2(x + h)^2 - 3(x + h) - (2x^2 - 3x)}{h}$$

$$= \frac{2(x^2 + 2xh + h^2) - 3x - 3h - 2x^2 + 3x}{h} \quad \text{Square } x + h; \text{ use the distributive property.}$$

$$= \frac{2x^2 + 4xh + 2h^2 - 3x - 3h - 2x^2 + 3x}{h}$$

$$= \frac{4xh + 2h^2 - 3h}{h} \quad \text{Combine terms.}$$

$$= \frac{h(4x + 2h - 3)}{h} \quad \text{Factor out } h.$$

$$= 4x + 2h - 3. \quad \text{Divide.} \quad ■$$

---

[*]In our work with functions, we restrict all variables to *real numbers*.

Notice that $f(x + h)$ is not the same as $f(x) + f(h)$. For $f(x) = 2x^2 - 3x$, as shown in Example 3,

$$f(x + h) = 2(x + h)^2 - 3(x + h) = 2x^2 + 4xh + 2h^2 - 3x - 3h$$

but

$$f(x) + f(h) = (2x^2 - 3x) + (2h^2 - 3h) = 2x^2 - 3x + 2h^2 - 3h.$$

These expressions differ by $4xh$.

DOMAIN AND RANGE Throughout this book, if the domain for a function specified by an algebraic formula is not given, it will be assumed to be the largest possible set of real numbers for which the formula is meaningful. For example, if

$$f(x) = \frac{-4x}{2x - 3},$$

then any real number can be used for x except $x = 3/2$, which makes the denominator equal to 0. Based on our assumption, the domain of this function must be $\{x | x \neq 3/2\}$ or $(-\infty, 3/2) \cup (3/2, \infty)$ in interval form.

■ **Example 4**
FINDING DOMAIN AND RANGE

Give the domain and range of each of the following functions.

(a) $f(x) = 4 - 3x$

Here x can be any real number. Multiplying by -3 and adding 4 will give a real number $f(x)$ for each value of x. Thus, both the domain and range are the set of all real numbers. In interval notation, both the domain and range are $(-\infty, \infty)$.

(b) $f(x) = x^2 + 4$

Since any real number can be squared, the domain is the set of all real numbers, $(-\infty, \infty)$. The square of any number is nonnegative, so that $x^2 \geq 0$ and $x^2 + 4 \geq 4$. The range is the interval $[4, \infty)$. We can verify this by noting that the graph of f is a parabola that opens upward with vertex at $(0, 4)$.

(c) $g(x) = \sqrt{x - 2}$

If $\sqrt{x - 2}$ is to be a real number, then

$$x - 2 \geq 0 \quad \text{or} \quad x \geq 2,$$

so the domain of the function is given by the interval $[2, \infty)$. Since $\sqrt{x - 2}$ is nonnegative, the range is $[0, \infty)$.

(d) $h(x) = -\sqrt{100 - x^2}$

For $-\sqrt{100 - x^2}$ to be a real number,

$$100 - x^2 \geq 0.$$

Since $100 - x^2$ factors as $(10 - x)(10 + x)$, use a sign graph to verify that

$$-10 \leq x \leq 10,$$

making the domain of the function $[-10, 10]$. As x takes values from -10 to 10, $h(x)$ (or y) goes from 0 to -10 and back to 0. (Verify this by substituting -10, 0, 10, and some values in between for x.) Thus, the range is $[-10, 0]$.

(e) $k(x) = \dfrac{2}{x - 5}$

The quotient is undefined if the denominator is 0, so x cannot equal 5. Therefore, the domain is written in interval notation as $(-\infty, 5) \cup (5, \infty)$. As x takes on all real-number values except 5, y will take on all real-number values except 0. The range is $(-\infty, 0) \cup (0, \infty)$. ■

■ **FIGURE 4.1**

Example 4 suggests some generalizations about the types of functions where the domain is restricted. As shown in parts (c) and (d), for functions involving even roots, the domain includes only numbers that make a radicand nonnegative, and in part (e) the domain includes only numbers that make a denominator nonzero.

For most of the functions discussed in this book, the domain can be found with the methods already presented. As Example 4 suggests, one reason for studying inequalities is to find domains. The range, however, often must be found by using graphing, more involved algebra, or calculus. For example, the graph in Figure 4.1 suggests that the function has domain $[-3, 3]$ and range $[-1, 3]$.

VERTICAL LINE TEST There is a quick way to tell whether a given graph is the graph of a function. Figure 4.2 shows two graphs. In the graph for part (a), each value of x leads to only one value of y, so that this is the graph of a function. On the other hand, the graph in part (b) is not the graph of a function. For exam-

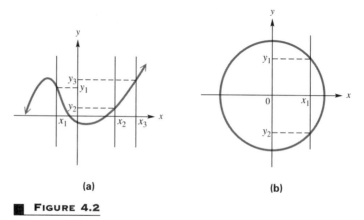

(a)   (b)

■ **FIGURE 4.2**

ple, if $x = x_1$, the vertical line through x_1 intersects the graph at two points, showing that there are two values of y that correspond to this x-value. This idea is known as the *vertical line test* for a function.

VERTICAL LINE TEST
If each vertical line intersects a graph in no more than one point, the graph is the graph of a function.

NOTE By the vertical line test, lines and vertical parabolas are graphs of functions; horizontal parabolas, circles, ellipses, and hyperbolas are not.

APPLYING FUNCTIONS In manufacturing, the cost of making a product usually consists of two parts. One part is a *fixed cost* for designing the product, setting up a factory, training workers, and so on. Usually the fixed cost is constant for a particular product and does not change as more items are made. The other part of the cost is a *variable cost* per item for labor, materials, packaging, shipping, and so on. The variable cost is often the same per item, so that the total amount of variable cost increases as more items are produced. A *linear cost function* has the form $C(x) = mx + b$, where m represents the variable cost per item and b represents the fixed cost. The revenue from selling a product depends on the price per item and the number of items sold, as given by the *revenue function*, $R(x) = px$, where p is the price per item and $R(x)$ is the revenue from the sale of x items. The profit is described by the *profit function* given by $P(x) = R(x) - C(x)$.

■ **Example 5**
WRITING LINEAR COST FUNCTIONS

(a) Assume that the cost to produce an item is a linear function. If the fixed cost is \$1500 and the variable cost is \$100, write a cost function for the product.

Since the cost function is linear it will have the form $C(x) = mx + b$, with $m = 100$ and $b = 1500$. That is,

$$C(x) = 100x + 1500.$$

(b) Find the revenue function if the item in part (a) sells for \$125.

The revenue function is

$$R(x) = px = 125x. \qquad \text{Let } p = 125.$$

(c) Give the profit function for the item in part (a).

The profit function is given by

$$\begin{aligned}
P(x) &= R(x) - C(x) \\
&= 125x - (100x + 1500) \\
&= 125x - 100x - 1500 \\
&= 25x - 1500.
\end{aligned}$$

(d) How many items must be produced and sold before the company makes a profit?

To make a profit, $P(x)$ must be positive. Set $P(x) = 25x - 1500 > 0$ and solve for x.

$$25x - 1500 > 0$$
$$25x > 1500 \qquad \text{Add 1500 to each side.}$$
$$x > 60 \qquad \text{Divide by 25.}$$

At least 61 items must be sold for the company to make any profit. ■

4.1 EXERCISES ■

For each of the following, find the indicated function values: (a) $f(-2)$, (b) $f(0)$, (c) $f(1)$, *and* (d) $f(4)$.

1.

2.

3.

4.

Let $f(x) = 3x - 1$ and $g(x) = x^2$. Find each of the following function values. See Example 2.

5. $f(-1)$ **6.** $f(-3)$ **7.** $f(4)$ **8.** $g(2)$ **9.** $f(a)$

10. $g(b)$ **11.** $f(1) + g(1)$ **12.** $f(-2) - g(-2)$ **13.** $f(3) \cdot g(3)$ **14.** $\dfrac{g(-1)}{f(-1)}$

15. $f(-2m)$ **16.** $f(-11p)$ **17.** $f(5a - 2)$ **18.** $f(3 + 2k)$

19. Complete the following sentence. If the ordered pair $(2, 5)$ belongs to function g, then $g(\underline{\hspace{1cm}}) = \underline{\hspace{1cm}}$.

Give the domain and range of each function defined as follows. See Example 4.

20.

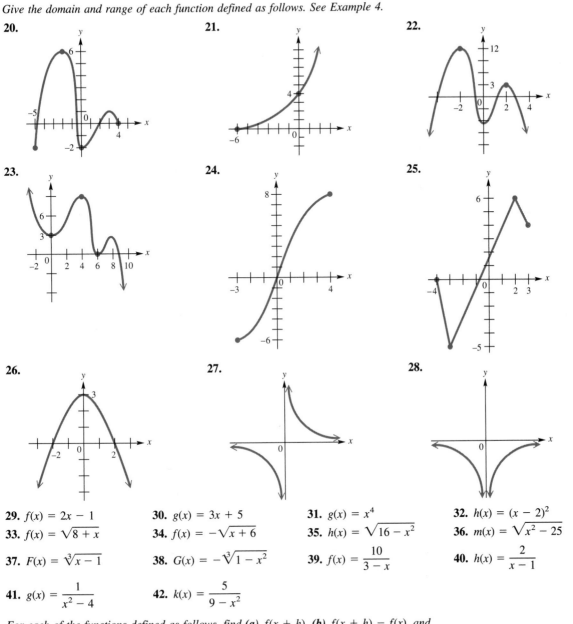

21.

22.

23.

24.

25.

26.

27.

28.

29. $f(x) = 2x - 1$

30. $g(x) = 3x + 5$

31. $g(x) = x^4$

32. $h(x) = (x - 2)^2$

33. $f(x) = \sqrt{8 + x}$

34. $f(x) = -\sqrt{x + 6}$

35. $h(x) = \sqrt{16 - x^2}$

36. $m(x) = \sqrt{x^2 - 25}$

37. $F(x) = \sqrt[3]{x - 1}$

38. $G(x) = -\sqrt[3]{1 - x^2}$

39. $f(x) = \dfrac{10}{3 - x}$

40. $h(x) = \dfrac{2}{x - 1}$

41. $g(x) = \dfrac{1}{x^2 - 4}$

42. $k(x) = \dfrac{5}{9 - x^2}$

*For each of the functions defined as follows, find (**a**) $f(x + h)$, (**b**) $f(x + h) - f(x)$, and*

(**c**) $\dfrac{f(x + h) - f(x)}{h}$. *See Example 3.*

43. $f(x) = 6x + 2$

44. $f(x) = 4x + 11$

45. $f(x) = -2x + 5$

46. $f(x) = 8 - 3x$

47. $f(x) = 1 - x^2$

48. $f(x) = x^2 - 4$

49. $f(x) = 8 - 3x^2$

Let $f(x) = 2^x$ for all rational numbers x. Find each of the following function values. Assume that m and r represent rational numbers.

50. $f(2)$ **51.** $f(4)$ **52.** $f(-2)$ **53.** $f(-4)$

54. $f(m)$ **55.** $f(-5r)$ **56.** $f(1/2)$ **57.** $f(1/4)$

58. Give an example of a function from everyday life. (*Hint:* Fill in the blanks: _____ depends on _____, so _____ is a function of _____.)

A firm will break even (*no profit and no loss*) *as long as revenue just equals cost. The value of x (the number of items produced and sold) where $C(x) = R(x)$ is called the* break-even point. *Assume that each of the following can be expressed as a linear cost function. Find (a) the cost function, (b) the revenue function, and (c) the profit function. (d) Find the break-even point; then decide whether the product should be produced. See Example 5.*

	Fixed Cost	**Variable Cost**	**Price of Item**	
59.	$500	$10	$35	No more than 18 units can be sold.
60.	$180	$11	$20	No more than 30 units can be sold.
61.	$2700	$150	$280	No more than 25 units can be sold.
62.	$1650	$400	$305	All units produced can be sold.

63. The manager of a small company that produces roof tile has determined that his total cost in dollars, $C(x)$, of producing x units of tile is given by $C(x) = 200x + 1000$, while the revenue in dollars, $R(x)$, from the sale of x units of tile is given by $R(x) = 240x$.

(a) Find the break-even point.

(b) Graph functions R and C on the same axes. Identify the break-even point. Identify any regions of profit or loss.

(c) What is the cost/revenue at the break-even point?

64. Suppose the manager of the company in Exercise 63 finds that he has miscalculated his variable cost, and that it is actually $220 per unit, instead of $200. How does this affect the break-even point? Is he better off or not?

4.2 ——————— **VARIATION: APPLICATIONS OF FUNCTIONS**

In many applications of mathematics, it is necessary to express relationships between quantities. For example, in chemistry, the ideal gas law shows how temperature, pressure, and volume are related. In physics, various formulas in optics show how the focal length of a lens and the size of an image are related. All of these relations are functions.

DIRECT VARIATION If the quotient of two variables is constant, then one variable *varies directly* as or is *directly proportional* to the other. This idea can be stated in a different way as follows.

DIRECT VARIATION y **varies directly** as x, or y is **directly proportional** to x, if a nonzero real number k, called the **constant of variation,** exists such that

$$y = kx.$$

| NOTE | If y varies directly as x, then y is a linear function of x. |

■ *Example 1*
SOLVING A DIRECT
VARIATION PROBLEM

Suppose the area of a certain rectangle varies directly as the length. If the area is 50 square meters when the length is 10 meters, find the area when the length is 25 meters.

Since the area varies directly as the length,

$$A = kl,$$

where A represents the area of the rectangle, l is the length, and k is a nonzero constant that must be found. When $A = 50$ and $l = 10$, the equation $A = kl$ becomes

$$50 = 10k \quad \text{or} \quad k = 5.$$

Using this value of k, the relationship between the area and the length can be expressed as

$$A = 5l.$$

To find the area when the length is 25, replace l with 25 to get

$$A = 5l = 5(25) = 125.$$

The area of the rectangle is 125 square meters when the length is 25 meters. ■

Sometimes y varies as a power of x. In this case y is a polynomial function of x.

DIRECT VARIATION
AS nTH POWER

Let n be a positive real number. Then y **varies directly as the nth power** of x, or y is **directly proportional to the nth power** of x, if a nonzero real number k exists such that

$$y = kx^n.$$

The phrase "directly proportional" is sometimes abbreviated to just "proportional."

For example, the area of a square of side x is given by the formula $A = x^2$, so that the area varies directly as the square of the length of a side. Here $k = 1$.

INVERSE VARIATION   The case where y increases as x decreases is an example of inverse variation, where y is a rational function of x. In this case, the product of the variables is constant.

INVERSE
VARIATION

Let n be a positive real number. Then y **varies inversely as the nth power** of x, or y is **inversely proportional** to the nth power of x, if a nonzero real number k exists such that

$$y = \frac{k}{x^n}.$$

If $n = 1$, then $y = k/x$, and y **varies inversely** as x.

■ *Example 2*

SOLVING AN
INVERSE VARIATION
PROBLEM

In a certain manufacturing process, the cost of producing a single item varies inversely as the square of the number of items produced. If 100 items are produced, each costs $2. Find the cost per item if 400 items are produced.

Let x represent the number of items produced and y the cost per item, and write

$$y = \frac{k}{x^2}$$

for some nonzero constant k. Since $y = 2$ when $x = 100$,

$$2 = \frac{k}{100^2} \quad \text{or} \quad k = 20{,}000.$$

Thus, the relationship between x and y is given by

$$y = \frac{20{,}000}{x^2}.$$

When 400 items are produced, the cost per item is

$$y = \frac{20{,}000}{400^2} = .125, \text{ or } 12.5¢. \quad ■$$

The steps involved in solving a problem in variation are summarized here.

SOLVING
VARIATION
PROBLEMS

1. Write the general relationship among the variables as a function. Use the constant k.
2. Substitute given values of the variables and find the value of k.
3. Substitute this value of k into the equation from Step 1, obtaining a specific formula.
4. Solve for the required unknown.

COMBINED AND JOINT VARIATION One variable may depend on more than one other variable. Such variation is called *combined variation*. More specifically, when a variable depends on the *product* of two or more other variables, it is referred to as *joint variation*.

JOINT VARIATION

Let m and n be real numbers. Then y **varies jointly** as the nth power of x and the mth power of z if a nonzero real number k exists such that

$$y = kx^n z^m.$$

■ *Example 3*

SOLVING A JOINT
VARIATION PROBLEM

The area of a triangle varies jointly as the lengths of the base and the height. A triangle with a base of 10 feet and a height of 4 feet has an area of 20 square feet. Find the area of a triangle with a base of 3 centimeters and a height of 8 centimeters.

Let A represent the area, b the base, and h the height of the triangle. Then

$$A = kbh$$

for some number k. Since A is 20 when b is 10 and h is 4,

$$20 = k(\mathbf{10})(\mathbf{4})$$

$$\frac{1}{2} = k.$$

Then
$$A = \frac{1}{2} bh,$$

which is the familiar formula for the area of a triangle. When $b = 3$ centimeters and $h = 8$ centimeters,

$$A = \frac{1}{2} (\mathbf{3})(\mathbf{8}) = 12 \text{ square centimeters.} \quad ■$$

The final example in this section shows combined variation.

■ **Example 4**

SOLVING A COMBINED VARIATION PROBLEM

The number of vibrations per second (the pitch) of a steel guitar string varies directly as the square root of the tension and inversely as the length of the string. If the number of vibrations per second is 5 when the tension is 225 kilograms and the length is .60 meter, find the number of vibrations per second when the tension is 196 kilograms and the length is .65 meter.

Let n represent the number of vibrations per second, T represent the tension, and L represent the length of the string. Then, from the information in the problem,

$$n = \frac{k\sqrt{T}}{L}.$$

Substitute the given values for n, T, and L to find k.

$$5 = \frac{k\sqrt{225}}{.60} \qquad \text{Let } n = 5, T = 225, L = .60.$$

$$3 = k\sqrt{225} \qquad \text{Multiply by .60.}$$

$$3 = 15k \qquad \sqrt{225} = 15$$

$$k = \frac{1}{5} = .2 \qquad \text{Divide by 15.}$$

Now substitute for k and use the second set of values for T and L to find n.

$$n = \frac{.2\sqrt{196}}{.65} \qquad \text{Let } k = .2, T = 196, L = .65.$$

$$n = 4.3$$

The number of vibrations per second is 4.3. ■

222 ■ **IN SIMPLEST TERMS** ─────────────────────────────── ■

15

Variation can be seen extensively in the field of photography. The formula $L = \dfrac{25F^2}{st}$ represents a combined variation. The luminance, L, varies directly as the square of the F-stop, F. It also varies inversely as the product of the film ASA number, s, and the shutter speed, t. The constant of variation is 25.

Suppose we want to use 200 ASA film and a shutter speed of 1/250 when 500 footcandles of light are available. What would be an appropriate F-stop?

$$L = \frac{25F^2}{st}$$

$$500 = \frac{25F^2}{200\left(\dfrac{1}{250}\right)}$$

$$400 = 25F^2$$

$$16 = F^2$$

$$4 = F$$

An F-stop of 4 would be appropriate.

SOLVE EACH PROBLEM

A. Determine the luminance needed when a photographer is using 400 ASA film, a shutter speed of 1/60 of a second, and an F-stop of 5.6.

B. If 125 footcandles of light are available and an F-stop of 2 is used with 200 ASA film, what shutter speed should be used?

ANSWERS   A. About 118 footcandles of light are needed.   B. The shutter speed should be 1/250 (or .004) of a second.

■ ─── ■

4.2 EXERCISES ■ ──────────────────────────────────────

Express each of the following statements as an equation.

1. a varies directly as b.

2. m is proportional to n.

3. x is inversely proportional to y.

4. p varies inversely as y.

5. r varies jointly as s and t.

6. R is proportional to m and p.

7. w is proportional to x^2 and inversely proportional to y.

8. c varies directly as d and inversely as f^2 and g.

Solve each of the following problems. See Examples 1–4.

9. If m varies directly as x and y, and $m = 10$ when $x = 4$ and $y = 7$, find m when $x = 11$ and $y = 8$.

10. Suppose m varies directly as z and p. If $m = 10$ when $z = 3$ and $p = 5$, find m when $z = 5$ and $p = 7$.

11. Suppose r varies directly as the square of m, and inversely as s. If $r = 12$ when $m = 6$ and $s = 4$, find r when $m = 4$ and $s = 10$.

12. Suppose p varies directly as the square of z, and inversely as r. If $p = 32/5$ when $z = 4$ and $r = 10$, find p when $z = 2$ and $r = 16$.

13. Let a be proportional to m and n^2, and inversely proportional to y^3. If $a = 9$ when $m = 4$, $n = 9$, and $y = 3$, find a if $m = 6$, $n = 2$, and $y = 5$.

14. If y varies directly as x, and inversely as m^2 and r^2, and $y = 5/3$ when $x = 1$, $m = 2$, and $r = 3$, find y if $x = 3$, $m = 1$, and $r = 8$.

15. For $k > 0$, if y varies directly as x, when x increases, y _____, and when x decreases, y _____.

16. For $k > 0$, if y varies inversely as x, when x increases, y _____, and when x decreases, y _____.

In Exercises 17–22, tell whether each equation represents direct, inverse, joint, or combined variation.

17. $y = \dfrac{4}{x}$ **18.** $y = \dfrac{x}{zw}$ **19.** $y = 6xz^2$ **20.** $y = 3x$ **21.** $y = \dfrac{x^2}{zw^3}$ **22.** $y = \dfrac{10}{x}$

Solve each of the following problems. See Examples 1–4. A calculator will be helpful for many of these problems.

23. Hooke's law for an elastic spring states that the distance a spring stretches varies directly as the force applied. If a force of 15 lb stretches a certain spring 8 in, how much will a force of 30 lb stretch the spring?

24. In electric current flow, it is found that the resistance (measured in units called ohms) offered by a fixed length of wire of a given material varies inversely as the square of the diameter of the wire. If a wire .01 inch in diameter has a resistance of .4 ohm, what is the resistance of a wire of the same length and material with a diameter of .03 inch?

25. The illumination produced by a light source varies inversely as the square of the distance from the source. The illumination of a light source at 5 m is 70 candela. What is the illumination 12 m from the source?

26. The pressure exerted by a certain liquid at a given point is proportional to the depth of the point below the surface of the liquid. If the pressure 20 m below the surface is 70 kg per cm^2, what pressure is exerted 40 m below the surface?

27. The distance that a person can see to the horizon from a point above the surface of the earth varies directly as the square root of the height of the point above the earth. A person on a hill 121 m high can see for 15 km to the horizon. How far can a person see to the horizon from a hill 900 m high?

28. Simple interest varies jointly as principal and time. If $1000 left at interest for 2 yr earned $110, find the amount of interest earned by $5000 for 5 yr.

29. The volume of a right circular cylinder is jointly proportional to the square of the radius of the circular base and to the height. If the volume is 300 cu cm when the height is 10.62 cm and the radius is 3 cm, find the volume for a cylinder with a radius of 4 cm and a height of 15.92 cm.

30. The roof of a new sports arena rests on round concrete pillars. The maximum load a cylindrical column of circular cross section can hold varies directly as the fourth power of the diameter and inversely as the square of the height. The arena has 9 m tall columns that are 1 m in diameter and will support a load of 8 metric tons. How many metric tons will be supported by a column 12 m high and 2/3 m in diameter?

31. The sports arena in Exercise 30 requires a beam 16 m long, 24 cm wide, and 8 cm high. The maximum load of a horizontal beam that is supported at both ends varies directly as the width and square of the height and inversely as the length between supports. If a beam of the same material 8 m long, 12 cm wide, and 15 cm high can support a maximum of 400 kg, what is the maximum load the beam in the arena will support?

32. The period of a pendulum varies directly as the square root of the length of the pendulum and inversely as the square root of the acceleration due to gravity. Find the period when the length is 121 cm and the acceleration due to gravity is 980 cm per sec squared, if the period is 6π sec when the length is 289 cm and the acceleration due to gravity is 980 cm per sec squared.

33. The force needed to keep a car from skidding on a curve varies inversely as the radius of the curve and

224

jointly as the weight of the car and the square of the speed. It takes 3000 lb of force to keep a 2000-lb car from skidding on a curve of radius 500 ft at 30 mph. What force is needed to keep the same car from skidding on a curve of radius 800 ft at 60 mph?

34. The pressure on a point in a liquid is directly proportional to the distance from the surface to the point. In a certain liquid the pressure at a depth of 4 m is 60 kg per m^2. Find the pressure at a depth of 10 m.

35. The volume V of a gas varies directly as the temperature T and inversely as the pressure P. If V is 10 when T is 280 and P is 6, find V if T is 300 and P is 10.

36. Under certain conditions, the length of time that it takes for fruit to ripen during the growing season varies inversely as the average maximum temperature during the season. If it takes 25 days for fruit to ripen with an average maximum temperature of 80°, find the number of days it would take at 75°.

37. The number of long-distance phone calls between two cities in a certain time period varies directly as the populations p_1 and p_2 of the cities, and inversely as the distance between them. If 10,000 calls are made between two cities 500 mi apart, having populations of 50,000 and 125,000, find the number of calls between two cities 800 mi apart having populations of 20,000 and 80,000.

38. The horsepower needed to run a boat through water varies directly as the cube of the speed. If 80 horsepower are needed to go 15 km/hr in a certain boat, how many horsepower would be needed to go 30 km/hr?

39. According to Poiseuille's law, the resistance to flow of a blood vessel, R, is directly proportional to the length, l, and inversely proportional to the fourth power of the radius, r. If $R = 25$ when $l = 12$ and $r = .2$, find R as r increases to .3, while l is unchanged.

40. The Stefan-Boltzmann law says that the radiation of heat R from an object is directly proportional to the fourth power of the Kelvin temperature of the object. For a certain object, $R = 213.73$ at room temperature (293° Kelvin). Find R if the temperature increases to 335° Kelvin.

41. Suppose a nuclear bomb is detonated at a certain site. The effects of the bomb will be felt over a distance from the point of detonation that is directly proportional to the cube root of the yield of the bomb. Suppose a 100-kiloton bomb has certain effects to a radius of 3 km from the point of detonation. Find the distance that the effects would be felt for a 1500-kiloton bomb.

42. The maximum speed possible on a length of railroad track is directly proportional to the cube root of the amount of money spent on maintaining the track. Suppose that a maximum speed of 25 km/hr is possible on a stretch of track for which $450,000 was spent on maintenance. Find the maximum speed if the amount spent on maintenance is increased to $1,750,000.

43. A measure of malnutrition, called the *pelidisi,* varies directly as the cube root of a person's weight in grams and inversely as the person's sitting height in centimeters. A person with a pelidisi below 100 is considered to be undernourished, while a pelidisi greater than 100 indicates overfeeding. A person who weighs 48,820 g with a sitting height of 78.7 cm has a pelidisi of 100. Find the pelidisi (to the nearest whole number) of a person whose weight is 54,430 g and whose sitting height is 88.9 cm. Is this individual undernourished or overfed?

44. The cost of a pizza varies directly as the square of its radius. If a pizza with a radius of 15 in costs $7, find the cost of a pizza having a radius of 12.5 in. (You might want to do some research at a nearby pizza establishment and see if this assumption is reasonable.)

4.3 — COMBINING FUNCTIONS: ALGEBRA OF FUNCTIONS

When a company accountant sits down to estimate the firm's overhead, the first step might be to find functions representing the cost of materials, labor charges, equipment maintenance, and so on. The sum of these various functions could then be used to find the total overhead for the company. Such methods of combining functions are discussed in this section.

Given two functions f and g, their *sum*, written $f + g$, is defined as

$$(f + g)(x) = f(x) + g(x),$$

for all x such that both $f(x)$ and $g(x)$ exist. Similar definitions can be given for the difference, $f - g$, product, fg, and quotient, f/g, of functions; however, the quotient,

$$\left(\frac{f}{g}\right)(x) = \frac{f(x)}{g(x)},$$

is defined only for those values of x where both $f(x)$ and $g(x)$ exist, and in addition, $g(x) \neq 0$. The various operations on functions are defined as follows.

OPERATIONS ON FUNCTIONS

If f and g are functions, then for all values of x for which both $f(x)$ and $g(x)$ exist, the sum of f and g is defined by

$$(f + g)(x) = f(x) + g(x),$$

the difference of f and g is defined by

$$(f - g)(x) = f(x) - g(x),$$

the product of f and g is defined by

$$(fg)(x) = f(x) \cdot g(x),$$

and the quotient of f and g is defined by

$$\left(\frac{f}{g}\right)(x) = \frac{f(x)}{g(x)}, \text{ where } g(x) \neq 0.$$

The domains of $f + g$, $f - g$, fg, and f/g are summarized below. (Recall that the intersection of two sets is the set of all elements belonging to *both* sets.)

DOMAINS OF $f + g$, $f - g$, fg, f/g

For functions f and g, the domains of $f + g$, $f - g$, and fg include all real numbers in the intersection of the domains of f and g, while the domain of f/g includes those real numbers in the intersection of the domains of f and g for which $g(x) \neq 0$.

■ **Example 1**

USING THE OPERATIONS ON FUNCTIONS

Let $f(x) = x^2 + 1$, and $g(x) = 3x + 5$. Find each of the following.

(a) $(f + g)(1)$

Since $f(1) = 2$ and $g(1) = 8$, use the definition above to get

$$(f + g)(1) = f(1) + g(1) = 2 + 8 = 10.$$

(b) $(f - g)(-3) = f(-3) - g(-3) = 10 - (-4) = 14$

(c) $(fg)(5) = f(5) \cdot g(5) = 26 \cdot 20 = 520$

(d) $\left(\dfrac{f}{g}\right)(0) = \dfrac{f(0)}{g(0)} = \dfrac{1}{5}$ ■

■ **Example 2**

USING THE OPERATIONS ON FUNCTIONS

Let $f(x) = 8x - 9$ and $g(x) = \sqrt{2x - 1}$.

(a) $(f + g)(x) = f(x) + g(x) = 8x - 9 + \sqrt{2x - 1}$

(b) $(f - g)(x) = f(x) - g(x) = 8x - 9 - \sqrt{2x - 1}$

(c) $(fg)(x) = f(x) \cdot g(x) = (8x - 9)\sqrt{2x - 1}$

(d) $\left(\dfrac{f}{g}\right)(x) = \dfrac{f(x)}{g(x)} = \dfrac{8x - 9}{\sqrt{2x - 1}}$

(e) Find the domains of f, g, $f + g$, $f - g$, fg, and f/g.

The domain of f is the set of all real numbers, while the domain of g for $g(x) = \sqrt{2x - 1}$ includes just those real numbers that make $2x - 1 \geq 0$; the domain of g is the interval $[1/2, \infty)$. The domain of $f + g$, $f - g$, and fg is thus $[1/2, \infty)$. With f/g, the denominator cannot be zero, so the value $1/2$ is excluded from the domain. The domain of f/g is $(1/2, \infty)$. ■

COMPOSITION OF FUNCTIONS The sketch in Figure 4.3 shows a function f that assigns to each element x of set X some element y of set Y. Suppose also that a function g takes each element of set Y and assigns a value z of set Z. Using both f and g, then, an element x in X is assigned to an element z in Z. The result of this process is a new function h, that takes an element x in X and assigns an element z in Z. This function h is called the *composition* of functions g and f, written $g \circ f$, and is defined as follows.

COMPOSITION OF FUNCTIONS

If f and g are functions, then the **composite function**, or **composition**, of g and f is

$$(g \circ f)(x) = g[f(x)]$$

for all x in the domain of f such that $f(x)$ is in the domain of g.

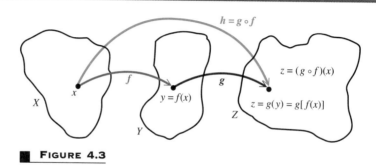

■ **FIGURE 4.3**

Suppose an oil well off the California coast is leaking, with the leak spreading oil in a circular layer over the surface. At any time t, in minutes, after the beginning of the leak, the radius of the circular oil slick is $r(t) = 5t$ feet. Since $A(r) = \pi r^2$ gives the area of a circle of radius r, the area can be expressed as a function of time by substituting $5t$ for r in $A(r) = \pi r^2$ to get

$$A(r) = \pi r^2$$
$$A[r(t)] = \pi(5t)^2 = 25\pi t^2.$$

The function $A[r(t)]$ is a composite function of the functions A and r.

■ **Example 3**

FINDING COMPOSITE FUNCTIONS

Let $f(x) = 4x + 1$ and $g(x) = 2x^2 + 5x$. Find each of the following.

(a) $(g \circ f)(x)$

By definition, $(g \circ f)(x) = g[f(x)]$. Using the given functions,

$$
\begin{aligned}
(g \circ f)(x) = g[f(x)] &= g(4x + 1) && f(x) = 4x + 1 \\
&= 2(4x + 1)^2 + 5(4x + 1) && g(x) = 2x^2 + 5x \\
&= 2(16x^2 + 8x + 1) + 20x + 5 && \text{Square } 4x + 1. \\
&= 32x^2 + 16x + 2 + 20x + 5 && \text{Multiply.} \\
&= 32x^2 + 36x + 7. && \text{Combine terms.}
\end{aligned}
$$

(b) $(f \circ g)(x)$

If we use the definition above with f and g interchanged, $(f \circ g)(x)$ becomes $f[g(x)]$, with

$$
\begin{aligned}
(f \circ g)(x) = f[g(x)] & \\
&= f(2x^2 + 5x) && g(x) = 2x^2 + 5x \\
&= 4(2x^2 + 5x) + 1 && f(x) = 4x + 1 \\
&= 8x^2 + 20x + 1. && \text{Multiply.} \quad ■
\end{aligned}
$$

As this example shows, it is not always true that $f \circ g = g \circ f$. In fact, two composite functions are equal only for a special class of functions, discussed in Section 4.5. In Example 3, the domain of both composite functions is the set of all real numbers.

CAUTION

In general, the composite function $f \circ g$ is not the same as the product fg. For example, with f and g defined as in Example 3,

$$(f \circ g)(x) = 8x^2 + 20x + 1$$

but

$$(fg)(x) = (4x + 1)(2x^2 + 5x) = 8x^3 + 22x^2 + 5x.$$

■ *Example 4*

FINDING COMPOSITE
FUNCTIONS AND
THEIR DOMAINS

Let $f(x) = 1/x$ and $g(x) = \sqrt{3-x}$. Find $f \circ g$ and $g \circ f$. Give the domain of each.

First find $f \circ g$.

$$(f \circ g)(x) = f[g(x)]$$

$$= f(\sqrt{3-x}) \qquad g(x) = \sqrt{3-x}$$

$$= \frac{1}{\sqrt{3-x}} \qquad f(x) = \frac{1}{x}$$

The radical $\sqrt{3-x}$ is a nonzero real number only when $3 - x > 0$ or $x < 3$, so the domain of $f \circ g$ is the interval $(-\infty, 3)$.

Use the same functions to find $g \circ f$, as follows.

$$(g \circ f)(x) = g[f(x)]$$

$$= g\left(\frac{1}{x}\right) \qquad f(x) = \frac{1}{x}$$

$$= \sqrt{3 - \frac{1}{x}} \qquad g(x) = \sqrt{3-x}$$

$$= \sqrt{\frac{3}{1} \cdot \frac{x}{x} - \frac{1}{x}} \qquad \text{Get a common denominator.}$$

$$= \sqrt{\frac{3x-1}{x}} \qquad \text{Combine terms.}$$

The domain of $g \circ f$ is the set of all real numbers x such that $x \neq 0$ and $3 - f(x) \geq 0$. As shown above,

$$3 - f(x) = \frac{3x-1}{x}.$$

We need to solve the inequality

$$\frac{3x-1}{x} \geq 0.$$

By the methods of Section 2.7, find values of x that make the numerator and denominator each zero. The required numbers are 0 and 1/3. Use a sign graph to verify that the domain of $g \circ f$ is the set $(-\infty, 0) \cup [1/3, \infty)$. ■

■ *Example 5*

FINDING COMPOSITE
FUNCTIONS AND
THEIR DOMAINS

Given $f(x) = \sqrt{x-2}$ and $g(x) = x^2 + 2$, find $f \circ g$ and $g \circ f$ and their domains.

$$(f \circ g)(x) = \sqrt{(x^2+2) - 2} \qquad \text{Substitute } g(x) \text{ for } x \text{ in } f(x).$$

$$= \sqrt{x^2} \qquad \text{Combine terms.}$$

$$= |x| \qquad \sqrt{x^2} \text{ is the principal square root.}$$

$$(g \circ f)(x) = (\sqrt{x-2})^2 + 2 \qquad \text{Substitute } f(x) \text{ for } x \text{ in } g(x).$$

$$= |x-2| + 2 \qquad (\sqrt{x-2})^2 \text{ must be positive.}$$

The domain of g is $(-\infty, \infty)$ and, since $g(x) \geq 2$ for all x, the domain of $f \circ g$ is $(-\infty, \infty)$. The domain of f is $[2, \infty)$ and $f(x) \geq 0$ for x in $[2, \infty)$. Thus, the domain of $g \circ f$ is $[2, \infty)$. Since $|x - 2| = x - 2$ for x in $[2, \infty)$,

$$(g \circ f)(x) = x - 2 + 2 = x. \quad ■$$

| NOTE | Comparing the domains of $f \circ g$ and $g \circ f$ in Examples 4 and 5 shows that we cannot always look just at the equation of a composite function to determine the domain. |

In calculus it is sometimes useful to treat a function as a composition of two functions. The next example shows how this can be done.

■ **Example 6**

FINDING FUNCTIONS
THAT FORM A
GIVEN COMPOSITE

Suppose $h(x) = \sqrt{2x + 3}$. Find functions f and g, so that $(f \circ g)(x) = h(x)$.

Since there is a quantity, $2x + 3$, under a radical, one possibility is to choose $f(x) = \sqrt{x}$ and $g(x) = 2x + 3$. Then $(f \circ g)(x) = \sqrt{2x + 3}$, as required. Other combinations are possible. For example, we could choose $f(x) = \sqrt{x + 3}$ and $g(x) = 2x$. ■

4.3 EXERCISES ■

For each pair of functions defined as follows, find (a) $f + g$, (b) $f - g$, (c) fg, and (d) f/g. Give the domain of each. See Example 2.

1. $f(x) = 4x - 1$, $g(x) = 6x + 3$

2. $f(x) = 9 - 2x$, $g(x) = -5x + 2$

3. $f(x) = 3x^2 - 2x$, $g(x) = x^2 - 2x + 1$

4. $f(x) = 6x^2 - 11x$, $g(x) = x^2 - 4x - 5$

5. $f(x) = \sqrt{2x + 5}$, $g(x) = \sqrt{4x - 9}$

6. $f(x) = \sqrt{11x - 3}$, $g(x) = \sqrt{2x - 15}$

Let $f(x) = 4x^2 - 2x$ and let $g(x) = 8x + 1$. Find each of the following. See Examples 1–3.

7. $(f + g)(3)$

8. $(f + g)(-5)$

9. $(fg)(4)$

10. $(fg)(-3)$

11. $\left(\dfrac{f}{g}\right)(-1)$

12. $\left(\dfrac{f}{g}\right)(4)$

13. $(f - g)(m)$

14. $(f - g)(2k)$

15. $(f \circ g)(2)$

16. $(f \circ g)(-5)$

17. $(g \circ f)(2)$

18. $(g \circ f)(-5)$

19. $(f \circ g)(k)$

20. $(g \circ f)(5z)$

Find $f \circ g$ and $g \circ f$ and their domains for each pair of functions defined as follows. See Examples 3–5.

21. $f(x) = 8x + 12$, $g(x) = 3x - 1$

22. $f(x) = -6x + 9$, $g(x) = 5x + 7$

23. $f(x) = -x^3 + 2$, $g(x) = 4x$

24. $f(x) = 2x$, $g(x) = 6x^2 - x^3$

25. $f(x) = \dfrac{1}{x}$, $g(x) = x^2$

26. $f(x) = \dfrac{2}{x^4}$, $g(x) = 2 - x$

27. $f(x) = \sqrt{x + 2}$, $g(x) = 8x - 6$

28. $f(x) = 9x - 11$, $g(x) = 2\sqrt{x + 2}$

29. $f(x) = \dfrac{1}{x - 5}$, $g(x) = \dfrac{2}{x}$

30. $f(x) = \dfrac{8}{x - 6}$, $g(x) = \dfrac{4}{3x}$

31. $f(x) = \sqrt{x + 1}$, $g(x) = -\dfrac{1}{x}$

32. $f(x) = \dfrac{8}{x}$, $g(x) = \sqrt{3 - x}$

For each pair of functions defined as follows, show that $(f \circ g)(x) = x$ *and* $(g \circ f)(x) = x$.

33. $f(x) = 8x$, $g(x) = \dfrac{1}{8}x$

34. $f(x) = \dfrac{3}{4}x$, $g(x) = \dfrac{4}{3}x$

35. $f(x) = 8x - 11$, $g(x) = \dfrac{x + 11}{8}$

36. $f(x) = \dfrac{x - 3}{4}$, $g(x) = 4x + 3$

37. $f(x) = x^3 + 6$, $g(x) = \sqrt[3]{x - 6}$

38. $f(x) = \sqrt[5]{x - 9}$, $g(x) = x^5 + 9$

39. Explain how to find the domain of $(f \circ g)(x) = \dfrac{1}{x^2 - 1}$, where $f(x) = \dfrac{1}{x}$ and $g(x) = x^2 - 1$.

40. For each of the following functions f, decide what operation was used with $g(x)$ and $m(x)$ to get $f(x)$.
(a) $f(x) = ax^2$; $g(x) = a$, $m(x) = x^2$
(b) $f(x) = x^2 + k$; $m(x) = x^2$, $g(x) = k$
(c) $f(x) = (x - h)^2$; $m(x) = x^2$, $g(x) = x - h$
(d) $f(x) = a(x - h)^2 + k$; $g(x) = ax^2 + k$, $m(x) = x - h$

In each of the following exercises, a function h is defined. Find f(x) and g(x) such that $h(x) = (f \circ g)(x)$. *Many such pairs of functions exist. See Example 6.*

41. $h(x) = (6x - 2)^2$

42. $h(x) = (11x^2 + 12x)^2$

43. $h(x) = \sqrt{x^2 - 1}$

44. $h(x) = \dfrac{1}{x^2 + 2}$

45. $h(x) = \dfrac{(x - 2)^2 + 1}{5 - (x - 2)^2}$

46. $h(x) = (x + 2)^3 - 3(x + 2)^2$

47. A couple planning their wedding has found that the cost to hire a caterer for the reception depends on the number of guests attending. If 100 people attend, the cost per person will be $2. For each person less than 100, the cost will increase by $.20. Assume that no more than 100 people will attend. Let x represent the number less than 100 that attend. For example, if 95 attend, $x = 5$.
(a) Write a function $N(x)$ for the possible number of guests.
(b) Write a function $G(x)$ for the possible cost per guest.
(c) Write the function $N(x) \cdot G(x)$ for the total cost, $C(x)$.

48. The manager of a music store has found that the price $p(x)$, in dollars, of a compact disc depends on the demand for the disc, x. For one particular disc,

$$p(x) = -.001x + 20.$$

Express the revenue as $R(x)$, where $R(x)$ is the product of price and demand, that is, $p(x) \cdot x$.

49. Suppose the population P of a certain species of fish depends on the number x (in hundreds) of a smaller kind of fish which serves as its food supply, so that

$$P(x) = 2x^2 + 1.$$

Suppose, also, that the number x (in hundreds) of the smaller species of fish depends upon the amount a (in

appropriate units) of its food supply, a kind of plankton. Suppose

$$x = f(a) = 3a + 2.$$

A biologist wants to find the relationship between the population P of the large fish and the amount a of plankton available, that is, $(P \circ f)(a)$. What is the relationship? If the amount of plankton decreases, what will happen to the fish population?

50. Suppose the demand for a certain brand of graphing calculator is given by

$$D(p) = \dfrac{-p^2}{100} + 500,$$

where p is the price in dollars. If the price, in terms of the cost, c, is expressed as

$$p(c) = 2c - 10,$$

find $D(c)$, the demand in terms of the cost. What happens to the demand for the calculator as the cost goes down?

51. When a thermal inversion layer is over a city, pollutants cannot rise vertically but are trapped below the layer and must disperse horizontally. Assume that a factory smokestack begins emitting a pollutant at 8 A.M. Assume that the pollutant disperses horizontally over a circular area. If t represents the time, in hours, since the factory began emitting pollutants ($t = 0$ represents

8 A.M.), assume that the radius of the circle of pollution is $r(t) = 2t$ mi. Let $A(r) = \pi r^2$ represent the area of a circle of radius r.

(a) Find $(A \circ r)(t)$.

(b) Interpret $(A \circ r)(t)$.

52. An oil well off the Gulf Coast is leaking, with the leak spreading oil over the surface as a circle. At any time t,

in minutes, after the beginning of the leak, the radius of the circular oil slick on the surface is $r(t) = 4t$ ft. Let $A(r) = \pi r^2$ represent the area of a circle of radius r.

(a) Find $(A \circ r)(t)$.

(b) Interpret $(A \circ r)(t)$.

4.4 ——— GRAPHING BASIC FUNCTIONS AND THEIR VARIATIONS

Many of the graphs discussed in Chapter 3 are the graphs of functions. By the vertical line test, any straight line that is not vertical is the graph of a function, as is the graph of any vertical parabola. Because of this, linear and quadratic functions can be defined as follows.

LINEAR AND QUADRATIC FUNCTIONS

If a, b, and c are real numbers, then the function defined by

$$f(x) = ax + b$$

is a **linear function,** and the function defined by

$$f(x) = ax^2 + bx + c \ (a \neq 0)$$

is a **quadratic function.**

■ **Example 1**

GRAPHING LINEAR AND QUADRATIC FUNCTIONS

Graph each function defined as follows.

(a) $f(x) = -4x + 5$

This defines a linear function with a slope of -4 and y-intercept 5. The graph is shown in Figure 4.4.

(b) $f(x) = -(x - 2)^2 + 5$

The graph of this quadratic function, a parabola with vertex at $(2, 5)$ opening downward, is shown in Figure 4.5. ■

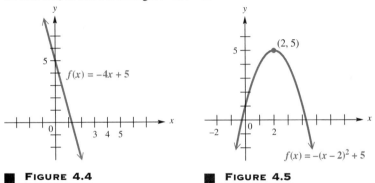

■ **FIGURE 4.4**   ■ **FIGURE 4.5**

NOTE The function in Example 1(b) is a good illustration of the result of combining simple functions to get a more complicated function. If $g(x) = -x^2 + 5$ and $h(x) = x - 2$, then $f(x) = -(x - 2)^2 + 5 = (g \circ h)(x)$.

In the rest of this section we discuss the graphs of several basic functions that are neither linear nor quadratic. The graphs of these functions are made up of portions of different straight lines or of straight lines and curves.

ABSOLUTE VALUE FUNCTIONS One common function that is not linear and not quadratic is the **absolute value function,** defined by $f(x) = |x|$, or $y = |x|$. Since $|x|$ can be found for any real number x, the domain is $(-\infty, \infty)$. Also, $|x| \geq 0$ for any real number x, so the range is $[0, \infty)$. When $x \geq 0$, then $y = |x| = x$, so $y = x$ is graphed for nonnegative values of x. On the other hand, if $x < 0$, then $y = |x| = -x$, and $y = -x$ is graphed for negative values of x. The graph is the union of two rays. The final graph is shown in Figure 4.6. By the vertical line test, the graph is that of a function. Notice that the graph of $f(x) = |x|$ is symmetric with respect to the y-axis.

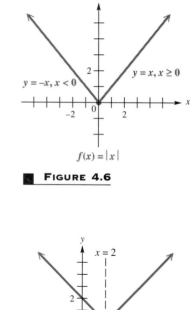

■ **FIGURE 4.6**

■ *Example 2*
GRAPHING AN ABSOLUTE VALUE FUNCTION

Graph $f(x) = |x - 2|$.

Notice that f is the composite function $g \circ h$, where $g(x) = |x|$ and $h(x) = x - 2$. The domain of f is $(-\infty, \infty)$, and the range is $[0, \infty)$. This graph has the same shape as that of $y = |x|$, but the "vertex" point is translated 2 units to the right, from $(0, 0)$ to $(2, 0)$. See Figure 4.7. The axis of symmetry is the vertical line $x = 2$, through $(2, 0)$. ■

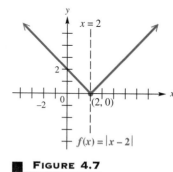

■ **FIGURE 4.7**

■ *Example 3*
GRAPHING AN ABSOLUTE VALUE FUNCTION

Graph $f(x) = |3x + 4| + 1$.

The value of $f(x) = y$ is always greater than or equal to 1, since $|3x + 4| \geq 0$. Thus, the y-value of the "vertex" is 1, and the range is $[1, \infty)$. The domain is $(-\infty, \infty)$. The x-value of the "vertex" can be found by substituting 1 for y in the equation.

$$y = |3x + 4| + 1$$

$$1 = |3x + 4| + 1 \qquad \text{Let } y = 1.$$

$$0 = |3x + 4| \qquad \text{Subtract 1.}$$

$$x = -\frac{4}{3} \qquad \text{Solve } 3x + 4 = 0.$$

Plotting a few other ordered pairs leads to the graph in Figure 4.8. The graph shows the "vertex" translated to $(-4/3, 1)$. The axis of symmetry is $x = -4/3$. The coefficient of x, 3, determines the slopes of the two rays that form the graph. One has slope 3, and the other has slope -3. Because the absolute value of the slopes is greater than 1, the rays are steeper than the rays that form the graph of $y = |x|$. ■

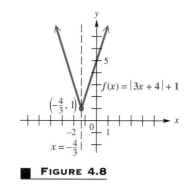

■ **FIGURE 4.8**

FUNCTIONS DEFINED PIECEWISE The graphs of the absolute value functions of Examples 2 and 3 are made up of portions of two different straight lines. Such functions, called **functions defined piecewise,** are often defined with different equations for different parts of the domain.

■ *Example 4*

GRAPHING A FUNCTION DEFINED PIECEWISE

Graph the function

$$f(x) = \begin{cases} x + 1 \text{ if } x \le 2 \\ -2x + 7 \text{ if } x > 2. \end{cases}$$

We must graph each part of the domain separately. If $x \le 2$, this portion of the graph has an endpoint at $x = 2$. Find the y-value by substituting 2 for x in $y = x + 1$ to get $y = 3$. Another point is needed to graph this part of the graph. Choose an x-value less than 2. Choosing $x = -1$ gives $y = -1 + 1 = 0$. Draw the graph through $(2, 3)$ and $(-1, 0)$ as a ray with an endpoint at $(2, 3)$. Graph the ray for $x > 2$ similarly. This ray will have an open endpoint when $x = 2$ and $y = -2(2) + 7 = 3$. Choosing $x = 4$ gives $y = -2(4) + 7 = -1$. The ray through $(2, 3)$ and $(4, -1)$ completes the graph. In this example the two rays meet at $(2, 3)$, although this is not always the case. The graph is shown in Figure 4.9. ■

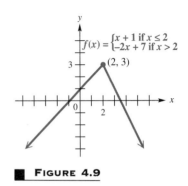

■ **FIGURE 4.9**

234   ■ *Example 5*
GRAPHING A
FUNCTION DEFINED
PIECEWISE

Graph

$$f(x) = \begin{cases} x + 2 \text{ if } x \leq 0 \\ \dfrac{1}{2}x^2 \text{ if } x > 0. \end{cases}$$

Graph the ray $y = x + 2$, choosing x so that $x \leq 0$, with a solid endpoint at $(0, 2)$. The ray has slope 1 and y-intercept 2. Then graph $y = \dfrac{1}{2}x^2$ for $x > 0$. This graph will be half of a parabola with an open endpoint at $(0, 0)$. See Figure 4.10. Notice that, in this example, the two portions do not meet.   ■

■ **FIGURE 4.10**

GREATEST INTEGER FUNCTIONS   Another type of function with a graph composed of line segments is the **greatest integer function,** written $y = [\![x]\!]$, and defined as follows:

$[\![x]\!]$ is the greatest integer less than or equal to x.

For example, $[\![8]\!] = 8$, $[\![-5]\!] = -5$, $[\![\pi]\!] = 3$, $[\![12\ 1/9]\!] = 12$, $[\![-2.001]\!] = -3$, and so on.

■ *Example 6*
GRAPHING THE
GREATEST INTEGER
FUNCTION

Graph $y = [\![x]\!]$.

For any value of x in the interval $[0, 1)$, $[\![x]\!] = 0$. Also, for x in $[1, 2)$, $[\![x]\!] = 1$. This process continues; for x in $[2, 3)$, the value of $[\![x]\!]$ is 2. The values of y are constant between integers, but they jump at integer values of x. This makes the graph, shown in Figure 4.11, a series of line segments. In each case, the left endpoint of the segment is included, and the right endpoint is excluded. The domain of the function is $(-\infty, \infty)$, while the range is the set of integers, $\{\ldots, -2, -1, 0, 1, 2, \ldots\}$.   ■

■ **FIGURE 4.11**

Each of the graphs in Examples 6, 7, and 8 is made up of a series of horizontal line segments. (See Figures 4.11, 4.12, and 4.13.) The functions producing these graphs are called **step functions.**

IN SIMPLEST TERMS ━━━━━━━━━━━━━━

13 ▭

Special functions called step functions are used when the range value is expected to remain constant for a given interval of domain values. For example, parking fees are often calculated by charging a certain flat fee for the first hour and a different fee for each additional hour (or fraction of an hour).

Suppose Downtown Daily Parking charges a $5 base fee for the first hour and $1 for each additional hour or fraction thereof. The maximum fee per day is $15. It would cost $5 + $1(6) = $11 to park for 7 hours.

SOLVE EACH PROBLEM

A. How much would it cost to park at Downtown Daily Parking for 3.5 hours?

B. Melissa and her children plan to spend the entire day touring the city. How much would it cost her to park at Downtown Daily Parking for 12 hours?

ANSWERS   A. It would cost $8.   B. It would cost $15, the maximum charge per day.

▪ **Example 7**

GRAPHING A STEP FUNCTION

Graph $y = \left[\!\left[\dfrac{1}{2}x + 1 \right]\!\right]$.

Try some values of x in the equation to see how the values of y behave. Some sample ordered pairs are given here.

x	0	$\frac{1}{2}$	1	2	3	4	-1	-2	-3
y	1	1	1	2	2	3	0	0	-1

These ordered pairs suggest that if x is in the interval $[0, 2)$, then $y = 1$. For x in $[2, 4)$, $y = 2$, and so on. The graph is shown in Figure 4.12. Again, the domain is $(-\infty, \infty)$. The range is $\{\ldots, -1, 0, 1, 2, \ldots\}$. ▪

▪ **FIGURE 4.12**

The greatest integer function can be used to describe many common pricing practices encountered in everyday life, as shown in the next example.

▪ **Example 8**

APPLYING THE GREATEST INTEGER FUNCTION

An express mail company charges $10 for a package weighing 1 pound or less. Each additional pound or part of a pound costs $3 more. Find the cost to send a package weighing 2 pounds; 2.5 pounds; 5.8 pounds. Graph the ordered pairs (pounds, cost). Is this the graph of a function?

The cost for a package weighing 2 pounds is $10 for the first pound and $3 for the second pound, for a total of $13. For a 2.5-pound package, the cost will be the same as for 3 pounds: $10 + 2(3) = 16$, or $16. A 5.8-pound package will cost the same as a 6-pound package: $10 + 5(3) = 25$, or $25. The graph of this step function is shown in Figure 4.13. Notice that the right endpoints are included in this case, instead of the left endpoints. ■

■ **FIGURE 4.13**

4.4 EXERCISES ■

Graph each of the following. Give the domain and range of each function. See Examples 1–3.

1. $f(x) = \dfrac{3}{4}x - 1$ **2.** $f(x) = (x - 3)^2 + 1$ **3.** $f(x) = |x + 1|$ **4.** $y = |-3 - x|$

5. $y = |x| + 4$ **6.** $y = 2|x| - 1$ **7.** $y = 3|x - 2| + 1$ **8.** $y = \dfrac{1}{2}|x + 3| + 1$

9. $f(x) = -|x + 1| + 2$ **10.** $f(x) = -|x - 3| - 2$ **11.** $f(x) = |x| + x$ **12.** $f(x) = |x| - x$

13. Explain why the graph of an absolute value function typically has a V-shape.

14. When functions are defined piecewise, what two points *must* be plotted?

For each of the following, find (a) $f(-5)$, (b) $f(-1)$, (c) $f(0)$, (d) $f(3)$, and (e) $f(5)$.

15. $f(x) = \begin{cases} 2x \text{ if } x \le -1 \\ x - 1 \text{ if } x > -1 \end{cases}$ **16.** $f(x) = \begin{cases} x - 2 \text{ if } x < 3 \\ 4 - x \text{ if } x \ge 3 \end{cases}$

17. $f(x) = \begin{cases} 3x + 5 \text{ if } x \le 0 \\ 4 - 2x \text{ if } 0 < x < 2 \\ x \text{ if } x \ge 2 \end{cases}$ **18.** $f(x) = \begin{cases} 4x + 1 \text{ if } x < 2 \\ 3x \text{ if } 2 \le x \le 5 \\ 3 - 2x \text{ if } x > 5 \end{cases}$

Graph each of the functions defined as follows. See Examples 4 and 5.

19. $f(x) = \begin{cases} x - 1 \text{ if } x \le 3 \\ 2 \text{ if } x > 3 \end{cases}$ **20.** $f(x) = \begin{cases} 6 - x \text{ if } x \le 3 \\ 3x - 6 \text{ if } x > 3 \end{cases}$ **21.** $f(x) = \begin{cases} 4 - x \text{ if } x < 2 \\ 1 + 2x \text{ if } x \ge 2 \end{cases}$

22. $f(x) = \begin{cases} 2x + 1 \text{ if } x \ge 0 \\ x \text{ if } x < 0 \end{cases}$ **23.** $f(x) = \begin{cases} 2 + x \text{ if } x < -4 \\ -x \text{ if } -4 \le x \le 5 \\ 3x \text{ if } x > 5 \end{cases}$ **24.** $f(x) = \begin{cases} -2x \text{ if } x < -3 \\ 3x - 1 \text{ if } -3 \le x \le 2 \\ -4x \text{ if } x > 2 \end{cases}$

25. $f(x) = \begin{cases} |x| \text{ if } x > -2 \\ x \text{ if } x \le -2 \end{cases}$ **26.** $f(x) = \begin{cases} |x| - 1 \text{ if } x > -1 \\ x - 1 \text{ if } x \le -1 \end{cases}$

27. Explain how to find the *y*-values of the greatest integer function for negative *x*-values.

28. The graph of the greatest integer function defined by $f(x) = [\![x - 2]\!] + 1$ has each line segment translated _____ units _____ and _____ unit(s)
                                    (left/right)
_____ compared to the graph of $f(x) = [\![x]\!]$.
     (up/down)

Graph each of the following. See Examples 6 and 7.

29. $f(x) = [\![-x]\!]$

30. $f(x) = [\![2x - 1]\!]$

31. $f(x) = [\![3x + 1]\!]$

32. $f(x) = [\![3x]\!]$

33. $f(x) = [\![3x]\!] + 1$

34. $f(x) = [\![3x]\!] - 1$

Work each problem. See Example 8.

35. A mail order firm charges 30¢ to mail a package weighing one ounce or less, and then 27¢ for each additional ounce or fraction of an ounce. Let $M(x)$ be the cost of mailing a package weighing x oz. Find
(a) $M(.75)$,   (b) $M(1.6)$, and   (c) $M(4)$.
(d) Graph $y = M(x)$.
(e) Give the domain and range of M.

36. Use the greatest integer function and write an expression for the number of ounces for which postage will be charged on a package weighing x oz (see Exercise 35).

37. A car rental costs $37 for one day, which includes 50 free miles. Each additional 25 mi or portion costs $10. Graph the ordered pairs (miles, cost).

38. For a lift truck rental of no more than three days, the charge is $300. An additional charge of $75 is made for each day or portion of a day after three. Graph the ordered pairs (number of days, cost).

39. Montreal taxi rates in a recent year were $1.80 for the first 1/9 mi and $.20 for each additional 1/9 mi or fraction of 1/9. Let $C(x)$ be the cost for a taxi ride of x 1/9 mi. Find (a) $C(1)$,   (b) $C(2.3)$, and   (c) $C(8)$.
(d) Graph $y = C(x)$.
(e) Give the domain and range of C.

40. When a diabetic takes long-acting insulin, the insulin reaches its peak effect on the blood sugar level in about 3 hr. This effect remains fairly constant for 5 hr, then declines, and is very low until the next injection. In a typical patient, the level of insulin might be given by the following function.

$$i(t) = \begin{cases} 40t + 100 & \text{if } 0 \le t \le 3 \\ 220 & \text{if } 3 < t \le 8 \\ -80t & \text{if } 8 < t \le 10 \\ 60 & \text{if } 10 < t \le 24 \end{cases}$$

Here $i(t)$ is the blood sugar level, in appropriate units, at time t measured in hours from the time of the injection. Chuck takes his insulin at 6 A.M. Find the blood sugar level at each of the following times.
(a) 7 A.M.   (b) 9 A.M.   (c) 10 A.M.
(d) noon   (e) 2 P.M.   (f) 5 P.M.
(g) midnight   (h) Graph $y = i(t)$.

41. The snow depth in Michigan's Isle Royale National Park varies throughout the winter. In a typical winter, the snow depth in inches is approximated by the following function.

$$f(x) = \begin{cases} 6.5x & \text{if } 0 \le x \le 4 \\ -5.5x + 48 & \text{if } 4 < x \le 6 \\ -30x + 195 & \text{if } 6 < x \le 6.5 \end{cases}$$

Here, x represents the time in months with $x = 0$ representing the beginning of October, $x = 1$ representing the beginning of November, and so on.
(a) Graph $f(x)$.
(b) In what month is the snow deepest? What is the deepest snow depth?
(c) In what months does the snow begin and end?

4.5 —————— INVERSE FUNCTIONS

Addition and subtraction are inverse operations: starting with a number x, adding 5, and subtracting 5 gives x back as a result. Similarly, some functions are inverses of each other. For example, the functions

$$f(x) = 8x \qquad \text{and} \qquad g(x) = \frac{1}{8}x$$

are inverses of each other. This means that if a value of x such as $x = 12$ is chosen, so that

$$f(12) = 8 \cdot 12 = 96,$$

calculating $g(96)$ gives

$$g(96) = \frac{1}{8} \cdot 96 = 12.$$

Thus, $$g[f(12)] = 12.$$

Also, $f[g(12)] = 12$. For these functions f and g, it can be shown that

$$f[g(x)] = x \qquad \text{and} \qquad g[f(x)] = x$$

for any value of x.

This section will show how to start with a function such as $f(x) = 8x$ and obtain the inverse function $g(x) = (1/8)x$. Not all functions have inverse functions. The only functions that do have inverse functions are *one-to-one functions*.

ONE-TO-ONE FUNCTIONS   For the function $y = 5x - 8$, any two different values of x produce two different values of y. On the other hand, for the function $y = x^2$, two different values of x can lead to the *same* value of y; for example, both $x = 4$ and $x = -4$ give $y = 4^2 = (-4)^2 = 16$. A function such as $y = 5x - 8$, where different elements from the domain always lead to different elements from the range, is called a *one-to-one function*.

ONE-TO-ONE FUNCTION	A function f is a **one-to-one function** if, for elements a and b from the domain of f, $$a \neq b \qquad \text{implies} \qquad f(a) \neq f(b).$$

■ *Example 1*
DECIDING WHETHER A FUNCTION IS ONE-TO-ONE

Decide whether each of the following functions is one-to-one.

(a) $f(x) = -4x + 12$

Suppose that $a \neq b$. Then $-4a \neq -4b$, and $-4a + 12 \neq -4b + 12$. Thus, the fact that $a \neq b$ implies that $f(a) \neq f(b)$, so f is one-to-one.

(b) $f(x) = \sqrt{25 - x^2}$

If $a = 3$ and $b = -3$, then $3 \neq -3$, but

$$f(3) = \sqrt{25 - 3^2} = \sqrt{25 - 9} = \sqrt{16} = 4$$

and $$f(-3) = \sqrt{25 - (-3)^2} = \sqrt{25 - 9} = 4.$$

Here, even though $3 \neq -3$, $f(3) = f(-3)$. By definition, this is not a one-to-one function. ■

As shown in Example 1(b), a way to show that a function is *not* one-to-one is to produce a pair of unequal numbers that lead to the same function value. There is also a useful graphical test that tells whether or not a function is one-to-one. This *horizontal line test* for one-to-one functions can be summarized as follows.

HORIZONTAL LINE TEST	If each horizontal line intersects the graph of a function in no more than one point, then the function is one-to-one.

NOTE   In Example 1(b), the graph of the function is a semicircle. There are infinitely many horizontal lines that cut the graph of a semicircle in two points, so the horizontal line test shows that the function is not one-to-one.

▪ *Example 2*
USING THE HORIZONTAL LINE TEST

Use the horizontal line test to determine whether the graphs in Figures 4.14 and 4.15 are graphs of one-to-one functions.

(a)

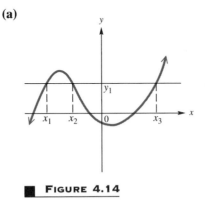

▪ **FIGURE 4.14**

Each point where the horizontal line intersects the graph has the same value of y but a different value of x. Since more than one (here three) different values of x lead to the same value of y, the function is not one-to-one.

(b)

▪ **FIGURE 4.15**

Every horizontal line will intersect the graph in Figure 4.15 in exactly one point. This function is one-to-one.   ▪

INVERSE FUNCTIONS As mentioned earlier, certain pairs of one-to-one functions "undo" one another. For example, if

■ **FIGURE 4.16**

$$f(x) = 8x + 5 \qquad \text{and} \qquad g(x) = \frac{x - 5}{8}$$

$$f(10) = 8 \cdot 10 + 5 = \mathbf{85} \qquad \text{and} \qquad g(85) = \frac{85 - 5}{8} = 10.$$

Starting with 10, we "applied" function f and then "applied" function g to the result, which gave back the number 10. See Figure 4.16. Similarly, for these same functions, check that

$$f(3) = \mathbf{29} \qquad \text{and} \qquad g(29) = 3,$$
$$f(-5) = -35 \qquad \text{and} \qquad g(-35) = -5,$$
$$g(2) = -\frac{3}{8} \qquad \text{and} \qquad f\left(-\frac{3}{8}\right) = 2.$$

In particular, for these functions,

$$f[g(2)] = 2 \qquad \text{and} \qquad g[f(2)] = 2.$$

In fact for *any* value of x,

$$f[g(x)] = x \qquad \text{and} \qquad g[f(x)] = x,$$
or $\qquad (f \circ g)(x) = x \qquad \text{and} \qquad (g \circ f)(x) = x.$

Because of this property, g is called the *inverse* of f.

INVERSE FUNCTION Let f be a one-to-one function. Then g is the **inverse function** of f if

$$(f \circ g)(x) = x \quad \text{for every } x \text{ in the domain of } g,$$

and $\qquad (g \circ f)(x) = x \quad \text{for every } x \text{ in the domain of } f.$

■ *Example 3*

DECIDING WHETHER
TWO FUNCTIONS
ARE INVERSES

Let $f(x) = x^3 - 1$, and let $g(x) = \sqrt[3]{x + 1}$. Is g the inverse function of f? Use the definition to get

$$(f \circ g)(x) = f[g(x)] = (\sqrt[3]{x + 1})^3 - 1$$
$$= x + 1 - 1 = x$$
$$(g \circ f)(x) = g[f(x)] = \sqrt[3]{(x^3 - 1) + 1} = \sqrt[3]{x^3} = x.$$

Since $f[g(x)] = x$ and $g[f(x)] = x$, function g is the inverse of function f. Also, f is the inverse of function g. ■

A special notation is often used for inverse functions: if g is the inverse function of f, then g can be written as f^{-1} (read "f-inverse"). In Example 3,

$$f^{-1}(x) = \sqrt[3]{x + 1}.$$

CAUTION	Do not confuse the -1 in f^{-1} with a negative exponent. The symbol $f^{-1}(x)$ does not represent $1/f(x)$; it represents the inverse function of f. Keep in mind that a function f can have an inverse function f^{-1} if and only if f is one-to-one.

The definition of inverse function can be used to show that the domain of f equals the range of f^{-1}, and the range of f equals the domain of f^{-1}. See Figure 4.17.

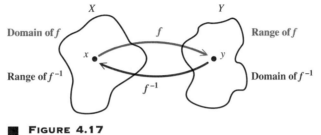

■ **FIGURE 4.17**

For the inverse functions f and g discussed at the beginning of this section, $f(12) = 96$ and $g(96) = 12$; that is $(12, 96)$ belonged to f and $(96, 12)$ belonged to g. The inverse of any one-to-one function f can be found by exchanging the components of the ordered pairs of f. The equation of the inverse of a function defined by $y = f(x)$ also is found by exchanging x and y. For example, if $f(x) = 7x - 2$, then $y = 7x - 2$. The function f is one-to-one, so that f^{-1} exists. The ordered pairs in f^{-1} have the form (y, x), where $x = f^{-1}(y)$. Thus, we can solve the equation $y = 7x - 2$ for x to find $f^{-1}(y)$. Since it is customary to use x as the independent variable, we then exchange x and y to get $y = f^{-1}(x)$.

$$y = 7x - 2$$
$$7x = y + 2 \qquad \text{Add 2.}$$
$$x = \frac{y + 2}{7} = f^{-1}(y) \qquad \text{Divide by 7.}$$
$$y = \frac{x + 2}{7} = f^{-1}(x) \qquad \text{Exchange } x \text{ and } y.$$

Check that $(f \circ f^{-1})(x) = x$ and $(f^{-1} \circ f)(x) = x$, so that f^{-1} is the correct inverse of f.

In summary, finding the equation of an inverse function involves the following steps.

FINDING AN EQUATION FOR f^{-1}	1. Verify that f is a one-to-one function. 2. Replace $f(x)$ with y. 3. Solve for x. Let $x = f^{-1}(y)$. 4. Exchange x and y to get $y = f^{-1}(x)$. 5. Verify that $(f \circ f^{-1})(x) = x$ and $(f^{-1} \circ f)(x) = x$.

242

■ *Example 4*

FINDING THE
INVERSE OF A
FUNCTION

For each of the following functions, find the inverse function, if it exists.

(a) $f(x) = \dfrac{4x + 6}{5}$

This function is one-to-one and thus has an inverse. Let $f(x) = y$, and solve for x, getting

$$y = \frac{4x + 6}{5}$$

$$5y = 4x + 6 \qquad\qquad \text{Multiply by 5.}$$

$$5y - 6 = 4x \qquad\qquad \text{Subtract 6.}$$

$$x = \frac{5y - 6}{4} = f^{-1}(y). \qquad \text{Divide by 4.}$$

Finally, exchange x and y to get

$$y = \frac{5x - 6}{4} = f^{-1}(x)$$

or $\qquad\qquad f^{-1}(x) = \dfrac{5x - 6}{4}.$

The domain and range of both f and f^{-1} are the set of real numbers. In function f, the value of y is found by multiplying x by 4, adding 6 to the product, then dividing that sum by 5. In the equation for the inverse, x is *multiplied* by 5, then 6 is *subtracted*, and the result is *divided* by 4. This shows how an inverse function is used to "undo" what a function does to the variable x.

(b) $f(x) = x^3 - 1$

Two different values of x will produce two different values of $x^3 - 1$, so the function is one-to-one and has an inverse. To find the inverse, first solve $y = x^3 - 1$ for x, as follows.

$$y = x^3 - 1$$

$$y + 1 = x^3 \qquad\qquad \text{Add 1.}$$

$$\sqrt[3]{y + 1} = x \qquad\qquad \text{Take cube roots.}$$

Exchange x and y, giving

$$\sqrt[3]{x + 1} = y,$$

or $\qquad\qquad f^{-1}(x) = \sqrt[3]{x + 1}.$

In Example 3, we verified that $\sqrt[3]{x + 1}$ is the inverse of $x^3 - 1$.

(c) $f(x) = x^2$

We can find two different values of x that give the same value of y. For example, both $x = 4$ and $x = -4$ give $y = 16$. Therefore, the function is not one-to-one and thus has no inverse function. ■

Suppose f and f^{-1} are inverse functions, and $f(a) = b$ for real numbers a and b. Then, by the definition of inverse, $f^{-1}(b) = a$. This shows that if a point (a, b) is on the graph of f, then (b, a) will belong to the graph of f^{-1}. As shown in Figure 4.18, the points (a, b) and (b, a) are symmetric with respect to the line $y = x$. Thus, the graph of f^{-1} can be obtained from the graph of f by reflecting the graph of f about the line $y = x$.

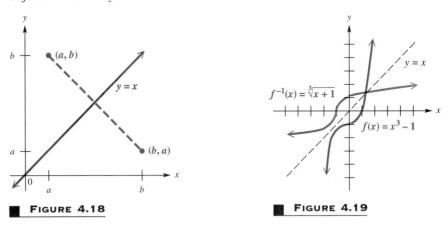

■ **FIGURE 4.18**　　　　　　　　■ **FIGURE 4.19**

For example, Figure 4.19 shows the graph of $f(x) = x^3 - 1$ in blue and the graph of $f^{-1}(x) = \sqrt[3]{x + 1}$ in red. These graphs are symmetric with respect to the line $y = x$.

■ ***Example 5***

FINDING THE INVERSE OF A FUNCTION WITH A RESTRICTED DOMAIN

Let $f(x) = \sqrt{x + 5}$ with domain $[-5, \infty)$. Find $f^{-1}(x)$.

The function f is one-to-one and has an inverse function. To find this inverse function, start with

$$y = \sqrt{x + 5}$$

and solve for x, to get

$$y^2 = x + 5 \qquad \text{Square both sides.}$$
$$y^2 - 5 = x. \qquad \text{Subtract 5.}$$

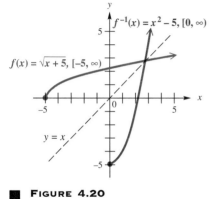

■ **FIGURE 4.20**

Exchanging x and y gives

$$x^2 - 5 = y.$$

However, $x^2 - 5$ is not $f^{-1}(x)$. In the definition of f above, the domain was given as $[-5, \infty)$. The range of f is $[0, \infty)$. As mentioned above, the range of f equals the domain of f^{-1}, so f^{-1} must be given as

$$f^{-1}(x) = x^2 - 5, \qquad \text{domain } [0, \infty).$$

As a check, the range of f^{-1}, $[-5, \infty)$, equals the domain of f. Graphs of f and f^{-1} are shown in Figure 4.20. The line $y = x$ is included on the graph to show that the graphs of f and f^{-1} are mirror images with respect to this line. ■

■ **IN SIMPLEST TERMS**

14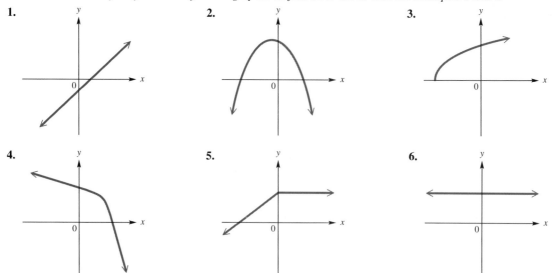

Inverse functions are used by government agencies and other businesses to send and receive coded information. The functions they use are usually very complicated. A simple example involves the function $f(x) = 2x + 5$. If each letter of the alphabet is assigned a numerical value according to its position ($a = 1, \ldots, z = 26$), the word ALGEBRA would be encoded as 7 29 19 15 9 41 7. The "message" can be decoded using the inverse function $f^{-1}(x) = \dfrac{x - 5}{2}$.

SOLVE EACH PROBLEM   Use the alphabet assignment given above.

A. The function $f(x) = 3x - 2$ was used to encode the following message:

37 25 19 61 13 34   22 1 55   1 52 52 25 64 13 10.

Find the inverse function and decode the message.

B. Encode the message SEND HELP using the one-to-one function $f(x) = x^3 - 1$. Give the inverse function that the decoder would need when the message is received.

ANSWERS   A. $f^{-1}(x) = \dfrac{x + 2}{3}$; the message reads MIGUEL HAS ARRIVED.

B. 6858   124   2743   63   511   124   1727   4095; $f^{-1}(x) = \sqrt[3]{x + 1}$

4.5 EXERCISES ■

Decide whether each of the functions defined or graphed as follows is one-to-one. See Examples 1 and 2.

1.

2.

3.

4.

5.

6.

7. $y = 4x - 5$

8. $y = -x^2$

9. $y = (x - 2)^2$

10. $y = -(x + 3)^2 - 8$

11. $y = \sqrt{36 - x^2}$

12. $y = -\sqrt{100 - x^2}$

13. $y = 2x^3 + 1$

14. $y = -\sqrt[3]{x + 5}$

15. $y = \dfrac{1}{x + 2}$

16. $y = \dfrac{-4}{x - 8}$

17. $y = 9$

18. $y = -4$

In Exercises 19–24, an everyday activity is described. Keeping in mind that an inverse operation "undoes" what an operation does, describe the inverse activity.

19. Tying your shoelaces

20. Starting a car

21. Entering a room

22. Climbing the stairs

23. Taking off in an airplane

24. Filling a cup

25. Explain why the function $f(x) = x^4$ does not have an inverse. Give examples of ordered pairs to illustrate your explanation.

26. Explain why the function $f(x) = x^5$ is one-to-one.

Decide whether the functions described in each pair are inverses of each other. See Example 3.

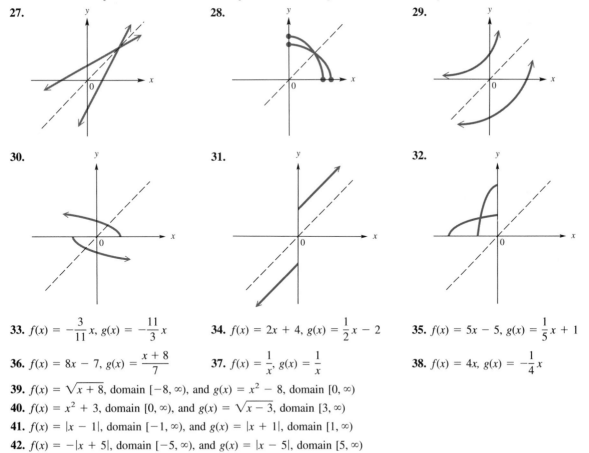

27.

28.

29.

30.

31.

32.

33. $f(x) = -\dfrac{3}{11}x$, $g(x) = -\dfrac{11}{3}x$

34. $f(x) = 2x + 4$, $g(x) = \dfrac{1}{2}x - 2$

35. $f(x) = 5x - 5$, $g(x) = \dfrac{1}{5}x + 1$

36. $f(x) = 8x - 7$, $g(x) = \dfrac{x + 8}{7}$

37. $f(x) = \dfrac{1}{x}$, $g(x) = \dfrac{1}{x}$

38. $f(x) = 4x$, $g(x) = -\dfrac{1}{4}x$

39. $f(x) = \sqrt{x + 8}$, domain $[-8, \infty)$, and $g(x) = x^2 - 8$, domain $[0, \infty)$

40. $f(x) = x^2 + 3$, domain $[0, \infty)$, and $g(x) = \sqrt{x - 3}$, domain $[3, \infty)$

41. $f(x) = |x - 1|$, domain $[-1, \infty)$, and $g(x) = |x + 1|$, domain $[1, \infty)$

42. $f(x) = -|x + 5|$, domain $[-5, \infty)$, and $g(x) = |x - 5|$, domain $[5, \infty)$

Graph the inverse of each one-to-one function.

43.

44.

45.

46.

47.

48.

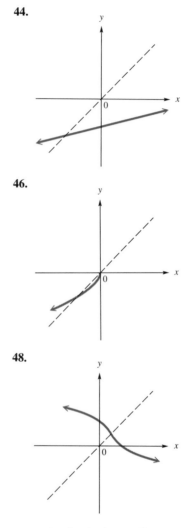

For each function defined as follows that is one-to-one, write an equation for the inverse function in the form of $y = f^{-1}(x)$, and then graph f and f^{-1} on the same axes. See Examples 4 and 5.

49. $y = 3x - 4$

50. $y = 4x - 5$

51. $y = \dfrac{1}{3}x$

52. $y = -\dfrac{2}{5}x$

53. $y = x^3 + 1$

54. $y = -x^3 - 2$

55. $y = x^2$

56. $y = -x^2 + 2$

57. $y = \dfrac{1}{x}$

58. $y = \dfrac{4}{x}$

59. $f(x) = 4 - x^2$, domain $(-\infty, 0]$

60. $f(x) = \sqrt{6 + x}$, domain $[-6, \infty)$

61. $f(x) = -\sqrt{x^2 - 16}$, domain $[4, \infty)$

62. $f(x) = (x - 1)^2$, domain $(-\infty, 1]$

SECTION	TERMS	KEY IDEAS
4.1 Functions	independent variable dependent variable	A function is a relation in which each element in the domain corresponds to exactly one element in the range. **Vertical Line Test** If each vertical line intersects a graph in no more than one point, the graph is the graph of a function.
4.2 Variation: **Applications of** **Functions**	constant of variation	**Direct Variation** y varies directly as the nth power of x if a nonzero real number k exists such that $y = kx^n$. **Inverse Variation** y varies inversely as the nth power of x if a nonzero real number k exists such that $y = \dfrac{k}{x^n}$. **Joint Variation** For real numbers m and n, y varies jointly as the nth power of x and the mth power of z if a nonzero real number k exists such that $y = kx^n z^m$.
4.3 Combining **Functions:** **Algebra of** **Functions**		**Operations on Functions** If f and g are functions, then for all values of x for which both $f(x)$ and $g(x)$ exist, the sum of f and g is defined by $$(f + g)(x) = f(x) + g(x),$$ the difference of f and g is defined by $$(f - g)(x) = f(x) - g(x),$$ the product of f and g is defined by $$(fg)(x) = f(x) \cdot g(x),$$ and the quotient of f and g is defined by $$\left(\frac{f}{g}\right)(x) = \frac{f(x)}{g(x)}, \quad \text{where } g(x) \neq 0.$$ **Composition of Functions** If f and g are functions, then the composite function, or composition, of g and f is $$(g \circ f)(x) = g[f(x)]$$ for all x in the domain of f such that $f(x)$ is in the domain of g.

SECTION	TERMS	KEY IDEAS
4.4 Graphing Basic Functions and Their Variations	linear function quadratic function absolute value function function defined piecewise greatest integer function step function	
4.5 Inverse Functions		**One-to-One Function** A function f is a one-to-one function if, for elements a and b from the domain of f, $$a \neq b \quad \text{implies} \quad f(a) \neq f(b).$$ **Horizontal Line Test** If each horizontal line intersects the graph of a function in no more than one point, then the function is one-to-one. **Inverse Functions** Let f be a one-to-one function. Then g is the inverse function of f if $$(f \circ g)(x) = x \text{ for every } x \text{ in the domain of } g,$$ and $(g \circ f)(x) = x$ for every x in the domain of f.

CHAPTER 4 REVIEW EXERCISES ■

Give the domain and range of each of the following. Give just the domain in Exercises 11 and 12.

1. $f(x) = -x$

2. $f(x) = 6 - 4x$

3. $f(x) = |x| - 4$

4. $f(x) = -3|x - 4|$

5. $f(x) = -(x + 3)^2$

6. $f(x) = \sqrt{5 - x}$

7. $f(x) = \sqrt{49 - x^2}$

8. $f(x) = -\sqrt{x^2 - 4}$

9. $f(x) = \sqrt{\dfrac{1}{x^2 + 9}}$

10. $f(x) = \sqrt{x^2 + 3x - 10}$

11. $f(x) = \dfrac{x + 2}{x - 7}$

12. $f(x) = \dfrac{3x - 8}{4x - 7}$

13. What ordered pair corresponds to the notation $f(-2) = 3$?

Let $f(x) = -x^2 + 4x + 2$ and $g(x) = 3x + 5$. Find each of the following.

14. $f(-3)$ **15.** $f(4)$ **16.** $f(0)$ **17.** $g(3)$

18. $f[g(-1)]$ **19.** $g[f(-2)]$ **20.** $f(3 + z)$ **21.** $g(5y - 4)$

For each of the following, find $\dfrac{f(x + h) - f(x)}{h}$, where $h \neq 0$.

22. $f(x) = 12x - 5$ **23.** $f(x) = -3$ **24.** $f(x) = -2x^2 + 4x - 3$ **25.** $f(x) = -x^3 + 2x^2$

Write each of the following statements as an equation.

26. m varies directly as the square of z.

27. y varies inversely as r and directly as the cube of p.

28. Y varies jointly as M and the square of N and inversely as the cube of X.

29. A varies jointly as the third power of t and the fourth power of s, and inversely as p and the square of h.

Solve each of the following problems.

30. Suppose r varies directly as x and inversely as the square of y. If r is 10 when x is 5 and y is 3, find r when x is 12 and y is 4.

31. Suppose m varies jointly as n and the square of p, and inversely as q. If m is 20 when n is 5, p is 6, and q is 18, find m when n is 7, p is 11, and q is 2.

32. Suppose Z varies jointly as the square of J and the cube of M, and inversely as the fourth power of W. If Z is 125 when J is 3, M is 5 and W is 1, find Z if J is 2, M is 7, and W is 3.

33. The power a windmill obtains from the wind varies directly as the cube of the wind velocity. If a 10 km/hr wind produces 10,000 units of power, how much power is produced by a wind of 15 km/hr?

34. Hooke's law for an elastic spring states that the distance a spring stretches varies directly as the force applied. If a force of 32 lb stretches a certain spring 48 in, how much will a force of 24 lb stretch the spring?

35. The weight w of an object varies inversely as the square of the distance d between the object and the center of the earth. If a man weighs 90 kg on the surface of the earth, how much would he weigh 800 km above the surface? (The radius of the earth is about 6400 km.)

Let $f(x) = 3x^2 - 4$ and $g(x) = x^2 - 3x - 4$. Find each of the following values.

36. $(f + g)(x)$ **37.** $(fg)(x)$ **38.** $(f - g)(4)$ **39.** $(f + g)(-4)$

40. $(f + g)(2k)$ **41.** $(fg)(1 + r)$ **42.** $\left(\dfrac{f}{g}\right)(3)$ **43.** $\left(\dfrac{f}{g}\right)(-1)$

44. Give the domain of $(fg)(x)$.   **45.** Give the domain of $\left(\dfrac{f}{g}\right)(x)$.

46. Which of the following is *not* equal to $(f \circ g)(x)$ for $f(x) = 1/x$ and $g(x) = x^2 + 1$? (*Hint:* There may be more than one.)

   (a) $f[g(x)]$ **(b)** $\dfrac{1}{x^2 + 1}$ **(c)** $\dfrac{1}{x^2}$ **(d)** $(g \circ f)(x)$

Let $f(x) = \sqrt{x - 2}$ and $g(x) = x^2$. Find each of the following.

47. $(f \circ g)(x)$ **48.** $(g \circ f)(x)$ **49.** $(f \circ g)(-6)$ **50.** $(f \circ g)(2)$ **51.** $(g \circ f)(3)$ **52.** $(g \circ f)(24)$

53. Explain what is meant by a "function defined piecewise."

Graph each function defined as follows.

54. $f(x) = -|x|$

55. $f(x) = |x| - 3$

56. $f(x) = -|x| - 2$

57. $f(x) = -|x + 1| + 3$

58. $f(x) = 2|x - 3| - 4$

59. $f(x) = [\![x - 3]\!]$

60. $f(x) = [\![\frac{1}{2}x - 2]\!]$

61. $f(x) = \begin{cases} -4x + 2 \text{ if } x \le 1 \\ 3x - 5 \text{ if } x > 1 \end{cases}$

62. $f(x) = \begin{cases} 3x + 1 \text{ if } x < 2 \\ -x + 4 \text{ if } x \ge 2 \end{cases}$

63. $f(x) = \begin{cases} |x| \text{ if } x < 3 \\ 6 - x \text{ if } x \ge 3 \end{cases}$

64. Let f be a function that gives the cost to rent a floor polisher for x days. The cost is a flat $3 for cleaning the polisher plus $4 per day or fraction of a day for using the polisher.
  (a) Graph f.
  (b) Give the domain and range of f.

65. A trailer hauling service charges $45, plus $2 per mile or part of a mile. Graph the ordered pairs (miles, cost). Give the domain and range of the function.

Which of the functions graphed or defined as follows are one-to-one?

66.

67.

68.

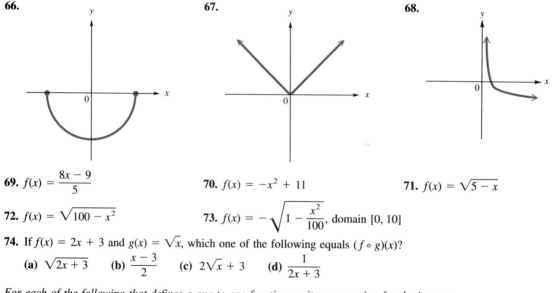

69. $f(x) = \dfrac{8x - 9}{5}$

70. $f(x) = -x^2 + 11$

71. $f(x) = \sqrt{5 - x}$

72. $f(x) = \sqrt{100 - x^2}$

73. $f(x) = -\sqrt{1 - \dfrac{x^2}{100}}$, domain $[0, 10]$

74. If $f(x) = 2x + 3$ and $g(x) = \sqrt{x}$, which one of the following equals $(f \circ g)(x)$?
  (a) $\sqrt{2x + 3}$ **(b)** $\dfrac{x - 3}{2}$ **(c)** $2\sqrt{x} + 3$ **(d)** $\dfrac{1}{2x + 3}$

For each of the following that defines a one-to-one function, write an equation for the inverse function in the form $y = f^{-1}(x)$ and then graph f and f^{-1}.

75. $f(x) = 12x + 3$

76. $f(x) = 2 - x^2$

77. $f(x) = x^3 - 3$

78. $f(x) = \sqrt{x^2 - 6}$, domain $(-\infty, -\sqrt{6}]$

79. $f(x) = \sqrt{25 - x^2}$, domain $[0, 5]$

80. $f(x) = -\sqrt{x - 3}$

Give the domain of each of the following. Give the range in Exercises 1 and 2.

1. $f(x) = 3 + |x + 8|$

2. $f(x) = \sqrt{x^2 + 7x + 12}$

3. $f(x) = \dfrac{x - 1}{2x + 8}$

Let $f(x) = 6 - 3x - x^2$ and $g(x) = 2x + 3$. Find each of the following.

4. $f(-2)$

5. $(f + g)(2a)$

6. $\left(\dfrac{f}{g}\right)(1)$

7. $g\left(\dfrac{1}{2}y - 4\right)$

8. $g[f(2)]$

9. $(g \circ f)(p)$

Graph each function defined as follows.

10. $f(x) = |x - 2| - 1$

11. $f(x) = [\![x + 1]\!]$

12. $f(x) = \begin{cases} 3 \text{ if } x < -2 \\ 2 - \dfrac{1}{2}x \text{ if } x \geq -2 \end{cases}$

Which of the functions graphed or defined as follows are one-to-one?

13.

14. $f(x) = \sqrt{3x + 2}$

15. $f(x) = (x - 2)^2$, domain $[2, \infty)$

For each of the following that defines a one-to-one function, write an equation for the inverse in the form $y = f^{-1}(x)$.

16. $f(x) = 2x + 8$

17. $f(x) = 2 + x^2$, domain $(-\infty, 0]$

18. Are the following two functions inverses of each other? Explain why or why not.

$$f(x) = \frac{x}{x + 2} \quad \text{and} \quad g(x) = \frac{2x}{x - 1}$$

19. Suppose y varies directly as the square root of x, and y is 10 when x is 25/9.
   (a) Write an equation for y in terms of x that defines this functional relationship.
   (b) Find y when x is 144.

20. The force of wind on a sail varies jointly as the area of the sail and the square of the wind velocity. If the force is 8 lb when the velocity is 15 mph and the area is 3 sq ft, find the force when the area is 6 sq ft and the velocity is 22.5 mph.

■ THE GRAPHING CALCULATOR ■

With a graphing calculator in graphing mode, it is possible to get a wide variety of graphs of *functions* quickly and easily. Most graphing calculators have some built-in graphs for basic functions. One of these is the graph of x^2. By touching the x^2 key and the "execute", =, or "graph" key, the graph appears automatically. For user-defined graphs, it is necessary to first use the range key to set the domain and range that will be shown on the screen. The user may choose the minimum and maximum x-values and the scale (the number of units between tick marks) for the x-axis, and the minimum and maximum y-values and the scale for the y-axis. The equations (formulas) are input using the variable x (see your instruction booklet for how this is done), the operation keys, and the numerals. It is necessary to clear the screen with the appropriate key before entering a new graph, unless you want to superimpose graphs.

Functions defined piecewise can be graphed by superimposing the graphs of the different functions over the entire domain. To determine the appropriate "pieces," match the domain of each part with the graph of the function for that part. Some calculators (for example, the TI-81) have special methods for inputting piecewise functions.

In Chapter 3 we discussed graphing relations that are not functions by superimposing the graphs of two functions. For example, to graph $x = y^2$, we solve for y to get $y = \pm\sqrt{x}$, which represents the two functions $y = \sqrt{x}$ and $y = -\sqrt{x}$. Graph $y = \sqrt{x}$ using the $\sqrt{\ }$ key, then graph $y = -\sqrt{x}$. The two graphs will both show and, combined, they give the graph of $x = y^2$, a horizontal parabola.

One way to decide whether two functions are inverses is to graph them both on the same grid, using the same domain and range with the same x-scale and y-scale. Recall, if they are inverse functions, their graphs will be symmetric to the line $y = x$. Remember that *both* expressions must define *functions*, if they are to be inverse functions.

You may want to use the following functions and relations to experiment with your calculator. Most of these functions and relations have been selected from examples in the text, so you can compare your calculator graphs with those given in the example figures. The figure number of the graph in the text is given with each exercise. There are a few with no corresponding figure.

---

1. $y = -(x - 2)^2 + 5$ (Figure 4.5)

2. $y = (x + 1)^2 - 3$

3. $y = |x|$ (Figure 4.6)

4. $y = -|x|$

5. $y = |x + 3|$

6. $y = |x - 2|$ (Figure 4.7)

7. $y = 2|x| - 1$

8. $y = .3|x| + 1$

9. $y = |3x + 4| + 1$ (Figure 4.8)

10. $y = |-x + 2| - 3$

11. $f(x) = \begin{cases} x + 1 \text{ if } x \le 2 \\ -2x + 7 \text{ if } x > 2 \end{cases}$ (Figure 4.9)

12. $f(x) = \begin{cases} x + 2 \text{ if } x \le 0 \\ \dfrac{1}{2}x^2 \text{ if } x > 0 \end{cases}$ (Figure 4.10)

13. $y = x^3 - 1$ and $y = \sqrt[3]{x + 1}$ (Figure 4.19)

14. $y = \sqrt{x + 5}$ and $y = x^2 - 5$ (Figure 4.20)

15. $\dfrac{x^2}{9} + \dfrac{y^2}{4} = 1$ (Figure 3.38)

16. $\dfrac{x^2}{9} + \dfrac{y^2}{16} = 1.$

17. $\dfrac{x^2}{16} - \dfrac{y^2}{9} = 1$ (Figure 3.40)

18. $\dfrac{x^2}{25} - \dfrac{y^2}{49} = 1$ (Figure 3.41)

19. $25y^2 - 4x^2 = 100.$

20. $y = -\sqrt{1 + 4x^2}$ (Figure 3.49)

CHAPTER ■ FIVE

EXPONENTIAL AND LOGARITHMIC FUNCTIONS

5.1 EXPONENTIAL FUNCTIONS

5.2 LOGARITHMIC FUNCTIONS

5.3 EVALUATING LOGARITHMS; CHANGE OF BASE

5.4 EXPONENTIAL AND LOGARITHMIC EQUATIONS

5.5 EXPONENTIAL GROWTH AND DECAY

In this chapter we introduce two kinds of functions that are quite different from the functions discussed earlier. Until now, we have worked with *algebraic functions* that involve only the basic operations of addition, subtraction, multiplication, division, and taking roots. The *exponential* and *logarithmic* functions in this chapter go beyond (transcend) these basic operations, so they are called **transcendental functions.*** Many applications of mathematics, particularly those involving growth and decay of quantities, require the closely interrelated exponential and logarithmic functions introduced in this chapter.

5.1 ———— **EXPONENTIAL FUNCTIONS**

Recall from Chapter 1 the definition of a^r, where r is a rational number: if $r = m/n$, then for appropriate values of m and n,

$$a^{m/n} = (\sqrt[n]{a})^m.$$

For example,

$$16^{3/4} = (\sqrt[4]{16})^3 = 2^3 = 8,$$

$$27^{-1/3} = \frac{1}{27^{1/3}} = \frac{1}{\sqrt[3]{27}} = \frac{1}{3},$$

and

$$64^{-1/2} = \frac{1}{64^{1/2}} = \frac{1}{\sqrt{64}} = \frac{1}{8}.$$

In this section the definition of a^r is extended to include all real (not just rational) values of the exponent r. For example, the new symbol $2^{\sqrt{3}}$ might be evaluated by approximating the exponent $\sqrt{3}$ by the numbers 1.7, 1.73, 1.732, and so on. Since these decimals approach the value of $\sqrt{3}$ more and more closely, it seems reasonable that $2^{\sqrt{3}}$ should be approximated more and more closely by the numbers $2^{1.7}$, $2^{1.73}$, $2^{1.732}$, and so on. (Recall, for example, that $2^{1.7} = 2^{17/10} = \sqrt[10]{2^{17}}$.) In fact, this is exactly how $2^{\sqrt{3}}$ is defined (in a more advanced course).

With this interpretation of real exponents, all rules and theorems for exponents are valid for real-number exponents as well as rational ones. In addition to the rules for exponents presented earlier, several new properties are used in this chapter. For example, if $y = 2^x$, then each real value of x leads to exactly one value of y, and therefore, $y = 2^x$ defines a function. Furthermore,

$$\text{if } 3^x = 3^4, \text{ then } x = 4,$$

and for $p > 0$,

$$\text{if } p^2 = 3^2, \text{ then } p = 3.$$

Also, $\qquad 4^2 < 4^3 \qquad \text{but} \qquad \left(\frac{1}{2}\right)^2 > \left(\frac{1}{2}\right)^3,$

so that when $a > 1$, increasing the exponent on a leads to a *larger* number, but if $0 < a < 1$, increasing the exponent on a leads to a *smaller* number.

———

*The trigonometric functions studied in trigonometry are also transcendental functions.

These properties are generalized below. Proofs of the properties are not given here, as they require more advanced mathematics.

ADDITIONAL PROPERTIES OF EXPONENTS	**(a)** If $a > 0$ and $a \neq 1$, then a^x is a unique real number for all real numbers x. **(b)** If $a > 0$ and $a \neq 1$, then $a^b = a^c$ if and only if $b = c$. **(c)** If $a > 1$ and $m < n$, then $a^m < a^n$. **(d)** If $0 < a < 1$ and $m < n$, then $a^m > a^n$.

Properties (a) and (b) require $a > 0$ so that a^x is always defined. For example, $(-6)^x$ is not a real number if $x = 1/2$. This means that a^x will always be positive, since a is positive. In part (a), $a \neq 1$ because $1^x = 1$ for every real-number value of x, so that each value of x does not lead to a distinct real number. For Property (b) to hold, a must not equal 1 since, for example, $1^4 = 1^5$, even though $4 \neq 5$.

EXPONENTIAL EQUATIONS The properties given above are useful in solving equations, as shown by the next examples.

∎ *Example 1*

USING A PROPERTY OF EXPONENTS TO SOLVE AN EQUATION

Solve $\left(\dfrac{1}{3}\right)^x = 81$.

First, write $1/3$ as 3^{-1}, so that $(1/3)^x = (3^{-1})^x = 3^{-x}$. Since $81 = 3^4$,

$$\left(\frac{1}{3}\right)^x = 81$$

becomes

$$3^{-x} = 3^4.$$

By the second property above,

$$-x = 4, \quad \text{or} \quad x = -4.$$

The solution set of the given equation is $\{-4\}$. ∎

In Section 5.4 we describe a more general method for solving exponential equations where the approach used in Example 1 is not possible. For instance, this method could not be used to solve an equation like $7^x = 12$, since it is not easy to express both sides as exponential expressions with the same base.

∎ *Example 2*

USING A PROPERTY OF EXPONENTS TO SOLVE AN EQUATION

Solve $81 = b^{4/3}$.

Begin by writing $b^{4/3}$ as $(\sqrt[3]{b})^4$.

$$81 = b^{4/3}$$
$$81 = (\sqrt[3]{b})^4$$
$$\pm 3 = \sqrt[3]{b} \qquad \text{Take fourth roots on both sides.}$$
$$\pm 27 = b \qquad \text{Cube both sides.}$$

Check both solutions in the original equation. Since both solutions check, the solution set is $\{-27, 27\}$. ∎

GRAPHING EXPONENTIAL FUNCTIONS As mentioned above, the expression a^x satisfies all the properties of exponents from Chapter 1. We can now define a function $f(x) = a^x$ whose domain is the set of all real numbers (and not just the rationals).

EXPONENTIAL FUNCTION	If $a > 0$ and $a \neq 1$, then
	$$f(x) = a^x$$
	defines the **exponential function** with base a.

NOTE If $a = 1$, the function is the constant function $f(x) = 1$, and not an exponential function.

■ ***Example 3***

EVALUATING AN EXPONENTIAL EXPRESSION

If $f(x) = 2^x$, find each of the following.

(a) $f(-1)$

Replace x with -1.

$$f(-1) = 2^{-1} = \frac{1}{2}$$

(b) $f(3) = 2^3 = 8$

(c) $f(5/2) = 2^{5/2} = (2^5)^{1/2} = 32^{1/2} = \sqrt{32} = 4\sqrt{2}$ ■

Figure 5.1 shows the graph of $f(x) = 2^x$. The graph was found by obtaining a number of ordered pairs belonging to the function and then drawing a smooth curve through them. As we choose smaller and smaller negative values of x, the y-values get closer and closer to 0, as shown in the table below.

x	0	-1	-2	-3	-4
y	1	$\frac{1}{2}$	$\frac{1}{4}$	$\frac{1}{8}$	$\frac{1}{16}$

Because 2^x is always positive, the values of y will never become 0. The line $y = 0$, which the graph gets closer and closer to, is called a **horizontal asymptote.** Asymptotes will be discussed in more detail in Chapter 6. By Property (c), as x increases, so does y, making $f(x) = 2^x$ an **increasing function.** As suggested by the graph in Figure 5.1, the domain of the function is $(-\infty, \infty)$, and the range is $(0, \infty)$.

■ **FIGURE 5.1**

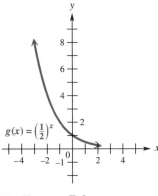

FIGURE 5.2

In Figure 5.2, the graph of $g(x) = (1/2)^x$ was sketched in a similar way. The domain and range are the same as those of $f(x) = 2^x$. However, here, as the values of x increase, the values of y *decrease,* so $g(x) = (1/2)^x$ is a **decreasing function.** The graph of $g(x) = (1/2)^x$ is the reflection of the graph of $f(x) = 2^x$ across the y-axis, because $g(x) = f(-x)$.

As the graphs suggest, by the horizontal line test, $f(x) = 2^x$ and $g(x) = (1/2)^x$ are one-to-one functions.

The graph of $f(x) = 2^x$ is typical of graphs of $f(x) = a^x$, where $a > 1$. For larger values of a, the graphs rise more steeply, but the general shape is similar to the graph in Figure 5.1. Exponential functions with $0 < a < 1$ have graphs similar to that of $g(x) = (1/2)^x$. Based on our work above, the following generalizations can be made about the graphs of exponential functions defined by $f(x) = a^x$.

GRAPH OF
$f(x) = a^x$

1. The point $(0, 1)$ is on the graph.
2. If $a > 1$, f is an increasing function; if $0 < a < 1$, f is a decreasing function.
3. The x-axis is a horizontal asymptote.
4. The domain is $(-\infty, \infty)$ and the range is $(0, \infty)$.

We can use the composition of functions to produce more general exponential functions. If $h(u) = ka^u$, where k is a constant and $u = g(x)$, and $f(x) = h[g(x)]$, then

$$f(x) = h[g(x)] = ka^{g(x)}.$$

For example, if $a = 7$, $g(x) = 3x - 1$, and $k = 4$, then

$$f(x) = 4 \cdot 7^{3x-1}.$$

■ *Example 4*
GRAPHING A
COMPOSITE
EXPONENTIAL
FUNCTION

Graph $f(x) = 2^{-x+2}$.

The graph will have the same shape as the graph of $g(x) = 2^{-x} = (1/2)^x$. Because of the 2 added to $-x$, the graph will be translated 2 units to the right, compared with the graph of $g(x) = 2^{-x}$. This means that the point $(2, 1)$ is on the graph instead of $(0, 1)$. When $x = 0$, $y = 2^2 = 4$, so the point $(0, 4)$ is on the graph. Plotting a few additional points, such as $(-1, 8)$ and $(1, 2)$, gives the graph in Figure 5.3. The graph of $g(x) = 2^{-x}$ is also shown for comparison. ■

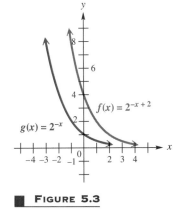

FIGURE 5.3

■ **IN SIMPLEST TERMS**

18 📼

The exponential growth of deer in Massachusetts can be calculated using the equation $T = 50{,}000(1 + .06)^n$, where 50,000 is the initial deer population and .06 is the rate of growth. T is the total population after n years have passed.

Given the initial population and growth rate above, we could predict the total population after 4 years by using $n = 4$.

$$T = 50{,}000(1 + .06)^4$$
$$\approx 50{,}000(1.26)*$$
$$= 63{,}000$$

We can expect a total population of about 63,000, or an increase of about 13,000 deer.

SOLVE EACH PROBLEM

A. If the initial population were 30,000 and the growth rate was .12, approximately how many deer would be present after 3 years?

B. How many additional deer can we expect in 5 years if the initial population is 45,000 and the current growth rate is .08?

ANSWERS   A. About 42,148 deer would be present after 3 years.   B. The population would increase by about 21,120 deer.

■ *Example 5*

GRAPHING A COMPOSITE EXPONENTIAL FUNCTION

Graph $f(x) = -2^x + 3$.

The graph of $y = -2^x$ is a reflection across the x-axis of the graph of $y = 2^x$. The 3 indicates that the graph should be translated up 3 units, as compared to the graph of $y = -2^x$. Find some ordered pairs. Since $y = -2^x$ would have y-intercept -1, this function has y-intercept 2, which is up 3 units from the y-intercept of $y = -2^x$. Some other ordered pairs are $(1, 1)$, $(2, -1)$, and $(3, -5)$. For negative values of x, the graph approaches the line $y = 3$ as a horizontal asymptote. The graph is shown in Figure 5.4. ■

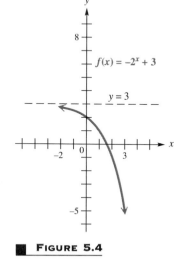

■ **FIGURE 5.4**

---

*The symbol $\approx$ means "is approximately equal to."

■ **Example 6**

GRAPHING A
COMPOSITE
EXPONENTIAL
FUNCTION

Graph $f(x) = 2^{-x^2}$.

Write $f(x) = 2^{-x^2}$ as $f(x) = 1/(2^{x^2})$ to find ordered pairs that belong to the function. Some ordered pairs are shown in the chart below.

x	-2	-1	0	1	2
y	$\frac{1}{16}$	$\frac{1}{2}$	1	$\frac{1}{2}$	$\frac{1}{16}$

As the chart suggests, $0 < y \le 1$ for all values of x. Plotting these points and drawing a smooth curve through them gives the graph in Figure 5.5. This graph is symmetric with respect to the y-axis and has the x-axis as a horizontal asymptote. ■

The important normal curve in probability theory has a graph very similar to the one in Figure 5.5.

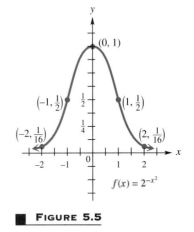

■ **FIGURE 5.5**

COMPOUND INTEREST   The formula for *compound interest* (interest paid on both principal and interest) is an important application of exponential functions. You may recall the formula for simple interest, $I = Prt$, where P is the amount left at interest, r is the rate of interest expressed as a decimal, and t is time in years that the principal earns interest. Suppose $t = 1$ year. Then at the end of the year the amount has grown to

$$P + Pr = P(1 + r),$$

the original principal plus the interest. If this amount is left at the same interest rate for another year, the total amount becomes

$$[P(1 + r)] + [P(1 + r)]r = [P(1 + r)](1 + r)$$
$$= P(1 + r)^2.$$

After the third year, this will grow to

$$[P(1 + r)^2] + [P(1 + r)^2]r = [P(1 + r)^2](1 + r)$$
$$= P(1 + r)^3.$$

Continuing in this way produces the following formula for compound interest.

COMPOUND
INTEREST

If P dollars is deposited in an account paying an annual rate of interest r compounded (paid) m times per year, then after t years the account will contain A dollars, where

$$A = P\left(1 + \frac{r}{m}\right)^{tm}.$$

For example, let $1000 be deposited in an account paying 8% per year compounded quarterly, or four times per year. After 10 years the account will contain

$$P\left(1 + \frac{r}{m}\right)^{tm} = 1000\left(1 + \frac{.08}{4}\right)^{10(4)}$$

$$= 1000(1 + .02)^{40}$$

$$= 1000(1.02)^{40}$$

dollars. The number $(1.02)^{40}$ can be found by using a calculator with a y^x key. To five decimal places, $(1.02)^{40} = 2.20804$. The amount on deposit after 10 years is

$$1000(1.02)^{40} = 1000(2.20804) = 2208.04,$$

or $2208.04.

In the formula for compound interest, A is sometimes called the **future value** and P the **present value.**

■ *Example 7*
FINDING PRESENT VALUE

An accountant wants to buy a new computer in three years that will cost $20,000.

(a) How much should be deposited now, at 6% interest compounded annually, to give the required $20,000 in three years?

Since the money deposited should amount to $20,000 in three years, $20,000 is the future value of the money. To find the present value P of $20,000 (the amount to deposit now), use the compound interest formula with $A = 20,000$, $r = .06$, $m = 1$, and $t = 3$.

$$A = P\left(1 + \frac{r}{m}\right)^{tm}$$

$$20,000 = P\left(1 + \frac{.06}{1}\right)^{3(1)} = P(1.06)^3$$

$$\frac{20,000}{(1.06)^3} = P$$

$$P = 16,792.39$$

The accountant must deposit $16,792.39.

(b) If only $15,000 is available to deposit now, what annual interest rate is required for it to increase to $20,000 in three years?

Here $P = 15,000$, $A = 20,000$, $m = 1$, $t = 3$, and r is unknown. Substitute the known values into the compound interest formula and solve for r.

$$A = P\left(1 + \frac{r}{m}\right)^{tm}$$

$$20,000 = 15,000\left(1 + \frac{r}{1}\right)^3$$

$$\frac{4}{3} = (1 + r)^3 \qquad \text{Divide both sides by 15,000.}$$

$$\left(\frac{4}{3}\right)^{1/3} = 1 + r \qquad \text{Take the cube root on both sides.}$$

$$\left(\frac{4}{3}\right)^{1/3} - 1 = r \qquad \text{Subtract 1 on both sides.}$$

$$r \approx .10 \qquad \text{Use a calculator.}$$

An interest rate of 10% will produce enough interest to increase the $15,000 deposit to the $20,000 needed at the end of three years. ■

Perhaps the single most useful exponential function is the function defined by $f(x) = e^x$, where e is an irrational number that occurs often in practical applications. The number e comes up in a natural way when using the formula for compound interest.

Suppose that a lucky investment produces an annual interest of 100%, so that $r = 1.00$, or $r = 1$. Suppose also that only $1 can be deposited at this rate, and for only one year. Then $P = 1$ and $t = 1$. Substitute into the formula for compound interest:

m	$\left(1 + \dfrac{1}{m}\right)^m$
1	2
2	2.25
5	2.48832
10	2.59374
25	2.66584
50	2.69159
100	2.70481
500	2.71557
1000	2.71692
10,000	2.71815
1,000,000	2.71828

$$P\left(1 + \frac{r}{m}\right)^{tm} = 1\left(1 + \frac{1}{m}\right)^{1(m)} = \left(1 + \frac{1}{m}\right)^m$$

As interest is compounded more and more often, the value of this expression will increase. If interest is compounded annually, making $m = 1$, the total amount on deposit is

$$\left(1 + \frac{1}{m}\right)^m = \left(1 + \frac{1}{1}\right)^1 = 2^1 = 2,$$

so an investment of $1 becomes $2 in one year.

A calculator with a y^x key gives the results in the table at the left. These results have been rounded to five decimal places. The table suggests that, as m increases, the value of $(1 + 1/m)^m$ gets closer and closer to some fixed number. It turns out that this is indeed the case. This fixed number is called e.

VALUE OF e   To nine decimal places,

$$e \approx 2.718281828.$$

NOTE   Values of e^x can be found with a calculator that has a key marked e^x or by using a combination of keys marked INV and ln x. See your instruction booklet for details or ask your instructor for assistance.

In Figure 5.6 the functions $y = 2^x$, $y = e^x$, and $y = 3^x$ are graphed for comparison.

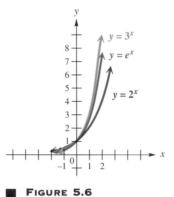

■ **FIGURE 5.6**

EXPONENTIAL GROWTH AND DECAY As mentioned above, the number e is important as the base of an exponential function because many practical applications require an exponential function with base e. For example, it can be shown that in situations involving growth or decay of a quantity, the amount or number present at time t often can be closely approximated by a function defined by

$$A = A_0 e^{kt},$$

where A_0 is the amount or number present at time $t = 0$ and k is a constant.

The next example illustrates exponential growth.

■ *Example 8*

SOLVING AN
EXPONENTIAL
GROWTH PROBLEM

The U. S. Consumer Price Index (CPI, or cost of living index) has risen exponentially over the years. From 1960 to 1990, the CPI is approximated by

$$A(t) = 34e^{.04t},$$

where t is time in years, with $t = 0$ corresponding to 1960. The index in 1960, at $t = 0$, was

$$A(0) = 34e^{(.04)(0)}$$
$$= 34e^0$$
$$= 34. \qquad e^0 = 1$$

To find the CPI for 1990, let $t = 1990 - 1960 = 30$, and find $A(30)$.

$$A(30) = 34e^{(.04)(30)}$$
$$= 34e^{1.2}$$
$$= 113 \qquad e^{1.2} \approx 3.3201$$

The index measures the average change in prices relative to the base year 1983 (1983 corresponds to 100) of a common group of goods and services. Our result of 113 means that prices increased an average of $113 - 34 = 79$ percent over the 30-year period from 1960 to 1990. ■

5.1 EXERCISES ∎ ——————————————————————

Solve each of the following equations. See Examples 1 and 2.

1. $4^x = 2$ **2.** $125^r = 5$ **3.** $\left(\dfrac{1}{2}\right)^k = 4$ **4.** $\left(\dfrac{2}{3}\right)^x = \dfrac{9}{4}$

5. $2^{3-y} = 8$ **6.** $5^{2p+1} = 25$ **7.** $\dfrac{1}{27} = b^{-3}$ **8.** $\dfrac{1}{81} = k^{-4}$

9. $4 = r^{2/3}$ **10.** $z^{5/2} = 32$ **11.** $27^{4z} = 9^{z+1}$ **12.** $32^t = 16^{1-t}$

13. $\left(\dfrac{1}{8}\right)^{-2p} = 2^{p+3}$ **14.** $3^{-h} = \left(\dfrac{1}{27}\right)^{1-2h}$ **15.** $\left(\dfrac{1}{2}\right)^{-x} = \left(\dfrac{1}{4}\right)^{x+1}$ **16.** $\left(\dfrac{2}{3}\right)^{k-1} = \left(\dfrac{81}{16}\right)^{k+1}$

17. Graph f for each of the following. Compare the graphs to that of $f(x) = 2^x$. See Examples 3–5.
  (a) $f(x) = 2^x + 1$ **(b)** $f(x) = 2^x - 4$ **(c)** $f(x) = 2^{x+1}$ **(d)** $f(x) = 2^{x-4}$

18. Graph f for each of the following. See Examples 3–5.
  (a) $f(x) = 3^{-x} - 2$ **(b)** $f(x) = 3^{-x} + 4$ **(c)** $f(x) = 3^{-x-2}$ **(d)** $f(x) = 3^{-x+4}$

19. Explain how you could use the graph of $y = 4^x$ to graph $y = -4^x$.

Graph f for each of the following. See Examples 4 – 6.

20. $f(x) = 3^x$ **21.** $f(x) = 4^x$ **22.** $f(x) = (3/2)^x$ **23.** $f(x) = e^{-x}$

24. $f(x) = e^{3x}$ **25.** $f(x) = e^{x+1}$ **26.** $f(x) = 2^{|x|}$ **27.** $f(x) = 2^{-|x|}$

28. Explain why the exponential equation $3^x = 12$ cannot be solved by using the properties of exponentials given in this section.

29. For $a > 1$, how does the value of $f(x) = a^x$ change as x increases? What if $0 < a < 1$?

30. If $f(x) = a^x$ and $f(3) = 27$, find the following values of $f(x)$.
  (a) $f(1)$ **(b)** $f(-1)$ **(c)** $f(2)$ **(d)** $f(0)$

31. What two points on the graph of $f(x) = a^x$ can be found with no computation?

Use a calculator to help graph f in Exercises 32–37.

32. $f(x) = \dfrac{e^x - e^{-x}}{2}$ **33.** $f(x) = \dfrac{e^x + e^{-x}}{2}$ **34.** $f(x) = x \cdot 2^x$

35. $f(x) = x^2 \cdot 2^{-x}$ **36.** $f(x) = (1 - x)e^x$ **37.** $f(x) = xe^x$

Use the formula for compound interest and a calculator to find the future value of each amount.

38. \$4292 at 6% compounded annually for 10 years

39. \$8906.54 at 5% compounded semiannually for 9 years

40. \$56,780 at 5.3% compounded quarterly for 23 quarters

41. \$45,788 at 6% compounded daily (ignoring leap years) for 11 years of 365 days

Find the present value for the following future values. See Example 7(a).

42. \$10,000, if interest is 12% compounded semiannually for 5 years

43. \$25,000, if interest is 6% compounded quarterly for 11 quarters

44. \$45,678.93, if interest is 9.6% compounded monthly for 11 months.

45. \$123,788, if interest is 8.7% compounded daily for 195 days

264 📖 *Find the required annual interest rate to the nearest tenth in each of the following. See Example 7(b).*

46. $25,000 compounded annually for 2 years to yield $31,360

47. $65,000 compounded monthly for 6 months to yield $65,325

48. $1200 compounded quarterly for 5 years to yield $1780

49. $15,000 compounded seminnually for 4 semiannual periods to yield $19,000

📖 *An annual interest rate of 6.50%, for example, compounded more than once per year produces an annual yield of more than 6.50%. With daily compounding (365 times per year), the annual yield is found using the compound interest formula for P = 1 and t = 1 as follows.*

$$A = \left(1 + \frac{.065}{365}\right)^{365} = 1.06715$$

Subtract the 1 to get the annual yield, .06715 or 6.715%.

50. Find the annual yield for an annual rate of 6.50% compounded as follows.
 (a) quarterly   **(b)** monthly

51. Find the annual yield for an annual rate of 7.00% compounded as follows.
 (a) monthly   **(b)** daily

📖 *Solve each of the following problems. See Example 8.*

52. Since 1950, the growth in the world population in millions closely fits the exponential function defined by

$$A(t) = 2600e^{.018t},$$

where t is the number of years since 1950.
 (a) The world population was about 3700 million in 1970. How closely does the function approximate this value?
 (b) Use the function to approximate the population in 1990. (The actual 1990 population was about 5320 million.)
 (c) Estimate the population in the year 2000.

53. A sample of 500 g of lead 210 decays to polonium 210 according to the function given by

$$A(t) = 500e^{-.032t},$$

where t is time in years. Find the amount of the sample after each of the following times.
 (a) 4 years   **(b)** 8 years
 (c) 20 years   **(d)** Graph $y = A(t)$.

54. Vehicle theft in the United States has been rising exponentially since 1972. The number of stolen vehicles, in millions, is given by

$$f(x) = .88(1.03)^x,$$

where $x = 0$ represents the year 1972. Find the number of vehicles stolen in the following years.
 (a) 1975   **(b)** 1980   **(c)** 1985   **(d)** 1990

55. Experiments have shown that the sales of a product, under relatively stable market conditions, but in the absence of promotional activities such as advertising, tend to decline at a constant yearly rate. This rate of sales decline varies considerably from product to product, but seems to remain the same for any particular product. The sales decline can be expressed by a function of the form

$$S(t) = S_0 e^{-at},$$

where $S(t)$ is the rate of sales at time t measured in years, S_0 is the rate of sales at time $t = 0$, and a is the sales decay constant.
 (a) Suppose the sales decay constant for a particular product is $a = .10$. Let $S_0 = 50,000$ and find $S(1)$ and $S(3)$.
 (b) Find $S(2)$ and $S(10)$ if $S_0 = 80,000$ and $a = .05$.

56. The number of nuclear warheads in a certain major superpower arsenal from 1965 to 1980 is approximated by

$$A(t) = 830e^{.15t},$$

where t is the number of years since 1965. Find the number of nuclear warheads in the following years.
 (a) 1965   **(b)** 1975   **(c)** 1980
 (d) In 1985, the number of warheads was actually 10,012. Find the corresponding number using the given function. Is the discrepancy significant? If so, what might have caused it?

In calculus, it is shown that

$$e^x = 1 + x + \frac{x^2}{2 \cdot 1} + \frac{x^3}{3 \cdot 2 \cdot 1} + \frac{x^4}{4 \cdot 3 \cdot 2 \cdot 1} + \frac{x^5}{5 \cdot 4 \cdot 3 \cdot 2 \cdot 1} + \cdots .$$

57. Use the terms shown here and replace x with 1 to approximate $e^1 = e$ to three decimal places. Then check your results with a calculator.

58. Use the terms shown here and replace x with $-.05$ to approximate $e^{-.05}$ to four decimal places. Check your results with a calculator.

5.2 ———— LOGARITHMIC FUNCTIONS

The previous section dealt with exponential functions of the form $y = a^x$ for all positive values of a, where $a \neq 1$. As mentioned there, the horizontal line test shows that exponential functions are one-to-one, and thus have inverse functions. In this section we discuss the inverses of exponential functions. The equation defining the inverse of a function is found by exchanging x and y in the equation that defines the function. Doing so with $y = a^x$ gives

$$x = a^y$$

as the equation of the inverse function of the exponential function defined by $y = a^x$. This equation can be solved for y by using the following definition.

LOGARITHM For all real numbers y, and all positive numbers a and x, where $a \neq 1$:

$$y = \log_a x \qquad \text{if and only if} \qquad x = a^y.$$

The "log" in the definition above is an abbreviation for *logarithm.* Read $\log_a x$ as "the logarithm to the base a of x." Intuitively, the logarithm to the base a of x is the power to which a must be raised to yield x.

In working with logarithms, it is helpful to remember the following.

MEANING OF $\log_a x$ A **logarithm** is an exponent; $\log_a x$ is the exponent on the base a that yields the number x.

CAUTION The "log" in $y = \log_a x$ is the notation for a particular function and there must be a replacement for x following it, as in $\log_a 3$, $\log_a (2x - 1)$, or $\log_a x^2$. Avoid writing meaningless notation such as $y = \log$ or $y = \log_a$.

■ *Example 1*

CONVERTING
BETWEEN
EXPONENTIAL AND
LOGARITHMIC
STATEMENTS

The chart below shows several pairs of equivalent statements. The same statement is written in both exponential and logarithmic forms.

Exponential Form	Logarithmic Form
$2^3 = 8$	$\log_2 8 = 3$
$(1/2)^{-4} = 16$	$\log_{1/2} 16 = -4$
$10^5 = 100{,}000$	$\log_{10} 100{,}000 = 5$
$3^{-4} = 1/81$	$\log_3 (1/81) = -4$
$5^1 = 5$	$\log_5 5 = 1$
$(3/4)^0 = 1$	$\log_{3/4} 1 = 0$

■

LOGARITHMIC EQUATIONS The definition of logarithm can be used to solve logarithmic equations, as shown in the next example.

■ *Example 2*

SOLVING
LOGARITHMIC
EQUATIONS

Solve each equation.

(a) $\log_x \dfrac{8}{27} = 3$

First, write the expression in exponential form.

$$x^3 = \frac{8}{27}$$

$$x^3 = \left(\frac{2}{3}\right)^3 \qquad \frac{8}{27} = \left(\frac{2}{3}\right)^3$$

$$x = \frac{2}{3} \qquad \text{Property (b) of exponents}$$

The solution set is $\{2/3\}$.

(b) $\log_4 x = 5/2$

In exponential form, the given statement becomes

$$4^{5/2} = x$$
$$(4^{1/2})^5 = x$$
$$2^5 = x$$
$$32 = x.$$

The solution set is $\{32\}$. ■

LOGARITHMIC FUNCTIONS The logarithmic function with base a is defined as follows.

LOGARITHMIC
FUNCTION

If $a > 0$, $a \neq 1$, and $x > 0$, then

$$f(x) = \log_a x$$

defines the **logarithmic function** with base a.

Exponential and logarithmic functions are inverses of each other. Since the domain of an exponential function is the set of all real numbers, the range of a logarithmic function also will be the set of all real numbers. In the same way, both the range of an exponential function and the domain of a logarithmic function are the set of all positive real numbers, so logarithms can be found for positive numbers only.

The graph of $y = 2^x$ is shown in red in Figure 5.7. The graph of its inverse is found by reflecting the graph of $y = 2^x$ about the line $y = x$. The graph of the inverse function, defined by $y = \log_2 x$, shown in blue, has the y-axis as a **vertical asymptote.**

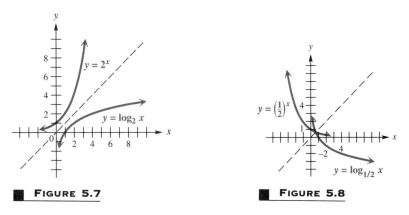

■ **FIGURE 5.7**   ■ **FIGURE 5.8**

The graph of $y = (1/2)^x$ is shown in red in Figure 5.8. The graph of its inverse, defined by $y = \log_{1/2} x$, in blue, is found by reflecting the graph of $y = (1/2)^x$ about the line $y = x$. As Figure 5.8 suggests, the graph of $y = \log_{1/2} x$ also has the y-axis for a vertical asymptote.

The graphs of $y = \log_2 x$ in Figure 5.7 and $y = \log_{1/2} x$ in Figure 5.8 suggest the following generalizations about the graphs of logarithmic functions of the form $f(x) = \log_a x$.

GRAPH OF $f(x) = \log_a x$	**1.** The point $(1, 0)$ is on the graph.
	2. If $a > 1$, f is an increasing function; if $0 < a < 1$, f is a decreasing function.
	3. The y-axis is a vertical asymptote.
	4. The domain is $(0, \infty)$ and the range is $(-\infty, \infty)$.

Compare these generalizations to those for exponential functions discussed in Section 5.1.

More general logarithmic functions can be obtained by forming the composition of $h(x) = \log_a x$ with a function $g(x)$ to get

$$f(x) = h[g(x)] = \log_a [g(x)].$$

The next examples illustrate some composite functions of this type.

268

■ **Example 3**

GRAPHING A
COMPOSITE
LOGARITHMIC
FUNCTION

Graph $f(x) = \log_2 (x - 1)$.

   The graph of this function will be the same as that of $f(x) = \log_2 x$, but shifted 1 unit to the right because $x - 1$ is given instead of x. This makes the domain $(1, \infty)$ instead of $(0, \infty)$. The line $x = 1$ is a vertical asymptote. The range is $(-\infty, \infty)$. See Figure 5.9. ■

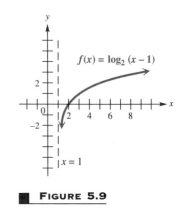

■ **FIGURE 5.9**

■ **Example 4**

GRAPHING A
TRANSLATED
LOGARITHMIC
FUNCTION

Graph $f(x) = (\log_3 x) - 1$.

   This function will have the same graph as that of $g(x) = \log_3 x$ translated down 1 unit. A table of values is given below for both $g(x) = \log_3 x$ and $f(x) = (\log_3 x) - 1$.

x	$\frac{1}{3}$	1	3	9
$g(x)$	-1	0	1	2
$f(x)$	-2	-1	0	1

The graph is shown in Figure 5.10. ■

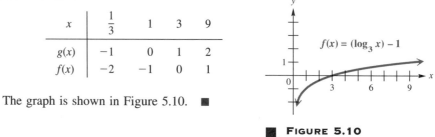

■ **FIGURE 5.10**

■ **Example 5**

GRAPHING A
COMPOSITE
LOGARITHMIC
FUNCTION

Graph $y = \log_3 |x|$.

   Write $y = \log_3 |x|$ in exponential form as $3^y = |x|$ to help identify some ordered pairs that satisfy the equation. (This is usually a good idea when graphing a logarithmic function.) Here, it is easier to choose y-values and find the corresponding x-values. Doing so gives the following ordered pairs.

x	-3	-1	$-\frac{1}{3}$	$\frac{1}{3}$	1	3
y	1	0	-1	-1	0	1

Plotting these points and connecting them with a smooth curve gives the graph in Figure 5.11. The y-axis is a vertical asymptote. Notice that, since $y = \log_3 |-x| = \log_3 |x|$, the graph is symmetric with respect to the y-axis. ■

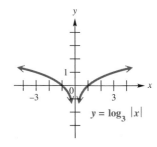

■ **FIGURE 5.11**

<table>
<tr><td>CAUTION</td><td>If you write a logarithmic function in exponential form, choosing y-values to calculate x-values as we did in Example 5, be careful to get the ordered pairs in the correct order.</td></tr>
</table>

PROPERTIES OF LOGARITHMS Logarithms originally were important as an aid for numerical calculations, but the availability of inexpensive calculators has made this application of logarithms obsolete. Yet the principles behind the use of logarithms for calculation are important; these principles are based on the properties listed below.

PROPERTIES OF LOGARITHMS

For any positive real numbers x and y, real number r, and any positive real number a, $a \neq 1$:

(a) $\log_a xy = \log_a x + \log_a y$ **(b)** $\log_a \dfrac{x}{y} = \log_a x - \log_a y$

(c) $\log_a x^r = r \log_a x$ **(d)** $\log_a a = 1$

(e) $\log_a 1 = 0$

PROOF

To prove Property (a), let

$$m = \log_a x \qquad \text{and} \qquad n = \log_a y.$$

By the definition of logarithm,

$$a^m = x \qquad \text{and} \qquad a^n = y.$$

Multiplication gives

$$a^m \cdot a^n = xy.$$

By a property of exponents,

$$a^{m+n} = xy.$$

Now use the definition of logarithm to write

$$\log_a xy = m + n.$$

Since $m = \log_a x$ and $n = \log_a y$,

$$\log_a xy = \log_a x + \log_a y. \quad ■$$

Properties (b) and (c) are proven in a similar way. (See Exercises 68 and 69.) Properties (d) and (e) follow directly from the definition of logarithm since $a^1 = a$ and $a^0 = 1$.

The properties of logarithms are useful for rewriting expressions with logarithms in different forms, as shown in the next examples.

270 ■ *Example 6*
USING THE
PROPERTIES OF
LOGARITHMS

Assuming that all variables represent positive real numbers, use the properties of logarithms to rewrite each of the following expressions.

(a) $\log_6 7 \cdot 9$

$$\log_6 7 \cdot 9 = \log_6 7 + \log_6 \mathbf{9}$$

(b) $\log_9 \dfrac{15}{7}$

$$\log_9 \frac{15}{7} = \log_9 15 - \log_9 7$$

(c) $\log_5 \sqrt{8}$

$$\log_5 \sqrt{8} = \log_5 8^{1/2} = \frac{1}{2} \log_5 8$$

(d) $\log_a \dfrac{mnq}{p^2} = \log_a m + \log_a n + \log_a q - 2 \log_a p$

(e) $\log_a \sqrt[3]{m^2} = \dfrac{2}{3} \log_a m$

(f) $\log_b \sqrt[n]{\dfrac{x^3 y^5}{z^m}} = \dfrac{1}{n} \log_b \dfrac{x^3 y^5}{z^m}$

$$= \frac{1}{n} (\log_b x^3 + \log_b y^5 - \log_b z^m)$$

$$= \frac{1}{n} (3 \log_b x + 5 \log_b y - m \log_b z)$$

$$= \frac{3}{n} \log_b x + \frac{5}{n} \log_b y - \frac{m}{n} \log_b z$$

Notice the use of parentheses in the second step. The factor $1/n$ applies to each term. ■

■ *Example 7*
USING THE
PROPERTIES OF
LOGARITHMS

Use the properties of logarithms to write each of the following as a single logarithm with a coefficient of 1. Assume that all variables represent positive real numbers.

(a) $\log_3 (x + 2) + \log_3 x - \log_3 2$
Using Properties (a) and (b),

$$\log_3 (x + 2) + \log_3 x - \log_3 2 = \log_3 \frac{(x + 2)x}{2}.$$

(b) $2 \log_a m - 3 \log_a n = \log_a m^2 - \log_a n^3 = \log_a \dfrac{m^2}{n^3}$

Here we used Property (c), then Property (b).

(c) $\dfrac{1}{2} \log_b m + \dfrac{3}{2} \log_b 2n - \log_b m^2 n$

$\quad = \log_b m^{1/2} + \log_b (2n)^{3/2} - \log_b m^2 n$ Property (c)

$\quad = \log_b \dfrac{m^{1/2}(2n)^{3/2}}{m^2 n}$ Properties (a) and (b)

$\quad = \log_b \dfrac{2^{3/2} n^{1/2}}{m^{3/2}}$ Rules for exponents

$\quad = \log_b \left(\dfrac{2^3 n}{m^3} \right)^{1/2}$ Rules for exponents

$\quad = \log_b \sqrt{\dfrac{8n}{m^3}}$ Definition of $a^{1/n}$ ■

CAUTION

There is no property of logarithms to rewrite a logarithm of a *sum* or *difference*. That is why, in Example 7(a), $\log_3 (x + 2)$ was not written as $\log_3 x + \log_3 2$. Remember, $\log_3 x + \log_3 2 = \log_3 (x \cdot 2)$.

■ **Example 8**

USING THE PROPERTIES OF LOGARITHMS WITH NUMERICAL VALUES

Assume that $\log_{10} 2 = .3010$. Find the base 10 logarithms of 4 and 5.

By the properties of logarithms,

$$\log_{10} 4 = \log_{10} 2^2 = 2 \log_{10} 2 = 2(.3010) = .6020$$

$$\log_{10} 5 = \log_{10} \frac{10}{2} = \log_{10} 10 - \log_{10} 2 = 1 - .3010 = .6990.$$

We used Property (d) to replace $\log_{10} 10$ with 1. ■

Compositions of the exponential and logarithmic functions can be used to get two more useful properties. If $f(x) = a^x$ and $g(x) = \log_a x$, then

$$f[g(x)] = a^{\log_a x}$$

and
$$g[f(x)] = \log_a a^x.$$

THEOREM ON INVERSES

For $a > 0$, $a \neq 1$:

$$a^{\log_a x} = x \qquad \text{and} \qquad \log_a a^x = x.$$

PROOF

Exponential and logarithmic functions are inverses of each other, so $f[g(x)] = x$ and $g[f(x)] = x$. Letting $f(x) = a^x$ and $g(x) = \log_a x$ gives both results. ■

By the results of the last theorem,

$$\log_5 5^3 = 3, \qquad 7^{\log_7 10} = 10, \qquad \text{and} \qquad \log_r r^{k+1} = k + 1.$$

The second statement in the theorem will be useful in Sections 5.4 and 5.5 when solving logarithmic or exponential equations.

5.2 EXERCISES ■

For each of the following statements, write an equivalent statement in logarithmic form. See Example 1.

1. $3^4 = 81$　　　　　　**2.** $2^5 = 32$　　　　　　**3.** $(1/2)^{-4} = 16$

4. $(2/3)^{-3} = 27/8$　　　**5.** $10^{-4} = .0001$　　　**6.** $(1/100)^{-2} = 10,000$

For each of the following statements, write an equivalent statement in exponential form. See Example 1.

7. $\log_6 36 = 2$　　　　　**8.** $\log_5 5 = 1$　　　　　**9.** $\log_{\sqrt{3}} 81 = 8$

10. $\log_4 \dfrac{1}{64} = -3$　　　**11.** $\log_{10} .0001 = -4$　　　**12.** $\log_3 \sqrt[3]{9} = \dfrac{2}{3}$

Find the value of each of the following expressions. (Hint: In Exercises 13–22, let the expression equal y, and write in exponential form.) See Example 2(a).

13. $\log_5 25$　　　**14.** $\log_3 81$　　　**15.** $\log_8 8$　　　**16.** $\log_7 1$

17. $\log_{10} .001$　　　**18.** $\log_6 \dfrac{1}{216}$　　　**19.** $\log_4 \dfrac{\sqrt[3]{4}}{2}$　　　**20.** $\log_9 \dfrac{\sqrt[4]{27}}{3}$

21. $\log_{1/3} \dfrac{9^{-4}}{3}$　　　**22.** $\log_{1/4} \dfrac{16^2}{2^{-3}}$　　　**23.** $2^{\log_2 9}$　　　**24.** $8^{\log_8 11}$

Solve each of the following equations. See Example 2.

25. $x = \log_2 32$　　　**26.** $x = \log_2 128$　　　**27.** $\log_x 25 = -2$　　　**28.** $\log_x \dfrac{1}{16} = -2$

29. Explain why logarithms of negative numbers are not defined.

30. Compare the summary of facts about the graph of $f(x) = \log_a x$ with the similar summary about the graph of $y = a^x$ in Section 5.1. Make a list of the facts that reinforce the idea that these are inverse functions.

31. Graph each of the following functions. Compare the graphs to that of $f(x) = \log_2 x$. See Examples 3–5.
　　(a) $f(x) = (\log_2 x) + 3$　　**(b)** $f(x) = \log_2 (x + 3)$　　**(c)** $f(x) = |\log_2 (x + 3)|$

32. Graph each of the following functions. Compare the graphs to that of $f(x) = \log_{1/2} x$. See Examples 3–5.
　　(a) $f(x) = (\log_{1/2} x) - 2$　　**(b)** $f(x) = \log_{1/2} (x - 2)$　　**(c)** $f(x) = |\log_{1/2} (x - 2)|$

Graph each of the following functions. See Examples 3–5.

33. $f(x) = \log_3 x$　　　**34.** $f(x) = \log_{10} x$　　　**35.** $f(x) = \log_{1/2} (1 - x)$　　　**36.** $f(x) = \log_{1/3} (3 - x)$

37. $f(x) = \log_2 x^2$　　　**38.** $f(x) = \log_3 (x - 1)$　　　**39.** $f(x) = x \log_{10} x$　　　**40.** $f(x) = x^2 \log_{10} x$

Write each of the following expressions as a sum, difference, or product of logarithms. Simplify the result if possible. Assume that all variables represent positive real numbers. See Example 6.

41. $\log_3 \dfrac{2}{5}$

42. $\log_4 \dfrac{6}{7}$

43. $\log_2 \dfrac{6x}{y}$

44. $\log_3 \dfrac{4p}{q}$

45. $\log_5 \dfrac{5\sqrt{7}}{3}$

46. $\log_2 \dfrac{2\sqrt{3}}{5}$

47. $\log_4 (2x + 5y)$

48. $\log_6 (7m + 3q)$

49. $\log_k \dfrac{pq^2}{m}$

50. $\log_z \dfrac{x^5 y^3}{3}$

51. $\log_m \sqrt{\dfrac{5r^3}{z^5}}$

52. $\log_p \sqrt[3]{\dfrac{m^5 n^4}{t^2}}$

Write each of the following expressions as a single logarithm. Assume that all variables represent positive real numbers. See Example 7.

53. $\log_a x + \log_a y - \log_a m$

54. $(\log_b k - \log_b m) - \log_b a$

55. $2 \log_m a - 3 \log_m b^2$

56. $\dfrac{1}{2} \log_y p^3 q^4 - \dfrac{2}{3} \log_y p^4 q^3$

57. $2 \log_a (z - 1) + \log_a (3z + 2), z > 1$

58. $\log_b (2y + 5) - \dfrac{1}{2} \log_b (y + 3)$

59. $-\dfrac{2}{3} \log_5 5m^2 + \dfrac{1}{2} \log_5 25m^2$

60. $-\dfrac{3}{4} \log_3 16p^4 - \dfrac{2}{3} \log_3 8p^3$

61. Why does $\log_a 1$ always equal 0 for any valid base a?

Given $\log_{10} 2 = .3010$ and $\log_{10} 3 = .4771$, find each of the following logarithms without using a calculator. See Example 8.

62. $\log_{10} 6$

63. $\log_{10} 12$

64. $\log_{10} (9/4)$

65. $\log_{10} (20/27)$

66. $\log_{10} \sqrt{30}$

67. $\log_{10} (36)^{1/3}$

Prove the following properties of logarithms.

68. $\log_a (x/y) = \log_a x - \log_a y$

69. $\log_a x^r = r \log_a x$

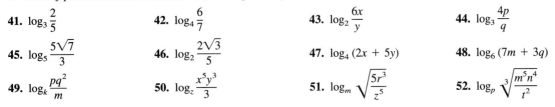

5.3 ——— **EVALUATING LOGARITHMS; CHANGE OF BASE**

COMMON LOGARITHMS Base 10 logarithms are called **common logarithms.** The common logarithm of the number x, or $\log_{10} x$, is often abbreviated as just log x, and we will use that convention from now on. A calculator with a log key can be used to find base 10 logarithms of any positive number.

■ *Example 1*
EVALUATING
COMMON
LOGARITHMS

Use a calculator to evaluate the following logarithms.

(a) log 142
    Enter 142 and press the log key. This may be a second function key on some calculators. With other calculators, these steps may be reversed. Consult your owner's manual if you have any problem using this key. The result should be 2.152 to the nearest thousandth.

(b) log .005832
    A calculator gives

$$\log .005832 \approx -2.234. \quad ■$$

NOTE Logarithms of numbers less than 1 are always negative, as suggested by the graphs in Section 5.2.

In chemistry, the pH of a solution is defined as

$$pH = -\log [H_3O^+],$$

where $[H_3O^+]$ is the hydronium ion concentration in moles per liter.* The pH value is a measure of the acidity or alkalinity of solutions. Pure water has a pH of 7.0, substances with pH values greater than 7.0 are alkaline, and substances with pH values less than 7.0 are acidic.

■ *Example 2*
FINDING pH

(a) Find the pH of a solution with $[H_3O^+] = 2.5 \times 10^{-4}$

$$pH = -\log [H_3O^+]$$
$$pH = -\log (2.5 \times 10^{-4}) \qquad \text{Substitute.}$$
$$= -(\log 2.5 + \log 10^{-4}) \qquad \text{Property (a) of logarithms}$$
$$= -(.3979 - 4)$$
$$= -.3979 + 4$$
$$= 3.6$$

It is customary to round pH values to the nearest tenth.

(b) Find the hydronium ion concentration of a solution with pH $= 7.1$.

$$pH = -\log [H_3O^+]$$
$$7.1 = -\log [H_3O^+] \qquad \text{Substitute.}$$
$$-7.1 = \log [H_3O^+] \qquad \text{Multiply by } -1.$$
$$[H_3O^+] = 10^{-7.1} \qquad \text{Write in exponential form.}$$

Evaluate $10^{-7.1}$ with a calculator to get

$$[H_3O^+] \approx 7.9 \times 10^{-8}. \quad ■$$

■ *Example 3*
SOLVING AN
APPLICATION OF
BASE 10
LOGARITHMS

The loudness of sounds is measured in a unit called a *decibel*. To measure with this unit, we first assign an intensity of I_0 to a very faint sound, called the *threshold sound*. If a particular sound has intensity I, then the decibel rating of this louder sound is

$$d = 10 \log \frac{I}{I_0}.$$

*A *mole* is the amount of a substance that contains the same number of molecules as the number of atoms in exactly 12 grams of carbon 12.

Find the decibel rating of a sound with intensity $10,000I_0$.

Let $I = 10,000I_0$ and find d.

$$d = 10 \log \frac{10,000I_0}{I_0}$$

$$= 10 \log 10,000$$

$$= 10(4) \qquad \text{log } 10,000 = 4$$

$$= 40$$

The sound has a decibel rating of 40. ▪

NATURAL LOGARITHMS In most practical applications of logarithms, the number $e \approx 2.718281828$ is used as base. The number e is irrational, like π. Logarithms to base e are called **natural logarithms,** since they occur in the life sciences and economics in natural situations that involve growth and decay. The base e logarithm of x is written ln x (read "el-en x"). A graph of the natural logarithm function defined by $f(x) = \ln x$ is given in Figure 5.12.

Natural logarithms can be found with a calculator that has an ln key.

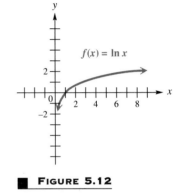

▪ **FIGURE 5.12**

▪ *Example 4*

EVALUATING NATURAL LOGARITHMS

Use a calculator to find the following logarithms.

(a) ln 85

With a calculator, enter 85, press the ln key, and read the result, 4.4427. The steps may be reversed with some calculators. If your calculator has an e^x key, but not a key labeled ln x, natural logarithms can be found by entering the number, pressing the INV key and then the e^x key. This works because $y = e^x$ is the inverse function of $y = \ln x$ (or $y = \log_e x$).

(b) ln 127.8 = 4.850

(c) ln .049 = −3.02

As with common logarithms, natural logarithms of numbers between 0 and 1 are negative. ▪

▪ *Example 5*

APPLYING NATURAL LOGARITHMS

Geologists sometimes measure the age of rocks by using "atomic clocks." By measuring the amounts of potassium 40 and argon 40 in a rock, the age t of the specimen in years is found with the formula

$$t = (1.26 \times 10^9) \frac{\ln [1 + 8.33(A/K)]}{\ln 2}.$$

A and K are respectively the numbers of atoms of argon 40 and potassium 40 in the specimen.

(a) How old is a rock in which $A = 0$ and $K > 0$?

If $A = 0$, $A/K = 0$ and the equation becomes

$$t = (1.26 \times 10^9) \frac{\ln 1}{\ln 2} = (1.26 \times 10^9)(0) = 0.$$

The rock is 0 years old or new.

(b) The ratio A/K for a sample of granite from New Hampshire is .212. How old is the sample?

Since A/K is .212, we have

$$t = (1.26 \times 10^9) \frac{\ln [1 + 8.33 (.212)]}{\ln 2} = 1.85 \times 10^9.$$

The granite is about 1.85 billion years old. ■

LOGARITHMS TO OTHER BASES A calculator can be used to find the values of either natural logarithms (base e) or common logarithms (base 10). However, sometimes it is convenient to use logarithms to other bases. The following theorem can be used to convert logarithms from one base to another.

CHANGE OF BASE THEOREM

For any positive real numbers x, a, and b, where $a \neq 1$ and $b \neq 1$:

$$\log_a x = \frac{\log_b x}{\log_b a}.$$

This theorem is proved by using the definition of logarithm to write $y = \log_a x$ in exponential form.

PROOF

Let

$$y = \log_a x.$$

$$a^y = x \qquad \text{Change to exponential form.}$$

$$\log_b a^y = \log_b x \qquad \text{Take logarithms on both sides.}$$

$$y \log_b a = \log_b x \qquad \text{Property (c) of logarithms}$$

$$y = \frac{\log_b x}{\log_b a} \qquad \text{Divide both sides by } \log_b a.$$

$$\log_a x = \frac{\log_b x}{\log_b a} \qquad \text{Substitute } \log_a x \text{ for } y. \quad ■$$

Any positive number other than 1 can be used for base b in the change of base rule, but usually the only practical bases are e and 10, since calculators give logarithms only for these two bases.

NOTE Some calculators have only a log key or an ln key. In that case, the change of base rule can be used to find logarithms to the missing base.

The next example shows how the change of base rule is used to find logarithms to bases other than 10 or e with a calculator.

■ **Example 6**
USING THE CHANGE OF BASE RULE

Use natural logarithms to find each of the following. Round to the nearest hundredth.

(a) $\log_5 17$

Use natural logarithms and the change of base theorem.

$$\log_5 17 = \frac{\log_e 17}{\log_e 5}$$

$$= \frac{\ln 17}{\ln 5}$$

Now use a calculator to evaluate this quotient.

$$\log_5 17 \approx \frac{2.8332}{1.6094}$$

$$\approx 1.76$$

To check, use a calculator with a y^x key, along with the definition of logarithm, to verify that $5^{1.76} \approx 17$.

(b) $\log_2 .1$

$$\log_2 .1 = \frac{\ln .1}{\ln 2} \approx \frac{-2.3026}{.6931} = -3.32 \quad ■$$

NOTE

In Example 6, logarithms that were evaluated in the intermediate steps, such as ln 17 and ln 5, were shown to four decimal places. However, the final answers were obtained *without* rounding off these intermediate values, using all the digits obtained with the calculator. In general, it is best to wait until the final step to round off the answer; otherwise, a build-up of round-off error may cause the final answer to have an incorrect final decimal place digit.

■ **Example 7**
SOLVING AN APPLICATION WITH BASE 2 LOGARITHMS

One measure of the diversity of the species in an ecological community is given by the formula

$$H = -[P_1 \log_2 P_1 + P_2 \log_2 P_2 + \cdots + P_n \log_2 P_n],$$

where $P_1, P_2, \ldots, P_n$ are the proportions of a sample belonging to each of n species found in the sample. For example, in a community with two species, where there are 90 of one species and 10 of the other, $P_1 = 90/100 = .9$ and $P_2 = 10/100 = .1$. Thus,

$$H = -[.9 \log_2 .9 + .1 \log_2 .1].$$

In Example 6(b), $\log_2 .1$ was found to be -3.32. Now find $\log_2 .9$.

$$\log_2 .9 = \frac{\ln .9}{\ln 2}$$

$$\approx \frac{-.1054}{.6931}$$

$$\approx -.152$$

Therefore,

$$H \approx -[(.9)(-.152) + (.1)(-3.32)] \approx .469.$$

If the number in each species is the same, the measure of diversity is 1, representing "perfect" diversity. In a community with little diversity, H is close to 0. In this example, since $H \approx .5$, there is neither great nor little diversity. ■

5.3 EXERCISES ■

▦ *A calculator will be necessary for most of the exercises in this section.*

Use a calculator to evaluate each of the following logarithms to four decimal places. See Examples 1 and 4.

1. log 43   **2.** log 1247   **3.** log 783   **4.** log .014

5. log .0069   **6.** log 60,000   **7.** ln 580   **8.** ln .08

9. ln .7   **10.** ln 81,000   **11.** ln 121,000   **12.** ln 350

Use the change of base rule to find each of the following logarithms to the nearest hundredth. See Example 6.

13. $\log_5 10$ **14.** $\log_9 12$ **15.** $\log_{15} 5$ **16.** $\log_{1/2} 3$

17. $\log_{100} 83$ **18.** $\log_{200} 175$ **19.** $\log_{2.9} 7.5$ **20.** $\log_{5.8} 12.7$

21. Is the logarithm to the base 3 of 4 written as $\log_4 3$ or $\log_3 4$?

22. What is the relationship between e^x and $\ln x$?

For each of the following substances, find the pH from the given hydronium ion concentration. See Example 2(a).

23. Grapefruit, 6.3×10^{-4} **24.** Crackers, 3.9×10^{-9}

25. Limes, 1.6×10^{-2} **26.** Sodium hydroxide (lye), 3.2×10^{-14}

Find the $[H_3O^+]$ for each of the following substances from the given pH. See Example 2(b).

27. Soda pop, 2.7   **28.** Wine, 3.4   **29.** Beer, 4.8   **30.** Drinking water, 6.5

Solve each problem. See Example 3.

31. Find the decibel ratings of sounds having the following intensities:

    (a) $100I_0$ **(b)** $1000I_0$

    (c) $100,000I_0$ **(d)** $1,000,000I_0$

32. Find the decibel ratings of the following sounds, having intensities as given. (You will need a calculator with a log key.) Round each answer to the nearest whole number.

    (a) whisper, $115I_0$

    (b) busy street, $9,500,000I_0$

    (c) heavy truck, 20 m away, $1,200,000,000I_0$

    (d) rock music, $895,000,000,000I_0$

    (e) jetliner at takeoff, $109,000,000,000,000I_0$

33. The intensity of an earthquake, measured on the *Richter scale,* is $\log_{10}(I/I_0)$, where I_0 is the intensity of an earthquake of a certain (small) size. Find the Richter scale ratings of earthquakes having the following intensities.
    (a) $1000I_0$
    (b) $1,000,000I_0$
    (c) $100,000,000I_0$

34. On July 14, 1991, Peshawar, Pakistan, was shaken by an earthquake that measured 6.6 on the Richter scale.
    (a) Express this reading in terms of I_0. See Exercise 33.
    (b) In February of the same year a quake measuring 6.5 on the Richter scale killed about 900 people in the mountains of Pakistan and Afghanistan. Express the intensity of a 6.5 reading in terms of I_0.
    (c) How much greater was the force of the earthquake with a measure of 6.6?

35. (a) The San Francisco earthquake of 1906 had a Richter scale rating of 8.3. Express the intensity of this earthquake as a multiple of I_0.
    (b) In 1989, the San Francisco region experienced an earthquake with a Richter scale rating of 7.1. Express the intensity of this earthquake as a multiple of I_0.
    (c) Compare the intensity of the two San Francisco earthquakes discussed above.

Solve each problem. See Examples 3 and 5.

36. The number of years, n, since two independently evolving languages split off from a common ancestral language is approximated by $n \approx -7600 \log r$, where r is the proportion of words from the ancestral language common to both languages.
    (a) Find n if $r = .9$.
    (b) Find n if $r = .3$.

(c) How many years have elapsed since the split if half of the words of the ancestral language are common to both languages?

37. The number of species in a sample is given by

$$S(n) = a \ln \left(1 + \frac{n}{a} \right).$$

Here n is the number of individuals in the sample and a is a constant that indicates the diversity of species in the community. If $a = .36$, find $S(n)$ for the following values of n.
    (a) 100   (b) 200   (c) 150   (d) 10

38. In Exercise 37, find $S(n)$ if a changes to .88. Use the following values of n.
    (a) 50   (b) 100   (c) 250

In the central Sierra Nevada mountains of California, the percent of moisture that falls as snow rather than rain is approximated reasonably well by

$$p = 86.3 \ln h - 680,$$

where p is the percent of snow at an altitude h in feet. (Assume $h \geq 3000$.)

39. Find the percent of moisture that falls as snow at the following altitudes.
    (a) 3000 ft   (b) 4000 ft
    (c) 7000 ft   (d) Graph p.

40. Suppose a sample of a small community shows two species with 50 individuals each. Find the index of diversity H. See Example 7.

41. A virgin forest in northwestern Pennsylvania has 4 species of large trees with the following proportions of each: hemlock, .521; beech, .324; birch, .081; maple, .074. Find the index of diversity H. See Example 7.

Graph f for each of the following.

42. $f(x) = |\ln x|$ 43. $f(x) = x \ln x$ 44. $f(x) = x^2 \ln x$

45. Given $g(x) = e^x$, evaluate the following.
    (a) $g(\ln 3)$ (b) $g(\ln 5^2)$ (c) $g\left(\ln \dfrac{1}{e} \right)$

46. Given $f(x) = 3^x$, evaluate the following.
    (a) $f(\log_3 7)$ (b) $f[\log_3 (\ln 3)]$ (c) $f[\log_3 (2 \ln 3)]$

47. Given $f(x) = \ln x$, evaluate the following.
    (a) $f(e^5)$ (b) $f(e^{\ln 3})$ (c) $f(e^{2 \ln 3})$

48. Given $f(x) = \log_2 x$, evaluate the following.
    (a) $f(2^3)$ (b) $f(2^{\log_2 2})$ (c) $f(2^{2 \log_2 2})$

5.4 ——————— **EXPONENTIAL AND LOGARITHMIC EQUATIONS**

As mentioned at the beginning of this chapter, exponential and logarithmic functions are important in many useful applications of mathematics. Using these functions in applications often requires solving exponential and logarithmic equations. Some simple equations were solved in the first two sections of this chapter. More general methods for solving these equations depend on the properties below. These properties follow from the fact that exponential and logarithmic functions are one-to-one. Property 1 was given and used to solve exponential equations in Section 5.1.

PROPERTIES OF LOGARITHMIC AND EXPONENTIAL FUNCTIONS	For $b > 0$ and $b \neq 1$: **1.** $b^x = b^y$ if and only if $x = y$. **2.** If $x > 0$ and $y > 0$, $\qquad \log_b x = \log_b y$ if and only if $x = y$.

EXPONENTIAL EQUATIONS   The first examples illustrate a general method, using Property 2, for solving exponential equations.

■ *Example 1*

SOLVING AN
EXPONENTIAL
EQUATION

Solve the equation $7^x = 12$.

In Section 5.1, we saw that Property 1 cannot be used to solve this equation, so we apply Property 2. While any appropriate base b can be used to apply Property 2, the best practical base to use is base 10 or base e. Taking base e (natural) logarithms of both sides gives

$$7^x = 12$$
$$\ln 7^x = \ln 12$$
$$x \ln 7 = \ln 12 \qquad \text{Property (c) of logarithms}$$
$$x = \frac{\ln 12}{\ln 7}. \qquad \text{Divide by ln 7.}$$

A decimal approximation for x can be found using a calculator:

$$x = \frac{\ln 12}{\ln 7} \approx \frac{2.4849}{1.9459} \approx 1.277.$$

A calculator with a y^x key can be used to check this answer. Evaluate $7^{1.277}$; the result should be approximately 12. This step verifies that, to the nearest thousandth, the solution set is $\{1.277\}$.   ■

CAUTION

> Be careful when evaluating a quotient like $\dfrac{\ln 12}{\ln 7}$ in Example 1. Do not confuse this quotient with $\ln \dfrac{12}{7}$ which can be written as $\ln 12 - \ln 7$. You *cannot* change the quotient of two *logarithms* to a difference of logarithms.
>
> $$\frac{\ln 12}{\ln 7} \neq \ln \frac{12}{7}$$

■ ***Example 2***

SOLVING AN
EXPONENTIAL
EQUATION

Solve $e^{-2 \ln x} = \dfrac{1}{16}$.

Use a property of logarithms to rewrite the exponent on the left side of the equation.

$$e^{-2 \ln x} = \frac{1}{16}$$

$$e^{\ln x^{-2}} = \frac{1}{16} \qquad \text{Property (c) of logarithms}$$

$$x^{-2} = \frac{1}{16} \qquad \text{Theorem on inverses: } e^{\ln k} = k$$

$$x^{-2} = 4^{-2} \qquad \frac{1}{16} = \frac{1}{4^2} = 4^{-2}$$

$$x = 4 \qquad \text{Property 1 given above}$$

Check this answer by substituting in the original equation to see that the solution set is $\{4\}$. ■

LOGARITHMIC EQUATIONS   The next examples show some ways to solve logarithmic equations. The properties of logarithms given in Section 5.2 are useful here, as is Property 2.

■ ***Example 3***

SOLVING A
LOGARITHMIC
EQUATION

Solve $\log_a (x + 6) - \log_a (x + 2) = \log_a x$.

Using a property of logarithms, rewrite the equation as

$$\log_a \frac{x + 6}{x + 2} = \log_a x. \qquad \text{Property (b) of logarithms}$$

Now the equation is in the proper form to use Property 2.

$$\frac{x + 6}{x + 2} = x \qquad \text{Property 2}$$

$$x + 6 = x(x + 2) \qquad \text{Multiply by } x + 2.$$

$$x + 6 = x^2 + 2x \qquad \text{Distributive property}$$

$$x^2 + x - 6 = 0 \qquad \text{Get 0 on one side.}$$

$$(x + 3)(x - 2) = 0 \qquad \text{Use the zero-factor property.}$$

$$x = -3 \qquad \text{or} \qquad x = 2.$$

The negative solution ($x = -3$) cannot be used since it is not in the domain of $\log_a x$ in the original equation. For this reason, the only valid solution is the positive number 2, giving the solution set $\{2\}$. ■

CAUTION

> Recall that the domain of $y = \log_b x$ is $(0, \infty)$. For this reason, it is always necessary to check that the solution of a logarithmic equation results in the logarithms of positive numbers in the original equation.

■ IN SIMPLEST TERMS

23

When physicians prescribe medication they must consider how the drug's effectiveness decreases over time. If, each hour, a drug is only 90% as effective as the previous hour, at some point the patient will not be receiving enough medication and must receive another dose. This situation can be modeled with a geometric sequence (see Section 9.2). If the initial dose was 200 mg and the drug was administered 3 hours ago, the expression $200(.90)^2$ represents the amount of effective medication still available. Thus, $200(.90)^2 = 162$ mg are still in the system. To determine how long it would take for the medication to reach the dangerously low level of 50 mg, we consider the equation $200(.90)^x = 50$, which is solved using logarithms.

$$200(.90)^x = 50$$
$$(.90)^x = .25$$
$$\log (.90)^x = \log .25$$
$$x \log .90 = \log .25$$
$$x = \frac{\log .25}{\log .90} \approx 13.16$$

Since x represents $n - 1$, the drug will reach a level of 50 mg in about 14 hours.

SOLVE EACH PROBLEM

A. If 250 mg of a drug are administered, and the drug is only 75% as effective each subsequent hour, how much effective medicine will remain in the person's system after 6 hours?

B. A new drug has been introduced which is 80% as effective each hour as the previous hour. A minimum of 20 mg must remain in the patient's bloodstream during the course of treatment. If 100 mg are administered, how many hours may elapse before another dose is necessary?

ANSWERS   A. About 59 mg will remain.   B. A new dose should be given before 8 hours elapse.

■ *Example 4*

SOLVING A
LOGARITHMIC
EQUATION

Solve $\log (3x + 2) + \log (x - 1) = 1$.

Since $\log x$ is an abbreviation for $\log_{10} x$, and $1 = \log_{10} 10$, the properties of logarithms give

$$\log (3x + 2)(x - 1) = \log 10 \qquad \text{Property (a) of logarithms}$$
$$(3x + 2)(x - 1) = 10 \qquad \text{Property 2}$$
$$3x^2 - x - 2 = 10$$
$$3x^2 - x - 12 = 0.$$

Now use the quadratic formula to get

$$x = \frac{1 \pm \sqrt{1 + 144}}{6}.$$

If $x = (1 - \sqrt{145})/6$, then $x - 1 < 0$; therefore, $\log (x - 1)$ is not defined and this proposed solution must be discarded, giving the solution set

$$\left\{ \frac{1 + \sqrt{145}}{6} \right\}. \quad ■$$

The definition of logarithm could have been used in Example 4 by first writing

$$\log (3x + 2) + \log (x - 1) = 1$$
$$\log_{10} (3x + 2)(x - 1) = 1 \qquad \text{Property (a)}$$
$$(3x - 2)(x - 1) = 10^1, \qquad \text{Definition of logarithm}$$

then continuing as shown above.

■ *Example 5*

SOLVING A
LOGARITHMIC
EQUATION

Solve $\ln e^{\ln x} - \ln (x - 3) = \ln 2$.

On the left, $\ln e^{\ln x}$ can be written as $\ln x$ using the theorem on inverses at the end of Section 5.2. The equation becomes

$$\ln x - \ln (x - 3) = \ln 2$$
$$\ln \frac{x}{x - 3} = \ln 2 \qquad \text{Property (b)}$$
$$\frac{x}{x - 3} = 2 \qquad \text{Property 2}$$
$$x = 2x - 6 \qquad \text{Multiply by } x - 3.$$
$$6 = x.$$

Verify that the solution set is $\{6\}$. ■

A summary of the methods used for solving equations in this section follows.

> **SOLVING EXPONENTIAL AND LOGARITHMIC EQUATIONS**
>
> An exponential or logarithmic equation may be solved by changing the equation into one of the following forms, where a and b are real numbers, $a > 0$, and $a \neq 1$.
>
> **1.** $a^{f(x)} = b$
> Solve by taking logarithms of each side. (Natural logarithms are often a good choice.)
> **2.** $\log_a f(x) = \log_a g(x)$
> From the given equation, $f(x) = g(x)$, which is solved algebraically.
> **3.** $\log_a f(x) = b$
> Solve by using the definition of logarithm to write the expression in exponential form as $f(x) = a^b$.

The next examples show applications of exponential and logarithmic equations.

■ *Example 6*

SOLVING A COMPOSITE EXPONENTIAL EQUATION

The strength of a habit is a function of the number of times the habit is repeated. If N is the number of repetitions and H is the strength of the habit, then, according to psychologist C. L. Hull,

$$H = 1000(1 - e^{-kN}),$$

where k is a constant. Solve this equation for k.

We must first solve the equation for e^{-kN}.

$$\frac{H}{1000} = 1 - e^{-kN} \qquad \text{Divide by 1000.}$$

$$\frac{H}{1000} - 1 = -e^{-kN} \qquad \text{Subtract 1.}$$

$$e^{-kN} = 1 - \frac{H}{1000} \qquad \text{Multiply by } -1.$$

Now solve for k. As shown earlier, we take logarithms on each side of the equation and use the fact that $\ln e^x = x$.

$$\ln e^{-kN} = \ln\left(1 - \frac{H}{1000}\right)$$

$$-kN = \ln\left(1 - \frac{H}{1000}\right) \qquad \ln e^x = x$$

$$k = -\frac{1}{N} \ln\left(1 - \frac{H}{1000}\right) \qquad \text{Multiply by } -\frac{1}{N}.$$

With the last equation, if one pair of values for H and N is known, k can be found, and the equation can then be used to find either H or N, for given values of the other variable. ■

■ *Example 7*

SOLVING A
COMPOSITE
LOGARITHMIC
EQUATION

In the exercises for Section 5.3, we saw that the number of species in a sample is given by S, where

$$S = a \ln \left(1 + \frac{n}{a} \right),$$

n is the number of individuals in the sample, and a is a constant. Solve this equation for n.

We begin by solving for $\ln \left(1 + \frac{n}{a} \right)$. Then we can change to exponential form and solve the resulting equation for n.

$$\frac{S}{a} = \ln \left(1 + \frac{n}{a} \right) \qquad \text{Divide by } a.$$

$$e^{S/a} = 1 + \frac{n}{a} \qquad \text{Write in exponential form.}$$

$$e^{S/a} - 1 = \frac{n}{a} \qquad \text{Subtract 1.}$$

$$n = a(e^{S/a} - 1) \qquad \text{Multiply by } a.$$

Using this equation and given values of S and a, the number of species in a sample can be found. ■

5.4 EXERCISES ■ ──────────────────────────

▦ *A calculator will be necessary for many of these exercises.*

Solve the following equations. When necessary, give answers as decimals rounded to the nearest thousandth. See Examples 1–5.

1. $3^x = 6$
2. $4^x = 12$
3. $3^{a+2} = 5$
4. $5^{2-x} = 12$

5. $6^{1-2k} = 8$
6. $3^{2m-5} = 13$
7. $e^{k-1} = 4$
8. $e^{2-y} = 12$

9. $2e^{5a+2} = 8$
10. $10e^{3z-7} = 5$
11. $2^x = -3$
12. $(1/4)^p = -4$

13. $e^{2x} \cdot e^{5x} = e^{14}$
14. $e^{\ln x} = 3$
15. $e^{-\ln x} = 2$
16. $e^{3 \ln x} = 1/8$

17. $e^{\ln x + \ln (x - 2)} = 8$
18. $e^{2 \ln x - \ln (x + 2)} = 8/3$
19. $100(1 + .02)^{3+n} = 150$

20. $500(1 + .05)^{p/4} = 200$
21. $\log (t - 1) = 1$
22. $\log q^2 = 1$

23. $\log (x - 3) = 1 - \log x$
24. $\log (z - 6) = 2 - \log (z + 15)$
25. $\ln (y + 2) = \ln (y - 7) + \ln 4$

26. $\ln p - \ln (p + 1) = \ln 5$
27. $\ln (5 + 4y) - \ln (3 + y) = \ln 3$
28. $\ln m + \ln (2m + 5) = \ln 7$

29. $\ln x + 1 = \ln (x - 4)$
30. $\ln (4x - 2) = \ln 4 - \ln (x - 2)$
31. $2 \ln (x - 3) = \ln (x + 5) + \ln 4$

32. $\ln (k + 5) + \ln (k + 2) = \ln 14k$
33. $\log_5 (r + 2) + \log_5 (r - 2) = 1$
34. $\log_4 (z + 3) + \log_4 (z - 3) = 1$

35. $\log_3 (a - 3) = 1 + \log_3 (a + 1)$
36. $\log w + \log (3w - 13) = 1$
37. $\ln e^x - \ln e^3 = \ln e^5$

38. $\ln e^x - 2 \ln e = \ln e^4$
39. $\log_2 \sqrt{2y^2 - 1} = 1/2$
40. $\log_2 (\log_2 x) = 1$

41. $\log z = \sqrt{\log z}$
42. $\log x^2 = (\log x)^2$

43. Suppose you overhear the following statement: "I must reject any negative answer when I solve an equation involving logarithms." Is this correct? Write an explanation of why it is or is not correct.

44. What values of x could not possibly be solutions of the following equation?
$$\log_a (4x - 7) + \log_a (x^2 + 4) = 0$$

Solve each of the following equations for the indicated variables. Use logarithms to the appropriate bases. See Examples 6 and 7.

45. $P = P_0 e^{kt/1000}$ for t

46. $I = \dfrac{E}{R} (1 - e^{-Rt/2})$ for t

47. $r = p - k \ln t$ for t

48. $p = a + \dfrac{k}{\ln x}$ for x

49. $T = T_0 + (T_1 - T_0) 10^{-kt}$ for t

50. $A = \dfrac{Pi}{1 - (1 + i)^{-n}}$ for n

51. Recall (from Section 5.3) the formula for the decibel rating of the loudness of a sound is
$$d = 10 \log \frac{I}{I_0}.$$
Solve this formula for I.

52. A few years ago, there was a controversy about a proposed government limit on factory noise. One group wanted a maximum of 89 decibels, while another group wanted 86. This difference seemed very small to many people. Find the percent by which the 89-decibel intensity exceeds that for 86 decibels. (See Exercise 51.)

The formula for compound interest,
$$A = P\left(1 + \frac{r}{m}\right)^{tm},$$
was given in Section 5.1. Use natural logarithms and solve for t to the nearest tenth in Exercises 53 and 54.

53. $A = \$10,000$, $P = \$7500$, $r = 6\%$, $m = 2$

54. $A = \$8400$, $P = \$5000$, $r = 8\%$, $m = 4$

55. George Tom wants to buy a $30,000 car. He has saved $27,000. Find the number of years (to the nearest tenth) it will take for his $27,000 to grow to $30,000 at 6% interest compounded quarterly.

56. Find t to the nearest hundredth if $1786 becomes $2063.40 at 11.6%, with interest compounded monthly.

57. Find the number of years for $2000 to amount to $2500 at 6% compounded semianually.

58. Find the number of years for $16,000 to become $25,000 at 8% compounded quarterly.

59. The growth of bacteria in food products makes it necessary to time-date some products (such as milk) so that they will be sold and consumed before the bacteria count becomes too high. Suppose for a certain product that the number of bacteria present is given by
$$f(t) = 500e^{.1t},$$
under certain storage conditions, where t is time in days after packing of the product and the value of $f(t)$ is in millions.

(a) If the product cannot be safely eaten after the bacteria count reaches 3,000,000,000, how long will this take?

(b) If $t = 0$ corresponds to January 1, what date should be placed on the product?

60. In Example 8 of Section 5.1, the U.S. Consumer Price Index was approximated by $A(t) = 34e^{.04t}$, where t represents the number of years after 1960. Assuming the same equation continues to apply, find the year in which costs will be 50% higher than in 1983, that is, when the CPI equals 150.

61. The population of an animal species that is introduced into a certain area may grow rapidly at first but then grow more slowly as time goes on. A logarithmic function can provide an excellent description of such growth. Suppose that the population of foxes in an area t months after the foxes were first introduced there is
$$F = 500 \log (2t + 3).$$
Solve the equation for t. Then find t to the nearest tenth for the following values of F.

(a) 600   **(b)** 1000

62. Recall from Section 5.3 that the number of years, n, since two independently evolving languages split off from a common language is approximated by $n = -7600 \log r$, where r is the proportion of words from the ancestral language still common to the two languages. Solve the formula for r. Then find r if the languages split

(a) 1000 years ago,   **(b)** 2500 years ago.

In many cases, quantities grow or decay according to a function defined by

$$A(t) = A_0 e^{kt}.$$

As mentioned in Section 5.1, when k is positive, the result is a growth function; when k is negative, it is a decay function. This section gives several examples of applications of this function.

CONTINUOUS COMPOUNDING   The compound interest formula

$$A = P\left(1 + \frac{r}{m}\right)^{tm}$$

was discussed in Section 5.1. The table presented there shows that increasing the frequency of compounding makes smaller and smaller differences in the amount of interest earned. In fact, it can be shown that even if interest is compounded at intervals of time as small as one chooses (such as each hour, each minute, or each second), the total amount of interest earned will be only slightly more than for daily compounding. This is true even for a process called **continuous compounding.** As suggested in Section 5.1, the value of the expression $(1 + 1/m)^m$ approaches e as m gets larger. Because of this, the formula for continuous compounding involves the number e.

CONTINUOUS COMPOUNDING   If P dollars is deposited at a rate of interest r compounded continuously for t years, the final amount on deposit is

$$A = Pe^{rt}$$

dollars.

∎ **Example 1**
SOLVING A CONTINUOUS COMPOUNDING PROBLEM

Suppose $5000 is deposited in an account paying 8% compounded continuously for five years. Find the total amount on deposit at the end of five years.

Let $P = 5000$, $t = 5$, and $r = .08$. Then

$$A = 5000 e^{.08(5)} = 5000 e^{.4}.$$

Using a calculator, we find that $e^{.4} \approx 1.49182$, and

$$A = 5000 e^{.4} = 7459.12,$$

or $7459.12. As a comparison, the compound interest formula with daily compounding gives

$$A = P\left(1 + \frac{r}{m}\right)^{tm}$$

$$= 5000\left(1 + \frac{.08}{365}\right)^{5(365)} = 7458.80,$$

about 32¢ less.   ∎

■ *Example 2*

SOLVING A
CONTINUOUS
COMPOUNDING
PROBLEM

How long will it take for the money in an account that is compounded continuously at 8% interest to double?

Use the formula for continuous compounding, $A = Pe^{rt}$, to find the time t that makes $A = 2P$. Substitute $2P$ for A and .08 for r; then solve for t.

$$A = Pe^{rt}$$
$$2P = Pe^{.08t} \qquad \text{Substitute.}$$
$$2 = e^{.08t} \qquad \text{Divide by } P.$$

Taking natural logarithms on both sides gives

$$\ln 2 = \ln e^{.08t}.$$

Use the property $\ln e^x = x$ to get $\ln e^{.08t} = .08t$.

$$\ln 2 = .08t \qquad \text{Substitute.}$$
$$\frac{\ln 2}{.08} = t \qquad \text{Divide by .08.}$$
$$8.664 = t$$

It will take about 8 2/3 years for the amount to double. ■

GROWTH AND DECAY The next examples illustrate applications of exponential growth and decay.

■ *Example 3*

SOLVING AN
EXPONENTIAL
DECAY PROBLEM

Nuclear energy derived from radioactive isotopes can be used to supply power to space vehicles. The output of the radioactive power supply for a certain satellite is given by the function

$$y = 40e^{-.004t},$$

where y is in watts and t is the time in days.

(a) How much power will be available at the end of 180 days?
Let $t = 180$ in the formula.

$$y = 40e^{-.004(180)}$$
$$y \approx 19.5 \qquad \text{Use a calculator.}$$

About 19.5 watts will be left.

(b) How long will it take for the amount of power to be half of its original strength?
The original amount of power is 40 watts. (Why?) Since half of 40 is 20, replace y with 20 in the formula, and solve for t.

$$20 = 40e^{-.004t}$$

$$.5 = e^{-.004t} \qquad \text{Divide by 40.}$$

$$\ln .5 = \ln e^{-.004t}$$

$$\ln .5 = -.004t \qquad \ln e^x = x$$

$$t = \frac{\ln .5}{-.004}$$

$$t \approx 173 \qquad \text{Use a calculator.}$$

After about 173 days, the amount of available power will be half of its original amount. ■

In Examples 2 and 3(b), we found the amount of time that it would take for an amount to double and to become half of its original amount. These are examples of *doubling time* and *half-life*. The **doubling time** of a quantity that grows exponentially is the amount of time that it takes for any initial amount to grow to twice its value. Similarly, the **half-life** of a quantity that decays exponentially is the amount of time that it takes for any initial amount to decay to half its value.

■ *Example 4*

SOLVING AN
EXPONENTIAL
GROWTH PROBLEM

Carbon 14 is a radioactive form of carbon that is found in all living plants and animals. After a plant or animal dies, the radiocarbon disintegrates. Scientists determine the age of the remains by comparing the amount of carbon 14 present with the amount found in living plants and animals. The amount of carbon 14 present after t years is given by the exponential equation

$$A(t) = A_0 e^{kt}$$

with $k \approx -(\ln 2)(1/5700)$.

(a) Find the half-life.
   Let $A(t) = (1/2)A_0$ and $k = -(\ln 2)(1/5700)$.

$$\frac{1}{2} A_0 = A_0 e^{-(\ln 2)\,(1/5700)t}$$

$$\frac{1}{2} = e^{-(\ln 2)(1/5700)t} \qquad \text{Divide by } A_0.$$

$$\ln \frac{1}{2} = \ln e^{-(\ln 2)(1/5700)t} \qquad \text{Take logarithms on both sides.}$$

$$\ln \frac{1}{2} = -\frac{\ln 2}{5700} t \qquad \ln e^x = x$$

$$-\frac{5700}{\ln 2} \ln \frac{1}{2} = t \qquad \text{Multiply by } -\frac{5700}{\ln 2}.$$

$$-\frac{5700}{\ln 2} (\ln 1 - \ln 2) = t \qquad \text{Property (b)}$$

$$-\frac{5700}{\ln 2} (-\ln 2) = t \qquad \ln 1 = 0$$

$$5700 = t$$

The half-life is 5700 years.

(b) Charcoal from an ancient fire pit on Java contained 1/4 the carbon 14 of a living sample of the same size. Estimate the age of the charcoal.

Let $A(t) = \dfrac{1}{4} A_0$ and $k = -(\ln 2)(1/5700)$.

$$\frac{1}{4} A_0 = A_0 e^{-(\ln 2)(1/5700)t}$$

$$\frac{1}{4} = e^{-(\ln 2)(1/5700)t}$$

$$\ln \frac{1}{4} = \ln e^{-(\ln 2)(1/5700)t}$$

$$\ln \frac{1}{4} = -\frac{\ln 2}{5700} t$$

$$-\frac{5700}{\ln 2} \ln \frac{1}{4} = t$$

$$t = 11{,}400$$

The charcoal is about 11,400 years old.   ■

5.5 EXERCISES ■ ─────────────────────────────

⊞ *A calculator will be necessary for many of these exercises.*

Work the following problems. See Examples 1 and 2.

1. Assuming continuous compounding, what will it cost to buy a $10 item in 3 years at the following inflation rates?
   (a) 3%   **(b)** 4%   **(c)** 5%

2. Bert Bezzone invests a $25,000 inheritance in a fund paying 5% per year compounded continuously. What will be the amount on deposit after each of the following time periods?
   (a) 1 year   **(b)** 5 years   **(c)** 10 years

3. Linda Youngman, who is self-employed, wants to invest $60,000 in a pension plan. One investment offers 7% compounded quarterly. Another offers 6.75% compounded continuously. Which investment will earn more interest in 5 years? How much more will the better plan earn?

4. If Ms. Youngman (see Exercise 3) chooses the plan with continuous compounding, how long will it take for her $60,000 to grow to $80,000?

5. Assume the cost of a loaf of bread is $1. With continuous compounding, find the doubling time at an annual inflation rate of 6%.

6. Suppose the annual rate of inflation is 5%. How long will it take for prices to double?

7. Find the interest rate that will cause $5000 to grow to $7250 in 4 years if the money is compounded continuously.

Work the following problems. See Example 3.

8. In the exercises for Section 5.1, we gave the sales decline function as $S(t) = S_0 e^{-at}$. If $a = .1$, $S_0 = 50{,}000$, and t is time measured in years, find the number of years it will take for sales to fall to half the initial sales.

9. One measure of living standards in the United States is given by $L = 9 + 2e^{.15t}$, where t is the number of years since 1982. Find L for the following years.
   (a) 1982   (b) 1986   (c) 1992
   (d) Graph L.
   (e) What can be said about the growth of living standards in the U.S. according to this equation?

10. The quantity in grams of a radioactive substance present at time t is $A(t) = 500e^{-.05t}$, where t is time measured in days.
    (a) Find the half-life of the substance.
    (b) Find the amount present after 10 days.

Find the half-life of the following radioactive substances to three significant digits. (t is measured in years.) See Examples 3 and 4.

11. Plutonium 241; $A(t) = A_0e^{-.053t}$

12. Radium 226; $A(t) = A_0e^{-.00043t}$

13. Iodine 131; $A(t) = A_0e^{-.087t}$

14. Suppose the number of rabbits in a colony is $y = y_0e^{.4t}$, where t represents time in months and y_0 is the rabbit population when $t = 0$.
    (a) If $y_0 = 100$, find the number of rabbits present at time $t = 4$.
    (b) How long will it take for the number of rabbits to triple?

15. A midwestern city finds its residents moving to the suburbs. Its population is declining according to the relationship $P = P_0e^{-.04t}$, where t is time measured in years and P_0 is the population at time $t = 0$. Assume that $P_0 = 1,000,000$.
    (a) Find the population at time $t = 1$.
    (b) Estimate the time it will take for the population to be reduced to 750,000.
    (c) How long will it take for the population to be cut in half?

For Exercises 16–21, refer to Example 4.

16. Suppose an Egyptian mummy is discovered in which the amount of carbon 14 present is only about one-third the amount found in the atmosphere. About how long ago did the Egyptian die?

17. A sample from a refuse deposit near the Strait of Magellan had 60% of the carbon 14 of a contemporary living sample. How old was the sample?

18. Paint from the Lascaux caves of France contains 15% of the normal amount of carbon 14. Estimate the age of the caves.

19. Solve the formula $A = A_0e^{kt}$ for t.

20. Suppose a specimen is found in which there is about 2/3 the carbon 14 of a comparable living specimen. Estimate the age of the specimen.

21. Estimate the age of a specimen that contains 20% of the carbon 14 of a comparable living specimen.

A large cloud of radioactive debris from a nuclear explosion has floated over the Pacific Northwest, contaminating much of the hay supply. Consequently, farmers in the area are concerned that the cows who eat this hay will give contaminated milk. (The tolerance level for radioactive iodine in milk is 0.) The percent of the initial amount of radioactive iodine still present in the hay after t days is approximated by $P(t) = 100e^{-.1t}$.

22. Some scientists feel that the hay is safe after the percent of radioactive iodine has declined to 10% of the original amount. Find the number of days before the hay can be used.

23. Other scientists believe that the hay is not safe until the level of radioactive iodine has declined to only 1% of the original level. Find the number of days this would take.

24. The amount of a chemical that will dissolve in a solution increases exponentially as the (Celsius) temperature t is increased according to the equation $A(t) = 10e^{.0095t}$. At what temperature will 15 g dissolve?

25. By Newton's law of cooling, the temperature of a body at time t after being introduced into an environment having constant temperature T_0 is

$$A(t) = T_0 + Ce^{-kt},$$

where C and k are constants. If $C = 100$, $k = .1$, and t is time measured in minutes, how long will it take a hot cup of coffee to cool to a temperature of $25°$ C in a room at $20°$ C?

CHAPTER 5 SUMMARY ■ ————————————————

SECTION	TERMS	KEY IDEAS
5.1 Exponential Functions	exponential function horizontal asymptote increasing function decreasing function compound interest future value present value	**Additional Properties of Exponents** **(a)** If $a > 0$ and $a \neq 1$, then a^x is a unique real number for all real numbers x. **(b)** If $a > 0$ and $a \neq 1$, then $a^b = a^c$ if and only if $b = c$. **(c)** If $a > 1$ and $m < n$, then $a^m < a^n$. **(d)** If $0 < a < 1$ and $m < n$, then $a^m > a^n$.
5.2 Logarithmic Functions	logarithm logarithmic function vertical asymptote	**Properties of Logarithms** For any positive real numbers x and y, real number r, and positive real number a, $a \neq 1$: **(a)** $\log_a xy = \log_a x + \log_a y$ **(b)** $\log_a \dfrac{x}{y} = \log_a x - \log_a y$ **(c)** $\log_a x^r = r \cdot \log_a x$ **(d)** $\log_a a = 1$ **(e)** $\log_a 1 = 0$.
5.3 Evaluating Logarithms; Change of Base	common logarithm natural logarithm	**Change of Base Theorem** For any positive real numbers x, a, and b, where $a \neq 1$ and $b \neq 1$: $$\log_a x = \frac{\log_b x}{\log_b a}.$$
5.4 Exponential and Logarithmic Equations		**Properties of Exponential and Logarithmic Functions** For $b > 0$ and $b \neq 1$: **1.** $b^x = b^y$ if and only if $x = y$. **2.** If $x > 0$ and $y > 0$, $$\log_b x = \log_b y \quad \text{if and only if} \quad x = y.$$
5.5 Exponential Growth and Decay	continuous compounding doubling time half-life	**Growth or Decay Function** $$A = A_0 e^{kt},$$ for constants A_0 and k.

Solve each of the following equations.

1. $8^p = 32$

2. $9^{2y-1} = 27^y$

3. $\dfrac{8}{27} = b^{-3}$

4. $\dfrac{1}{2} = \left(\dfrac{b}{4}\right)^{1/4}$

5. The amount of a certain radioactive material, in grams, present after t days is given by

$$A(t) = 800e^{-.04t}.$$

How much is present after 5 days?

Match each equation with one of the graphs below.

6. $y = \log_{.3} x$

7. $y = e^x$

8. $y = \ln x$

9. $y = (.3)^x$

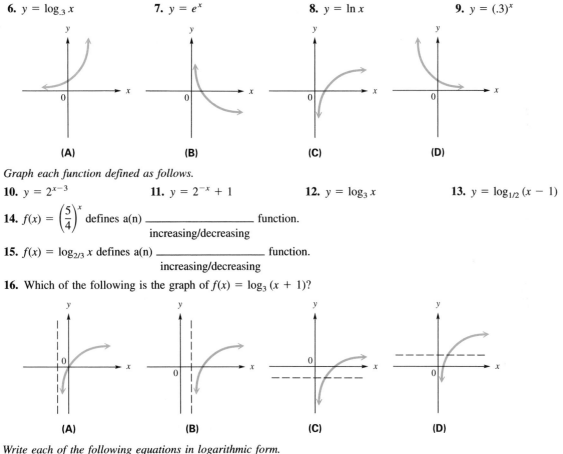

(A)   (B)   (C)   (D)

Graph each function defined as follows.

10. $y = 2^{x-3}$

11. $y = 2^{-x} + 1$

12. $y = \log_3 x$

13. $y = \log_{1/2}(x - 1)$

14. $f(x) = \left(\dfrac{5}{4}\right)^x$ defines a(n) ——————————— function.
increasing/decreasing

15. $f(x) = \log_{2/3} x$ defines a(n) ——————————— function.
increasing/decreasing

16. Which of the following is the graph of $f(x) = \log_3(x + 1)$?

(A)   (B)   (C)   (D)

Write each of the following equations in logarithmic form.

17. $2^5 = 32$ **18.** $100^{1/2} = 10$ **19.** $(1/16)^{1/4} = 1/2$ **20.** $(3/4)^{-1} = 4/3$ **21.** $10^{.4771} = 3$ **22.** $e^{2.4849} = 12$

Write each of the following equations in exponential form.

23. $\log_{10} .001 = -3$ **24.** $\log_2 \sqrt{32} = 5/2$ **25.** $\log 3.45 = .537819$ **26.** $\ln 45 = 3.806662$

Use properties of logarithms to write each of the following logarithms as a sum, difference, or product of logarithms.

27. $\log_3 \dfrac{mn}{5r}$

28. $\log_2 \dfrac{\sqrt{7}}{15}$

29. $\log_5 x^2 y^4 \sqrt[5]{m^3 p}$

30. $\log_7 (7k + 5r^2)$

▦ *A calculator will be necessary for many of these expressions.*

Find each of the following logarithms. Round to the nearest thousandth.

31. log 45.6

32. log 1,230,000

33. log .00056

34. log .0411

35. ln 35

36. ln 470

37. ln 144,000

38. ln 98,000

39. $\log_{3.4} 15.8$

40. $\log_{1/2} 9.45$

41. $\log_3 769$

42. $\log_{2/3} 5/8$

The height in meters of the members of a certain tribe is approximated by $h = .5 + \log t$, where t is the tribe member's age in years, and $1 \le t \le 20$. Estimate the heights of tribe members of the following ages.

43. 2 yr

44. 5 yr

45. 10 yr

46. 20 yr

Use a formula for compound interest to find the total amount on deposit if the following amounts are placed on deposit at the given interest rates for the given numbers of years.

47. $1000 at 8% compounded annually for 9 years

48. $312.45 at 6% compounded semiannually for 16 years

In Exercises 49 and 50, find the present value of each of the following future values.

49. $2000 at 4% compounded semiannually for 11 years

50. $2000 at 6% compounded quarterly for 8 years

51. Manuel deposits $10,000 for 12 years in an account paying 12% compounded annually. He then puts this total amount on deposit in another account paying 10% compounded semiannually for another 9 years. Find the total amount on deposit after the entire 21-year period.

52. Anne Kelly deposits $12,000 for 8 years in an account paying 5% compounded annually. She then leaves the money alone with no further deposits at 6% compounded annually for an additional 6 years. Find the total amount on deposit after the entire 14-year period.

Solve each of the following equations. Round to the nearest thousandth if necessary.

53. $5^r = 11$

54. $10^{2r-3} = 17$

55. $e^{p+1} = 10$

56. $(1/2)^{3k+1} = 3$

57. $\log_{64} y = 1/3$

58. $\ln 6x - \ln (x + 1) = \ln 4$

59. $\log_{16} \sqrt{x + 1} = 1/4$

60. $\ln x + 3 \ln 2 = \ln \dfrac{2}{x}$

61. $\ln (\ln e^{-x}) = \ln 3$

62. What annual interest rate, to the nearest tenth, will produce $8780 if $3500 is left at interest for 10 years?

63. Find the annual interest rate to the nearest tenth that will produce $27,208.60 if $12,700 is left at interest compounded semiannually for 10 years.

64. How many years (to the nearest tenth) will be needed for $7800 to increase to $14,088 at 6% compounded quarterly?

65. Find the number of years (to the nearest tenth) needed for $48,000 to become $58,344 at 5% interest compounded semiannually.

66. If the inflation rate were 10%, use the formula for continuous compounding to find the number of years for a $1 item to cost $2.

67. In Exercise 66, find the number of years if the rate of inflation were 4%.

How much would $1200 amount to at 10% compounded continuously for the following numbers of years?

68. 4 yr

69. 10 yr

70. Historically, the consumption of electricity has increased at a continuous rate of 6% per year. If it continued to increase at this rate, find the number of years before exactly twice as much electricity would be needed.

71. Suppose a conservation compaign together with higher rates caused demand for electricity to increase at only 2% per year. (See Exercise 70.) Find the number of years before twice as much electricity would be needed.

72. *Escherichia coli* is a strain of bacteria that occurs naturally in many different organisms. Under certain conditions, the number of bacteria present in a colony is

$$E(t) = E_0 2^{t/30},$$

where $E(t)$ is the number of bacteria present t min after the beginning of an experiment, and E_0 is the number present when $t = 0$. Let $E_0 = 2{,}400{,}000$ and find the number of bacteria at the following times.

(a) 5 min **(b)** 10 min **(c)** 60 min **(d)** 120 min

73. Use the function defined by

$$t = T \frac{\ln[1 + 8.33(A/K)]}{\ln 2}$$

from Section 5.3 to estimate the age of a rock sample, if tests show that A/K is .103 for the sample. Let $T = 1.26 \times 10^9$.

74. The function defined by

$$A(t) = (5 \times 10^{12})e^{-.04t}$$

gives the known coal reserves in the world in year t (in tons), where $t = 0$ corresponds to 1970, and $-.04$ indicates the rate of consumption.

(a) Find the amount of coal available in 1990.

(b) When will the coal reserves be half of what they were in 1970?

CHAPTER 5 TEST ■

Solve each of the following equations.

1. $25^{2x-1} = 125^{x+1}$

2. $\left(\dfrac{a}{2}\right)^{-1/3} = 4$

Write each of the following in logarithmic form.

3. $a^2 = b$

4. $e^c = 4.82$

Write each of the following in exponential form.

5. $\log_3 \sqrt{27} = \dfrac{3}{2}$

6. $\ln 5 = a$

7. What two points on the graph of $y = \log_a x$ can be found without computation?

Graph each function defined as follows.

8. $y = (1.5)^{x+2}$

9. $y = \log_{1/2} x$

10. Use properties of logarithms to write the following as a sum, difference, or product of logarithms.

$$\log_7 \frac{x^2 \sqrt[4]{y}}{z^3}$$

A calculator will be necessary for most of the remaining exercises in this test.

Find each of the following logarithms. Round to the nearest thousandth.

11. $\ln 2300$

12. $\log_{2.7} 94.6$

13. What values of x cannot possibly be solutions of the equation $\log_a (2x - 3) = -1$?

Solve the following equations. Round answers to the nearest thousandth.

14. $8^{2w-4} = 100$

15. $\log_3 (m + 2) = 2$

16. $\ln x - 4 \ln 3 = \ln \dfrac{5}{x}$

17. The amount of radioactive material, in grams, present after t days is given by $A(t) = 600e^{-.05t}$.
(a) Find the amount present after 12 days.
(b) Find the half-life of the material.

18. How much will \$2500 amount to at 8% compounded continuously for 12 years?

19. If a population increases at a continuous rate of 4% per year, how many years will it take for the population to double in size?

20. How many years, to the nearest tenth, will be needed for \$5000 to increase to \$18,000 at 12% compounded monthly?

■ THE GRAPHING CALCULATOR ■

The functions $f(x) = 10^x$, $f(x) = e^x$, $f(x) = \log_{10} x$, and $f(x) = \log_e x = \ln x$ are built into graphing calculators and can be graphed by touching the appropriate key after touching the graph key. Variations of these functions can be graphed as follows. To graph an exponential function with any base other than 10 or e, first use the range key to set the domain, range, and scales. Then touch the graph key and the required keys to input the function. For example, to graph $y = 3^{x^2}$ with the TI-81 calculator, touch the following sequence of keys.

$$y = \quad 3 \quad \wedge \quad x \quad x^2 \quad \text{graph}$$

(There may be slight differences with some calculators.) To graph $y = 3^{-x^2}$, we can use the following sequence of keystrokes:

$$y = \quad 3 \quad \wedge \quad - \quad x \quad x^2 \quad \text{graph}$$

Logarithmic functions to bases other than 10 or e can be graphed by using the change of base theorem. For example, to graph

$$y = \log_3 (x - 1) = \frac{\ln (x - 1)}{\ln 3}$$

with base e logarithms, use the following sequence.

$$y = \quad \ln \quad (\quad x \quad - \quad 1 \quad) \quad \div \quad \ln \quad 3 \quad \text{graph}$$

You may want to use the following functions to experiment with your calculator. Most of these functions have been selected from examples in the text, so you can compare your calculator graphs with those given in the example figures. The figure number of the graph in the text is given when applicable.

---

1. $f(x) = 2^x$ (Figure 5.1)

2. $g(x) = \left(\dfrac{1}{2}\right)^x$ (Figure 5.2)

3. $f(x) = 2^{-x+2}$ (Figure 5.3)

4. $f(x) = -2^x + 3$ (Figure 5.4)

5. $f(x) = 2^{-x^2}$ (Figure 5.5)

6. $y = e^x$ (Figure 5.6)

7. $y = \log_2 x$ (Figure 5.7)

8. $y = \log_{1/2} x$ (Figure 5.8)

9. $f(x) = \log_2 (x - 1)$ (Figure 5.9)

10. $f(x) = (\log_3 x) - 1$ (Figure 5.10)

11. $y = \log_3 |x|$ (Figure 5.11)

12. $f(x) = \ln x$ (Figure 5.12)

13. $h(x) = \ln (2x + 1)$

14. $k(x) = 2 + \ln x$

POLYNOMIAL AND RATIONAL FUNCTIONS

6.1 GRAPHING POLYNOMIAL FUNCTIONS

6.2 POLYNOMIAL DIVISION

6.3 ZEROS OF POLYNOMIAL FUNCTIONS

6.4 RATIONAL ZEROS OF POLYNOMIAL FUNCTIONS

6.5 REAL ZEROS OF POLYNOMIAL FUNCTIONS

6.6 GRAPHING RATIONAL FUNCTIONS

298

In Chapter 4 we studied linear and quadratic functions. Recall that these are functions defined by equations of the form $f(x) = mx + b$ and $f(x) = ax^2 + bx + c$, and are the simplest examples of a larger group of functions known as *polynomial functions*. In this chapter we will study polynomial functions, although a more complete analysis of polynomial functions requires the concepts of calculus.

Rational expressions were introduced in Chapter 1. In this chapter we will also look at rational functions which are defined by such expressions. They, too, are studied in more detail in calculus.

Polynomial and rational functions can be used to predict outcomes of various activities, as illustrated in some of the exercises in this chapter.

6.1 ——— GRAPHING POLYNOMIAL FUNCTIONS

Polynomial functions are defined as follows.

POLYNOMIAL FUNCTION	A **polynomial function of degree n** is a function defined by $$P(x) = a_n x^n + a_{n-1} x^{n-1} + \cdots + a_1 x + a_0,$$ for complex numbers $a_n, a_{n-1}, \ldots, a_1$, and a_0, where n is a whole number and $a_n \neq 0$.

$P(x)$ is used instead of $f(x)$ in this chapter to emphasize that the functions are polynomial functions. The polynomials discussed in Chapter 1 had only real-number coefficients. In this chapter the domain of the coefficients is extended to include all complex numbers.* The number a_n is the **leading coefficient** of P.

We begin the discussion of graphing polynomial functions with the graphs of polynomial functions defined by equations of the form $P(x) = ax^n$.

■ **Example 1**
GRAPHING
FUNCTIONS OF THE
FORM $P(x) = ax^n$

Graph each function P defined as follows.

(a) $P(x) = x^3$

Choose several values for x, and find the corresponding values of $P(x)$, or y, as shown in the upper table in Figure 6.1. Plot the resulting ordered pairs and connect the points with a smooth curve. The graph of $P(x) = x^3$ is shown in blue in Figure 6.1.

(b) $P(x) = x^5$

Work as in part (a) of this example to get the graph shown in red in Figure 6.1. Notice that the graphs of $P(x) = x^3$ and $P(x) = x^5$ are both symmetric with respect to the origin.

$P(x) = x^3$

x	$P(x)$
-2	-8
-1	-1
0	0
1	1
2	8

$P(x) = x^5$

x	$P(x)$
-1.5	-7.6
-1	-1
0	0
1	1
1.5	7.6

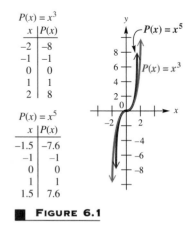

■ **FIGURE 6.1**

---

*Recall that the set of complex numbers includes both real numbers and imaginary numbers such as $5i$ or $6 + i\sqrt{7}$.

(c) $P(x) = x^4$, $P(x) = x^6$

Some typical ordered pairs for the graphs of $P(x) = x^4$ and $P(x) = x^6$ are given in the tables in Figure 6.2. These graphs are symmetric with respect to the y-axis, as is the graph of $P(x) = ax^2$ for a nonzero real number a. ∎

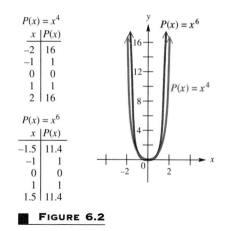

$P(x) = x^4$

x	P(x)
-2	16
-1	1
0	0
1	1
2	16

$P(x) = x^6$

x	P(x)
-1.5	11.4
-1	1
0	0
1	1
1.5	11.4

∎ **FIGURE 6.2**

As with the graph of $y = ax^2$ in Chapter 3, the value of a in $P(x) = ax^n$ affects the width of the graph. When $|a| > 1$, the graph is narrower than the graph of $P(x) = x^n$, when $0 < |a| < 1$, the graph is broader. The graph of $P(x) = -ax^n$ is reflected about the x-axis as compared to the graph of $P(x) = ax^n$.

∎ **Example 2**

EXAMINING THE
EFFECT OF a FOR
$P(x) = ax^n$

Graph each function P defined as follows.

(a) $P(x) = \frac{1}{2}x^3$

The graph is broader than that of $P(x) = x^3$ but has the same general shape. It includes the points $(-2, -4)$, $(-1, -1/2)$, $(0, 0)$, $(1, 1/2)$, and $(2, 4)$. See Figure 6.3.

(b) $P(x) = -\frac{3}{2}x^4$

The following table gives some ordered pairs for this function.

x	-2	-1	0	1	2
P(x)	-24	$-\frac{3}{2}$	0	$-\frac{3}{2}$	-24

The graph is shown in Figure 6.4. This graph is narrower than that of $P(x) = x^4$, since $|-3/2| > 1$, and opens downward instead of upward, since $-3/2 < 0$. ∎

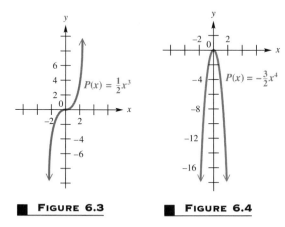

∎ **FIGURE 6.3**   ∎ **FIGURE 6.4**

Compared with the graph of $P(x) = ax^n$, the graph of $P(x) = ax^n + k$ is translated (shifted) k units up if $k > 0$ and $|k|$ units down if $k < 0$. Also, the graph of $P(x) = a(x - h)^n$ is translated h units to the right if $h > 0$ and $|h|$ units to the left if $h < 0$, when compared with the graph of $P(x) = ax^n$.

The graph of $P(x) = a(x - h)^n + k$ shows a combination of these translations. The effects here are the same as those we saw earlier with quadratic functions.

300 ■ *Example 3*

EXAMINING
VERTICAL AND
HORIZONTAL
TRANSLATIONS

Graph each of the following.

(a) $P(x) = x^5 - 2$

The graph will be the same as that of $P(x) = x^5$, but translated down 2 units. See Figure 6.5.

(b) $P(x) = (x + 1)^6$

This function P has a graph like that of $P(x) = x^6$, but since $x + 1 = x - (-1)$, it is translated one unit to the left as shown in Figure 6.6.

(c) $P(x) = -(x - 1)^3 + 3$

The negative sign causes the graph to be reflected about the x-axis when compared with the graph of $P(x) = x^3$. As shown in Figure 6.7, the graph is also translated 1 unit to the right and 3 units up. ■

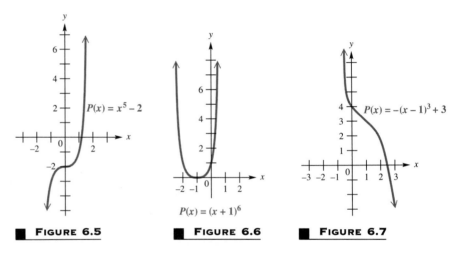

■ **FIGURE 6.5**   ■ **FIGURE 6.6**   ■ **FIGURE 6.7**

Generalizing from the graphs in Examples 1–3, the domain of a polynomial function is the set of all real numbers. The range of a polynomial function of odd degree is also the set of all real numbers. Some typical graphs of polynomial functions of odd degree are shown in Figure 6.8.

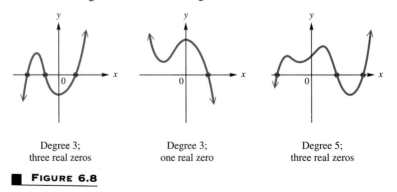

Degree 3;   Degree 3;   Degree 5;
three real zeros   one real zero   three real zeros

■ **FIGURE 6.8**

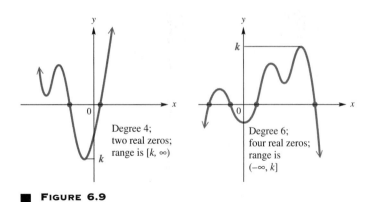

■ **FIGURE 6.9**

A polynomial function of even degree will have a range that takes the form $(-\infty, k]$ or $[k, \infty)$ for some real number k. Figure 6.9 shows two typical graphs of polynomial functions of even degree.

The values of x that satisfy $P(x) = 0$ are called the **zeros** of the function P. Methods for finding zeros of first- and second-degree polynomial functions were given in earlier chapters. In later sections of this chapter, several theorems are presented that help to find, or at least approximate, the zeros of polynomials of higher degree. The graphs in Figure 6.8 suggest that every polynomial function P of odd degree has at least one real zero.

It is difficult to accurately graph most polynomial functions without using calculus or a graphing calculator. A large number of points must be plotted to get a reasonably accurate graph. If a polynomial function P can be factored into linear factors, however, its graph can be approximated by plotting only a few points.

■ *Example 4*

GRAPHING A POLYNOMIAL FUNCTION GIVEN IN FACTORED FORM

Graph the function P defined as follows:

$$P(x) = (2x + 3)(x - 1)(x + 2).$$

Multiplying out the expression on the right would show that it is a third-degree polynomial, also called a *cubic* polynomial. Sketch the graph of P by first setting each of the three factors equal to 0 and solving the resulting equations to find the zeros of the function.

$$2x + 3 = 0 \qquad \text{or} \qquad x - 1 = 0 \qquad \text{or} \qquad x + 2 = 0$$

$$x = -\frac{3}{2} \qquad\qquad x = 1 \qquad\qquad x = -2$$

The three zeros, $-3/2$, 1, and -2, divide the x-axis into four intervals:

$$(-\infty, -2), \left(-2, -\frac{3}{2}\right), \left(-\frac{3}{2}, 1\right), (1, \infty).$$

These intervals are shown in Figure 6.10.

$(-\infty, -2)$ $\left(-2, -\frac{3}{2}\right)$ $\left(-\frac{3}{2}, 1\right)$ $(1, \infty)$

-2 $-\frac{3}{2}$ 1

■ **FIGURE 6.10**

Since the three zeros give the only x-intercepts, the values of $P(x)$ in the intervals shown in Figure 6.10 are either always positive or always negative. To find the sign of $P(x)$ in each interval, select a value of x in the interval and determine by substitution whether the function values are positive or negative in that interval. A typical selection of test points and the results of the tests are shown below. (When 0 lies in an interval, it is a good idea to use it as a test point. This will give the y-intercept.)

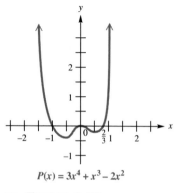

$P(x) = (2x + 3)(x - 1)(x + 2)$

■ **FIGURE 6.11**

Interval	Test Point	Value of $P(x)$	Sign of $P(x)$
$(-\infty, -2)$	-3	-12	Negative
$(-2, -3/2)$	$-7/4$	$11/32$	Positive
$(-3/2, 1)$	0	-6	Negative
$(1, \infty)$	2	28	Positive

When the values of $P(x)$ are negative, the graph is below the x-axis, and when $P(x)$ takes on positive values, the graph is above the x-axis. The y-intercept is -6. These results suggest that the graph looks like the sketch in Figure 6.11. The sketch could be improved by plotting additional points in each region. ■

■ ***Example 5***

GRAPHING A
POLYNOMIAL
FUNCTION
REQUIRING
FACTORING

Sketch the graph of the function P, where $P(x) = 3x^4 + x^3 - 2x^2$.

The polynomial can be factored as follows:

$$3x^4 + x^3 - 2x^2 = x^2(3x^2 + x - 2)$$
$$= x^2(3x - 2)(x + 1).$$

The zeros, 0, 2/3, and -1, divide the x-axis into four intervals:

$$(-\infty, -1), \quad (-1, 0), \quad \left(0, \frac{2}{3}\right), \quad \text{and} \quad \left(\frac{2}{3}, \infty\right).$$

Determine the sign of $P(x)$ (and thus the location of the graph relative to the x-axis) in each interval by substituting a value for x from each interval to get the following information. Since $P(0) = 0$, the y-intercept is 0.

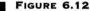

$P(x) = 3x^4 + x^3 - 2x^2$

■ **FIGURE 6.12**

Interval	Test Point	Value of $P(x)$	Location Relative to x-Axis
$(-\infty, -1)$	-2	32	Above
$(-1, 0)$	$-1/2$	$-7/16$	Below
$(0, 2/3)$	$1/3$	$-4/27$	Below
$(2/3, \infty)$	1	2	Above

With the values of x used for the test points and the corresponding values of $P(x)$, sketch the graph as shown in Figure 6.12. ■

A general procedure that can be used to graph polynomial functions is given below.

GRAPHING POLYNOMIAL FUNCTIONS

Let $y = P(x)$ define a polynomial function. To sketch its graph, use the following steps.

1. Find x-intercepts by solving $P(x) = 0$.
2. Find the y-intercept by evaluating $P(0)$.
3. Use test points from the intervals formed by the x-intercepts to determine whether the graph lies above or below the x-axis in each interval.
4. Plot any additional points, as necessary, and join the points with a smooth, unbroken curve.

NOTE

When graphing polynomial functions using only the methods of college algebra, determining the exact location of the "turning points" of the graph may be difficult or impossible. Therefore, we sometimes just approximate them. Methods of calculus or the use of graphing calculators allow us to graph polynomials more accurately.

There are important relationships among the following ideas:

1. the **x-intercepts of the graph** of $y = P(x)$;
2. the **zeros of the function** P; and
3. the **solutions of the equation** $P(x) = 0$.

Consider the function from Example 4, with $P(x) = (2x + 3)(x - 1)(x + 2)$. When the factors are multiplied out, $P(x) = 2x^3 + 5x^2 - x - 6$. The graph of this function, shown in Figure 6.11, has x-intercepts -2, $-3/2$, and 1. Since -2, $-3/2$, and 1 are the x-values for which the function is 0, they are the zeros of P.

Furthermore, since $P(-2) = P(-3/2) = P(1) = 0$, they are also solutions of the equation $P(x) = 2x^3 + 5x^2 - x - 6 = 0$. The discussion above can be summarized as follows.

---

x-INTERCEPTS, ZEROS, AND SOLUTIONS

If a is an x-intercept of the graph of the function defined by $y = P(x)$, then a is a zero of P and a is a solution of the equation $P(x) = 0$.

■ **Example 6**

DESCRIBING RELATIONSHIPS AMONG x-INTERCEPTS, ZEROS, AND SOLUTIONS

Describe the relationships among the x-intercepts of the graph of $P(x) = 3x^4 + x^3 - 2x^2$, the zeros of P, and the solutions of the equation $3x^4 + x^3 - 2x^2 = 0$.

This function was graphed in Example 5. See Figure 6.12. From the graph, we see that the x-intercepts are -1, 0, and 2/3. Therefore, -1, 0, and 2/3 are zeros of P, and -1, 0, and 2/3 are solutions of the equation $3x^4 + x^3 - 2x^2 = 0$.   ■

■ IN SIMPLEST TERMS ─────────────────────────────────────── ■

16 ▱

AIDS researchers have discovered that the cumulative number of AIDS cases, measured over time, can be accurately described by the cubic polynomial $f(t) = 174.6(t - 1981.2)^3 + 340$, where t is the calendar year in question.

To approximate the cumulative number of cases through 1985, let $t = 1985$.

$$f(1985) = 174.6(1985 - 1981.2)^3 + 340$$
$$= 174.6(3.8)^3 + 340$$
$$= 174.6(54.872) + 340$$
$$\approx 9581 + 340$$
$$= 9921$$

There were about 9921 AIDS cases reported through 1985.

SOLVE EACH PROBLEM

A. Approximate the cumulative number of AIDS cases present in 1989.

B. Assuming that the same polynomial will be an accurate model in 1993, how many AIDS cases will have developed through that year?

───

ANSWERS   A. There were about 83,197 cases.   B. Approximately 287,213 cases would have developed through 1993.

■ ── ■

6.1 EXERCISES ■ ────────────────────────────────────

Sketch the graph of each polynomial function P as defined in Exercises 1–10. See Examples 1–3.

1. $P(x) = \frac{1}{4}x^6$ **2.** $P(x) = -\frac{2}{3}x^5$ **3.** $P(x) = -\frac{5}{4}x^5$ **4.** $P(x) = 2x^4$

5. $P(x) = \frac{1}{2}x^3 + 1$ **6.** $P(x) = -x^4 + 2$ **7.** $P(x) = -(x + 1)^3$ **8.** $P(x) = \frac{1}{3}(x + 3)^4$

9. $P(x) = (x - 1)^4 + 2$ **10.** $P(x) = (x + 2)^3 - 1$

Graph each of the following. See Examples 4 and 5.

11. $P(x) = 2x(x - 3)(x + 2)$ **12.** $P(x) = x^2(x + 1)(x - 1)$

13. $P(x) = x^2(x - 2)(x + 3)^2$ **14.** $P(x) = x^2(x - 5)(x + 3)(x - 1)$

15. $P(x) = x^3 - x^2 - 2x$ **16.** $P(x) = -x^3 - 4x^2 - 3x$

17. $P(x) = (x + 2)(x - 1)(x + 1)$ **18.** $P(x) = (x - 4)(x + 2)(x - 1)$

19. $P(x) = (3x - 1)(x + 2)^2$ **20.** $P(x) = (4x + 3)(x + 2)^2$

21. $P(x) = x^3 + 5x^2 - x - 5$ (*Hint:* Factor by grouping.)   **22.** $P(x) = x^3 + x^2 - 36x - 36$

23. Which one of the following does not define a polynomial function?
(a) $P(x) = x^2$ (b) $P(x) = (x + 1)^3$
(c) $P(x) = \frac{1}{x}$ (d) $P(x) = 2x^5$

24. Write a short explanation of how the values of a, h, and k affect the graph of $P(x) = a(x - h)^n + k$ in comparison to the graph of $P(x) = x^n$.

▦ **25.** During the early part of the twentieth century, the deer population of the Kaibab Plateau in Arizona experienced a rapid increase, because hunters had reduced the number of natural predators. The increase in population depleted the food resources and eventually caused the population to decline. For the period from 1905 to 1930, the deer population was approximated by the function D, where

$$D(x) = -.125x^5 + 3.125x^4 + 4000,$$

and x is time in years from 1905.
(a) Use a calculator to find enough points to graph the function.
(b) From the graph, over what period of time (from 1905 to 1930) was the population increasing? relatively stable? decreasing?

26. The pressure of the oil in a reservoir tends to drop with time. By taking sample pressure readings for a particular oil reservoir, petroleum engineers have found that the change in pressure is given by the function P, where

$$P(t) = t^3 - 25t^2 + 200t,$$

and t is time in years from the date of the first reading.
(a) Graph the function.
(b) For what time period is the change in pressure (drop) increasing? decreasing?

For each of the graphs in Exercises 27 and 28, state the relationships among the x-intercepts of the graph of P(x), the zeros of the function P, and the solutions of the equation P(x) = 0. See Example 6.

27.

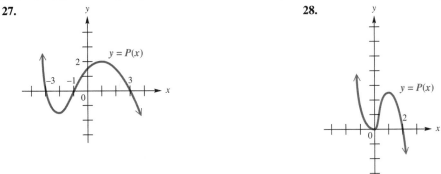

28.

29. Give an example (if possible) of a polynomial function whose graph is symmetric with respect to the y-axis.

30. Give an example (if possible) of a polynomial function whose graph is symmetric with respect to both the x-axis and the origin. (Do not consider $P(x) = 0$.)

Approximate maximum or minimum values of polynomial functions can be found for given intervals by first evaluating the function at the left endpoint of the interval, then adding .1 to the value of x and reevaluating the polynomial. Keep doing this until the right endpoint of the interval is reached. Then identify the approximate maximum and minimum value for the polynomial on the interval.

31. $P(x) = x^3 + 4x^2 - 8x - 8, [-3.8, -3]$

32. $P(x) = x^3 + 4x^2 - 8x - 8, [.3, 1]$

33. $P(x) = 2x^3 - 5x^2 - x + 1, [-1, 0]$

34. $P(x) = 2x^3 - 5x^2 - x + 1, [1.4, 2]$

35. $P(x) = x^4 - 7x^3 + 13x^2 + 6x - 28, [-2, -1]$

36. $P(x) = x^4 - 7x^3 + 13x^2 + 6x - 28, [2, 3]$

37. The graphs of the functions defined by $P(x) = x$, $P(x) = x^3$, and $P(x) = x^5$ all pass through the points $(-1, -1)$, $(0, 0)$, and $(1, 1)$. Draw a graph showing these three functions for the domain $[-1, 1]$. Use your result to decide how the graph of $P(x) = x^{11}$ would look for the same interval.

38. The graphs of the functions defined by $P(x) = x^2$, $P(x) = x^4$, and $P(x) = x^6$ all pass through the points $(-1, 1)$, $(0, 0)$, and $(1, 1)$. Draw a graph showing these three functions for the domain $[-1, 1]$. Use your result to decide how the graph of $P(x) = x^{12}$ would look for the same interval.

◼ **6.2** ─────── **POLYNOMIAL DIVISION**

In Chapter 1 we reviewed how to add, subtract, and multiply polynomials. In this section we will introduce several methods of dividing polynomials. Polynomial division is useful for evaluating polynomial functions, as we shall soon see.

DIVIDING A POLYNOMIAL BY A MONOMIAL The quotient $\dfrac{P}{Q}$ is found by multiplying P by the reciprocal of Q. That is,

$$\frac{P}{Q} = P \cdot \frac{1}{Q}, \quad \text{if } Q \neq 0.$$

■ *Example 1*
| DIVIDING BY A
| MONOMIAL

Simplify each quotient.

(a) $\dfrac{2m^5 - 6m^3}{2m^3} = (2m^5 - 6m^3) \cdot \dfrac{1}{2m^3} = \dfrac{2m^5}{2m^3} - \dfrac{6m^3}{2m^3} = m^2 - 3$

(b) $\dfrac{3y^6x^3 - 6y^3x^6 + 8y^5x}{3y^3x^3} = \dfrac{3y^6x^3}{3y^3x^3} - \dfrac{6y^3x^6}{3y^3x^3} + \dfrac{8y^5x}{3y^3x^3} = y^3 - 2x^3 + \dfrac{8y^2}{3x^2}$ ◼

DIVIDING A POLYNOMIAL BY ANOTHER POLYNOMIAL To divide one polynomial by another, use a **division algorithm.** (An **algorithm** is an orderly procedure for performing an operation.) The division algorithm for polynomials is very similar to that used for dividing whole numbers, as the following examples show.

■ *Example 2*
| USING THE DIVISION
| ALGORITHM

Divide $6q^3 - 17q^2 + 22q - 23$ by $2q - 3$.

Begin by writing both polynomials with exponents in descending order.

$$2q - 3)\overline{6q^3 - 17q^2 + 22q - 23}$$

Step 1 The quotient $6q^3/2q$ is $3q^2$. Multiply $3q^2$ and $2q - 3$, getting $6q^3 - 9q^2$.

$$
\begin{array}{r}
3q^2 \leftarrow \\
2q - 3)\overline{6q^3 - 17q^2 + 22q - 23} \\
6q^3 - 9q^2 \leftarrow
\end{array}
$$

$\dfrac{6q^3}{2q} = 3q^2$

$3q^2(2q - 3) = 6q^3 - 9q^2$

Step 2 Subtract $6q^3 - 9q^2$ from $6q^3 - 17q^2$ by changing the signs on $6q^3 - 9q^2$ and adding. Then bring down $22q$.

$$
\begin{array}{r}
3q^2 \\
2q - 3)\overline{6q^3 - 17q^2 + 22q - 23} \\
\underline{-6q^3 + 9q^2} \\
-8q^2 + 22q
\end{array}
$$

Change signs and add.

Step 3 The quotient of $-8q^2$ and $2q$ is $-4q$. Multiply $-4q$ and $2q - 3$.

$$
\begin{array}{r}
3q^2 - 4q \\
2q - 3\overline{)6q^3 - 17q^2 + 22q - 23} \\
\underline{6q^3 - 9q^2} \\
-8q^2 + 22q \\
\underline{-8q^2 + 12q}
\end{array}
$$

$$\frac{-8q^2}{2q} = -4q$$

$$-4q(2q - 3) = -8q^2 + 12q$$

Step 4 Subtract $-8q^2 + 12q$ from $-8q^2 + 22q$. Bring down -23.

$$
\begin{array}{r}
3q^2 - 4q \\
2q - 3\overline{)6q^3 - 17q^2 + 22q - 23} \\
\underline{6q^3 - 9q^2} \\
-8q^2 + 22q \\
\underline{8q^2 - 12q} \\
10q - 23
\end{array}
$$

Change signs and add.

Step 5 The quotient of $10q$ and $2q$ is 5. Multiply 5 and $2q - 3$.

$$
\begin{array}{r}
3q^2 - 4q + 5 \\
2q - 3\overline{)6q^3 - 17q^2 + 22q - 23} \\
\underline{6q^3 - 9q^2} \\
-8q^2 + 22q \\
\underline{-8q^2 + 12q} \\
10q - 23 \\
\underline{10q - 15}
\end{array}
$$

$$\frac{10q}{2q} = 5$$

$$5(2q - 3) = 10q - 15$$

Step 6 Subtract $10q - 15$ from $10q - 23$, getting -8.

$$
\begin{array}{r}
3q^2 - 4q + 5 \\
2q - 3\overline{)6q^3 - 17q^2 + 22q - 23} \\
\underline{6q^3 - 9q^2} \\
-8q^2 + 22q \\
\underline{-8q^2 + 12q} \\
10q - 23 \\
\underline{-10q + 15} \\
-8
\end{array}
$$

Change signs and add.

The quotient is $3q^2 - 4q + 5$, with remainder -8. Write this result as

$$\frac{6q^3 - 17q^2 + 22q - 23}{2q - 3} = 3q^2 - 4q + 5 + \frac{-8}{2q - 3}. \quad ■$$

■ ***Example 3***
USING THE DIVISION
ALGORITHM

Divide $3x^3 - 2x^2 - 150$ by $x - 4$.

The polynomial $3x^3 - 2x^2 - 150$ has a missing term, the term where the variable is x. When dividing a polynomial with a missing term, it is convenient to insert the term with a 0 coefficient, to act as a placeholder. Since $3x^3 - 2x^2 - 150$ is missing a first-degree term, insert $0x$ as shown on the next page.

$$
\begin{array}{r}
3x^2 + 10x\ \ + 40 \\
x - 4 \overline{)\ 3x^3 - \ \ 2x^2 + \ \ 0x - 150} \\
\underline{3x^3 - 12x^2} \\
10x^2 + \ \ 0x \\
\underline{10x^2 - 40x} \\
40x - 150 \\
\underline{40x - 160} \\
10
\end{array}
$$

This result is written as follows:

$$
\frac{3x^3 - 2x^2 - 150}{x - 4} = 3x^2 + 10x + 40 + \frac{10}{x - 4}. \quad \blacksquare
$$

By the process of division shown above, if a polynomial P is divided by a polynomial D of lower degree, the result is a **quotient polynomial** Q and a **remainder polynomial** R such that

$$
P = D \cdot Q + R,
$$

where the degree of R is less than the degree of D. To see this, note that dividing P by D gives

$$
\begin{array}{r}
Q + R/D \\
D \overline{)\, P}
\end{array}.
$$

As a check, the product of D and $Q + R/D$ should give P:

$$
D\left(Q + \frac{R}{D}\right) = D \cdot Q + R = P.
$$

SYNTHETIC DIVISION In many polynomial division problems the divisor is a first-degree binomial of the form $x - k$, where the coefficient of x is 1. A shortcut for this type of division problem is developed by omitting all variables and writing only coefficients. Zero is used to represent the coefficient of any missing powers of the variables. Since the coefficient of x in the divisor is always 1 in problems of this type, the 1 can be left out also. Using these omissions, the division problem on the left below is written in the shortened form shown on the right.

$$
\begin{array}{r}
3x^2 + 10x\ \ + 40 \\
x - 4 \overline{)\ 3x^3 - \ \ 2x^2 + \ \ 0x - 150} \\
\underline{3x^3 - 12x^2} \\
10x^2 \\
\underline{10x^2 - 40x} \\
40x \\
40x - 160 \\
\underline{} \\
10
\end{array}
\qquad
\begin{array}{r}
3\quad 10\quad 40 \\
-4 \overline{)\ 3 - \ \ 2\quad \ \ 0 - 150} \\
\underline{3 - 12} \\
10 \\
\underline{10 - 40} \\
40 \\
40 - 160 \\
\underline{} \\
10
\end{array}
$$

The numbers in color are repetitions of the numbers directly above them and can be omitted also.

$$\begin{array}{r} 3 \quad 10 \quad 40 \\ -4\overline{)3 \ - \ 2 \quad\ \ 0 \ - \ 150} \\ \underline{- \ 12} \\ 10 \\ \underline{- \ 40} \\ 40 \\ \underline{- \ 160} \\ 10 \end{array}$$

The entire problem can now be condensed, with the top row of numbers omitted, since it duplicates the bottom row if we bring down the 3.

$$\begin{array}{r} -4\overline{)3 \ - \ 2 \quad\ \ 0 \ - \ 150} \\ \underline{- \ 12 \ - \ 40 \ - \ 160} \\ 3 \quad\ \ 10 \quad\ \ 40 \quad\ \ 10 \end{array}$$

The bottom row was obtained by subtracting -12, -40, and -160 from the corresponding terms above them. For reasons that will become clear later, change the -4 at the left to 4, which also changes the signs of the numbers in the second row, and then *add*.

$$\begin{array}{r} 4\overline{)3 \ - \ 2 \quad\ \ 0 \ - \ 150} \\ \underline{12 \quad\ 40 \quad\ 160} \\ 3 \quad\ 10 \quad\ 40 \quad\ 10 \end{array}$$

Read the quotient and remainder from the bottom row. The quotient will have degree one less than the polynomial being divided.

$$\frac{3x^3 - 2x^2 - 150}{x - 4} = 3x^2 + 10x + 40 + \frac{10}{x - 4}$$

This shortcut process is called **synthetic division.**

▪ *Example 4*
USING SYNTHETIC
DIVISION

Use synthetic division to divide $5m^3 - 6m^2 - 28m - 2$ by $m + 2$.
    Begin by writing

$$-2\overline{)5 \quad -6 \quad -28 \quad -2.}$$

The 2 is changed to -2 since k is found by writing $m + 2$ as $m - (-2)$. Next, bring down the 5.

$$\begin{array}{r} -2\overline{)5 \quad -6 \quad -28 \quad -2} \\ \\ \hline 5 \end{array}$$

Now, multiply -2 by 5 to get -10, and add it to -6 from the first row. The result is -16.

$$\begin{array}{r} -2\overline{)5 \quad\ -6 \quad -28 \quad -2} \\ \underline{-10 } \\ 5 \quad -16 \end{array}$$

Next, multiply -2 by -16 to get 32. Add this to -28 from the first row.

$$
\begin{array}{r}
-2\overline{)\,5\quad\ -6\quad -28\quad -2} \\
-10\quad\ \ 32 \\
\hline
5\quad -16\quad\ \ \ 4
\end{array}
$$

Finally, $(-2)(4) = -8$. Add this result to -2 to get -10.

$$
\begin{array}{r}
-2\overline{)\,5\quad\ -6\quad -28\quad\ -2} \\
-10\quad\ \ 32\quad\ -8 \\
\hline
\mathbf{5}\quad \mathbf{-16}\quad\ \ \mathbf{4}\quad -10
\end{array}
$$

The coefficients of the quotient polynomial and the remainder are read directly from the bottom row. Since the degree of the quotient will always be one less than the degree of the polynomial to be divided,

$$\frac{5m^3 - 6m^2 - 28m - 2}{m + 2} = \mathbf{5m^2 - 16m + 4} + \frac{-10}{m + 2}. \quad ■$$

The result of the division in Example 4 can be written as

$$5m^3 - 6m^2 - 28m - 2 = (m + 2)(5m^2 - 16m + 4) + (-10)$$

by multiplying both sides by the denominator $m + 2$. The following theorem is a generalization of the division process illustrated above.

DIVISION ALGORITHM For any polynomial $P(x)$ and any complex number k, there exists a unique polynomial $Q(x)$ and number r such that

$$P(x) = (x - k)Q(x) + r.$$

For example, in the synthetic division above,

$$5m^3 - 6m^2 - 28m - 2 = (m + 2)(5m^2 - 16m + 4) + (-10)$$
$$P(m) \qquad\qquad = (m - k) \cdot \qquad Q(m) \qquad + \quad r.$$

By the division algorithm, $P(x) = (x - k)Q(x) + r.$ This equality is true for all complex values of x, so it is true for $x = k$. Thus,

$$P(k) = (k - k)Q(k) + r$$
$$P(k) = r.$$

This proves the following theorem.

REMAINDER THEOREM If the polynomial $P(x)$ is divided by $x - k$, then the remainder is $P(k)$.

As an example of this theorem, the polynomial $P(m) = 5m^3 - 6m^2 - 28m - 2$ was written above as

$$P(m) = (m + 2)(5m^2 - 16m + 4) + (-10).$$

To find $P(-2)$, substitute -2 for m.

$$\begin{aligned} P(-2) &= (-2 + 2)[5(-2)^2 - 16(-2) + 4] + (-10) \\ &= 0[5(-2)^2 - 16(-2) + 4] + (-10) \\ &= -10 \end{aligned}$$

By the remainder theorem, instead of replacing m with -2 to find $P(-2)$, divide $P(m)$ by $m + 2$ using synthetic division. Then $P(-2)$ is the remainder, -10.

■ **Example 5**

USING THE
REMAINDER
THEOREM

Let $P(x) = -x^4 + 3x^2 - 4x - 5$. Find $P(-2)$.
Use the remainder theorem and synthetic division.

$$\begin{array}{r|rrrrr} -2) & -1 & 0 & 3 & -4 & -5 \\ & & 2 & -4 & 2 & 4 \\ \hline & -1 & 2 & -1 & -2 & -1 \end{array}$$

The remainder when $P(x)$ is divided by $x - (-2) = x + 2$ is -1, so $P(-2) = -1$. ∎

CAUTION

As seen in Example 5, it is very important to use a zero for any missing terms (including the constant term) when setting up the synthetic division.

■ **Example 6**

USING THE
REMAINDER
THEOREM

Find $P(-2 + i)$ if $P(x) = x^3 - 4x^2 + 2x - 29i$.
Use synthetic division and addition and multiplication of complex numbers. (Recall that the product of two complex numbers can be found in the same way as the product of two binomials.)

$$\begin{array}{r|rrrr} -2 + i) & 1 & -4 & 2 & -29i \\ & & -2 + i & 11 - 8i & -18 + 29i \\ \hline & 1 & -6 + i & 13 - 8i & -18 \end{array}$$

Since the remainder is -18, $P(-2 + i) = -18$. ∎

The remainder theorem gives a quick way to decide if a number k is a zero of a polynomial function. Use synthetic division to find $P(x)$; if the remainder is zero, then $P(x) = 0$ and k is a zero of P.

■ *Example 7*

DECIDING WHETHER A NUMBER IS A ZERO OF A POLYNOMIAL FUNCTION

Decide whether the given number is a zero of the function defined by the given polynomial.

(a) 2; $P(x) = x^3 - 4x^2 + 9x - 10$

Use synthetic division.

$$
\begin{array}{r|rrrr}
2 & 1 & -4 & 9 & -10 \\
  &   & 2 & -4 & 10 \\
\hline
  & 1 & -2 & 5 & 0
\end{array}
$$

Since the remainder is 0, $P(2) = 0$, and 2 is a zero of the polynomial function with $P(x) = x^3 - 4x^2 + 9x - 10$.

(b) -2; $P(x) = 3x^3 - 2x^2 + 4x$

Remember to use a coefficient of 0 for the missing constant term in the synthetic division.

$$
\begin{array}{r|rrrr}
-2 & 3 & -2 & 4 & 0 \\
   &   & -6 & 16 & -40 \\
\hline
   & 3 & -8 & 20 & -40
\end{array}
$$

The remainder is not zero, so -2 is not a zero of P, where $P(x) = 3x^3 - 2x^2 + 4x$. In fact, $P(-2) = -40$. ■

6.2 EXERCISES ■

Perform each of the following divisions. See Examples 1–3.

1. $\dfrac{4m^3 - 8m^2 + 16m}{2m}$

2. $\dfrac{30k^5 - 12k^3 + 18k^2}{6k^2}$

3. $\dfrac{25x^2y^4 - 15x^3y^3 + 40x^4y^2}{5x^2y^2}$

4. $\dfrac{-8r^3s - 12r^2s^2 + 20rs^3}{4rs}$

5. $\dfrac{6y^2 + y - 2}{2y - 1}$

6. $\dfrac{16r^2 + 2r - 3}{2r + 1}$

7. $\dfrac{8z^2 + 14z - 20}{2z + 5}$

8. $\dfrac{6a^2 - 13a - 18}{2a + 1}$

9. $\dfrac{2x^3 - 11x^2 + 19x - 10}{2x - 5}$

10. $\dfrac{3p^3 - 11p^2 + 5p + 3}{3p + 1}$

11. $\dfrac{15x^3 + 11x^2 + 20}{3x + 4}$

12. $\dfrac{4r^3 + 2r^2 - 14r + 15}{2r + 5}$

13. $\dfrac{x^4 + 2x^3 + 2x^2 - 2x - 3}{x^2 - 1}$

14. $\dfrac{2y^5 + y^3 - 2y^2 - 1}{2y^2 + 1}$

15. $\dfrac{4z^5 - 4z^2 - 5z + 3}{2z^2 + z + 1}$

16. $\dfrac{12z^4 - 25z^3 + 35z^2 - 26z + 10}{4z^2 - 3z + 5}$

17. When a second-degree polynomial in x is divided by a third-degree polynomial in x, is the quotient a polynomial? Explain.

18. If polynomial P is divided by polynomial Q and the remainder is 0, then Q is a _____ of P.

Use synthetic division to perform each of the following divisions. See Example 4.

19. $\dfrac{x^3 + 2x^2 - 17x - 10}{x + 5}$

20. $\dfrac{a^4 + 4a^3 + 2a^2 + 9a + 4}{a + 4}$

21. $\dfrac{m^4 - 3m^3 - 4m^2 + 12m}{m - 2}$

22. $\dfrac{p^4 - 3p^3 - 5p^2 + 2p - 16}{p + 2}$

23. $\dfrac{3x^3 - 11x^2 - 20x + 3}{x - 5}$

24. $\dfrac{4p^3 + 8p^2 - 16p - 9}{p + 3}$

25. $\dfrac{x^5 + 3x^4 + 2x^3 + 2x^2 + 3x + 1}{x + 2}$

26. $\dfrac{m^6 - 3m^4 + 2m^3 - 6m^2 - 5m + 3}{m + 2}$

27. $\dfrac{\frac{1}{3}x^3 - \frac{2}{9}x^2 + \frac{1}{27}x + 1}{x - \frac{1}{3}}$

28. $\dfrac{x^3 + x^2 + \frac{1}{2}x + \frac{1}{8}}{x + \frac{1}{2}}$

29. $\dfrac{y^3 - 1}{y - 1}$

30. $\dfrac{r^5 - 1}{r - 1}$

31. $\dfrac{x^4 - 1}{x - 1}$

32. $\dfrac{x^7 + 1}{x + 1}$

Express each polynomial in the form $P(x) = (x - k)Q(x) + r$ for the given value of k.

33. $P(x) = x^3 + x^2 + x - 8; \quad k = 1$

34. $P(x) = 2x^3 + 3x^2 + 4x - 10; \quad k = -1$

35. $P(x) = -x^3 + 2x^2 + 4; \quad k = -2$

36. $P(x) = -4x^3 + 2x^2 - 3x - 10; \quad k = 2$

37. $P(x) = x^4 - 3x^3 + 2x^2 - x + 5; \quad k = 3$

38. $P(x) = 2x^4 + x^3 - 15x^2 + 3x; \quad k = -3$

For each of the following polynomials, use the remainder theorem to find $P(k)$. See Examples 5 and 6.

39. $k = 3; \quad P(x) = x^2 - 4x + 5$

40. $k = -2; \quad P(x) = x^2 + 5x + 6$

41. $k = -2; \quad P(x) = 5x^3 + 2x^2 - x$

42. $k = 2; \quad P(x) = 2x^3 - 3x^2 - 5x + 4$

43. $k = 2 + i; \quad P(x) = x^2 - 5x + 1$

44. $k = 3 - 2i; \quad P(x) = x^2 - x + 3$

45. $k = 1 - i; \quad P(x) = x^3 + x^2 - x + 1$

46. $k = 2 - 3i; \quad P(x) = x^3 + 2x^2 + x - 5$

Use synthetic division to decide whether the given number is a zero of the given polynomial. See Example 7.

47. $2; \quad P(x) = x^2 + 2x - 8$

48. $-1; \quad P(m) = m^2 + 4m - 5$

49. $2; \quad P(g) = g^3 - 3g^2 + 4g - 4$

50. $-3; \quad P(m) = m^3 + 2m^2 - m + 6$

51. $4; \quad P(r) = 2r^3 - 6r^2 - 9r + 6$

52. $-4; \quad P(y) = 9y^3 + 39y^2 + 12y$

53. $2 + i; \quad P(k) = k^2 + 3k + 4$

54. $1 - 2i; \quad P(z) = z^2 - 3z + 5$

55. $i; \quad P(x) = x^3 + 2ix^2 + 2x + i$

56. $-i; \quad P(p) = p^3 - ip^2 + 3p + 5i$

6.3 ZEROS OF POLYNOMIAL FUNCTIONS

In this section we will build upon some of the ideas presented in Section 6.2 to learn more about finding zeros of polynomial functions.

By the remainder theorem, if $P(k) = 0$, then the remainder when $P(x)$ is divided by $x - k$ is zero. This means that $x - k$ is a factor of $P(x)$. Conversely, if $x - k$ is a factor of $P(x)$, then $P(k)$ must equal 0. This is summarized in the following theorem.

FACTOR THEOREM The polynomial $x - k$ is a factor of the polynomial $P(x)$ if and only if $P(k) = 0$.

■ ***Example 1***

DECIDING WHETHER
$x - k$ IS A FACTOR
OF $P(x)$ (REAL
COEFFICIENTS)

Is $x - 1$ a factor of $P(x) = 2x^4 + 3x^2 - 5x + 7$?

By the factor theorem, $x - 1$ will be a factor of $P(x)$ only if $P(1) = 0$. Use synthetic division and the remainder theorem to decide.

$$
\begin{array}{r|rrrrr}
1) & 2 & 0 & 3 & -5 & 7 \\
  &   & 2 & 2 & 5 & 0 \\
\hline
  & 2 & 2 & 5 & 0 & 7
\end{array}
$$

Since the remainder is 7, $P(1) = 7$, not 0, so $x - 1$ is not a factor of $P(x)$. ■

■ ***Example 2***

DECIDING WHETHER
$x - k$ IS A FACTOR
OF $P(x)$ (COMPLEX
COEFFICIENTS)

Is $x - i$ a factor of $P(x) = 3x^3 + (-4 - 3i)x^2 + (5 + 4i)x - 5i$?

The only way $x - i$ can be a factor of $P(x)$ is for $P(i)$ to be 0. Decide if this is the case by using synthetic division.

$$
\begin{array}{r|rrrr}
i) & 3 & -4 - 3i & 5 + 4i & -5i \\
  &   & 3i & -4i & 5i \\
\hline
  & 3 & -4 & 5 & 0
\end{array}
$$

Since the remainder is 0, $P(i) = 0$, and $x - i$ is a factor of $P(x)$. The other factor is the quotient $3x^2 - 4x + 5$, so $P(x)$ can be factored as

$$P(x) = (x - i)(3x^2 - 4x + 5). \quad ■$$

The next theorem says that every polynomial of degree 1 or more has a zero, so that every such polynomial can be factored. This theorem was first proved by the mathematician Carl F. Gauss in his doctoral thesis in 1799 when he was 22 years old. Although many proofs of this result have been given, all of them involve mathematics beyond the algebra in this book, so no proof is included here.

FUNDAMENTAL THEOREM OF ALGEBRA	Every function defined by a polynomial of degree 1 or more has at least one complex zero.

From the fundamental theorem, if $P(x)$ is of degree 1 or more then there is some number k such that $P(k) = 0$. By the factor theorem, then

$$P(x) = (x - k) \cdot Q(x)$$

for some polynomial $Q(x)$. The fundamental theorem and the factor theorem can be used to factor $Q(x)$ in the same way. Assuming that $P(x)$ has degree n, repeating this process n times gives

$$P(x) = a(x - k_1)(x - k_2) \ldots (x - k_n),$$

where a is the leading coefficient of $P(x)$. Each of these factors leads to a zero of $P(x)$, so $P(x)$ has the n zeros $k_1, k_2, k_3, \ldots, k_n$. This result can be used to prove the next theorem. The proof is left for the exercises.

ZEROS OF A POLYNOMIAL FUNCTION	A function defined by a polynomial of degree n has at most n distinct zeros.

The theorem says that there exist *at most n* distinct zeros. For example, the polynomial function $P(x) = x^3 + 3x^2 + 3x + 1 = (x + 1)^3$ is of degree 3 but has only one zero, -1. Actually, the zero -1 occurs three times, since there are three factors of $x + 1$; this zero is called a *zero of* **multiplicity 3.**

■ *Example 3*

DETERMINING THE MULTIPLICITY OF ZEROS

Determine the zeros of the polynomial function P defined by

$$P(x) = (x - 5)^4(x + 1)^2 x^5$$

and give their multiplicities.

Since $(x - 5)^4$ is the largest power of $(x - 5)$ that appears in the factored form of $P(x)$, 5 is a zero of multiplicity 4. Because $(x + 1)^2$ can be written $[x - (-1)]^2$, -1 is a zero of multiplicity 2. Finally, because $x^5 = (x - 0)^5$, 0 is a zero of multiplicity 5. ■

The fundamental theorem of algebra and the results concerning zeros and their multiplicities can be used to find a polynomial, given its zeros and a particular function value.

■ *Example 4*

DETERMINING POLYNOMIALS THAT SATISFY GIVEN CONDITIONS (REAL ZEROS)

Find a polynomial function P of degree 3 that satisfies the following conditions.

(a) Zeros of -1, 2, and 4; $P(1) = 3$

These three zeros give $x - (-1) = x + 1$, $x - 2$, and $x - 4$ as factors of $P(x)$. Since $P(x)$ is to be of degree 3, these are the only possible factors that can be found by using the theorem above. Therefore, $P(x)$ has the form

$$P(x) = a(x + 1)(x - 2)(x - 4)$$

for some nonzero real number a. To find a, use the fact that $P(1) = 3$.

$$P(1) = a(1 + 1)(1 - 2)(1 - 4) = 3$$
$$a(2)(-1)(-3) = 3$$
$$6a = 3$$
$$a = \frac{1}{2}$$

Thus, $$P(x) = \frac{1}{2}(x + 1)(x - 2)(x - 4),$$

or $$P(x) = \frac{1}{2}x^3 - \frac{5}{2}x^2 + x + 4.$$

(b) -2 is a zero of multiplicity 3; $P(-1) = 4$

The polynomial $P(x)$ has the form

$$P(x) = a(x + 2)^3.$$

Since $P(-1) = 4$,

$$P(-1) = a(-1 + 2)^3 = 4$$
$$a(1)^3 = 4$$
$$a = 4,$$

and $P(x) = 4(x + 2)^3 = 4x^3 + 24x^2 + 48x + 32.$ ■

| **CAUTION** | In Example 4(a), it would be *wrong* to clear the polynomial of fractions by multiplying through by 2. The result would then be $2 \cdot P(x)$ and not $P(x)$. |

The remainder theorem can be used to show that both $2 + i$ and $2 - i$ are zeros of $P(x) = x^3 - x^2 - 7x + 15$. It is not a coincidence that both $2 + i$ and its conjugate $2 - i$ are zeros of this polynomial. If $a + bi$ is a zero of a polynomial function with *real* coefficients, then so is $a - bi$. This is given as the next theorem. The proof is left for the exercises.

| **CONJUGATE ZEROS THEOREM** | If $P(x)$ is a polynomial having only real coefficients and if $a + bi$ is a zero of P then the conjugate $a - bi$ is also a zero of P. |

| **CAUTION** | The requirement that the polynomial have only real coefficients is very important. For example, $P(x) = x - (1 + i)$ has $1 + i$ as a zero, but the conjugate $1 - i$ is not a zero. |

■ ***Example 5***

DETERMINING A POLYNOMIAL SATISFYING GIVEN CONDITIONS (COMPLEX ZEROS)

Find a polynomial of lowest degree having real coefficients and defining a function with zeros 3 and $2 + i$.

The complex number $2 - i$ also must be a zero, so there are at least three zeros, 3, $2 + i$, and $2 - i$. For the polynomial to be of lowest degree these must be the only zeros. Then by the factor theorem there must be three factors, $x - 3$, $x - (2 + i)$, and $x - (2 - i)$. A polynomial of lowest degree is

$$P(x) = (x - 3)[x - (2 + i)][x - (2 - i)]$$
$$= (x - 3)(x - 2 - i)(x - 2 + i)$$
$$= x^3 - 7x^2 + 17x - 15.$$

Other polynomials, such as $2(x^3 - 7x^2 + 17x - 15)$ or $\sqrt{5}(x^3 - 7x^2 + 17x - 15)$, for example, also satisfy the given conditions on zeros. There is not enough information on zeros given in the problem to give a specific value for the leading coefficient. ■

The conjugate zeros theorem helps predict the number of real zeros of functions defined by polynomials with real coefficients. A function defined by a polynomial of odd degree n, where $n \geq 1$, with real coefficients must have at least one real zero (since zeros of the form $a + bi$, where $b \neq 0$, occur in conjugate pairs). On the other hand, a function defined by a polynomial of even degree n with real coefficients need have no real zeros but may have up to n real zeros.

■ *Example 6*

FINDING ALL ZEROS OF A POLYNOMIAL FUNCTION GIVEN ONE ZERO (COMPLEX ZEROS)

Find all zeros of the function P defined by $P(x) = x^4 - 7x^3 + 18x^2 - 22x + 12$, given that $1 - i$ is a zero.

Since $1 - i$ is a zero and the coefficients are real numbers, by the conjugate zeros theorem $1 + i$ is also a zero. The remaining zeros are found by first dividing the original polynomial by $x - (1 - i)$.

$$
\begin{array}{r|rrrrr}
1 - i) & 1 & -7 & 18 & -22 & 12 \\
 & & 1 - i & -7 + 5i & 16 - 6i & -12 \\
\hline
 & 1 & -6 - i & 11 + 5i & -6 - 6i & 0
\end{array}
$$

Rather than go back to the original polynomial, divide the quotient from the first division by $x - (1 + i)$ as follows.

$$
\begin{array}{r|rrrr}
1 + i) & 1 & -6 - i & 11 + 5i & -6 - 6i \\
 & & 1 + i & -5 - 5i & 6 + 6i \\
\hline
 & 1 & -5 & 6 & 0
\end{array}
$$

Find the zeros of the function defined by the quadratic polynomial $x^2 - 5x + 6$ by solving the equation $x^2 - 5x + 6 = 0$. Factoring the polynomial shows that the zeros are 2 and 3, so the four zeros of P are $1 - i$, $1 + i$, 2, and 3. ■

The theorems given in this section can be used to factor a polynomial $P(x)$ into linear factors (factors of the form $ax - b$) when one or more of the zeros of P are known.

■ *Example 7*

FACTORING A POLYNOMIAL GIVEN INFORMATION ABOUT ITS ZEROS

Factor $P(x)$ into linear factors, given that k is a zero of P.

(a) $P(x) = 6x^3 + 19x^2 + 2x - 3$; $k = -3$

Since $k = -3$ is a zero of P, $x - (-3) = x + 3$ is a factor. Use synthetic division to divide $P(x)$ by $x + 3$.

$$
\begin{array}{r|rrrr}
-3) & 6 & 19 & 2 & -3 \\
 & & -18 & -3 & 3 \\
\hline
 & 6 & 1 & -1 & 0
\end{array}
$$

The quotient is $6x^2 + x - 1$, so

$$P(x) = (x + 3)(6x^2 + x - 1).$$

Factor $6x^2 + x - 1$ as $(2x + 1)(3x - 1)$ to get

$$P(x) = (x + 3)(2x + 1)(3x - 1),$$

where all factors are linear.

(b) $P(x) = 3x^3 + (-1 + 3i)x^2 + (-12 + 5i)x + 4 - 2i;\ k = 2 - i$

One factor is $x - (2 - i)$ or $x - 2 + i$. Divide $P(x)$ by $x - (2 - i)$.

$$
\begin{array}{r|rrrr}
2 - i) & 3 & -1 + 3i & -12 + 5i & 4 - 2i \\
& & 6 - 3i & 10 - 5i & -4 + 2i \\
\hline
& 3 & 5 & -2 & 0
\end{array}
$$

By the division algorithm,

$$P(x) = (x - 2 + i)(3x^2 + 5x - 2).$$

Factor $3x^2 + 5x - 2$ as $(3x - 1)(x + 2)$; then a linear factored form of $P(x)$ is

$$P(x) = (x - 2 + i)(3x - 1)(x + 2). \quad ■$$

NOTE	In Example 7(b), the conjugate $2 + i$ is not also a zero, because $P(x)$ has some imaginary coefficients.

6.3 EXERCISES ■

Use the factor theorem to decide whether the second polynomial is a factor of the first. See Examples 1 and 2.

1. $4x^2 + 2x + 42;\quad x - 3$ **2.** $-3x^2 - 4x + 2;\quad x + 2$ **3.** $x^3 + 2x^2 - 3;\quad x - 1$ **4.** $2x^3 + x + 2;\quad x + 1$

5. $3x^3 - 12x^2 - 11x - 20;\quad x - 5$ **6.** $4x^3 + 6x^2 - 5x - 2;\quad x + 2$

7. $2x^4 + 5x^3 - 2x^2 + 5x + 3;\quad x + 3$ **8.** $5x^4 + 16x^3 - 15x^2 + 8x + 16;\quad x + 4$

For each of the following polynomial functions, give all zeros and their multiplicities. See Example 3.

9. $P(x) = x^4(x + 3)^5(x - 8)^2$ **10.** $P(x) = x^9(x - 4)^6(x + 7)^4$ **11.** $P(x) = (4x - 7)^3(x - 5)$

12. $P(x) = (5x + 1)^3(x + 1)$ **13.** $P(x) = (x - i)^4(x + i)^4$ **14.** $P(x) = (x + 3i)^9(x - 3i)^9$

For each of the following, find a function P defined by a polynomial of degree 3 with real coefficients that satisfies the given conditions. See Example 4.

15. Zeros of -3, -1, and 4; $P(2) = 5$ **16.** Zeros of 1, -1, and 0; $P(2) = -3$

17. Zeros of -2, 1, and 0; $P(-1) = -1$ **18.** Zeros of 2, 5, and -3; $P(1) = -4$

19. Zeros of 3, i, and $-i$; $P(2) = 50$ **20.** Zeros of -2, i, and $-i$; $P(-3) = 30$

For each of the following, find a function P defined by a polynomial of lowest degree with real coefficients having the given zeros. See Example 5.

21. $5 + i$ and $5 - i$ **22.** $3 - 2i$ and $3 + 2i$ **23.** 2, $1 - i$, and $1 + i$

24. -3, $2 - i$, and $2 + i$ **25.** $1 + \sqrt{2}$, $1 - \sqrt{2}$, and 1   **26.** $1 - \sqrt{3}$, $1 + \sqrt{3}$, and -2

27. $2 + i$, $2 - i$, 3, and -1 **28.** $3 + 2i$, $3 - 2i$, -1, and 2   **29.** 2 and $3 + i$

30. -1 and $4 - 2i$ **31.** $2 - i$ and $3 + 2i$ **32.** $5 + i$ and $4 - i$

33. 4, $1 - 2i$, and $3 + 4i$ **34.** -1, $1 + \sqrt{2}$, $1 - \sqrt{2}$, and $1 + 4i$ **35.** $1 + 2i$ and 2 (multiplicity 2)

36. $2 + i$ and -3 (multiplicity 2)

For each of the following polynomials, one zero is given. Find all others. See Example 6.

37. $P(x) = x^3 - x^2 - 4x - 6;\quad 3$ **38.** $P(x) = x^3 - 5x^2 + 17x - 13;\quad 1$

39. $P(x) = 2x^3 - 2x^2 - x - 6;\quad 2$ **40.** $P(x) = 2x^3 - 5x^2 + 6x - 2;\quad 1 + i$

41. $P(x) = x^4 + 5x^2 + 4;\quad -i$

42. $P(x) = x^4 + 10x^3 + 27x^2 + 10x + 26;\quad i$

43. $P(x) = x^4 - 3x^3 + 6x^2 + 2x - 60;\quad 1 + 3i$

44. $P(x) = x^4 - 6x^3 - x^2 + 86x + 170;\quad 5 + 3i$

Factor P(x) into linear factors given that k is a zero of P. See Example 7.

45. $P(x) = 2x^3 - 3x^2 - 17x + 30;\quad k = 2$

46. $P(x) = 2x^3 - 3x^2 - 5x + 6;\quad k = 1$

47. $P(x) = 6x^3 + 25x^2 + 3x - 4;\quad k = -4$

48. $P(x) = 8x^3 + 50x^2 + 47x - 15;\quad k = -5$

49. $P(x) = x^3 + (7 - 3i)x^2 + (12 - 21i)x - 36i;\quad k = 3i$

50. $P(x) = 2x^3 + (3 + 2i)x^2 + (1 + 3i)x + i;\quad k = -i$

51. $P(x) = 2x^3 + (3 - 2i)x^2 + (-8 - 5i)x + 3 + 3i;\quad k = 1 + i$

52. $P(x) = 6x^3 + (19 - 6i)x^2 + (16 - 7i)x + 4 - 2i;\quad k = -2 + i$

53. Show that -2 is a zero of multiplicity 2 of P, where $P(x) = x^4 + 2x^3 - 7x^2 - 20x - 12$, and find all other complex zeros. Then write $P(x)$ in factored form.

54. Show that -1 is a zero of multiplicity 3 of P, where $P(x) = x^5 + 9x^4 + 33x^3 + 55x^2 + 42x + 12$, and find all other complex zeros. Then write $P(x)$ in factored form.

55. What are the possible numbers of real zeros (counting multiplicities) for a polynomial function with real coefficients of degree five?

56. Explain why a function defined by a polynomial of degree four with real coefficients has either 0, 2, or 4 real zeros.

57. Explain why it is not possible for a function defined by a polynomial of degree 3 with real coefficients to have zeros of 1, 2, and $1 + i$.

58. Show that the zeros of the function P defined by the polynomial $P(x) = x^3 + ix^2 - (7 - i)x + (6 - 6i)$

are $1 - i$, 2, and -3. Does the conjugate zeros theorem apply? Why or why not?

59. The displacement at time t of a particle moving along a straight line is given by

$$s(t) = t^3 - 2t^2 - 5t + 6,$$

where t is in seconds and s is measured in centimeters. The displacement is 0 after 1 second has elapsed. At what other times (positive) is the displacement 0?

60. For headphone radios, the cost function (in thousands of dollars) is given by

$$C(x) = 2x^3 - 9x^2 + 17x - 4,$$

and the revenue function (in thousands of dollars) is given by $R(x) = 5x$, where x is the number of items (in hundred thousands) produced. Cost equals revenue if 200,000 items are produced ($x = 2$). Find all other break-even points.

If c and d are complex numbers, prove each of the following statements. These statements are used in Exercise 66. (Hint: Let $c = a + bi$ and $d = m + ni$ and form the conjugates, the sums, and the products. The notation $\bar{c}$ represents the conjugate of c.)

61. $\overline{c + d} = \bar{c} + \bar{d}$

62. $\overline{cd} = \bar{c} \cdot \bar{d}$

63. $\bar{a} = a$ for any real number a

64. $\overline{c^n} = (\bar{c})^n$, n is a positive integer

Use the theorems presented in this section to prove each of the following statements.

65. A polynomial of degree n has at most n distinct zeros. (*Hint:* Use an indirect proof, where you assume the opposite of the statement is true and show that it leads to a contradiction.)

66. Complete the proof of the conjugate zeros theorem, outlined below. Assume that

$$P(x) = a_n x^n + a_{n-1}x^{n-1} + \cdots + a_1 x + a_0,$$

where all coefficients are real numbers.

(a) If the complex number z is a zero of P, find $P(z)$.

(b) Take the conjugate of both sides of the result from part (a).

(c) Use generalizations of the properties given in Exercises 61–64 on the result of part (b) to show that $a_n(\bar{z})^n + a_{n-1}(\bar{z})^{n-1} + \cdots + a_1(\bar{z}) + a_0 = 0$.

(d) Why does the result in part (c) mean that $\bar{z}$ is a zero of P?

6.4 ———————— RATIONAL ZEROS OF POLYNOMIAL FUNCTIONS

By the fundamental theorem of algebra, every function defined by a polynomial of degree 1 or more has a zero. However, the fundamental theorem merely says that such a zero exists. It gives no help at all in identifying zeros. Other theorems can be used to find any rational zeros of polynomial functions with rational coefficients or to find decimal approximations of any irrational zeros.

The next theorem gives a useful method for finding a set of possible zeros of a polynomial function with integer coefficients.

RATIONAL ZEROS THEOREM

Let $P(x) = a_n x^n + a_{n-1} x^{n-1} + \cdots + a_1 x + a_0$, where $a_n \neq 0$, define a polynomial function with integer coefficients. If p/q is a rational number written in lowest terms and if p/q is a zero of P, then p is a factor of the constant term a_0 and q is a factor of the leading coefficient a_n.

PROOF

$P(p/q) = 0$ since p/q is a zero of P, so

$$a_n \left(\frac{p}{q}\right)^n + a_{n-1}\left(\frac{p}{q}\right)^{n-1} + \cdots + a_1\left(\frac{p}{q}\right) + a_0 = 0.$$

This also can be written as

$$a_n \left(\frac{p^n}{q^n}\right) + a_{n-1}\left(\frac{p^{n-1}}{q^{n-1}}\right) + \cdots + a_1\left(\frac{p}{q}\right) + a_0 = 0.$$

Multiply both sides of this last result by q^n and add $-a_0 q^n$ to both sides.

$$a_n p^n + a_{n-1} p^{n-1} q + \cdots + a_1 p q^{n-1} = -a_0 q^n$$

Factoring out p gives

$$p(a_n p^{n-1} + a_{n-1} p^{n-2} q + \cdots + a_1 q^{n-1}) = -a_0 q^n.$$

This result shows that $-a_0 q^n$ equals the product of the two factors, p and $(a_n p^{n-1} + \cdots + a_1 q^{n-1})$. For this reason, p must be a factor of $-a_0 q^n$. Since it was assumed that p/q is written in lowest terms, p and q have no common factor other than 1, so p is not a factor of q^n. Thus p must be a factor of a_0. In a similar way it can be shown that q is a factor of a_n. ■

In the following example we use the rational zeros theorem to find all rational zeros of a function defined by a polynomial with integer coefficients.

■ *Example 1*

FINDING THE RATIONAL ZEROS OF A FUNCTION DEFINED BY A POLYNOMIAL WITH INTEGER COEFFICIENTS

Find all rational zeros of P, if $P(x) = 2x^4 - 11x^3 + 14x^2 - 11x + 12$.

If p/q is to be a rational zero of P, by the rational zeros theorem p must be a factor of $a_0 = 12$ and q must be a factor of $a_4 = 2$. The possible values of p are $\pm 1, \pm 2, \pm 3, \pm 4, \pm 6$, or ± 12, while q must be ± 1 or ± 2. The possible rational zeros are found by forming all possible quotients of the form p/q; any rational zero of P will come from the list

$$\pm 1, \quad \pm \frac{1}{2}, \quad \pm 2, \quad \pm 3, \quad \pm \frac{3}{2}, \quad \pm 4, \quad \pm 6, \quad \text{or} \quad \pm 12$$

Though none of these numbers may be zeros, if P has any rational zeros, they will be in the list above. These proposed zeros can be checked by synthetic division. Doing so shows that 4 is a zero.

$$
\begin{array}{r|rrrrr}
4 & 2 & -11 & 14 & -11 & 12 \\
  &   & 8 & -12 & 8 & -12 \\
\hline
  & 2 & -3 & 2 & -3 & 0 \leftarrow P(4) = 0
\end{array}
$$

As a fringe benefit of this calculation, the simpler polynomial $Q(x) = 2x^3 - 3x^2 + 2x - 3$ can be used to find the remaining zeros. Any rational zero of Q will have a numerator of ± 3 or ± 1 and a denominator of ± 1 or ± 2 and so will come from the list

$$
\pm 3, \quad \pm\frac{3}{2}, \quad \pm 1, \quad \pm\frac{1}{2}.
$$

Again use synthetic division and trial and error to find that 3/2 is a zero.

$$
\begin{array}{r|rrrr}
\frac{3}{2} & 2 & -3 & 2 & -3 \\
  &   & 3 & 0 & 3 \\
\hline
  & 2 & 0 & 2 & 0 \leftarrow Q\left(\frac{3}{2}\right) = 0
\end{array}
$$

The quotient is $2x^2 + 2$, which, by the quadratic formula, has i and $-i$ as zeros. They are imaginary zeros, however. The rational zeros of the polynomial function defined by $P(x) = 2x^4 - 11x^3 + 14x^2 - 11x + 12$ are 4 and 3/2. ■

■ **Example 2**

FINDING RATIONAL
ZEROS AND
FACTORING A
POLYNOMIAL

Find all rational zeros of P, if $P(x) = 6x^4 + 7x^3 - 12x^2 - 3x + 2$, and factor the polynomial.

For a rational number p/q to be a zero of P, p must be a factor of $a_0 = 2$ and q must be a factor of $a_4 = 6$. Thus, p can be ± 1 or ± 2 and q can be ± 1, ± 2, ± 3, or ± 6. The rational zeros, p/q, must come from the following list.

$$
\pm 1, \quad \pm 2, \quad \pm\frac{1}{2}, \quad \pm\frac{1}{3}, \quad \pm\frac{1}{6}, \quad \pm\frac{2}{3}
$$

Check 1 first because it is easy.

$$
\begin{array}{r|rrrrr}
1 & 6 & 7 & -12 & -3 & 2 \\
  &   & 6 & 13 & 1 & -2 \\
\hline
  & 6 & 13 & 1 & -2 & 0
\end{array}
$$

The 0 remainder shows that 1 is a zero. Now use the quotient polynomial $6x^3 + 13x^2 + x - 2$ and synthetic division to find that -2 is also a zero.

$$
\begin{array}{r|rrrr}
-2 & 6 & 13 & 1 & -2 \\
  &   & -12 & -2 & 2 \\
\hline
  & 6 & 1 & -1 & 0
\end{array}
$$

The new quotient polynomial is $6x^2 + x - 1$. Use the quadratic formula or factor to solve the equation $6x^2 + x - 1 = 0$. The remaining two zeros are 1/3 and $-1/2$.

Factor the polynomial $P(x)$ in the following way. Since the four zeros of P are 1, -2, 1/3, and $-1/2$, the corresponding factors are $x - 1$, $x + 2$, $x - 1/3$, and $x + 1/2$, and

$$P(x) = a_4(x - 1)(x + 2)\left(x - \frac{1}{3}\right)\left(x + \frac{1}{2}\right).$$

Since $a_4 = 6 = 2 \cdot 3$, fractions can be cleared in the last two factors by writing the product as

$$P(x) = 6(x - 1)(x + 2)\left(x - \frac{1}{3}\right)\left(x + \frac{1}{2}\right)$$

$$= (x - 1)(x + 2)\left[(3)\left(x - \frac{1}{3}\right)\right]\left[(2)\left(x + \frac{1}{2}\right)\right]$$

$$= (x - 1)(x + 2)(3x - 1)(2x + 1). \quad ■$$

| NOTE | We found by using synthetic division that 1 is a zero of P, where $P(x) = 6x^4 + 7x^3 - 12x^2 - 3x + 2$ in Example 2. An easy way to determine whether 1 is a zero of a polynomial function is as follows: *If the sum of the coefficients of a polynomial is 0, then 1 is a zero of the function defined by the polynomial.* Since any power of 1 is 1, when each term of the polynomial is evaluated, the answer is the coefficient of the term; the result above follows. Verify this in Example 2. (This does not help in determining other zeros, however, as synthetic division does.) |

Rational zeros of a polynomial function with rational coefficients can be found by first multiplying the polynomial by a number that will clear it of all fractions, then using the rational zeros theorem.

■ **Example 3**

FINDING RATIONAL ZEROS OF A FUNCTION DEFINED BY A POLYNOMIAL WITH FRACTIONS AS COEFFICIENTS

Find all rational zeros of P, if $P(x) = x^4 - \frac{1}{6}x^3 + \frac{2}{3}x^2 - \frac{1}{6}x - \frac{1}{3}$.

Find the values of x that make $P(x) = 0$;

$$x^4 - \frac{1}{6}x^3 + \frac{2}{3}x^2 - \frac{1}{6}x - \frac{1}{3} = 0.$$

Multiply both sides by 6 to eliminate all fractions.

$$6x^4 - x^3 + 4x^2 - x - 2 = 0$$

The solutions of this equation are the zeros of P. The possible rational zeros are of the form p/q where p is ± 1 or ± 2, and q is ± 1, ± 2, ± 3, or ± 6. Then p/q may be

$$\pm 1, \quad \pm 2, \quad \pm \frac{1}{2}, \quad \pm \frac{1}{3}, \quad \pm \frac{1}{6}, \quad \text{or} \quad \pm \frac{2}{3}.$$

Use synthetic division to find that $-1/2$ and $2/3$ are zeros.

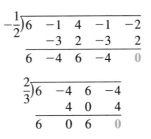

The final quotient is $Q(x) = 6x^2 + 6 = 6(x^2 + 1)$. By setting $x^2 + 1$ equal to 0, we find that $x = \pm i$. Since these zeros are imaginary numbers, there are just two rational zeros: $-1/2$ and $2/3$. ∎

CAUTION

Remember, the rational zeros theorem can be used only if the coefficients of the polynomial are integers. Functions defined by polynomials with rational coefficients can be rewritten with integer coefficients in order to use the theorem, but the theorem cannot be used with functions having irrational or imaginary coefficients in the defining polynomials.

6.4 EXERCISES ∎

Give all possible rational zeros for each function P defined as follows.

1. $P(x) = 6x^3 + 17x^2 - 31x - 1$

2. $P(x) = 15x^3 + 61x^2 + 2x - 1$

3. $P(x) = 12x^3 + 20x^2 - x - 2$

4. $P(x) = 12x^3 + 40x^2 + 41x + 3$

5. $P(x) = 2x^3 + 7x^2 + 12x - 8$

6. $P(x) = 2x^3 + 20x^2 + 68x - 40$

7. $P(x) = x^4 + 4x^3 + 3x^2 - 10x + 50$

8. $P(x) = x^4 - 2x^3 + x^2 + 18$

9. Discuss a major drawback of the rational zeros theorem.

10. Can the rational zeros theorem be used for a function defined by a polynomial with irrational coefficients?

Find all rational zeros of each function P defined as follows. See Example 1.

11. $P(x) = x^3 - 2x^2 - 13x - 10$

12. $P(x) = x^3 + 5x^2 + 2x - 8$

13. $P(x) = x^3 + 6x^2 - x - 30$

14. $P(x) = x^3 - x^2 - 10x - 8$

15. $P(x) = x^3 + 9x^2 - 14x - 24$

16. $P(x) = x^3 + 3x^2 - 4x - 12$

17. $P(x) = x^4 + 9x^3 + 21x^2 - x - 30$

18. $P(x) = x^4 + 4x^3 - 7x^2 - 34x - 24$

Find the rational zeros of each function P defined as follows. Then write each polynomial in factored form. See Example 2.

19. $P(x) = 6x^3 + 17x^2 - 31x - 12$

20. $P(x) = 15x^3 + 61x^2 + 2x - 8$

21. $P(x) = 12x^3 + 20x^2 - x - 6$

22. $P(x) = 12x^3 + 40x^2 + 41x + 12$

23. $P(x) = 2x^3 + 7x^2 + 12x - 8$

24. $P(x) = 2x^3 + 20x^2 + 68x - 40$

25. $P(x) = 2x^4 + 3x^3 - 4x^2 - 3x + 2$

26. $P(x) = x^4 - 2x^3 + x^2 + 18$

27. $P(x) = 3x^4 + 5x^3 - 10x^2 - 20x - 8$

28. $P(x) = 6x^4 + x^3 - 7x^2 - x + 1$

Find all rational zeros of each function P defined as follows. See Example 3.

29. $P(x) = x^3 - \dfrac{4}{3}x^2 - \dfrac{13}{3}x - 2$

30. $P(x) = x^3 + x^2 - \dfrac{16}{9}x + \dfrac{4}{9}$

31. $P(x) = x^4 + \dfrac{1}{4}x^3 + \dfrac{11}{4}x^2 + x - 5$

32. $P(x) = \dfrac{10}{7}x^4 - x^3 - 7x^2 + 5x - \dfrac{5}{7}$

33. $P(x) = \dfrac{1}{3}x^5 + x^4 - \dfrac{5}{3}x^3 - \dfrac{11}{3}x^2 + 4$

34. $P(x) = x^5 + x^4 - \dfrac{37}{4}x^2 + \dfrac{9}{4}x + \dfrac{9}{4}$

For each of the polynomial functions defined as follows, find all rational zeros and factor the polynomial. Then graph the function using the method described in Section 6.1.

35. $P(x) = 2x^3 - 5x^2 - x + 6$

36. $P(x) = 3x^3 + x^2 - 10x - 8$

37. $P(x) = x^3 + x^2 - 8x - 12$

38. $P(x) = x^3 + 6x^2 - 32$

39. $P(x) = -x^3 - x^2 + 8x + 12$

40. $P(x) = -x^3 + 10x^2 - 33x + 36$

41. $P(x) = x^4 - 18x^2 + 81$

42. $P(x) = x^4 - 8x^2 + 16$

43. $P(x) = 2x^4 + x^3 - 6x^2 - 7x - 2$

44. $P(x) = 3x^4 - 7x^3 - 6x^2 + 12x + 8$

45. Show that $P(x) = x^2 - 2$ has no rational zeros, so $\sqrt{2}$ must be irrational.

46. Show that $P(x) = x^2 - 5$ has no rational zeros, so $\sqrt{5}$ must be irrational.

47. Show that $P(x) = x^4 + 5x^2 + 4$ has no rational zeros.

48. Show that $P(x) = x^5 - 3x^3 + 5$ has no rational zeros.

49. Show that any integer zeros of a polynomial function must be factors of the constant term a_0 of the defining polynomial.

50. If k is a zero of the polynomial function P, then $P(k) =$ _____, a factor of $P(x)$ is _____, and an x-intercept of the graph of P is _____.

6.5 ——— REAL ZEROS OF POLYNOMIAL FUNCTIONS

Every function defined by a polynomial of degree 1 or more has a zero. However, the fundamental theorem of algebra does not say whether a polynomial function has *real* zeros. Even if it does have real zeros, often their exact values cannot be determined. This section discusses methods of determining how many real zeros a polynomial function may have, and explains methods for approximating any real zeros.

The recent strides made in computer technology and graphing calculators have made some of the material in this section of limited value. However, the concepts presented here allow students to understand the ideas of finding real zeros of polynomial functions. Once these ideas are mastered, students may then wish to investigate the use of computers and graphing calculators to find real zeros of polynomial functions. Learning the methods of this section first will help students realize the power of today's technology.

Descartes' rule of signs, stated below, gives a test for finding the number of positive or negative real zeros of a given polynomial function.

DESCARTES' RULE OF SIGNS	Let $P(x)$ define a polynomial function with real coefficients and terms in descending powers of x.

(a) The number of positive real zeros of P either equals the number of variations in sign occurring in the coefficients of $P(x)$, or is less than the number of variations by a positive even integer.

(b) The number of negative real zeros of P either equals the number of variations in sign occurring in the coefficients of $P(-x)$, or is less than the number of variations by a positive even integer.

In the theorem, variation in sign is a change from positive to negative or negative to positive in successive terms of the polynomial. Missing terms (those with 0 coefficients) are counted as no change in sign and can be ignored.

For the purposes of this theorem, zeros of multiplicity k count as k zeros. For example.

$$P(x) = (x - 1)^4 = + x^4 - 4x^3 + 6x^2 - 4x + 1$$

$$\underset{1 \quad\quad 2 \quad\quad 3 \quad\quad 4}{}$$

has 4 changes of sign. By Descartes' rule of signs, P has either 4, 2, or 0 positive real zeros. In this case there are 4, and each of the 4 positive real zeros is 1.

■ *Example 1*

USING DESCARTES' RULE OF SIGNS

Find the number of positive and negative real zeros possible for the polynomial function P defined by

$$P(x) = x^4 - 6x^3 + 8x^2 + 2x - 1.$$

$P(x)$ has 3 variations in sign:

$$+ x^4 - 6x^3 + 8x^2 + 2x - 1.$$

$$\underset{1 \quad\quad 2 \quad\quad\quad\quad 3}{}$$

Thus, by Descartes' rule of signs, P has either 3 or $3 - 2 = 1$ positive real zeros. Since

$$P(-x) = (-x)^4 - 6(-x)^3 + 8(-x)^2 + 2(-x) - 1$$
$$= x^4 + 6x^3 + 8x^2 - 2x - 1$$

has only one variation in sign, P has only one negative real zero. ■

■ *Example 2*

USING DESCARTES' RULE OF SIGNS

Find the number of positive and negative real zeros possible for the function Q defined by

$$Q(x) = x^5 + 5x^4 + 3x^2 + 2x + 1.$$

The polynomial $Q(x)$ has no variations in sign and so Q has no positive real zeros. Here

$$Q(-x) = -x^5 + 5x^4 + 3x^2 - 2x + 1$$

has three variations in sign, so Q has either 3 or 1 negative real zeros. The other zeros are imaginary numbers. ■

| NOTE | Every polynomial function of odd degree having real coefficients must have at least one real zero. This is convenient to remember when using Descartes' rule of signs. |

Much of our work in locating real zeros uses the following result, which is related to the fact that graphs of polynomial functions are unbroken curves, with no gaps or sudden jumps. The proof requires advanced methods, so it is not given here.

| INTERMEDIATE VALUE THEOREM FOR POLYNOMIAL FUNCTIONS | If $P(x)$ defines a polynomial function with *only real coefficients,* and if for real numbers a and b, $P(a)$ and $P(b)$ are opposite in sign, then there exists at least one real zero between a and b. |

This theorem helps to identify intervals where zeros of polynomials are located. For example, in Figure 6.13 $P(a)$ and $P(b)$ are opposite in sign, so 0 is between $P(a)$ and $P(b)$. Then, by the intermediate value theorem, there must be a number c between a and b such that $P(c) = 0$.

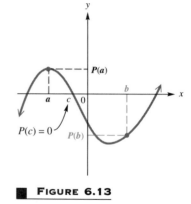

■ **FIGURE 6.13**

■ **Example 3**

USING THE INTERMEDIATE VALUE THEOREM

Does the function P defined by $P(x) = x^3 - 2x^2 - x + 1$ have any real zeros between 2 and 3?

Use synthetic division to find $P(2)$ and $P(3)$.

$$
\begin{array}{r}
2)\overline{1 \quad -2 \quad -1 \quad 1} \\
2 \quad 0 \quad -2 \\
\hline
1 \quad 0 \quad -1 \quad -1
\end{array}
\qquad
\begin{array}{r}
3)\overline{1 \quad -2 \quad -1 \quad 1} \\
3 \quad 3 \quad 6 \\
\hline
1 \quad 1 \quad 2 \quad 7
\end{array}
$$

The results show that $P(2) = -1$ and $P(3) = 7$. Since $P(2)$ is negative but $P(3)$ is positive, by the intermediate value theorem, there must be a real zero between 2 and 3. ■

| CAUTION | Be careful how you interpret the intermediate value theorem. If $P(a)$ and $P(b)$ are *not* opposite in sign, it does not necessarily mean that there is no zero between a and b. For example, in Figure 6.14, $P(a)$ and $P(b)$ are both negative, but -3 and -1, which are between a and b, are zeros of P. |

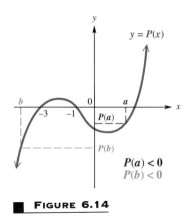

$y = P(x)$

$P(a) < 0$
$P(b) < 0$

∎ **FIGURE 6.14**

The intermediate value theorem for polynomial functions is helpful in limiting the search for real zeros to a smaller and smaller interval. In Example 3 the theorem was used to show that the polynomial function P defined by $P(x) = x^3 - 2x^2 - x + 1$ has a real zero between 2 and 3. The theorem then could be used repeatedly to express the zero more accurately.

As suggested by the graphs of Figure 6.15, if the values of $|x|$ in a polynomial get larger and larger, then so will the values of $|y|$. This is used in the next theorem, the boundedness theorem, which shows how the bottom row of a synthetic division can be used to place upper and lower bounds on the possible real zeros of a polynomial.

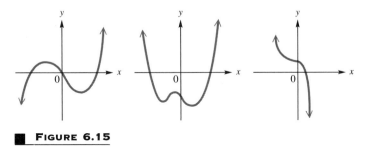

∎ **FIGURE 6.15**

BOUNDEDNESS THEOREM   Let $P(x)$ define a polynomial function such that $P(x)$ has *real* coefficients and a *positive* leading coefficient. If $P(x)$ is divided synthetically by $x - c$, and

(a) if $c > 0$ and all numbers in the bottom row of the synthetic division are nonnegative, then P has no zero greater than c;

(b) if $c < 0$ and the numbers in the bottom row of the synthetic division alternate in sign (with 0 considered positive or negative, as needed), then P has no zero less than c.

■ *Example 4*

USING THE
BOUNDEDNESS
THEOREM

Approximate the real zeros of the polynomial function P defined by $P(x) = x^4 - 6x^3 + 8x^2 + 2x - 1$.

Use Descartes' rule of signs first. From Example 1, P has either three or one positive real zeros and one negative real zero. Next, check for rational zeros. The only possible rational zeros are ± 1.

$$
\begin{array}{r|rrrrl}
1) & 1 & -6 & 8 & 2 & -1 \\
  &   & 1 & -5 & 3 & 5 \\
\hline
  & 1 & -5 & 3 & 5 & 4 \leftarrow P(1) = 4
\end{array}
$$

$$
\begin{array}{r|rrrrl}
-1) & 1 & -6 & 8 & 2 & -1 \\
   &   & -1 & 7 & -15 & 13 \\
\hline
   & 1 & -7 & 15 & -13 & 12 \leftarrow P(-1) = 12
\end{array}
$$

Neither 1 nor -1 is a zero.

We now use the two theorems in this section to search in some consistent way for the location of irrational real zeros. The leading coefficient of $P(x)$ is positive and the numbers in the last row of the second synthetic division above alternate in sign. Since $-1 < 0$, by the boundedness theorem -1 is less than or equal to any zero of P. By the synthetic division above, -1 is not a zero of P. Also, $P(-1) = 12 > 0$. By substitution, or synthetic division, $P(0) = -1 < 0$. Thus the one negative real zero is between -1 and 0.

Try $c = -.5$. Divide $P(x)$ by $x + .5$.

$$
\begin{array}{r|rrrrl}
-.5) & 1 & -6 & 8 & 2 & -1 \\
    &   & -.5 & 3.25 & -5.625 & 1.8125 \\
\hline
    & 1 & -6.5 & 11.25 & -3.625 & .8125
\end{array}
$$

Since $P(-.5) = .8125 > 0$ and $P(0) = -1 < 0$, there is a real zero between $-.5$ and 0.

Now try $c = -.4$.

$$
\begin{array}{r|rrrrl}
-.4) & 1 & -6 & 8 & 2 & -1 \\
    &   & -.4 & 2.56 & -4.224 & .8896 \\
\hline
    & 1 & -6.4 & 10.56 & -2.224 & -.1104
\end{array}
$$

Since $P(-.5)$ is positive, but $P(-.4)$ is negative, there is a zero between $-.5$ and $-.4$. The value of $P(-.4)$ is closer to zero than $P(-.5)$, so it is probably safe to say that, to one decimal place of accuracy, $-.4$ is a real zero of P. A more accurate result can be found, if desired, by continuing this process.

Find the remaining real zeros of P by using synthetic division to find $P(1)$, $P(2)$, $P(3)$, and so on, until a change in sign is noted. It is helpful to use the shortened form of synthetic division shown below. Only the last row of the synthetic division is shown for each division. The first row of the chart is used for each division and the work in the second row of the division is done mentally.

x					$P(x)$
	1	−6	8	2	−1
−1	1	−7	15	−13	12
0	1	−6	8	2	−1
1	1	−5	3	5	4
2	1	−4	0	2	3
3	1	−3	−1	−1	−4
4	1	−2	0	2	7

⟵ Zero between −1 and 0

⟵ Zero between 0 and 1

⟵ Zero between 2 and 3

⟵ Zero between 3 and 4

Since the polynomial is degree 4, there are no more than four zeros. Expand the table to find the real zeros to the nearest tenth. For example, for the zero between 0 and 1, work as follows. Start halfway between 0 and 1 with $x = .5$. Since $P(.5) > 0$ and $P(0) < 0$, try $x = .4$ next, and so on.

x					$P(x)$
	1	−6	8	2	−1
.5	1	−5.5	5.25	4.63	1.31
.4	1	−5.6	5.76	4.30	.72
.3	1	−5.7	6.29	3.89	.17
.2	1	−5.8	6.84	3.37	−.33

⟵ Zero between .3 and .2

The value $P(.3) = .17$ is closer to 0 than $P(.2) = -.33$, so to the nearest tenth, the zero is .3. Use synthetic division to verify that the remaining two zeros are approximately 2.4 and 3.7. ∎

Many of today's scientific calculators have programming capabilities. It would be fairly simple to program the function of Example 4 into one of these calculators, and evaluate the polynomial for successive integer values. When it is found, for example, that a zero lies between −1 and 0, this interval could be subdivided and a closer approximation then obtained. By repeating this process, zeros can be found to whatever accuracy is required.

With many polynomial functions, in order to graph the function without the use of a graphing calculator you will need to plot a large number of points. The theorems studied in this chapter are helpful in deciding which points to plot and in finding the ordered pairs.

∎ **_Example 5_**

GRAPHING A POLYNOMIAL FUNCTION USING VARIOUS RULES AND THEOREMS

Graph the polynomial function P defined by $P(x) = 8x^3 - 12x^2 + 2x + 1$.

By Descartes' rule of signs, P has two or zero positive real zeros and one negative real zero. Locate the real zeros using synthetic division. Start with $x = 0$ and then find $P(1)$, $P(2)$, $P(3)$, and so on until a row of the synthetic division is all positive, indicating there are no zeros greater than the x-value for that row. Then do the same thing in the negative direction; find $P(-1)$, $P(-2)$, and so on until alternating signs in the last row of the synthetic division indicate the number that is less than or equal to all the real zeros. As shown in the shortened form of synthetic division below, a sign change in the value of $P(x)$ indicates a zero.

x				$P(x)$	Ordered Pair	
	8	−12	2	1		
2	8	4	10	21	(2, 21)	←— Zero
1	8	−4	−2	−1	(1, −1)	←— Zero
0	8	−12	2	1	(0, 1)	←— Zero
−1	8	−20	22	−21	(−1, −21)	

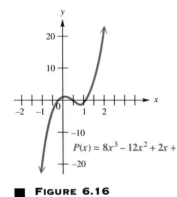

$$P(x) = 8x^3 - 12x^2 + 2x + 1$$

■ **FIGURE 6.16**

All three real zeros have been located. As expected, two are positive and one is negative. Since the numbers in the row for $x = 2$ are all positive and $2 > 0$, 2 is greater than any real zero of P. Also, −1 is less than any real zero of P.

By the intermediate value theorem, there is a zero between 0 and 1 and between −1 and 0, as well as between 1 and 2. The polynomial function is graphed by plotting the points from the chart and then drawing a continuous curve through them, as shown in Figure 6.16. ■

■ *Example 6*

GRAPHING A
POLYNOMIAL
FUNCTION USING
VARIOUS RULES
AND THEOREMS

Graph the polynomial function P defined by $P(x) = 3x^4 - 14x^3 + 54x - 3$.

Use Descartes' rule of signs to see that there are either three positive real zeros or one positive real zero, and there is one negative real zero. The points to plot are found using synthetic division to make a table like the one shown below. Start with $x = 0$ and work up through the positive integers until a row with all nonnegative numbers is found. Then work down through the negative integers until a row with alternating signs is found.

x					$P(x)$	Ordered Pair	
	3	−14	0	54	−3		
5	3	1	5	79	392	(5, 392)	← All positive
4	3	−2	−8	22	85	(4, 85)	
3	3	−5	−15	9	24	(3, 24)	
2	3	−8	−16	22	41	(2, 41)	
1	3	−11	−11	43	40	(1, 40)	
0	3	−14	0	54	−3	(0, −3)	
−1	3	−17	17	37	−40	(−1, −40)	
−2	3	−20	40	−26	49	(−2, 49)	← Alternating signs

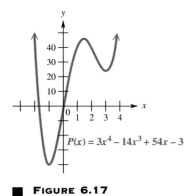

$P(x) = 3x^4 - 14x^3 + 54x - 3$

■ **FIGURE 6.17**

Since the row in the chart for $x = 5$ contains all positive numbers, the function has no zero greater than 5. Also, since the row for $x = -2$ has numbers that alternate in sign, there is no zero less than -2. By the changes in sign of $P(x)$, the function has zeros between 0 and 1 and between -2 and -1. Plotting the points found above and drawing a continuous curve through them gives the graph in Figure 6.17. ■

6.5 EXERCISES ■

A calculator will be helpful in working many of the exercises in this section.

Use Descartes' rule of signs to find the number of positive and negative real zeros possible for each polynomial function defined as follows. See Examples 1 and 2.

1. $P(x) = 2x^3 - 4x^2 + 2x + 7$

2. $P(x) = x^3 + 2x^2 + x - 10$

3. $P(x) = 5x^4 + 3x^2 + 2x - 9$

4. $P(x) = 3x^4 + 2x^3 - 8x^2 - 10x - 1$

5. $P(x) = x^5 + 3x^4 - x^3 + 2x + 3$

6. $P(x) = 2x^5 - x^4 + x^3 - x^2 + x + 5$

Use the intermediate value theorem for polynomial functions to show that each function defined as follows has a real zero between the numbers given. See Example 3.

7. $P(x) = 3x^2 - 2x - 6$;   1 and 2

8. $P(x) = x^3 + x^2 - 5x - 5$;   2 and 3

9. $P(x) = 2x^3 - 8x^2 + x + 16$;   2 and 2.5

10. $P(x) = 3x^3 + 7x^2 - 4$;   1/2 and 1

11. $P(x) = 2x^4 - 4x^2 + 3x - 6$;   2 and 1.5

12. $P(x) = x^4 - 4x^3 - x + 1$;   1 and .3

13. Suppose that a polynomial function P is defined in such a way that $P(2) = -4$ and $P(2.5) = 2$. What conclusion does the intermediate value theorem allow you to make?

14. Suppose that a polynomial function P is defined in such a way that $P(3) = -4$ and $P(4) = -10$. Can we be certain that there is no zero between 3 and 4? Explain.

Show that the real zeros of the polynomial functions defined as follows satisfy the given conditions.

15. $P(x) = x^4 - x^3 + 3x^2 - 8x + 8$;   no real zero greater than 2

16. $P(x) = 2x^5 - x^4 + 2x^3 - 2x^2 + 4x - 4$;   no real zero greater than 1

17. $P(x) = x^4 + x^3 - x^2 + 3$;   no real zero less than -2

18. $P(x) = x^5 + 2x^3 - 2x^2 + 5x + 5$;   no real zero less than -1

19. $P(x) = 3x^4 + 2x^3 - 4x^2 + x - 1$;   no real zero greater than 1

20. $P(x) = 3x^4 + 2x^3 - 4x^2 + x - 1$;   no real zero less than -2

21. $P(x) = x^5 - 3x^3 + x + 2$;   no real zero greater than 2

22. $P(x) = x^5 - 3x^3 + x + 2$;   no real zero less than -3

*For each of the polynomial functions P defined as follows: (**a**) Find the number of positive and negative real zeros that are possible. See Examples 1 and 2. (**b**) Approximate each zero as a decimal to the nearest tenth. See Example 4.*

23. $P(x) = x^3 + 3x^2 - 2x - 6$

24. $P(x) = x^3 + x^2 - 5x - 5$

25. $P(x) = x^3 - 4x^2 - 5x + 14$

26. $P(x) = x^3 + 9x^2 + 34x + 13$

27. $P(x) = x^3 + 6x - 13$

28. $P(x) = 4x^3 - 3x^2 + 4x - 5$

29. $P(x) = 4x^4 - 8x^3 + 17x^2 - 2x - 14$

30. $P(x) = 3x^4 - 4x^3 - x^2 + 8x - 2$

31. $P(x) = -x^4 + 2x^3 + 3x^2 + 6$

32. $P(x) = -2x^4 - x^2 + x - 5$

Graph each of the polynomial functions defined as follows. See Examples 5 and 6.

33. $P(x) = x^3 - 7x - 6$

34. $P(x) = x^3 + x^2 - 4x - 4$

35. $P(x) = x^4 - 5x^2 + 6$

36. $P(x) = x^3 - 3x^2 - x + 3$

37. $P(x) = 6x^3 + 11x^2 - x - 6$

38. $P(x) = x^4 - 2x^2 - 8$

39. $P(x) = -x^3 + 6x^2 - x - 14$

40. $P(x) = 6x^4 - x^3 - 23x^2 - 4x + 12$

The following polynomials define functions that have zeros in the given intervals. Approximate these zeros to the nearest hundredth.

41. $P(x) = x^4 + x^3 - 6x^2 - 20x - 16$; [3.2, 3.3] and [−1.4, −1.1]

42. $P(x) = x^4 - 2x^3 - 2x^2 - 18x + 5$; [.2, .4] and [3.7, 3.8]

43. $P(x) = x^4 - 4x^3 - 20x^2 + 32x + 12$; [−4, −3], [−1, 0], [1, 2], and [6, 7]

44. $P(x) = x^4 - 4x^3 - 44x^2 + 160x - 80$; [−7, −6], [0, 1], [2, 3], and [7, 8]

45. A technique for measuring cardiac output depends on the concentration of a dye in the bloodstream after a known amount is injected into a vein near the heart. For a normal heart, the concentration of dye in the bloodstream at time x (in sec) is given by the function defined as follows.

$$g(x) = -.006x^4 + .140x^3 - .053x^2 + 1.79x$$

(**a**) Find $g(20)$.

(**b**) Graph g.

46. The polynomial function defined by

$$A(x) = -.015x^3 + 1.058x$$

gives the approximate alcohol concentration (in tenths of a percent) in an average person's bloodstream x hours after drinking about 8 oz of 100-proof whiskey. The function is approximately valid for x in the interval [0, 8].

(**a**) Graph A.

(**b**) Using the graph you drew for part (a), estimate the time of maximum alcohol concentration.

(**c**) In one state, a person is legally drunk if the blood alcohol concentration exceeds .08 percent. Use the graph from part (a) to estimate the period in which the average person is legally drunk.

47. Give an example of a polynomial function that is never negative and has −3 and 2 as zeros.

48. Give an example of a polynomial function that has −3 and 2 as zeros and is positive only between −3 and 2.

49. Explain why a seventh-degree polynomial with some terms missing, that is, with one or more zero coefficients, cannot have seven positive zeros.

50. Show that a polynomial of the form $x^4 \ldots + 1$ must have an even number of positive zeros and that a polynomial of the form $x^4 \ldots -1$ must have an odd number of positive zeros.

6.6 ———— GRAPHING RATIONAL FUNCTIONS

We begin with a definition of *rational function*.

RATIONAL FUNCTION	A function f of the form p/q defined by $$f(x) = \frac{p(x)}{q(x)},$$ where $p(x)$ and $q(x)$ are polynomials, is called a **rational function.**

Since any values of x such that $q(x) = 0$ are excluded from the domain of a rational function, this type of function often has a graph that has one or more breaks in it.

The simplest rational function with a variable denominator is defined by

$$f(x) = \frac{1}{x}.$$

The domain of this function is the set of all real numbers except 0. The number 0 cannot be used as a value of x, but for graphing it is helpful to find the values of $f(x)$ for some values of x close to 0. The following table shows what happens to $f(x)$ as x gets closer and closer to 0 from either side.

x approaches 0.

x	-1	$-.1$	$-.01$	$-.001$	$.001$	$.01$	$.1$	1
$f(x)$	-1	-10	-100	-1000	1000	100	10	1

$|f(x)|$ gets larger and larger.

The table suggests that $|f(x)|$ gets larger and larger as x gets closer and closer to 0, which is written in symbols as

$$|f(x)| \to \infty \text{ as } x \to 0.$$

(The symbol $x \to 0$ means that x approaches 0, without necessarily ever being equal to 0.) Since x cannot equal 0, the graph of $f(x) = 1/x$ will never intersect the vertical line $x = 0$. This line is called a *vertical asymptote*.

On the other hand, as $|x|$ gets larger and larger, the values of $f(x) = 1/x$ get closer and closer to 0, as shown in the table below.

x	$-10{,}000$	-1000	-100	-10	10	100	1000	$10{,}000$
$f(x)$	$-.0001$	$-.001$	$-.01$	$-.1$	$.1$	$.01$	$.001$	$.0001$

Letting $|x|$ get larger and larger without bound (written $|x| \to \infty$) causes the graph of $f(x) = 1/x$ to move closer and closer to the horizontal line $y = 0$. This line is called a *horizontal asymptote*.

If the point (a, b) lies on the graph of $f(x) = 1/x$, then so does the point $(-a, -b)$. Therefore, the graph of f is symmetric with respect to the origin. Choosing some positive values of x and finding the corresponding values of $f(x)$ gives the first-quadrant part of the graph shown in Figure 6.18. The other part of the graph (in the third quadrant) can be found by symmetry.

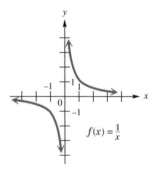

■ **FIGURE 6.18**

■ *Example 1*

GRAPHING A RATIONAL FUNCTION USING REFLECTION

Graph $f(x) = -\dfrac{2}{x}$.

The expression on the right side of the equation can be rewritten so that

$$f(x) = -2 \cdot \frac{1}{x}.$$

Compared to $f(x) = 1/x$, the graph will be reflected about the x-axis (because of the negative sign), and each point will be twice as far from the x-axis. See Figure 6.19. ■

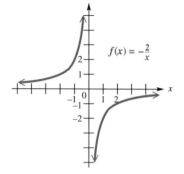

■ **FIGURE 6.19**

■ *Example 2*

GRAPHING A RATIONAL FUNCTION USING TRANSLATION

Graph $f(x) = \dfrac{2}{1 + x}$.

The domain of this function is the set of all real numbers except -1. As shown in Figure 6.20, the graph is that of $f(x) = 1/x$, translated 1 unit to the left, with each y-value doubled. This can be seen by writing the expression as

$$f(x) = 2 \cdot \frac{1}{x - (-1)}.$$ ■

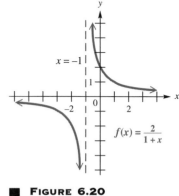

■ **FIGURE 6.20**

Earlier we observed the presence of vertical and horizontal asymptotes for the graph of $f(x) = 1/x$. We now give a formal definition for such asymptotes.

DEFINITIONS OF VERTICAL AND HORIZONTAL ASYMPTOTES	For the rational function f with $f(x) = \dfrac{p(x)}{q(x)}$, written in lowest terms, if $|f(x)| \to \infty$ as $x \to a$, then the line $x = a$ is a **vertical asymptote**; and if $f(x) \to a$ as $|x| \to \infty$, then the line $y = a$ is a **horizontal asymptote**.

Locating asymptotes is an important part of sketching the graphs of rational functions. Vertical asymptotes are found by determining the values of x which make the denominator equal to 0 but do not make the numerator equal to 0. Horizontal asymptotes (and, in some cases, oblique asymptotes) are found by considering what happens to $f(x)$ as $|x| \to \infty$. The next example shows how to find asymptotes.

■ *Example 3*

FINDING ASYMPTOTES OF GRAPHS OF RATIONAL FUNCTIONS

For each rational function f, find all asymptotes.

(a) $f(x) = \dfrac{x + 1}{(2x - 1)(x + 3)}$

To find the vertical asymptotes, set the denominator equal to zero and solve.

$$(2x - 1)(x + 3) = 0$$

$$2x - 1 = 0 \quad \text{or} \quad x + 3 = 0 \qquad \text{Zero-factor property}$$

$$x = \frac{1}{2} \quad \text{or} \quad x = -3$$

The equations of the vertical asymptotes are $x = 1/2$ and $x = -3$.

To find the equation of the horizontal asymptote, we divide each term by the largest power of x in the expression. Begin by multiplying the factors in the denominator to get

$$f(x) = \frac{x + 1}{(2x - 1)(x + 3)} = \frac{x + 1}{2x^2 + 5x - 3}.$$

Now divide each term in the numerator and denominator by x^2, since 2 is the largest exponent on x. This gives

$$f(x) = \frac{\dfrac{x}{x^2} + \dfrac{1}{x^2}}{\dfrac{2x^2}{x^2} + \dfrac{5x}{x^2} - \dfrac{3}{x^2}} = \frac{\dfrac{1}{x} + \dfrac{1}{x^2}}{2 + \dfrac{5}{x} - \dfrac{3}{x^2}}.$$

As $|x|$ gets larger and larger, the quotients $1/x$, $1/x^2$, $5/x$, and $3/x^2$ all approach 0, and the value of $f(x)$ approaches

$$\frac{0 + 0}{2 + 0 - 0} = \frac{0}{2} = 0.$$

The line $y = 0$ (that is, the x-axis) is therefore the horizontal asymptote.

(b) $f(x) = \dfrac{2x + 1}{x - 3}$

Set the denominator equal to zero to find that the vertical asymptote has the equation $x = 3$. To find the horizontal asymptote, divide each term in the rational expression by x, since the greatest power of x in the expression is 1.

$$f(x) = \frac{2x + 1}{x - 3} = \frac{\dfrac{2x}{x} + \dfrac{1}{x}}{\dfrac{x}{x} - \dfrac{3}{x}} = \frac{2 + \dfrac{1}{x}}{1 - \dfrac{3}{x}}$$

As $|x|$ gets larger and larger, both $1/x$ and $3/x$ approach 0, and $f(x)$ approaches

$$\frac{2 + 0}{1 - 0} = \frac{2}{1} = 2,$$

so the line $y = 2$ is the horizontal asymptote.

(c) $f(x) = \dfrac{x^2 + 1}{x - 2}$

Setting the denominator equal to zero shows that the vertical asymptote has the equation $x = 2$. If we divide by the largest power of x as before (x^2 in this case), we see that there is no horizontal asymptote because

$$f(x) = \frac{\dfrac{x^2}{x^2} + \dfrac{1}{x^2}}{\dfrac{x}{x^2} - \dfrac{2}{x^2}} = \frac{1 + \dfrac{1}{x^2}}{\dfrac{1}{x} - \dfrac{2}{x^2}}$$

does not approach any real number as $|x| \to \infty$, since $1/0$ is undefined. This will happen whenever the degree of the numerator is greater than the degree of the denominator. In such cases, divide the denominator into the numerator to write the expression in another form. Using synthetic division gives

$$2)\overline{\begin{array}{ccc} 1 & 0 & 1 \\ & 2 & 4 \\ \hline 1 & 2 & 5. \end{array}}$$

The function can now be written as

$$f(x) = \frac{x^2 + 1}{x - 2} = x + 2 + \frac{5}{x - 2}.$$

For very large values of $|x|$, $5/(x - 2)$ is close to 0, and the graph approaches the line $y = x + 2$. This line is an **oblique asymptote** (neither vertical nor horizontal) for the graph of the function.

In general, if the degree of the numerator is exactly one more than the degree of the denominator, a rational function may have an oblique asymptote. The equation of this asymptote is found by dividing the numerator by the denominator and disregarding the remainder. ■

The results of Example 3 can be summarized as follows.

DETERMINING ASYMPTOTES

In order to find asymptotes of a rational function defined by a rational expression *in lowest terms,* use the following procedures.

1. Vertical Asymptotes

   Find any vertical asymptotes by setting the denominator equal to 0 and solving for *x.* If *a* is a zero of the denominator, then the line $x = a$ is a vertical asymptote.

2. Other Asymptotes

   Determine any other asymptotes. We consider three possibilities:

   (a) If the numerator has lower degree than the denominator, there is a horizontal asymptote, $y = 0$ (the *x*-axis).

   (b) If the numerator and denominator have the same degree, and the function is of the form

   $$f(x) = \frac{a_n x^n + \cdots + a_0}{b_n x^n + \cdots + b_0}, \quad \text{where } b_n \neq 0,$$

   dividing by x^n in the numerator and denominator produces the horizontal asymptote

   $$y = \frac{a_n}{b_n}.$$

   (c) If the numerator is of degree exactly one more than the denominator, there may be an oblique asymptote. To find it, divide the numerator by the denominator and disregard any remainder. Set the rest of the quotient equal to *y* to get the equation of the asymptote.

NOTE The graph of a rational function may have more than one vertical asymptote, or it may have none at all. The graph cannot intersect any vertical asymptote. There can be only one other (non-vertical) asymptote, and the graph *may* intersect that asymptote. This will be seen in Example 6. The method of graphing a rational function having common variable factors in the numerator and denominator of the defining expression will be covered in Example 8.

The following procedure can be used to graph functions defined by rational expressions reduced to lowest terms.

| GRAPHING RATIONAL FUNCTIONS | Let $f(x) = \dfrac{p(x)}{q(x)}$ define a function where the rational expression is written in lowest terms. To sketch its graph, follow the steps below. |

1. Find any vertical asymptotes.
2. Find any horizontal or oblique asymptote.
3. Find the y-intercept by evaluating $f(0)$.
4. Find the x-intercepts, if any, by solving $f(x) = 0$. (These will be the zeros of the numerator, p.)
5. Determine whether the graph will intersect its non-vertical asymptote by solving $f(x) = k$, where k (or $mx + b$) is the y-value of the non-vertical asymptote.
6. Plot a few selected points, as necessary. Choose an x-value in each interval of the domain as determined by the vertical asymptotes and x-intercepts.
7. Complete the sketch.

The next example shows how the above guidelines can be used to graph a rational function.

■ *Example 4*

GRAPHING A RATIONAL FUNCTION DEFINED BY AN EXPRESSION WITH DEGREE OF NUMERATOR LESS THAN DEGREE OF DENOMINATOR

Graph $f(x) = \dfrac{x + 1}{(2x - 1)(x + 3)}$.

Step 1 As shown in Example 3(a), the vertical asymptotes have equations $x = 1/2$ and $x = -3$.

Step 2 Again, as shown in Example 3(a), the horizontal asymptote is the x-axis.

Step 3 Since $f(0) = \dfrac{0 + 1}{(2(0) - 1)((0) + 3)} = -\dfrac{1}{3}$, the y-intercept is $-1/3$.

Step 4 The x-intercept is found by solving $f(x) = 0$.

$$\frac{x + 1}{(2x - 1)(x + 3)} = 0$$

$$x + 1 = 0 \qquad \text{Multiply by } (2x - 1)(x + 3).$$

$$x = -1$$

The x-intercept is -1.

Step 5 To determine whether the graph intersects its horizontal asymptote, solve

$$f(x) = 0.$$
$$\underset{\text{\textit{y}-value of horizontal asymptote}}{\uparrow}$$

Since the horizontal asymptote is the x-axis, the solution of this equation was found in Step 4. The graph intersects its horizontal asymptote at $(-1, 0)$.

Step 6 Plot a point in each of the intervals determined by the *x*-intercepts and vertical asymptotes, $(-\infty, -3)$, $(-3, -1)$, $(-1, 1/2)$, and $(1/2, \infty)$ to get an idea of how the graph behaves in each region.

Step 7 Complete the sketch. Keep in mind that the graph approaches its asymptotes as the points on the graph become farther away from the origin. The graph is shown in Figure 6.21. ∎

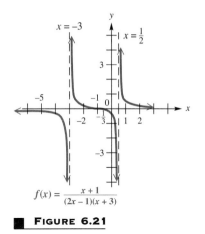

$$f(x) = \frac{x+1}{(2x-1)(x+3)}$$

∎ **FIGURE 6.21**

In the remaining examples, we will not specifically number the steps.

∎ ***Example 5***

GRAPHING A RATIONAL FUNCTION DEFINED BY AN EXPRESSION WITH DEGREE OF NUMERATOR EQUAL TO DEGREE OF DENOMINATOR

Graph $f(x) = \dfrac{2x+1}{x-3}$.

As shown in Example 3(b), the equation of the vertical asymptote is $x = 3$ and the equation of the horizontal asymptote is $y = 2$. Since $f(0) = -1/3$, the *y*-intercept is $-1/3$. The solution of $f(x) = 0$ is $-1/2$, so the only *x*-intercept is $-1/2$. The graph does not intersect its horizontal asymptote, since $f(x) = 2$ has no solution. (Verify this.) The points $(-4, 1)$ and $(6, 13/3)$ are on the graph and can be used to complete the sketch, as shown in Figure 6.22. ∎

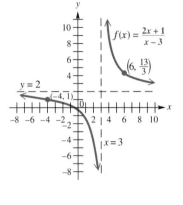

∎ **FIGURE 6.22**

∎ ***Example 6***

GRAPHING A RATIONAL FUNCTION DEFINED BY AN EXPRESSION WITH DEGREE OF NUMERATOR EQUAL TO DEGREE OF DENOMINATOR

Graph $f(x) = \dfrac{3(x+1)(x-2)}{(x+4)^2}$.

The only vertical asymptote is the line $x = -4$. To find any horizontal asymptotes, multiply the factors in the numerator and denominator.

$$f(x) = \frac{3x^2 - 3x - 6}{x^2 + 8x + 16}$$

As explained in the guidelines above, the equation of the horizontal asymptote can be shown to be

$$y = \frac{3}{1} \begin{array}{l} \leftarrow \text{Leading coefficient of numerator} \\ \leftarrow \text{Leading coefficient of denominator} \end{array}$$

or $y = 3$. The *y*-intercept is $-3/8$ and the *x*-intercepts are -1 and 2. By setting $f(x) = 3$ and solving, we can find the point where the graph intersects the horizontal asymptote.

$$f(x) = \frac{3x^2 - 3x - 6}{x^2 + 8x + 16}$$

$$3 = \frac{3x^2 - 3x - 6}{x^2 + 8x + 16}$$

$$3x^2 - 3x - 6 = 3x^2 + 24x + 48 \qquad \text{Multiply by } x^2 + 8x + 16.$$

$$-3x - 6 = 24x + 48 \qquad \text{Subtract } 3x^2.$$

$$-27x = 54$$

$$x = -2$$

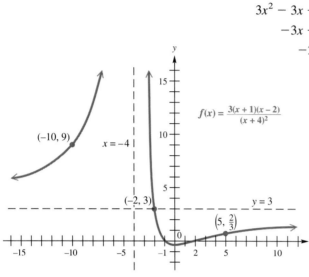

$$f(x) = \frac{3(x+1)(x-2)}{(x+4)^2}$$

The graph intersects its horizontal asymptote at $(-2, 3)$.

Some other points that lie on the graph are $(-10, 9)$, $(-3, 30)$, and $(5, 2/3)$. These can be used to complete the graph, shown in Figure 6.23. ■

FIGURE 6.23

The next example discusses a rational function defined by an expression having the degree of its numerator greater than the degree of its denominator.

■ *Example 7*

GRAPHING A
RATIONAL
FUNCTION DEFINED
BY AN EXPRESSION
WITH DEGREE OF
NUMERATOR
GREATER THAN
DEGREE OF
DENOMINATOR

Graph $f(x) = \dfrac{x^2 + 1}{x - 2}$.

As shown in Example 3(c), the vertical asymptote has the equation $x = 2$, and the graph has an oblique asymptote with the equation $y = x + 2$. The y-intercept is $-1/2$, and the graph has no x-intercepts, since the numerator, $x^2 + 1$, has no real zeros. It can be shown that the graph does not intersect its oblique asymptote. Using the intercepts, asymptotes, the points $(4, 17/2)$ and $(-1, -2/3)$, and the general behavior of the graph near its asymptotes, we obtain the graph shown in Figure 6.24. ■

FIGURE 6.24

As mentioned earlier, a rational function must be defined by an expression in lowest terms before we can use the methods discussed in this section to determine the graph. The next example shows a typical rational function defined by an expression that is not in lowest terms.

∎ *Example 8*

GRAPHING A RATIONAL FUNCTION DEFINED BY AN EXPRESSION THAT IS NOT IN LOWEST TERMS

Graph $f(x) = \dfrac{x^2 - 4}{x - 2}$.

Start by noticing that the domain of this function cannot contain 2. The rational expression $(x^2 - 4)/(x - 2)$ can be reduced to lowest terms by factoring the numerator, and using the fundamental principle.

$$f(x) = \frac{x^2 - 4}{x - 2} = \frac{(x + 2)(x - 2)}{x - 2} = x + 2 \quad (x \neq 2)$$

Therefore, the graph of this function will be the same as the graph of $y = x + 2$ (a straight line), with the exception of the point with x-value 2. A "hole" appears in the graph at $(2, 4)$. See Figure 6.25. ∎

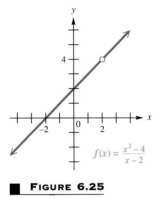

$f(x) = \dfrac{x^2 - 4}{x - 2}$

∎ **FIGURE 6.25**

6.6 EXERCISES ∎ ───────────────────────────────

Use reflections and translations to graph the rational functions defined as follows. See Examples 1 and 2.

1. $f(x) = \dfrac{2}{x}$

2. $f(x) = -\dfrac{3}{x}$

3. $f(x) = \dfrac{1}{x + 2}$

4. $f(x) = \dfrac{1}{x - 3}$

5. $f(x) = \dfrac{1}{x} + 1$

6. $f(x) = \dfrac{1}{x} - 2$

7. Sketch the following graphs and compare them with the graph of $f(x) = 1/x^2$.

(a) $f(x) = \dfrac{1}{(x - 3)^2}$ (b) $f(x) = -\dfrac{2}{x^2}$ (c) $f(x) = \dfrac{-2}{(x - 3)^2}$

8. Describe in your own words what is meant by an *asymptote* of the graph of a rational function.

Give the equations of any vertical, horizontal, or oblique asymptotes for the graphs of the rational functions in Exercises 9–22. See Example 3.

9. $f(x) = \dfrac{2}{x - 5}$

10. $f(x) = \dfrac{-1}{x + 2}$

11. $f(x) = \dfrac{-8}{3x - 7}$

12. $f(x) = \dfrac{5}{4x - 9}$

13. $f(x) = \dfrac{2 - x}{x + 2}$

14. $f(x) = \dfrac{x - 4}{5 - x}$

15. $f(x) = \dfrac{3x - 5}{2x + 9}$

16. $f(x) = \dfrac{4x + 3}{3x - 7}$

17. $f(x) = \dfrac{2}{x^2 - 4x + 3}$

18. $f(x) = \dfrac{-5}{x^2 - 3x - 10}$

19. $f(x) = \dfrac{x^2 - 1}{x + 3}$

20. $f(x) = \dfrac{x^2 + 4}{x - 1}$

21. $f(x) = \dfrac{(x - 3)(x + 1)}{(x + 2)(2x - 5)}$

22. $f(x) = \dfrac{3(x + 2)(x - 4)}{(5x - 1)(x - 5)}$

23. Which one of the following has a graph that does not have a vertical asymptote?

   (a) $f(x) = \dfrac{1}{x^2 + 2}$ **(b)** $f(x) = \dfrac{1}{x^2 - 2}$ **(c)** $f(x) = \dfrac{3}{x^2}$ **(d)** $f(x) = \dfrac{2x + 1}{x - 8}$

24. Which one of the following has a graph that does not have a horizontal asymptote?

   (a) $f(x) = \dfrac{2x - 7}{x + 3}$ **(b)** $f(x) = \dfrac{3x}{x^2 - 9}$ **(c)** $f(x) = \dfrac{x^2 - 9}{x + 3}$ **(d)** $f(x) = \dfrac{x + 5}{(x + 2)(x - 3)}$

Graph each of the following. See Examples 4–8.

25. $f(x) = \dfrac{4}{5 + 3x}$

26. $f(x) = \dfrac{1}{(x - 2)(x + 4)}$

27. $f(x) = \dfrac{3}{(x + 4)^2}$

28. $f(x) = \dfrac{3x}{(x + 1)(x - 2)}$

29. $f(x) = \dfrac{2x + 1}{(x + 2)(x + 4)}$

30. $f(x) = \dfrac{5x}{x^2 - 1}$

31. $f(x) = \dfrac{-x}{x^2 - 4}$

32. $f(x) = \dfrac{3x}{x - 1}$

33. $f(x) = \dfrac{4x}{1 - 3x}$

34. $f(x) = \dfrac{x + 1}{x - 4}$

35. $f(x) = \dfrac{x - 5}{x + 3}$

36. $f(x) = \dfrac{x}{x^2 - 9}$

37. $f(x) = \dfrac{3x}{x^2 - 16}$

38. $f(x) = \dfrac{2x^2 + 3}{x - 4}$

39. $f(x) = \dfrac{x^2 + 1}{x + 3}$

40. $f(x) = \dfrac{x^2 + 2x}{2x - 1}$

41. $f(x) = \dfrac{x^2 - x}{x + 2}$

42. $f(x) = \dfrac{(x - 3)(x + 1)}{(x - 1)^2}$

43. $f(x) = \dfrac{x(x - 2)}{(x + 3)^2}$

44. $f(x) = \dfrac{(x - 5)(x - 2)}{x^2 + 9}$

45. $f(x) = \dfrac{1}{x^2 + 1}$

46. $f(x) = \dfrac{x^2 - 9}{x + 3}$

47. $f(x) = \dfrac{x^2 - 16}{x + 4}$

48. The figures below show the four ways that the graph of a rational function can approach the vertical line $x = 2$ as an asymptote. Identify the graph of each of the following rational functions defined as follows.

   (a) $f(x) = \dfrac{1}{(x - 2)^2}$ **(b)** $f(x) = \dfrac{1}{x - 2}$ **(c)** $f(x) = \dfrac{-1}{x - 2}$ **(d)** $f(x) = \dfrac{-1}{(x - 2)^2}$

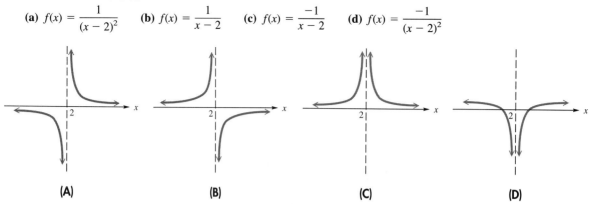

(A)   **(B)**   **(C)**   **(D)**

Work the following problems.

49. Antique-car owners often enter their cars in a *concours d'elegance* in which a maximum of 100 points can be awarded to a particular car. Points are awarded for the general attractiveness of the car. The function defined by

$$C(x) = \frac{10x}{49(101 - x)}$$

expresses the cost, in thousands of dollars, of restoring a car so that it will win x points. Graph the function.

50. Suppose the average cost per unit in thousands of dollars, $C(x)$, to produce x units of margarine is given by

$$C(x) = \frac{500}{x + 30}.$$

(a) Find $C(10)$, $C(20)$, $C(50)$, $C(75)$, and $C(100)$.
(b) Graph the function.

51. In situations involving environmental pollution, a cost-benefit model expresses cost as a function of the percentage of pollutant removed from the environment. Suppose a cost-benefit model is expressed as

$$C(x) = \frac{6.7x}{100 - x},$$

where y is the cost in thousands of dollars of removing x percent of a certain pollutant.
(a) Graph the function.
(b) Is it possible, according to this function, to remove all of the pollutant?

52. In a recent year, the cost per ton, y, to build an oil tanker of x thousand deadweight tons was approximated by

$$C(x) = \frac{110,000}{x + 225}.$$

(a) Find y for $x = 25$, $x = 50$, $x = 100$, $x = 200$, $x = 300$, and $x = 400$.
(b) Graph the function.

*In recent years the economist Arthur Laffer has been a center of controversy because of his **Laffer curve**, an idealized version of which is shown here.*

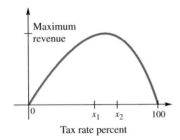

Tax rate percent

According to this curve, increasing a tax rate, say from x_1 percent to x_2 percent on the graph above, can actually lead to a decrease in government revenue. All economists agree on the endpoints, 0 revenue at tax rates of both 0% and 100%, but there is much disagreement on the location of the rate x_1 that produces maximum revenue.

53. Suppose an economist studying the Laffer curve produces the rational function defined by

$$R(x) = \frac{80x - 8000}{x - 110},$$

with y giving government revenue in tens of millions of dollars for a tax rate of x percent, with the function valid for $55 \le x \le 100$. Find the revenue for the following tax rates.
(a) 55%   (b) 60%   (c) 70%
(d) 90%   (e) 100%
(f) Graph the function.

54. Suppose an economist studies a different tax, this time producing

$$R(x) = \frac{60x - 6000}{x - 120},$$

where y is government revenue in millions of dollars from a tax rate of x percent, with $y = R(x)$ valid for $50 \le x \le 100$. Find the revenue from the following tax rates.
(a) 50%   (b) 60%   (c) 80%   (d) 100%
(e) Graph the function.

55. Let $f(x) = p(x)/q(x)$ define a rational function where the expression is reduced to lowest terms. Suppose that the degree of $p(x)$ is m and the degree of $q(x)$ is n. Write an explanation of how you would determine the nonvertical asymptote in each of the following situations.
(a) $m < n$ (b) $m = n$ (c) $m > n$

56. Suppose that a friend tells you that the graph of

$$f(x) = \frac{x^2 - 25}{x + 5}$$

has a vertical asymptote with equation $x = -5$. Is this correct? If not, describe the behavior of the graph at $x = -5$.

CHAPTER 6 SUMMARY ■ ────────────

SECTION	TERMS	KEY IDEAS
6.1 Graphing Polynomial Functions	polynomial function of degree n leading coefficient zero of a polynomial function	**Graphing Polynomial Functions** To graph a polynomial function in factored form, find x-intercepts and y-intercepts. Choose a test point in each interval determined by the x-intercepts to find out whether the graph is above or below the x-axis. Plot a few points, as necessary, and sketch the graph as a smooth, unbroken curve.
6.2 Polynomial Division	division algorithm quotient polynomial remainder polynomial synthetic division	**Division Algorithm** (See page 310.) **Remainder Theorem** (See page 310.)
6.3 Zeros of Polynomial Functions	multiplicity	**Factor Theorem** (See page 313.) **Fundamental Theorem of Algebra** (See page 314.) **Conjugate Zeros Theorem** (See page 316.)
6.4 Rational Zeros of Polynomial Functions		**Rational Zeros Theorem** (See page 320.)
6.5 Real Zeros of Polynomial Functions		**Descartes' Rule of Signs** (See page 325.) **Intermediate Value Theorem for Polynomial Functions** (See page 326.) **Boundedness Theorem** (See page 327.)
6.6 Graphing Rational Functions	rational function vertical asymptote horizontal asymptote oblique asymptote	**Graphing Rational Functions** To graph a rational function defined by an expression in lowest terms, find asymptotes and intercepts. Determine whether the graph intersects the nonvertical asymptote. Plot a few points, as necessary, to complete the sketch. If a rational function is defined by an expression not reduced to lowest terms, there may be a "hole" in the graph.

Sketch the graph of each polynomial function P as defined in Exercises 1–6.

1. $P(x) = x^3 + 5$

2. $P(x) = 1 - x^4$

3. $P(x) = x^2(2x + 1)(x - 2)$

4. $P(x) = (4x - 3)(3x + 2)(x - 1)$

5. $P(x) = 2x^3 + 13x^2 + 15x$

6. $P(x) = x(x - 1)(x + 2)(x - 3)$

7. For the polynomial function defined by $P(x) = x^3 - 3x^2 - 7x + 12$, $P(4) = 0$. Therefore, we can say that 4 is a(n) _____ of the function, 4 is a(n) _____ of the equation $x^3 - 3x^2 - 7x + 12 = 0$, and that 4 is a(n) _____ of the graph of the function.

8. Which one of the following is not a polynomial function?
    (a) $f(x) = x^2$ **(b)** $f(x) = 2x + 5$ **(c)** $f(x) = 1/x$ **(d)** $f(x) = x^{100}$

Perform each division.

9. $\dfrac{15m^4n^5 - 10m^3n^7 + 20m^7n^3}{10mn^4}$

10. $\dfrac{10r^6m^3 + 8r^4m - 3r^{10}m^2}{12r^3m^4}$

11. $\dfrac{6p^2 - 5p - 56}{2p - 7}$

12. $\dfrac{72r^2 + 59r + 12}{8r + 3}$

13. $\dfrac{5m^3 - 7m^2 - 10m + 14}{m^2 - 2}$

14. $\dfrac{3b^3 - 8b^2 + 12b - 30}{b^2 + 4}$

Use synthetic division to perform each division.

15. $(2x^3 - x^2 - x - 6) \div (x + 1)$

16. $(3t^3 + 10t^2 - 9t - 4) \div (t + 4)$

17. $(4m^3 + m^2 - 12) \div (m - 2)$

18. $(5y^3 + 24y^2 + 25y - 6) \div (y + 3)$

Use synthetic division to find P(2) for each of the following.

19. $P(x) = x^3 + 3x^2 - 5x + 1$

20. $P(x) = 2x^3 - 4x^2 + 3x - 10$

21. $P(x) = 5x^4 - 12x^2 + 2x - 8$

22. $P(x) = x^5 - 3x^2 + 2x - 4$

Find a polynomial function P of lowest degree having the given zeros.

23. $-1, 4, 7$

24. $8, 2, 3$

25. $-\sqrt{7}, \sqrt{7}, 2, -1$

26. $1 + \sqrt{5}, 1 - \sqrt{5}, -4, 1$

27. Is -1 a zero of P, if $P(x) = 2x^4 + x^3 - 4x^2 + 3x + 1$?

28. Is -2 a zero of P, if $P(x) = 2x^4 + x^3 - 4x^2 + 3x + 1$?

29. Is $x + 1$ a factor of $P(x) = x^3 + 2x^2 + 3x - 1$?

30. Is $x + 1$ a factor of $P(x) = 2x^3 - x^2 + x + 4$?

31. Find a function defined by a polynomial $P(x)$ of degree 3 with real coefficients, having $-2, 1$, and 4 as zeros, and $P(2) = 16$.

32. Find a function defined by a polynomial $P(x)$ of degree 4 with real coefficients having $1, -1$, and $3i$ as zeros, and $P(2) = 39$.

33. Find a lowest-degree polynomial $P(x)$ with real coefficients defining a function having zeros $2, -2$, and $-i$.

34. Find a lowest-degree polynomial $P(x)$ with real coefficients defining a function having zeros $2, -3$, and $5i$.

35. Find a polynomial $P(x)$ of lowest degree with real coefficients defining a function having -3 and $1 - i$ as zeros.

36. Find all zeros of P, if $P(x) = x^4 - 3x^3 - 8x^2 + 22x - 24$, given that $1 - i$ is a zero, and factor $P(x)$.

37. Find all zeros of P, if $P(x) = x^4 - 6x^3 + 14x^2 - 24x + 40$, given that $3 + i$ is a zero, and factor $P(x)$.

38. Find all zeros of P, if $P(x) = x^4 + x^3 - x^2 + x - 2$, given that 1 and -2 are zeros, and factor $P(x)$.

Find all rational zeros of each polynomial function P, if $P(x)$ is defined as follows.

39. $P(x) = 2x^3 - 9x^2 - 6x + 5$

40. $P(x) = 3x^3 - 10x^2 - 27x + 10$

41. $P(x) = x^3 - \dfrac{17}{6}x^2 - \dfrac{13}{3}x - \dfrac{4}{3}$

42. $P(x) = 8x^4 - 14x^3 - 29x^2 - 4x + 3$

For the polynomial functions defined in Exercises 43 and 44, find all rational zeros and factor the polynomial. Then graph the function.

43. $P(x) = x^3 - 5x^2 - 13x - 7$

44. $P(x) = 3x^3 - 2x^2 - 19x - 6$

45. Use a polynomial to show that $\sqrt{11}$ is irrational.

46. Show that the function defined by $P(x) = x^3 - 9x^2 + 2x - 5$ has no rational zeros.

Show that each polynomial P defined as follows has real zeros satisfying the given conditions.

47. $P(x) = 3x^3 - 8x^2 + x + 2$, zero in $[-1, 0]$ and $[2, 3]$

48. $P(x) = 4x^3 - 37x^2 + 50x + 60$, zero in $[2, 3]$ and $[7, 8]$

49. $P(x) = x^3 + 2x^2 - 22x - 8$, zero in $[-1, 0]$ and $[-6, -5]$

50. $P(x) = 2x^4 - x^3 - 21x^2 + 51x - 36$, has no real zero greater than 4

51. $P(x) = 6x^4 + 13x^3 - 11x^2 - 3x + 5$, has no zero greater than 1 or less than -3

In Exercises 52 and 53, approximate the real zeros of each polynomial function P as defined. *Give answers as decimals to the nearest tenth.*

52. $P(x) = 2x^3 - 11x^2 - 2x + 2$

53. $P(x) = x^4 - 4x^3 - 5x^2 + 14x - 15$

54. (a) Find the possible numbers of positive and negative zeros of the polynomial function P defined by $P(x) = x^3 + 3x^2 - 4x - 2$.

  (b) Show that the polynomial function P has a zero between -4 and -3. Approximate this zero to the nearest tenth.

Graph each polynomial function defined as follows.

55. $P(x) = 2x^3 - 11x^2 - 2x + 2$ (See Exercise 52.)

56. $P(x) = x^4 - 4x^3 - 5x^2 + 14x - 15$ (See Exercise 53.)

57. $P(x) = x^3 + 3x^2 - 4x - 2$ (See Exercise 54.)

58. $P(x) = 2x^4 - 3x^3 + 4x^2 + 5x - 1$

Graph each rational function defined as follows.

59. $f(x) = \dfrac{8}{x}$

60. $f(x) = \dfrac{2}{3x - 1}$

61. $f(x) = \dfrac{4x - 2}{3x + 1}$

62. $f(x) = \dfrac{6x}{(x - 1)(x + 2)}$

63. $f(x) = \dfrac{2x}{x^2 - 1}$

64. $f(x) = \dfrac{x^2 + 4}{x + 2}$

65. $f(x) = \dfrac{x^2 - 1}{x}$

66. $f(x) = \dfrac{-2}{x^2 + 1}$

67. $f(x) = \dfrac{4x^2 - 9}{2x + 3}$

68. Under what conditions will the graph of a rational function defined by an expression reduced to lowest terms have an oblique asymptote?

Graph each polynomial function P as defined in Exercises 1 and 2.

1. $P(x) = (1 - x)^4$

2. $P(x) = (x + 1)(x - 2)x$

Use synthetic division to perform each division.

3. $(3x^3 + 4x^2 - 9x + 6) \div (x + 2)$

4. $(2x^3 - 11x^2 + 28) \div (x - 5)$

5. Use synthetic division to find $P(3)$ for $P(x) = 2x^3 - 9x^2 + 4x + 8$.

6. Find a polynomial function P of lowest degree with -1, -2, and 4 as zeros.

7. Is 3 a zero of the function P defined by $P(x) = 6x^4 - 11x^3 - 35x^2 + 34x + 24$? Why or why not?

8. Is $x + 2$ a factor of $P(x) = 6x^4 - 11x^3 - 35x^2 + 34x + 24$? Why or why not?

9. Find a function P defined by a polynomial of degree 4 with real coefficients having 2, -1, and $-i$ as zeros, and with $P(3) = 80$.

10. Factor $P(x) = 2x^3 - x^2 - 13x - 6$, given that -2 is a zero of P.

11. For the polynomial function defined by $P(x) = 6x^3 - 25x^2 + 12x + 7$,
  (a) list all rational numbers that can possibly be zeros of P;
  (b) find all rational zeros of P.

12. Explain why the function P defined by the polynomial $P(x) = x^4 + x^2 + 1$ cannot have any negative zeros. (*Hint:* Look at the exponents.)

13. Show that the polynomial function P defined by $P(x) = 2x^4 - 3x^3 + 4x^2 - 5x - 1$ has no real zeros greater than 2 or less than -1.

14. Find the possible numbers of positive and negative real zeros of the polynomial function P defined by $P(x) = x^4 + 3x^3 - x^2 - 2x + 1$.

15. Graph the polynomial function defined by $P(x) = x^3 - 5x^2 + 8x - 4$.

Graph the rational functions defined as follows.

16. $f(x) = \dfrac{-2}{x + 3}$

17. $f(x) = \dfrac{3x - 1}{x - 2}$

18. $f(x) = \dfrac{x^2 - 1}{x^2 - 9}$

19. What is the equation of the oblique asymptote of the graph of $f(x) = \dfrac{2x^2 + x - 6}{x - 1}$?

20. Which one of the functions defined below has a graph with no x-intercepts?

  (a) $f(x) = (x - 2)(x + 3)^2$ 　**(b)** $f(x) = \dfrac{x + 7}{x - 2}$ 　**(c)** $f(x) = x^3 - x$ 　**(d)** $f(x) = \dfrac{1}{x^2 + 4}$

■ THE GRAPHING CALCULATOR ■

Graphing calculators can easily be used to graph polynomial and rational functions. In Example 4 in Section 6.1, we graphed the polynomial function defined by

$$P(x) = (2x + 3)(x - 1)(x + 2)$$

by using intercepts and test points in each interval formed by the x-intercepts. The graph is shown in Figure 6.11. Use a graphing calculator to graph this function, with the minimum and maximum x-values of -4 and 2, and minimum and maximum y-values of -6 and 6. Compare the result to the figure in the text.

The rational function defined by

$$f(x) = \frac{3(x + 1)(x - 2)}{(x + 4)^2}$$

was analyzed in Example 6 in Section 6.6. Program your graphing calculator to graph this function, using minimum and maximum x-values of -15 and 12, and minimum and maximum y-values of -1 and 15. Compare the display to Figure 6.23.

Graphing calculators have the capability of "zooming in" on a particular portion of the graph of a function. Zoom in on the portion of the graph of this rational function in the vicinity of the origin to see how the function changes from decreasing to increasing. These calculators also have the capability to "trace" the graph while displaying the coordinates of the points on the graph.

In calculus we are often required to find the highest and lowest points on a graph in a particular region of the domain. By experimenting with your graphing calculator, you can begin to appreciate just how useful it can be in helping us to find excellent approximations of the coordinates of these points.

You may want to use the functions defined below to experiment with your calculator. These functions have been selected from examples in the text, so you can compare your calculator graphs with those given in the example figures. The figure number of the graph in the text is given with each equation.

1. $P(x) = x^3$ (Figure 6.1)

2. $P(x) = x^6$ (Figure 6.2)

3. $P(x) = \frac{1}{2}x^3$ (Figure 6.3)

4. $P(x) = -\frac{3}{2}x^4$ (Figure 6.4)

5. $P(x) = x^5 - 2$ (Figure 6.5)

6. $P(x) = (x + 1)^6$ (Figure 6.6)

7. $P(x) = -(x - 1)^3 + 3$ (Figure 6.7)

8. $P(x) = 3x^4 + x^3 - 2x^2$ (Figure 6.12)

9. $P(x) = 8x^3 - 12x^2 + 2x + 1$ (Figure 6.16)

10. $P(x) = 3x^4 - 14x^3 + 54x - 3$ (Figure 6.17)

11. $f(x) = \frac{1}{x}$ (Figure 6.18)

12. $f(x) = -\frac{2}{x}$ (Figure 6.19)

13. $f(x) = \frac{2}{1 + x}$ (Figure 6.20)

14. $f(x) = \frac{x + 1}{(2x - 1)(x + 3)}$ (Figure 6.21)

15. $f(x) = \frac{2x + 1}{x - 3}$ (Figure 6.22)

16. $f(x) = \frac{x^2 + 1}{x - 2}$ (Figure 6.24)

17. Graph $f(x) = \frac{x^2 - 4}{x - 2}$ on your graphing calculator and compare your result with Figure 6.25.

18. On the same screen, graph $y = \frac{1}{x^2}$ and $y = -\frac{1}{x^2}$. Describe in words how they are alike and how they are different.

SYSTEMS OF EQUATIONS AND INEQUALITIES

7.1 LINEAR SYSTEMS WITH TWO VARIABLES

7.2 LINEAR SYSTEMS WITH THREE VARIABLES

7.3 SOLUTION OF LINEAR SYSTEMS BY MATRICES

7.4 NONLINEAR SYSTEMS OF EQUATIONS

7.5 SYSTEMS OF INEQUALITIES; LINEAR PROGRAMMING

350

When an applied problem requires that more than one unknown quantity must be found, it is often helpful (or, in some cases, absolutely necessary) to write several equations in several variables, and then solve this resulting *system of equations.* At the college algebra level, we usually restrict the number of unknowns to two or three. After studying this chapter, it will become clear how the methods of solving systems can be extended to a larger number of unknowns. Graphing calculators and computers are able to solve systems of equations, eliminating the need for time-consuming computations. The methods of this chapter (especially those in Section 7.3) can help the student appreciate the capabilities of today's electronic marvels.

7.1 ———— # LINEAR SYSTEMS WITH TWO VARIABLES

Many applications of mathematics require the simultaneous solution of a large number of equations or inequalities having many variables. A group of equations that place restrictions on the same variables is called a **system of equations.** The solution set of a system of equations is the intersection of the solution sets of the individual equations. It is customary to write a system by listing its equations. For example, the system of equations $2x + y = 4$ and $x - y = 6$ is written as

$$2x + y = 4$$
$$x - y = 6.$$

In general, a **first-degree equation in n unknowns** is any equation of the form

$$a_1x_1 + a_2x_2 + \cdots + a_nx_n = k,$$

where $a_1, a_2, \ldots, a_n,$ and k are constants and $x_1, x_2, \ldots, x_n$ are variables. Such equations are also called *linear equations.* Generally, only systems of linear equations with two or three variables will be discussed in this book, although the methods used can be extended to systems with more variables.

The solution set of a linear equation in two variables is an infinite set of ordered pairs. Since the graph of such an equation is a straight line, there are three possibilities for the solution set of a system of two linear equations in two variables. An example of each possibility is shown in Figure 7.1.

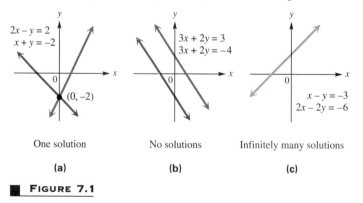

One solution | No solutions | Infinitely many solutions

(a) | (b) | (c)

■ **FIGURE 7.1**

POSSIBLE GRAPHS OF A LINEAR SYSTEM WITH TWO EQUATIONS AND TWO VARIABLES	1. The graphs of the two equations intersect in a single point. The coordinates of this point give the solution of the system. This is the most common case. See Figure 7.1(a). 2. The graphs are distinct parallel lines. In this case, the system is said to be **inconsistent.** That is, there is no solution common to both equations. The solution set of the linear system is empty. See Figure 7.1(b). 3. The graphs are the same line. In this case, the equations are said to be **dependent,** and any solution of one equation is also a solution of the other. Thus, there are infinitely many solutions. See Figure 7.1(c).

Although the *number* of solutions of a linear system can often be seen from the graph of the equations of the system, it is usually difficult to determine an exact solution from the graph. A general algebraic method of finding the solution of a system of two linear equations, called the **substitution method,** is illustrated in the following example and used again in Section 7.4.

The substitution method involves substituting an expression for one variable in terms of the other in another equation of a system. For example, to solve the system

$$3x + 2y = 11$$
$$y = x + 3$$

by substitution, replace y in the first equation with $x + 3$ to get

$$3x + 2(x + 3) = 11,$$

noticing that parentheses are required. Then solve this equation for x.

$$3x + 2x + 6 = 11$$
$$5x + 6 = 11$$
$$5x = 5$$
$$x = 1$$

Replace x with 1 in $y = x + 3$ to find that $y = 1 + 3 = 4$. The solution set for this system is $\{(1, 4)\}$.

■ *Example 1*
SOLVING A SYSTEM
BY SUBSTITUTION

Solve the system

$$2x + 5y = 21 \qquad \qquad (1)$$
$$3x - 2y = -16. \qquad \qquad (2)$$

Begin by solving one equation for one of the variables in terms of the other. For example, solving equation (1) for x gives

$$2x + 5y = 21 \qquad \qquad (1)$$
$$2x = 21 - 5y$$
$$x = \frac{21 - 5y}{2}. \qquad \qquad (3)$$

Now substitute this result for x into equation (2).

$$3x - 2y = -16 \qquad (2)$$

$$3\left(\frac{21 - 5y}{2}\right) - 2y = -16$$

To eliminate the fraction on the left, multiply both sides of the equation by 2 and then solve for y.

$$2 \cdot 3\left(\frac{21 - 5y}{2}\right) - 2 \cdot 2y = 2(-16)$$

$$3(21 - 5y) - 4y = -32$$

$$63 - 15y - 4y = -32 \qquad \text{Distributive property}$$

$$-19y = -95 \qquad \text{Combine terms; subtract 63}$$

$$y = 5 \qquad \text{Divide by } -19.$$

Substitute $y = 5$ back into equation (3) to find x.

$$x = \frac{21 - 5y}{2} \qquad (3)$$

$$x = \frac{21 - 5 \cdot 5}{2}$$

$$x = -2$$

The solution set for the system is $\{(-2, 5)\}$. Check by substituting -2 for x and 5 for y in each of the equations of the system. ■

Another method of solving systems of two equations is the **addition method.** With this method, we first multiply the equations on both sides by suitable numbers, so that when they are added, one variable is eliminated. The result is an equation in one variable that can be solved by methods from Chapter 2. The solution is then substituted into one of the original equations, making it possible to solve for the other variable. In this process the given system is replaced by new systems that have the same solution set as the original system. Systems that have the same solution set are called **equivalent systems.** The addition method is illustrated by the following examples.

■ *Example 2*

SOLVING A SYSTEM BY ADDITION

Solve the system

$$3x - 4y = 1 \qquad \textbf{(4)}$$

$$2x + 3y = 12. \qquad \textbf{(5)}$$

To eliminate x, multiply both sides of equation (4) by -2 and both sides of equation (5) by 3 to get equations (6) and (7).

$$-6x + 8y = -2 \qquad \textbf{(6)}$$

$$6x + 9y = 36 \qquad \textbf{(7)}$$

Although this new system is not the same as the given system, it will have the same solution set.

Now add the two equations to eliminate x, and then solve the result for y.

$$-6x + 8y = -2$$
$$\underline{6x + 9y = 36}$$
$$17y = 34$$
$$y = 2$$

Substitute 2 for y in equation (4) or (5). Choosing equation (4) gives

$$3x - 4(2) = 1$$
$$3x = 9$$
$$x = 3.$$

The solution set of the given system is $\{(3, 2)\}$, which can be checked by substituting 3 for x and 2 for y in equation (5). ■

Since the addition method of solution results in the elimination of one variable from the system, it is also called the **elimination method.**

■ *Example 3*

SOLVING A SYSTEM
BY ELIMINATION
(INCONSISTENT
SYSTEM)

Solve the system

$$3x - 2y = 4$$
$$-6x + 4y = 7.$$

The variable x can be eliminated by multiplying both sides of the first equation by 2 and then adding.

$$6x - 4y = 8$$
$$\underline{-6x + 4y = 7}$$
$$0 = 15 \qquad \text{False}$$

Both variables were eliminated here, leaving the false statement $0 = 15$, a signal that these two equations have no solutions in common. The system is inconsistent, and the solution set is $\emptyset$. In slope-intercept form, the equations would show that the graphs are parallel lines because they have the same slope but different y-intercepts. ■

■ *Example 4*

SOLVING A SYSTEM
BY ELIMINATION
(DEPENDENT
EQUATIONS)

Solve the system

$$-4x + y = 2$$
$$8x - 2y = -4.$$

Eliminate x by multiplying both sides of the first equation by 2 and then adding it to the second equation.

$$-8x + 2y = 4$$
$$\underline{8x - 2y = -4}$$
$$0 = 0 \qquad \text{True}$$

This true statement, $0 = 0$, indicates that a solution of one equation is also a solution of the other, so the solution set is an infinite set of ordered pairs. The graphs of the equations are the same line, since the slopes and y-intercepts are equal. The two equations are dependent.

We will write the solution of a system of dependent equations as an ordered pair by expressing x in terms of y as follows. Choose either equation and solve for x. Choosing the first equation gives

$$-4x + y = 2$$

$$x = \frac{2 - y}{-4} = \frac{y - 2}{4}.$$

We write the solution set as

$$\left\{ \left(\frac{y - 2}{4}, y \right) \right\}.$$

By selecting values for y and calculating the corresponding values for x, individual ordered pairs of the solution set can be found. For example, if $y = -2$, $x = (-2 - 2)/4 = -1$ and the ordered pair $(-1, -2)$ is a solution. ■

NOTE In Example 4 we wrote the solution set in a form with the variable y arbitrary. However, it would be acceptable to write the ordered pair with x arbitrary. In this case, the solution set would be written

$$\{(x, 4x + 2)\}.$$

By selecting values for x and solving for y in the ordered pair above, individual solutions can be found. Verify that $(-1, -2)$ is a solution.

Examples 3 and 4 suggest the following summary.

SPECIAL CASES FOR SYSTEMS

Consider the system of equations

$$a_1 x + b_1 y = c_1$$
$$a_2 x + b_2 y = c_2,$$

with $a_2 \neq 0$, $b_2 \neq 0$, and $c_2 \neq 0$. The graphs of the equations are **parallel lines** and the system is **inconsistent** if

$$\frac{a_1}{a_2} = \frac{b_1}{b_2} \neq \frac{c_1}{c_2}.$$

(See Figure 7.1(b).) The graphs of the equations are the **same line** and the equations are **dependent** if

$$\frac{a_1}{a_2} = \frac{b_1}{b_2} = \frac{c_1}{c_2}.$$

(See Figure 7.1(c).)

PROBLEM SOLVING
Many applied problems involve more than one unknown quantity. Although some problems with two unknowns can be solved using just one variable, many times it is easier to use two variables. To solve a problem using a system, determine the unknown quantities you are asked to find, and let different variables represent each of these quantities. Then write a system of equations, and solve it using one of the methods of this section. Be sure that you answer the question(s) posed in the problem, and check to see that your answer is reasonable. ■

■ *Example 5*
SOLVING AN
APPLICATION USING
A SYSTEM

The manager of a shoe store purchased 10 pairs of style 1501 shoes and 12 pairs of style 1470 shoes for $551. A later purchase of 4 pairs of style 1501 and 3 pairs of style 1470 cost $170. Find the cost per pair for each style.

A system of linear equations can be written from the information in the problem. To begin, let

$$x = \text{cost of a pair of style 1501 shoes}$$
$$y = \text{cost of a pair of style 1470 shoes.}$$

Since 10 pairs of 1501 and 12 pairs of 1470 shoes cost $551,

$$10x + 12y = 551. \tag{8}$$

Also, $\qquad\qquad\qquad 4x + 3y = 170. \tag{9}$

To solve this system by the addition method, multiply both sides of equation (9) by -4 and add the result to equation (8).

$$\begin{array}{r} 10x + 12y = 551 \\ -16x - 12y = -680 \\ \hline -6x = -129 \\ x = 21.5 \end{array}$$

Substitute 21.5 for x in equation (8) or (9). Using (8),

$$10(21.5) + 12y = 551$$
$$215 + 12y = 551$$
$$12y = 336$$
$$y = 28.$$

Style 1501 shoes cost $21.50 a pair and style 1470 shoes cost $28 a pair.
Check by using the information of the original problem: 10 pairs of style 1501 and 12 pairs of style 1470 cost

$$10(21.50) + 12(28) = 215 + 336 = 551 \text{ dollars,}$$

while 4 pairs of style 1501 and 3 pairs of style 1470 cost

$$4(21.50) + 3(28) = 86 + 84 = 170 \text{ dollars,}$$

as required. ■

356

Usually, as the price of an item goes up, demand for the item goes down and the supply of the item goes up. Changes in gasoline prices illustrate this situation. The price where supply and demand are equal is called the *equilibrium price,* and the resulting supply or demand is called the *equilibrium supply* or *equilibrium demand.*

■ ***Example 6***
SOLVING AN
APPLICATION USING
A SYSTEM

(a) Suppose that the supply of a product is related to its price by the equation

$$p = \frac{2}{3}q,$$

where p is price in dollars and q is supply in appropriate units. (Here, q stands for quantity.) Find the price for the supply levels $q = 9$ and $q = 18$.
When $q = 9$,

$$p = \frac{2}{3}q = \frac{2}{3}(9) = 6.$$

When $q = 18$,

$$p = \frac{2}{3}q = \frac{2}{3}(18) = 12.$$

(b) Suppose demand and price for the same product are related by

$$p = -\frac{1}{3}q + 18,$$

where p is price and q is demand. Find the price for the demand levels $q = 6$ and $q = 18$.
When $q = 6$,

$$p = -\frac{1}{3}q + 18 = -\frac{1}{3}(6) + 18 = 16,$$

and when $q = 18$,

$$p = -\frac{1}{3}q + 18 = -\frac{1}{3}(18) + 18 = 12.$$

(c) Graph both functions on the same axes.

Use the ordered pairs found in parts (a) and (b) and the p-intercepts to get the graphs shown in Figure 7.2.

(d) Find the equilibrium price, supply, and demand.

Solve the system

$$p = \frac{2}{3}q$$

$$p = -\frac{1}{3}q + 18$$

using substitution.

■ **FIGURE 7.2**

$$\frac{2}{3}q = -\frac{1}{3}q + 18$$

$$q = 18$$

This gives 18 units as the equilibrium supply or demand. Find the equilibrium price by substituting 18 for q in either equation. Using $p = (2/3)q$ gives

$$p = \frac{2}{3}(18) = 12$$

or \$12, the equilibrium price. The point (18, 12) that gives the equilibrium values is shown in Figure 7.2. ■

7.1 EXERCISES ■

Decide whether the systems in Exercises 1–8 have a single solution, no solution, or infinitely many solutions.

1. One line has positive slope and one line has negative slope.

2. One line has slope 0 and one line has undefined slope.

3. Both lines have slope 0 and have the same y-intercept.

4. Both lines have undefined slope and have the same x-intercept.

5. $y = 3x + 6$
$y = 3x + 9$

6. $y = 4x + 8$
$y = 4x + 12$

7. $x + y = 4$
$kx + ky = 4k$ $(k \neq 0)$

8. $x + y = 10$
$kx + ky = 9k$ $(k \neq 0)$

Use the substitution method to solve each system. Check your answers. (Hint: In Exercises 13 and 14, clear fractions first by multiplying through by the least common denominator.) See Example 1.

9. $y = 2x + 3$
$3x + 4y = 78$

10. $y = 4x - 6$
$2x + 5y = -8$

11. $3x - 2y = 12$
$5x = 4 - 2y$

12. $8x + 3y = 2$
$5x = 17 + 6y$

13. $\dfrac{x}{2} = \dfrac{7}{6} - \dfrac{y}{3}$

$\dfrac{2x}{3} = \dfrac{3y}{2} - \dfrac{7}{3}$

14. $\dfrac{3x}{4} = 4 + \dfrac{y}{2}$

$\dfrac{x}{3} + \dfrac{5y}{4} = -\dfrac{7}{6}$

15. Refer to Example 2 in this section. If we began solving the system by eliminating y, by what numbers might we have multiplied equations (4) and (5)?

16. Explain how one can determine whether a system is inconsistent or has dependent equations when using the substitution or elimination method.

Use the addition method to solve each system. Check your answers. (Hint for Exercises 29–32: Let $1/x = t$, $1/y = u$.) See Examples 2–4.

17. $5x + 3y = 7$
$7x - 3y = -19$

18. $2x + 7y = -8$
$-2x + 3y = -12$

19. $3x + 2y = 5$
$6x + 4y = 8$

20. $9x - 5y = 1$
$-18x + 10y = 1$

21. $2x - 3y = -7$
$5x + 4y = 17$

22. $4x + 3y = -1$
$2x + 5y = 3$

23. $5x + 7y = 6$
$10x - 3y = 46$

24. $12x - 5y = 9$
$3x - 8y = -18$

25. $4x - y = 9$
$-8x + 2y = -18$

26. $3x + 5y + 2 = 0$
$9x + 15y + 6 = 0$

27. $\dfrac{x}{2} + \dfrac{y}{3} = 8$

$\dfrac{2x}{3} + \dfrac{3y}{2} = 17$

28. $\dfrac{x}{5} + 3y = 31$

$2x - \dfrac{y}{5} = 8$

29. $\dfrac{2}{x} + \dfrac{1}{y} = \dfrac{3}{2}$

$\dfrac{3}{x} - \dfrac{1}{y} = 1$

30. $\dfrac{1}{x} + \dfrac{3}{y} = \dfrac{16}{5}$

$\dfrac{5}{x} + \dfrac{4}{y} = 5$

31. $\dfrac{2}{x} + \dfrac{1}{y} = 11$

$\dfrac{3}{x} - \dfrac{5}{y} = 10$

32. $\dfrac{2}{x} + \dfrac{3}{y} = 18$

$\dfrac{4}{x} - \dfrac{5}{y} = -8$

33. Find values of a and b so that the line with equation $ax + by = 5$ passes through the points $(-2, 1)$ and $(-1, -2)$.

34. Find m and b so that the line $y = mx + b$ passes through $(4, 6)$ and $(-5, -3)$.

35. Find m and b so that the line $y = mx + b$ passes through $(2, 5)$ and $(-1, 4)$.

36. Find values of a and b so that the line with equation $ax + by = 23$ passes through $(-3, 2)$ and $(-4, -5)$.

Write each problem as a system of equations and then solve. See Example 5.

37. To start a new business, Shannon borrowed money from two financial institutions. One loan was at 7% interest and the other was for one-third as much money at 8% interest. How much did she borrow at each rate if the total amount of annual interest was $1160?

38. Mark LeBeau has invested a total of $20,000 in two ways. Part of the money is in certificates of deposit paying 3.75% interest, while the rest is in municipal bonds that pay 8.2% interest. How much is there in each account if the total annual interest is $1417.50?

39. Alexis is a botanist who has patented a successful type of plant food that contains two chemicals, X and Y. Eight hundred kilograms of these chemicals will be used to make a batch of the food, and the ratio of X to Y must be 3 to 2. How much of each chemical should be used?

40. A manufacturer of portable compact disc players shipped 200 of the players to its two Quebec warehouses. It costs $3 per unit to ship to Warehouse A, and $2.50 per unit to ship to Warehouse B. If the total shipping cost was $537.50, how many were shipped to each warehouse?

41. Octane ratings show the percent of isooctane in gasoline. An octane rating of 98, for example, indicates a gasoline that is 98% isooctane. How many gallons of 98-octane gasoline should be mixed with 92-octane gasoline to produce 40 gallons of 94-octane gasoline?

42. A chemist needs 10 liters of a 24% alcohol solution. She has on hand a 30% alcohol solution and an 18% alcohol solution. How many liters of each should be mixed to get the required solution?

43. A supplier of poultry sells 20 turkeys and 8 chickens for $74. He also sells 15 turkeys and 24 chickens for $87. Find the cost of a turkey and the cost of a chicken.

44. Jose Ortega is a building contractor. If he hires 8 brick layers and 2 roofers, his daily payroll is $960, while 10 brick layers and 5 roofers require a daily payroll of $1500. What is the daily wage of a brick layer and the daily wage of a roofer?

45. During summer vacation Hector and Ann earned a total of $6496. Hector worked 8 days less than Ann and earned $4 per day less. Find the number of days he worked and the daily wage made, if the total number of days worked by both was 72.

46. A bank teller has ten-dollar bills and twenty-dollar bills. He has 25 more twenties than tens. The value of the bills is $2900. How many of each kind does he have?

47. Mike Karelius plans to invest $30,000 he won in a lottery. With part of the money he buys a mutual fund, paying 4.5% a year. The rest he invests in utility bonds paying 5% per year. The first year his investments bring a return of $1410. How much is invested at each rate?

48. How much milk that is 3% butterfat should be mixed with milk that is 18% butterfat to get 25 gal of milk that is 4.8% butterfat?

The break-even point for a company is the point where its costs equal its revenues. If both cost and revenue are expressed as linear equations, the break-even point is the solution of a linear system. In each of the following exercises, C represents the cost to produce x items, and R represents the revenue from the sale of x items. Use the substitution method to find the break-even point in each case, that is, the point where C = R.

49. $C = 1.5x + 252$
$R = 5.5x$

50. $C = 2.5x + 90$
$R = 3x$

51. $C = 20x + 10,000$
$R = 30x - 11,000$

52. $C = 4x + 125$
$R = 9x - 200$

In each of the following exercises, p is the price of an item, while q represents the supply in one equation and the demand in the other. Find the equilibrium price and the equilibrium supply/demand. See Example 6.

53. $p = 80 - \dfrac{3}{5}q$

$p = \dfrac{2}{5}q$

54. $p = 630 - \dfrac{3}{4}q$

$p = \dfrac{3}{4}q$

55. $3p = 84 - 2q$
$3p - q = 0$

56. $4p + q = 80$
$3p - 2q = 5$

57. Suppose that the demand and price for a certain model of electric can opener are related by

$$p = 16 - \frac{5}{4}q,$$

where p is price and q is demand, in appropriate units. Find the price when the demand is at the following levels.

(a) 0 units   **(b)** 4 units   **(c)** 8 units

Find the demand for the electric can opener at the following prices.

(d) \$6   **(e)** \$11   **(f)** \$16
(g) Graph $p = 16 - (5/4)q$.

Suppose that the price and supply of the item above are related by

$$p = \frac{3}{4}q,$$

where q represents the supply and p the price. Find the supply at the following prices.

(h) \$0   **(i)** \$10   **(j)** \$20
(k) Graph $p = (3/4)q$ on the same axes used for part (g).
(l) Find the equilibrium supply.
(m) Find the equilibrium price.

58. Let the supply and demand equations for banana smoothies be

$$\text{supply: } p = \frac{3}{2}q \quad \text{and} \quad \text{demand: } p = 81 - \frac{3}{4}q.$$

(a) Graph these on the same axes.
(b) Find the equilibrium demand.
(c) Find the equilibrium price.

59. Let the supply and demand equations for chocolate frozen yogurt be given by

$$\text{supply: } p = \frac{2}{5}q \quad \text{and} \quad \text{demand: } p = 100 - \frac{2}{5}q.$$

(a) Graph these on the same axes.
(b) Find the equilibrium demand.
(c) Find the equilibrium price.

60. Let the supply and demand equations for onions be given by

$$\text{supply: } p = 1.4q - .6 \quad \text{and demand: } p = -2q + 3.2.$$

(a) Graph these on the same axes.
(b) Find the equilibrium demand.
(c) Find the equilibrium price.

7.2 ─────── LINEAR SYSTEMS WITH THREE VARIABLES

A solution of a linear equation $ax + by + cz = k$ with three variables is an **ordered triple** (x, y, z). For example, $(1, 2, -4)$ is a solution of $2x + 5y - 3z = 24$. The solution set of such an equation is an infinite set of ordered triples. In geometry the graph of a linear equation in three variables is a plane in three-dimensional space. Considering the possible intersections of the planes representing three equations in three unknowns shows that the solution set of such a system may be either a single ordered triple (x, y, z), an infinite set of ordered triples (dependent equations), or the empty set (an inconsistent system). See Figure 7.3.

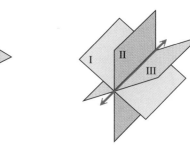

One solution

(a)

Infinitely many solutions

(b)

Infinitely many solutions

(c)

No solution

(d)

No solution

(e)

■ **FIGURE 7.3**

A system of equations with three or more variables can be solved with repeated use of the addition method, as shown in the next example.

■ *Example 1*

SOLVING A SYSTEM OF THREE EQUATIONS IN THREE VARIABLES

Solve the system

$$2x + y - z = 2 \tag{1}$$
$$x + 3y + 2z = 1 \tag{2}$$
$$x + y + z = 2. \tag{3}$$

Several steps are needed to obtain an equation in one variable. First, eliminate the same variable from each of two pairs of equations. Then eliminate another variable from the two resulting equations. For example, to eliminate x, multiply both sides of equation (2) by -2 and add to equation (1).

$$\begin{array}{r} 2x + y - z = 2 \\ -2x - 6y - 4z = -2 \\ \hline -5y - 5z = 0 \end{array} \tag{4}$$

The variable x must be eliminated again from a different pair of equations, say (2) and (3). Multiply both sides of (2) by -1, then add the result to equation (3).

$$\begin{array}{r} -x - 3y - 2z = -1 \\ x + y + z = 2 \\ \hline -2y - z = 1 \end{array} \tag{5}$$

Now solve the system formed by the two equations (4) and (5). One way to do this is to eliminate y by multiplying equation (4) by 2 and equation (5) by -5.

$$
\begin{array}{ll}
-10y - 10z = 0 & \text{2 times equation (4)} \\
\underline{10y + 5z = -5} & \text{-5 times equation (5)} \\
-5z = -5 & \\
z = 1 &
\end{array}
$$

Substitute 1 for z in equation (5) to find y. (Equation (4) could also have been used.)

$$
\begin{aligned}
-2y - z &= 1 \\
-2y - 1 &= 1 \\
-2y &= 2 \\
y &= -1
\end{aligned}
$$

Now use equation (3) and the values for y and z to find x. (Either equation (1) or (2) could also have been used.)

$$
\begin{aligned}
x + y + z &= 2 \\
x - 1 + 1 &= 2 \\
x &= 2
\end{aligned}
$$

The solution set of the system is $\{(2, -1, 1)\}$. ■

NOTE

In Example 1 we used a process that yielded a system of equations that is equivalent to the original system.

$$
\begin{aligned}
x + y + z &= 2 \quad \text{(3)} \\
-2y - z &= 1 \quad \text{(5)} \\
z &= 1
\end{aligned}
$$

Notice the "triangular" form here. While it is not absolutely necessary to eliminate x first and then y to find z, it is a good preparation for the matrix method of solving systems, introduced in the next section.

■ *Example 2*

Solving a System Having Two Equations with Three Variables

Solve the system

$$
\begin{aligned}
x + 2y + z &= 4 \qquad\qquad &\textbf{(6)} \\
3x - y - 4z &= -9. &\textbf{(7)}
\end{aligned}
$$

Geometrically, the solution is the intersection of the two planes given by equations (6) and (7). The intersection of two different nonparallel planes is a line. Thus there will be an infinite number of ordered triples in the solution set, representing the points on the line of intersection. To describe these ordered triples, proceed as follows.

To eliminate x, multiply both sides of equation (6) by -3 and add to equation (7). (Either y or z could have been eliminated instead.)

$$\begin{array}{r} -3x - 6y - 3z = -12 \\ 3x - y - 4z = -9 \\ \hline -7y - 7z = -21 \end{array} \qquad (8)$$

Now solve equation (8) for z.

$$-7y - 7z = -21$$
$$-7z = 7y - 21$$
$$z = -y + 3$$

This gives z in terms of y. Now, express x also in terms of y by solving equation (6) for x and substituting $-y + 3$ for z in the result.

$$x + 2y + z = 4$$
$$x = -2y - z + 4$$
$$x = -2y - (-y + 3) + 4$$
$$x = -y + 1$$

The system has an infinite number of solutions. For any value of y, the value of z is given by $-y + 3$ and x equals $-y + 1$. For example, if $y = 1$, then $x = -1 + 1 = 0$ and $z = -1 + 3 = 2$, giving the solution $(0, 1, 2)$. Verify that another solution is $(-1, 2, 1)$.

With y arbitrary, the solution set is of the form $\{(-y + 1, y, -y + 3)\}$. Had equation (8) been solved for y instead of z, the solution would have had a different form but would have led to the same set of solutions. In that case we would have z arbitrary, and the solution set would be of the form $\{(-2 + z, 3 - z, z)\}$. By choosing $z = 2$, one solution would be $(0, 1, 2)$, which was verified above. ■

A system like the one in Example 2 occurs when two of the equations in a system of three equations with three variables are dependent. In such a case, there are really only two equations in three variables, and Example 2 illustrates the method of solution. On the other hand, an inconsistent system is indicated by a false statement at some point in the solution, as in Example 3 of Section 7.1.

Applications with three unknowns often require solving systems of three linear equations. As an example, the equation of the parabola $y = ax^2 + bx + c$ that passes through three given points can be found by solving a system of three equations with three unknowns.

■ *Example 3*

SOLVING A
GEOMETRIC
APPLICATION USING
A SYSTEM

Find the equation of the parabola $y = ax^2 + bx + c$ that passes through $(2, 4)$, $(-1, 1)$, and $(-2, 5)$.

Since the three points lie on the graph of the equation $y = ax^2 + bx + c$, they must satisfy the equation. Substituting each ordered pair into the equation gives three equations with three variables.

$$4 = a(2)^2 + b(2) + c \qquad \text{or} \qquad 4 = 4a + 2b + c \qquad \textbf{(9)}$$
$$1 = a(-1)^2 + b(-1) + c \qquad \text{or} \qquad 1 = a - b + c \qquad \textbf{(10)}$$
$$5 = a(-2)^2 + b(-2) + c \qquad \text{or} \qquad 5 = 4a - 2b + c \qquad \textbf{(11)}$$

This system can be solved by the addition method. First eliminate c using equations (9) and (10).

$$
\begin{array}{ll}
4 = 4a + 2b + c & \\
\underline{-1 = -a + b - c} & -1 \text{ times equation (10)} \\
3 = 3a + 3b & \textbf{(12)}
\end{array}
$$

Now, use equations (10) and (11) to also eliminate c.

$$
\begin{array}{ll}
1 = a - b + c & \\
\underline{-5 = -4a + 2b - c} & -1 \text{ times equation (11)} \\
-4 = -3a + b & \textbf{(13)}
\end{array}
$$

Solve the system of equations (12) and (13) in two variables by eliminating a.

$$
\begin{array}{l}
3 = 3a + 3b \\
\underline{-4 = -3a + b} \\
-1 = 4b \\
-\dfrac{1}{4} = b
\end{array}
$$

Find a by substituting $-1/4$ for b in equation (12), which is equivalent to $1 = a + b$.

$$
\begin{array}{ll}
1 = a + b & \text{Equation (12) divided by 3} \\
1 = a - \dfrac{1}{4} & \text{Let } b = -\dfrac{1}{4}. \\
\dfrac{5}{4} = a &
\end{array}
$$

Finally, find c by substituting $a = 5/4$ and $b = -1/4$ in equation (10).

$$
\begin{array}{ll}
1 = a - b + c & \\
1 = \dfrac{5}{4} - \left(-\dfrac{1}{4}\right) + c & \quad a = \dfrac{5}{4}, b = -\dfrac{1}{4} \\
1 = \dfrac{6}{4} + c & \\
-\dfrac{1}{2} = c &
\end{array}
$$

An equation of the parabola is $y = \dfrac{5}{4}x^2 - \dfrac{1}{4}x - \dfrac{1}{2}$. ■

PROBLEM SOLVING

In Section 7.1 we discussed solving problems with systems of two equations in two unknowns. Those ideas can be extended to problems requiring three unknowns. With three unknowns, we must write a system of three equations, as shown in the next example. ■

■ *Example 4*

SOLVING AN APPLICATION USING A SYSTEM OF THREE EQUATIONS

An animal feed is made from three ingredients: corn, soybeans, and cottonseed. One unit of each ingredient provides units of protein, fat, and fiber as shown in the table below. How many units of each ingredient should be used to make a feed that contains 22 units of protein, 28 units of fat, and 18 units of fiber?

	Corn	Soybeans	Cottonseed	Total
Protein	.25	.4	.2	22
Fat	.4	.2	.3	28
Fiber	.3	.2	.1	18

Let x represent the number of units of corn, y, the number of units of soybeans, and z, the number of units of cottonseed that are required. Since the total amount of protein is to be 22 units,

$$.25x + .4y + .2z = 22.$$

Also, for the 28 units of fat,

$$.4x + .2y + .3z = 28,$$

and, for the 18 units of fiber,

$$.3x + .2y + .1z = 18.$$

Multiply the first equation on both sides by 100, and the second and third equations by 10 to get the system

$$25x + 40y + 20z = 2200$$
$$4x + 2y + 3z = 280$$
$$3x + 2y + z = 180.$$

Using the methods described earlier in this section, we can show that $x = 40$, $y = 15$, and $z = 30$. The feed should contain 40 units of corn, 15 units of soybeans, and 30 units of cottonseed to fulfill the given requirements. ■

NOTE

The table shown in Example 4 is useful in setting up the equations of the system, since the coefficients in each equation can be read from left to right. This idea is extended in the next section, where we introduce solution of systems by matrices.

Use the addition method to solve each of the following systems of three equations in three un-knowns. Check your answers. See Example 1. (Hint: In Exercises 15–18, let $t = 1/x$, $u = 1/y$, and $v = 1/z$. Solve for t, u, and v, and then for x, y, and z.)

1. $x + y + z = 2$
$2x + y - z = 5$
$x - y + z = -2$

2. $2x + y + z = 9$
$-x - y + z = 1$
$3x - y + z = 9$

3. $x + 3y + 4z = 14$
$2x - 3y + 2z = 10$
$3x - y + z = 9$

4. $4x - y + 3z = -2$
$3x + 5y - z = 15$
$-2x + y + 4z = 14$

5. $x + 2y + 3z = 8$
$3x - y + 2z = 5$
$-2x - 4y - 6z = 5$

6. $3x - 2y - 8z = 1$
$9x - 6y - 24z = -2$
$x - y + z = 1$

7. $x + 4y - z = 6$
$2x - y + z = 3$
$3x + 2y + 3z = 16$

8. $4x - 3y + z = 9$
$3x + 2y - 2z = 4$
$x - y + 3z = 5$

9. $5x + y - 3z = -6$
$2x + 3y + z = 5$
$-3x - 2y + 4z = 3$

10. $2x - 5y + 4z = -35$
$5x + 3y - z = 1$
$x + y + z = 1$

11. $x - 3y - 2z = -3$
$3x + 2y - z = 12$
$-x - y + 4z = 3$

12. $x + y + z = 3$
$3x - 3y - 4z = -1$
$x + y + 3z = 11$

13. $2x + 6y - z = 6$
$4x - 3y + 5z = -5$
$6x + 9y - 2z = 11$

14. $8x - 3y + 6z = -2$
$4x + 9y + 4z = 18$
$12x - 3y + 8z = -2$

15. $\dfrac{1}{x} + \dfrac{1}{y} - \dfrac{1}{z} = \dfrac{1}{4}$
$\dfrac{2}{x} - \dfrac{1}{y} + \dfrac{3}{z} = \dfrac{9}{4}$
$-\dfrac{1}{x} - \dfrac{2}{y} + \dfrac{4}{z} = 1$

16. $\dfrac{3}{x} + \dfrac{2}{y} - \dfrac{1}{z} = \dfrac{11}{6}$
$\dfrac{1}{x} - \dfrac{1}{y} + \dfrac{3}{z} = -\dfrac{11}{12}$
$\dfrac{2}{x} + \dfrac{1}{y} + \dfrac{1}{z} = \dfrac{7}{12}$

17. $\dfrac{2}{x} - \dfrac{2}{y} + \dfrac{1}{z} = -1$
$\dfrac{4}{x} + \dfrac{1}{y} - \dfrac{2}{z} = -9$
$\dfrac{1}{x} + \dfrac{1}{y} - \dfrac{3}{z} = -9$

18. $\dfrac{5}{x} - \dfrac{1}{y} - \dfrac{2}{z} = -6$
$-\dfrac{1}{x} + \dfrac{3}{y} - \dfrac{3}{z} = -12$
$\dfrac{2}{x} - \dfrac{1}{y} - \dfrac{1}{z} = 6$

19. Consider the linear equation in three variables $x + y + z = 4$. Find a pair of linear equations that, when considered together with the given equation, will form a system having the following.

(a) Exactly one solution   (b) No solution   (c) Infinitely many solutions

20. Refer to Example 2 in this section. Write the solution set with x arbitrary.

21. Give an example using your immediate surroundings of three planes that intersect in a single point.

22. Give an example using your immediate surroundings of three planes that intersect in a line.

Solve each of the following systems in terms of the arbitrary variable x. (Hint: Begin by eliminating either y or z.) See Example 2.

23. $x - 2y + 3z = 6$
$2x - y + 2z = 5$

24. $3x + 4y - z = 13$
$x + y + 2z = 15$

25. $5x - 4y + z = 9$
$x + y = 15$

26. $x - y + z = -6$
$4x + y + z = 7$

27. $3x - 5y - 4z = -7$
$y - z = -13$

28. $3x - 2y + z = 15$
$x + 4y - z = 11$

Solve the following problems. See Examples 3 and 4.

29. Find a, b, and c so that the graph of the equation $y = ax^2 + bx + c$ passes through the points $(2, 3)$, $(-1, 0)$, and $(-2, 2)$.

30. Find a, b, and c so that $(2, 14)$, $(0, 0)$, and $(-1, -1)$ lie on the graph of $y = ax^2 + bx + c$.

31. Find the equation of the circle $x^2 + y^2 + ax + by + c = 0$ that passes through the points $(2, 1)$, $(-1, 0)$, and $(3, 3)$.

32. Find the equation of the circle $x^2 + y^2 + ax + by + c = 0$ that passes through the points $(4, 2)$, $(-5, -2)$, and $(0, 3)$.

33. A coin collection contains a total of 29 coins, made up of cents, nickels, and quarters. The number of quarters is 8 less than the number of cents. The total face value of the coins is $1.77. How many of each denomination are there?

34. A sparkling water distributor wants to make up 300 gallons of sparkling water to sell for $6.00 per gallon. She wishes to mix three grades of water selling for $9.00, $3.00, and $4.50 per gallon, respectively. She must use twice as much of the $4.50 water as the $3.00 water. How many gallons of each should she use?

35. A glue company needs to make some glue that it can sell for $120 per barrel. It wants to use 150 barrels of glue worth $100 per barrel, along with some glue worth $150 per barrel, and glue worth $190 per barrel. It must use the same number of barrels of $150 and $190 glue. How much of the $150 and $190 glue will be needed? How many barrels of $120 glue will be produced?

36. Billy Dixon and the Topics sell three kinds of concert tickets, "up close," "middle," and "farther back." "Up close" tickets cost $6 more than "middle" tickets, while "middle" tickets cost $3 more than "farther back" tickets. Twice the cost of an "up close" ticket is $3 more than 3 times the cost of a "farther back" seat. Find the price of each kind of ticket.

37. The perimeter of a triangle is 59 inches. The longest side is 11 inches longer than the medium side, and the medium side is 3 inches more than the shortest side. Find the length of each side of the triangle.

38. The sum of the measures of the angles of any triangle is 180°. In a certain triangle, the largest angle measures 55° less than twice the medium angle, and the smallest measures 25° less than the medium angle. Find the measures of each of the three angles.

39. Sam Abo-zahrah wins $100,000 in the Louisiana state lottery. He invests part of the money in real estate with an annual return of 5% and another part in a money market account at 4.5% interest. He invests the rest, which amounts to $20,000 less than the sum of the other two parts, in certificates of deposit that pay 3.75%. If the total annual interest on the money is $4450, how much was invested at each rate?

40. Ellen Keith invests $10,000 received in an inheritance in three parts. With one part she buys mutual funds which offer a return of 4% per year. The second part, which amounts to twice the first, is used to buy government bonds paying 4.5% per year. She puts the rest into a savings account that pays 2.5% annual interest. During the first year, the total interest is $415. How much did she invest at each rate?

41. Write an application similar to the one in Exercise 33. (Do this by first starting with an answer, and writing the exercise to fit the answer.) Then solve your own problem using a system of equations.

42. Explain why your solution to the problem in Exercise 41 cannot have fractional answers. Can it have negative answers? Explain.

7.3 ———— **SOLUTION OF LINEAR SYSTEMS BY MATRICES**

The elimination method used to solve linear systems of equations in the last two sections can be streamlined to a systematic method by using *matrices* (singular, *matrix*). In this section we describe one way matrices are used to solve linear systems. Matrix methods are particularly suitable for computer solutions of large systems of equations having many unknowns.

To begin, consider a system of three equations and three unknowns such as

$$a_1 x + b_1 y + c_1 z = d_1$$
$$a_2 x + b_2 y + c_2 z = d_2$$
$$a_3 x + b_3 y + c_3 z = d_3.$$

This system can be written in an abbreviated form as

$$\begin{bmatrix} a_1 & b_1 & c_1 & d_1 \\ a_2 & b_2 & c_2 & d_2 \\ a_3 & b_3 & c_3 & d_3 \end{bmatrix}.$$

Such a rectangular array of numbers enclosed by brackets is called a **matrix.** Each number in the array is an **element** or **entry.** The constants in the last column of the matrix can be set apart from the coefficients of the variables by using a vertical line, as shown in the following **augmented matrix.** (Because the matrix of coefficients has an extra column determined by the constants of the system, the coefficient matrix is *augmented.*)

As an example, the system

$$x + 3y + 2z = 1$$
$$2x + y - z = 2$$
$$x + y + z = 2$$

has the augmented matrix

$$\begin{bmatrix} 1 & 3 & 2 & 1 \\ 2 & 1 & -1 & 2 \\ 1 & 1 & 1 & 2 \end{bmatrix}.$$

This matrix has 3 rows (horizontal) and 4 columns (vertical). To refer to a number in the matrix, use its row and column numbers. For example, the number 3 is in the first row, second column position.

The rows of this augmented matrix can be treated just like the equations of the system of linear equations. Since the augmented matrix is nothing more than a short form of the system, any transformation of the matrix that results in an equivalent system of equations can be performed. Operations that produce such transformations are given below.

MATRIX ROW TRANSFORMATIONS

For any augmented matrix of a system of linear equations, the following row transformations will result in the matrix of an equivalent system.

1. Any two rows may be interchanged.
2. The elements of any row may be multiplied by a nonzero real number.
3. Any row may be changed by adding to its elements a multiple of the corresponding elements of another row.

■ *Example 1*
USING THE ROW
TRANSFORMATIONS

(a) The first row transformation is used to change the matrix

$$\begin{bmatrix} 1 & 3 & 5 \\ 0 & 1 & 2 \\ 1 & -1 & -2 \end{bmatrix} \quad \text{to} \quad \begin{bmatrix} 0 & 1 & 2 \\ 1 & 3 & 5 \\ 1 & -1 & -2 \end{bmatrix}$$

by interchanging the first two rows.

(b) Using the second row transformation with $k = -2$ changes

$$\begin{bmatrix} 1 & 3 & 5 \\ 0 & 1 & 2 \\ 1 & -1 & -2 \end{bmatrix} \quad \text{to} \quad \begin{bmatrix} -2 & -6 & -10 \\ 0 & 1 & 2 \\ 1 & -1 & -2 \end{bmatrix},$$

where the elements of the first row of the original matrix were multiplied by -2.

(c) The third row transformation is used to change

$$\begin{bmatrix} 1 & 3 & 5 \\ 0 & 1 & 2 \\ 1 & -1 & -2 \end{bmatrix} \quad \text{to} \quad \begin{bmatrix} 0 & 4 & 7 \\ 0 & 1 & 2 \\ 1 & -1 & -2 \end{bmatrix},$$

by multiplying each element in the third row of the original matrix by -1 and adding the results to the corresponding elements in the first row of that matrix. That is, the elements in the new first row were found as follows.

$$\begin{bmatrix} 1 + 1(-1) & 3 + (-1)(-1) & 5 + (-2)(-1) \\ 0 & 1 & 2 \\ 1 & -1 & -2 \end{bmatrix} = \begin{bmatrix} 0 & 4 & 7 \\ 0 & 1 & 2 \\ 1 & -1 & -2 \end{bmatrix}$$

Rows two and three were left unchanged. ■

If the word "row" is replaced by "equation," it can be seen that the three row transformations also apply to a system of equations, so that a system of equations can be solved by transforming its corresponding matrix into the matrix of an equivalent, simpler system. The goal is a matrix in the form

$$\begin{bmatrix} 1 & 0 & | & a \\ 0 & 1 & | & b \end{bmatrix} \quad \text{or} \quad \begin{bmatrix} 1 & 0 & 0 & | & a \\ 0 & 1 & 0 & | & b \\ 0 & 0 & 1 & | & c \end{bmatrix}$$

for systems with two or three equations respectively. Notice that on the left of the vertical bar there are ones down the diagonal from upper left to lower right and zeros elsewhere in the matrices. When these matrices are rewritten as systems of equations, the values of the variables are known. The **Gauss-Jordan method** (first developed by Carl F. Gauss (1777–1855)) is a systematic way of using the matrix row transformations to change the augmented matrix of a system into the form that shows its solution. The following examples will illustrate this method.

■ *Example 2*

USING THE
GAUSS-JORDAN
METHOD

Solve the linear system

$$3x - 4y = 1$$
$$5x + 2y = 19.$$

The equations should all be in the same form, with the variable terms in the same order on the left, and the constant term on the right. Begin by writing the augmented matrix.

$$\begin{bmatrix} 3 & -4 & | & 1 \\ 5 & 2 & | & 19 \end{bmatrix}$$

The goal is to transform this augmented matrix into one in which the value of the variables will be easy to see. That is, since each column in the matrix represents the coefficients of one variable, the augmented matrix should be transformed so that it is of the form

$$\begin{bmatrix} 1 & 0 & | & k \\ 0 & 1 & | & j \end{bmatrix}$$

for real numbers k and j. Once the augmented matrix is in this form, the matrix can be rewritten as a linear system to get

$$x = k$$
$$y = j.$$

The necessary transformations are performed as follows. It is best to work in columns beginning in each column with the element that is to become 1. In the augmented matrix,

$$\begin{bmatrix} 3 & -4 & | & 1 \\ 5 & 2 & | & 19 \end{bmatrix},$$

there is a 3 in the first row, first column position. Use transformation 2, multiplying each entry in the first row by 1/3 to get a 1 in this position. (This step is abbreviated as (1/3)R1.)

$$\begin{bmatrix} 1 & -4/3 & | & 1/3 \\ 5 & 2 & | & 19 \end{bmatrix} \quad \frac{1}{3}R1$$

Get 0 in the second row, first column by multiplying each element of the first row by -5 and adding the result to the corresponding element in the second row, using transformation 3.

$$\begin{bmatrix} 1 & -4/3 & | & 1/3 \\ 0 & 26/3 & | & 52/3 \end{bmatrix} \quad -5R1 + R2$$

Get 1 in the second row, second column by multiplying each element of the second row by 3/26, using transformation 2.

$$\begin{bmatrix} 1 & -4/3 & | & 1/3 \\ 0 & 1 & | & 2 \end{bmatrix} \quad \frac{3}{26}R2$$

Finally, get 0 in the first row, second column by multiplying each element of the second row by 4/3 and adding the result to the corresponding element in the first row.

$$\begin{bmatrix} 1 & 0 & | & 3 \\ 0 & 1 & | & 2 \end{bmatrix} \qquad \tfrac{4}{3}\,R2 + R1$$

This last matrix corresponds to the system

$$x = 3$$
$$y = 2,$$

that has the solution set $\{(3, 2)\}$. This solution could have been read directly from the third column of the final matrix. ■

A linear system with three equations is solved in a similar way. Row transformations are used to get 1s down the diagonal from left to right and 0s above and below each 1.

■ *Example 3*

USING THE
GAUSS-JORDAN
METHOD

Use the Gauss-Jordan method to solve the system

$$x - y + 5z = -6$$
$$3x + 3y - z = 10$$
$$x + 3y + 2z = 5.$$

Since the system is in proper form, begin by writing the augmented matrix of the linear system.

$$\begin{bmatrix} 1 & -1 & 5 & | & -6 \\ 3 & 3 & -1 & | & 10 \\ 1 & 3 & 2 & | & 5 \end{bmatrix}$$

The final matrix is to be of the form

$$\begin{bmatrix} 1 & 0 & 0 & | & m \\ 0 & 1 & 0 & | & n \\ 0 & 0 & 1 & | & p \end{bmatrix},$$

where m, n, and p are real numbers. This final form of the matrix gives the system $x = m$, $y = n$, and $z = p$, so the solution set is $\{(m, n, p)\}$.

There is already a 1 in the first row, first column. Get a 0 in the second row of the first column by multiplying each element in the first row by -3 and adding the result to the corresponding element in the second row, using transformation 3.

$$\begin{bmatrix} 1 & -1 & 5 & | & -6 \\ 0 & 6 & -16 & | & 28 \\ 1 & 3 & 2 & | & 5 \end{bmatrix} \qquad -3R1 + R2$$

Now, to change the last element in the first column to 0, use transformation 3 and multiply each element of the first row by -1, then add the results to the corresponding elements of the third row.

$$\begin{bmatrix} 1 & -1 & 5 & | & -6 \\ 0 & 6 & -16 & | & 28 \\ 0 & 4 & -3 & | & 11 \end{bmatrix} \quad -1R1 + R3$$

The same procedure is used to transform the second and third columns. For both of these columns perform the additional step of getting 1 in the appropriate position of each column. Do this by multiplying the elements of the row by the reciprocal of the number in that position.

$$\begin{bmatrix} 1 & -1 & 5 & | & -6 \\ 0 & 1 & -8/3 & | & 14/3 \\ 0 & 4 & -3 & | & 11 \end{bmatrix} \quad \frac{1}{6}R2$$

$$\begin{bmatrix} 1 & 0 & 7/3 & | & -4/3 \\ 0 & 1 & -8/3 & | & 14/3 \\ 0 & 4 & -3 & | & 11 \end{bmatrix} \quad R2 + R1$$

$$\begin{bmatrix} 1 & 0 & 7/3 & | & -4/3 \\ 0 & 1 & -8/3 & | & 14/3 \\ 0 & 0 & 23/3 & | & -23/3 \end{bmatrix} \quad -4R2 + R3$$

$$\begin{bmatrix} 1 & 0 & 7/3 & | & -4/3 \\ 0 & 1 & -8/3 & | & 14/3 \\ 0 & 0 & 1 & | & -1 \end{bmatrix} \quad \frac{3}{23}R3$$

$$\begin{bmatrix} 1 & 0 & 0 & | & 1 \\ 0 & 1 & -8/3 & | & 14/3 \\ 0 & 0 & 1 & | & -1 \end{bmatrix} \quad -\frac{7}{3}R3 + R1$$

$$\begin{bmatrix} 1 & 0 & 0 & | & 1 \\ 0 & 1 & 0 & | & 2 \\ 0 & 0 & 1 & | & -1 \end{bmatrix} \quad \frac{8}{3}R3 + R2$$

The linear system associated with this final matrix is

$$x = 1$$
$$y = 2$$
$$z = -1,$$

and the solution set is $\{(1, 2, -1)\}$. ■

The next two examples show how to recognize inconsistent systems or systems with dependent equations when solving such systems using row transformations.

■ *Example 4*

RECOGNIZING AN INCONSISTENT SYSTEM

Use the Gauss-Jordan method to solve the system

$$x + y = 2$$
$$2x + 2y = 5.$$

Write the augmented matrix

$$\begin{bmatrix} 1 & 1 & | & 2 \\ 2 & 2 & | & 5 \end{bmatrix}.$$

Multiply the elements in the first row by -2 and add the result to the corresponding elements in the second row.

$$\begin{bmatrix} 1 & 1 & | & 2 \\ 0 & 0 & | & 1 \end{bmatrix} \quad -2R1 + R2$$

The next step would be to get a 1 in the second row, second column. Because of the zeros, it is impossible to go further. Since the second row corresponds to the equation

$$0x + 0y = 1,$$

which has no solution, the system is inconsistent, and the solution set is $\emptyset$. ■

■ *Example 5*

SOLVING A SYSTEM WITH DEPENDENT EQUATIONS

Use the Gauss-Jordan method to solve the system

$$2x - 5y + 3z = 1$$
$$x - 2y - 2z = 8.$$

Recall from Section 7.2 that a system with two equations and three variables has an infinite number of solutions. The Gauss-Jordan method can be used to give the solution with z arbitrary. Start with the augmented matrix

$$\begin{bmatrix} 2 & -5 & 3 & | & 1 \\ 1 & -2 & -2 & | & 8 \end{bmatrix}.$$

Exchange rows to get a 1 in the first row, first column position.

$$\begin{bmatrix} 1 & -2 & -2 & | & 8 \\ 2 & -5 & 3 & | & 1 \end{bmatrix}$$

Now multiply each element in the first row by -2 and add to the corresponding element in the second row.

$$\begin{bmatrix} 1 & -2 & -2 & | & 8 \\ 0 & -1 & 7 & | & -15 \end{bmatrix} \quad -2R1 + R2$$

Multiply each element in the second row by -1.

$$\begin{bmatrix} 1 & -2 & -2 & \bigm| & 8 \\ 0 & 1 & -7 & \bigm| & 15 \end{bmatrix} \quad -1R2$$

Multiply each element in the second row by 2 and add to the corresponding element in the first row.

$$\begin{bmatrix} 1 & 0 & -16 & \bigm| & 38 \\ 0 & 1 & -7 & \bigm| & 15 \end{bmatrix} \quad 2R2 + R1$$

It is not possible to go further with the Gauss-Jordan method. The equations that correspond to the final matrix are

$$x - 16z = 38 \qquad \text{and} \qquad y - 7z = 15.$$

Solve these equations for x and y, respectively.

$$x - 16z = 38 \qquad\qquad y - 7z = 15$$
$$x = 16z + 38 \qquad\qquad y = 7z + 15$$

The solution can now be written with z arbitrary, as

$$\{(16z + 38, 7z + 15, z)\}. \quad \blacksquare$$

The cases that might occur when matrix methods are used to solve a system of linear equations are summarized below.

SUMMARY OF POSSIBLE CASES USING MATRIX METHODS	**1.** If the number of rows with nonzero elements to the left of the vertical line is equal to the number of variables in the system, then the system has a single solution. **2.** If one of the rows has the form $[0 \quad 0 \ldots 0 \mid a]$ with $a \neq 0$, then the system has no solution. **3.** If there are fewer rows in the matrix containing nonzero elements than the number of variables, then there are infinitely many solutions for the system. These solutions should be given in terms of an arbitrary variable. See Example 5.

Although only examples with two variables or three variables have been shown, the Gauss-Jordan method can be extended to solve linear systems with more variables. As the number of variables increases, the process quickly becomes very tedious and the opportunity for error increases. The method, however, is suitable for use by computers and a fairly large system can be solved quickly in that way.

7.3 EXERCISES ■ ─────────────────────────────

Use the third row transformation to change each of the following matrices as indicated. See Example 1.

1. $\begin{bmatrix} 2 & 4 \\ 4 & 7 \end{bmatrix}$; -2 times row 1 added to row 2

2. $\begin{bmatrix} -1 & 4 \\ 7 & 0 \end{bmatrix}$; 7 times row 1 added to row 2

3. $\begin{bmatrix} 1 & 5 & 6 \\ -2 & 3 & -1 \\ 4 & 7 & 0 \end{bmatrix}$; 2 times row 1 added to row 2

4. $\begin{bmatrix} 2 & 5 & 6 \\ 4 & -1 & 2 \\ 3 & 7 & 1 \end{bmatrix}$; -6 times row 3 added to row 1

5. $\begin{bmatrix} -3 & 1 & -4 \\ 2 & 1 & 3 \\ -7 & 5 & 2 \end{bmatrix}$; -5 times row 2 added to row 3

6. $\begin{bmatrix} 4 & 10 & -8 \\ 7 & 4 & 3 \\ -1 & 1 & 0 \end{bmatrix}$; -4 times row 3 added to row 2

Write the augmented matrix for each of the following systems. Do not solve the system.

7. $2x + 3y = 11$
$x + 2y = 8$

8. $3x + 5y = -13$
$2x + 3y = -9$

9. $x + 5y = 6$
$x + 2y = 8$

10. $2x + 7y = 1$
$5x = -15$

11. $2x + y + z = 3$
$3x - 4y + 2z = -7$
$x + y + z = 2$

12. $4x - 2y + 3z = 4$
$3x + 5y + z = 7$
$5x - y + 4z = 7$

13. $x + y = 2$
$2y + z = -4$
$z = 2$

14. $x = 6$
$y + 2z = 2$
$x - 3z = 6$

Write the system of equations associated with each of the following augmented matrices. Do not try to solve.

15. $\begin{bmatrix} 2 & 1 & | & 1 \\ 3 & -2 & | & -9 \end{bmatrix}$

16. $\begin{bmatrix} 1 & -5 & | & -18 \\ 6 & 2 & | & 20 \end{bmatrix}$

17. $\begin{bmatrix} 1 & 0 & 0 & | & 2 \\ 0 & 1 & 0 & | & 3 \\ 0 & 0 & 1 & | & -2 \end{bmatrix}$

18. $\begin{bmatrix} 1 & 0 & 1 & | & 4 \\ 0 & 1 & 0 & | & 2 \\ 0 & 0 & 1 & | & 3 \end{bmatrix}$

19. $\begin{bmatrix} 3 & 2 & 1 & | & 1 \\ 0 & 2 & 4 & | & 22 \\ -1 & -2 & 3 & | & 15 \end{bmatrix}$

20. $\begin{bmatrix} 2 & 1 & 3 & | & 12 \\ 4 & -3 & 0 & | & 10 \\ 5 & 0 & -4 & | & -11 \end{bmatrix}$

Use the Gauss-Jordan method to solve each of the following systems of equations. See Examples 2–5.

21. $x + y = 5$
$x - y = -1$

22. $x + 2y = 5$
$2x + y = -2$

23. $x + y = -3$
$2x - 5y = -6$

24. $3x - 2y = 4$
$3x + y = -2$

25. $2x - 3y = 10$
$2x + 2y = 5$

26. $4x + y = 5$
$2x + y = 3$

27. $2x - 3y = 2$
$4x - 6y = 1$

28. $x + 2y = 1$
$2x + 4y = 3$

29. $6x - 3y = 1$
$-12x + 6y = -2$

30. $x - y = 1$
$-x + y = -1$

31. $x + y = -1$
$y + z = 4$
$x + z = 1$

32. $x - z = -3$
$y + z = 9$
$x + z = 7$

33. $x + y - z = 6$
$2x - y + z = -9$
$x - 2y + 3z = 1$

34. $x + 3y - 6z = 7$
$2x - y + 2z = 0$
$x + y + 2z = -1$

35. $-x + y = -1$
$y - z = 6$
$x + z = -1$

36. $x + y = 1$
$2x - z = 0$
$y + 2z = -2$

37. $2x - y + 3z = 0$
$x + 2y - z = 5$
$2y + z = 1$

38. $4x + 2y - 3z = 6$
$x - 4y + z = -4$
$-x + 2z = 2$

39. Compare the use of an augmented matrix as a shorthand way of writing a system of linear equations and the use of synthetic division as a shorthand way to divide polynomials.

40. Compare the use of the third row transformation on a matrix and the elimination method of solving a system of linear equations.

Use the Gauss-Jordan method to solve each of the following systems.

41.
$$3x + 2y - w = 0$$
$$2x + z + 2w = 5$$
$$x + 2y - z = -2$$
$$2x - y + z + w = 2$$

42.
$$x + 3y - 2z - w = 9$$
$$4x + y + z + 2w = 2$$
$$-3x - y + z - w = -5$$
$$x - y - 3z - 2w = 2$$

Solve each of the following problems by first setting up a system of equations. Use the Gauss-Jordan method.

43. George Esquibel deposits some money in a bank account paying 3% per year. He uses some additional money, amounting to 1/3 the amount placed in the bank, to buy bonds paying 4% per year. With the balance of his funds he buys a 4.5% certificate of deposit. The first year his investments bring a return of $400. If the total of the investments is $10,000, how much is invested at each rate?

44. To get necessary funds for a planned expansion, a small company took out three loans totaling $25,000. The company was able to borrow some of the money at 8%. It borrowed $2000 more than 1/2 the amount of the 8% loan at 10%, and the rest at 9%. The total annual interest was $2220. How much did the company borrow at each rate?

45. In one day, a service station sold 400 gallons of premium gasoline, 150 gallons of regular gasoline, and 130 gallons of super gasoline for a total of $909. The next day 380 gallons of premium, 170 gallons of regular, and 150 gallons of super were sold for $931. The difference in price per gallon between super and regular is one-half the difference in price per gallon between premium and regular. How much does this station charge for each type of gasoline?

46. A biologist has three salt solutions: some 5% solution, some 15% solution, and some 25% solution. She needs to mix some of each to get 50 liters of 20% solution. She wants to use twice as much of the 5% solution as the 15% solution. How much of each solution should she use?

Solve each of the systems in Exercises 47–52 by the Gauss-Jordan method. Let z be the arbitrary variable if necessary. See Example 5.

47.
$$x - 3y + 2z = 10$$
$$2x - y - z = 8$$

48.
$$3x + y - z = 12$$
$$x + 2y + z = 10$$

49.
$$x + 2y - z = 0$$
$$3x - y + z = 6$$
$$-2x - 4y + 2z = 0$$

50.
$$3x + 5y - z = 0$$
$$4x - y + 2z = 1$$
$$-6x - 10y + 2z = 0$$

51.
$$x - 2y + z = 5$$
$$-2x + 4y - 2z = 2$$
$$2x + y - z = 2$$

52.
$$3x + 6y - 3z = 12$$
$$-x - 2y + z = 16$$
$$x + y - 2z = 20$$

53. At rush hours, substantial traffic is encountered at the traffic intersections shown in the figure. The city wishes to improve the signals at these corners so as to speed the flow of traffic. The traffic engineers first gather data. As the figure shows, 700 cars per hour come down M Street to intersection A; 300 cars per hour come

to intersection A on 10th Street. A total of x_1 of these cars leave A on M Street, while x_4 cars leave A on 10th Street. The number of cars entering A must equal the number leaving, so that

$$x_1 + x_4 = 700 + 300$$
or
$$x_1 + x_4 = 1000. \qquad \textbf{(1)}$$

For intersection B, x_1 cars enter B on M Street, and x_2 cars enter B on 11th Street. The figure shows that 900 cars leave B on 11th while 200 leave on M. We have

$$x_1 + x_2 = 900 + 200$$
or
$$x_1 + x_2 = 1100. \qquad \textbf{(2)}$$

At intersection C, 400 cars enter on N Street, 300 on 11th Street, while x_2 leave on 11th Street and x_3 leave on N Street. This gives

$$x_2 + x_3 = 400 + 300$$

or $\quad x_2 + x_3 = 700.$ **(3)**

Finally, intersection D has x_3 cars entering on N and x_4 entering on 10th. There are 400 leaving D on 10th and 200 leaving on N.

(a) Set up an equation for intersection D.

(b) Use equations (1), (2), (3), and your answer to part (a) to set up a system of equations. Solve the system by the Gauss-Jordan method.

(c) Since you got a row of zeros, the system of equations is dependent and does not have a unique solu-

tion. Solve equation (1) for x_1 and substitute the result into equation (2).

(d) Solve equation (1) for x_4. What is the largest possible value of x_1 so that x_4 is not negative?

(e) Your answer to part (c) should be $x_2 - x_4 = 100$. Solve the equation for x_4 and then find the smallest possible value of x_2 so that x_4 is not negative.

(f) Your answer to part (a) should be $x_3 + x_4 = 600$. Solve the equation for x_4 and then find the largest possible values of x_3 and x_4 so that neither is negative.

(g) From your answers to parts (d)–(f), what is the maximum value of x_4 so that all the equations are satisfied and all variables are nonnegative? Of x_3? Of x_2? Of x_1?

7.4 ———————— ## NONLINEAR SYSTEMS OF EQUATIONS

A **nonlinear system of equations** is one in which at least one of the equations is not a first-degree equation. Since nonlinear systems vary, depending upon the type of equations in the system, different solution methods are required for different systems. Although the substitution method discussed in Section 7.1 can almost always be used, it may not be the most efficient approach to the solution. This section illustrates some of the methods best suited to solving certain types of nonlinear systems.

It is often helpful to visualize the types of graphs involved in a nonlinear system to get an idea of the possible numbers of ordered pairs of real numbers that may be in the solution set of the system. (Graphs of nonlinear equations were studied in Chapters 3 and 4.) For example, a line and a parabola may have 0, 1, or 2 points of intersection, as shown in Figure 7.4. A parabola and an ellipse may have 0, 1, 2, 3, or 4 points of intersection, as shown in Figure 7.5.

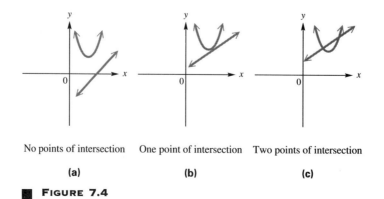

No points of intersection   One point of intersection   Two points of intersection

(a)   **(b)**   **(c)**

■ **FIGURE 7.4**

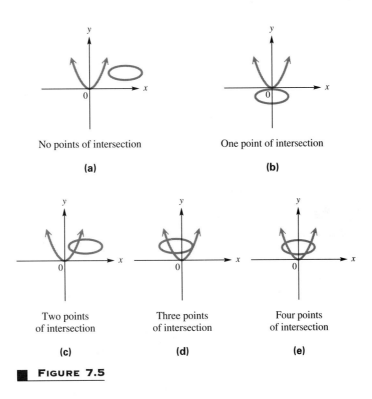

No points of intersection

(a)

One point of intersection

(b)

Two points
of intersection

(c)

Three points
of intersection

(d)

Four points
of intersection

(e)

■ **Figure 7.5**

Some nonlinear systems can be solved by the addition method. This method works well if we can eliminate completely one variable from the system, as shown in the following example.

■ *Example 1*

**Solving a
Nonlinear System
by Addition**

Solve the system

$$x^2 + y^2 = 4 \tag{1}$$
$$2x^2 - y^2 = 8. \tag{2}$$

The graph of equation (1) is a circle and that of equation (2) is a hyperbola. Visualizing these suggests that there may be 0, 1, 2, 3, or 4 points of intersection.

Add the two equations to eliminate y^2.

$$\begin{aligned} x^2 + y^2 &= 4 \\ 2x^2 - y^2 &= 8 \\ \hline 3x^2 \quad\;\; &= 12 \\ x^2 &= 4 \end{aligned}$$

$$x = 2 \quad \text{or} \quad x = -2$$

Substituting into equation (1) gives the corresponding values of y.

$$2^2 + y^2 = 4 \qquad \text{or} \qquad (-2)^2 + y^2 = 4$$
$$y^2 = 0 \qquad\qquad\qquad y^2 = 0$$
$$y = 0 \qquad\qquad\qquad y = 0$$

The solution set of the system is $\{(2, 0), (-2, 0)\}$. Check these solutions in the original system. The graph shown in Figure 7.6 also verifies the result. ■

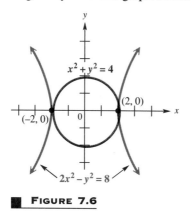

■ **FIGURE 7.6**

The substitution method is particularly useful in solving a nonlinear system that includes a linear equation. This method is illustrated by the next example.

■ *Example 2*
SOLVING A
NONLINEAR SYSTEM
BY SUBSTITUTION

Use the substitution method to solve the system

$$3x^2 - 2y = 5 \qquad (3)$$
$$x + 3y = -4. \qquad (4)$$

The graph of equation (3) is a parabola, and that of equation (4) is a line. There may be 0, 1, or 2 points of intersection of these graphs.

Although either equation could be solved for either variable, it is best to use the linear equation. Here, it is simpler to solve equation (4) for y, since x is squared in equation (3).

$$x + 3y = -4$$
$$3y = -4 - x$$
$$\frac{-4 - x}{3} \qquad (5)$$

Substituting this value of y into equation (3) gives

$$3x^2 - 2\left(\frac{-4 - x}{3}\right) = 5.$$

Multiply both sides by 3 to eliminate the denominator.

$$9x^2 - 2(-4 - x) = 15$$
$$9x^2 + 8 + 2x = 15$$
$$9x^2 + 2x - 7 = 0$$
$$(9x - 7)(x + 1) = 0$$

$$x = \frac{7}{9} \quad \text{or} \quad x = -1$$

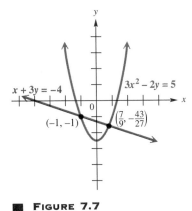

■ **FIGURE 7.7**

Substitute both values of x into equation (5) to find the corresponding y-values.

$$y = \frac{-4 - \left(\dfrac{7}{9}\right)}{3} \quad \text{or} \quad y = \frac{-4 - (-1)}{3}$$

$$y = -\frac{43}{27} \qquad\qquad y = -1$$

Check in the original system that the solution set is $\{(7/9, -43/27), (-1, -1)\}$. Figure 7.7 shows the graphs of the equations in the system. ■

Sometimes a combination of the elimination method and the substitution method is effective in solving a system, as illustrated in Example 3.

■ ***Example 3***

SOLVING A NONLINEAR SYSTEM BY A COMBINATION OF METHODS

Solve the system

$$x^2 + 3xy + y^2 = 22 \tag{6}$$
$$x^2 - xy + y^2 = 6. \tag{7}$$

Multiply both sides of equation (7) by -1, and then add the result to equation (6), as follows.

$$4xy = 16 \tag{8}$$

Now solve equation (8) for either x or y and substitute the result into one of the given equations. Solving for y gives

$$y = \frac{4}{x}, \quad \text{if } x \neq 0. \tag{9}$$

(The restriction $x \neq 0$ is included since if $x = 0$ there is no value of y that satisfies the system.) Substituting for y in equation (7) (equation (6) could have been used) and simplifying gives

$$x^2 - x\left(\frac{4}{x}\right) + \left(\frac{4}{x}\right)^2 = 6.$$

Now solve for x.

$$x^2 - 4 + \frac{16}{x^2} = 6$$

$$x^4 - 4x^2 + 16 = 6x^2$$

$$x^4 - 10x^2 + 16 = 0$$

$$(x^2 - 2)(x^2 - 8) = 0$$

$$x^2 = 2 \quad \text{or} \quad x^2 = 8$$

$$x = \sqrt{2} \quad \text{or} \quad x = -\sqrt{2} \quad \text{or} \quad x = 2\sqrt{2} \quad \text{or} \quad x = -2\sqrt{2}$$

Substitute these x values into equation (9) to find corresponding values of y.

$$\text{If } x = \sqrt{2}, y = \frac{4}{\sqrt{2}} = 2\sqrt{2}.$$

$$\text{If } x = -\sqrt{2}, y = \frac{4}{-\sqrt{2}} = -2\sqrt{2}.$$

$$\text{If } x = 2\sqrt{2}, y = \frac{4}{2\sqrt{2}} = \sqrt{2}.$$

$$\text{If } x = -2\sqrt{2}, y = \frac{4}{-2\sqrt{2}} = -\sqrt{2}.$$

The solution set of the system is:

$$\{(\sqrt{2}, 2\sqrt{2}), (-\sqrt{2}, -2\sqrt{2}), (2\sqrt{2}, \sqrt{2}), (-2\sqrt{2}, -\sqrt{2})\}.$$

Verify these solutions by substitution in the original system. ■

NOTE   Since we have not graphed equations of the type found in Example 3, it was not possible (at this level) to analyze the possible numbers of points of intersection. This step is not an essential part of the solution process, but it is useful when applicable.

■ *Example 4*

**SOLVING A
NONLINEAR SYSTEM
WITH AN ABSOLUTE
VALUE EQUATION**

Solve the system

$$x^2 + y^2 = 16 \tag{10}$$

$$|x| + y = 4. \tag{11}$$

The substitution method is required here. Equation (11) can be rewritten as $|x| = 4 - y$, then the definition of absolute value can be used to get

$$x = 4 - y \quad \text{or} \quad x = -(4 - y) = y - 4. \tag{12}$$

(Since $|x| \geq 0$ for all real x, $4 - y \geq 0$, or $4 \geq y$.) Substituting from either part of equation (12) into equation (10) gives the same result.

$$(4 - y)^2 + y^2 = 16 \quad \text{or} \quad (y - 4)^2 + y^2 = 16$$

Since $(4 - y)^2 = (y - 4)^2 = 16 - 8y + y^2$, either equation becomes

$$(16 - 8y + y^2) + y^2 = 16$$
$$2y^2 - 8y = 0$$
$$2y(y - 4) = 0$$
$$y = 0 \quad \text{or} \quad y = 4.$$

From equation (12),

If $y = 0$, then $x = 4 - 0$ or $x = 0 - 4$.
$$x = 4 \qquad\qquad x = -4$$

If $y = 4$, then $x = 4 - 4 = 0$.

The solution set, $\{(4, 0), (-4, 0), (0, 4)\}$, includes the points of intersection shown in Figure 7.8. Be sure to check the solutions in the original system. ■

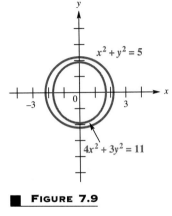

■ **FIGURE 7.8**

Nonlinear systems sometimes lead to solutions that are imaginary numbers.

■ *Example 5*

SOLVING A
NONLINEAR SYSTEM
WITH IMAGINARY
NUMBERS IN ITS
SOLUTIONS

Solve the system

$$x^2 + y^2 = 5 \tag{13}$$
$$4x^2 + 3y^2 = 11. \tag{14}$$

A circle and an ellipse (equations (13) and (14), respectively) may intersect in 0, 1, 2, 3, or 4 points.

Multiplying equation (13) on both sides by -3 and adding the result to equation (14) gives

$$-3x^2 - 3y^2 = -15$$
$$\underline{4x^2 + 3y^2 = 11}$$
$$x^2 = -4.$$

By the square root property,

$$x = \pm\sqrt{-4}$$
$$x = 2i \quad \text{or} \quad x = -2i.$$

Find y by substitution. Using equation (13) gives

$$-4 + y^2 = 5$$
$$y^2 = 9$$
$$y = 3 \quad \text{or} \quad y = -3,$$

for either $x = 2i$ or $x = -2i$.

Checking the solutions in the given system shows that the solution set is $\{(2i, 3), (2i, -3), (-2i, 3), (-2i, -3)\}$. As the graph in Figure 7.9 suggests, imaginary solutions may occur when the graphs of the equations do not intersect. ■

■ **FIGURE 7.9**

For each of Exercises 1–8, draw a sketch of the two graphs described, with the indicated number of points of intersection. (There may be more than one way to do this.) See Figures 7.4 and 7.5.

1. a line and an ellipse; two points

2. a line and a hyperbola; one point

3. a circle and an ellipse; one point

4. a circle and a hyperbola; two points

5. a line and a hyperbola; two points

6. a circle and an ellipse; four points

7. a parabola and an ellipse; four points

8. a parabola and a hyperbola; four points

9. In Example 3, we used a combination of methods to get the equation

$$x^4 - 10x^2 + 16 = 0$$

in order to find the solutions of the system. Explain why this equation cannot possibly have more than four solutions.

10. In Example 5, there are four solutions to the system, but there are no points of intersection of the graphs. When will the number of solutions of a system be less than the number of points of intersection of the graphs of the equations of the system?

Give all solutions of the following nonlinear systems of equations, including those with imaginary values. See Examples 1–5.

11. $y = x^2$
$x + y = 2$

12. $y = -x^2 + 2$
$x - y = 0$

13. $y = (x - 1)^2$
$x - 3y = -1$

14. $y = (x + 3)^2$
$x + 2y = -2$

15. $y = x^2 + 4x$
$2x - y = -8$

16. $y = 6x + x^2$
$3x - 2y = 10$

17. $3x^2 + 2y^2 = 5$
$x - y = -2$

18. $x^2 + y^2 = 5$
$-3x + 4y = 2$

19. $x^2 + y^2 = 8$
$x^2 - y^2 = 0$

20. $x^2 + y^2 = 10$
$2x^2 - y^2 = 17$

21. $5x^2 - y^2 = 0$
$3x^2 + 4y^2 = 0$

22. $x^2 + y^2 = 4$
$2x^2 - 3y^2 = -12$

23. $3x^2 + y^2 = 3$
$4x^2 + 5y^2 = 26$

24. $x^2 + 2y^2 = 9$
$3x^2 - 4y^2 = 27$

25. $2x^2 + 3y^2 = 5$
$3x^2 - 4y^2 = -1$

26. $3x^2 + 5y^2 = 17$
$2x^2 - 3y^2 = 5$

27. $2x^2 + 2y^2 = 20$
$3x^2 + 3y^2 = 30$

28. $x^2 + y^2 = 4$
$5x^2 + 5y^2 = 28$

29. $9x^2 + 4y^2 = 1$
$x^2 + y^2 = 1$

30. $2x^2 - 3y^2 = 8$
$6x^2 + 5y^2 = 24$

31. $xy = -15$
$4x + 3y = 3$

32. $xy = 8$
$3x + 2y = -16$

33. $2xy + 1 = 0$
$x + 16y = 2$

34. $-5xy + 2 = 0$
$x - 15y = 5$

35. $x^2 + 4y^2 = 25$
$xy = 6$

36. $5x^2 - 2y^2 = 6$
$xy = 2$

37. $x^2 + 2xy - y^2 = 14$
$x^2 - y^2 = -16$

38. $3x^2 + xy + 3y^2 = 7$
$x^2 + y^2 = 2$

39. $x^2 - xy + y^2 = 5$
$2x^2 + xy - y^2 = 10$

40. $3x^2 + 2xy - y^2 = 9$
$x^2 - xy + y^2 = 9$

41. $x = |y|$
$x^2 + y^2 = 18$

42. $2x + |y| = 4$
$x^2 + y^2 = 5$

43. $y = |x - 1|$
$y = x^2 - 4$

44. $2x^2 - y^2 = 4$
$|x| = |y|$

Solve each problem using a system of equations in two variables.

45. Find two numbers whose ratio is 9 to 2 and whose product is 162.

46. Find two numbers whose ratio is 4 to 3 such that the sum of their squares is 100.

47. Does the straight line $3x - 2y = 9$ intersect the circle $x^2 + y^2 = 25$? (To find out, solve the system made up of these two equations.)

48. Do the parabola $y = x^2 + 4$ and the ellipse $2x^2 + y^2 - 4x - 4y = 0$ have any points in common?

49. Find the equation of the straight line through $(2, 4)$ that touches the parabola $y = x^2$ at only one point. (*Note:* Recall that a quadratic equation has a unique solution when the discriminant is 0.)

50. For what value of b will the line $x + 2y = b$ touch the circle $x^2 + y^2 = 9$ in only one point?

51. For what nonzero values of a do the graphs of $x^2 + y^2 = 25$ and $x^2/a^2 + y^2/25 = 1$ have exactly two points in common?

52. Find the equation of the line passing through the points of intersection of the graphs of $y = x^2$ and $x^2 + y^2 = 90$.

53. Suppose that you are given the equations of two circles that are known to intersect in exactly two points. Explain how you would find the equation of the only chord common to these circles.

54. In electronics, circuit gain is given by

$$G = \frac{Bt}{R + R_t}$$

where R is the value of a resistor, t is temperature, and B is a constant. The sensitivity of the circuit to temperature is given by

$$S = \frac{BR}{(R + R_t)^2}$$

If $B = 3.7$ and t is 90K (Kelvin), find the values of R and R_t that will make $G = .4$ and $S = .001$.

7.5 SYSTEMS OF INEQUALITIES; LINEAR PROGRAMMING

As shown in Chapter 3, the solution set of an inequality in two variables is usually an infinite set whose graph is one or more regions of the coordinate plane. The solution set of a **system of inequalities,** such as

$$x + y < 3$$
$$x^2 < 2y,$$

is the intersection of the solution sets of its members. The solution is best visualized by its graph. To graph the solution set of the system, graph both inequalities (using the method described in Section 3.7) on the same axes and identify the solution by shading heavily the region common to both graphs.

■ *Example 1*

GRAPHING A
SYSTEM OF
INEQUALITIES

Graph the solution set of the system

$$x + y < 3$$
$$x^2 < 2y.$$

Figure 7.10 shows the graph of $x + y < 3$, and Figure 7.11 shows the graph of $x^2 < 2y$. The solution set of the system consists of all points in the intersection of the two regions, as shown in Figure 7.12. Because the points on the boundaries of $x + y < 3$ and $x^2 < 2y$ do not belong to the graph of the solution, the boundaries are dashed. ■

■ **FIGURE 7.10**

■ **FIGURE 7.11**

■ **FIGURE 7.12**

NOTE

While we illustrated three graphs in the solution of Example 1, in practice it is customary to give only the final graph (Figure 7.12). The two individual inequalities were shown simply to illustrate the procedure.

■ *Example 2*

GRAPHING A
SYSTEM OF
INEQUALITIES

Graph the solution set of the system

$$y \geq 2^x$$
$$9x^2 + 4y^2 \leq 36$$
$$2x + y < 1.$$

Graph the three inequalities on the same axes and shade the region common to all three, as shown in Figure 7.13. Two boundary lines are solid and one is dashed. The points where the dashed line intersects the solid lines are shown as open circles, since they do not belong to the solution set. ■

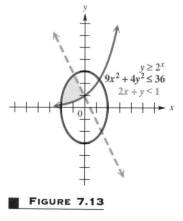

■ **FIGURE 7.13**

■ *Example 3*

GRAPHING A SYSTEM OF SEVERAL LINEAR INEQUALITIES

Graph the solution set of the system

$$2x + 3y \geq 12$$
$$7x + 4y \geq 28$$
$$y \leq 6$$
$$x \leq 5.$$

The graph is obtained by graphing the four inequalities on the same axes and shading the region common to all four as shown in Figure 7.14. As the graph shows, the boundary lines are all solid. ■

$$2x + 3y \geq 12$$
$$7x + 4y \geq 28$$
$$y \leq 6$$
$$x \leq 5$$

■ **FIGURE 7.14**

LINEAR PROGRAMMING   The type of system shown in Example 3 is used in many applications from business and social science in conjunction with a procedure known as **linear programming.**

PROBLEM SOLVING

This type of system can be used to find such things as minimum cost and maximum profit. The basic ideas of this technique are shown in the following example. ■

The Smith Company makes two products, tape decks and amplifiers. Each tape deck gives a profit of $3, while each amplifier produces $7 profit. The company must manufacture at least one tape deck per day to satisfy one of its customers, but no more than five because of production problems. Also, the number of amplifiers produced cannot exceed six per day. As a further requirement, the number of tape decks cannot exceed the number of amplifiers. How many of each should the company manufacture in order to obtain the maximum profit?

To begin, translate the statement of the problem into symbols by assuming

$x =$ number of tape decks to be produced daily

$y =$ number of amplifiers to be produced daily.

According to the statement of the problem given above, the company must produce at least one tape deck (one or more), so

$$x \geq 1.$$

Since no more than 5 tape decks may be produced,

$$x \leq 5.$$

Since no more than 6 amplifiers may be made in one day,

$$y \leq 6.$$

The requirement that the number of tape decks may not exceed the number of amplifiers translates as

$$x \leq y.$$

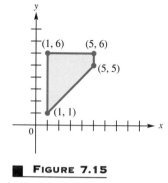

■ **FIGURE 7.15**

The number of tape decks and of amplifiers cannot be negative, so

$$x \geq 0 \quad \text{and} \quad y \geq 0.$$

These restrictions, or **constraints,** that are placed on production form the system of inequalities

$$x \geq 1, \quad x \leq 5, \quad y \leq 6, \quad x \leq y, \quad x \geq 0, \quad y \geq 0.$$

The maximum possible profit that the company can make, subject to these constraints, is found by sketching the graph of the solution of the system. See Figure 7.15. The only feasible values of x and y are those that satisfy all constraints. These values correspond to points that lie on the boundary or in the shaded region, called the **region of feasible solutions.**

Since each tape deck gives a profit of $3, the daily profit from the production of x tape decks is $3x$ dollars. Also, the profit from the production of y amplifiers will be $7y$ dollars per day. The total daily profit is thus given by the following **objective function:**

$$\text{Profit} = 3x + 7y.$$

The problem of the Smith Company may now be stated as follows: find values of x and y in the region of feasible solutions as shown in Figure 7.15 that will produce the maximum possible value of $3x + 7y$.

It can be shown that any optimum value (maximum or minimum) will always occur at a **vertex** (or **corner point**) of the region of feasible solutions. Locate the point (x, y) that gives the maximum profit by checking the coordinates of the vertex points, shown in Figure 7.15 and listed below. Find the profit that corresponds to each coordinate pair and choose the one that gives the maximum profit.

Point	Profit $= 3x + 7y$
(1, 1)	$3(1) + 7(1) = 10$
(1, 6)	$3(1) + 7(6) = 45$
(5, 6)	$3(5) + 7(6) = 57$ ← Maximum
(5, 5)	$3(5) + 7(5) = 50$

The maximum profit of $57 is obtained when 5 tape decks and 6 amplifiers are produced each day.

■ *Example 4*
SOLVING A
PROBLEM USING
LINEAR
PROGRAMMING

Robin, who is ill, takes vitamin pills. Each day, she must have at least 16 units of vitamin A, at least 5 units of vitamin B_1, and at least 20 units of vitamin C. She can choose between red pills, costing 10¢ each, which contain 8 units of A, 1 of B_1, and 2 of C, and blue pills, costing 20¢ each, which contain 2 units of A, 1 of B_1, and 7 of C. How many of each pill should she buy in order to minimize her cost yet fulfill her daily requirements?

Let x represent the number of red pills to buy, and let y represent the number of blue pills to buy. Then the cost in pennies per day is given by

$$\text{Cost} = 10x + 20y.$$

Since Robin buys x of the 10¢ pills and y of the 20¢ pills, she gets vitamin A as follows: 8 units from each red pill and 2 units from each blue pill. Altogether, she gets $8x + 2y$ units of A per day. Since she needs at least 16 units,

$$8x + 2y \geq 16.$$

Each red pill or each blue pill supplies 1 unit of vitamin B_1. Robin needs at least 5 units per day, so

$$x + y \geq 5.$$

For vitamin C, the inequality is

$$2x + 7y \geq 20.$$

Also, $x \geq 0$ and $y \geq 0$, since Robin cannot buy negative numbers of the pills.

The total cost of the pills can be minimized by finding the solution of the system of inequalities formed by the constraints. (See Figure 7.16.) The solution to this minimizing problem will also occur at a vertex point. Check the coordinates of the vertex points in the cost function to find the lowest cost.

■ **FIGURE 7.16**

Point	Cost = $10x + 20y$	
(10, 0)	$10(10) + 20(0) = 100$	
(3, 2)	$10(3) + 20(2) = 70$	← Minimum
(1, 4)	$10(1) + 20(4) = 90$	
(0, 8)	$10(0) + 20(8) = 160$	

Robin's solution is to buy 3 red pills and 2 blue ones, for a total cost of 70¢ per day. She receives minimum amounts of vitamins B_1 and C but an excess of vitamin A. Even with an excess of A, this is still the best buy. ■

PROBLEM SOLVING

To solve a linear programming problem in general, use the following steps.

SOLVING A LINEAR PROGRAMMING PROBLEM	**1.** Write the objective function and all necessary constraints. **2.** Graph the feasible region. **3.** Identify all vertices or corner points. **4.** Find the value of the objective function at each vertex. **5.** The solution is given by the vertex producing the optimum value of the objective function. ■

■ **IN SIMPLEST TERMS**

21 Linear programming was used during the Berlin Airlift after WW II to determine which combination of goods to pack on each plane. Many different constraints were involved, including volume and weight.

Suppose we want to ship food and clothing to hurricane victims in Mexico. Commercial carriers have volunteered to transport the packages, provided they fit in the available cargo space. Each 20-cubic-foot box of food weighs 40 pounds and each 30-cubic-foot box of clothing weighs 10 pounds. The total weight cannot exceed 16,000 pounds and the total volume must be less than 18,000 cubic feet. Each carton of food will feed 10 people, while each carton of clothing will help 8 people.

Let

$$F = \text{the number of cartons of food to send, and}$$
$$C = \text{the number of cartons of clothing to send.}$$

We have the following inequalities:

$$40F + 10C \leq 16{,}000$$
$$20F + 30C \leq 18{,}000.$$

We want to maximize the number of people we can help, $10F + 8C$. With C on the vertical axis, the corners of the feasible region are (0, 600), (300, 400), and (400, 0).

$$10(0) + 8(600) = 4800$$
$$10(300) + 8(400) = 6200$$
$$10(400) + 8(0) = 4000$$

We should send 300 cartons of food and 400 cartons of clothes.

SOLVE EACH PROBLEM

A. Earthquake victims in China need medical supplies and bottled water. Each medical kit measures 1 cubic foot and weighs 10 pounds. Each container of water is also 1 cubic foot but weighs 20 pounds. The plane can only carry 80,000 pounds with a total volume of 6000 cubic feet. Each medical kit will aid 4 people, while each container of water will serve 10 people. How many of each should be sent?

B. If each medical kit could aid 6 people instead of 4, how would the results above change?

A. Answers   B. Ship no medical kits and 4000 containers of water.   C. Ship 4000 medical kits and 2000 containers of water.

Graph the solution set of each system of inequalities. See Examples 1–3.

1. $x + y \leq 4$
$x - 2y \geq 6$

2. $2x + y > 2$
$x - 3y < 6$

3. $4x + 3y < 12$
$y + 4x > -4$

4. $3x + 5y \leq 15$
$x - 3y \geq 9$

5. $x + y \leq 6$
$2x + 2y \geq 12$

6. $3x + 4y < 15$
$6x + 8y > 30$

7. $x + 2y \leq 4$
$y \geq x^2 - 1$

8. $4x - 3y \leq 12$
$y \leq x^2$

9. $y \leq -x^2$
$y \geq x^2 - 6$

10. $x^2 + y^2 \leq 9$
$x \leq -y^2$

11. $x^2 - y^2 < 1$
$-1 < y < 1$

12. $x^2 + y^2 \leq 36$
$-4 \leq x \leq 4$

13. $2x^2 - y^2 > 4$
$2y^2 - x^2 > 4$

14. $y \geq x^2 + 4x + 4$
$y < -x^2$

15. $\dfrac{x^2}{16} + \dfrac{y^2}{9} \leq 1$
$\dfrac{x^2}{4} - \dfrac{y^2}{16} \geq 1$

16. $\dfrac{x^2}{36} - \dfrac{y^2}{9} \geq 1$
$\dfrac{x^2}{81} + y^2 \leq 1$

17. $x + y \leq 4$
$x - y \leq 5$
$4x + y \leq -4$

18. $3x - 2y \geq 6$
$x + y \leq -5$
$y \leq 4$

19. $2y + x \leq -5$
$y \geq 3 + x$
$x \geq 0$
$y \geq 0$

20. $2x + 3y \leq 12$
$2x + 3y > -6$
$3x + y < 4$
$x \geq 0$
$y \geq 0$

21. $\dfrac{x^2}{4} + \dfrac{y^2}{9} > 1$
$x^2 - y^2 \geq 1$
$-4 \leq x \leq 4$

22. $2x - 3y < 6$
$4x^2 + 9y^2 < 36$
$x \geq -1$

23. $y \geq 3^x$
$y \geq 2$

24. $y \leq \left(\dfrac{1}{2}\right)^x$
$y \geq 4$

25. $|x| \geq 2$
$|y| \geq 4$
$y < x^2$

26. $|x| + 2 \geq 4$
$|y| \leq 1$
$\dfrac{x^2}{9} + \dfrac{y^2}{16} \leq 1$

27. $y \leq |x + 2|$
$\dfrac{x^2}{16} - \dfrac{y^2}{9} \leq 1$

28. $y \leq \log x$
$y \geq |x - 2|$

29. Which one of the following is a description of the solution set of the system below?
$$x^2 + 4y^2 < 36$$
$$y < x$$
 (a) all points outside the ellipse $x^2 + 4y^2 = 36$ and above the line $y = x$
 (b) all points outside the ellipse $x^2 + 4y^2 = 36$ and below the line $y = x$
 (c) all points inside the ellipse $x^2 + 4y^2 = 36$ and above the line $y = x$
 (d) all points inside the ellipse $x^2 + 4y^2 = 36$ and below the line $y = x$

30. Fill in the blanks with the appropriate responses. The graph of the system
$$y > x^2 + 2$$
$$x^2 + y^2 < 16$$
$$y < 7$$
consists of all points _____ the parabola $y = x^2 + 2$, _____
 (above/below)   (inside/outside)
the circle $x^2 + y^2 = 16$, and _____ the line $y = 7$.
 (above/below)

In Exercises 31–34, use graphical methods to find values of x and y satisfying the given conditions. Find the value of the maximum or minimum.

31. Find $x \geq 0$ and $y \geq 0$ such that
$$2x + 3y \leq 6$$
$$4x + y \leq 6$$
and $5x + 2y$ is maximized.

32. Find $x \geq 0$ and $y \geq 0$ such that
$$x + y \leq 10$$
$$5x + 2y \geq 20$$
$$2y \geq x$$
and $x + 3y$ is minimized.

33. Find $x \geq 2$ and $y \geq 5$ such that
$$3x - y \geq 12$$
$$x + y \leq 15$$
and $2x + y$ is minimized.

34. Find $x \geq 10$ and $y \geq 20$ such that
$$2x + 3y \leq 100$$
$$5x + 4y \leq 200$$
and $x + 3y$ is maximized.

Solve each of the following linear programming problems. See Example 4.

35. Farmer Jones raises only pigs and geese. She wants to raise no more than 16 animals with no more than 12 geese. She spends $50 to raise a pig and $20 to raise a goose. She has $500 available for this purpose. Find the maximum profit she can make if she makes a profit of $80 per goose and $40 per pig.

36. A wholesaler of party goods wishes to display her products at a convention of social secretaries in such a way that she gets the maximum number of inquiries about her whistles and hats. Her booth at the convention has 12 sq m of floor space to be used for display purposes. A display unit for hats requires 2 sq m, and for whistles, 4 sq m. Experience tells the wholesaler that she should never have more than a total of 5 units of whistles and hats on display at one time. If she receives three inquiries for each unit of hats and two inquiries for each unit of whistles on display, how many of each should she display in order to get the maximum number of inquiries?

37. An office manager wants to buy some filing cabinets. She knows that cabinet #1 costs $10 each, requires 6 sq ft of floor space, and holds 8 cu ft of files. Cabinet #2 costs $20 each, requires 8 sq ft of floor space, and holds 12 cu ft. She can spend no more than $140 due to budget limitations, while her office has room for no more than 72 sq ft of cabinets. She wants the maximum storage capacity within the limits imposed by funds and space. How many of each type of cabinet should she buy?

38. In a small town in South Carolina, zoning rules require that the window space (in square feet) in a house be at least one-sixth of the space used up by solid walls. The cost to heat the house is 20¢ for each square foot of solid walls and 80¢ for each square foot of windows. Find the maximum total area (windows plus walls) if $160 is available to pay for heat.

39. The manufacturing process requires that oil refineries manufacture at least 2 gal of gasoline for each gallon of fuel oil. To meet the winter demand for fuel oil, at least 3 million gal a day must be produced. The demand for gasoline is no more than 6.4 million gal per day. If the price of gasoline is $1.90 and the price of fuel oil is $1.50 per gal, how much of each should be produced to maximize revenue?

40. Seall Manufacturing Company makes color television sets. It produces a bargain set that sells for $100 profit and a deluxe set that sells for $150 profit. On the assembly line the bargain set requires 3 hr, while the deluxe set takes 5 hr. The cabinet shop spends 1 hr on the cabinet for the bargain set and 3 hr on the cabinet for the deluxe set. Both sets require 2 hr of time for testing and packing. On a particular production run the Seall Company has available 3900 work hours on the assembly line, 2100 work hours in the cabinet shop, and 2200 work hours in the testing and packing department. How many sets of each type should it produce to make maximum profit? What is the maximum profit?

CHAPTER 7 SUMMARY ■ ———————————————

SECTION	TERMS	KEY IDEAS
7.1 Linear Systems with Two Variables	system of equations first-degree equation in n unknowns inconsistent system dependent equations substitution method addition method equivalent systems elimination method	**Possible Graphs of a Linear System with Two Equations and Two Variables** **1.** The graphs of the two equations intersect in a single point. The coordinates of this point give the solution of the system. **2.** The graphs are distinct parallel lines. In this case, the system is said to be inconsistent. That is, there is no solution common to both equations. The solution set of the linear system is empty. **3.** The graphs are the same line. In this case, the equations are said to be dependent, and any solution of one equation is also a solution of the other. Thus, there are infinitely many solutions.
7.2 Linear Systems with Three Variables	ordered triple	**Solution Set of a System with Three Equations and Three Variables** The solution set of a system of three equations in three unknowns may be either a single ordered triple, an infinite set of ordered triples (dependent equations), or the empty set (an inconsistent system).
7.3 Solution of Linear Systems by Matrices	matrix, matrices element (entry) augmented matrix Gauss-Jordan method	**Matrix Row Transformations** For an augmented matrix of a system of linear equations, the following row transformations will result in the matrix of an equivalent system. **1.** Any two rows may be interchanged. **2.** The elements of any row may be multiplied by a nonzero real number. **3.** Any row may be changed by adding to its elements a multiple of the corresponding elements of another row.
7.4 Nonlinear Systems of Equations	nonlinear system of equations	

SECTION	TERMS	KEY IDEAS
7.5 Systems of Inequalities; Linear Programming	system of inequalities linear programming constraints region of feasible solutions objective function vertex (corner point)	**Solving a Linear Programming Problem** **1.** Write the objective function and all necessary constraints. **2.** Graph the feasible region. **3.** Identify all vertex points. **4.** Find the value of the objective function at each vertex point. **5.** The solution is given by the vertex producing the optimum value of the objective function.

CHAPTER 7 REVIEW EXERCISES ■

Use the substitution method to solve each of the following linear systems. Identify any systems with dependent equations or any inconsistent systems.

1. $4x - 3y = -1$
$3x + 5y = 50$

2. $10x + 3y = 8$
$5x - 4y = 26$

3. $7x - 10y = 11$
$\dfrac{3x}{2} - 5y = 8$

4. $\dfrac{x}{2} + \dfrac{2y}{3} = -8$
$\dfrac{3x}{4} + \dfrac{y}{3} = 0$

5. $\dfrac{x}{2} - \dfrac{y}{5} = \dfrac{11}{10}$
$2x - \dfrac{4y}{5} = \dfrac{22}{5}$

6. $4x + 5y = 5$
$3x + 7y = -6$

Use the addition method to solve each of the following linear systems. Identify any systems with dependent equations or any inconsistent systems.

7. $3x - 5y = -18$
$2x + 7y = 19$

8. $6x + 5y = 53$
$4x - 3y = 29$

9. $\dfrac{2}{3}x - \dfrac{3}{4}y = 13$
$\dfrac{1}{2}x + \dfrac{2}{3}y = -5$

10. $3x + 7y = 10$
$18x + 42y = 50$

11. $\dfrac{1}{x} + \dfrac{1}{y} = \dfrac{7}{10}$
$\dfrac{3}{x} - \dfrac{5}{y} = \dfrac{1}{2}$

12. A student solves the system

$$x + y = 3$$
$$2x + 2y = 6$$

and obtains the result $0 = 0$. Then the student gives the solution set as $\{(0, 0)\}$. Is this correct? If not, explain.

Solve each of the following problems by writing a system of equations and then solving it.

13. A student bought some candy bars, paying 25¢ each for some and 50¢ each for others. The student bought a total of 22 bars, paying a total of $8.50. How many of each kind of bar did he buy?

14. Ink worth $25 a bottle is to be mixed with ink worth $18 per bottle to get 12 bottles of ink worth $20 each. How much of each type of ink should be used?

15. Mai Ling wins $50,000 in a lottery. She invests part of the money at 3%, twice as much at 3.5%, with $10,000 more than the amount invested at 3% invested at 4.5%. The total annual interest is $1900. How much is invested at each rate?

16. The sum of three numbers is 23. The second number is 3 more than the first. The sum of the first and twice the third is 4. Find the three numbers.

17. A gold merchant has some 12-carat gold (12/24 pure gold), and some 22-carat gold (22/24 pure). How many grams of each should be mixed to get 25 g of 15-carat gold?

18. A chemist has some 40% acid solution and some 60% solution. How many liters of each should be used to get 40 liters of a 45% solution?

Use the addition method to solve each of the following linear systems.

19. $2x - 3y + z = -5$
$x + 4y + 2z = 13$
$5x + 5y + 3z = 14$

20. $x - 3y = 12$
$2y + 5z = 1$
$4x + z = 25$

21. $x + y - z = 5$
$2x + y + 3z = 2$
$4x - y + 2z = -1$

22. $5x - 3y + 2z = -5$
$2x + 2y - z = 4$
$4x - y + z = -1$

Find solutions for the following systems in terms of the arbitrary variable z.

23. $3x - 4y + z = 2$
$2x + y - 4z = 1$

24. $2x + 3y - z = 5$
$-x + 2y + 4z = 8$

Solve each of the following problems by writing a system of equations and then solving it.

25. Three kinds of tea worth $4.60, $5.75, and $6.50 per pound are to be mixed to get 20 lb of tea worth $5.25 per pound. The amount of $4.60 tea used is to be equal to the total amount of the other two kinds together. How many pounds of each tea should be used?

26. A 5% solution of a drug is to be mixed with some 15% solution and some 10% solution to get 20 ml of 8% solution. The amount of 5% solution used must be

2 ml more than the sum of the other two solutions. How many milliliters of each solution should be used?

27. The cashier at an amusement park has a total of $2480, made up of fives, tens, and twenties. The total number of bills is 290, and the value of the tens is $60 more than the value of the twenties. How many of each type of bill does the cashier have?

28. Can a system consisting of two equations in three variables have a unique solution? Explain.

Use the Gauss-Jordan method to solve each of the following systems.

29. $2x + 3y = 10$
$-3x + y = 18$

30. $5x + 2y = -10$
$3x - 5y = -6$

31. $3x + y = -7$
$x - y = -5$

32. $x - z = -3$
$y + z = 6$
$2x - 3z = -9$

33. $2x - y + 4z = -1$
$-3x + 5y - z = 5$
$2x + 3y + 2z = 3$

34. $5x - 8y + z = 1$
$3x - 2y + 4z = 3$
$10x - 16y + 2z = 3$

Solve each of the following nonlinear systems of equations.

35. $y = x^2 - 1$
$x + y = 1$

36. $x^2 + y^2 = 2$
$3x + y = 4$

37. $x^2 + 2y^2 = 22$
$2x^2 - y^2 = -1$

38. $x^2 - 4y^2 = 19$
$x^2 + y^2 = 29$

39. $xy = 4$
$x - 6y = 2$

40. $x^2 + 2xy + y^2 = 4$
$x = 3y - 2$

41. Do the circle $x^2 + y^2 = 144$ and the line $x + 2y = 8$ have any points in common? If so, what are they?

42. Find a value of b so that the straight line $3x - y = b$ intersects the circle $x^2 + y^2 = 25$ in only one point.

43. Sketch a line and a hyperbola on the same set of axes so that their intersection contains each of the following numbers of points.
   (a) 0   **(b)** 1   **(c)** 2

Graph the solution of each of the following systems of inequalities.

44. $x + y \leq 6$
     $2x - y \geq 3$

45. $x - 3y \geq 6$
     $y^2 \leq 16 - x^2$

46. $9x^2 + 16y^2 \geq 144$
     $x^2 - y^2 \geq 16$

47. Find $x \geq 0$ and $y \geq 0$ such that

$$3x + 2y \leq 12$$
$$5x + y \geq 5$$

and $2x + 4y$ is maximized.

48. Find $x \geq 0$ and $y \geq 0$ such that

$$x + y \leq 50$$
$$2x + y \geq 20$$
$$x + 2y \geq 30$$

and $4x + 2y$ is minimized.

Solve the following linear programming problems.

49. A bakery makes both cakes and cookies. Each batch of cakes requires 2 hr in the oven and 3 hr in the decorating room. Each batch of cookies needs 1 1/2 hr in the oven and 2/3 hr in the decorating room. The oven is available no more than 16 hr a day, while the decorating room can be used no more than 12 hr per day. A batch of cookies produces a profit of $20; the profit on a batch of cakes is $30. Find the number of batches of each item that will maximize profit.

50. A candy company has 100 kg of chocolate-covered nuts and 125 kg of chocolate-covered raisins to be sold as two different mixtures. One mix will contain 1/2 nuts and 1/2 raisins, while the other mix will contain 1/3 nuts and 2/3 raisins. How much of each mixture should be made to maximize revenue if the first mix sells for $6.00 per kilogram and the second mix sells for $4.80 per kilogram?

CHAPTER 7   TEST   ■

Use the substitution method to solve each of the following systems. Identify any systems with dependent equations or any inconsistent systems. If a system has dependent equations, express the solution set with y arbitrary.

1. $x - 3y = -5$
     $2x + y = 4$

2. $3m - n = 9$
     $m + 2n = 10$

3. $6x + 9y = -21$
     $4x + 6y = -14$

Use the addition method to solve each of the following systems. Identify any systems with dependent equations or any inconsistent systems. If a system has dependent equations, express the solution set with y arbitrary.

4. $4a + 6b = 31$
     $3a + 4b = 22$

5. $\dfrac{1}{4}x - \dfrac{1}{3}y = -\dfrac{5}{12}$
     $\dfrac{1}{10}x + \dfrac{1}{5}y = \dfrac{1}{2}$

6. $x - 2y = 4$
     $-2x + 4y = 6$

7. Can a system of two linear equations in two variables have exactly two solutions? Explain.

Solve the following problem using a system of equations.

8. Christopher Michael has $1000 more invested at 5% than he has invested at 4%. If his annual income from the two investments together is $698, how much does Christopher have invested at each rate?

Use the addition method to solve the following linear systems.

9. $3a - 4b + 2c = 15$
 $2a - b + c = 13$
 $a + 2b - c = 5$

10. $2x + y + z = 3$
 $x + 2y - z = 3$
 $3x - y + z = 5$

11. Find a solution for the following system in terms of the arbitrary variable z.

$$x - 2y + 3z = 2$$
$$4x + y - z = 1$$

12. The sum of three numbers is 2. The first number is equal to the sum of the other two, and the third number is the result of subtracting the first from the second. Find the numbers by solving a system of equations.

Use the Gauss-Jordan method to solve each of the following systems.

13. $3a - 2b = 13$
 $4a - b = 19$

14. $2x + 3y - 6z = 1$
 $x - y + 2z = 3$
 $4x + y - 2z = 7$

Solve each of the following nonlinear systems of equations.

15. $2x^2 + y^2 = 6$
 $x^2 - 4y^2 = -15$

16. $x^2 + y^2 = 25$
 $x + y = 7$

17. If a system of two nonlinear equations contains one equation whose graph is a circle and another equation whose graph is a line, can the system have exactly one solution? If so, draw a sketch to indicate this situation.

18. Find two numbers such that their sum is -1 and the sum of their squares is 61.

19. Graph the solution of the following system of inequalities.

$$9x^2 + 4y^2 \geq 36$$
$$y < x^2$$

20. Gwen, who is dieting, requires two food supplements, I and II. She can get these supplements from two different products, A and B. Product A provides 3 grams per serving of supplement I and 2 grams per serving of supplement II. Product B provides 2 grams per serving of supplement I and 4 grams per serving of supplement II. Her dietician, Dr. Shoemake, has recommended that she include at least 15 grams of each supplement in her daily diet. If product A costs 25¢ per serving and product B costs 40¢ per serving, how can she satisfy her requirements most economically?

■ **THE GRAPHING CALCULATOR** ■

Graphing calculators are capable of graphing more than one equation on the same set of axes. Thus, we can use them to graph systems of equations. In Example 1 of Section 7.1, we found algebraically that the solution of the system

$$2x + 5y = 21$$
$$3x - 2y = -16$$

is the ordered pair $(-2, 5)$. This result can be verified on a graphing calculator by graphing both lines on the same set of axes, and then zooming in on the point of intersection. By using the tracing capability of the calculator, we can find the coordinates of the point.

However, since graphing calculators require that equations be entered in function form, we must first write each equation in the form $y = mx + b$ (since they are both linear), and then enter them into the calculator. The system above is equivalent to

$$y = -\frac{2}{5}x + \frac{21}{5}$$

$$y = \frac{3}{2}x + 8.$$

Thus, we see an important use of the function concept.

Suppose that we wish to solve the nonlinear system

$$x^2 + y^2 = 4$$
$$2x^2 - y^2 = 8$$

as shown in Example 1 of Section 7.4. Since neither of these equations defines a function, we must rewrite each of them as two equations involving square roots. The first equation consists of the two functions

$$y = \sqrt{4 - x^2}$$
$$y = -\sqrt{4 - x^2},$$

and the second consists of the two functions

$$y = \sqrt{2x^2 - 8}$$
$$y = -\sqrt{2x^2 - 8}.$$

The system can now be solved by entering these four functions into the calculator. By zooming and tracing, you should be able to identify the solutions as $(-2, 0)$ and $(2, 0)$, as shown in Figure 7.6.

The following systems can be used to experiment with your calculator. Most have been selected from examples and figures in the text, so you can compare your calculator results with those given there. The example or figure number is given with each system.

---

1. $2x - y = 2$ (Figure 7.1(a))
$x + y = -2$

2. $3x - 4y = 1$ (Example 2, Section 7.1)
$2x + 3y = 12$

3. $3x - 4y = 1$ (Example 2, Section 7.3)
$5x + 2y = 19$

4. $3x^2 - 2y = 5$ (Figure 7.7)
$x + 3y = -4$

5. $x^2 + y^2 = 16$ (Figure 7.8)
$|x| + y = 4$

6. $x^2 + y^2 = 5$ (Figure 7.9)
$4x^2 + 3y^2 = 11$

7. Use a graphing calculator to show that the system

$$x^2 + y^2 = 4$$
$$x^2 + y^2 = 25$$

consists of two concentric (having the same center) circles, and thus has no solutions.

8. Use a graphing calculator to graph these four functions on the same set of axes.

$$y = x^2$$
$$y = x - 4$$
$$y = x - \frac{1}{4}$$
$$y = x + 2$$

This will show how a line may intersect a parabola in either 0, 1, or 2 points.

MATRICES

8.1 BASIC PROPERTIES OF MATRICES

8.2 MATRIX PRODUCTS

8.3 MULTIPLICATIVE INVERSES OF MATRICES

8.4 DETERMINANTS

8.5 PROPERTIES OF DETERMINANTS

8.6 SOLUTION OF LINEAR SYSTEMS BY CRAMER'S RULE

In Chapter 7, we saw matrices used to solve systems of equations. Although this is an important application of matrices, they also have many other uses because they provide an efficient way to handle an array of numbers.

8.1 ——— BASIC PROPERTIES OF MATRICES

Suppose you are the manager of a health food store and you receive the following shipments of vitamins from two suppliers: from Dexter, 2 cartons of vitamin A pills, 7 cartons of vitamin E pills, and 5 cartons of vitamin K pills; from Sullivan, 4 cartons of vitamin A pills, 6 cartons of vitamin E pills, and 9 cartons of vitamin K pills. It might be helpful to rewrite the information in a chart to make it more understandable.

	Cartons of Vitamins		
Manufacturer	**A**	**E**	**K**
Dexter	2	7	5
Sullivan	4	6	9

The information is clearer when presented this way. In fact, as long as you remember what each number represents, you can remove all the labels and write the numbers as a matrix.

$$\begin{bmatrix} 2 & 7 & 5 \\ 4 & 6 & 9 \end{bmatrix}$$

This array of numbers gives all the information needed.

Matrices are classified by their size, that is, by the number of rows and columns that they contain. For example, the matrix

$$\begin{bmatrix} 2 & 7 & 5 \\ 4 & 6 & 9 \end{bmatrix}$$

has two rows and three columns and is called a 2×3 (read "2 by 3") matrix. A matrix with m rows and n columns is an $m \times n$ matrix. The number of rows is always given first.

■ *Example 1*
CLASSIFYING
MATRICES BY SIZE

(a) The matrix $\begin{bmatrix} 6 & 5 \\ 3 & 4 \\ 5 & -1 \end{bmatrix}$ is a 3×2 matrix.

(b) $\begin{bmatrix} 5 & 8 & 9 \\ 0 & 5 & -3 \\ -4 & 0 & 5 \end{bmatrix}$ is a 3×3 matrix.

(c) $[1 \quad 6 \quad 5 \quad -2 \quad 5]$ is a 1×5 matrix.

(d) $\begin{bmatrix} 3 \\ -5 \\ 0 \\ 2 \end{bmatrix}$ is a 4×1 matrix. ■

A matrix having the same number of rows as columns is called a **square matrix.** The matrix given in Example 1(b) above is a square matrix, as are

$$\begin{bmatrix} -5 & 6 \\ 8 & 3 \end{bmatrix} \quad \text{and} \quad \begin{bmatrix} 0 & 0 & 0 & 0 \\ -2 & 4 & 1 & 3 \\ 0 & 0 & 0 & 0 \\ -5 & -4 & 1 & 8 \end{bmatrix}.$$

A matrix containing only one row is called a **row matrix.** The matrix in Example 1(c) is a row matrix, as are

$$[5 \quad 8], \quad [6 \quad -9 \quad 2], \quad \text{and} \quad [-4 \quad 0 \quad 0 \quad 0].$$

Finally, a matrix of only one column, as in part (d) of Example 1, is a **column matrix.**

It is customary to use capital letters to name matrices. Subscript notation is used to name the elements of a matrix, as follows.

$$A = \begin{bmatrix} a_{11} & a_{12} & a_{13} & \cdots & a_{1n} \\ a_{21} & a_{22} & a_{23} & \cdots & a_{2n} \\ a_{31} & a_{32} & a_{33} & \cdots & a_{3n} \\ \cdot & \cdot & \cdot & & \cdot \\ \cdot & \cdot & \cdot & & \cdot \\ \cdot & \cdot & \cdot & & \cdot \\ a_{m1} & a_{m2} & a_{m3} & \cdots & a_{mn} \end{bmatrix}$$

Using this notation, the first row, first column element is denoted a_{11}, the second row, third column element is denoted a_{23}, and the ith row, jth column element is denoted a_{ij}.

Two matrices are **equal** if they are the same size and if each corresponding element, position by position, is equal. Using this definition, the matrices

$$\begin{bmatrix} 2 & 1 \\ 3 & -5 \end{bmatrix} \quad \text{and} \quad \begin{bmatrix} 1 & 2 \\ -5 & 3 \end{bmatrix}$$

are *not* equal (even though they contain the same elements and are the same size), since the corresponding elements differ.

■ *Example 2*

DECIDING WHETHER
TWO MATRICES ARE
EQUAL

From the definition of equality given above, the only way that the statement

$$\begin{bmatrix} 2 & 1 \\ p & q \end{bmatrix} = \begin{bmatrix} x & y \\ -1 & 0 \end{bmatrix}$$

can be true is if $2 = x$, $1 = y$, $p = -1$, and $q = 0$. ■

■ *Example 3*

DECIDING WHETHER
TWO MATRICES ARE
EQUAL

The statement

$$\begin{bmatrix} x \\ y \end{bmatrix} = \begin{bmatrix} 1 \\ 4 \\ 0 \end{bmatrix}$$

can never be true, since the two matrices are different sizes. (One is 2×1 and the other is 3×1.) ■

ADDING MATRICES   At the beginning of this section we used the matrix

$$\begin{bmatrix} 2 & 7 & 5 \\ 4 & 6 & 9 \end{bmatrix},$$

where the columns represent the numbers of cartons of three different types of vitamins (A, E, and K, respectively), and the rows represent two different manufacturers (Dexter and Sullivan, respectively). For example, the element 7 represents 7 cartons of vitamin E pills from Dexter, and so on. Suppose another shipment from these two suppliers is described by the following matrix.

$$\begin{bmatrix} 3 & 12 & 10 \\ 15 & 11 & 8 \end{bmatrix}$$

Here, for example, 8 cartons of vitamin K pills arrived from Sullivan. The number of cartons of each kind of pill that were received from these two shipments can be found from these two matrices.

In the first shipment, 2 cartons of vitamin A pills were received from Dexter, and in the second shipment, 3 cartons of vitamin A pills were received. Altogether, $2 + 3$, or 5, cartons of these pills were received. Corresponding elements can be added to find the total number of cartons of each type of pill received.

$$\begin{bmatrix} 2 & 7 & 5 \\ 4 & 6 & 9 \end{bmatrix} + \begin{bmatrix} 3 & 12 & 10 \\ 15 & 11 & 8 \end{bmatrix} = \begin{bmatrix} 2+3 & 7+12 & 5+10 \\ 4+15 & 6+11 & 9+8 \end{bmatrix}$$

$$= \begin{bmatrix} 5 & 19 & 15 \\ 19 & 17 & 17 \end{bmatrix}$$

The last matrix gives the total number of cartons of each type of pill that were received. For example, 15 cartons of vitamin K pills were received from Dexter. Generalizing from this example leads to the following definition.

MATRIX ADDITION   The sum of two $m \times n$ matrices A and B is the $m \times n$ matrix $A + B$ in which each element is the sum of the corresponding elements of A and B.

CAUTION   Only matrices of the same size can be added.

■ *Example 4*
| ADDING MATRICES

Find each of the following sums.

(a) $\begin{bmatrix} 5 & -6 \\ 8 & 9 \end{bmatrix} + \begin{bmatrix} -4 & 6 \\ 8 & -3 \end{bmatrix} = \begin{bmatrix} 5+(-4) & -6+6 \\ 8+8 & 9+(-3) \end{bmatrix} = \begin{bmatrix} 1 & 0 \\ 16 & 6 \end{bmatrix}$

(b) $\begin{bmatrix} 2 \\ 5 \\ 8 \end{bmatrix} + \begin{bmatrix} -6 \\ 3 \\ 12 \end{bmatrix} = \begin{bmatrix} -4 \\ 8 \\ 20 \end{bmatrix}$

(c) The matrices

$$A = \begin{bmatrix} 5 & 8 \\ 6 & 2 \end{bmatrix} \quad \text{and} \quad B = \begin{bmatrix} 3 & 9 & 1 \\ 4 & 2 & 5 \end{bmatrix}$$

are different sizes. Therefore, the sum $A + B$ does not exist. ■

A matrix containing only zero elements is called a **zero matrix.** For example, [0 0 0] is the 1×3 zero matrix, while

$$\begin{bmatrix} 0 & 0 & 0 \\ 0 & 0 & 0 \end{bmatrix}$$

is the 2×3 zero matrix. A zero matrix can be written of any size.

In Chapter 1 the additive inverse of a real number a was defined as the real number $-a$ such that $a + (-a) = 0$ and $-a + a = 0$. Given a matrix A, a matrix $-A$ can be found such that $A + (-A) = O$, where O is the appropriate zero matrix, and $-A + A = O$. For example, if

$$A = \begin{bmatrix} -5 & 2 & -1 \\ 3 & 4 & -6 \end{bmatrix},$$

then the elements of matrix $-A$ are the additive inverses of the corresponding elements of A. (Remember that each element of A is a real number and thus has an additive inverse.)

$$-A = \begin{bmatrix} 5 & -2 & 1 \\ -3 & -4 & 6 \end{bmatrix}$$

To check, first test that $A + (-A)$ equals O, the appropriate zero matrix.

$$A + (-A) = \begin{bmatrix} -5 & 2 & -1 \\ 3 & 4 & -6 \end{bmatrix} + \begin{bmatrix} 5 & -2 & 1 \\ -3 & -4 & 6 \end{bmatrix} = \begin{bmatrix} 0 & 0 & 0 \\ 0 & 0 & 0 \end{bmatrix} = O$$

Then test that $-A + A$ is also O. Matrix $-A$ is the **additive inverse,** or **negative,** of matrix A. Every matrix has a unique additive inverse.

SUBTRACTING MATRICES Subtraction of real numbers was defined in Chapter 1 by saying that $a - b = a + (-b)$. The same definition is used for subtraction of matrices.

MATRIX SUBTRACTION	If A and B are matrices of the same size, then
	$$A - B = A + (-B).$$

■ *Example 5*

SUBTRACTING
MATRICES

Find each of the following differences.

(a) $\begin{bmatrix} -5 & 6 \\ 2 & 4 \end{bmatrix} - \begin{bmatrix} -3 & 2 \\ 5 & -8 \end{bmatrix} = \begin{bmatrix} -5 & 6 \\ 2 & 4 \end{bmatrix} + \begin{bmatrix} 3 & -2 \\ -5 & 8 \end{bmatrix} = \begin{bmatrix} -2 & 4 \\ -3 & 12 \end{bmatrix}$

(b) $[8 \quad 6 \quad -4] - [3 \quad 5 \quad -8] = [5 \quad 1 \quad 4]$

(c) The matrices

$$\begin{bmatrix} -2 & 5 \\ 0 & 1 \end{bmatrix} \quad \text{and} \quad \begin{bmatrix} 3 \\ 5 \end{bmatrix}$$

are different sizes and cannot be subtracted. ■

If a matrix A is added to itself, each element in the sum is twice as large as the corresponding element of A. For example,

$$\begin{bmatrix} 2 & 5 \\ 1 & 3 \\ 4 & 6 \end{bmatrix} + \begin{bmatrix} 2 & 5 \\ 1 & 3 \\ 4 & 6 \end{bmatrix} = \begin{bmatrix} 4 & 10 \\ 2 & 6 \\ 8 & 12 \end{bmatrix} = 2\begin{bmatrix} 2 & 5 \\ 1 & 3 \\ 4 & 6 \end{bmatrix}.$$

In the last expression, the number 2 in front of the matrix is called a **scalar** to distinguish it from a matrix. The example above suggests the following definition of multiplication of a matrix by a scalar.

MULTIPLICATION OF A MATRIX BY A SCALAR

The product of a scalar k and a matrix X is the matrix kX, each of whose elements is k times the corresponding element of X.

■ *Example 6*

MULTIPLYING A
MATRIX BY A
SCALAR

Find each product.

(a) $5\begin{bmatrix} 2 & -3 \\ 0 & 4 \end{bmatrix} = \begin{bmatrix} 10 & -15 \\ 0 & 20 \end{bmatrix}$

(b) $\dfrac{3}{4}\begin{bmatrix} 20 & 36 \\ 12 & -16 \end{bmatrix} = \begin{bmatrix} 15 & 27 \\ 9 & -12 \end{bmatrix}$ ■

8.1 EXERCISES ■

1. A 3×8 matrix has _____ columns and _____ rows.

Find the size of each of the following matrices. Identify any square, column, or row matrices. See Example 1.

2. $\begin{bmatrix} -4 & 8 \\ 2 & 3 \end{bmatrix}$

3. $\begin{bmatrix} -9 & 6 & 2 \\ 4 & 1 & 8 \end{bmatrix}$

4. $\begin{bmatrix} -6 & 8 & 0 & 0 \\ 4 & 1 & 9 & 2 \\ 3 & -5 & 7 & 1 \end{bmatrix}$

5. $[8 \quad -2 \quad 4 \quad 6 \quad 3]$

6. $\begin{bmatrix} 2 \\ 4 \end{bmatrix}$

7. $[-9]$

8. $\begin{bmatrix} -4 & 2 & 3 \\ -8 & 2 & 1 \\ 4 & 6 & 8 \end{bmatrix}$

9. $\begin{bmatrix} -4 & 2 \\ 3 & 5 \end{bmatrix}$

Find the values of the variables in each of the following matrices. See Examples 2 and 3.

10. $\begin{bmatrix} 2 & 1 \\ 4 & 8 \end{bmatrix} = \begin{bmatrix} x & 1 \\ y & z \end{bmatrix}$

11. $\begin{bmatrix} -5 \\ y \end{bmatrix} = \begin{bmatrix} -5 \\ 8 \end{bmatrix}$

12. $\begin{bmatrix} x+6 & y+2 \\ 8 & 3 \end{bmatrix} = \begin{bmatrix} -9 & 7 \\ 8 & k \end{bmatrix}$

13. $\begin{bmatrix} 9 & 7 \\ r & 0 \end{bmatrix} = \begin{bmatrix} m-3 & n+5 \\ 8 & 0 \end{bmatrix}$

14. $\begin{bmatrix} 3 & 5 \\ 8 & 9 \end{bmatrix} + \begin{bmatrix} m & 3 \\ 5 & n \end{bmatrix} = \begin{bmatrix} 9 & 8 \\ 13 & 0 \end{bmatrix}$

15. $[8 \quad p+9 \quad q+5] + [9 \quad -3 \quad 12] = [k-2 \quad 12 \quad 2q]$

16. $\begin{bmatrix} -7+z & 4r & 8s \\ 6p & 2 & 5 \end{bmatrix} + \begin{bmatrix} -9 & 8r & 3 \\ 2 & 5 & 4 \end{bmatrix} = \begin{bmatrix} 2 & 36 & 27 \\ 20 & 7 & 12a \end{bmatrix}$

17. $\begin{bmatrix} a+2 & 3z+1 & 5m \\ 4k & 0 & 3 \end{bmatrix} + \begin{bmatrix} 3a & 2z & 5m \\ 2k & 5 & 6 \end{bmatrix} = \begin{bmatrix} 10 & -14 & 80 \\ 10 & 5 & 9 \end{bmatrix}$

18. Your friend missed the lecture on adding matrices. In your own words, explain to him how to add two matrices.

19. Explain to a friend in your own words how to subtract two matrices.

20. What is a scalar?

Perform each of the following operations, whenever possible. See Examples 4 and 5.

21. $3\begin{bmatrix} 6 & -1 & 4 \\ 2 & 8 & -3 \\ -4 & 5 & 6 \end{bmatrix} + 5\begin{bmatrix} -2 & -8 & -6 \\ 4 & 1 & 3 \\ 2 & -1 & 5 \end{bmatrix}$

22. $4\begin{bmatrix} 1 & -4 \\ 2 & -3 \\ -8 & 4 \end{bmatrix} - 3\begin{bmatrix} -6 & 9 \\ -2 & 5 \\ -7 & -12 \end{bmatrix}$

23. $\begin{bmatrix} -8 & 4 & 0 \\ 2 & 5 & 0 \end{bmatrix} + \begin{bmatrix} 6 & 3 \\ 8 & 9 \end{bmatrix}$

24. $\begin{bmatrix} 2 \\ 3 \end{bmatrix} - \begin{bmatrix} 8 & 1 \\ 9 & 4 \end{bmatrix}$

25. $\begin{bmatrix} 9 & 4 & 1 & -2 \\ 5 & -6 & 3 & 4 \\ 2 & -5 & 1 & 2 \end{bmatrix} - \begin{bmatrix} -2 & 5 & 1 & 3 \\ 0 & 1 & 0 & 2 \\ -8 & 3 & 2 & 1 \end{bmatrix} + \begin{bmatrix} 2 & 4 & 0 & 3 \\ 4 & -5 & 1 & 6 \\ 2 & -3 & 0 & 8 \end{bmatrix}$

26. $\begin{bmatrix} 6 & -2 & 4 \\ -2 & 5 & 8 \\ 1 & 0 & 2 \end{bmatrix} + \begin{bmatrix} 3 & 0 & 8 \\ 1 & -2 & 4 \\ 6 & 9 & -2 \end{bmatrix} - \begin{bmatrix} -4 & 2 & 1 \\ 0 & 3 & -2 \\ 4 & 2 & 0 \end{bmatrix}$

27. $\begin{bmatrix} -4x+2y & -3x+y \\ 6x-3y & 2x-5y \end{bmatrix} + \begin{bmatrix} -8x+6y & 2x \\ 3y-5x & 6x+4y \end{bmatrix}$

28. $\begin{bmatrix} 4k-8y \\ 6z-3x \\ 2k+5a \\ -4m+2n \end{bmatrix} - \begin{bmatrix} 5k+6y \\ 2z+5x \\ 4k+6a \\ 4m-2n \end{bmatrix}$

Let $A = \begin{bmatrix} -2 & 4 \\ 0 & 3 \end{bmatrix}$ *and* $B = \begin{bmatrix} -6 & 2 \\ 4 & 0 \end{bmatrix}$. *Find each of the following matrices.*
See Example 6.

29. $2A$

30. $-3B$

31. $-4A$

32. $5B$

33. $2A - B$

34. $-4A + 5B$

35. $3A - 11B$

36. $-2A + 4B$

Work each of the following problems.

37. Richard Maciasz bought 7 shares of Sears stock, 9 shares of IBM stock, and 8 shares of Chrysler stock. The following month, he bought 2 shares of Sears stock, no IBM, and 6 shares of Chrysler. Write this information first as a 3 × 2 matrix and then as a 2 × 3 matrix.

38. Margie Bezzone works in a computer store. The first week she sold 5 computers, 3 printers, 4 disk drives, and 6 monitors. The next week she sold 4 computers, 2 printers, 6 disk drives, and 5 monitors. Write this information first as a 2 × 4 matrix and then as a 4 × 2 matrix.

39. A recent study revealed that the average number of miles driven in the United States has been increasing steadily, with miles driven by women increasing much faster than miles driven by men. In 1969, 5411 thousand miles were driven by women and 11,352 thousand miles were driven by men. In 1990, 9371 thousand miles were driven by women and 15,956 thousand miles by men. Write this information as a 2 × 2 matrix in two ways.

40. The proportion of the population of China living in cities was slowed for a time by government-imposed birth-control policies. The urban population proportion has increased again in recent years. In 1952, the proportion was 12.5%; in 1960, 19.7%; in 1975, 12.1%; and in 1985, 19.7%. Write this information as a row matrix and as a column matrix.

Addition of the real numbers satisfies certain properties, such as the commutative, associative, identity, and inverse properties. In the following exercises, you can check to see which of these properties are satisfied by addition of matrices.

Let $\quad A = \begin{bmatrix} a & b \\ c & d \end{bmatrix}, \quad B = \begin{bmatrix} e & f \\ g & h \end{bmatrix}, \quad$ *and* $\quad C = \begin{bmatrix} j & m \\ k & n \end{bmatrix}.$ *Decide which of the follow-*

ing statements are true for these square matrices. Then decide if a similar property holds for any matrices of the same size.

41. $A + B = B + A$ (commutative property)

42. $A + (B + C) = (A + B) + C$ (associative property)

43. There exists a matrix O such that $A + O = A$ and $O + A = A$. (identity property)

44. There exists a matrix $-A$ such that $A + (-A) = O$ and $-A + A = O$. (inverse property)

45. Are these properties valid for matrices that are not square?

46. Does $kM = Mk$ for any scalar k and matrix M? Explain.

8.2 ——— MATRIX PRODUCTS

In the last section we multiplied a matrix by a scalar. Multiplication of two matrices is more complicated, but it is important in solving practical problems. To show the reasoning behind matrix multiplication, the example about the vitamin pills from Section 8.1 is continued here.

The matrix below, from the previous section, shows the number of cartons of each type of vitamin received from Dexter and Sullivan, respectively.

$$\begin{bmatrix} 2 & 7 & 5 \\ 4 & 6 & 9 \end{bmatrix}$$

Now suppose that each carton of vitamin A pills costs the store $12, each carton of vitamin E pills costs $18, and each carton of vitamin K costs $9.

To find the total cost of the pills from Dexter, multiply as follows.

Vitamin	Number of Cartons	Cost Per Carton	Total Cost
A	2	$12	$ 24
E	7	$18	$126
K	5	$ 9	$ 45
			$195 Total from Dexter

The Dexter pills cost a total of $195.

This result is the sum of three products:

$$2(\$12) + 7(\$18) + 5(\$9) = \$195.$$

In the same way, using the second row of the matrix and the three costs, the total cost of the Sullivan pills is

$$4(\$12) + 6(\$18) + 9(\$9) = \$237.$$

The costs, $12, $18, and $9, can be written as a column matrix.

$$\begin{bmatrix} 12 \\ 18 \\ 9 \end{bmatrix}$$

The total costs for each supplier, $195 and $237, also can be written as a column matrix.

$$\begin{bmatrix} 195 \\ 237 \end{bmatrix}$$

The product of the matrices

$$\begin{bmatrix} 2 & 7 & 5 \\ 4 & 6 & 9 \end{bmatrix} \quad \text{and} \quad \begin{bmatrix} 12 \\ 18 \\ 9 \end{bmatrix}$$

can be written as follows.

$$\begin{bmatrix} 2 & 7 & 5 \\ 4 & 6 & 9 \end{bmatrix} \begin{bmatrix} 12 \\ 18 \\ 9 \end{bmatrix} = \begin{bmatrix} 2 \cdot 12 + 7 \cdot 18 + 5 \cdot 9 \\ 4 \cdot 12 + 6 \cdot 18 + 9 \cdot 9 \end{bmatrix} = \begin{bmatrix} 195 \\ 237 \end{bmatrix}$$

Each element of the product was found by multiplying the elements of the *rows* of the matrix on the left and the corresponding elements of the *columns* of the matrix on the right, and then finding the sum of these products. Notice that the product of a 2×3 matrix and a 3×1 matrix is a 2×1 matrix.

Generalizing from this example gives the following definition of matrix multiplication.

MATRIX **MULTIPLICATION**	The product AB of an $m \times n$ matrix A and an $n \times k$ matrix B is found as follows. To get the ith row, jth column element of AB, multiply each element in the ith row of A by the corresponding element in the jth column of B. The sum of these products will give the element of row i, column j of AB.

■ **Example 1**

FINDING THE
PRODUCT OF TWO
MATRICES

Find the product AB, where

$$A = \begin{bmatrix} 2 & 4 \\ 5 & 6 \end{bmatrix} \quad \text{and} \quad B = \begin{bmatrix} -3 & 5 \\ 4 & -2 \end{bmatrix}.$$

Step 1 Multiply the elements of the first row of A and the corresponding elements of the first column of B, and add these products.

$$AB = \begin{bmatrix} 2 & 4 \\ 5 & 6 \end{bmatrix}\begin{bmatrix} -3 & 5 \\ 4 & -2 \end{bmatrix} \qquad 2(-3) + 4(4) = -6 + 16 = 10$$

The first-row, first-column entry of the product matrix AB is 10.

Step 2 Multiply the elements of the first row of A and the second column of B and then add the products to get the first-row, second-column entry of the product matrix.

$$AB = \begin{bmatrix} 2 & 4 \\ 5 & 6 \end{bmatrix}\begin{bmatrix} -3 & 5 \\ 4 & -2 \end{bmatrix} \qquad 2(5) + 4(-2) = 10 + (-8) = 2$$

Step 3

$$AB = \begin{bmatrix} 2 & 4 \\ 5 & 6 \end{bmatrix}\begin{bmatrix} -3 & 5 \\ 4 & -2 \end{bmatrix} \qquad 5(-3) + 6(4) = -15 + 24 = 9$$

The second-row, first-column entry of the product matrix is 9.

Step 4

$$AB = \begin{bmatrix} 2 & 4 \\ 5 & 6 \end{bmatrix}\begin{bmatrix} -3 & 5 \\ 4 & -2 \end{bmatrix} \qquad 5(5) + 6(-2) = 25 + (-12) = 13$$

Finally, 13 is the second-row, second-column entry.

Step 5 Write the product. The four entries in the product matrix come from the four steps above.

$$AB = \begin{bmatrix} 2 & 4 \\ 5 & 6 \end{bmatrix}\begin{bmatrix} -3 & 5 \\ 4 & -2 \end{bmatrix} = \begin{bmatrix} 10 & 2 \\ 9 & 13 \end{bmatrix} \qquad ■$$

■ **Example 2**

FINDING THE
PRODUCT OF TWO
MATRICES

Find the product.

$$\begin{bmatrix} -3 & 4 & 2 \\ 5 & 0 & 4 \end{bmatrix}\begin{bmatrix} -6 & 4 \\ 2 & 3 \\ 3 & -2 \end{bmatrix}$$

Step 1 $\begin{bmatrix} -3 & 4 & 2 \\ 5 & 0 & 4 \end{bmatrix}\begin{bmatrix} -6 & 4 \\ 2 & 3 \\ 3 & -2 \end{bmatrix}$ $\qquad (-3)(-6) + 4(2) + 2(3) = 32$

Step 2 $\begin{bmatrix} -3 & 4 & 2 \\ 5 & 0 & 4 \end{bmatrix}\begin{bmatrix} -6 & 4 \\ 2 & 3 \\ 3 & -2 \end{bmatrix}$ $\qquad (-3)(4) + 4(3) + 2(-2) = -4$

Step 3 $\begin{bmatrix} -3 & 4 & 2 \\ 5 & 0 & 4 \end{bmatrix}\begin{bmatrix} -6 & 4 \\ 2 & 3 \\ 3 & -2 \end{bmatrix}$ $\qquad 5(-6) + 0(2) + 4(3) = -18$

Step 4 $\begin{bmatrix} -3 & 4 & 2 \\ 5 & 0 & 4 \end{bmatrix}\begin{bmatrix} -6 & 4 \\ 2 & 3 \\ 3 & -2 \end{bmatrix}$ $\qquad 5(4) + 0(3) + 4(-2) = 12$

Step 5 Write the product.

$$\begin{bmatrix} -3 & 4 & 2 \\ 5 & 0 & 4 \end{bmatrix}\begin{bmatrix} -6 & 4 \\ 2 & 3 \\ 3 & -2 \end{bmatrix} = \begin{bmatrix} 32 & -4 \\ -18 & 12 \end{bmatrix}$$

As this example shows, the product of a 2 × 3 matrix and a 3 × 2 matrix is a 2 × 2 matrix. ■

The examples suggest the following restriction on matrix multiplication.

RESTRICTION ON MATRIX MULTIPLICATION The product *AB* of two matrices *A* and *B* can be found only if the number of *columns* of *A* is the same as the number of *rows* of *B*. The final product will have as many rows as *A* and as many columns as *B*.

■ ***Example 3***
DECIDING WHETHER TWO MATRICES CAN BE MULTIPLIED

Suppose matrix *A* is 2 × 2, while matrix *B* is 2 × 4. Can the product *AB* be calculated? What is the size of the product?

The following diagram helps answer these questions.

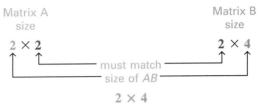

The product *AB* can be calculated because *A* has two columns and *B* has two rows. The size of the product is 2 × 4. (However, the product *BA* could not be found.) ■

■ *Example 4*

MULTIPLYING
MATRICES IN
DIFFERENT ORDERS

Find AB and BA if

$$A = \begin{bmatrix} 1 & -3 \\ 7 & 2 \end{bmatrix} \quad \text{and} \quad B = \begin{bmatrix} 1 & 0 & -1 & 2 \\ 3 & 1 & 4 & -1 \end{bmatrix}.$$

Use the definition of matrix multiplication to find AB.

$$AB = \begin{bmatrix} 1 & -3 \\ 7 & 2 \end{bmatrix}\begin{bmatrix} 1 & 0 & -1 & 2 \\ 3 & 1 & 4 & -1 \end{bmatrix}$$

$$= \begin{bmatrix} 1(1) + (-3)(3) & 1(0) + (-3)1 & 1(-1) + (-3)4 & 1(2) + (-3)(-1) \\ 7(1) + 2(3) & 7(0) + 2(1) & 7(-1) + 2(4) & 7(2) + 2(-1) \end{bmatrix}$$

$$= \begin{bmatrix} -8 & -3 & -13 & 5 \\ 13 & 2 & 1 & 12 \end{bmatrix}$$

Since B is a 2×4 matrix, and A is a 2×2 matrix, the product BA cannot be found. ■

■ *Example 5*

MULTIPLYING
SQUARE MATRICES
IN DIFFERENT
ORDERS

Find MN and NM, given

$$M = \begin{bmatrix} 1 & 3 \\ -2 & 4 \end{bmatrix} \quad \text{and} \quad N = \begin{bmatrix} 2 & 5 \\ 10 & -3 \end{bmatrix}.$$

By the definition of matrix multiplication,

$$MN = \begin{bmatrix} 1 & 3 \\ -2 & 4 \end{bmatrix}\begin{bmatrix} 2 & 5 \\ 10 & -3 \end{bmatrix}$$

$$= \begin{bmatrix} 2 + 30 & 5 - 9 \\ -4 + 40 & -10 - 12 \end{bmatrix}$$

$$= \begin{bmatrix} 32 & -4 \\ 36 & -22 \end{bmatrix}.$$

Similarly,

$$NM = \begin{bmatrix} 2 & 5 \\ 10 & -3 \end{bmatrix}\begin{bmatrix} 1 & 3 \\ -2 & 4 \end{bmatrix}$$

$$= \begin{bmatrix} 2 - 10 & 6 + 20 \\ 10 + 6 & 30 - 12 \end{bmatrix}$$

$$= \begin{bmatrix} -8 & 26 \\ 16 & 18 \end{bmatrix}. \quad ■$$

In Example 4 the product AB could be found, but not BA. In Example 5, although both MN and NM could be found, they were not equal, showing that multiplication of matrices is not commutative. This fact distinguishes matrix arithmetic from the arithmetic of real numbers. Other properties of matrix arithmetic are given as Exercises 40–43 at the end of this section.

Since multiplication of square matrices is not commutative, be careful about the ordering when you multiply matrices.

■ **Example 6**
APPLYING MATRIX
MULTIPLICATION

A contractor builds three kinds of houses, models A, B, and C, with a choice of two styles, colonial or ranch. Matrix P below shows the number of each kind of house the contractor is planning to build for a new 100-home subdivision. The amounts for each of the main materials used depend on the style of the house. These amounts are shown in matrix Q below, while matrix R gives the cost in dollars for each kind of material. Concrete is measured here in cubic yards, lumber in 1000 board feet, brick in 1000's, and shingles in 100 square feet.

$$
\begin{array}{c}
 \\
\text{Model A} \\
\text{Model B} \\
\text{Model C}
\end{array}
\begin{array}{c}
\text{Colonial} \quad \text{Ranch} \\
\begin{bmatrix}
0 & 30 \\
10 & 20 \\
20 & 20
\end{bmatrix} = P
\end{array}
$$

(a) What is the total cost of materials for all houses of each model?

To find the materials cost for each model, first find matrix PQ, which will show the total amount of each material needed for all houses of each model.

$$
PQ = \begin{bmatrix} 0 & 30 \\ 10 & 20 \\ 20 & 20 \end{bmatrix}\begin{bmatrix} 10 & 2 & 0 & 2 \\ 50 & 1 & 20 & 2 \end{bmatrix} = \begin{array}{c} \text{Concrete} \ \text{Lumber} \ \text{Brick} \ \text{Shingles} \\ \begin{bmatrix} 1500 & 30 & 600 & 60 \\ 1100 & 40 & 400 & 60 \\ 1200 & 60 & 400 & 80 \end{bmatrix} \begin{array}{l} \text{Model A} \\ \text{Model B} \\ \text{Model C} \end{array} \end{array}
$$

Multiplying PQ and the cost matrix R gives the total cost of materials for each model.

$$
(PQ)R = \begin{bmatrix} 1500 & 30 & 600 & 60 \\ 1100 & 40 & 400 & 60 \\ 1200 & 60 & 400 & 80 \end{bmatrix}\begin{bmatrix} 20 \\ 180 \\ 60 \\ 25 \end{bmatrix} = \begin{array}{c} \text{Cost} \\ \begin{bmatrix} 72,900 \\ 54,700 \\ 60,800 \end{bmatrix} \begin{array}{l} \text{Model A} \\ \text{Model B} \\ \text{Model C} \end{array} \end{array}
$$

(b) How much of each of the four kinds of material must be ordered?

The totals of the columns of matrix PQ will give a matrix whose elements represent the total amounts of each material needed for the subdivision. Call this matrix T and write it as a row matrix.

$$
T = \begin{bmatrix} 3800 & 130 & 1400 & 200 \end{bmatrix}
$$

(c) What is the total cost of the materials?

The total cost of all the materials is given by the product of matrix R, the cost matrix, and matrix T, the total amounts matrix. To multiply these and get a 1×1 matrix, representing the total cost, requires multiplying a 1×4 matrix and a 4×1 matrix. This is why in (b) a row matrix was written rather than a column matrix. The total materials cost is given by TR, so

$$TR = [3800 \quad 130 \quad 1400 \quad 200] \begin{bmatrix} 20 \\ 180 \\ 60 \\ 25 \end{bmatrix} = [188,400].$$

The total cost of the materials is $188,400. ■

To help keep track of the quantities a matrix represents, let matrix P, from Example 6, represent models/styles, matrix Q represent styles/materials, and matrix R represent materials/cost. In each case the meaning of the rows is written first and that of the columns second. When the product PQ was found in Example 6, the rows of the matrix represented models and the columns represented materials. Therefore, the matrix product PQ represents models/materials. The common quantity, styles, in both P and Q was eliminated in the product PQ. Do you see that the product $(PQ)R$ represents models/cost?

In practical problems this notation helps to identify the order in which two matrices should be multiplied so that the results are meaningful. In Example 6(c), either product RT or product TR could have been found. However, since T represents subdivisions/materials and R represents materials/cost, only TR gave the required matrix representing subdivisions/cost.

8.2 EXERCISES ■

In each of the following exercises, the sizes of the two matrices A and B are given. Find the size of the product AB and the product BA, whenever these products exist. See Example 3.

1. A is 2×2, B is 2×2

2. A is 3×3, B is 3×3

3. A is 4×2, B is 2×4

4. A is 3×1, B is 1×3

5. A is 3×5, B is 5×2

6. A is 4×3, B is 3×6

7. A is 4×2, B is 3×4

8. A is 7×3, B is 2×7

9. A is 4×3, B is 2×5

10. A is 1×6, B is 2×4

11. The product MN of two matrices can be found only if the number of _____ of M equals the number of _____ of N.

12. True or false: For matrices A and B, if AB can be found, then BA can always be found, too.

13. To find the product AB of matrices A and B, the first row, second column entry is found by multiplying the _____ elements in A and the _____ elements in B and then _____ these products.

Find each of the following matrix products, whenever possible. See Examples 1 and 2.

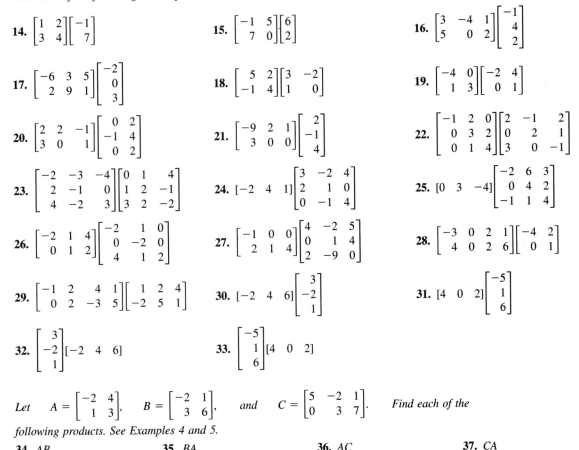

14. $\begin{bmatrix} 1 & 2 \\ 3 & 4 \end{bmatrix}\begin{bmatrix} -1 \\ 7 \end{bmatrix}$

15. $\begin{bmatrix} -1 & 5 \\ 7 & 0 \end{bmatrix}\begin{bmatrix} 6 \\ 2 \end{bmatrix}$

16. $\begin{bmatrix} 3 & -4 & 1 \\ 5 & 0 & 2 \end{bmatrix}\begin{bmatrix} -1 \\ 4 \\ 2 \end{bmatrix}$

17. $\begin{bmatrix} -6 & 3 & 5 \\ 2 & 9 & 1 \end{bmatrix}\begin{bmatrix} -2 \\ 0 \\ 3 \end{bmatrix}$

18. $\begin{bmatrix} 5 & 2 \\ -1 & 4 \end{bmatrix}\begin{bmatrix} 3 & -2 \\ 1 & 0 \end{bmatrix}$

19. $\begin{bmatrix} -4 & 0 \\ 1 & 3 \end{bmatrix}\begin{bmatrix} -2 & 4 \\ 0 & 1 \end{bmatrix}$

20. $\begin{bmatrix} 2 & 2 & -1 \\ 3 & 0 & 1 \end{bmatrix}\begin{bmatrix} 0 & 2 \\ -1 & 4 \\ 0 & 2 \end{bmatrix}$

21. $\begin{bmatrix} -9 & 2 & 1 \\ 3 & 0 & 0 \end{bmatrix}\begin{bmatrix} 2 \\ -1 \\ 4 \end{bmatrix}$

22. $\begin{bmatrix} -1 & 2 & 0 \\ 0 & 3 & 2 \\ 0 & 1 & 4 \end{bmatrix}\begin{bmatrix} 2 & -1 & 2 \\ 0 & 2 & 1 \\ 3 & 0 & -1 \end{bmatrix}$

23. $\begin{bmatrix} -2 & -3 & -4 \\ 2 & -1 & 0 \\ 4 & -2 & 3 \end{bmatrix}\begin{bmatrix} 0 & 1 & 4 \\ 1 & 2 & -1 \\ 3 & 2 & -2 \end{bmatrix}$

24. $\begin{bmatrix} -2 & 4 & 1 \end{bmatrix}\begin{bmatrix} 3 & -2 & 4 \\ 2 & 1 & 0 \\ 0 & -1 & 4 \end{bmatrix}$

25. $\begin{bmatrix} 0 & 3 & -4 \end{bmatrix}\begin{bmatrix} -2 & 6 & 3 \\ 0 & 4 & 2 \\ -1 & 1 & 4 \end{bmatrix}$

26. $\begin{bmatrix} -2 & 1 & 4 \\ 0 & 1 & 2 \end{bmatrix}\begin{bmatrix} -2 & 1 & 0 \\ 0 & -2 & 0 \\ 4 & 1 & 2 \end{bmatrix}$

27. $\begin{bmatrix} -1 & 0 & 0 \\ 2 & 1 & 4 \end{bmatrix}\begin{bmatrix} 4 & -2 & 5 \\ 0 & 1 & 4 \\ 2 & -9 & 0 \end{bmatrix}$

28. $\begin{bmatrix} -3 & 0 & 2 & 1 \\ 4 & 0 & 2 & 6 \end{bmatrix}\begin{bmatrix} -4 & 2 \\ 0 & 1 \end{bmatrix}$

29. $\begin{bmatrix} -1 & 2 & 4 & 1 \\ 0 & 2 & -3 & 5 \end{bmatrix}\begin{bmatrix} 1 & 2 & 4 \\ -2 & 5 & 1 \end{bmatrix}$

30. $\begin{bmatrix} -2 & 4 & 6 \end{bmatrix}\begin{bmatrix} 3 \\ -2 \\ 1 \end{bmatrix}$

31. $\begin{bmatrix} 4 & 0 & 2 \end{bmatrix}\begin{bmatrix} -5 \\ 1 \\ 6 \end{bmatrix}$

32. $\begin{bmatrix} 3 \\ -2 \\ 1 \end{bmatrix}\begin{bmatrix} -2 & 4 & 6 \end{bmatrix}$

33. $\begin{bmatrix} -5 \\ 1 \\ 6 \end{bmatrix}\begin{bmatrix} 4 & 0 & 2 \end{bmatrix}$

Let $A = \begin{bmatrix} -2 & 4 \\ 1 & 3 \end{bmatrix}$, $B = \begin{bmatrix} -2 & 1 \\ 3 & 6 \end{bmatrix}$, and $C = \begin{bmatrix} 5 & -2 & 1 \\ 0 & 3 & 7 \end{bmatrix}$. Find each of the following products. See Examples 4 and 5.

34. AB

35. BA

36. AC

37. CA

38. Did you get the same answer in Exercises 34 and 35? What about Exercises 36 and 37? Do you think that matrix multiplication is commutative?

39. For any matrices P and Q, what must be true for both PQ and QP to exist?

For the following exercises, let $A = \begin{bmatrix} a & b \\ c & d \end{bmatrix}$, $B = \begin{bmatrix} e & f \\ g & h \end{bmatrix}$, and $C = \begin{bmatrix} j & m \\ k & n \end{bmatrix}$.

Decide which of the following statements are true for these three matrices. Then decide if a similar property holds for any square matrices of the same size.

40. $(AB)C = A(BC)$ (associative property)

41. $A(B + C) = AB + AC$ (distributive property)

42. $k(A + B) = kA + kB$ for any real number k

43. $(k + p)A = kA + pA$ for any real numbers k and p

44. If matrix A represents costs/models and matrix B represents materials/costs, in what order should A and B be multiplied to get a meaningful product? What would that product represent?

45. Yummy Yogurt sells three types of yogurt: nonfat, regular, and super creamy at three locations. Location I sells 50 gallons of nonfat, 100 gallons of regular, and 30 gallons of super creamy each day. Location II sells 10 gallons of nonfat and Location III sells 60 gallons of nonfat each day. Daily sales of regular yogurt are 90 gallons at Location II and 120 gallons at Location III. At Location II, 50 gallons of super creamy are sold each day, and 40 gallons of super creamy are sold each day at Location III.

(a) Write a 3×3 matrix that shows the sales figures for the three locations.

(b) The income per gallon for nonfat, regular, and super creamy is $12, $10, and $15, respectively. Write a 1×3 or 3×1 matrix displaying the income.

(c) Find a matrix product that gives the daily income at each of the three locations.

(d) What is Yummy Yogurt's total daily income from the three locations?

46. The Bread Box, a small neighborhood bakery, sells four main items: sweet rolls, bread, cakes, and pies. The amount of each ingredient (in cups, except for eggs) required for these items is given by matrix A.

$$
\begin{array}{cc}
 & \begin{array}{ccccc} \text{Eggs} & \text{Flour} & \text{Sugar} & \text{Shortening} & \text{Milk} \end{array} \\
\begin{array}{c} \text{Rolls} \\ \text{(doz)} \\ \text{Bread} \\ \text{(loaves)} \\ \text{Cakes} \\ \\ \text{Pies} \\ \text{(crust)} \end{array} &
\begin{bmatrix} 1 & 4 & 1/4 & 1/4 & 1 \\ 0 & 3 & 0 & 1/4 & 0 \\ 4 & 3 & 2 & 1 & 1 \\ 0 & 1 & 0 & 1/3 & 0 \end{bmatrix} = A
\end{array}
$$

The cost (in cents) for each ingredient when purchased in large lots or small lots is given in matrix B.

$$
\begin{array}{cc}
 & \begin{array}{cc} \text{Cost} \\ \text{Large lot} \quad \text{Small lot} \end{array} \\
\begin{array}{c} \text{Eggs} \\ \text{Flour} \\ \text{Sugar} \\ \text{Shortening} \\ \text{Milk} \end{array} &
\begin{bmatrix} 5 & 5 \\ 8 & 10 \\ 10 & 12 \\ 12 & 15 \\ 5 & 6 \end{bmatrix} = B
\end{array}
$$

(a) Use matrix multiplication to find a matrix giving the comparative cost per item for the two purchase options.

Suppose a day's orders consist of 20 dozen sweet rolls, 200 loaves of bread, 50 cakes, and 60 pies.

(b) Write the orders as a 1×4 matrix and, using matrix multiplication, write as a matrix the amount of each ingredient needed to fill the day's orders.

(c) Use matrix multiplication to find a matrix giving the costs under the two purchase options to fill the day's orders.

8.3 ————— **MULTIPLICATIVE INVERSES OF MATRICES**

In this section multiplicative identity elements and multiplicative inverses are introduced and used to solve matrix equations. This leads to another method for solving systems of equations.

IDENTITY MATRICES The identity property for real numbers says that $a \cdot 1 = a$ and $1 \cdot a = a$ for any real number a. If there is to be a multiplicative identity matrix I, such that

$$AI = A \quad \text{and} \quad IA = A,$$

for any matrix A, then A and I must be square matrices of the same size. Otherwise it would not be possible to find both products. For example, let A be the 2×2 matrix

$$A = \begin{bmatrix} a_{11} & a_{12} \\ a_{21} & a_{22} \end{bmatrix},$$

and let
$$I = \begin{bmatrix} x_{11} & x_{12} \\ x_{21} & x_{22} \end{bmatrix}$$

represent the 2×2 identity matrix. To find I, use the fact that $IA = A$, or

$$\begin{bmatrix} x_{11} & x_{12} \\ x_{21} & x_{22} \end{bmatrix} \begin{bmatrix} a_{11} & a_{12} \\ a_{21} & a_{22} \end{bmatrix} = \begin{bmatrix} a_{11} & a_{12} \\ a_{21} & a_{22} \end{bmatrix}.$$

Multiplying the two matrices on the left side of this equation and setting the elements of the product matrix equal to the corresponding elements of A gives the following system of equations with variables x_{11}, x_{12}, x_{21}, and x_{22}.

$$x_{11}a_{11} + x_{12}a_{21} = a_{11}$$
$$x_{11}a_{12} + x_{12}a_{22} = a_{12}$$
$$x_{21}a_{11} + x_{22}a_{21} = a_{21}$$
$$x_{21}a_{12} + x_{22}a_{22} = a_{22}$$

Notice that this is really two systems of equations in two variables. Use one of the methods of Chapter 7 to find the solution of this system: $x_{11} = 1$, $x_{12} = x_{21} = 0$, and $x_{22} = 1$. From the solution of the system, the 2×2 identity matrix is

$$I = \begin{bmatrix} 1 & 0 \\ 0 & 1 \end{bmatrix}.$$

Check that with this definition of I, both $AI = A$ and $IA = A$.

■ **Example 1**
VERIFYING THE IDENTITY PROPERTY

Let $M = \begin{bmatrix} -2 & 6 \\ 3 & 5 \end{bmatrix}$. Verify that $MI = M$ and $IM = M$.

$$MI = \begin{bmatrix} -2 & 6 \\ 3 & 5 \end{bmatrix} \begin{bmatrix} 1 & 0 \\ 0 & 1 \end{bmatrix} \qquad\qquad IM = \begin{bmatrix} 1 & 0 \\ 0 & 1 \end{bmatrix} \begin{bmatrix} -2 & 6 \\ 3 & 5 \end{bmatrix}$$

$$= \begin{bmatrix} -2 & 6 \\ 3 & 5 \end{bmatrix} = M \qquad\qquad = \begin{bmatrix} -2 & 6 \\ 3 & 5 \end{bmatrix} = M \quad ■$$

The 2×2 identity matrix found above suggests the following generalization.

$n \times n$ IDENTITY MATRIX

For any value of n there is an $n \times n$ identity matrix having 1's down the diagonal and 0's elsewhere. The **$n \times n$ identity matrix** is given by I, where

$$I = \begin{bmatrix} 1 & 0 & \cdots & 0 \\ 0 & 1 & \cdots & 0 \\ \cdot & \cdot & & \cdot \\ \cdot & \cdot & a_{ij} & \cdot \\ \cdot & \cdot & & \cdot \\ 0 & 0 & \cdots & 1 \end{bmatrix}.$$

Here $a_{ij} = 1$ when $i = j$ (the diagonal elements) and $a_{ij} = 0$ otherwise.

■ *Example 2*
STATING AND
VERIFYING THE
3 × 3 IDENTITY
MATRIX

Let $K = \begin{bmatrix} -2 & 4 & 0 \\ 3 & 5 & 9 \\ 0 & 8 & -6 \end{bmatrix}$. Give the 3 × 3 identity matrix I and show that $KI = K$.

The 3 × 3 identity matrix is

$$I = \begin{bmatrix} 1 & 0 & 0 \\ 0 & 1 & 0 \\ 0 & 0 & 1 \end{bmatrix}.$$

By the definition of matrix multiplication,

$$KI = \begin{bmatrix} -2 & 4 & 0 \\ 3 & 5 & 9 \\ 0 & 8 & -6 \end{bmatrix}\begin{bmatrix} 1 & 0 & 0 \\ 0 & 1 & 0 \\ 0 & 0 & 1 \end{bmatrix} = \begin{bmatrix} -2 & 4 & 0 \\ 3 & 5 & 9 \\ 0 & 8 & -6 \end{bmatrix} = K. \quad ■$$

MULTIPLICATIVE INVERSES For every nonzero real number a, there is a multiplicative inverse $1/a$ such that

$$a \cdot \frac{1}{a} = 1 \quad \text{and} \quad \frac{1}{a} \cdot a = 1.$$

(Recall that $1/a$ can also be written a^{-1}.) In the rest of this section, a method is developed for finding a multiplicative inverse for square matrices. The **multiplicative inverse** of a matrix A is written A^{-1}. This matrix must satisfy the statements

$$AA^{-1} = I \quad \text{and} \quad A^{-1}A = I.$$

The multiplicative inverse of a matrix can be found using the matrix row transformations given in Section 7.3 and repeated here for convenience.

MATRIX ROW TRANSFORMATIONS

The **matrix row transformations** are:

1. interchanging any two rows of a matrix;
2. multiplying the elements of any row of a matrix by the same nonzero scalar k; and
3. adding a multiple of the elements of one row to the elements of another row.

As an example, let us find the inverse of

$$A = \begin{bmatrix} 2 & 4 \\ 1 & -1 \end{bmatrix}.$$

Let the unknown inverse matrix be

$$A^{-1} = \begin{bmatrix} x & y \\ z & w \end{bmatrix}.$$

By the definition of matrix inverse, $AA^{-1} = I$, or

$$AA^{-1} = \begin{bmatrix} 2 & 4 \\ 1 & -1 \end{bmatrix}\begin{bmatrix} x & y \\ z & w \end{bmatrix} = \begin{bmatrix} 1 & 0 \\ 0 & 1 \end{bmatrix}.$$

By matrix multiplication,

$$\begin{bmatrix} 2x + 4z & 2y + 4w \\ x - z & y - w \end{bmatrix} = \begin{bmatrix} 1 & 0 \\ 0 & 1 \end{bmatrix}.$$

Setting corresponding elements equal gives the system of equations

$$2x + 4z = 1 \tag{1}$$
$$2y + 4w = 0 \tag{2}$$
$$x - z = 0 \tag{3}$$
$$y - w = 1. \tag{4}$$

Since equations (1) and (3) involve only x and z, while equations (2) and (4) involve only y and w, these four equations lead to two systems of equations,

$$\begin{array}{ll} 2x + 4z = 1 & \\ & \text{and} \\ x - z = 0 & \end{array} \qquad \begin{array}{l} 2y + 4w = 0 \\ \\ y - w = 1. \end{array}$$

Writing the two systems as augmented matrices gives

$$\begin{bmatrix} 2 & 4 & | & 1 \\ 1 & -1 & | & 0 \end{bmatrix} \quad \text{and} \quad \begin{bmatrix} 2 & 4 & | & 0 \\ 1 & -1 & | & 1 \end{bmatrix}.$$

Each of these systems can be solved by the Gauss-Jordan method. However, since the elements to the left of the vertical bar are identical, the two systems can be combined into the one augmented matrix,

$$\begin{bmatrix} 2 & 4 & | & 1 & 0 \\ 1 & -1 & | & 0 & 1 \end{bmatrix},$$

and solved simultaneously as follows. Exchange the two rows to get a 1 in the upper left corner.

$$\begin{bmatrix} 1 & -1 & | & 0 & 1 \\ 2 & 4 & | & 1 & 0 \end{bmatrix}$$

Multiply the first row by -2 and add the results to the second row to get

$$\begin{bmatrix} 1 & -1 & | & 0 & 1 \\ 0 & 6 & | & 1 & -2 \end{bmatrix}. \qquad -2\,R_1 + R_2$$

Now, to get a 1 in the second-row, second-column position, multiply the second row by 1/6.

$$\begin{bmatrix} 1 & -1 & | & 0 & 1 \\ 0 & 1 & | & 1/6 & -1/3 \end{bmatrix} \qquad \tfrac{1}{6}R_2$$

Finally, add the second row to the first row to get a 0 in the second column above the 1.

$$\left[\begin{array}{cc|cc} 1 & 0 & 1/6 & 2/3 \\ 0 & 1 & 1/6 & -1/3 \end{array}\right] \quad R_2 + R_1$$

The numbers in the first column to the right of the vertical bar give the values of x and z. The second column gives the values of y and w. That is,

$$\left[\begin{array}{cc|cc} 1 & 0 & x & y \\ 0 & 1 & z & w \end{array}\right] = \left[\begin{array}{cc|cc} 1 & 0 & 1/6 & 2/3 \\ 0 & 1 & 1/6 & -1/3 \end{array}\right]$$

so that

$$A^{-1} = \left[\begin{array}{cc} x & y \\ z & w \end{array}\right] = \left[\begin{array}{cc} 1/6 & 2/3 \\ 1/6 & -1/3 \end{array}\right].$$

To check, multiply A by A^{-1}. The result should be I.

$$AA^{-1} = \left[\begin{array}{cc} 2 & 4 \\ 1 & -1 \end{array}\right]\left[\begin{array}{cc} 1/6 & 2/3 \\ 1/6 & -1/3 \end{array}\right] = \left[\begin{array}{cc} 1/3 + 2/3 & 4/3 - 4/3 \\ 1/6 - 1/6 & 2/3 + 1/3 \end{array}\right] = \left[\begin{array}{cc} 1 & 0 \\ 0 & 1 \end{array}\right] = I$$

Verify that $A^{-1}A = I$, also. Finally,

$$A^{-1} = \left[\begin{array}{cc} 1/6 & 2/3 \\ 1/6 & -1/3 \end{array}\right].$$

The process for finding the multiplicative inverse A^{-1} for any $n \times n$ matrix A that has an inverse is summarized below.

FINDING AN INVERSE MATRIX

To obtain A^{-1} for any $n \times n$ matrix A for which A^{-1} exists, follow these steps.

1. Form the augmented matrix $[A|I]$, where I is the $n \times n$ identity matrix.
2. Perform row transformations on $[A|I]$ to get a matrix of the form $[I|B]$.
3. Matrix B is A^{-1}.
4. Verify by showing that $BA = AB = I$.

CAUTION Only square matrices have inverses, but not every square matrix has an inverse. If an inverse exists, it is unique. That is, any given square matrix has no more than one inverse. Note that the symbol A^{-1} does not mean $1/A$; the symbol A^{-1} is just the notation for the inverse of matrix A.

■ *Example 3*

FINDING THE
INVERSE OF A 2 × 2
MATRIX

Given

$$A = \begin{bmatrix} 2 & 3 \\ 1 & -1 \end{bmatrix},$$

find A^{-1}.

Find A^{-1} by going through the following steps.

Step 1 Form the augmented matrix $[A|I]$.

$$[A|I] = \begin{bmatrix} 2 & 3 & | & 1 & 0 \\ 1 & -1 & | & 0 & 1 \end{bmatrix}$$

Perform row operations on $[A|I]$ until a new matrix of the form $[I|B]$ is obtained, as follows.

Step 2 Interchanging the first and second rows gives a 1 in the upper left-hand corner.

$$\begin{bmatrix} 1 & -1 & | & 0 & 1 \\ 2 & 3 & | & 1 & 0 \end{bmatrix}$$

Step 3 Multiplying the elements of the first row by -2 and adding the results to the second row gives a 0 in the lower left-hand corner.

$$\begin{bmatrix} 1 & -1 & | & 0 & 1 \\ 0 & 5 & | & 1 & -2 \end{bmatrix} \qquad -2\,R_1 + R_2$$

Step 4 Multiplying the elements of the second row by 1/5 gives a 1 for the second element in that row.

$$\begin{bmatrix} 1 & -1 & | & 0 & 1 \\ 0 & 1 & | & 1/5 & -2/5 \end{bmatrix} \qquad \tfrac{1}{5}R_2$$

Step 5 Replacing the first row by the sum of the first row and the second row gives a 0 as the second element in the first row.

$$\begin{bmatrix} 1 & 0 & | & 1/5 & 3/5 \\ 0 & 1 & | & 1/5 & -2/5 \end{bmatrix} \qquad R_2 + R_1$$

This last matrix is in the form $[I|B]$, where

$$B = \begin{bmatrix} 1/5 & 3/5 \\ 1/5 & -2/5 \end{bmatrix},$$

which should equal A^{-1}. To check, multiply B and A. The result should be I.

$$BA = \begin{bmatrix} 1/5 & 3/5 \\ 1/5 & -2/5 \end{bmatrix}\begin{bmatrix} 2 & 3 \\ 1 & -1 \end{bmatrix} = \begin{bmatrix} 2/5 + 3/5 & 3/5 - 3/5 \\ 2/5 - 2/5 & 3/5 + 2/5 \end{bmatrix} = \begin{bmatrix} 1 & 0 \\ 0 & 1 \end{bmatrix} = I$$

Verify that AB also equals I, so that

$$A^{-1} = \begin{bmatrix} 1/5 & 3/5 \\ 1/5 & -2/5 \end{bmatrix}. \qquad ■$$

■ *Example 4*
FINDING THE
INVERSE OF A 3 × 3
MATRIX

Find A^{-1} if $A = \begin{bmatrix} 1 & 0 & 1 \\ 2 & -2 & -1 \\ 3 & 0 & 0 \end{bmatrix}$.

Use row transformations as follows.

Step 1   Write the augmented matrix $[A|I]$.

$$\left[\begin{array}{ccc|ccc} 1 & 0 & 1 & 1 & 0 & 0 \\ 2 & -2 & -1 & 0 & 1 & 0 \\ 3 & 0 & 0 & 0 & 0 & 1 \end{array}\right]$$

Step 2   Since 1 is already in the upper left-hand corner as desired, begin by using the row transformation that will result in a 0 for the first element in the second row. Multiply the elements of the first row by −2, and add the results to the second row.

$$\left[\begin{array}{ccc|ccc} 1 & 0 & 1 & 1 & 0 & 0 \\ 0 & -2 & -3 & -2 & 1 & 0 \\ 3 & 0 & 0 & 0 & 0 & 1 \end{array}\right] \quad -2\,R_1 + R_2$$

Step 3   To get 0 for the first element in the third row, multiply the elements of the first row by −3 and add to the third row.

$$\left[\begin{array}{ccc|ccc} 1 & 0 & 1 & 1 & 0 & 0 \\ 0 & -2 & -3 & -2 & 1 & 0 \\ 0 & 0 & -3 & -3 & 0 & 1 \end{array}\right] \quad -3\,R_1 + R_3$$

Step 4   To get 1 for the second element in the second row, multiply the elements of the second row by −1/2.

$$\left[\begin{array}{ccc|ccc} 1 & 0 & 1 & 1 & 0 & 0 \\ 0 & 1 & 3/2 & 1 & -1/2 & 0 \\ 0 & 0 & -3 & -3 & 0 & 1 \end{array}\right] \quad -\frac{1}{2}R_2$$

Step 5   To get 1 for the third element in the third row, multiply the elements of the third row by −1/3.

$$\left[\begin{array}{ccc|ccc} 1 & 0 & 1 & 1 & 0 & 0 \\ 0 & 1 & 3/2 & 1 & -1/2 & 0 \\ 0 & 0 & 1 & 1 & 0 & -1/3 \end{array}\right] \quad -\frac{1}{3}R_3$$

Step 6   To get 0 for the third element in the first row, multiply the elements of the third row by −1 and add to the first row.

$$\left[\begin{array}{ccc|ccc} 1 & 0 & 0 & 0 & 0 & 1/3 \\ 0 & 1 & 3/2 & 1 & -1/2 & 0 \\ 0 & 0 & 1 & 1 & 0 & -1/3 \end{array}\right] \quad -1\,R_3 + R_1$$

Step 7   To get 0 for the third element in the second row, multiply the elements of the third row by $-3/2$ and add to the second row.

$$\begin{bmatrix} 1 & 0 & 0 & \bigm| & 0 & 0 & 1/3 \\ 0 & 1 & 0 & \bigm| & -1/2 & -1/2 & 1/2 \\ 0 & 0 & 1 & \bigm| & 1 & 0 & -1/3 \end{bmatrix} \qquad -\frac{3}{2}R_3 + R_2$$

The last transformation shows that the inverse is

$$A^{-1} = \begin{bmatrix} 0 & 0 & 1/3 \\ -1/2 & -1/2 & 1/2 \\ 1 & 0 & -1/3 \end{bmatrix}.$$

Confirm this by forming the products $A^{-1}A$ and AA^{-1}, each of which should equal the matrix *I*.   ■

As illustrated by the examples, the most efficient order of steps is to make the changes column by column from left to right, so that for each column the required 1 is the result of the first change. Next, perform the steps that obtain the zeros in that column. Then proceed to another column. Since it is tedious to find an inverse with paper and pencil, these same steps can be adapted for a computer program. A computer can produce the inverse of a large matrix, even a 40 × 40 matrix, in a few seconds.

■ *Example 5*

IDENTIFYING A
MATRIX WITH NO
INVERSE

Find A^{-1} given $A = \begin{bmatrix} 2 & -4 \\ 1 & -2 \end{bmatrix}$.

Using row transformations to change the first column of the augmented matrix

$$\begin{bmatrix} 2 & -4 & \bigm| & 1 & 0 \\ 1 & -2 & \bigm| & 0 & 1 \end{bmatrix}$$

results in the following matrices:

$$\begin{bmatrix} 1 & -2 & \bigm| & 1/2 & 0 \\ 1 & -2 & \bigm| & 0 & 1 \end{bmatrix} \qquad \text{and} \qquad \begin{bmatrix} 1 & -2 & \bigm| & 1/2 & 0 \\ 0 & 0 & \bigm| & -1/2 & 1 \end{bmatrix}.$$

(We multiplied the elements in row one by 1/2 in the first step.) At this point, the matrix should be changed so that the second-row, second-column element will be 1. Since that element is now 0, there is no way to complete the desired transformation, so A^{-1} does not exist for this matrix *A*. What is wrong? Just as there is no multiplicative inverse for the real number 0, not every matrix has a multiplicative inverse. Matrix *A* is an example of such a matrix.   ■

MATRIX EQUATIONS   Matrix equations are solved in much the same way as real-number equations. Properties of matrices are used, together with a multiplication property of equality for matrices that allows both sides of a matrix equation to be multiplied by any appropriate matrix. The next example shows how to solve a matrix equation.

420

■ ***Example 6***

SOLVING A MATRIX
EQUATION

Given

$$A = \begin{bmatrix} 2 & 2 \\ -1 & -2 \end{bmatrix} \quad \text{and} \quad B = \begin{bmatrix} 2 \\ 3 \end{bmatrix},$$

find a matrix X so that $AX = B$.

Since A is 2×2 and B is 2×1, matrix X will be a 2×1 matrix, like B. The matrix equation $AX = B$ can be solved as a linear equation is solved, with the properties of matrices (given in 8.2 Exercises) and the facts that if A^{-1} exists, $A^{-1}A = I$ and $IX = X$, as follows.

$$\begin{array}{ll} AX = B & \text{Given} \\ A^{-1}(AX) = A^{-1}B & \text{Multiplication property of equality} \\ (A^{-1}A)X = A^{-1}B & \text{Associative property of multiplication} \\ IX = A^{-1}B & \text{Multiplicative inverse property} \\ X = A^{-1}B & \text{Multiplicative identity property} \end{array}$$

To find X, first find A^{-1}. Verify that

$$A^{-1} = \begin{bmatrix} 1 & 1 \\ -1/2 & -1 \end{bmatrix}.$$

As shown above,

$$X = A^{-1}B = \begin{bmatrix} 1 & 1 \\ -1/2 & -1 \end{bmatrix}\begin{bmatrix} 2 \\ 3 \end{bmatrix} = \begin{bmatrix} 5 \\ -4 \end{bmatrix}.$$

Check:

$$AX = \begin{bmatrix} 2 & 2 \\ -1 & -2 \end{bmatrix}\begin{bmatrix} 5 \\ -4 \end{bmatrix} = \begin{bmatrix} 2 \\ 3 \end{bmatrix} = B. \quad ■$$

CAUTION

In using the multiplication property of equality for matrices, be careful to multiply on the left on both sides of the equation. Unlike multiplication of real numbers, multiplication of matrices is not commutative.

The result from Example 6 can be used to solve linear systems by first writing the system as a matrix equation $AX = B$, where X is a matrix of the variables of the system. Then the solution is $X = A^{-1}B$.

■ ***Example 7***

SOLVING A SYSTEM
OF EQUATIONS
USING A MATRIX
INVERSE

Use the inverse of the coefficient matrix to solve the system

$$2x - 3y = 4$$
$$x + 5y = 2.$$

To represent the system as a matrix equation, use one matrix for the coefficients, one for the variables, and one for the constants, as follows.

$$A = \begin{bmatrix} 2 & -3 \\ 1 & 5 \end{bmatrix}, \qquad X = \begin{bmatrix} x \\ y \end{bmatrix}, \qquad \text{and} \qquad B = \begin{bmatrix} 4 \\ 2 \end{bmatrix}$$

The system can then be written in matrix form as the equation $AX = B$, since

$$AX = \begin{bmatrix} 2 & -3 \\ 1 & 5 \end{bmatrix}\begin{bmatrix} x \\ y \end{bmatrix} = \begin{bmatrix} 2x - 3y \\ x + 5y \end{bmatrix} = \begin{bmatrix} 4 \\ 2 \end{bmatrix} = B.$$

To solve the system, first find A^{-1}.

$$A^{-1} = \begin{bmatrix} 5/13 & 3/13 \\ -1/13 & 2/13 \end{bmatrix}$$

Next, find the product $A^{-1}B$.

$$A^{-1}B = \begin{bmatrix} 5/13 & 3/13 \\ -1/13 & 2/13 \end{bmatrix}\begin{bmatrix} 4 \\ 2 \end{bmatrix} = \begin{bmatrix} 2 \\ 0 \end{bmatrix}$$

Since $X = A^{-1}B$,

$$X = \begin{bmatrix} x \\ y \end{bmatrix} = \begin{bmatrix} 2 \\ 0 \end{bmatrix},$$

and the solution set of the system is $\{(2, 0)\}$. ■

8.3 EXERCISES ■

1. What is the product when an $n \times n$ matrix A is multiplied by the $n \times n$ identity matrix?
2. When an $n \times n$ matrix A is multiplied by the inverse matrix A^{-1}, the product is what special matrix?

Decide whether the given matrices are inverses of each other.

3. $\begin{bmatrix} 2 & 3 \\ 1 & 1 \end{bmatrix}, \begin{bmatrix} -1 & 3 \\ 1 & -2 \end{bmatrix}$

4. $\begin{bmatrix} 5 & 7 \\ 2 & 3 \end{bmatrix}, \begin{bmatrix} 3 & -7 \\ -2 & 5 \end{bmatrix}$

5. $\begin{bmatrix} 2 & 1 \\ 3 & 2 \end{bmatrix}, \begin{bmatrix} 2 & 1 \\ -3 & 2 \end{bmatrix}$

6. $\begin{bmatrix} -1 & 2 \\ 3 & -5 \end{bmatrix}, \begin{bmatrix} -5 & -2 \\ -3 & -1 \end{bmatrix}$

7. $\begin{bmatrix} 1 & -2 & -3 \\ 2 & -2 & -5 \\ -1 & 1 & 4 \end{bmatrix}, \begin{bmatrix} -1 & 5/3 & 4/3 \\ -1 & 1/3 & -1/3 \\ 0 & 1/3 & 2/3 \end{bmatrix}$

8. $\begin{bmatrix} 1 & 2 & -1 \\ 2 & -1 & 3 \\ 3 & -2 & 3 \end{bmatrix}, \begin{bmatrix} 3/10 & -2/5 & 1/2 \\ 3/10 & 3/5 & -1/2 \\ -1/10 & 4/5 & -1/2 \end{bmatrix}$

9. $\begin{bmatrix} 1 & 2 & -1 \\ 0 & 1 & 3 \\ 2 & 1 & -2 \end{bmatrix}, \begin{bmatrix} 1 & 1 & 2 \\ 1 & 1 & 1 \\ 2 & 3 & 4 \end{bmatrix}$

10. $\begin{bmatrix} 2 & -1 & 4 \\ 0 & 5 & 0 \\ 3 & 2 & -1 \end{bmatrix}, \begin{bmatrix} 1 & 0 & 1 \\ 6 & 4 & 2 \\ 1 & 1 & 0 \end{bmatrix}$

Use row transformations to find any inverses that exist for the following matrices. See Examples 3–5.

11. $\begin{bmatrix} 1 & -1 \\ 2 & 0 \end{bmatrix}$

12. $\begin{bmatrix} 3 & -1 \\ -5 & 2 \end{bmatrix}$

13. $\begin{bmatrix} -6 & 4 \\ -3 & 2 \end{bmatrix}$

14. $\begin{bmatrix} -1 & 2 \\ -2 & -1 \end{bmatrix}$

15. $\begin{bmatrix} -1 & -2 \\ 3 & 4 \end{bmatrix}$ 　　**16.** $\begin{bmatrix} 5 & 10 \\ -3 & -6 \end{bmatrix}$ 　　**17.** $\begin{bmatrix} .6 & .2 \\ .5 & .1 \end{bmatrix}$ 　　**18.** $\begin{bmatrix} .8 & -.3 \\ .5 & -.2 \end{bmatrix}$

19. $\begin{bmatrix} 1 & 0 & 0 \\ 0 & -1 & 0 \\ 1 & 0 & 1 \end{bmatrix}$ 　**20.** $\begin{bmatrix} 1 & 3 & 3 \\ -1 & 0 & 0 \\ -4 & -4 & -3 \end{bmatrix}$ 　**21.** $\begin{bmatrix} 3 & 6 & 3 \\ 6 & 4 & -2 \\ 0 & 1 & -1 \end{bmatrix}$ 　**22.** $\begin{bmatrix} 2 & 6 & 0 \\ 1 & 5 & 3 \\ 0 & 0 & 1 \end{bmatrix}$

23. $\begin{bmatrix} -1 & -1 & -1 \\ 4 & 5 & 0 \\ 0 & 1 & -3 \end{bmatrix}$ 　**24.** $\begin{bmatrix} 2 & 0 & 4 \\ 3 & 1 & 5 \\ -1 & 1 & -2 \end{bmatrix}$ 　**25.** $\begin{bmatrix} -.4 & .1 & .2 \\ 0 & .6 & .8 \\ .3 & 0 & -.2 \end{bmatrix}$ 　**26.** $\begin{bmatrix} .8 & .2 & .1 \\ -.2 & 0 & .3 \\ 0 & 0 & .5 \end{bmatrix}$

Write the matrix of coefficients, A, the matrix of variables, X, and the matrix of constants, B, for each of the following systems. Do not solve.

27. $x + y = 8$
　　$2x - y = 4$

28. $2x + y = 9$
　　$x + 3y = 17$

29. $4x + 5y = 7$
　　$2x + 3y = 5$

30. $2x + 3y = -2$
　　$5x + 4y = -12$

31. $x + y + z = 9$
　　$2x + y + 3z = 17$
　　$5x - 2y + 2z = 16$

32. $3x + y - 5z = 8$
　　$2x + 6y - 3z = 6$
　　$x + 5y - 9z = -12$

33. A system of linear equations can be solved by using the matrix equation $X = A^{-1}B$. How are the matrices X, A^{-1}, and B related to the linear system?

Solve each of the following systems by using the inverse of the coefficient matrix. See Example 7.

34. $-x + y = 1$
　　$2x - y = 1$

35. $x + y = 5$
　　$x - y = -1$

36. $2x - y = -8$
　　$3x + y = -2$

37. $x + 3y = -12$
　　$2x - y = 11$

38. $2x + 3y = -10$
　　$3x + 4y = -12$

39. $2x - 3y = 10$
　　$2x + 2y = 5$

40. $2x - 5y = 10$
　　$4x - 5y = 15$

41. $2x - 3y = 2$
　　$4x - 6y = 1$

Use the inverse of the coefficient matrix to solve each of the following systems of equations. The inverses of the coefficient matrices for the first four problems are found in the text and exercises of this section.

42. 　$x + 3y + 3z = 11$
　　　　$-x = -2$
　　$-4x - 4y - 3z = -10$

43. 　　$x + z = 3$
　　$2x - 2y - z = -2$
　　　　$3x = 3$

44. 　$2x + 4z = 14$
　　$3x + y + 5z = 19$
　　$-x + y - 2z = -7$

45. $3x + 6y + 3z = 12$
　　$6x + 4y - 2z = -4$
　　　　$y - z = -3$

46. $x + 3y + z = 2$
　　$x - 2y + 3z = -3$
　　$2x - 3y - z = 34$

47. $x + y - z = 6$
　　$2x - y + z = -9$
　　$x - 2y + 3z = 1$

In Exercises 48 and 49, write a system of equations and solve it by using the inverse of the coefficient matrix.

48. Midtown Manufacturing Company makes two products, plastic plates and plastic cups. Both require time on two machines: plates require 1 hr on machine A and 2 hr on machine B; cups require 3 hr on machine A and 1 hr on machine B. Both machines operate 15 hr a day. How many of each product can be produced in a day under these conditions?

49. A company produces two models of bicycles, model 201 and model 301. Model 201 requires 2 hr of assembly time and model 301 requires 3 hr of assembly time. The parts of model 201 cost \$25 per bike and the parts for model 301 cost \$30 per bike. If the company has a total of 34 hr of assembly time and \$365 available per day for these two models, how many of each can be made in a day?

50. List the methods presented so far in this text to solve a linear system. Which one do you prefer? Explain why.

Let $A = \begin{bmatrix} a & b \\ c & d \end{bmatrix}$. *Show that the following are true.*

51. $AA^{-1} = I$ and $A^{-1}A = I$ 　　**52.** $IA = A$ 　　**53.** $AI = A$ 　　**54.** $A \cdot O = O$

———— **DETERMINANTS**

In Chapter 7, we saw that not every system of linear equations has a single solution. Sometimes a system of n equations in n variables has no solution or an infinite set of solutions. In this section, we introduce the *determinant* of a matrix. We shall see in Section 8.6 that the determinant can be used to determine whether a system of equations has a single solution. (Also, see Exercise 55 at the end of this section.)

Every square matrix A is associated with a real number called the determinant of A, written $|A|$.

DETERMINANT OF A 2 × 2 MATRIX

The **determinant of a 2 × 2 matrix** A,

$$A = \begin{bmatrix} a_{11} & a_{12} \\ a_{21} & a_{22} \end{bmatrix},$$

is defined as

$$|A| = \begin{vmatrix} a_{11} & a_{12} \\ a_{21} & a_{22} \end{vmatrix} = a_{11}a_{22} - a_{21}a_{12}.$$

NOTE Notice that matrices are enclosed with square brackets, while determinants are denoted with vertical bars. Also, the matrix is an *array* of numbers, but its determinant is a *single number.*

■ *Example 1*
EVALUATING A 2 × 2 DETERMINANT

If $P = \begin{bmatrix} -3 & 4 \\ 6 & 8 \end{bmatrix}$, then

$$|P| = \begin{vmatrix} -3 & 4 \\ 6 & 8 \end{vmatrix} = -3(8) - 6(4) = -48. \quad ■$$

The definition of a determinant can be extended to a 3 × 3 matrix as follows.

DETERMINANT OF A 3 × 3 MATRIX

The **determinant of a 3 × 3 matrix** A,

$$A = \begin{bmatrix} a_{11} & a_{12} & a_{13} \\ a_{21} & a_{22} & a_{23} \\ a_{31} & a_{32} & a_{33} \end{bmatrix},$$

is defined as

$$|A| = \begin{vmatrix} a_{11} & a_{12} & a_{13} \\ a_{21} & a_{22} & a_{23} \\ a_{31} & a_{32} & a_{33} \end{vmatrix} = (a_{11}a_{22}a_{33} + a_{12}a_{23}a_{31} + a_{13}a_{21}a_{32}) - (a_{31}a_{22}a_{13} + a_{32}a_{23}a_{11} + a_{33}a_{21}a_{12}).$$

An easy method for calculating 3×3 determinants is found by rearranging and factoring the terms given above to get

$$\begin{vmatrix} a_{11} & a_{12} & a_{13} \\ a_{21} & a_{22} & a_{23} \\ a_{31} & a_{32} & a_{33} \end{vmatrix} = a_{11}(a_{22}a_{33} - a_{32}a_{23}) - a_{21}(a_{12}a_{33} - a_{32}a_{13}) + a_{31}(a_{12}a_{23} - a_{22}a_{13}).$$

Each of the quantities in parentheses represents the determinant of a 2×2 matrix that is the part of the 3×3 matrix remaining when the row and column of the multiplier are eliminated, as shown below.

$$a_{11}(a_{22}a_{33} - a_{32}a_{23}) \qquad \begin{bmatrix} a_{11} & a_{12} & a_{13} \\ a_{21} & a_{22} & a_{23} \\ a_{31} & a_{32} & a_{33} \end{bmatrix}$$

$$a_{21}(a_{12}a_{33} - a_{32}a_{13}) \qquad \begin{bmatrix} a_{11} & a_{12} & a_{13} \\ a_{21} & a_{22} & a_{23} \\ a_{31} & a_{32} & a_{33} \end{bmatrix}$$

$$a_{31}(a_{12}a_{23} - a_{22}a_{13}) \qquad \begin{bmatrix} a_{11} & a_{12} & a_{13} \\ a_{21} & a_{22} & a_{23} \\ a_{31} & a_{32} & a_{33} \end{bmatrix}$$

These determinants of the 2×2 matrices are called **minors** of an element in the 3×3 matrix. The symbol M_{ij} represents the determinant of the matrix that results when row i and column j are eliminated. The following list gives some of the minors from the matrix above.

Element	Minor	Element	Minor
a_{11}	$M_{11} = \begin{vmatrix} a_{22} & a_{23} \\ a_{32} & a_{33} \end{vmatrix}$	a_{22}	$M_{22} = \begin{vmatrix} a_{11} & a_{13} \\ a_{31} & a_{33} \end{vmatrix}$
a_{21}	$M_{21} = \begin{vmatrix} a_{12} & a_{13} \\ a_{32} & a_{33} \end{vmatrix}$	a_{23}	$M_{23} = \begin{vmatrix} a_{11} & a_{12} \\ a_{31} & a_{32} \end{vmatrix}$
a_{31}	$M_{31} = \begin{vmatrix} a_{12} & a_{13} \\ a_{22} & a_{23} \end{vmatrix}$	a_{33}	$M_{33} = \begin{vmatrix} a_{11} & a_{12} \\ a_{21} & a_{22} \end{vmatrix}$

In a 4×4 matrix, the minors are determinants of 3×3 matrices, and an $n \times n$ matrix has minors that are determinants of $(n - 1) \times (n - 1)$ matrices.

To find the determinant of a 3×3 or larger matrix, first choose any row or column. Then the minor of each element in that row or column must be multiplied by $+1$ or -1, depending on whether the sum of the row numbers and column numbers is even or odd. The product of a minor and the number $+1$ or -1 is called a *cofactor.*

COFACTOR	Let M_{ij} be the minor for element a_{ij} in an $n \times n$ matrix. The **cofactor** of a_{ij}, written A_{ij}, is

$$A_{ij} = (-1)^{i+j} \cdot M_{ij}.$$

Finally, the determinant of an $n \times n$ matrix is found as follows.

FINDING THE DETERMINANT OF A MATRIX	Multiply each element in any row or column of the matrix by its cofactor. The sum of these products gives the value of the determinant.

The process of forming this sum of products is called **expansion by a given row or column.** (See Exercises 51–52.)

■ *Example 2*

FINDING THE COFACTOR OF AN ELEMENT

For the matrix

$$\begin{bmatrix} 6 & 2 & 4 \\ 8 & 9 & 3 \\ 1 & 2 & 0 \end{bmatrix},$$

find the cofactor of each of the following elements.

(a) 6

Since 6 is in the first row and first column of the matrix, $i = 1$ and $j = 1$.

$$M_{11} = \begin{vmatrix} 9 & 3 \\ 2 & 0 \end{vmatrix} = -6$$

The cofactor is $(-1)^{1+1} \cdot -6 = 1 \cdot -6 = -6$.

(b) 3

Here $i = 2$ and $j = 3$.

$$M_{23} = \begin{vmatrix} 6 & 2 \\ 1 & 2 \end{vmatrix} = 10$$

The cofactor is $(-1)^{2+3} \cdot 10 = -1 \cdot 10 = -10$.

(c) 8

We have $i = 2$ and $j = 1$.

$$M_{21} = \begin{vmatrix} 2 & 4 \\ 2 & 0 \end{vmatrix} = -8$$

The cofactor is $(-1)^{2+1} \cdot -8 = -1 \cdot -8 = 8$. ■

■ *Example 3*

EVALUATING A 3 × 3
DETERMINANT

Evaluate $\begin{vmatrix} 2 & -3 & -2 \\ -1 & -4 & -3 \\ -1 & 0 & 2 \end{vmatrix}$, expanding by the second column.

To find this determinant, first get the minors of each element in the second column.

$$M_{12} = \begin{vmatrix} -1 & -3 \\ -1 & 2 \end{vmatrix} = -1(2) - (-1)(-3) = -5$$

$$M_{22} = \begin{vmatrix} 2 & -2 \\ -1 & 2 \end{vmatrix} = 2(2) - (-1)(-2) = 2$$

$$M_{32} = \begin{vmatrix} 2 & -2 \\ -1 & -3 \end{vmatrix} = 2(-3) - (-1)(-2) = -8$$

Now find the cofactor of each of these minors.

$$A_{12} = (-1)^{1+2} \cdot M_{12} = (-1)^3 \cdot (-5) = (-1)(-5) = 5$$
$$A_{22} = (-1)^{2+2} \cdot M_{22} = (-1)^4 \cdot (2) = 1 \cdot 2 = 2$$
$$A_{32} = (-1)^{3+2} \cdot M_{32} = (-1)^5 \cdot (-8) = (-1)(-8) = 8$$

The determinant is found by multiplying each cofactor by its corresponding element in the matrix and finding the sum of these products.

$$\begin{vmatrix} 2 & -3 & -2 \\ -1 & -4 & -3 \\ -1 & 0 & 2 \end{vmatrix} = a_{12} \cdot A_{12} + a_{22} \cdot A_{22} + a_{32} \cdot A_{32}$$

$$= -3(5) + (-4)(2) + (0)(8)$$
$$= -15 + (-8) + 0 = -23 \quad ■$$

CAUTION

Be very careful to keep track of all negative signs when evaluating determinants. Work carefully, writing down each step as in the examples. Skipping steps frequently leads to errors in these computations.

Exactly the same answer would be found using any row or column of the matrix. One reason that column 2 was used in Example 3 is that it contains a 0 element, so that it was not really necessary to calculate M_{32} and A_{32} above. One learns quickly that zeros can be very useful in working with determinants.

Instead of calculating $(-1)^{i+j}$ for a given element, the following sign checkerboards can be used.

ARRAY OF SIGNS	For 3 × 3 Matrices	For 4 × 4 Matrices
	+ − +	+ − + −
	− + −	− + − +
	+ − +	+ − + −
		− + − +

The signs alternate for each row and column, beginning with + in the first row, first column position. Thus, these arrays of signs can be reproduced as needed. If we expand a 3 × 3 matrix about row 3, for example, the first minor would have a + sign associated with it, the second minor a − sign, and the third minor a + sign. These arrays of signs can be extended in this way for determinants of 5 × 5, 6 × 6, and larger matrices.

■ *Example 4*

EVALUATING A 4 × 4 DETERMINANT

Evaluate

$$\begin{vmatrix} -1 & -2 & 3 & 2 \\ 0 & 1 & 4 & -2 \\ 3 & -1 & 4 & 0 \\ 2 & 1 & 0 & 3 \end{vmatrix}.$$

Expanding by minors about the fourth row gives

$$-2\begin{vmatrix} -2 & 3 & 2 \\ 1 & 4 & -2 \\ -1 & 4 & 0 \end{vmatrix} + 1\begin{vmatrix} -1 & 3 & 2 \\ 0 & 4 & -2 \\ 3 & 4 & 0 \end{vmatrix} - 0\begin{vmatrix} -1 & -2 & 2 \\ 0 & 1 & -2 \\ 3 & -1 & 0 \end{vmatrix} + 3\begin{vmatrix} -1 & -2 & 3 \\ 0 & 1 & 4 \\ 3 & -1 & 4 \end{vmatrix}$$

$$= -2(6) + 1(-50) - 0 + 3(-41)$$

$$= -185. \quad ■$$

Each of the four determinants in Example 4 must be evaluated by expansion of three minors, requiring much work to get the final value. Always look for the row or column with the most zeros to simplify the work. In the next section we introduce several properties that make it easier to calculate determinants. Fortunately, determinants of large matrices can be evaluated quickly and easily with the aid of a computer or with certain calculators.

8.4 EXERCISES ■ ─────────────────────────────

Evaluate each of the following determinants. See Example 1.

1. $\begin{vmatrix} 5 & 8 \\ 2 & -4 \end{vmatrix}$
2. $\begin{vmatrix} -3 & 0 \\ 0 & 9 \end{vmatrix}$
3. $\begin{vmatrix} -1 & -2 \\ 5 & 3 \end{vmatrix}$
4. $\begin{vmatrix} 6 & -4 \\ 0 & -1 \end{vmatrix}$
5. $\begin{vmatrix} 9 & 3 \\ -3 & -1 \end{vmatrix}$

6. $\begin{vmatrix} 0 & 2 \\ 1 & 5 \end{vmatrix}$
7. $\begin{vmatrix} 3 & 4 \\ 5 & -2 \end{vmatrix}$
8. $\begin{vmatrix} -9 & 7 \\ 2 & 6 \end{vmatrix}$
9. $\begin{vmatrix} 0 & 4 \\ 4 & 0 \end{vmatrix}$
10. $\begin{vmatrix} 1 & 0 \\ 0 & 2 \end{vmatrix}$

11. $\begin{vmatrix} 8 & 3 \\ 8 & 3 \end{vmatrix}$
12. $\begin{vmatrix} 9 & -4 \\ -4 & 9 \end{vmatrix}$
13. $\begin{vmatrix} x & 4 \\ 8 & 2 \end{vmatrix}$
14. $\begin{vmatrix} k & 3 \\ 0 & 4 \end{vmatrix}$
15. $\begin{vmatrix} y & 2 \\ 8 & y \end{vmatrix}$

16. $\begin{vmatrix} 3 & 8 \\ m & n \end{vmatrix}$
17. $\begin{vmatrix} x & y \\ y & x \end{vmatrix}$
18. $\begin{vmatrix} 2m & 8n \\ 8n & 2m \end{vmatrix}$
19. $\begin{vmatrix} 1.4 & 2.5 \\ 3.7 & 6.2 \end{vmatrix}$
20. $\begin{vmatrix} .123 & .054 \\ .691 & .302 \end{vmatrix}$

21. Explain how to evaluate a 2 × 2 determinant.

22. In your own words, define the minor of an element of a matrix. Then define the cofactor of the element.

For the following determinants, find the cofactor of each element in the second row of the corresponding matrix. See Example 2.

23. $\begin{vmatrix} -2 & 0 & 1 \\ 3 & 2 & -1 \\ 1 & 0 & 2 \end{vmatrix}$
24. $\begin{vmatrix} 0 & -1 & 2 \\ 1 & 0 & 2 \\ 0 & -3 & 1 \end{vmatrix}$
25. $\begin{vmatrix} 1 & 2 & -1 \\ 2 & 3 & -2 \\ -1 & 4 & 1 \end{vmatrix}$
26. $\begin{vmatrix} 2 & -1 & 4 \\ 3 & 0 & 1 \\ -2 & 1 & 4 \end{vmatrix}$

Evaluate each of the following determinants. See Examples 2 and 3.

27. $\begin{vmatrix} 1 & 0 & 0 \\ 0 & -1 & 0 \\ 1 & 0 & 1 \end{vmatrix}$
28. $\begin{vmatrix} -2 & 0 & 1 \\ 0 & 1 & 0 \\ 0 & 0 & -1 \end{vmatrix}$
29. $\begin{vmatrix} -2 & 0 & 0 \\ 4 & 0 & 1 \\ 3 & 4 & 2 \end{vmatrix}$
30. $\begin{vmatrix} 3 & -2 & 0 \\ 0 & -1 & 1 \\ 4 & 0 & 2 \end{vmatrix}$
31. $\begin{vmatrix} 1 & 2 & 0 \\ -1 & 2 & -1 \\ 0 & 1 & 4 \end{vmatrix}$

32. $\begin{vmatrix} 2 & 1 & -1 \\ 4 & 7 & -2 \\ 2 & 4 & 0 \end{vmatrix}$
33. $\begin{vmatrix} 10 & 2 & 1 \\ -1 & 4 & 3 \\ -3 & 8 & 10 \end{vmatrix}$
34. $\begin{vmatrix} 7 & -1 & 1 \\ 1 & -7 & 2 \\ -2 & 1 & 1 \end{vmatrix}$
35. $\begin{vmatrix} 1 & -2 & 3 \\ 0 & 0 & 0 \\ 1 & 10 & -12 \end{vmatrix}$
36. $\begin{vmatrix} 2 & 3 & 0 \\ 1 & 9 & 0 \\ -1 & -2 & 0 \end{vmatrix}$

37. $\begin{vmatrix} 3 & 3 & -1 \\ 2 & 6 & 0 \\ -6 & -6 & 2 \end{vmatrix}$
38. $\begin{vmatrix} 5 & -3 & 2 \\ -5 & 3 & -2 \\ 1 & 0 & 1 \end{vmatrix}$
39. $\begin{vmatrix} 3 & 2 & 0 \\ 0 & 1 & x \\ 2 & 0 & 0 \end{vmatrix}$
40. $\begin{vmatrix} 0 & 3 & y \\ 0 & 4 & 2 \\ 1 & 0 & 1 \end{vmatrix}$

41. Explain why a determinant with a row or column of zeros has a value of zero.

42. Does a determinant with two identical rows have a value of zero?

Solve each of the following equations for x.

43. $\begin{vmatrix} -2 & 0 & 1 \\ -1 & 3 & x \\ 5 & -2 & 0 \end{vmatrix} = 3$
44. $\begin{vmatrix} 4 & 3 & 0 \\ 2 & 0 & 1 \\ -3 & x & -1 \end{vmatrix} = 5$
45. $\begin{vmatrix} 5 & 3x & -3 \\ 0 & 2 & -1 \\ 4 & -1 & x \end{vmatrix} = -7$
46. $\begin{vmatrix} 2x & 1 & -1 \\ 0 & 4 & x \\ 3 & 0 & 2 \end{vmatrix} = x$

Find the value of each of the following determinants. See Example 4.

47. $\begin{vmatrix} 4 & 0 & 0 & 2 \\ -1 & 0 & 3 & 0 \\ 2 & 4 & 0 & 1 \\ 0 & 0 & 1 & 2 \end{vmatrix}$
48. $\begin{vmatrix} -2 & 0 & 4 & 2 \\ 3 & 6 & 0 & 4 \\ 0 & 0 & 0 & 3 \\ 9 & 0 & 2 & -1 \end{vmatrix}$
49. $\begin{vmatrix} 1 & 1 & 0 & 1 \\ 2 & 1 & 0 & 2 \\ 0 & 1 & -1 & 1 \\ 1 & -1 & 1 & 1 \end{vmatrix}$
50. $\begin{vmatrix} 2 & 7 & 0 & -1 \\ 1 & 0 & 1 & 3 \\ 2 & 4 & -1 & -1 \\ -1 & 1 & 0 & 8 \end{vmatrix}$

Let $A = \begin{bmatrix} a_{11} & a_{12} & a_{13} \\ a_{21} & a_{22} & a_{23} \\ a_{31} & a_{32} & a_{33} \end{bmatrix}$ for Exercises 51–52.

51. Find $|A|$ by expansion about row 3 of the matrix. Show that your result is really equal to $|A|$ as given in the definition of the determinant of a 3×3 matrix.

52. Repeat Exercise 51 for column 3.

Determinants can be used to find the equation of a line passing through two given points. The next two exercises show how to do this.

53. Expand the determinant below and show that the result is the equation of the line through $(2, 3)$, and $(-1, 4)$.

$$\begin{vmatrix} x & y & 1 \\ 2 & 3 & 1 \\ -1 & 4 & 1 \end{vmatrix} = 0$$

54. (a) Write the equation of the line through the points (x_1, y_1) and (x_2, y_2) using the point-slope formula.

(b) Expanding the determinant in the equation below, show that the equation is equivalent to the equation in part (a).

$$\begin{vmatrix} x & y & 1 \\ x_1 & y_1 & 1 \\ x_2 & y_2 & 1 \end{vmatrix} = 0$$

55. The inverse of matrix $A = \begin{bmatrix} a & b \\ c & d \end{bmatrix}$ is

$$A^{-1} = \begin{bmatrix} \dfrac{d}{ad-bc} & \dfrac{-b}{ad-bc} \\ \dfrac{-c}{ad-bc} & \dfrac{a}{ad-bc} \end{bmatrix}.$$

Show that A^{-1} does not exist if $|A| = 0$. What does this tell you about the solution of a linear system with A as the coefficient matrix?

56. Show that if $A = \begin{bmatrix} a & b \\ c & d \end{bmatrix}$ and $|A| \neq 0$, then

$$A^{-1} = \begin{bmatrix} \dfrac{d}{|A|} & \dfrac{-b}{|A|} \\ \dfrac{-c}{|A|} & \dfrac{a}{|A|} \end{bmatrix}.$$

This is an alternative method for finding the inverse of a 2×2 matrix A, where $|A| \neq 0$. Unfortunately, the method does not generalize to any $n \times n$ matrix.

8.5 ——— PROPERTIES OF DETERMINANTS

Row transformations can be used to show that for the linear system

$$ax + by = m$$
$$cx + dy = n,$$

the matrix of coefficients

$$A = \begin{bmatrix} a & b \\ c & d \end{bmatrix}$$

has inverse matrix

$$A^{-1} = \begin{bmatrix} \dfrac{d}{ad-bc} & \dfrac{-b}{ad-bc} \\ \dfrac{-c}{ad-bc} & \dfrac{a}{ad-bc} \end{bmatrix} = \begin{bmatrix} \dfrac{d}{|A|} & \dfrac{-b}{|A|} \\ \dfrac{-c}{|A|} & \dfrac{a}{|A|} \end{bmatrix}.$$

Thus, if $|A| = 0$, A^{-1} does not exist. It can be shown that this idea extends to $n \times n$ matrices in general. This provides a way to determine whether a system of equations has a single solution without actually solving the system. The properties of determinants given in this section make it easier to decide quickly if a determinant is zero or to evaluate a nonzero determinant more quickly.

The value of

$$\begin{vmatrix} 3 & 0 \\ 5 & 0 \end{vmatrix}$$

is $3 \cdot 0 - 5 \cdot 0 = 0$. Also, expanded about the second row,

$$\begin{vmatrix} 1 & 2 & 3 \\ 0 & 0 & 0 \\ 4 & 6 & 8 \end{vmatrix} = -0 \begin{vmatrix} 2 & 3 \\ 6 & 8 \end{vmatrix} + 0 \begin{vmatrix} 1 & 3 \\ 4 & 8 \end{vmatrix} - 0 \begin{vmatrix} 1 & 2 \\ 4 & 6 \end{vmatrix} = 0.$$

The fact that the matrix for each of these determinants has a row or column of zeros leads to a determinant of 0. These examples suggest the following result.

DETERMINANT PROPERTY 1 If every element in a row or column of a matrix is 0, then the determinant equals 0.

To prove Property 1, expand the determinant about the 0 row or column. Each term of this expansion has a 0 factor, so each term is 0. For example,

$$\begin{vmatrix} a & b & c \\ 0 & 0 & 0 \\ d & e & f \end{vmatrix} = -0 \begin{vmatrix} b & c \\ e & f \end{vmatrix} + 0 \begin{vmatrix} a & c \\ d & f \end{vmatrix} - 0 \begin{vmatrix} a & b \\ d & e \end{vmatrix} = 0.$$

■ *Example 1*
APPLYING
PROPERTY 1

$$\begin{vmatrix} 2 & 4 \\ 0 & 0 \end{vmatrix} = 0 \qquad \text{and} \qquad \begin{vmatrix} -3 & 7 & 0 \\ 4 & 9 & 0 \\ -6 & 8 & 0 \end{vmatrix} = 0 \quad ■$$

Suppose the matrix

$$\begin{bmatrix} 1 & 5 \\ 3 & 7 \end{bmatrix}$$

is rewritten as

$$\begin{bmatrix} 1 & 3 \\ 5 & 7 \end{bmatrix},$$

so the rows of the first matrix are the columns of the second matrix. Evaluating the determinant of each matrix gives

$$\begin{vmatrix} 1 & 5 \\ 3 & 7 \end{vmatrix} = 1 \cdot 7 - 3 \cdot 5 = -8$$

$$\begin{vmatrix} 1 & 3 \\ 5 & 7 \end{vmatrix} = 1 \cdot 7 - 5 \cdot 3 = -8.$$

The two determinants are the same, suggesting the next property of determinants.

| **DETERMINANT** | If corresponding rows and columns of a matrix are interchanged, the determinant |
| **PROPERTY 2** | is not changed. |

A proof of Property 2 is requested in Exercise 37.

■ *Example 2*
APPLYING
PROPERTY 2

Given

$$A = \begin{bmatrix} 2 & 1 & 6 \\ 3 & 0 & 5 \\ -4 & 6 & 9 \end{bmatrix} \quad \text{and} \quad B = \begin{bmatrix} 2 & 3 & -4 \\ 1 & 0 & 6 \\ 6 & 5 & 9 \end{bmatrix},$$

by Property 2, since the rows of the first matrix are the columns of the second,

$$\begin{vmatrix} 2 & 1 & 6 \\ 3 & 0 & 5 \\ -4 & 6 & 9 \end{vmatrix} = \begin{vmatrix} 2 & 3 & -4 \\ 1 & 0 & 6 \\ 6 & 5 & 9 \end{vmatrix}. \quad ■$$

Exchanging the rows of the matrix

$$\begin{bmatrix} 1 & 6 \\ 5 & 3 \end{bmatrix} \quad \text{gives} \quad \begin{bmatrix} 5 & 3 \\ 1 & 6 \end{bmatrix}.$$

While the first determinant is

$$\begin{vmatrix} 1 & 6 \\ 5 & 3 \end{vmatrix} = 1 \cdot 3 - 5 \cdot 6 = -27,$$

the second determinant is

$$\begin{vmatrix} 5 & 3 \\ 1 & 6 \end{vmatrix} = 5 \cdot 6 - 1 \cdot 3 = 27.$$

The second determinant is the negative of the first. This suggests the third property of determinants.

| **DETERMINANT** | Interchanging two rows (or columns) of a matrix reverses the sign of the deter- |
| **PROPERTY 3** | minant. |

Exercise 38 asks for a proof of Property 3.

■ *Example 3*
APPLYING
PROPERTY 3

(a) Interchange the two columns of

$$\begin{bmatrix} 2 & 5 \\ 3 & 4 \end{bmatrix} \qquad \text{to get} \qquad \begin{bmatrix} 5 & 2 \\ 4 & 3 \end{bmatrix}.$$

By Property 3, since

$$\begin{vmatrix} 2 & 5 \\ 3 & 4 \end{vmatrix} = -7,$$

then

$$\begin{vmatrix} 5 & 2 \\ 4 & 3 \end{vmatrix} = 7.$$

(b) $$\begin{vmatrix} 2 & 1 & 6 \\ 3 & 0 & 5 \\ -4 & 6 & 9 \end{vmatrix} = - \begin{vmatrix} -4 & 6 & 9 \\ 3 & 0 & 5 \\ 2 & 1 & 6 \end{vmatrix} \quad ▩$$

Multiplying each element of the second row of the matrix

$$\begin{bmatrix} 2 & -3 \\ 4 & 1 \end{bmatrix}$$

by -5 gives the new matrix

$$\begin{bmatrix} 2 & -3 \\ 4(-5) & 1(-5) \end{bmatrix} = \begin{bmatrix} 2 & -3 \\ -20 & -5 \end{bmatrix}.$$

The determinants of these two matrices are

$$\begin{vmatrix} 2 & -3 \\ 4 & 1 \end{vmatrix} = 2 \cdot 1 - 4(-3) = 14$$

and

$$\begin{vmatrix} 2 & -3 \\ -20 & -5 \end{vmatrix} = 2(-5) - (-20)(-3) = -70.$$

Since $-70 = -5(14)$, the new determinant is -5 times the original determinant. The next property generalizes this idea. A proof is requested in Exercise 39.

DETERMINANT
PROPERTY 4

If every element of a row (or column) of a matrix is multiplied by the real number k, then the determinant of the new matrix is k times the determinant of the original matrix.

■ *Example 4*
APPLYING
PROPERTY 4

By Property 4, if

$$A = \begin{bmatrix} -2 & 3 & 7 \\ 0 & 5 & 2 \\ -16 & 6 & 4 \end{bmatrix} \qquad \text{and} \qquad B = \begin{bmatrix} 1 & 3 & 7 \\ 0 & 5 & 2 \\ 8 & 6 & 4 \end{bmatrix},$$

then $|A| = -2 \cdot |B|$, since the first column of matrix A is -2 times the first column of matrix B. ■

The matrix

$$\begin{bmatrix} 2 & 1 & 2 \\ 5 & 10 & 5 \\ 3 & 6 & 3 \end{bmatrix}$$

has first and third columns that are identical. Expanding the determinant about the second column gives

$$\begin{vmatrix} 2 & 1 & 2 \\ 5 & 10 & 5 \\ 3 & 6 & 3 \end{vmatrix} = -1 \begin{vmatrix} 5 & 5 \\ 3 & 3 \end{vmatrix} + 10 \begin{vmatrix} 2 & 2 \\ 3 & 3 \end{vmatrix} - 6 \begin{vmatrix} 2 & 2 \\ 5 & 5 \end{vmatrix}$$

$$= -1(15 - 15) + 10(6 - 6) - 6(10 - 10)$$

$$= 0.$$

This result suggests the next property.

DETERMINANT PROPERTY 5	The determinant of a matrix with two identical rows (or columns) equals 0.

∎ *Example 5*
APPLYING
PROPERTY 5

Since two rows are identical, the determinant of

$$\begin{bmatrix} -4 & 2 & 3 \\ 0 & 1 & 6 \\ -4 & 2 & 3 \end{bmatrix}$$

equals 0. ∎

If we multiply each element of the second row of the matrix

$$\begin{bmatrix} -3 & 5 \\ 1 & 2 \end{bmatrix}$$

by 3 and add the result to the first row, we get

$$\begin{bmatrix} -3 + 1(3) & 5 + 2(3) \\ 1 & 2 \end{bmatrix} = \begin{bmatrix} -3 + 3 & 5 + 6 \\ 1 & 2 \end{bmatrix} = \begin{bmatrix} 0 & 11 \\ 1 & 2 \end{bmatrix}.$$

Verify that the determinant of the new matrix is the same as the determinant of the original matrix, -11. This idea, which is generalized below, is perhaps the most useful property of determinants presented in this section.

DETERMINANT PROPERTY 6	The determinant of a matrix is unchanged if a multiple of a row (or column) of the matrix is added to the corresponding elements of another row (or column).

■ *Example 6*

APPLYING
PROPERTY 6

Multiply each element of the first column of the matrix

$$\begin{bmatrix} -2 & 4 & 1 \\ 2 & 1 & 5 \\ 3 & 0 & 2 \end{bmatrix}$$

by 3, and add the results to the third column to get the new matrix

$$\begin{bmatrix} -2 & 4 & 1 + 3(-2) \\ 2 & 1 & 5 + 3(2) \\ 3 & 0 & 2 + 3(3) \end{bmatrix} = \begin{bmatrix} -2 & 4 & -5 \\ 2 & 1 & 11 \\ 3 & 0 & 11 \end{bmatrix}.$$

By Property 6, the determinants of these two matrices are equal. Verify that each determinant equals 37. ■

The following examples show how the properties of determinants can be used to simplify the calculation of determinants.

■ *Example 7*

USING THE
PROPERTIES TO
EVALUATE A
DETERMINANT

Without expanding, show that the determinant of the following matrix is zero.

$$\begin{bmatrix} 2 & 5 & -1 \\ 1 & -15 & 3 \\ -2 & 10 & -2 \end{bmatrix}$$

Each element in the second column equals -5 times the corresponding element in the third column. Multiply the elements of the third column by 5 and then add the results to the corresponding elements in the second column to get a matrix with an equivalent determinant,

$$\begin{bmatrix} 2 & 0 & -1 \\ 1 & 0 & 3 \\ -2 & 0 & -2 \end{bmatrix}.$$

By Property 1, the determinant of this matrix is zero. ■

■ *Example 8*

USING THE
PROPERTIES TO
EVALUATE A
DETERMINANT

Evaluate

$$\begin{vmatrix} 4 & 2 & 1 & 0 \\ -2 & 4 & -1 & 7 \\ -5 & 2 & 3 & 1 \\ 6 & 4 & -3 & 2 \end{vmatrix}.$$

Use Property 6 to change the first row (any row or column could be used) of the matrix to a row in which every element but one is zero. Begin by multiplying the elements of the second column by -2 and adding the results to the first column, replacing the first column with this new column. The result is a 0 for the first element in the first row.

$$\begin{bmatrix} 0 & 2 & 1 & 0 \\ -10 & 4 & -1 & 7 \\ -9 & 2 & 3 & 1 \\ -2 & 4 & -3 & 2 \end{bmatrix}$$

The first column replaced by the sum of -2 times the second column and the first column

To get a zero for the second element in the first row, multiply the elements of the third column by -2 and add to the second column.

$$\begin{bmatrix} 0 & 0 & 1 & 0 \\ -10 & 6 & -1 & 7 \\ -9 & -4 & 3 & 1 \\ -2 & 10 & -3 & 2 \end{bmatrix}$$

The second column replaced by the sum of -2 times the third column and the second column

Since the first row now has only one nonzero number, expand the determinant about the first row to get

$$1 \begin{vmatrix} -10 & 6 & 7 \\ -9 & -4 & 1 \\ -2 & 10 & 2 \end{vmatrix} = \begin{vmatrix} -10 & 6 & 7 \\ -9 & -4 & 1 \\ -2 & 10 & 2 \end{vmatrix}.$$

Repeat the process with the 3×3 matrix

$$\begin{bmatrix} -10 & 6 & 7 \\ -9 & -4 & 1 \\ -2 & 10 & 2 \end{bmatrix}.$$

Change the third column to a column with two zeros, as shown below.

$$\begin{bmatrix} 53 & 34 & 0 \\ -9 & -4 & 1 \\ -2 & 10 & 2 \end{bmatrix}$$

The first row replaced by the sum of -7 times the second row and the first row

$$\begin{bmatrix} 53 & 34 & 0 \\ -9 & -4 & 1 \\ 16 & 18 & 0 \end{bmatrix}$$

The third row replaced by the sum of -2 times the second row and the third row

Expand the determinant about the third column to evaluate.

$$-1 \begin{vmatrix} 53 & 34 \\ 16 & 18 \end{vmatrix} = -1(954 - 544) = -410 \quad ■$$

> **NOTE** When applying Property 6, work with sums of *rows* to get a *column* with only one nonzero number or with sums of *columns* to get a *row* with one nonzero number.

The following summary lists the properties of determinants presented in this section.

436

> **PROPERTIES OF DETERMINANTS**
>
> **1.** If every element in a row or column of a matrix is 0, then the determinant equals 0.
> **2.** If corresponding rows and columns of a matrix are interchanged, the determinant is not changed.
> **3.** Interchanging two rows (or columns) of a matrix reverses the sign of the determinant.
> **4.** If every element of a row (or column) of a matrix is multiplied by the real number k, then the determinant of the new matrix is k times the determinant of the original matrix.
> **5.** The determinant of a matrix with two identical rows (or columns) equals 0.
> **6.** The determinant of a matrix is unchanged if a multiple of a row (or column) of the matrix is added to the corresponding elements of another row (or column).

8.5 EXERCISES ■

Give the property that tells why each of the following determinants equals 0. (More than one property may be used.)

1. $\begin{vmatrix} 2 & 3 \\ 2 & 3 \end{vmatrix}$

2. $\begin{vmatrix} -5 & -5 \\ 6 & 6 \end{vmatrix}$

3. $\begin{vmatrix} 2 & 0 \\ 3 & 0 \end{vmatrix}$

4. $\begin{vmatrix} -8 & 0 \\ -6 & 0 \end{vmatrix}$

5. $\begin{vmatrix} -1 & 2 & 4 \\ 4 & -8 & -16 \\ 3 & 0 & 5 \end{vmatrix}$

6. $\begin{vmatrix} 2 & -8 & 3 \\ 0 & 2 & -1 \\ -6 & 24 & -9 \end{vmatrix}$

7. $\begin{vmatrix} 3 & 6 & 6 \\ 2 & 0 & 4 \\ 1 & 4 & 2 \end{vmatrix}$

8. $\begin{vmatrix} 1 & 0 & 0 \\ 1 & 0 & 1 \\ 3 & 0 & 0 \end{vmatrix}$

9. $\begin{vmatrix} m & 2 & 2m \\ 3n & 1 & 6n \\ 5p & 6 & 10p \end{vmatrix}$

10. $\begin{vmatrix} 7z & 8x & 2y \\ z & x & y \\ 7z & 7x & 7y \end{vmatrix}$

11. Explain why the determinant of a matrix with two identical rows (or columns) equals 0.

Use the appropriate properties from this section to tell why each of the following is true. Do not evaluate the determinants. See Examples 1–6.

12. $\begin{vmatrix} 2 & 1 & 6 \\ 3 & 0 & 2 \\ 4 & 1 & 8 \end{vmatrix} = \begin{vmatrix} 2 & 3 & 4 \\ 1 & 0 & 1 \\ 6 & 2 & 8 \end{vmatrix}$

13. $\begin{vmatrix} 4 & -2 \\ 3 & 8 \end{vmatrix} = \begin{vmatrix} 4 & 3 \\ -2 & 8 \end{vmatrix}$

14. $\begin{vmatrix} 2 & 6 \\ 3 & 5 \end{vmatrix} = -\begin{vmatrix} 3 & 5 \\ 2 & 6 \end{vmatrix}$

15. $\begin{vmatrix} -1 & 8 & 9 \\ 0 & 2 & 1 \\ 3 & 2 & 0 \end{vmatrix} = -\begin{vmatrix} 8 & -1 & 9 \\ 2 & 0 & 1 \\ 2 & 3 & 0 \end{vmatrix}$

16. $3\begin{vmatrix} 6 & 0 & 2 \\ 4 & 1 & 3 \\ 2 & 8 & 6 \end{vmatrix} = \begin{vmatrix} 6 & 0 & 2 \\ 4 & 3 & 3 \\ 2 & 24 & 6 \end{vmatrix}$

17. $-\dfrac{1}{2}\begin{vmatrix} 5 & -8 & 2 \\ 3 & -6 & 9 \\ 2 & 4 & 4 \end{vmatrix} = \begin{vmatrix} 5 & 4 & 2 \\ 3 & 3 & 9 \\ 2 & -2 & 4 \end{vmatrix}$

18. $\begin{vmatrix} 3 & -4 \\ 2 & 5 \end{vmatrix} = \begin{vmatrix} 3 & -4 \\ 5 & 1 \end{vmatrix}$

19. $\begin{vmatrix} -1 & 6 \\ 3 & -5 \end{vmatrix} = \begin{vmatrix} -1 & 6 \\ 2 & 1 \end{vmatrix}$

20. $\begin{vmatrix} 5 & 8 \\ 2 & -1 \end{vmatrix} = \begin{vmatrix} 5 & -2 \\ 2 & -5 \end{vmatrix}$

21. $\begin{vmatrix} 13 & 5 \\ 6 & 1 \end{vmatrix} = \begin{vmatrix} -2 & 5 \\ 3 & 1 \end{vmatrix}$

22. $\begin{vmatrix} 2 & 5 & 8 \\ 1 & 0 & 2 \\ 4 & 3 & 5 \end{vmatrix} = \begin{vmatrix} 2 & 5 & 8 \\ 1 & 0 & 2 \\ 7 & 3 & 11 \end{vmatrix}$

23. $2 \begin{vmatrix} 4 & 2 & -1 \\ m & 2n & 3p \\ 5 & 1 & 0 \end{vmatrix} = \begin{vmatrix} 4 & 2 & -1 \\ 2m & 4n & 6p \\ 5 & 1 & 0 \end{vmatrix}$

24. $\begin{vmatrix} 3 & 5 & 0 \\ 2 & 1 & 3 \\ -5 & 1 & 6 \end{vmatrix} = \begin{vmatrix} 3 & 5 & 0 \\ 2+3k & 1+5k & 3+0k \\ -5 & 1 & 6 \end{vmatrix}$

25. $\begin{vmatrix} -4 & 2 & 1 \\ 3 & 0 & 5 \\ -1 & 4 & -2 \end{vmatrix} = \begin{vmatrix} -4 & 2 & 1+(-4)k \\ 3 & 0 & 5+3k \\ -1 & 4 & -2+(-1)k \end{vmatrix}$

Use Property 6 to evaluate each of the following determinants. See Examples 7 and 8.

26. $\begin{vmatrix} 2 & 4 \\ 3 & 6 \end{vmatrix}$

27. $\begin{vmatrix} -5 & 10 \\ 6 & -12 \end{vmatrix}$

28. $\begin{vmatrix} 4 & 8 & 0 \\ -1 & -2 & 1 \\ 2 & 4 & 3 \end{vmatrix}$

29. $\begin{vmatrix} 6 & 8 & -12 \\ -1 & 16 & 2 \\ 4 & 0 & -8 \end{vmatrix}$

30. $\begin{vmatrix} 3 & 1 & 2 \\ 2 & 3 & 1 \\ 1 & 0 & -2 \end{vmatrix}$

31. $\begin{vmatrix} -2 & 2 & 3 \\ 0 & 2 & 1 \\ -1 & 4 & 0 \end{vmatrix}$

32. $\begin{vmatrix} -4 & 2 & 3 \\ 2 & 0 & 1 \\ 0 & 4 & 2 \end{vmatrix}$

33. $\begin{vmatrix} 6 & 3 & 2 \\ 1 & 0 & 2 \\ -1 & 4 & 1 \end{vmatrix}$

34. $\begin{vmatrix} 1 & 0 & 2 & 2 \\ 2 & 4 & 1 & -1 \\ 1 & -3 & 1 & 0 \\ 1 & 1 & 0 & 1 \end{vmatrix}$

35. $\begin{vmatrix} 2 & -1 & 1 & 0 \\ 1 & 1 & 0 & 1 \\ 0 & -1 & 1 & 1 \\ 1 & 2 & 1 & 2 \end{vmatrix}$

36. The note following Example 8 advises working with sums of rows to get a column with only one nonzero number. Why is this good advice? What happens if you work with columns to get a column with just one nonzero number?

Use the determinant

$$\begin{vmatrix} a & b & c \\ d & e & f \\ g & h & j \end{vmatrix}$$

to prove the following properties of determinants by finding the values of the original determinant and the new determinant and comparing them.

37. Property 2

38. Property 3

39. Property 4

8.6 SOLUTION OF LINEAR SYSTEMS BY CRAMER'S RULE

We have now seen several methods for solving linear systems of equations in Chapter 7 and in this chapter. Cramer's rule is a method of solving a linear system of equations using determinants. An advantage of this method is that by finding the determinant of the matrix of coefficients, we can decide whether a single solution exists before actually solving the system.

To derive Cramer's rule, we use the elimination method to solve the general system of two equations with two variables,

$$a_1x + b_1y = c_1 \tag{1}$$
$$a_2x + b_2y = c_2. \tag{2}$$

To begin, eliminate y and solve for x by first multiplying both sides of equation (1) by b_2 and both sides of equation (2) by $-b_1$. Then add these results and solve for x.

$$a_1b_2x + b_1b_2y = c_1b_2$$
$$\underline{-a_2b_1x - b_1b_2y = -c_2b_1}$$
$$(a_1b_2 - a_2b_1)x \qquad = c_1b_2 - c_2b_1$$
$$x = \frac{c_1b_2 - c_2b_1}{a_1b_2 - a_2b_1}, \text{ if } a_1b_2 - a_2b_1 \neq 0$$

Solve for y by multiplying both sides of equation (1) by $-a_2$ and equation (2) by a_1 and then adding the two equations.

$$-a_1a_2x - a_2b_1y = -a_2c_1$$
$$\underline{a_1a_2x + a_1b_2y = a_1c_2}$$
$$(a_1b_2 - a_2b_1)y = a_1c_2 - a_2c_1$$
$$y = \frac{a_1c_2 - a_2c_1}{a_1b_2 - a_2b_1}, \text{ if } a_1b_2 - a_2b_1 \neq 0$$

Both numerators and the common denominator of these values for x and y can be written as determinants, since

$$c_1b_2 - c_2b_1 = \begin{vmatrix} c_1 & b_1 \\ c_2 & b_2 \end{vmatrix}, \qquad a_1c_2 - a_2c_1 = \begin{vmatrix} a_1 & c_1 \\ a_2 & c_2 \end{vmatrix},$$
$$\text{and} \qquad a_1b_2 - a_2b_1 = \begin{vmatrix} a_1 & b_1 \\ a_2 & b_2 \end{vmatrix}.$$

Using these determinants, the solutions for x and y become

$$x = \frac{\begin{vmatrix} c_1 & b_1 \\ c_2 & b_2 \end{vmatrix}}{\begin{vmatrix} a_1 & b_1 \\ a_2 & b_2 \end{vmatrix}} \qquad \text{and} \qquad y = \frac{\begin{vmatrix} a_1 & c_1 \\ a_2 & c_2 \end{vmatrix}}{\begin{vmatrix} a_1 & b_1 \\ a_2 & b_2 \end{vmatrix}}, \qquad \text{if } \begin{vmatrix} a_1 & b_1 \\ a_2 & b_2 \end{vmatrix} \neq 0.$$

For convenience, denote the three determinants in the solution as

$$\begin{vmatrix} a_1 & b_1 \\ a_2 & b_2 \end{vmatrix} = D, \qquad \begin{vmatrix} c_1 & b_1 \\ c_2 & b_2 \end{vmatrix} = D_x, \qquad \text{and} \qquad \begin{vmatrix} a_1 & c_1 \\ a_2 & c_2 \end{vmatrix} = D_y.$$

NOTE The elements of D are the four coefficients of the variables in the given system, the elements of D_x are obtained by replacing the coefficients of x in D by the respective constants, and the elements of D_y are obtained by replacing the coefficients of y in D by the respective constants.

These results are summarized as **Cramer's rule.**

CRAMER'S RULE FOR 2 × 2 SYSTEMS

Given the system

$$a_1x + b_1y = c_1$$
$$a_2x + b_2y = c_2,$$

with $a_1b_2 - a_2b_1 \neq 0$,

then

$$x = \frac{D_x}{D} \quad \text{and} \quad y = \frac{D_y}{D},$$

where

$$D_x = \begin{vmatrix} c_1 & b_1 \\ c_2 & b_2 \end{vmatrix}, \quad D_y = \begin{vmatrix} a_1 & c_1 \\ a_2 & c_2 \end{vmatrix}, \quad \text{and} \quad D = \begin{vmatrix} a_1 & b_1 \\ a_2 & b_2 \end{vmatrix} \neq 0.$$

Although this theorem is well-known as Cramer's rule, it was probably first discovered by Colin Maclaurin (1698–1746) as early as 1729 and was published under his name in 1748, two years earlier than Cramer's first publication of the rule in 1750.

Cramer's rule is used to solve a system of linear equations by evaluating the three determinants D, D_x, and D_y and then writing the appropriate quotients for x and y.

CAUTION As indicated above, Cramer's rule does not apply if $D = 0$. When $D = 0$, the system is inconsistent or has dependent equations. For this reason, it is a good idea to evaluate D first.

■ *Example 1*
APPLYING CRAMER'S RULE TO A 2 × 2 SYSTEM

Use Cramer's rule to solve the system

$$5x + 7y = -1$$
$$6x + 8y = 1.$$

By Cramer's rule, $x = D_x/D$ and $y = D_y/D$. As mentioned above, it is a good idea to find D first, since if $D = 0$, Cramer's rule does not apply. If $D \neq 0$, then find D_x and D_y.

$$D = \begin{vmatrix} 5 & 7 \\ 6 & 8 \end{vmatrix} = 5(8) - 6(7) = -2$$

$$D_x = \begin{vmatrix} -1 & 7 \\ 1 & 8 \end{vmatrix} = (-1)(8) - (1)(7) = -15$$

$$D_y = \begin{vmatrix} 5 & -1 \\ 6 & 1 \end{vmatrix} = 5(1) - (6)(-1) = 11$$

From Cramer's rule,

$$x = \frac{D_x}{D} = \frac{-15}{-2} = \frac{15}{2}$$

and

$$y = \frac{D_y}{D} = \frac{11}{-2} = -\frac{11}{2}.$$

The solution set is $\{(15/2, -11/2)\}$, as can be verified by substituting in the given system. ■

Cramer's rule can be generalized to systems of three equations in three variables (or n equations in n variables).

CRAMER'S RULE FOR 3 × 3 SYSTEMS

Given the system

$$a_1x + b_1y + c_1z = d_1$$
$$a_2x + b_2y + c_2z = d_2$$
$$a_3x + b_3y + c_3z = d_3,$$

then

$$x = \frac{D_x}{D}, \qquad y = \frac{D_y}{D}, \qquad \text{and} \qquad z = \frac{D_z}{D},$$

where

$$D_x = \begin{vmatrix} d_1 & b_1 & c_1 \\ d_2 & b_2 & c_2 \\ d_3 & b_3 & c_3 \end{vmatrix}, \qquad D_y = \begin{vmatrix} a_1 & d_1 & c_1 \\ a_2 & d_2 & c_2 \\ a_3 & d_3 & c_3 \end{vmatrix},$$

$$D_z = \begin{vmatrix} a_1 & b_1 & d_1 \\ a_2 & b_2 & d_2 \\ a_3 & b_3 & d_3 \end{vmatrix}, \qquad \text{and} \qquad D = \begin{vmatrix} a_1 & b_1 & c_1 \\ a_2 & b_2 & c_2 \\ a_3 & b_3 & c_3 \end{vmatrix} \neq 0.$$

■ **Example 2**

APPLYING CRAMER'S RULE TO A 3 × 3 SYSTEM

Use Cramer's rule to solve the system

$$x + y - z + 2 = 0$$
$$2x - y + z + 5 = 0$$
$$x - 2y + 3z - 4 = 0.$$

For Cramer's rule, the system must be rewritten in the form

$$x + y - z = -2$$
$$2x - y + z = -5$$
$$x - 2y + 3z = 4.$$

Verify that the required determinants are as follows:

$$D = \begin{vmatrix} 1 & 1 & -1 \\ 2 & -1 & 1 \\ 1 & -2 & 3 \end{vmatrix} = -3, \qquad D_x = \begin{vmatrix} -2 & 1 & -1 \\ -5 & -1 & 1 \\ 4 & -2 & 3 \end{vmatrix} = 7,$$

$$D_y = \begin{vmatrix} 1 & -2 & -1 \\ 2 & -5 & 1 \\ 1 & 4 & 3 \end{vmatrix} = -22, \qquad D_z = \begin{vmatrix} 1 & 1 & -2 \\ 2 & -1 & -5 \\ 1 & -2 & 4 \end{vmatrix} = -21.$$

Thus

$$x = \frac{D_x}{D} = \frac{7}{-3} = -\frac{7}{3}, \qquad y = \frac{D_y}{D} = \frac{-22}{-3} = \frac{22}{3},$$

and

$$z = \frac{D_z}{D} = \frac{-21}{-3} = 7,$$

so the solution set is $\{(-7/3, 22/3, 7)\}$. ■

CAUTION   As shown in Example 2, each equation in the system must be written in the form $ax + by + cz + \cdots = k$ before using Cramer's rule.

■ *Example 3*

APPLYING CRAMER'S RULE WHEN $D = 0$

Use Cramer's rule to solve the system

$$2x - 3y + 4z = 10$$
$$6x - 9y + 12z = 24$$
$$x + 2y - 3z = 5.$$

First find D by expanding about column 1.

$$D = \begin{vmatrix} 2 & -3 & 4 \\ 6 & -9 & 12 \\ 1 & 2 & -3 \end{vmatrix} = 2\begin{vmatrix} -9 & 12 \\ 2 & -3 \end{vmatrix} - 6\begin{vmatrix} -3 & 4 \\ 2 & -3 \end{vmatrix} + 1\begin{vmatrix} -3 & 4 \\ -9 & 12 \end{vmatrix}$$
$$= 2(3) - 6(1) + 1(0)$$
$$= 0$$

As mentioned above, Cramer's rule does not apply if $D = 0$. When $D = 0$, the system either is inconsistent or contains dependent equations. Use the elimination method or the Gauss-Jordan method to tell which is the case. Verify that this system is inconsistent. ■

8.6 EXERCISES ■

Use Cramer's rule to solve each of the following systems of equations. If $D = 0$, use another method to determine the solution. See Example 1.

1. $x + y = 4$
$2x - y = 2$

2. $3x + 2y = -4$
$2x - y = -5$

3. $4x + 3y = -7$
$2x + 3y = -11$

4. $4x - y = 0$
$2x + 3y = 14$

5. $5x + 4y = 10$
$3x - 7y = 6$

6. $3x + 2y = -4$
$5x - y = 2$

7. $2x - 3y = -5$
$x + 5y = 17$

8. $x + 9y = -15$
$3x + 2y = 5$

9. $3x + 2y = 4$
$6x + 4y = 8$

10. $1.5x + 3y = 5$
$2x + 4y = 3$

11. $12x + 8y = 3$
$15x + 10y = 9$

12. $15x - 10y = 5$
$9x + 6y = 3$

Use Cramer's rule to solve each of the following systems of equations. If $D = 0$, use another method to complete the solution. See Examples 2 and 3.

13. $4x - y + 3z = -3$
$3x + y + z = 0$
$2x - y + 4z = 0$

14. $5x + 2y + z = 15$
$2x - y + z = 9$
$4x + 3y + 2z = 13$

15. $2x - y + 4z = -2$
$3x + 2y - z = -3$
$x + 4y + 2z = 17$

16. $x + y + z = 4$
$2x - y + 3z = 4$
$4x + 2y - z = -15$

17. $4x - 3y + z = -1$
$5x + 7y + 2z = -2$
$3x - 5y - z = 1$

18. $2x - 3y + z = 8$
$-x - 5y + z = -4$
$3x - 5y + 2z = 12$

19. $x + 2y + 3z = 4$
$4x + 3y + 2z = 1$
$-x - 2y - 3z = 0$

20. $2x - y + 3z = 1$
$-2x + y - 3z = 2$
$5x - y + z = 2$

21. $-2x - 2y + 3z = 4$
$5x + 7y - z = 2$
$2x + 2y - 3z = -4$

22. $-3x + 2y - 2z = 4$
$4x + y + z = 5$
$3x - 2y + 2z = 1$

23. $2x + 3y = 13$
$2y - z = 5$
$x + 2z = 4$

24. $3x - z = -10$
$y + 4z = 8$
$x + 2z = -1$

25. $5x - y = -4$
$3x + 2z = 4$
$4y + 3z = 22$

26. $3x + 5y = -7$
$2x + 7z = 2$
$4y + 3z = -8$

27. $x + 2y = 10$
$3x + 4z = 7$
$-y - z = 1$

28. $5x - 2y = 3$
$4y + z = 8$
$x + 2z = 4$

29. In your own words, explain what it means in applying Cramer's rule if $D = 0$.

30. Describe D_x, D_y, and D_z in terms of the coefficients and constants in the given system of equations.

Use Cramer's rule to solve each of the following systems.

31. $x + 3y - 2z - w = 9$
$4x + y + z + 2w = 2$
$-3x - y + z - w = -5$
$x - y - 3z - 2w = 2$

32. $3x + 2y - w = 0$
$2x + z + 2w = 5$
$x + 2y - z = -2$
$2x - y + z + w = 2$

33. $x + y - z + w = 2$
$x - y + z + w = 4$
$-2x + y + 2z - w = -5$
$x + 3z + 2w = 5$

34. $x + 2y - z + w = 8$
$2x - y + 2w = 8$
$y + 3z = 5$
$x - z = 4$

Solve each system for x and y using Cramer's rule. Assume a and b are nonzero constants.

35. $bx + y = a^2$
$ax + y = b^2$

36. $ax + by = \dfrac{b}{a}$
$x + y = \dfrac{1}{b}$

37. $b^2x + a^2y = b^2$
$ax + by = a$

38. $x + \dfrac{1}{b}y = b$
$\dfrac{1}{a}x + y = a$

For the following two exercises, use the system of equations

$$a_1x + b_1y = c_1$$
$$a_2x + b_2y = c_2.$$

39. Assume $D_x = 0$ and $D_y = 0$. Show that if $c_1c_2 \neq 0$, then $D = 0$, and the equations are dependent.

40. Assume $D = 0$, $D_x = 0$, and $b_1b_2 \neq 0$. Show that $D_y = 0$.

CHAPTER 8 SUMMARY ■ ———————————

SECTION	TERMS	KEY IDEAS
8.1 Basic Properties of Matrices	square matrix row matrix column matrix zero matrix scalar	**Addition of Matrices** The sum of two $m \times n$ matrices A and B is the $m \times n$ matrix $A + B$ in which each element is the sum of the corresponding elements of A and B. **Subtraction of Matrices** If A and B are matrices of the same size, then $$A - B = A + (-B).$$
8.2 Matrix Products		**Multiplication of Matrices** The product AB of an $m \times n$ matrix A and an $n \times k$ matrix B is found as follows. To get the ith row, jth column element of AB, multiply each element in the ith row of A by the corresponding element in the jth column of B. The sum of these products will give the element of row i, column j of AB.
8.3 Multiplicative Inverses of Matrices	identity matrix inverse of a matrix	**Finding an Inverse Matrix** To obtain A^{-1} for any $n \times n$ matrix A for which A^{-1} exists, follow these steps. **1.** Form the augmented matrix $[A\|I]$, where I is the $n \times n$ identity matrix. **2.** Perform row transformations on $[A\|I]$ to get a matrix of the form $[I\|B]$. **3.** Matrix B is A^{-1}. **4.** Verify by showing that $BA = AB = I$.

SECTION	TERMS	KEY IDEAS
8.4 Determinants	determinant minor cofactor expansion by a row or column	**Determinant of a 2 × 2 Matrix** The determinant of a 2 × 2 matrix A, $$A = \begin{bmatrix} a_{11} & a_{12} \\ a_{21} & a_{22} \end{bmatrix},$$ is defined as $$|A| = \begin{vmatrix} a_{11} & a_{12} \\ a_{21} & a_{22} \end{vmatrix} = a_{11}a_{22} - a_{21}a_{12}.$$ **Determinant of a 3 × 3 Matrix** The determinant of a 3 × 3 matrix A, $$A = \begin{bmatrix} a_{11} & a_{12} & a_{13} \\ a_{21} & a_{22} & a_{23} \\ a_{31} & a_{32} & a_{33} \end{bmatrix},$$ is defined as $$|A| = \begin{vmatrix} a_{11} & a_{12} & a_{13} \\ a_{21} & a_{22} & a_{23} \\ a_{31} & a_{32} & a_{33} \end{vmatrix}$$ $$= (a_{11}a_{22}a_{33} + a_{12}a_{23}a_{31} + a_{13}a_{21}a_{32})$$ $$- (a_{31}a_{22}a_{13} + a_{32}a_{23}a_{11} + a_{33}a_{21}a_{12}).$$
8.5 Properties of Determinants		**Properties of Determinants** 1. If every element in a row or column of a matrix is 0, then the determinant equals 0. 2. If corresponding rows and columns of a matrix are interchanged, the determinant is not changed. 3. Interchanging two rows (or columns) of a matrix reverses the sign of the determinant. 4. If every element of a row (or column) of a matrix is multiplied by the real number k, then the determinant of the new matrix is k times the determinant of the original matrix. 5. The determinant of a matrix with two identical rows (or columns) equals 0. 6. The determinant of a matrix is unchanged if a multiple of a row (or column) of the matrix is added to the corresponding elements of another row (or column).

SECTION	TERMS	KEY IDEAS
8.6 Solution of Linear Systems by Cramer's Rule		**Cramer's Rule for 2 × 2 Systems** Given the system $$a_1x + b_1y = c_1$$ $$a_2x + b_2y = c_2,$$ then $$x = \frac{D_x}{D} \quad \text{and} \quad y = \frac{D_y}{D},$$ where $$D_x = \begin{vmatrix} c_1 & b_1 \\ c_2 & b_2 \end{vmatrix},$$ $$D_y = \begin{vmatrix} a_1 & c_1 \\ a_2 & c_2 \end{vmatrix},$$ and $$D = \begin{vmatrix} a_1 & b_1 \\ a_2 & b_2 \end{vmatrix} \neq 0.$$ **Cramer's Rule for 3 × 3 Systems** Given the system $$a_1x + b_1y + c_1z = d_1$$ $$a_2x + b_2y + c_2z = d_2$$ $$a_3x + b_3y + c_3z = d_3,$$ then $$x = \frac{D_x}{D}, \quad y = \frac{D_y}{D}, \quad \text{and} \quad z = \frac{D_z}{D},$$ where $$D_x = \begin{vmatrix} d_1 & b_1 & c_1 \\ d_2 & b_2 & c_2 \\ d_3 & b_3 & c_3 \end{vmatrix},$$ $$D_y = \begin{vmatrix} a_1 & d_1 & c_1 \\ a_2 & d_2 & c_2 \\ a_3 & d_3 & c_3 \end{vmatrix},$$ $$D_z = \begin{vmatrix} a_1 & b_1 & d_1 \\ a_2 & b_2 & d_2 \\ a_3 & b_3 & d_3 \end{vmatrix},$$ and $$D = \begin{vmatrix} a_1 & b_1 & c_1 \\ a_2 & b_2 & c_2 \\ a_3 & b_3 & c_3 \end{vmatrix} \neq 0.$$

Find the values of all variables in the following statements.

1. $\begin{bmatrix} 2 & z & 1 \\ m & 9 & -7 \end{bmatrix} = \begin{bmatrix} x & 5 & 1 \\ -8 & y & p \end{bmatrix}$

2. $\begin{bmatrix} 5 & -10 \\ a & 0 \end{bmatrix} = \begin{bmatrix} b-3 & c+2 \\ 4 & 0 \end{bmatrix}$

3. $\begin{bmatrix} 5 & x+2 \\ -6y & z \end{bmatrix} = \begin{bmatrix} a & 3x-1 \\ 5y & 9 \end{bmatrix}$

4. $\begin{bmatrix} 6+k & 2 & a+3 \\ -2+m & 3p & 2r \end{bmatrix} + \begin{bmatrix} 3-2k & 5 & 7 \\ 5 & 8p & 5r \end{bmatrix} = \begin{bmatrix} 5 & y & 6a \\ 2m & 11 & -35 \end{bmatrix}$

Perform each of the following operations whenever possible.

5. $\begin{bmatrix} 3 & -4 & 2 \\ 5 & -1 & 6 \end{bmatrix} + \begin{bmatrix} -3 & 2 & 5 \\ 1 & 0 & 4 \end{bmatrix}$

6. $\begin{bmatrix} 3 \\ 2 \\ 5 \end{bmatrix} - \begin{bmatrix} 8 \\ -4 \\ 6 \end{bmatrix} + \begin{bmatrix} 1 \\ 0 \\ 2 \end{bmatrix}$

7. $\begin{bmatrix} 2 & 5 & 8 \\ 1 & 9 & 2 \end{bmatrix} - \begin{bmatrix} 3 & 4 \\ 7 & 1 \end{bmatrix}$

8. $3\begin{bmatrix} 2 & 4 \\ -1 & 4 \end{bmatrix} - 2\begin{bmatrix} 5 & 8 \\ 2 & -2 \end{bmatrix}$

9. $-1\begin{bmatrix} 3 & -5 & 2 \\ 1 & 7 & -4 \end{bmatrix} + 5\begin{bmatrix} 0 & 2 \\ -1 & 3 \end{bmatrix}$

10. $10\begin{bmatrix} 2x+3y & 4x+y \\ x-5y & 6x+2y \end{bmatrix} + 2\begin{bmatrix} -3x-y & x+6y \\ 4x+2y & 5x-y \end{bmatrix}$

11. Complete the following sentence. The sum of two $m \times n$ matrices A and B is found _____.

12. The speed limits in Italy, Britain, and the U.S. are 87 mph, 70 mph, and 55 mph, respectively. The corresponding fatalities per 100 million miles driven in a recent year were 6.4, 4.0, and 3.3. Write this information as a 3 × 2 matrix and then as a 2 × 3 matrix.

13. In a recent year, Dan Marino, quarterback for the Miami Dolphins, attempted 606 passes and completed 354 for 4434 yd and 28 touchdowns. Joe Montana, San Francisco 49'ers quarterback, attempted 397 passes and completed 238 for 2981 yd and 18 touchdowns. Phil Simms, quarterback for the New York Giants, attempted 479 passes and completed 263 for 3359 yd and 21 touchdowns. Write this information as a matrix in two ways.

Find each of the following matrix products, whenever possible.

14. $\begin{bmatrix} -3 & 4 \\ 2 & 8 \end{bmatrix}\begin{bmatrix} -1 & 0 \\ 2 & 5 \end{bmatrix}$

15. $\begin{bmatrix} 3 & 2 & -1 \\ 4 & 0 & 6 \end{bmatrix}\begin{bmatrix} -2 & 0 \\ 0 & 2 \\ 3 & 1 \end{bmatrix}$

16. $\begin{bmatrix} 1 & -2 & 4 & 2 \\ 0 & 1 & -1 & 8 \end{bmatrix}\begin{bmatrix} -1 \\ 2 \\ 0 \\ 1 \end{bmatrix}$

17. $\begin{bmatrix} 1 & 2 & 5 \\ -3 & 4 & 7 \\ 0 & 2 & -1 \end{bmatrix}\begin{bmatrix} 4 & 2 & 3 \\ 10 & -5 & 6 \end{bmatrix}$

18. $\begin{bmatrix} 4 & 2 & 3 \\ 10 & -5 & 6 \end{bmatrix}\begin{bmatrix} 1 & 2 & 5 \\ -3 & 4 & 7 \\ 0 & 2 & -1 \end{bmatrix}$

19. $\begin{bmatrix} 3 & -1 & 0 \end{bmatrix}\begin{bmatrix} 1 & 3 & 2 \\ 2 & -4 & 0 \\ 5 & 7 & 3 \end{bmatrix}$

20. What must be true of two matrices to find their product?

Decide whether or not each of the following pairs of matrices are inverses.

21. $\begin{bmatrix} 2 & -3 \\ 1 & -2 \end{bmatrix}, \begin{bmatrix} 2 & -3 \\ 1 & -2 \end{bmatrix}$

22. $\begin{bmatrix} 1 & 0 \\ 2 & -3 \end{bmatrix}, \begin{bmatrix} 1 & 0 \\ 2/3 & -1/3 \end{bmatrix}$

23. $\begin{bmatrix} 2 & 0 & 6 \\ 0 & 1 & 0 \\ 1 & 0 & 1 \end{bmatrix}, \begin{bmatrix} -1 & 0 & 3/2 \\ 0 & 1 & 0 \\ 1/4 & 0 & -1 \end{bmatrix}$

24. $\begin{bmatrix} 1 & 0 & 2 \\ 0 & 2 & 4 \\ 0 & 0 & 1 \end{bmatrix}, \begin{bmatrix} 1 & 0 & -2 \\ 0 & 1/2 & -2 \\ 0 & 0 & 1 \end{bmatrix}$

Find the inverse, if it exists, for each of the following matrices.

25. $\begin{bmatrix} 2 & 1 \\ 5 & 3 \end{bmatrix}$

26. $\begin{bmatrix} -4 & 2 \\ 0 & 3 \end{bmatrix}$

27. $\begin{bmatrix} 2 & 0 \\ -1 & 5 \end{bmatrix}$

28. $\begin{bmatrix} 2 & 0 & 4 \\ 1 & -1 & 0 \\ 0 & 1 & -2 \end{bmatrix}$

29. $\begin{bmatrix} 2 & -1 & 0 \\ 1 & 0 & 1 \\ 1 & -2 & 0 \end{bmatrix}$

30. $\begin{bmatrix} 2 & 3 & 5 \\ -2 & -3 & -5 \\ 1 & 4 & 2 \end{bmatrix}$

Use matrix inverses to solve each of the following. Identify any inconsistent systems or systems with dependent equations.

31. $x + y = 4$
$2x + 3y = 10$

32. $5x - 3y = -2$
$2x + 7y = -9$

33. $2x + y = 5$
$3x - 2y = 4$

34. $x - 2y = 7$
$3x + y = 7$

35. $3x - 2y + 4z = 1$
$4x + y - 5z = 2$
$-6x + 4y - 8z = -2$

36. $x + 2y = -1$
$3y - z = -5$
$x + 2y - z = -3$

37. $x + y + z = 1$
$2x - y = -2$
$3y + z = 2$

38. $x = -3$
$y + z = 6$
$2x - 3z = -9$

Evaluate each of the following determinants.

39. $\begin{vmatrix} -1 & 8 \\ 2 & 9 \end{vmatrix}$

40. $\begin{vmatrix} -2 & 4 \\ 0 & 3 \end{vmatrix}$

41. $\begin{vmatrix} -2 & 4 & 1 \\ 3 & 0 & 2 \\ -1 & 0 & 3 \end{vmatrix}$

42. $\begin{vmatrix} -1 & 2 & 3 \\ 4 & 0 & 3 \\ 5 & -1 & 2 \end{vmatrix}$

Solve each of the following determinant equations for x.

43. $\begin{vmatrix} -3 & 2 \\ 1 & x \end{vmatrix} = 5$

44. $\begin{vmatrix} 3x & 7 \\ -x & 4 \end{vmatrix} = 8$

45. $\begin{vmatrix} 2 & 5 & 0 \\ 1 & 3x & -1 \\ 0 & 2 & 0 \end{vmatrix} = 4$

46. $\begin{vmatrix} 6x & 2 & 0 \\ 1 & 5 & 3 \\ x & 2 & -1 \end{vmatrix} = 2x$

47. What are the determinant properties used for?

Give the property (or properties) that justifies each of the following statements.

48. $\begin{vmatrix} 8 & 9 & 2 \\ 0 & 0 & 0 \\ 3 & 1 & 4 \end{vmatrix} = 0$

49. $\begin{vmatrix} 4 & 6 \\ 3 & 5 \end{vmatrix} = \begin{vmatrix} 4 & 3 \\ 6 & 5 \end{vmatrix}$

50. $\begin{vmatrix} 8 & 2 \\ 4 & 3 \end{vmatrix} = 2\begin{vmatrix} 4 & 1 \\ 4 & 3 \end{vmatrix}$

51. $\begin{vmatrix} 4 & 6 & 2 \\ -3 & 8 & -5 \\ 4 & 6 & 2 \end{vmatrix} = 0$

52. $\begin{vmatrix} 5 & -1 & 2 \\ 3 & -2 & 0 \\ -4 & 1 & 2 \end{vmatrix} = \begin{vmatrix} 5 & -1 & 2 \\ 8 & -3 & 2 \\ -4 & 1 & 2 \end{vmatrix}$

53. $\begin{vmatrix} 8 & 2 & -5 \\ -3 & 1 & 4 \\ 2 & 0 & 5 \end{vmatrix} = -\begin{vmatrix} 8 & -5 & 2 \\ -3 & 4 & 1 \\ 2 & 5 & 0 \end{vmatrix}$

Use Determinant Property 6 to evaluate the following.

54. $\begin{vmatrix} 5 & -1 & 2 \\ 3 & -2 & 0 \\ -4 & 1 & 2 \end{vmatrix}$

55. $\begin{vmatrix} 8 & 2 & -5 \\ -3 & 1 & 4 \\ 2 & 0 & 5 \end{vmatrix}$

56. Cramer's rule has the condition that $D \neq 0$. Why is this necessary? What is true of the system when $D = 0$?

Solve each of the following systems by Cramer's rule. Identify any systems with dependent equations or any inconsistent systems.

57. $3x + y = -1$
$5x + 4y = 10$

58. $3x + 7y = 2$
$5x - y = -22$

59. $2x - 5y = 8$
$3x + 4y = 10$

60. $3x + 2y + z = 2$
$4x - y + 3z = -16$
$x + 3y - z = 12$

61. $5x - 2y - z = 8$
$-5x + 2y + z = -8$
$x - 4y - 2z = 0$

62. $-x + 3y - 4z = 2$
$2x + 4y + z = 3$
$3x - z = 9$

1. Find the values of all variables in the following equation. $\begin{bmatrix} 5 & x+6 \\ 0 & 4 \end{bmatrix} = \begin{bmatrix} y-2 & 4-x \\ 0 & w+7 \end{bmatrix}$

Perform each of the following operations whenever possible.

2. $3\begin{bmatrix} 2 & 3 \\ 1 & -4 \\ 5 & 9 \end{bmatrix} - \begin{bmatrix} -2 & 6 \\ 3 & -1 \\ 0 & 8 \end{bmatrix}$

3. $\begin{bmatrix} 1 \\ 2 \end{bmatrix} + \begin{bmatrix} 4 \\ -6 \end{bmatrix} + \begin{bmatrix} 2 & 8 \\ -7 & 5 \end{bmatrix}$

4. At a small college, the freshman class consists of 108 women and 142 men. The sophomore class consists of 112 women and 98 men. The junior class consists of 123 women and 117 men. The senior class consists of 110 women and 130 men. Write this information as a 2 × 4 matrix.

Find each of the following matrix products, whenever possible.

5. $\begin{bmatrix} 2 & 1 & -3 \\ 4 & 0 & 5 \end{bmatrix} \begin{bmatrix} 1 & 3 \\ 2 & 4 \\ 3 & -2 \end{bmatrix}$

6. $\begin{bmatrix} 2 & -4 \\ 3 & 5 \end{bmatrix} \begin{bmatrix} 4 \\ 2 \\ 7 \end{bmatrix}$

7. Which of the following properties does not apply to multiplication of matrices?
   (a) commutative   **(b)** associative   **(c)** distributive   **(d)** identity

8. Decide whether both of the following matrices are inverses. $\begin{bmatrix} 1 & 0 & 1 \\ -1 & 0 & 1 \\ 0 & 1 & 0 \end{bmatrix}$, $\begin{bmatrix} 1/2 & -1/2 & 0 \\ 0 & 0 & 1 \\ 1/2 & 1/2 & 0 \end{bmatrix}$

Find the inverse, if it exists, for each of the following matrices.

9. $\begin{bmatrix} -8 & 5 \\ 3 & -2 \end{bmatrix}$

10. $\begin{bmatrix} 4 & 12 \\ 2 & 6 \end{bmatrix}$

11. $\begin{bmatrix} 1 & 3 & 4 \\ 2 & 7 & 8 \\ -2 & -5 & -7 \end{bmatrix}$

Use matrix inverses to solve each of the following.

12. $2x + y = -6$
    $3x - y = -29$

13. $x + y = 5$
    $y - 2z = 23$
    $x + 3z = -27$

Evaluate each of the following determinants.

14. $\begin{vmatrix} 6 & 8 \\ 2 & -7 \end{vmatrix}$

15. $\begin{vmatrix} 2 & 0 & 8 \\ -1 & 7 & 9 \\ 12 & 5 & -3 \end{vmatrix}$

16. Solve the following determinant equation for x. $\begin{vmatrix} 1 & 3x & 3 \\ -2 & 0 & 7 \\ 4 & x & 5 \end{vmatrix} = -202$

Give the property (or properties) that justifies the following statements.

17. $\begin{vmatrix} 6 & 7 \\ -5 & 2 \end{vmatrix} = -\begin{vmatrix} -5 & 2 \\ 6 & 7 \end{vmatrix}$

18. $\begin{vmatrix} 7 & 2 & -1 \\ 5 & -4 & 3 \\ 6 & -2 & 1 \end{vmatrix} = \begin{vmatrix} 7 & 2 & -1 \\ -7 & 0 & 1 \\ 13 & 0 & 0 \end{vmatrix}$

Solve each of the following systems by Cramer's rule. Identify any systems with dependent equations or any inconsistent systems.

19. $2x - 3y = -33$
    $4x + 5y = 11$

20. $x + y - z = -4$
    $2x - 3y - z = 5$
    $x + 2y + 2z = 3$

While the most obvious feature of graphing calculators is their capability to graph numerous types of relations and functions, several current models are also capable of working with matrices. One popular model, Texas Instrument's TI-81, has a matrix "menu", designated MATRX. The menu indicates the capabilities of row transformations (see Section 7.3) and evaluation of determinants ("det"). The EDIT feature allows the user to enter the dimensions and elements of up to three matrices (designated [A], [B], and [C]), and each matrix may have up to 6 rows and 6 columns.

Suppose that we wish to use the TI-81 to evaluate the determinant of the matrix

$$\begin{bmatrix} 1 & 3 & -2 \\ -1 & -2 & -3 \\ 1 & 1 & 2 \end{bmatrix}.$$

We simply use the EDIT feature of the MATRX menu to designate matrix [A] as having 3 rows and 3 columns and then enter the elements of this matrix into the calculator. (For this particular calculator, the prompt "1, 1" represents row 1, column 1, "1, 2" represents row 1, column 2, and so on.) When this is completed, return to the MATRX menu and choose the "det" function. Enter matrix [A] after the "det" prompt using the 2nd function key, and press ENTER to get the display −6, the correct determinant for this matrix. Determinants found in this way can then be used with Cramer's rule to solve a system of equations.

Solving systems using matrix inverses, as explained in Section 8.3, is easily performed with the TI-81. To solve the system

$$2x - 3y = 4$$
$$x + 5y = 2,$$

first solved in Example 7, Section 8.3, we enter the coefficient matrix and the constant matrix

$$[A] = \begin{bmatrix} 2 & -3 \\ 1 & 5 \end{bmatrix} \quad \text{and} \quad [B] = \begin{bmatrix} 4 \\ 2 \end{bmatrix},$$

using the EDIT feature of the MATRX menu. This calculator can perform matrix multiplication and can find the inverse of a matrix using the standard $\boxed{x^{-1}}$ key. Therefore, enter the product $[A]^{-1}[B]$ and press ENTER. The display is the matrix

$$\begin{bmatrix} 2 \\ 1 & E & -13 \end{bmatrix}.$$

The entry in row 2, column 1, in decimal form, is .0000000000001, which is how the internal routine for this calculation displays the number 0. The product is therefore

$$\begin{bmatrix} 2 \\ 0 \end{bmatrix}$$

and thus the solution of the system is (2, 0).

As seen above, the user of the graphing calculator must be aware that reading *exactly* what is seen on the screen can lead to incorrect conclusions. For an excellent treatment of the perils and pitfalls associated with graphing calculators, see "The Graphics Calculator: A Tool for Critical Thinking," by Gloria Dion, in the October, 1990 issue of *Mathematics Teacher.*

You may want to experiment with your graphing calculator by using the following exercises, taken from Sections 8.3, 8.4, and 8.6. They are chosen from odd-numbered exercises, so you can check your answers with those given at the end of this text. The exercise number is given with each item.

Use a graphing calculator to find each determinant.

1. $\begin{vmatrix} 3 & 4 \\ 5 & -2 \end{vmatrix}$ (8.4 Exercise 7)

2. $\begin{vmatrix} 1.4 & 2.5 \\ 3.7 & 6.2 \end{vmatrix}$ (8.4 Exercise 19)

3. $\begin{vmatrix} 10 & 2 & 1 \\ -1 & 4 & 3 \\ -3 & 8 & 10 \end{vmatrix}$ (8.4 Exercise 33)

4. $\begin{vmatrix} 3 & 3 & -1 \\ 2 & 6 & 0 \\ -6 & -6 & 2 \end{vmatrix}$ (8.4 Exercise 37)

Use a graphing calculator to solve each of the following systems by using matrix inverses.

5. $x + 3y = -12$
$2x - y = 11$ (8.3 Exercise 37)

6. $2x - 3y = 10$
$2x + 2y = 5$ (8.3 Exercise 39)

7. $x + z = 3$
$2x - 2y - z = -2$
$3x = 3$ (8.3 Exercise 43)

8. $x + y - z = 6$
$2x - y + z = -9$
$x - 2y + 3z = 1$ (8.3 Exercise 47)

Use a graphing calculator with Cramer's rule to solve the following systems.

9. $x + 3y - 2z - w = 9$
$4x + y + z + 2w = 2$
$-3x - y + z - w = -5$
$x - y - 3z - 2w = 2$ (8.6 Exercise 31)

10. $x + y - z + w = 2$
$x - y + z + w = 4$
$-2x + y + 2z - w = -5$
$x + 3z + 2w = 5$ (8.6 Exercise 33)

■ NINE

SEQUENCES AND SERIES; PROBABILITY

9.1 SEQUENCES: ARITHMETIC SEQUENCES

9.2 GEOMETRIC SEQUENCES

9.3 SERIES; APPLICATIONS OF SEQUENCES AND SERIES

9.4 SUMS OF INFINITE GEOMETRIC SEQUENCES

9.5 MATHEMATICAL INDUCTION

9.6 PERMUTATIONS

9.7 COMBINATIONS

9.8 AN INTRODUCTION TO PROBABILITY

This chapter concludes our study of college algebra by discussing sequences, series, and probability. A *sequence* is a list of terms in a specific order. A *series* is the sum of the terms of a sequence. *Mathematical induction* is a powerful method used to prove that a statement is true for all positive integers. We will use it to prove some facts about sequences.

Counting principles, *permutations* and *combinations,* are methods of finding the number of possible arrangements or groupings of the elements of a set. This information is used in *probability theory* to find the likelihood that an event will occur.

———— ## SEQUENCES; ARITHMETIC SEQUENCES

If the domain of a function is the set of positive integers, the range elements can be *ordered,* as $f(1), f(2), f(3)$, and so on. This ordered list of numbers is called a **sequence.** Since the letter x has been used to suggest real numbers, the variable n is used instead with sequences to suggest the positive integer domain. Although a sequence may be defined by $f(n) = 2n + 3$, for example, it is customary to use a_n instead of $f(n)$ and write $a_n = 2n + 3$.

The elements of a sequence, called the **terms of the sequence,** are written in order as $a_1, a_2, a_3, \ldots$. The **general term,** or ***n*th term,** of the sequence is a_n. The general term of a sequence is used to find any term of the sequence. For example, if $a_n = n + 1$, then $a_2 = 2 + 1 = 3$.

A sequence is a **finite sequence** if the domain is the set $\{1, 2, 3, 4, \ldots, n\}$, where n is a positive integer. An **infinite sequence** has the set of all positive integers as its domain. For example, the sequence of positive even integers,

$$2, 4, 6, 8, 10, 12, 14, \ldots,$$

is infinite, but the sequence of dates in June,

$$1, 2, 3, 4, \ldots, 29, 30,$$

is finite.

■ *Example 1*

FINDING THE TERMS OF A SEQUENCE FROM THE GENERAL TERM

Write the first five terms for each of the following sequences.

(a) $a_n = \dfrac{n + 1}{n + 2}$

Replacing n, in turn, with 1, 2, 3, 4, and 5 gives

$$\frac{2}{3}, \frac{3}{4}, \frac{4}{5}, \frac{5}{6}, \frac{6}{7}.$$

(b) $a_n = (-1)^n \cdot n$

Replace n, in turn, with 1, 2, 3, 4, and 5, to get

$$a_1 = (-1)^1 \cdot 1 = -1$$
$$a_2 = (-1)^2 \cdot 2 = 2$$
$$a_3 = (-1)^3 \cdot 3 = -3$$
$$a_4 = (-1)^4 \cdot 4 = 4$$
$$a_5 = (-1)^5 \cdot 5 = -5.$$

(c) $b_n = \dfrac{(-1)^n}{2^n}$

Here $b_1 = -1/2$, $b_2 = 1/4$, $b_3 = -1/8$, $b_4 = 1/16$, and $b_5 = -1/32$. ■

Sequences can be defined using recursion formulas. A **recursion formula** defines the nth term of a sequence in terms of the previous term. For example, if the first term of a sequence is $a_1 = 2$ and the nth term is $a_n = a_{n-1} + 3$, then $a_2 = a_1 + 3 = 2 + 3 = 5$, $a_3 = a_2 + 3 = 5 + 3 = 8$, and so on.

■ *Example 2*

USING A
RECURSION
FORMULA

Find the first four terms for the sequences defined as follows.

(a) $a_1 = 4$; for $n \geq 2$, $a_n = 2a_{n-1} + 1$

We have $a_1 = 4$, so

$$a_2 = 2 \cdot a_1 + 1 = 2 \cdot 4 + 1 = 9,$$
$$a_3 = 2 \cdot a_2 + 1 = 2 \cdot 9 + 1 = 19,$$

and

$$a_4 = 2 \cdot a_3 + 1 = 2 \cdot 19 + 1 = 39.$$

(b) $a_1 = 2$; for $n \geq 2$, $a_n = a_{n-1} + n - 1$

$$a_1 = 2$$
$$a_2 = a_1 + 2 - 1 = 2 + 1 = 3$$
$$a_3 = a_2 + 3 - 1 = 3 + 2 = 5$$
$$a_4 = a_3 + 4 - 1 = 5 + 3 = 8$$ ■

ARITHMETIC SEQUENCES   A sequence in which each term after the first is obtained by adding a fixed number to the preceding term is called an **arithmetic sequence** (or **arithmetic progression**). The fixed number that is added is called the **common difference.** The sequence

$$5, 9, 13, 17, 21, \ldots$$

is an arithmetic sequence since each term after the first is obtained by adding 4 to the previous term. That is,

$$9 = 5 + 4$$
$$13 = 9 + 4$$
$$17 = 13 + 4$$
$$21 = 17 + 4,$$

and so on. The common difference is 4.

If the common difference of an arithmetic sequence is d, then by the definition of an arithmetic sequence,

$$d = a_{n+1} - a_n$$

for every positive integer n in its domain.

■ *Example 3*
FINDING THE
COMMON
DIFFERENCE

Find the common difference, d, for the arithmetic sequence

$$-9, -7, -5, -3, -1, \ldots.$$

Since this sequence is arithmetic, d can be found by choosing any two adjacent terms and subtracting the first from the second. Choosing -7 and -5 gives

$$d = -5 - (-7) = 2.$$

Choosing -9 and -7 would give $d = -7 - (-9) = 2$, the same result. ■

If a_1 and d are known, then all the terms of an arithmetic sequence can be found.

■ *Example 4*
FINDING THE TERMS
GIVEN a_1 AND d

Write the first five terms for each of the following arithmetic sequences.

(a) The first term is 7, and the common difference is -3.

Here
$$a_1 = 7 \quad \text{and} \quad d = -3.$$
$$a_2 = 7 + (-3) = 4,$$
$$a_3 = 4 + (-3) = 1,$$
$$a_4 = 1 + (-3) = -2,$$
$$a_5 = -2 + (-3) = -5.$$

(b) $a_1 = -12, d = 5$

Starting with a_1, add d to each term to get the next term.

$$a_1 = -12$$
$$a_2 = -12 + d = -12 + 5 = -7$$
$$a_3 = -7 + d = -7 + 5 = -2$$
$$a_4 = -2 + d = -2 + 5 = 3$$
$$a_5 = 3 + d = 3 + 5 = 8 \quad ■$$

If a_1 is the first term of an arithmetic sequence and d is the common difference, then the terms of the sequence are given by

$$a_1 = a_1$$
$$a_2 = a_1 + d$$
$$a_3 = a_2 + d = a_1 + d + d = a_1 + 2d$$
$$a_4 = a_3 + d = a_1 + 2d + d = a_1 + 3d$$
$$a_5 = a_1 + 4d$$
$$a_6 = a_1 + 5d,$$

and, by this pattern $a_n = a_1 + (n - 1)d.$

nTH TERM OF AN ARITHMETIC SEQUENCE	In an arithmetic sequence with first term a_1 and common difference d, the nth term, a_n, is given by $$a_n = a_1 + (n - 1)d.$$

■ *Example 5*

USING THE FORMULA FOR THE nTH TERM

Find a_{13} and a_n for the arithmetic sequence

$$-3, 1, 5, 9, \ldots .$$

Here $a_1 = -3$ and $d = 1 - (-3) = 4$. To find a_{13}, substitute 13 for n in the preceding formula.

$$a_{13} = a_1 + (13 - 1)d$$
$$a_{13} = -3 + (12)4$$
$$a_{13} = -3 + 48$$
$$a_{13} = 45$$

Find a_n by substituting values for a_1 and d in the formula for a_n.

$$a_n = -3 + (n - 1) \cdot 4$$
$$a_n = -3 + 4n - 4 \qquad \text{Distributive property}$$
$$a_n = 4n - 7 \quad ■$$

■ *Example 6*

USING THE FORMULA FOR THE nTH TERM

Find a_{18} and a_n for the arithmetic sequence having $a_2 = 9$ and $a_3 = 15$.

Find d first; $d = a_3 - a_2 = 15 - 9 = 6$.

Since
$$a_2 = a_1 + d,$$
$$9 = a_1 + 6 \qquad \text{Let } a_2 = 9, d = 6.$$
$$a_1 = 3.$$

Then
$$a_{18} = 3 + (18 - 1) \cdot 6 \qquad \text{Formula for } a_n; \quad n = 18$$
$$a_{18} = 105,$$

and
$$a_n = 3 + (n - 1) \cdot 6$$
$$a_n = 3 + 6n - 6 \qquad \text{Distributive property}$$
$$a_n = 6n - 3. \quad ■$$

■ **IN SIMPLEST TERMS** ━━━━━━━━━━━━━━━━━━━━━━━━━━━━━

22 ▭

The normal growth pattern for children aged 3–11 follows that of an arithmetic sequence. An increase in height of about 6 cm per year is expected. Thus, 6 would be the common difference of the sequence.

A child who measures 96 cm at age 3 would have his expected height in subsequent years represented by the sequence 102, 108, 114, 120, 126, 132, 138, 144. Each term differs from the adjacent terms by the common difference, 6.

SOLVE EACH PROBLEM

A. If a child measures 98.2 cm at age 3 and 109.8 cm at age 5, what would be the common difference of the arithmetic sequence describing her yearly height?

B. What would we expect her height to be at age 8?

ANSWERS   A. The common difference is 5.8.   B. We would expect the child to be 127.2 cm at age 8.

■ *Example 7*

USING THE
FORMULA FOR THE
*n*TH TERM

Suppose that an arithmetic sequence has $a_8 = -16$ and $a_{16} = -40$. Find a_1.

We must find d first. Since $a_8 = a_1 + (8 - 1)d$, replacing a_8 with -16 gives $-16 = a_1 + 7d$ or $a_1 = -16 - 7d$. Similarly, $-40 = a_1 + 15d$ or $a_1 = -40 - 15d$. From these two equations, using the substitution method from Chapter 7,

$$-16 - 7d = -40 - 15d,$$

so $d = -3$. To find a_1, substitute -3 for d in $-16 = a_1 + 7d$:

$$-16 = a_1 + 7d$$
$$-16 = a_1 + 7(-3) \quad \text{Let } d = -3.$$
$$a_1 = 5. \quad ■$$

9.1 EXERCISES ■ ─────────────────────────────────────

Write the first five terms of each of the following sequences. See Example 1.

1. $a_n = 6n + 4$ **2.** $a_n = 3n - 2$ **3.** $a_n = 2^n$ **4.** $a_n = 3^n$

5. $a_n = (-1)^{n+1}$ **6.** $a_n = (-1)^n(n + 2)$ **7.** $a_n = \dfrac{2n}{n + 3}$ **8.** $a_n = \dfrac{-4}{n + 5}$

9. $a_n = \dfrac{8n - 4}{2n + 1}$ **10.** $a_n = \dfrac{-3n + 6}{n + 1}$ **11.** $a_n = (-2)^n(n)$

12. $a_n = (-1/2)^n(n^{-1})$ **13.** $a_n = x^n$ **14.** $a_n = n \cdot x^{-n}$

Find the first ten terms for the sequences defined as follows. See Example 2.

15. $a_1 = 4$; for $n \geq 2$, $a_n = a_{n-1} + 5$

16. $a_1 = -8$; for $n \geq 2$, $a_n = a_{n-1} - 7$

17. $a_1 = 2$; for $n \geq 2$, $a_n = 2 \cdot a_{n-1}$

18. $a_1 = -3$; for $n \geq 2$, $a_n = 2 \cdot a_{n-1}$

19. $a_1 = 1$, $a_2 = 1$; for $n \geq 3$, $a_n = a_{n-1} + a_{n-2*}$

20. $a_1 = 3$, $a_2 = 2$; for $n \geq 3$, $a_n = a_{n-1} - a_{n-2}$

For each of the following arithmetic sequences, write the indicated number of terms. See Example 4.

21. $a_1 = 4$, $d = 2$, $n = 5$

22. $a_1 = 6$, $d = 8$, $n = 4$

23. $a_2 = 9$, $d = -2$, $n = 4$

24. $a_3 = 7$, $d = -4$, $n = 4$

25. $a_3 = -2$, $d = -4$, $n = 4$

26. $a_2 = -12$, $d = -6$, $n = 5$

27. $a_3 = 6$, $a_4 = 12$, $n = 6$

28. $a_5 = 8$, $a_6 = 5$, $n = 6$

For each of the following sequences that are arithmetic, find d and a_n. See Examples 3 and 5.

29. 12, 17, 22, 27, 32, 37, . . .

30. 8, 17, 26, 35, 44, 53, . . .

31. $-19, -12, -5, 2, 9, \ldots$

32. $-30, -20, -12, -6, -2, \ldots$

33. $x, x + m, x + 2m, x + 3m, x + 4m, \ldots$

34. $k + p, k + 2p, k + 3p, k + 4p, \ldots$

35. $2z + m, 2z, 2z - m, 2z - 2m, 2z - 3m, \ldots$

36. $3r - 4z, 3r - 3z, 3r - 2z, 3r - z, 3r, 3r + z, \ldots$

Find a_8 and a_n for each of the following arithmetic sequences. See Examples 6 and 7.

37. $a_1 = 5$, $d = 2$

38. $a_1 = -3$, $d = -4$

39. $a_3 = 2$, $d = 1$

40. $a_4 = 5$, $d = -2$

41. $a_1 = 8$, $a_2 = 6$

42. $a_1 = 6$, $a_2 = 3$

43. $a_{10} = 6$, $a_{12} = 15$

44. $a_{15} = 8$, $a_{17} = 2$

45. $a_1 = x$, $a_2 = x + 3$

46. $a_2 = y + 1$, $d = -3$

47. $a_6 = 2m$, $a_7 = 3m$

48. $a_5 = 4p + 1$, $a_7 = 6p + 7$

49. Give an example of an arithmetic sequence with $a_4 = 12$.

50. Explain in your own words what is meant by an *arithmetic sequence*.

51. Which one of the following is not an arithmetic sequence?
   (a) 4, 6, 8, 10, . . .   **(b)** $-2, 6, 14, 22, \ldots$
   (c) 1/2, 1, 3/2, 2, . . .   **(d)** 5, 10, 20, 40, . . .

52. Refer to the sequence in Exercise 51 that is not arithmetic. Explain in your own words how each term after the first is determined by using the previous term of the sequence.

Find a_1 for each of the following arithmetic sequences. See Example 7.

53. $a_9 = 47$, $a_{15} = 77$

54. $a_{10} = 50$, $a_{20} = 110$

55. $a_{15} = 168$, $a_{16} = 180$

56. $a_{10} = -54$, $a_{17} = -89$

Explain why each of the following sequences is arithmetic.

57. log 2, log 4, log 8, log 16, log 32, . . .

58. log 12, log 36, log 108, log 324, . . .

---

*This is the Fibonacci sequence, a well-known sequence that occurs widely in nature.

458 | 9.2 —————— **GEOMETRIC SEQUENCES**

A **geometric sequence** (or **geometric progression**) is a sequence in which each term after the first is obtained by multiplying the preceding term by a constant nonzero real number, called the **common ratio.** An example of a geometric sequence is 2, 8, 32, 128, . . . in which the first term is 2 and the common ratio is 4.

If the common ratio of a geometric sequence is r, then by the definition of a geometric sequence,

$$\frac{a_{n+1}}{a_n} = r$$

for every positive integer n in its domain. By this definition, the common ratio can be found by choosing any term except the first and dividing it by the preceding term.

■ *Example 1*
FINDING THE COMMON RATIO

Find the common ratio, r, for the geometric sequence

$$15, \frac{15}{2}, \frac{15}{4}, \frac{15}{8}, \ldots$$

Find r by choosing *any* two adjacent terms and dividing the second one by the first. Choosing the second and third terms of the sequence gives

$$r = \frac{15}{4} \div \frac{15}{2} = \frac{1}{2}.$$

Additional terms of the sequence can be found by multiplying each successive term by 1/2. ■

As mentioned above, the geometric sequence, 2, 8, 32, 128, . . . has $r = 4$. Notice that

$$8 = 2 \cdot 4$$
$$32 = 8 \cdot 4 = (2 \cdot 4) \cdot 4 = 2 \cdot 4^2$$
$$128 = 32 \cdot 4 = (2 \cdot 4^2) \cdot 4 = 2 \cdot 4^3.$$

To generalize this result, assume that a geometric sequence has first term a_1 and common ratio r. The second term can be written as $a_2 = a_1 r$, the third as $a_3 = a_2 r = (a_1 r)r = a_1 r^2$, and so on. Following this pattern, the nth term is $a_n = a_1 r^{n-1}$.

---

*n*TH **TERM OF A** **GEOMETRIC** **SEQUENCE**	In the geometric sequence with first term a_1 and common ratio r, the nth term is $$a_n = a_1 r^{n-1}.$$

■ *Example 2*
USING THE
FORMULA FOR THE
*n*TH TERM

Find the sixth term of the geometric sequence with first term 8, and common ratio $-1/2$.

In this sequence, we have $a_1 = 8$, $r = -1/2$, and $n = 6$. Substituting into the formula, we get

$$a_n = a_1 r^{n-1}$$

$$a_6 = 8\left(-\frac{1}{2}\right)^{6-1}$$

$$= 8\left(-\frac{1}{2}\right)^5 = 8\left(-\frac{1}{32}\right) = -\frac{1}{4}. \quad ■$$

■ *Example 3*
USING THE
FORMULA FOR THE
*n*TH TERM

Find a_5 and a_n for the following geometric sequence.

$$4, 12, 36, 108, \ldots$$

The first term, a_1, is 4. Find r by choosing any term except the first and dividing it by the preceding term. For example,

$$r = \frac{36}{12} = 3.$$

Since $a_4 = 108$, $a_5 = 3 \cdot 108 = 324$. The fifth term also could be found using the formula for a_n, $a_n = a_1 r^{n-1}$, and replacing n with 5, r with 3, and a_1 with 4.

$$a_5 = 4 \cdot (3)^{5-1} = 4 \cdot 3^4 = 324$$

By the formula, $a_n = 4 \cdot 3^{n-1}.$ ■

■ *Example 4*
USING THE
FORMULA FOR THE
*n*TH TERM

Find a_1 and r for each of the following geometric sequences.

(a) The third term is 20 and the sixth term is -160.

Use the formula for the *n*th term of a geometric sequence.

$$\text{For } n = 3, \quad a_3 = a_1 r^2 = 20;$$

$$\text{for } n = 6, \quad a_6 = a_1 r^5 = -160.$$

We have $a_1 r^2 = 20$, so that $a_1 = 20/r^2$. Substitute this in the second equation and solve for r.

$$a_1 r^5 = -160$$

$$\left(\frac{20}{r^2}\right) r^5 = -160 \qquad \text{Let } a_1 = \frac{20}{r^2}.$$

$$20 r^3 = -160$$

$$r^3 = -8$$

$$r = -2$$

Since $a_1 r^2 = 20$ and $r = -2$, we have $a_1 = 5$.

(b) $a_5 = 15$ and $a_7 = 375$

First substitute $n = 5$ and then $n = 7$ into $a_n = a_1 r^{n-1}$.

$$a_5 = a_1 r^4 = 15 \qquad \text{and} \qquad a_7 = a_1 r^6 = 375$$

Solve the first equation for a_1 to get $a_1 = 15/r^4$. Then substitute for a_1 in the second equation.

$$a_1 r^6 = 375$$

$$\frac{15}{r^4} \cdot r^6 = 375 \qquad \text{Let } a_1 = \frac{15}{r^4}.$$

$$15 r^2 = 375$$

$$r^2 = 25$$

$$r = \pm 5 \qquad \text{Square root property}$$

Either 5 or -5 can be used for r. To find a_1, use

$$a_1 = \frac{15}{r^4}.$$

Replace r with ± 5.

$$a_1 = \frac{15}{(\pm 5)^4} = \frac{15}{625} = \frac{3}{125}$$

There are two sequences that satisfy the given conditions: one with $a_1 = 3/125$ and $r = 5$ and the other with $a_1 = 3/125$ and $r = -5$. ■

9.2 EXERCISES ■

For each of the following, write the first n terms of the geometric sequence that satisfies the given conditions.

1. $a_1 = 2$, $r = 3$, $n = 4$ **2.** $a_1 = 4$, $r = 2$, $n = 5$ **3.** $a_1 = 1/2$, $r = 4$, $n = 4$

4. $a_1 = 2/3$, $r = 6$, $n = 3$ **5.** $a_1 = -2$, $r = -3$, $n = 4$ **6.** $a_1 = -4$, $r = 2$, $n = 5$

7. $a_1 = 3125$, $r = 1/5$, $n = 7$ **8.** $a_1 = 729$, $r = 2/3$, $n = 5$ **9.** $a_1 = -1$, $r = -1$, $n = 6$

10. $a_1 = -1$, $r = 1$, $n = 5$ **11.** $a_3 = 6$, $a_4 = 12$, $n = 5$ **12.** $a_2 = 9$, $a_3 = 3$, $n = 4$

Find a_5 and a_n for each of the following geometric sequences. See Examples 2 and 3.

13. $a_1 = 4$, $r = 3$ **14.** $a_1 = 8$, $r = 4$ **15.** $a_1 = -2$, $r = 3$ **16.** $a_1 = -5$, $r = 4$

17. $a_1 = -3$, $r = -5$ **18.** $a_1 = -4$, $r = -2$ **19.** $a_2 = 3$, $r = 2$ **20.** $a_3 = 6$, $r = 3$

21. $a_4 = 64$, $r = -4$ **22.** $a_4 = 81$, $r = -3$

For each of the following sequences that are geometric, find a_n. See Examples 1 and 3.

23. 6, 12, 24, 48, . . .   **24.** 4, 16, 64, 256, . . .   **25.** 3/4, 3/2, 3, 6, 12, . . .

26. 5/6, 5/3, 10/3, 20/3, 40/3, . . .   **27.** -4, 2, -1, 1/2, . . .   **28.** 49, -7, 1, $-1/7$, . . .

29. 18, 20, 24, 32, 48, . . .   **30.** -7, -5, -3, -1, 1, 3, . . .   **31.** $a_3 = 9$, $r = 2$

32. $a_5 = 6$, $r = 1/2$ **33.** $a_3 = -2$, $r = 3$ **34.** $a_2 = 5$, $r = 1/2$

Work the following problems.

35. John Vasquez contracts to work for thirty days, receiving $.01 on the first day of work, $.02 on the second, $.04 on the third, $.08 on the fourth, and so on, with each day's pay double that of the previous day. What would Vasquez earn on the twentieth day?

36. In Exercise 35, what would Vasquez earn on the thirtieth day?

37. Give an example of a geometric sequence with $a_3 = 15$.

38. Explain in your own words what is meant by a *geometric sequence.*

39. Which one of the following is not a geometric sequence?
(a) $1, -2, 4, -8, \ldots$ (c) $3, 3/2, 3/4, 3/8, \ldots$
(b) $1, 10, 100, 1000, \ldots$ (d) $1, 1, 2, 3, 5, 8, \ldots$

40. Refer to the sequence in Exercise 39 that is not geometric. Explain in your own words how each term after the first two terms is determined. (This is a famous sequence known as the Fibonacci sequence.)

Find a_1 and r for each of the following geometric sequences. See Example 4.

41. $a_2 = 6, \quad a_6 = 486$ **42.** $a_3 = -12, \quad a_6 = 96$ **43.** $a_2 = 64, \quad a_8 = 1$ **44.** $a_2 = 100, \quad a_5 = 1/10$

Explain why the following sequences are geometric.

45. $\log 6, \log 6^2, \log 6^4, \log 6^8, \ldots$

46. $\log 2, \log 4, \log 16, \log 256, \ldots$

9.3 ——— SERIES; APPLICATIONS OF SEQUENCES AND SERIES

The sum of the terms of a sequence is called a **series.** A compact shorthand notation can be used to write a series. For example,

$$1 + 5 + 9 + 13 + 17 + 21$$

is the sum of the first six terms in the sequence with general term $a_n = 4n - 3$. This series can be written as

$$\sum_{i=1}^{6} (4i - 3)$$

(read "the sum from $i = 1$ to 6 of $4i - 3$").

The Greek letter sigma, Σ, is used to mean "sum." To evaluate this sum, replace i in $4i - 3$ first by 1, then 2, then 3, until finally i is replaced with 6.

$$\sum_{i=1}^{6}(4i - 3) = (4 \cdot 1 - 3) + (4 \cdot 2 - 3) + (4 \cdot 3 - 3)$$
$$+ (4 \cdot 4 - 3) + (4 \cdot 5 - 3) + (4 \cdot 6 - 3)$$
$$= 1 + 5 + 9 + 13 + 17 + 21 = 66$$

In this notation, i is called the **index of summation.**

CAUTION | Do not confuse this use of i with the use of i to represent an imaginary number. Other letters may be used for the index of summation.

■ *Example 1*

Usɪɴɢ Sᴜᴍᴍᴀᴛɪᴏɴ
Nᴏᴛᴀᴛɪᴏɴ

Evaluate each of the following sums.

(a) $\displaystyle\sum_{i=1}^{4} i^2(i + 1) = 1^2(1 + 1) + 2^2(2 + 1) + 3^2(3 + 1) + 4^2(4 + 1)$

$$= 1 \cdot 2 + 4 \cdot 3 + 9 \cdot 4 + 16 \cdot 5$$

$$= 2 + 12 + 36 + 80$$

$$= 130$$

(b) $\displaystyle\sum_{j=3}^{6} \frac{j + 1}{j - 2} = \frac{3 + 1}{3 - 2} + \frac{4 + 1}{4 - 2} + \frac{5 + 1}{5 - 2} + \frac{6 + 1}{6 - 2}$

$$= \frac{4}{1} + \frac{5}{2} + \frac{6}{3} + \frac{7}{4} = \frac{41}{4} \quad ■$$

SUM OF AN ARITHMETIC SEQUENCE Suppose someone borrows $3000 and agrees to pay $100 per month plus interest of 1% per month on the unpaid balance until the loan is paid off. The first month he pays $100 to reduce the loan, plus interest of $(.01)3000 = 30$ dollars. The second month he pays another $100 toward the loan and interest of $(.01)2900 = 29$ dollars. Since the loan is reduced by $100 each month, his interest payments decrease by $(.01)100 = 1$ dollar each month, forming the arithmetic sequence

$$30, 29, 28, \ldots, 3, 2, 1.$$

The total amount of interest paid is given by the sum of the terms of this sequence. A formula can be developed to find this sum without adding all thirty numbers directly.

If an arithmetic sequence has terms $a_1, a_2, a_3, a_4, \ldots, a_n$, and S_n is defined as the sum of the first n terms of the sequence, then

$$S_n = a_1 + a_2 + a_3 + \cdots + a_n.$$

A formula for S_n can be found by writing the sum of the first n terms as follows.

$$S_n = a_1 + [a_1 + d] + [a_1 + 2d] + \cdots + [a_1 + (n - 1)d]$$

Next, write this same sum in reversed order.

$$S_n = [a_1 + (n - 1)d] + [a_1 + (n - 2)d] + \cdots + [a_1 + d] + a_1$$

Now add the corresponding sides of these two equations.

$$S_n + S_n = (a_1 + [a_1 + (n - 1)d]) + ([a_1 + d] + [a_1 + (n - 2)d])$$
$$+ \cdots + ([a_1 + (n - 1)d] + a_1)$$

From this,

$$2S_n = [2a_1 + (n - 1)d] + [2a_1 + (n - 1)d] + \cdots + [2a_1 + (n - 1)d].$$

Since there are n of the $[2a_1 + (n - 1)d]$ terms on the right,

$$2S_n = n[2a_1 + (n - 1)d]$$

$$S_n = \frac{n}{2}[2a_1 + (n - 1)d].$$

Using the formula $a_n = a_1 + (n - 1)d$, S_n can also be written as

$$S_n = \frac{n}{2}[a_1 + a_1 + (n - 1)d],$$

or

$$S_n = \frac{n}{2}(a_1 + a_n).$$

A summary of this work with sums of arithmetic sequences follows.

SUM OF THE FIRST
n TERMS OF AN
ARITHMETIC
SEQUENCE

If an arithmetic sequence has first term a_1 and common difference d, then the sum of the first n terms is given by

$$S_n = \frac{n}{2}(a_1 + a_n)$$

or

$$S_n = \frac{n}{2}[2a_1 + (n - 1)d].$$

The first formula is used when the first and last terms are known; otherwise the second formula is used.

Either one of these formulas can be used to find the total interest on the $3000 loan discussed above. In the sequence of interest payments, $a_1 = 30$, $d = -1$, $n = 30$, and $a_n = 1$. Choosing the first formula,

$$S_n = \frac{n}{2}(a_1 + a_n),$$

gives

$$S_{30} = \frac{30}{2}(30 + 1) = 15(31) = 465,$$

so a total of $465 interest will be paid over the 30 months.

■ **Example 2**
USING THE SUM
FORMULA
(ARITHMETIC
SEQUENCE)

Find S_{12} for the arithmetic sequence

$$-9, -5, -1, 3, 7, \ldots .$$

We want the sum of the first twelve terms. Using $a_1 = -9$, $n = 12$, and $d = 4$ in the second formula,

$$S_n = \frac{n}{2}[2a_1 + (n - 1)d],$$

gives

$$S_{12} = \frac{12}{2}[2(-9) + 11(4)] = 6(-18 + 44) = 156. \quad ■$$

■ *Example 3*

USING THE SUM
FORMULA
(ARITHMETIC
SEQUENCE)

Find the sum of the first 60 positive integers.

Here $n = 60$, $a_1 = 1$, and $a_{60} = 60$. Use the first formula for S_n.

$$S_n = \frac{n}{2}(a_1 + a_n)$$

$$S_{60} = \frac{60}{2}(1 + 60) = 30 \cdot 61 = 1830 \quad ■$$

■ *Example 4*

USING THE SUM
FORMULA
(ARITHMETIC
SEQUENCE)

The sum of the first 17 terms of an arithmetic sequence is 187. If $a_{17} = -13$, find a_1 and d.

Use the first formula for S_n, with $n = 17$, to find a_1.

$$S_{17} = \frac{17}{2}(a_1 + a_{17}) \qquad \text{Let } n = 17.$$

$$187 = \frac{17}{2}(a_1 - 13) \qquad \text{Let } S_{17} = 187, a_{17} = -13.$$

$$22 = a_1 - 13 \qquad \text{Multiply by } \frac{2}{17}.$$

$$a_1 = 35$$

Since $a_{17} = a_1 + (17 - 1)d$,

$$-13 = 35 + 16d \qquad \text{Let } a_{17} = -13, a_1 = 35.$$

$$-48 = 16d$$

$$d = -3. \quad ■$$

Any sum of the form

$$\sum_{i=1}^{n}(mi + p),$$

where m and p are real numbers, represents the sum of the terms of an arithmetic sequence having first term

$$a_1 = m(1) + p = m + p$$

and common difference $d = m$. These sums can be evaluated by the formulas in this section, as shown by the next example.

■ *Example 5*

USING SUMMATION
NOTATION

Find the following sum.

$$\sum_{i=1}^{10}(4i + 8)$$

This sum represents the sum of the first ten terms of the arithmetic sequence having

$$a_1 = 4 \cdot 1 + 8 = 12,$$
$$n = 10,$$

and
$$a_n = a_{10} = 4 \cdot 10 + 8 = 48.$$

Thus
$$\sum_{i=1}^{10} (4i + 8) = S_{10} = \frac{10}{2}(12 + 48) = 5(60) = 300. \quad \blacksquare$$

SUM OF A GEOMETRIC SEQUENCE Just as formulas were developed to find the sum of the first n terms of an arithmetic sequence, the same can be done for a geometric sequence. We begin by writing the sum S_n as

$$S_n = a_1 + a_2 + a_3 + \cdots + a_n.$$

This can also be written as

$$S_n = a_1 + a_1 r + a_1 r^2 + \cdots + a_1 r^{n-1}. \tag{1}$$

If $r = 1$, then $S_n = na_1$, which is a correct formula for this case. If $r \neq 1$, multiply both sides of (1) by r, obtaining

$$rS_n = a_1 r + a_1 r^2 + a_1 r^3 + \cdots + a_1 r^n. \tag{2}$$

If equation (2) is subtracted from equation (1), the result is

$$S_n = a_1 + a_1 r + a_1 r^2 + \cdots + a_1 r^{n-1}$$
$$\underline{rS_n = \qquad a_1 r + a_1 r^2 + \cdots + a_1 r^{n-1} + a_1 r^n}$$
$$S_n - rS_n = a_1 \qquad\qquad\qquad\qquad\qquad - a_1 r^n \qquad \text{Subtract.}$$

or
$$S_n(1 - r) = a_1(1 - r^n), \qquad\qquad\qquad\qquad \text{Factor.}$$

which finally gives

$$S_n = \frac{a_1(1 - r^n)}{1 - r}, \quad \text{where } r \neq 1. \qquad \text{Divide by } 1 - r.$$

This discussion is summarized below.

SUM OF THE FIRST **n TERMS OF A** **GEOMETRIC** **SEQUENCE**	If a geometric sequence has first term a_1 and common ratio r, then the sum of the first n terms is given by $$S_n = \frac{a_1(1 - r^n)}{1 - r}, \quad \text{where } r \neq 1.$$

■ *Example 6*

**USING THE SUM
FORMULA
(GEOMETRIC
SEQUENCE)**

(a) Find S_5 for the geometric sequence

$$3, 6, 12, 24, 48.$$

Here $a_1 = 3$ and $r = 2$. Using the formula above,

$$S_5 = \frac{3(1 - 2^5)}{1 - 2} = \frac{3(1 - 32)}{-1} = \frac{3(-31)}{-1} = 93.$$

(b) Find S_4 for the geometric sequence

$$10, 2, \frac{2}{5}, \ldots$$

Here $a_1 = 10$, $r = 1/5$, and $n = 4$. Substitute these values into the formula for S_n.

$$S_4 = \frac{10\left[1 - \left(\frac{1}{5}\right)^4\right]}{1 - \frac{1}{5}}$$

$$= \frac{10\left(1 - \frac{1}{625}\right)}{\frac{4}{5}}$$

$$= 10\left(\frac{624}{625}\right) \cdot \frac{5}{4}$$

$$S_4 = \frac{312}{25} \quad ■$$

A sum of the form

$$\sum_{i=1}^{n} m \cdot p^i$$

represents the sum of the terms of a geometric sequence having first term

$$a_1 = m \cdot p^1 = mp$$

and common ratio $r = p$. These sums can be found by using the formula for S_n given above.

■ *Example 7*

**USING SUMMATION
NOTATION**

Find each of the following sums.

(a) $\displaystyle\sum_{i=1}^{7} 2 \cdot 3^i$

In this sum, $a_1 = 2 \cdot 3^1 = 6$ and $r = 3$. Thus,

$$\sum_{i=1}^{7} 2 \cdot 3^i = S_7 = \frac{6(1 - 3^7)}{1 - 3} = \frac{6(1 - 2187)}{-2} = \frac{6(-2186)}{-2} = 6558.$$

(b) $\displaystyle\sum_{i=1}^{5} \left(\frac{3}{4}\right)^{i} = S_5 = \frac{\frac{3}{4}\left[1 - \left(\frac{3}{4}\right)^5\right]}{1 - \frac{3}{4}}$

$$= \frac{\frac{3}{4}\left(1 - \frac{243}{1024}\right)}{\frac{1}{4}}$$

$$= 3\left(\frac{781}{1024}\right) = \frac{2343}{1024} \quad ■$$

APPLICATIONS The formulas developed so far in this chapter will help us to solve applied problems that lead to sequences.

PROBLEM SOLVING
In some cases, we might be asked to find a particular term of an arithmetic or geometric sequence that appears in an applied problem. We would then use the appropriate formula for finding a_n (developed in the earlier sections). If the problem requires finding the sum of a specified number of terms of an arithmetic or geometric sequence, we then use one of the formulas developed in this section. It is important to read carefully to determine whether we want to find a specific term or a sum of terms. ■

■ *Example 8*
SOLVING AN
APPLIED PROBLEM
INVOLVING AN
ARITHMETIC
SEQUENCE

A child building a tower with blocks uses 15 for the bottom row. Each row has 2 fewer blocks than the previous row. Suppose that there are 8 rows in the tower.

(a) How many blocks are used for the top row?
    The number of blocks in each row forms an arithmetic sequence with $a_1 = 15$ and $d = -2$. Find a_n for $n = 8$ by using the formula $a_n = a_1 + (n - 1)d$.

$$a_8 = 15 + (8 - 1)(-2) = 1$$

There is just one block in the top row.

(b) What is the total number of blocks in the tower?
    Here we must find the sum of the terms of the arithmetic sequence formed. We have $a_1 = 15$, $n = 8$, and $a_8 = 1$ (as found in part (a)). Use the formula

$$S_n = \frac{n}{2}(a_1 + a_n).$$

Then $\qquad\qquad S_8 = \dfrac{8}{2}(15 + 1) = 4(16) = 64.$

There are 64 blocks in the tower. ■

NOTE   In Example 8(b), we could have used the other form of the formula for the sum of the first n terms of an arithmetic sequence,

$$S_n = \frac{n}{2}[2a_1 + (n - 1)d],$$

with $n = 8$, $a_1 = 15$, and $d = -2$.

■ ***Example 9***

Sᴏʟᴠɪɴɢ ᴀɴ
Aᴘᴘʟɪᴇᴅ Pʀᴏʙʟᴇᴍ
Iɴᴠᴏʟᴠɪɴɢ ᴀ
Gᴇᴏᴍᴇᴛʀɪᴄ
Sᴇǫᴜᴇɴᴄᴇ

An insect population is growing in such a way that each generation is 1.5 times as large as the previous generation. Suppose there are 100 insects in the first generation.

(a) How many will there be in the fourth generation?

The population can be written as a geometric sequence with a_1 as the first-generation population, a_2 the second-generation population, and so on. Then the fourth-generation population will be a_4. Using the formula for a_n, where $n = 4$, $r = 1.5$, and $a_1 = 100$, gives

$$a_4 = a_1 r^3 = 100(1.5)^3 = 100(3.375) = 337.5.$$

In the fourth generation, the population will number about 338 insects.

(b) What will be the total number of insects in the first four generations?

Since we must find the sum of the first four terms of this geometric sequence, we use the formula

$$S_n = \frac{a_1(1 - r^n)}{1 - r},$$

with $n = 4$, $a_1 = 100$, and $r = 1.5$. Use a calculator to get

$$S_4 = \frac{100(1 - 1.5^4)}{1 - 1.5} = \frac{100(1 - 5.0625)}{-.5} = 812.5.$$

The total population for the four generations will amount to about 813 insects.   ■

A sequence of equal payments made at equal periods of time is called an **annuity.** The sum of the payments and interest on the payments is the **future value** of the annuity.

■ ***Example 10***

Fɪɴᴅɪɴɢ ᴛʜᴇ
Fᴜᴛᴜʀᴇ Vᴀʟᴜᴇ ᴏꜰ
ᴀɴ Aɴɴᴜɪᴛʏ

To save money for a trip to Europe, Marge deposited $1000 at the end of each year for four years in an account paying 6% interest compounded annually. What is the future value of this annuity?

Recall the formula for compound interest from Section 5.1: $A = P(1 + r)^t$. The first payment will earn interest for 3 years, the second payment for 2 years, and the third payment for 1 year. The last payment earns no interest. The total amount is

$$1000(1.06)^3 + 1000(1.06)^2 + 1000(1.06) + 1000.$$

This is the sum of a geometric sequence with first term (starting at the end of the sum as written above) $a_1 = 1000$ and common ratio $r = 1.06$. Using the formula for S_4, the sum of four terms, gives

$$S_4 = \frac{1000[1 - (1.06)^4]}{1 - 1.06}$$

$$= 4374.62.$$

The future value of the annuity is $4374.62. ■

9.3 EXERCISES ■ ————————————————————

Evaluate each of the following sums. See Example 1.

1. $\displaystyle\sum_{i=1}^{5} (2i + 1)$ **2.** $\displaystyle\sum_{i=1}^{6} (3i - 2)$ **3.** $\displaystyle\sum_{j=1}^{4} \frac{1}{j}$ **4.** $\displaystyle\sum_{i=1}^{5} (i + 1)^{-1}$

5. $\displaystyle\sum_{i=1}^{4} i^i$ **6.** $\displaystyle\sum_{k=1}^{4} (k + 1)^2$ **7.** $\displaystyle\sum_{k=1}^{6} (-1)^k \cdot k$ **8.** $\displaystyle\sum_{i=1}^{7} (-1)^{i+1} \cdot i^2$

Find the sum of the first ten terms for each of the following arithmetic sequences. See Examples 2 and 3.

9. $a_1 = 8, d = 3$ **10.** $a_1 = -9, d = 4$ **11.** $a_3 = 5, a_4 = 8$ **12.** $a_2 = 9, a_4 = 13$

13. $5, 9, 13, \ldots$ **14.** $8, 6, 4, \ldots$ **15.** $a_1 = 10, a_{10} = 5\ 1/2$ **16.** $a_1 = -8, a_{10} = -5/4$

Find a_1 and d for each of the following arithmetic sequences. See Example 4.

17. $S_{20} = 1090, \quad a_{20} = 102$ **18.** $S_{31} = 5580, \quad a_{31} = 360$

19. $S_{12} = -108, \quad a_{12} = -19$ **20.** $S_{25} = 650, \quad a_{25} = 62$

Evaluate each of the following sums. See Example 5.

21. $\displaystyle\sum_{i=1}^{3} (i + 4)$ **22.** $\displaystyle\sum_{i=1}^{5} (i - 8)$ **23.** $\displaystyle\sum_{j=1}^{10} (2j + 3)$ **24.** $\displaystyle\sum_{j=1}^{15} (5j - 9)$

25. $\displaystyle\sum_{i=1}^{12} (-5 - 8i)$ **26.** $\displaystyle\sum_{k=1}^{19} (-3 - 4k)$ **27.** $\displaystyle\sum_{i=1}^{1000} i$ **28.** $\displaystyle\sum_{k=1}^{2000} k$

Find the sum of the first five terms for each of the following geometric sequences. See Example 6.

29. $3, 6, 12, 24, \ldots$ **30.** $5, 20, 80, 320, \ldots$ **31.** $12, -6, 3, -3/2, \ldots$ **32.** $18, -3, 1/2, -1/12, \ldots$

33. $a_1 = 4, \quad r = 2$ **34.** $a_1 = 3, \quad r = 3$ **35.** $a_2 = 1/3, \quad r = 3$ **36.** $a_2 = -1, \quad r = 2$

Find each of the following sums. (Hint: In Exercises 43–44 note the beginning value of i.) See Example 7.

37. $\displaystyle\sum_{i=1}^{4} 2^i$ **38.** $\displaystyle\sum_{j=1}^{6} 3^j$ **39.** $\displaystyle\sum_{i=1}^{4} (-3)^i$ **40.** $\displaystyle\sum_{i=1}^{4} (-3^i)$ **41.** $\displaystyle\sum_{i=1}^{6} 81(2/3)^i$

42. $\displaystyle\sum_{k=1}^{5} 9(5/3)^k$ **43.** $\displaystyle\sum_{i=3}^{6} 2^i$ **44.** $\displaystyle\sum_{i=4}^{7} 3^i$ **45.** $\displaystyle\sum_{i=1}^{4} 5(3/5)^{i-1}$ **46.** $\displaystyle\sum_{i=1}^{5} 256(-3/4)^{i-1}$

Solve each problem. See Examples 3 and 8.

47. Find the sum of all the integers from 51 to 71.

48. Find the sum of all the integers from −8 to 30.

49. If a clock strikes the proper number of chimes each hour on the hour, how many times will it chime in a month of 30 days?

50. A stack of telephone poles has 30 in the bottom row, 29 in the next, and so on, with one pole in the top row. How many poles are in the stack?

51. A sky diver falls 10 m during the first second, 20 m during the second, 30 m during the third, and so on. How many meters will the diver fall during the tenth second? During the first ten seconds?

52. Deepwell Drilling Company charges a flat $100 set-up charge, plus $5 for the first foot, $6 for the second, $7 for the third, and so on. Find the total charge for a 70-foot well.

53. An object falling under the force of gravity falls about 16 ft the first second, 48 ft during the next second, 80 ft during the third second, and so on. How far would the object fall during the eighth second? What is the total distance the object would fall in 8 seconds?

54. The population of a city was 49,000 five years ago. Each year the zoning commission permits an increase of 580 in the population. What will the maximum population be five years from now?

55. A super slide of uniform slope is to be built on a level piece of land. There are to be twenty equally spaced supports, with the longest support 15 m long and the shortest 2 m long. Find the total length of all the supports.

56. How much material would be needed for the rungs of a ladder of 31 rungs, if the rungs taper uniformly from 18 in to 28 in?

Solve each problem. See Example 9.

57. Suppose you could save $1 on January 1, $2 on January 2, $4 on January 3, and so on. What amount would you save on January 31?

58. See Exercise 57. What would be the total amount of your savings during January?

59. Richland Oil has a well that produced $4,000,000 of income its first year. Each year thereafter, the well produced half as much as it did the previous year. What total amount of income would be produced by the well in 6 years?

60. Fruit and vegetable dealer Olive Greene paid 10¢ per lb for 10,000 lb of onions. Each week the price she charges increases by .1¢ per lb, while her onions lose 5% of their weight. If she sells all the onions after 6 weeks, does she make or lose money? How much?

61. The final step in processing a black and white photographic print is to immerse the print in a chemical called "fixer." The print is then washed in running water. Under certain conditions, 98% of the fixer in a print will be removed with 15 min of washing. How much of the original fixer would then be left after 1 hour?

62. A scientist has a vat containing 100 liters of a pure chemical. Twenty liters are drained and replaced with water. After complete mixing, 20 liters of the mixture are drained and replaced with water. What will be the strength of the mixture after nine such drainings?

63. The half-life of a radioactive substance is the time it takes for half the substance to decay. Suppose that the half-life of a substance is 3 yr and that 10^{15} molecules of the substance are present initially. How many molecules will be present after 15 yr?

64. Each year a machine loses 20% of the value it had at the beginning of the year. Find the value of a machine at the end of 6 yr if it cost $100,000 new.

Find the future value of each of the following annuities. See Example 10.

65. Payments of $1000 at the end of each year for 9 years at 8% interest compounded annually

66. Payments of $800 at the end of each year for 12 years at 10% interest compounded annually

67. Payments of $2430 at the end of each year for 10 years at 11% interest compounded annually

68. Payments of $1500 at the end of each year for 6 years at 12% interest compounded annually

In the preceding section the sum of the first n terms of a geometric sequence was given as

$$S_n = \sum_{i=1}^{n} a_i = \frac{a_1(1 - r^n)}{1 - r},$$

where a_1 is the first term and r $(r \neq 1)$ is the common ratio. Now consider an infinite geometric sequence such as

$$2, 1, \frac{1}{2}, \frac{1}{4}, \frac{1}{8}, \frac{1}{16}, \cdots$$

with first term 2 and common ratio 1/2. Using the formula for S_n gives the following sequence.

$$S_1 = 2, \quad S_2 = 3, \quad S_3 = \frac{7}{2}, \quad S_4 = \frac{15}{4}, \quad S_5 = \frac{31}{8}, \quad S_6 = \frac{63}{16}$$

These sums seem to be getting closer and closer to the number 4. In fact, by selecting a value of n large enough, it is possible to make S_n as close as desired to 4. This idea is expressed as

$$\lim_{n \to \infty} S_n = 4.$$

(Read: "the limit of S_n as n increases without bound is 4.") For no value of n is $S_n = 4$. However, if n is large enough, then S_n is as close to 4 as desired.*

Since

$$\lim_{n \to \infty} S_n = 4,$$

the number 4 is called the *sum* of the infinite geometric sequence

$$2, 1, \frac{1}{2}, \frac{1}{4}, \cdots$$

and

$$2 + 1 + \frac{1}{2} + \frac{1}{4} + \frac{1}{8} + \cdots = 4.$$

■ **Example 1**

FINDING THE SUM OF AN INFINITE GEOMETRIC SEQUENCE

Find $1 + \frac{1}{3} + \frac{1}{9} + \frac{1}{27} + \cdots$

Use the formula for the first n terms of a geometric sequence to get

$$S_1 = 1, \quad S_2 = \frac{4}{3}, \quad S_3 = \frac{13}{9}, \quad S_4 = \frac{40}{27},$$

and in general

$$S_n = \frac{1\left[1 - \left(\frac{1}{3}\right)^n\right]}{1 - \frac{1}{3}}. \qquad \text{Let } a_1 = 1, r = \frac{1}{3}.$$

────────────

*These phrases "large enough" and "as close as desired" are not nearly precise enough for mathematicians; much of a standard calculus course is devoted to making them more precise.

The chart below shows the value of $(1/3)^n$ for larger and larger values of n.

n	1	10	100	200
$\left(\dfrac{1}{3}\right)^n$	$\dfrac{1}{3}$	1.69×10^{-5}	1.94×10^{-48}	3.76×10^{-96}

As n gets larger and larger, $(1/3)^n$ gets closer and closer to 0. That is,

$$\lim_{n \to \infty} \left(\frac{1}{3}\right)^n = 0,$$

making it reasonable that

$$\lim_{n \to \infty} S_n = \lim_{n \to \infty} \frac{1\left[1 - \left(\dfrac{1}{3}\right)^n\right]}{1 - \dfrac{1}{3}} = \frac{1(1 - 0)}{1 - \dfrac{1}{3}} = \frac{1}{\dfrac{2}{3}} = \frac{3}{2}.$$

Hence,

$$1 + \frac{1}{3} + \frac{1}{9} + \frac{1}{27} + \cdots = \frac{3}{2}. \quad ■$$

If a geometric sequence has a first term a_1 and a common ratio r, then

$$S_n = \frac{a_1(1 - r^n)}{1 - r} \quad (r \neq 1)$$

for every positive integer n. If $-1 < r < 1$, then $\lim_{n \to \infty} r^n = 0$, and

$$\lim_{n \to \infty} S_n = \frac{a_1(1 - 0)}{1 - r} = \frac{a_1}{1 - r}.$$

This quotient, $a_1/(1 - r)$, is called the **sum of an infinite geometric sequence.**

The limit $\lim_{n \to \infty} S_n$ is often expressed as S_∞ or $\sum_{i=1}^{\infty} a_i$. These results lead to the following definition.

SUM OF AN INFINITE GEOMETRIC SEQUENCE

The sum of an infinite geometric sequence with first term a_1 and common ratio r, where $-1 < r < 1$, is given by

$$S_\infty = \frac{a_1}{1 - r}.$$

If $|r| > 1$, the terms get larger and larger, so there is no limit as $n \to \infty$. Hence the sequence will not have a sum.

■ ***Example 2***
USING THE SUM
FORMULA

(a) Find the sum

$$-\frac{3}{4} + \frac{3}{8} - \frac{3}{16} + \frac{3}{32} - \frac{3}{64} + \cdots.$$

The first term is $a_1 = -3/4$. To find r, divide any two adjacent terms. For example,

$$r = \frac{-\dfrac{3}{16}}{\dfrac{3}{8}} = -\frac{1}{2}.$$

Since $-1 < r < 1$, the formula in this section applies, and

$$S_\infty = \frac{a_1}{1-r} = \frac{-\dfrac{3}{4}}{1 - \left(-\dfrac{1}{2}\right)} = -\frac{1}{2}.$$

(b) $\displaystyle\sum_{i=1}^{\infty} \left(\frac{3}{5}\right)^i = \frac{\dfrac{3}{5}}{1 - \dfrac{3}{5}} = \frac{3}{2}$

Notice that the upper limit is ∞, indicating that this is an infinite geometric sequence. ■

The formula in this section can be used to convert repeating decimals (which represent rational numbers) to fractions of the form p/q, where p and q are integers, with $q \neq 0$.

■ ***Example 3***

**WRITING A
REPEATING DECIMAL
AS A QUOTIENT OF
INTEGERS**

Write each repeating decimal in the form p/q, where p and q are integers.

(a) .090909 . . .

This decimal can be written as

$$.090909 \ldots = .09 + .0009 + .000009 + \ldots,$$

which is the sum of the terms of an infinite geometric sequence having $a_1 = .09$ and $r = .01$. The sum of this sequence is given by

$$S_\infty = \frac{a_1}{1-r} = \frac{.09}{1 - .01} = \frac{.09}{.99} = \frac{1}{11}.$$

Thus, .090909 . . . = 1/11.

(b) 2.5121212 . . .

Write the number as

$$2.5121212 \ldots = 2.5 + .012 + .00012 + .0000012 + \ldots$$
$$= 2.5 + (.012 + .00012 + .0000012 + \ldots).$$

Beginning with the second term, this is the sum of the terms of an infinite geometric sequence with $a_1 = .012$ and $r = .01$, so 2.5121212 . . . can be written as

$$2.5 + \frac{.012}{1 - .01} = 2.5 + \frac{.012}{.990} = 2.5 + \frac{2}{165} = \frac{829}{330}. \quad ■$$

9.4 EXERCISES ■ ───────────────────────────────

Find r for each of the following infinite geometric sequences. Identify any whose sums would exist. See Example 2.

1. 9, 18, 36, 72, 144, . . .

2. 3, 9, 27, 81, . . .

3. 10, 100, 1000, 10,000, . . .

4. −8, −4, −2, −1, −1/2, . . .

5. 12, 6, 3, 3/2, . . .

6. −8, −16, −32, −64, . . .

7. 1, 1.1, 1.21, 1.331, . . .

8. −1, 1.2, −1.44, 1.728, . . .

9. Under what conditions will an infinite geometric sequence have a sum?

10. The sum of the terms of

$$.9 + .09 + .009 + .0009 + \ldots$$

may be written as .9999 . . . . The sum does exist, since $r = .1$. Use the formula to find this sum, and then complete this statement: The value of .9999 . . . is _____.

Find the sum of each geometric sequence by using the formula of this section where it applies. See Example 2.

11. $16 + 4 + 1 + \ldots$

12. $81 + 27 + 9 + 3 + 1 + \ldots$

13. $100 + 10 + 1 + \ldots$

14. $128 + 64 + 32 + \ldots$

15. $90 + 30 + 10 + \ldots$

16. $25 + 5 + 1 + \ldots$

17. $256 - 128 + 64 - 32 + 16 - \ldots$

18. $120 - 60 + 30 - 15 + \ldots$

19. $108 - 36 + 12 - 4 + \ldots$

20. $10{,}000 - 1000 + 100 - 10 + 1 - \ldots$

21. $\dfrac{3}{4} + \dfrac{3}{8} + \dfrac{3}{16} + \ldots$

22. $\dfrac{4}{5} + \dfrac{2}{5} + \dfrac{1}{5} + \ldots$

23. $3 - \dfrac{3}{2} + \dfrac{3}{4} - \ldots$

24. $9 - 3 + 1 - \ldots$

25. $\dfrac{1}{3} - \dfrac{2}{9} + \dfrac{4}{27} - \dfrac{8}{81} + \ldots$

26. $1 + \dfrac{1}{1.01} + \dfrac{1}{(1.01)^2} + \ldots$

27. $\dfrac{1}{36} + \dfrac{1}{30} + \dfrac{1}{25} + \ldots$

28. $1 + \dfrac{1}{2^2} + \dfrac{1}{2^4} + \ldots$

29. $\displaystyle\sum_{i=1}^{\infty} (1/4)^i$

30. $\displaystyle\sum_{i=1}^{\infty} (9/10)^i$

31. $\displaystyle\sum_{i=1}^{\infty} (1.2)^i$

32. $\displaystyle\sum_{i=1}^{\infty} (1.001)^i$

33. $\displaystyle\sum_{i=1}^{\infty} (-1/4)^i$

34. $\displaystyle\sum_{i=1}^{\infty} (.3)^i$

35. $\displaystyle\sum_{i=1}^{\infty} 1/5^i$

36. $\displaystyle\sum_{i=1}^{\infty} (-1/2)^i$

37. $\displaystyle\sum_{i=1}^{\infty} 10^{-i}$

38. $\displaystyle\sum_{i=1}^{\infty} 4^{-i}$

39. $\displaystyle\sum_{i=1}^{\infty} (1/2)^{-i}$

40. $\displaystyle\sum_{i=1}^{\infty} (3/4)^{-i}$

41. Which one of the following does not have a sum?

(a) $\displaystyle\sum_{i=1}^{10} (1/2)^i$ **(b)** $\displaystyle\sum_{i=1}^{\infty} (3/2)^i$ **(c)** $\displaystyle\sum_{i=1}^{\infty} (1/2)^i$ **(d)** $\displaystyle\sum_{i=1}^{1000} 3 \cdot 4^i$

42. Refer to Exercise 10. The same result can be obtained a different way. Use the fact that $1/3 = .3333 \ldots$ to develop an argument that leads to the same conclusion.

Express each repeating decimal in the form p/q, where p and q are integers. See Example 3.

43. .55555 . . .   **44.** .77777 . . .   **45.** .121212 . . .   **46.** .858585 . . .

47. .313131 . . .   **48.** .909090 . . .   **49.** 1.508508508 . . .   **50.** 2.613613613 . . .

Use the formula of this section to solve each problem.

51. Mitzi drops a ball from a height of 10 m and notices that on each bounce the ball returns to about 3/4 of its previous height. About how far will the ball travel before it comes to rest? (*Hint:* Consider the sum of two sequences.)

52. A sugar factory receives an order for 1000 units of sugar. The production manager thus orders production of 1000 units of sugar. He forgets, however, that the production of sugar requires some sugar (to prime the machines, for example), and so he ends up with only 900 units of sugar. He then orders an additional 100 units, and receives only 90 units. A further order for 10 units produces 9 units. Finally seeing his mistake, the manager decides to try mathematics. He views the production process as an infinite geometric progression with $a_1 = 1000$ and $r = .1$. Using this, find the number of units of sugar that he should have ordered originally.

53. After a person pedaling a bicycle removes his or her feet from the pedals, the wheel rotates 400 times the first minute. As it continues to slow down, each minute it rotates only 3/4 as many times as in the previous minute. How many times will the wheel rotate before coming to a complete stop?

54. A pendulum bob swings through an arc 40 cm long on its first swing. Each swing thereafter, it swings only 80% as far as on the previous swing. How far will it swing altogether before coming to a complete stop?

55. A sequence of equilateral triangles is constructed. The first triangle has sides 2 m in length. To get the second triangle, midpoints of the sides of the original triangle are connected. If this process could be continued indefinitely, what would be the total perimeter of all the triangles?

56. What would be the total area of all the triangles in Exercise 55, disregarding the overlapping?

9.5 ———— MATHEMATICAL INDUCTION

Many results in mathematics are claimed true for any positive integer. Any of these results could be checked for $n = 1$, $n = 2$, $n = 3$, and so on, but since the set of positive integers is infinite it would be impossible to check every possible case. For example, let S_n represent the statement that the sum of the first n positive integers is $n(n + 1)/2$,

$$S_n: 1 + 2 + 3 + \cdots + n = \frac{n(n + 1)}{2}.$$

The truth of this statement can be checked quickly for the first few values of n.

If $n = 1$, S_1 is $\qquad 1 = \dfrac{1(1 + 1)}{2}$, a true statement, since $1 = 1$.

If $n = 2$, S_2 is $\qquad 1 + 2 = \dfrac{2(2 + 1)}{2}$, a true statement, since $3 = 3$.

If $n = 3$, S_3 is $\quad 1 + 2 + 3 = \dfrac{3(3 + 1)}{2}$, a true statement, since $6 = 6$.

If $n = 4$, S_4 is $1 + 2 + 3 + 4 = \dfrac{4(4 + 1)}{2}$, a true statement, since $10 = 10$.

Since the statement is true for $n = 1, 2, 3,$ and 4, can we conclude that the statement is true for all positive integers by observing this finite number of examples? The answer is no. However, we have an idea that it *may* be true for all positive integers.

To prove that such a statement is true for every positive integer, we use the following principle.

PRINCIPLE OF MATHEMATICAL INDUCTION

Let S_n be a statement concerning the positive integer n. Suppose that

1. S_1 is true;

2. for any positive integer k, $k \leq n$, if S_k is true, then S_{k+1} is also true.

Then S_n is true for every positive integer value of n.

A proof by mathematical induction can be explained as follows. By assumption (1) above, the statement is true when $n = 1$. By (2) above, the fact that the statement is true for $n = 1$ implies that it is true for $n = 1 + 1 = 2$. Using (2) again, the statement is thus true for $2 + 1 = 3$, for $3 + 1 = 4$, for $4 + 1 = 5$, and so on. Continuing in this way shows that the statement must be true for *every* positive integer, no matter how large.

The situation is similar to that of a number of dominoes lined up as shown in Figure 9.1. If the first domino is pushed over, it pushes the next, which pushes the next, and so on until all are down.

■ FIGURE 9.1

Another example of the principle of mathematical induction might be an infinite ladder. Suppose the rungs are spaced so that, whenever you are on a rung, you know you can move to the next rung. Then *if* you can get to the first rung, you can go as high up the ladder as you wish.

Two separate steps are required for a proof by mathematical induction.

PROOF BY MATHEMATICAL INDUCTION

Step 1 Prove that the statement is true for $n = 1$.

Step 2 Show that, for any positive integer k, $k \leq n$, if S_k is true, then S_{k+1} is also true.

Mathematical induction is used in the next example to prove the statement discussed above: $S_n = 1 + 2 + 3 + \cdots + n = \dfrac{n(n + 1)}{2}$.

■ *Example 1*
PROVING A
STATEMENT BY
MATHEMATICAL
INDUCTION

Let S_n represent the statement

$$1 + 2 + 3 + \cdots + n = \frac{n(n + 1)}{2}.$$

Prove that S_n is true for every positive integer n.

PROOF

The proof by mathematical induction is as follows.

Step 1 Show that the statement is true when $n = 1$. If $n = 1$, S_1 becomes

$$1 = \frac{1(1 + 1)}{2},$$

which is true.

Step 2 Show that S_k implies S_{k+1}, where S_k is the statement

$$1 + 2 + 3 + \cdots + k = \frac{k(k + 1)}{2},$$

and S_{k+1} is the statement

$$1 + 2 + 3 + \cdots + k + (k + 1) = \frac{(k + 1)[(k + 1) + 1]}{2}.$$

Start with S_k.

$$1 + 2 + 3 + \cdots + k = \frac{k(k + 1)}{2}$$

Add $k + 1$ to both sides of this equation.

$$1 + 2 + 3 + \cdots + k + (k + 1) = \frac{k(k + 1)}{2} + (k + 1)$$

Now factor out the common factor $k + 1$ on the right to get

$$= (k + 1)\left(\frac{k}{2} + 1\right)$$

$$= (k + 1)\left(\frac{k + 2}{2}\right)$$

$$1 + 2 + 3 + \cdots + k + (k + 1) = \frac{(k + 1)[(k + 1) + 1]}{2}. \quad ∎$$

This final result is the statement for $n = k + 1$; it has been shown that if S_k is true, then S_{k+1} is also true. The two steps required for a proof by mathematical induction have now been completed, so the statement S_n is true for every positive integer value of n. ∎

Notice that the left side of the statement always includes *all* the terms up to the *n*th term, as well as the *n*th term.

■ *Example 2*

PROVING A STATEMENT BY MATHEMATICAL INDUCTION

Prove: $4 + 7 + 10 + \cdots + (3n + 1) = \dfrac{n(3n + 5)}{2}$ for all positive integers *n*.

PROOF

Step 1 Show that the statement is true for S_1. S_1 is

$$4 = \frac{1(3 \cdot 1 + 5)}{2}.$$

Since the right side equals 4, S_1 is a true statement.

Step 2 Show that if S_k is true, then S_{k+1} is true, where S_k is

$$4 + 7 + 10 + \cdots + (3k + 1) = \frac{k(3k + 5)}{2},$$

and S_{k+1} is

$$4 + 7 + 10 + \cdots + (3k + 1) + [3(k + 1) + 1]$$
$$= \frac{(k + 1)[3(k + 1) + 5]}{2}.$$

Start with S_k:

$$4 + 7 + 10 + \cdots + (3k + 1) = \frac{k(3k + 5)}{2}.$$

To get the left side of the equation S_k to be the left side of the equation S_{k+1}, we must add the $(k + 1)$th term. Now we try to algebraically change the right side of S_k to look like the right side of S_{k+1}. Adding $[3(k + 1) + 1]$ to both sides of S_k gives

$$4 + 7 + 10 + \cdots + (3k + 1) + [3(k + 1) + 1]$$
$$= \frac{k(3k + 5)}{2} + [3(k + 1) + 1].$$

Clear the parentheses in the new term on the right side of the equals sign and simplify.

$$= \frac{k(3k + 5)}{2} + 3k + 3 + 1$$

$$= \frac{k(3k + 5)}{2} + 3k + 4$$

Now combine the two terms on the right.

$$= \frac{k(3k + 5)}{2} + \frac{2(3k + 4)}{2}$$

$$= \frac{k(3k + 5) + 2(3k + 4)}{2}$$

$$= \frac{3k^2 + 5k + 6k + 8}{2}$$

$$= \frac{3k^2 + 11k + 8}{2}$$

$$= \frac{(k + 1)(3k + 8)}{2}$$

Since $3k + 8$ can be written as $3(k + 1) + 5$,

$$4 + 7 + 10 + \cdots + (3k + 1) + [3(k + 1) + 1]$$

$$= \frac{(k + 1)[3(k + 1) + 5]}{2}. \quad ∎$$

The final result is the statement S_{k+1}. Therefore, if S_k is true, then S_{k+1} is true. The two steps required for a proof by mathematical induction are completed, so the general statement S_n is true for every positive integer value of n. ∎

■ *Example 3*

PROVING A
STATEMENT BY
MATHEMATICAL
INDUCTION

Prove that if x is a real number between 0 and 1, then for every positive integer n,

$$0 < x^n < 1.$$

PROOF

Here S_1 is the statement

$$\text{if } 0 < x < 1, \text{ then } 0 < x^1 < 1,$$

which is true. S_k is the statement

$$\text{if } 0 < x < 1, \text{ then } 0 < x^k < 1.$$

To show that this implies that S_{k+1} is true, multiply all expressions of $0 < x^k < 1$ by x to get

$$x \cdot 0 < x \cdot x^k < x \cdot 1.$$

(Here the fact that $0 < x$ is used.) Simplify to get

$$0 < x^{k+1} < x.$$

Since $x < 1$,

$$x^{k+1} < x < 1$$

and

$$0 < x^{k+1} < 1. \quad ∎$$

By this work, if S_k is true, then S_{k+1} is true. Since both conditions for a proof have been satisfied, the given statement is true for every positive integer n. ∎

9.5 EXERCISES ■

Use the method of mathematical induction to prove the following statements. Assume that n is a positive integer. See Examples 1–3.

1. $2 + 4 + 6 + \cdots + 2n = n(n + 1)$

2. $1 + 3 + 5 + \cdots + (2n - 1) = n^2$

3. $3 + 6 + 9 + \cdots + 3n = \dfrac{3n(n + 1)}{2}$

4. $5 + 10 + 15 + \cdots + 5n = \dfrac{5n(n + 1)}{2}$

5. $2 + 4 + 8 + \cdots + 2^n = 2^{n+1} - 2$

6. $3 + 3^2 + 3^3 + \cdots + 3^n = \dfrac{3(3^n - 1)}{2}$

7. $1^2 + 2^2 + 3^2 + \cdots + n^2 = \dfrac{n(n + 1)(2n + 1)}{6}$

8. $1^3 + 2^3 + 3^3 + \cdots + n^3 = \dfrac{n^2(n + 1)^2}{4}$

9. $5 \cdot 6 + 5 \cdot 6^2 + 5 \cdot 6^3 + \cdots + 5 \cdot 6^n = 6(6^n - 1)$

10. $7 \cdot 8 + 7 \cdot 8^2 + 7 \cdot 8^3 + \cdots + 7 \cdot 8^n = 8(8^n - 1)$

11. $\dfrac{1}{1 \cdot 2} + \dfrac{1}{2 \cdot 3} + \dfrac{1}{3 \cdot 4} + \cdots + \dfrac{1}{n(n + 1)} = \dfrac{n}{n + 1}$

12. $\dfrac{1}{1 \cdot 4} + \dfrac{1}{4 \cdot 7} + \dfrac{1}{7 \cdot 10} + \cdots + \dfrac{1}{(3n - 2)(3n + 1)} = \dfrac{n}{3n + 1}$

13. $\dfrac{1}{2} + \dfrac{1}{2^2} + \dfrac{1}{2^3} + \cdots + \dfrac{1}{2^n} = 1 - \dfrac{1}{2^n}$

14. $\dfrac{4}{5} + \dfrac{4}{5^2} + \dfrac{4}{5^3} + \cdots + \dfrac{4}{5^n} = 1 - \dfrac{1}{5^n}$

15. $x^{2n} + x^{2n-1}y + \cdots + xy^{2n-1} + y^{2n} = \dfrac{x^{2n+1} - y^{2n+1}}{x - y}$

16. $x^{2n-1} + x^{2n-2}y + \cdots + xy^{2n-2} + y^{2n-1} = \dfrac{x^{2n} - y^{2n}}{x - y}$

17. $(a^m)^n = a^{mn}$ (Assume that a and m are constant.)

18. $(ab)^n = a^n b^n$ (Assume that a and b are constant.)

19. If $a > 1$, then $a^n > 1$.

20. If $a > 1$, then $a^n > a^{n-1}$.

21. If $0 < a < 1$, then $a^n < a^{n-1}$.

22. A proof by mathematical induction allows us to prove that a statement is true for all _____.

23. Suppose that Step 2 in a proof by mathematical induction can be satisfied, but Step 1 cannot. May we conclude that the proof is complete? Explain.

24. What is wrong with the following proof by mathematical induction?

Prove: Any natural number equals the next natural number; that is, $n = n + 1$.

Proof. To begin, we assume the statement true for some natural number $n = k$:

$$k = k + 1.$$

We must now show that the statement is true for $n = k + 1$. If we add 1 to both sides, we have

$$k + 1 = k + 1 + 1$$
$$k + 1 = k + 2.$$

Hence, if the statement is true for $n = k$, it is also true for $n = k + 1$. Thus, the theorem is proved.

25. Suppose that n straight lines (with $n \geq 2$) are drawn in a plane, where no two lines are parallel and no three lines pass through the same point. Show that the number of points of intersection of the lines is $(n^2 - n)/2$.

26. The series of sketches below starts with an equilateral triangle having sides of length 1. In the following steps, equilateral triangles are constructed on each side of the preceding figure. The lengths of the sides of these new triangles are 1/3 the length of the sides of the preceding triangles. Develop a formula for the number of sides of the nth figure. Use mathematical induction to prove your answer.

27. Find the perimeter of the nth figure in Exercise 26.

28. Show that the area of the nth figure in Exercise 26 is

$$\sqrt{3}\left[\frac{2}{5} - \frac{3}{20}\left(\frac{4}{9}\right)^{n-1}\right].$$

9.6 ———— PERMUTATIONS

If there are 3 roads from Albany to Baker and 2 roads from Baker to Creswich, in how many ways can one travel from Albany to Creswich by way of Baker? For each of the 3 roads from Albany to Baker, there are 2 different roads from Baker to Creswich, so that there are $3 \cdot 2 = 6$ different ways to make the trip, as shown in the **tree diagram** in Figure 9.2. This example illustrates the following property.

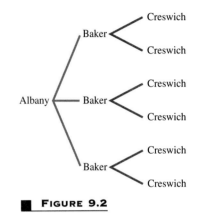

■ **FIGURE 9.2**

MULTIPLICATION PRINCIPLE OF COUNTING	If one event can occur in m ways and a second event can occur in n ways, then both events can occur in mn ways, provided the outcome of the first event does not influence the outcome of the second.

The multiplication principle of counting can be extended to any number of events, provided the outcome of no one event influences the outcome of another. Such events are called **independent events.**

■ *Example 1*
USING THE
MULTIPLICATION
PRINCIPLE

A restaurant offers a choice of 3 salads, 5 main dishes, and 2 desserts. Use the multiplication principle of counting to find the number of different 3-course meals that can be selected.

Three independent events are involved: selecting a salad, selecting a main dish, and selecting a dessert. The first event can occur in 3 ways, the second event can occur in 5 ways, and the third event can occur in 2 ways; thus there are

$$3 \cdot 5 \cdot 2 = 30 \text{ possible meals.} \quad ■$$

■ *Example 2*

USING THE
MULTIPLICATION
PRINCIPLE

Janet Branson has 5 different books that she wishes to arrange on her desk. How many different arrangements are possible?

Five events are involved: selecting a book for the first spot, selecting a book for the second spot, and so on. Here the outcome of the first event *does* influence the outcome of the other events (since one book has already been chosen). For the first spot Branson has 5 choices, for the second spot 4 choices, for the third spot 3 choices, and so on. Now use the multiplication principle of counting to find that there are

$$5 \cdot 4 \cdot 3 \cdot 2 \cdot 1 \text{ or } 120 \text{ different arrangements.} \quad ■$$

In using the multiplication principle of counting we often encounter such products as $5 \cdot 4 \cdot 3 \cdot 2 \cdot 1$ from Example 2. For convenience in writing these products, use the symbol $n!$ (read "n factorial"), which was defined in Section 1.4 and is repeated here.

---

n-FACTORIAL ($n!$) For any positive integer n,

$$n! = n(n - 1)(n - 2)(n - 3) \cdots (2)(1)$$

and

$$0! = 1.$$

By the definition, $5 \cdot 4 \cdot 3 \cdot 2 \cdot 1$ is written as $5!$. Also, $3! = 3 \cdot 2 \cdot 1 = 6$. The definition of $n!$ means that $n[(n - 1)!] = n!$ for all natural numbers $n \geq 2$. It is useful to have this relation hold also for $n = 1$, so, by definition, $0! = 1$.

■ *Example 3*

USING THE
MULTIPLICATION
PRINCIPLE

Suppose Branson (from Example 2) wishes to place only 3 of the 5 books on her desk. How many arrangements of 3 books are possible?

She still has 5 ways to fill the first spot, 4 ways to fill the second spot, and 3 ways to fill the third. Since she wants to use 3 books, there are only 3 spots to be filled (3 events) instead of 5, so there are

$$5 \cdot 4 \cdot 3 = 60 \text{ arrangements.} \quad ■$$

PERMUTATIONS The number 60 in the example above is called the number of *permutations* of 5 things taken 3 at a time, written $P(5, 3) = 60$. Example 2 showed that the number of ways of arranging 5 elements from a set of 5 elements, written $P(5, 5)$, is 120.

A **permutation** of n elements taken r at a time is one of the ways of arranging r elements taken from a set of n elements ($r \leq n$). Generalizing from the examples above, the number of permutations of n elements, taken r at a time, denoted by $P(n, r)$, is

$$
\begin{aligned}
P(n, r) &= n(n - 1)(n - 2) \cdots (n - r + 1) \\
&= \frac{n(n - 1)(n - 2) \cdots (n - r + 1)(n - r)(n - r - 1) \cdots (2)(1)}{(n - r)(n - r - 1) \cdots (2)(1)} \\
&= \frac{n!}{(n - r)!}.
\end{aligned}
$$

This derivation gives the following permutations formula.

PERMUTATIONS OF n ELEMENTS r AT A TIME	If $P(n, r)$ denotes the number of permutations of n elements taken r at a time, $r \le n$, then $$P(n, r) = \frac{n!}{(n - r)!}.$$

Some other symbols used for permutations of n things taken r at a time are $_nP_r$ and P^n_r.

> **NOTE** Some scientific calculators have a key for permutations. Consult the owner's manual for instructions on how to use it.

■ IN SIMPLEST TERMS

Telephone numbers are examples of permutations. The number 269-8057 is not the same as the number 269-5870. To determine how many telephone numbers are available with the 269 prefix, we need only consider the last 4 digits. Each place can hold any digit from 0–9, so there are 10 choices for each: $10 \cdot 10 \cdot 10 \cdot 10 = 10,000$ possible numbers with a 269 prefix.

SOLVE EACH PROBLEM

A. Area codes always have a middle digit of 0 or 1. If we assume that, in addition to this, the first digit of the area code is not 0, how many codes can be formed?

B. Recently, the city of Boston had to add a new area code since the population demanded more phone numbers than were possible with one area code. If we assume that the first digit of any prefix is not 0 and that the second digit is neither 0 nor 1, how many numbers can be assigned to one area code?

ANSWERS   A. With the given restrictions, 180 different area codes are possible.
B. With the given restrictions, one area code can have 7,200,000 phone numbers.

■ *Example 4*

USING THE
PERMUTATIONS
FORMULA

Suppose 8 people enter an event in a swim meet. Assuming there are no ties, in how many ways could the gold, silver, and bronze prizes be awarded?

Using the multiplication principle of counting, there are 3 choices to be made giving $8 \cdot 7 \cdot 6 = 336$. However, we can also use the formula for $P(n, r)$ to get the same result.

$$P(8, 3) = \frac{8!}{(8 - 3)!}$$

$$= \frac{8!}{5!} = \frac{8 \cdot 7 \cdot 6 \cdot 5 \cdot 4 \cdot 3 \cdot 2 \cdot 1}{5 \cdot 4 \cdot 3 \cdot 2 \cdot 1}$$

$$= 8 \cdot 7 \cdot 6 = 336 \quad ■$$

■ *Example 5*

USING THE
PERMUTATIONS
FORMULA AND THE
MULTIPLICATION
PRINCIPLE

A televised talk show will include 4 women and 3 men as panelists.

(a) In how many ways can the panelists be seated in a row of 7 chairs?

Find $P(7, 7)$, the total number of ways to seat 7 panelists in 7 chairs.

$$P(7, 7)$$

$$= \frac{7!}{(7 - 7)!}$$

$$= \frac{7!}{0!} = \frac{7!}{1}$$

$$= 7 \cdot 6 \cdot 5 \cdot 4 \cdot 3 \cdot 2 \cdot 1 = 5040$$

There are 5040 ways to seat the 7 panelists.

(b) In how many ways can the panelists be seated if the men and women are to be alternated?

Use the multiplication principle. In order to alternate men and women, a woman must be seated in the first chair (since there are 4 women and only 3 men), any of the men next, and so on. Thus, there are 4 ways to fill the first seat, 3 ways to fill the second seat, 3 ways to fill the third seat (with any of the 3 remaining women), and so on. This gives

$$4 \cdot 3 \cdot 3 \cdot 2 \cdot 2 \cdot 1 \cdot 1 = 144$$

ways to seat the panelists.　■

■ *Example 6*

USING THE
MULTIPLICATION
PRINCIPLE WITH
RESTRICTIONS

In how many ways can three letters of the alphabet be arranged if a vowel cannot be used in the middle position, and repetitions of the letters are allowed?

We cannot use $P(26, 3)$ here, because of the restriction for the middle position. In the first and third positions, we can use any of the 26 letters of the alphabet, but in the middle position, we can only use one of $26 - 5 = 21$ letters (since there are 5 vowels). Now, using the multiplication counting principle, there are $26 \cdot 21 \cdot 26 = 14,196$ ways to arrange the letters according to the problem.　■

Evaluate each of the following. See Examples 4 and 5.

1. $P(7, 7)$ **2.** $P(5, 3)$ **3.** $P(6, 5)$ **4.** $P(4, 2)$ **5.** $P(10, 2)$ **6.** $P(8, 2)$

7. $P(8, 3)$ **8.** $P(11, 4)$ **9.** $P(7, 1)$ **10.** $P(18, 0)$ **11.** $P(9, 0)$ **12.** $P(14, 1)$

13. Explain in your own words what is meant by a *permutation*.

14. Explain why the restriction $r \leq n$ is needed in the formula for $P(n, r)$.

Use the multiplication principle to solve the following problems. (See Examples 1–3, 5 and 6.)

15. How many different types of homes are available if a builder offers a choice of 5 basic plans, 3 roof styles, and 2 exterior finishes?

16. An auto manufacturer produces 7 models, each available in 6 different colors, with 4 different upholstery fabrics, and 5 interior colors. How many varieties of the auto are available?

17. How many different 4-letter radio station call letters can be made
(a) if the first letter must be K or W and no letter may be repeated;
(b) if repeats are allowed (but the first letter is K or W)?
(c) How many of the 4-letter call letters (starting with K or W) with no repeats end in R?

18. A menu offers a choice of 3 salads, 8 main dishes, and 5 desserts. How many different 3-course meals (salad, main dish, dessert) are possible?

19. A couple has narrowed down the choice of a name for their new baby to 3 first names and 5 middle names. How many different first- and middle-name arrangements are possible?

20. A concert to raise money for an economics prize is to consist of 5 works: 2 overtures, 2 sonatas, and a piano concerto. In how many ways can a program with these 5 works be arranged?

21. For many years, the state of California used three letters followed by three digits on its automobile license plates.
(a) How many different license plates are possible with this arrangement?
(b) When the state ran out of new plates, the order was reversed to three digits followed by three letters. How many additional plates were then possible?

(c) Several years ago, the plates described in (b) were also used up. The state then issued plates with one letter followed by three digits and then three letters. How many plates does this scheme provide?

22. How many 7-digit telephone numbers are possible if the first digit cannot be zero and
(a) only odd digits may be used;
(b) the telephone number must be a multiple of 10 (that is, it must end in zero);
(c) the telephone number must be a multiple of 100;
(d) the first 3 digits are 481;
(e) no repetitions are allowed?

Use permutations to solve each of the following problems. (See Examples 4–5.)

23. In an experiment on social interaction, 6 people will sit in 6 seats in a row. In how many ways can this be done?

24. In how many ways can 7 of 10 monkeys be arranged in a row for a genetics experiment?

25. A business school offers courses in typing, shorthand, transcription, business English, technical writing, and accounting. In how many ways can a student arrange a schedule if 3 courses are taken?

26. If your college offers 400 courses, 20 of which are in mathematics, and your counselor arranges your schedule of 4 courses by random selection, how many schedules are possible that do not include a math course?

27. In a club with 15 members, how many ways can a slate of 3 officers consisting of president, vice-president, and secretary/treasurer be chosen?

28. A baseball team has 20 players. How many 9-player batting orders are possible?

29. In how many ways can 5 players be assigned to the 5 positions on a basketball team, assuming that any player can play any position? In how many ways can 10 players be assigned to the 5 positions?

30. Show that $P(n, n - 1) = P(n, n)$.

9.7 ——————— **COMBINATIONS**

In the preceding section we discussed a method for finding the number of ways to arrange r elements taken from a set of n elements. Sometimes, however, the arrangement (or order) of the elements is not important.

For example, suppose three people (Ms. Opelka, Mr. Adams, and Ms. Jacobs) apply for 2 identical jobs. Ignoring all other factors, in how many ways can the personnel officer select 2 people from the 3 applicants? Here the arrangement or order of the people is unimportant. Selecting Ms. Opelka and Mr. Adams is the same as selecting Mr. Adams and Ms. Opelka. Therefore, there are only 3 ways to select 2 of the 3 applicants:

Ms. Opelka and Mr. Adams;

Ms. Opelka and Ms. Jacobs;

Mr. Adams and Ms. Jacobs.

These three choices are called the *combinations* of 3 elements taken 2 at a time. A **combination** of n elements taken r at a time is one of the ways in which r elements can be chosen from n elements.

In the example above, each combination of 2 applicants forms 2! permutations (Ms. Opelka and Mr. Adams and Mr. Adams and Ms. Opelka, for example). So the number of combinations in the example could be found by dividing the number of *permutations* of 3 things taken 2 at a time by 2! to get

$$\frac{P(3, 2)}{P(2, 2)} = \frac{\frac{3!}{(3-2)!}}{2!} = \frac{3 \cdot 2}{2 \cdot 1} = 3.$$

This agrees with the answer we found by writing out the different groups of two applicants. Similarly, the number of combinations of n elements taken r at a time is found by dividing the number of permutations, $P(n, r)$, by $r!$ to get

$$\frac{P(n, r)}{r!}$$

combinations. This expression can be rewritten as follows.

$$\frac{P(n, r)}{r!} = \frac{\frac{n!}{(n-r)!}}{r!} = \frac{n!}{(n-r)!r!}$$

The symbol $\binom{n}{r}$ is used to represent the number of combinations of n things taken r at a time. (Recall from Section 1.4 that the quantity expressed by $\binom{n}{r}$ also gives the coefficients in the binomial theorem.) With this symbol we have the following combinations formula.

COMBINATIONS OF n ELEMENTS r AT A TIME If $\binom{n}{r}$ represents the number of combinations of n elements taken r at a time, $r \le n$, then

$$\binom{n}{r} = \frac{n!}{(n-r)!r!}.$$

Other symbols used for $\binom{n}{r}$ are $C(n, r)$, $_nC_r$, and C_r^n.

■ **Example 1**

USING THE
COMBINATIONS
FORMULA

How many different committees of 3 people can be chosen from a group of 8 people?

Since the order in which the members of the committee are chosen does not affect the result, use combinations to get

$$\binom{8}{3} = \frac{8!}{5!3!} = \frac{8 \cdot 7 \cdot 6 \cdot 5 \cdot 4 \cdot 3 \cdot 2 \cdot 1}{5 \cdot 4 \cdot 3 \cdot 2 \cdot 1 \cdot 3 \cdot 2 \cdot 1} = 56. \quad ■$$

■ **Example 2**

USING THE
COMBINATIONS
FORMULA

From a group of 30 employees, 3 are to be selected to work on a special project.

(a) In how many different ways can the employees be selected?

The number of 3-element combinations from a set of 30 elements must be found. (Use combinations, not permutations, because order within the group of 3 does not affect the result.) Using the formula gives

$$\binom{30}{3} = \frac{30!}{27!3!} = 4060.$$

There are 4060 ways to select the project group.

(b) In how many different ways can the group of 3 be selected if it has already been decided that a certain employee must work on the project?

Since one employee has already been selected to work on the project, the problem is reduced to selecting 2 more employees from the 29 employees that are left:

$$\binom{29}{2} = \frac{29!}{27!2!} = 406.$$

In this case, the project group can be selected in 406 different ways.

(c) In how many ways can a (nonempty) group of *at most* 3 employees be selected from the group of 30?

Here, "at most 3" means "exactly 1 or exactly 2 or exactly 3." We shall find the number of ways to select employees for each case.

Case	Number of Ways
1	$\binom{30}{1} = \frac{30!}{29!1!} = \frac{30 \cdot 29!}{29! \cdot 1} = 30$
2	$\binom{30}{2} = \frac{30 \cdot 29 \cdot 28!}{28! \cdot 2 \cdot 1} = 435$
3	$\binom{30}{3} = \frac{30 \cdot 29 \cdot 28 \cdot 27!}{27! \cdot 3 \cdot 2 \cdot 1} = 4060$

The total number of ways to select at most 3 employees will be the sum $30 + 435 + 4060 = 4525.$ ■

PROBLEM SOLVING

The formulas for permutations and combinations given in this section are very useful in solving probability problems. Any difficulty in using these formulas usually comes from being unable to differentiate between them. Both permutations and combinations give the number of ways to choose r objects from a set of n objects. The differences between permutations and combinations are outlined below.

Permutations	Combinations
Different orderings or arrangements of the r objects are different permutations.	Each choice or subset of r objects gives 1 combination. Order within the r objects does not matter.
$$P(n, r) = \frac{n!}{(n - r)!}$$	$$\binom{n}{r} = \frac{n!}{(n - r)!r!}$$
Clue words: Arrangement, Schedule, Order	Clue words: Group, Committee, Sample

■ **Example 3**

DISTINGUISHING BETWEEN PERMUTATIONS AND COMBINATIONS

For each of the following problems, tell whether permutations or combinations should be used to solve the problem.

(a) How many 4-digit code numbers are possible if no digits are repeated?

Since changing the order of the 4 digits results in a different code, use permutations.

(b) A sample of 3 light bulbs is randomly selected from a batch of 15 items. How many different samples are possible?

The order in which the 3 light bulbs are selected is not important. The sample is unchanged if the items are rearranged, so combinations should be used.

(c) In a basketball tournament with 8 teams, how many games must be played so that each team plays every other team exactly once?

Selection of 2 teams for a game is an *unordered* subset of 2 from the set of 8 teams. Use combinations again.

(d) In how many ways can 4 patients be assigned to 6 hospital rooms so that each patient has a private room?

The room assignments are an *ordered* selection of 4 rooms from the 6 rooms. Exchanging the rooms of any 2 patients within a selection of 4 rooms gives a different assignment, so permutations should be used. ■

■ **Example 4**

DISTINGUISHING BETWEEN PERMUTATIONS AND COMBINATIONS

To illustrate the differences between permutations and combinations in another way, suppose 2 cans of soup are to be selected from 4 cans on a shelf: noodle (N), bean (B), mushroom (M), and tomato (T). As shown in Figure 9.3(a), there are 12 ways to select 2 cans from the 4 cans if the order matters (if noodle first and bean second is considered different from bean, then noodle, for example). On the other hand, if order is unimportant, then there are 6 ways to choose 2 cans of soup from the 4, as illustrated in Figure 9.3(b). ■

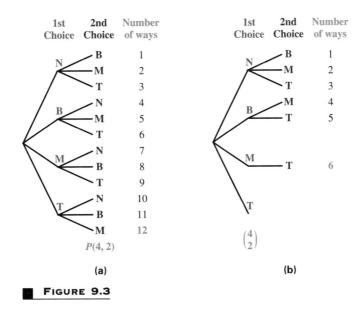

FIGURE 9.3

CAUTION It should be stressed that not all counting problems lend themselves to either permutations or combinations. Whenever a tree diagram or the multiplication principle can be used directly, as in Example 4, then use it.

9.7 EXERCISES ■

Evaluate each of the following. See Examples 1 and 2.

1. $\binom{6}{5}$ **2.** $\binom{4}{2}$ **3.** $\binom{8}{5}$ **4.** $\binom{10}{2}$ **5.** $\binom{15}{4}$ **6.** $\binom{9}{3}$

7. $\binom{10}{7}$ **8.** $\binom{10}{3}$ **9.** $\binom{14}{1}$ **10.** $\binom{20}{2}$ **11.** $\binom{18}{0}$ **12.** $\binom{13}{0}$

13. Explain the difference between a permutation and a combination.

14. Is a telephone number an example of a permutation of digits or a combination of digits? Explain.

15. Padlocks with digit dials are often referred to as "combination locks." According to the mathematical definition of combination, is this an accurate description? Explain.

16. Describe each of the following as either a permutation or a combination.
(a) Your social security number
(b) A particular five-card hand of playing cards in the game of poker
(c) A committee of school board members
(d) An automobile license plate number

Work each problem using combinations. See Examples 1 and 2.

17. A club has 30 members. If a committee of 4 is selected at random, how many committees are possible?

18. How many different samples of 3 apples can be drawn from a crate of 25 apples?

19. Hal's Hamburger Heaven sells hamburgers with cheese, relish, lettuce, tomato, mustard, or ketchup. How many different hamburgers can be made using any 3 of the extras?

20. Three students are to be selected from a group of 12 students to participate in a special class. In how many ways can this be done? In how many ways can the group that will not participate be selected?

21. Five cards are marked with the numbers 1, 2, 3, 4, and 5, shuffled, and 2 cards are then drawn. How many different 2-card combinations are possible?

22. If a bag contains 15 marbles, how many samples of 2 marbles can be drawn from it? How many samples of 4 marbles?

23. In Exercise 22, if the bag contains 3 yellow, 4 white, and 8 blue marbles, how many samples of 2 can be drawn in which both marbles are blue?

24. In Exercise 18, if it is known that there are 5 rotten apples in the crate:
(a) How many samples of 3 could be drawn in which all 3 are rotten?
(b) How many samples of 3 could be drawn in which there are 1 rotten apple and 2 good apples?

25. A city council is composed of 5 liberals and 4 conservatives. Three members are to be selected randomly as delegates to a convention.
(a) How many delegations are possible?
(b) How many delegations could have all liberals?
(c) How many delegations could have 2 liberals and 1 conservative?
(d) If 1 member of the council serves as mayor, how many delegations are possible which include the mayor?

26. Seven workers decide to send a delegation of 2 to their supervisor to discuss their grievances.
(a) How many different delegations are possible?
(b) If it is decided that a certain employee must be in the delegation, how many different delegations are possible?
(c) If there are 2 women and 5 men in the group, how many delegations would include at least one woman?

Solve the following problems by using either combinations or permutations. See Examples 3 and 4.

27. How many ways can the letters of the word TOUGH be arranged?

28. If Matthew has 8 courses to choose from, how many ways can he arrange his schedule if he must pick 4 of them?

29. How many samples of 3 pineapples can be drawn from a crate of 12?

30. Velma specializes in making different vegetable soups with carrots, celery, beans, peas, mushrooms, and potatoes. How many different soups can she make using any 4 ingredients?

31. From a pool of 7 secretaries, 3 are selected to be assigned to 3 managers, 1 secretary to each manager. In how many ways can this be done?

32. In a game of musical chairs, 12 children will sit in 11 chairs (1 will be left out). How many seatings are possible?

33. In an experiment on plant hardiness, a researcher gathers 6 wheat plants, 3 barley plants, and 2 rye plants. She wishes to select 4 plants at random.
(a) In how many ways can this be done?
(b) In how many ways can this be done if exactly 2 wheat plants must be included?

34. In a club with 8 men and 11 women members, how many 5-member committees can be chosen that have the following?
(a) all men
(b) all women
(c) 3 men and 2 women
(d) no more than 3 women

35. From 10 names on a ballot, 4 will be elected to a political party committee. In how many ways can the committee of 4 be formed if each person will have a different responsibility?

36. In how many ways can 5 out of 9 plants be arranged in a row on a window sill?

9.8 ————— AN INTRODUCTION TO PROBABILITY

The study of probability has become increasingly popular because it has a wide range of practical applications. The basic ideas of probability are introduced in this section.

Consider an experiment that has one or more possible **outcomes,** each of which is equally likely to occur. For example, the experiment of tossing a fair coin has two equally likely possible outcomes: landing heads up *(H)* or landing tails up *(T)*. Also, the experiment of rolling a fair die has 6 equally likely outcomes: landing so the face that is up shows 1, 2, 3, 4, 5, or 6 points.

The set *S* of all possible outcomes of a given experiment is called the **sample space** for the experiment. (In this text all sample spaces are finite.) One sample space for the experiment of tossing a coin could consist of the outcomes *H* and *T*. This sample space can be written in set notation as

$$S = \{H, T\}.$$

Similarly, a sample space for the experiment of rolling a single die might be

$$S = \{1, 2, 3, 4, 5, 6\}.$$

Any subset of the sample space is called an **event.** In the experiment with the die, for example, "the number showing is a three" is an event, say E_1, such that $E_1 = \{3\}$. "The number showing is greater than three" is also an event, say E_2, such that $E_2 = \{4, 5, 6\}$. To represent the number of outcomes that belong to event E, the notation $n(E)$ is used. Then $n(E_1) = 1$ and $n(E_2) = 3$.

Another sample space might give as outcomes the number of heads in one toss of a fair coin. This sample space would be written

$$S = \{0, 1\}.$$

Are these outcomes equally likely?

PROBABILITY OF EVENT *E*

In a sample space with equally likely outcomes, the **probability** of an event *E*, written $P(E)$, is the ratio of the number of outcomes in sample space *S* that belong to event *E*, $n(E)$, to the total number of outcomes in sample space *S*, $n(S)$. That is,

$$P(E) = \frac{n(E)}{n(S)}.$$

This definition is used to find the probability of the event E_1 given above, by starting with the sample space for the experiment, $S = \{1, 2, 3, 4, 5, 6\}$, and the desired event, $E_1 = \{3\}$. Since $n(E_1) = 1$ and since there are 6 outcomes in the sample space,

$$P(E_1) = \frac{n(E_1)}{n(S)} = \frac{1}{6}.$$

■ **Example 1**
FINDING
PROBABILITIES
OF EVENTS

A single die is rolled. Write the following events in set notation and give the probability for each event.

(a) E_3: the number showing is even
Use the definition above. Since $E_3 = \{2, 4, 6\}$, $n(E_3) = 3$. As shown above, $n(S) = 6$, so

$$P(E_3) = \frac{3}{6} = \frac{1}{2}.$$

(b) E_4: the number showing is greater than 4
Again $n(S) = 6$. Event $E_4 = \{5, 6\}$, with $n(E_4) = 2$. By the definition,

$$P(E_4) = \frac{2}{6} = \frac{1}{3}.$$

(c) E_5: the number showing is less than 7

$$E_5 = \{1, 2, 3, 4, 5, 6\} \quad \text{and} \quad P(E_5) = \frac{6}{6} = 1$$

(d) E_6: the number showing is 7

$$E_6 = \emptyset \quad \text{and} \quad P(E_6) = \frac{0}{6} = 0 \quad ■$$

In Example 1(c), $E_5 = S$. Therefore, the event E_5 is certain to occur every time the experiment is performed. An event that is certain to occur, such as E_5, always has a probability of 1. On the other hand, $E_6 = \emptyset$ and $P(E_6)$ is 0. The probability of an impossible event, such as E_6, is always 0, since none of the outcomes in the sample space satisfy the event. For any event E, $P(E)$ is between 0 and 1 inclusive.

The set of all outcomes in the sample space that do *not* belong to event E is called the **complement** of E, written E'. For example, in the experiment of drawing a single card from a standard deck of 52 cards, let E be the event "the card is an ace." Then E' is the event "the card is not an ace." From the definition of E', for an event E,

$$E \cup E' = S \quad \text{and} \quad E \cap E' = \emptyset.$$

A standard deck of 52 cards has four suits: hearts, clubs, diamonds, and spades, with thirteen cards of each suit. Each suit has an ace, king, queen, jack, and cards numbered from 2 to 10. The hearts and diamonds are red and the spades and clubs are black. We will refer to this standard deck of cards in this section.

Probability concepts can be illustrated using **Venn diagrams,** as shown in Figure 9.4. The rectangle in Figure 9.4 represents the sample space in an experiment. The area inside the circle represents event E, while the area inside the rectangle, but outside the circle, represents event E'.

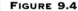

■ **FIGURE 9.4**

▪ *Example 2*

USING THE COMPLEMENT IN A PROBABILITY PROBLEM

In the experiment of drawing a card from a well-shuffled deck, find the probability of event E, the card is an ace, and event E'.

Since there are four aces in the deck of 52 cards, $n(E) = 4$ and $n(S) = 52$. Therefore,

$$P(E) = \frac{n(E)}{n(S)} = \frac{4}{52} = \frac{1}{13}.$$

Of the 52 cards, 48 are not aces, so

$$P(E') = \frac{n(E')}{n(S)} = \frac{48}{52} = \frac{12}{13}. \quad ▪$$

In Example 2, $P(E) + P(E') = (1/13) + (12/13) = 1$. This is always true for any event E and its complement E'. That is,

$$P(E) + P(E') = 1.$$

This can be restated as

$$P(E) = 1 - P(E') \quad \text{or} \quad P(E') = 1 - P(E).$$

These two equations suggest an alternative way to compute the probability of an event. For example, if it is known that $P(E) = 1/10$, then

$$P(E') = 1 - \frac{1}{10} = \frac{9}{10}.$$

ODDS Sometimes probability statements are expressed in terms of odds, a comparison of $P(E)$ with $P(E')$. The **odds** in favor of an event E are expressed as the ratio of $P(E)$ to $P(E')$ or as the fraction $P(E)/P(E')$. For example, if the probability of rain can be established as 1/3, the odds that it will rain are

$$P(\text{rain}) \text{ to } P(\text{no rain}) = \frac{1}{3} \text{ to } \frac{2}{3} = \frac{1/3}{2/3} = \frac{1}{2} \quad \text{or} \quad 1 \text{ to } 2.$$

On the other hand, the odds that it will not rain are 2 to 1 (or 2/3 to 1/3). If the odds in favor of an event are, say, 3 to 5, then the probability of the event is 3/8, while the probability of the complement of the event is 5/8. If the odds favoring event E are m to n, then

$$P(E) = \frac{m}{m + n} \quad \text{and} \quad P(E') = \frac{n}{m + n}.$$

▪ *Example 3*

FINDING ODDS IN FAVOR OF AN EVENT

A shirt is selected at random from a dark closet containing 6 blue shirts and 4 shirts that are not blue. Find the odds in favor of a blue shirt being selected.

Let E represent "a blue shirt is selected." Then $P(E) = 6/10$ or 3/5, Also, $P(E') = 1 - (3/5) = 2/5$. Therefore, the odds in favor of a blue shirt being selected are

$$P(E) \text{ to } P(E') = \frac{3}{5} \text{ to } \frac{2}{5} = \frac{3/5}{2/5} = \frac{3}{2} \quad \text{or} \quad 3 \text{ to } 2. \quad ▪$$

THE UNION OF TWO EVENTS   We now extend the rules for probability to more complex events. Since events are sets, we can use set operations to find the union of two events. (The *union* of sets A and B includes all elements of set A in addition to all elements of set B.)

Suppose a fair die is tossed. Let H be the event "the result is a 3," and K the event "the result is an even number." From the results earlier in this section,

$$H = \{3\} \qquad\qquad P(H) = \frac{1}{6}$$

$$K = \{2, 4, 6\} \qquad\qquad P(K) = \frac{3}{6} = \frac{1}{2}$$

$$H \cup K = \{2, 3, 4, 6\} \qquad P(H \cup K) = \frac{4}{6} = \frac{2}{3}.$$

Notice that $P(H) + P(K) = P(H \cup K)$.

Before assuming that this relationship is true in general, consider another event for this experiment, "the result is a 2," event G.

$$G = \{2\} \qquad\qquad P(G) = \frac{1}{6}$$

$$K = \{2, 4, 6\} \qquad\qquad P(K) = \frac{3}{6} = \frac{1}{2}$$

$$K \cup G = \{2, 4, 6\} \qquad P(K \cup G) = \frac{3}{6} = \frac{1}{2}$$

In this case $P(K) + P(G) \neq P(K \cup G)$. See Figure 9.5.

As Figure 9.5 suggests, the difference in the two examples above comes from the fact that events H and K cannot occur simultaneously. Such events are called **mutually exclusive events.** In fact, $H \cap K = \emptyset$, which is true for any two mutually exclusive events. Events K and G, however, can occur simultaneously. Both are satisfied if the result of the roll is a 2, the element in their intersection ($K \cap G = \{2\}$). This example suggests the following property.

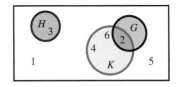

■ **FIGURE 9.5**

PROBABILITY OF THE UNION OF TWO EVENTS	For any events E and F: $P(E \text{ or } F) = P(E \cup F) = P(E) + P(F) - P(E \cap F)$

■ *Example 4*

FINDING THE
PROBABILITY OF A
UNION

One card is drawn from a well-shuffled deck of 52 cards. What is the probability of the following outcomes?

(a) The card is an ace or a spade.

The events "drawing an ace" and "drawing a spade" are not mutually exclusive since it is possible to draw the ace of spades, an outcome satisfying both events. The probability is

$$P(\text{ace or spade}) = P(\text{ace}) + P(\text{spade}) - P(\text{ace and spade})$$

$$= \frac{4}{52} + \frac{13}{52} - \frac{1}{52} = \frac{16}{52} = \frac{4}{13}.$$

(b) The card is a three or a king.

"Drawing a 3" and "drawing a king" are mutually exclusive events because it is impossible to draw one card that is both a 3 and a king. Using the rule given above,

$$P(3 \text{ or } K) = P(3) + P(K) - P(3 \text{ and } K)$$

$$= \frac{4}{52} + \frac{4}{52} - 0 = \frac{8}{52} = \frac{2}{13}. \quad ■$$

■ *Example 5*

FINDING THE
PROBABILITY OF A
UNION

For the experiment consisting of one roll of a pair of dice, find the probability that the sum of the points showing is at most 4.

The statement "at most 4" can be rewritten as "exactly 2 or exactly 3 or exactly 4." (A sum of 1 is meaningless here.) Then

$$P(\text{at most 4}) = P(2 \text{ or } 3 \text{ or } 4)$$

$$= P(2) + P(3) + P(4), \tag{1}$$

since the events represented by "2," "3," and "4" are mutually exclusive.

The sample space for this experiment includes the 36 possible pairs of numbers from 1 to 6: (1, 1), (1, 2), (1, 3), (1, 4), (1, 5), (1, 6), (2, 1), (2, 2), and so on. The pair (1, 1) is the only one with a sum of 2, so $P(2) = 1/36$. Also $P(3) = 2/36$ since both (1, 2) and (2, 1) give a sum of 3. The pairs, (1, 3), (2, 2), and (3, 1) have a sum of 4, so $P(4) = 3/36$. Substituting into equation (1) above gives

$$P(\text{at most 4}) = \frac{1}{36} + \frac{2}{36} + \frac{3}{36} = \frac{6}{36} = \frac{1}{6}. \quad ■$$

The properties of probability discussed in this section are summarized as follows.

PROPERTIES OF PROBABILITY	For any events E and F:
	1. $0 \leq P(E) \leq 1$
	2. P(a certain event) $= 1$
	3. P(an impossible event) $= 0$
	4. $P(E') = 1 - P(E)$
	5. $P(E$ or $F) = P(E \cup F) = P(E) + P(F) - P(E \cap F)$.

CAUTION When finding the probability of a union, don't forget to subtract the probability of the intersection from the sum of the probabilities of the individual events.

9.8 EXERCISES ■

Write a sample space with equally likely outcomes for each of the following experiments.

1. A two-headed coin is tossed once.

2. Two ordinary coins are tossed.

3. Three ordinary coins are tossed.

4. Slips of paper marked with the numbers 1, 2, 3, 4, and 5 are placed in a box. After mixing well, two slips are drawn.

5. An unprepared student takes a three-question true/false quiz in which he guesses the answer to all three questions.

6. A die is rolled and then a coin is tossed.

Write the following events in set notation and give the probability of each event. See Example 1.

7. In Exercise 1:
   (a) the result of the toss is heads
   (b) the result of the toss is tails.

8. In the experiment described in Exercise 2:
   (a) both coins show the same face;
   (b) at least one coin turns up heads.

9. In Exercise 5:
   (a) the student gets all three answers correct;
   (b) he gets all three answers wrong;
   (c) he gets exactly two answers correct;
   (d) he gets at least one answer correct.

10. In Exercise 4:
   (a) both slips are marked with even numbers;
   (b) both slips are marked with odd numbers;
   (c) both slips are marked with the same number;
   (d) one slip is marked with an odd number, the other with an even number.

11. A student gives the answer to a probability problem as 6/5. Explain why this answer must be incorrect.

12. If the probability of an event is .857, what is the probability that the event will not occur?

Work the following problems. See Examples 1–5.

13. A marble is drawn at random from a box containing 3 yellow, 4 white, and 8 blue marbles. Find the probabilities in (a)–(c).

(a) A yellow marble is drawn.

(b) A black marble is drawn.

(c) The marble is yellow or white.

(d) What are the odds in favor of drawing a yellow marble?

(e) What are the odds against drawing a blue marble?

14. A baseball player with a batting average of .300 comes to bat. What are the odds in favor of his getting a hit?

15. In Exercise 4, what are the odds that the sum of the numbers on the two slips of paper is 5?

16. If the odds that it will rain are 4 to 5, what is the probability of rain?

17. If the odds that a candidate will win an election are 3 to 2, what is the probability that the candidate will lose?

18. A card is drawn from a well-shuffled deck of 52 cards. Find the probability that the card is

(a) a 9;   (b) black;   (c) a black 9;

(d) a heart;

(e) a face card (K, Q, J of any suit);

(f) red or a 3;

(g) less than a 4 (consider aces as 1s).

19. Mrs. Elliott invites 10 relatives to a party: her mother, two uncles, three brothers, and four cousins. If the chances of any one guest arriving first are equally likely, find the following probabilities.

(a) The first guest is an uncle or a brother.

(b) The first guest is a brother or cousin.

(c) The first guest is a brother or her mother.

20. Two dice are rolled. Find the probability of the following events.

(a) The sum of the points is at least 10.

(b) The sum of the points is either 7 or at least 10.

(c) The sum of the points is 2 or the dice both show the same number.

21. The table shows the probability that a customer of a department store will make a purchase in the indicated price range.

Cost	Probability
Below $5	.25
$5–$19.99	.37
$20–$39.99	.11
$40–$69.99	.09
$70–$99.99	.07
$100–$149.99	.08
$150 or more	.03

Find the probability that a customer makes a purchase that is

(a) less than $20;   (b) $40 or more;

(c) more than $99.99;   (d) less than $100.

22. One game in a state lottery requires you to pick one heart, one club, one diamond, and one spade, in that order, from the thirteen cards in each suit. What is the probability of getting all four picks correct and winning $5000?

23. If three of the four selections in Exercise 22 are correct, the player wins $200. Find the probability of this outcome.

24. The law firm of Alam, Bartolini, Chinn, Dickinson, and Ellsberg has two senior partners, Alam and Bartolini. Two of the attorneys are to be selected to attend a conference. Assuming that all are equally likely to be selected, find the following probabilities.

(a) Chinn is selected.

(b) Alam and Dickinson are selected.

(c) At least one senior partner is selected.

25. The management of a firm wishes to survey the opinions of its workers, classified as follows for the purpose of an interview:

30% have worked for the company more than 5 years, 28% are female,

65% contribute to a voluntary retirement plan, and

1/2 of the female workers contribute to the retirement plan.

Find the following probabilities.

(a) A male worker is selected.

(b) A worker is selected who has worked for the company less than 5 years.

(c) A worker is selected who contributes to the retirement plan or is female.

CHAPTER 9 SUMMARY ■ ────────────

SECTION	TERMS	KEY IDEAS
9.1 Sequences; Arithmetic Sequences	sequence terms of a sequence nth term finite sequence infinite sequence recursion formula arithmetic sequence common difference	**nth Term of an Arithmetic Sequence** In an arithmetic sequence with first term a_1 and common difference d, the nth term, a_n, is given by $$a_n = a_1 + (n-1)d.$$
9.2 Geometric Sequences	geometric sequence common ratio	**nth Term of a Geometric Sequence** In the geometric sequence with first term a_1 and common ratio r, the nth term is $$a_n = a_1 r^{n-1}.$$
9.3 Series; Applications of Sequences and Series	series index of summation annuity future value	**Sum of the First n Terms of an Arithmetic Sequence** If an arithmetic sequence has first term a_1 and common difference d, then the sum of the first n terms is given by $$S_n = \frac{n}{2}(a_1 + a_n)$$ or $$S_n = \frac{n}{2}[2a_1 + (n-1)d].$$ **Sum of the First n Terms of a Geometric Sequence** If a geometric sequence has first term a_1 and common ratio r, then the sum of the first n terms is given by $$S_n = \frac{a_1(1-r^n)}{1-r}, \quad \text{where } r \neq 1.$$
9.4 Sums of Infinite Geometric Sequences		**Sum of an Infinite Geometric Sequence** The sum of an infinite geometric sequence with first term a_1 and common ratio r, where $-1 < r < 1$, is given by $$S_\infty = \frac{a_1}{1-r}.$$

SECTION	TERMS	KEY IDEAS
9.5 Mathematical Induction		**Principle of Mathematical Induction** Let S_n be a statement concerning the positive integer n. Suppose that **1.** S_1 is true; **2.** for any positive integer k, $k \leq n$, if S_k is true, then S_{k+1} is also true. Then S_n is true for every positive integer value of n.
9.6 Permutations	tree diagram independent events permutation	**Multiplication Principle of Counting** If one event can occur in m ways and a second event can occur in n ways, then both events can occur in mn ways, provided the outcome of the first event does not influence the outcome of the second. **Permutations of n Elements, r at a Time** If $P(n, r)$ denotes the number of permutations of n elements taken r at a time, $r \leq n$, then $$P(n, r) = \frac{n!}{(n - r)!}.$$
9.7 Combinations	combination	**Combinations of n Elements, r at a Time** If $\binom{n}{r}$ represents the number of combinations of n elements taken r at a time, $r \leq n$, then $$\binom{n}{r} = \frac{n!}{(n - r)!r!}.$$
9.8 An Introduction to Probability	outcome sample space event complement Venn diagram odds mutually exclusive events	**Probability of Event E** In a sample space S with equally likely outcomes, the probability of an event E is $$P(E) = \frac{n(E)}{n(S)}.$$

Write the first five terms for each of the following sequences.

1. $a_n = \dfrac{n}{n+1}$ 　　　　　　　　　　**2.** $a_n = (-2)^n$

3. $a_n = 2(n+3)$ 　　　　　　　　　　**4.** $a_n = n(n+1)$

5. $a_1 = 5$; for $n \geq 2$, $a_n = a_{n-1} - 3$ 　　　**6.** $a_1 = -2$; for $n \geq 2$, $a_n = 3a_{n-1}$

7. $a_1 = 5$, $a_2 = 3$; for $n \geq 3$, $a_n = a_{n-1} - a_{n-2}$

8. $b_1 = -2$, $b_2 = 2$, $b_3 = -4$; for $n \geq 4$, $b_n = -2 \cdot b_{n-2}$ if n is even, and $b_n = 2 \cdot b_{n-2}$ if n is odd.

9. Arithmetic, $a_1 = 6$, $d = -4$ 　　　　　**10.** Arithmetic, $a_3 = 9$, $a_4 = 7$

11. Arithmetic, $a_1 = 3 - \sqrt{5}$, $a_2 = 4$ 　　**12.** Arithmetic, $a_3 = \pi$, $a_4 = 0$

13. A certain arithmetic sequence has $a_6 = -4$ and $a_{17} = 51$. Find a_1 and a_{20}.

Find a_8 for each of the following arithmetic sequences.

14. $a_1 = 6$, $d = 2$ 　　　　　　　　　　**15.** $a_1 = -4$, $d = 3$

16. $a_1 = 6x - 9$, $a_2 = 5x + 1$ 　　　　　**17.** $a_3 = 11m$, $a_5 = 7m - 4$

Write the first five terms for the geometric sequences in Exercises 18–21.

18. $a_1 = 4$, $r = 2$ 　　**19.** $a_4 = 8$, $r = 1/2$ 　　**20.** $a_1 = -3$, $a_2 = 4$ 　　**21.** $a_3 = 8$, $a_5 = 72$

22. For a given geometric sequence, $a_1 = 4$ and $a_5 = 324$. Find a_6.

Find a_5 for each of the following geometric sequences.

23. $a_1 = 3$, $r = 2$ 　　**24.** $a_2 = 3125$, $r = 1/5$ 　　**25.** $a_1 = 5x$, $a_2 = x^2$ 　　**26.** $a_2 = \sqrt{6}$, $a_4 = 6\sqrt{6}$

27. Explain the difference between an arithmetic sequence and a geometric sequence.

Find S_{12} for each of the following arithmetic sequences.

28. $a_1 = 2$, $d = 3$ 　　　　**29.** $a_2 = 6$, $d = 10$ 　　　　**30.** $a_1 = -4k$, $d = 2k$

Find S_4 for each of the following geometric sequences.

31. $a_1 = 1$, $r = 2$ 　　　　**32.** $a_1 = 3$, $r = 3$ 　　　　**33.** $a_1 = 2k$, $a_2 = -4k$

Evaluate each of the following sums.

34. $\displaystyle\sum_{i=1}^{4} \dfrac{2}{i}$ 　　　　**35.** $\displaystyle\sum_{i=1}^{7} (-1)^{i+1} \cdot 6$ 　　　　**36.** $\displaystyle\sum_{i=4}^{8} 3i(2i-5)$

37. $\displaystyle\sum_{i=1}^{6} i(i+2)$ 　　　　**38.** $\displaystyle\sum_{i=1}^{6} 4 \cdot 3^i$ 　　　　**39.** $\displaystyle\sum_{i=1}^{4} 8 \cdot 2^i$

Solve each problem.

40. A stack of canned goods in a market display requires 15 cans on the bottom, 13 in the next layer, 11 in the next layer, and so on. How many cans are needed for the display?

41. Gale Stockdale borrows $6000 at simple interest of 12% per year. He will repay the loan and interest in monthly payments of $260, $258, $256, and so on. If he makes 30 payments, what is the total amount required to pay off the loan plus the interest?

42. On a certain production line, new employees during their first week turn out 5/4 as many items each day as on the previous day. If a new employee produces 48 items the first day, how many will she produce on the fifth day of work?

43. The half-life of a radioactive substance is 20 years. If 600 g are present at the start, how much will be left after 100 years?

Evaluate each of the following sums that exists.

44. $18 + 9 + 9/2 + 9/4 + \cdots$

45. $20 + 15 + 45/4 + 135/16 + \cdots$

46. $-5/6 + 5/9 - 10/27 + \cdots$

47. $.8 + .08 + .008 + .0008 + \cdots$

48. $\displaystyle\sum_{i=1}^{\infty} \left(\frac{5}{8}\right)^i$

49. $\displaystyle\sum_{i=1}^{\infty} -10\left(\frac{5}{2}\right)^i$

Convert each of the following repeating decimals to a quotient of integers.

50. $.6666 \ldots$

51. $.512512512 \ldots$

52. Explain the difference between a sequence and a series.

Use mathematical induction to prove that each of the following is true for every positive integer n.

53. $2 + 6 + 10 + 14 + \cdots + (4n - 2) = 2n^2$

54. $2^2 + 4^2 + 6^2 + \cdots + (2n)^2 = \dfrac{2n(n + 1)(2n + 1)}{3}$

55. $2 + 2^2 + 2^3 + \cdots + 2^n = 2(2^n - 1)$

56. $1^3 + 3^3 + 5^3 + \cdots + (2n - 1)^3 = n^2(2n^2 - 1)$

Evaluate each of the following.

57. $P(5, 5)$ **58.** $P(9, 2)$ **59.** $P(6, 0)$ **60.** $\dbinom{8}{3}$ **61.** $\dbinom{10}{5}$ **62.** $\dbinom{6}{0}$

Solve each problem.

63. Two people are planning their wedding. They can select from 2 different chapels, 4 soloists, 3 organists, and 2 ministers. How many different wedding arrangements are possible?

64. John Jacobs, who is furnishing his apartment, wants to buy a new couch. He can select from 5 different styles, each available in 3 different fabrics, with 6 color choices. How many different couches are available?

65. Four students are to be assigned to 4 different summer jobs. Each student is qualified for all 4 jobs. In how many ways can the jobs be assigned?

66. A student body council consists of a president, vice-president, secretary-treasurer, and 3 representatives at large. Three members are to be selected to attend a conference.
(a) How many different such delegations are possible?
(b) How many are possible if the president must attend?

67. Nine football teams are competing for first-, second-, and third-place titles in a statewide tournament. In how many ways can the winners be determined?

68. How many different license plates can be formed with a letter followed by 3 digits and then 3 letters? How many such license plates have no repeats?

Solve each problem.

69. A marble is drawn at random from a box containing 4 green, 5 black, and 6 white marbles. Find the following probabilities.
(a) A green marble is drawn.
(b) A marble that is not black is drawn.
(c) A blue marble is drawn.

70. Refer to Exercise 69 and answer each question.
(a) What are the odds in favor of drawing a green marble?
(b) What are the odds against drawing a white marble?

A card is drawn from a standard deck of 52 cards. Find the probability that each of the following is drawn.

71. A black king   **72.** A face card or an ace   **73.** An ace or a diamond

74. A card that is not a diamond   **75.** A card that is not a diamond or not black

A sample shipment of 5 swimming pool filters is chosen. The probability of exactly 0, 1, 2, 3, 4, or 5 filters being defective is given in the following table.

Number defective	0	1	2	3	4	5
Probability	.31	.25	.18	.12	.08	.06

Find the probability that the following numbers of filters are defective.

76. No more than 3   **77.** At least 2   **78.** More than 5

Write the first five terms for each of the following sequences.

1. $a_n = (-1)^n(n^2 + 1)$

2. $a_1 = 5$; for $n \geq 2$, $a_n = n + a_{n-1}$

3. $a_n = n + 1$ if n is odd and $a_n = a_{n-1} + 2$ if n is even.

4. Arithmetic, $a_2 = 1$, $a_4 = 25$

5. Geometric, $a_1 = 81$, $r = 2/3$

6. A certain arithmetic sequence has $a_7 = -6$ and $a_{15} = -2$. Find a_{31}.

7. Find S_{10} for the arithmetic sequence with $a_1 = 37$ and $d = 13$.

8. For a given geometric sequence, $a_1 = 12$ and $a_6 = -3/8$. Find a_3.

9. Find a_7 for the geometric sequence with $a_1 = 2x^3$ and $a_3 = 18x^7$.

10. Find S_4 for the geometric sequence with $a_1 = 4$ and $r = 1/2$.

Evaluate each of the following sums that exists.

11. $\sum_{i=1}^{30} (5i + 2)$

12. $\sum_{i=1}^{5} (-3 \cdot 2^i)$

13. $75 + 30 + 12 + \dfrac{24}{5} + \ldots$

14. $\sum_{i=1}^{\infty} 54(2/9)^i$

Work each problem.

15. Fred Meyers deposited $50 in a new savings account on February 1. On the first of each month thereafter he deposited $5 more than the previous month's deposit. Find the total amount of money he deposited after twenty $5 payments.

16. The number of bacteria in a certain culture doubles every 30 minutes. If 50 are present at noon, how many will be present at 4:30 P.M.?

17. Use mathematical induction to prove that

$$8 + 14 + 20 + 26 + \cdots + (6n + 2) = 3n^2 + 5n$$

for every positive integer n.

Find each of the following.

18. $P(11, 3)$

19. $P(7, 7)$

20. $\dbinom{10}{2}$

21. $\dbinom{12}{0}$

22. A clothing manufacturer makes women's coats in four different styles. Each coat can be made from one of three fabrics. Each fabric comes in five different colors. How many different coats can be made?

23. A club with 30 members is to elect a president, secretary, and treasurer from its membership. If a member can hold at most one position, in how many ways can the offices be filled?

24. If there are 14 women and 16 men in the club in Exercise 23, in how many ways can 2 women and 2 men be chosen to attend a conference?

25. Discuss the similarities and differences between permutations and combinations.

A card is drawn from a standard deck of 52 cards. Find the probability that each of the following is drawn.

26. A red three

27. A card that is not a face card

28. A king or a spade

29. In the card-drawing experiment above, what are the odds in favor of drawing a face card?

30. A sample of 4 transistors is chosen. The probability of exactly 0, 1, 2, 3, or 4 transistors being defective is given in the following table.

Number defective	0	1	2	3	4
Probability	.19	.43	.30	.07	.01

Find the probability that at most two are defective.

APPENDICES

Think of a **set** as a collection of objects. The objects that belong to a set are called the **elements** or **members** of the set. In algebra, the elements of a set are usually numbers. Sets are commonly written using **set braces,** { }. For example, the set containing the elements 1, 2, 3, and 4 is written

$$\{1, 2, 3, 4\}.$$

Since the order in which the elements are listed is not important, this same set can also be written as $\{4, 3, 2, 1\}$ or with any other arrangement of the four numbers.

To show that 4 is an element of the set $\{1, 2, 3, 4\}$, we use the symbol $\in$ and write

$$4 \in \{1, 2, 3, 4\}.$$

Also, $2 \in \{1, 2, 3, 4\}$. To show that 5 is *not* an element of this set, we place a slash through the symbol:

$$5 \notin \{1, 2, 3, 4\}.$$

It is customary to name sets with capital letters. If S is used to name the set above, then

$$S = \{1, 2, 3, 4\}.$$

Set S was written by listing its elements. It is sometimes easier to describe a set in words. For example, set S might be described as "the set containing the first four counting numbers." In this example, the notation $\{1, 2, 3, 4\}$, with the elements listed between set braces, is briefer than the verbal description. However, the set F, consisting of all fractions between 0 and 1, could not be described by listing its elements. (Try it.)

Set F is an example of an *infinite set,* one that has an unending list of distinct elements. A *finite set* is one that has a limited number of elements. Some infinite sets, unlike F, can be described by a listing process. For example, the set of numbers used for counting, called the **natural numbers,** or the **counting numbers,** can be written as

$$N = \{1, 2, 3, 4, \ldots\},$$

where the three dots show that the list of the elements of the set continues in the same way.

Sets are often written using a **variable,** a letter that is used to represent one element from a set of numbers. For example,

$$\{x | x \text{ is a natural number between 2 and 7}\}$$

(read "the set of all elements x such that x is a natural number between 2 and 7") represents the set $\{3, 4, 5, 6\}$. The numbers 2 and 7 are *not* between 2 and 7. The notation used here, $\{x | x \text{ is a natural number between 2 and 7}\}$, is called **set-builder notation.**

■ *Example 1*
**LISTING THE
ELEMENTS OF A SET**

Write the elements belonging to each of the following sets.

(a) $\{x | x \text{ is a counting number less than 5}\}$
The counting numbers less than 5 make up the set $\{1, 2, 3, 4\}$.

(b) $\{x | x \text{ is a state that borders Florida}\}$
The states bordering Florida make up the set $\{\text{Alabama, Georgia}\}$. ■

When discussing a particular situation or problem, we can usually identify a **universal set** (whether expressed or implied) that contains all the elements appearing in any set used in the given problem. The letter U is used to represent the universal set.

At the other extreme from the universal set is the **null set,** or **empty set,** the set containing no elements. The set of all people twelve feet tall is an example of the null set. We write the null set in either of two ways, using the special symbol $\emptyset$ or else writing set braces enclosing no elements, $\{\ \}$.

| CAUTION | Be careful not to combine these symbols; $\{\emptyset\}$ is *not* the null set. |

Every element of the set $S = \{1, 2, 3, 4\}$ is a natural number. Because of this, set S is a *subset* of the set N of natural numbers, written $S \subseteq N$. By definition, set A is a **subset** of set B if every element of set A is also an element of set B. For example, if $A = \{2, 5, 9\}$ and $B = \{2, 3, 5, 6, 9, 10\}$, then $A \subseteq B$. However, there are some elements of B that are not in A, so B is not a subset of A, written $B \not\subseteq A$. By the definition, every set is a subset of itself. Also, by definition, $\emptyset$ is a subset of every set.

If A is any set, then $\emptyset \subseteq A$.

Figure 1 shows a set A that is a subset of set B. The rectangle of the drawing represents the universal set U. Such diagrams are called **Venn diagrams.** These diagrams are used as an aid in clarifying and discussing the relationships among sets.

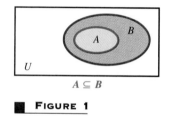

$A \subseteq B$

■ **FIGURE 1**

Two sets A and B are **equal** whenever $A \subseteq B$ and $B \subseteq A$. In other words, $A = B$ if the two sets contain exactly the same elements. For example,

$$\{1, 2, 3\} = \{3, 1, 2\},$$

since both sets contain exactly the same elements. However,

$$\{1, 2, 3\} \neq \{0, 1, 2, 3\},$$

since the set $\{0, 1, 2, 3\}$ contains the element 0, which is not an element of $\{1, 2, 3\}$.

OPERATIONS ON SETS Given a set A and a universal set U, the set of all elements of U that do not belong to set A is called the **complement** of set A. For example, if set A is the set of all the female students in your class, and U is the set of all students in the class, then the complement of A would be the set of all the male students in the class. The complement of set A is written A' (read "A-prime"). The Venn diagram in Figure 2 shows a set B. Its complement, B', is colored.

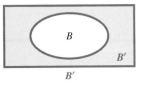

■ **FIGURE 2**

■ *Example 2*

FINDING THE COMPLEMENT OF A SET

Let $U = \{1, 2, 3, 4, 5, 6, 7\}$, $A = \{1, 3, 5, 7\}$, and $B = \{3, 4, 6\}$. Find each of the following sets.

(a) A'

Set A' contains the elements of U that are not in A.

$$A' = \{2, 4, 6\}$$

(b) $B' = \{1, 2, 5, 7\}$

(c) $\emptyset' = U$ and $U' = \emptyset$ ■

Given two sets A and B, the set of all elements belonging both to set A and to set B is called the **intersection** of the two sets, written $A \cap B$. For example, the elements that belong to both $A = \{1, 2, 4, 5, 7\}$ and $B = \{2, 4, 5, 7, 9, 11\}$ are 2, 4, 5, and 7, so

$$A \cap B = \{1, 2, 4, 5, 7\} \cap \{2, 4, 5, 7, 9, 11\} = \{2, 4, 5, 7\}.$$

The Venn diagram in Figure 3 shows two sets A and B; their intersection, $A \cap B$, is colored.

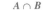

■ **FIGURE 3**

■ *Example 3*

FINDING THE INTERSECTION OF TWO SETS

(a) $\{9, 15, 25, 36\} \cap \{15, 20, 25, 30, 35\} = \{15, 25\}$

The elements 15 and 25 are the only ones belonging to both sets.

(b) $\{2, 3, 4, 5, 6\} \cap \{1, 2, 3, 4\} = \{2, 3, 4\}$ ■

Two sets that have no elements in common are called **disjoint sets.** For example, there are no elements common to both $\{50, 51, 54\}$ and $\{52, 53, 55, 56\}$, so these two sets are disjoint, and

$$\{50, 51, 54\} \cap \{52, 53, 55, 56\} = \emptyset.$$

This result can be generalized.

DISJOINT SETS   If A and B are any two disjoint sets, then $A \cap B = \emptyset$.

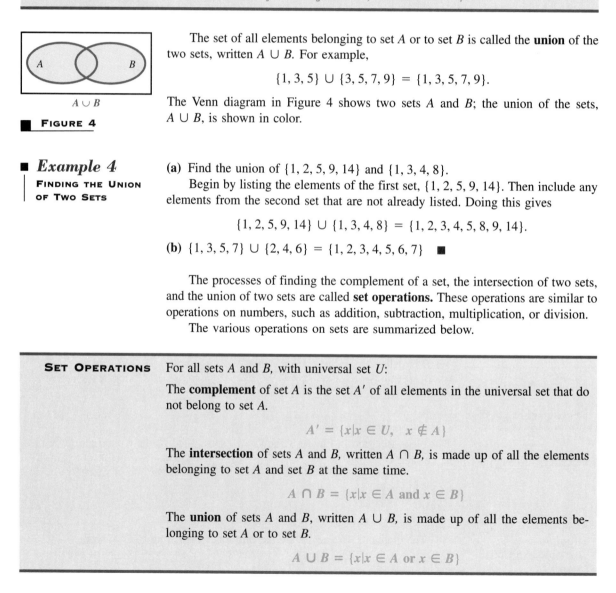

$A \cup B$

■ **FIGURE 4**

The set of all elements belonging to set A or to set B is called the **union** of the two sets, written $A \cup B$. For example,

$$\{1, 3, 5\} \cup \{3, 5, 7, 9\} = \{1, 3, 5, 7, 9\}.$$

The Venn diagram in Figure 4 shows two sets A and B; the union of the sets, $A \cup B$, is shown in color.

■ ***Example 4***

FINDING THE UNION
OF TWO SETS

(a) Find the union of $\{1, 2, 5, 9, 14\}$ and $\{1, 3, 4, 8\}$.

Begin by listing the elements of the first set, $\{1, 2, 5, 9, 14\}$. Then include any elements from the second set that are not already listed. Doing this gives

$$\{1, 2, 5, 9, 14\} \cup \{1, 3, 4, 8\} = \{1, 2, 3, 4, 5, 8, 9, 14\}.$$

(b) $\{1, 3, 5, 7\} \cup \{2, 4, 6\} = \{1, 2, 3, 4, 5, 6, 7\}$ ■

The processes of finding the complement of a set, the intersection of two sets, and the union of two sets are called **set operations.** These operations are similar to operations on numbers, such as addition, subtraction, multiplication, or division. The various operations on sets are summarized below.

SET OPERATIONS   For all sets A and B, with universal set U:

The **complement** of set A is the set A' of all elements in the universal set that do not belong to set A.

$$A' = \{x | x \in U, \ x \notin A\}$$

The **intersection** of sets A and B, written $A \cap B$, is made up of all the elements belonging to set A and set B at the same time.

$$A \cap B = \{x | x \in A \text{ and } x \in B\}$$

The **union** of sets A and B, written $A \cup B$, is made up of all the elements belonging to set A or to set B.

$$A \cup B = \{x | x \in A \text{ or } x \in B\}$$

APPENDIX A EXERCISES ■

List all the elements of each of the following sets. See Example 1.

1. $\{12, 13, 14, \ldots, 20\}$

2. $\{8, 9, 10, \ldots, 17\}$

3. $\{1, 1/2, 1/4, \ldots, 1/32\}$

4. $\{3, 9, 27, \ldots, 729\}$

5. $\{17, 22, 27, \ldots, 47\}$

6. $\{74, 68, 62, \ldots, 38\}$

7. $\{$all natural numbers greater than 7 and less than 15$\}$

8. $\{$all natural numbers not greater than 4$\}$

Identify the sets in Exercises 9–16 as finite *or* infinite.

9. $\{4, 5, 6, \ldots, 15\}$

10. $\{4, 5, 6, \ldots\}$

11. $\{1, 1/2, 1/4, 1/8, \ldots\}$

12. $\{0, 1, 2, 3, 4, 5, \ldots, 75\}$

13. $\{x|x$ is a natural number larger than 5$\}$

14. $\{x|x$ is a person alive on the earth now$\}$

15. $\{x|x$ is a fraction between 0 and 1$\}$

16. $\{x|x$ is an even natural number$\}$

Complete the blanks with either $\in$ *or* $\notin$ *so that the resulting statement is true.*

17. 6 _____ $\{3, 4, 5, 6\}$

18. 9 _____ $\{3, 2, 5, 9, 8\}$

19. -4 _____ $\{4, 6, 8, 10\}$

20. -12 _____ $\{3, 5, 12, 14\}$

21. 0 _____ $\{2, 0, 3, 4\}$

22. 0 _____ $\{5, 6, 7, 8, 10\}$

23. $\{3\}$ _____ $\{2, 3, 4, 5\}$

24. $\{5\}$ _____ $\{3, 4, 5, 6, 7\}$

25. $\{0\}$ _____ $\{0, 1, 2, 5\}$

26. $\{2\}$ _____ $\{2, 4, 6, 8\}$

27. 0 _____ $\emptyset$

28. $\emptyset$ _____ $\emptyset$

Tell whether each statement is true *or* false.

29. $3 \in \{2, 5, 6, 8\}$

30. $6 \in \{-2, 5, 8, 9\}$

31. $1 \in \{3, 4, 5, 11, 1\}$

32. $12 \in \{18, 17, 15, 13, 12\}$

33. $9 \notin \{2, 1, 5, 8\}$

34. $3 \notin \{7, 6, 5, 4\}$

35. $\{2, 5, 8, 9\} = \{2, 5, 9, 8\}$

36. $\{3, 0, 9, 6, 2\} = \{2, 9, 0, 3, 6\}$

37. $\{5, 8, 9\} = \{5, 8, 9, 0\}$

38. $\{3, 7, 12, 14\} = \{3, 7, 12, 14, 0\}$

39. $\{x|x$ is a natural number less than 3$\} = \{1, 2\}$

40. $\{x|x$ is a natural number greater than 10$\} = \{11, 12, 13, \ldots\}$

41. $\{5, 7, 9, 19\} \cap \{7, 9, 11, 15\} = \{7, 9\}$

42. $\{8, 11, 15\} \cap \{8, 11, 19, 20\} = \{8, 11\}$

43. $\{2, 1, 7\} \cup \{1, 5, 9\} = \{1\}$

44. $\{6, 12, 14, 16\} \cup \{6, 14, 19\} = \{6, 14\}$

45. $\{3, 2, 5, 9\} \cap \{2, 7, 8, 10\} = \{2\}$

46. $\{8, 9, 6\} \cup \{9, 8, 6\} = \{8, 9\}$

47. $\{3, 5, 9, 10\} \cap \emptyset = \{3, 5, 9, 10\}$

48. $\{3, 5, 9, 10\} \cup \emptyset = \{3, 5, 9, 10\}$

49. $\{1, 2, 4\} \cup \{1, 2, 4\} = \{1, 2, 4\}$

50. $\{1, 2, 4\} \cap \{1, 2, 4\} = \emptyset$

51. $\emptyset \cup \emptyset = \emptyset$

52. $\emptyset \cap \emptyset = \emptyset$

Let $A = \{2, 4, 6, 8, 10, 12\}$, $B = \{2, 4, 8, 10\}$, $C = \{4, 10, 12\}$, $D = \{2, 10\}$, *and* $U = \{2, 4, 6, 8, 10, 12, 14\}$.

Tell whether each statement is true *or* false.

53. $A \subseteq U$

54. $C \subseteq U$

55. $D \subseteq B$

56. $D \subseteq A$

57. $A \subseteq B$

58. $B \subseteq C$

59. $\emptyset \subseteq A$

60. $\emptyset \subseteq \emptyset$

61. $\{4, 8, 10\} \subseteq B$

62. $\{0, 2\} \subseteq D$

63. $B \subseteq D$

64. $A \not\subseteq C$

Insert ⊆ or ⊄ in each blank to make the resulting statement true.

65. {2, 4, 6} _____ {3, 2, 5, 4, 6}   **66.** {1, 5} _____ {0, −1, 2, 3, 1, 5}

67. {0, 1, 2} _____ {1, 2, 3, 4, 5}   **68.** {5, 6, 7, 8} _____ {1, 2, 3, 4, 5, 6, 7}

69. ∅ _____ {1, 4, 6, 8}   **70.** ∅ _____ ∅

Let $U = \{0, 1, 2, 3, 4, 5, 6, 7, 8, 9, 10, 11, 12, 13\}$, $M = \{0, 2, 4, 6, 8\}$,
$N = \{1, 3, 5, 7, 9, 11, 13\}$, $Q = \{0, 2, 4, 6, 8, 10, 12\}$, *and* $R = \{0, 1, 2, 3, 4\}$.

Use these sets to find each of the following. Identify any disjoint sets. See Examples 2–4.

71. $M \cap R$	**72.** $M \cup R$	**73.** $M \cup N$	**74.** $M \cap N$
75. $M \cap U$	**76.** $M \cup Q$	**77.** $N \cup R$	**78.** $U \cap N$
79. N'	**80.** Q'	**81.** $M' \cap Q$	**82.** $Q \cap R'$
83. $\emptyset \cap R$	**84.** $\emptyset \cap Q$	**85.** $N \cup \emptyset$	**86.** $R \cup \emptyset$
87. $(M \cap N) \cup R$	**88.** $(N \cup R) \cap M$	**89.** $(Q \cap M) \cup R$	**90.** $(R \cup N) \cap M'$
91. $(M' \cup Q) \cap R$	**92.** $Q \cap (M \cup N)$	**93.** $Q' \cap (N' \cap U)$	**94.** $(U \cap \emptyset') \cup R$

Let $U = \{$all students in this school$\}$, $M = \{$all students taking this course$\}$,
$N = \{$all students taking calculus$\}$, *and* $P = \{$all students taking history$\}$.

Describe each of the following sets in words.

95. M' **96.** $M \cup N$ **97.** $N \cap P$ **98.** $N' \cap P'$ **99.** $M \cup P$ **100.** $P' \cup M'$

In the study of probability, the set of all possible outcomes for an experiment is called the sample space for the experiment. For example, the sample space for tossing an honest coin is the set
$\{h, t\}$. *Find the sample space for each of these experiments.*

101. Rolling an honest die   **102.** Tossing a coin twice

103. Tossing a coin three times   **104.** Having three children

State the conditions on sets X and Y for which the following are true.

105. $X \cap Y = \emptyset$ **106.** $X \cup Y = \emptyset$ **107.** $X \cap Y = X$

108. $X \cup Y = X$ **109.** $X \cap \emptyset = X$ **110.** $X \cup \emptyset = X$

▬ ——————— APPENDIX B GEOMETRY REVIEW AND FORMULAS

SPECIAL ANGLES ■ ———————————————————————		
NAME	**CHARACTERISTIC**	**EXAMPLES**
Right Angle	Measure is 90°.	

SPECIAL ANGLES ■

NAME	CHARACTERISTIC	EXAMPLES
Straight Angle	Measure is 180°.	$180°$
Complementary Angles	The sum of the measures of two complementary angles is 90°.	Angle 1 and angle 2 are complementary.
Supplementary Angles	The sum of the measures of two supplementary angles is 180°.	Angle 3 and angle 4 are supplementary.
Vertical Angles	Vertical angles have equal measures.	Angle 2 = Angle 4 Angle 1 = Angle 3
Angles Formed by Parallel Lines and a Transversal		m and n are parallel.
Alternate Interior Angles	Measures are equal.	Angle 3 = Angle 6
Alternate Exterior Angles	Measures are equal.	Angle 1 = Angle 8
Interior Angles on the Same Side	Angles are supplements.	Angles 3 and 5 are supplementary.

SPECIAL TRIANGLES ■ ————————————————————

NAME	CHARACTERISTIC	EXAMPLES
Right Triangle	Triangle has a right angle.	90°
Isosceles Triangle	Triangle has two equal sides.	$AB = BC$
Equilateral Triangle	Triangle has three equal sides.	$AB = BC = CA$
Similar Triangles	Corresponding angles are equal; corresponding sides are proportional.	$A = D,\ B = E,\ C = F$ $$\frac{AB}{DE} = \frac{AC}{DF} = \frac{BC}{EF}$$

FORMULAS ∎

FIGURE	FORMULAS	EXAMPLES
Square	Perimeter: $P = 4s$ Area: $A = s^2$	
Rectangle	Perimeter: $P = 2L + 2W$ Area: $A = LW$	
Triangle	Perimeter: $P = a + b + c$ Area: $A = \dfrac{1}{2}bh$	
Pythagorean Theorem (for Right Triangles)	$c^2 = a^2 + b^2$	
Sum of the Angles of a Triangle	$A + B + C = 180°$	
Circle	Diameter: $d = 2r$ Circumference: $C = 2\pi r = \pi d$ Area: $A = \pi r^2$	

FORMULAS ■ ──────────────────────

FIGURE	FORMULAS	EXAMPLES
Parallelogram	Area: $A = bh$ Perimeter: $P = 2a + 2b$	
Trapezoid	Area: $A = \dfrac{1}{2}(B + b)h$ Perimeter: $P = a + b + c + B$	
Sphere	Volume: $V = \dfrac{4}{3}\pi r^3$ Surface area: $S = 4\pi r^2$	
Cone	Volume: $V = \dfrac{1}{3}\pi r^2 h$ Surface area: $S = \pi r\sqrt{r^2 + h^2}$	
Cube	Volume: $V = e^3$ Surface area: $S = 6e^2$	

FORMULAS ∎

FIGURE	FORMULAS	EXAMPLES
Rectangular Solid	Volume: $V = LWH$ Surface area: $S = 2HW + 2LW + 2LH$	
Right Circular Cylinder	Volume: $V = \pi r^2 h$ Surface area: $S = 2\pi rh + 2\pi r^2$	
Right Pyramid	Volume: $V = \dfrac{1}{3} Bh$ $B = $ area of the base	

APPENDIX C SYMBOLS USED IN THIS BOOK

$\{a, b\}$	set containing the elements a and b
$\emptyset$	empty set
$\{x \mid x \text{ has property } P\}$	set of all elements x such that x has property P
$=$	equals
$\neq$	is not equal to
$\lvert x \rvert$	absolute value of x
$\cap$	set intersection
$\cup$	set union
$<$	is less than
$\leq$	is less than or equal to
$>$	is greater than
$\geq$	is greater than or equal to

$\not<$ is not less than

$\not>$ is not greater than

a^m a to the power m

a^{-n} $1/a^n$

$\sqrt[n]{a}$ nth root of a

$a^{1/n}$ $\sqrt[n]{a}$

$\sqrt{a}$ square root of a

$\approx$ is approximately equal to

$f(x)$ f of x

i $\sqrt{-1}$

$a + bi$ complex number

$\pm$ plus or minus

(a, b) ordered pair of numbers

m slope

f^{-1} inverse of function f

$f \circ g$ composite function

$\begin{vmatrix} a & b \\ c & d \end{vmatrix}$ two-by-two determinant

$\begin{bmatrix} a & b & c \\ d & e & f \end{bmatrix}$ matrix of two rows, three columns

$\log_a x$ logarithm of x to the base a

$\log x$ common, or base 10, logarithm of x

$\ln x$ natural (base e) logarithm of x

e a number, approximately 2.7182818

a_n nth term of a sequence

$\sum_{i=1}^{n} x_i$ summation

$n \rightarrow \infty$ n increases without bound

$\lim_{x \to a} f(x)$ limit of $f(x)$ as x approaches a

S_n sum of the first n terms of a sequence

S_∞ sum of the terms of an infinite sequence

$n!$ n factorial

$P(n, r)$ number of permutations of n elements r at a time

$\binom{n}{r}$ number of combinations of n elements r at a time

$n(E)$ number of elements in set E

$P(E)$ probability of event E

E' complement of set E

ANSWERS TO SELECTED EXERCISES

TO THE STUDENT

If you need further help with algebra, you may want to obtain a copy of the *Student's Solution Manual* that goes with this book. It contains solutions to all the odd-numbered exercises and all the chapter test exercises. You also may want the *Student's Study Guide*. It has extra examples and exercises to complete, corresponding to each section of the book. In addition, there is a practice test for each chapter. Your college bookstore either has the *Manual* or *Guide* or can order them for you.

In this section we provide the answers that we think most students will obtain when they work the exercises using the methods explained in the text. If your answer does not look exactly like the one given here, it is not necessarily wrong. In many cases there are equivalent forms of the answer. For example, if the answer section shows 3/4 and your answer is .75, you have obtained the correct answer but written it in a different (yet equivalent) form. Unless the directions specify otherwise, .75 is just as valid an answer as 3/4. In general, if your answer does not agree with the one given in the text, see whether it can be transformed into the other form. If it can, then it is the correct answer. If you still have doubts, talk with your instructor.

■ CHAPTER 1 ALGEBRAIC EXPRESSIONS

1.1 EXERCISES (PAGE 12)

1. 1, 3   **3.** $-6, -\frac{12}{4}$ (or -3), 0, 1, 3   **5.** $-\sqrt{3}, 2\pi, \sqrt{12}$ **7.** natural, whole, integer, rational, real
9. rational, real   **11.** irrational, real   **15.** -243 **17.** 81   **19.** -243 **21.** negative, positive   **23.** 79
25. 9   **27.** 23   **29.** 18   **31.** 28   **33.** $-\frac{5}{8}$ **35.** $\frac{6}{5}$ **37.** 18   **39.** $-\frac{1}{2}$ **41.** $\frac{1}{10}$ **43.** $\frac{7}{4}$
45. distributive   **47.** inverse   **49.** commutative and inverse   **51.** distributive   **53.** $(8 - 14)p = -6p$
55. $6(3y + 1)$ **57.** $9r - 9s$ **59.** $-2m - 2n$ **61.** $2 - 5m$ **63.** $pq - pw + px$ **65.** $20z$ **67.** $15p$
69. $-5m - 2y + 8z$ **71.** $\frac{3}{2}a + \frac{5}{7}b - \frac{9}{10}$

1.2 EXERCISES (PAGE 19)

1. $-|9|, -|-6|, |-8|$ **3.** $-5, -4, -2, -\sqrt{3}, \sqrt{6}, \sqrt{8}, 3$ **5.** $\frac{3}{4}, \frac{7}{5}, \sqrt{2}, \frac{22}{15}, \frac{8}{5}$ **9.** $-6 < 15$
11. $1 \le 2$ **13.** $x < 3$ **15.** $k < -7$ **17.** 8   **19.** 6   **21.** 4   **23.** 16   **25.** 8   **29.** -1
31. $5 - \sqrt{7}$ **33.** $\pi - 3$ **35.** $x - 4$ **37.** $8 - 2k$ **39.** $8 + 4m$ **41.** $y - x$ **43.** $3 + x^2$
45. $1 + p^2$ **47.** addition property of order   **49.** transitive property of order   **51.** addition property of order
53. triangle inequality, $|a + b| \le |a| + |b|$ **55.** property of absolute value, $|a| \ge 0$ **57.** property of absolute
value, $\left|\frac{a}{b}\right| = \frac{|a|}{|b|}$ **59.** if x and y both have the same sign or if either x or y equals 0   **61.** if $x = 0$
63. always true   **65.** 1

1.3 EXERCISES (PAGE 26)

1. $(-4)^5$ **3.** 1 **5.** 1 **7.** 2^{10} **9.** $2^3 x^{15} y^{12}$ **11.** $-\dfrac{p^8}{q^2}$ **13.** polynomial; degree 11; monomial

15. polynomial; degree 6; binomial **17.** polynomial; degree 6; binomial **19.** polynomial;
degree 6; trinomial **21.** not a polynomial **23.** not a polynomial **25.** $x^2 - x + 3$ **27.** $9y^2 - 4y + 4$

29. $6m^4 - 2m^3 - 7m^2 - 4m$ **31.** $28r^2 + r - 2$ **33.** $15x^2 - \dfrac{7}{3}x - \dfrac{2}{9}$ **35.** $12x^5 + 8x^4 - 20x^3 + 4x^2$

37. $-2z^3 + 7z^2 - 11z + 4$ **39.** $m^2 + mn - 2n^2 - 2km + 5kn - 3k^2$ **41.** $a^2 - 2ab + b^2 + 4ac - 4bc + 4c^2$
45. $4m^2 - 9$ **47.** $16m^2 + 16mn + 4n^2$ **49.** $25r^2 + 30rt + 9t^4$ **51.** $4p^2 - 12p + 9 + 4pq - 6q + q^2$
53. $9q^2 + 30q + 25 - p^2$ **55.** $a^2 - b^2 - 2bc - c^2$ **57.** $9a^2 + 6ab + b^2 - 6a - 2b + 1$
59. $18p^2 - 27pq - 35q^2$ **61.** $27x^3 - 108x^2 y + 144xy^2 - 64y^3$ **63.** $36k^2 - 36k + 9$
65. $p^3 - 7p^2 - p - 7$ **67.** $49m^2 - 4n^2$ **69.** $-14q^2 + 11q - 14$ **71.** $4p^2 - 16$ **73.** $11y^3 - 18y^2 + 4y$
75. -8 **77.** 1 **79.** -24 **81.** $k^{2m} - 4$ **83.** $b^{2r} + b^r - 6$ **85.** $3p^{2x} - 5p^x - 2$
87. $m^{2x} - 4m^x + 4$ **89.** $q^{2p} - 10q^p p^q + 25p^{2q}$ **91.** $27k^{3a} - 54k^{2a} + 36k^a - 8$ **93.** m **95.** $m + n$

1.4 EXERCISES (PAGE 32)

1. 20 **3.** 35 **5.** 56 **7.** 45 **9.** n **13.** they alternate, starting with positive
15. $x^6 + 6x^5 y + 15x^4 y^2 + 20x^3 y^3 + 15x^2 y^4 + 6xy^5 + y^6$ **17.** $p^5 - 5p^4 q + 10p^3 q^2 - 10p^2 q^3 + 5pq^4 - q^5$
19. $r^{10} + 5r^8 s + 10r^6 s^2 + 10r^4 s^3 + 5r^2 s^4 + s^5$ **21.** $p^4 + 8p^3 q + 24p^2 q^2 + 32pq^3 + 16q^4$
23. $2401p^4 + 2744p^3 q + 1176p^2 q^2 + 224pq^3 + 16q^4$
25. $729x^6 - 2916x^5 y + 4860x^4 y^2 - 4320x^3 y^3 + 2160x^2 y^4 - 576xy^5 + 64y^6$
27. $\dfrac{m^6}{64} - \dfrac{3m^5}{16} + \dfrac{15m^4}{16} - \dfrac{5m^3}{2} + \dfrac{15m^2}{4} - 3m + 1$ **29.** $4r^4 + \dfrac{8\sqrt{2}r^3}{m} + \dfrac{12r^2}{m^2} + \dfrac{4\sqrt{2}r}{m^3} + \dfrac{1}{m^4}$ **31.** $7920m^8 p^4$
33. $126x^4 y^5$ **35.** $180m^2 n^8$ **37.** $4845p^8 q^{16}$ **39.** $439{,}296x^{21} y^7$ **41.** 1.002; 5.010 **43.** .942

1.5 EXERCISES (PAGE 39)

1. $4k^2 m^3(1 + 2k^2 - 3m)$ **3.** $2(a + b)(1 + 2m)$ **5.** $(y + 2)(3y + 2)$ **7.** $(r + 3)(3r - 5)$
9. $(m - 1)(2m^2 - 7m + 7)$ **11.** $(2s + 3)(3t - 5)$ **13.** $(t^3 + s^2)(r - p)$ **15.** $(3p - 7)(2p + 5)$
17. $(5z - 2x)(4z - 9x)$ **19.** $(3 - m^2)(5 - r^2)$ **21.** $8(h - 8)(h + 5)$ **23.** $9y^2(y - 5)(y - 1)$
25. $(7m - 5r)(2m + 3r)$ **27.** $(4s + 5t)(3s - t)$ **29.** $(5a + m)(6a - m)$ **31.** $3x^3(3x - 5z)(2x + 5z)$
33. $(3m - 2)^2$ **35.** $2(4a - 3b)^2$ **37.** $(2xy + 7)^2$ **39.** $(a - 3b - 3)^2$ **41.** $(3a + 4)(3a - 4)$
43. $(5s^2 + 3t)(5s^2 - 3t)$ **45.** $(a + b + 4)(a + b - 4)$ **47.** $(p^2 + 25)(p + 5)(p - 5)$ **49.** (b)
51. $(2 - a)(4 + 2a + a^2)$ **53.** $(5x - 3)(25x^2 + 15x + 9)$ **55.** $(3y^3 + 5z^2)(9y^6 - 15y^3 z^2 + 25z^4)$
57. $r(r^2 + 18r + 108)$ **59.** $(3 - m - 2n)(9 + 3m + 6n + m^2 + 4mn + 4n^2)$ **63.** $(a^2 - 8)(a^2 + 6)$
65. $2(8z - 3)(6z - 5)$ **67.** $(18 - 5p)(17 - 4p)$ **69.** $(2b + c + 4)(2b + c - 4)$ **71.** $(x + y)(x - 5)$
73. $(m - 2n)(p^4 + q)$ **75.** $(2z + 7)^2$ **77.** $(10x + 7y)(100x^2 - 70xy + 49y^2)$
79. $(5m^2 - 6)(25m^4 + 30m^2 + 36)$ **81.** $(6m - 7n)(2m + 5n)$ **83.** $(4p - 1)(p + 1)$ **85.** prime **87.** $4xy$
89. $(r + 3s^q)(r - 2s^q)$ **91.** $(3a^{2k} + b^{4k})(3a^{2k} - b^{4k})$ **93.** $(2y^a - 3)^2$ **95.** $[3(m + p)^k + 5][2(m + p)^k - 3]$
97. ± 36 **99.** 9

1.6 EXERCISES (PAGE 46)

1. $x \neq -6$ **3.** $x \neq \dfrac{3}{5}$ **5.** no restrictions **7.** (a) **9.** $\dfrac{5p}{2}$ **11.** $\dfrac{8}{9}$ **13.** $\dfrac{3}{t - 3}$ **15.** $\dfrac{2x + 4}{x}$

17. $\dfrac{m - 2}{m + 3}$ **19.** $\dfrac{2m + 3}{4m + 3}$ **21.** $\dfrac{25p^2}{9}$ **23.** $\dfrac{2}{9}$ **25.** $\dfrac{5x}{y}$ **27.** $\dfrac{2(a + 4)}{a - 3}$ **29.** 1 **31.** $\dfrac{m + 6}{m + 3}$

33. $\dfrac{m-3}{2m-3}$ **35.** $\dfrac{x^2-1}{x^2}$ **37.** $\dfrac{x+y}{x-y}$ **39.** $\dfrac{x^2-xy+y^2}{x^2+xy+y^2}$ **41.** (b) and (c) **43.** $\dfrac{19}{6k}$ **45.** 1 **47.** $\dfrac{6+p}{2p}$

49. $\dfrac{137}{30m}$ **51.** $\dfrac{-2}{(a+1)(a-1)}$ **53.** $\dfrac{2m^2+2}{(m-1)(m+1)}$ **55.** $\dfrac{4}{a-2}$ or $\dfrac{-4}{2-a}$ **57.** $\dfrac{3x+y}{2x-y}$ or $\dfrac{-3x-y}{y-2x}$

59. $\dfrac{5}{(a-2)(a-3)(a+2)}$ **61.** $\dfrac{x-11}{(x+4)(x-4)(x-3)}$ **63.** $\dfrac{a^2+5a}{(a+6)(a+1)(a-1)}$ **65.** $\dfrac{x+1}{x-1}$ **67.** $\dfrac{-1}{x+1}$

69. $\dfrac{(2-b)(1+b)}{b(1-b)}$ **71.** $\dfrac{m^3-4m-1}{m-2}$ **73.** $\dfrac{p+5}{p(p+1)}$ **75.** $\dfrac{-1}{x(x+h)}$

1.7 EXERCISES (PAGE 55)

1. $-\dfrac{1}{64}$ **3.** 8 **5.** -2 **7.** 4 **9.** $\dfrac{1}{9}$ **11.** not defined **13.** $\dfrac{256}{81}$ **15.** .0024 **17.** $4p^2$

19. $9x^4$ **23.** (d) **25.** $\dfrac{1}{2^7}$ **27.** $\dfrac{1}{27^3}$ **29.** 1 **31.** $m^{7/3}$ **33.** $(1+n)^{5/4}$ **35.** $\dfrac{6z^{2/3}}{y^{5/4}}$ **37.** $2^6 a^{1/4} b^{37/2}$

39. $\dfrac{r^6}{s^{15}}$ **41.** $-\dfrac{1}{ab^3}$ **43.** $12^{9/4}y$ **45.** $\dfrac{1}{2p^2}$ **47.** $\dfrac{m^3 p}{n}$ **49.** $-4a^{5/3}$ **51.** $\dfrac{1}{(k+5)^{1/2}}$ **53.** $\dfrac{4z^2}{x^5 y}$

55. r^{6+p} **57.** $m^{3/2}$ **59.** $x^{(2n^2-1)/n}$ **61.** $p^{(m+n+m^2)/(mn)}$ **63.** $y-10y^2$ **65.** $-4k^{10/3}+24k^{4/3}$

67. x^2-x **69.** $r-2+r^{-1}$ or $r-2+\dfrac{1}{r}$ **71.** $k^{-2}(4k+1)$ or $\dfrac{4k+1}{k^2}$ **73.** $z^{-1/2}(9+2z)$ or

$\dfrac{9+2z}{z^{1/2}}$ **75.** $p^{-7/4}(p-2)$ or $\dfrac{p-2}{p^{7/4}}$ **77.** $(p+4)^{-3/2}(p^2+9p+21)$ or $\dfrac{p^2+9p+21}{(p+4)^{3/2}}$ **79.** $64

81. $64,000,000 **83.** about $10,000,000 **85.** 29 **87.** 177 **89.** $b+a$ **91.** -1 **93.** $\dfrac{y(xy-9)}{x^2 y^2-9}$

1.8 EXERCISES (PAGE 64)

1. $\sqrt[3]{(-m)^2}$ or $(\sqrt[3]{-m})^2$ **3.** $5\sqrt[5]{m^4}$ or $5(\sqrt[5]{m})^4$ **5.** $\dfrac{-4}{\sqrt[3]{z}}$ **7.** $\sqrt[3]{(2m+p)^2}$ or $(\sqrt[3]{2m+p})^2$ **9.** $k^{2/5}$

11. $-a^{2/3}$ **13.** $-3\cdot 5^{1/2} p^{3/2}$ **15.** $18mn^{3/2} p^{1/2}$ **21.** 5 **23.** -5 **25.** $5\sqrt{2}$ **27.** $3\sqrt[3]{3}$ **29.** $-2\sqrt[4]{2}$

31. $-\dfrac{3\sqrt{5}}{5}$ **33.** $-\dfrac{\sqrt[3]{100}}{5}$ **35.** $32\sqrt[3]{2}$ **37.** $2x^2 z^4 \sqrt{2x}$ **39.** $2zx^2 y\sqrt[3]{2z^2 x^2 y}$ **41.** $np^2\sqrt[4]{m^2 n^3}$

43. cannot simplify further **45.** $\dfrac{\sqrt{6x}}{3x}$ **47.** $\dfrac{x^2 y\sqrt{xy}}{z}$ **49.** $\dfrac{2\sqrt[3]{x}}{x}$ **51.** $\dfrac{h\sqrt[4]{9g^3 hr^2}}{3r^2}$ **53.** $\dfrac{m\sqrt[4]{n^2}}{n}$

55. $2\sqrt[4]{x^3 y^3}$ **57.** $\sqrt[3]{2}$ **59.** $\sqrt[18]{x}$ **61.** $9\sqrt{3}$ **63.** $-2\sqrt{7p}$ **65.** $7\sqrt[3]{3}$ **67.** $2\sqrt{3}$

69. $\dfrac{13\sqrt[3]{4}}{6}$ **71.** -7 **73.** 10 **75.** $11+4\sqrt{6}$ **77.** $5\sqrt{6}$ **79.** $2\sqrt[3]{9}-7\sqrt[3]{3}-4$ **81.** $\dfrac{\sqrt{15}-3}{2}$

83. $\dfrac{3\sqrt{5}+3\sqrt{15}-2\sqrt{3}-6}{33}$ **85.** $\dfrac{p(\sqrt{p}-2)}{p-4}$ **87.** $\dfrac{a(\sqrt{a+b}+1)}{a+b-1}$ **89.** $\dfrac{-1}{2(1-\sqrt{2})}$ **91.** $\dfrac{x}{\sqrt{x}+x}$

93. $\dfrac{-1}{2x-2\sqrt{x(x+1)}+1}$ **95.** $|m+n|$ **97.** $|z-3x|$

CHAPTER 1 REVIEW EXERCISES (PAGE 68)

1. $-12, -6, -\sqrt{4}$ (or -2), $0, 6$ **3.** $-\sqrt{7}, \dfrac{\pi}{4}, \sqrt{11}$ **5.** integers, rational, real **7.** irrational, real **9.** 9

11. 31 **13.** $-\dfrac{37}{20}$ **15.** $-\dfrac{19}{42}$ **17.** -32 **21.** distributive **23.** distributive **25.** distributive

27. $kr + ks - kt$ **29.** $-|3 - (-2)|, -|-2|, |6 - 4|, |8 + 1|$ **31.** -3 **33.** $3 - \sqrt{8}$ **35.** $m - 3$
37. $4 - \pi$ **39.** $7q^3 - 9q^2 - 8q + 9$ **41.** $16y^2 + 42y - 49$ **43.** $9k^2 - 30km + 25m^2$
45. $15w^3 - 22w^2 + 11w - 2$ **47.** $p^{2q} - 2p^q - 3$ **49.** $x^4 + 8x^3y + 24x^2y^2 + 32xy^3 + 16y^4$ **51.** $2160x^2y^4$
53. $3^{16} + 16 \cdot 3^{15}x + 120 \cdot 3^{14}x^2 + 560 \cdot 3^{13}x^3$ **55.** $z(7z - 9z^2 + 1)$ **57.** $(r + 7p)(r - 6p)$
59. $(3m + 1)(2m - 5)$ **61.** $(13y^2 + 1)(13y^2 - 1)$ **63.** $8(y - 5z^2)(y^2 + 5yz^2 + 25z^4)$ **65.** $(r - 3s)(a + 5b)$
67. $(4m - 7 + 5a)(4m - 7 - 5a)$ **69.** (a)   **71.** $\dfrac{3}{8r}$ **73.** $(3m + n)(3m - n)$ **75.** $\dfrac{37}{20y}$
77. $\dfrac{x + 9}{(x - 3)(x - 1)(x + 1)}$ **79.** $\dfrac{q + p}{pq - 1}$ **81.** $\dfrac{1}{64}$ **83.** $\dfrac{16}{25}$ **85.** $-10z^8$ **87.** 1 **89.** $-8y^{11}p$
91. $\dfrac{1}{(p + q)^5}$ **93.** $\dfrac{r^2}{s^4t^4}$ **95.** $p^{1/3}$ **97.** $-14r^{17/12}$ **99.** $y^{1/2}$ **101.** k^{2-9p} **103.** $10z^{7/3} - 4z^{1/3}$
105. $3p^2 + 3p^{3/2} - 5p - 5p^{1/2}$ **107.** $10\sqrt{2}$ **109.** $5\sqrt[4]{2}$ **111.** $-\dfrac{\sqrt[3]{50p}}{5p}$ **113.** $\sqrt[12]{m}$ **115.** 66
117. $-9m\sqrt{2m} + 5m\sqrt{m}$ or $m(-9\sqrt{2m} + 5\sqrt{m})$ **119.** $\dfrac{6(3 + \sqrt{2})}{7}$ **121.** $\dfrac{k(\sqrt{k} + 3)}{k - 9}$

CHAPTER 1 TEST   (PAGE 70)

[1.1] 1. $\dfrac{4}{\pi}$ **2.** $-10, -\dfrac{9}{3}$ (or -3), $0, \sqrt{25}$ **[1.2] 3.** $\dfrac{17}{9}$ **4.** commutative   **5.** distributive
[1.3] 6. $2a^2 - a - 3$ **7.** $4x^{2n} - 12x^n + 9$ **8.** $3y^3 + 5y^2 + 2y + 8$
[1.4] 9. $16x^4 - 96x^3y + 216x^2y^2 - 216xy^3 + 81y^4$ **10.** $60w^4y^2$ **[1.5] 11.** $(x^2 + 4)(x + 2)(x - 2)$
12. $2m(4m + 3)(3m - 4)$ **13.** $(x - 2)(x^2 + 2x + 4)(y + 3)(y - 3)$ **[1.6] 14.** The first step is wrong.
The numerator must be in factored form before cancelling factors of $x - 1$. As shown, the numerator is a difference,
not a product.   **15.** $\dfrac{x^4(x + 1)}{3(x^2 + 1)}$ **16.** $\dfrac{4x^2 + x}{(x + 2)(x + 1)(2x - 3)}$ **17.** $\dfrac{2a}{2a - 3}$ or $\dfrac{-2a}{3 - 2a}$ **18.** $\dfrac{y}{y + 2}$
[1.7] 19. $\dfrac{4}{x^{14}}$ **20.** $\dfrac{yz}{x}$ **[1.8] 21.** $3x^2y^4\sqrt{2x}$ **22.** $\dfrac{\sqrt[5]{8xy^3}}{2y}$ **23.** $2\sqrt{2x}$ **24.** $3\sqrt{2} + 3$
25. $x - y$

■ CHAPTER 2   EQUATIONS AND INEQUALITIES

2.1 EXERCISES   (PAGE 77)

1. identity;   {all real numbers}   **3.** conditional; $\left\{-\dfrac{17}{2}\right\}$ **5.** identity;   {all real numbers except 0}
7. contradiction; $\emptyset$ **9.** equivalent   **11.** not equivalent   **13.** not equivalent   **15.** (b)   **19.** {12}
21. $\left\{-\dfrac{2}{7}\right\}$ **23.** $\left\{-\dfrac{7}{8}\right\}$ **25.** $\{-1\}$ **27.** $\{3\}$ **29.** $\left\{\dfrac{3}{4}\right\}$ **31.** $\left\{-\dfrac{12}{5}\right\}$ **33.** $\emptyset$ **35.** $\left\{\dfrac{27}{7}\right\}$
37. $\left\{-\dfrac{59}{6}\right\}$ **39.** $\emptyset$ **41.** {0}   **43.** {50}   **45.** $\left\{-\dfrac{19}{75}\right\}$ **47.** $x = -3a + b$ **49.** $x = \dfrac{3a + b}{3 - a}$
51. $x = \dfrac{3 - 3a}{a^2 - a - 1}$ **53.** $x = \dfrac{2a^2}{a^2 + 3}$ **55.** $x = \dfrac{b - yd}{yc - a}$ **57.** earned runs allowed: 88   **59.** E.R.A.: 3.37
61. innings pitched: 241   **63.** 68° F   **65.** 15° C   **67.** 37.8° C   **69.** 13%   **71.** \$432   **73.** \$1500
75. \$205.41   **77.** \$66.50   **79.** Since 3 is the solution of the equation, dividing by $x - 3$ is dividing by zero, which is not allowed.   **81.** 6   **83.** -2

2.2 EXERCISES (PAGE 88)

1. $p = \dfrac{i}{rt}$ **3.** $w = \dfrac{P - 2l}{2}$ or $w = \dfrac{P}{2} - l$ **5.** $h = \dfrac{2A}{B + b}$ **7.** $h = \dfrac{S - 2lw}{2w + 2l}$ **9.** $g = \dfrac{2s}{t^2}$

11. $R = \dfrac{r_1 r_2}{r_2 + r_1}$ **13.** $f = \dfrac{AB(p + 1)}{24}$ **15.** $m = \dfrac{Pi}{A - P}$ **21.** 6 cm **23.** 96 **25.** 2 liters **27.** 4 ft

29. $\dfrac{400}{3}$ liters **31.** about 840 mi **33.** 15 min **35.** 35 km per hr **37.** 2.7 mi **39.** 3.6 hr **41.** 78 hr

43. $\dfrac{40}{3}$ hr **45.** $350 **47.** 9.1 yr **49.** $60,000 at 8.5% and $80,000 at 7% **51.** $70,000 for land that

makes a profit and $50,000 for land that produces a loss **53.** $800 **55.** $3000 in passbook accounts and
$17,000 in long-term deposits **57.** $20,000 at 4.5% and $40,000 at 5% **59.** (b) and (c)

2.3 EXERCISES (PAGE 97)

1. imaginary **3.** real **5.** imaginary **7.** imaginary **9.** $10i$ **11.** $-20i$ **13.** $-i\sqrt{39}$ **15.** $5 + 2i$
17. $9 - 5i\sqrt{2}$ **19.** -5 **21.** 2 **23.** -2 **25.** $7 - i$ **27.** $2 + 0i$ **29.** $1 - 10i$ **31.** $8 - i$
33. $-14 + 2i$ **35.** $5 - 12i$ **37.** $13 + 0i$ **39.** $7 + 0i$ **41.** $0 + 25i$ **43.** $12 + 9i$ **45.** i **47.** i
49. 1 **51.** $-i$ **53.** 1 **55.** i **59.** $0 + i$ **61.** $\dfrac{7}{25} - \dfrac{24}{25}i$ **63.** $\dfrac{26}{29} + \dfrac{7}{29}i$ **65.** $-2 + i$
67. $0 - 2i$ **69.** $\dfrac{3 - 2\sqrt{5}}{13} + \dfrac{-2 - 3\sqrt{5}}{13}i$ **71.** $\dfrac{5}{2} + i$ **73.** $-\dfrac{16}{65} - \dfrac{37}{65}i$ **75.** $\dfrac{27}{10} + \dfrac{11}{10}i$
77. $a = 0$ or $b = 0$ **81.** $4 + 6i$

2.4 EXERCISES (PAGE 106)

1. $\{\pm 4\}$ **3.** $\{\pm 3\sqrt{3}\}$ **5.** $\{\pm 4i\}$ **7.** $\{\pm 3i\sqrt{2}\}$ **9.** $\left\{\dfrac{1 \pm 2\sqrt{3}}{3}\right\}$ **11.** $\{2, 3\}$ **13.** $\left\{-\dfrac{5}{2}, \dfrac{10}{3}\right\}$

15. $\left\{\dfrac{3}{5} \pm \dfrac{\sqrt{3}}{5}i\right\}$ **17.** $\{3, 5\}$ **19.** $\{1 \pm \sqrt{5}\}$ **21.** $\left\{-\dfrac{1}{2} \pm \dfrac{1}{2}i\right\}$ **25.** $\left\{\dfrac{3 \pm \sqrt{17}}{2}\right\}$ **27.** $\left\{\dfrac{7 \pm \sqrt{5}}{22}\right\}$

29. $\{1 \pm 2i\}$ **31.** $\left\{\dfrac{-3 \pm 3\sqrt{129}}{16}\right\}$ **33.** $\left\{\dfrac{-3 \pm \sqrt{41}}{8}\right\}$ **35.** $\left\{\dfrac{2 \pm \sqrt{10}}{3}\right\}$ **37.** (d) **39.** (a)

41. $\left\{1, -\dfrac{1}{2} \pm \dfrac{\sqrt{3}}{2}i\right\}$ **43.** $\left\{-3, \dfrac{3}{2} \pm \dfrac{3\sqrt{3}}{2}i\right\}$ **45.** $\left\{\dfrac{1}{2}, 2\right\}$ **47.** $\left\{\dfrac{1 \pm \sqrt{13}}{2}\right\}$ **49.** $\left\{\dfrac{7}{4}, -\dfrac{7}{8} \pm \dfrac{7\sqrt{3}}{8}i\right\}$

51. $\left\{-\dfrac{1}{3} \pm \dfrac{\sqrt{7}}{3}i\right\}$ **53.** $\left\{\dfrac{\sqrt{2} \pm \sqrt{6}}{2}\right\}$ **55.** $\left\{\sqrt{2}, \dfrac{\sqrt{2}}{2}\right\}$ **57.** $\left\{\dfrac{-\sqrt{5} \pm 1}{2}\right\}$ **59.** $t = \dfrac{\pm\sqrt{2sg}}{g}$

61. $v = \dfrac{\pm\sqrt[4]{Frk^3M^3}}{kM}$ **63.** $R = \dfrac{E^2 - 2Pr \pm E\sqrt{E^2 - 4Pr}}{2P}$ **65. (a)** $x = \dfrac{y \pm \sqrt{8 - 11y^2}}{4}$

(b) $y = \dfrac{x \pm \sqrt{6 - 11x^2}}{3}$ **67.** 0; one rational solution **69.** 1; two different rational solutions

71. 84; two different irrational solutions **73.** -23; two different imaginary solutions

75. 2304; two different rational solutions **77.** 1 **81.** $x \neq 7, x \neq -\dfrac{2}{3}$ **83.** $y \neq -\dfrac{1}{4}$ **85.** no restrictions

2.5 EXERCISES (PAGE 113)

1. 100 yd by 400 yd **3.** 10 in by 13 in **5.** 2 cm **7.** Felipe: 36 hr; Felix: 45 hr **9.** 4 hr **11.** 50 mph
13. 2.5 hr or 16 hr (16 is not realistic) **15.** (d) **17.** 40 ft **19.** 61 min **21. (a)** 1 sec, 5 sec **(b)** 6 sec
23. (a) it will not reach 80 ft **(b)** 2 sec **27.** 4 **29.** 80

2.6 EXERCISES (PAGE 120)

1. $\{\pm\sqrt{3}, \pm i\sqrt{5}\}$ **3.** $\left\{\pm 1, \pm\dfrac{\sqrt{10}}{2}\right\}$ **5.** $\{4, 6\}$ **7.** $\left\{\dfrac{-5 \pm \sqrt{21}}{2}\right\}$ **9.** $\left\{\dfrac{-6 \pm 2\sqrt{3}}{3}, \dfrac{-4 \pm \sqrt{2}}{2}\right\}$

11. $\left\{\dfrac{7}{2}, -\dfrac{1}{3}\right\}$ **13.** $\{-63, 28\}$ **15.** $\left\{\pm i\sqrt{3}, \pm i\dfrac{\sqrt{6}}{3}\right\}$ **19.** $\{-1\}$ **21.** $\{5\}$ **23.** $\{9\}$ **25.** $\{9\}$

27. $\emptyset$ **29.** $\{8\}$ **31.** $\{-2\}$ **33.** $\left\{2, -\dfrac{2}{9}\right\}$ **35.** $\{-2\}$ **37.** $\{2\}$ **39.** $\left\{\dfrac{3}{2}\right\}$ **41.** $\{31\}$

43. $\{-3, 1\}$ **45.** $\{-27, 3\}$ **47.** $\left\{\dfrac{1}{4}, 1\right\}$ **51.** $h = \dfrac{d^2}{k^2}$ **53.** $L = \dfrac{P^2 g}{4}$ **55.** $y = (a^{2/3} - x^{2/3})^{3/2}$

2.7 EXERCISES (PAGE 130)

1. $(-1, 4)$

3. $(-\infty, 0)$

5. $[1, 2)$

7. $(-\infty, -9)$

9. $\{x \mid -4 < x < 3\}$ **11.** $\{x \mid x \le -1\}$ **13.** $\{x \mid -2 \le x < 6\}$ **15.** $\{x \mid x \le -4\}$

19. $[-1, \infty)$

21. $(-\infty, 6]$

23. $(-\infty, 4)$

25. $\left[-\dfrac{11}{5}, \infty\right)$

27. $[1, 4]$

29. $(-6, -4)$

31. $(-16, 19]$

33. $[500, \infty)$ **35.** $[45, \infty)$

37. $[-3, 3]$

39. $(-\infty, -3] \cup [-1, \infty)$

41. $[-2, 3]$

43. $\left(-\infty, \dfrac{1}{2}\right) \cup (4, \infty)$

45. $(-\infty, 0) \cup (0, \infty)$

47. $\left(\dfrac{-5 - \sqrt{33}}{2}, \dfrac{-5 + \sqrt{33}}{2}\right)$

49. $[1 - \sqrt{2}, 1 + \sqrt{2}]$ **51.** $(-5, 3]$ **53.** $(-\infty, -2)$

55. $(-\infty, 6) \cup \left[\dfrac{15}{2}, \infty\right)$ **57.** $(-\infty, 1) \cup \left(\dfrac{9}{5}, \infty\right)$ **59.** $\left(-\infty, -\dfrac{3}{2}\right) \cup \left[-\dfrac{1}{2}, \infty\right)$ **61.** $(-2, \infty)$

63. $\left(0, \dfrac{4}{11}\right) \cup \left(\dfrac{1}{2}, \infty\right)$ **65.** $(-\infty, -2] \cup (1, 2)$ **67.** $(-\infty, 5)$ **69.** both included   **71.** both excluded

73. 2 included, 5 excluded   **75.** $\left[-2, \dfrac{3}{2}\right] \cup [3, \infty)$ **77.** $(-\infty, -2] \cup [0, 2]$ **79.** $(-2, 0) \cup \left(\dfrac{1}{4}, \infty\right)$

81. (a), (e)   **85.** $\left(0, \dfrac{5}{4}\right) \cup (6, \infty)$ **87.** for values of t in $\left[4, \dfrac{39}{4}\right]$ **89.** for values of x in $(100, 150)$

2.8 EXERCISES   (PAGE 136)

1. $\{1, 3\}$ **3.** $\left\{-\dfrac{1}{3}, 1\right\}$ **5.** $\left\{\dfrac{2}{3}, \dfrac{8}{3}\right\}$ **7.** $\{-6, 14\}$ **9.** $\left\{\dfrac{5}{2}, \dfrac{7}{2}\right\}$ **11.** $\left\{\dfrac{3}{2}, -3\right\}$ **13.** $\left\{-\dfrac{3}{2}\right\}$

15. $\left\{-\dfrac{4}{3}, \dfrac{2}{9}\right\}$ **17.** $\left\{-\dfrac{7}{3}, -\dfrac{1}{7}\right\}$ **19.** $\left\{-\dfrac{3}{5}, 11\right\}$ **21.** $\left\{-\dfrac{7}{4}, -\dfrac{7}{12}\right\}$ **23.** $\left\{-\dfrac{1}{2}\right\}$ **27.** $[-3, 3]$

29. $(-\infty, -1) \cup (1, \infty)$ **31.** $\emptyset$ **33.** $[-10, 10]$ **35.** $(-4, -1)$ **37.** $\left(-\infty, -\dfrac{2}{3}\right) \cup (2, \infty)$

39. $\left(-\infty, -\dfrac{8}{3}\right] \cup [2, \infty)$ **41.** $\left[-1, -\dfrac{1}{2}\right]$ **43.** $\left(-\dfrac{3}{2}, \dfrac{13}{10}\right)$ **45.** $(-\infty, \infty)$ **47.** $(-\infty, -12) \cup (-12, \infty)$

49. $|p - q| = 5$ (or $|q - p| = 5$)   **53.** $|x - 2| \le 4$ **55.** $|z - 12| \ge 2$ **57.** $|k - 1| = 6$

59. If $|x - 2| \le .0004$, then $|y - 7| \le .00001$.   **61.** $[-140, -28]$ **63.** $|F - 730| \le 50$

65. choose any $m \le 2$ and any $n \ge 20$

CHAPTER 2 REVIEW EXERCISES   (PAGE 139)

1. $\{6\}$ **3.** $\left\{-\dfrac{11}{3}\right\}$ **5.** $\emptyset$ **7.** $\left\{\dfrac{1}{60}\right\}$ **9.** $x = -6b - a - 6$ **11.** $x = \dfrac{6 - 3m}{1 + 2k - km}$ **13.** $P = \dfrac{A}{1 + i}$

15. $f = \dfrac{Ab(p + 1)}{24}$ **17.** $x = \dfrac{-4}{y^2 - 5y - 6p}$ **19.** \$500   **21.** 8 mi   **23.** \$55,000 at 11.5% and \$35,000 at 12%

25. $10 - 3i$ **27.** $-8 + 13i$ **29.** $19 + 17i$ **31.** $146 + 0i$ **33.** $7 - 24i$ **35.** $\dfrac{5}{2} + \dfrac{7}{2}i$ **37.** real

39. $-i$ **41.** i **43.** $\{-7 \pm \sqrt{5}\}$ **45.** $\left\{\dfrac{5}{2}, -3\right\}$ **47.** $\left\{7, -\dfrac{3}{2}\right\}$ **49.** $\left\{3, -\dfrac{1}{2}\right\}$ **51.** $\{\sqrt{2} \pm 1\}$

53. -188; two different imaginary solutions   **55.** 484; two different rational solutions   **57.** 0; one rational

solution   **59.** $\dfrac{1}{2}$ ft   **61.** 6.5 hr   **63.** 10 mph   **65.** $\left\{\pm i, \pm\dfrac{1}{2}\right\}$ **67.** $\left\{\dfrac{5}{2}, -15\right\}$ **69.** $\{3\}$

71. $\{-2, -1\}$ **73.** $\emptyset$ **75.** $\{-1\}$ **77.** $\left(-\dfrac{7}{13}, \infty\right)$ **79.** $(-\infty, 1]$ **81.** $(1, \infty)$ **83.** $[4, 5]$ **85.** $[-4, 1]$

87. $\left(-\dfrac{2}{3}, \dfrac{5}{2}\right)$ **89.** $\{3\}$ **91.** $(-2, 0)$ **93.** $(-3, 1) \cup [7, \infty)$ **95.** positive on the intervals $(-\infty, a)$ and

(b, ∞); negative on the interval (a, b); zero at the numbers a and b **97.** $[300, \infty)$ **99.** $\{3, -11\}$

101. $\left\{\dfrac{11}{27}, \dfrac{25}{27}\right\}$ **103.** $\left\{\dfrac{4}{3}, -\dfrac{2}{7}\right\}$ **105.** $[-7, 7]$ **107.** $(-\infty, -3) \cup (3, \infty)$ **109.** $[-6, -3]$ **111.** $\left(-\dfrac{2}{7}, \dfrac{8}{7}\right)$

113. $(-\infty, -4) \cup \left(-\dfrac{2}{3}, \infty\right)$

A-8　**CHAPTER 2 TEST**　(PAGE 142)

[2.1]　**1.** $\left\{\dfrac{5}{2}\right\}$　**2.** $\emptyset$　**3.** $c = \dfrac{ab}{a+b}$　**[2.2]**　**4.** $13\dfrac{1}{3}$ qt　**5.** 225 mi　**[2.3]**　**6.** $1 + 9i$

7. $-26 - 7i$　**8.** $2 + i$　**9.** i　**[2.4]**　**10.** $\left\{1 \pm \dfrac{\sqrt{6}}{6}i\right\}$　**11.** $\left\{\dfrac{6 \pm \sqrt{42}}{3}\right\}$　**12.** 1;　2 different rational

solutions　**[2.5]**　**13.** 6 ft by 8 ft　**[2.6]**　**14.** $\{2\}$　**15.** $\{0, 4\}$　**16.** $\{\pm\sqrt{5}, \pm i\sqrt{2}\}$　**17.** $\{-4, 2\}$

[2.7]　**18.** $[-3, \infty)$　**19.** $[-10, 2]$　**20.** $(-\infty, -1] \cup \left[\dfrac{3}{2}, \infty\right)$　**21.** $\left(-\infty, \dfrac{5}{2}\right) \cup [4, \infty)$

[2.8]　**23.** $\left\{-\dfrac{1}{3}, -3\right\}$　**24.** $(-2, 7)$　**25.** $(-\infty, -6] \cup [5, \infty)$

■ CHAPTER 3　RELATIONS AND THEIR GRAPHS

3.1 EXERCISES　(PAGE 149)

Other correct ordered pairs may be given in Exercises 1–12.
1. $(-3, 5), (-2, 4), (-1, 6)$; $\{-3, -2, -1, 0, 1\}$; $\{5, 4, 6, -8, 2\}$　**3.** $(0, 2), (1, 11), (2, 20)$; $(-\infty, \infty)$; $(-\infty, \infty)$
5. $(-1, -1), (0, 0), (1, 1)$; $(-\infty, \infty)$; $(-\infty, \infty)$　**7.** $(0, 0), (1, 1), (4, 2)$; $[0, \infty)$; $[0, \infty)$
9. $(0, 0), (1, 1), (2, 4)$; $(-\infty, \infty)$; $[0, \infty)$　**11.** $(-3, 0), (-2, 1), (1, 2)$; $[-3, \infty)$; $[0, \infty)$　**15.** $8\sqrt{2}$; $(9, 3)$
17. $\sqrt{34}$; $(-11/2, -7/2)$　**19.** $2\sqrt{185}$; $(-1, 6)$　**21.** $\sqrt{4a^2 + 25b^2}$; $(2a, -3b/2)$
23. $\sqrt{133}$; $(2\sqrt{2}, 3\sqrt{5}/2)$　**25.** yes　**27.** no　**29.** no　**31.** yes　**33.** no　**35.** yes　**37.** $(13, 10)$
39. $(19, 16)$　**41.** $(2c - a, 2d - b)$　**45.** quadrants I and IV　**47.** quadrants II and IV
49. quadrants III and IV

3.2 EXERCISES　(PAGE 160)

1. domain and range: $(-\infty, \infty)$　**3.** domain and range: $(-\infty, \infty)$　**5.** domain and range: $(-\infty, \infty)$

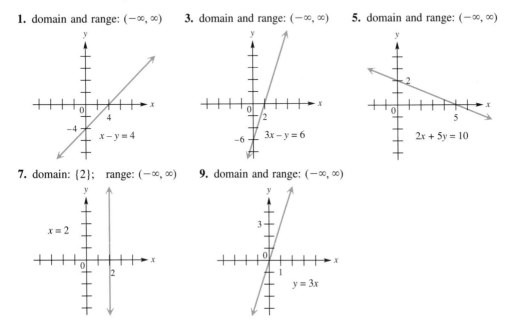

7. domain: $\{2\}$;　range: $(-\infty, \infty)$　**9.** domain and range: $(-\infty, \infty)$

11. $1/5$　**13.** $7/9$　**15.** $-3/4$　**17.** 0　**19.** **(a)** D　**(b)** B　**(c)** A　**(d)** C

23. 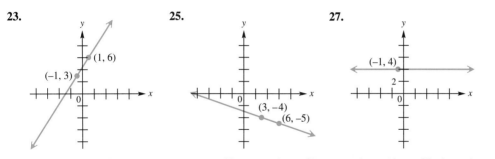 **25.** **27.**

29. $2x + y = 5$ **31.** $3x + 2y = -7$ **33.** $x = -8$ **35.** $-x + 4y = 13$ **37.** $2x - 3y = 6$ **39.** $x = -6$
41. -2, does not, undefined, 1/2, does not, zero   **43.** $x = 0$ **45.** slope 3; y-intercept -1 **47.** slope 4;
y-intercept -7 **49.** slope $-3/4$; y-intercept 0   **51.** $x + 3y = 11$ **53.** $5x - 3y = -13$ **55.** $x = -5$
57. yes   **59.** no   **61. (a)** $-1/2$ **(b)** $-7/2$ **67.** $y = 640x + 1100$ **69.** $y = -1000x + 40,000$

71. $y = 2.5x - 70$ **73.** $F = \dfrac{9}{5}C + 32$ **75.** $-40°$

3.3 EXERCISES   (PAGE 173)

1. **3.** **5. (a)** III   **(b)** II   **(c)** IV   **(d)** I

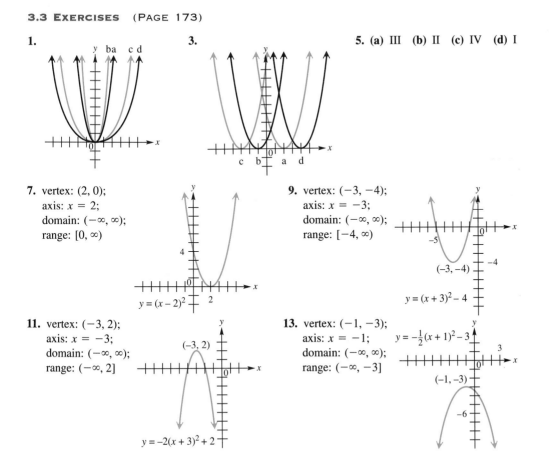

7. vertex: $(2, 0)$;
   axis: $x = 2$;
   domain: $(-\infty, \infty)$;
   range: $[0, \infty)$

$y = (x - 2)^2$

9. vertex: $(-3, -4)$;
   axis: $x = -3$;
   domain: $(-\infty, \infty)$;
   range: $[-4, \infty)$

$y = (x + 3)^2 - 4$

11. vertex: $(-3, 2)$;
   axis: $x = -3$;
   domain: $(-\infty, \infty)$;
   range: $(-\infty, 2]$

$y = -2(x + 3)^2 + 2$

13. vertex: $(-1, -3)$;
   axis: $x = -1$;
   domain: $(-\infty, \infty)$;
   range: $(-\infty, -3]$

$y = -\frac{1}{2}(x + 1)^2 - 3$

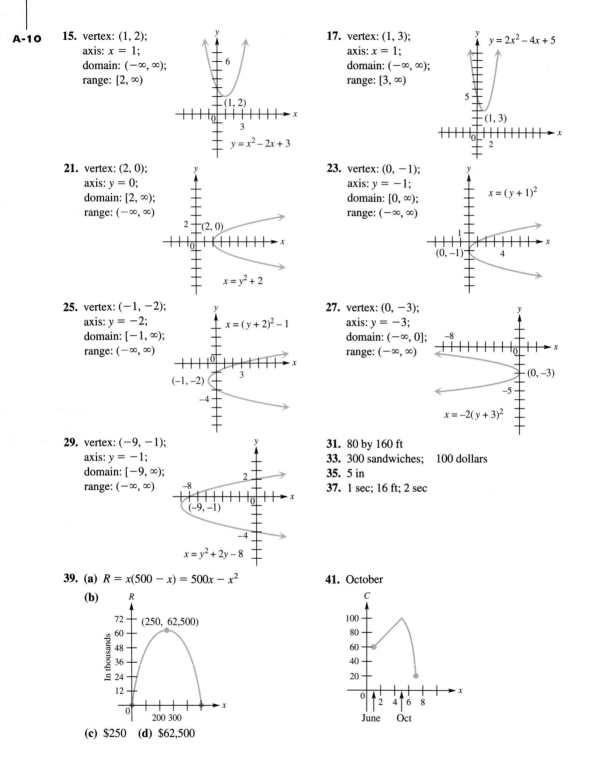

15. vertex: (1, 2);
axis: $x = 1$;
domain: $(-\infty, \infty)$;
range: $[2, \infty)$

17. vertex: (1, 3);
axis: $x = 1$;
domain: $(-\infty, \infty)$;
range: $[3, \infty)$

21. vertex: (2, 0);
axis: $y = 0$;
domain: $[2, \infty)$;
range: $(-\infty, \infty)$

23. vertex: (0, −1);
axis: $y = -1$;
domain: $[0, \infty)$;
range: $(-\infty, \infty)$

25. vertex: (−1, −2);
axis: $y = -2$;
domain: $[-1, \infty)$;
range: $(-\infty, \infty)$

27. vertex: (0, −3);
axis: $y = -3$;
domain: $(-\infty, 0]$;
range: $(-\infty, \infty)$

29. vertex: (−9, −1);
axis: $y = -1$;
domain: $[-9, \infty)$;
range: $(-\infty, \infty)$

31. 80 by 160 ft
33. 300 sandwiches; 100 dollars
35. 5 in
37. 1 sec; 16 ft; 2 sec

39. **(a)** $R = x(500 - x) = 500x - x^2$

(b)

(c) \$250 **(d)** \$62,500

41. October

43. $x^2 = 12y$ **45.** $y^2 = -8x$ **47.** $(x - 3)^2 = 8(y - 4)$ **49.** 25   **51.** E   **53.** C

55. $y = 2x^2 + 5x - 3$

(a) $(-3, 1/2)$
(b) $(-\infty, -3) \cup (1/2, \infty)$
(c) $-3, 1/2$

3.4 EXERCISES (PAGE 182)

1. $(x - 1)^2 + (y - 4)^2 = 9$ **3.** $x^2 + y^2 = 1$ **5.** $(x - 2/3)^2 + (y + 4/5)^2 = 9/49$ **7.** $(x + 1)^2 + (y - 2)^2 = 25$
9. $(x + 3)^2 + (y + 2)^2 = 4$

11. domain: $[-6, 6]$;
range: $[-6, 6]$

$x^2 + y^2 = 36$

13. domain: $[-4, 8]$;
range: $[-6, 6]$

$(x - 2)^2 + y^2 = 36$

15. domain: $[-6, 2]$;
range: $[1, 9]$

$(x + 2)^2 + (y - 5)^2 = 16$

17. domain: $[-9, 3]$;
range: $[-8, 4]$

$(x + 3)^2 + (y + 2)^2 = 36$

23. $(-3, -4)$; $r = 4$ **25.** $(6, -5)$; $r = 6$ **27.** $(-4, 7)$; $r = 1$ **29.** $(0, 1)$; $r = 7$ **31.** y-axis
33. x-axis   **35.** origin   **37.**   **39.**

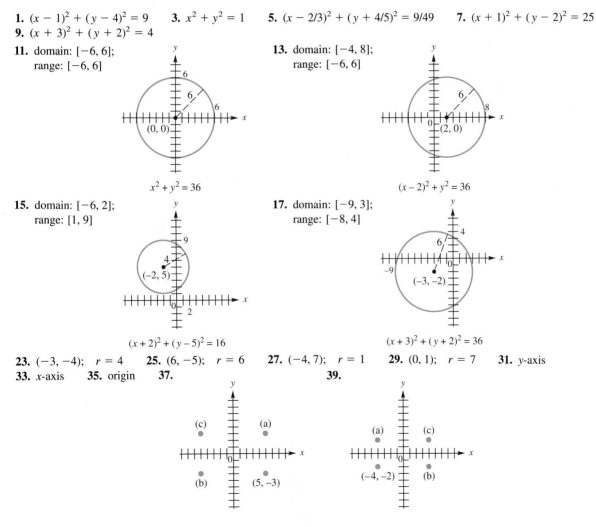

41. $(s, -t)$ **43.** $(-s, -t)$ **45.** x-axis, y-axis, origin   **47.** none   **49.** origin   **51.** y-axis
Many answers are possible in Exercises 53 and 55.   **53.**   **55.**

57. $(x - 2)^2 + (y + 3)^2 = 45$ **59.** $(2 + \sqrt{7}, 2 + \sqrt{7}), (2 - \sqrt{7}, 2 - \sqrt{7})$ **61.** It is reflected about the x-axis.
63. yes

3.5 EXERCISES   (PAGE 191)

1. domain: $[-3, 3]$;
range: $[-2, 2]$

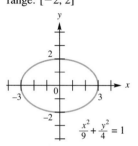

$$\frac{x^2}{9} + \frac{y^2}{4} = 1$$

3. domain; $[-\sqrt{6}, \sqrt{6}]$;
range: $[-3, 3]$

$$\frac{x^2}{6} + \frac{y^2}{9} = 1$$

5. domain: $[-1/3, 1/3]$;
range: $[-1/4, 1/4]$

$$\frac{x^2}{1/9} + \frac{y^2}{1/16} = 1$$

7. domain: $[-3/8, 3/8]$;
range: $[-6/5, 6/5]$

$$\frac{64x^2}{9} + \frac{25y^2}{36} = 1$$

9. domain: $[-2, 4]$;
range: $[-8, 2]$

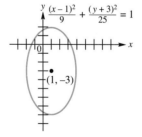

$$\frac{(x-1)^2}{9} + \frac{(y+3)^2}{25} = 1$$

11. domain: $[-2, 6]$;
range: $[-2, 4]$

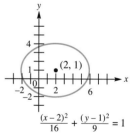

$$\frac{(x-2)^2}{16} + \frac{(y-1)^2}{9} = 1$$

13. domain: $(-\infty, -4] \cup [4, \infty)$;
range: $(-\infty, \infty)$

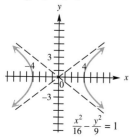

$$\frac{x^2}{16} - \frac{y^2}{9} = 1$$

15. domain: $(-\infty, \infty)$;
range: $(-\infty, -6] \cup [6, \infty)$

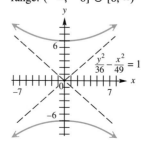

$$\frac{y^2}{36} - \frac{x^2}{49} = 1$$

17. domain: $(-\infty, -3/2] \cup [3/2, \infty)$;
range: $(-\infty, \infty)$

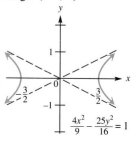

$$\frac{4x^2}{9} - \frac{25y^2}{16} = 1$$

19. domain: $(-\infty, -2] \cup [4, \infty)$; range: $(-\infty, \infty)$

21. domain: $(-\infty, -1] \cup [7, \infty)$; range: $(-\infty, \infty)$

23. domain: $(-\infty, \infty)$; range: $(-\infty, -6] \cup [4, \infty)$

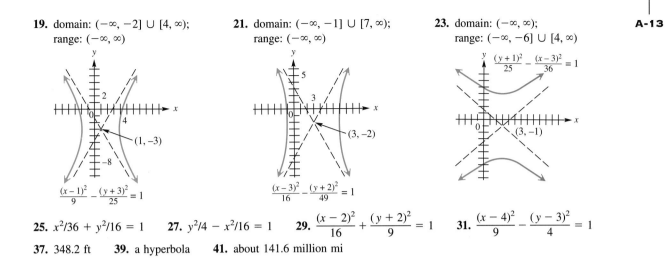

25. $x^2/36 + y^2/16 = 1$ **27.** $y^2/4 - x^2/16 = 1$ **29.** $\dfrac{(x-2)^2}{16} + \dfrac{(y+2)^2}{9} = 1$ **31.** $\dfrac{(x-4)^2}{9} - \dfrac{(y-3)^2}{4} = 1$

37. 348.2 ft   **39.** a hyperbola   **41.** about 141.6 million mi

3.6 EXERCISES (PAGE 198)

1. circle   **3.** parabola   **5.** parabola   **7.** ellipse   **9.** hyperbola   **11.** hyperbola   **13.** ellipse   **15.** circle
17. line   **19.** hyperbola   **21.** ellipse   **23.** circle   **25.** parabola   **27.** no graph   **29.** circle
31. parabola   **33.** hyperbola   **35.** ellipse   **37.** no graph   **39.** ellipse
41. (a)   **(b)**

43. domain: $(-\infty, 4]$; range: $[0, \infty)$

45. domain: $(-\infty, -3] \cup [3, \infty)$; range: $[0, \infty)$

$y = \sqrt{4 - x}$

$y = \sqrt{x^2 - 9}$

47. domain: $(-\infty, \infty)$; range: $(-\infty, -3]$

$\dfrac{y}{3} = -\sqrt{1 + \dfrac{x^2}{16}}$

49. domain: $[-8, 8]$; range: $[0, 1]$

$y = \sqrt{1 - \dfrac{x^2}{64}}$

1.

$x \le 3$

3.

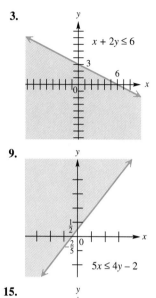

$x + 2y \le 6$

5.

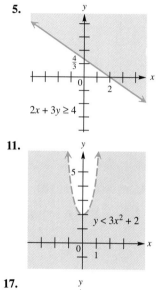

$\frac{4}{3}$ 2

$2x + 3y \ge 4$

7.

$3x - 5y > 6$

2

$-\frac{6}{5}$

9.

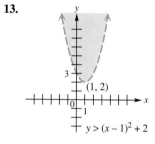

$\frac{1}{2}$

$-\frac{2}{5}$

$5x \le 4y - 2$

11.

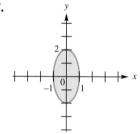

5

$y < 3x^2 + 2$

1

13.

3

$(1, 2)$

1

$y > (x - 1)^2 + 2$

15.

$(0, -3)$

$x^2 + (y + 3)^2 \le 16$

17.

2

-1 1

$4x^2 \le 4 - y^2$

19.

3

-4 4

-3

$9x^2 - 16y^2 > 144$

21. (b)   **23.** (d)

25.

4

$x - 3y < 4,\ x \le 0$

27.

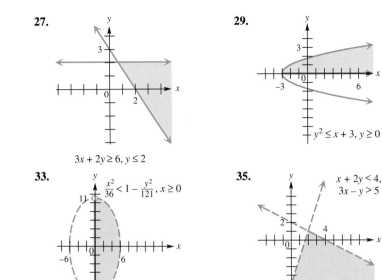

$3x + 2y \geq 6,\ y \leq 2$

29.

$y^2 \leq x + 3,\ y \geq 0$

31.

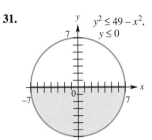

$y^2 \leq 49 - x^2,$
$y \leq 0$

33.

$\dfrac{x^2}{36} < 1 - \dfrac{y^2}{121},\ x \geq 0$

35.

$x + 2y < 4,$
$3x - y > 5$

37.

$2x + 3y < 6,\ x - 5y \geq 10$

CHAPTER 3 REVIEW EXERCISES (PAGE 205)

1. domain: $\{-3, -1, 8\}$;   range: $\{6, 4, 5\}$ **3.** $\sqrt{85}$; $(-1/2, 2)$ **5.** 5; $(-6, 11/2)$ **7.** -7; -1; 8; 23
9. 6/5   **11.** undefined slope   **13.** 9/4   **15.** 1/5   **17.** 0
19. domain and range: $(-\infty, \infty)$ **21.** domain and range: $(-\infty, \infty)$ **23.** domain: $\{-5\}$;   range: $(-\infty, \infty)$

$3x + 7y = 14$

$3y = x$

$x = -5$

25. $x + 3y = 10$ **27.** $2x + y = 1$ **29.** $15x + 30y = 13$ **31.** $y = 3/4$ **33.** $5x - 8y = -40$ **35.** $y = -5$
37.   **39.**   **41.** domain: $(-\infty, \infty)$;   range: $[-4, \infty)$

$(6, -1)$
$(2, -4)$

$(-3, 4)$
$(-4, 1)$

$y = x^2 - 4$
$(0, -4)$

43. domain: $(-\infty, \infty)$;
range: $[-5, \infty)$

45. domain: $(-\infty, \infty)$;
range: $[-2, \infty)$

47. domain: $[-2, \infty)$;
range: $(-\infty, \infty)$

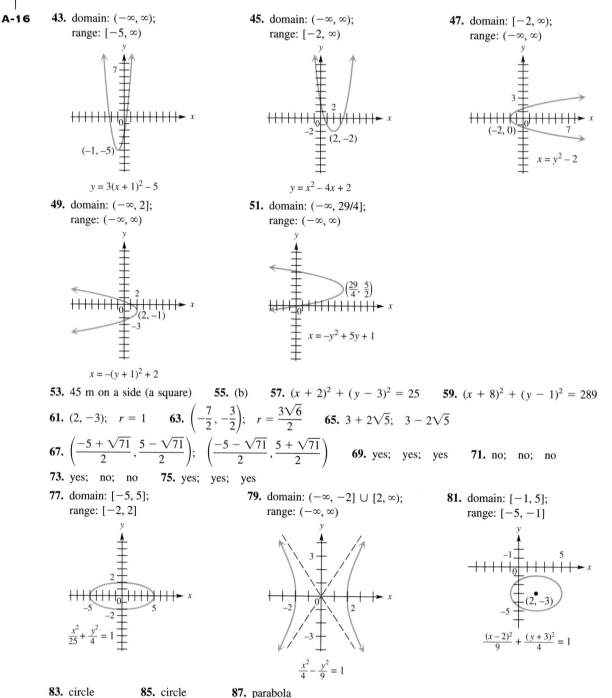

$y = 3(x + 1)^2 - 5$ $y = x^2 - 4x + 2$

$x = y^2 - 2$

49. domain: $(-\infty, 2]$;
range: $(-\infty, \infty)$

51. domain: $(-\infty, 29/4]$;
range: $(-\infty, \infty)$

$x = -(y + 1)^2 + 2$ $x = -y^2 + 5y + 1$

53. 45 m on a side (a square)   **55.** (b)   **57.** $(x + 2)^2 + (y - 3)^2 = 25$ **59.** $(x + 8)^2 + (y - 1)^2 = 289$

61. $(2, -3)$; $r = 1$ **63.** $\left(-\dfrac{7}{2}, -\dfrac{3}{2}\right)$; $r = \dfrac{3\sqrt{6}}{2}$ **65.** $3 + 2\sqrt{5}$; $3 - 2\sqrt{5}$

67. $\left(\dfrac{-5 + \sqrt{71}}{2}, \dfrac{5 - \sqrt{71}}{2}\right)$; $\left(\dfrac{-5 - \sqrt{71}}{2}, \dfrac{5 + \sqrt{71}}{2}\right)$ **69.** yes;   yes;   yes   **71.** no;   no;   no

73. yes;   no;   no   **75.** yes;   yes;   yes

77. domain: $[-5, 5]$;
range: $[-2, 2]$

79. domain: $(-\infty, -2] \cup [2, \infty)$;
range: $(-\infty, \infty)$

81. domain: $[-1, 5]$;
range: $[-5, -1]$

$\dfrac{x^2}{25} + \dfrac{y^2}{4} = 1$ $\dfrac{x^2}{4} - \dfrac{y^2}{9} = 1$ $\dfrac{(x - 2)^2}{9} + \dfrac{(y + 3)^2}{4} = 1$

83. circle   **85.** circle   **87.** parabola

89. domain: $[-10, 10]$;
range: $[0, 10]$

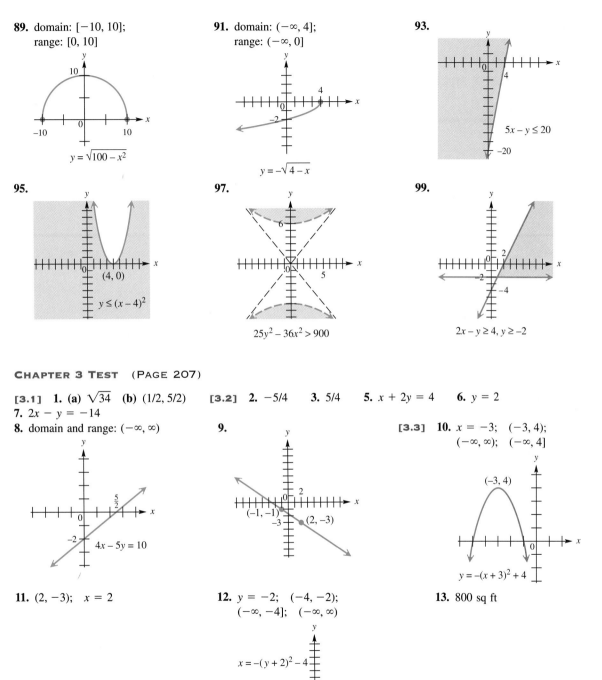

91. domain: $(-\infty, 4]$;
range: $(-\infty, 0]$

93.

$$y = \sqrt{100 - x^2}$$

$$y = -\sqrt{4 - x}$$

$$5x - y \leq 20$$

95.

$$y \leq (x - 4)^2$$

$(4, 0)$

97.

$$25y^2 - 36x^2 > 900$$

99.

$$2x - y \geq 4, y \geq -2$$

CHAPTER 3 TEST (PAGE 207)

[3.1] **1. (a)** $\sqrt{34}$ **(b)** $(1/2, 5/2)$ **[3.2]** **2.** $-5/4$ **3.** $5/4$ **5.** $x + 2y = 4$ **6.** $y = 2$
7. $2x - y = -14$

8. domain and range: $(-\infty, \infty)$

9.

[3.3] **10.** $x = -3$; $(-3, 4)$;
$(-\infty, \infty)$; $(-\infty, 4]$

$$4x - 5y = 10$$

$(-1, -1)$ $(2, -3)$

$(-3, 4)$

$$y = -(x + 3)^2 + 4$$

11. $(2, -3)$; $x = 2$

12. $y = -2$; $(-4, -2)$;
$(-\infty, -4]$; $(-\infty, \infty)$

13. 800 sq ft

$$x = -(y + 2)^2 - 4$$

$(-4, -2)$

[3.4] **14.** $(x - 5)^2 + (y + 1)^2 = 16$;
[1, 9];   [−5, 3]

15. center: $(−2, 3)$;   radius: 3
16. origin only

[3.5] **17.** domain: $[−5, −1]$;
range: $[−1, 5]$

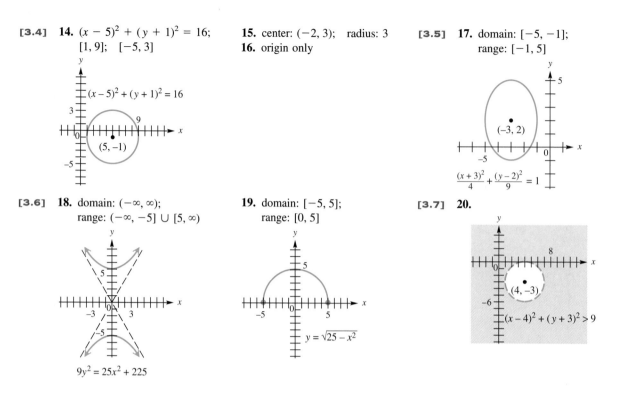

[3.6] **18.** domain: $(−∞, ∞)$;
range: $(−∞, −5] \cup [5, ∞)$

19. domain: $[−5, 5]$;
range: $[0, 5]$

[3.7] **20.**

■ CHAPTER 4   FUNCTIONS

4.1 EXERCISES   (PAGE 216)

1. (a) 0 (b) 4 (c) 2 (d) 4   **3.** (a) −3 (b) −2 (c) 0 (d) 2   **5.** −4   **7.** 11   **9.** $3a − 1$ **11.** 3
13. 72   **15.** $−6m − 1$ **17.** $15a − 7$ **19.** 2, 5   **21.** $[−6, ∞)$; $[0, ∞)$ **23.** $(−∞, ∞)$; $(−∞, ∞)$
25. $[−4, 3]$; $[−5, 6]$ **27.** $(−∞, 0) \cup (0, ∞)$; $(−∞, 0) \cup (0, ∞)$ **29.** $(−∞, ∞)$; $(−∞, ∞)$
31. $(−∞, ∞)$; $[0, ∞)$ **33.** $[−8, ∞)$; $[0, ∞)$ **35.** $[−4, 4]$; $[0, 4]$ **37.** $(−∞, ∞)$; $(−∞, ∞)$
39. $(−∞, 3) \cup (3, ∞)$; $(−∞, 0) \cup (0, ∞)$ **41.** $(−∞, −2) \cup (−2, 2) \cup (2, ∞)$; $(−∞, 0) \cup (0, ∞)$
43. (a) $6x + 6h + 2$ (b) $6h$ (c) 6   **45.** (a) $−2x − 2h + 5$ (b) $−2h$ (c) −2   **47.** (a) $1 − x^2 − 2xh − h^2$
(b) $−2xh − h^2$ (c) $−2x − h$ **49.** (a) $8 − 3x^2 − 6xh − 3h^2$ (b) $−6xh − 3h^2$ (c) $−6x − 3h$ **51.** 16
53. 1/16   **55.** $2^{−5r}$ or $1/2^{5r}$ **57.** $2^{1/4}$ or $\sqrt[4]{2}$ **59.** (a) $C(x) = 10x + 500$ (b) $R(x) = 35x$
(c) $P(x) = 25x − 500$ (d) 20 units; do not produce   **61.** (a) $C(x) = 150x + 2700$ (b) $R(x) = 280x$
(c) $P(x) = 130x − 2700$ (d) 20.77 or 21 units; produce   **63.** (a) 25 units (b)
(c) $6000

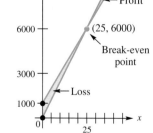

4.2 EXERCISES (PAGE 222)

1. $a = kb$ **3.** $x = k/y$ **5.** $r = kst$ **7.** $w = kx^2/y$ **9.** 220/7 **11.** 32/15 **13.** 18/125
15. increases, decreases **17.** inverse **19.** joint **21.** combined **23.** 16 in **25.** 875/72 candela
27. 450/11 km **29.** 799.5 cm^3 **31.** 1024/9 kg **33.** 7500 lb **35.** 45/7 **37.** 1600 calls **39.** 4.94
41. 7.4 km **43.** 92, undernourished

4.3 EXERCISES (PAGE 229)

1. (a) $10x + 2$ **(b)** $-2x - 4$ **(c)** $24x^2 + 6x - 3$ **(d)** $\dfrac{4x - 1}{6x + 3}$ **(e)** All domains are $(-\infty, \infty)$, except for f/g which

is $(-\infty, -1/2) \cup (-1/2, \infty)$. **3. (a)** $4x^2 - 4x + 1$ **(b)** $2x^2 - 1$ **(c)** $(3x^2 - 2x)(x^2 - 2x + 1)$ **(d)** $\dfrac{3x^2 - 2x}{x^2 - 2x + 1}$

(e) All domains are $(-\infty, \infty)$, except for f/g which is $(-\infty, 1) \cup (1, \infty)$. **5. (a)** $\sqrt{2x + 5} + \sqrt{4x - 9}$

(b) $\sqrt{2x + 5} - \sqrt{4x - 9}$ **(c)** $\sqrt{(2x + 5)(4x - 9)}$ **(d)** $\sqrt{\dfrac{2x + 5}{4x - 9}}$ **(e)** All domains are $[9/4, \infty)$, except for f/g which

is $(9/4, \infty)$. **7.** 55 **9.** 1848 **11.** $-6/7$ **13.** $4m^2 - 10m - 1$ **15.** 1122 **17.** 97
19. $256k^2 + 48k + 2$ **21.** $24x + 4$; $24x + 35$; both domains are $(-\infty, \infty)$ **23.** $-64x^3 + 2$; $-4x^3 + 8$;
both domains are $(-\infty, \infty)$ **25.** $1/x^2$; $1/x^2$; both domains are $(-\infty, 0) \cup (0, \infty)$ **27.** $2\sqrt{2x - 1}$;

$8\sqrt{x + 2} - 6$; domain of $f \circ g$ $[1/2, \infty)$; domain of $g \circ f$ $[-2, \infty)$ **29.** $\dfrac{x}{2 - 5x}$; $2(x - 5)$;

domain of $f \circ g$ $(-\infty, 0) \cup \left(0, \dfrac{2}{5}\right) \cup \left(\dfrac{2}{5}, \infty\right)$; domain of $g \circ f$ $(-\infty, 5) \cup (5, \infty)$ **31.** $\sqrt{\dfrac{x - 1}{x}}$; $-\dfrac{1}{\sqrt{x + 1}}$;

domain of $f \circ g$ $(-\infty, 0) \cup [1, \infty)$; domain of $g \circ f$ $(-1, \infty)$ Other correct answers are possible for Exercises

41–45. **41.** $f(x) = x^2$; $g(x) = 6x - 2$ **43.** $f(x) = \sqrt{x}$; $g(x) = x^2 - 1$ **45.** $f(x) = \dfrac{x^2 + 1}{5 - x^2}$;

$g(x) = x - 2$ **47. (a)** $N(x) = 100 - x$ **(b)** $G(x) = 2 + .2x$ **(c)** $C(x) = (100 - x)(2 + .2x)$
49. $18a^2 + 24a + 9$; It will decrease. **51. (a)** $4\pi t^2$

4.4 EXERCISES (PAGE 236)

1. domain $(-\infty, \infty)$; range $(-\infty, \infty)$ **3.** domain $(-\infty, \infty)$; range $[0, \infty)$ **5.** domain $(-\infty, \infty)$; range $[4, \infty)$

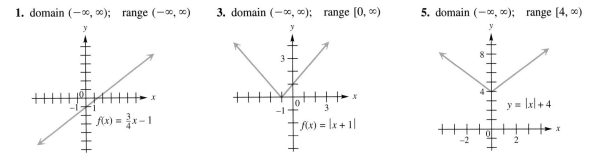

7. domain $(-\infty, \infty)$; range $[1, \infty)$

$y = 3|x - 2| + 1$

9. domain $(-\infty, \infty)$; range $(-\infty, 2]$

$f(x) = -|x + 1| + 2$

11. domain $(-\infty, \infty)$; range $[0, \infty)$

$f(x) = |x| + x$

15. (a) -10 **(b)** -2 **(c)** -1 **(d)** 2 **(e)** 4   **17. (a)** -10 **(b)** 2 **(c)** 5 **(d)** 3 **(e)** 5

19.

$f(x) = \begin{cases} x - 1 \text{ if } x \le 3 \\ 2 \text{ if } x > 3 \end{cases}$

21.

$f(x) = \begin{cases} 4 - x \text{ if } x < 2 \\ 1 + 2x \text{ if } x \ge 2 \end{cases}$

23.

$f(x) = \begin{cases} 2 + x \text{ if } x < -4 \\ -x \text{ if } -4 \le x \le 5 \\ 3x \text{ if } x > 5 \end{cases}$

25.

$f(x) = \begin{cases} |x| \text{ if } x > -2 \\ x \text{ if } x \le -2 \end{cases}$

29.

$f(x) = [\![-x]\!]$

31.

$f(x) = [\![3x + 1]\!]$

33.

$f(x) = [\![3x]\!] + 1$

35. (a) 30¢ **(b)** 57¢ **(c)** 111¢ **(d)**
(e) domain $(0, \infty)$;
range $\{30, 57, 84, 111, \ldots\}$

37.

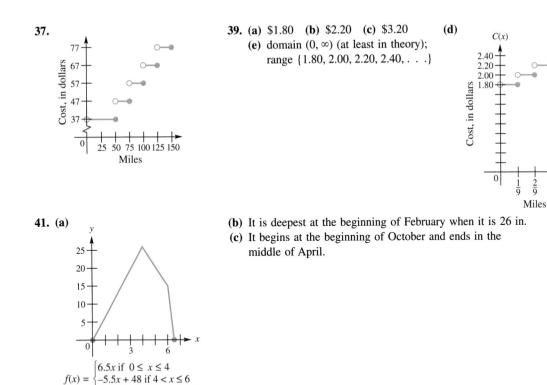

39. (a) $1.80 **(b)** $2.20 **(c)** $3.20 **(d)**
(e) domain $(0, \infty)$ (at least in theory);
range $\{1.80, 2.00, 2.20, 2.40, \ldots\}$

41. (a)

$$f(x) = \begin{cases} 6.5x \text{ if } 0 \le x \le 4 \\ -5.5x + 48 \text{ if } 4 < x \le 6 \\ -30x + 195 \text{ if } 6 < x \le 6.5 \end{cases}$$

(b) It is deepest at the beginning of February when it is 26 in.
(c) It begins at the beginning of October and ends in the middle of April.

4.5 EXERCISES (PAGE 244)

1. one-to-one   **3.** one-to-one   **5.** not one-to-one   **7.** one-to-one   **9.** not one-to-one   **11.** not one-to-one
13. one-to-one   **15.** one-to-one   **17.** not one-to-one   **19.** untying your shoelaces   **21.** leaving a room
23. landing in an airplane   **27.** inverses   **29.** not inverses   **31.** not inverses   **33.** inverses   **35.** inverses
37. inverses   **39.** inverses   **41.** not inverses

43.   **45.**   **47.**

A-22

49. $f^{-1}(x) = \dfrac{x+4}{3}$

51. $f^{-1}(x) = 3x$

53. $f^{-1}(x) = \sqrt[3]{x-1}$

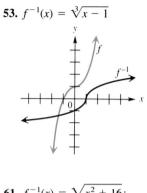

55. not one-to-one

57. $f^{-1}(x) = 1/x$

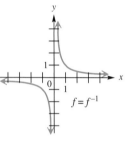

59. $f^{-1}(x) = -\sqrt{4-x}$;
domain $(-\infty, 4]$

61. $f^{-1}(x) = \sqrt{x^2 + 16}$;
domain $(-\infty, 0]$

CHAPTER 4 REVIEW EXERCISES (PAGE 248)

1. $(-\infty, \infty)$; $(-\infty, \infty)$ **3.** $(-\infty, \infty)$; $[-4, \infty)$ **5.** $(-\infty, \infty)$; $(-\infty, 0]$ **7.** $[-7, 7]$; $[0, 7]$
9. $(-\infty, \infty)$; $(0, 1/3]$ **11.** $(-\infty, 7) \cup (7, \infty)$ **13.** $(-2, 3)$ **15.** 2   **17.** 14   **19.** -25

21. $15y - 7$ **23.** 0   **25.** $-3x^2 - 3xh - h^2 + 4x + 2h$ **27.** $y = \dfrac{kp^3}{r}$ **29.** $A = \dfrac{kt^3s^4}{ph^2}$

31. 847   **33.** 33,750 units   **35.** 640/9 kg   **37.** $3x^4 - 9x^3 - 16x^2 + 12x + 16$ **39.** 68
41. $3(1+r)^4 - 9(1+r)^3 - 16(1+r)^2 + 12(1+r) + 16 = 3r^4 + 3r^3 - 25r^2 - 35r + 6$ **43.** undefined
45. $(-\infty, -1) \cup (-1, 4) \cup (4, \infty)$ **47.** $\sqrt{x^2 - 2}$ **49.** $\sqrt{34}$ **51.** 1

55.

57.

59.

61.

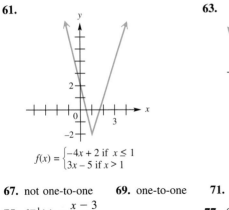

$$f(x) = \begin{cases} -4x + 2 \text{ if } x \le 1 \\ 3x - 5 \text{ if } x > 1 \end{cases}$$

63.

$$f(x) = \begin{cases} |x| \text{ if } x < 3 \\ 6 - x \text{ if } x \ge 3 \end{cases}$$

65. domain $(0, \infty)$;
range $\{47, 49, 51, \ldots\}$

67. not one-to-one   **69.** one-to-one   **71.** one-to-one   **73.** one-to-one

75. $f^{-1}(x) = \dfrac{x - 3}{12}$

77. $f^{-1}(x) = \sqrt[3]{x + 3}$

79. $f^{-1}(x) = \sqrt{25 - x^2}$;
domain $[0, 5]$

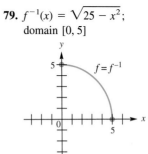

CHAPTER 4 TEST (PAGE 251)

[4.1] **1.** $(-\infty, \infty)$; $[3, \infty)$ **2.** $(-\infty, -4] \cup [-3, \infty)$; $[0, \infty)$ **3.** $(-\infty, -4) \cup (-4, \infty)$ **[4.3]** **4.** 8
5. $9 - 2a - 4a^2$ **6.** 2/5   **7.** $y - 5$ **8.** -5 **9.** $15 - 6p - 2p^2$

[4.4] **10.**

11.

12.

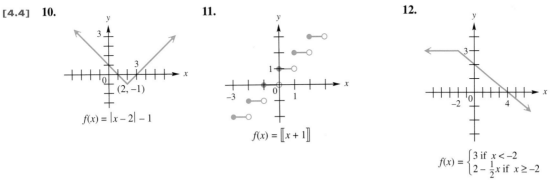

$f(x) = |x - 2| - 1$

$f(x) = [\![x + 1]\!]$

$$f(x) = \begin{cases} 3 \text{ if } x < -2 \\ 2 - \frac{1}{2}x \text{ if } x \ge -2 \end{cases}$$

[4.5] **13.** not one-to-one   **14.** one-to-one   **15.** one-to-one   **16.** $f^{-1}(x) = \dfrac{x - 8}{2}$
17. $f^{-1}(x) = -\sqrt{x - 2}$; domain $[2, \infty)$ **[4.2]** **19. (a)** $y = 6\sqrt{x}$ **(b)** 72   **20.** 36 lb

5.1 EXERCISES (PAGE 263)

1. $\{1/2\}$ **3.** $\{-2\}$ **5.** $\{0\}$ **7.** $\{3\}$ **9.** $\{-8, 8\}$ **11.** $\{1/5\}$ **13.** $\{3/5\}$ **15.** $\{-2/3\}$

17. (a) **(b)** **(c)**

(d) **21.** **23.**

25. **27.** **29.** If $a > 1$, the function value increases. If $0 < a < 1$, the function value decreases.
31. $(0, 1)$ and $(1, a)$

33. **35.** **37.**

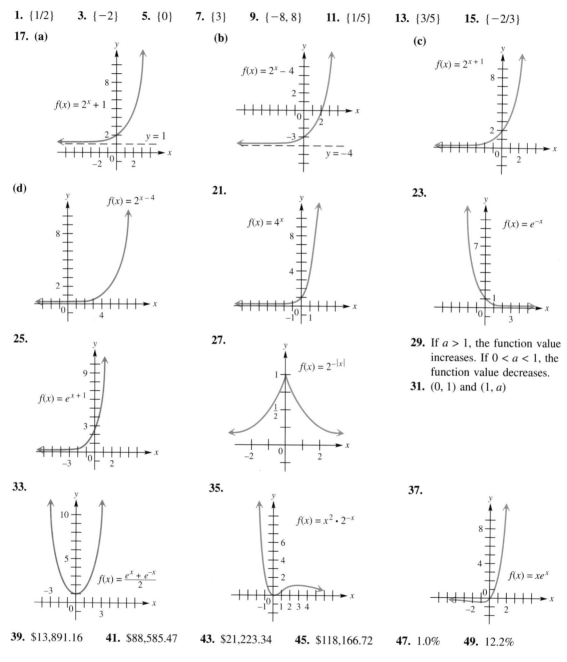

39. \$13,891.16 **41.** \$88,585.47 **43.** \$21,223.34 **45.** \$118,166.72 **47.** 1.0% **49.** 12.2%

51. (a) 7.229% **(b)** 7.250% **53. (a)** 440 g **(b)** 387 g **(c)** 264 g **(d)**

$A(t) = 500e^{-.032t}$

55. (a) about 45,200; about 37,000 **(b)** about 72,400; about 48,500 **57.** 2.717

5.2 EXERCISES (PAGE 272)

1. $\log_3 81 = 4$ **3.** $\log_{1/2} 16 = -4$ **5.** $\log_{10} .0001 = -4$ **7.** $6^2 = 36$ **9.** $(\sqrt{3})^8 = 81$
11. $10^{-4} = .0001$ **13.** 2 **15.** 1 **17.** -3 **19.** $-1/6$ **21.** 9 **23.** 9 **25.** {5} **27.** {1/5}

31. (a) **(b)** **(c)**

33. **35.** **37.**

39. **41.** $\log_3 2 - \log_3 5$ **43.** $\log_2 6 + \log_2 x - \log_2 y$

45. $1 + (1/2) \log_5 7 - \log_5 3$ **47.** no change possible **49.** $\log_k p + 2 \log_k q - \log_k m$
51. $(1/2)(\log_m 5 + 3 \log_m r - 5 \log_m z)$ **53.** $\log_a \dfrac{xy}{m}$ **55.** $\log_m \dfrac{a^2}{b^6}$ **57.** $\log_a [(z - 1)^2(3z + 2)]$
59. $\log_5 \dfrac{5^{1/3}}{m^{1/3}}$ **63.** 1.0791 **65.** $-.1303$ **67.** .5187

5.3 EXERCISES (PAGE 278)

1. 1.6335   **3.** 2.8938   **5.** −2.1612   **7.** 6.3630   **9.** −.3567   **11.** 11.7035   **13.** 1.43   **15.** .59
17. .96   **19.** 1.89   **21.** $\log_3 4$ **23.** 3.2   **25.** 1.8   **27.** 2×10^{-3} **29.** 1.6×10^{-5} **31.** (a) 20
(b) 30   (c) 50   (d) 60   **33.** (a) 3   (b) 6   (c) 8   **35.** (a) about $200,000,000 I_0$ (b) about $13,000,000 I_0$
(c) The 1906 earthquake was more than 15 times as intense as the 1989 earthquake.   **37.** (a) 2.03   (b) 2.28
(c) 2.17   (d) 1.21   **39.** (a) 11%   (b) 36%   (c) 84%   (d)

41. 1.59   **43.**

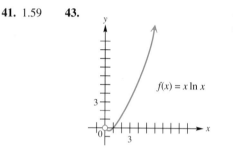

45. (a) 3   (b) $5^2 = 25$ (c) $1/e$
47. (a) 5   (b) ln 3   (c) 2 ln 3   or   ln 9

5.4 EXERCISES (PAGE 285)

1. {1.631}   **3.** {−.535}   **5.** {−.080}   **7.** {2.386}   **9.** {−.123}   **11.** ∅   **13.** {2}   **15.** {.5}
17. {4}   **19.** {17.475}   **21.** {11}   **23.** {5}   **25.** {10}   **27.** {4}   **29.** ∅   **31.** {11}   **33.** {3}
35. ∅   **37.** {8}   **39.** {−2, 2}   **41.** {1, 10}   **45.** $t = \dfrac{1000}{k} \ln \dfrac{P}{P_0}$ **47.** $t = e^{(p-r)/k}$

49. $t = -\dfrac{1}{k} \log \dfrac{T - T_0}{T_1 - T_0}$ **51.** $I = I_0 \cdot 10^{d/10}$ **53.** 4.9 yr   **55.** 1.8 yr   **57.** 3.8 yr   **59.** (a) about 18 days

(b) January 19th   **61.** $t = \dfrac{10^{F/500} - 3}{2}$ (a) 6.4 mo   (b) 48.5 mo

5.5 EXERCISES (PAGE 290)

1. (a) $10.94   (b) $11.27   (c) $11.62   **3.** 7% compounded quarterly; $800.31   **5.** 11.6 yr   **7.** 9.3%
9. (a) 11   (b) 12.6   (c) 18   (d)   (e) They are growing exponentially, but at a slow rate.

11. 13.1 yr   **13.** 7.97 yr   **15. (a)** about 961,000   **(b)** about 7.2 years   **(c)** about 17.3 yr   **17.** about 4200 yr

19. $t = \dfrac{\ln (A/A_0)}{k}$ **21.** about 13,000 yr   **23.** 46 days   **25.** 30 min

CHAPTER 5 REVIEW EXERCISES (PAGE 293)

1. {5/3}   **3.** {3/2}   **5.** 655 g   **7. (a)**   **9. (d)**

11.   **13.**

15. decreasing   **17.** $\log_2 32 = 5$ **19.** $\log_{1/16} 1/2 = 1/4$ **21.** $\log_{10} 3 = .4771$ **23.** $10^{-3} = .001$
25. $10^{.537819} = 3.45$ **27.** $\log_3 m + \log_3 n - \log_3 5 - \log_3 r$ **29.** $2 \log_5 x + 4 \log_5 y + (1/5)(3 \log_5 m + \log_5 p)$
31. 1.659   **33.** −3.252   **35.** 3.555   **37.** 11.878   **39.** 2.255   **41.** 6.049   **43.** .8 m   **45.** 1.5 m
47. $1999   **49.** $1293.68   **51.** $93,761.31   **53.** {1.490}   **55.** {1.303}   **57.** {4}   **59.** {3}
61. {−3}   **63.** 7.8%   **65.** 4.0 yr   **67.** about 17.3 yr   **69.** $3261.94   **71.** about 35 yr
73. about 1.126 billion yr

CHAPTER 5 TEST (PAGE 295)

[5.1] **1.** {5}   **2.** {1/32}   **[5.2]** **3.** $\log_a b = 2$ **4.** $\ln 4.82 = c$ **5.** $3^{3/2} = \sqrt{27}$ **6.** $e^a = 5$

7. $(a, 1)$ and $(1, 0)$ **[5.1]** **8.**   **[5.2]** **9.**

10. $2 \log_7 x + \dfrac{1}{4} \log_7 y - 3 \log_7 z$ **[5.3]** **11.** 7.741   **12.** 4.581   **[5.4]** **13.** $(-\infty, 3/2]$ **14.** {3.107}
15. {7}   **16.** {20.125}   **[5.5]** **17. (a)** 329.3 g   **(b)** 13.9 days   **18.** $6529.24   **19.** 17.3 yr   **20.** 10.7 yr

6.1 EXERCISES (PAGE 304)

1.

$P(x) = \frac{1}{4}x^6$

3.

$P(x) = -\frac{5}{4}x^5$

5.

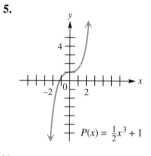

$P(x) = \frac{1}{2}x^3 + 1$

7.

$P(x) = -(x+1)^3$

9.

$P(x) = (x-1)^4 + 2$

11.

$P(x) = 2x(x-3)(x+2)$

13.

$P(x) = x^2(x-2)(x+3)^2$

15.

$P(x) = x^3 - x^2 - 2x$

17.

$P(x) = (x+2)(x-1)(x+1)$

19.

$P(x) = (3x-1)(x+2)^2$

21.

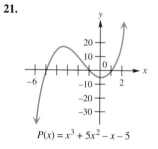

$P(x) = x^3 + 5x^2 - x - 5$

23. (c)

25. (a)

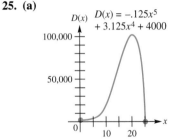

$D(x)$ $D(x) = -.125x^5 + 3.125x^4 + 4000$

(b) 1910 to 1925;   1905 to 1910;   1925 to 1930

29. $P(x) = x^2$ (There are many others.)   **31.** maximum is 26.136 when $x = -3.4$;   minimum is 25 when $x = -3$
33. maximum is 1.048 when $x = -.1$;   minimum is -5 when $x = -1$
35. maximum is 84 when $x = -2$;   minimum is -13 when $x = -1$ **37.**

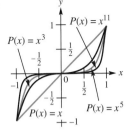

6.2 EXERCISES   (PAGE 312)

1. $2m^2 - 4m + 8$ **3.** $5y^2 - 3xy + 8x^2$ **5.** $3y + 2$ **7.** $4z - 3 + \dfrac{-5}{2z + 5}$ **9.** $x^2 - 3x + 2$

11. $5x^2 - 3x + 4 + \dfrac{4}{3x + 4}$ **13.** $x^2 + 2x + 3$ **15.** $2z^3 - z^2 - \dfrac{1}{2}z - \dfrac{5}{4} + \dfrac{-\frac{13}{4}z + \frac{17}{4}}{2z^2 + z + 1}$ **19.** $x^2 - 3x - 2$

21. $m^3 - m^2 - 6m$ **23.** $3x^2 + 4x + \dfrac{3}{x - 5}$ **25.** $x^4 + x^3 + 2x - 1 + \dfrac{3}{x + 2}$ **27.** $\dfrac{1}{3}x^2 - \dfrac{1}{9}x + \dfrac{1}{x - \frac{1}{3}}$

29. $y^2 + y + 1$ **31.** $x^3 + x^2 + x + 1$ **33.** $P(x) = (x - 1)(x^2 + 2x + 3) + (-5)$
35. $P(x) = (x + 2)(-x^2 + 4x - 8) + 20$ **37.** $P(x) = (x - 3)(x^3 + 2x + 5) + 20$ **39.** 2   **41.** -30
43. $-6 - i$ **45.** $-2 - 3i$ **47.** yes   **49.** yes   **51.** no   **53.** no   **55.** yes

6.3 EXERCISES   (PAGE 318)

1. no   **3.** yes   **5.** yes   **7.** no   **9.** 0: multiplicity 4; -3: multiplicity 5;   8: multiplicity 2

11. 7/4: multiplicity 3;   5: multiplicity 1   **13.** i: multiplicity 4; $-i$: multiplicity 4   **15.** $P(x) = -\dfrac{1}{6}x^3 + \dfrac{13}{6}x + 2$

17. $P(x) = -\dfrac{1}{2}x^3 - \dfrac{1}{2}x^2 + x$ **19.** $P(x) = -10x^3 + 30x^2 - 10x + 30$ **21.** $P(x) = x^2 - 10x + 26$

23. $P(x) = x^3 - 4x^2 + 6x - 4$ **25.** $P(x) = x^3 - 3x^2 + x + 1$ **27.** $P(x) = x^4 - 6x^3 + 10x^2 + 2x - 15$
29. $P(x) = x^3 - 8x^2 + 22x - 20$ **31.** $P(x) = x^4 - 10x^3 + 42x^2 - 82x + 65$
33. $P(x) = x^5 - 12x^4 + 74x^3 - 248x^2 + 445x - 500$ **35.** $P(x) = x^4 - 6x^3 + 17x^2 - 28x + 20$
37. $-1 + i, -1 - i$ **39.** $-\dfrac{1}{2} + \dfrac{\sqrt{5}}{2}i, -\dfrac{1}{2} - \dfrac{\sqrt{5}}{2}i$ **41.** $i, 2i, -2i$ **43.** $3, -2, 1 - 3i$
45. $P(x) = (x - 2)(2x - 5)(x + 3)$ **47.** $P(x) = (x + 4)(3x - 1)(2x + 1)$ **49.** $P(x) = (x - 3i)(x + 4)(x + 3)$
51. $P(x) = (x - 1 - i)(2x - 1)(x + 3)$ **53.** zeros are $-2, -1, 3$; $P(x) = (x + 2)^2(x + 1)(x - 3)$
55. 1, 3, or 5   **59.** at 3 seconds

A-30 **6.4 EXERCISES** (PAGE 323)

1. $\pm 1, \pm 1/2, \pm 1/3, \pm 1/6$ **3.** $\pm 1, \pm 1/2, \pm 1/3, \pm 1/4, \pm 1/6, \pm 1/12, \pm 2, \pm 2/3$ **5.** $\pm 1, \pm 1/2, \pm 2, \pm 4, \pm 8$
7. $\pm 1, \pm 2, \pm 5, \pm 10, \pm 25, \pm 50$ **11.** $-1, -2, 5$ **13.** $2, -3, -5$ **15.** no rational zeros
17. $1, -2, -3, -5$ **19.** $-4, 3/2, -1/3$; $P(x) = (x + 4)(2x - 3)(3x + 1)$ **21.** $-3/2, -2/3, 1/2$;
$P(x) = (2x - 1)(3x + 2)(2x + 3)$ **23.** $1/2$; $P(x) = (2x - 1)(x^2 + 4x + 8)$ **25.** $-2, -1, 1/2, 1$;
$P(x) = (x + 2)(x + 1)(2x - 1)(x - 1)$ **27.** $-1, -2, -2/3, 2$; $P(x) = (x + 1)(x + 2)(3x + 2)(x - 2)$
29. $-2/3, -1, 3$ **31.** $1, -5/4$ **33.** 1 **35.** $-1, 3/2, 2$; $P(x) = (x + 1)(2x - 3)(x - 2)$

$P(x) = 2x^3 - 5x^2 - x + 6$

37. -2 (multiplicity 2), 3; $P(x) = (x + 2)^2(x - 3)$ **39.** -2 (multiplicity 2), 3; $P(x) = (x + 2)^2(-x + 3)$

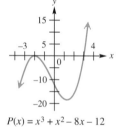

$P(x) = x^3 + x^2 - 8x - 12$

$P(x) = -x^3 - x^2 + 8x + 12$

41. 3 (multiplicity 2), -3 (multiplicity 2);   **43.** $-1/2, 2, -1$ (multiplicity 2);
     $P(x) = (x - 3)^2(x + 3)^2$ $P(x) = (2x + 1)(x - 2)(x + 1)^2$

$P(x) = x^4 - 18x^2 + 81$

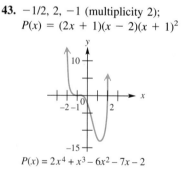

$P(x) = 2x^4 + x^3 - 6x^2 - 7x - 2$

6.5 EXERCISES (PAGE 331)

1. 2 or 0 positive real zeros;   1 negative real zero   **3.** 1 positive real zero;   1 negative real zero
5. 2 or 0 positive real zeros;   3 or 1 negative real zeros   **13.** P has a zero between 2 and 2.5.
23. (a) positive: 1;   negative: 2 or 0   **(b)** $-3, -1.4, 1.4$ **25. (a)** positive: 2 or 0;   negative: 1
(b) $-2, 1.6, 4.4$ **27. (a)** positive: 1;   negative: 0   **(b)** 1.5   **29. (a)** positive: 3 or 1;   negative: 1
(b) $1.1, -.7$ **31. (a)** positive: 1;   negative: 1   **(b)** $-1.5, 3.1$

33.

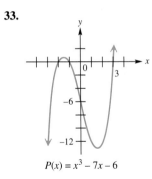

$P(x) = x^3 - 7x - 6$

35.

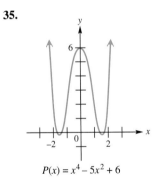

$P(x) = x^4 - 5x^2 + 6$

37.

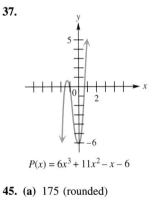

$P(x) = 6x^3 + 11x^2 - x - 6$

39.

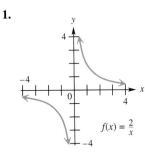

$P(x) = -x^3 + 6x^2 - x - 14$

41. 3.24, -1.24
43. -3.65, $-.32$, 1.65, 6.32

45. (a) 175 (rounded)
(b)

$g(x) = -.006x^4 + .140x^3 - .053x^2 + 1.79x$

47. for example, $P(x) = (x + 3)^2 (x - 2)^2 = x^4 + 2x^3 - 11x^2 - 12x + 36$

6.6 EXERCISES (PAGE 341)

1.

$f(x) = \frac{2}{x}$

3.

$f(x) = \frac{1}{x + 2}$

$x = -2$

5.

$y = 1$

$f(x) = \frac{1}{x} + 1$

7. (a)

$f(x) = \frac{1}{x^2}$

$f(x) = \frac{1}{(x - 3)^2}$

(b)

$f(x) = \frac{1}{x^2}$

$f(x) = -\frac{2}{x^2}$

(c)

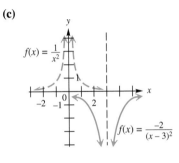

$f(x) = \frac{1}{x^2}$

$f(x) = \frac{-2}{(x - 3)^2}$

9. vertical asymptote: $x = 5$;   horizontal asymptote: $y = 0$ **11.** vertical asymptote: $x = 7/3$; horizontal asymptote: $y = 0$ **13.** vertical asymptote: $x = -2$;   horizontal asymptote: $y = -1$ **15.** vertical asymptote: $x = -9/2$;   horizontal asymptote: $y = 3/2$ **17.** vertical asymptotes: $x = 3, x = 1$; horizontal asymptote: $y = 0$ **19.** vertical asymptote: $x = -3$;   oblique asymptote: $y = x - 3$ **21.** vertical asymptotes: $x = -2, x = 5/2$;   horizontal asymptote: $y = 1/2$ **23.** (a)

25.

27.

29.

31.

33.

35.

37.

39.

41.

43.

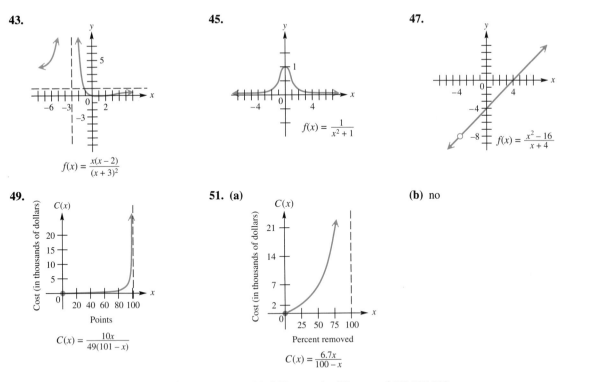

$$f(x) = \frac{x(x-2)}{(x+3)^2}$$

45.

$$f(x) = \frac{1}{x^2+1}$$

47.

$$f(x) = \frac{x^2-16}{x+4}$$

49.

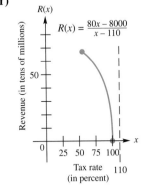

$$C(x) = \frac{10x}{49(101-x)}$$

51. (a)

$$C(x) = \frac{6.7x}{100-x}$$

(b) no

53. (a) \$65.5 tens of millions, or \$655,000,000   **(b)** \$64 tens of millions, or \$640,000,000
(c) \$60 tens of millions, or \$600,000,000   **(d)** \$40 tens of millions, or \$400,000,000   **(e)** \$0
(f)

$$R(x) = \frac{80x - 8000}{x - 110}$$

1.

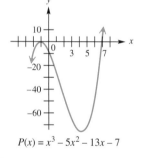

$P(x) = x^3 + 5$

3.

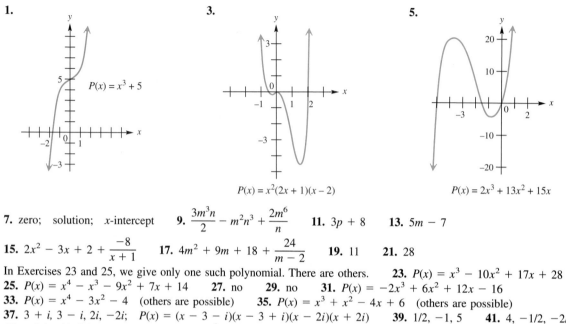

$P(x) = x^2(2x + 1)(x - 2)$

5.

$P(x) = 2x^3 + 13x^2 + 15x$

7. zero; solution; x-intercept   **9.** $\dfrac{3m^3 n}{2} - m^2 n^3 + \dfrac{2m^6}{n}$ **11.** $3p + 8$ **13.** $5m - 7$

15. $2x^2 - 3x + 2 + \dfrac{-8}{x + 1}$ **17.** $4m^2 + 9m + 18 + \dfrac{24}{m - 2}$ **19.** 11   **21.** 28

In Exercises 23 and 25, we give only one such polynomial. There are others.   **23.** $P(x) = x^3 - 10x^2 + 17x + 28$
25. $P(x) = x^4 - x^3 - 9x^2 + 7x + 14$ **27.** no   **29.** no   **31.** $P(x) = -2x^3 + 6x^2 + 12x - 16$
33. $P(x) = x^4 - 3x^2 - 4$ (others are possible)   **35.** $P(x) = x^3 + x^2 - 4x + 6$ (others are possible)
37. $3 + i,\ 3 - i,\ 2i,\ -2i;$ $P(x) = (x - 3 - i)(x - 3 + i)(x - 2i)(x + 2i)$ **39.** $1/2,\ -1,\ 5$ **41.** $4,\ -1/2,\ -2/3$
43. -1 (multiplicity 2), 7; $P(x) = (x + 1)^2\ (x - 7)$ **53.** $-2.3,\ 4.6$ **55.**

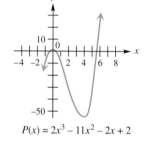

$P(x) = x^3 - 5x^2 - 13x - 7$

$P(x) = 2x^3 - 11x^2 - 2x + 2$

57.

$P(x) = x^3 + 3x^2 - 4x - 2$

59.

$f(x) = \dfrac{8}{x}$

61.

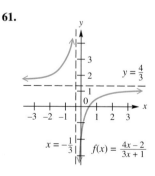

$y = \dfrac{4}{3}$

$x = -\dfrac{1}{3}$ $f(x) = \dfrac{4x - 2}{3x + 1}$

63.

$$f(x) = \frac{2x}{x^2 - 1}$$

65.

$$f(x) = \frac{x^2 - 1}{x}$$

67.

$$f(x) = \frac{4x^2 - 9}{2x + 3}$$

CHAPTER 6 TEST (PAGE 347)

[6.1] **1.**

2.

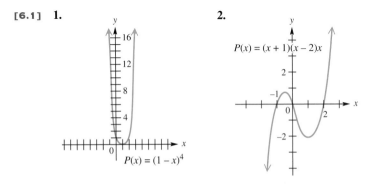

$$P(x) = (1 - x)^4$$

$$P(x) = (x + 1)(x - 2)x$$

[6.2] **3.** $3x^2 - 2x - 5 + \dfrac{16}{x + 2}$ **4.** $2x^2 - x - 5 + \dfrac{3}{x - 5}$ **5.** -7

[6.3] **6.** $P(x) = x^3 - x^2 - 10x - 8$ (others are possible)   **7.** Yes, because when $P(x)$ is divided by $x - 3$, the remainder is 0.   **8.** Yes, because $P(-2) = 0$.   **9.** $P(x) = 2x^4 - 2x^3 - 2x^2 - 2x - 4$

10. $P(x) = (x + 2)(2x + 1)(x - 3)$ **[6.4]** **11.** **(a)** $\pm 1, \pm 1/2, \pm 1/3, \pm 1/6, \pm 7, \pm 7/2, \pm 7/3, \pm 7/6$

(b) $-1/3, 1, 7/2$ **[6.5]** **14.** positive: 2 or 0;   negative: 2 or 0

15.

$P(x) = x^3 - 5x^2 + 8x - 4$

[6.6] **16.**

$$f(x) = \frac{-2}{x + 3}$$

17.

$$f(x) = \frac{3x - 1}{x - 2}$$

18.

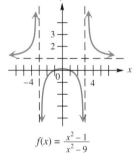

$$f(x) = \frac{x^2 - 1}{x^2 - 9}$$

19. $y = 2x + 3$ **[6.1, 6.5]** **20.** (d)

■ CHAPTER 7   SYSTEMS OF EQUATIONS AND INEQUALITIES

7.1 EXERCISES (PAGE 357)

1. single solution **3.** infinitely many solutions **5.** no solution **7.** infinitely many solutions **9.** $\{(6, 15)\}$
11. $\{(2, -3)\}$ **13.** $\{(1, 2)\}$ **15.** Multiply equation (4) by 3 and equation (5) by 4. **17.** $\{(-1, 4)\}$ **19.** $\emptyset$
21. $\{(1, 3)\}$ **23.** $\{(4, -2)\}$ **25.** $\left\{\left(\dfrac{y + 9}{4}, y\right)\right\}$ **27.** $\{(12, 6)\}$ **29.** $\{(2, 2)\}$ **31.** $\{(1/5, 1)\}$
33. $a = -3$; $b = -1$ **35.** $m = 1/3$; $b = 13/3$ **37.** \$12,000 at 7\%; \$4000 at 8\% **39.** 480 kg of X;
320 kg of Y **41.** 40/3 gal of 98 octane; 80/3 gal of 92 octane **43.** \$3 for a turkey; \$1.75 for a chicken
45. 32 days at \$88 per day **47.** \$18,000 at 4.5\%; \$12,000 at 5\% **49.** $x = 63$; $R = C = 346.5$
51. $x = 2100$; $R = C = 52,000$ **53.** 32; 80 **55.** 28/3; 28 **57.** (a) 16 (b) 11 (c) 6 (d) 8 (e) 4

(f) 0 **(g), (k)**   **(h)** 0   **59. (a)**   **(b)** 125 **(c)** 50
(i) 40/3
(j) 80/3
(l) 8
(m) 6

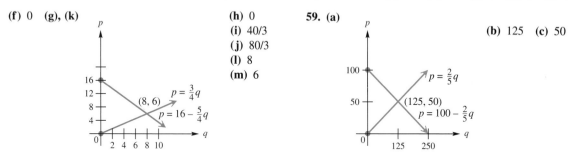

7.2 EXERCISES (PAGE 365)

1. $\{(1, 2, -1)\}$ **3.** $\{(2, 0, 3)\}$ **5.** $\emptyset$ **7.** $\{(1, 2, 3)\}$ **9.** $\{(-1, 2, 1)\}$ **11.** $\{(4, 1, 2)\}$
13. $\{(1/2, 2/3, -1)\}$ **15.** $\{(2, 4, 2)\}$ **17.** $\{(-1, 1, 1/3)\}$ **19. (a)** for example: $x + 2y + z = 5$,
$2x - y + 3z = 4$ (There are others.) **(b)** for example: $x + y + z = 5$, $2x - y + 3z = 4$ (There are others.)
(c) for example: $2x + 2y + 2z = 8$, $2x - y + 3z = 4$ (There are others.) **21.** For example, the ceiling and two
perpendicular walls of a standard room meet in a point. **23.** $\{(x, -4x + 3, -3x + 4)\}$
25. $\{(x, -x + 15, -9x + 69)\}$ **27.** $\left\{\left(x, \dfrac{-15 + x}{3}, \dfrac{24 + x}{3}\right)\right\}$ **29.** $a = 3/4$; $b = 1/4$; $c = -1/2$
31. $x^2 + y^2 + x - 7y = 0$ **33.** 12 cents; 13 nickels; 4 quarters **35.** 30 barrels each of \$150 and \$190 glue;
210 barrels of \$120 glue **37.** 28 in; 17 in; 14 in **39.** \$50,000 at 5\%; \$10,000 at 4.5\%; \$40,000 at 3.75\%

7.3 EXERCISES (PAGE 374)

1. $\begin{bmatrix} 2 & 4 \\ 0 & -1 \end{bmatrix}$ **3.** $\begin{bmatrix} 1 & 5 & 6 \\ 0 & 13 & 11 \\ 4 & 7 & 0 \end{bmatrix}$ **5.** $\begin{bmatrix} -3 & 1 & -4 \\ 2 & 1 & 3 \\ -17 & 0 & -13 \end{bmatrix}$ **7.** $\left[\begin{array}{cc|c} 2 & 3 & 11 \\ 1 & 2 & 8 \end{array}\right]$ **9.** $\left[\begin{array}{cc|c} 1 & 5 & 6 \\ 1 & 2 & 8 \end{array}\right]$

11. $\left[\begin{array}{ccc|c} 2 & 1 & 1 & 3 \\ 3 & -4 & 2 & -7 \\ 1 & 1 & 1 & 2 \end{array}\right]$ **13.** $\left[\begin{array}{ccc|c} 1 & 1 & 0 & 2 \\ 0 & 2 & 1 & -4 \\ 0 & 0 & 1 & 2 \end{array}\right]$ **15.** $2x + y = 1;\quad 3x - 2y = -9$ **17.** $x = 2;\quad y = 3;$

$z = -2$ **19.** $3x + 2y + z = 1;\quad 2y + 4z = 22;\quad -x - 2y + 3z = 15$ **21.** $\{(2, 3)\}$ **23.** $\{(-3, 0)\}$

25. $\{(7/2, -1)\}$ **27.** $\emptyset$ **29.** $\left\{\left(\dfrac{3y + 1}{6}, y\right)\right\}$ **31.** $\{(-2, 1, 3)\}$ **33.** $\{(-1, 23, 16)\}$ **35.** $\{(3, 2, -4)\}$

37. $\{(2, 1, -1)\}$ **41.** $\{(-1, 2, 5, 1)\}$ **43.** \$3000 at 3%; \$1000 at 4%; \$6000 at 4.5% **45.** \$1.40 for

premium; \$1.20 for regular; \$1.30 for super **47.** $\left\{\left(\dfrac{5z + 14}{5}, \dfrac{5z - 12}{5}, z\right)\right\}$ **49.** $\left\{\left(\dfrac{12 - z}{7}, \dfrac{4z - 6}{7}, z\right)\right\}$

51. $\emptyset$ **53. (a)** $x_3 + x_4 = 600$ **(b)** The system has no unique solution; equations $x_2 + x_3 = 700$ and
$x_3 + x_4 = 600$ are dependent. **(c)** $x_2 - x_4 = 100$ **(d)** $x_4 = 1000 - x_1$; 1000 **(e)** $x_4 = x_2 - 100$; 100
(f) $x_4 = 600 - x_3$; 600, 600 **(g)** 600; 600; 700; 1000

7.4 EXERCISES (PAGE 382)

1.

3.

5.

7.

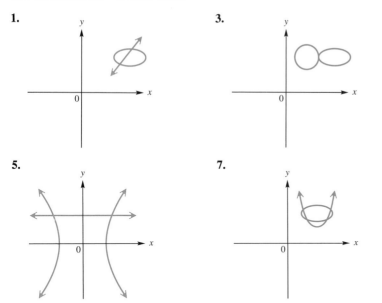

11. $\{(1, 1), (-2, 4)\}$ **13.** $\{(2, 1), (1/3, 4/9)\}$ **15.** $\{(2, 12), (-4, 0)\}$ **17.** $\{(-3/5, 7/5), (-1, 1)\}$
19. $\{(2, 2), (2, -2), (-2, 2), (-2, -2)\}$ **21.** $\{(0, 0)\}$ **23.** $\{(i, \sqrt{6}), (-i, \sqrt{6}), (i, -\sqrt{6}), (-i, -\sqrt{6})\}$
25. $\{(1, -1), (-1, 1), (1, 1), (-1, -1)\}$ **27.** $\{(x, \pm\sqrt{10 - x^2})\}$
29. $\{(i\sqrt{15}/5, 2\sqrt{10}/5), (i\sqrt{15}/5, -2\sqrt{10}/5), (-i\sqrt{15}/5, 2\sqrt{10}/5), (-i\sqrt{15}/5, -2\sqrt{10}/5)\}$
31. $\{(-3, 5), (15/4, -4)\}$ **33.** $\{(4, -1/8), (-2, 1/4)\}$ **35.** $\{(3, 2), (-3, -2), (4, 3/2), (-4, -3/2)\}$
37. $\{(3, 5), (-3, -5), (5i, -3i), (-5i, 3i)\}$ **39.** $\{(\sqrt{5}, 0), (-\sqrt{5}, 0), (\sqrt{5}, \sqrt{5}), (-\sqrt{5}, -\sqrt{5})\}$
41. $\{(3, -3), (3, 3)\}$ **43.** $\left\{\left(\dfrac{1 + \sqrt{13}}{2}, \dfrac{-1 + \sqrt{13}}{2}\right), \left(\dfrac{-1 - \sqrt{21}}{2}, \dfrac{3 + \sqrt{21}}{2}\right)\right\}$ **45.** 27 and 6 or -27 and -6
47. yes **49.** $y = 4x - 4$ **51.** $a \neq 5, a \neq -5$

1.

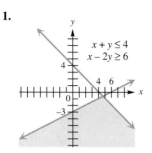

$$x + y \le 4$$
$$x - 2y \ge 6$$

3.

$$4x + 3y < 12$$
$$y + 4x > -4$$

5.

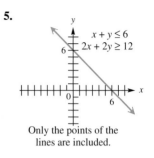

$$x + y \le 6$$
$$2x + 2y \ge 12$$

Only the points of the
lines are included.

7.

$$x + 2y \le 4$$
$$y \ge x^2 - 1$$

9.

$$y \le -x^2$$
$$y \ge x^2 - 6$$

11.

$$x^2 - y^2 < 1$$
$$-1 < y < 1$$

13.

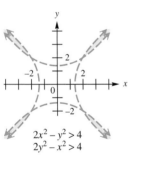

$$2x^2 - y^2 > 4$$
$$2y^2 - x^2 > 4$$

15.

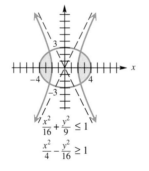

$$\frac{x^2}{16} + \frac{y^2}{9} \le 1$$
$$\frac{x^2}{4} - \frac{y^2}{16} \ge 1$$

17.

$$x + y \le 4$$
$$x - y \le 5$$
$$4x + y \le -4$$

19.

$$2y + x \le -5$$
$$y \ge 3 + x$$
$$x \ge 0$$
$$y \ge 0$$

No solution

21.

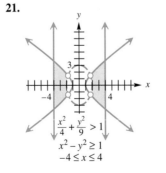

$$\frac{x^2}{4} + \frac{y^2}{9} > 1$$
$$x^2 - y^2 \ge 1$$
$$-4 \le x \le 4$$

23.

$$y \ge 3^x$$
$$y \ge 2$$

25.

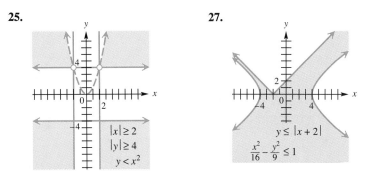

27.

29. (d)   **31.** (6/5, 6/5);   42/5   **33.** (17/3, 5);   49/3   **35.** \$1120, with 4 pigs, 12 geese   **37.** 8 of #1 and 3 of #2, for 100 cu ft of storage   **39.** 6.4 million gallons of gasoline and 3.2 million gallons of fuel oil, for maximum revenue of \$16,960,000

CHAPTER 7 REVIEW EXERCISES   (PAGE 392)

1. $\{(5, 7)\}$ **3.** $\{(-5/4, -79/40)\}$ **5.** $\left\{\left(\dfrac{2y + 11}{5}, y\right)\right\}$;   dependent   **7.** $\{(-1, 3)\}$ **9.** $\{(6, -12)\}$

11. $\{(2, 5)\}$ **13.** 10 at 25¢;   12 at 50¢   **15.** \$10,000 at 3%;   \$20,000 at 3.5%;   \$20,000 at 4.5%

17. 17.5 g of 12-carat;   7.5 g of 22-carat   **19.** $\{(-1, 2, 3)\}$ **21.** $\{(1, 3, -1)\}$ **23.** $\left\{\left(\dfrac{15z + 6}{11}, \dfrac{14z - 1}{11}, z\right)\right\}$

25. 10 lb of \$4.60 tea;   8 lb of \$5.75 tea; 2 lb of \$6.50 tea   **27.** 164 fives;   86 tens;   40 twenties

29. $\{(-4, 6)\}$ **31.** $\{(-3, 2)\}$ **33.** $\{(0, 1, 0)\}$ **35.** $\{(-2, 3), (1, 0)\}$ **37.** $\{(2, 3), (-2, 3), (2, -3), (-2, -3)\}$

39. $\{(6, 2/3), (-4, -1)\}$ **41.** yes; $\left\{\left(\dfrac{8 - 8\sqrt{41}}{5}, \dfrac{16 + 4\sqrt{41}}{5}\right), \left(\dfrac{8 + 8\sqrt{41}}{5}, \dfrac{16 - 4\sqrt{41}}{5}\right)\right\}$

43. (a)   **(b)**   **(c)**

45.

47. maximum of 24 at (0, 6)   **49.** $44/19 \approx 2$ batches of cakes and $144/19 \approx 8$ batches of cookies, for a maximum profit of approximately \$220

CHAPTER 7 TEST (PAGE 394)

[7.1] **1.** $\{(1, 2)\}$ **2.** $\{(4, 3)\}$ **3.** $\left\{\left(\dfrac{-3y - 7}{2}, y\right)\right\}$; dependent   **4.** $\{(4, 5/2)\}$ **5.** $\{(1, 2)\}$

6. $\emptyset$; inconsistent   **8.** \$7200 at 4%; \$8200 at 5%   **[7.2]** **9.** $\{(5, 3, 6)\}$ **10.** $\{(2, 0, -1)\}$

11. $\left\{\left(\dfrac{4 - z}{9}, \dfrac{-7 + 13z}{9}, z\right)\right\}$ **12.** 1, 1, 0   **[7.3]** **13.** $\{(5, 1)\}$ **14.** $\{(2, 2z - 1, z)\}$

[7.4] **15.** $\{(1, 2), (-1, 2), (1, -2), (-1, -2)\}$ **16.** $\{(3, 4), (4, 3)\}$

17. yes   **18.** 5 and -6 **[7.5]** **19.**

$9x^2 + 4y^2 \geq 36$
$y < x^2$

20. 3 3/4 servings of A and 1 7/8 servings of B, for a minimum cost of \$1.69

■ CHAPTER 8   MATRICES

8.1 EXERCISES (PAGE 402)

1. eight; three   **3.** 2×3 **5.** 1×5; row   **7.** 1×1; square; row; column   **9.** 2×2; square

11. $y = 8$ **13.** $r = 8$; $m = 12$; $n = 2$ **15.** $k = 19$; $p = 6$; $q = 17$ **17.** $a = 2$; $z = -3$; $m = 8$;

$k = 5/3$ **21.** $\begin{bmatrix} 8 & -43 & -18 \\ 26 & 29 & 6 \\ -2 & 10 & 43 \end{bmatrix}$ **23.** not possible   **25.** $\begin{bmatrix} 13 & 3 & 0 & -2 \\ 9 & -12 & 4 & 8 \\ 12 & -11 & -1 & 9 \end{bmatrix}$

27. $\begin{bmatrix} -12x + 8y & -x + y \\ x & 8x - y \end{bmatrix}$ **29.** $\begin{bmatrix} -4 & 8 \\ 0 & 6 \end{bmatrix}$ **31.** $\begin{bmatrix} 8 & -16 \\ 0 & -12 \end{bmatrix}$ **33.** $\begin{bmatrix} 2 & 6 \\ -4 & 6 \end{bmatrix}$ **35.** $\begin{bmatrix} 60 & -10 \\ -44 & 9 \end{bmatrix}$

37. $\begin{bmatrix} 7 & 2 \\ 9 & 0 \\ 8 & 6 \end{bmatrix}$; $\begin{bmatrix} 7 & 9 & 8 \\ 2 & 0 & 6 \end{bmatrix}$ **39.** $\begin{bmatrix} 5411 & 11{,}352 \\ 9371 & 15{,}956 \end{bmatrix}$; $\begin{bmatrix} 5411 & 9371 \\ 11{,}352 & 15{,}956 \end{bmatrix}$ **41.** yes, always true

43. yes, always true   **45.** yes

8.2 EXERCISES (PAGE 410)

1. 2×2; 2×2 **3.** 4×4; 2×2 **5.** 3×2; BA cannot be found   **7.** AB cannot be found; 3×2

9. AB cannot be found; BA cannot be found   **11.** columns; rows   **13.** first row; second column; adding

15. $\begin{bmatrix} 4 \\ 42 \end{bmatrix}$ **17.** $\begin{bmatrix} 27 \\ -1 \end{bmatrix}$ **19.** $\begin{bmatrix} 8 & -16 \\ -2 & 7 \end{bmatrix}$ **21.** $\begin{bmatrix} -16 \\ 6 \end{bmatrix}$ **23.** $\begin{bmatrix} -15 & -16 & 3 \\ -1 & 0 & 9 \\ 7 & 6 & 12 \end{bmatrix}$ **25.** $[4 \quad 8 \quad -10]$

27. $\begin{bmatrix} -4 & 2 & -5 \\ 16 & -39 & 14 \end{bmatrix}$ **29.** not possible   **31.** $[-8]$ **33.** $\begin{bmatrix} -20 & 0 & -10 \\ 4 & 0 & 2 \\ 24 & 0 & 12 \end{bmatrix}$ **35.** $\begin{bmatrix} 5 & -5 \\ 0 & 30 \end{bmatrix}$

37. not possible   **39.** Both products will exist if P is $m \times n$ and Q is $n \times m$.   **41.** always true   **43.** always true

45. (a) $\begin{bmatrix} 50 & 100 & 30 \\ 10 & 90 & 50 \\ 60 & 120 & 40 \end{bmatrix}$ **(b)** $\begin{bmatrix} 12 \\ 10 \\ 15 \end{bmatrix}$ (If the rows and columns are interchanged in part (a), this should be a 1×3

matrix.) **(c)** $\begin{bmatrix} 2050 \\ 1770 \\ 2520 \end{bmatrix}$ (This may be a 1×3 matrix.) **(d)** $6340

8.3 Exercises (Page 421)

1. matrix A **3.** yes **5.** no **7.** yes **9.** no **11.** $\begin{bmatrix} 0 & 1/2 \\ -1 & 1/2 \end{bmatrix}$ **13.** does not exist

15. $\begin{bmatrix} 2 & 1 \\ -3/2 & -1/2 \end{bmatrix}$ **17.** $\begin{bmatrix} -2.5 & 5 \\ 12.5 & -15 \end{bmatrix}$ **19.** $\begin{bmatrix} 1 & 0 & 0 \\ 0 & -1 & 0 \\ -1 & 0 & 1 \end{bmatrix}$ **21.** $\begin{bmatrix} -1/24 & 3/16 & -1/2 \\ 1/8 & -1/16 & 1/2 \\ 1/8 & -1/16 & -1/2 \end{bmatrix}$

23. $\begin{bmatrix} 15 & 4 & -5 \\ -12 & -3 & 4 \\ -4 & -1 & 1 \end{bmatrix}$ **25.** $\begin{bmatrix} -10/3 & 5/9 & -10/9 \\ 20/3 & 5/9 & 80/9 \\ -5 & 5/6 & -20/3 \end{bmatrix}$ **27.** $A = \begin{bmatrix} 1 & 1 \\ 2 & -1 \end{bmatrix}$; $X = \begin{bmatrix} x \\ y \end{bmatrix}$; $B = \begin{bmatrix} 8 \\ 4 \end{bmatrix}$

29. $A = \begin{bmatrix} 4 & 5 \\ 2 & 3 \end{bmatrix}$; $X = \begin{bmatrix} x \\ y \end{bmatrix}$; $B = \begin{bmatrix} 7 \\ 5 \end{bmatrix}$ **31.** $A = \begin{bmatrix} 1 & 1 & 1 \\ 2 & 1 & 3 \\ 5 & -2 & 2 \end{bmatrix}$; $X = \begin{bmatrix} x \\ y \\ z \end{bmatrix}$; $B = \begin{bmatrix} 9 \\ 17 \\ 16 \end{bmatrix}$ **33.** X is a matrix of the

variables of the system; A^{-1} is the inverse of matrix A, which is a matrix of the coefficients of the variables of the system; B is a matrix of the constants in the system. **35.** {(2, 3)} **37.** {(3, −5)} **39.** {(7/2, −1)} **41.** $\emptyset$
43. {(1, 1, 2)} **45.** {(1/4, 1/4, 13/4)} **47.** {(−1, 23, 16)} **49.** 5 model 201; 8 model 301

8.4 Exercises (Page 427)

1. −36 **3.** 7 **5.** 0 **7.** −26 **9.** −16 **11.** 0 **13.** $2x − 32$ **15.** $y^2 − 16$ **17.** $x^2 − y^2$
19. −.57 **23.** 0, −5, 0 **25.** −6, 0, −6 **27.** −1 **29.** 8 **31.** 17 **33.** 166 **35.** 0 **37.** 0
39. $4x$ **43.** {−4} **45.** {13} **47.** −88 **49.** 1 **53.** $x + 3y − 11 = 0$ **55.** The system has either no
solution or an infinite number of solutions if $|A| = 0$.

8.5 Exercises (Page 436)

1. Property 5 **3.** Property 1 **5.** By Property 4, multiply each element in the second row by −1/4. The
determinant of the resulting matrix is 0, by Property 5. **7.** Property 4 and Property 5 **9.** Property 4 and
Property 5 **13.** Property 2 **15.** Property 3 **17.** Property 4 **19.** Property 6 (first row added to second row)
21. Property 6 (−3 times the elements of the second column added to the elements of the first column)
23. Property 4 **25.** Property 6 **27.** 0 **29.** 0 **31.** 12 **33.** −49 **35.** −6

8.6 Exercises (Page 441)

1. {(2, 2)} **3.** {(2, −5)} **5.** {(2, 0)} **7.** {(2, 3)} **9.** can't use Cramer's rule, $D = 0$; $\left\{ \left(\dfrac{4 - 2y}{3}, y \right) \right\}$

11. can't use Cramer's rule, $D = 0$; $\emptyset$ **13.** {(−1, 2, 1)} **15.** {(−3, 4, 2)} **17.** {(0, 0, −1)} **19.** can't use

Cramer's rule, $D = 0$; $\emptyset$ **21.** can't use Cramer's rule, $D = 0$; $\left\{ \left(\dfrac{-32 + 19z}{4}, \dfrac{24 - 13z}{4}, z \right) \right\}$ **23.** {(2, 3, 1)}

25. {(0, 4, 2)} **27.** {(31/5, 19/10, −29/10)} **31.** {(0, 2, −2, 1)} **33.** can't use Cramer's rule, $D = 0$
35. {(−a − b, $a^2 + ab + b^2$)} **37.** {(1, 0)}

CHAPTER 8 REVIEW EXERCISES   (PAGE 446)

1. $m = -8$; $p = -7$; $x = 2$; $y = 9$; $z = 5$ **3.** $a = 5$; $x = 3/2$; $y = 0$; $z = 9$ **5.** $\begin{bmatrix} 0 & -2 & 7 \\ 6 & -1 & 10 \end{bmatrix}$

7. not possible   **9.** not possible   **11.** by adding corresponding elements   **13.** $\begin{bmatrix} 606 & 354 & 4434 & 28 \\ 397 & 238 & 2981 & 18 \\ 479 & 263 & 3359 & 21 \end{bmatrix}$;

$\begin{bmatrix} 606 & 397 & 479 \\ 354 & 238 & 263 \\ 4434 & 2981 & 3359 \\ 28 & 18 & 21 \end{bmatrix}$ **15.** $\begin{bmatrix} -9 & 3 \\ 10 & 6 \end{bmatrix}$ **17.** not possible   **19.** $[1 \quad 13 \quad 6]$ **21.** yes   **23.** no

25. $\begin{bmatrix} 3 & -1 \\ -5 & 2 \end{bmatrix}$ **27.** $\begin{bmatrix} 1/2 & 0 \\ 1/10 & 1/5 \end{bmatrix}$ **29.** $\begin{bmatrix} 2/3 & 0 & -1/3 \\ 1/3 & 0 & -2/3 \\ -2/3 & 1 & 1/3 \end{bmatrix}$ **31.** $\{(2, 2)\}$ **33.** $\{(2, 1)\}$ **35.** not possible;
dependent equations   **37.** $\{(-1, 0, 2)\}$ **39.** -25 **41.** -44 **43.** $\{-7/3\}$ **45.** {all real numbers}
49. Property 2 (Rows and columns are interchanged, so the determinants are equal.)   **51.** Property 5 (Two identical rows indicate the determinant equals zero.)   **53.** Property 3 (Exchanging columns 2 and 3 produces a matrix with determinant equal to -1 times the value of the first determinant.)   **55.** 96   **57.** $\{(-2, 5)\}$
59. $\{(82/23, -4/23)\}$ **61.** can't use Cramer's rule, $D = 0$;   dependent equations

CHAPTER 8 TEST   (PAGE 448)

[8.1]   **1.** $w = -3$; $x = -1$; $y = 7$ **2.** $\begin{bmatrix} 8 & 3 \\ 0 & -11 \\ 15 & 19 \end{bmatrix}$ **3.** not possible   **4.** $\begin{bmatrix} 108 & 112 & 123 & 110 \\ 142 & 98 & 117 & 130 \end{bmatrix}$

[8.2]   **5.** $\begin{bmatrix} -5 & 16 \\ 19 & 2 \end{bmatrix}$ **6.** not possible   **7.** (a)   **[8.3]**   **8.** yes   **9.** $\begin{bmatrix} -2 & -5 \\ -3 & -8 \end{bmatrix}$ **10.** does not exist

11. $\begin{bmatrix} -9 & 1 & -4 \\ -2 & 1 & 0 \\ 4 & -1 & 1 \end{bmatrix}$ **12.** $\{(-7, 8)\}$ **13.** $\{(0, 5, -9)\}$ **[8.4]**   **14.** -58 **15.** -844 **16.** $\{-2\}$

[8.5]   **17.** Property 3 (The rows are interchanged.)   **18.** Property 6 (-2 times row 3 was added to row 2 and row 1 was added to row 3.)   **[8.6]**   **19.** $\{(-6, 7)\}$ **20.** $\{(1, -2, 3)\}$

■ CHAPTER 9   SEQUENCES AND SERIES; PROBABILITY

9.1 EXERCISES   (PAGE 456)

1. 10, 16, 22, 28, 34   **3.** 2, 4, 8, 16, 32   **5.** 1, -1, 1, -1, 1   **7.** 1/2, 4/5, 1, 8/7, 5/4   **9.** 4/3, 12/5, 20/7, 28/9, 36/11   **11.** -2, 8, -24, 64, -160 **13.** x, x^2, x^3, x^4, x^5 **15.** 4, 9, 14, 19, 24, 29, 34, 39, 44, 49
17. 2, 4, 8, 16, 32, 64, 128, 256, 512, 1024   **19.** 1, 1, 2, 3, 5, 8, 13, 21, 34, 55   **21.** 4, 6, 8, 10, 12
23. 11, 9, 7, 5   **25.** 6, 2, -2, -6 **27.** -6, 0, 6, 12, 18, 24   **29.** $d = 5$, $a_n = 7 + 5n$ **31.** $d = 7$, $a_n = -26 + 7n$ **33.** $d = m$, $a_n = x + nm - m$ **35.** $d = -m$, $a_n = 2z + 2m - mn$ **37.** $a_8 = 19$; $a_n = 3 + 2n$ **39.** $a_8 = 7$; $a_n = n - 1$ **41.** $a_8 = -6$; $a_n = 10 - 2n$ **43.** $a_8 = -3$; $a_n = -39 + 9n/2$
45. $a_8 = x + 21$; $a_n = x + 3n - 3$ **47.** $a_8 = 4m$; $a_n = mn - 4m$ **49.** One example is the arithmetic sequence with $a_1 = 6$ and $d = 2$.   **51.** (d)   **53.** $a_1 = 7$ **55.** $a_1 = 0$

9.2 EXERCISES (PAGE 460)

1. 2, 6, 18, 54 **3.** 1/2, 2, 8, 32 **5.** $-2, 6, -18, 54$ **7.** 3125, 625, 125, 25, 5, 1, 1/5 **9.** $-1, 1, -1, 1, -1, 1$
11. 3/2, 3, 6, 12, 24 **13.** $a_5 = 324$; $a_n = 4 \cdot 3^{n-1}$ **15.** $a_5 = -162$; $a_n = -2 \cdot 3^{n-1}$ **17.** $a_5 = -1875$;
$a_n = -3(-5)^{n-1}$ **19.** $a_5 = 24$; $a_n = (3/2)2^{n-1}$ **21.** $a_5 = -256$; $a_n = -1(-4)^{n-1}$ **23.** $a_n = 6 \cdot 2^{n-1}$
25. $a_n = (3/4)2^{n-1}$ **27.** $a_n = -4(-1/2)^{n-1}$ **29.** not geometric **31.** $a_n = (9/4)2^{n-1}$ **33.** $a_n = (-2/9)3^{n-1}$
35. \$5242.88 **37.** One example is the geometric sequence with $a_1 = 5/3$ and $r = 3$. **39.** (d) **41.** $a_1 = 2, r = 3$
or $a_1 = -2, r = -3$ **43.** $a_1 = 128, r = 1/2$ or $a_1 = -128, r = -1/2$

9.3 EXERCISES (PAGE 469)

1. 35 **3.** 25/12 **5.** 288 **7.** 3 **9.** 215 **11.** 125 **13.** 230 **15.** 155/2 **17.** $a_1 = 7, d = 5$
19. $a_1 = 1, d = -20/11$ **21.** 18 **23.** 140 **25.** -684 **27.** 500,500 **29.** 93 **31.** 33/4 **33.** 124
35. 121/9 **37.** 30 **39.** 60 **41.** 1330/9 **43.** 120 **45.** 272/25 **47.** 1281 **49.** 4680 **51.** 100 m;
550 m **53.** 240 ft; 1024 ft **55.** 170 m **57.** $\$2^{30}$ or \$1,073,741,824 **59.** \$7,875,000 **61.** .00002%
63. $(1/32) \times 10^{15}$ molecules **65.** \$12,487.56 **67.** \$40,634.48

9.4 EXERCISES (PAGE 474)

1. 2 **3.** 10 **5.** 1/2; sum would exist **7.** 1.1 **9.** $-1 < r < 1$ **11.** 64/3 **13.** 1000/9 **15.** 135
17. 512/3 **19.** 81 **21.** 3/2 **23.** 2 **25.** 1/5 **27.** The sum does not exist. **29.** 1/3 **31.** The sum
does not exist. **33.** $-1/5$ **35.** 1/4 **37.** 1/9 **39.** The sum does not exist. **41.** (b) **43.** 5/9
45. 4/33 **47.** 31/99 **49.** 1507/999 **51.** 70 m **53.** 1600 **55.** 12 m

9.5 EXERCISES (PAGE 480)

Exercises 1–25 are proofs, so no answers are supplied. **27.** $\dfrac{4^{n-1}}{3^{n-2}}$ or $3\left(\dfrac{4}{3}\right)^{n-1}$

9.6 EXERCISES (PAGE 485)

1. 5040 **3.** 720 **5.** 90 **7.** 336 **9.** 7 **11.** 1 **15.** 30 **17.** (a) 27,600 (b) 35,152 (c) 1104
19. 15 **21.** (a) 17,576,000 (b) 17,576,000 (c) 456,976,000 **23.** 720 **25.** 120 **27.** 2730
29. 120; 30,240

9.7 EXERCISES (PAGE 489)

1. 6 **3.** 56 **5.** 1365 **7.** 120 **9.** 14 **11.** 1 **17.** 27,405 **19.** 20 **21.** 10 **23.** 28
25. (a) 84 (b) 10 (c) 40 (d) 28 **27.** 120 **29.** 220 **31.** 210 **33.** (a) 330 (b) 150 **35.** 5040

9.8 EXERCISES (PAGE 496)

1. Let h = heads, t = tails. $S = \{h\}$ **3.** $S = \{hhh, hht, hth, thh, htt, tht, tth, ttt\}$ **5.** Let c = correct, w = wrong.
$S = \{ccc, ccw, cwc, wcc, wwc, wcw, cww, www\}$ **7.** (a) $\{h\}$; 1 (b) $\emptyset$; 0 **9.** (a) $\{ccc\}$; 1/8
(b) $\{www\}$; 1/8 (c) $\{ccw, cwc, wcc\}$; 3/8 (d) $\{cww, wcw, wwc, ccw, cwc, wcc, ccc\}$; 7/8 **13.** (a) 1/5 (b) 0
(c) 7/15 (d) 1 to 4 (e) 7 to 8 **15.** 1 to 4 **17.** 2/5 **19.** (a) 1/2 (b) 7/10 (c) 2/5 **21.** (a) .62
(b) .27 (c) .11 (d) .89 **23.** $48/28,561 \approx .001681$ **25.** (a) .72 (b) .70 (c) .79

CHAPTER 9 REVIEW EXERCISES (PAGE 500)

1. 1/2, 2/3, 3/4, 4/5, 5/6   **3.** 8, 10, 12, 14, 16   **5.** 5, 2, −1, −4, −7   **7.** 5, 3, −2, −5, −3
9. 6, 2, −2, −6, −10   **11.** $3 - \sqrt{5}, 4, 5 + \sqrt{5}, 6 + 2\sqrt{5}, 7 + 3\sqrt{5}$ **13.** $a_1 = -29$; $a_{20} = 66$ **15.** 17
17. $m - 10$ **19.** 64, 32, 16, 8, 4   **21.** 8/9, ±8/3, 8, ±24, 72   **23.** 48   **25.** $x^5/125$ **29.** 612   **31.** 15
33. $-10k$ **35.** 6   **37.** 133   **39.** 240   **41.** \$6930   **43.** 18.75 g   **45.** 80   **47.** 8/9   **49.** Sum does
not exist.   **51.** 512/999   **57.** 120   **59.** 1   **61.** 252   **63.** 48   **65.** 24   **67.** 504   **69. (a)** 4/15
(b) 2/3   **(c)** 0   **71.** 1/26   **73.** 4/13   **75.** 1   **77.** .44

CHAPTER 9 TEST (PAGE 502)

[9.1] **1.** −2, 5, −10, 17, −26   **2.** 5, 7, 10, 14, 19   **3.** 2, 4, 4, 6, 6   **4.** −11, 1, 13, 25, 37
[9.2] **5.** 81, 54, 36, 24, 16   **[9.1]** **6.** 6   **7.** 955   **[9.2]** **8.** 3   **9.** $1458x^{15}$ **10.** 15/2
[9.3] **11.** 2385   **12.** −186   **[9.4]** **13.** 125   **14.** 108/7   **[9.3]** **15.** \$2100   **16.** 25,600
[9.6] **18.** 990   **19.** 5040   **[9.7]** **20.** 45   **21.** 1   **[9.6]** **22.** 60   **23.** 24,360   **[9.7]** **24.** 10,920
[9.8] **26.** 1/26   **27.** 10/13   **28.** 4/13   **29.** 3 to 10   **30.** .92

APPENDIX A EXERCISES (PAGE 507)

1. 12, 13, 14, 15, 16, 17, 18, 19, 20   **3.** 1, 1/2, 1/4, 1/8, 1/16, 1/32   **5.** 17, 22, 27, 32, 37, 42, 47
7. 8, 9, 10, 11, 12, 13, 14   **9.** finite   **11.** infinite   **13.** infinite   **15.** infinite   **17.** ∈   **19.** ∉   **21.** ∈
23. ∉   **25.** ∉   **27.** ∉   **29.** false   **31.** true   **33.** true   **35.** true   **37.** false   **39.** true   **41.** true
43. false   **45.** true   **47.** false   **49.** true   **51.** true   **53.** true   **55.** true   **57.** false   **59.** true
61. true   **63.** false   **65.** ⊆   **67.** ⊄   **69.** ⊆   **71.** {0, 2, 4}   **73.** {0, 1, 2, 3, 4, 5, 6, 7, 8, 9, 11, 13}
75. M or {0, 2, 4, 6, 8}   **77.** {0, 1, 2, 3, 4, 5, 7, 9, 11, 13}   **79.** Q or {0, 2, 4, 6, 8, 10, 12}   **81.** {10, 12}
83. ∅;   ∅ and R are disjoint   **85.** N or {1, 3, 5, 7, 9, 11, 13}   **87.** R or {0, 1, 2, 3, 4}   **89.** {0, 1, 2, 3, 4, 6, 8}
91. R or   {0, 1, 2, 3, 4}   **93.** ∅; Q' and $(N' \cap U)$ are disjoint   **95.** all students in this school who are not
taking this course   **97.** all students in this school who are taking calculus and history   **99.** all students in this
school who are taking this course or history or both   **101.** {1, 2, 3, 4, 5, 6}   **103.** {$hhh, hht, hth, thh, htt, tht, tth, ttt$}
105. X and Y are disjoint.   **107.** $X \subseteq Y$ **109.** $X = \emptyset$

INDEX

A

Absolute value, 16
  definition of, 17
  equations, 133
  functions, 232
  inequalities, 134
  properties of, 17
Addition
  of complex numbers, 93
  identity element for, 9
  of matrices, 400
  of polynomials, 23
  of rational expressions, 43
Addition method for systems, 352
Addition property
  of equality, 74
  of order, 15
Additive identity
  of a complex number, 93
  of a real number, 9
Additive inverse, 9
  of a complex number, 93
  of a matrix, 401
Algebra of functions, 225
Algebraic expression, 22
Algorithm, division, 306, 310
Annuity, 468
Arithmetic progression, 453
Arithmetic sequence, 453
  nth term of, 455
  sum of n terms, 463
Array of signs for a matrix, 426
Associative property, 8
Asymptote, 189
  horizontal, 256, 335
  of a hyperbola, 189
  oblique, 336
  summary of, 337
  vertical, 267, 335
Augmented matrix, 367
Axis, of a parabola, 163
Axis, symmetry with respect to, 180

B

Base, 5
Binomial, 23
  expansion of, 30
  square of, 25
Binomial coefficient, 29
Binomial theorem, 30
  rth term, 32
Blood pressure, 18
Boundedness theorem, 327
Break-even point, 124

C

Carbon 14 dating, 289
Cartesian coordinate system, 146
Celsius, 179
Center
  of a circle, 176

of an ellipse, 184
of a hyperbola, 188
Center-radius form, 177
Change in x, 153
Change in y, 153
Change of base, 276
Change of base theorem, 276
Circle, 176
  equation of, 177
  translation of, 179
Closed interval, 123
Coefficient, 22
  binomial, 29
  leading, 298
Coefficient matrix, 420
Cofactor of an element of a matrix, 425
Column matrix, 399
Combinations, 486
Combined variation, 220
Common denominator, 43
Common difference, 453
Common factor, 33
Common logarithm, 273
Common ratio, 458
Commutative property, 8
Complement
  of an event, 492
  of a set, 505
Completing the square, 100, 166
Complex fraction, 45
Complex numbers, 91
  addition, 93
  additive identity, 93
  additive inverse of, 93
  conjugate of, 95
  division, 96
  multiplication, 94
  negative of, 93
  operations on, 93
  standard form, 91
  subtraction, 94
Composition of functions, 226
Compound interest, 259
Compounding continuously, 287
Conditional equation, 73
Conic sections, 193
  summary, 194
Conjugate
  of a complex number, 95
  of a radical, 63
Conjugate zeros theorem, 316
Constant of variation, 218
Constraints for linear programming, 386
Continuous compounding, 287
Contradiction equation, 73
Coordinate, 13
Coordinate plane, 146
Coordinate system, 13, 146
Coordinates of a point, 146
Corner point of region, 386
Cost function, 215
Counting numbers, 503

Cramer's rule, 439
Cubes, sum and difference of, 37
Cubic equation, 103
Cubic polynomial, 301

D

Data points, 159
Decibel, 274
Decreasing function, 257
Degree, of a polynomial, 22
Delta x, 153
Dependent equations, 351, 354
Dependent variable, 210
Descartes' rule of signs, 325
Descending order, 23
Determinant, 423
  of matrices, 423
  properties of, 429
Difference of cubes, 37
Difference of squares, 25, 36
Dimension, 398
Direct variation, 218
  as nth power, 219
Directly proportional, 218
Directrix, 169
Discriminant, 104
Disjoint sets, 506
Distance, 16, 132
Distance formula, 147
Distance, rate and time problems, 84, 110
Distributive property, 10
Division
  algorithm, 306, 310
  of a complex number, 96
  of polynomials, 306
  of rational expressions, 42
  synthetic, 308
Domain
  of a function, 210
  of a relation, 144
Doubling time, 289

E

e, 261
Element
  cofactor, 425
  of a matrix, 367
  minor of, 424
  of a set, 503
Elimination method for systems, 353
Ellipse, 184
  equation of, 185
  translation of, 187
Empty set, 74, 504
Entry of a matrix, 367
Equal matrices, 399
Equality, addition property of, 74
Equation, 73
  absolute value, 133
  of a circle, 177
  conditional, 73
  contradiction, 73

A-46

cubic, 103
dependent, 351, 354
of an ellipse, 185
equivalent, 74
exponential, 255, 280
first-degree, 350
of a horizontal line, 152, 157
of a hyperbola, 187
identity, 73
linear, 75, 350
literal, 77
logarithmic, 266, 280
matrix, 419
of a parabola, 162
quadratic, 98
quadratic in form, 116
radical, 118
of a relation, 144
solution set, 73
system, 350, 359
of a vertical line, 152, 157
Equilibrium demand, 356
Equilibrium price, 356
Equilibrium supply, 356
Equivalent equations, 74
Equivalent inequalities, 121
Equivalent systems, 352
Event, 491
  complement of, 492
  mutually exclusive, 494
Expansion of a binomial, 30
Expansion of a matrix, 425
Exponent, 5
  definition of, 5
  integer, 5, 20
  negative, 48
  power rule, 21
  product rule, 20
  properties of, 20, 255
  quotient rule, 49
  rational, 51
  rules for, 20, 53
  zero, 21
Exponential equation, 255, 280
Exponential function, 256
  properties of, 280
Expression
  algebraic, 22
  fractional, 40
  rational, 40

F

Factor theorem, 313
Factored form, 33
Factorial, 28, 482
Factoring, 33
  difference of cubes, 37
  difference of squares, 36
  greatest common factor, 33
  by grouping, 34
  method of substitution, 38
  perfect square trinomial, 36
  sum of cubes, 37
  trinomial, 35
Factors, 33
Fahrenheit, 79
Finite sequence, 452

Finite set, 503
First-degree equation, 350
Fixed cost, 215
Foci
  of an ellipse, 184
  of a hyperbola, 187
Focus of a parabola, 169
FOIL, 24
Fraction, complex, 45
Fractional expression, 40
Fractions, fundamental principle of, 41
Function, 210
  absolute value, 232
  algebra of, 225
  composition, 226
  cost, 215
  decreasing, 257
  domain of, 210
  exponential, 256
  graphing, 231
  greatest integer, 234
  increasing, 256
  inverse, 240
  linear, 231
  logarithmic, 266
  notation, 211
  objective, 386
  one-to-one, 238
  operations on, 225
  piecewise, 233
  polynomial, 298
  profit, 215
  quadratic, 231
  range of, 210
  rational, 333
  revenue, 215
  sequence, 452
  step, 234
  transcendental, 254
  vertical line test for, 215
Fundamental principle of fractions, 41
Fundamental rectangle, 189
Fundamental theorem of algebra, 314
Future value, 260, 468
$f(x)$, 211

G

Gauss-Jordan method, 368
General term, 452
Geometric progression, 458
Geometric sequence, 458
  infinite, 471
  nth term of, 458
  sum of infinite, 472
  sum of n terms, 465
Geometry problems, 82, 108
Geometry review, 500
Greater than, 14
Greatest integer function, 234
Grouping, factoring by, 34
Growth and decay, 262, 288

H

Half-life, 289
Half-plane, 199
Horizontal asymptote, 256, 335
Horizontal line, 152
  equation, 152, 157

slope, 154
test, 238
Horizontal parabola, 170
Horizontal translation, 300
Hyperbola, 187
  asymptote, 189
  equation of, 187
  translation of, 190

I

i, 91
  powers of, 95
Identity element
  for addition, 9, 93
  for multiplication, 9
Identity equation, 73
Identity matrix, 413
Identity property, 9
Imaginary numbers, 91
Inconsistent system, 351, 354
Increasing function, 256
Independent events, 481
Independent variable, 210
Index of a radical, 57
Index of summation, 461
Induction, 475
Inequality, 14, 121, 199
  absolute value, 134
  equivalent, 121
  linear, 122, 199
  properties of, 15, 121
  quadratic, 124, 200
  rational, 127
  second-degree, 200
  system of, 383
  three-part, 124
  triangle, 17
Infinite geometric sequence, 471
  sum of terms, 472
Infinite sequence, 452
Infinite set, 503
Integer exponent, 5, 20
Integers, 2
Intercepts of a linear relation, 151
Interest, compound, 259
Interest problems, 87
Intermediate value theorem, 326
Intersection of sets, 505
Interval, 123
  notation, 123
Inverse
  additive, 9, 93
  multiplicative, 10
Inverse functions, 240
  theorem for, 271
Inverse variation, 219
Inversely proportional, 219
Irrational numbers, 3
Irreducible polynomial, 33

J

Joint variation, 220

L

Leading coefficient, 298
Less than, 14
Like radicals, 60

Like terms, 22
Linear equation, 75, 350
   applications, 80
Linear function, 231
Linear inequality, 122, 199
Linear programming, 385
   constraints for, 386
Linear relation in two variables, 151
   standard form, 151
Linear system
   Cramer's rule for, 439
   with three variables, 359
   with two variables, 350
Literal equation, 77
Logarithm, 265
   change of base, 276
   common, 273
   natural, 275
   properties of, 269
Logarithmic equation, 266, 280
Logarithmic function, 266
   properties of, 280

M

Magic number, 7
Major axis of an ellipse, 184
Mathematical induction, 475
Matrix, 367, 398
   addition, 400
   additive inverse, 401
   array of signs, 426
   augmented, 367
   coefficient, 420
   cofactor of an element, 425
   column, 399
   determinant of, 423
   element, 367
   entry, 367
   equal, 399
   equation, 419
   expansion of, 425
   identity, 413
   multiplication, 406
   multiplicative inverse, 416
   negative of, 401
   row, 399
   row transformations, 367, 414
   scalar, 402
   scalar multiplication, 402
   size, 398
   square, 399
   subtraction, 401
   zero, 401
Members of a set, 503
Midpoint formula, 149
Minor axis of an ellipse, 184
Minor of a matrix element, 424
Mixture problems, 86
Monomial, 23
Multiplication
   of complex numbers, 94
   identity element for, 9
   of matrices, 406
   of polynomials, 24
   of rational expressions, 42
Multiplication principle of counting, 481
Multiplication property of equality, 74

Multiplication property of order, 15
Multiplicative identity, 9
Multiplicative inverse, 10
   for a matrix, 416
Multiplicity of zeros, 315
Mutually exclusive events, 494

N

Natural logarithm, 275
Natural numbers, 2, 503
Negative
   of a complex number, 93
   of a matrix, 401
   of a real number, 9
Negative exponent, 48
n-factorial, 28, 482
Nonlinear system of equations, 376
nth root, 57
nth term, 452
   of an arithmetic sequence, 455
   of a geometric sequence, 452
Null set, 74, 504
Number line, 2
Numbers
   complex, 91
   counting, 503
   imaginary, 91
   integers, 2
   irrational, 3
   magic, 7
   natural, 2, 503
   rational, 2
   real, 3
   sets of, 4
   whole, 2
Numerical coefficient, 22

O

Objective function, 386
Oblique asymptote, 336
Odds of an event, 493
One-to-one function, 238
   horizontal line test for, 238
Open interval, 123
Operations
   on complex numbers, 93
   on functions, 225
Order, properties of, 15
Order of operations, 5
Ordered pair, 144
Ordered triple, 359
Origin, 145
   symmetry with respect to, 181
Outcomes, 491

P

Parabola, 162
   applications, 168
   axis of, 163
   equation of, 162
   focus of, 169
   geometric definition, 169
   graphs of, 163
   horizontal, 170
Parallel lines, 157, 354
Pascal's triangle, 28

Perfect square trinomial, 36
Permutations, 482
Perpendicular lines, 157
pH, 274
Piecewise functions, 233
Point-slope form, 155
Polynomial, 22
   addition, 23
   binomial, 23
   cubic, 301
   degree, 22
   descending order, 23
   division, 306
   factoring, 33
   function, 298
   in x, 22
   irreducible, 33
   monomial, 23
   multiplication, 24
   prime, 33
   quotient, 308
   remainder, 308
   subtraction, 23
   term, 22
   trinomial, 23
Polynomial function, 298
   zeros of, 301, 315
Polynomial in x, 22
Power rules for exponents, 21
Powers of i, 95
Present value, 260
Prime polynomial, 33
Principle nth root, 57
Principle of mathematical induction, 475
Probability of an event, 491
   properties of, 496
Product
   of complex numbers, 95
   of the sum and difference of two terms, 25
Product rule for exponents, 20
Profit function, 215
Progression
   arithmetic, 453
   geometric, 458
Property
   of absolute value, 17
   associative, 8
   commutative, 8
   of determinants, 429
   distributive, 10
   of equality, 74
   of exponents, 20, 255
   identity, 9
   of inequalities, 15, 121
   of logarithms, 269
   of order, 15
   of probability, 496
   of real numbers, 8
   square-root, 99
   transitive, 15
   zero-factor, 98
Proportional, 218
Pythagorean theorem, 111

Q

Quadrants, 146
Quadratic equation, 98
   applications, 108

A-48

discriminant of, 104
  standard form, 98
Quadratic formula, 102
Quadratic function, 231
Quadratic inequalities, 124, 200
Quadratic in form, 116
Quadratic relation, 162
  applications, 168
Quotient polynomial, 308
Quotient rule for exponents, 49

R

Radical equation, 118
Radical notation, 57
Radical sign, 57
Radicals, 57
  conjugate of, 63
  like, 60
  rationalizing the denominator, 62
  rules for, 59
  simplifying, 60
Radicand, 57
Radius, 176
Range
  of a function, 210
  of a relation, 144
Rational zeros theorem, 320
Rationalizing the denominator, 62
Real numbers, 3
  negative of, 9
  properties of, 8, 11
Reciprocal, 10
Rectangular coordinate system, 146
Recursive formula, 453
Region of feasible solutions, 386
  vertex of, 386
Relation, 144
  domain of, 144
  equation of, 144
  graph of, 146
  linear, 151
  quadratic, 162
  range of, 144
Remainder polynomial, 308
Remainder theorem, 310
Revenue function, 215
Richter scale, 279
Row matrix, 399
Row transformations, 367, 414
rth term, 32
  of a binomial expansion, 32
Rules for exponents, 20, 53
Rules for radicals, 59

S

Sample space, 491
Satisfying an equation, 73
Scalar, 402
Scalar multiplication, 402
Second-degree inequality, 200
Sequence, 452
  arithmetic, 453
  finite, 452
  general term, 452
  geometric, 458
  infinite 452
  nth term of, 452

Series, 461
Set, 503
  complement of, 505
  element of, 503
Set braces, 503
Set-builder notation, 2, 504
Sets of numbers, summary of, 4
Sigma notation, 461
Sign graph, 126
Simplifying radicals, 60
Slope, 153
  horizontal line, 154
  parallel lines, 157
  perpendicular lines, 157
  vertical line, 153
Slope-intercept form, 156
Solution of an equation, 73
Solution of an inequality, 122
Solution set, 73
Solving for a specified variable, 77, 80
Square matrix, 399
Square of a binomial, 25
Square-root property, 99
Squares, difference of, 25, 26
Standard form
  of a complex number, 91
  of a quadratic equation, 98
Step function, 234
Subset, 504
Substitution method for systems, 351
Subtraction
  of complex numbers, 94
  of matrices, 401
  of polynomials, 23
  of rational expressions, 43
Sum
  of an arithmetic sequence, 463
  of cubes, 37
  of a geometric sequence, 465
  of an infinite geometric sequence, 472
Supply and demand, 56
Symmetry, 180
  origin, 181
  x-axis, 180
  y-axis, 180
Synthetic division, 308
System
  coordinate, 13
  of dependent equations, 351, 354
  of equations, 350, 359
  equivalent, 352
  inconsistent, 351, 354
  of inequalities, 383
  linear, 350, 359
  nonlinear, 376
  solution of, 351

T

Term, 22
  coefficient of, 22
Tests for symmetry, 182
Theorem on inverses, 271
Three-part inequality, 124
Threshold sound, 274
Transcendental function, 254
Transitive property, 15
Translation, 165, 299

of a circle, 179
of an ellipse, 187
horizontal, 300
of a hyperbola, 190
vertical, 300
Transverse axis of a hyperbola, 188
Tree diagram, 481
Triangle inequality, 17
Trinomial, 23
  factoring of, 35
  perfect square, 36

U

Union of events, 494
Union of intervals, 127
Union of sets, 506
Universal set, 504

V

Variable, 504
  dependent, 210
  independent, 210
  solving for, 77, 80
Variable cost, 215
Variation, 218
  combined, 220
  constant of, 218
  direct, 218
  inverse, 219
  joint, 220
Venn diagram, 492, 504
Vertex
  of a parabola, 163, 167, 171
  of a region, 386
Vertical asymptote, 267, 335
Vertical line, 152
  equation, 152, 157
  slope, 153
  test, 215
Vertical translation, 300
Vertices
  of an ellipse, 184
  of a hyperbola, 188

W

Whole numbers, 2
Work problems, 109

X

x-axis, 145
  symmetry with respect to, 180
x-intercept, 151
xy-plane, 146

Y

y-axis, 145
  symmetry with respect to, 180
y-intercept, 151

Z

Zero exponent, 21
Zero matrix, 401
Zero-factor property, 98
Zeros
  multiplicity of, 315
  of polynomial functions, 301, 315

INSTRUCTOR'S ANSWER SECTION

For the instructor's convenience, all of the exercises from the student's edition are reprinted in this section. Answers are printed in place for all exercises *except* those for which many correct answers are possible (i.e., proofs or writing exercises). Each exercise set is cross-referenced to the pages on which the exercises appear in the student text, and two bound-in ribbon markers provide for ease of movement between chapter text and answer section pages.

Conceptual and writing exercises are identified for the instructor by the symbols ◎ and ✐, respectively. Challenging exercises, which will require most students to stretch beyond the concepts discussed in the text, are marked with the symbol ▲. Annotated sample pages from the Instructor's Answer Section can be found immediately following the Preface of this book.

In addition to the exercises and answers, each section begins with a complete list of the print and nonprint supplement resources for that section of the text, with cross-references to the relevant portions of each supplement.

■ CONTENTS

CHAPTER 1   ALGEBRAIC EXPRESSIONS   I-2

CHAPTER 2   EQUATIONS AND INEQUALITIES   I-25

CHAPTER 3   RELATIONS AND THEIR GRAPHS   I-52

CHAPTER 4   FUNCTIONS   I-87

CHAPTER 5   EXPONENTIAL AND LOGARITHMIC FUNCTIONS   I-112

CHAPTER 6   POLYNOMIAL AND RATIONAL FUNCTIONS   I-129

CHAPTER 7   SYSTEMS OF EQUATIONS AND INEQUALITIES   I-156

CHAPTER 8   MATRICES   I-179

CHAPTER 9   SEQUENCES AND SERIES; PROBABILITY   I-198

APPENDIX A   SETS   I-215

■ CHAPTER 1 ALGEBRAIC EXPRESSIONS

1.1 RESOURCES

Instructor's Test Manual
Test Items, p. 299–300
Instructor's Solution Manual
pp. 1–3, 423–425
Study Guide
pp. 6–12
Student's Solution Manual
pp. 1–3

Test Generator
IBM Disks
5.25″ Disk A
3.5″ Disk A
Mac Disk A
Tutorials
IBM Disks
5.25″ Disk A
3.5″ Disk A
Mac Disk A

HarperCollins Videotape 1
"College Algebra: In Simplest Terms" Program 2

1.1 EXERCISES (TEXT PAGE 12) ■

Let set B = {−6, −12/4, −5/8, −$\sqrt{3}$, 0, 1/4, 1, 2π, 3, $\sqrt{12}$}. List all the elements of B that belong to each of the following sets. See Example 1.

1. Natural numbers $1, 3$

2. Whole numbers $0, 1, 3$

3. Integers $-6, -\dfrac{12}{4}$ (or -3), $0, 1, 3$

4. Rational numbers $-6, -\dfrac{12}{4}$ (or -3), $-\dfrac{5}{8}, 0, \dfrac{1}{4}, 1, 3$

5. Irrational numbers $-\sqrt{3}, 2\pi, \sqrt{12}$

6. Real numbers $-6, -\dfrac{12}{4}$ (or -3), $-\dfrac{5}{8}, -\sqrt{3}, 0, \dfrac{1}{4}, 1, 2\pi, 3, \sqrt{12}$

For Exercises 7–12, choose all words from the following list that apply: (a) natural number, (b) whole number, (c) integer, (d) rational number, (e) irrational number, (f) real number, (g) not defined. See Example 1.

7. 12 natural, whole, integer, rational, real

8. 0 whole, integer, rational, real

9. $-\dfrac{3}{4}$ rational, real

10. $\dfrac{5}{9}$ rational, real

11. $\sqrt{8}$ irrational, real

12. $-\sqrt{2}$ irrational, real

✐ **13.** Explain why division by zero is not defined.

Simplify each of the following. See Example 2.

14. 3^4 81

15. -3^5 -243

16. -2^6 -64

17. $(-3)^4$ 81

18. $(-2)^5$ -32

19. $(-3)^5$ -243

20. $(-3)^6$ 729

◎ **21.** Based on your answers to Exercises 14–20, complete the following statements. A negative base raised to an odd exponent is _____. A negative base raised to an even
positive/negative
exponent is _____. negative, positive
positive/negative

◎ CONCEPTUAL   ✐ WRITING   ▲ CHALLENGING

Use the order of operations to evaluate each of the following. See Example 3.

22. $9 - 4^2 - (-12)$ 5   **23.** $8^2 - (-4) + 11$ 79   **24.** $16(-9) - 4$ -148 **25.** $-15 - 3(-8)$ 9

26. $-2 \cdot 5 + 12 \div 3$ -6 **27.** $9 \cdot 3 - 16 \div 4$ 23   **28.** $-4(9 - 8) + (-7)(2)^3$ -60

29. $6(-5) - (-3)(2)^4$ 18   **30.** $(4 - 2^3)(-2 + \sqrt{25})$ -12 **31.** $[-3^2 - (-2)] [\sqrt{16} - 2^3]$ 28

32. $\left(-\dfrac{2}{9} - \dfrac{1}{4}\right) - \left[-\dfrac{5}{18} - \left(-\dfrac{1}{2}\right)\right]$ $-\dfrac{25}{36}$ **33.** $\left[-\dfrac{5}{8} - \left(-\dfrac{2}{5}\right)\right] - \left(\dfrac{3}{2} - \dfrac{11}{10}\right)$ $-\dfrac{5}{8}$

34. $\dfrac{-8 + (-4)(-6) \div 12}{4 - (-3)}$ $-\dfrac{6}{7}$ **35.** $\dfrac{15 \div 5 \cdot 4 \div 6 - 8}{-6 - (-5) - 8 \div 2}$ $\dfrac{6}{5}$

▦ *Evaluate each of the following if $p = -2$, $q = 4$, and $r = -5$. See Example 4.*

36. $-3(p + 5q)$ -54 **37.** $2(q - r)$ 18   **38.** $\dfrac{p}{q} + \dfrac{3}{r}$ $-\dfrac{11}{10}$ **39.** $\dfrac{q + r}{q + p}$ $-\dfrac{1}{2}$

40. $\dfrac{3q}{3p - 2r}$ 3   **41.** $\dfrac{3q}{r} - \dfrac{5}{p}$ $\dfrac{1}{10}$ **42.** $\dfrac{\dfrac{q}{4} - \dfrac{r}{5}}{\dfrac{p}{2} + \dfrac{q}{2}}$ 2   **43.** $\dfrac{\dfrac{3r}{10} - \dfrac{5p}{2}}{q + \dfrac{2r}{5}}$ $\dfrac{7}{4}$

Identify the properties that are illustrated in each of the following statements. Some will require more than one property. Assume that all variables represent real numbers. See Examples 5–8.

44. $6 \cdot 12 + 6 \cdot 15 = 6(12 + 15)$ distributive   **45.** $8(m + 4) = 8m + 8 \cdot 4$ distributive

46. $(x + 6) \cdot \left(\dfrac{1}{x + 6}\right) = 1$,   if $x + 6 \neq 0$ inverse   **47.** $\dfrac{2 + m}{2 - m} \cdot \dfrac{2 - m}{2 + m} = 1$,   if $m \neq 2$ or -2 inverse

48. $(7 - y) + 0 = 7 - y$ identity   **49.** $[9 + (-9)] \cdot 5 = 5 \cdot 0$ commutative and inverse

50. $(-5 + 7)(3 + 4) = (-5 + 7) \cdot 3 + (-5 + 7) \cdot 4$ distributive

51. $x \cdot \dfrac{1}{x} + x \cdot \dfrac{1}{x} = x\left(\dfrac{1}{x} + \dfrac{1}{x}\right)$,   if $x \neq 0$ distributive

Use the distributive property to rewrite sums as products and products as sums. See Example 8.

52. $10k + 3k$ $(10 + 3)k = 13k$ **53.** $8p - 14p$ $(8 - 14)p = -6p$ **54.** $15 - 10x$ $5(3 - 2x)$

55. $18y + 6$ $6(3y + 1)$ **56.** $6(x + y)$ $6x + 6y$ **57.** $9(r - s)$ $9r - 9s$

58. $-3(z - y)$ $-3z + 3y$ **59.** $-2(m + n)$ $-2m - 2n$ **60.** $-(8r - k)$ $-8r + k$

61. $-(-2 + 5m)$ $2 - 5m$ **62.** $a(r + s - t)$ $ar + as - at$ **63.** $p(q - w + x)$ $pq - pw + px$

Use the various properties of real numbers to simplify each of the following expressions. See Example 7.

64. $\dfrac{5}{3}(12y)$ $20y$ **65.** $\dfrac{10}{11}(22z)$ $20z$ **66.** $\left(\dfrac{3}{4}r\right)(-12)$ $-9r$ **67.** $\left(-\dfrac{5}{8}p\right)(-24)$ $15p$

68. $\dfrac{2}{3}(12y - 6z + 18q)$ $8y - 4z + 12q$ **69.** $-\dfrac{1}{4}(20m + 8y - 32z)$ $-5m - 2y + 8z$

70. $\dfrac{3}{8}\left(\dfrac{16}{9}y + \dfrac{32}{27}z - \dfrac{40}{9}\right)$ $\dfrac{2}{3}y + \dfrac{4}{9}z - \dfrac{5}{3}$ **71.** $\dfrac{2}{3}\left(\dfrac{9}{4}a + \dfrac{15}{14}b - \dfrac{27}{20}\right)$ $\dfrac{3}{2}a + \dfrac{5}{7}b - \dfrac{9}{10}$

The symbol ▦ identifies problems that can be solved more quickly with the help of a calculator.

72. Is there a commutative property for subtraction? That is, does $a - b = b - a$? Support your answer with examples.

73. Is there an associative property for subtraction? Does $(a - b) - c = a - (b - c)$? Support your answer with examples.

74. This exercise is designed to show that $\sqrt{2}$ is irrational. Give a reason for each of steps (a) through (h).

There are two possibilities:

(1) A rational number a/b exists such that $(a/b)^2 = 2$.

(2) There is no such rational number.

We work with assumption (1). If it leads to a contradiction, then we will know that (2) must be correct. Start by assuming that a rational number a/b exists with $(a/b)^2 = 2$. Assume also that a/b is written in lowest terms.

(a) Since $(a/b)^2 = 2$, we must have $a^2/b^2 = 2$, or $a^2 = 2b^2$.

Multiply both sides of $\dfrac{a^2}{b^2} = 2$ by b^2.

(b) $2b^2$ is an even number.

$2b^2$ is a multiple of 2.

(c) Therefore, a^2, and a itself, must be even numbers.

a^2 is even because a^2 equals the even number $2b^2$; if a perfect square is even, then its square root is even; therefore, a must be even.

(d) Since a is an even number, it must be a multiple of 2. That is, we can find a natural number c such that $a = 2c$. This changes $a^2 = 2b^2$ into $(2c)^2 = 2b^2$.

Replace a with $2c$.

(e) Therefore, $4c^2 = 2b^2$ or $2c^2 = b^2$.

$(2c)^2 = 4c^2$;   divide both sides by 2.

(f) $2c^2$ is an even number.

$2c^2$ is a multiple of 2.

(g) This makes b^2 an even number, so that b must be even.

If $2c^2$ is even, b^2 is even, since $b^2 = 2c^2$. If a perfect square is even, then its square root is even.

(h) We have reached a contradiction. Show where the contradiction occurs.

We have both a and b even, but we claimed that the fraction $\dfrac{a}{b}$ was in lowest terms.

(i) Since assumption (1) leads to a contradiction, we are forced to accept assumption (2), which says that $\sqrt{2}$ is irrational.

1.2   RESOURCES

Instructor's Test Manual
Test Items, p. 300–301
Instructor's Solution Manual
pp. 3–5, 425–427
Study Guide
pp. 12–16
Student's Solution Manual
pp. 3–5

Test Generator
IBM Disks
5.25″ Disk A
3.5″ Disk A
Mac Disk A
Tutorials
IBM Disks
5.25″ Disk A
3.5″ Disk A
Mac Disk A

HarperCollins Videotape   1
"College Algebra: In Simplest Terms" Program   9

◎ CONCEPTUAL   ✐ WRITING   ▲ CHALLENGING

1.2 EXERCISES (TEXT PAGE 19) ∎

Write the following numbers in numerical order, from smallest to largest. Use a calculator as necessary. See Example 1.

1. $|-8|, -|9|, -|-6|$ $-|9|, -|-6|, |-8|$

2. $-|-9|, -|7|, -|-2|$ $-|-9|, -|7|, -|-2|$

3. $\sqrt{8}, -4, -\sqrt{3}, -2, -5, \sqrt{6}, 3$ $-5, -4, -2, -\sqrt{3}, \sqrt{6}, \sqrt{8}, 3$

4. $\sqrt{2}, -1, 4, 3, \sqrt{8}, -\sqrt{6}, \sqrt{7}$ $-\sqrt{6}, -1, \sqrt{2}, \sqrt{7}, \sqrt{8}, 3, 4$

5. $\dfrac{3}{4}, \sqrt{2}, \dfrac{7}{5}, \dfrac{8}{5}, \dfrac{22}{15}$ $\dfrac{3}{4}, \dfrac{7}{5}, \sqrt{2}, \dfrac{22}{15}, \dfrac{8}{5}$

6. $-\dfrac{9}{8}, -3, -\sqrt{3}, -\sqrt{5}, -\dfrac{9}{5}, -\dfrac{8}{5}$ $-3, -\sqrt{5}, -\dfrac{9}{5}, -\sqrt{3}, -\dfrac{8}{5}, -\dfrac{9}{8}$

7. Explain why it is necessary to reverse the direction of the inequality symbol when multiplying an inequality by a negative number. Give examples to illustrate your explanation.

8. What is wrong with writing the statement "$x < 2$ or $x > 5$" as $5 < x < 2$? An "or" statement cannot be written as a combined inequality. Note that $5 < x < 2$ indicates that $5 < 2$, which is false.

For each of the given inequalities, multiply both sides by the indicated number to get an equivalent inequality. See Example 3(c).

9. $-2 < 5, 3$ $-6 < 15$

10. $-7 < 1, 2$ $-14 < 2$

11. $-11 \geq -22, -\dfrac{1}{11}$ $1 \leq 2$

12. $-2 \geq -14, -\dfrac{1}{2}$ $1 \leq 7$

13. $3x < 9, \dfrac{1}{3}$ $x < 3$

14. $5x > 45, \dfrac{1}{5}$ $x > 9$

15. $-9k > 63, -\dfrac{1}{9}$ $k < -7$

16. $-3p \leq 24, -\dfrac{1}{3}$ $p \geq -8$

Let $x = -4$ and $y = 2$. Evaluate each of the following. See Examples 4, 5, and 7.

17. $|2x|$ 8

18. $|-3y|$ 6

19. $|x - y|$ 6

20. $|2x + 5y|$ 2

21. $|3x + 4y|$ 4

22. $|-5y + x|$ 14

23. $|-4x + y| - |y|$ 16

24. $\dfrac{|-8y + x|}{-|x|}$ -5

25. $\dfrac{|x| + 2|y|}{5 + x}$ 8

26. Explain how to remove the absolute value bars from the expression $|1 - \sqrt{2}|$.

27. Explain the steps necessary to write $|-x^2 - 4|$ without absolute value bars.

Write each of the following without absolute value bars. See Examples 4 and 5.

28. $-|-8| + |-2|$ -6

29. $3 - |-4|$ -1

30. $|2 - \sqrt{3}|$ $2 - \sqrt{3}$

31. $|\sqrt{7} - 5|$ $5 - \sqrt{7}$

32. $|\sqrt{2} - 3|$ $3 - \sqrt{2}$

33. $|\pi - 3|$ $\pi - 3$

34. $|\pi - 5|$ $5 - \pi$

35. $|x - 4|$, if $x > 4$ $x - 4$

36. $|y - 3|$, if $y < 3$ $3 - y$

37. $|2k - 8|$, if $k < 4$ $8 - 2k$

38. $|3r - 15|$, if $r > 5$ $3r - 15$

39. $|-8 - 4m|$, if $m > -2$ $8 + 4m$

40. $|6 - 5r|$, if $r < -2$ $6 - 5r$

41. $|x - y|$, if $x < y$ $y - x$

42. $|x - y|$, if $x > y$ $x - y$

43. $|3 + x^2|$ $3 + x^2$

44. $|x^2 + 4|$ $x^2 + 4$

45. $|-1 - p^2|$ $1 + p^2$

Justify each of the following statements by giving the correct property from this section. Assume that all variables represent real numbers. See Examples 3 and 6.

46. If $2k < 8$, then $k < 4$.   first part of multiplication property of order

47. If $x + 8 < 15$, then $x < 7$.   addition property of order

48. If $-4x < 24$, then $x > -6$.   second part of multiplication property of order

49. If $x < 5$ and $5 < m$, then $x < m$.   transitive property of order

50. If $m > 0$, then $9m > 0$.   first part of multiplication property of order

51. If $k > 0$, then $8 + k > 8$.   addition property of order

52. $|8 + m| \leq |8| + |m|$ triangle inequality, $|a + b| \leq |a| + |b|$

53. $|k - m| \leq |k| + |-m|$ triangle inequality, $|a + b| \leq |a| + |b|$

54. $|8| \cdot |-4| = |-32|$ property of absolute value, $|a| \cdot |b| = |ab|$

55. $|12 + 11r| \geq 0$ property of absolute value, $|a| \geq 0$

56. $\left| \dfrac{-12}{5} \right| = \dfrac{|-12|}{|5|}$ property of absolute value, $\left| \dfrac{a}{b} \right| = \dfrac{|a|}{|b|}$

57. $\left| \dfrac{6}{5} \right| = \dfrac{|6|}{|5|}$ property of absolute value, $\left| \dfrac{a}{b} \right| = \dfrac{|a|}{|b|}$

▲ ◎ *Under what conditions are the following statements true?*

58. $|x| = |y|$ if $x = y$ or $x = -y$

59. $|x + y| = |x| + |y|$ if x and y both have the same sign or if either x or y equals 0

60. $|x + y| = |x| - |y|$ if $y = 0$ or if $|x| \geq |y|$ and x and y have opposite signs

61. $|x| \leq 0$ if $x = 0$

62. $|x - y| = |x| - |y|$ if $|x| \geq |y|$ and either both are positive or both are negative

63. $||x + y|| = |x + y|$ always true

▲ ◎ *Evaluate each of the following if x is a nonzero real number.*

64. $\dfrac{|x|}{x}$ -1 if $x < 0$ and 1 if $x > 0$

65. $\left| \dfrac{|x|}{x} \right|$ 1

1.3 RESOURCES

Instructor's Test Manual
  Test Items, p. 301–302
Instructor's Solution Manual
  pp. 5–9, 427–431
Study Guide
  pp. 16–21
Student's Solution Manual
  pp. 5–9

Test Generator
  IBM Disks
  5.25″ Disk A
  3.5″ Disk A
  Mac Disk A
Tutorials
  IBM Disks
  5.25″ Disk A
  3.5″ Disk A
  Mac Disk A

HarperCollins Videotape 1

1.3 EXERCISES   (TEXT PAGE 26) ■

Use the properties of exponents to simplify each exponential expression. Leave answers with exponents. See Examples 1–3.

1. $(-4)^3 \cdot (-4)^2$ $(-4)^5$ **2.** $(-5)^2 \cdot (-5)^6$ $(-5)^8$ **3.** 2^0 1   **4.** -2^0 -1

5. $(5m)^0$ 1   **6.** $(-4z)^0$ 1   **7.** $(2^2)^5$ 2^{10} **8.** $(6^4)^3$ 6^{12}

◎ CONCEPTUAL   ✎ WRITING   ▲ CHALLENGING

9. $(2x^5y^4)^3$ $2^3x^{15}y^{12}$ **10.** $(-4m^3n^9)^2$ $(-4)^2m^6n^{18}$ **11.** $-\left(\dfrac{p^4}{q}\right)^2$ $-\dfrac{p^8}{q^2}$ **12.** $\left(\dfrac{r^8}{s^2}\right)^3$ $\dfrac{r^{24}}{s^6}$

Identify each of the following as a polynomial *or* not a polynomial. *For each polynomial, give the degree and identify as* monomial, binomial, trinomial, *or* none of these. *See Example 4.*

13. $5x^{11}$ polynomial;   degree 11;   monomial

14. $-9y^{12} + y^2$ polynomial; degree 12; binomial

15. $8p^5q + 6pq$ polynomial;   degree 6;   binomial

16. $2a^6 - 5a^2 + 4a$ polynomial; degree 6; trinomial

17. $\sqrt{2}x + \sqrt{3}x^6$ polynomial;   degree 6;   binomial

18. $-\sqrt{7}m^5n^2 + 2\sqrt{3}m^4n^2$ polynomial;   degree 7;   binomial

19. $\dfrac{1}{3}r^2s - \dfrac{3}{5}r^4s^2 + rs^3$ polynomial;   degree 6;   trinomial

20. $\dfrac{3}{10}p^7 - \dfrac{2}{7}p^5$ polynomial;   degree 7;   binomial   **21.** $\dfrac{5}{p} + \dfrac{2}{p^2} - \dfrac{5}{p^3}$ not a polynomial

22. $\dfrac{q^2}{6} - \dfrac{q^5}{9} + \dfrac{q^8}{10}$ polynomial;   degree 8;   trinomial   **23.** $5\sqrt{z} + 2\sqrt{z^3} - 5\sqrt{z^5}$ not a polynomial

24. $\dfrac{8}{\sqrt{y}} - \dfrac{6}{\sqrt{y^3}} + \dfrac{3}{\sqrt{y^5}}$ not a polynomial

Find each of the following sums and differences. See Example 5.

25. $(3x^2 - 4x + 5) + (-2x^2 + 3x - 2)$ $x^2 - x + 3$

26. $(4m^3 - 3m^2 + 5) + (-3m^3 - m^2 + 5)$ $m^3 - 4m^2 + 10$

27. $(12y^2 - 8y + 6) - (3y^2 - 4y + 2)$ $9y^2 - 4y + 4$ **28.** $(8p^2 - 5p) - (3p^2 - 2p + 4)$ $5p^2 - 3p - 4$

29. $(6m^4 - 3m^2 + m) - (2m^3 + 5m^2 + 4m) + (m^2 - m)$ $6m^4 - 2m^3 - 7m^2 - 4m$

30. $-(8x^3 + x - 3) + (2x^3 + x^2) - (4x^2 + 3x - 1)$ $-6x^3 - 3x^2 - 4x + 4$

Find each of the following products. See Examples 6–7.

31. $(4r - 1)(7r + 2)$ $28r^2 + r - 2$ **32.** $(5m - 6)(3m + 4)$ $15m^2 + 2m - 24$

33. $\left(3x - \dfrac{2}{3}\right)\left(5x + \dfrac{1}{3}\right)$ $15x^2 - \dfrac{7}{3}x - \dfrac{2}{9}$ **34.** $\left(2m - \dfrac{1}{4}\right)\left(3m + \dfrac{1}{2}\right)$ $6m^2 + \dfrac{1}{4}m - \dfrac{1}{8}$

35. $4x^2(3x^3 + 2x^2 - 5x + 1)$ $12x^5 + 8x^4 - 20x^3 + 4x^2$

36. $2b^3(b^2 - 4b + 3)$ $2b^5 - 8b^4 + 6b^3$ **37.** $(2z - 1)(-z^2 + 3z - 4)$ $-2z^3 + 7z^2 - 11z + 4$

38. $(k + 2)(12k^3 - 3k^2 + k + 1)$ $12k^4 + 21k^3 - 5k^2 + 3k + 2$

39. $(m - n + k)(m + 2n - 3k)$ $m^2 + mn - 2n^2 - 2km + 5kn - 3k^2$

40. $(r - 3s + t)(2r - s + t)$ $2r^2 - 7rs + 3s^2 + 3rt - 4st + t^2$

41. $(a - b + 2c)^2$ $a^2 - 2ab + b^2 + 4ac - 4bc + 4c^2$

42. $(k - y + 3m)^2$ $k^2 - 2ky + y^2 + 6km - 6ym + 9m^2$

✐ **43.** State the formula for the square of a binomial in words.

✐ **44.** State the formula for the product of the sum and difference of two terms in words.

Use the special products formulas for each of the following. See Examples 8 and 9.

45. $(2m + 3)(2m - 3)$ $4m^2 - 9$ **46.** $(8s - 3t)(8s + 3t)$ $64s^2 - 9t^2$

47. $(4m + 2n)^2$ $16m^2 + 16mn + 4n^2$ **48.** $(a - 6b)^2$ $a^2 - 12ab + 36b^2$

49. $(5r + 3t^2)^2$ $25r^2 + 30rt^2 + 9t^4$ **50.** $(2z^4 - 3y)^2$ $4z^8 - 12z^4y + 9y^2$

51. $[(2p - 3) + q]^2$ $4p^2 - 12p + 9 + 4pq - 6q + q^2$

52. $[(4y - 1) + z]^2$ $16y^2 - 8y + 1 + 8yz - 2z + z^2$

53. $[(3q + 5) - p][(3q + 5) + p]$ $9q^2 + 30q + 25 - p^2$

54. $[(9r - s) + 2][(9r - s) - 2]$ $81r^2 - 18rs + s^2 - 4$

▲ **55.** $[a + (b + c)][a - (b + c)]$ $a^2 - b^2 - 2bc - c^2$

▲ **56.** $[(k + 2) + r][(k + 2) - r]$ $k^2 + 4k + 4 - r^2$

▲ **57.** $[(3a + b) - 1]^2$ $9a^2 + 6ab + b^2 - 6a - 2b + 1$

▲ **58.** $[(2m + 7) - n]^2$ $4m^2 + 28m + 49 - 4mn - 14n + n^2$

Perform the indicated operations.

59. $(6p + 5q)(3p - 7q)$ $18p^2 - 27pq - 35q^2$

60. $(2p - 1)(3p^2 - 4p + 5)$ $6p^3 - 11p^2 + 14p - 5$

61. $(3x - 4y)^3$ $27x^3 - 108x^2y + 144xy^2 - 64y^3$

62. $(r^5 - r^3 + r) + (3r^5 - 4r^4 + r^3 + 2r)$ $4r^5 - 4r^4 + 3r$

63. $(6k - 3)^2$ $36k^2 - 36k + 9$

64. $(4x + 3y)(4x - 3y)$ $16x^2 - 9y^2$

65. $(p^3 - 4p^2 + p) - (3p^2 + 2p + 7)$ $p^3 - 7p^2 - p - 7$

66. $(2z + y)(3z - 4y)$ $6z^2 - 5zy - 4y^2$

67. $(7m + 2n)(7m - 2n)$ $49m^2 - 4n^2$

68. $(3p + 5)^2$ $9p^2 + 30p + 25$

69. $-3(4q^2 - 3q + 2) + 2(-q^2 + q - 4)$ $-14q^2 + 11q - 14$

70. $2(3r^2 + 4r + 2) - 3(-r^2 + 4r - 5)$ $9r^2 - 4r + 19$

71. $p(4p - 6) + 2(3p - 8)$ $4p^2 - 16$

72. $m(5m - 2) + 9(5 - m)$ $5m^2 - 11m + 45$

73. $-y(y^2 - 4) + 6y^2(2y - 3)$ $11y^3 - 18y^2 + 4y$

74. $-z^3(9 - z) + 4z(2 + 3z)$ $z^4 - 9z^3 + 12z^2 + 8z$

◎ *In each of the following, find the coefficient of x^3 without finding the entire product.*

75. $(x^2 + 4x)(-3x^2 + 4x - 1)$ -8

76. $(4x^3 - 2x + 5)(x^3 + 2)$ 13

77. $(1 + x^2)(1 + x)$ 1

78. $(3 - x)(2 - x^2)$ 1

79. $x^2(4 - 3x)^2$ -24

80. $-4x^2(2 - x)(2 + x)$ 0

▲ *Find each of the following products. Assume that all variables used in exponents represent non-negative integers.*

81. $(k^m + 2)(k^m - 2)$ $k^{2m} - 4$

82. $(y^x - 4)(y^x + 4)$ $y^{2x} - 16$

83. $(b^r + 3)(b^r - 2)$ $b^{2r} + b^r - 6$

84. $(q^y + 4)(q^y + 3)$ $q^{2y} + 7q^y + 12$

85. $(3p^x + 1)(p^x - 2)$ $3p^{2x} - 5p^x - 2$

86. $(2^a + 5)(2^a + 3)$ $2^{2a} + 8 \cdot 2^a + 15$

87. $(m^x - 2)^2$ $m^{2x} - 4m^x + 4$

88. $(z^r + 5)^2$ $z^{2r} + 10z^r + 25$

89. $(q^p - 5p^q)^2$ $q^{2p} - 10q^p p^q + 25p^{2q}$

90. $(3y^x - 2x^y)^2$ $9y^{2x} - 12y^x x^y + 4x^{2y}$

91. $(3k^a - 2)^3$ $27k^{3a} - 54k^{2a} + 36k^a - 8$

92. $(r^x - 4)^3$ $r^{3x} - 12r^{2x} + 48r^x - 64$

◎ *Suppose one polynomial has degree m and another has degree n, where m and n are natural numbers with $n < m$. Find the degree of the following for the polynomials.*

93. Sum m

94. Difference m

95. Product $m + n$

96. What would be the degree of the square of the polynomial of degree m? $2m$

1.4 RESOURCES

Instructor's Test Manual
Test Items, p. 303
Instructor's Solution Manual
pp. 9–12, 431–434
Study Guide
pp. 22–24
Student's Solution Manual
pp. 9–12

Test Generator
IBM Disks
5.25" Disk A
3.5" Disk A
Mac Disk A
Tutorials
IBM Disks
5.25" Disk A
3.5" Disk A
Mac Disk A

HarperCollins Videotape 1

1.4 EXERCISES (TEXT PAGE 32) ■

Evaluate each of the following. See Examples 1 and 2.

1. $\dfrac{6!}{3!3!}$ 20

2. $\dfrac{5!}{2!3!}$ 10

3. $\dfrac{7!}{3!4!}$ 35

4. $\dfrac{8!}{5!3!}$ 56

5. $\dbinom{8}{3}$ 56

6. $\dbinom{7}{4}$ 35

7. $\dbinom{10}{8}$ 45

8. $\dbinom{9}{6}$ 84

9. $\dbinom{n}{n-1}$ n

10. $\dbinom{n}{n-2}$ $\dfrac{n(n-1)}{2}$

11. Describe in your own words how you would determine the binomial coefficient for the fifth term in the expansion of $(x + y)^8$.

12. How many terms are there in the expansion of $(x + y)^{10}$? 11

13. What is true of the signs (positive/negative) of the terms in the expansion of $(x - y)^n$?
they alternate, starting with positive

14. In the expansion of $(a + 5b)^n$, what quantity replaces y in the binomial theorem, 5 or $5b$? $5b$

Write out the binomial expansion for each of the following. See Examples 3 and 4.

15. $(x + y)^6$ $x^6 + 6x^5y + 15x^4y^2 + 20x^3y^3 + 15x^2y^4 + 6xy^5 + y^6$

16. $(m + n)^4$ $m^4 + 4m^3n + 6m^2n^2 + 4mn^3 + n^4$

17. $(p - q)^5$ $p^5 - 5p^4q + 10p^3q^2 - 10p^2q^3 + 5pq^4 - q^5$

18. $(a - b)^7$ $a^7 - 7a^6b + 21a^5b^2 - 35a^4b^3 + 35a^3b^4 - 21a^2b^5 + 7ab^6 - b^7$

19. $(r^2 + s)^5$ $r^{10} + 5r^8s + 10r^6s^2 + 10r^4s^3 + 5r^2s^4 + s^5$

20. $(m + n^2)^4$ $m^4 + 4m^3n^2 + 6m^2n^4 + 4mn^6 + n^8$ **21.** $(p + 2q)^4$ $p^4 + 8p^3q + 24p^2q^2 + 32pq^3 + 16q^4$

22. $(3r - s)^6$ $729r^6 - 1458r^5s + 1215r^4s^2 - 540r^3s^3 + 135r^2s^4 - 18rs^5 + s^6$

23. $(7p + 2q)^4$ $2401p^4 + 2744p^3q + 1176p^2q^2 + 224pq^3 + 16q^4$

24. $(4a - 5b)^5$ $1024a^5 - 6400a^4b + 16{,}000a^3b^2 - 20{,}000a^2b^3 + 12{,}500ab^4 - 3125b^5$

25. $(3x - 2y)^6$ $729x^6 - 2916x^5y + 4860x^4y^2 - 4320x^3y^3 + 2160x^2y^4 - 576xy^5 + 64y^6$

26. $(7k - 9j)^4$ $2401k^4 - 12{,}348k^3j + 23{,}814k^2j^2 - 20{,}412kj^3 + 6561j^4$

27. $\left(\dfrac{m}{2} - 1\right)^6$ $\dfrac{m^6}{64} - \dfrac{3m^5}{16} + \dfrac{15m^4}{16} - \dfrac{5m^3}{2} + \dfrac{15m^2}{4} - 3m + 1$

28. $\left(3 + \dfrac{y}{3}\right)^5$ $243 + 135y + 30y^2 + \dfrac{10y^3}{3} + \dfrac{5y^4}{27} + \dfrac{y^5}{243}$

29. $\left(\sqrt{2}r + \dfrac{1}{m}\right)^4$ $4r^4 + \dfrac{8\sqrt{2}r^3}{m} + \dfrac{12r^2}{m^2} + \dfrac{4\sqrt{2}r}{m^3} + \dfrac{1}{m^4}$

30. $\left(\dfrac{1}{k} - \sqrt{3}p\right)^3$ $\dfrac{1}{k^3} - \dfrac{3\sqrt{3}p}{k^2} + \dfrac{9p^2}{k} - 3\sqrt{3}p^3$

For each of the following, write the indicated term of the binomial expansion. See Example 5.

31. Fifth term of $(m - 2p)^{12}$ $7920m^8p^4$

32. Fourth term of $(3x + y)^6$ $540x^3y^3$

33. Sixth term of $(x + y)^9$ $126x^4y^5$

34. Twelfth term of $(a - b)^{15}$ $-1365a^4b^{11}$

35. Ninth term of $(2m + n)^{10}$ $180m^2n^8$

36. Seventh term of $(3r - 5s)^{12}$ $10,524,937,500r^6s^6$

37. Seventeenth term of $(p^2 + q)^{20}$ $4845p^8q^{16}$

38. Tenth term of $(2x^2 + y)^{14}$ $64,064x^{10}y^9$

39. Eighth term of $(x^3 + 2y)^{14}$ $439,296x^{21}y^7$

40. Thirteenth term of $(a + 2b^3)^{12}$ $4096b^{36}$

▲ 🖩 *In calculus it is shown that*

$$(1 + x)^n = 1 + nx + \frac{n(n - 1)}{2!}x^2 + \frac{n(n - 1)(n - 2)}{3!}x^3 + \cdots$$

for any real number n (not just positive integers) and any real number x, where $|x| < 1$. This result, a generalized binomial theorem, may be used to find approximate values of powers or roots. For example,

$$\sqrt[4]{630} = (625 + 5)^{1/4} = \left[625\left(1 + \frac{5}{625}\right)\right]^{1/4} = 625^{1/4}\left(1 + \frac{5}{625}\right)^{1/4}.$$

41. Use the expression given above for $(1 + x)^n$ to approximate $\left(1 + \dfrac{5}{625}\right)^{1/4}$ to the nearest thousandth. Then approximate $\sqrt[4]{630}$. 1.002; 5.010

42. Approximate $\sqrt[3]{9.42}$, using this method. 2.112

43. Approximate $(1.02)^{-3}$. $.942$

▲ **44.** Show that $\dbinom{n}{2} = \dfrac{n(n - 1)}{2!}$ and $\dbinom{n}{3} = \dfrac{n(n - 1)(n - 2)}{3!}$.

1.5 RESOURCES

Instructor's Test Manual
   Test Items, p. 303–305

Instructor's Solution Manual
   pp. 12–17, 434–439

Study Guide
   pp. 24–30

Student's Solution Manual
   pp. 12–17

Test Generator
   IBM Disks
   5.25" Disk A
   3.5" Disk A
   Mac Disk A

Tutorials
   IBM Disks
   5.25" Disk A
   3.5" Disk A
   Mac Disk A

HarperCollins Videotape   1

© CONCEPTUAL   ✎ WRITING   ▲ CHALLENGING

1.5 EXERCISES (TEXT PAGE 39) ■

Factor the greatest common factor from each polynomial. See Example 1.

1. $4k^2m^3 + 8k^4m^3 - 12k^2m^4$ $4k^2m^3(1 + 2k^2 - 3m)$

2. $28r^4s^2 + 7r^3s - 35r^4s^3$ $7r^3s(4rs + 1 - 5rs^2)$

3. $2(a + b) + 4m(a + b)$ $2(a + b)(1 + 2m)$

4. $4(y - 2)^2 + 3(y - 2)$ $(y - 2)(4y - 5)$

5. $(2y - 3)(y + 2) + (y + 5)(y + 2)$ $(y + 2)(3y + 2)$

6. $(6a - 1)(a + 2) + (6a - 1)(3a - 1)$ $(6a - 1)(4a + 1)$

7. $(5r - 6)(r + 3) - (2r - 1)(r + 3)$ $(r + 3)(3r - 5)$

8. $(3z + 2)(z + 4) - (z + 6)(z + 4)$ $2(z + 4)(z - 2)$

9. $2(m - 1) - 3(m - 1)^2 + 2(m - 1)^3$ $(m - 1)(2m^2 - 7m + 7)$

10. $5(a + 3)^3 - 2(a + 3) + (a + 3)^2$ $(a + 3)(5a^2 + 31a + 46)$

Factor each of the following by grouping. See Example 2.

11. $6st + 9t - 10s - 15$ $(2s + 3)(3t - 5)$

12. $10ab - 6b + 35a - 21$ $(5a - 3)(2b + 7)$

13. $rt^3 + rs^2 - pt^3 - ps^2$ $(t^3 + s^2)(r - p)$

14. $2m^4 + 6 - am^4 - 3a$ $(m^4 + 3)(2 - a)$

15. $6p^2 - 14p + 15p - 35$ $(3p - 7)(2p + 5)$

16. $8r^2 - 10r + 12r - 15$ $(4r - 5)(2r + 3)$

17. $20z^2 - 8zx - 45zx + 18x^2$ $(5z - 2x)(4z - 9x)$

18. $16a^2 + 10ab - 24ab - 15b^2$ $(8a + 5b)(2a - 3b)$

19. $15 - 5m^2 - 3r^2 + m^2r^2$ $(3 - m^2)(5 - r^2)$

Factor each trinomial. See Example 3.

20. $6a^2 - 48a - 120$ $6(a - 10)(a + 2)$

21. $8h^2 - 24h - 320$ $8(h - 8)(h + 5)$

22. $3m^3 + 12m^2 + 9m$ $3m(m + 1)(m + 3)$

23. $9y^4 - 54y^3 + 45y^2$ $9y^2(y - 5)(y - 1)$

24. $6k^2 + 5kp - 6p^2$ $(3k - 2p)(2k + 3p)$

25. $14m^2 + 11mr - 15r^2$ $(7m - 5r)(2m + 3r)$

26. $5a^2 - 7ab - 6b^2$ $(5a + 3b)(a - 2b)$

27. $12s^2 + 11st - 5t^2$ $(4s + 5t)(3s - t)$

28. $9x^2 - 6x^3 + x^4$ $x^2(3 - x)^2$

29. $30a^2 + am - m^2$ $(5a + m)(6a - m)$

30. $24a^4 + 10a^3b - 4a^2b^2$ $2a^2(4a - b)(3a + 2b)$

31. $18x^5 + 15x^4z - 75x^3z^2$ $3x^3(3x - 5z)(2x + 5z)$

🖉 **32.** When asked to factor $6x^4 - 3x^2 - 3$ completely, a student gave the following result:
$6x^4 - 3x^2 - 3 = (2x^2 + 1)(3x^2 - 3)$. Is this answer correct? Explain why.

Factor each perfect square trinomial. See Example 4.

33. $9m^2 - 12m + 4$ $(3m - 2)^2$

34. $16p^2 - 40p + 25$ $(4p - 5)^2$

35. $32a^2 - 48ab + 18b^2$ $2(4a - 3b)^2$

36. $20p^2 - 100pq + 125q^2$ $5(2p - 5q)^2$

37. $4x^2y^2 + 28xy + 49$ $(2xy + 7)^2$

38. $9m^2n^2 - 12mn + 4$ $(3mn - 2)^2$

39. $(a - 3b)^2 - 6(a - 3b) + 9$ $(a - 3b - 3)^2$

40. $(2p + q)^2 - 10(2p + q) + 25$ $(2p + q - 5)^2$

Factor each difference of two squares. See Example 5.

41. $9a^2 - 16$ $(3a + 4)(3a - 4)$

42. $16q^2 - 25$ $(4q + 5)(4q - 5)$

43. $25s^4 - 9t^2$ $(5s^2 + 3t)(5s^2 - 3t)$

44. $36z^2 - 81y^4$ $9(2z + 3y^2)(2z - 3y^2)$

45. $(a + b)^2 - 16$ $(a + b + 4)(a + b - 4)$

46. $(p - 2q)^2 - 100$ $(p - 2q + 10)(p - 2q - 10)$

47. $p^4 - 625$ $(p^2 + 25)(p + 5)(p - 5)$

48. $m^4 - 81$ $(m^2 + 9)(m + 3)(m - 3)$

◎ **49.** Which of the following is the correct complete factorization of $x^4 - 1$?
   (a) $(x^2 - 1)(x^2 + 1)$ **(b)** $(x^2 + 1)(x + 1)(x - 1)$ **(c)** $(x^2 - 1)^2$ **(d)** $(x - 1)^2(x + 1)^2$ (b)

◎ **50.** Which of the following is the correct factorization of $x^3 + 8$?
   (a) $(x + 2)^3$ **(b)** $(x + 2)(x^2 + 2x + 4)$ **(c)** $(x + 2)(x^2 - 2x + 4)$ **(d)** $(x + 2)(x^2 - 4x + 4)$ (c)

Factor each sum or difference of cubes. See Example 6.

51. $8 - a^3$ $(2 - a)(4 + 2a + a^2)$

52. $r^3 + 27$ $(r + 3)(r^2 - 3r + 9)$

53. $125x^3 - 27$ $(5x - 3)(25x^2 + 15x + 9)$

54. $8m^3 - 27n^3$ $(2m - 3n)(4m^2 + 6mn + 9n^2)$

55. $27y^9 + 125z^6$ $(3y^3 + 5z^2)(9y^6 - 15y^3z^2 + 25z^4)$

56. $27z^3 + 729y^3$ $27(z + 3y)(z^2 - 3zy + 9y^2)$

57. $(r + 6)^3 - 216$ $r(r^2 + 18r + 108)$

58. $(b + 3)^3 - 27$ $b(b^2 + 9b + 27)$

59. $27 - (m + 2n)^3$ $(3 - m - 2n)(9 + 3m + 6n + m^2 + 4mn + 4n^2)$

60. $125 - (4a - b)^3$ $(5 - 4a + b)(25 + 20a - 5b + 16a^2 - 8ab + b^2)$

🖊 **61.** Is the following factorization of $3a^4 + 14a^2 - 5$ correct? Explain. If it is incorrect, give the correct factors.

$$3a^4 + 14a^2 - 5 = 3u^2 + 14u - 5 \quad \text{Let } u = a^2.$$
$$= (3u - 1)(u + 5)$$

Factor each of the following, using the method of substitution. See Example 7.

62. $m^4 - 3m^2 - 10$ $(m^2 - 5)(m^2 + 2)$

63. $a^4 - 2a^2 - 48$ $(a^2 - 8)(a^2 + 6)$

64. $7(3k - 1)^2 + 26(3k - 1) - 8$ $9(7k - 3)(k + 1)$

65. $6(4z - 3)^2 + 7(4z - 3) - 3$ $2(8z - 3)(6z - 5)$

66. $9(a - 4)^2 + 30(a - 4) + 25$ $(3a - 7)^2$

67. $20(4 - p)^2 - 3(4 - p) - 2$ $(18 - 5p)(17 - 4p)$

Factor by any method.

68. $a^3(r + s) + b^2(r + s)$ $(r + s)(a^3 + b^2)$

69. $4b^2 + 4bc + c^2 - 16$ $(2b + c + 4)(2b + c - 4)$

70. $(2y - 1)^2 - 4(2y - 1) + 4$ $(2y - 3)^2$

71. $x^2 + xy - 5x - 5y$ $(x + y)(x - 5)$

72. $8r^2 - 3rs + 10s^2$ prime

73. $p^4(m - 2n) + q(m - 2n)$ $(m - 2n)(p^4 + q)$

74. $36a^2 + 60a + 25$ $(6a + 5)^2$

75. $4z^2 + 28z + 49$ $(2z + 7)^2$

76. $6p^4 + 7p^2 - 3$ $(3p^2 - 1)(2p^2 + 3)$

77. $1000x^3 + 343y^3$ $(10x + 7y)(100x^2 - 70xy + 49y^2)$

78. $b^2 + 8b + 16 - a^2$ $(b + 4 + a)(b + 4 - a)$

79. $125m^6 - 216$ $(5m^2 - 6)(25m^4 + 30m^2 + 36)$

80. $q^2 + 6q + 9 - p^2$ $(q + 3 + p)(q + 3 - p)$

81. $12m^2 + 16mn - 35n^2$ $(6m - 7n)(2m + 5n)$

82. $216p^3 + 125q^3$ $(6p + 5q)(36p^2 - 30pq + 25q^2)$

83. $4p^2 + 3p - 1$ $(4p - 1)(p + 1)$

84. $100r^2 - 169s^2$ $(10r + 13s)(10r - 13s)$

85. $144z^2 + 121$ prime

86. $(3a + 5)^2 - 18(3a + 5) + 81$ $(3a - 4)^2$

87. $(x + y)^2 - (x - y)^2$ $4xy$

88. $4z^4 - 7z^2 - 15$ $(4z^2 + 5)(z^2 - 3)$

▲ *Factor each of the following polynomials. Assume that all variables used in exponents represent positive integers. (See Example 5(d).)*

89. $r^2 + rs^q - 6s^{2q}$ $(r + 3s^q)(r - 2s^q)$

90. $6z^{2a} - z^a x^b - x^{2b}$ $(3z^a + x^b)(2z^a - x^b)$

91. $9a^{4k} - b^{8k}$ $(3a^{2k} + b^{4k})(3a^{2k} - b^{4k})$

92. $16y^{2c} - 25x^{4c}$ $(4y^c + 5x^{2c})(4y^c - 5x^{2c})$

93. $4y^{2a} - 12y^a + 9$ $(2y^a - 3)^2$

94. $25x^{4c} - 20x^{2c} + 4$ $(5x^{2c} - 2)^2$

95. $6(m + p)^{2k} + (m + p)^k - 15$ $[3(m + p)^k + 5][2(m + p)^k - 3]$

96. $8(2k + q)^{4z} - 2(2k + q)^{2z} - 3$ $[4(2k + q)^{2z} - 3][2(2k + q)^{2z} + 1]$

▲ *Find a value of b or c that will make the following expressions perfect square trinomials.*

97. $4z^2 + bz + 81$ ±36

98. $9p^2 + bp + 25$ ±30

99. $100r^2 - 60r + c$ 9

100. $49x^2 + 70x + c$ 25

1.6 RESOURCES

Instructor's Test Manual
  Test Items, p. 305–306
Instructor's Solution Manual
  pp. 17–23, 439–445
Study Guide
  pp. 31–35
Student's Solution Manual
  pp. 17–23

Test Generator
  IBM Disks
    5.25″ Disk A
    3.5″ Disk A
    Mac Disk A
Tutorials
  IBM Disks
    5.25″ Disk A
    3.5″ Disk A
    Mac Disk A

HarperCollins Videotape 1

1.6 EXERCISES (TEXT PAGE 46) ∎

For each of the following, give the restrictions on the variable.

1. $\dfrac{x-2}{x+6}$ $x \neq -6$

2. $\dfrac{x+5}{x-3}$ $x \neq 3$

3. $\dfrac{2x}{5x-3}$ $x \neq \dfrac{3}{5}$

4. $\dfrac{6x}{2x-1}$ $x \neq \dfrac{1}{2}$

5. $\dfrac{-8}{x^2+1}$ no restrictions

6. $\dfrac{3x}{3x^2+7}$ no restrictions

7. Which one of the following expressions is equivalent to $\dfrac{x^2+4x+3}{x+1}$?

  (a) $x+3$ **(b)** $x+7$ **(c)** $5x+3$ **(d)** x^2+7 (a)

8. Explain why $\dfrac{2x+3}{2x+5} = \dfrac{3}{5}$ is false for all real numbers x.

Write each of the following in lowest terms. See Example 1.

9. $\dfrac{25p^3}{10p^2}$ $\dfrac{5p}{2}$

10. $\dfrac{14z^3}{6z^2}$ $\dfrac{7z}{3}$

11. $\dfrac{8k+16}{9k+18}$ $\dfrac{8}{9}$

12. $\dfrac{20r+10}{30r+15}$ $\dfrac{2}{3}$

13. $\dfrac{3(t+5)}{(t+5)(t-3)}$ $\dfrac{3}{t-3}$

14. $\dfrac{-8(y-4)}{(y+2)(y-4)}$ $\dfrac{-8}{y+2}$

15. $\dfrac{8x^2+16x}{4x^2}$ $\dfrac{2x+4}{x}$

16. $\dfrac{36y^2+72y}{9y}$ $4y+8$

17. $\dfrac{m^2-4m+4}{m^2+m-6}$ $\dfrac{m-2}{m+3}$

18. $\dfrac{r^2-r-6}{r^2+r-12}$ $\dfrac{r+2}{r+4}$

19. $\dfrac{8m^2+6m-9}{16m^2-9}$ $\dfrac{2m+3}{4m+3}$

20. $\dfrac{6y^2+11y+4}{3y^2+7y+4}$ $\dfrac{2y+1}{y+1}$

Find each of the following products or quotients. See Example 2.

21. $\dfrac{15p^3}{9p^2} \div \dfrac{6p}{10p^2}$ $\dfrac{25p^2}{9}$

22. $\dfrac{3r^2}{9r^3} \div \dfrac{8r^3}{6r}$ $\dfrac{1}{4r^3}$

23. $\dfrac{2k+8}{6} \div \dfrac{3k+12}{2}$ $\dfrac{2}{9}$

24. $\dfrac{5m+25}{10} \cdot \dfrac{12}{6m+30}$ 1

25. $\dfrac{x^2+x}{5} \cdot \dfrac{25}{xy+y}$ $\dfrac{5x}{y}$

26. $\dfrac{3m-15}{4m-20} \cdot \dfrac{m^2-10m+25}{12m-60}$ $\dfrac{m-5}{16}$

27. $\dfrac{4a+12}{2a-10} \div \dfrac{a^2-9}{a^2-a-20}$ $\dfrac{2(a+4)}{a-3}$

28. $\dfrac{6r-18}{9r^2+6r-24} \cdot \dfrac{12r-16}{4r-12}$ $\dfrac{2}{r+2}$

29. $\dfrac{p^2 - p - 12}{p^2 - 2p - 15} \cdot \dfrac{p^2 - 9p + 20}{p^2 - 8p + 16}$ 1

30. $\dfrac{x^2 + 2x - 15}{x^2 + 11x + 30} \cdot \dfrac{x^2 + 2x - 24}{x^2 - 8x + 15}$ $\dfrac{x - 4}{x - 5}$

31. $\dfrac{m^2 + 3m + 2}{m^2 + 5m + 4} \div \dfrac{m^2 + 5m + 6}{m^2 + 10m + 24}$ $\dfrac{m + 6}{m + 3}$

32. $\dfrac{y^2 + y - 2}{y^2 + 3y - 4} \div \dfrac{y^2 + 3y + 2}{y^2 + 4y + 3}$ $\dfrac{y + 3}{y + 4}$

33. $\dfrac{2m^2 - 5m - 12}{m^2 - 10m + 24} \div \dfrac{4m^2 - 9}{m^2 - 9m + 18}$ $\dfrac{m - 3}{2m - 3}$

34. $\dfrac{6n^2 - 5n - 6}{6n^2 + 5n - 6} \cdot \dfrac{12n^2 - 17n + 6}{12n^2 - n - 6}$ $\dfrac{2n - 3}{2n + 3}$

35. $\left(1 + \dfrac{1}{x}\right)\left(1 - \dfrac{1}{x}\right)$ $\dfrac{x^2 - 1}{x^2}$

36. $\left(3 + \dfrac{2}{y}\right)\left(3 - \dfrac{2}{y}\right)$ $\dfrac{9y^2 - 4}{y^2}$

37. $\dfrac{x^3 + y^3}{x^2 - y^2} \cdot \dfrac{x + y}{x^2 - xy + y^2}$ $\dfrac{x + y}{x - y}$

38. $\dfrac{8y^3 - 125}{4y^2 - 20y + 25} \cdot \dfrac{2y - 5}{y}$ $\dfrac{4y^2 + 10y + 25}{y}$

39. $\dfrac{x^3 + y^3}{x^3 - y^3} \cdot \dfrac{x^2 - y^2}{x^2 + 2xy + y^2}$ $\dfrac{x^2 - xy + y^2}{x^2 + xy + y^2}$

40. $\dfrac{x^2 - y^2}{(x - y)^2} \cdot \dfrac{x^2 - xy + y^2}{x^2 - 2xy + y^2} \div \dfrac{x^3 + y^3}{(x - y)^4}$ $x - y$

◉ **41.** Which of the following rational expressions equals -1?

   (a) $\dfrac{x - 4}{x + 4}$ **(b)** $\dfrac{-x - 4}{x + 4}$ **(c)** $\dfrac{x - 4}{4 - x}$ **(d)** $\dfrac{x - 4}{-x - 4}$ **(b) and (c)**

✐ **42.** In your own words, explain how to find the least common denominator for two fractions.

Perform each of the following additions or subtractions. See Example 3.

43. $\dfrac{3}{2k} + \dfrac{5}{3k}$ $\dfrac{19}{6k}$

44. $\dfrac{8}{5p} + \dfrac{3}{4p}$ $\dfrac{47}{20p}$

45. $\dfrac{a + 1}{2} - \dfrac{a - 1}{2}$ 1

46. $\dfrac{y + 6}{5} - \dfrac{y - 6}{5}$ $\dfrac{12}{5}$

47. $\dfrac{3}{p} + \dfrac{1}{2}$ $\dfrac{6 + p}{2p}$

48. $\dfrac{9}{r} - \dfrac{2}{3}$ $\dfrac{27 - 2r}{3r}$

49. $\dfrac{1}{6m} + \dfrac{2}{5m} + \dfrac{4}{m}$ $\dfrac{137}{30m}$

50. $\dfrac{8}{3p} + \dfrac{5}{4p} + \dfrac{9}{2p}$ $\dfrac{101}{12p}$

51. $\dfrac{1}{a + 1} - \dfrac{1}{a - 1}$ $\dfrac{-2}{(a + 1)(a - 1)}$

52. $\dfrac{1}{x + z} + \dfrac{1}{x - z}$ $\dfrac{2x}{(x + z)(x - z)}$

53. $\dfrac{m + 1}{m - 1} + \dfrac{m - 1}{m + 1}$ $\dfrac{2m^2 + 2}{(m - 1)(m + 1)}$

54. $\dfrac{2}{x - 1} + \dfrac{1}{1 - x}$ $\dfrac{1}{x - 1}$ or $\dfrac{-1}{1 - x}$

55. $\dfrac{3}{a - 2} - \dfrac{1}{2 - a}$ $\dfrac{4}{a - 2}$ or $\dfrac{-4}{2 - a}$

56. $\dfrac{q}{p - q} - \dfrac{q}{q - p}$ $\dfrac{2q}{p - q}$ or $\dfrac{-2q}{q - p}$

57. $\dfrac{x + y}{2x - y} - \dfrac{2x}{y - 2x}$ $\dfrac{3x + y}{2x - y}$ or $\dfrac{-3x - y}{y - 2x}$

58. $\dfrac{m - 4}{3m - 4} + \dfrac{3m + 2}{4 - 3m}$ $\dfrac{-2m - 6}{3m - 4}$ or $\dfrac{2m + 6}{4 - 3m}$

59. $\dfrac{1}{a^2 - 5a + 6} - \dfrac{1}{a^2 - 4}$ $\dfrac{5}{(a - 2)(a - 3)(a + 2)}$

60. $\dfrac{-3}{m^2 - m - 2} - \dfrac{1}{m^2 + 3m + 2}$ $\dfrac{-4}{(m - 2)(m + 2)}$

61. $\dfrac{1}{x^2 + x - 12} - \dfrac{1}{x^2 - 7x + 12} + \dfrac{1}{x^2 - 16}$ $\dfrac{x - 11}{(x + 4)(x - 4)(x - 3)}$

62. $\dfrac{2}{2p^2 - 9p - 5} + \dfrac{p}{3p^2 - 17p + 10} - \dfrac{2p}{6p^2 - p - 2}$ $\dfrac{17p - 4}{(2p + 1)(p - 5)(3p - 2)}$

63. $\dfrac{3a}{a^2 + 5a - 6} - \dfrac{2a}{a^2 + 7a + 6}$ $\dfrac{a^2 + 5a}{(a + 6)(a + 1)(a - 1)}$

64. $\dfrac{2k}{k^2 + 4k + 3} + \dfrac{3k}{k^2 + 5k + 6}$ $\dfrac{5k^2 + 7k}{(k + 3)(k + 1)(k + 2)}$

Perform the indicated operations. See Example 4.

65. $\dfrac{1 + \dfrac{1}{x}}{1 - \dfrac{1}{x}}$ $\dfrac{x + 1}{x - 1}$

66. $\dfrac{2 - \dfrac{2}{y}}{2 + \dfrac{2}{y}}$ $\dfrac{y - 1}{y + 1}$

67. $\dfrac{\dfrac{1}{x + 1} - \dfrac{1}{x}}{\dfrac{1}{x}}$ $\dfrac{-1}{x + 1}$

68. $\dfrac{\dfrac{1}{y + 3} - \dfrac{1}{y}}{\dfrac{1}{y}}$ $\dfrac{-3}{y + 3}$

◉ CONCEPTUAL   ✐ WRITING   ▲ CHALLENGING

69. $\dfrac{1 + \dfrac{1}{1-b}}{1 - \dfrac{1}{1+b}}$ $\dfrac{(2-b)(1+b)}{b(1-b)}$

70. $m - \dfrac{m}{m + \dfrac{1}{2}}$ $\dfrac{m(2m-1)}{2m+1}$

71. $\dfrac{m - \dfrac{1}{m^2-4}}{\dfrac{1}{m+2}}$ $\dfrac{m^3 - 4m - 1}{m-2}$

72. $\dfrac{\dfrac{3}{p^2-16} + p}{\dfrac{1}{p-4}}$ $\dfrac{p^3 - 16p + 3}{p+4}$

73. $\left(\dfrac{3}{p-1} - \dfrac{2}{p+1}\right)\left(\dfrac{p-1}{p}\right)$ $\dfrac{p+5}{p(p+1)}$

74. $\left(\dfrac{y}{y^2-1} - \dfrac{y}{y^2-2y+1}\right)\left(\dfrac{y-1}{y+1}\right)$ $\dfrac{-2y}{(y+1)^2(y-1)}$

75. $\dfrac{\dfrac{1}{x+h} - \dfrac{1}{x}}{h}$ $\dfrac{-1}{x(x+h)}$

76. $\dfrac{1}{h}\left(\dfrac{1}{(x+h)^2+9} - \dfrac{1}{x^2+9}\right)$ $-\dfrac{2x+h}{(x^2+9)[(x+h)^2+9]}$

1.7 RESOURCES

Instructor's Test Manual
  Test Items, p. 307
Instructor's Solution Manual
  pp. 23–27, 445–449
Study Guide
  pp. 35–40
Student's Solution Manual
  pp. 23–27

Test Generator
  IBM Disks
    5.25″ Disk A
    3.5″ Disk A
    Mac Disk A
Tutorials
  IBM Disks
    5.25″ Disk A
    3.5″ Disk A
    Mac Disk A

HarperCollins Videotape 2
"College Algebra: In Simplest Terms" Program 3

1.7 EXERCISES (TEXT PAGE 55) ■

Simplify each of the following. Assume that all variables represent positive real numbers. See Examples 1, 4, and 5.

1. $(-4)^{-3}$ $-\dfrac{1}{64}$

2. $(-5)^{-2}$ $\dfrac{1}{25}$

3. $\left(\dfrac{1}{2}\right)^{-3}$ 8

4. $\left(\dfrac{2}{3}\right)^{-2}$ $\dfrac{9}{4}$

5. $-4^{1/2}$ -2

6. $25^{1/2}$ 5

7. $8^{2/3}$ 4

8. $-81^{3/4}$ -27

9. $27^{-2/3}$ $\dfrac{1}{9}$

10. $(-32)^{-4/5}$ $\dfrac{1}{16}$

11. $\left(-\dfrac{4}{9}\right)^{-3/2}$ not defined

12. $\left(\dfrac{1}{8}\right)^{-5/3}$ 32

13. $\left(\dfrac{27}{64}\right)^{-4/3}$ $\dfrac{256}{81}$

14. $\left(\dfrac{121}{100}\right)^{-3/2}$ $\dfrac{1000}{1331}$

🖩 **15.** $7^{-3.1}$ $.0024$

🖩 **16.** $2^{5.41}$ 42.5179

17. $(16p^4)^{1/2}$ $4p^2$

18. $(36r^6)^{1/2}$ $6r^3$

19. $(27x^6)^{2/3}$ $9x^4$

20. $(64a^{12})^{5/6}$ $32a^{10}$

📝 **21.** Why is $a^{1/n}$ defined to be the nth root of a (with appropriate restrictions)?

📝 **22.** Explain why a must be positive if n is even for $a^{1/n}$ to be defined.

◉ **23.** Which of the following expressions is equivalent to $(2x^{-3/2})^2$?

   (a) $2x^{-3}$ **(b)** 2^{-3} **(c)** $2^2x^{-3/4}$ **(d)** $\dfrac{2^2}{x^3}$ (d)

🖉 **24.** Explain how you would evaluate $(-27)^{5/3}$.

Perform the indicated operations. Write answers using only positive exponents. Assume that all variables represent positive real numbers and that variables used as exponents represent rational numbers. See Examples 2, 3, and 6.

25. $2^{-3} \cdot 2^{-4}$ $\dfrac{1}{2^7}$ **26.** $5^{-2} \cdot 5^{-6}$ $\dfrac{1}{5^8}$ **27.** $27^{-2} \cdot 27^{-1}$ $\dfrac{1}{27^3}$ **28.** $9^{-4} \cdot 9^{-1}$ $\dfrac{1}{9^5}$

29. $\dfrac{4^{-2} \cdot 4^{-1}}{4^{-3}}$ 1 **30.** $\dfrac{3^{-1} \cdot 3^{-4}}{3^2 \cdot 3^{-2}}$ $\dfrac{1}{3^5}$ **31.** $(m^{2/3})(m^{5/3})$ $m^{7/3}$ **32.** $(x^{4/5})(x^{2/5})$ $x^{6/5}$

33. $(1+n)^{1/2}(1+n)^{3/4}$ $(1+n)^{5/4}$ **34.** $(m+7)^{-1/6}(m+7)^{-2/3}$ $\dfrac{1}{(m+7)^{5/6}}$

35. $(2y^{3/4}z)(3y^{-2}z^{-1/3})$ $\dfrac{6z^{2/3}}{y^{5/4}}$ **36.** $(4a^{-1}b^{2/3})(a^{3/2}b^{-3})$ $\dfrac{4a^{1/2}}{b^{7/3}}$

37. $(4a^{-2}b^7)^{1/2} \cdot (2a^{1/4}b^3)^5$ $2^6a^{1/4}b^{37/2}$ **38.** $(x^{-2}y^{1/3})^5 \cdot (8x^2y^{-2})^{-1/3}$ $\dfrac{y^{7/3}}{2x^{32/3}}$

39. $\left(\dfrac{r^{-2}}{s^{-5}}\right)^{-3}$ $\dfrac{r^6}{s^{15}}$ **40.** $\left(\dfrac{p^{-1}}{q^{-5}}\right)^{-2}$ $\dfrac{p^2}{q^{10}}$ **41.** $\left(\dfrac{-a}{b^{-3}}\right)^{-1}$ $-\dfrac{1}{ab^3}$ **42.** $\dfrac{7^{-1/3}7r^{-3}}{7^{2/3}r^{-2}}$ $\dfrac{1}{r}$

43. $\dfrac{12^{5/4}y^{-2}}{12^{-1}y^{-3}}$ $12^{9/4}y$ **44.** $\dfrac{6k^{-4}(3k^{-1})^{-2}}{2^3k^{1/2}}$ $\dfrac{1}{12k^{5/2}}$ **45.** $\dfrac{8p^{-3}(4p^2)^{-2}}{p^{-5}}$ $\dfrac{1}{2p^2}$ **46.** $\dfrac{k^{-3/5}h^{-1/3}t^{2/5}}{k^{-1/5}h^{-2/3}t^{1/5}}$ $\dfrac{h^{1/3}t^{1/5}}{k^{2/5}}$

47. $\dfrac{m^{7/3}n^{-2/5}p^{3/8}}{m^{-2/3}n^{3/5}p^{-5/8}}$ $\dfrac{m^3p}{n}$ **48.** $\dfrac{m^{2/5}m^{3/5}m^{-4/5}}{m^{1/5}m^{-6/5}}$ $m^{6/5}$ **49.** $\dfrac{-4a^{-1}a^{2/3}}{a^{-2}}$ $-4a^{5/3}$ **50.** $\dfrac{8y^{2/3}y^{-1}}{2^{-1}y^{3/4}y^{-1/6}}$ $\dfrac{16}{y^{11/12}}$

51. $\dfrac{(k+5)^{1/2}(k+5)^{-1/4}}{(k+5)^{3/4}}$ $\dfrac{1}{(k+5)^{1/2}}$ **52.** $\dfrac{(x+y)^{-5/8}(x+y)^{3/8}}{(x+y)^{1/8}(x+y)^{-1/8}}$ $\dfrac{1}{(x+y)^{1/4}}$

53. $\left(\dfrac{x^4y^3z}{16x^{-6}yz^5}\right)^{-1/2}$ $\dfrac{4z^2}{x^5y}$ **54.** $\left(\dfrac{p^3r^9}{27p^{-3}r^{-6}}\right)^{-1/3}$ $\dfrac{3}{p^2r^5}$

▲ *Perform the indicated operations. Write answers without denominators. Assume that all variables used in denominators are not zero and all variables used as exponents represent rational numbers. See Examples 2(e) and 6(f).*

55. $(r^{3/p})^{2p}(r^{1/p})^{p^2}$ r^{6+p} **56.** $(m^{2/x})^{x/3}(m^{x/4})^{2/x}$ $m^{7/6}$ **57.** $\dfrac{m^{1-a}m^a}{m^{-1/2}}$ $m^{3/2}$

58. $\dfrac{(y^{3-b})(y^{2b-1})}{y^{1/2}}$ $y^{(2b+3)/2}$ **59.** $\dfrac{(x^{n/2})(x^{3n})^{1/2}}{x^{1/n}}$ $x^{(2n^2-1)/n}$ **60.** $\dfrac{(a^{2/3})(a^{1/x})}{(a^{x/3})^{-2}}$ $a^{(2x^2+2x+3)/(3x)}$

61. $\dfrac{(p^{1/n})(p^{1/m})}{p^{-m/n}}$ $p^{(m+n+m^2)/(mn)}$ **62.** $\dfrac{(q^{2r/3})(q^r)^{-1/3}}{(q^{4/3})^{1/r}}$ $q^{(r^2-4)/(3r)}$

▲ *Find each of the following products. Assume that all variables represent positive real numbers. See Example 6(e). (Hint: Use the special binomial product formulas in Exercises 67, 69, and 70.)*

63. $y^{5/8}(y^{3/8} - 10y^{11/8})$ $y - 10y^2$ **64.** $p^{11/5}(3p^{4/5} + 9p^{19/5})$ $3p^3 + 9p^6$

65. $-4k(k^{7/3} - 6k^{1/3})$ $-4k^{10/3} + 24k^{4/3}$ **66.** $-5y(3y^{9/10} + 4y^{3/10})$ $-15y^{19/10} - 20y^{13/10}$

67. $(x + x^{1/2})(x - x^{1/2})$ $x^2 - x$ **68.** $(2z^{1/2} + z)(z^{1/2} - z)$ $2z - z^{3/2} - z^2$

◉ CONCEPTUAL   🖉 WRITING   ▲ CHALLENGING

69. $(r^{1/2} - r^{-1/2})^2$ $r - 2 + r^{-1}$ or $r - 2 + \dfrac{1}{r}$

70. $(p^{1/2} - p^{-1/2})(p^{1/2} + p^{-1/2})$ $p - p^{-1}$ or $p - \dfrac{1}{p}$

Factor, using the given common factor. Assume all variables represent positive real numbers. See Example 7.

71. $4k^{-1} + k^{-2}$; k^{-2} $k^{-2}(4k + 1)$ or $\dfrac{4k + 1}{k^2}$

72. $y^{-5} - 3y^{-3}$; y^{-5} $y^{-5}(1 - 3y^2)$ or $\dfrac{1 - 3y^2}{y^5}$

73. $9z^{-1/2} + 2z^{1/2}$; $z^{-1/2}$ $z^{-1/2}(9 + 2z)$ or $\dfrac{9 + 2z}{z^{1/2}}$

74. $3m^{2/3} - 4m^{-1/3}$; $m^{-1/3}$ $m^{-1/3}(3m - 4)$ or $\dfrac{3m - 4}{m^{1/3}}$

75. $p^{-3/4} - 2p^{-7/4}$; $p^{-7/4}$ $p^{-7/4}(p - 2)$ or $\dfrac{p - 2}{p^{7/4}}$

76. $6r^{-2/3} - 5r^{-5/3}$; $r^{-5/3}$ $r^{-5/3}(6r - 5)$ or $\dfrac{6r - 5}{r^{5/3}}$

▲ **77.** $(p + 4)^{-3/2} + (p + 4)^{-1/2} + (p + 4)^{1/2}$; $(p + 4)^{-3/2}$ $(p + 4)^{-3/2}(p^2 + 9p + 21)$ or $\dfrac{p^2 + 9p + 21}{(p + 4)^{3/2}}$

▲ **78.** $(3r + 1)^{-2/3} + (3r + 1)^{1/3} + (3r + 1)^{4/3}$; $(3r + 1)^{-2/3}$ $3(3r + 1)^{-2/3}(3r^2 + 3r + 1)$ or $\dfrac{3(3r^2 + 3r + 1)}{(3r + 1)^{2/3}}$

One important application of mathematics to business and management concerns supply and demand. Usually, as the price of an item increases, the supply increases and the demand decreases. By studying past records of supply and demand at different prices, economists can construct an equation that describes (approximately) supply and demand for a given item. The next two exercises show examples of this.

79. The price (in dollars) of a certain type of solar heater is approximated by p, where

$$p = 2x^{1/2} + 3x^{2/3}$$

and x is the number of units supplied. Find the price when the supply is 64 units. $64

80. For a certain commodity the demand and the price (in dollars) are related by the equation

$$p = 1000 - 200x^{-2/3}   (x > 0),$$

where x is the number of units of the product demanded. Find the price when the demand is 27.

$977.78

In our system of government, the president is elected by the electoral college, and not by individual voters. Because of this, smaller states have a greater voice in the selection of a president than they otherwise would have. Two political scientists have studied the problems of campaigning for president under the current system and have concluded that candidates should allot their money according to the formula

$$\text{Amount for} \atop \text{large state} = \left(\dfrac{E_{\text{large}}}{E_{\text{small}}}\right)^{3/2} \times \text{amount for} \atop \text{small state}.$$

Here E_{large} represents the electoral vote of the large state, and E_{small} represents the electoral vote of the small state. Find the amount that should be spent in each of the following larger states if $1,000,000 is spent in the small state and the following statements are true.

81. The large state has 48 electoral votes, and the small state has 3. $64,000,000

82. The large state has 36 electoral votes, and the small state has 4. $27,000,000

83. 6 votes in a small state;   28 in a large

about $10,000,000

84. 9 votes in a small state;   32 in a large

about $6,700,000

The Galapagos Islands are a chain of islands ranging in size from 2 to 2249 square miles. A biologist has shown that the number of different land-plant species on an island in this chain is related to the size of the island approximately by

$$S = 28.6A^{.32},$$

where A is the area of the island in square miles and S is the number of different plant species on that island. Estimate S (rounding to the nearest whole number) for islands of the following areas.

85. 1 square mile   29

86. 25 square miles   80

87. 300 square miles   177

88. 2000 square miles   326

Perform all indicated operations and write all answers with positive integer exponents. See Example 8.

89. $\dfrac{a^{-1}+b^{-1}}{(ab)^{-1}}$ $b+a$

90. $\dfrac{p^{-1}-q^{-1}}{(pq)^{-1}}$ $q-p$

91. $\dfrac{r^{-1}+q^{-1}}{r^{-1}-q^{-1}}\cdot\dfrac{r-q}{r+q}$ -1

92. $\dfrac{xy^{-1}+yx^{-1}}{x^2+y^2}$ $\dfrac{1}{xy}$

93. $\dfrac{x-9y^{-1}}{(x-3y^{-1})(x+3y^{-1})}$ $\dfrac{y(xy-9)}{x^2y^2-9}$

94. $\dfrac{(m+n)^{-1}}{m^{-2}-n^{-2}}$ $\dfrac{m^2n^2}{(n-m)(n+m)^2}$

1.8   RESOURCES

Instructor's Test Manual
  Test Items, p. 308
Instructor's Solution Manual
  pp. 28–31, 449–453
Study Guide
  pp. 40–45
Student's Solution Manual
  pp. 28–31

Test Generator
  IBM Disks
  5.25″ Disk A
  3.5″ Disk A
  Mac Disk A
Tutorials
  IBM Disks
  5.25″ Disk A
  3.5″ Disk A
  Mac Disk A

HarperCollins Videotape   2

1.8   EXERCISES   (TEXT PAGE 64) ■

Write in radical form. Assume all variables are positive real numbers. See Example 2.

1. $(-m)^{2/3}$ $\sqrt[3]{(-m)^2}$ or $(\sqrt[3]{-m})^2$

2. $p^{5/4}$ $\sqrt[4]{p^5}$ or $(\sqrt[4]{p})^5$

3. $5m^{4/5}$ $5\sqrt[5]{m^4}$ or $5(\sqrt[5]{m})^4$

4. $-2k^{1/6}$ $-2\sqrt[6]{k}$

5. $-4z^{-1/3}$ $\dfrac{-4}{\sqrt[3]{z}}$

6. $10a^{-3/4}$ $\dfrac{10}{\sqrt[4]{a^3}}$ or $\dfrac{10}{(\sqrt[4]{a})^3}$

7. $(2m+p)^{2/3}$ $\sqrt[3]{(2m+p)^2}$ or $(\sqrt[3]{2m+p})^2$

8. $(5r+3t)^{4/7}$ $\sqrt[7]{(5r+3t)^4}$ or $(\sqrt[7]{5r+3t})^4$

Write in exponential form. Assume all variables are nonnegative real numbers. See Example 3.

9. $\sqrt[5]{k^2}$ $k^{2/5}$

10. $-\sqrt[4]{z^5}$ $-z^{5/4}$

11. $-\sqrt[3]{a^2}$ $-a^{2/3}$

12. $2\sqrt{m^9}$ $2m^{9/2}$

13. $-3\sqrt{5p^3}$ $-3\cdot5^{1/2}p^{3/2}$

14. $m\sqrt{2y^5}$ $m\cdot2^{1/2}y^{5/2}$

15. $18\sqrt{m^2n^3p}$ $18mn^{3/2}p^{1/2}$

16. $-12\sqrt[5]{x^3y^2z^4}$ $-12x^{3/5}y^{2/5}z^{4/5}$

✐ **17.** What is wrong with the statement $\sqrt[3]{4}\cdot\sqrt[3]{4}=4$?

◉ **18.** Which of the following expressions is *not* simplified? Give the simplified form.

  (a) $\sqrt[3]{2y}$ **(b)** $\dfrac{\sqrt{5}}{2}$ **(c)** $\sqrt[4]{m^3}$ **(d)** $\sqrt{\dfrac{3}{4}}$ (d), $\dfrac{\sqrt{3}}{2}$

✐ **19.** Explain how to rationalize the denominator of $\sqrt[3]{\dfrac{3}{2}}$.

✐ **20.** How can we multiply $\sqrt{2}$ and $\sqrt[3]{2}$?

Simplify each of the following. Assume that all variables represent positive real numbers. See Examples 1, 5, 6, 8, 10, and 11(a).

21. $\sqrt[3]{125}$ 5   **22.** $\sqrt[4]{81}$ 3   **23.** $\sqrt[5]{-3125}$ -5 **24.** $\sqrt[3]{343}$ 7   **25.** $\sqrt{50}$ $5\sqrt{2}$

26. $\sqrt{45}$ $3\sqrt{5}$ **27.** $\sqrt[3]{81}$ $3\sqrt[3]{3}$ **28.** $\sqrt[3]{250}$ $5\sqrt[3]{2}$ **29.** $-\sqrt[4]{32}$ $-2\sqrt[4]{2}$ **30.** $-\sqrt[4]{243}$ $-3\sqrt[4]{3}$

31. $-\sqrt{\dfrac{9}{5}}$ $-\dfrac{3\sqrt{5}}{5}$ **32.** $-\sqrt[3]{\dfrac{3}{2}}$ $-\dfrac{\sqrt[3]{12}}{2}$ **33.** $-\sqrt[3]{\dfrac{4}{5}}$ $-\dfrac{\sqrt[3]{100}}{5}$ **34.** $\sqrt[4]{\dfrac{3}{2}}$ $\dfrac{\sqrt[4]{24}}{2}$

35. $\sqrt[3]{16(-2)^4(2)^8}$ $32\sqrt[3]{2}$ **36.** $\sqrt[3]{25(3)^4(5)^3}$ $15\sqrt[3]{75}$ **37.** $\sqrt{8x^5z^8}$ $2x^2z^4\sqrt{2x}$

38. $\sqrt{24m^6n^5}$ $2m^3n^2\sqrt{6n}$ **39.** $\sqrt[3]{16z^5x^8y^4}$ $2zx^2y\sqrt[3]{2z^2x^2y}$ **40.** $-\sqrt[6]{64a^{12}b^8}$ $-2a^2b\sqrt[3]{b}$

41. $\sqrt[4]{m^2n^7p^8}$ $np^2\sqrt[4]{m^2n^3}$ **42.** $\sqrt[4]{x^8y^7z^9}$ $x^2yz^2\sqrt[4]{y^3z}$

43. $\sqrt[4]{x^4 + y^4}$ cannot simplify further   **44.** $\sqrt[3]{27 + a^3}$ cannot simplify further

45. $\sqrt{\dfrac{2}{3x}}$ $\dfrac{\sqrt{6x}}{3x}$ **46.** $\sqrt{\dfrac{5}{3p}}$ $\dfrac{\sqrt{15p}}{3p}$ **47.** $\sqrt{\dfrac{x^5y^3}{z^2}}$ $\dfrac{x^2y\sqrt{xy}}{z}$ **48.** $\sqrt{\dfrac{g^3h^5}{r^3}}$ $\dfrac{gh^2\sqrt{ghr}}{r^2}$

49. $\sqrt[3]{\dfrac{8}{x^2}}$ $\dfrac{2\sqrt[3]{x}}{x}$ **50.** $\sqrt[3]{\dfrac{9}{16p^4}}$ $\dfrac{\sqrt[3]{36p^2}}{4p^2}$ **51.** $\sqrt[4]{\dfrac{g^3h^5}{9r^6}}$ $\dfrac{h\sqrt[4]{9g^3hr^2}}{3r^2}$ **52.** $\sqrt[4]{\dfrac{32x^5}{y^5}}$ $\dfrac{2x\sqrt[4]{2xy^3}}{y^2}$

53. $\dfrac{\sqrt[3]{mn}\cdot\sqrt[3]{m^2}}{\sqrt[3]{n^2}}$ $\dfrac{m\sqrt[3]{n^2}}{n}$ **54.** $\dfrac{\sqrt[3]{8m^2n^3}\cdot\sqrt[3]{2m^2}}{\sqrt[3]{32m^4n^3}}$ $\dfrac{\sqrt[3]{4}}{2}$ **55.** $\dfrac{\sqrt[4]{32x^5y}\cdot\sqrt[4]{2xy^4}}{\sqrt[4]{4x^3y^2}}$ $2\sqrt[4]{x^3y^3}$

56. $\dfrac{\sqrt[4]{rs^2t^3}\cdot\sqrt[4]{r^3s^2t}}{\sqrt[4]{r^2t^3}}$ $s\sqrt[4]{r^2t}$ **57.** $\sqrt[3]{\sqrt{4}}$ $\sqrt[3]{2}$ **58.** $\sqrt[4]{\sqrt[3]{2}}$ $\sqrt[12]{2}$

59. $\sqrt[6]{\sqrt[3]{x}}$ $\sqrt[18]{x}$ **60.** $\sqrt[8]{\sqrt[4]{y}}$ $\sqrt[32]{y}$

Simplify each of the following, assuming that all variables represent nonnegative numbers. See Examples 7, 9, and 11(b).

61. $4\sqrt{3} - 5\sqrt{12} + 3\sqrt{75}$ $9\sqrt{3}$ **62.** $2\sqrt{5} - 3\sqrt{20} + 2\sqrt{45}$ $2\sqrt{5}$

63. $3\sqrt{28p} - 4\sqrt{63p} + \sqrt{112p}$ $-2\sqrt{7p}$ **64.** $9\sqrt{8k} + 3\sqrt{18k} - \sqrt{32k}$ $23\sqrt{2k}$

65. $2\sqrt[3]{3} + 4\sqrt[3]{24} - \sqrt[3]{81}$ $7\sqrt[3]{3}$ **66.** $\sqrt[3]{32} - 5\sqrt[3]{4} + 2\sqrt[3]{108}$ $3\sqrt[3]{4}$ **67.** $\dfrac{1}{\sqrt{3}} - \dfrac{2}{\sqrt{12}} + 2\sqrt{3}$ $2\sqrt{3}$

68. $\dfrac{1}{\sqrt{2}} + \dfrac{3}{\sqrt{8}} + \dfrac{1}{\sqrt{32}}$ $\dfrac{11\sqrt{2}}{8}$ **69.** $\dfrac{5}{\sqrt[3]{2}} - \dfrac{2}{\sqrt[3]{16}} + \dfrac{1}{\sqrt[3]{54}}$ $\dfrac{13\sqrt[3]{4}}{6}$ **70.** $\dfrac{-4}{\sqrt[3]{3}} + \dfrac{1}{\sqrt[3]{24}} - \dfrac{2}{\sqrt[3]{81}}$ $\dfrac{-25\sqrt[3]{9}}{18}$

71. $(\sqrt{2} + 3)(\sqrt{2} - 3)$ -7 **72.** $(\sqrt{5} + \sqrt{2})(\sqrt{5} - \sqrt{2})$ 3

73. $(\sqrt[3]{11} - 1)(\sqrt[3]{11^2} + \sqrt[3]{11} + 1)$ 10   **74.** $(\sqrt[3]{7} + 3)(\sqrt[3]{7^2} - 3\sqrt[3]{7} + 9)$ 34

75. $(\sqrt{3} + \sqrt{8})^2$ $11 + 4\sqrt{6}$ **76.** $(\sqrt{2} - 1)^2$ $3 - 2\sqrt{2}$

77. $(3\sqrt{2} + \sqrt{3})(2\sqrt{3} - \sqrt{2})$ $5\sqrt{6}$ **78.** $(4\sqrt{5} - 1)(3\sqrt{5} + 2)$ $58 + 5\sqrt{5}$

79. $(2\sqrt[3]{3} + 1)(\sqrt[3]{3} - 4)$ $2\sqrt[3]{9} - 7\sqrt[3]{3} - 4$ **80.** $(\sqrt[3]{4} + 3)(5\sqrt[3]{4} + 1)$ $10\sqrt[3]{2} + 16\sqrt[3]{4} + 3$

Rationalize the denominator of each of the following. Assume that all variables represent nonnegative numbers and that no denominators are zero. See Example 12.

81. $\dfrac{\sqrt{3}}{\sqrt{5} + \sqrt{3}}$ $\dfrac{\sqrt{15} - 3}{2}$ **82.** $\dfrac{\sqrt{7}}{\sqrt{3} - \sqrt{7}}$ $\dfrac{-(\sqrt{21} + 7)}{4}$

83. $\dfrac{1 + \sqrt{3}}{3\sqrt{5} + 2\sqrt{3}}$ $\dfrac{3\sqrt{5} + 3\sqrt{15} - 2\sqrt{3} - 6}{33}$ **84.** $\dfrac{\sqrt{7} - 1}{2\sqrt{7} + 4\sqrt{2}}$ $\dfrac{-7 + \sqrt{7} + 2\sqrt{14} - 2\sqrt{2}}{2}$

85. $\dfrac{p}{\sqrt{p}+2}$ $\dfrac{p(\sqrt{p}-2)}{p-4}$

86. $\dfrac{\sqrt{r}}{3-\sqrt{r}}$ $\dfrac{\sqrt{r}\,(3+\sqrt{r})}{9-r}$

87. $\dfrac{a}{\sqrt{a+b}-1}$ $\dfrac{a(\sqrt{a+b}+1)}{a+b-1}$

88. $\dfrac{3m}{2+\sqrt{m+n}}$ $\dfrac{3m(2-\sqrt{m+n})}{4-m-n}$

▲ *In advanced mathematics it is sometimes useful to write a radical expression with a rational numerator. The procedure is similar to rationalizing the denominator. Rationalize the numerator of each of the following expressions.*

89. $\dfrac{1+\sqrt{2}}{2}$ $\dfrac{-1}{2(1-\sqrt{2})}$

90. $\dfrac{1-\sqrt{3}}{3}$ $\dfrac{-2}{3(1+\sqrt{3})}$

91. $\dfrac{\sqrt{x}}{1+\sqrt{x}}$ $\dfrac{x}{\sqrt{x}+x}$

92. $\dfrac{\sqrt{p}}{1-\sqrt{p}}$ $\dfrac{p}{\sqrt{p}-p}$

93. $\dfrac{\sqrt{x}+\sqrt{x+1}}{\sqrt{x}-\sqrt{x+1}}$ $\dfrac{-1}{2x-2\sqrt{x(x+1)}+1}$

94. $\dfrac{\sqrt{p}+\sqrt{p^2-1}}{\sqrt{p}-\sqrt{p^2-1}}$ $\dfrac{-p^2+p+1}{p+p^2-2\sqrt{p(p^2-1)}-1}$

Write the following without radicals. Use absolute value if necessary. See Example 4.

95. $\sqrt{(m+n)^2}$ $|m+n|$

96. $\sqrt[4]{(a+2b)^4}$ $|a+2b|$

97. $\sqrt{z^2-6zx+9x^2}$ $|z-3x|$

98. $\sqrt[3]{(r+2s)(r^2+4rs+4s^2)}$ $r+2s$

RESOURCES FOR REVIEW

Instructor's Test Manual
  Chapter Test Forms, pp. 15–36
Instructor's Solution Manual
  pp. 31–39, 453–459

Study Guide
  pp. 46–48
Student's Solution Manual
  pp. 31–39

Test Generator
  IBM Disks
  5.25" Disk A
  3.5" Disk A
  Mac Disk A

CHAPTER 1 REVIEW EXERCISES   (TEXT PAGE 68) ■

Let set $K = \{-12, -6, -.9, -\sqrt{7}, -\sqrt{4}, 0, 1/8, \pi/4, 6, \sqrt{11}\}$. List the elements of K that are members of the following sets.

1. Integers $-12, -6, -\sqrt{4}$ (or -2), $0, 6$

2. Rational numbers $-12, -6, -.9, -\sqrt{4}, 0, \dfrac{1}{8}, 6$

3. Irrational numbers $-\sqrt{7}, \dfrac{\pi}{4}, \sqrt{11}$

For Exercises 4–7 choose all words from the following list that apply.

(a) natural numbers   **(b)** whole numbers   **(c)** integers   **(d)** rational numbers
(e) irrational numbers   **(f)** real numbers

4. 0   whole, integers, rational, real

5. $-\sqrt{25}$ integers, rational, real

6. $\dfrac{5}{8}$ rational, real

7. $\dfrac{3\pi}{4}$ irrational, real

Use the order of operations to simplify each of the following.

8. $-4 - [2 - (-3^2)]$ -15

9. $[2^3 - (-5)] - 2^2$ 9

10. $(-4 - 1)(-3 - 5) - 2^3$ 32

◎ CONCEPTUAL   ✎ WRITING   ▲ CHALLENGING

11. $(6 - 9)(-2 - 7) - (-4)$ 31

12. $\left(-\dfrac{5}{9} - \dfrac{2}{3}\right) - \dfrac{5}{6}$ $-\dfrac{37}{18}$

13. $\left(-\dfrac{2^3}{5} - \dfrac{3}{4}\right) - \left(-\dfrac{1}{2}\right)$ $-\dfrac{37}{20}$

14. $\dfrac{6(-4) - 3^2(-2)^3}{-5[-2 - (-6)]}$ $-\dfrac{12}{5}$

15. $\dfrac{(-7)(-3) - (-2^3)(-5)}{(-2^2 - 2)(-1 - 6)}$ $-\dfrac{19}{42}$

Evaluate each of the following if $a = -1$, $b = -2$, and $c = 4$.

16. $9a - 5b + 4c$ 17

17. $-4(2a - 5b)$ -32

18. $\dfrac{9a + 2b}{a + b + c}$ -13

✏ **19.** Suppose your friend missed class the day order of operations was discussed. Explain to her how to work the following problem. Evaluate $\dfrac{4a - 5c}{2a + 3b}$ if $a = -1$, $b = -2$, and $c = 4$.

Identify the properties illustrated in each of the following.

20. $8(5 + 9) = (5 + 9)8$ commutative

21. $4 \cdot 6 + 4 \cdot 12 = 4(6 + 12)$ distributive

22. $3 \cdot (4 \cdot 2) = (3 \cdot 4) \cdot 2$ associative

23. $-(r - 2) = -r + 2$ distributive

24. $(9 + p) + 0 = 9 + p$ identity

25. $[6 \cdot 5 + 8 \cdot 5] \cdot 2 = [(6 + 8) \cdot 5] \cdot 2$ distributive

Use the distributive property to rewrite sums as products and products as sums.

26. $-11(m - n)$ $-11m + 11n$

27. $k(r + s - t)$ $kr + ks - kt$

28. $mn - 9m$ $m(n - 9)$

Write the following numbers in numerical order from smallest to largest.

29. $|6 - 4|, -|-2|, |8 + 1|, -|3 - (-2)|$ $-|3 - (-2)|, -|-2|, |6 - 4|, |8 + 1|$

30. $\sqrt{7}, -\sqrt{8}, -|\sqrt{16}|, |-\sqrt{12}|$ $-|\sqrt{16}|, -\sqrt{8}, \sqrt{7}, |-\sqrt{12}|$

Write without absolute value bars.

31. $-|-6| + |3|$ -3

32. $7 - |-8|$ -1

33. $|\sqrt{8} - 3|$ $3 - \sqrt{8}$

34. $|\sqrt{52} - 8|$ $8 - \sqrt{52}$

35. $|m - 3|$ if $m > 3$ $m - 3$

▲ **36.** $|-6 - x^2|$ $6 + x^2$

37. $|\pi - 4|$ $4 - \pi$

38. $|3 + 5k|$ if $k < -3/5$ $-3 - 5k$

Perform each of the following operations. Assume that all variables appearing as exponents represent positive integers.

39. $(3q^3 - 9q^2 + 6) + (4q^3 - 8q + 3)$ $7q^3 - 9q^2 - 8q + 9$

40. $2(3y^6 - 9y^2 + 2y) - (5y^6 - 10y^2 - 4y)$ $y^6 - 8y^2 + 8y$

41. $(8y - 7)(2y + 7)$ $16y^2 + 42y - 49$

42. $(2r + 11s)(4r - 9s)$ $8r^2 + 26rs - 99s^2$

43. $(3k - 5m)^2$ $9k^2 - 30km + 25m^2$

44. $(4a - 3b)^2$ $16a^2 - 24ab + 9b^2$

45. $(3w - 2)(5w^2 - 4w + 1)$ $15w^3 - 22w^2 + 11w - 2$

46. $(2k + 5)(3k^3 - 4k^2 + 8k - 2)$ $6k^4 + 7k^3 - 4k^2 + 36k - 10$

47. $(p^q + 1)(p^q - 3)$ $p^{2q} - 2p^q - 3$

48. $(a^y + 6)^2$ $a^{2y} + 12a^y + 36$

Use the binomial theorem to expand each of the following.

49. $(x + 2y)^4$ $x^4 + 8x^3y + 24x^2y^2 + 32xy^3 + 16y^4$

50. $\left(\dfrac{k}{2} - g\right)^5$ $\dfrac{k^5}{32} - \dfrac{5k^4g}{16} + \dfrac{5k^3g^2}{4} - \dfrac{5k^2g^3}{2} + \dfrac{5kg^4}{2} - g^5$

Find the indicated term or terms for each of the following expansions.

51. Fifth term of $(3x - 2y)^6$ $2160x^2y^4$

52. Eighth term of $(2m + n^2)^{12}$ $25{,}344m^5n^{14}$

53. First four terms of $(3 + x)^{16}$ $3^{16} + 16 \cdot 3^{15}x + 120 \cdot 3^{14}x^2 + 560 \cdot 3^{13}x^3$

54. Last three terms of $(2m - 3n)^{15}$ $-420 \cdot 3^{13}m^2n^{13} + 30 \cdot 3^{14}mn^{14} - 3^{15}n^{15}$

Factor as completely as possible.

55. $7z^2 - 9z^3 + z$ $z(7z - 9z^2 + 1)$

56. $3(z - 4)^2 + 9(z - 4)^3$ $3(z - 4)^2(3z - 11)$

57. $r^2 + rp - 42p^2$ $(r + 7p)(r - 6p)$

58. $z^2 - 6zk - 16k^2$ $(z - 8k)(z + 2k)$

59. $6m^2 - 13m - 5$ $(3m + 1)(2m - 5)$

60. $48a^8 - 12a^7b - 90a^6b^2$ $6a^6(4a + 5b)(2a - 3b)$

61. $169y^4 - 1$ $(13y^2 + 1)(13y^2 - 1)$

62. $49m^8 - 9n^2$ $(7m^4 + 3n)(7m^4 - 3n)$

63. $8y^3 - 1000z^6$ $8(y - 5z^2)(y^2 + 5yz^2 + 25z^4)$

64. $6(3r - 1)^2 + (3r - 1) - 35$ $3(9r - 10)(2r + 1)$

65. $ar - 3as + 5rb - 15sb$ $(r - 3s)(a + 5b)$

66. $15mp + 9mq - 10np - 6nq$ $(5p + 3q)(3m - 2n)$

67. $(16m^2 - 56m + 49) - 25a^2$ $(4m - 7 + 5a)(4m - 7 - 5a)$

68. A student factored $64a^3 + 8b^3$ as $(4a + 2b)(16a^2 - 8ab + 4b^2)$. Is the polynomial factored completely? Explain.

69. Which one of the following is equal to $\dfrac{2a + b}{4a^2 - b^2}$?

(a) $\dfrac{1}{2a - b}$ **(b)** $\dfrac{2}{2a - b}$ **(c)** $\dfrac{1}{2ab}$ **(d)** $-\dfrac{1}{2ab}$ (a)

Perform each of the following operations.

70. $\dfrac{k^2 + k}{8k^3} \cdot \dfrac{4}{k^2 - 1}$ $\dfrac{1}{2k^2(k - 1)}$

71. $\dfrac{3r^3 - 9r^2}{r^2 - 9} \div \dfrac{8r^3}{r + 3}$ $\dfrac{3}{8r}$

72. $\dfrac{x^2 + x - 2}{x^2 + 5x + 6} \div \dfrac{x^2 + 3x - 4}{x^2 + 4x + 3}$ $\dfrac{x + 1}{x + 4}$

73. $\dfrac{27m^3 - n^3}{3m - n} \div \dfrac{9m^2 + 3mn + n^2}{9m^2 - n^2}$ $(3m + n)(3m - n)$

74. $\dfrac{p^2 - 36q^2}{(p - 6q)^2} \cdot \dfrac{p^2 - 5pq - 6q^2}{p^2 - 6pq + 36q^2} \div \dfrac{5p}{p^3 + 216q^3}$ $\dfrac{(p + q)(p + 6q)^2}{5p}$

75. $\dfrac{1}{4y} + \dfrac{8}{5y}$ $\dfrac{37}{20y}$

76. $\dfrac{m}{4 - m} + \dfrac{3m}{m - 4}$ $\dfrac{2m}{m - 4}$ or $\dfrac{-2m}{4 - m}$

77. $\dfrac{3}{x^2 - 4x + 3} - \dfrac{2}{x^2 - 1}$ $\dfrac{x + 9}{(x - 3)(x - 1)(x + 1)}$

78. $\left[\dfrac{1}{(x + h)^2 + 16} - \dfrac{1}{x^2 + 16}\right] \div h$ $\dfrac{-2x - h}{[(x + h)^2 + 16](x^2 + 16)}$

79. $\dfrac{\dfrac{1}{p} + \dfrac{1}{q}}{1 - \dfrac{1}{pq}}$ $\dfrac{q + p}{pq - 1}$

80. $\dfrac{3 + \dfrac{2m}{m^2 - 4}}{\dfrac{5}{m - 2}}$ $\dfrac{3m^2 + 2m - 12}{5(m + 2)}$

Simplify each of the following. Write results with only positive exponents. Assume that all variables represent positive real numbers.

81. 2^{-6} $\dfrac{1}{64}$

82. -3^{-2} $-\dfrac{1}{9}$

83. $\left(-\dfrac{5}{4}\right)^{-2}$ $\dfrac{16}{25}$

84. $3^{-1} - 4^{-1}$ $\dfrac{1}{12}$

85. $(5z^3)(-2z^5)$ $-10z^8$

86. $(8p^2q^3)(-2p^5q^{-4})$ $-\dfrac{16p^7}{q}$

87. $(-6p^5w^4m^{12})^0$ 1

88. $(-6x^2y^{-3}z^2)^{-2}$ $\dfrac{y^6}{36x^4z^4}$

89. $\dfrac{-8y^7p^{-2}}{y^{-4}p^{-3}}$ $-8y^{11}p$

90. $\dfrac{a^{-6}(a^{-8})}{a^{-2}(a^{11})}$ $\dfrac{1}{a^{23}}$

91. $\dfrac{(p + q)^4(p + q)^{-3}}{(p + q)^6}$ $\dfrac{1}{(p + q)^5}$

92. $\dfrac{[p^2(m + n)^3]^{-2}}{p^{-2}(m + n)^{-5}}$ $\dfrac{1}{p^2(m + n)}$

93. $\left(\dfrac{r^{-1}s^2}{t^{-2}}\right)^{-2}$ $\dfrac{r^2}{s^4t^4}$

94. $\dfrac{(2x^{-3})^2(3x^2)^{-2}}{6(x^2y^3)}$ $\dfrac{2}{27x^{12}y^3}$

95. $\dfrac{p^4(p^{-2})}{p^{5/3}}$ $p^{1/3}$

96. $(2 + p)^{5/4}(2 + p)^{-9/4}$ $\dfrac{1}{2 + p}$

97. $(7r^{1/2})(2r^{3/4})(-r^{1/6})$ $-14r^{17/12}$

98. $(a^{3/4}b^{2/3})(a^{5/8}b^{-5/6})$ $\dfrac{a^{11/8}}{b^{1/6}}$

99. $\dfrac{y^{5/3} \cdot y^{-2}}{y^{-5/6}}$ $y^{1/2}$

100. $\left(\dfrac{25m^3n^5}{m^{-2}n^6}\right)^{-1/2}$ $\dfrac{n^{1/2}}{5m^{5/2}}$

Simplify. Assume that all variables represent positive real numbers. Write without denominators.

101. $\dfrac{k^{2+p} \cdot k^{-4p}}{k^{6p}}$ k^{2-9p}

102. $\dfrac{(r^{3+2z})(r^{-5-z})}{(r^{-z/2})^{-4}}$ r^{-2-z}

Find each product. Assume that all variables represent positive real numbers.

103. $2z^{1/3}(5z^2 - 2)$ $10z^{7/3} - 4z^{1/3}$

104. $-m^{3/4}(8m^{1/2} + 4m^{-3/2})$ $-8m^{5/4} - 4m^{-3/4}$

105. $(p + p^{1/2})(3p - 5)$ $3p^2 + 3p^{3/2} - 5p - 5p^{1/2}$

106. $(m^{1/2} - 4m^{-1/2})^2$ $m - 8 + 16m^{-1}$ or $m - 8 + \dfrac{16}{m}$

Simplify. Assume that all variables represent positive numbers.

107. $\sqrt{200}$ $10\sqrt{2}$

108. $\sqrt[3]{16}$ $2\sqrt[3]{2}$

109. $\sqrt[4]{1250}$ $5\sqrt[4]{2}$

110. $-\sqrt{\dfrac{16}{3}}$ $-\dfrac{4\sqrt{3}}{3}$

111. $-\sqrt[3]{\dfrac{2}{5p^2}}$ $-\dfrac{\sqrt[3]{50p}}{5p}$

112. $\sqrt{\dfrac{2^7y^8}{m^3}}$ $\dfrac{2^3y^4\sqrt{2m}}{m^2}$ or $\dfrac{8y^4\sqrt{2m}}{m^2}$

113. $\sqrt[4]{\sqrt[3]{m}}$ $\sqrt[12]{m}$

114. $\dfrac{\sqrt[4]{8p^2q^5} \cdot \sqrt[4]{2p^3q}}{\sqrt[4]{p^5q^2}}$ $2q$

115. $(\sqrt[3]{2} + 4)(\sqrt[3]{2^2} - 4\sqrt[3]{2} + 16)$ 66

116. $\dfrac{3}{\sqrt{5}} - \dfrac{2}{\sqrt{45}} + \dfrac{6}{\sqrt{80}}$ $\dfrac{23\sqrt{5}}{30}$

117. $\sqrt{18m^3} - 3m\sqrt{32m} + 5\sqrt{m^3}$ $-9m\sqrt{2m} + 5m\sqrt{m}$ or $m(-9\sqrt{2m} + 5\sqrt{m})$

118. $\dfrac{2}{7 - \sqrt{3}}$ $\dfrac{7 + \sqrt{3}}{23}$

119. $\dfrac{6}{3 - \sqrt{2}}$ $\dfrac{6(3 + \sqrt{2})}{7}$

120. $\dfrac{z}{\sqrt{z + 1}}$ $\dfrac{z\sqrt{z + 1}}{z + 1}$

121. $\dfrac{k}{\sqrt{k} - 3}$ $\dfrac{k(\sqrt{k} + 3)}{k - 9}$

122. $\dfrac{\sqrt{x} - \sqrt{x - 2}}{\sqrt{x} + \sqrt{x - 2}}$ $x - 1 - \sqrt{x(x - 2)}$

CHAPTER 1 TEST (TEXT PAGE 70) ■ ——————

Let set $A = \left\{-10, -\dfrac{9}{3}, 0, \dfrac{2}{3}, \dfrac{4}{\pi}, \sqrt{25}\right\}$. List the elements of set A that are members of the following sets.

1. Irrational numbers $\dfrac{4}{\pi}$

2. Integers $-10, -\dfrac{9}{3}$ (or -3), $0, \sqrt{25}$

3. Evaluate the following when $x = -2$, $y = -3$, and $z = -4$.

$$\dfrac{2x - |3y| - |z|}{-|x + y| + z}$$ $\dfrac{17}{9}$

Identify the properties illustrated in each of the following.

4. $2(x + 3) = 2(3 + x)$ commutative

5. $3(x + 4) = 3x + 12$ distributive

In Exercises 6–8, perform each of the following operations. Assume that all variables appearing as exponents represent positive integers.

6. $(a^2 - 3a + 1) - (a - 2a^2) + (-4 + 3a - a^2)$ $2a^2 - a - 3$

7. $(2x^n - 3)^2$ $4x^{2n} - 12x^n + 9$

8. $(3y^2 - y + 4)(y + 2)$ $3y^3 + 5y^2 + 2y + 8$

9. Use the binomial theorem to expand $(2x - 3y)^4$. $16x^4 - 96x^3y + 216x^2y^2 - 216xy^3 + 81y^4$

10. Find the third term in the expansion of $(w - 2y)^6$. $60w^4y^2$

Factor as completely as possible.

11. $x^4 - 16$ $(x^2 + 4)(x + 2)(x - 2)$

12. $24m^3 - 14m^2 - 24m$ $2m(4m + 3)(3m - 4)$

13. $x^3y^2 - 9x^3 - 8y^2 + 72$ $(x - 2)(x^2 + 2x + 4)(y + 3)(y - 3)$

◉ **14.** What is wrong with the following simplification?

$$\frac{4(x - 1) - (2x + 3)}{x(x - 1)} = \frac{4 - (2x + 3)}{x}$$

$$= \frac{4 - 2x - 3}{x}$$

$$= \frac{1 - 2x}{x}$$

The first step is wrong. The numerator must be in factored form before cancelling factors of $x - 1$. As shown, the numerator is a difference, not a product.

Perform each of the following operations.

15. $\dfrac{5x^2 - 9x - 2}{30x^3 + 6x^2} \cdot \dfrac{2x^8 + 6x^7 + 4x^6}{x^4 - 3x^2 - 4}$ $\dfrac{x^4(x + 1)}{3(x^2 + 1)}$

16. $\dfrac{x}{x^2 + 3x + 2} + \dfrac{2x}{2x^2 - x - 3}$ $\dfrac{4x^2 + x}{(x + 2)(x + 1)(2x - 3)}$

17. $\dfrac{a + b}{2a - 3} - \dfrac{a - b}{3 - 2a}$ $\dfrac{2a}{2a - 3}$ or $\dfrac{-2a}{3 - 2a}$

18. $\dfrac{y - 2}{y - \dfrac{4}{y}}$ $\dfrac{y}{y + 2}$

Simplify each of the following. Write results without negative exponents. Assume that all variables represent positive numbers.

19. $\left(\dfrac{x^{-1}x^3}{2x^0x^{-5}}\right)^{-2}$ $\dfrac{4}{x^{14}}$

20. $\left(\dfrac{x^{-2}y^{-1/3}z}{x^{-5/3}y^{-2/3}z^{2/3}}\right)^3$ $\dfrac{yz}{x}$

Simplify. Assume that all variables represent positive numbers.

21. $\sqrt{18x^5y^8}$ $3x^2y^4\sqrt{2x}$

22. $\sqrt[5]{\dfrac{x}{4y^2}}$ $\dfrac{\sqrt[5]{8xy^3}}{2y}$

23. $\sqrt{32x} + \sqrt{2x} - \sqrt{18x}$ $2\sqrt{2x}$

24. $\dfrac{3}{\sqrt{2} - 1}$ $3\sqrt{2} + 3$

25. $(\sqrt{x} - \sqrt{y})(\sqrt{x} + \sqrt{y})$ $x - y$

■ **CHAPTER 2 EQUATIONS AND INEQUALITIES**

2.1 RESOURCES

Instructor's Test Manual
Test Items, p. 309
Instructor's Solution Manual
pp. 40–46, 460–466
Study Guide
pp. 52–56
Student's Solution Manual
pp. 40–46

Test Generator
IBM Disks
5.25″ Disk A
3.5″ Disk A
Mac Disk A
Tutorials
IBM Disks
5.25″ Disk B
3.5″ Disk A
Mac Disk A
Apple II Module 2, Lessons 1–2

HarperCollins Videotape 2

2.1 EXERCISES (TEXT PAGE 77) ■

Decide whether each of the following equations is an identity, a conditional equation, or a contradiction. Give the solution set. See Example 1.

1. $x^2 + 5x = x(x + 5)$ identity; {all real numbers}

2. $3y + 4 = 5(y - 2)$ conditional; {7}

3. $2(x - 7) = 5x + 3 - x$ conditional; $\left\{-\dfrac{17}{2}\right\}$

4. $2x - 4 = 2(x + 2)$ contradiction; $\emptyset$

5. $\dfrac{m + 3}{m} = 1 + \dfrac{3}{m}$ identity; {all real numbers except 0}

6. $\dfrac{p}{2 - p} = \dfrac{2}{p} - 1$ conditional; {1}

7. $4q + 20q - 25 = 24q + 5$ contradiction; $\emptyset$

8. $3(k + 2) - 5(k + 2) = -2k - 4$ identity; {all real numbers}

Decide which of the following pairs of equations are equivalent. See Example 2.

9. $\dfrac{3x}{x - 1} = \dfrac{2}{x - 1}$ equivalent
$3x = 2$

10. $\dfrac{x + 1}{12} = \dfrac{5}{12}$ equivalent
$x + 1 = 5$

11. $\dfrac{x}{x - 2} = \dfrac{2}{x - 2}$ not equivalent
$x = 2$

12. $\dfrac{x + 3}{x + 1} = \dfrac{2}{x + 1}$ not equivalent
$x = -1$

13. $x = 4$ not equivalent
$x^2 = 16$

14. $z^2 = 9$ not equivalent
$z = 3$

15. Which one of the following is not a linear equation?
  (a) $5x + 7(x - 1) = -3x$ **(b)** $8x^2 - 4x + 3 = 0$
  (c) $7y + 8y = 13y$ **(d)** $.04t - .08t = .40$ (b)

16. True or false: The solution set of the equation in Exercise 15(c) is $\emptyset$. false (It is {0}.)

17. In solving the equation $3(2t - 4) = 6t - 12$, a student obtains the result $0 = 0$, and gives the solution set {0}. Is this correct? Explain.

18. Write the steps you would use to solve the equation in Exercise 15(a).

Solve each of the following equations. See Examples 3–5.

19. $2m - 5 = m + 7$ $\{12\}$

20. $.01p + 3.1 = 2.03p - 2.96$ $\{3\}$

21. $\dfrac{5}{6}k - 2k + \dfrac{1}{3} = \dfrac{2}{3}$ $\left\{-\dfrac{2}{7}\right\}$

22. $\dfrac{3}{4} + \dfrac{1}{5}r - \dfrac{1}{2} = \dfrac{4}{5}r$ $\left\{\dfrac{5}{12}\right\}$

23. $3r + 2 - 5(r + 1) = 6r + 4$ $\left\{-\dfrac{7}{8}\right\}$

24. $5(a + 3) + 4a - 5 = -(2a - 4)$ $\left\{-\dfrac{6}{11}\right\}$

25. $2[m - (4 + 2m) + 3] = 2m + 2$ $\{-1\}$

26. $4[2p - (3 - p) + 5] = -7p - 2$ $\left\{-\dfrac{10}{19}\right\}$

27. $\dfrac{3x - 2}{7} = \dfrac{x + 2}{5}$ $\{3\}$

28. $\dfrac{2p + 5}{5} = \dfrac{p + 2}{3}$ $\{-5\}$

29. $\dfrac{1}{4p} + \dfrac{2}{p} = 3$ $\left\{\dfrac{3}{4}\right\}$

30. $\dfrac{2}{t} + 6 = \dfrac{5}{2t}$ $\left\{\dfrac{1}{12}\right\}$

31. $\dfrac{m}{2} - \dfrac{1}{m} = \dfrac{6m + 5}{12}$ $\left\{-\dfrac{12}{5}\right\}$

32. $\dfrac{-3k}{2} + \dfrac{9k - 5}{6} = \dfrac{11k + 8}{k}$ $\left\{-\dfrac{48}{71}\right\}$

33. $\dfrac{2r}{r - 1} = 5 + \dfrac{2}{r - 1}$ $\emptyset$

34. $\dfrac{3x}{x + 2} = \dfrac{1}{x + 2} - 4$ $\{-1\}$

35. $\dfrac{5}{2a + 3} + \dfrac{1}{a - 6} = 0$ $\left\{\dfrac{27}{7}\right\}$

36. $\dfrac{2}{x + 1} = \dfrac{3}{5x + 5}$ $\emptyset$

37. $\dfrac{4}{x - 3} - \dfrac{8}{2x + 5} + \dfrac{3}{x - 3} = 0$ $\left\{-\dfrac{59}{6}\right\}$

38. $\dfrac{5}{2p + 3} - \dfrac{3}{p - 2} = \dfrac{4}{2p + 3}$ $\left\{-\dfrac{11}{5}\right\}$

39. $\dfrac{2p}{p - 2} = 3 + \dfrac{4}{p - 2}$ $\emptyset$

40. $\dfrac{5k}{k + 4} = 3 - \dfrac{20}{k + 4}$ $\emptyset$

41. $\dfrac{3}{y - 2} + \dfrac{1}{y + 1} = \dfrac{1}{y^2 - y - 2}$ $\{0\}$

42. $\dfrac{2}{p + 3} - \dfrac{5}{p - 1} = \dfrac{1}{3 - 2p - p^2}$ $\left\{-\dfrac{16}{3}\right\}$

43. $.08w + .06(w + 12) = 7.72$ $\{50\}$

44. $.04(x - 12) + .06x = 1.52$ $\{20\}$

45. $(3x - 4)^2 - 5 = 3(x + 5)(3x + 2)$ $\left\{-\dfrac{19}{75}\right\}$

46. $(2x + 5)^2 = 3x^2 + (x + 3)^2$ $\left\{-\dfrac{8}{7}\right\}$

Solve each of the following equations for x. See Example 6.

47. $2(x - a) + b = 3x + a$ $x = -3a + b$

48. $5x - (2a + c) = a(x + 1)$ $x = \dfrac{3a + c}{5 - a}$

49. $ax + b = 3(x - a)$ $x = \dfrac{3a + b}{3 - a}$

50. $4a - ax = 3b + bx$ $x = \dfrac{4a - 3b}{b + a}$

51. $\dfrac{x}{a - 1} = ax + 3$ $x = \dfrac{3 - 3a}{a^2 - a - 1}$

52. $\dfrac{2a}{x - 1} = a - b$ $x = \dfrac{3a - b}{a - b}$

53. $a^2x + 3x = 2a^2$ $x = \dfrac{2a^2}{a^2 + 3}$

54. $ax + b^2 = bx - a^2$ $x = \dfrac{a^2 + b^2}{b - a}$

55. $y = \dfrac{ax + b}{cx + d}$ $x = \dfrac{b - yd}{yc - a}$

56. $y = \dfrac{px - q}{rx - s}$ $x = \dfrac{ys - q}{yr - p}$

▦ *A baseball pitcher's earned run average (E.R.A.) gives the average number of earned runs given up per 9 innings. An equation that gives this average is*

$$\text{E.R.A.} = \dfrac{9(\text{number of earned runs allowed})}{\text{number of innings pitched}}.$$

In each of the following exercises, two of the three values in this equation are given for a particular major league pitcher during the 1990 season. Find the remaining value.

◎ CONCEPTUAL   ✍ WRITING   ▲ CHALLENGING

57. Roger Clemens;   E.R.A.: 3.13;   innings pitched: $253\frac{1}{3}$ earned runs allowed: 88

58. Dwight Gooden;   E.R.A.: 2.89;   innings pitched: $118\frac{1}{3}$ earned runs allowed: 38

59. Greg Swindell;   innings pitched: $184\frac{1}{3}$;   earned runs allowed: 69   E.R.A.: 3.37

60. Frank Wills;   innings pitched: $71\frac{1}{3}$;   earned runs allowed: 29   E.R.A.: 3.66

61. Bert Blyleven;   E.R.A.: 2.73;   earned runs allowed: 73   innings pitched: 241

62. Tim Belcher;   E.R.A.: 2.82;   earned runs allowed: 72   innings pitched: 230

In the metric system of weights and measures, temperature is measured in degrees Celsius (° C) instead of degrees Fahrenheit (° F). To convert back and forth between the two systems, use the equations

$$C = \frac{5(F - 32)}{9} \qquad and \qquad F = \frac{9}{5}C + 32.$$

In each of the following exercises, convert to the other system. Round answers to the nearest tenth of a degree if necessary.

63. 20° C   68° F   **64.** 100° C   212° F   **65.** 59° F   15° C

66. 86° F   30° C   **67.** 100° F   37.8° C   **68.** 350° F   176.7° C

When a consumer borrows money, the lender must tell the consumer the true annual interest rate of the loan. The method of finding the exact true annual interest rate requires special tables available from the government, but a quick approximate rate can be found by using the equation

$$A = \frac{2pf}{b(q + 1)},$$

where p is the number of payments made in one year, f is the finance charge, b is the balance owed on the loan, and q is the total number of payments. Find the value of the variables not given in Exercises 69–74. Round A to the nearest percent and round other variables to the nearest whole number. (This formula is not *accurate enough for the requirements of federal law.)*

69. $p = 12, f = \$800, b = \$4000, q = 36$;   find A 13%

70. $p = 12, f = \$60, b = \$740, q = 12$;   find A 15%

71. $A = 14\%$ (or .14), $p = 12, b = \$2000, q = 36$;   find f \$432

72. $A = 11\%, p = 12, b = \$1500, q = 24$;   find f \$172

73. $A = 16\%, p = 12, f = \$370, q = 36$;   find b \$1500

74. $A = 10\%, p = 12, f = \$490, q = 48$;   find b \$2400

When a loan is paid off early, a portion of the finance charge must be returned to the borrower. By one method of calculating finance charge (called the rule of 78), *the amount of unearned interest (finance charge to be returned) is given by*

$$u = f \cdot \frac{n(n + 1)}{q(q + 1)},$$

where u represents unearned interest, f is the original finance charge, n is the number of payments remaining when the loan is paid off, and q is the original number of payments. Find the amount of the unearned interest in each of the following.

75. Original finance charge = \$800, loan scheduled to run 36 months, paid off with 18 payments remaining   \$205.41

76. Original finance charge = \$1400, loan scheduled to run 48 months, paid off with 12 payments remaining $92.86

77. Original finance charge = \$950, loan scheduled to run 24 months, paid off with 6 payments remaining $66.50

78. Original finance charge = \$175, loan scheduled to run 12 months, paid off with 3 payments remaining $13.46

79. Find the error in the following.

$$x^2 + 2x - 15 = x^2 - 3x$$

$$(x + 5)(x - 3) = x(x - 3)$$

$$x + 5 = x$$ Since 3 is the solution of the equation, dividing by $x - 3$
is dividing by zero, which is not allowed.

$$5 = 0$$

Solution set: $\emptyset$

▲ *Find the value of k that will make each equation equivalent to x = 2.*

80. $9x - 7 = k$ 11

81. $-5x + 11x - 2 = k + 4$ 6

82. $\dfrac{8}{k + x} = 4$ 0

83. $\sqrt{x + k} = 0$ -2

84. $\sqrt{3x - 2k} = 4$ -5

2.2 RESOURCES

Instructor's Test Manual
Test Items, p. 310–311
Instructor's Solution Manual
pp. 46–53, 466–472
Study Guide
pp. 57–62
Student's Solution Manual
pp. 46–53

Test Generator
IBM Disks
5.25" Disk A
3.5" Disk A
Mac Disk A
Tutorials
IBM Disks
5.25" Disk B
3.5" Disk A
Mac Disk A
Apple II Module 2, Lessons 1–2

HarperCollins Videotape   2
"College Algebra: In Simplest Terms" Program   5

2.2 EXERCISES   (TEXT PAGE 88) ■

Solve each formula for the indicated variable. Assume that all denominators are nonzero. See Example 1.

1. $i = prt$ for p (simple interest) $p = \dfrac{i}{rt}$

2. $V = lwh$ for l (volume of a rectangular box) $l = \dfrac{V}{wh}$

3. $P = 2l + 2w$ for w (perimeter of a rectangle) $w = \dfrac{P - 2l}{2}$ or $w = \dfrac{P}{2} - l$

4. $P = a + b + c$ for c (perimeter of a triangle) $c = P - a - b$

5. $A = \dfrac{1}{2}(B + b)h$ for h (area of a trapezoid) $h = \dfrac{2A}{B + b}$

6. $A = \dfrac{1}{2}(B + b)h$ for B (area of a trapezoid) $B = \dfrac{2A}{h} - b$ or $B = \dfrac{2A - hb}{h}$

7. $S = 2lw + 2wh + 2hl$ for h (surface area of a rectangular box) $h = \dfrac{S - 2lw}{2w + 2l}$

◎ CONCEPTUAL   ✍ WRITING   ▲ CHALLENGING

8. $S = 2\pi rh + 2\pi r^2$ for h (surface area of a right circular cylinder) $h = \dfrac{S - 2\pi r^2}{2\pi r}$ or $h = \dfrac{S}{2\pi r} - r$

9. $s = \dfrac{1}{2}gt^2$ for g (distance traveled by a falling object) $g = \dfrac{2s}{t^2}$

10. $C = \dfrac{5}{9}(F - 32)$ for F (Fahrenheit to Celsius) $F = \dfrac{9C}{5} + 32$

11. $\dfrac{1}{R} = \dfrac{1}{r_1} + \dfrac{1}{r_2}$ for R (electricity) $R = \dfrac{r_1 r_2}{r_2 + r_1}$ **12.** $u = f \cdot \dfrac{k(k + 1)}{n(n + 1)}$ for f (unearned interest) $f = \dfrac{un(n + 1)}{k(k + 1)}$

13. $A = \dfrac{24f}{B(p + 1)}$ for f (approximate annual interest rate) $f = \dfrac{AB(p + 1)}{24}$

14. $A = \dfrac{24f}{B(p + 1)}$ for B (approximate annual interest rate) $B = \dfrac{24f}{A(p + 1)}$

15. $A = P\left(1 + \dfrac{i}{m}\right)$ for m (compound interest) $m = \dfrac{Pi}{A - P}$

16. $V = \dfrac{1}{3}\pi r^2 h$ for h (volume of a right circular cone) $h = \dfrac{3V}{\pi r^2}$

17. Refer to Exercise 16. Why is it not possible, using the methods of this section, to solve this formula for r?

18. Refer to Example 1. Suppose that someone tells you that there is no reason to solve for x, since the right side of the formula is already equal to x. How would you respond?

19. Suppose two acid solutions are mixed. One is 26% acid and the other is 32% acid. Which one of the following concentrations cannot possibly be the concentration of the mixture? Explain.
(a) 36%   (b) 28%   (c) 30%   (d) 31%

20. Suppose that a computer that originally sells for x dollars has been discounted 30%. Which one of the following expressions does not represent its sale price?
(a) $x - .30x$ (b) $.70x$
(c) $\dfrac{7}{10}x$ (d) $x - .30$ **(d)**

Solve each of the following problems. See Examples 2–7.

21. A puzzle piece in the shape of a triangle has a perimeter of 30 cm. Two sides of the triangle are each twice as long as the shortest side. Find the length of the shortest side. (In the figure, let $x =$ length of the shortest side. Then $2x =$ length of the other sides.) **6 cm**

Side lengths are in centimeters.

22. The length of a rectangular label is 3 cm less than twice the width. The perimeter is 54 cm. Find the width. (In the figure, let $w =$ width. Then $2w - 3$ $=$ length.) **10 cm**

Side lengths are in centimeters.

23. Hien has grades of 84, 88, and 92 on his first three calculus tests. What grade on his fourth test will give him an average of 90? **96**

24. Doug scored 78, 94, and 60 on his three trigonometry tests. If his final exam score is to be counted as two test grades in determining his course average, what grade must he make on his final exam to give him an average of 80? **84**

25. A pharmacist wishes to strengthen a mixture that is 10% alcohol to one that is 30% alcohol. How much pure alcohol should be added to 7 liters of the 10% mixture? **2 liters**

26. A student needs 10% hydrochloric acid for a chemistry experiment. How much 5% acid should be mixed with 60 ml of 20% acid to get a 10% solution? **120 ml**

27. A recycling bin is in the shape of a rectangular box. Find the height of the box if its length is 18 ft, its width is 8 ft, and its surface area is 496 sq ft. (In the figure, let $h =$ height.) **4 ft**

h is in feet.

28. A right circular cylinder has radius 6 in and volume 144π cu in. What is its height? (In the figure, let h = height.) **4 in**

h is in inches.

Exercises 29 and 30 depend on the idea of the octane rating of gasoline, a measure of its antiknock qualities. In one measure of octane, a standard fuel is made with only two ingredients: heptane and isooctane. For this fuel, the octane rating is the percent of isooctane. An actual gasoline blend is then compared to a standard fuel. For example, a gasoline with an octane rating of 98 has the same antiknock properties as a standard fuel that is 98% isooctane.

29. How many liters of 94-octane gasoline should be mixed with 200 liters of 99-octane gasoline to get a mixture that is 97-octane? $\dfrac{400}{3}$ **liters**

30. A service station has 92-octane and 98-octane gasoline. How many liters of each should be mixed to provide 12 liters of 96-octane gasoline needed for chemical research?

4 liters of 92-octane and 8 liters of 98-octane

31. On a vacation trip, Jose averaged 50 mph traveling from Denver to Minneapolis. Returning by a different route that covered the same number of miles, he averaged 55 mph. What is the distance between the two cities if his total traveling time was 32 hr?

about 840 mi

32. Cindy left by plane to visit her mother in Hartford, 420 km away. Fifteen minutes later, her mother left to meet her at the airport. She drove the 20 km to the airport at 40 km/hr, arriving just as the plane taxied in. What was the speed of the plane?

560 km per hr

33. Russ and Janet are running in the Apple Hill Fun Run. Russ runs at 7 mph, Janet at 5 mph. If they start at the same time, how long will it be before they are 1/2 mi apart? **15 min**

34. If the run in Exercise 33 has a staggered start, and Janet starts first, with Russ starting 10 min later, how long will it be before he catches up with her?

25 min

35. Joann took 20 min to drive her boat upstream to water-ski at her favorite spot. Coming back later in the day, at the same boat speed, took her 15 min. If the current in that part of the river is 5 km/hr, what was her boat speed? **35 km per hr**

36. Joe traveled against the wind in a small plane for 3 hr. The return trip with the wind took 2.8 hr. Find the speed of the wind if the speed of the plane in still air is 180 mph. **about 6.2 mph**

37. Johnny gets to work in 20 min when he drives his car. Riding his bike (by the same route) takes him 45 min. His average driving speed is 4.5 mph greater than his average speed on his bike. How far does he travel to work? **2.7 mi**

38. In the morning, Marge drove to a business appointment at 50 mph. Her average speed on the return trip in the afternoon was 40 mph. The return trip took 1/4 hr longer because of heavy traffic. How far did she travel to the appointment? **50 mi**

39. Le can clean the house in 9 hr, while Tran needs 6 hr. How long will it take them to clean the house if they work together? **3.6 hr**

40. Helen can paint a room in 5 hr. Jay can paint the same room in 4 hr. How long will it take them to paint the room together? $2\dfrac{2}{9}$ **hr**

41. Two chemical plants are polluting a river. If plant A produces a predetermined maximum amount of pollutant twice as fast as plant B, and together they produce the maximum pollutant in 26 hr, how long will it take plant B alone? **78 hr**

42. A sewage treatment plant has two inlet pipes to its settling pond. One can fill the pond in 10 hours, the other in 12 hours. If the first pipe is open for 5 hours and then the second pipe is opened, how long will it take to fill the pond?

$\dfrac{30}{11}$ **hr after the second pipe is opened**

43. An inlet pipe can fill Dominic's pool in 5 hr, while an outlet pipe can empty it in 8 hr. In his haste to watch television, Dominic left both pipes open. How long did it take to fill the pool? $\dfrac{40}{3}$ **hr**

44. Suppose Dominic discovered his error (see Exercise 43) after an hour-long program. If he then closed the outlet pipe, how much more time would be needed to fill the pool? $\dfrac{37}{8}$ hr

45. A VCR is on sale for $245. If the sale price is 30% less than the regular price, what was the regular price?

$350

46. A jeweler prices his items 60% over their wholesale price. If a watch sells for $152, what is its wholesale price? $95

A person's intelligence quotient (IQ) is found by multiplying the mental age by 100 *and dividing by the chronological age. Use this to solve Exercises 47 and 48.*

47. Jack is 7 years old. His IQ is 130. Find his mental age.

9.1 yr

48. If a person is 16 years old with a mental age of 20, what is the person's IQ? 125

49. Bill Cornett won $200,000 in a state lottery. He first paid income tax of 30% on the winnings. Of the rest, he invested some at 8.5% and some at 7%, making $10,700 interest per year. How much is invested at each rate? $60,000 at 8.5% and $80,000 at 7%

50. Marjorie Williams earned $48,000 from royalties on her cookbook. She paid a 28% income tax on these royalties. The balance was invested in two ways, some of it at 6.5% and some at 6.25%. The investments produced $2210 interest per year. Find the amount invested at each rate.

$20,000 at 6.5% and $14,560 at 6.25%

51. Janet Tilden bought two plots of land for a total of $120,000. When she sold the first plot, she made a profit of 15%. When she sold the second, she lost 10%. Her total profit was $5500. How much did she pay for each piece of land?

$70,000 for land that makes a profit and $50,000 for land that produces a loss

52. Suppose $10,000 is invested at 6%. How much additional money must be invested at 8% to produce a yield of 7.2% on the entire amount invested?

$15,000

53. Cathy Wacaser earns take-home pay of $592 a week. If her deductions for taxes, retirement, union dues, and medical plan amount to 26% of her wages, what is her weekly pay before deductions? $800

54. George Duda gives 10% of his net income to charity. This amounts to $167.20 per month. In addition, his paycheck deductions are 24% of his gross monthly income. What is his gross monthly income? $2200

55. A bank pays 5% interest on passbook accounts and 7% interest on long-term deposits. Suppose a depositor divides $20,000 among the two types of deposits. Find the amount deposited at each rate if the total annual income from interest is $1340.

$3000 in passbook accounts and $17,000 in long-term deposits

56. Adam Bryer wishes to sell a piece of property for $125,000. He wishes the money to be paid off in two ways—a short-term note at 12% and a long-term note at 10%. Find the amount of each note if the total annual interest paid is $13,700.

short-term note is for $60,000 and long-term note is for $65,000

57. In planning her retirement, Louise Howe deposits some money at 4.5% with twice as much deposited at 5%. Find the amount deposited at each rate if the total annual interest income is $2900.

$20,000 at 4.5% and $40,000 at 5%

58. A church building fund has invested $75,000 in two ways: part of the money at 7% and four times as much at 11%. Find the amount invested at each rate if the total annual income from interest is $7650.

$15,000 at 7% and $60,000 at 11%

◎ **59.** Which of the following cannot be a correct equation to solve a geometry problem, if x represents the length of a rectangle? (*Hint:* Solve each equation and consider the solutions.)
   (a) $2x + 2(x - 1) = 14$
   (b) $-2x + 7(5 - x) = 62$
   (c) $4(x + 2) + 4x = 8$
   (d) $2x + 2(x - 3) = 22$ (b) and (c)

◎ **60.** If x represents the number of pennies in a jar in an applied problem, which of the following equations cannot be a correct equation for finding x? (*Hint:* Solve each equation and consider the solutions.)
   (a) $5x + 3 = 9$ **(b)** $12x + 3 = -4$
   (c) $100x = 50(x + 3)$ **(d)** $6(x + 4) = x + 24$

   (a) and (b)

2.3 RESOURCES

Instructor's Test Manual
  Test Items, p. 311–312
Instructor's Solution Manual
  pp. 53–56, 473–476
Study Guide
  pp. 62–66
Student's Solution Manual
  pp. 53–56

Test Generator
  IBM Disks
    5.25" Disk A
    3.5" Disk A
    Mac Disk A
Tutorials
  IBM Disks
    5.25" Disk B
    3.5" Disk A
    Mac Disk A
    Apple II Module 2, Lessons 1–2

HarperCollins Videotape 2
Transparency T-1

2.3 EXERCISES (TEXT PAGE 97) ■

Identify each complex number as real or imaginary. See Example 1.

1. $-9i$ imaginary   **2.** 6 real   **3.** π real   **4.** $-\sqrt{7}$ real

5. $i\sqrt{6}$ imaginary   **6.** $-3i$ imaginary   **7.** $2 + 5i$ imaginary   **8.** $-7 - 6i$ imaginary

Write each of the following without negative radicands. See Examples 3 and 4.

9. $\sqrt{-100}$ $10i$ **10.** $\sqrt{-169}$ $13i$ **11.** $-\sqrt{-400}$ $-20i$ **12.** $-\sqrt{-225}$ $-15i$

13. $-\sqrt{-39}$ $-i\sqrt{39}$ **14.** $-\sqrt{-95}$ $-i\sqrt{95}$ **15.** $5 + \sqrt{-4}$ $5 + 2i$

16. $-7 + \sqrt{-100}$ $-7 + 10i$ **17.** $9 - \sqrt{-50}$ $9 - 5i\sqrt{2}$ **18.** $-11 - \sqrt{-24}$ $-11 - 2i\sqrt{6}$

19. $\sqrt{-5} \cdot \sqrt{-5}$ -5 **20.** $\sqrt{-20} \cdot \sqrt{-20}$ -20 **21.** $\dfrac{\sqrt{-40}}{\sqrt{-10}}$ 2

22. $\dfrac{\sqrt{-190}}{\sqrt{-19}}$ $\sqrt{10}$ **23.** $\dfrac{\sqrt{-6} \cdot \sqrt{-2}}{\sqrt{3}}$ -2 **24.** $\dfrac{\sqrt{-12} \cdot \sqrt{-6}}{\sqrt{8}}$ -3

Add or subtract. Write each result in standard form. See Examples 5 and 6.

25. $(3 + 2i) + (4 - 3i)$ $7 - i$ **26.** $(4 - i) + (2 + 5i)$ $6 + 4i$

27. $(-2 + 3i) - (-4 + 3i)$ $2 + 0i$ **28.** $(-3 + 5i) - (-4 + 3i)$ $1 + 2i$

29. $(2 - 5i) - (3 + 4i) - (-2 + i)$ $1 - 10i$ **30.** $(-4 - i) - (2 + 3i) + (-4 + 5i)$ $-10 + i$

Multiply. Write each result in standard form. See Example 7.

31. $(2 + i)(3 - 2i)$ $8 - i$ **32.** $(-2 + 3i)(4 - 2i)$ $-2 + 16i$ **33.** $(2 + 4i)(-1 + 3i)$ $-14 + 2i$

34. $(1 + 3i)(2 - 5i)$ $17 + i$ **35.** $(-3 + 2i)^2$ $5 - 12i$ **36.** $(2 + i)^2$ $3 + 4i$

37. $(2 + 3i)(2 - 3i)$ $13 + 0i$ **38.** $(6 - 4i)(6 + 4i)$ $52 + 0i$ **39.** $(\sqrt{6} + i)(\sqrt{6} - i)$ $7 + 0i$

40. $(\sqrt{2} - 4i)(\sqrt{2} + 4i)$ $18 + 0i$ **41.** $i(3 - 4i)(3 + 4i)$ $0 + 25i$ **42.** $i(2 + 7i)(2 - 7i)$ $0 + 53i$

43. $3i(2 - i)^2$ $12 + 9i$ **44.** $-5i(4 - 3i)^2$ $-120 - 35i$

Find each of the following powers of i. See Example 8.

45. i^5 i **46.** i^8 1 **47.** i^9 i **48.** i^{11} $-i$ **49.** i^{12} 1 **50.** i^{25} i

51. i^{43} $-i$ **52.** $\dfrac{1}{i^9}$ $-i$ **53.** $\dfrac{1}{i^{12}}$ 1 **54.** i^{-6} -1 **55.** i^{-15} i **56.** i^{-49} $-i$

© CONCEPTUAL   ✎ WRITING   ▲ CHALLENGING

57. Suppose that your friend, Susan Katz, tells you that she has discovered a method of simplifying a positive power of i. "Just divide the exponent by 4," she says, "and then look at the remainder. Then refer to the table of powers of i in this section. The large power of i is equal to i to the power indicated by the remainder.

And if the remainder is 0, the result is $i^0 = 1$." Explain why Susan's method works.

58. Explain why the method of dividing complex numbers (that is, multiplying both the numerator and the denominator by the conjugate of the denominator) works. That is, what property justifies this process?

Divide. Write each result in standard form. See Example 10.

59. $\dfrac{1+i}{1-i}$ $0 + i$

60. $\dfrac{2-i}{2+i}$ $\dfrac{3}{5} - \dfrac{4}{5}i$

61. $\dfrac{4-3i}{4+3i}$ $\dfrac{7}{25} - \dfrac{24}{25}i$

62. $\dfrac{5-2i}{6-i}$ $\dfrac{32}{37} - \dfrac{7}{37}i$

63. $\dfrac{3-4i}{2-5i}$ $\dfrac{26}{29} + \dfrac{7}{29}i$

64. $\dfrac{1-3i}{1+i}$ $-1 - 2i$

65. $\dfrac{-3+4i}{2-i}$ $-2 + i$

66. $\dfrac{5+6i}{5-6i}$ $-\dfrac{11}{61} + \dfrac{60}{61}i$

67. $\dfrac{2}{i}$ $0 - 2i$

68. $\dfrac{-7}{3i}$ $0 + \dfrac{7}{3}i$

▲ **69.** $\dfrac{1-\sqrt{-5}}{3+\sqrt{-4}}$ $\dfrac{3-2\sqrt{5}}{13} + \dfrac{-2-3\sqrt{5}}{13}i$

▲ **70.** $\dfrac{2+\sqrt{-3}}{1-\sqrt{-9}}$ $\dfrac{2-3\sqrt{3}}{10} + \dfrac{6+\sqrt{3}}{10}i$

▲ *Perform the indicated operations. Write answers in standard form.*

71. $\dfrac{2+i}{3-i} \cdot \dfrac{5+2i}{1+i}$ $\dfrac{5}{2} + i$

72. $\dfrac{1-i}{2+i} \cdot \dfrac{4+3i}{1+i}$ $\dfrac{2}{5} - \dfrac{11}{5}i$

73. $\dfrac{6+2i}{5-i} \cdot \dfrac{1-3i}{2+6i}$ $-\dfrac{16}{65} - \dfrac{37}{65}i$

74. $\dfrac{5-3i}{1+2i} \cdot \dfrac{2-4i}{1+i}$ $-\dfrac{38}{5} + \dfrac{16}{5}i$

75. $\dfrac{5-i}{3+i} + \dfrac{2+7i}{3+i}$ $\dfrac{27}{10} + \dfrac{11}{10}i$

76. $\dfrac{4-3i}{2+5i} + \dfrac{8-i}{2+5i}$ $\dfrac{4}{29} - \dfrac{68}{29}i$

▲ ◉ *Give the necessary conditions for a and b so that the square $(a + bi)^2$ is.*

77. Real. $a = 0$ or $b = 0$

78. Imaginary. $a \neq 0$ and $b \neq 0$

79. Show that $\dfrac{\sqrt{2}}{2} + \dfrac{\sqrt{2}}{2}i$ is a square root of i.

80. Show that $\dfrac{\sqrt{3}}{2} + \dfrac{1}{2}i$ is a cube root of i.

81. Evaluate $3z - z^2$ if $z = 3 - 2i$. $4 + 6i$

82. Evaluate $-2z + z^3$ if $z = -6i$. $228i$

2.4   RESOURCES

Instructor's Test Manual
Test Items, p. 312–313

Instructor's Solution Manual
pp. 56–64, 476–484

Study Guide
pp. 64–74

Student's Solution Manual
pp. 56–64

Test Generator
IBM Disks
5.25" Disk A
3.5" Disk A
Mac Disk A

Tutorials
IBM Disks
5.25" Disk B
3.5" Disk A
Mac Disk A
Apple II Module 2, Lessons 1–2

HarperCollins Videotape   2

2.4 EXERCISES   (TEXT PAGE 106) ■ ─────────────────────────

Solve the following equations by factoring or by using the square root property. See Examples 1 and 2.

1. $p^2 = 16$ $\{\pm 4\}$ 　　　**2.** $k^2 = 25$ $\{\pm 5\}$ 　　　**3.** $x^2 = 27$ $\{\pm 3\sqrt{3}\}$ 　　　**4.** $r^2 = 48$ $\{\pm 4\sqrt{3}\}$

5. $t^2 = -16$ $\{\pm 4i\}$ 　　　**6.** $y^2 = -100$ $\{\pm 10i\}$ 　　　**7.** $x^2 = -18$ $\{\pm 3i\sqrt{2}\}$

8. $(p + 2)^2 = 7$ $\{-2 \pm \sqrt{7}\}$ 　　　**9.** $(3k - 1)^2 = 12$ $\left\{\dfrac{1 \pm 2\sqrt{3}}{3}\right\}$ 　　　**10.** $(4t + 1)^2 = 20$ $\left\{\dfrac{-1 \pm 2\sqrt{5}}{4}\right\}$

11. $p^2 - 5p + 6 = 0$ $\{2, 3\}$ 　　　　　　　**12.** $q^2 + 2q - 8 = 0$ $\{-4, 2\}$

13. $6z^2 - 5z - 50 = 0$ $\left\{-\dfrac{5}{2}, \dfrac{10}{3}\right\}$ 　　　　　**14.** $21p^2 = 10 - 29p$ $\left\{-\dfrac{5}{3}, \dfrac{2}{7}\right\}$

15. $(5r - 3)^2 = -3$ $\left\{\dfrac{3}{5} \pm \dfrac{\sqrt{3}}{5}i\right\}$ 　　　　**16.** $(-2w + 5)^2 = -8$ $\left\{\dfrac{5}{2} \pm i\sqrt{2}\right\}$

Solve the following equations by completing the square. See Examples 3 and 4.

17. $p^2 - 8p + 15 = 0$ $\{3, 5\}$ 　　　　　**18.** $m^2 + 5m = 6$ $\{-6, 1\}$

19. $x^2 - 2x - 4 = 0$ $\{1 \pm \sqrt{5}\}$ 　　　　**20.** $r^2 + 8r + 13 = 0$ $\{-4 \pm \sqrt{3}\}$

21. $2p^2 + 2p + 1 = 0$ $\left\{-\dfrac{1}{2} \pm \dfrac{1}{2}i\right\}$ 　　　**22.** $9z^2 - 12z + 8 = 0$ $\left\{\dfrac{2}{3} \pm \dfrac{2}{3}i\right\}$

✍ ◉ **23.** Erin solved Exercise 16 correctly and wrote the solution as $\dfrac{5}{2} \pm i\sqrt{2}$. Paul wrote his solution

　　as $\dfrac{-5}{-2} \pm i\sqrt{2}$. Was Paul correct? Explain.

Solve the following equations by using the quadratic formula. See Examples 5 and 6.

24. $m^2 - m - 1 = 0$ $\left\{\dfrac{1 \pm \sqrt{5}}{2}\right\}$ 　　　　**25.** $y^2 - 3y - 2 = 0$ $\left\{\dfrac{3 \pm \sqrt{17}}{2}\right\}$

26. $x^2 - 6x + 7 = 0$ $\{3 \pm \sqrt{2}\}$ 　　　　**27.** $11p^2 - 7p + 1 = 0$ $\left\{\dfrac{7 \pm \sqrt{5}}{22}\right\}$

28. $4z^2 - 12z + 11 = 0$ $\left\{\dfrac{3}{2} \pm \dfrac{\sqrt{2}}{2}i\right\}$ 　　　**29.** $x^2 = 2x - 5$ $\{1 \pm 2i\}$

30. $\dfrac{1}{2}t^2 + \dfrac{1}{4}t - 3 = 0$ $\left\{\dfrac{-1 \pm \sqrt{97}}{4}\right\}$ 　　　**31.** $\dfrac{2}{3}x^2 + \dfrac{1}{4}x = 3$ $\left\{\dfrac{-3 \pm 3\sqrt{129}}{16}\right\}$

32. $.2x^2 + .4x - .3 = 0$ $\left\{\dfrac{-2 \pm \sqrt{10}}{2}\right\}$ 　　　**33.** $4 + \dfrac{3}{x} - \dfrac{2}{x^2} = 0$ $\left\{\dfrac{-3 \pm \sqrt{41}}{8}\right\}$

34. $4 - \dfrac{11}{x} - \dfrac{3}{x^2} = 0$ $\left\{-\dfrac{1}{4}, 3\right\}$ 　　　**35.** $3 - \dfrac{4}{p} = \dfrac{2}{p^2}$ $\left\{\dfrac{2 \pm \sqrt{10}}{3}\right\}$

✍ ◉ **36.** Why is the restriction $a \neq 0$ necessary in the statement of the quadratic formula?

◉ **37.** Which one of the following equations has two real, distinct solutions? Do not actually solve.
   　(a) $(3x - 4)^2 = -4$ 　　　　**(b)** $(4 + 7x)^2 = 0$
   　(c) $(5x + 9)(5x + 9) = 0$ 　　**(d)** $(7x + 4)^2 = 11$ **(d)**

◉ **38.** Which equations in Exercise 37 have only one distinct, real solution?   **(b) and (c)**

◉ **39.** Which one of the equations in Exercise 37 has two imaginary solutions?   **(a)**

✍ ◉ **40.** Discuss the advantages and disadvantages of the three methods of solving quadratic equations discussed in this section: using the square root property, completing the square, and using the quadratic formula.

◉ CONCEPTUAL 　　✍ WRITING 　　▲ CHALLENGING

Solve each of the following equations by factoring first. See Example 7.

41. $x^3 - 1 = 0$ $\left\{1, -\dfrac{1}{2} \pm \dfrac{\sqrt{3}}{2}i\right\}$

42. $x^3 + 64 = 0$ $\{-4, 2 \pm 2i\sqrt{3}\}$

43. $x^3 + 27 = 0$ $\left\{-3, \dfrac{3}{2} \pm \dfrac{3\sqrt{3}}{2}i\right\}$

Solve the following equations by any method.

44. $8p^3 + 125 = 0$ $\left\{-\dfrac{5}{2}, \dfrac{5}{4} \pm \dfrac{5\sqrt{3}}{4}i\right\}$

45. $2 - \dfrac{5}{k} + \dfrac{2}{k^2} = 0$ $\left\{\dfrac{1}{2}, 2\right\}$

46. $(m - 3)^2 = 5$ $\{3 \pm \sqrt{5}\}$

47. $t^2 - t = 3$ $\left\{\dfrac{1 \pm \sqrt{13}}{2}\right\}$

48. $x^2 + x = -1$ $\left\{-\dfrac{1}{2} \pm \dfrac{\sqrt{3}}{2}i\right\}$

49. $64r^3 - 343 = 0$ $\left\{\dfrac{7}{4}, -\dfrac{7}{8} \pm \dfrac{7\sqrt{3}}{8}i\right\}$

50. $2s^2 + 2s = 3$ $\left\{\dfrac{-1 \pm \sqrt{7}}{2}\right\}$

51. $(3y + 1)^2 = -7$ $\left\{-\dfrac{1}{3} \pm \dfrac{\sqrt{7}}{3}i\right\}$

52. $\dfrac{1}{3}x^2 + \dfrac{1}{6}x + \dfrac{1}{9} = 0$ $\left\{-\dfrac{1}{4} \pm \dfrac{\sqrt{39}}{12}i\right\}$

▲ *Each of the following quadratic equations has at least one coefficient that is an irrational number. To solve for example,*

$$\sqrt{2}m^2 + 5m - 3\sqrt{2} = 0,$$

we can use the quadratic formula with $a = \sqrt{2}$, $b = 5$, and $c = -3\sqrt{2}$. Solve each equation using the quadratic formula.

53. $m^2 - \sqrt{2}m - 1 = 0$ $\left\{\dfrac{\sqrt{2} \pm \sqrt{6}}{2}\right\}$

54. $z^2 - \sqrt{3}z - 2 = 0$ $\left\{\dfrac{\sqrt{3} \pm \sqrt{11}}{2}\right\}$

55. $\sqrt{2}p^2 - 3p + \sqrt{2} = 0$ $\left\{\sqrt{2}, \dfrac{\sqrt{2}}{2}\right\}$

56. $-\sqrt{6}k^2 - 2k + \sqrt{6} = 0$ $\left\{\dfrac{-\sqrt{6} \pm \sqrt{42}}{6}\right\}$

57. $x^2 + \sqrt{5}x + 1 = 0$ $\left\{\dfrac{-\sqrt{5} \pm 1}{2}\right\}$

58. $3\sqrt{5}m^2 - 2m = \sqrt{5}$ $\left\{\dfrac{\sqrt{5}}{3}, -\dfrac{\sqrt{5}}{5}\right\}$

Solve each equation for the indicated variable. Assume that no denominators are zero. See Example 8.

59. $s = \dfrac{1}{2}gt^2$ for t $t = \dfrac{\pm\sqrt{2sg}}{g}$

60. $A = \pi r^2$ for r $r = \dfrac{\pm\sqrt{A\pi}}{\pi}$

61. $F = \dfrac{kMv^4}{r}$ for v $v = \dfrac{\pm\sqrt[4]{Frk^3M^3}}{kM}$

62. $s = s_0 + gt^2 + k$ for t $t = \dfrac{\pm\sqrt{(s - s_0 - k)g}}{g}$

63. $P = \dfrac{E^2R}{(r + R)^2}$ for R $R = \dfrac{E^2 - 2Pr \pm E\sqrt{E^2 - 4Pr}}{2P}$

64. $S = 2\pi rh + 2\pi r^2$ for r $r = \dfrac{-\pi h \pm \sqrt{\pi^2h^2 + 2\pi S}}{2\pi}$

▲ *For each of the following equations, (**a**) solve for x in terms of y, and (**b**) solve for y in terms of x. See Example 8.*

65. $4x^2 - 2xy + 3y^2 = 2$ (a) $x = \dfrac{y \pm \sqrt{8 - 11y^2}}{4}$ (b) $y = \dfrac{x \pm \sqrt{6 - 11x^2}}{3}$

66. $3y^2 + 4xy - 9x^2 = -1$ (a) $x = \dfrac{2y \pm \sqrt{31y^2 + 9}}{9}$ (b) $y = \dfrac{-2x \pm \sqrt{31x^2 - 3}}{3}$

Identify the values of a, b, and c for each of the following: then evaluate the discriminant
$b^2 - 4ac$ *and use it to predict the type of solutions. Do not solve the equations. See Example 9.*

67. $x^2 + 8x + 16 = 0$ 0;   one rational solution

68. $x^2 - 5x + 4 = 0$ 9;   two different rational solutions

69. $3m^2 - 5m + 2 = 0$ 1;   two different rational solutions

70. $8y^2 = 14y - 3$ 100;   two different rational solutions

71. $4p^2 = 6p + 3$ 84;   two different irrational solutions

72. $2r^2 - 4r + 1 = 0$ 8;   two different irrational solutions

73. $9k^2 + 11k + 4 = 0$ −23;   two different imaginary solutions

74. $3z^2 = 4z - 5$ −44;   two different imaginary solutions

75. $8x^2 - 72 = 0$ 2304;   two different rational solutions

Find all values of k for which each of the following equations has exactly one solution.
See Example 10.

76. $9x^2 + kx + 4 = 0$ ±12   **77.** $25m^2 - 10m + k = 0$ 1   **78.** $y^2 + 11y + k = 0$ $\dfrac{121}{4}$

☑◎ **79.** Show that the discriminant for the equation $\sqrt{2}m^2 + 5m - 3\sqrt{2} = 0$ is 49. If this equation is completely solved, it can be shown that the solution set is $\{-3\sqrt{2}, \sqrt{2}/2\}$. Here we have a discriminant that is positive and a perfect square, yet the two solutions are irrational. Does this contradict the discussion in this section? Explain.

80. Solve the equation in Exercise 79. $\left\{-3\sqrt{2}, \dfrac{\sqrt{2}}{2}\right\}$

Give the real number restrictions on the variable for each of the following rational expressions.
See Example 11.

81. $\dfrac{3 - 2x}{3x^2 - 19x - 14}$ $x \neq 7, x \neq -\dfrac{2}{3}$ **82.** $\dfrac{7 - 3x}{2x^2 + 7x - 15}$ $x \neq \dfrac{3}{2}, x \neq -5$

83. $\dfrac{-3}{16y^2 + 8y + 1}$ $y \neq -\dfrac{1}{4}$ **84.** $\dfrac{18}{25y^2 - 30y + 9}$ $y \neq \dfrac{3}{5}$

85. $\dfrac{8 + 7x}{x^2 + x + 1}$ no restrictions   **86.** $\dfrac{-9x + 4}{7x^2 + 2x + 2}$ no restrictions

▲ **87.** Let r_1 and r_2 be the solutions of the quadratic equation $ax^2 + bx + c = 0$. Show that the following statements are correct.

   (a) $r_1 + r_2 = -\dfrac{b}{a}$ **(b)** $r_1 \cdot r_2 = \dfrac{c}{a}$

☑◎ **88.** State in words the result of Exercise 87(a).

☑◎ **89.** State in words the result of Exercise 87(b).

◎ **90.** Is it possible for the solution set of a quadratic equation with real coefficients to consist of one real and one imaginary solution?   no

◎ CONCEPTUAL   ☑ WRITING   ▲ CHALLENGING

2.5 RESOURCES

Instructor's Test Manual
Test Items, p. 314–315
Instructor's Solution Manual
pp. 64–70, 484–490
Study Guide
pp. 74–77
Student's Solution Manual
pp. 64–70

Test Generator
IBM Disks
5.25″ Disk A
3.5″ Disk A
Mac Disk A
Tutorials
IBM Disks
5.25″ Disk B
3.5″ Disk A
Mac Disk A
Apple II Module 2, Lessons 1–2

HarperCollins Videotape 3
"College Algebra: In Simplest Terms" Program 7

2.5 EXERCISES (TEXT PAGE 113) ■

Solve each of the following problems. See Examples 1–4.

1. A shopping center has a rectangular area of 40,000 sq yd enclosed on three sides for a parking lot. The length is 200 yd more than twice the width. What are the dimensions of the lot? **100 yd by 400 yd**

$2x + 200$

x is in yards.

2. An ecology center wants to set up an experimental garden using 300 m of fencing to enclose a rectangular area of 5000 sq m. Find the dimensions of the rectangle. **50 m by 100 m**

$150 - x$

x

x is in meters.

3. Pat Kelley went into a frame-it-yourself shop. He wanted a frame 3 in longer than it was wide. The frame he chose extended 1.5 in beyond the picture on each side. Find the outside dimensions of the frame if the area of the unframed picture is 70 sq in.

   10 in by 13 in

4. Kathy Monaghan wants to buy a rug for a room that is 12 ft wide and 15 ft long. She wants to leave a uniform strip of floor around the rug. She can afford to buy 108 sq ft of carpeting. What dimensions should the rug have? **9 ft by 12 ft**

5. A rectangular page in a book is to have an 18 cm by 23 cm illustration in the center with equal margins on all four sides. How wide should the margins be if the page has an area of 594 sq cm? **2 cm**

6. A landscape architect has included a rectangular flower bed measuring 9 ft by 5 ft in her plans for a new building. She wants to use two colors of flowers in the bed, one in the center and the other for a border of the same width on all four sides. If she has enough plants to cover 24 sq ft for the border, how wide can the border be? **1 ft**

7. Felipe Santiago can clean a garage in 9 hr less time than his brother Felix. Working together, they can do the job in 20 hr. How long would it take each one to do the job alone? **Felipe: 36 hr; Felix: 45 hr**

8. An experienced roofer can do a complete roof in a housing development in half the time compared to an inexperienced roofer. If the two work together on a roof, they complete the job in 2 2/3 hr. How long would it take the experienced roofer to do a roof by himself? **4 hr**

9. Two typists are working on a special project. The experienced typist could complete the project in 2 hr less time than the new typist. Together they complete the project in 2.4 hr. How long would it have taken the experienced typist working alone? **4 hr**

10. It takes two copy machines 6/5 hr to make the copies for a company newsletter. One copy machine would take 1 hr longer than the other to do the job alone. How long would it take the faster machine to complete the job alone? **2 hr**

11. Paula drives 10 mph faster than Steve. Both start at the same time for Atlanta from Chattanooga, a distance of about 100 mi. It takes Steve 1/3 hr longer than Paula to make the trip. What is Steve's average speed?

50 mph

12. Marjorie Seachrist walks 1 mph faster than her daughter Sibyl. In a walk for charity, both walked the full distance of 24 mi. Sibyl took 2 hr longer than Marjorie. What was Sibyl's average speed?   3 mph

13. A plane flew 1000 mi. It later took off and flew 2025 mi at an average speed of 50 mph faster than the first trip. The second trip took 2 hr more time than the first. Find the time for the first trip.

2.5 hr   or   16 hr (16 is not realistic)

14. The Branson family traveled 100 mi to a lake for their vacation. On the return trip their average speed was 50 mph faster. The total time for the round trip was 11/3 hr. What was the family's average speed on their trip to the lake?   39.27 mph

◉ 15. To solve for the lengths of the sides of the right triangle shown, which equation is correct?

(a) $x^2 = (2x - 2)^2 + (x + 4)^2$
(b) $x^2 + (x + 4)^2 = (2x - 2)^2$
(c) $x^2 = (2x - 2)^2 - (x + 4)^2$
(d) $x^2 + (2x - 2)^2 = (x + 4)^2$

(d)

◉ 16. If a rectangle is r feet long and s feet wide, which one of the following is the length of its diagonal in terms of r and s?   (a) $\sqrt{rs}$ (b) $r + s$
(c) $\sqrt{r^2 + s^2}$ (d) $r^2 + s^2$ (c)

17. A kite is flying on 50 ft of string. How high is it above the ground if its height is 10 ft more than the horizontal distance from the person flying it? Assume the string is being released at ground level.   40 ft

18. A boat is being pulled into a dock with a rope attached at water level. When the boat is 12 ft from the dock, the length of the rope from the boat to the dock is 3 ft longer than twice the height of the dock above the water. Find the height of the dock.   5 ft

▦ 19. Chris and Josh have received walkie-talkies for Christmas. If they leave from the same point at the same time, Chris walking north at 2.5 mph and Josh walking east at 3 mph, how long will they be able to talk to each other if the range of the walkie-talkies is 4 miles? Round your answer to the nearest minute.   61 min

20. There is a bamboo 10 ft high, the upper end of which, being broken, reaches the ground 3 ft from the stem. Find the height of the break. (Adapted from an ancient Chinese work, *Arithmetic in Nine Sections*.)   4.55 ft

▦ *A projectile is launched from ground level with an initial velocity of v_0 feet per second. Neglecting air resistance, its height t seconds after launch in feet is given by*

$$s = -16t^2 + v_0 t.$$

In Exercises 21–24, find the time(s) that the projectile will (a) reach a height of 80 ft and (b) return to the ground for the given value of v_0. Round answers to the nearest hundredth if necessary. See Example 5.

21. $v_0 = 96$ (a) 1 sec, 5 sec   (b) 6 sec

22. $v_0 = 128$ (a) .68 sec, 7.32 sec   (b) 8 sec

23. $v_0 = 32$ (a) it will not reach 80 ft   (b) 2 sec

24. $v_0 = 16$ (a) it will not reach 80 ft   (b) 1 sec

✒ 25. Exercise 21(a) has two answers. A student was once asked why such a problem can have two answers, and his response was "What goes up must come down." Explain in more detail than the student why such a problem may have two answers, one answer, or no answers.

▲ 26. Consider the following problem.

The manager of an 80-unit apartment complex knows from experience that at a rent of $300, all the units will be full. On the average, one additional unit will remain vacant for each $20 increase in rent over $300. Currently, the revenue from the complex is $35,000. How many apartments are rented?

If x represents the number of $20 increases over $300, which one of the following equations indicates that the revenue generated will be $35,000?
(a) $(80 - x)(300 + 20x) = 35{,}000$
(b) $(80 + x)(300 + 20x) = 35{,}000$
(c) $(80 - x)(300 - 20x) = 35{,}000$
(d) $(80 + x)(300 - 20x) = 35{,}000$ (a)

▲ 27. The manager of a cherry orchard is trying to decide when to schedule the annual harvest. If the cherries are picked now, the average yield per tree will be 100 lb, and the cherries can be sold for 40 cents per pound. Past experience shows that the yield per tree will increase about 5 lb per week, while the price will decrease about 2 cents per pound per week. How many weeks should the manager wait to get an average revenue of $38.40 per tree?   4

▲ **28.** A local group of scouts has been collecting old aluminum cans for recycling. The group has already collected 12,000 lb of cans, for which they could currently receive $4 per hundred pounds. The group can continue to collect cans at the rate of 400 lb per day. However, a glut in the old-can market has caused the recycling company to announce that it will lower its price, starting immediately, by $.10 per hundred pounds per day. The scouts can make only one trip to the recycling center. How many days should they wait in order to get $490 for their cans? **5**

▲ **29.** A local club is arranging a charter flight to Miami. The cost of the trip is $225 each for 75 passengers, with a refund of $5 per passenger for each passenger in excess of 75. How many passengers must take the flight to produce revenue of $16,000? **80**

▲ **30.** Suppose one solution of the equation $km^2 + 10m = 8$ is -4. Find the value of k, and the other solution.

$k = 3$;   the other solution is 2/3

2.6 RESOURCES

Instructor's Test Manual
  Test Items, p. 315
Instructor's Solution Manual
  pp. 70–76, 490–496
Study Guide
  pp. 77–80
Student's Solution Manual
  pp. 70–76

Test Generator
  IBM Disks
    5.25" Disk A
    3.5" Disk A
    Mac Disk A
Tutorials
  IBM Disks
    5.25" Disk B
    3.5" Disk A
    Mac Disk A
    Apple II Module 2, Lessons 1–2

HarperCollins Videotape 3

2.6 EXERCISES (TEXT PAGE 120) ■

Solve each of the following equations. Use substitution to first write each as a quadratic equation. See Examples 1–2.

1. $m^4 + 2m^2 - 15 = 0$ $\{\pm\sqrt{3}, \pm i\sqrt{5}\}$

2. $3k^4 + 10k^2 - 25 = 0$ $\left\{\pm i\sqrt{5}, \pm\dfrac{\sqrt{15}}{3}\right\}$

3. $2r^4 - 7r^2 + 5 = 0$ $\left\{\pm 1, \pm\dfrac{\sqrt{10}}{2}\right\}$

4. $4x^4 - 8x^2 + 3 = 0$ $\left\{\pm\dfrac{\sqrt{6}}{2}, \pm\dfrac{\sqrt{2}}{2}\right\}$

5. $(g - 2)^2 - 6(g - 2) + 8 = 0$ $\{4, 6\}$

6. $(p + 2)^2 - 2(p + 2) - 15 = 0$ $\{3, -5\}$

7. $-(r + 1)^2 - 3(r + 1) + 3 = 0$ $\left\{\dfrac{-5 \pm \sqrt{21}}{2}\right\}$

8. $-2(z - 4)^2 + 2(z - 4) + 3 = 0$ $\left\{\dfrac{9 \pm \sqrt{7}}{2}\right\}$

9. $6(k + 2)^4 - 11(k + 2)^2 + 4 = 0$ $\left\{\dfrac{-6 \pm 2\sqrt{3}}{3}, \dfrac{-4 \pm \sqrt{2}}{2}\right\}$

10. $8(m - 4)^4 - 10(m - 4)^2 + 3 = 0$ $\left\{\dfrac{8 \pm \sqrt{2}}{2}, \dfrac{8 \pm \sqrt{3}}{2}\right\}$

11. $7p^{-2} + 19p^{-1} = 6$ $\left\{\dfrac{7}{2}, -\dfrac{1}{3}\right\}$

12. $5k^{-2} - 43k^{-1} = 18$ $\left\{-\dfrac{5}{2}, \dfrac{1}{9}\right\}$

13. $(r - 1)^{2/3} + (r - 1)^{1/3} = 12$ $\{-63, 28\}$

14. $(y + 3)^{2/3} - 2(y + 3)^{1/3} - 3 = 0$ $\{24, -4\}$

15. $3 + \dfrac{5}{p^2 + 1} = \dfrac{2}{(p^2 + 1)^2}$ $\left\{\pm i\sqrt{3}, \pm i\dfrac{\sqrt{6}}{3}\right\}$

16. $6 = \dfrac{7}{2y - 3} + \dfrac{3}{(2y - 3)^2}$ $\left\{\dfrac{9}{4}, \dfrac{4}{3}\right\}$

What is wrong with each solution in Exercises 17 and 18?

17. Solve $4x^4 - 11x^2 - 3 = 0$.
    Let $t = x^2$.

$$4t^2 - 11t - 3 = 0$$
$$(4t + 1)(t - 3) = 0$$
$$4t + 1 = 0 \quad \text{or} \quad t - 3 = 0$$
$$t = -\frac{1}{4} \quad \text{or} \quad t = 3$$

    The solution set is $\{-1/4, 3\}$.

18. Solve $x = \sqrt{3x + 4}$.
    Square both sides to get

$$x^2 = 3x + 4$$
$$x^2 - 3x - 4 = 0$$
$$(x - 4)(x + 1) = 0$$
$$x - 4 = 0 \quad \text{or} \quad x + 1 = 0$$
$$x = 4 \qquad\qquad x = -1.$$

    The solution set is $\{4, -1\}$.

Solve the following equations. See Examples 3–5.

19. $\sqrt{3z + 7} = 3z + 5$ $\{-1\}$

20. $\sqrt{4r + 13} = 2r - 1$ $\{3\}$

21. $\sqrt{4k + 5} - 2 = 2k - 7$ $\{5\}$

22. $\sqrt{6m + 7} - 1 = m + 1$ $\{-1, 3\}$

23. $\sqrt{4x} - x + 3 = 0$ $\{9\}$

24. $\sqrt{2t} - t + 4 = 0$ $\{8\}$

25. $\sqrt{y} = \sqrt{y - 5} + 1$ $\{9\}$

26. $\sqrt{2m} = \sqrt{m + 7} - 1$ $\{2\}$

27. $\sqrt{m + 7} + 3 = \sqrt{m - 4}$ $\emptyset$

28. $\sqrt{r + 5} - 2 = \sqrt{r - 1}$ $\left\{\dfrac{5}{4}\right\}$

29. $\sqrt{2z} = \sqrt{3z + 12} - 2$ $\{8\}$

30. $\sqrt{5k + 1} - \sqrt{3k} = 1$ $\{0, 3\}$

31. $\sqrt{r + 2} = 1 - \sqrt{3r + 7}$ $\{-2\}$

32. $\sqrt{2p - 5} - 2 = \sqrt{p - 2}$ $\{27\}$

33. $\sqrt{2\sqrt{7x + 2}} = \sqrt{3x + 2}$ $\left\{2, -\dfrac{2}{9}\right\}$

34. $\sqrt{3\sqrt{2m + 3}} = \sqrt{5m - 6}$ $\{3\}$

35. $\sqrt[3]{4n + 3} = \sqrt[3]{2n - 1}$ $\{-2\}$

36. $\sqrt[3]{2z} = \sqrt[3]{5z + 2}$ $\left\{-\dfrac{2}{3}\right\}$

37. $\sqrt[3]{t^2 + 2t - 1} = \sqrt[3]{t^2 + 3}$ $\{2\}$

38. $\sqrt[3]{2x^2 - 5x + 4} = \sqrt[3]{2x^2}$ $\left\{\dfrac{4}{5}\right\}$

39. $(2r + 5)^{1/3} = (6r - 1)^{1/3}$ $\left\{\dfrac{3}{2}\right\}$

40. $(3m + 7)^{1/3} = (4m + 2)^{1/3}$ $\{5\}$

41. $\sqrt[4]{q - 15} = 2$ $\{31\}$

42. $\sqrt[4]{3x + 1} = 1$ $\{0\}$

43. $\sqrt[4]{y^2 + 2y} = \sqrt[4]{3}$ $\{-3, 1\}$

44. $\sqrt[4]{k^2 + 6k} = 2$ $\{2, -8\}$

45. $(z^2 + 24z)^{1/4} = 3$ $\{-27, 3\}$

46. $(3t^2 + 52t)^{1/4} = 4$ $\left\{4, -\dfrac{64}{3}\right\}$

47. $(2r - 1)^{2/3} = r^{1/3}$ $\left\{\dfrac{1}{4}, 1\right\}$

48. $(z - 3)^{2/5} = (4z)^{1/5}$ $\{1, 9\}$

49. How can we tell that the equation $x^{1/4} = -2$ has no real solution without actually going through a solution process?

What is wrong with the solution in Exercise 50?

50. Solve $x^4 - x^2 = 0$.
    Since x^2 is a common factor, divide both sides by x^2 to get

$$x^2 - 1 = 0$$
$$(x - 1)(x + 1) = 0$$
$$x = 1 \quad \text{or} \quad x = -1.$$

    The solution set is $\{-1, 1\}$.

◎ CONCEPTUAL   ✐ WRITING   ▲ CHALLENGING

Solve each equation for the indicated variable. Assume that all denominators are nonzero.

51. $d = k\sqrt{h}$ for h $\quad h = \dfrac{d^2}{k^2}$

52. $v = \dfrac{k}{\sqrt{d}}$ for d $\quad d = \dfrac{k^2}{v^2}$

53. $P = 2\sqrt{\dfrac{L}{g}}$ for L $\quad L = \dfrac{P^2 g}{4}$

54. $c = \sqrt{a^2 + b^2}$ for a $\quad a = \pm\sqrt{c^2 - b^2}$

55. $x^{2/3} + y^{2/3} = a^{2/3}$ for y $\quad y = (a^{2/3} - x^{2/3})^{3/2}$

56. $m^{3/4} + n^{3/4} = 1$ for m $\quad m = (1 - n^{3/4})^{4/3}$

2.7 RESOURCES

Instructor's Test Manual
Test Items, p. 316
Instructor's Solution Manual
pp. 76–86, 497–506
Study Guide
pp. 81–87
Student's Solution Manual
pp. 76–86

Test Generator
IBM Disks
5.25" Disk A
3.5" Disk A
Mac Disk A
Tutorials
IBM Disks
5.25" Disk B
3.5" Disk A
Mac Disk A
Apple II Module 2, Lessons 1–2

HarperCollins Videotape 3
"College Algebra: In Simplest Terms" Program 8
Transparency T-2

2.7 EXERCISES (TEXT PAGE 130) ■ ─────────────

Write each of the following in interval notation. Graph each interval.

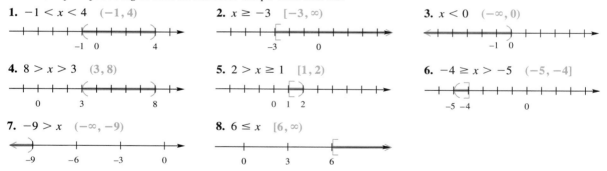

1. $-1 < x < 4$ $\quad (-1, 4)$

2. $x \geq -3$ $\quad [-3, \infty)$

3. $x < 0$ $\quad (-\infty, 0)$

4. $8 > x > 3$ $\quad (3, 8)$

5. $2 > x \geq 1$ $\quad [1, 2)$

6. $-4 \geq x > -5$ $\quad (-5, -4]$

7. $-9 > x$ $\quad (-\infty, -9)$

8. $6 \leq x$ $\quad [6, \infty)$

Using the variable x, write each of the following intervals as an inequality with set-builder notation.

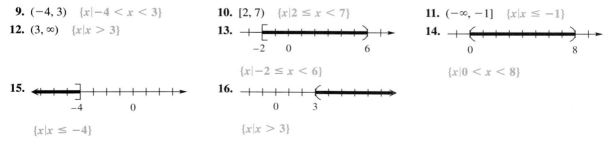

9. $(-4, 3)$ $\quad \{x | -4 < x < 3\}$

10. $[2, 7)$ $\quad \{x | 2 \leq x < 7\}$

11. $(-\infty, -1]$ $\quad \{x | x \leq -1\}$

12. $(3, \infty)$ $\quad \{x | x > 3\}$

13. $\{x | -2 \leq x < 6\}$

14. $\{x | 0 < x < 8\}$

15. $\{x | x \leq -4\}$

16. $\{x | x > 3\}$

📝 ◉ **17.** Explain how to determine whether a parenthesis or a square bracket is used when graphing the solution set of a linear inequality.

◉ **18.** The three-part inequality $a < x < b$ means "a is less than x and x is less than b." Which one of the following inequalities is not satisfied by some real number x?
   (a) $-3 < x < 5$ **(b)** $0 < x < 4$ **(c)** $-3 < x < -2$ **(d)** $-7 < x < -10$ **(d)**

Solve the following inequalities. Write the solutions in interval notation. Graph each solution. See Examples 1–3.

19. $-3p - 2 \le 1$ $[-1, \infty)$

20. $-5r + 3 \ge -2$ $(-\infty, 1]$

21. $2(m + 5) - 3m + 1 \ge 5$ $(-\infty, 6]$

22. $6m - (2m + 3) \ge 4m - 5$ $(-\infty, \infty)$

23. $8k - 3k + 2 < 2(k + 7)$ $(-\infty, 4)$

24. $2 - 4x + 5(x - 1) < -6(x - 2)$ $\left(-\infty, \dfrac{15}{7}\right)$

25. $\dfrac{4x + 7}{-3} \le 2x + 5$ $\left[-\dfrac{11}{5}, \infty\right)$

26. $\dfrac{2z - 5}{-8} \ge 1 - z$ $\left[\dfrac{1}{2}, \infty\right)$

27. $2 \le y + 1 \le 5$ $[1, 4]$

28. $-3 \le 2t \le 6$ $\left[-\dfrac{3}{2}, 3\right]$

29. $-10 > 3r + 2 > -16$ $(-6, -4)$

30. $4 > 6a + 5 > -1$ $\left(-1, -\dfrac{1}{6}\right)$

31. $-3 \le \dfrac{x - 4}{-5} < 4$ $(-16, 19]$

32. $1 < \dfrac{4m - 5}{-2} < 9$ $\left(-\dfrac{13}{4}, \dfrac{3}{4}\right)$

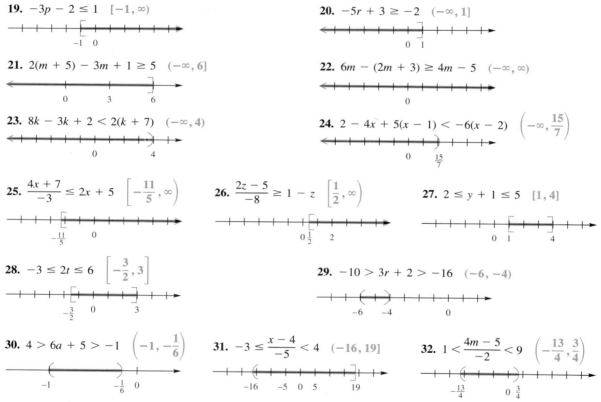

Find all intervals where the following products will at least break even. See Example 4.

33. The cost to produce x units of wire is $C = 50x + 5000$, while the revenue is $R = 60x$. $[500, \infty)$

34. The cost to produce x units of squash is $C = 100x + 6000$, while the revenue is $R = 500x$. $[15, \infty)$

35. $C = 85x + 900;$ $R = 105x$ $[45, \infty)$

36. $C = 70x + 500;$ $R = 60x$ $R < C$ for all positive x;   this product will never break even

Solve the following quadratic inequalities. Write the solutions in interval notation. Graph each solution. (Hint: In Exercises 47–50, use the quadratic formula.) See Examples 5 and 6.

37. $x^2 \le 9$ $[-3, 3]$

38. $p^2 > 16$ $(-\infty, -4) \cup (4, \infty)$

◉ **CONCEPTUAL**   📝 **WRITING**   ▲ **CHALLENGING**

39. $r^2 + 4r + 6 \geq 3$ $(-\infty, -3] \cup [-1, \infty)$

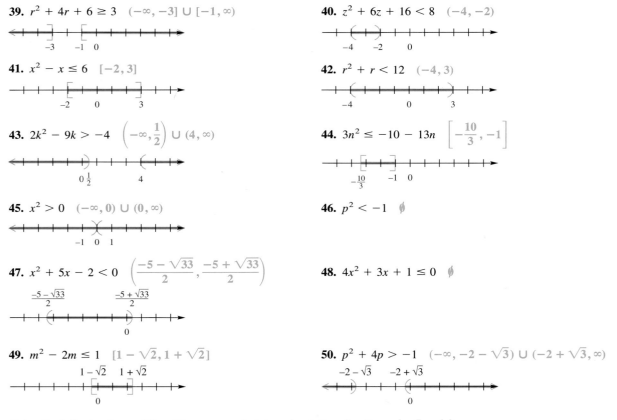

40. $z^2 + 6z + 16 < 8$ $(-4, -2)$

41. $x^2 - x \leq 6$ $[-2, 3]$

42. $r^2 + r < 12$ $(-4, 3)$

43. $2k^2 - 9k > -4$ $\left(-\infty, \dfrac{1}{2}\right) \cup (4, \infty)$

44. $3n^2 \leq -10 - 13n$ $\left[-\dfrac{10}{3}, -1\right]$

45. $x^2 > 0$ $(-\infty, 0) \cup (0, \infty)$

46. $p^2 < -1$ $\emptyset$

47. $x^2 + 5x - 2 < 0$ $\left(\dfrac{-5 - \sqrt{33}}{2}, \dfrac{-5 + \sqrt{33}}{2}\right)$

48. $4x^2 + 3x + 1 \leq 0$ $\emptyset$

49. $m^2 - 2m \leq 1$ $[1 - \sqrt{2}, 1 + \sqrt{2}]$

50. $p^2 + 4p > -1$ $(-\infty, -2 - \sqrt{3}) \cup (-2 + \sqrt{3}, \infty)$

Solve the following inequalities. Give answers in interval notation. See Examples 7 and 8.

51. $\dfrac{m - 3}{m + 5} \leq 0$ $(-5, 3]$

52. $\dfrac{r + 1}{r - 4} > 0$ $(-\infty, -1) \cup (4, \infty)$

53. $\dfrac{k - 1}{k + 2} > 1$ $(-\infty, -2)$

54. $\dfrac{a - 6}{a + 2} < -1$ $(-2, 2)$

55. $\dfrac{3}{x - 6} \leq 2$ $(-\infty, 6) \cup \left[\dfrac{15}{2}, \infty\right)$

56. $\dfrac{1}{k - 2} < \dfrac{1}{3}$ $(-\infty, 2) \cup (5, \infty)$

57. $\dfrac{1}{m - 1} < \dfrac{5}{4}$ $(-\infty, 1) \cup \left(\dfrac{9}{5}, \infty\right)$

58. $\dfrac{6}{5 - 3x} \leq 2$ $\left(-\infty, \dfrac{2}{3}\right] \cup \left(\dfrac{5}{3}, \infty\right)$

59. $\dfrac{10}{3 + 2x} \leq 5$ $\left(-\infty, -\dfrac{3}{2}\right) \cup \left[-\dfrac{1}{2}, \infty\right)$

60. $\dfrac{1}{x + 2} \geq 3$ $\left(-2, -\dfrac{5}{3}\right]$

61. $\dfrac{7}{k + 2} \geq \dfrac{1}{k + 2}$ $(-2, \infty)$

62. $\dfrac{5}{p + 1} > \dfrac{12}{p + 1}$ $(-\infty, -1)$

63. $\dfrac{3}{2r - 1} > -\dfrac{4}{r}$ $\left(0, \dfrac{4}{11}\right) \cup \left(\dfrac{1}{2}, \infty\right)$

64. $-\dfrac{5}{3h + 2} \geq \dfrac{5}{h}$ $\left(-\infty, -\dfrac{2}{3}\right) \cup \left[-\dfrac{1}{2}, 0\right)$

65. $\dfrac{4}{y - 2} \leq \dfrac{3}{y - 1}$ $(-\infty, -2] \cup (1, 2)$

66. $\dfrac{4}{n + 1} < \dfrac{2}{n + 3}$ $(-\infty, -5) \cup (-3, -1)$

67. $\dfrac{y + 3}{y - 5} \leq 1$ $(-\infty, 5)$

68. $\dfrac{a + 2}{3 + 2a} \leq 5$ $\left(-\infty, -\dfrac{3}{2}\right) \cup \left[-\dfrac{13}{9}, \infty\right)$

◎ *In each of the following inequalities, the intervals of the sign graph are*

$$(-\infty, 2), \quad (2, 5), \quad \text{and} \quad (5, \infty).$$

Without actually solving the inequality, state whether the numbers 2 and 5 will be included in or excluded from the solution set of the inequality.

69. $(x - 2)(x - 5) \geq 0$ both included

70. $(x - 2)(x - 5) < 0$ both excluded

71. $\dfrac{x - 2}{x - 5} > 0$ both excluded

72. $\dfrac{x - 2}{x - 5} < 0$ both excluded

73. $\dfrac{x - 2}{x - 5} \geq 0$ 2 included, 5 excluded

74. $\dfrac{x - 5}{x - 2} \leq 0$ 5 included, 2 excluded

▲ *The following inequalities are not quadratic inequalities, but they still may be solved in much the same way (with four regions). Solve each inequality. Write solutions in interval notation.*

75. $(2r - 3)(r + 2)(r - 3) \geq 0$ $\left[-2, \dfrac{3}{2}\right] \cup [3, \infty)$

76. $(y + 5)(3y - 4)(y + 2) > 0$ $(-5, -2) \cup \left(\dfrac{4}{3}, \infty\right)$

77. $x^3 - 4x \leq 0$ $(-\infty, -2] \cup [0, 2]$

78. $r^3 - 9r \geq 0$ $[-3, 0] \cup [3, \infty)$

79. $4m^3 + 7m^2 - 2m > 0$ $(-2, 0) \cup \left(\dfrac{1}{4}, \infty\right)$

80. $6p^3 - 11p^2 + 3p > 0$ $\left(0, \dfrac{1}{3}\right) \cup \left(\dfrac{3}{2}, \infty\right)$

◎ **81.** Which of the following inequalities have solution set $(-\infty, \infty)$?
  (a) $(x + 3)^2 \geq 0$
  (b) $(5x - 6)^2 \leq 0$
  (c) $(6y + 4)^2 > 0$
  (d) $(8p - 7)^2 < 0$
  (e) $\dfrac{x^2 + 7}{2x^2 + 4} \geq 0$
  (f) $\dfrac{2x^2 + 8}{x^2 + 9} < 0$ (a), (e)

◎ **82.** Which of the inequalities in Exercise 81 have solution set $\emptyset$?   (d), (f)

📝◎ **83.** A student attempted to solve the inequality

$$\frac{2x - 1}{x + 2} \leq 0$$

by multiplying both sides by $x + 2$ to get

$$2x - 1 \leq 0$$

$$x \leq \frac{1}{2}.$$

He wrote the solution set as $(-\infty, 1/2]$. Is his solution correct? Explain.

📝◎ **84.** A student solved the inequality $p^2 \leq 16$ by taking the square root of both sides to get $p \leq 4$. She wrote the solution set as $(-\infty, 4]$. Is her solution correct? Explain.

85. The commodity market is very unstable; money can be made or lost quickly when investing in soybeans, wheat, pork bellies, and the like. Suppose that an investor kept track of her total profit, P, at time t, measured in months, after she began investing, and found that

$$P = 4t^2 - 29t + 30.$$

Find the time intervals when she has been ahead. (*Hint:* $t > 0$ in this case.) $\left(0, \dfrac{5}{4}\right) \cup (6, \infty)$

86. Suppose the velocity of an object is given by

$$v = 2t^2 - 5t - 12,$$

where t is time in seconds. (Here t can be positive or negative.) Find the intervals where the velocity is negative. $\left(-\dfrac{3}{2}, 4\right)$

87. A projectile is fired from ground level. After t sec its height above the ground is $220t - 16t^2$ ft. For what time period is the projectile at least 624 ft above the ground?   for values of t in $\left[4, \dfrac{39}{4}\right]$

88. An analyst has found that his company's profits, in hundreds of thousands of dollars, are given by

$$P = 3x^2 - 35x + 50,$$

where x is the amount, in hundreds of dollars, spent on advertising. For what values of x does the company make a profit?   for values of x in $\left[0, \dfrac{5}{3}\right) \cup (10, \infty)$

89. The manager of a large apartment complex has found that the profit is given by

$$P = -x^2 + 250x - 15{,}000,$$

where x is the number of units rented. For what values of x does the complex produce a profit?

for values of x in $(100, 150)$

90. If $a > b$, is it always true that $1/a < 1/b$? Explain.

91. If $b > 0$, when is it true that $b^2 > b$? Explain.

92. If $a > b$, is it always true that $a^2 > b^2$? Explain.

2.8 RESOURCES

Instructor's Test Manual
  Test Items, p. 316–317
Instructor's Solution Manual
  pp. 86–90, 506–510
Study Guide
  pp. 87–89
Student's Solution Manual
  pp. 86–90

Test Generator
  IBM Disks
    5.25″ Disk A
    3.5″ Disk A
    Mac Disk A
Tutorials
  IBM Disks
    5.25″ Disk B
    3.5″ Disk A
    Mac Disk A
    Apple II Module 2, Lessons 1–2

HarperCollins Videotape 3

2.8 EXERCISES (TEXT PAGE 136) ▪

Solve each of the following equations. See Examples 1, 2, and 7(c).

1. $|a - 2| = 1$ $\{1, 3\}$

2. $|x - 3| = 2$ $\{1, 5\}$

3. $|3m - 1| = 2$ $\left\{-\dfrac{1}{3}, 1\right\}$

4. $|4p + 2| = 5$ $\left\{\dfrac{3}{4}, -\dfrac{7}{4}\right\}$

5. $|5 - 3x| = 3$ $\left\{\dfrac{2}{3}, \dfrac{8}{3}\right\}$

6. $|-3a + 7| = 3$ $\left\{\dfrac{4}{3}, \dfrac{10}{3}\right\}$

7. $\left|\dfrac{z - 4}{2}\right| = 5$ $\{-6, 14\}$

8. $\left|\dfrac{m + 2}{2}\right| = 7$ $\{12, -16\}$

9. $\left|\dfrac{5}{r - 3}\right| = 10$ $\left\{\dfrac{5}{2}, \dfrac{7}{2}\right\}$

10. $\left|\dfrac{3}{2h - 1}\right| = 4$ $\left\{\dfrac{7}{8}, \dfrac{1}{8}\right\}$

11. $|4w + 3| - 2 = 7$ $\left\{\dfrac{3}{2}, -3\right\}$

12. $|8 - 3t| - 3 = -2$ $\left\{\dfrac{7}{3}, 3\right\}$

13. $|6x + 9| = 0$ $\left\{-\dfrac{3}{2}\right\}$

14. $|12t - 3| = 0$ $\left\{\dfrac{1}{4}\right\}$

15. $\left|\dfrac{6y + 1}{y - 1}\right| = 3$ $\left\{-\dfrac{4}{3}, \dfrac{2}{9}\right\}$

16. $\left|\dfrac{3a - 4}{2a + 3}\right| = 1$ $\left\{7, \dfrac{1}{5}\right\}$

17. $|2k - 3| = |5k + 4|$ $\left\{-\dfrac{7}{3}, -\dfrac{1}{7}\right\}$

18. $|p + 1| = |3p - 1|$ $\{1, 0\}$

19. $|4 - 3y| = |7 + 2y|$ $\left\{-\dfrac{3}{5}, 11\right\}$

20. $|5k + 16| = |3k|$ $\{-8, -2\}$

21. $|8b + 7| = |4b|$ $\left\{-\dfrac{7}{4}, -\dfrac{7}{12}\right\}$

22. $|2 + 5a| = |4 - 6a|$ $\left\{\dfrac{2}{11}, 6\right\}$

23. $|x + 2| = |x - 1|$ $\left\{-\dfrac{1}{2}\right\}$

24. $|y - 5| = -8$ $\emptyset$

25. Without actually going through the solution process, we can say that the equation $|5x - 6| = 6x$ cannot have a negative solution. Why is this true?

26. Determine by inspection the solution set of each of the following absolute value equations.

   (a) $-|x| = |x|$ $\{0\}$ **(b)** $|-x| = |x|$ $(-\infty, \infty)$ **(c)** $|x^2| = |x|$ $\{-1, 0, 1\}$ **(d)** $-|x| = 3$ $\emptyset$

Solve each of the following inequalities. Give the solution in interval notation. See Examples 3–7.

27. $|x| \le 3$ $[-3, 3]$ **28.** $|y| \le 10$ $[-10, 10]$ **29.** $|m| > 1$ $(-\infty, -1) \cup (1, \infty)$

30. $|z| > 5$ $(-\infty, -5) \cup (5, \infty)$ **31.** $|a| < -2$ $\emptyset$ **32.** $|b| > -5$ $(-\infty, \infty)$

33. $|x| - 3 \le 7$ $[-10, 10]$ **34.** $|r| + 3 \le 10$ $[-7, 7]$ **35.** $|2x + 5| < 3$ $(-4, -1)$

36. $\left|x - \dfrac{1}{2}\right| < 2$ $\left(-\dfrac{3}{2}, \dfrac{5}{2}\right)$ **37.** $|3m - 2| > 4$ $\left(-\infty, -\dfrac{2}{3}\right) \cup (2, \infty)$

38. $|4x - 6| > 10$ $(-\infty, -1) \cup (4, \infty)$ **39.** $|3z + 1| \ge 7$ $\left(-\infty, -\dfrac{8}{3}\right] \cup [2, \infty)$

40. $|8b + 5| \ge 7$ $\left(-\infty, -\dfrac{3}{2}\right] \cup \left[\dfrac{1}{4}, \infty\right)$ **41.** $\left|\dfrac{2}{3}t + \dfrac{1}{2}\right| \le \dfrac{1}{6}$ $\left[-1, -\dfrac{1}{2}\right]$

42. $\left|\dfrac{5}{3} - \dfrac{1}{2}x\right| > \dfrac{2}{9}$ $\left(-\infty, \dfrac{26}{9}\right) \cup \left(\dfrac{34}{9}, \infty\right)$ **43.** $\left|5x + \dfrac{1}{2}\right| - 2 < 5$ $\left(-\dfrac{3}{2}, \dfrac{13}{10}\right)$

44. $\left|x + \dfrac{2}{3}\right| + 1 < 4$ $\left(-\dfrac{11}{3}, \dfrac{7}{3}\right)$ **45.** $|6x + 3| \ge -2$ $(-\infty, \infty)$ **46.** $|7 - 8x| < -4$ $\emptyset$

47. $\left|\dfrac{1}{2}x + 6\right| > 0$ $(-\infty, -12) \cup (-12, \infty)$ **48.** $\left|\dfrac{2}{3}r - 4\right| \le 0$ $\{6\}$

49. Write an equation involving absolute value that says that the distance between p and q is 5 units. $|p - q| = 5$ (or $|q - p| = 5$)

50. Write an inequality involving absolute value that says that the distance between r and s is less than 9 units. $|r - s| < 9$ (or $|s - r| < 9$)

51. Suppose that you hear someone say "The absolute value of a number is always positive." How might you politely correct this person's misconception?

52. The following triangle inequality holds for all real numbers x and y:

$$|x| + |y| \ge |x + y|.$$

Illustrate cases for this inequality by choosing values of x and y that satisfy the following.
   (a) x and y are both positive   **(b)** x and y are both negative
   (c) x is positive and y is negative   **(d)** x is negative and y is positive

Write each of the statements in Exercises 53–60 as an absolute value equation or inequality. See Example 8.

53. x is within 4 units of 2 $|x - 2| \le 4$ **54.** m is no more than 8 units from 9 $|m - 9| \le 8$

55. z is no less than 2 units from 12 $|z - 12| \ge 2$ **56.** p is at least 5 units from 9 $|p - 9| \ge 5$

57. k is 6 units from 1 $|k - 1| = 6$ **58.** r is 5 units from 3 $|r - 3| = 5$

59. If x is within .0004 unit of 2, then y is within .00001 unit of 7.

   If $|x - 2| \le .0004$, then $|y - 7| \le .00001$.

60. y is within 10^{-6} unit of 10 whenever x is within 2×10^{-4} unit of 5.

   If $|x - 5| \le 2 \times 10^{-4}$, then $|y - 10| \le 10^{-6}$.

◉ CONCEPTUAL   ✐ WRITING   ▲ CHALLENGING

Solve each problem.

61. The temperatures on the surface of Mars in degrees Celsius approximately satisfy the inequality

$$|C + 84| \leq 56.$$

What range of temperatures corresponds to this inequality? $[-140, -28]$

62. Dr. Tydings has found that, over the years, 95% of the babies that he has delivered have weighed y pounds, where $|y - 8.0| \leq 1.5$. What range of weights corresponds to this inequality? $[6.5, 9.5]$

63. The industrial process that is used to convert methanol to gasoline is carried out at a temperature range of 680° F to 780° F. Using F as the variable, write an absolute value inequality that corresponds to this range. $|F - 730| \leq 50$

64. When a model kite was flown across the wind in tests to determine its limits of power extraction, it attained speeds of 98 to 148 feet per second in winds of 16 to 26 feet per second. Using x as the variable in each case, write absolute value inequalities that correspond to these ranges. $|x - 123| \leq 25$; $|x - 21| \leq 5$

▲ **65.** If $|x - 2| < 3$, find values of m and n such that $m < 3x + 5 < n$.   choose any $m \leq 2$ and any $n \geq 20$

▲ **66.** If $|x + 8| < 16$, find values of p and q such that $p < 2x - 1 < q$.   choose any $p \leq -49$ and any $q \geq 15$

✎ ◎ **67.** Explain why it is *incorrect* to write the absolute value inequality $|x| > 6$ as any of these:
$-6 > x > 6$, $-6 > x < 6$, or $-6 < x > 6$.

✎ ◎ **68.** Is $|a - b|^2$ always equal to $(b - a)^2$? Why?

RESOURCES FOR REVIEW

Instructor's Test Manual
Chapter Test Forms, pp. 37–54
Instructor's Solution Manual
pp. 90–105, 510–519

Study Guide
pp. 90–92
Student's Solution Manual
pp. 90–105

Test Generator
IBM Disks
5.25″ Disk A
3.5″ Disk A
Mac Disk A

CHAPTER 2 REVIEW EXERCISES   (TEXT PAGE 139) ∎ ——————

Solve each of the following equations.

1. $2m + 7 = 3m + 1$ $\{6\}$

2. $4k - 2(k - 1) = 12$ $\{5\}$

3. $5y - 2(y + 4) = 3(2y + 1)$ $\left\{ -\dfrac{11}{3} \right\}$

4. $\dfrac{2}{x - 3} = \dfrac{3}{x + 3}$ $\{15\}$

5. $\dfrac{10}{4z - 4} = \dfrac{1}{1 - z}$ $\emptyset$

6. $\dfrac{2}{p} - \dfrac{4}{3p} = 8 + \dfrac{3}{p}$ $\left\{ -\dfrac{7}{24} \right\}$

7. $\dfrac{5}{3r} - 10 = \dfrac{3}{2r}$ $\left\{ \dfrac{1}{60} \right\}$

8. $\dfrac{p}{p - 2} - \dfrac{3}{5} = \dfrac{2}{p - 2}$ $\emptyset$

Solve for x.

9. $3(x + 2b) + a = 2x - 6$ $x = -6b - a - 6$

10. $9x - 11(k + p) = x(a - 1)$ $x = \dfrac{11(k + p)}{10 - a}$

11. $\dfrac{x}{m - 2} = kx - 3$ $x = \dfrac{6 - 3m}{1 + 2k - km}$

12. $r^2x - 5x = 3r^2$ $x = \dfrac{3r^2}{r^2 - 5}$

Solve each of the following for the indicated variable.

13. $A = P + Pi$ for P $P = \dfrac{A}{1 + i}$

14. $A = I\left(1 - \dfrac{j}{n}\right)$ for j $j = n - \dfrac{nA}{I}$

15. $A = \dfrac{24f}{b(p + 1)}$ for f $f = \dfrac{Ab(p + 1)}{24}$

16. $P(r + R)^2 = E^2R$ for P $P = \dfrac{E^2R}{(r + R)^2}$

17. $\dfrac{xy^2 - 5xy + 4}{3x} = 2p$ for x $x = \dfrac{-4}{y^2 - 5y - 6p}$

18. $\dfrac{zx^2 - 5x + z}{z + 1} = 9$ for z $z = \dfrac{5x + 9}{x^2 - 8}$

Solve each of the following problems.

19. A computer printer is on sale for 15% off. The sale price is $425. What was the original price? $500

20. To make a special candy mix for Valentine's Day, the owner of a candy store wants to combine chocolate hearts that sell for $5 per lb with candy kisses that sell for $3.50 per lb. How many pounds of each kind should be used to get 30 lb of a mix which can be sold for $4.50 per pound?

20 lb of hearts and 10 lb of kisses

21. Alison can ride her bike to the university library in 20 min. The trip home, which is all uphill, takes her 30 min. If her rate is 8 mph slower on the return trip, how far does she live from the library?   8 mi

22. Ed and Faye Clement are stuffing envelopes for a political campaign. Working together, they can stuff 5000 envelopes in 4 hr. When Faye worked at the job alone, it took her 6 hr to stuff the 5000 envelopes. How long would it take Ed working alone to stuff 5000 envelopes?   12 hr

23. A realtor borrowed $90,000 to develop some property. He was able to borrow part of the money at 11.5% interest and the rest at 12%. The annual interest on the two loans amounts to $10,525. How much was borrowed at each rate?

$55,000 at 11.5% and $35,000 at 12%

24. An excursion boat travels upriver to a landing and then returns to its starting point. The trip upriver takes 1.2 hr, and the trip back takes .9 hr. If the average speed on the return trip is 5 mph faster than on the trip upriver, what is the boat's speed upriver?

15 mph

Perform each operation. Write each result in standard form.

25. $(6 - i) + (4 - 2i)$ $10 - 3i$

26. $(-11 + 2i) - (5 - 7i)$ $-16 + 9i$

27. $15i - (3 + 2i) - 5$ $-8 + 13i$

28. $-6 + 4i - (5i - 2)$ $-4 - i$

29. $(5 - i)(3 + 4i)$ $19 + 17i$

30. $(8 - 2i)(1 - i)$ $6 - 10i$

31. $(5 - 11i)(5 + 11i)$ $146 + 0i$

32. $(7 + 6i)(7 - 6i)$ $85 + 0i$

33. $(4 - 3i)^2$ $7 - 24i$

34. $(1 + 7i)^2$ $-48 + 14i$

35. $\dfrac{6 + i}{1 - i}$ $\dfrac{5}{2} + \dfrac{7}{2}i$

36. $\dfrac{5 + 2i}{5 - 2i}$ $\dfrac{21}{29} + \dfrac{20}{29}i$

◉ **37.** The product of a complex number and its conjugate is always a _____ number.   real

◉ **38.** True or false: A real number is a complex number.   true

Find each of the following powers of i.

39. i^7 $-i$

40. i^{150} -1

41. i^{-35} i

42. i^{-20} 1

Solve each equation.

43. $(b + 7)^2 = 5$ $\{-7 \pm \sqrt{5}\}$

44. $(3y - 2)^2 = 8$ $\left\{\dfrac{2 \pm 2\sqrt{2}}{3}\right\}$

45. $2a^2 + a - 15 = 0$ $\left\{\dfrac{5}{2}, -3\right\}$

46. $12x^2 = 8x - 1$ $\left\{\dfrac{1}{2}, \dfrac{1}{6}\right\}$

47. $2q^2 - 11q = 21$ $\left\{7, -\dfrac{3}{2}\right\}$

48. $3x^2 + 2x = 16$ $\left\{2, -\dfrac{8}{3}\right\}$

49. $2 - \dfrac{5}{p} = \dfrac{3}{p^2}$ $\left\{3, -\dfrac{1}{2}\right\}$

50. $\dfrac{4}{m^2} = 2 + \dfrac{7}{m}$ $\left\{\dfrac{1}{2}, -4\right\}$

51. $\sqrt{2}x^2 - 4x + \sqrt{2} = 0$ $\{\sqrt{2} \pm 1\}$

◉ **CONCEPTUAL**   ✎ **WRITING**   ▲ **CHALLENGING**

52. Find all values of k for which $4x^2 + 3kx + 9 = 0$ will have one real solution (a double solution).   4, −4

Evaluate the discriminant for each of the following, and then use it to predict the type of solutions for the equation.

53. $8y^2 = 2y - 6$ −188;   two different imaginary solutions

54. $6k^2 - 2k = 3$ 76;   two different irrational solutions

55. $16r^2 + 3 = 26r$ 484;   two different rational solutions

56. $8p^2 + 10p = 7$ 324;   two different rational solutions

57. $25z^2 - 110z + 121 = 0$ 0;   one rational solution

58. $4y^2 - 8y + 17 = 0$ −208;   two different imaginary solutions

Work each problem.

59. Paula Story plans to replace the vinyl floor covering in her 10- by 12-ft kitchen. She wants to have a border of even width of a special material. She can afford only 21 sq ft of this material. How wide a border can she have? $\dfrac{1}{2}$ ft

60. Calvin wants to fence off a rectangular playground beside an apartment building. The building forms one boundary, so he needs to fence only the other three sides. The area of the playground is to be 11,250 sq m. He has enough material to build 325 m of fence. Find the length and width of the playground.

   50 m by 225 m;   or 112.5 m by 100 m

61. It takes two gardeners 3 hr (working together) to mow the lawns in a city park. One gardener could do the entire job in 1 hr less time than the other. How long would it take the slower gardener to complete the work alone? Give the answer to the nearest tenth.

   6.5 hr

62. Steve and Paula sell pies. It takes Paula 1 hr longer than Steve to bake a day's supply of pies. Working together, it takes them 1 1/5 hr to bake the pies. How long would it take Steve working alone?   2 hr

63. In a marathon (a 26-mile run), the winner finished 2/5 hr before his friend. If the difference in their rates was 4/3 mph, what was the winner's average speed in the race?   10 mph

64. Suppose that one solution of the equation $km^2 - 11m = 3$ is 3. Find the value of k and the other solution. $k = 4$;   the other solution is $-\dfrac{1}{4}$

Solve each equation.

65. $4a^4 + 3a^2 - 1 = 0$ $\left\{\pm i, \pm \dfrac{1}{2}\right\}$

66. $2x^4 - x^2 = 0$ $\left\{0, \pm \dfrac{\sqrt{2}}{2}\right\}$

67. $(2z + 3)^{2/3} + (2z + 3)^{1/3} = 6$ $\left\{\dfrac{5}{2}, -15\right\}$

68. $5\sqrt{m} = \sqrt{3m + 2}$ $\left\{\dfrac{1}{11}\right\}$

69. $\sqrt{4y - 2} = \sqrt{3y + 1}$ {3}

70. $\sqrt{2x + 3} = x + 2$ {−1}

71. $\sqrt{p + 2} = 2 + p$ {−2, −1}

72. $\sqrt{k} = \sqrt{k + 3} - 1$ {1}

73. $\sqrt{x + 3} - \sqrt{3x + 10} = 1$ ∅

74. $\sqrt{5x - 15} - \sqrt{x + 1} = 2$ {8}

75. $\sqrt[3]{6y + 2} = \sqrt[3]{4y}$ {−1}

76. $(x - 2)^{2/3} = x^{1/3}$ {4, 1}

Solve the following inequalities. Write answers in interval notation.

77. $-9x < 4x + 7$ $\left(-\dfrac{7}{13}, \infty\right)$

78. $11y \geq 2y - 8$ $\left[-\dfrac{8}{9}, \infty\right)$

79. $-5z - 4 \geq 3(2z - 5)$ $(-\infty, 1]$

80. $-(4a + 6) < 3a - 2$ $\left(-\dfrac{4}{7}, \infty\right)$

81. $3r - 4 + r > 2(r - 1)$ $(1, \infty)$

82. $7p - 2(p - 3) \leq 5(2 - p)$ $\left(-\infty, \dfrac{2}{5}\right]$

83. $5 \leq 2x - 3 \leq 7$ [4, 5]

84. $-8 > 3a - 5 > -12$ $\left(-\dfrac{7}{3}, -1\right)$

85. $x^2 + 3x - 4 \leq 0$ [−4, 1]

86. $p^2 + 4p > 21$ $(-\infty, -7) \cup (3, \infty)$

87. $6m^2 - 11m - 10 < 0$ $\left(-\frac{2}{3}, \frac{5}{2}\right)$

88. $k^2 - 3k - 5 \geq 0$ $\left(-\infty, \frac{3 - \sqrt{29}}{2}\right] \cup \left[\frac{3 + \sqrt{29}}{2}, \infty\right)$

89. $x^2 - 6x + 9 \leq 0$ $\{3\}$

90. $(x - 2)(x - 4)(x - 3) > 0$ $(2, 3) \cup (4, \infty)$

91. $\frac{3a - 2}{a} > 4$ $(-2, 0)$

92. $\frac{5p + 2}{p} < -1$ $\left(-\frac{1}{3}, 0\right)$

93. $\frac{3}{r - 1} \leq \frac{5}{r + 3}$ $(-3, 1) \cup [7, \infty)$

94. $\frac{3}{x + 2} > \frac{2}{x - 4}$ $(-2, 4) \cup (16, \infty)$

◉ 95. If $0 < a < b$, on what interval is $(x - a)(x - b)$ positive? negative? zero?
positive on the intervals $(-\infty, a)$ and (b, ∞);   negative on the interval (a, b);
zero at the numbers a and b

✎ ◉ 96. Without actually solving the inequality, explain why 3 cannot be in the solution set of $\frac{2x + 5}{x - 3} < 0$.

Work the following problems.

97. A company produces videotapes. The revenue from the sale of x units of tapes is $R = 8x$. The cost to produce x units of tapes is $C = 3x + 1500$. In what interval will the company at least break even? $[300, \infty)$

98. A projectile is launched upward. Its height in feet above the ground after t seconds is $320t - 16t^2$.   **(a)** After how many seconds in the air will it hit the ground?   20 sec
(b) During what time interval is the projectile more than 576 ft above the ground?   between 2 and 18 sec

Solve each equation.

99. $|a + 4| = 7$ $\{3, -11\}$

100. $|-y + 2| = -4$ $\emptyset$

101. $\left|\frac{7}{2 - 3a}\right| = 9$ $\left\{\frac{11}{27}, \frac{25}{27}\right\}$

102. $|5 - 8x| + 1 = 3$ $\left\{\frac{3}{8}, \frac{7}{8}\right\}$

103. $|5r - 1| = |2r + 3|$ $\left\{\frac{4}{3}, -\frac{2}{7}\right\}$

104. $|k + 7| = 2k$ $\{7\}$

Solve each inequality. Write solutions with interval notation.

105. $|m| \leq 7$ $[-7, 7]$

106. $|r| < 2$ $(-2, 2)$

107. $|p| > 3$ $(-\infty, -3) \cup (3, \infty)$

108. $|z| > -1$ $(-\infty, \infty)$

109. $|2z + 9| \leq 3$ $[-6, -3]$

110. $|5m - 8| \leq 2$ $\left[\frac{6}{5}, 2\right]$

111. $|7k - 3| < 5$ $\left(-\frac{2}{7}, \frac{8}{7}\right)$

112. $|2p - 1| > 2$ $\left(-\infty, -\frac{1}{2}\right) \cup \left(\frac{3}{2}, \infty\right)$

113. $|3r + 7| - 5 > 0$ $(-\infty, -4) \cup \left(-\frac{2}{3}, \infty\right)$

114. Write as an absolute value inequality: k is at least 4 units from 1. $|k - 1| \geq 4$

CHAPTER 2 TEST   (TEXT PAGE 142) ■

Solve each of the following equations.

1. $x - (2x + 1) = 7 - 3(x + 1)$ $\left\{\frac{5}{2}\right\}$

2. $\frac{x}{x - 3} = \frac{3}{x - 3} + 4$ $\emptyset$

3. $\frac{1}{c} - \frac{1}{a} = \frac{1}{b}$ for c $c = \frac{ab}{a + b}$

◉ CONCEPTUAL   ✎ WRITING   ▲ CHALLENGING

Solve each of the following problems.

4. How many quarts of a 60% alcohol solution must be added to 40 qt of a 20% alcohol solution to obtain a mixture that is 30% alcohol? $13\frac{1}{3}$ **qt**

5. Fred and Wilma start from the same point and travel on a straight road. Fred travels at 30 mph, while Wilma travels at 50 mph. If Wilma starts 3 hr after Fred, find the distance they travel before Wilma catches up with Fred.   **225 mi**

Perform each operation. Write each result in standard form.

6. $(7 - i) - (6 - 10i)$ $1 + 9i$

7. $(4 + 3i)(-5 + 2i)$ $-26 - 7i$

8. $\dfrac{5 - 5i}{1 - 3i}$ $2 + i$

9. Is i^{297} equal to i, -1, $-i$, or 1? i

Solve each equation.

10. $6x(2 - x) = 7$ $\left\{1 \pm \dfrac{\sqrt{6}}{6}i\right\}$

11. $\dfrac{3x - 2}{3x + 2} = \dfrac{2x + 3}{4x - 1}$ $\left\{\dfrac{6 \pm \sqrt{42}}{3}\right\}$

12. Evaluate the discriminant, and use it to predict the type of solutions for the equation $3x^2 = 5x - 2$.
  1; 2 different rational solutions

13. Tony Lally has a rectangular-shaped flower box that measures 4 ft by 6 ft. He wants to double the available area by increasing the length and width by the same amount. What should the new dimensions be?   **6 ft by 8 ft**

Solve each equation.

14. $\sqrt{5 + 2x} = x + 1$ $\{2\}$

15. $\sqrt{2x + 1} - \sqrt{x} = 1$ $\{0, 4\}$

16. $x^4 - 3x^2 - 10 = 0$ $\{\pm\sqrt{5}, \pm i\sqrt{2}\}$

17. $2 - \sqrt[3]{2x + x^2} = 0$ $\{-4, 2\}$

Solve each of the following inequalities. Write each answer in interval notation.

18. $-2(x - 1) - 10 \le 2(2 + x)$ $[-3, \infty)$

19. $4 \ge 3 + \dfrac{x}{2} \ge -2$ $[-10, 2]$

20. $2x^2 - x - 3 \ge 0$ $(-\infty, -1] \cup \left[\dfrac{3}{2}, \infty\right)$

21. $\dfrac{6}{2x - 5} \le 2$ $\left(-\infty, \dfrac{5}{2}\right) \cup [4, \infty)$

✐ ◉ **22.** What is wrong with the following "solution" of $\dfrac{1}{x - 3} \ge 2$?

$$\dfrac{1}{x - 3} \ge 2$$

$$(x - 3)\left(\dfrac{1}{x - 3}\right) \ge (x - 3)(2) \qquad \text{Multiply by } x - 3.$$

$$1 \ge 2x - 6 \qquad \text{Distributive property}$$

$$7 \ge 2x \qquad \text{Add 6.}$$

$$\dfrac{7}{2} \ge x \qquad \text{Divide by 2.}$$

The solution set is $\left(-\infty, \dfrac{7}{2}\right]$.

23. Solve $|-5 - 3x| = 4$. $\left\{-\dfrac{1}{3}, -3\right\}$

Solve each inequality. Write solutions in interval notation.

24. $|2x - 5| < 9$ $(-2, 7)$

25. $|2x + 1| - 11 \ge 0$ $(-\infty, -6] \cup [5, \infty)$

■ CHAPTER 3 RELATIONS AND THEIR GRAPHS

3.1 RESOURCES

Instructor's Test Manual
  Test Items, p. 318
Instructor's Solution Manual
  pp. 106–110, 520–524
Study Guide
  pp. 97–101
Student's Solution Manual
  pp. 106–110

Test Generator
  IBM Disks
    5.25″ Disk A
    3.5″ Disk A
    Mac Disk A
Tutorials
  IBM Disks
    5.25″ Disk C
    3.5″ Disk B
    Mac Disk B

HarperCollins Videotape 3
Transparency T-3

3.1 EXERCISES (TEXT PAGE 149) ■ ——————————————

For each relation, give three ordered pairs that belong to the relation and state the domain and the range of the relation. See Example 1.

*1. $\{(-3, 5), (-2, 4), (-1, 6), (0, -8), (1, 2)\}$ $(-3, 5), (-2, 4), (-1, 6)$; $\{-3, -2, -1, 0, 1\}$; $\{5, 4, 6, -8, 2\}$

2. $\{(5, 12), (4, 12), (3, 9), (-1, 6), (-5, 5)\}$ $(5, 12), (4, 12), (3, 9)$; $\{5, 4, 3, -1, -5\}$; $\{12, 9, 6, 5\}$

3. $y = 9x + 2$ $(0, 2), (1, 11), (2, 20)$; $(-\infty, \infty)$; $(-\infty, \infty)$

4. $y = -2x + 7$ $(0, 7), (1, 5), (2, 3)$; $(-\infty, \infty)$; $(-\infty, \infty)$

5. $x = y$ $(-1, -1), (0, 0), (1, 1)$; $(-\infty, \infty)$; $(-\infty, \infty)$

6. $x = -y$ $(-1, 1), (0, 0), (1, -1)$; $(-\infty, \infty)$; $(-\infty, \infty)$

7. $y = \sqrt{x}$ $(0, 0), (1, 1), (4, 2)$; $[0, \infty)$; $[0, \infty)$ **8.** $x = \sqrt{y}$ $(0, 0), (1, 1), (2, 4)$; $[0, \infty)$; $[0, \infty)$

9. $y = x^2$ $(0, 0), (1, 1), (2, 4)$; $(-\infty, \infty)$; $[0, \infty)$

10. $x = \sqrt{y + 3}$ $(0, -3), (1, -2), (2, 1)$; $[0, \infty)$; $[-3, \infty)$

11. $y = \sqrt{x + 3}$ $(-3, 0), (-2, 1), (1, 2)$; $[-3, \infty)$; $[0, \infty)$

12. $y = |x|$ $(0, 0), (-1, 1), (1, 1)$; $(-\infty, \infty)$; $[0, \infty)$

13. Explain in your own words the terms *relation, domain,* and *range.*

14. Fill in the blank with the appropriate number: The relation in Exercise 2 has _____ elements. There are _____ elements in its domain and _____ elements in its range. 5; 5; 4

Find the distance d(P, Q) and the midpoint of segment PQ. See Examples 2 and 5.

15. $P(5, 7), Q(13, -1)$ $8\sqrt{2}$; $(9, 3)$ **16.** $P(-2, 5), Q(4, -3)$ 10; $(1, 1)$

17. $P(-8, -2), Q(-3, -5)$ $\sqrt{34}$; $(-11/2, -7/2)$ **18.** $P(-6, -10), Q(6, 5)$ $3\sqrt{41}$; $(0, -5/2)$

19. $P(3, -7), Q(-5, 19)$ $2\sqrt{185}$; $(-1, 6)$ **20.** $P(-4, 6), Q(8, -5)$ $\sqrt{265}$; $(2, 1/2)$

▲21. $P(a, b), Q(3a, -4b)$ $\sqrt{4a^2 + 25b^2}$; $(2a, -3b/2)$ **▲22.** $P(4x, 6y), Q(-3x, 2y)$ $\sqrt{49x^2 + 16y^2}$; $(x/2, 4y)$

▲23. $P(\sqrt{2}, -\sqrt{5}), Q(3\sqrt{2}, 4\sqrt{5})$ $\sqrt{133}$; $(2\sqrt{2}, 3\sqrt{5}/2)$

▲24. $P(5\sqrt{7}, -\sqrt{3}), Q(-\sqrt{7}, 8\sqrt{3})$ $3\sqrt{55}$; $(2\sqrt{7}, 7\sqrt{3}/2)$

—————————

*Other correct ordered pairs may be given in Exercises 1–12.

◎ CONCEPTUAL   ▨ WRITING   ▲ CHALLENGING

Decide whether or not the given points are the vertices of a right triangle. See Example 3.

25. $(2, 8)$, $(0, 4)$, $(4, 7)$ yes

26. $(6, 4)$, $(0, 2)$, $(10, -8)$ yes

27. $(-4, 0)$, $(1, 3)$, $(-6, -2)$ no

28. $(-8, 2)$, $(5, -7)$, $(3, -9)$ yes

▲ **29.** $(\sqrt{3}, 2\sqrt{3} + 3)$, $(\sqrt{3} + 4, -\sqrt{3} + 3)$, $(2\sqrt{3}, 2\sqrt{3} + 4)$ no

▲ **30.** $(4 - \sqrt{3}, -2\sqrt{3})$, $(2 - \sqrt{3}, -\sqrt{3})$, $(3 - \sqrt{3}, -2\sqrt{3})$ no

Decide whether or not the given points are collinear. See Example 4.

31. $(0, 7)$, $(3, -5)$, $(-2, 15)$ yes

32. $(1, -4)$, $(2, 1)$, $(-1, -14)$ yes

33. $(0, -9)$, $(3, 7)$, $(-2, -19)$ no

34. $(1, 3)$, $(5, -12)$, $(-1, 11)$ no

35. $(2, 7)$, $(-4, -2)$, $(10, 19)$ yes

36. $(-2, -13)$, $(0, 1)$, $(5, 36)$ yes

Find the other endpoint of each segment having endpoint and midpoint as given. See Example 6.

37. Endpoint $(-3, 6)$, midpoint $(5, 8)$ $(13, 10)$

38. Endpoint $(-5, 3)$, midpoint $(-7, 6)$ $(-9, 9)$

39. Endpoint $(5, -4)$, midpoint $(12, 6)$ $(19, 16)$

40. Endpoint $(-2, 7)$, midpoint $(-9, 8)$ $(-16, 9)$

41. Endpoint (a, b), midpoint (c, d) $(2c - a, 2d - b)$

42. Endpoint $(-r, -s)$, midpoint (r, s) $(3r, 3s)$

▲ **43.** Show that if M is the midpoint of the segment with endpoints $P(x_1, y_1)$ and $Q(x_2, y_2)$, then $d(P, M) + d(M, Q) = d(P, Q)$, and $d(P, M) = d(M, Q)$.

◎ **44.** The distance formula as given in the text involves a square root radical. Write the distance formula using rational exponents. $d = [(x_2 - x_1)^2 + (y_2 - y_1)^2]^{1/2}$

◎ *Give all quadrants in which the set of points satisfying the following conditions for ordered pairs (x, y) are located.*

45. $x > 0$ quadrants I and IV

46. $y < 0$ quadrants III and IV

47. $xy < 0$ quadrants II and IV

48. $\dfrac{x}{y} > 0$ quadrants I and III

49. $|x| < 3$, $y < -2$ quadrants III and IV

50. $|y| < 2$, $x > 1$ quadrants I and IV

▲ **51.** Show that the points $(-2, 2)$, $(13, 10)$, $(21, -5)$, and $(6, -13)$ are the vertices of a rhombus (all sides equal in length).

▲ **52.** Are the points $A(1, 1)$, $B(5, 2)$, $C(3, 4)$, $D(-1, 3)$ the vertices of a parallelogram? Of a rhombus (all sides equal in length)? yes; no

✐ ◎ **53.** "If p then q and if q then p" is called a *biconditional,* and is often stated *"p if and only if q."* State the Pythagorean theorem and its converse using this language.

3.2 RESOURCES

Instructor's Test Manual
Test Items, p. 318–321

Instructor's Solution Manual
pp. 111–116, 524–531

Study Guide
pp. 101–108

Student's Solution Manual
pp. 111–116

Test Generator
IBM Disks
5.25" Disk A
3.5" Disk A
Mac Disk A

Tutorials
IBM Disks
5.25" Disk C
3.5" Disk B
Mac Disk B

HarperCollins Videotape 3
"College Algebra: In Simplest Terms" Program 10

3.2 EXERCISES   (TEXT PAGE 160) ■ ──────────────────

Graph each of the following. Give the domain and range. See Examples 1–3.

1. $x - y = 4$
domain and range: $(-\infty, \infty)$

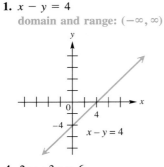

2. $x + y = 4$
domain and range: $(-\infty, \infty)$

3. $3x - y = 6$
domain and range: $(-\infty, \infty)$

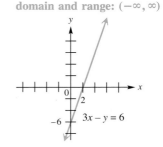

4. $2x - 3y = 6$
domain and range: $(-\infty, \infty)$

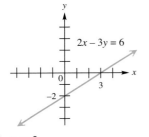

5. $2x + 5y = 10$
domain and range: $(-\infty, \infty)$

6. $4x - 3y = 9$
domain and range: $(-\infty, \infty)$

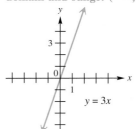

7. $x = 2$
domain: $\{2\}$;   range: $(-\infty, \infty)$

8. $y = -3$
domain: $(-\infty, \infty)$;   range: $\{-3\}$

9. $y = 3x$
domain and range: $(-\infty, \infty)$

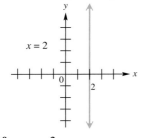

10. $x = -2y$
domain and range: $(-\infty, \infty)$

Find the slope of each of the following lines. See Examples 4 and 8.

11. Through $(-2, 1)$ and $(3, 2)$ **1/5**

12. Through $(-2, 3)$ and $(-1, 2)$ **−1**

13. Through $(8, 4)$ and $(-1, -3)$ **7/9**

14. Through $(-4, -3)$ and $(5, 0)$ **1/3**

15. $3x + 4y = 6$ **−3/4**

16. $2x + y = 8$ **−2**

17. $y = 4$ **0**

18. $x = -6$ **undefined slope**

19. Match each equation with the line that would most closely resemble its graph. (*Hint:* Consider the signs of m and b in the slope-intercept form.)

   (a) $y = 3x + 2$ **D**   **(b)** $y = -3x + 2$ **B**   **(c)** $y = 3x - 2$ **A**   **(d)** $y = -3x - 2$ **C**

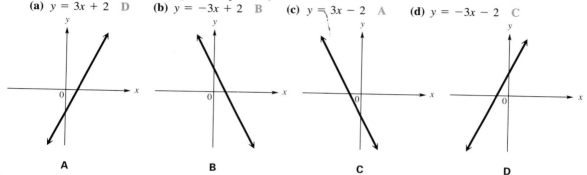

20. Match each equation with the line that would most closely resemble its graph.

   (a) $y = 2$ **A**   **(b)** $y = -2$ **C**   **(c)** $x = 2$ **D**   **(d)** $x = -2$ **B**

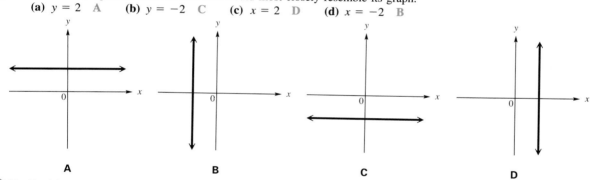

21. Explain in your own words what is meant by the slope of a line.

22. Explain how to graph a line using a point on the line and the slope of the line.

Graph the line passing through the given point and having the indicated slope. Indicate two points on the line. See Example 5.

23. Through $(-1, 3)$, $m = 3/2$

24. Through $(-2, 8)$, $m = -1$

25. Through $(3, -4)$, $m = -1/3$

26. Through $(-2, -3)$, $m = -3/4$

27. Through $(-1, 4)$, $m = 0$

28. Through $(9/4, 2)$, undefined slope

Write an equation in standard form for each of the following lines. See Examples 6–7.

29. Through $(1, 3)$, $m = -2$ $2x + y = 5$

30. Through $(2, 4)$, $m = -1$ $x + y = 6$

31. Through $(-5, 4)$, $m = -3/2$ $3x + 2y = -7$

32. Through $(-4, 3)$, $m = 3/4$ $-3x + 4y = 24$

33. Through $(-8, 1)$, undefined slope $x = -8$

34. Through $(6, 1)$, $m = 0$ $y = 1$

35. Through $(-1, 3)$ and $(3, 4)$ $-x + 4y = 13$

36. Through $(8, -1)$ and $(4, 3)$ $x + y = 7$

37. x-intercept 3, y-intercept -2 $2x - 3y = 6$

38. x-intercept -2, y-intercept 4 $-2x + y = 4$

39. Vertical, through $(-6, 5)$ $x = -6$

40. Horizontal, through $(8, 7)$ $y = 7$

◉ **41.** Fill in each blank with the appropriate response: The line $x + 2 = 0$ has x-intercept
_____. It _____ have a y-intercept. The slope of this line is
(does/does not)
_____. The line $4y = 2$ has y-intercept _____. It _____ have an
(zero/undefined)   (does/does not)
x-intercept. The slope of this line is _____. -2, **does not**, **undefined**, **1/2**, **does not**, **zero**
(zero/undefined)

◉ **42.** What is the equation of the x-axis? $y = 0$

◉ **43.** What is the equation of the y-axis? $x = 0$

◉ **44.** What can be said about the sign of the slope of a line perpendicular to a line whose slope is a
positive number?   **It must be negative.**

Give the slope and y-intercept of each of the following lines. See Example 8.

45. $y = 3x - 1$ slope 3; y-intercept -1

46. $y = -2x + 7$ slope -2; y-intercept 7

47. $4x - y = 7$ slope 4; y-intercept -7

48. $2x + 3y = 16$ slope $-2/3$; y-intercept 16/3

49. $4y = -3x$ slope $-3/4$; y-intercept 0

50. $2y - x = 0$ slope 1/2; y-intercept 0

Write an equation in standard form for each of the following lines. See Example 9.

51. Through $(-1, 4)$, parallel to $x + 3y = 5$ $x + 3y = 11$

52. Through $(3, -2)$, parallel to $2x - y = 5$ $2x - y = 8$

53. Through $(1, 6)$, perpendicular to $3x + 5y = 1$ $5x - 3y = -13$

54. Through $(-2, 0)$, perpendicular to $8x - 3y = 7$ $3x + 8y = -6$

55. Through $(-5, 7)$, perpendicular to $y = -2$ $x = -5$

56. Through $(1, -4)$, perpendicular to $x = 4$ $y = -4$

▲ *Use slopes to decide whether or not the given points lie on a straight line. (Hint: In each exercise, first find the slope of the line through the first and second points and then the slope of the
line through the second and third points. If these slopes are the same, the points lie on a straight
line.)*

57. $M(1, -2)$, $N(3, -18)$, $P(-2, 22)$ yes

58. $A(0, -2)$, $B(3, 7)$, $C(-4, -14)$ yes

59. $X(1, 3)$, $Y(-4, 73)$, $Z(5, -50)$ no

60. $R(0, -7)$, $S(8, -11)$, $T(-9, -25)$ no

◎ CONCEPTUAL   ▨ WRITING   ▲ CHALLENGING

▲ **61.** Find k so that the line through $(4, -1)$ and $(k, 2)$ is
   (a) parallel to $3y + 2x = 6$; $-1/2$
   (b) perpendicular to $2y - 5x = 1$. $-7/2$

▲ **62.** Find r so that the line through $(2, 6)$ and $(-4, r)$ is
   (a) parallel to $2x - 3y = 4$; 2
   (b) perpendicular to $x + 2y = 1$. -6

▲ *To prove that two perpendicular lines, neither of which is vertical, have slopes with a product of* -1, *go through the following steps. Let line* L_1 *have equation* $y = m_1x + b_1$, *and let line* L_2 *have equation* $y = m_2x + b_2$. *Assume that* L_1 *and* L_2 *are perpendicular, and complete right triangle MPN as shown in the figure.* $\overline{PQ}$ *is horizontal and* $\overline{MN}$ *is vertical.*

63. Show that MQ has length m_1.

64. Show that QN has length $-m_2$.

65. Show that triangles MPQ and PNQ are similar.

66. Show that $m_1/1 = 1/-m_2$ and that $m_1m_2 = -1$.

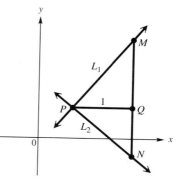

In each of the following problems, assume that the data can be approximated fairly closely by a straight line. Find the equation of the line. See Example 10.

67. A company finds that it can make a total of 20 solar heaters for $13,900, while 10 heaters cost $7500. Let y be the total cost to produce x solar heaters.

$y = 640x + 1100$

68. Consumer prices in the United States at the beginning of each year, measured as a percent of the 1967 average, have produced a graph that is approximately linear. In 1984 the consumer price index was 300%, and in 1987 it was 333%. Let y represent the consumer price index in year x, where $x = 0$ corresponds to 1980. $y = 11x + 256$

69. When a certain industrial pollutant is introduced into a river, the reproduction of catfish declines. In a given period of time, dumping three tons of the pollutant results in a fish population of 37,000. Also, 12 tons of pollutant produce a fish population of 28,000. Let y be the fish population when x tons of pollutant are introduced into the river. $y = -1000x + 40,000$

70. In the snake *Lampropelbis polyzona*, total length y is related to tail length x in the domain 30 mm $\leq x \leq$ 200 mm by a linear relation. Find such a linear relation, if a snake 455 mm long has a 60-mm tail, and a 1050-mm snake has a 140-mm tail. $16y = 119x + 140$

71. According to research done by the political scientist James March, if the Democrats win 45% of the two-party vote for the House of Representatives, they win 42.5% of the seats. If the Democrats win 55% of the vote, they win 67.5% of the seats. Let y be the percent of seats won, and x the percent of the two-party vote. $y = 2.5x - 70$

72. If the Republicans win 45% of the two-party vote, they win 32.5% of the seats (see Exercise 71). If they win 60% of the vote, they get 70% of the seats. Let y represent the percent of the seats, and x the percent of the vote. $y = 2.5x - 80$

73. When the Celsius temperature is 0°, the corresponding Fahrenheit temperature is 32°. When the Celsius temperature is 100°, the corresponding Fahrenheit temperature is 212°. Let C represent the Celsius temperature and F the Fahrenheit temperature. (Solve for F in terms of C; this is an exact linear relationship.)

$F = \dfrac{9}{5}C + 32$

74. Solve the equation found in Exercise 73 for C in terms of F. $C = \dfrac{5}{9}(F - 32)$

75. For what temperature does Celsius equal Fahrenheit? See the results of Exercises 73–74. $-40°$

▲ **76.** Show that the line $y = x$ is the perpendicular bisector of the segment with endpoints (a, b) and (b, a).

3.3 RESOURCES

Instructor's Test Manual
Test Items, p. 321–322
Instructor's Solution Manual
pp. 116–122, 531–539
Study Guide
pp. 109–116
Student's Solution Manual
pp. 116–122

Test Generator
IBM Disks
5.25" Disk A
3.5" Disk A
Mac Disk A
Tutorials
IBM Disks
5.25" Disk C
3.5" Disk B
Mac Disk B

HarperCollins Videotape 4
Transparency T-4–T-10

3.3 EXERCISES   (TEXT PAGE 173) ■ ────────────

1. Graph the following on the same coordinate system.

   (a) $y = 2x^2$ (b) $y = 3x^2$ (c) $y = \dfrac{1}{2}x^2$ (d) $y = \dfrac{1}{3}x^2$

   ⬜◉ (e) How does the coefficient of x^2 affect the shape of the graph?

2. Graph the following on the same coordinate system.
   (a) $y = x^2 + 2$ (b) $y = x^2 - 1$ (c) $y = x^2 + 1$ (d) $y = x^2 - 2$
   ⬜◉ (e) How do these graphs differ from the graph of $y = x^2$?

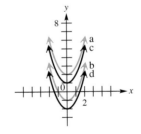

3. Graph the following on the same coordinate system.
   (a) $y = (x - 2)^2$ (b) $y = (x + 1)^2$ (c) $y = (x + 3)^2$ (d) $y = (x - 4)^2$
   ⬜◉ (e) How do these graphs differ from the graph of $y = x^2$?

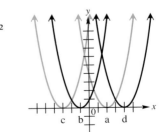

◉ CONCEPTUAL   ⬜ WRITING   ▲ CHALLENGING

◉ **4.** Match each equation with the description of the parabola that is its graph.

(**a**) $y = (x - 4)^2 - 2$ **D**   **A.** vertex $(2, -4)$, opens down

(**b**) $y = (x - 2)^2 - 4$ **B**   **B.** vertex $(2, -4)$, opens up

(**c**) $y = -(x - 4)^2 - 2$ **C**   **C.** vertex $(4, -2)$, opens down

(**d**) $y = -(x - 2)^2 - 4$ **A**   **D.** vertex $(4, -2)$, opens up

(**e**) $x = (y - 4)^2 - 2$ **F**   **E.** vertex $(-2, 4)$, opens left

(**f**) $x = (y - 2)^2 - 4$ **H**   **F.** vertex $(-2, 4)$, opens right

(**g**) $x = -(y - 4)^2 - 2$ **E**   **G.** vertex $(-4, 2)$, opens left

(**h**) $x = -(y - 2)^2 - 4$ **G**   **H.** vertex $(-4, 2)$, opens right

◉ **5.** For the graph of $y = a(x - h)^2 + k$, in what quadrant is the vertex if:

(**a**) $h < 0$, $k < 0$;   **III**   (**b**) $h < 0$, $k > 0$;   **II**   (**c**) $h > 0$, $k < 0$;   **IV**   (**d**) $h > 0$, $k > 0$?   **I**

◉ **6.** Repeat parts (a)–(d) of Exercise 5 for the graph of $x = a(y - k)^2 + h$.

(a) III   (b) II   (c) IV   (d) I

Graph each of the following parabolas. Give the vertex, axis, domain, and range of each. See Examples 1–7.

7. $y = (x - 2)^2$
vertex: $(2, 0)$;
axis: $x = 2$;
domain: $(-\infty, \infty)$;
range: $[0, \infty)$

8. $y = (x + 4)^2$
vertex: $(-4, 0)$;
axis: $x = -4$;
domain: $(-\infty, \infty)$;
range: $[0, \infty)$

9. $y = (x + 3)^2 - 4$
vertex: $(-3, -4)$;
axis: $x = -3$;
domain: $(-\infty, \infty)$;
range: $[-4, \infty)$

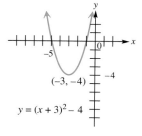

10. $y = (x - 5)^2 - 4$
vertex: $(5, -4)$;
axis: $x = 5$;
domain: $(-\infty, \infty)$;
range: $[-4, \infty)$

11. $y = -2(x + 3)^2 + 2$
vertex: $(-3, 2)$;
axis: $x = -3$;
domain: $(-\infty, \infty)$;
range: $(-\infty, 2]$

12. $y = -3(x - 2)^2 + 1$
vertex: $(2, 1)$;
axis: $x = 2$;
domain: $(-\infty, \infty)$;
range: $(-\infty, 1]$

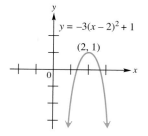

13. $y = -\dfrac{1}{2}(x + 1)^2 - 3$

vertex: $(-1, -3)$;
axis: $x = -1$;
domain: $(-\infty, \infty)$;
range: $(-\infty, -3]$

14. $y = \dfrac{2}{3}(x - 2)^2 - 1$

vertex: $(2, -1)$;
axis: $x = 2$;
domain: $(-\infty, \infty)$;
range: $[-1, \infty)$

15. $y = x^2 - 2x + 3$

vertex: $(1, 2)$;
axis: $x = 1$;
domain: $(-\infty, \infty)$;
range: $[2, \infty)$

16. $y = x^2 + 6x + 5$

vertex: $(-3, -4)$;
axis: $x = -3$;
domain: $(-\infty, \infty)$;
range: $[-4, \infty)$

17. $y = 2x^2 - 4x + 5$

vertex: $(1, 3)$;
axis: $x = 1$;
domain: $(-\infty, \infty)$;
range: $[3, \infty)$

18. $y = -3x^2 + 24x - 46$

vertex: $(4, 2)$;
axis: $x = 4$;
domain: $(-\infty, \infty)$;
range: $(-\infty, 2]$

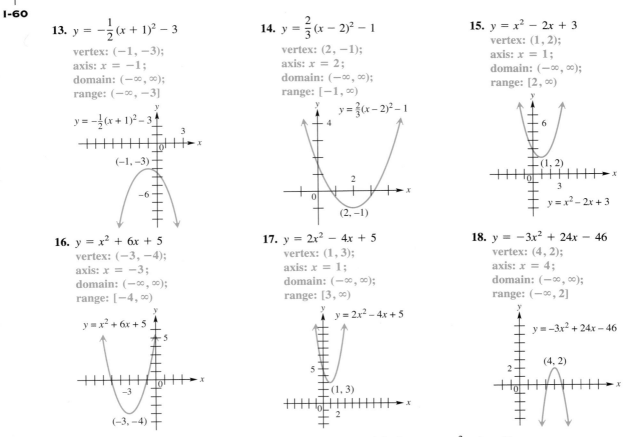

19. Explain what causes the graph of a parabola with equation of the form $y = ax^2$ to be wider or narrower than the graph of $y = x^2$.

20. Explain what causes the graph of a parabola with equation of the form $x = ay^2$ to open to the left or the right.

Graph each horizontal parabola. Give the vertex, axis, domain, and range of each. See Examples 9 and 10.

21. $x = y^2 + 2$ vertex: $(2, 0)$;   axis: $y = 0$;
domain: $[2, \infty)$;   range: $(-\infty, \infty)$

22. $x = -y^2$ vertex: $(0, 0)$;   axis: $y = 0$;
domain: $(-\infty, 0]$;   range: $(-\infty, \infty)$

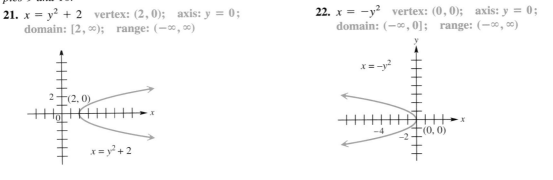

23. $x = (y + 1)^2$
vertex: $(0, -1)$;
axis: $y = -1$;
domain: $[0, \infty)$;
range: $(-\infty, \infty)$

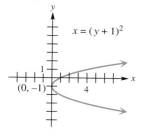

24. $x = (y - 3)^2$
vertex: $(0, 3)$;
axis: $y = 3$;
domain: $[0, \infty)$;
range: $(-\infty, \infty)$

25. $x = (y + 2)^2 - 1$
vertex: $(-1, -2)$;
axis: $y = -2$;
domain: $[-1, \infty)$;
range: $(-\infty, \infty)$

26. $x = (y - 4)^2 + 2$
vertex: $(2, 4)$;
axis: $y = 4$;
domain: $[2, \infty)$;
range: $(-\infty, \infty)$

27. $x = -2(y + 3)^2$
vertex: $(0, -3)$;
axis: $y = -3$;
domain: $(-\infty, 0]$;
range: $(-\infty, \infty)$

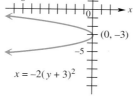

28. $x = \dfrac{2}{3}(y - 3)^2 + 2$
vertex: $(2, 3)$;
axis: $y = 3$;
domain: $[2, \infty)$;
range: $(-\infty, \infty)$

29. $x = y^2 + 2y - 8$
vertex: $(-9, -1)$;
axis: $y = -1$;
domain: $[-9, \infty)$;
range: $(-\infty, \infty)$

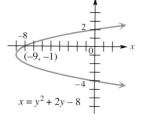

30. $x = -4y^2 - 4y - 3$
vertex: $(-2, -1/2)$;
axis: $y = -1/2$;
domain: $(-\infty, -2]$;
range: $(-\infty, \infty)$

Solve each application of quadratic relations.
See Example 8.

31. Glenview Community College wants to construct a rectangular parking lot on land bordered on one side by a highway. It has 320 ft of fencing with which to fence off the other three sides. What should be the dimensions of the lot if the enclosed area is to be a maximum? (*Hint:* Let x represent the width of the lot and let $320 - 2x$ represent the length. Graph the area parabola, $A = x(320 - 2x)$, and investigate the vertex.)

80 by 160 ft

32. What would be the maximum area that could be enclosed by the college's 320 ft of fencing if it decided to close the entrance by enclosing all four sides of the lot? (See Exercise 31.)   6400 sq ft

33. Ed Doskey owns a sandwich company. By studying data concerning his past costs, he has found that the cost of operating his company is given by

$$C = 2x^2 - 1200x + 180,100,$$

where C is the daily cost in dollars to make x sandwiches. Find the number of sandwiches Ed must sell to minimize the cost. What is the minimum cost?

300 sandwiches;   100 dollars

34. The revenue of a charter bus company depends on the number of unsold seats. If the revenue R, in dollars, is given by

$$R = 5000 + 50x - x^2,$$

where x is the number of unsold seats, find the maximum revenue and the number of unsold seats which produce maximum revenue.

25 unsold seats;   maximum revenue is $5625

35. The number of mosquitoes M, in millions, in a certain area of Alabama depends on the July rainfall x, in inches, approximately as follows.

$$M = 10x - x^2$$

Find the rainfall that will produce the maximum number of mosquitoes.   5 in

36. Refer to Example 5 in Section 2.5. After how many seconds will the projectile reach its maximum height? What is the maximum height?   3.125 sec, 156.25 ft

37. If an object is thrown upward with an initial velocity of 32 ft per sec, then its height h in ft after t sec is given by

$$h = 32t - 16t^2.$$

After how many seconds does it reach its maximum height? Find the maximum height attained by the object. Find the number of seconds it takes the object to hit the ground.   1 sec, 16 ft, 2 sec

38. A charter flight charges a fare of $200 per person plus $4 per person for each unsold seat on the plane. If the plane holds 100 passengers, and if x represents the number of unsold seats, find the following.
(a) an expression for the total revenue R, in dollars, received for the flight (*Hint:* Multiply the number of people flying, $100 - x$, by the price per ticket.)

$R = (100 - x)(200 + 4x) = 20,000 + 200x - 4x^2$

(b) the graph for the expression of part (a)

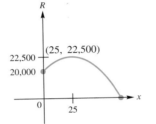

(c) the number of unsold seats that will produce the maximum revenue   25
(d) the maximum revenue $22,500

39. The demand for a certain type of cosmetic is given by

$$p = 500 - x,$$

where p is the price when x units are demanded.
(a) Find the revenue R, in dollars, that would be obtained at a price of x. (*Hint:* Revenue = Demand × Price.)

$R = x(500 - x) = 500x - x^2$

(b) Graph the revenue relation R.

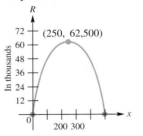

(c) From the graph of the revenue relation, estimate the price that will produce maximum revenue. $250
(d) What is the maximum revenue? $62,500

40. For the months of June through October, the percent of maximum possible chlorophyll production in a leaf is approximated by C, where

$$C = 10x + 50.$$

Here x is time in months with $x = 1$ representing June. From October through December, C is approximated by

$$C = -20(x - 5)^2 + 100,$$

with x as above. Find the percent of maximum possible chlorophyll production in each of the following months:

(a) June   60   **(b)** July   70
(c) September   90   **(d)** October   100
(e) November   80   **(f)** December.   20

41. Use your results from Exercise 40 to sketch a graph of $y = C$ from June through December. In what month is chlorophyll production at a maximum?   October

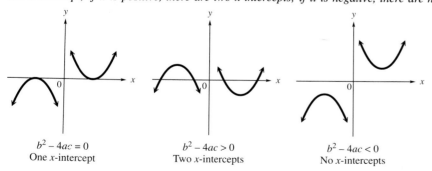

June   Oct

42. The temperature T at which water boils varies with the elevation E, where T is in degrees Celsius and E is in meters above sea level, according to the equation

$$E = 580(100 - T)^2 + 1000(100 - T).$$

(a) Determine the boiling point at each of the following elevations. (*Hint:* Let $X = 100 - T$. Solve for X, and then solve for T.)

1. Mount McKinley (Alaska), 6194 meters   97.5° C
2. Mount Everest (Nepal-Tibet), 8848 meters 96.9° C
3. Mount Andrew Jackson (Antarctica), 4191 meters 98.0° C
4. Mount Mauna Loa (Hawaii), 4170 meters   98.0° C

(b) Discuss the interpretation of letting $T = 100$ in the formula.

(c) What is the boiling point of water in Denver, elevation 1600 meters?   99.0° C

Use the geometric definition of parabola given in the text to find the equation of each parabola having the given point as focus.

43. $(0, 3)$, directrix $y = -3$ $x^2 = 12y$
44. $(0, -5)$, directrix $y = 5$ $x^2 = -20y$
45. $(-2, 0)$, directrix $x = 2$ $y^2 = -8x$
46. $(8, 0)$, directrix $x = -8$ $y^2 = 32x$
47. $(3, 6)$, vertex $(3, 4)$ $(x - 3)^2 = 8(y - 4)$
48. $(-5, 2)$, vertex $(-5, 5)$ $(x + 5)^2 = -12(y - 5)$

Just as the discriminant of the quadratic equation $ax^2 + bx + c = 0$ ($a \neq 0$) determines the number of real solutions of the equation (see Section 2.4), the discriminant $b^2 - 4ac$ for the vertical parabola $y = ax^2 + bx + c$ determines the number of x-intercepts. If the discriminant is 0, there is one x-intercept; if it is positive, there are two x-intercepts; if it is negative, there are no x-intercepts.

$b^2 - 4ac = 0$	$b^2 - 4ac > 0$	$b^2 - 4ac < 0$
One x-intercept	Two x-intercepts	No x-intercepts

49. Find a value of c so that $y = x^2 - 10x + c$ has exactly one x-intercept.   25
50. Find b so that $y = x^2 + bx + 9$ has exactly one x-intercept. ± 6

▲ *The figures below show several possible graphs of* $y = ax^2 + bx + c$. *For the restrictions on a, b, and c in Exercises 51–54, select the possible corresponding graph.*

51. $a < 0$, $b^2 - 4ac = 0$ **E**

52. $a > 0$, $b^2 - 4ac < 0$ **A**

53. $a > 0$, $b^2 - 4ac > 0$ **C**

54. $a > 0$, $b^2 - 4ac = 0$ **B**

A   B   C   D   E   F

▲ **55.** Graph the parabola $y = 2x^2 + 5x - 3$.

(a) Use the graph to find the solution of the quadratic inequality
$2x^2 + 5x - 3 < 0$. $(-3, 1/2)$

(b) Use the graph to find the solution of the quadratic inequality
$2x^2 + 5x - 3 > 0$. $(-\infty, -3) \cup (1/2, \infty)$

(c) Use the graph to find the solutions of the quadratic equation
$2x^2 + 5x - 3 = 0$. $-3, 1/2$

$y = 2x^2 + 5x - 3$

$\left(-\frac{5}{4}, -\frac{49}{8}\right)$

▲ **56.** Find the equation of the vertical parabola having x-intercepts 2 and 5, and y-intercept 5.

$y = \frac{1}{2}x^2 - \frac{7}{2}x + 5$

3.4 RESOURCES

Instructor's Test Manual
Test Items, p. 322–323

Instructor's Solution Manual
pp. 122–127, 539–544

Study Guide
pp. 117–119

Student's Solution Manual
pp. 122–127

Test Generator
IBM Disks
5.25" Disk A
3.5" Disk A
Mac Disk A

Tutorials
IBM Disks
5.25" Disk C
3.5" Disk B
Mac Disk B

HarperCollins Videotape   4
"College Algebra: In Simplest Terms" Program   11
Transparency   T-11–T-14

3.4 EXERCISES   (TEXT PAGE 182)   ■

Find the center-radius form of the equation for each of the following circles. See Example 1.

1. Center $(1, 4)$, radius 3 $(x - 1)^2 + (y - 4)^2 = 9$

2. Center $(-2, 5)$, radius 4 $(x + 2)^2 + (y - 5)^2 = 16$

3. Center $(0, 0)$, radius 1 $x^2 + y^2 = 1$

4. Center $(0, 0)$, radius 5 $x^2 + y^2 = 25$

◎ **CONCEPTUAL**   ✎ **WRITING**   ▲ **CHALLENGING**

5. Center $(2/3, -4/5)$, radius $3/7$ $(x - 2/3)^2 + (y + 4/5)^2 = 9/49$

6. Center $(-1/2, -1/4)$, radius $12/5$ $(x + 1/2)^2 + (y + 1/4)^2 = 144/25$

7. Center $(-1, 2)$, passing through $(2, 6)$ $(x + 1)^2 + (y - 2)^2 = 25$

8. Center $(2, -7)$, passing through $(-2, -4)$ $(x - 2)^2 + (y + 7)^2 = 25$

9. Center $(-3, -2)$, tangent to the x-axis (*Hint: Tangent to* means touching at one point.) $(x + 3)^2 + (y + 2)^2 = 4$

10. Center $(5, -1)$, tangent to the y-axis $(x - 5)^2 + (y + 1)^2 = 25$

Graph each of the following circles. Give the domain and the range. See Examples 1 and 3.

11. $x^2 + y^2 = 36$
domain: $[-6, 6]$;
range: $[-6, 6]$

12. $x^2 + y^2 = 81$
domain: $[-9, 9]$;
range: $[-9, 9]$

13. $(x - 2)^2 + y^2 = 36$
domain: $[-4, 8]$;
range: $[-6, 6]$

14. $x^2 + (y + 3)^2 = 49$
domain: $[-7, 7]$;
range: $[-10, 4]$

15. $(x + 2)^2 + (y - 5)^2 = 16$
domain: $[-6, 2]$;
range: $[1, 9]$

16. $(x - 4)^2 + (y - 3)^2 = 25$
domain: $[-1, 9]$;
range: $[-2, 8]$

17. $(x + 3)^2 + (y + 2)^2 = 36$
domain: $[-9, 3]$;
range: $[-8, 4]$

18. $(x - 5)^2 + (y + 4)^2 = 49$
domain: $[-2, 12]$;
range: $[-11, 3]$

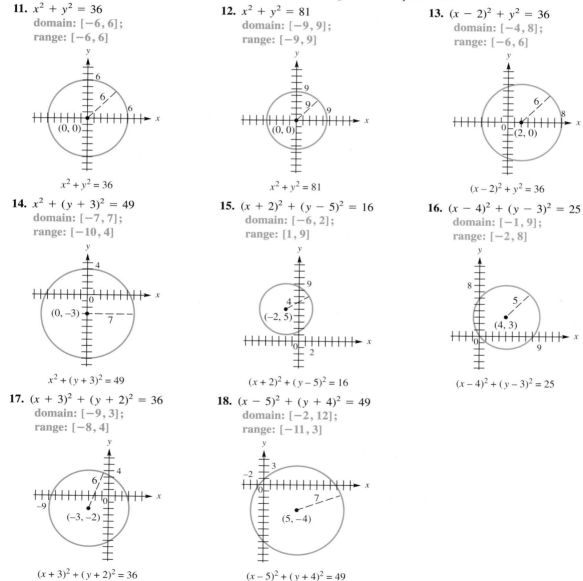

🖉 ◉ **19.** Describe the graph of the equation $(x - 3)^2 + (y - 3)^2 = 0$.

🖉 ◉ **20.** Describe the graph of the equation $(x - 3)^2 + (y - 3)^2 = -1$.

🖉 ◉ **21.** Without actually graphing, state whether or not the graphs of $x^2 + y^2 = 4$ and $x^2 + y^2 = 25$ will intersect. Explain your answer.

🖉 ◉ **22.** Can a circle have its center at $(2, 4)$ and be tangent to both axes? Explain.

Find the center and the radius of each of the following that are circles. See Example 2.

23. $x^2 + 6x + y^2 + 8y = -9$ $(-3, -4)$; $r = 4$

24. $x^2 - 4x + y^2 + 12y + 4 = 0$ $(2, -6)$; $r = 6$

25. $x^2 - 12x + y^2 + 10y + 25 = 0$ $(6, -5)$; $r = 6$

26. $x^2 + 8x + y^2 - 6y = -16$ $(-4, 3)$; $r = 3$

27. $x^2 + 8x + y^2 - 14y + 64 = 0$ $(-4, 7)$; $r = 1$

28. $x^2 - 8x + y^2 + 7 = 0$ $(4, 0)$; $r = 3$

29. $x^2 + y^2 = 2y + 48$ $(0, 1)$; $r = 7$

30. $x^2 + 4x + y^2 = 21$ $(-2, 0)$; $r = 5$

◉ *State whether each of the following graphs is symmetric to any of the following: x-axis, y-axis, origin.*

31. *y*-axis   **32.** *y*-axis   **33.** *x*-axis

34. *x*-axis   **35.** origin   **36.** *x*-axis, *y*-axis, origin

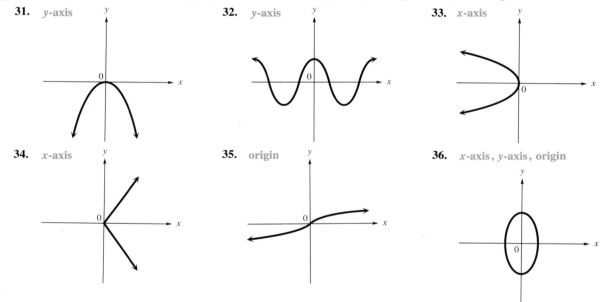

*Plot the following points, and then use the same axes to plot the points that are symmetric to the given point with respect to the following elements: (**a**) x-axis, (**b**) y-axis, (**c**) origin.*

37. $(5, -3)$ **38.** $(-6, 1)$ **39.** $(-4, -2)$ **40.** $(-8, 3)$

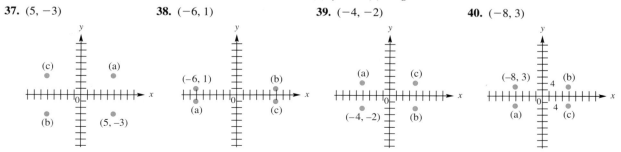

◉ **CONCEPTUAL**   🖉 **WRITING**   ▲ **CHALLENGING**

◉ *For Exercises 41–43, suppose that the point (s, t) lies on the graph of a relation R.*

41. If R is symmetric with respect to the x-axis, the point _____ must lie on the graph of R.

$(s, -t)$

42. If R is symmetric with respect to the y-axis, the point _____ must lie on the graph of R.

$(-s, t)$

43. If R is symmetric with respect to the origin, the point _____ must lie on the graph of R.

$(-s, -t)$

☑ ◉ **44.** If a relation is graphed on a sheet of paper, the concepts of symmetry with respect to the axes or the origin can be illustrated using paper-folding. Explain how this can be done.

Use the tests for symmetry to decide whether the graph of each relation is symmetric with respect to the x-axis, the y-axis, or the origin. See Examples 4 and 5.

45. $x^2 + y^2 = 5$ *x*-axis, *y*-axis, origin   **46.** $y^2 = 4 - x^2$ *x*-axis, *y*-axis, origin

47. $y = x^2 - 8x$ none   **48.** $y = 4x - x^2$ none   **49.** $y = x^3$ origin

50. $y = -x^3$ origin   **51.** $y = \dfrac{1}{1 + x^2}$ *y*-axis   **52.** $x = \dfrac{-1}{y^2 + 9}$ *x*-axis

◉ *Sketch examples of graphs that satisfy the following conditions.*   *Many answers are possible in Exercises 53–55.*

53. Symmetric with respect to the x-axis but not to the y-axis

54. Symmetric with respect to the y-axis but not to the x-axis

55. Symmetric with respect to the origin but to neither the x-axis nor the y-axis

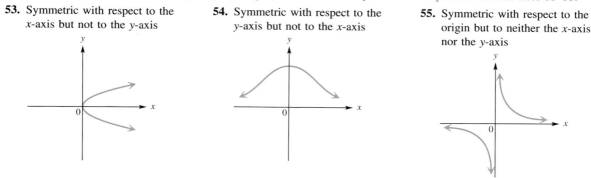

▲ *In Exercises 56–58, find the equation of the circle having the given points as endpoints of a diameter. (Hint: Use the midpoint formula to find the center, use the distance formula to find the radius, and then use the center-radius form of the equation of a circle to find the equation.)*

56. $(4, 3), (4, -3)$ $(x - 4)^2 + y^2 = 9$ **57.** $(-1, 3), (5, -9)$ $(x - 2)^2 + (y + 3)^2 = 45$

58. $(3, -5), (-7, 3)$ $(x + 2)^2 + (y + 1)^2 = 41$

▲ **59.** Find all points (x, y) with $x = y$ that are 4 units from $(1, 3)$.

$(2 + \sqrt{7}, 2 + \sqrt{7}), (2 - \sqrt{7}, 2 - \sqrt{7})$

▲ **60.** Find all points satisfying $x + y = 0$ that are 8 units from $(-2, 3)$.

$\left(\dfrac{-5 + \sqrt{127}}{2}, \dfrac{5 - \sqrt{127}}{2} \right), \left(\dfrac{-5 - \sqrt{127}}{2}, \dfrac{5 + \sqrt{127}}{2} \right)$

◉ **61.** Let F be some algebraic expression involving x as the only variable. Suppose the equation $y = F$ is changed to $y = -F$. What is the effect on the graph of $y = F$?

It is reflected about the x-axis.

▲ ◎ **62.** Suppose a circle is tangent to both axes, is in the third quadrant, and has a radius of $\sqrt{2}$. Find its equation. $(x + \sqrt{2})^2 + (y + \sqrt{2})^2 = 2$

▲ ◎ **63.** One circle has center at $(3, 4)$ and radius 5. A second circle has center at $(-1, -3)$ and radius 4. Do the circles intersect?   yes

▲ ◎ **64.** Does a circle with radius 6 and center at $(0, 5)$ intersect a circle with center at $(-5, -4)$ and radius 4?   no

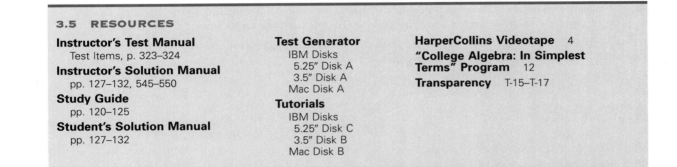

3.5 RESOURCES

Instructor's Test Manual
Test Items, p. 323–324
Instructor's Solution Manual
pp. 127–132, 545–550
Study Guide
pp. 120–125
Student's Solution Manual
pp. 127–132

Test Generator
IBM Disks
5.25″ Disk A
3.5″ Disk A
Mac Disk A
Tutorials
IBM Disks
5.25″ Disk C
3.5″ Disk B
Mac Disk B

HarperCollins Videotape   4
"College Algebra: In Simplest Terms" Program   12
Transparency   T-15–T-17

3.5 EXERCISES   (TEXT PAGE 191) ■ ────────────────

Graph each ellipse. Give the domain and range. See Examples 1 and 3.

1. $\dfrac{x^2}{9} + \dfrac{y^2}{4} = 1$
domain: $[-3, 3]$;
range: $[-2, 2]$

2. $\dfrac{x^2}{16} + \dfrac{y^2}{36} = 1$
domain: $[-4, 4]$;
range: $[-6, 6]$

3. $\dfrac{x^2}{6} + \dfrac{y^2}{9} = 1$
domain: $[-\sqrt{6}, \sqrt{6}]$;
range: $[-3, 3]$

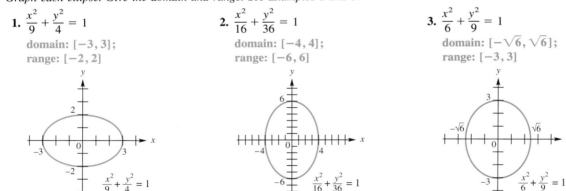

4. $\dfrac{x^2}{8} + \dfrac{y^2}{12} = 1$

domain: $[-2\sqrt{2}, 2\sqrt{2}]$;
range: $[-2\sqrt{3}, 2\sqrt{3}]$

5. $\dfrac{x^2}{1/9} + \dfrac{y^2}{1/16} = 1$

domain: $[-1/3, 1/3]$;
range: $[-1/4, 1/4]$

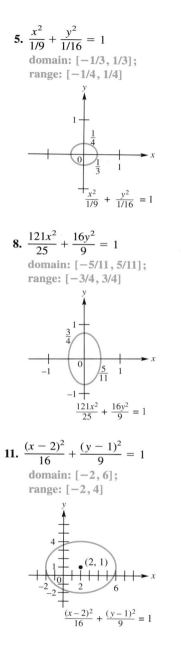

6. $\dfrac{x^2}{4/25} + \dfrac{y^2}{9/49} = 1$

domain: $[-2/5, 2/5]$;
range: $[-3/7, 3/7]$

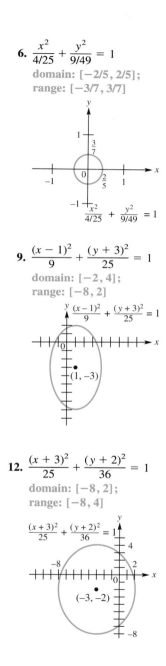

7. $\dfrac{64x^2}{9} + \dfrac{25y^2}{36} = 1$

domain: $[-3/8, 3/8]$;
range: $[-6/5, 6/5]$

8. $\dfrac{121x^2}{25} + \dfrac{16y^2}{9} = 1$

domain: $[-5/11, 5/11]$;
range: $[-3/4, 3/4]$

9. $\dfrac{(x-1)^2}{9} + \dfrac{(y+3)^2}{25} = 1$

domain: $[-2, 4]$;
range: $[-8, 2]$

10. $\dfrac{(x+3)^2}{16} + \dfrac{(y-2)^2}{36} = 1$

domain: $[-7, 1]$;
range: $[-4, 8]$

11. $\dfrac{(x-2)^2}{16} + \dfrac{(y-1)^2}{9} = 1$

domain: $[-2, 6]$;
range: $[-2, 4]$

12. $\dfrac{(x+3)^2}{25} + \dfrac{(y+2)^2}{36} = 1$

domain: $[-8, 2]$;
range: $[-8, 4]$

Graph each hyperbola. Give the domain and range. See Examples 4–6.

13. $\dfrac{x^2}{16} - \dfrac{y^2}{9} = 1$

domain: $(-\infty, -4] \cup [4, \infty)$;
range: $(-\infty, \infty)$

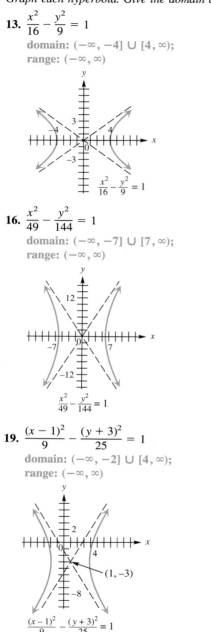

14. $\dfrac{y^2}{9} - \dfrac{x^2}{9} = 1$

domain: $(-\infty, \infty)$;
range: $(-\infty, -3] \cup [3, \infty)$

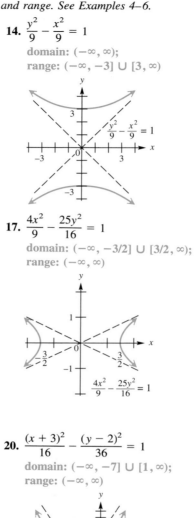

15. $\dfrac{y^2}{36} - \dfrac{x^2}{49} = 1$

domain: $(-\infty, \infty)$;
range: $(-\infty, -6] \cup [6, \infty)$

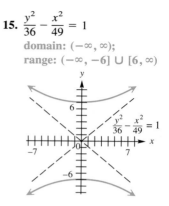

16. $\dfrac{x^2}{49} - \dfrac{y^2}{144} = 1$

domain: $(-\infty, -7] \cup [7, \infty)$;
range: $(-\infty, \infty)$

17. $\dfrac{4x^2}{9} - \dfrac{25y^2}{16} = 1$

domain: $(-\infty, -3/2] \cup [3/2, \infty)$;
range: $(-\infty, \infty)$

18. $x^2 - y^2 = 1$

domain: $(-\infty, -1] \cup [1, \infty)$
range: $(-\infty, \infty)$

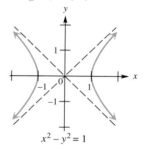

19. $\dfrac{(x-1)^2}{9} - \dfrac{(y+3)^2}{25} = 1$

domain: $(-\infty, -2] \cup [4, \infty)$;
range: $(-\infty, \infty)$

20. $\dfrac{(x+3)^2}{16} - \dfrac{(y-2)^2}{36} = 1$

domain: $(-\infty, -7] \cup [1, \infty)$;
range: $(-\infty, \infty)$

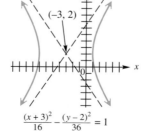

21. $\dfrac{(x-3)^2}{16} - \dfrac{(y+2)^2}{49} = 1$

domain: $(-\infty, -1] \cup [7, \infty)$;
range: $(-\infty, \infty)$

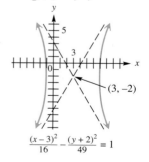

Ⓒ **CONCEPTUAL**   ✍ **WRITING**   ▲ **CHALLENGING**

22. $\dfrac{(y-5)^2}{4} - \dfrac{(x+1)^2}{9} = 1$

domain: $(-\infty, \infty)$;
range: $(-\infty, 3] \cup [7, \infty)$

23. $\dfrac{(y+1)^2}{25} - \dfrac{(x-3)^2}{36} = 1$

domain: $(-\infty, \infty)$;
range: $(-\infty, -6] \cup [4, \infty)$

24. $\dfrac{(x+2)^2}{16} - \dfrac{(y+2)^2}{25} = 1$

domain: $(-\infty, -6] \cup [2, \infty)$;
range: $(-\infty, \infty)$

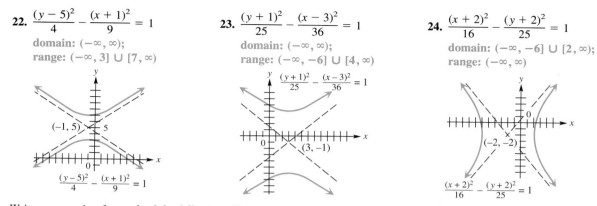

Write an equation for each of the following ellipses or hyperbolas. See Examples 2 and 7.

25. Ellipse, center at the origin, length of major axis $= 12$, endpoint of minor axis at $(0, 4)$ $x^2/36 + y^2/16 = 1$

26. Ellipse, center at the origin, vertex $(6, 0)$, length of minor axis $= 8$ $x^2/36 + y^2/16 = 1$

27. Hyperbola, center at the origin, vertex $(0, 2)$, $a = 2b$ $y^2/4 - x^2/16 = 1$

28. Hyperbola, center at the origin, length of transverse axis $= 8$, $b = 3a$ $x^2/16 - y^2/144 = 1$

29. Ellipse, center at $(2, -2)$, $a = 4$, $b = 3$, major axis parallel to the x-axis $\dfrac{(x-2)^2}{16} + \dfrac{(y+2)^2}{9} = 1$

30. Hyperbola, center at $(-3, 1)$, $a = 4$, $b = 2$, transverse axis parallel to the y-axis $\dfrac{(y-1)^2}{4} - \dfrac{(x+3)^2}{16} = 1$

31. Hyperbola, center at $(4, 3)$, vertex $(1, 3)$, $b = 2$ $\dfrac{(x-4)^2}{9} - \dfrac{(y-3)^2}{4} = 1$

32. Ellipse, center at $(-1, -2)$, length of minor axis $= 4$, vertex $(-1, 1)$ $\dfrac{(x+1)^2}{4} + \dfrac{(y+2)^2}{9} = 1$

33. Explain how it can be determined, before graphing, whether the branches of a hyperbola open up and down or right and left.

34. How can a circle be considered a special case of an ellipse?

35. Draftspeople often use the method shown in the sketch to draw an ellipse. Explain why the method works.

36. Explain how the method of Exercise 35 can be modified to draw a circle.

37. The Roman Coliseum is an ellipse with major axis 620 ft and minor axis 513 ft. Find the distance between the foci of this ellipse.   348.2 ft

38. A formula for the approximate circumference of an ellipse is

$$C \approx 2\pi \sqrt{\dfrac{a^2 + b^2}{2}},$$

where a and b are the lengths as shown in the figure. Use this formula to find the approximate circumference of the Roman Coliseum (see Exercise 37).

1787.6 ft

39. Ships and planes often use a location finding system called LORAN. With this system, a radio transmitter at M on the figure sends out a series of pulses. When each pulse is received at transmitter S, it then sends out a pulse. A ship at P receives pulses from both M and S. A receiver on the ship measures the difference in the arrival times of the pulses. The navigator then consults a special map, showing certain curves according to the differences in arrival times. In this way, the ship can be located as lying on a portion of which curve?   **a hyperbola**

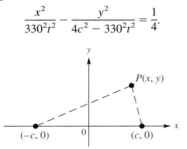

40. Microphones are placed at points $(-c, 0)$ and $(c, 0)$. An explosion occurs at point $P(x, y)$ having positive x-coordinate. (See the figure.) The sound is detected at the closer microphone t sec before being detected at the farther microphone. Assume that sound travels at a speed of 330 m per sec, and show that P must be on the hyperbola

$$\frac{x^2}{330^2 t^2} - \frac{y^2}{4c^2 - 330^2 t^2} = \frac{1}{4}.$$

41. The orbit of Venus is an ellipse with the sun at one focus. An approximate equation for its orbit is

$$\frac{x^2}{5013} + \frac{y^2}{4970} = 1,$$

where x and y are measured in millions of miles. Find the widest possible distance across the ellipse.
about 141.6 million mi

42. A one-way road passes under an overpass in the form of half of an ellipse, 15 ft high at the center and 20 ft wide. Assuming a truck is 12 ft wide, what is the tallest truck that can pass under the overpass?

12 ft tall

43. Suppose that $(c, 0)$ and $(-c, 0)$ are the foci of an ellipse. Suppose that the sum of the distances from any point (x, y) of the ellipse to the two foci is $2a$. See the accompanying figure.
(a) Use the distance formula to show that the equation of the resulting ellipse is

$$\frac{x^2}{a^2} + \frac{y^2}{a^2 - c^2} = 1.$$

(b) Show that a and $-a$ are the x-intercepts.
(c) Let $b^2 = a^2 - c^2$, and show that b and $-b$ are the y-intercepts.

44. Use the result of Exercise 43(a) to find an equation of an ellipse with foci $(3, 0)$ and $(-3, 0)$, where the sum of the distances from any point of the ellipse to the two foci is 10. $\dfrac{x^2}{25} + \dfrac{y^2}{16} = 1$

45. Suppose a hyperbola has center at the origin, foci at $F'(-c, 0)$ and $F(c, 0)$, and the value $|d(P, F') - d(P, F)| = 2a$. Let $b^2 = c^2 - a^2$, and show that the equation of the hyperbola is

$$\frac{x^2}{a^2} - \frac{y^2}{b^2} = 1.$$

46. Use the result of Exercise 45 to find an equation of a hyperbola with center at the origin, foci at $(-2, 0)$ and $(2, 0)$, and the absolute value of the difference of the distances from any point of the hyperbola to the two foci equal to 2. $x^2 - \dfrac{y^2}{3} = 1$

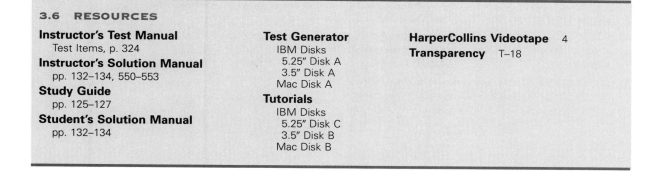

3.6 EXERCISES (TEXT PAGE 198) ▪ ─────────────────────────

In Exercises 1–12, the equation of a conic section is given in a familiar form. Identify the type of graph that the equation has, without actually graphing.

1. $x^2 + y^2 = 144$ circle

2. $(x - 2)^2 + (y + 3)^2 = 25$ circle

3. $y = 2x^2 + 3x - 4$ parabola

4. $x = 3y^2 + 5y - 6$ parabola

5. $x = -3(y - 4)^2 + 1$ parabola

6. $\dfrac{x^2}{25} + \dfrac{y^2}{36} = 1$ ellipse

7. $\dfrac{x^2}{49} + \dfrac{y^2}{100} = 1$ ellipse

8. $x^2 - y^2 = 1$ hyperbola

9. $\dfrac{x^2}{4} - \dfrac{y^2}{16} = 1$ hyperbola

10. $\dfrac{(x + 2)^2}{9} + \dfrac{(y - 4)^2}{16} = 1$ ellipse

11. $\dfrac{x^2}{25} - \dfrac{y^2}{25} = 1$ hyperbola

12. $y = 4(x + 3)^2 - 7$ parabola

For each of the following equations that has a graph, identify the corresponding graph. It may be necessary to transform the equation. See Examples 1 and 2.

13. $\dfrac{x^2}{4} = 1 - \dfrac{y^2}{9}$ ellipse

14. $\dfrac{x^2}{4} = 1 + \dfrac{y^2}{9}$ hyperbola

15. $\dfrac{x^2}{4} + \dfrac{y^2}{4} = 1$ circle

16. $\dfrac{x^2}{4} + \dfrac{y^2}{4} = -1$ no graph

17. $x^2 + 2x = x^2 + y - 6$ line

18. $y^2 - 4y = y^2 + 3 - x$ line

19. $x^2 = 25 + y^2$ hyperbola

20. $x^2 = 25 - y^2$ circle

21. $9x^2 + 36y^2 = 36$ ellipse

22. $x^2 = 4y - 8$ parabola

23. $\dfrac{(x + 3)^2}{16} + \dfrac{(y - 2)^2}{16} = 1$ circle

24. $\dfrac{(x - 4)^2}{8} + \dfrac{(y + 1)^2}{2} = 0$ point

25. $y^2 - 4y = x + 4$ parabola

26. $11 - 3x = 2y^2 - 8y$ parabola

27. $(x + 7)^2 + (y - 5)^2 + 4 = 0$ no graph

28. $4(x - 3)^2 + 3(y + 4)^2 = 0$ point

29. $3x^2 + 6x + 3y^2 - 12y = 12$ circle

30. $2x^2 - 8x + 2y^2 + 20y = 12$ circle

31. $x^2 - 6x + y = 0$ parabola

32. $x - 4y^2 - 8y = 0$ parabola

33. $4x^2 - 8x - y^2 - 6y = 6$ hyperbola

34. $x^2 + 2x = x^2 - 4y - 2$ line

35. $4x^2 - 8x + 9y^2 + 54y = -84$ ellipse

36. $3x^2 + 12x + 3y^2 = -11$ circle

37. $6x^2 - 12x + 6y^2 - 18y + 25 = 0$ no graph

38. $4x^2 - 24x + 5y^2 + 10y + 41 = 0$ point

▲ **39.** Identify the type of conic section consisting of the set of all points in the plane for which the sum of the distances from the points (5, 0) and (−5, 0) is 14.   **ellipse**

▲ **40.** Identify the type of conic section consisting of the set of all points in the plane for which the absolute value of the difference of the distances from the points (3, 0) and (−3, 0) is 2. **hyperbola**

◉ **41. (a)** Show by a sketch how a cone can be cut by a plane in exactly one point. (For this reason, a point is sometimes called a *degenerate circle* or *degenerate ellipse.*)

(b) Show by a sketch how a cone can be cut by a plane to produce exactly one straight line.(For this reason, a straight line is sometimes called a *degenerate parabola.*)

✎◉ **42.** Following the definitions in Exercise 41, what is a *degenerate hyperbola*?

Graph each of the following, and give the domain and range. See Example 3.

43. $y = \sqrt{4 - x}$
domain: $(-\infty, 4]$;
range: $[0, \infty)$

44. $y = \sqrt{16 - x}$
domain: $(-\infty, 16]$;
range: $[0, \infty)$

45. $y = \sqrt{x^2 - 9}$
domain: $(-\infty, -3] \cup [3, \infty)$;
range: $[0, \infty)$

46. $y = -\sqrt{16 - x^2}$
domain: $[-4, 4]$;
range: $[-4, 0]$

47. $\dfrac{y}{3} = -\sqrt{1 + \dfrac{x^2}{16}}$
domain: $(-\infty, \infty)$;
range: $(-\infty, -3]$

48. $\dfrac{y}{2} = \sqrt{1 - \dfrac{x^2}{25}}$
domain: $[-5, 5]$;
range: $[0, 2]$

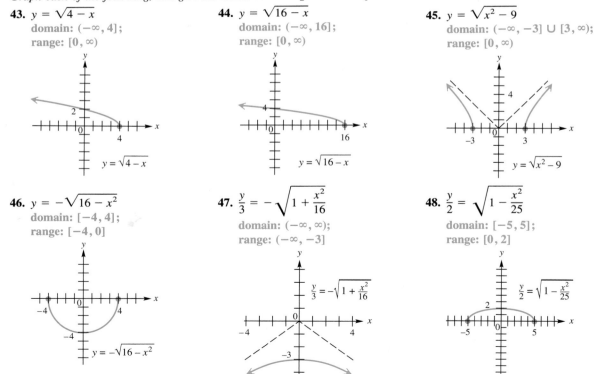

◉ CONCEPTUAL   ✎ WRITING   ▲ CHALLENGING

49. $y = \sqrt{1 - \dfrac{x^2}{64}}$

domain: $[-8, 8]$;
range: $[0, 1]$

50. $y = \sqrt{1 + \dfrac{x^2}{36}}$

domain: $(-\infty, \infty)$;
range: $[1, \infty)$

51. If you are given the graph of the relation $y = \sqrt{1 - x^2}$, how would you describe a procedure you can use to graph $y = -\sqrt{1 - x^2}$?

52. The graph of $y = \sqrt{x}$ is shown here.

Which one of the following is the graph of $y = -\sqrt{x}$?

(a)   **(b)**   **(c)**   **(d)** none of these   (a)

3.7 EXERCISES (TEXT PAGE 202) ■

Graph each of the following inequalities. See Examples 1–3.

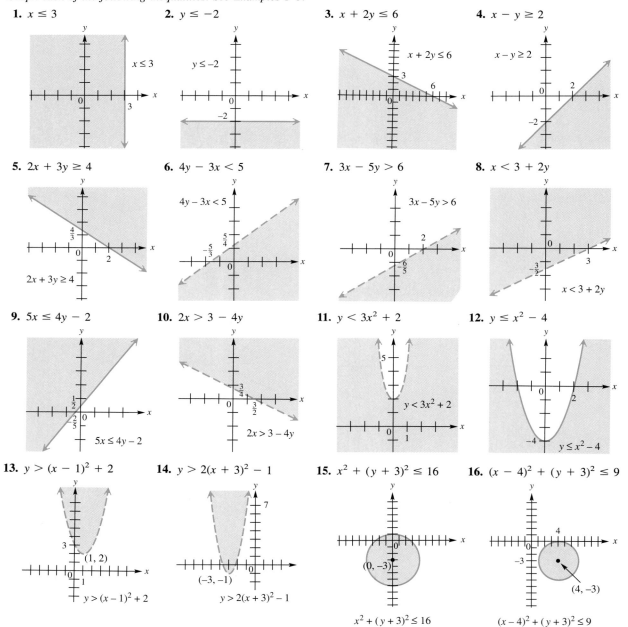

1. $x \leq 3$

2. $y \leq -2$

3. $x + 2y \leq 6$

4. $x - y \geq 2$

5. $2x + 3y \geq 4$

6. $4y - 3x < 5$

7. $3x - 5y > 6$

8. $x < 3 + 2y$

9. $5x \leq 4y - 2$

10. $2x > 3 - 4y$

11. $y < 3x^2 + 2$

12. $y \leq x^2 - 4$

13. $y > (x - 1)^2 + 2$

14. $y > 2(x + 3)^2 - 1$

15. $x^2 + (y + 3)^2 \leq 16$

16. $(x - 4)^2 + (y + 3)^2 \leq 9$

17. $4x^2 \leq 4 - y^2$ **18.** $x^2 + 9y^2 > 9$ **19.** $9x^2 - 16y^2 > 144$ **20.** $4x^2 \leq 36 + 9y^2$

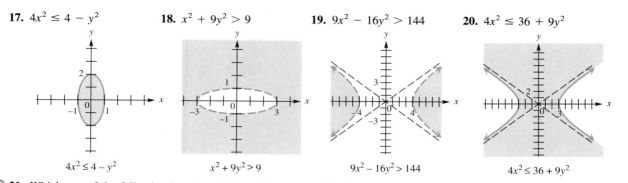

$4x^2 \leq 4 - y^2$ $x^2 + 9y^2 > 9$ $9x^2 - 16y^2 > 144$ $4x^2 \leq 36 + 9y^2$

◉ **21.** Which one of the following is a description of the graph of the inequality
$$(x - 5)^2 + (y - 2)^2 < 4?$$
  (a) the region inside a circle with center $(-5, -2)$ and radius 2
  (b) the region inside a circle with center $(5, 2)$ and radius 2
  (c) the region inside a circle with center $(-5, -2)$ and radius 4
  (d) the region outside a circle with center $(5, 2)$ and radius 4   **(b)**

✎ ◉ **22.** Without graphing, write a description of the graph of the inequality $y > 2(x - 3)^2 + 2$.

◉ **23.** Which one of the following inequalities satisfies the following description: the region outside
  an ellipse centered at the origin, with x-intercepts 4 and -4, and y-intercepts 9 and -9?
  (a) $\dfrac{x^2}{4} + \dfrac{y^2}{9} > 1$ **(b)** $\dfrac{x^2}{16} - \dfrac{y^2}{81} > 1$ **(c)** $\dfrac{x^2}{16} + \dfrac{y^2}{81} > 0$ **(d)** $\dfrac{x^2}{16} + \dfrac{y^2}{81} > 1$ **(d)**

✎ ◉ **24.** Explain how it is determined whether the boundary of an inequality is a solid line or a dashed line.

Graph each of the following inequalities. See Examples 4 and 5.

25. $x - 3y < 4$, where $x \leq 0$ **26.** $2x + y > 5$, where $y \geq 0$ **27.** $3x + 2y \geq 6$, where $y \leq 2$

$x - 3y < 4,\ x \leq 0$ $2x + y > 5,\ y \geq 0$ $3x + 2y \geq 6,\ y \leq 2$

28. $4x + 3y \leq 6$, where $x \leq -2$ **29.** $y^2 \leq x + 3$, where $y \geq 0$ **30.** $x^2 \leq 1 - y$, where $x \geq 0$

$4x + 3y \leq 6,$
$x \leq -2$

$y^2 \leq x + 3,\ y \geq 0$ $x^2 \leq 1 - y,\ x \geq 0$

31. $y^2 \leq 49 - x^2$, where $y \leq 0$

32. $x^2 > 16 - y^2$, where $x \leq 0$

33. $\dfrac{x^2}{36} < 1 - \dfrac{y^2}{121}$, where $x \geq 0$

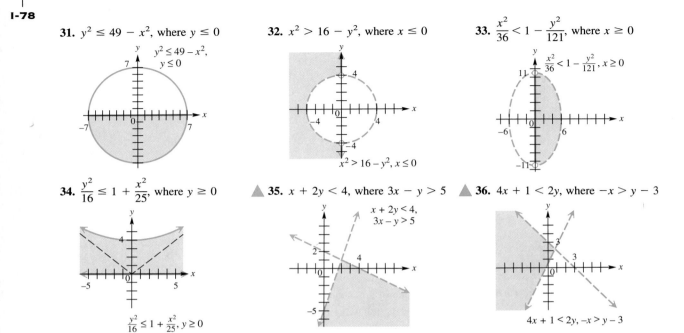

34. $\dfrac{y^2}{16} \leq 1 + \dfrac{x^2}{25}$, where $y \geq 0$

▲ **35.** $x + 2y < 4$, where $3x - y > 5$

▲ **36.** $4x + 1 < 2y$, where $-x > y - 3$

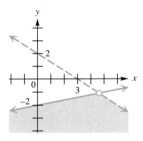

$\dfrac{y^2}{16} \leq 1 + \dfrac{x^2}{25}, y \geq 0$

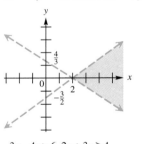

▲ **37.** $2x + 3y < 6$, where $x - 5y \geq 10$ ▲ **38.** $3x - 4y > 6$, where $2x + 3y > 4$

$2x + 3y < 6, x - 5y \geq 10$

$3x - 4y > 6, 2x + 3y > 4$

RESOURCES FOR REVIEW

Instructor's Test Manual
  Chapter Test Forms, pp. 55–84
Instructor's Solution Manual
  pp. 137–150, 558–568

Study Guide
  pp. 132–135
Student's Solution Manual
  pp. 137–150

Test Generator
  IBM Disks
  5.25″ Disk A
  3.5″ Disk A
  Mac Disk A

CHAPTER 3 REVIEW EXERCISES   (TEXT PAGE 205) ■

Give the domain and the range of each of the following relations.

  1. $\{(-3, 6), (-1, 4), (8, 5)\}$ domain: $\{-3, -1, 8\}$;   range: $\{6, 4, 5\}$

  2. $y = \sqrt{-x}$ domain: $(-\infty, 0]$;   range: $[0, \infty)$

◎ CONCEPTUAL   ✎ WRITING   ▲ CHALLENGING

Find the distance between each of the following pairs of points, and state the midpoint of each pair.

3. $P(3, -1)$, $Q(-4, 5)$ $\sqrt{85}$; $(-1/2, 2)$

4. $M(-8, 2)$, $N(3, -7)$ $\sqrt{202}$; $(-5/2, -5/2)$

5. $A(-6, 3)$, $B(-6, 8)$ 5; $(-6, 11/2)$

6. Are the points $(5, 7)$, $(3, 9)$, $(6, 8)$ the vertices of a right triangle?   yes

▲ **7.** Find all possible values of k so that $(-1, 2)$, $(-10, 5)$, and $(-4, k)$ are the vertices of a right triangle. -7; -1; 8; 23

8. Use the distance formula to determine whether the points $(-2, -5)$, $(1, 7)$, and $(3, 15)$ are collinear.   yes

Find the slope for each of the following lines that has a slope.

9. Through $(8, 7)$ and $(1/2, -2)$ $6/5$

10. Through $(2, -2)$ and $(3, -4)$ -2

11. Through $(5, 6)$ and $(5, -2)$ undefined slope

12. Through $(0, -7)$ and $(3, -7)$ 0

13. $9x - 4y = 2$ $9/4$

14. $11x + 2y = 3$ $-11/2$

15. $x - 5y = 0$ $1/5$

16. $x - 2 = 0$ undefined slope

17. $y + 6 = 0$ 0

18. $y = x$ 1

Graph each of the following. Give the domain and range.

19. $3x + 7y = 14$
domain and range: $(-\infty, \infty)$

20. $2x - 5y = 5$
domain and range: $(-\infty, \infty)$

21. $3y = x$
domain and range: $(-\infty, \infty)$

22. $y = 3$
domain: $(-\infty, \infty)$;   range: $\{3\}$

23. $x = -5$
domain: $\{-5\}$;   range: $(-\infty, \infty)$

24. $y = x$
domain and range: $(-\infty, \infty)$

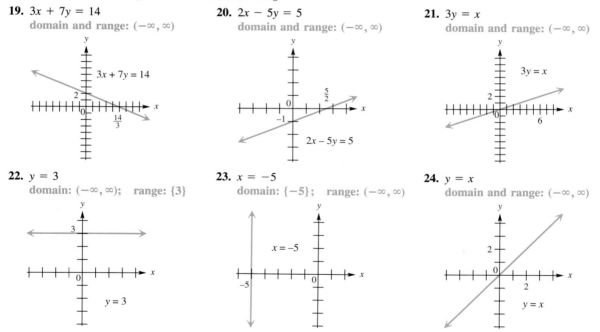

For each of the following lines, write the equation in standard form.

25. Through $(-2, 4)$ and $(1, 3)$ $x + 3y = 10$

26. Through $(-2/3, -1)$ and $(0, 4)$ $15x - 2y = -8$

27. Through $(3, -5)$ with slope -2 $2x + y = 1$

28. Through $(-4, 4)$ with slope $3/2$ $3x - 2y = -20$

29. Through $(1/5, 1/3)$ with slope $-1/2$ $15x + 30y = 13$

30. x-intercept -3, y-intercept 5 $5x - 3y = -15$

31. No x-intercept, y-intercept 3/4 $y = 3/4$

32. Through $(2, -1)$, parallel to $3x - y = 1$ $3x - y = 7$

33. Through $(0, 5)$, perpendicular to $8x + 5y = 3$ $5x - 8y = -40$

34. Through $(2, -10)$, perpendicular to a line with undefined slope $y = -10$

35. Through $(3, -5)$, parallel to $y = 4$ $y = -5$

36. Through $(-7, 4)$, perpendicular to $y = 8$ $x = -7$

Graph each line satisfying the given conditions.

37. Through $(2, -4)$, $m = 3/4$

38. Through $(0, 5)$, $m = -2/3$

39. Through $(-4, 1)$, $m = 3$

40. Through $(-3, -2)$, $m = -1$

Graph each of the following quadratic relations. Give the domain and range.

41. $y = x^2 - 4$
domain: $(-\infty, \infty)$;
range: $[-4, \infty)$

42. $y = 6 - x^2$
domain: $(-\infty, \infty)$;
range: $(-\infty, 6]$

43. $y = 3(x + 1)^2 - 5$
domain: $(-\infty, \infty)$;
range: $[-5, \infty)$

◎ CONCEPTUAL   ✎ WRITING   ▲ CHALLENGING

44. $y = -\dfrac{1}{4}(x-2)^2 + 3$

domain: $(-\infty, \infty)$;
range: $(-\infty, 3]$

45. $y = x^2 - 4x + 2$

domain: $(-\infty, \infty)$;
range: $[-2, \infty)$

46. $y = -3x^2 - 12x - 1$

domain: $(-\infty, \infty)$;
range: $(-\infty, 11]$

47. $x = y^2 - 2$
domain: $[-2, \infty)$;
range: $(-\infty, \infty)$

48. $x = (y+3)^2$
domain: $[0, \infty)$;
range: $(-\infty, \infty)$

49. $x = -(y+1)^2 + 2$
domain: $(-\infty, 2]$;
range: $(-\infty, \infty)$

50. $x = y^2 - 4y$
domain: $[-4, \infty)$;
range: $(-\infty, \infty)$

51. $x = -y^2 + 5y + 1$
domain: $(-\infty, 29/4]$;
range: $(-\infty, \infty)$

52. $x = -3y^2 + 6y - 2$
domain: $(-\infty, 1]$;
range: $(-\infty, \infty)$

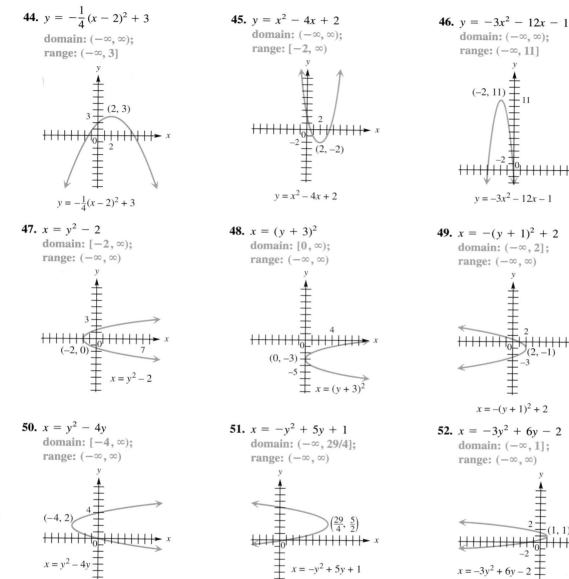

Solve each problem.

53. Find the dimensions of the rectangular region of maximum area that can be enclosed with 180 m of fencing.   **45 m on a side (a square)**

54. Find the dimensions of the rectangular region of maximum area that can be enclosed with 180 m of fencing, if no fencing is needed along one side of the region.   **45 m by 90 m**

◉ **55.** Which one of the following has a graph that is a parabola with a vertical axis and opens downward?
    (a) $y = 3x^2 - 4x + 2$ **(b)** $y = -3x^2 - 4x + 2$
    (c) $x = 3y^2 - 4y + 2$ **(d)** $x = -3y^2 - 4y + 2$ (b)

◉ **56.** If the graph of $y = ax^2 + bx + c$ (for some particular values of a, b, and c) has its vertex in the third quadrant and $a < 0$, how many real solutions does the equation $ax^2 + bx + c = 0$ have?   none

Find equations for each circle satisfying the given conditions.

57. Center $(-2, 3)$, radius 5 $(x + 2)^2 + (y - 3)^2 = 25$

58. Center $(\sqrt{5}, -\sqrt{7})$, radius $\sqrt{3}$ $(x - \sqrt{5})^2 + (y + \sqrt{7})^2 = 3$

59. Center $(-8, 1)$, passing through $(0, 16)$ $(x + 8)^2 + (y - 1)^2 = 289$

60. Center $(3, -6)$, tangent to the x-axis $(x - 3)^2 + (y + 6)^2 = 36$

Find the center and radius of each of the following circles.

61. $x^2 - 4x + y^2 + 6y + 12 = 0$ $(2, -3)$; $r = 1$ **62.** $x^2 - 6x + y^2 - 10y + 30 = 0$ $(3, 5)$; $r = 2$

63. $2x^2 + 14x + 2y^2 + 6y + 2 = 0$ $\left(-\dfrac{7}{2}, -\dfrac{3}{2}\right)$; $r = \dfrac{3\sqrt{6}}{2}$

64. $3x^2 + 33x + 3y^2 - 15y = 0$ $\left(-\dfrac{11}{2}, \dfrac{5}{2}\right)$; $r = \dfrac{\sqrt{146}}{2}$

▲ **65.** Find all possible values of x so that the distance between $(x, -9)$ and $(3, -5)$ is 6.
     $3 + 2\sqrt{5}$; $3 - 2\sqrt{5}$

▲ **66.** Find all points (x, y) with $x = 6$ so that (x, y) is 4 units from $(1, 3)$.
     No such points exist.

▲ **67.** Find all points (x, y) with $x + y = 0$ so that (x, y) is 6 units from $(-2, 3)$.
     $\left(\dfrac{-5 + \sqrt{71}}{2}, \dfrac{5 - \sqrt{71}}{2}\right)$; $\left(\dfrac{-5 - \sqrt{71}}{2}, \dfrac{5 + \sqrt{71}}{2}\right)$

✎◉ **68.** Describe the graph of $(x - 4)^2 + (y + 5)^2 = 0$.

Decide whether the relations defined below have graphs that are symmetric with respect to the x-axis, the y-axis, or the origin.

69. $3y^2 - 5x^2 = 15$ yes;   yes;   yes   **70.** $x + y^2 = 8$ yes;   no;   no

71. $y^3 = x + 1$ no;   no;   no   **72.** $x^2 = y^3$ no;   yes;   no

73. $|y| = -x$ yes;   no;   no   **74.** $|x + 2| = |y - 3|$ no;   no;   no

75. $|x| = |y|$ yes;   yes;   yes   **76.** $xy = 8$ no;   no;   yes

Graph each of the following relations. Give the domain and range.

77. $\dfrac{x^2}{25} + \dfrac{y^2}{4} = 1$ domain: $[-5, 5]$;   range: $[-2, 2]$ **78.** $\dfrac{x^2}{3} + \dfrac{y^2}{16} = 1$ domain: $[-\sqrt{3}, \sqrt{3}]$;   range: $[-4, 4]$

$\dfrac{x^2}{25} + \dfrac{y^2}{4} = 1$

$\dfrac{x^2}{3} + \dfrac{y^2}{16} = 1$

◉ CONCEPTUAL   ✎ WRITING   ▲ CHALLENGING

79. $\dfrac{x^2}{4} - \dfrac{y^2}{9} = 1$ domain: $(-\infty, -2] \cup [2, \infty)$;
range: $(-\infty, \infty)$

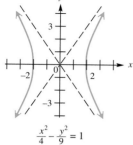

80. $\dfrac{y^2}{100} - \dfrac{x^2}{25} = 1$ domain: $(-\infty, \infty)$;
range: $(-\infty, -10] \cup [10, \infty)$

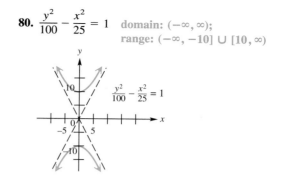

81. $\dfrac{(x-2)^2}{9} + \dfrac{(y+3)^2}{4} = 1$ domain: $[-1, 5]$;
range: $[-5, -1]$

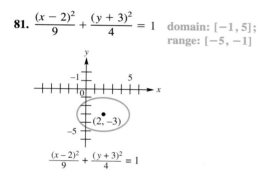

82. $\dfrac{(x+1)^2}{16} - \dfrac{(y-2)^2}{4} = 1$ domain: $(-\infty, 5] \cup [3, \infty)$;
range: $(-\infty, \infty)$

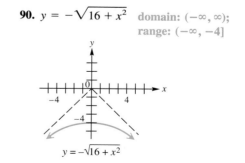

Identify the type of graph, if any, of each of the following.

83. $x^2 = 64 - y^2$ circle

84. $x^2 = 4y - 15$ parabola

85. $\dfrac{(x+3)^2}{9} + \dfrac{(y-2)^2}{9} = 1$ circle

86. $\dfrac{(x-4)^2}{9} - \dfrac{(y+1)^2}{16} = 1$ hyperbola

87. $y^2 - 3y = x + 4$ parabola

88. $(x+7)^2 + (y-5)^2 + 16 = 0$ no graph

Graph each of the following. Give the domain and range.

89. $y = \sqrt{100 - x^2}$ domain: $[-10, 10]$;
range: $[0, 10]$

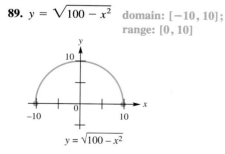

90. $y = -\sqrt{16 + x^2}$ domain: $(-\infty, \infty)$;
range: $(-\infty, -4]$

91. $y = -\sqrt{4 - x}$ domain: $(-\infty, 4]$;
range: $(-\infty, 0]$

$y = -\sqrt{4 - x}$

92. $y = \sqrt{1 + x}$ domain: $[-1, \infty)$;
range: $[0, \infty)$

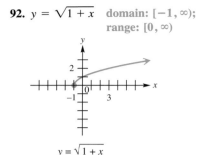

$y = \sqrt{1 + x}$

Graph each inequality.

93. $5x - y \le 20$

$5x - y \le 20$

94. $3x - 5y > 10$

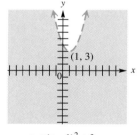

$3x - 5y > 10$

95. $y \le (x - 4)^2$

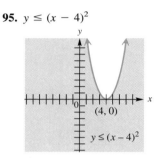

$y \le (x - 4)^2$

96. $y < (x - 1)^2 + 3$

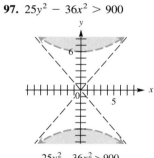

$y < (x - 1)^2 + 3$

97. $25y^2 - 36x^2 > 900$

$25y^2 - 36x^2 > 900$

98. $(x - 2)^2 + (y + 3)^2 \ge 9$

$(x - 2)^2 + (y + 3)^2 \ge 9$

99. $2x - y \ge 4$, where $y \ge -2$

$2x - y \ge 4$, $y \ge -2$

100. $y \le x^2$, where $x \ge 0$

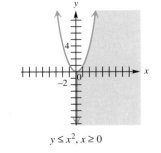

$y \le x^2$, $x \ge 0$

◎ CONCEPTUAL   ✍ WRITING   ▲ CHALLENGING

CHAPTER 3 TEST   (TEXT PAGE 207) ■ ──────────────────

1. For the points $(-2, 1)$ and $(3, 4)$:
   (a) Find the distance between them. $\sqrt{34}$
   (b) Find the midpoint of the line segment joining them.   (1/2, 5/2)

Find the slope of each of the following lines.

2. Through $(-3, 4)$ and $(5, -6)$ $-5/4$

3. $5x - 4y = 6$ 5/4

4. The line shown in the figure has a positive slope and a negative y-intercept. Explain why this is true.

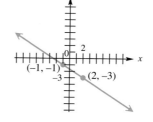

For each of the following lines, write the equation in standard form.

5. Through $(-2, 3)$ and $(6, -1)$ $x + 2y = 4$

6. Through $(-6, 2)$, perpendicular to $x = 4$ $y = 2$

7. Through $(-6, 2)$, parallel to $2x - y = 8$ $2x - y = -14$

8. Graph the line with the equation $4x - 5y = 10$. Give the domain and range.

   domain and range: $(-\infty, \infty)$

9. Graph the line through $(2, -3)$ with slope $-2/3$.

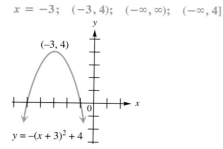

10. Graph $y = -(x + 3)^2 + 4$. Give the axis, vertex, domain, and range.

   $x = -3$; $(-3, 4)$; $(-\infty, \infty)$; $(-\infty, 4]$

11. Give the vertex and axis of the graph of $y = 3x^2 - 12x + 9$. $(2, -3)$; $x = 2$

12. Graph $x = -(y + 2)^2 - 4$. Give the axis, vertex, domain, and range.

$y = -2$; $(-4, -2)$; $(-\infty, -4]$; $(-\infty, \infty)$

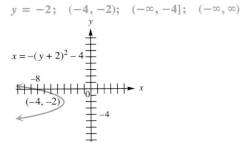

13. A developer has 80 ft of fencing. He wants to put a fence around a rectangular plot of land adjacent to a river. Use a quadratic relation to find the maximum area he can enclose. (Note: the side next to the river will not be fenced.)   800 sq ft

14. Find the equation of a circle centered at $(5, -1)$ with radius 4. Graph the circle, and give the domain and range. $(x - 5)^2 + (y + 1)^2 = 16$; $[1, 9]$; $[-5, 3]$

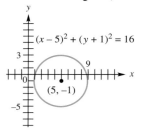

15. Find the center and radius of the circle with equation

$$x^2 + y^2 + 4x - 6y = -4.$$ center: $(-2, 3)$;   radius: 3

16. Decide whether the equation $y = 2x^3 - x$ has a graph that is symmetric with respect to the x-axis, y-axis, or the origin.   origin only

Graph each of the following relations. Give the domain and range.

17. $\dfrac{(x + 3)^2}{4} + \dfrac{(y - 2)^2}{9} = 1$ domain: $[-5, -1]$; range: $[-1, 5]$

18. $9y^2 = 25x^2 + 225$ domain: $(-\infty, \infty)$; range: $(-\infty, -5] \cup [5, \infty)$

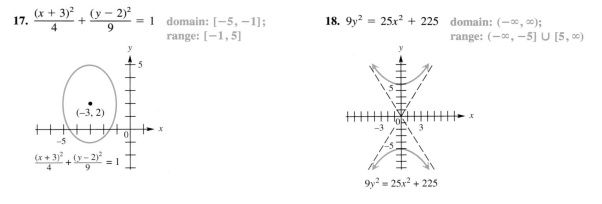

◎ CONCEPTUAL   ✎ WRITING   ▲ CHALLENGING

19. $y = \sqrt{25 - x^2}$ domain: $[-5, 5]$; range: $[0, 5]$

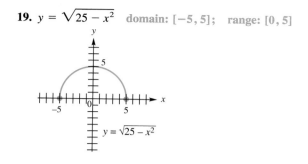

20. Graph the inequality $(x - 4)^2 + (y + 3)^2 > 9$.

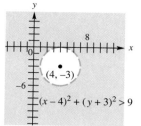

■ CHAPTER 4 FUNCTIONS

4.1 RESOURCES

Instructor's Test Manual
Test Items, p. 326–328
Instructor's Solution Manual
pp. 151–154, 569–572
Study Guide
pp. 139–144
Student's Solution Manual
pp. 151–154

Test Generator
IBM Disks
5.25" Disk A
3.5" Disk A
Mac Disk A
Tutorials
IBM Disks
5.25" Disk D
3.5" Disk B
Mac Disk B
Apple II Module 1, Lessons 1–5

HarperCollins Videotape 4
Transparency T-19

4.1 EXERCISES (TEXT PAGE 216) ■ ────────────────

◎ *For each of the following, find the indicated function values:* (***a***) $f(-2)$, (***b***) $f(0)$, (***c***) $f(1)$, *and*
(***d***) $f(4)$.

1.

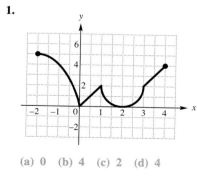

(a) 0 (b) 4 (c) 2 (d) 4

2.

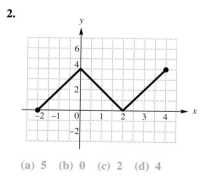

(a) 5 (b) 0 (c) 2 (d) 4

3.

(a) −3   (b) −2   (c) 0   (d) 2

4.

(a) 3   (b) 3   (c) 3   (d) 3

Let $f(x) = 3x − 1$ and $g(x) = x^2$. Find each of the following function values. See Example 2.

5. $f(−1)$ −4

6. $f(−3)$ −10

7. $f(4)$ 11

8. $g(2)$ 4

9. $f(a)$ $3a − 1$

10. $g(b)$ b^2

11. $f(1) + g(1)$ 3

12. $f(−2) − g(−2)$ −11

13. $f(3) \cdot g(3)$ 72

14. $\dfrac{g(−1)}{f(−1)}$ −1/4

15. $f(−2m)$ $−6m − 1$

16. $f(−11p)$ $−33p − 1$

17. $f(5a − 2)$ $15a − 7$

18. $f(3 + 2k)$ $8 + 6k$

◉ **19.** Complete the following sentence. If the ordered pair $(2, 5)$ belongs to function g, then $g(\rule{1cm}{0.4pt}) = \rule{1cm}{0.4pt}$.   2, 5

Give the domain and range of each function defined as follows. See Example 4.

20.

$[−5, 4]$; $[−2, 6]$

21.

$[−6, ∞)$; $[0, ∞)$

22.

$(−∞, ∞)$; $(−∞, 12]$

23.

$(−∞, ∞)$; $(−∞, ∞)$

24.

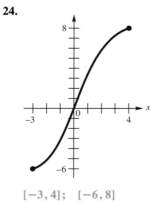

$[−3, 4]$; $[−6, 8]$

25.

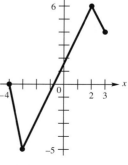

$[−4, 3]$; $[−5, 6]$

◉ **CONCEPTUAL**   ☑ **WRITING**   ▲ **CHALLENGING**

26.

$(-\infty, \infty)$; $(-\infty, 3]$

27.

$(-\infty, 0) \cup (0, \infty)$;
$(-\infty, 0) \cup (0, \infty)$

28.

$(-\infty, 0) \cup (0, \infty)$; $(-\infty, 0)$

29. $f(x) = 2x - 1$ $(-\infty, \infty)$; $(-\infty, \infty)$

30. $g(x) = 3x + 5$ $(-\infty, \infty)$; $(-\infty, \infty)$

31. $g(x) = x^4$ $(-\infty, \infty)$; $[0, \infty)$

32. $h(x) = (x - 2)^2$ $(-\infty, \infty)$; $[0, \infty)$

33. $f(x) = \sqrt{8 + x}$ $[-8, \infty)$; $[0, \infty)$

34. $f(x) = -\sqrt{x + 6}$ $[-6, \infty)$; $(-\infty, 0]$

35. $h(x) = \sqrt{16 - x^2}$ $[-4, 4]$; $[0, 4]$

36. $m(x) = \sqrt{x^2 - 25}$ $(-\infty, -5] \cup [5, \infty)$; $[0, \infty)$

37. $F(x) = \sqrt[3]{x - 1}$ $(-\infty, \infty)$; $(-\infty, \infty)$

38. $G(x) = -\sqrt[3]{1 - x^2}$ $(-\infty, \infty)$; $[-1, \infty)$

39. $f(x) = \dfrac{10}{3 - x}$ $(-\infty, 3) \cup (3, \infty)$; $(-\infty, 0) \cup (0, \infty)$

40. $h(x) = \dfrac{2}{x - 1}$ $(-\infty, 1) \cup (1, \infty)$; $(-\infty, 0) \cup (0, \infty)$

41. $g(x) = \dfrac{1}{x^2 - 4}$ $(-\infty, -2) \cup (-2, 2) \cup (2, \infty)$; $(-\infty, 0) \cup (0, \infty)$

42. $k(x) = \dfrac{5}{9 - x^2}$ $(-\infty, -3) \cup (-3, 3) \cup (3, \infty)$; $(-\infty, 0) \cup (0, \infty)$

For each of the functions defined as follows, find **(a)** $f(x + h)$, **(b)** $f(x + h) - f(x)$, *and*
(c) $\dfrac{f(x + h) - f(x)}{h}$. *See Example 3.*

43. $f(x) = 6x + 2$ (a) $6x + 6h + 2$ (b) $6h$ (c) 6 **44.** $f(x) = 4x + 11$ (a) $4x + 4h + 11$ (b) $4h$ (c) 4

45. $f(x) = -2x + 5$ (a) $-2x - 2h + 5$ (b) $-2h$ (c) -2

46. $f(x) = 8 - 3x$ (a) $8 - 3x - 3h$ (b) $-3h$ (c) -3

47. $f(x) = 1 - x^2$ (a) $1 - x^2 - 2xh - h^2$ (b) $-2xh - h^2$ (c) $-2x - h$

48. $f(x) = x^2 - 4$ (a) $x^2 + 2xh + h^2 - 4$ (b) $2xh + h^2$ (c) $2x + h$

49. $f(x) = 8 - 3x^2$ (a) $8 - 3x^2 - 6xh - 3h^2$ (b) $-6xh - 3h^2$ (c) $-6x - 3h$

▲ *Let $f(x) = 2^x$ for all rational numbers x. Find each of the following function values. Assume that
m and r represent rational numbers.*

50. $f(2)$ 4 **51.** $f(4)$ 16 **52.** $f(-2)$ $1/4$ **53.** $f(-4)$ $1/16$ **54.** $f(m)$ 2^m

55. $f(-5r)$ 2^{-5r} or $1/2^{5r}$ **56.** $f(1/2)$ $2^{1/2}$ or $\sqrt{2}$ **57.** $f(1/4)$ $2^{1/4}$ or $\sqrt[4]{2}$

◉ **58.** Give an example of a function from everyday life. (*Hint:* Fill in the blanks: _____ depends on _____, so _____ is a function of _____.)

Many answers are possible. One example is given here: getting good grades;   the time
spent studying;   getting good grades;   the time spent studying

A firm will break even (no profit and no loss) as long as revenue just equals cost. The value of x (the number of items produced and sold) where $C(x) = R(x)$ is called the break-even point. Assume that each of the following can be expressed as a linear cost function. Find (a) the cost function, (b) the revenue function, and (c) the profit function. (d) Find the break-even point; then decide whether the product should be produced. See Example 5.

	Fixed Cost	Variable Cost	Price of Item	
59.	$500	$10	$35	No more than 18 units can be sold.
60.	$180	$11	$20	No more than 30 units can be sold.
61.	$2700	$150	$280	No more than 25 units can be sold.
62.	$1650	$400	$305	All units produced can be sold.

59. (a) $C(x) = 10x + 500$ (b) $R(x) = 35x$ (c) $P(x) = 25x - 500$
(d) 20 units;   do not produce
60. (a) $C(x) = 11x + 180$ (b) $R(x) = 20x$ (c) $P(x) = 9x - 180$
(d) 20 units;   produce
61. (a) $C(x) = 150x + 2700$ (b) $R(x) = 280x$ (c) $P(x) = 130x - 2700$
(d) 20.77 or 21 units;   produce
62. (a) $C(x) = 400x + 1650$ (b) $R(x) = 305x$ (c) $P(x) = -95x - 1650$
(d) $R(x) \leq C(x)$ for all positive x;   don't produce, impossible to make a profit.

63. The manager of a small company that produces roof tile has determined that his total cost in dollars, $C(x)$, of producing x units of tile is given by $C(x) = 200x + 1000$, while the revenue in dollars, $R(x)$, from the sale of x units of tile is given by $R(x) = 240x$.
(a) Find the break-even point.   25 units
(b) Graph functions R and C on the same axes. Identify the break-even point. Identify any regions of profit or loss.

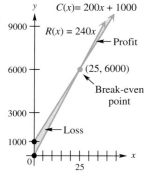

(c) What is the cost/revenue at the break-even point?
$6000

◉ 64. Suppose the manager of the company in Exercise 63 finds that he has miscalculated his variable cost, and that it is actually $220 per unit, instead of $200. How does this affect the break-even point? Is he better off or not?

The break-even point is 50 units instead of 25 units. He is not better off because he must sell twice as many units before beginning to show a profit.

4.2 RESOURCES

Instructor's Test Manual
Test Items, p. 328–329
Instructor's Solution Manual
pp. 155–159, 572–577
Study Guide
pp. 144–147
Student's Solution Manual
pp. 155–159

Test Generator
IBM Disks
5.25" Disk A
3.5" Disk A
Mac Disk A
Tutorials
IBM Disks
5.25" Disk D
3.5" Disk B
Mac Disk B
Apple II Module 1, Lessons 1–5

HarperCollins Videotape 5
"College Algebra: In Simplest Terms" Program 15

4.2 EXERCISES (TEXT PAGE 222) ■

Express each of the following statements as an equation.

1. *a* varies directly as *b*. $a = kb$

2. *m* is proportional to *n*. $m = kn$

3. *x* is inversely proportional to *y*. $x = k/y$

4. *p* varies inversely as *y*. $p = k/y$

5. *r* varies jointly as *s* and *t*. $r = kst$

6. *R* is proportional to *m* and *p*. $R = kmp$

7. *w* is proportional to x^2 and inversely proportional to *y*. $w = kx^2/y$

8. *c* varies directly as *d* and inversely as f^2 and *g*. $c = (kd)/(f^2g)$

Solve each of the following problems. See Examples 1–4.

9. If *m* varies directly as *x* and *y*, and *m* = 10 when *x* = 4 and *y* = 7, find *m* when *x* = 11 and *y* = 8. 220/7

10. Suppose *m* varies directly as *z* and *p*. If *m* = 10 when *z* = 3 and *p* = 5, find *m* when *z* = 5 and *p* = 7. 70/3

11. Suppose *r* varies directly as the square of *m*, and inversely as *s*. If *r* = 12 when *m* = 6 and *s* = 4, find *r* when *m* = 4 and *s* = 10. 32/15

12. Suppose *p* varies directly as the square of *z*, and inversely as *r*. If *p* = 32/5 when *z* = 4 and *r* = 10, find *p* when *z* = 2 and *r* = 16. 1

13. Let *a* be proportional to *m* and n^2, and inversely proportional to y^3. If *a* = 9 when *m* = 4, *n* = 9, and *y* = 3, find *a* if *m* = 6, *n* = 2, and *y* = 5. 18/125

14. If *y* varies directly as *x*, and inversely as m^2 and r^2, and *y* = 5/3 when *x* = 1, *m* = 2, and *r* = 3, find *y* if *x* = 3, *m* = 1, and *r* = 8. 45/16

15. For *k* > 0, if *y* varies directly as *x*, when *x* increases, *y* _____, and when *x* decreases, *y* _____. increases, decreases

16. For *k* > 0, if *y* varies inversely as *x*, when *x* increases, *y* _____, and when *x* decreases, *y* _____. decreases, increases

◉ *In Exercises 17–22, tell whether each equation represents direct, inverse, joint, or combined variation.*

17. $y = \dfrac{4}{x}$ inverse

18. $y = \dfrac{x}{zw}$ combined

19. $y = 6xz^2$ joint

20. $y = 3x$ direct

21. $y = \dfrac{x^2}{zw^3}$ combined

22. $y = \dfrac{10}{x}$ inverse

▦ *Solve each of the following problems. See Examples 1–4. A calculator will be helpful for many of these problems.*

23. Hooke's law for an elastic spring states that the distance a spring stretches varies directly as the force applied. If a force of 15 lb stretches a certain spring 8 in, how much will a force of 30 lb stretch the spring?

16 in

24. In electric current flow, it is found that the resistance (measured in units called ohms) offered by a fixed length of wire of a given material varies inversely as the square of the diameter of the wire. If a wire .01 inch in diameter has a resistance of .4 ohm, what is the resistance of a wire of the same length and material with a diameter of .03 inch?   .0444 ohm

25. The illumination produced by a light source varies inversely as the square of the distance from the source. The illumination of a light source at 5 m is 70 candela. What is the illumination 12 m from the source?

875/72 candela

26. The pressure exerted by a certain liquid at a given point is proportional to the depth of the point below the surface of the liquid. If the pressure 20 m below the surface is 70 kg per cm², what pressure is exerted 40 m below the surface?   140 kg per cm²

27. The distance that a person can see to the horizon from a point above the surface of the earth varies directly as the square root of the height of the point above the earth. A person on a hill 121 m high can see for 15 km to the horizon. How far can a person see to the horizon from a hill 900 m high?   450/11 km

28. Simple interest varies jointly as principal and time. If $1000 left at interest for 2 yr earned $110, find the amount of interest earned by $5000 for 5 yr. $1375

29. The volume of a right circular cylinder is jointly proportional to the square of the radius of the circular base and to the height. If the volume is 300 cu cm when the height is 10.62 cm and the radius is 3 cm, find the volume for a cylinder with a radius of 4 cm and a height of 15.92 cm.   799.5 cm³

30. The roof of a new sports arena rests on round concrete pillars. The maximum load a cylindrical column of circular cross section can hold varies directly as the fourth power of the diameter and inversely as the square of the height. The arena has 9 m tall columns that are 1 m in diameter and will support a load of 8 metric tons. How many metric tons will be supported by a column 12 m high and 2/3 m in diameter?

8/9 metric ton

31. The sports arena in Exercise 30 requires a beam 16 m long, 24 cm wide, and 8 cm high. The maximum load of a horizontal beam that is supported at both ends varies directly as the width and square of the height and inversely as the length between supports. If a beam of the same material 8 m long, 12 cm wide, and 15 cm high can support a maximum of 400 kg, what is the maximum load the beam in the arena will support?

1024/9 kg

32. The period of a pendulum varies directly as the square root of the length of the pendulum and inversely as the square root of the acceleration due to gravity. Find the period when the length is 121 cm and the acceleration due to gravity is 980 cm per sec squared, if the period is 6π sec when the length is 289 cm and the acceleration due to gravity is 980 cm per sec squared.

66π/17 sec

33. The force needed to keep a car from skidding on a curve varies inversely as the radius of the curve and jointly as the weight of the car and the square of the speed. It takes 3000 lb of force to keep a 2000-lb car from skidding on a curve of radius 500 ft at 30 mph. What force is needed to keep the same car from skidding on a curve of radius 800 ft at 60 mph?

7500 lb

34. The pressure on a point in a liquid is directly proportional to the distance from the surface to the point. In a certain liquid the pressure at a depth of 4 m is 60 kg per m². Find the pressure at a depth of 10 m.

150 kg per m²

◉ CONCEPTUAL   ▨ WRITING   ▲ CHALLENGING

35. The volume V of a gas varies directly as the temperature T and inversely as the pressure P. If V is 10 when T is 280 and P is 6, find V if T is 300 and P is 10.
45/7

36. Under certain conditions, the length of time that it takes for fruit to ripen during the growing season varies inversely as the average maximum temperature during the season. If it takes 25 days for fruit to ripen with an average maximum temperature of 80°, find the number of days it would take at 75°. 26 2/3 days

37. The number of long-distance phone calls between two cities in a certain time period varies directly as the populations p_1 and p_2 of the cities, and inversely as the distance between them. If 10,000 calls are made between two cities 500 mi apart, having populations of 50,000 and 125,000, find the number of calls between two cities 800 mi apart having populations of 20,000 and 80,000. 1600 calls

38. The horsepower needed to run a boat through water varies directly as the cube of the speed. If 80 horsepower are needed to go 15 km/hr in a certain boat, how many horsepower would be needed to go 30 km/hr?
640

39. According to Poiseuille's law, the resistance to flow of a blood vessel, R, is directly proportional to the length, l, and inversely proportional to the fourth power of the radius, r. If $R = 25$ when $l = 12$ and $r = .2$, find R as r increases to .3, while l is unchanged. 4.94

40. The Stefan-Boltzmann law says that the radiation of heat R from an object is directly proportional to the fourth power of the Kelvin temperature of the object. For a certain object, $R = 213.73$ at room temperature (293° Kelvin). Find R if the temperature increases to 335° Kelvin. 365.24

41. Suppose a nuclear bomb is detonated at a certain site. The effects of the bomb will be felt over a distance from the point of detonation that is directly proportional to the cube root of the yield of the bomb. Suppose a 100-kiloton bomb has certain effects to a radius of 3 km from the point of detonation. Find the distance that the effects would be felt for a 1500-kiloton bomb. 7.4 km

42. The maximum speed possible on a length of railroad track is directly proportional to the cube root of the amount of money spent on maintaining the track. Suppose that a maximum speed of 25 km/hr is possible on a stretch of track for which $450,000 was spent on maintenance. Find the maximum speed if the amount spent on maintenance is increased to $1,750,000.
39.3 km/hour

43. A measure of malnutrition, called the *pelidisi*, varies directly as the cube root of a person's weight in grams and inversely as the person's sitting height in centimeters. A person with a pelidisi below 100 is considered to be undernourished, while a pelidisi greater than 100 indicates overfeeding. A person who weighs 48,820 g with a sitting height of 78.7 cm has a pelidisi of 100. Find the pelidisi (to the nearest whole number) of a person whose weight is 54,430 g and whose sitting height is 88.9 cm. Is this individual undernourished or overfed? 92, undernourished

44. The cost of a pizza varies directly as the square of its radius. If a pizza with a radius of 15 in costs $7, find the cost of a pizza having a radius of 12.5 in. (You might want to do some research at a nearby pizza establishment and see if this assumption is reasonable.)
$4.86

4.3 RESOURCES

Instructor's Test Manual
Test Items, p. 330
Instructor's Solution Manual
pp. 159–164, 577–583
Study Guide
pp. 147–150
Student's Solution Manual
pp. 159–164

Test Generator
IBM Disks
5.25" Disk A
3.5" Disk A
Mac Disk A
Tutorials
IBM Disks
5.25" Disk D
3.5" Disk B
Mac Disk B
Apple II Module 1, Lessons 1–5

HarperCollins Videotape 5
Transparency T-20

4.3 EXERCISES   (TEXT PAGE 229) ■

For each pair of functions defined as follows, find (a) f + g, (b) f − g, (c) fg, and (d) f/g. Give the domain of each. See Example 2.

1. $f(x) = 4x - 1$, $g(x) = 6x + 3$

(a) $10x + 2$ (b) $-2x - 4$ (c) $24x^2 + 6x - 3$ (d) $\dfrac{4x - 1}{6x + 3}$

(e) All domains are $(-\infty, \infty)$, except for f/g which is $(-\infty, -1/2) \cup (-1/2, \infty)$.

2. $f(x) = 9 - 2x$, $g(x) = -5x + 2$

(a) $-7x + 11$ (b) $7 + 3x$ (c) $10x^2 - 49x + 18$ (d) $\dfrac{9 - 2x}{-5x + 2}$

(e) All domains are $(-\infty, \infty)$, except for f/g, which is $(-\infty, 2/5) \cup (2/5, \infty)$.

3. $f(x) = 3x^2 - 2x$, $g(x) = x^2 - 2x + 1$

(a) $4x^2 - 4x + 1$ (b) $2x^2 - 1$ (c) $(3x^2 - 2x)(x^2 - 2x + 1)$ (d) $\dfrac{3x^2 - 2x}{x^2 - 2x + 1}$

(e) All domains are $(-\infty, \infty)$, except for f/g which is $(-\infty, 1) \cup (1, \infty)$.

4. $f(x) = 6x^2 - 11x$, $g(x) = x^2 - 4x - 5$

(a) $7x^2 - 15x - 5$ (b) $5x^2 - 7x + 5$ (c) $(6x^2 - 11x)(x^2 - 4x - 5)$ (d) $\dfrac{6x^2 - 11x}{x^2 - 4x - 5}$

(e) All domains are $(-\infty, \infty)$, except for f/g, which is $(-\infty, -1) \cup (-1, 5) \cup (5, \infty)$.

5. $f(x) = \sqrt{2x + 5}$, $g(x) = \sqrt{4x - 9}$

(a) $\sqrt{2x + 5} + \sqrt{4x - 9}$ (b) $\sqrt{2x + 5} - \sqrt{4x - 9}$ (c) $\sqrt{(2x + 5)(4x - 9)}$

(d) $\sqrt{\dfrac{2x + 5}{4x - 9}}$ (e) All domains are $[9/4, \infty)$, except for f/g which is $(9/4, \infty)$.

6. $f(x) = \sqrt{11x - 3}$, $g(x) = \sqrt{2x - 15}$

(a) $\sqrt{11x - 3} + \sqrt{2x - 15}$ (b) $\sqrt{11x - 3} - \sqrt{2x - 15}$ (c) $\sqrt{(11x - 3)(2x - 15)}$

(d) $\sqrt{\dfrac{11x - 3}{2x - 15}}$ (e) All domains are $[15/2, \infty)$, except for f/g, which is $(15/2, \infty)$.

Let $f(x) = 4x^2 - 2x$ and let $g(x) = 8x + 1$. Find each of the following. See Examples 1–3.

7. $(f + g)(3)$ 55

8. $(f + g)(-5)$ 71

9. $(fg)(4)$ 1848

10. $(fg)(-3)$ −966

11. $\left(\dfrac{f}{g}\right)(-1)$ −6/7

12. $\left(\dfrac{f}{g}\right)(4)$ 56/33

13. $(f - g)(m)$ $4m^2 - 10m - 1$

14. $(f - g)(2k)$ $16k^2 - 20k - 1$

15. $(f \circ g)(2)$ 1122

16. $(f \circ g)(-5)$ 6162

17. $(g \circ f)(2)$ 97

18. $(g \circ f)(-5)$ 881

19. $(f \circ g)(k)$ $256k^2 + 48k + 2$

20. $(g \circ f)(5z)$ $800z^2 - 80z + 1$

Find $f \circ g$ and $g \circ f$ and their domains for each pair of functions defined as follows. See Examples 3–5.

21. $f(x) = 8x + 12$, $g(x) = 3x - 1$ $24x + 4$; $24x + 35$;   both domains are $(-\infty, \infty)$

22. $f(x) = -6x + 9$, $g(x) = 5x + 7$ $-30x - 33$; $-30x + 52$; both domains are $(-\infty, \infty)$

23. $f(x) = -x^3 + 2$, $g(x) = 4x$ $-64x^3 + 2$; $-4x^3 + 8$;   both domains are $(-\infty, \infty)$

24. $f(x) = 2x$, $g(x) = 6x^2 - x^3$ $12x^2 - 2x^3$; $24x^2 - 8x^3$;   both domains are $(-\infty, \infty)$

○ CONCEPTUAL   ✎ WRITING   ▲ CHALLENGING

25. $f(x) = \dfrac{1}{x}$, $g(x) = x^2$ $1/x^2$; $1/x^2$; both domains are $(-\infty, 0) \cup (0, \infty)$

26. $f(x) = \dfrac{2}{x^4}$, $g(x) = 2 - x$

$\dfrac{2}{(2-x)^4}$; $2 - \dfrac{2}{x^4}$; domain of $f \circ g$ $(-\infty, 2) \cup (2, \infty)$; domain of $g \circ f$ $(-\infty, 0) \cup (0, \infty)$

27. $f(x) = \sqrt{x + 2}$, $g(x) = 8x - 6$

$2\sqrt{2x - 1}$; $8\sqrt{x + 2} - 6$; domain of $f \circ g$ $[1/2, \infty)$; domain of $g \circ f$ $[-2, \infty)$

28. $f(x) = 9x - 11$, $g(x) = 2\sqrt{x + 2}$

$18\sqrt{x + 2} - 11$; $6\sqrt{x - 1}$; domain of $f \circ g$ $[-2, \infty)$; domain of $g \circ f$ $[1, \infty)$

29. $f(x) = \dfrac{1}{x - 5}$, $g(x) = \dfrac{2}{x}$

$\dfrac{x}{2 - 5x}$; $2(x - 5)$; domain of $f \circ g$ $(-\infty, 0) \cup \left(0, \dfrac{2}{5}\right) \cup \left(\dfrac{2}{5}, \infty\right)$;

domain of $g \circ f$ $(-\infty, 5) \cup (5, \infty)$

30. $f(x) = \dfrac{8}{x - 6}$, $g(x) = \dfrac{4}{3x}$

$\dfrac{12x}{2 - 9x}$; $\dfrac{x - 6}{6}$; domain of $f \circ g$ $(-\infty, 0) \cup \left(0, \dfrac{2}{9}\right) \cup \left(\dfrac{2}{9}, \infty\right)$;

domain of $g \circ f$ $(-\infty, 6) \cup (6, \infty)$

31. $f(x) = \sqrt{x + 1}$, $g(x) = -\dfrac{1}{x}$

$\sqrt{\dfrac{x - 1}{x}}$; $-\dfrac{1}{\sqrt{x + 1}}$; domain of $f \circ g$ $(-\infty, 0) \cup [1, \infty)$; domain of $g \circ f$ $(-1, \infty)$

32. $f(x) = \dfrac{8}{x}$, $g(x) = \sqrt{3 - x}$

$\dfrac{8}{\sqrt{3 - x}}$; $\sqrt{\dfrac{3x - 8}{x}}$; domain of $f \circ g$ $(-\infty, 3)$; domain of $g \circ f$ $(-\infty, 0) \cup \left[\dfrac{8}{3}, \infty\right)$

For each pair of functions defined as follows, show that $(f \circ g)(x) = x$ and $(g \circ f)(x) = x$.

33. $f(x) = 8x$, $g(x) = \dfrac{1}{8}x$

34. $f(x) = \dfrac{3}{4}x$, $g(x) = \dfrac{4}{3}x$

35. $f(x) = 8x - 11$, $g(x) = \dfrac{x + 11}{8}$

36. $f(x) = \dfrac{x - 3}{4}$, $g(x) = 4x + 3$

37. $f(x) = x^3 + 6$, $g(x) = \sqrt[3]{x - 6}$

38. $f(x) = \sqrt[5]{x - 9}$, $g(x) = x^5 + 9$

39. Explain how to find the domain of $(f \circ g)(x) = \dfrac{1}{x^2 - 1}$, where $f(x) = \dfrac{1}{x}$ and $g(x) = x^2 - 1$.

40. For each of the following functions f, decide what operation was used with $g(x)$ and $m(x)$ to get $f(x)$.

(a) $f(x) = ax^2$; $g(x) = a$, $m(x) = x^2$ multiplication

(b) $f(x) = x^2 + k$; $m(x) = x^2$, $g(x) = k$ addition

(c) $f(x) = (x - h)^2$; $m(x) = x^2$, $g(x) = x - h$ composition: $m[g(x)]$

(d) $f(x) = a(x - h)^2 + k$; $g(x) = ax^2 + k$, $m(x) = x - h$ composition: $g[m(x)]$

In each of the following exercises, a function h is defined. Find f(x) and g(x) such that
h(x) = (f ∘ g)(x). Many such pairs of functions exist. See Example 6.
Other correct answers are possible for Exercises 41–46.

41. $h(x) = (6x - 2)^2$ $f(x) = x^2$; $g(x) = 6x - 2$

42. $h(x) = (11x^2 + 12x)^2$ $f(x) = x^2$; $g(x) = 11x^2 + 12x$

43. $h(x) = \sqrt{x^2 - 1}$ $f(x) = \sqrt{x}$; $g(x) = x^2 - 1$

44. $h(x) = \dfrac{1}{x^2 + 2}$ $f(x) = \dfrac{1}{x + 2}$; $g(x) = x^2$

45. $h(x) = \dfrac{(x - 2)^2 + 1}{5 - (x - 2)^2}$ $f(x) = \dfrac{x^2 + 1}{5 - x^2}$; $g(x) = x - 2$

46. $h(x) = (x + 2)^3 - 3(x + 2)^2$ $f(x) = x^3 - 3x^2$; $g(x) = x + 2$

▲ **47.** A couple planning their wedding has found that the cost to hire a caterer for the reception depends on the number of guests attending. If 100 people attend, the cost per person will be $2. For each person less than 100, the cost will increase by $.20. Assume that no more than 100 people will attend. Let x represent the number less than 100 that attend. For example, if 95 attend, $x = 5$.
(a) Write a function $N(x)$ for the possible number of guests. $N(x) = 100 - x$
(b) Write a function $G(x)$ for the possible cost per guest. $G(x) = 2 + .2x$
(c) Write the function $N(x) \cdot G(x)$ for the total cost, $C(x)$. $C(x) = (100 - x)(2 + .2x)$

48. The manager of a music store has found that the price $p(x)$, in dollars, of a compact disc depends on the demand for the disc, x. For one particular disc,

$$p(x) = -.001x + 20.$$

Express the revenue as $R(x)$, where $R(x)$ is the product of price and demand, that is, $p(x) \cdot x$.

$R(x) = -.001x^2 + 20x$

49. Suppose the population P of a certain species of fish depends on the number x (in hundreds) of a smaller kind of fish which serves as its food supply, so that

$$P(x) = 2x^2 + 1.$$

Suppose, also, that the number x (in hundreds) of the smaller species of fish depends upon the amount a (in appropriate units) of its food supply, a kind of plankton. Suppose

$$x = f(a) = 3a + 2.$$

A biologist wants to find the relationship between the population P of the large fish and the amount a of plankton available, that is, $(P \circ f)(a)$. What is the relationship? If the amount of plankton decreases, what will happen to the fish population?

$18a^2 + 24a + 9$;   It will decrease.

50. Suppose the demand for a certain brand of graphing calculator is given by

$$D(p) = \frac{-p^2}{100} + 500,$$

where p is the price in dollars. If the price, in terms of the cost, $c,$ is expressed as

$$p(c) = 2c - 10,$$

find $D(c)$, the demand in terms of the cost. What happens to the demand for the calculator as the cost goes down? $\dfrac{-c^2 + 10c + 12{,}475}{25}$;   demand increases

51. When a thermal inversion layer is over a city, pollutants cannot rise vertically but are trapped below the layer and must disperse horizontally. Assume that a factory smokestack begins emitting a pollutant at 8 A.M. Assume that the pollutant disperses horizontally over a circular area. If t represents the time, in hours, since the factory began emitting pollutants ($t = 0$ represents 8 A.M.), assume that the radius of the circle of pollution is $r(t) = 2t$ mi. Let $A(r) = \pi r^2$ represent the area of a circle of radius r.
(a) Find $(A \circ r)(t)$. $4\pi t^2$
✐ (b) Interpret $(A \circ r)(t)$.

52. An oil well off the Gulf Coast is leaking, with the leak spreading oil over the surface as a circle. At any time t, in minutes, after the beginning of the leak, the radius of the circular oil slick on the surface is $r(t) = 4t$ ft. Let $A(r) = \pi r^2$ represent the area of a circle of radius r.
(a) Find $(A \circ r)(t)$. $16\pi t^2$
✐ (b) Interpret $(A \circ r)(t)$.

◎ **CONCEPTUAL**   ✐ **WRITING**   ▲ **CHALLENGING**

4.4   RESOURCES

Instructor's Test Manual
Test Items, p. 331
Instructor's Solution Manual
pp. 164–168, 583–589
Study Guide
pp. 150–156
Student's Solution Manual
pp. 164–168

Test Generator
IBM Disks
5.25" Disk A
3.5" Disk A
Mac Disk A
Tutorials
IBM Disks
5.25" Disk D
3.5" Disk B
Mac Disk B
Apple II Module 1, Lessons 1–5

HarperCollins Videotape   5
"College Algebra: In Simplest Terms" Program   13
Transparency   T-21–T-22

4.4 EXERCISES   (TEXT PAGE 236)   ∎

Graph each of the following. Give the domain and range of each function. See Examples 1–3.

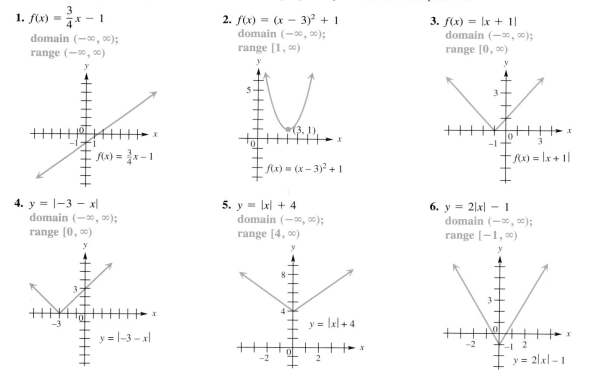

1. $f(x) = \dfrac{3}{4}x - 1$
domain $(-\infty, \infty)$;
range $(-\infty, \infty)$

2. $f(x) = (x - 3)^2 + 1$
domain $(-\infty, \infty)$;
range $[1, \infty)$

3. $f(x) = |x + 1|$
domain $(-\infty, \infty)$;
range $[0, \infty)$

4. $y = |-3 - x|$
domain $(-\infty, \infty)$;
range $[0, \infty)$

5. $y = |x| + 4$
domain $(-\infty, \infty)$;
range $[4, \infty)$

6. $y = 2|x| - 1$
domain $(-\infty, \infty)$;
range $[-1, \infty)$

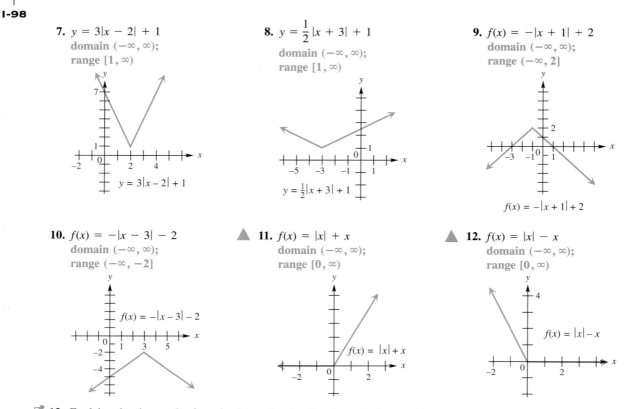

7. $y = 3|x - 2| + 1$
domain $(-\infty, \infty)$;
range $[1, \infty)$

$y = 3|x - 2| + 1$

8. $y = \dfrac{1}{2}|x + 3| + 1$
domain $(-\infty, \infty)$;
range $[1, \infty)$

$y = \frac{1}{2}|x + 3| + 1$

9. $f(x) = -|x + 1| + 2$
domain $(-\infty, \infty)$;
range $(-\infty, 2]$

$f(x) = -|x + 1| + 2$

10. $f(x) = -|x - 3| - 2$
domain $(-\infty, \infty)$;
range $(-\infty, -2]$

$f(x) = -|x - 3| - 2$

▲ 11. $f(x) = |x| + x$
domain $(-\infty, \infty)$;
range $[0, \infty)$

$f(x) = |x| + x$

▲ 12. $f(x) = |x| - x$
domain $(-\infty, \infty)$;
range $[0, \infty)$

$f(x) = |x| - x$

✎ **13.** Explain why the graph of an absolute value function typically has a V-shape.

◉ **14.** When functions are defined piecewise, what two points *must* be plotted?

The endpoints of each portion of the graph must be plotted. If they do not coincide, one will be an open endpoint.

*For each of the following, find (**a**) f(−5), (**b**) f(−1), (**c**) f(0), (**d**) f(3), and (**e**) f(5).*

15. $f(x) = \begin{cases} 2x \text{ if } x \le -1 \\ x - 1 \text{ if } x > -1 \end{cases}$ (a) −10   (b) −2   (c) −1   (d) 2   (e) 4

16. $f(x) = \begin{cases} x - 2 \text{ if } x < 3 \\ 4 - x \text{ if } x \ge 3 \end{cases}$ (a) −7   (b) −3   (c) −2   (d) 1   (e) −1

17. $f(x) = \begin{cases} 3x + 5 \text{ if } x \le 0 \\ 4 - 2x \text{ if } 0 < x < 2 \\ x \text{ if } x \ge 2 \end{cases}$ (a) −10   (b) 2   (c) 5   (d) 3   (e) 5

18. $f(x) = \begin{cases} 4x + 1 \text{ if } x < 2 \\ 3x \text{ if } 2 \le x \le 5 \\ 3 - 2x \text{ if } x > 5 \end{cases}$ (a) −19   (b) −3   (c) 1   (d) 9   (e) 15

◉ CONCEPTUAL   ✎ WRITING   ▲ CHALLENGING

Graph each of the functions defined as follows. See Examples 4 and 5.

19. $f(x) = \begin{cases} x - 1 \text{ if } x \le 3 \\ 2 \text{ if } x > 3 \end{cases}$

20. $f(x) = \begin{cases} 6 - x \text{ if } x \le 3 \\ 3x - 6 \text{ if } x > 3 \end{cases}$

21. $f(x) = \begin{cases} 4 - x \text{ if } x < 2 \\ 1 + 2x \text{ if } x \ge 2 \end{cases}$

22. $f(x) = \begin{cases} 2x + 1 \text{ if } x \ge 0 \\ x \text{ if } x < 0 \end{cases}$

23. $f(x) = \begin{cases} 2 + x \text{ if } x < -4 \\ -x \text{ if } -4 \le x \le 5 \\ 3x \text{ if } x > 5 \end{cases}$

24. $f(x) = \begin{cases} -2x \text{ if } x < -3 \\ 3x - 1 \text{ if } -3 \le x \le 2 \\ -4x \text{ if } x > 2 \end{cases}$

25. $f(x) = \begin{cases} |x| \text{ if } x > -2 \\ x \text{ if } x \le -2 \end{cases}$

26. $f(x) = \begin{cases} |x| - 1 \text{ if } x > -1 \\ x - 1 \text{ if } x \le -1 \end{cases}$

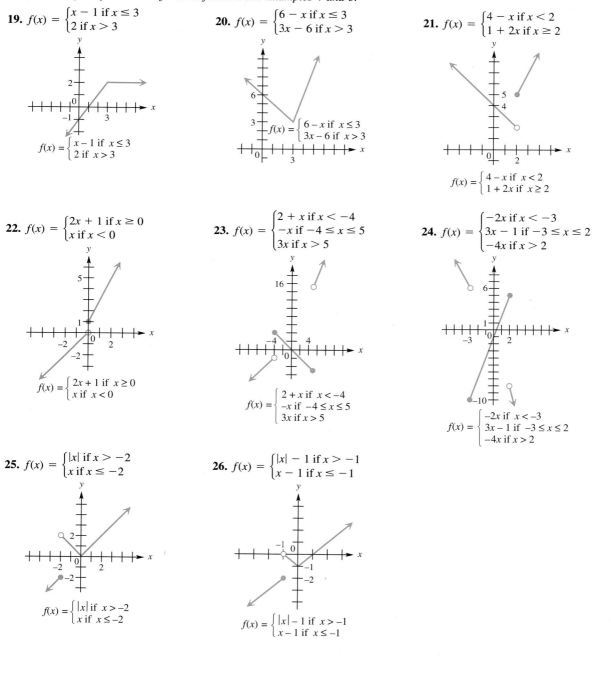

$f(x) = \begin{cases} x - 1 \text{ if } x \le 3 \\ 2 \text{ if } x > 3 \end{cases}$

$f(x) = \begin{cases} 6 - x \text{ if } x \le 3 \\ 3x - 6 \text{ if } x > 3 \end{cases}$

$f(x) = \begin{cases} 4 - x \text{ if } x < 2 \\ 1 + 2x \text{ if } x \ge 2 \end{cases}$

$f(x) = \begin{cases} 2x + 1 \text{ if } x \ge 0 \\ x \text{ if } x < 0 \end{cases}$

$f(x) = \begin{cases} 2 + x \text{ if } x < -4 \\ -x \text{ if } -4 \le x \le 5 \\ 3x \text{ if } x > 5 \end{cases}$

$f(x) = \begin{cases} -2x \text{ if } x < -3 \\ 3x - 1 \text{ if } -3 \le x \le 2 \\ -4x \text{ if } x > 2 \end{cases}$

$f(x) = \begin{cases} |x| \text{ if } x > -2 \\ x \text{ if } x \le -2 \end{cases}$

$f(x) = \begin{cases} |x| - 1 \text{ if } x > -1 \\ x - 1 \text{ if } x \le -1 \end{cases}$

27. Explain how to find the y-values of the greatest integer function for negative x-values.

28. The graph of the greatest integer function defined by $f(x) = [\![x - 2]\!] + 1$ has each line segment translated _____ units _____ and _____ unit(s)

(left/right)

_____ compared to the graph of $f(x) = [\![x]\!]$.   **2, right, 1, up**

(up/down)

Graph each of the following. See Examples 6 and 7.

29. $f(x) = [\![-x]\!]$ **30.** $f(x) = [\![2x - 1]\!]$ **31.** $f(x) = [\![3x + 1]\!]$

32. $f(x) = [\![3x]\!]$ **33.** $f(x) = [\![3x]\!] + 1$ **34.** $f(x) = [\![3x]\!] - 1$

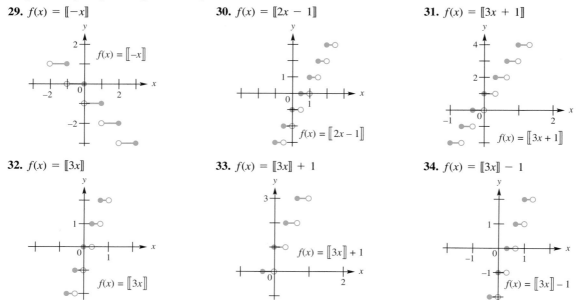

Work each problem. See Example 8.

35. A mail order firm charges 30¢ to mail a package weighing one ounce or less, and then 27¢ for each additional ounce or fraction of an ounce. Let M(x) be the cost of mailing a package weighing x oz. Find

(a) M(.75),   **(b)** M(1.6), and   **(c)** M(4).

(a) 30¢   (b) 57¢   (c) 111¢

(d) Graph y = M(x).

Ounces

(e) Give the domain and range of M.

domain $(0, \infty)$;   range $\{30, 57, 84, 111, \ldots\}$

36. Use the greatest integer function and write an expression for the number of ounces for which postage will be charged on a package weighing x oz (see Exercise 35). $-[\![-x]\!]$ oz

37. A car rental costs $37 for one day, which includes 50 free miles. Each additional 25 mi or portion costs $10. Graph the ordered pairs (miles, cost).

Miles

38. For a lift truck rental of no more than three days, the charge is $300. An additional charge of $75 is made for each day or portion of a day after three. Graph the ordered pairs (number of days, cost).

39. Montreal taxi rates in a recent year were $1.80 for the first 1/9 mi and $.20 for each additional 1/9 mi or fraction of 1/9. Let $C(x)$ be the cost for a taxi ride of x 1/9 mi. Find **(a)** $C(1)$, **(b)** $C(2.3)$, and **(c)** $C(8)$.

(a) $1.80   (b) $2.20   (c) $3.20

(d) Graph $y = C(x)$.

(e) Give the domain and range of C.

**domain $(0, \infty)$ (at least in theory);
range $\{1.80, 2.00, 2.20, 2.40, \ldots\}$**

40. When a diabetic takes long-acting insulin, the insulin reaches its peak effect on the blood sugar level in about 3 hr. This effect remains fairly constant for 5 hr, then declines, and is very low until the next injection. In a typical patient, the level of insulin might be given by the following function.

$$i(t) = \begin{cases} 40t + 100 & \text{if } 0 \le t \le 3 \\ 220 & \text{if } 3 < t \le 8 \\ -80t + 860 & \text{if } 8 < t \le 10 \\ 60 & \text{if } 10 < t \le 24 \end{cases}$$

Here $i(t)$ is the blood sugar level, in appropriate units, at time t measured in hours from the time of the injec-

tion. Chuck takes his insulin at 6 A.M. Find the blood sugar level at each of the following times.

(a) 7 A.M.   **140**   **(b)** 9 A.M.   **220**
(c) 10 A.M. **220**   **(d)** noon   **220**
(e) 2 P.M.   **220**   **(f)** 5 P.M.   **60**
(g) midnight **60**
(h) Graph $y = i(t)$.

41. The snow depth in Michigan's Isle Royale National Park varies throughout the winter. In a typical winter, the snow depth in inches is approximated by the following function.

$$f(x) = \begin{cases} 6.5x & \text{if } 0 \le x \le 4 \\ -5.5x + 48 & \text{if } 4 < x \le 6 \\ -30x + 195 & \text{if } 6 < x \le 6.5 \end{cases}$$

Here, x represents the time in months with $x = 0$ representing the beginning of October, $x = 1$ representing the beginning of November, and so on.

(a) Graph $f(x)$.

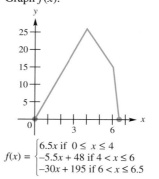

$$f(x) = \begin{cases} 6.5x & \text{if } 0 \le x \le 4 \\ -5.5x + 48 & \text{if } 4 < x \le 6 \\ -30x + 195 & \text{if } 6 < x \le 6.5 \end{cases}$$

(b) In what month is the snow deepest? What is the deepest snow depth?

It is deepest at the beginning of February when it is 26 in.

(c) In what months does the snow begin and end?

It begins at the beginning of October and ends in the middle of April.

4.5 RESOURCES

Instructor's Test Manual
    Test Items, p. 332–334
Instructor's Solution Manual
    pp. 168–172, 589–594
Study Guide
    pp. 156–158
Student's Solution Manual
    pp. 168–172

Test Generator
    IBM Disks
        5.25" Disk A
        3.5" Disk A
        Mac Disk A
Tutorials
    IBM Disks
        5.25" Disk D
        3.5" Disk B
        Mac Disk B
        Apple II Module 1, Lessons 1–5

HarperCollins Videotape 5
"College Algebra: In Simplest Terms" Program 14
Transparency T-23–T-25

4.5 EXERCISES (TEXT PAGE 244) ■

Decide whether each of the functions defined or graphed as follows is one-to-one. See Examples 1 and 2.

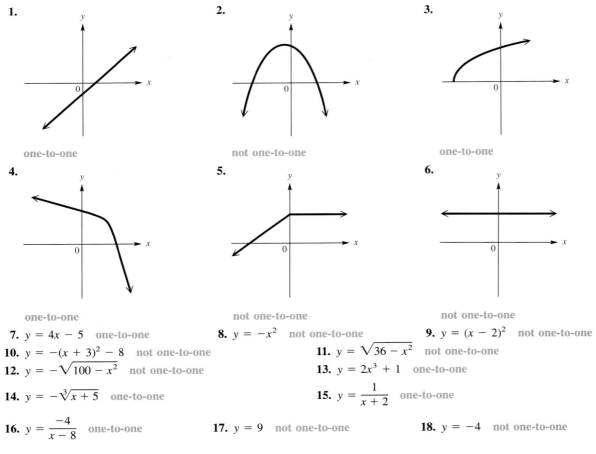

1.

one-to-one

2.

not one-to-one

3.

one-to-one

4.

one-to-one

5.

not one-to-one

6.

not one-to-one

7. $y = 4x - 5$ one-to-one

8. $y = -x^2$ not one-to-one

9. $y = (x - 2)^2$ not one-to-one

10. $y = -(x + 3)^2 - 8$ not one-to-one

11. $y = \sqrt{36 - x^2}$ not one-to-one

12. $y = -\sqrt{100 - x^2}$ not one-to-one

13. $y = 2x^3 + 1$ one-to-one

14. $y = -\sqrt[3]{x + 5}$ one-to-one

15. $y = \dfrac{1}{x + 2}$ one-to-one

16. $y = \dfrac{-4}{x - 8}$ one-to-one

17. $y = 9$ not one-to-one

18. $y = -4$ not one-to-one

ⓒ CONCEPTUAL   ✎ WRITING   ▲ CHALLENGING

⊙ *In Exercises 19–24, an everyday activity is described. Keeping in mind that an inverse operation "undoes" what an operation does, describe the inverse activity.*

19. Tying your shoelaces   untying your shoelaces

20. Starting a car   stopping a car

21. Entering a room   leaving a room

22. Climbing the stairs   descending the stairs

23. Taking off in an airplane   landing in an airplane

24. Filling a cup   emptying a cup

25. Explain why the function $f(x) = x^4$ does not have an inverse. Give examples of ordered pairs to illustrate your explanation.

26. Explain why the function $f(x) = x^5$ is one-to-one.

Decide whether the functions described in each pair are inverses of each other. See Example 3.

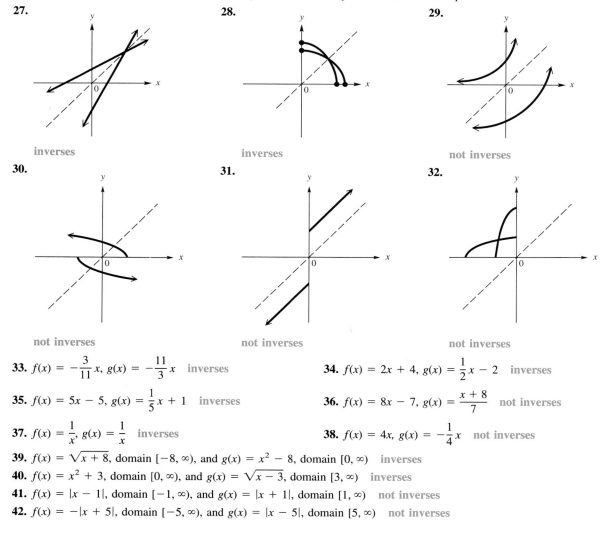

27.   inverses

28.   inverses

29.   not inverses

30.   not inverses

31.   not inverses

32.   not inverses

33. $f(x) = -\dfrac{3}{11}x$, $g(x) = -\dfrac{11}{3}x$ inverses

34. $f(x) = 2x + 4$, $g(x) = \dfrac{1}{2}x - 2$ inverses

35. $f(x) = 5x - 5$, $g(x) = \dfrac{1}{5}x + 1$ inverses

36. $f(x) = 8x - 7$, $g(x) = \dfrac{x + 8}{7}$ not inverses

37. $f(x) = \dfrac{1}{x}$, $g(x) = \dfrac{1}{x}$ inverses

38. $f(x) = 4x$, $g(x) = -\dfrac{1}{4}x$ not inverses

39. $f(x) = \sqrt{x + 8}$, domain $[-8, \infty)$, and $g(x) = x^2 - 8$, domain $[0, \infty)$ inverses

40. $f(x) = x^2 + 3$, domain $[0, \infty)$, and $g(x) = \sqrt{x - 3}$, domain $[3, \infty)$ inverses

41. $f(x) = |x - 1|$, domain $[-1, \infty)$, and $g(x) = |x + 1|$, domain $[1, \infty)$ not inverses

42. $f(x) = -|x + 5|$, domain $[-5, \infty)$, and $g(x) = |x - 5|$, domain $[5, \infty)$ not inverses

Graph the inverse of each one-to-one function.

43.   **44.**   **45.**

46.   **47.**   **48.**

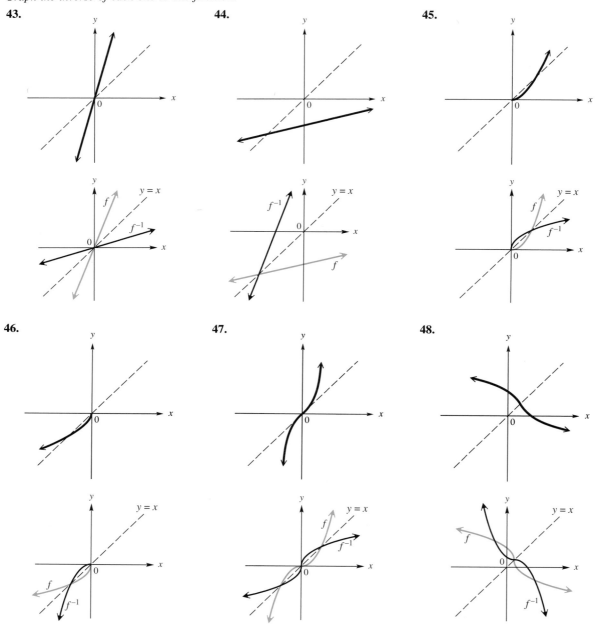

For each function defined as follows that is one-to-one, write an equation for the inverse function in the form of $y = f^{-1}(x)$, and then graph f and f^{-1} on the same axes. See Examples 4 and 5.

49. $y = 3x - 4$ $f^{-1}(x) = \dfrac{x + 4}{3}$

50. $y = 4x - 5$ $f^{-1}(x) = \dfrac{x + 5}{4}$

51. $y = \dfrac{1}{3}x$ $f^{-1}(x) = 3x$

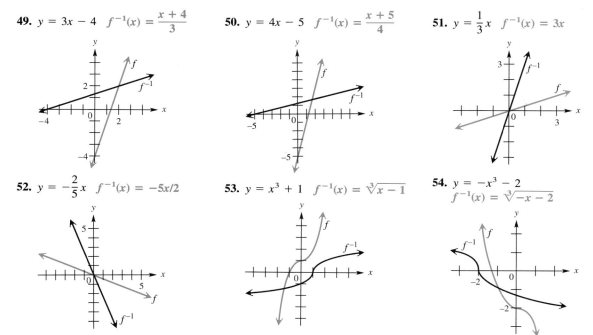

52. $y = -\dfrac{2}{5}x$ $f^{-1}(x) = -5x/2$

53. $y = x^3 + 1$ $f^{-1}(x) = \sqrt[3]{x - 1}$

54. $y = -x^3 - 2$
$f^{-1}(x) = \sqrt[3]{-x - 2}$

55. $y = x^2$ not one-to-one

56. $y = -x^2 + 2$ not one-to-one

57. $y = \dfrac{1}{x}$ $f^{-1}(x) = 1/x$

58. $y = \dfrac{4}{x}$ $f^{-1}(x) = 4/x$

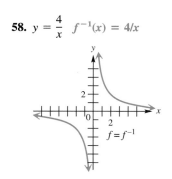

59. $f(x) = 4 - x^2$, domain $(-\infty, 0]$ $f^{-1}(x) = -\sqrt{4 - x}$;   domain $(-\infty, 4]$

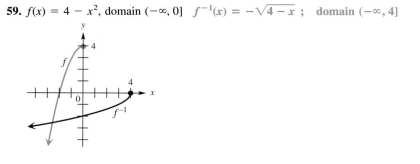

60. $f(x) = \sqrt{6 + x}$, domain $[-6, \infty)$
  $f^{-1}(x) = x^2 - 6$;   domain $[0, \infty)$

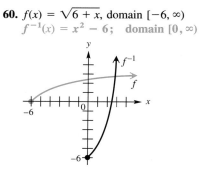

61. $f(x) = -\sqrt{x^2 - 16}$, domain $[4, \infty)$
  $f^{-1}(x) = \sqrt{x^2 + 16}$;   domain $(-\infty, 0]$

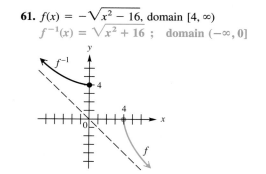

62. $f(x) = (x - 1)^2$, domain $(-\infty, 1]$
  $f^{-1}(x) = 1 - \sqrt{x}$;   domain $[0, \infty)$

RESOURCES FOR REVIEW

Instructor's Test Manual
   Chapter Test Forms, pp. 85–104
Instructor's Solution Manual
   pp. 172–181, 594–600

Study Guide
   pp. 159–161
Student's Solution Manual
   pp. 172–181

Test Generator
   IBM Disks
   5.25″ Disk A
   3.5″ Disk A
   Mac Disk A

CHAPTER 4 REVIEW EXERCISES   (TEXT PAGE 248) ■ ───────

*Give the domain and range of each of the following. Give just the domain in Exercises 11
and 12.*

1. $f(x) = -x$ $(-\infty, \infty)$; $(-\infty, \infty)$

2. $f(x) = 6 - 4x$ $(-\infty, \infty)$; $(-\infty, \infty)$

3. $f(x) = |x| - 4$ $(-\infty, \infty)$; $[-4, \infty)$

4. $f(x) = -3|x - 4|$ $(-\infty, \infty)$; $(-\infty, 0]$

5. $f(x) = -(x + 3)^2$ $(-\infty, \infty)$; $(-\infty, 0]$

6. $f(x) = \sqrt{5 - x}$ $(-\infty, 5]$; $[0, \infty)$

7. $f(x) = \sqrt{49 - x^2}$ $[-7, 7]$; $[0, 7]$

8. $f(x) = -\sqrt{x^2 - 4}$ $(-\infty, -2] \cup [2, \infty)$; $(-\infty, 0]$

9. $f(x) = \sqrt{\dfrac{1}{x^2 + 9}}$ $(-\infty, \infty)$; $(0, 1/3]$

◎ CONCEPTUAL   ✐ WRITING   ▲ CHALLENGING

10. $f(x) = \sqrt{x^2 + 3x - 10}$ $(-\infty, -5] \cup [2, \infty);$ $[0, \infty)$

11. $f(x) = \dfrac{x + 2}{x - 7}$ $(-\infty, 7) \cup (7, \infty)$

12. $f(x) = \dfrac{3x - 8}{4x - 7}$ $(-\infty, 7/4) \cup (7/4, \infty)$

◉ **13.** What ordered pair corresponds to the notation $f(-2) = 3$? $(-2, 3)$

Let $f(x) = -x^2 + 4x + 2$ and $g(x) = 3x + 5$. Find each of the following.

14. $f(-3)$ -19 **15.** $f(4)$ 2 **16.** $f(0)$ 2 **17.** $g(3)$ 14 **18.** $f[g(-1)]$ 6

19. $g[f(-2)]$ -25 **20.** $f(3 + z)$ $-z^2 - 2z + 5$ **21.** $g(5y - 4)$ $15y - 7$

For each of the following, find $\dfrac{f(x + h) - f(x)}{h}$, where $h \neq 0$.

22. $f(x) = 12x - 5$ 12

23. $f(x) = -3$ 0

24. $f(x) = -2x^2 + 4x - 3$ $-4x - 2h + 4$

25. $f(x) = -x^3 + 2x^2$ $-3x^2 - 3xh - h^2 + 4x + 2h$

Write each of the following statements as an equation.

26. m varies directly as the square of z. $m = kz^2$

27. y varies inversely as r and directly as the cube of p. $y = \dfrac{kp^3}{r}$

28. Y varies jointly as M and the square of N and inversely as the cube of X. $Y = \dfrac{kMN^2}{X^3}$

29. A varies jointly as the third power of t and the fourth power of s, and inversely as p and the square of h. $A = \dfrac{kt^3s^4}{ph^2}$

Solve each of the following problems.

30. Suppose r varies directly as x and inversely as the square of y. If r is 10 when x is 5 and y is 3, find r when x is 12 and y is 4. $27/2$

31. Suppose m varies jointly as n and the square of p, and inversely as q. If m is 20 when n is 5, p is 6, and q is 18, find m when n is 7, p is 11, and q is 2. 847

32. Suppose Z varies jointly as the square of J and the cube of M, and inversely as the fourth power of W. If Z is 125 when J is 3, M is 5 and W is 1, find Z if J is 2, M is 7, and W is 3. $1372/729$

33. The power a windmill obtains from the wind varies directly as the cube of the wind velocity. If a 10 km/hr wind produces 10,000 units of power, how much power is produced by a wind of 15 km/hr? $33{,}750$ units

34. Hooke's law for an elastic spring states that the distance a spring stretches varies directly as the force applied. If a force of 32 lb stretches a certain spring 48 in, how much will a force of 24 lb stretch the spring? 36 in

35. The weight w of an object varies inversely as the square of the distance d between the object and the center of the earth. If a man weighs 90 kg on the surface of the earth, how much would he weigh 800 km above the surface? (The radius of the earth is about 6400 km.) $640/9$ kg

Let $f(x) = 3x^2 - 4$ and $g(x) = x^2 - 3x - 4$. Find each of the following values.

36. $(f + g)(x)$ $4x^2 - 3x - 8$

37. $(fg)(x)$ $3x^4 - 9x^3 - 16x^2 + 12x + 16$

38. $(f - g)(4)$ 44

39. $(f + g)(-4)$ 68

40. $(f + g)(2k)$ $16k^2 - 6k - 8$

41. $(fg)(1 + r)$ $3(1 + r)^4 - 9(1 + r)^3 - 16(1 + r)^2 + 12(1 + r) + 16 = 3r^4 + 3r^3 - 25r^2 - 35r + 6$

42. $\left(\dfrac{f}{g}\right)(3)$ $-23/4$

43. $\left(\dfrac{f}{g}\right)(-1)$ undefined

44. Give the domain of $(fg)(x)$. $(-\infty, \infty)$

45. Give the domain of $\left(\dfrac{f}{g}\right)(x)$. $(-\infty, -1) \cup (-1, 4) \cup (4, \infty)$

◉ **46.** Which of the following is *not* equal to $(f \circ g)(x)$ for $f(x) = 1/x$ and $g(x) = x^2 + 1$? (*Hint:* There may be more than one.)

    (a) $f[g(x)]$ **(b)** $\dfrac{1}{x^2 + 1}$ **(c)** $\dfrac{1}{x^2}$ **(d)** $(g \circ f)(x)$ (c) and (d)

Let $f(x) = \sqrt{x - 2}$ and $g(x) = x^2$. Find each of the following.

47. $(f \circ g)(x)$ $\sqrt{x^2 - 2}$ **48.** $(g \circ f)(x)$ $x - 2$ **49.** $(f \circ g)(-6)$ $\sqrt{34}$

50. $(f \circ g)(2)$ $\sqrt{2}$ **51.** $(g \circ f)(3)$ 1 **52.** $(g \circ f)(24)$ 22

🗹 **53.** Explain what is meant by a "function defined piecewise."

Graph each function defined as follows.

54. $f(x) = -|x|$ **55.** $f(x) = |x| - 3$ **56.** $f(x) = -|x| - 2$

57. $f(x) = -|x + 1| + 3$ **58.** $f(x) = 2|x - 3| - 4$ **59.** $f(x) = [\![x - 3]\!]$

60. $f(x) = \left[\!\!\left[\dfrac{1}{2}x - 2\right]\!\!\right]$ **61.** $f(x) = \begin{cases} -4x + 2 \text{ if } x \le 1 \\ 3x - 5 \text{ if } x > 1 \end{cases}$

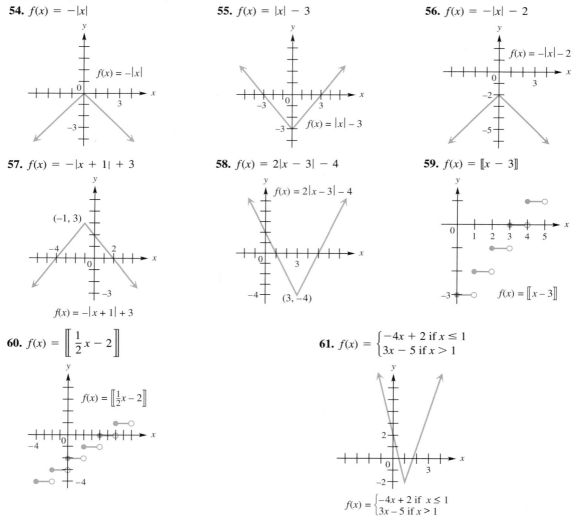

◉ CONCEPTUAL   🗹 WRITING   ▲ CHALLENGING

62. $f(x) = \begin{cases} 3x + 1 \text{ if } x < 2 \\ -x + 4 \text{ if } x \geq 2 \end{cases}$

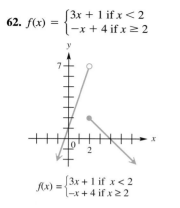

$f(x) = \begin{cases} 3x + 1 \text{ if } x < 2 \\ -x + 4 \text{ if } x \geq 2 \end{cases}$

63. $f(x) = \begin{cases} |x| \text{ if } x < 3 \\ 6 - x \text{ if } x \geq 3 \end{cases}$

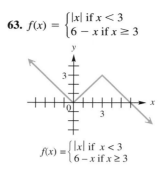

$f(x) = \begin{cases} |x| \text{ if } x < 3 \\ 6 - x \text{ if } x \geq 3 \end{cases}$

64. Let f be a function that gives the cost to rent a floor polisher for x days. The cost is a flat \$3 for cleaning the polisher plus \$4 per day or fraction of a day for using the polisher.

(a) Graph f.

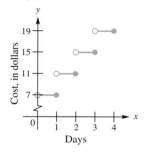

(b) Give the domain and range of f.   domain $(0, \infty)$;   range $\{7, 11, 15, 19, \ldots\}$

65. A trailer hauling service charges \$45, plus \$2 per mile or part of a mile. Graph the ordered pairs (miles, cost). Give the domain and range of the function.

domain $(0, \infty)$;   range $\{47, 49, 51, \ldots\}$

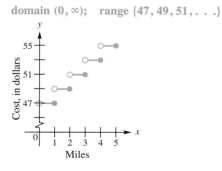

Which of the functions graphed or defined as follows are one-to-one?

66.

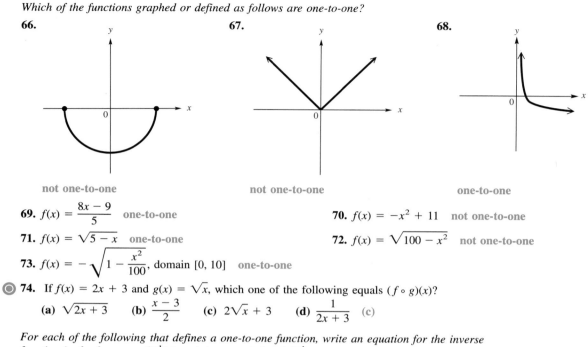

not one-to-one

67.

not one-to-one

68.

one-to-one

69. $f(x) = \dfrac{8x - 9}{5}$ one-to-one

70. $f(x) = -x^2 + 11$ not one-to-one

71. $f(x) = \sqrt{5 - x}$ one-to-one

72. $f(x) = \sqrt{100 - x^2}$ not one-to-one

73. $f(x) = -\sqrt{1 - \dfrac{x^2}{100}}$, domain $[0, 10]$ one-to-one

◉ **74.** If $f(x) = 2x + 3$ and $g(x) = \sqrt{x}$, which one of the following equals $(f \circ g)(x)$?

    (a) $\sqrt{2x + 3}$ **(b)** $\dfrac{x - 3}{2}$ **(c)** $2\sqrt{x} + 3$ **(d)** $\dfrac{1}{2x + 3}$ **(c)**

For each of the following that defines a one-to-one function, write an equation for the inverse function in the form $y = f^{-1}(x)$ and then graph f and f^{-1}.

75. $f(x) = 12x + 3$ $f^{-1}(x) = \dfrac{x - 3}{12}$

76. $f(x) = 2 - x^2$ not one-to-one

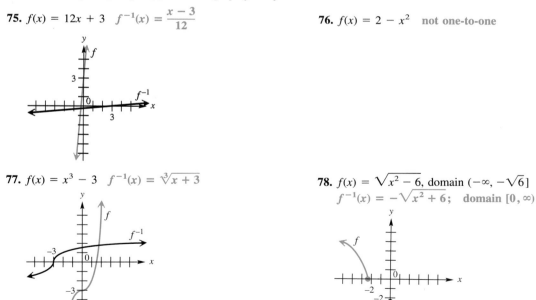

77. $f(x) = x^3 - 3$ $f^{-1}(x) = \sqrt[3]{x + 3}$

78. $f(x) = \sqrt{x^2 - 6}$, domain $(-\infty, -\sqrt{6}]$
$f^{-1}(x) = -\sqrt{x^2 + 6}$;   domain $[0, \infty)$

◎ CONCEPTUAL   ✐ WRITING   ▲ CHALLENGING

79. $f(x) = \sqrt{25 - x^2}$, domain [0, 5]
  $f^{-1}(x) = \sqrt{25 - x^2}$;   domain [0, 5]

80. $f(x) = -\sqrt{x - 3}$
  $f^{-1}(x) = x^2 + 3$;   domain $(-\infty, 0]$

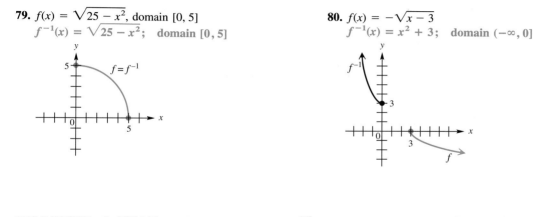

CHAPTER 4 TEST   (TEXT PAGE 251)   ■

Give the domain of each of the following. Give the range in Exercises 1 and 2.

1. $f(x) = 3 + |x + 8|$ $(-\infty, \infty)$; $[3, \infty)$

2. $f(x) = \sqrt{x^2 + 7x + 12}$ $(-\infty, -4] \cup [-3, \infty)$; $[0, \infty)$

3. $f(x) = \dfrac{x - 1}{2x + 8}$ $(-\infty, -4) \cup (-4, \infty)$

Let $f(x) = 6 - 3x - x^2$ and $g(x) = 2x + 3$. Find each of the following.

4. $f(-2)$ 8

5. $(f + g)(2a)$ $9 - 2a - 4a^2$

6. $\left(\dfrac{f}{g}\right)(1)$ 2/5

7. $g\left(\dfrac{1}{2}y - 4\right)$ $y - 5$

8. $g[f(2)]$ -5

9. $(g \circ f)(p)$ $15 - 6p - 2p^2$

Graph each function defined as follows.

10. $f(x) = |x - 2| - 1$

11. $f(x) = [\![x + 1]\!]$

12. $f(x) = \begin{cases} 3 \text{ if } x < -2 \\ 2 - \dfrac{1}{2}x \text{ if } x \geq -2 \end{cases}$

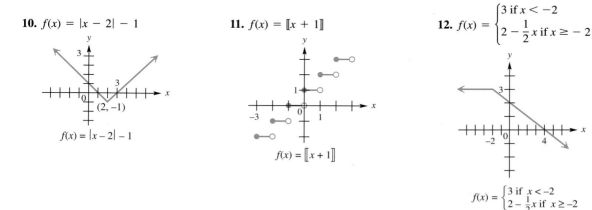

Which of the functions graphed or defined as follows are one-to-one?

13.

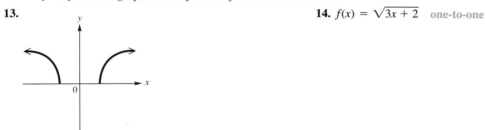

not one-to-one

14. $f(x) = \sqrt{3x + 2}$ one-to-one

15. $f(x) = (x - 2)^2$, domain $[2, \infty)$ one-to-one

For each of the following that defines a one-to-one function, write an equation for the inverse in the form $y = f^{-1}(x)$.

16. $f(x) = 2x + 8$ $f^{-1}(x) = \dfrac{x - 8}{2}$

17. $f(x) = 2 + x^2$, domain $(-\infty, 0]$ $f^{-1}(x) = -\sqrt{x - 2}$;   domain $[2, \infty)$

18. Are the following two functions inverses of each other? Explain why or why not.

$$f(x) = \frac{x}{x + 2} \quad \text{and} \quad g(x) = \frac{2x}{x - 1}$$

19. Suppose y varies directly as the square root of x, and y is 10 when x is 25/9.
  (a) Write an equation for y in terms of x that defines this functional relationship. $y = 6\sqrt{x}$
  (b) Find y when x is 144.   72

20. The force of wind on a sail varies jointly as the area of the sail and the square of the wind velocity. If the force is 8 lb when the velocity is 15 mph and the area is 3 sq ft, find the force when the area is 6 sq ft and the velocity is 22.5 mph.   36 lb

■ CHAPTER 5   EXPONENTIAL AND LOGARITHMIC FUNCTIONS

5.1   RESOURCES

Instructor's Test Manual
  Test Items, p. 335–336
Instructor's Solution Manual
  pp. 182–187, 601–608
Study Guide
  pp. 165–172
Student's Solution Manual
  pp. 182–187

Test Generator
  IBM Disks
    5.25" Disk B
    3.5" Disk A
    Mac Disk A
Tutorials
  IBM Disks
    5.25" Disk E
    3.5" Disk C
    Mac Disk C
    Apple II Module 4, Lessons 1–4

HarperCollins Videotape   5
"College Algebra: In Simplest Terms" Program   18
Transparency   T-26–T-27

◎ CONCEPTUAL   ✎ WRITING   ▲ CHALLENGING

5.1 EXERCISES (Text Page 263) ■

Solve each of the following equations. See Examples 1 and 2.

1. $4^x = 2$ {1/2}

2. $125^r = 5$ {1/3}

3. $\left(\dfrac{1}{2}\right)^k = 4$ {−2}

4. $\left(\dfrac{2}{3}\right)^x = \dfrac{9}{4}$ {−2}

5. $2^{3-y} = 8$ {0}

6. $5^{2p+1} = 25$ {1/2}

7. $\dfrac{1}{27} = b^{-3}$ {3}

8. $\dfrac{1}{81} = k^{-4}$ {−3, 3}

9. $4 = r^{2/3}$ {−8, 8}

10. $z^{5/2} = 32$ {4}

11. $27^{4z} = 9^{z+1}$ {1/5}

12. $32^t = 16^{1-t}$ {4/9}

13. $\left(\dfrac{1}{8}\right)^{-2p} = 2^{p+3}$ {3/5}

14. $3^{-h} = \left(\dfrac{1}{27}\right)^{1-2h}$ {3/7}

15. $\left(\dfrac{1}{2}\right)^{-x} = \left(\dfrac{1}{4}\right)^{x+1}$ {−2/3}

16. $\left(\dfrac{2}{3}\right)^{k-1} = \left(\dfrac{81}{16}\right)^{k+1}$ {−3/5}

17. Graph f for each of the following. Compare the graphs to that of $f(x) = 2^x$. See Examples 3–5.
  (a) $f(x) = 2^x + 1$ (b) $f(x) = 2^x - 4$ (c) $f(x) = 2^{x+1}$ (d) $f(x) = 2^{x-4}$

18. Graph f for each of the following. See Examples 3–5.
  (a) $f(x) = 3^{-x} - 2$ (b) $f(x) = 3^{-x} + 4$ (c) $f(x) = 3^{-x-2}$ (d) $f(x) = 3^{-x+4}$

✐ ◉ **19.** Explain how you could use the graph of $y = 4^x$ to graph $y = -4^x$.

Graph f for each of the following. See Examples 4–6.

20. $f(x) = 3^x$ **21.** $f(x) = 4^x$ **22.** $f(x) = (3/2)^x$ **23.** $f(x) = e^{-x}$

24. $f(x) = e^{3x}$

25. $f(x) = e^{x+1}$

26. $f(x) = 2^{|x|}$

27. $f(x) = 2^{-|x|}$

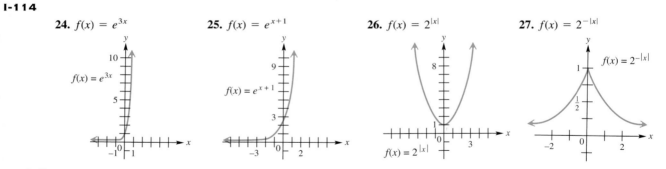

28. Explain why the exponential equation $3^x = 12$ cannot be solved by using the properties of exponentials given in this section.

29. For $a > 1$, how does the value of $f(x) = a^x$ change as x increases? What if $0 < a < 1$?

If $a > 1$, the function value increases. If $0 < a < 1$, the function value decreases.

30. If $f(x) = a^x$ and $f(3) = 27$, find the following values of $f(x)$.

(a) $f(1)$ 3   **(b)** $f(-1)$ 1/3   **(c)** $f(2)$ 9   **(d)** $f(0)$ 1

31. What two points on the graph of $f(x) = a^x$ can be found with no computation? $(0, 1)$ and $(1, a)$

Use a calculator to help graph f in Exercises 32–37

32. $f(x) = \dfrac{e^x - e^{-x}}{2}$

33. $f(x) = \dfrac{e^x + e^{-x}}{2}$

34. $f(x) = x \cdot 2^x$

35. $f(x) = x^2 \cdot 2^{-x}$

36. $f(x) = (1 - x)e^x$

37. $f(x) = xe^x$

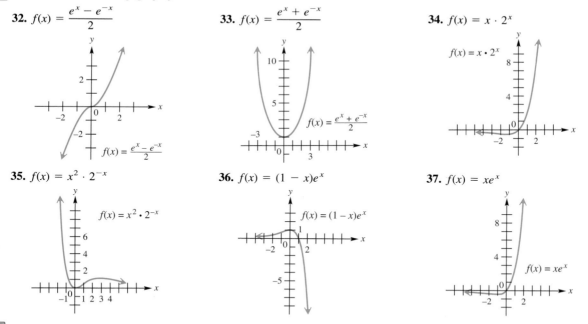

Use the formula for compound interest and a calculator to find the future value of each amount.

38. $4292 at 6% compounded annually for 10 years $7686.32

39. $8906.54 at 5% compounded semiannually for 9 years $13,891.16

40. $56,780 at 5.3% compounded quarterly for 23 quarters $76,855.95

41. $45,788 at 6% compounded daily (ignoring leap years) for 11 years of 365 days $88,585.47

○ CONCEPTUAL   ✎ WRITING   ▲ CHALLENGING

Find the present value for the following future values. See Example 7(a).

42. $10,000, if interest is 12% compounded semiannually for 5 years $5583.95

43. $25,000, if interest is 6% compounded quarterly for 11 quarters $21,223.34

44. $45,678.93, if interest is 9.6% compounded monthly for 11 months. $41,845.63

45. $123,788, if interest is 8.7% compounded daily for 195 days $118,166.72

Find the required annual interest rate to the nearest tenth in each of the following. See Example 7(b).

46. $25,000 compounded annually for 2 years to yield $31,360   12.0%

47. $65,000 compounded monthly for 6 months to yield $65,325   1.0%

48. $1200 compounded quarterly for 5 years to yield $1780   8.0%

49. $15,000 compounded seminnually for 4 semiannual periods to yield $19,000   12.2%

An annual interest rate of 6.50%, for example, compounded more than once per year produces an annual yield of more than 6.50%. With daily compounding (365 times per year), the annual yield is found using the compound interest formula for P = 1 and t = 1 as follows.

$$A = \left(1 + \frac{.065}{365}\right)^{365} = 1.06715$$

Subtract the 1 to get the annual yield, .06715 or 6.715%.

50. Find the annual yield for an annual rate of 6.50% compounded as follows.
   (a) quarterly   6.660%   **(b)** monthly   6.697%

51. Find the annual yield for an annual rate of 7.00% compounded as follows.
   (a) monthly   7.229%   **(b)** daily   7.250%

Solve each of the following problems. See Example 8.

52. Since 1950, the growth in the world population in millions closely fits the exponential function defined by

$$A(t) = 2600e^{.018t},$$

where t is the number of years since 1950.
(a) The world population was about 3700 million in 1970. How closely does the function approximate this value?

$A(20) = 3727$ million, which is off by 27 million

(b) Use the function to approximate the population in 1990. (The actual 1990 population was about 5320 million.)   5342 million

(c) Estimate the population in the year 2000.

6395 million

53. A sample of 500 g of lead 210 decays to polonium 210 according to the function given by

$$A(t) = 500e^{-.032t},$$

where t is time in years. Find the amount of the sample after each of the following times.

(a) 4 years   440 g   **(b)** 8 years   387 g
(c) 20 years   264 g
(d) Graph $y = A(t)$.

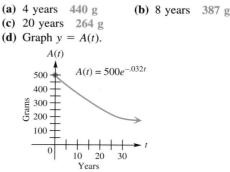

54. Vehicle theft in the United States has been rising exponentially since 1972. The number of stolen vehicles, in millions, is given by

$$f(x) = .88(1.03)^x,$$

where $x = 0$ represents the year 1972. Find the number of vehicles stolen in the following years.
(a) 1975   .96 million   **(b)** 1980   1.1 million
(c) 1985   1.3 million   **(d)** 1990   1.5 million

55. Experiments have shown that the sales of a product, under relatively stable market conditions, but in the absence of promotional activities such as advertising, tend to decline at a constant yearly rate. This rate of sales decline varies considerably from product to product, but seems to remain the same for any particular product. The sales decline can be expressed by a function of the form

$$S(t) = S_0 e^{-at},$$

where $S(t)$ is the rate of sales at time t measured in years, S_0 is the rate of sales at time $t = 0$, and a is the sales decay constant.

(a) Suppose the sales decay constant for a particular product is $a = .10$. Let $S_0 = 50,000$ and find $S(1)$ and $S(3)$.   about 45,200;   about 37,000

(b) Find $S(2)$ and $S(10)$ if $S_0 = 80,000$ and $a = .05$.
about 72,400;   about 48,500

56. The number of nuclear warheads in a certain major superpower arsenal from 1965 to 1980 is approximated by

$$A(t) = 830e^{.15t},$$

where t is the number of years since 1965. Find the number of nuclear warheads in the following years.

(a) 1965   830   **(b)** 1975   3700

(c) 1980   7900

✎ **(d)** In 1985, the number of warheads was actually 10,012. Find the corresponding number using the given function. Is the discrepancy significant? If so, what might have caused it?   17,000

▦ *In calculus, it is shown that*

$$e^x = 1 + x + \frac{x^2}{2 \cdot 1} + \frac{x^3}{3 \cdot 2 \cdot 1} + \frac{x^4}{4 \cdot 3 \cdot 2 \cdot 1} + \frac{x^5}{5 \cdot 4 \cdot 3 \cdot 2 \cdot 1} + \cdots$$

57. Use the terms shown here and replace x with 1 to approximate $e^1 = e$ to three decimal places. Then check your results with a calculator.   2.717

58. Use the terms shown here and replace x with $-.05$ to approximate $e^{-.05}$ to four decimal places. Check your results with a calculator.   .9512

5.2 RESOURCES

Instructor's Test Manual
Test Items, p. 336–337
Instructor's Solution Manual
pp. 187–191, 608–613
Study Guide
pp. 172–177
Student's Solution Manual
pp. 187–191

Test Generator
IBM Disks
5.25″ Disk B
3.5″ Disk A
Mac Disk A
Tutorials
IBM Disks
5.25″ Disk E
3.5″ Disk C
Mac Disk C
Apple II Module 4, Lessons 1–4

HarperCollins Videotape   5
Transparency   T-28

5.2 EXERCISES   (TEXT PAGE 272) ■ ———————

For each of the following statements, write an equivalent statement in logarithmic form. See Example 1.

1. $3^4 = 81$ $\log_3 81 = 4$

2. $2^5 = 32$ $\log_2 32 = 5$

3. $(1/2)^{-4} = 16$ $\log_{1/2} 16 = -4$

4. $(2/3)^{-3} = 27/8$ $\log_{2/3} 27/8 = -3$

5. $10^{-4} = .0001$ $\log_{10} .0001 = -4$

6. $(1/100)^{-2} = 10,000$ $\log_{1/100} 10,000 = -2$

◎ CONCEPTUAL   ✎ WRITING   ▲ CHALLENGING

For each of the following statements, write an equivalent statement in exponential form. See Example 1.

7. $\log_6 36 = 2$ $6^2 = 36$

8. $\log_5 5 = 1$ $5^1 = 5$

9. $\log_{\sqrt{3}} 81 = 8$ $(\sqrt{3})^8 = 81$

10. $\log_4 \dfrac{1}{64} = -3$ $4^{-3} = 1/64$

11. $\log_{10} .0001 = -4$ $10^{-4} = .0001$

12. $\log_3 \sqrt[3]{9} = \dfrac{2}{3}$ $3^{2/3} = \sqrt[3]{9}$

Find the value of each of the following expressions. (Hint: In Exercises 13–22, let the expression equal y, and write in exponential form.) See Example 2(a).

13. $\log_5 25$ 2

14. $\log_3 81$ 4

15. $\log_8 8$ 1

16. $\log_7 1$ 0

17. $\log_{10} .001$ -3

18. $\log_6 \dfrac{1}{216}$ -3

19. $\log_4 \dfrac{\sqrt[3]{4}}{2}$ $-1/6$

20. $\log_9 \dfrac{\sqrt[4]{27}}{3}$ $-1/8$

21. $\log_{1/3} \dfrac{9^{-4}}{3}$ 9

22. $\log_{1/4} \dfrac{16^2}{2^{-3}}$ $-11/2$

23. $2^{\log_2 9}$ 9

24. $8^{\log_8 11}$ 11

Solve each of the following equations. See Example 2.

25. $x = \log_2 32$ {5}

26. $x = \log_2 128$ {7}

27. $\log_x 25 = -2$ {1/5}

28. $\log_x \dfrac{1}{16} = -2$ {4}

29. Explain why logarithms of negative numbers are not defined.

30. Compare the summary of facts about the graph of $f(x) = \log_a x$ with the similar summary about the graph of $y = a^x$ in Section 5.1. Make a list of the facts that reinforce the idea that these are inverse functions.

31. Graph each of the following functions. Compare the graphs to that of $f(x) = \log_2 x$. See Examples 3–5.

(a) $f(x) = (\log_2 x) + 3$

(b) $f(x) = \log_2 (x + 3)$

(c) $f(x) = |\log_2 (x + 3)|$

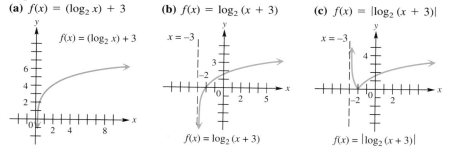

32. Graph each of the following functions. Compare the graphs to that of $f(x) = \log_{1/2} x$. See Examples 3–5.

(a) $f(x) = (\log_{1/2} x) - 2$

(b) $f(x) = \log_{1/2} (x - 2)$

(c) $f(x) = |\log_{1/2} (x - 2)|$

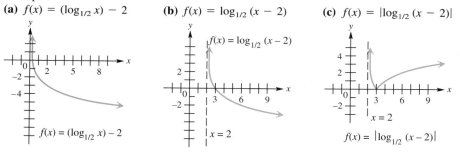

Graph each of the following functions. See Examples 3–5.

33. $f(x) = \log_3 x$ **34.** $f(x) = \log_{10} x$ **35.** $f(x) = \log_{1/2} (1 - x)$ **36.** $f(x) = \log_{1/3} (3 - x)$

37. $f(x) = \log_2 x^2$ **38.** $f(x) = \log_3 (x - 1)$ ▲ **39.** $f(x) = x \log_{10} x$ ▲ **40.** $f(x) = x^2 \log_{10} x$

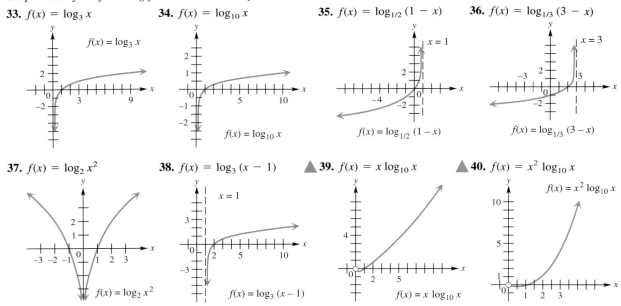

Write each of the following expressions as a sum, difference, or product of logarithms. Simplify the result if possible. Assume that all variables represent positive real numbers. See Example 6.

41. $\log_3 \dfrac{2}{5}$ $\log_3 2 - \log_3 5$

42. $\log_4 \dfrac{6}{7}$ $\log_4 6 - \log_4 7$

43. $\log_2 \dfrac{6x}{y}$ $\log_2 6 + \log_2 x - \log_2 y$

44. $\log_3 \dfrac{4p}{q}$ $\log_3 4 + \log_3 p - \log_3 q$

45. $\log_5 \dfrac{5\sqrt{7}}{3}$ $1 + (1/2) \log_5 7 - \log_5 3$

46. $\log_2 \dfrac{2\sqrt{3}}{5}$ $1 + (1/2) \log_2 3 - \log_2 5$

47. $\log_4 (2x + 5y)$ **no change possible**

48. $\log_6 (7m + 3q)$ **no change possible**

49. $\log_k \dfrac{pq^2}{m}$ $\log_k p + 2 \log_k q - \log_k m$

50. $\log_z \dfrac{x^5 y^3}{3}$ $5 \log_z x + 3 \log_z y - \log_z 3$

51. $\log_m \sqrt{\dfrac{5r^3}{z^5}}$ $(1/2)(\log_m 5 + 3 \log_m r - 5 \log_m z)$

52. $\log_p \sqrt[3]{\dfrac{m^5 n^4}{t^2}}$ $(1/3)(5 \log_p m + 4 \log_p n - 2 \log_p t)$

Write each of the following expressions as a single logarithm. Assume that all variables represent positive real numbers. See Example 7.

53. $\log_a x + \log_a y - \log_a m$ $\log_a \dfrac{xy}{m}$

54. $(\log_b k - \log_b m) - \log_b a$ $\log_b \dfrac{k}{ma}$

55. $2 \log_m a - 3 \log_m b^2$ $\log_m \dfrac{a^2}{b^6}$

56. $\dfrac{1}{2} \log_y p^3 q^4 - \dfrac{2}{3} \log_y p^4 q^3$ $-\log_y p^{7/6}$

57. $2 \log_a (z - 1) + \log_a (3z + 2), z > 1$ $\log_a [(z - 1)^2(3z + 2)]$

58. $\log_b (2y + 5) - \dfrac{1}{2} \log_b (y + 3)$ $\log_b \dfrac{2y + 5}{\sqrt{y + 3}}$

59. $-\dfrac{2}{3} \log_5 5m^2 + \dfrac{1}{2} \log_5 25m^2$ $\log_5 \dfrac{5^{1/3}}{m^{1/3}}$

60. $-\dfrac{3}{4} \log_3 16p^4 - \dfrac{2}{3} \log_3 8p^3$ $\log_3 \dfrac{1}{32p^5}$

◎ CONCEPTUAL   ✎ WRITING   ▲ CHALLENGING

61. Why does $\log_a 1$ always equal 0 for any valid base a?

Given $\log_{10} 2 = .3010$ *and* $\log_{10} 3 = .4771$, *find each of the following logarithms without using a calculator. See Example 8.*

62. $\log_{10} 6$ .7781

63. $\log_{10} 12$ 1.0791

64. $\log_{10} (9/4)$ .3522

65. $\log_{10} (20/27)$ $-.1303$

66. $\log_{10} \sqrt{30}$ .7386

67. $\log_{10} (36)^{1/3}$ .5187

▲ *Prove the following properties of logarithms.*

68. $\log_a (x/y) = \log_a x - \log_a y$

69. $\log_a x^r = r \log_a x$

5.3 RESOURCES

Instructor's Test Manual
Test Items, p. 337–339
Instructor's Solution Manual
pp. 191–194, 613–616
Study Guide
pp. 177–180
Student's Solution Manual
pp. 191–194

Test Generator
IBM Disks
5.25″ Disk B
3.5″ Disk A
Mac Disk A
Tutorials
IBM Disks
5.25″ Disk E
3.5″ Disk C
Mac Disk C
Apple II Module 4, Lessons 1–4

HarperCollins Videotape 6

5.3 EXERCISES (TEXT PAGE 278) ■

A calculator will be necessary for most of the exercises in this section.

Use a calculator to evaluate each of the following logarithms to four decimal places. See Examples 1 and 4.

1. log 43 1.6335

2. log 1247 3.0959

3. log 783 2.8938

4. log .014 -1.8539

5. log .0069 -2.1612

6. log 60,000 4.7782

7. ln 580 6.3630

8. ln .08 -2.5257

9. ln .7 $-.3567$

10. ln 81,000 11.3022

11. ln 121,000 11.7035

12. ln 350 5.8579

Use the change of base rule to find each of the following logarithms to the nearest hundredth. See Example 6.

13. $\log_5 10$ 1.43

14. $\log_9 12$ 1.13

15. $\log_{15} 5$ .59

16. $\log_{1/2} 3$ -1.58

17. $\log_{100} 83$ .96

18. $\log_{200} 175$ .97

19. $\log_{2.9} 7.5$ 1.89

20. $\log_{5.8} 12.7$ 1.45

21. Is the logarithm to the base 3 of 4 written as $\log_4 3$ or $\log_3 4$? $\log_3 4$

22. What is the relationship between e^x and ln x? They define a pair of inverse functions.

For each of the following substances, find the pH from the given hydronium ion concentration. See Example 2(a).

23. Grapefruit, 6.3×10^{-4} 3.2

24. Crackers, 3.9×10^{-9} 8.4

25. Limes, 1.6×10^{-2} 1.8

26. Sodium hydroxide (lye), 3.2×10^{-14} 13.5

Find the $[H_3O^+]$ for each of the following substances from the given pH. See Example 2(b).

27. Soda pop, 2.7 2×10^{-3}

28. Wine, 3.4 4×10^{-4}

29. Beer, 4.8 1.6×10^{-5}

30. Drinking water, 6.5 3.2×10^{-7}

Solve each problem. See Example 3.

31. Find the decibel ratings of sounds having the following intensities:
(a) $100I_0$ 20   (b) $1000I_0$ 30
(c) $100,000I_0$ 50   (d) $1,000,000I_0$ 60

32. Find the decibel ratings of the following sounds, having intensities as given. (You will need a calculator with a log key.) Round each answer to the nearest whole number.
(a) whisper, $115I_0$ 21
(b) busy street, $9,500,000I_0$ 70
(c) heavy truck, 20 m away, $1,200,000,000I_0$ 91
(d) rock music, $895,000,000,000I_0$ 120
(e) jetliner at takeoff, $109,000,000,000,000I_0$ 140

33. The intensity of an earthquake, measured on the *Richter scale*, is $\log_{10}(I/I_0)$, where I_0 is the intensity of an earthquake of a certain (small) size. Find the Richter scale ratings of earthquakes having the following intensities.
(a) $1000I_0$ 3   (b) $1,000,000I_0$ 6
(c) $100,000,000I_0$ 8

34. On July 14, 1991, Peshawar, Pakistan, was shaken by an earthquake that measured 6.6 on the Richter scale.
(a) Express this reading in terms of I_0. See Exercise 33.   about $4,000,000I_0$
(b) In February of the same year a quake measuring 6.5 on the Richter scale killed about 900 people in the mountains of Pakistan and Afghanistan. Express the intensity of a 6.5 reading in terms of I_0.

about $3,200,000I_0$

(c) How much greater was the force of the earthquake with a measure of 6.6?   about 1.25 times as great

35. (a) The San Francisco earthquake of 1906 had a Richter scale rating of 8.3. Express the intensity of this earthquake as a multiple of I_0.   about $200,000,000I_0$
(b) In 1989, the San Francisco region experienced an earthquake with a Richter scale rating of 7.1. Express the intensity of this earthquake as a multiple of I_0.

about $13,000,000I_0$

(c) Compare the intensity of the two San Francisco earthquakes discussed above.

The 1906 earthquake was more than 15 times as intense as the 1989 earthquake.

Solve each problem. See Examples 3 and 5.

36. The number of years, n, since two independently evolving languages split off from a common ancestral language is approximated by $n \approx -7600 \log r$, where r is the proportion of words from the ancestral language common to both languages.
(a) Find n if $r = .9$.   about 350 yr
(b) Find n if $r = .3$.   about 4000 yr
(c) How many years have elapsed since the split if half of the words of the ancestral language are common to both languages?   about 2300 yr

37. The number of species in a sample is given by

$$S(n) = a \ln \left(1 + \frac{n}{a} \right).$$

Here n is the number of individuals in the sample and a is a constant that indicates the diversity of species in the community. If $a = .36$, find $S(n)$ for the following values of n.
(a) 100   2.03   (b) 200   2.28
(c) 150   2.17   (d) 10   1.21

38. In Exercise 37, find $S(n)$ if a changes to .88. Use the following values of n.
(a) 50   3.57   (b) 100   4.17   (c) 250   4.97

In the central Sierra Nevada mountains of California, the percent of moisture that falls as snow rather than rain is approximated reasonably well by

$$p = 86.3 \ln h - 680,$$

where p is the percent of snow at an altitude h in feet. (Assume $h \geq 3000$.)

39. Find the percent of moisture that falls as snow at the following altitudes.
(a) 3000 ft   11%   (b) 4000 ft   36%
(c) 7000 ft   84%
(d) Graph p.

40. Suppose a sample of a small community shows two species with 50 individuals each. Find the index of diversity H. See Example 7.   1

41. A virgin forest in northwestern Pennsylvania has 4 species of large trees with the following proportions of each: hemlock, .521; beech, .324; birch, .081; maple, .074. Find the index of diversity H. See Example 7.

1.59

 Graph f for each of the following.

42. $f(x) = |\ln x|$

43. $f(x) = x \ln x$

44. $f(x) = x^2 \ln x$

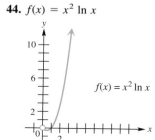

45. Given $g(x) = e^x$, evaluate the following.

    (a) $g(\ln 3)$ 3   **(b)** $g(\ln 5^2)$ $5^2 = 25$ **(c)** $g\left(\ln \dfrac{1}{e}\right)$ $1/e$

46. Given $f(x) = 3^x$, evaluate the following.

    (a) $f(\log_3 7)$ 7   **(b)** $f[\log_3 (\ln 3)]$ $\ln 3$

    (c) $f[\log_3 (2 \ln 3)]$ $\ln 9$

47. Given $f(x) = \ln x$, evaluate the following.

    (a) $f(e^5)$ 5   **(b)** $f(e^{\ln 3})$ $\ln 3$ **(c)** $f(e^{2 \ln 3})$ $2 \ln 3$ or $\ln 9$

48. Given $f(x) = \log_2 x$, evaluate the following.

    (a) $f(2^3)$ 3   **(b)** $f(2^{\log_2 2})$ 1   **(c)** $f(2^{2 \log_2 2})$ 2

5.4   RESOURCES

Instructor's Test Manual
Test Items, p. 339–340

Instructor's Solution Manual
pp. 195–200, 617–622

Study Guide
pp. 180–184

Student's Solution Manual
pp. 195–200

Test Generator
IBM Disks
  5.25" Disk B
  3.5" Disk A
  Mac Disk A

Tutorials
IBM Disks
  5.25" Disk E
  3.5" Disk C
  Mac Disk C
  Apple II Module 4, Lessons 1–4

HarperCollins Videotape   6
"College Algebra: In Simplest Terms" Program   23

5.4 EXERCISES (TEXT PAGE 285) ■

▦ *A calculator will be necessary for many of these exercises.*

Solve the following equations. When necessary, give answers as decimals rounded to the nearest thousandth. See Examples 1–5.

1. $3^x = 6$ {**1.631**}

2. $4^x = 12$ {**1.792**}

3. $3^{a+2} = 5$ {**−.535**}

4. $5^{2-x} = 12$ {**.456**}

5. $6^{1-2k} = 8$ {**−.080**}

6. $3^{2m-5} = 13$ {**3.667**}

7. $e^{k-1} = 4$ {**2.386**}

8. $e^{2-y} = 12$ {**−.485**}

9. $2e^{5a+2} = 8$ {**−.123**}

10. $10e^{3z-7} = 5$ {**2.102**}

11. $2^x = -3$ ∅

12. $(1/4)^P = -4$ ∅

13. $e^{2x} \cdot e^{5x} = e^{14}$ {**2**}

14. $e^{\ln x} = 3$ {**3**}

15. $e^{-\ln x} = 2$ {**.5**}

16. $e^{3 \ln x} = 1/8$ {**.5**}

17. $e^{\ln x + \ln (x - 2)} = 8$ {**4**}

18. $e^{2 \ln x - \ln (x + 2)} = 8/3$ {**4**}

19. $100(1 + .02)^{3+n} = 150$ {**17.475**}

20. $500(1 + .05)^{p/4} = 200$ {**−75.121**}

21. $\log (t - 1) = 1$ {**11**}

22. $\log q^2 = 1$ {**±3.162**}

23. $\log (x - 3) = 1 - \log x$ {**5**}

24. $\log (z - 6) = 2 - \log (z + 15)$ {**10**}

25. $\ln (y + 2) = \ln (y - 7) + \ln 4$ {**10**}

26. $\ln p - \ln (p + 1) = \ln 5$ ∅

27. $\ln (5 + 4y) - \ln (3 + y) = \ln 3$ {**4**}

28. $\ln m + \ln (2m + 5) = \ln 7$ {**1**}

29. $\ln x + 1 = \ln (x - 4)$ ∅

30. $\ln (4x - 2) = \ln 4 - \ln (x - 2)$ {**2.5**}

31. $2 \ln (x - 3) = \ln (x + 5) + \ln 4$ {**11**}

32. $\ln (k + 5) + \ln (k + 2) = \ln 14k$ {**2, 5**}

33. $\log_5 (r + 2) + \log_5 (r - 2) = 1$ {**3**}

34. $\log_4 (z + 3) + \log_4 (z - 3) = 1$ {**3.606**}

35. $\log_3 (a - 3) = 1 + \log_3 (a + 1)$ ∅

36. $\log w + \log (3w - 13) = 1$ {**5**}

37. $\ln e^x - \ln e^3 = \ln e^5$ {**8**}

38. $\ln e^x - 2 \ln e = \ln e^4$ {**6**}

39. $\log_2 \sqrt{2y^2} - 1 = 1/2$ {**−2, 2**}

40. $\log_2 (\log_2 x) = 1$ {**4**}

41. $\log z = \sqrt{\log z}$ {**1, 10**}

42. $\log x^2 = (\log x)^2$ {**1, 100**}

✐ **43.** Suppose you overhear the following statement: "I must reject any negative answer when I solve an equation involving logarithms." Is this correct? Write an explanation of why it is or is not correct.

◉ **44.** What values of x could not possibly be solutions of the following equation?

$$\log_a (4x - 7) + \log_a (x^2 + 4) = 0 \quad (-\infty, 7/4]$$

Solve each of the following equations for the indicated variables. Use logarithms to the appropriate bases. See Examples 6 and 7.

45. $P = P_0 e^{kt/1000}$ for t $\quad t = \dfrac{1000}{k} \ln \dfrac{P}{P_0}$

46. $I = \dfrac{E}{R} (1 - e^{-Rt/2})$ for t $\quad t = -\dfrac{2}{R} \ln \left(1 - \dfrac{RI}{E}\right)$

47. $r = p - k \ln t$ for t $\quad t = e^{(p-r)/k}$

48. $p = a + \dfrac{k}{\ln x}$ for x $\quad x = e^{k/(p-a)}$

49. $T = T_0 + (T_1 - T_0) 10^{-kt}$ for t $\quad t = -\dfrac{1}{k} \log \dfrac{T - T_0}{T_1 - T_0}$

50. $A = \dfrac{Pi}{1 - (1 + i)^{-n}}$ for n $\quad n = \dfrac{\ln \dfrac{A}{A - Pi}}{\ln (1 + i)}$

◉ CONCEPTUAL   ✐ WRITING   ▲ CHALLENGING

51. Recall (from Section 5.3) the formula for the decibel rating of the loudness of a sound is

$$d = 10 \log \frac{I}{I_0}.$$

Solve this formula for I.　$I = I_0 \cdot 10^{d/10}$

52. A few years ago, there was a controversy about a proposed government limit on factory noise. One group wanted a maximum of 89 decibels, while another group wanted 86. This difference seemed very small to many people. Find the percent by which the 89-decibel intensity exceeds that for 86 decibels. (See Exercise 51.)

89 decibels is about twice as loud as 86 decibels, for a 100% increase.

The formula for compound interest,

$$A = P\left(1 + \frac{r}{m}\right)^{tm},$$

was given in Section 5.1. Use natural logarithms and solve for t to the nearest tenth in Exercises 53 and 54.

53. $A = \$10,000$, $P = \$7500$, $r = 6\%$, $m = 2$　4.9 yr

54. $A = \$8400$, $P = \$5000$, $r = 8\%$, $m = 4$　6.5 yr

55. George Tom wants to buy a $30,000 car. He has saved $27,000. Find the number of years (to the nearest tenth) it will take for his $27,000 to grow to $30,000 at 6% interest compounded quarterly.　1.8 yr

56. Find t to the nearest hundredth if $1786 becomes $2063.40 at 11.6%, with interest compounded monthly.　1.25 yr

57. Find the number of years for $2000 to amount to $2500 at 6% compounded semianually.　3.8 yr

58. Find the number of years for $16,000 to become $25,000 at 8% compounded quarterly.　5.6 yr

59. The growth of bacteria in food products makes it necessary to time-date some products (such as milk) so that they will be sold and consumed before the bacteria count becomes too high. Suppose for a certain product that the number of bacteria present is given by

$$f(t) = 500e^{.1t},$$

under certain storage conditions, where t is time in days after packing of the product and the value of $f(t)$ is in millions.

(a) If the product cannot be safely eaten after the bacteria count reaches 3,000,000,000, how long will this take?　about 18 days

(b) If $t = 0$ corresponds to January 1, what date should be placed on the product?　January 19th

60. In Example 8 of Section 5.1, the U.S. Consumer Price Index was approximated by $A(t) = 34e^{.04t}$, where t represents the number of years after 1960. Assuming the same equation continues to apply, find the year in which costs will be 50% higher than in 1983, that is, when the CPI equals 150.　1997

61. The population of an animal species that is introduced into a certain area may grow rapidly at first but then grow more slowly as time goes on. A logarithmic function can provide an excellent description of such growth. Suppose that the population of foxes in an area t months after the foxes were first introduced there is

$$F = 500 \log (2t + 3).$$

Solve the equation for t. Then find t to the nearest tenth for the following values of F.

$$t = \frac{10^{F/500} - 3}{2}$$

(a) 600　6.4 mo　　**(b)** 1000　48.5 mo

62. Recall from Section 5.3 that the number of years, n, since two independently evolving languages split off from a common language is approximated by $n = -7600 \log r$, where r is the proportion of words from the ancestral language still common to the two languages. Solve the formula for r. Then find r if the languages split　$r = 10^{-n/7600}$

(a) 1000 years ago,　.74
(b) 2500 years ago.　.47

5.5 RESOURCES

Instructor's Test Manual
  Test Items, p. 340–343
Instructor's Solution Manual
  pp. 200–203, 622–624
Study Guide
  pp. 184–189
Student's Solution Manual
  pp. 200–203

Test Generator
  IBM Disks
   5.25" Disk B
   3.5" Disk A
   Mac Disk A
Tutorials
  IBM Disks
   5.25" Disk E
   3.5" Disk C
   Mac Disk C
   Apple II Module 4, Lessons 1–4

HarperCollins Videotape   6

5.5 EXERCISES   (TEXT PAGE 290) ■

▦ *A calculator will be necessary for many of these exercises.*

Work the following problems. See Examples 1 and 2.

1. Assuming continuous compounding, what will it cost to buy a $10 item in 3 years at the following inflation rates?
 (a) 3% $10.94   **(b)** 4% $11.27
 (c) 5% $11.62

2. Bert Bezzone invests a $25,000 inheritance in a fund paying 5% per year compounded continuously. What will be the amount on deposit after each of the following time periods?
 (a) 1 year $26,281.78   **(b)** 5 years $32,100.64
 (c) 10 years $41,218.03

3. Linda Youngman, who is self-employed, wants to invest $60,000 in a pension plan. One investment offers 7% compounded quarterly. Another offers 6.75% compounded continuously. Which investment will earn more interest in 5 years? How much more will the better plan earn?

 7% compounded quarterly; $800.31

4. If Ms. Youngman (see Exercise 3) chooses the plan with continuous compounding, how long will it take for her $60,000 to grow to $80,000? about 4.3 yr

5. Assume the cost of a loaf of bread is $1. With continuous compounding, find the doubling time at an annual inflation rate of 6%. 11.6 yr

6. Suppose the annual rate of inflation is 5%. How long will it take for prices to double? 13.9 yr

7. Find the interest rate that will cause $5000 to grow to $7250 in 4 years if the money is compounded continuously. 9.3%

Work the following problems. See Example 3.

8. In the exercises for Section 5.1, we gave the sales decline function as $S(t) = S_0 e^{-at}$. If $a = .1$, $S_0 = 50,000$, and t is time measured in years, find the number of years it will take for sales to fall to half the initial sales. 6.9 yr

9. One measure of living standards in the United States is given by $L = 9 + 2e^{.15t}$, where t is the number of years since 1982. Find L for the following years.
 (a) 1982 11   **(b)** 1986 12.6   **(c)** 1992 18
 (d) Graph L.

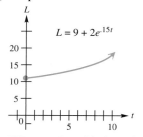

 (e) What can be said about the growth of living standards in the U.S. according to this equation?

 They are growing exponentially, but at a slow rate.

10. The quantity in grams of a radioactive substance present at time t is $A(t) = 500e^{-.05t}$, where t is time measured in days.
 (a) Find the half-life of the substance. 13.9 days
 (b) Find the amount present after 10 days. 303 g

Find the half-life of the following radioactive substances to three significant digits. (t is measured in years.) See Examples 3 and 4.

11. Plutonium 241; $A(t) = A_0 e^{-.053t}$ 13.1 yr

12. Radium 226; $A(t) = A_0 e^{-.00043t}$ 1610 yr

13. Iodine 131; $A(t) = A_0 e^{-.087t}$ 7.97 yr

14. Suppose the number of rabbits in a colony is $y = y_0 e^{.4t}$, where t represents time in months and y_0 is the rabbit population when $t = 0$.
  (a) If $y_0 = 100$, find the number of rabbits present at time $t = 4$.   about 495 rabbits
  (b) How long will it take for the number of rabbits to triple?   about 2.75 mo

15. A midwestern city finds its residents moving to the suburbs. Its population is declining according to the relationship $P = P_0 e^{-.04t}$, where t is time measured in years and P_0 is the population at time $t = 0$. Assume that $P_0 = 1,000,000$.
  (a) Find the population at time $t = 1$.

  about 961,000

  (b) Estimate the time it will take for the population to be reduced to 750,000.   about 7.2 yr
  (c) How long will it take for the population to be cut in half?   about 17.3 yr

For Exercises 16–21, refer to Example 4.

16. Suppose an Egyptian mummy is discovered in which the amount of carbon 14 present is only about one-third the amount found in the atmosphere. About how long ago did the Egyptian die?   9000 yr

17. A sample from a refuse deposit near the Strait of Magellan had 60% of the carbon 14 of a contemporary living sample. How old was the sample?

  about 4200 yr

18. Paint from the Lascaux caves of France contains 15% of the normal amount of carbon 14. Estimate the age of the caves.   about 16,000 yr

19. Solve the formula $A = A_0 e^{kt}$ for t. $t = \dfrac{\ln (A/A_0)}{k}$

20. Suppose a specimen is found in which there is about 2/3 the carbon 14 of a comparable living specimen. Estimate the age of the specimen.   about 3300 yr

21. Estimate the age of a specimen that contains 20% of the carbon 14 of a comparable living specimen.

  about 13,000 yr

A large cloud of radioactive debris from a nuclear explosion has floated over the Pacific Northwest, contaminating much of the hay supply. Consequently, farmers in the area are concerned that the cows who eat this hay will give contaminated milk. (The tolerance level for radioactive iodine in milk is 0.) The percent of the initial amount of radioactive iodine still present in the hay after t days is approximated by $P(t) = 100e^{-.1t}$.

22. Some scientists feel that the hay is safe after the percent of radioactive iodine has declined to 10% of the original amount. Find the number of days before the hay can be used.   23 days

23. Other scientists believe that the hay is not safe until the level of radioactive iodine has declined to only 1% of the original level. Find the number of days this would take.   46 days

24. The amount of a chemical that will dissolve in a solution increases exponentially as the (Celsius) temperature t is increased according to the equation $A(t) = 10e^{.0095t}$. At what temperature will 15 g dissolve?   43°C

25. By Newton's law of cooling, the temperature of a body at time t after being introduced into an environment having constant temperature T_0 is

$$A(t) = T_0 + Ce^{-kt},$$

where C and k are constants. If $C = 100$, $k = .1$, and t is time measured in minutes, how long will it take a hot cup of coffee to cool to a temperature of 25° C in a room at 20° C?   30 min

RESOURCES FOR REVIEW

Instructor's Test Manual
  Chapter Test Forms, pp. 105–124
Instructor's Solution Manual
  pp. 203–210, 625–630

Study Guide
  pp. 190–192
Student's Solution Manual
  pp. 203–210

Test Generator
  IBM Disks
  5.25″ Disk B
  3.5″ Disk A
  Mac Disk A

CHAPTER 5 REVIEW EXERCISES   (TEXT PAGE 293) ■ ————————————

Solve each of the following equations.

1. $8^p = 32$ {5/3}

2. $9^{2y-1} = 27^y$ {2}

3. $\dfrac{8}{27} = b^{-3}$ {3/2}

4. $\dfrac{1}{2} = \left(\dfrac{b}{4}\right)^{1/4}$ {1/4}

5. The amount of a certain radioactive material, in grams, present after t days is given by

$$A(t) = 800e^{-.04t}.$$

How much is present after 5 days?   655 g

◉ *Match each equation with one of the graphs below.*

6. $y = \log_{.3} x$ (B)

7. $y = e^x$ (A)

8. $y = \ln x$ (C)

9. $y = (.3)^x$ (D)

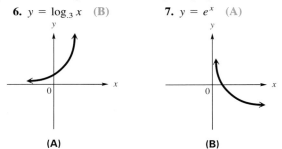

(A)   (B)   (C)   (D)

Graph each function defined as follows.

10. $y = 2^{x-3}$

11. $y = 2^{-x} + 1$

12. $y = \log_3 x$

13. $y = \log_{1/2}(x - 1)$

◉ **14.** $f(x) = \left(\dfrac{5}{4}\right)^x$ defines a(n) ——————— function.   increasing
 increasing/decreasing

◉ **15.** $f(x) = \log_{2/3} x$ defines a(n) ——————— function.   decreasing
 increasing/decreasing

◉ **16.** Which of the following is the graph of $f(x) = \log_3(x + 1)$?   (a)

(a)   (b)   (c)   (d)

(A)   (B)   (C)   (D)

◉ CONCEPTUAL   ✐ WRITING   ▲ CHALLENGING

Write each of the following equations in logarithmic form.

17. $2^5 = 32$ $\log_2 32 = 5$

18. $100^{1/2} = 10$ $\log_{100} 10 = 1/2$

19. $(1/16)^{1/4} = 1/2$ $\log_{1/16} 1/2 = 1/4$

20. $(3/4)^{-1} = 4/3$ $\log_{3/4} 4/3 = -1$

21. $10^{.4771} = 3$ $\log_{10} 3 = .4771$

22. $e^{2.4849} = 12$ $\log_e 12 = 2.4849$ or $\ln 12 = 2.4849$

Write each of the following equations in exponential form.

23. $\log_{10} .001 = -3$ $10^{-3} = .001$

24. $\log_2 \sqrt{32} = 5/2$ $2^{5/2} = \sqrt{32}$

25. $\log 3.45 = .537819$ $10^{.537819} = 3.45$

26. $\ln 45 = 3.806662$ $e^{3.806662} = 45$

Use properties of logarithms to write each of the following logarithms as a sum, difference, or product of logarithms.

27. $\log_3 \dfrac{mn}{5r}$ $\log_3 m + \log_3 n - \log_3 5 - \log_3 r$ $(1/2) \log_2 7 - \log_2 15$

29. $\log_5 x^2 y^4 \sqrt[5]{m^3 p}$ $2 \log_5 x + 4 \log_5 y + (1/5)(3 \log_5 m + \log_5 p)$

30. $\log_7 (7k + 5r^2)$ **cannot be simplified**

A calculator will be necessary for many of these exercises.

Find each of the following logarithms. Round to the nearest thousandth.

31. $\log 45.6$ **1.659**

32. $\log 1,230,000$ **6.090**

33. $\log .00056$ **−3.252**

34. $\log .0411$ **−1.386**

35. $\ln 35$ **3.555**

36. $\ln 470$ **6.153**

37. $\ln 144,000$ **11.878**

38. $\ln 98,000$ **11.493**

39. $\log_{3.4} 15.8$ **2.255**

40. $\log_{1/2} 9.45$ **−3.240**

41. $\log_3 769$ **6.049**

42. $\log_{2/3} 5/8$ **1.159**

The height in meters of the members of a certain tribe is approximated by

$$h = .5 + \log t,$$

where t is the tribe member's age in years, and $1 \le t \le 20$. Estimate the heights of tribe members of the following ages.

43. 2 yr   **.8 m**

44. 5 yr   **1.20 m**

45. 10 yr   **1.5 m**

46. 20 yr   **1.80 m**

Use a formula for compound interest to find the total amount on deposit if the following amounts are placed on deposit at the given interest rates for the given numbers of years.

47. $1000 at 8% compounded annually for 9 years   **$1999**

48. $312.45 at 6% compounded semiannually for 16 years   **$804.58**

Find the present value of each of the following future values.

49. $2000 at 4% compounded semiannually for 11 years   **$1293.68**

50. $2000 at 6% compounded quarterly for 8 years   **$1241.99**

51. Manuel deposits $10,000 for 12 years in an account paying 12% compounded annually. He then puts this total amount on deposit in another account paying 10% compounded semiannually for another 9 years. Find the total amount on deposit after the entire 21-year period.   **$93,761.31**

52. Anne Kelly deposits $12,000 for 8 years in an account paying 5% compounded annually. She then leaves the money alone with no further deposits at 6% compounded annually for an additional 6 years. Find the total amount on deposit after the entire 14-year period.
    $25,149.59

Solve each of the following equations. Round to the nearest thousandth if necessary.

53. $5^r = 11$ {1.490}

54. $10^{2r-3} = 17$ {2.115}

55. $e^{p+1} = 10$ {1.303}

56. $(1/2)^{3k+1} = 3$ {−.862}

57. $\log_{64} y = 1/3$ {4}

58. $\ln 6x - \ln (x + 1) = \ln 4$ {2}

59. $\log_{16} \sqrt{x + 1} = 1/4$ $\{3\}$

60. $\ln x + 3 \ln 2 = \ln \dfrac{2}{x}$ $\{1/2\}$

61. $\ln (\ln e^{-x}) = \ln 3$ $\{-3\}$

62. What annual interest rate, to the nearest tenth, will produce \$8780 if \$3500 is left at interest for 10 years?   9.6%

63. Find the annual interest rate to the nearest tenth that will produce \$27,208.60 if \$12,700 is left at interest compounded semiannually for 10 years.   7.8%

64. How many years (to the nearest tenth) will be needed for \$7800 to increase to \$14,088 at 6% compounded quarterly?   9.9 yr

65. Find the number of years (to the nearest tenth) needed for \$48,000 to become \$58,344 at 5% interest compounded semiannually.   4.0 yr

66. If the inflation rate were 10%, use the formula for continuous compounding to find the number of years for a \$1 item to cost \$2.   about 6.9 yr

67. In Exercise 66, find the number of years if the rate of inflation were 4%.   about 17.3 yr

How much would \$1200 *amount to at* 10% *compounded continuously for the following numbers of years?*

68. 4 yr   \$1790.19

69. 10 yr   \$3261.94

70. Historically, the consumption of electricity has increased at a continuous rate of 6% per year. If it continued to increase at this rate, find the number of years before exactly twice as much electricity would be needed.   about 12 yr

71. Suppose a conservation compaign together with higher rates caused demand for electricity to increase at only 2% per year. (See Exercise 70.) Find the number of years before twice as much electricity would be needed.   about 35 yr

72. *Escherichia coli* is a strain of bacteria that occurs naturally in many different organisms. Under certain conditions, the number of bacteria present in a colony is

$$E(t) = E_0 2^{t/30},$$

where $E(t)$ is the number of bacteria present t min after the beginning of an experiment, and E_0 is the number present when $t = 0$. Let $E_0 = 2,400,000$ and find the number of bacteria at the following times.
(a) 5 min   about 2,700,000
(b) 10 min   about 3,000,000

(c) 60 min   about 9,600,000
(d) 120 min   about 38,000,000

73. Use the function defined by

$$t = T \frac{\ln [1 + 8.33(A/K)]}{\ln 2}$$

from Section 5.3 to estimate the age of a rock sample, if tests show that A/K is .103 for the sample. Let $T = 1.26 \times 10^9$.   about 1.126 billion yr

74. The function defined by

$$A(t) = (5 \times 10^{12})e^{-.04t}$$

gives the known coal reserves in the world in year t (in tons), where $t = 0$ corresponds to 1970, and $-.04$ indicates the rate of consumption.
(a) Find the amount of coal available in 1990.
2.2 × 10¹² tons

(b) When will the coal reserves be half of what they were in 1970?   in 1987

CHAPTER 5 TEST   (TEXT PAGE 295) ■

Solve each of the following equations.

1. $25^{2x-1} = 125^{x+1}$ $\{5\}$

2. $\left(\dfrac{a}{2}\right)^{-1/3} = 4$ $\{1/32\}$

Write each of the following in logarithmic form.

3. $a^2 = b$ $\log_a b = 2$

4. $e^c = 4.82$ $\ln 4.82 = c$

Write each of the following in exponential form.

5. $\log_3 \sqrt{27} = \dfrac{3}{2}$ $3^{3/2} = \sqrt{27}$

6. $\ln 5 = a$ $e^a = 5$

◎ CONCEPTUAL   ✍ WRITING   ▲ CHALLENGING

7. What two points on the graph of $y = \log_a x$ can be found without computation?   (**a**, **1**) and (**1**, **0**)

Graph each function defined as follows.

8. $y = (1.5)^{x+2}$

9. $y = \log_{1/2} x$

10. Use properties of logarithms to write the following as a sum, difference, or product of logarithms.

$$\log_7 \frac{x^2\sqrt[4]{y}}{z^3} \qquad 2 \log_7 x + \frac{1}{4} \log_7 y - 3 \log_7 z$$

A calculator will be necessary for most of the remaining exercises in this test.

Find each of the following logarithms. Round to the nearest thousandth.

11. ln 2300   **7.741**

12. $\log_{2.7} 94.6$ **4.581**

13. What values of x cannot possibly be solutions of the equation $\log_a (2x - 3) = -1$? $(-\infty, 3/2]$

Solve the following equations. Round answers to the nearest thousandth.

14. $8^{2w-4} = 100$ {**3.107**}

15. $\log_3 (m + 2) = 2$ {**7**}

16. $\ln x - 4 \ln 3 = \ln \dfrac{5}{x}$ {**20.125**}

17. The amount of radioactive material, in grams, present after t days is given by $A(t) = 600e^{-.05t}$.
(a) Find the amount present after 12 days.   **329.3 g**
(b) Find the half-life of the material.   **13.9 days**

18. How much will $2500 amount to at 8% compounded continuously for 12 years?   **$6529.24**

19. If a population increases at a continuous rate of 4% per year, how many years will it take for the population to double in size?   **17.3 yr**

20. How many years, to the nearest tenth, will be needed for $5000 to increase to $18,000 at 12% compounded monthly?   **10.7 yr**

■ **CHAPTER 6   POLYNOMIAL AND RATIONAL FUNCTIONS**

6.1   RESOURCES

Instructor's Test Manual
Test Items, p. 344
Instructor's Solution Manual
pp. 211–214, 631–635
Study Guide
pp. 197–205
Student's Solution Manual
pp. 211–214

Test Generator
IBM Disks
5.25″ Disk B
3.5″ Disk A
Mac Disk A
Tutorials
IBM Disks
5.25″ Disk F
3.5″ Disk C
Mac Disk C
Apple II Module 3, Lessons 1–5

HarperCollins Videotape   6
"College Algebra: In Simplest Terms" Program   16

6.1 EXERCISES (TEXT PAGE 304) ■

Sketch the graph of each polynomial function P as defined in Exercises 1–10. See Examples 1–3.

1. $P(x) = \frac{1}{4}x^6$

2. $P(x) = -\frac{2}{3}x^5$

3. $P(x) = -\frac{5}{4}x^5$

4. $P(x) = 2x^4$

5. $P(x) = \frac{1}{2}x^3 + 1$

6. $P(x) = -x^4 + 2$

7. $P(x) = -(x + 1)^3$

8. $P(x) = \frac{1}{3}(x + 3)^4$

9. $P(x) = (x - 1)^4 + 2$

10. $P(x) = (x + 2)^3 - 1$

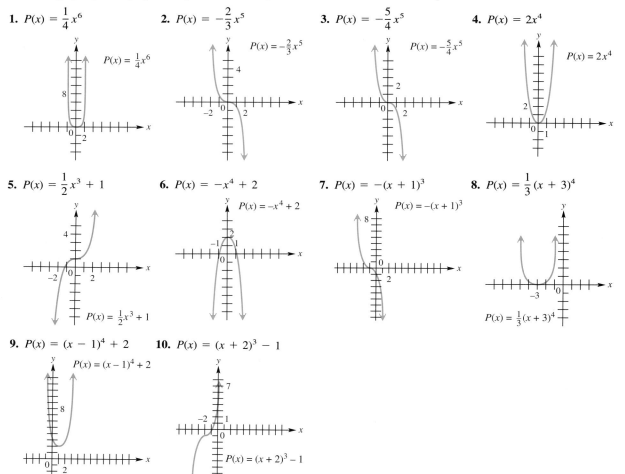

Graph each of the following. See Examples 4 and 5.

11. $P(x) = 2x(x - 3)(x + 2)$

12. $P(x) = x^2(x + 1)(x - 1)$

13. $P(x) = x^2(x - 2)(x + 3)^2$

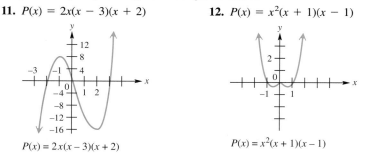

14. $P(x) = x^2(x - 5)(x + 3)(x - 1)$

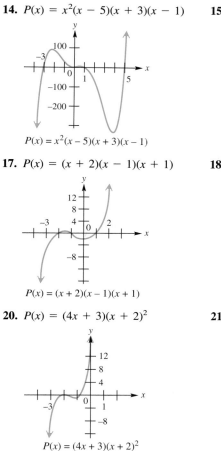

$P(x) = x^2(x - 5)(x + 3)(x - 1)$

15. $P(x) = x^3 - x^2 - 2x$

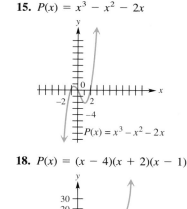

$P(x) = x^3 - x^2 - 2x$

16. $P(x) = -x^3 - 4x^2 - 3x$

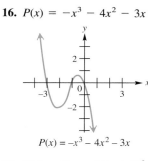

$P(x) = -x^3 - 4x^2 - 3x$

17. $P(x) = (x + 2)(x - 1)(x + 1)$

$P(x) = (x + 2)(x - 1)(x + 1)$

18. $P(x) = (x - 4)(x + 2)(x - 1)$

$P(x) = (x - 4)(x + 2)(x - 1)$

19. $P(x) = (3x - 1)(x + 2)^2$

$P(x) = (3x - 1)(x + 2)^2$

20. $P(x) = (4x + 3)(x + 2)^2$

$P(x) = (4x + 3)(x + 2)^2$

21. $P(x) = x^3 + 5x^2 - x - 5$
(*Hint:* Factor by grouping.)

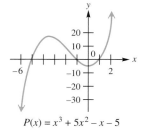

$P(x) = x^3 + 5x^2 - x - 5$

22. $P(x) = x^3 + x^2 - 36x - 36$

$P(x) = x^3 + x^2 - 36x - 36$

23. Which one of the following does not define a polynomial function?
   (a) $P(x) = x^2$ **(b)** $P(x) = (x + 1)^3$
   (c) $P(x) = \dfrac{1}{x}$ **(d)** $P(x) = 2x^5$ (c)

24. Write a short explanation of how the values of a, h, and k affect the graph of $P(x) = a(x - h)^n + k$ in comparison to the graph of $P(x) = x^n$.

25. During the early part of the twentieth century, the deer population of the Kaibab Plateau in Arizona experienced a rapid increase, because hunters had reduced the number of natural predators. The increase in population depleted the food resources and eventually caused the population to decline. For the period from 1905 to 1930, the deer population was approximated by the function D, where

$$D(x) = -.125x^5 + 3.125x^4 + 4000,$$

and x is time in years from 1905.
(a) Use a calculator to find enough points to graph the function.

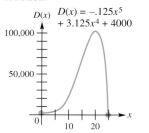

$D(x) = -.125x^5 + 3.125x^4 + 4000$

(b) From the graph, over what period of time (from 1905 to 1930) was the population increasing? relatively stable? decreasing?

1910 to 1925;   1905 to 1910;   1925 to 1930

26. The pressure of the oil in a reservoir tends to drop with time. By taking sample pressure readings for a particular oil reservoir, petroleum engineers have found that the change in pressure is given by the function P, where

$$P(t) = t^3 - 25t^2 + 200t,$$

and t is time in years from the date of the first reading.

(a) Graph the function.

(b) For what time period is the change in pressure (drop) increasing? decreasing?

increasing from $t = 0$ to $t = 6\ 2/3$ and from $t = 10$ on;   decreasing from $t = 6\ 2/3$ to $t = 10$

For each of the graphs in Exercises 27 and 28, state the relationships among the x-intercepts of the graph of $P(x)$, the zeros of the function P, and the solutions of the equation $P(x) = 0$. See Example 6.

27.   **28.**

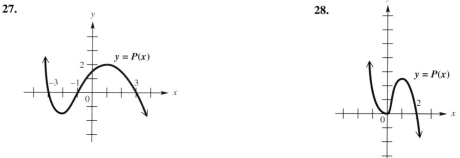

29. Give an example (if possible) of a polynomial function whose graph is symmetric with respect to the y-axis. $P(x) = x^2$ (There are many others.)

30. Give an example (if possible) of a polynomial function whose graph is symmetric with respect to both the x-axis and the origin. (Do not consider $P(x) = 0$.)   There is no such function.

Approximate maximum or minimum values of polynomial functions can be found for given intervals by first evaluating the function at the left endpoint of the interval, then adding .1 to the value of x and reevaluating the polynomial. Keep doing this until the right endpoint of the interval is reached. Then identify the approximate maximum and minimum value for the polynomial on the interval.

31. $P(x) = x^3 + 4x^2 - 8x - 8$, $[-3.8, -3]$

maximum is 26.136 when $x = -3.4$;   minimum is 25 when $x = -3$

32. $P(x) = x^3 + 4x^2 - 8x - 8$, $[.3, 1]$

maximum is -10.013 when $x = .3$;   minimum is -11.328 when $x = .8$

33. $P(x) = 2x^3 - 5x^2 - x + 1$, $[-1, 0]$

maximum is 1.048 when $x = -.1$;   minimum is -5 when $x = -1$

34. $P(x) = 2x^3 - 5x^2 - x + 1$, $[1.4, 2]$

maximum is -4.712 when $x = 1.4$;   minimum is -5.336 when $x = 1.8$

◉ CONCEPTUAL   ✎ WRITING   ▲ CHALLENGING

35. $P(x) = x^4 - 7x^3 + 13x^2 + 6x - 28$, $[-2, -1]$

maximum is 84 when $x = -2$; minimum is -13 when $x = -1$

36. $P(x) = x^4 - 7x^3 + 13x^2 + 6x - 28$, $[2, 3]$

maximum is -1 when $x = 3$; minimum is -4 when $x = 2$

◎ **37.** The graphs of the functions defined by $P(x) = x$, $P(x) = x^3$, and $P(x) = x^5$ all pass through the points $(-1, -1)$, $(0, 0)$, and $(1, 1)$. Draw a graph showing these three functions for the domain $[-1, 1]$. Use your result to decide how the graph of $P(x) = x^{11}$ would look for the same interval.

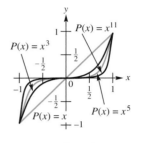

◎ **38.** The graphs of the functions defined by $P(x) = x^2$, $P(x) = x^4$, and $P(x) = x^6$ all pass through the points $(-1, 1)$, $(0, 0)$, and $(1, 1)$. Draw a graph showing these three functions for the domain $[-1, 1]$. Use your result to decide how the graph of $P(x) = x^{12}$ would look for the same interval.

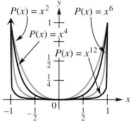

6.2 RESOURCES

Instructor's Test Manual
Test Items, p. 344–346
Instructor's Solution Manual
pp. 215–218, 635–638
Study Guide
pp. 205–210
Student's Solution Manual
pp. 215–218

Test Generator
IBM Disks
5.25" Disk B
3.5" Disk A
Mac Disk A
Tutorials
IBM Disks
5.25" Disk F
3.5" Disk C
Mac Disk C
Apple II Module 3, Lessons 1–5

HarperCollins Videotape 6
Transparency T-29, T-30(a)–(d)

6.2 EXERCISES (TEXT PAGE 312) ∎

Perform each of the following divisions. See Examples 1–3.

1. $\dfrac{4m^3 - 8m^2 + 16m}{2m}$ $\quad 2m^2 - 4m + 8$

2. $\dfrac{30k^5 - 12k^3 + 18k^2}{6k^2}$ $\quad 5k^3 - 2k + 3$

3. $\dfrac{25x^2y^4 - 15x^3y^3 + 40x^4y^2}{5x^2y^2}$ $\quad 5y^2 - 3xy + 8x^2$

4. $\dfrac{-8r^3s - 12r^2s^2 + 20rs^3}{4rs}$ $\quad -2r^2 - 3rs + 5s^2$

5. $\dfrac{6y^2 + y - 2}{2y - 1}$ $3y + 2$

6. $\dfrac{16r^2 + 2r - 3}{2r + 1}$ $8r - 3$

7. $\dfrac{8z^2 + 14z - 20}{2z + 5}$ $4z - 3 + \dfrac{-5}{2z + 5}$

8. $\dfrac{6a^2 - 13a - 18}{2a + 1}$ $3a - 8 + \dfrac{-10}{2a + 1}$

9. $\dfrac{2x^3 - 11x^2 + 19x - 10}{2x - 5}$ $x^2 - 3x + 2$

10. $\dfrac{3p^3 - 11p^2 + 5p + 3}{3p + 1}$ $p^2 - 4p + 3$

11. $\dfrac{15x^3 + 11x^2 + 20}{3x + 4}$ $5x^2 - 3x + 4 + \dfrac{4}{3x + 4}$

12. $\dfrac{4r^3 + 2r^2 - 14r + 15}{2r + 5}$ $2r^2 - 4r + 3$

13. $\dfrac{x^4 + 2x^3 + 2x^2 - 2x - 3}{x^2 - 1}$ $x^2 + 2x + 3$

14. $\dfrac{2y^5 + y^3 - 2y^2 - 1}{2y^2 + 1}$ $y^3 - 1$

15. $\dfrac{4z^5 - 4z^2 - 5z + 3}{2z^2 + z + 1}$ $2z^3 - z^2 - \dfrac{1}{2}z - \dfrac{5}{4} + \dfrac{-\frac{13}{4}z + \frac{17}{4}}{2z^2 + z + 1}$

16. $\dfrac{12z^4 - 25z^3 + 35z^2 - 26z + 10}{4z^2 - 3z + 5}$ $3z^2 - 4z + 2$

17. When a second-degree polynomial in x is divided by a third-degree polynomial in x, is the quotient a polynomial? Explain.

18. If polynomial P is divided by polynomial Q and the remainder is 0, then Q is a
_____ of P.   factor (or divisor)

Use synthetic division to perform each of the following divisions. See Example 4.

19. $\dfrac{x^3 + 2x^2 - 17x - 10}{x + 5}$ $x^2 - 3x - 2$

20. $\dfrac{a^4 + 4a^3 + 2a^2 + 9a + 4}{a + 4}$ $a^3 + 2a + 1$

21. $\dfrac{m^4 - 3m^3 - 4m^2 + 12m}{m - 2}$ $m^3 - m^2 - 6m$

22. $\dfrac{p^4 - 3p^3 - 5p^2 + 2p - 16}{p + 2}$ $p^3 - 5p^2 + 5p - 8$

23. $\dfrac{3x^3 - 11x^2 - 20x + 3}{x - 5}$ $3x^2 + 4x + \dfrac{3}{x - 5}$

24. $\dfrac{4p^3 + 8p^2 - 16p - 9}{p + 3}$ $4p^2 - 4p - 4 + \dfrac{3}{p + 3}$

25. $\dfrac{x^5 + 3x^4 + 2x^3 + 2x^2 + 3x + 1}{x + 2}$ $x^4 + x^3 + 2x - 1 + \dfrac{3}{x + 2}$

26. $\dfrac{m^6 - 3m^4 + 2m^3 - 6m^2 - 5m + 3}{m + 2}$ $m^5 - 2m^4 + m^3 - 6m + 7 + \dfrac{-11}{m + 2}$

27. $\dfrac{\frac{1}{3}x^3 - \frac{2}{9}x^2 + \frac{1}{27}x + 1}{x - \frac{1}{3}}$ $\dfrac{1}{3}x^2 - \dfrac{1}{9}x + \dfrac{1}{x - \frac{1}{3}}$

28. $\dfrac{x^3 + x^2 + \frac{1}{2}x + \frac{1}{8}}{x + \frac{1}{2}}$ $x^2 + \dfrac{1}{2}x + \dfrac{1}{4}$

29. $\dfrac{y^3 - 1}{y - 1}$ $y^2 + y + 1$

30. $\dfrac{r^5 - 1}{r - 1}$ $r^4 + r^3 + r^2 + r + 1$

31. $\dfrac{x^4 - 1}{x - 1}$ $x^3 + x^2 + x + 1$

32. $\dfrac{x^7 + 1}{x + 1}$ $x^6 - x^5 + x^4 - x^3 + x^2 - x + 1$

Express each polynomial in the form $P(x) = (x - k)Q(x) + r$ for the given value of k.

33. $P(x) = x^3 + x^2 + x - 8$; $k = 1$ $P(x) = (x - 1)(x^2 + 2x + 3) + (-5)$

34. $P(x) = 2x^3 + 3x^2 + 4x - 10$; $k = -1$ $P(x) = (x + 1)(2x^2 + x + 3) + (-13)$

35. $P(x) = -x^3 + 2x^2 + 4$; $k = -2$ $P(x) = (x + 2)(-x^2 + 4x - 8) + 20$

◎ **CONCEPTUAL**   ✎ **WRITING**   ▲ **CHALLENGING**

36. $P(x) = -4x^3 + 2x^2 - 3x - 10;$ $k = 2$ $P(x) = (x - 2)(-4x^2 - 6x - 15) + (-40)$

37. $P(x) = x^4 - 3x^3 + 2x^2 - x + 5;$ $k = 3$ $P(x) = (x - 3)(x^3 + 2x + 5) + 20$

38. $P(x) = 2x^4 + x^3 - 15x^2 + 3x;$ $k = -3$ $P(x) = (x + 3)(2x^3 - 5x^2 + 3) + (-9)$

For each of the following polynomials, use the remainder theorem to find P(k). See Examples 5 and 6.

39. $k = 3;$ $P(x) = x^2 - 4x + 5$ 2

40. $k = -2;$ $P(x) = x^2 + 5x + 6$ 0

41. $k = -2;$ $P(x) = 5x^3 + 2x^2 - x$ -30

42. $k = 2;$ $P(x) = 2x^3 - 3x^2 - 5x + 4$ -2

43. $k = 2 + i;$ $P(x) = x^2 - 5x + 1$ $-6 - i$

44. $k = 3 - 2i;$ $P(x) = x^2 - x + 3$ $5 - 10i$

45. $k = 1 - i;$ $P(x) = x^3 + x^2 - x + 1$ $-2 - 3i$

46. $k = 2 - 3i;$ $P(x) = x^3 + 2x^2 + x - 5$ $-59 - 36i$

Use synthetic division to decide whether the given number is a zero of the given polynomial. See Example 7.

47. $2;$ $P(x) = x^2 + 2x - 8$ yes

48. $-1;$ $P(m) = m^2 + 4m - 5$ no

49. $2;$ $P(g) = g^3 - 3g^2 + 4g - 4$ yes

50. $-3;$ $P(m) = m^3 + 2m^2 - m + 6$ yes

51. $4;$ $P(r) = 2r^3 - 6r^2 - 9r + 6$ no

52. $-4;$ $P(y) = 9y^3 + 39y^2 + 12y$ yes

53. $2 + i;$ $P(k) = k^2 + 3k + 4$ no

54. $1 - 2i;$ $P(z) = z^2 - 3z + 5$ no

55. $i;$ $P(x) = x^3 + 2ix^2 + 2x + i$ yes

56. $-i;$ $P(p) = p^3 - ip^2 + 3p + 5i$ no

6.3 RESOURCES

Instructor's Test Manual
Test Items, p. 346–347

Instructor's Solution Manual
pp. 219–225, 639–644

Study Guide
pp. 210–216

Student's Solution Manual
pp. 219–225

Test Generator
IBM Disks
5.25" Disk B
3.5" Disk A
Mac Disk A

Tutorials
IBM Disks
5.25" Disk F
3.5" Disk C
Mac Disk C
Apple II Module 3, Lessons 1–5

HarperCollins Videotape 6

6.3 EXERCISES (TEXT PAGE 318) ■

Use the factor theorem to decide whether the second polynomial is a factor of the first. See Examples 1 and 2.

1. $4x^2 + 2x + 42;$ $x - 3$ no

2. $-3x^2 - 4x + 2;$ $x + 2$ no

3. $x^3 + 2x^2 - 3;$ $x - 1$ yes

4. $2x^3 + x + 2;$ $x + 1$ no

5. $3x^3 - 12x^2 - 11x - 20;$ $x - 5$ yes

6. $4x^3 + 6x^2 - 5x - 2;$ $x + 2$ yes

7. $2x^4 + 5x^3 - 2x^2 + 5x + 3;$ $x + 3$ no

8. $5x^4 + 16x^3 - 15x^2 + 8x + 16;$ $x + 4$ yes

For each of the following polynomial functions, give all zeros and their multiplicities. See Example 3.

9. $P(x) = x^4(x + 3)^5(x - 8)^2$ 0: multiplicity 4;   −3: multiplicity 5;   8: multiplicity 2

10. $P(x) = x^9(x - 4)^6(x + 7)^4$ 0: multiplicity 9;   4: multiplicity 6;   −7: multiplicity 4

11. $P(x) = (4x - 7)^3(x - 5)$ 7/4: multiplicity 3;   5: multiplicity 1

12. $P(x) = (5x + 1)^3(x + 1)$ −1/5: multiplicity 3;   −1: multiplicity 1

13. $P(x) = (x - i)^4(x + i)^4$ i: multiplicity 4; $-i$: multiplicity 4

14. $P(x) = (x + 3i)^9(x - 3i)^9$ $-3i$: multiplicity 9; $3i$: multiplicity 9

For each of the following, find a function P defined by a polynomial of degree 3 with real coefficients that satisfies the given conditions. See Example 4.

15. Zeros of −3, −1, and 4; $P(2) = 5$ $P(x) = -\dfrac{1}{6}x^3 + \dfrac{13}{6}x + 2$

16. Zeros of 1, −1, and 0; $P(2) = -3$ $P(x) = -\dfrac{1}{2}x^3 + \dfrac{1}{2}x$

17. Zeros of −2, 1, and 0; $P(-1) = -1$ $P(x) = -\dfrac{1}{2}x^3 - \dfrac{1}{2}x^2 + x$

18. Zeros of 2, 5, and −3; $P(1) = -4$ $P(x) = -\dfrac{1}{4}x^3 + x^2 + \dfrac{11}{4}x - \dfrac{15}{2}$

19. Zeros of 3, i, and $-i$; $P(2) = 50$ $P(x) = -10x^3 + 30x^2 - 10x + 30$

20. Zeros of −2, i, and $-i$; $P(-3) = 30$ $P(x) = -3x^3 - 6x^2 - 3x - 6$

For each of the following, find a function P defined by a polynomial of lowest degree with real coefficients having the given zeros. See Example 5.

21. $5 + i$ and $5 - i$ $P(x) = x^2 - 10x + 26$ **22.** $3 - 2i$ and $3 + 2i$ $P(x) = x^2 - 6x + 13$

23. 2, $1 - i$, and $1 + i$ $P(x) = x^3 - 4x^2 + 6x - 4$ **24.** −3, $2 - i$, and $2 + i$ $P(x) = x^3 - x^2 - 7x + 15$

25. $1 + \sqrt{2}$, $1 - \sqrt{2}$, and 1 $P(x) = x^3 - 3x^2 + x + 1$

26. $1 - \sqrt{3}$, $1 + \sqrt{3}$, and −2 $P(x) = x^3 - 6x - 4$

27. $2 + i$, $2 - i$, 3, and −1 $P(x) = x^4 - 6x^3 + 10x^2 + 2x - 15$

28. $3 + 2i$, $3 - 2i$, −1, and 2 $P(x) = x^4 - 7x^3 + 17x^2 - x - 26$

29. 2 and $3 + i$ $P(x) = x^3 - 8x^2 + 22x - 20$ **30.** −1 and $4 - 2i$ $P(x) = x^3 - 7x^2 + 12x + 20$

31. $2 - i$ and $3 + 2i$ $P(x) = x^4 - 10x^3 + 42x^2 - 82x + 65$

32. $5 + i$ and $4 - i$ $P(x) = x^4 - 18x^3 + 123x^2 - 378x + 442$

33. 4, $1 - 2i$, and $3 + 4i$ $P(x) = x^5 - 12x^4 + 74x^3 - 248x^2 + 445x - 500$

34. −1, $1 + \sqrt{2}$, $1 - \sqrt{2}$, and $1 + 4i$ $P(x) = x^5 - 3x^4 + 16x^3 - 12x^2 - 49x - 17$

35. $1 + 2i$ and 2 (multiplicity 2) $P(x) = x^4 - 6x^3 + 17x^2 - 28x + 20$

36. $2 + i$ and −3 (multiplicity 2) $P(x) = x^4 + 2x^3 - 10x^2 - 6x + 45$

For each of the following polynomials, one zero is given. Find all others. See Example 6.

37. $P(x) = x^3 - x^2 - 4x - 6$;   3 $-1 + i$, $-1 - i$ **38.** $P(x) = x^3 - 5x^2 + 17x - 13$;   1 $2 + 3i$, $2 - 3i$

39. $P(x) = 2x^3 - 2x^2 - x - 6$;   2 $-\dfrac{1}{2} + \dfrac{\sqrt{5}}{2}i$, $-\dfrac{1}{2} - \dfrac{\sqrt{5}}{2}i$

◎ CONCEPTUAL   ✍ WRITING   ▲ CHALLENGING

40. $P(x) = 2x^3 - 5x^2 + 6x - 2$; $1 + i$ $1 - i, 1/2$ **41.** $P(x) = x^4 + 5x^2 + 4$; $-i$ $i, 2i, -2i$

42. $P(x) = x^4 + 10x^3 + 27x^2 + 10x + 26$; i $-i, -5 + i, -5 - i$

43. $P(x) = x^4 - 3x^3 + 6x^2 + 2x - 60$; $1 + 3i$ $3, -2, 1 - 3i$

44. $P(x) = x^4 - 6x^3 - x^2 + 86x + 170$; $5 + 3i$ $5 - 3i, -2 + i, -2 - i$

Factor P(x) into linear factors given that k is a zero of P. See Example 7.

45. $P(x) = 2x^3 - 3x^2 - 17x + 30$; $k = 2$ $P(x) = (x - 2)(2x - 5)(x + 3)$

46. $P(x) = 2x^3 - 3x^2 - 5x + 6$; $k = 1$ $P(x) = (x - 1)(2x + 3)(x - 2)$

47. $P(x) = 6x^3 + 25x^2 + 3x - 4$; $k = -4$ $P(x) = (x + 4)(3x - 1)(2x + 1)$

48. $P(x) = 8x^3 + 50x^2 + 47x - 15$; $k = -5$ $P(x) = (x + 5)(2x + 3)(4x - 1)$

49. $P(x) = x^3 + (7 - 3i)x^2 + (12 - 21i)x - 36i$; $k = 3i$ $P(x) = (x - 3i)(x + 4)(x + 3)$

50. $P(x) = 2x^3 + (3 + 2i)x^2 + (1 + 3i)x + i$; $k = -i$ $P(x) = (x + i)(2x + 1)(x + 1)$

51. $P(x) = 2x^3 + (3 - 2i)x^2 + (-8 - 5i)x + 3 + 3i$; $k = 1 + i$
    $P(x) = (x - 1 - i)(2x - 1)(x + 3)$

52. $P(x) = 6x^3 + (19 - 6i)x^2 + (16 - 7i)x + 4 - 2i$; $k = -2 + i$
    $P(x) = (x + 2 - i)(2x + 1)(3x + 2)$

53. Show that -2 is a zero of multiplicity 2 of P, where $P(x) = x^4 + 2x^3 - 7x^2 - 20x - 12$, and find all other complex zeros. Then write $P(x)$ in factored form.
zeros are $-2, -1, 3$;
$P(x) = (x + 2)^2(x + 1)(x - 3)$

54. Show that -1 is a zero of multiplicity 3 of P, where $P(x) = x^5 + 9x^4 + 33x^3 + 55x^2 + 42x + 12$, and find all other complex zeros. Then write $P(x)$ in factored form.
zeros are $-1, -3 + i\sqrt{3}, -3 - i\sqrt{3}$;
$P(x) = (x + 1)^3(x + 3 - i\sqrt{3})(x + 3 + i\sqrt{3})$

◉ **55.** What are the possible numbers of real zeros (counting multiplicities) for a polynomial function with real coefficients of degree five?   1, 3, or 5

◉ **56.** Explain why a function defined by a polynomial of degree four with real coefficients has either 0, 2, or 4 real zeros.

◉ **57.** Explain why it is not possible for a function defined by a polynomial of degree 3 with real coefficients to have zeros of 1, 2, and $1 + i$.

58. Show that the zeros of the function P defined by the polynomial $P(x) = x^3 + ix^2 - (7 - i)x + (6 - 6i)$ are $1 - i$, 2, and -3. Does the conjugate zeros theorem apply? Why or why not?

59. The displacement at time t of a particle moving along a straight line is given by
$$s(t) = t^3 - 2t^2 - 5t + 6,$$
where t is in seconds and s is measured in centimeters. The displacement is 0 after 1 second has elapsed. At what other times (positive) is the displacement 0?
at 3 seconds

60. For headphone radios, the cost function (in thousands of dollars) is given by
$$C(x) = 2x^3 - 9x^2 + 17x - 4,$$
and the revenue function (in thousands of dollars) is given by $R(x) = 5x$, where x is the number of items (in hundred thousands) produced. Cost equals revenue if 200,000 items are produced ($x = 2$). Find all other break-even points.   50,000 (when $x = 1/2$)

If c and d are complex numbers, prove each of the following statements. These statements are used in Exercise 66. (Hint: Let $c = a + bi$ and $d = m + ni$ and form the conjugates, the sums, and the products. The notation $\bar{c}$ represents the conjugate of c.)

61. $\overline{c + d} = \bar{c} + \bar{d}$

62. $\overline{cd} = \bar{c} \cdot \bar{d}$

63. $\bar{a} = a$ for any real number a

64. $\overline{c^n} = (\bar{c})^n$, n is a positive integer

▲ Use the theorems presented in this section to prove each of the following statements.

65. A polynomial of degree n has at most n distinct zeros. (*Hint:* Use an indirect proof, where you assume the opposite of the statement is true and show that it leads to a contradiction.)

66. Complete the proof of the conjugate zeros theorem, outlined below. Assume that

$$P(x) = a_n x^n + a_{n-1} x^{n-1} + \cdots + a_1 x + a_0,$$

where all coefficients are real numbers.

(a) If the complex number z is a zero of P, find $P(z)$.

(b) Take the conjugate of both sides of the result from part (a).

(c) Use generalizations of the properties given in Exercises 61–64 on the result of part (b) to show that $a_n(\bar{z})^n + a_{n-1}(\bar{z})^{n-1} + \cdots + a_1(\bar{z}) + a_0 = 0.$

(d) Why does the result in part (c) mean that $\bar{z}$ is a zero of P?

6.4   RESOURCES

Instructor's Test Manual
Test Items, p. 347–348

Instructor's Solution Manual
pp. 225–233, 645–651

Study Guide
pp. 217–221

Student's Solution Manual
pp. 225–233

Test Generator
IBM Disks
5.25" Disk B
3.5" Disk A
Mac Disk A

Tutorials
IBM Disks
5.25" Disk F
3.5" Disk C
Mac Disk C
Apple II Module 3, Lessons 1–5

HarperCollins Videotape   7

6.4 EXERCISES   (TEXT PAGE 323) ■

Give all possible rational zeros for each function P defined as follows.

1. $P(x) = 6x^3 + 17x^2 - 31x - 1$ $\pm 1, \pm 1/2, \pm 1/3, \pm 1/6$

2. $P(x) = 15x^3 + 61x^2 + 2x - 1$ $\pm 1, \pm 1/3, \pm 1/5, \pm 1/15$

3. $P(x) = 12x^3 + 20x^2 - x - 2$ $\pm 1, \pm 1/2, \pm 1/3, \pm 1/4, \pm 1/6, \pm 1/12, \pm 2, \pm 2/3$

4. $P(x) = 12x^3 + 40x^2 + 41x + 3$ $\pm 1, \pm 1/2, \pm 1/3, \pm 1/4, \pm 1/6, \pm 1/12, \pm 3, \pm 3/2, \pm 3/4$

5. $P(x) = 2x^3 + 7x^2 + 12x - 8$ $\pm 1, \pm 1/2, \pm 2, \pm 4, \pm 8$

6. $P(x) = 2x^3 + 20x^2 + 68x - 40$ $\pm 1, \pm 2, \pm 4, \pm 5, \pm 8, \pm 10, \pm 20, \pm 40, \pm 5/2, \pm 1/2$

7. $P(x) = x^4 + 4x^3 + 3x^2 - 10x + 50$ $\pm 1, \pm 2, \pm 5, \pm 10, \pm 25, \pm 50$

8. $P(x) = x^4 - 2x^3 + x^2 + 18$ $\pm 1, \pm 2, \pm 3, \pm 6, \pm 9, \pm 18$

9. Discuss a major drawback of the rational zeros theorem.

10. Can the rational zeros theorem be used for a function defined by a polynomial with irrational coefficients?   no

Find all rational zeros of each function P defined as follows. See Example 1.

11. $P(x) = x^3 - 2x^2 - 13x - 10$ $-1, -2, 5$

12. $P(x) = x^3 + 5x^2 + 2x - 8$ $1, -4, -2$

13. $P(x) = x^3 + 6x^2 - x - 30$ $2, -3, -5$

14. $P(x) = x^3 - x^2 - 10x - 8$ $-2, 4, -1$

◎ CONCEPTUAL   ▨ WRITING   ▲ CHALLENGING

15. $P(x) = x^3 + 9x^2 - 14x - 24$ no rational zeros   **16.** $P(x) = x^3 + 3x^2 - 4x - 12$ $2, -3, -2$

17. $P(x) = x^4 + 9x^3 + 21x^2 - x - 30$ $1, -2, -3, -5$ **18.** $P(x) = x^4 + 4x^3 - 7x^2 - 34x - 24$ $3, -2, -4, -1$

Find the rational zeros of each function P defined as follows. Then write each polynomial in factored form. See Example 2.

19. $P(x) = 6x^3 + 17x^2 - 31x - 12$ $-4, 3/2, -1/3$; $P(x) = (x + 4)(2x - 3)(3x + 1)$

20. $P(x) = 15x^3 + 61x^2 + 2x - 8$ $-4, -2/5, 1/3$; $P(x) = (x + 4)(5x + 2)(3x - 1)$

21. $P(x) = 12x^3 + 20x^2 - x - 6$ $-3/2, -2/3, 1/2$; $P(x) = (2x - 1)(3x + 2)(2x + 3)$

22. $P(x) = 12x^3 + 40x^2 + 41x + 12$ $-3/2, -4/3, -1/2$; $P(x) = (2x + 3)(3x + 4)(2x + 1)$

23. $P(x) = 2x^3 + 7x^2 + 12x - 8$ $1/2$; $P(x) = (2x - 1)(x^2 + 4x + 8)$

24. $P(x) = 2x^3 + 20x^2 + 68x - 40$ no rational zeros;   can't be factored

25. $P(x) = 2x^4 + 3x^3 - 4x^2 - 3x + 2$
  $-2, -1, 1/2, 1$; $P(x) = (x + 2)(x + 1)(2x - 1)(x - 1)$

26. $P(x) = x^4 - 2x^3 + x^2 + 18$ no rational zeros;   can't be factored

27. $P(x) = 3x^4 + 5x^3 - 10x^2 - 20x - 8$
  $-1, -2, -2/3, 2$; $P(x) = (x + 1)(x + 2)(3x + 2)(x - 2)$

28. $P(x) = 6x^4 + x^3 - 7x^2 - x + 1$
  $1, -1, 1/3, -1/2$; $P(x) = (x - 1)(x + 1)(3x - 1)(2x + 1)$

Find all rational zeros of each function P defined as follows. See Example 3.

29. $P(x) = x^3 - \dfrac{4}{3}x^2 - \dfrac{13}{3}x - 2$ $-2/3, -1, 3$ **30.** $P(x) = x^3 + x^2 - \dfrac{16}{9}x + \dfrac{4}{9}$ $-2, 1/3, 2/3$

31. $P(x) = x^4 + \dfrac{1}{4}x^3 + \dfrac{11}{4}x^2 + x - 5$ $1, -5/4$ **32.** $P(x) = \dfrac{10}{7}x^4 - x^3 - 7x^2 + 5x - \dfrac{5}{7}$ $1/2, 1/5$

33. $P(x) = \dfrac{1}{3}x^5 + x^4 - \dfrac{5}{3}x^3 - \dfrac{11}{3}x^2 + 4$ 1

34. $P(x) = x^5 + x^4 - \dfrac{37}{4}x^2 + \dfrac{9}{4}x + \dfrac{9}{4}$ no rational zeros

For each of the polynomial functions defined as follows, find all rational zeros and factor the polynomial. Then graph the function using the method described in Section 6.1.

35. $P(x) = 2x^3 - 5x^2 - x + 6$ **36.** $P(x) = 3x^3 + x^2 - 10x - 8$
  $-1, 3/2, 2$; $P(x) = (x + 1)(2x - 3)(x - 2)$ $2; -4/3, -1$; $P(x) = (x - 2)(3x + 4)(x + 1)$

$P(x) = 2x^3 - 5x^2 - x + 6$

$P(x) = 3x^3 + x^2 - 10x - 8$

37. $P(x) = x^3 + x^2 - 8x - 12$

-2 (multiplicity 2), 3; $P(x) = (x + 2)^2(x - 3)$

$P(x) = x^3 + x^2 - 8x - 12$

38. $P(x) = x^3 + 6x^2 - 32$

-4 (multiplicity 2), 2; $P(x) = (x + 4)^2(x - 2)$

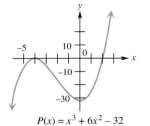

$P(x) = x^3 + 6x^2 - 32$

39. $P(x) = -x^3 - x^2 + 8x + 12$

-2 (multiplicity 2), 3; $P(x) = (x + 2)^2(-x + 3)$

$P(x) = -x^3 - x^2 + 8x + 12$

40. $P(x) = -x^3 + 10x^2 - 33x + 36$

3 (multiplicity 2), 4; $P(x) = (x - 3)^2(-x + 4)$

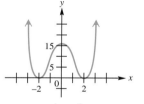

$P(x) = -x^3 + 10x^2 - 33x + 36$

41. $P(x) = x^4 - 18x^2 + 81$

3 (multiplicity 2), -3 (multiplicity 2);
$P(x) = (x - 3)^2(x + 3)^2$

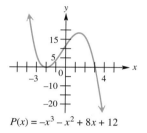

$P(x) = x^4 - 18x^2 + 81$

42. $P(x) = x^4 - 8x^2 + 16$

-2 (multiplicity 2), 2 (multiplicity 2);
$P(x) = (x + 2)^2(x - 2)^2$

$P(x) = x^4 - 8x^2 + 16$

43. $P(x) = 2x^4 + x^3 - 6x^2 - 7x - 2$

$-1/2$, 2, -1 (multiplicity 2);
$P(x) = (2x + 1)(x - 2)(x + 1)^2$

$P(x) = 2x^4 + x^3 - 6x^2 - 7x - 2$

44. $P(x) = 3x^4 - 7x^3 - 6x^2 + 12x + 8$

$-2/3$, -1, 2 (multiplicity 2);
$P(x) = (3x + 2)(x + 1)(x - 2)^2$

$P(x) =$
$3x^4 - 7x^3 - 6x^2 + 12x + 8$

ⓒ CONCEPTUAL   ✎ WRITING   ▲ CHALLENGING

▲ ◎ **45.** Show that $P(x) = x^2 - 2$ has no rational zeros, so $\sqrt{2}$ must be irrational.

▲ ◎ **46.** Show that $P(x) = x^2 - 5$ has no rational zeros, so $\sqrt{5}$ must be irrational.

47. Show that $P(x) = x^4 + 5x^2 + 4$ has no rational zeros.

48. Show that $P(x) = x^5 - 3x^3 + 5$ has no rational zeros.

▲ ◎ **49.** Show that any integer zeros of a polynomial function must be factors of the constant term a_0 of the defining polynomial.

◎ **50.** If k is a zero of the polynomial function P, then $P(k) = $ _____, a factor of $P(x)$ is _____, and an x-intercept of the graph of P is _____. 0; $(x - k)$; k

---

6.5 RESOURCES

Instructor's Test Manual
  Test Items, p. 348–349
Instructor's Solution Manual
  pp. 233–241, 651–660
Study Guide
  pp. 221–229
Student's Solution Manual
  pp. 233–241

Test Generator
  IBM Disks
    5.25" Disk B
    3.5" Disk A
    Mac Disk A
Tutorials
  IBM Disks
    5.25" Disk F
    3.5" Disk C
  Mac Disk C
  Apple II Module 3, Lessons 1–5

HarperCollins Videotape   7

6.5 EXERCISES   (TEXT PAGE 331)   ▪

▦ *A calculator will be helpful in working many of the exercises in this section.*

Use Descartes' rule of signs to find the number of positive and negative real zeros possible for each polynomial function defined as follows. See Examples 1 and 2.

1. $P(x) = 2x^3 - 4x^2 + 2x + 7$ **2 or 0 positive real zeros;   1 negative real zero**

2. $P(x) = x^3 + 2x^2 + x - 10$ **1 positive real zero;   2 or 0 negative real zeros**

3. $P(x) = 5x^4 + 3x^2 + 2x - 9$ **1 positive real zero;   1 negative real zero**

4. $P(x) = 3x^4 + 2x^3 - 8x^2 - 10x - 1$ **1 positive real zero;   3 or 1 negative real zeros**

5. $P(x) = x^5 + 3x^4 - x^3 + 2x + 3$ **2 or 0 positive real zeros;   3 or 1 negative real zeros**

6. $P(x) = 2x^5 - x^4 + x^3 - x^2 + x + 5$

  4, 2, or 0 positive real zeros;   1 negative real zero

Use the intermediate value theorem for polynomial functions to show that each function defined as follows has a real zero between the numbers given. See Example 3.

7. $P(x) = 3x^2 - 2x - 6$;   1 and 2

8. $P(x) = x^3 + x^2 - 5x - 5$;   2 and 3

9. $P(x) = 2x^3 - 8x^2 + x + 16$;   2 and 2.5

10. $P(x) = 3x^3 + 7x^2 - 4$;   1/2 and 1

11. $P(x) = 2x^4 - 4x^2 + 3x - 6$;   2 and 1.5

12. $P(x) = x^4 - 4x^3 - x + 1$;   1 and .3

◉ **13.** Suppose that a polynomial function P is defined in such a way that $P(2) = -4$ and $P(2.5) = 2$. What conclusion does the intermediate value theorem allow you to make? **P has a zero between 2 and 2.5.**

🖉 **14.** Suppose that a polynomial function P is defined in such a way that $P(3) = -4$ and $P(4) = -10$. Can we be certain that there is no zero between 3 and 4? Explain.
◉

Show that the real zeros of the polynomial functions defined as follows satisfy the given conditions.

15. $P(x) = x^4 - x^3 + 3x^2 - 8x + 8$; no real zero greater than 2

16. $P(x) = 2x^5 - x^4 + 2x^3 - 2x^2 + 4x - 4$; no real zero greater than 1

17. $P(x) = x^4 + x^3 - x^2 + 3$; no real zero less than -2

18. $P(x) = x^5 + 2x^3 - 2x^2 + 5x + 5$; no real zero less than -1

19. $P(x) = 3x^4 + 2x^3 - 4x^2 + x - 1$; no real zero greater than 1

20. $P(x) = 3x^4 + 2x^3 - 4x^2 + x - 1$; no real zero less than -2

21. $P(x) = x^5 - 3x^3 + x + 2$; no real zero greater than 2

22. $P(x) = x^5 - 3x^3 + x + 2$; no real zero less than -3

For each of the polynomial functions P defined as follows: **(a)** *Find the number of positive and negative real zeros that are possible. See Examples 1 and 2.* **(b)** *Approximate each zero as a decimal to the nearest tenth. See Example 4.*

23. $P(x) = x^3 + 3x^2 - 2x - 6$ (a) positive: 1; negative: 2 or 0 (b) $-3, -1.4, 1.4$

24. $P(x) = x^3 + x^2 - 5x - 5$ (a) positive: 1; negative: 2 or 0 (b) $-1, 2.2, -2.2$

25. $P(x) = x^3 - 4x^2 - 5x + 14$ (a) positive: 2 or 0; negative: 1 (b) $-2, 1.6, 4.4$

26. $P(x) = x^3 + 9x^2 + 34x + 13$ (a) positive: 0; negative: 3 or 1 (b) $-.4$

27. $P(x) = x^3 + 6x - 13$ (a) positive: 1; negative: 0 (b) 1.5

28. $P(x) = 4x^3 - 3x^2 + 4x - 5$ (a) positive: 3 or 1; negative: 0 (b) 1

29. $P(x) = 4x^4 - 8x^3 + 17x^2 - 2x - 14$ (a) positive: 3 or 1; negative: 1 (b) $1.1, -.7$

30. $P(x) = 3x^4 - 4x^3 - x^2 + 8x - 2$ (a) positive: 3 or 1; negative: 1 (b) $-1.2, .3$

31. $P(x) = -x^4 + 2x^3 + 3x^2 + 6$ (a) positive: 1; negative: 1 (b) $-1.5, 3.1$

32. $P(x) = -2x^4 - x^2 + x - 5$ (a) positive: 2 or 0; negative: 0 (b) no real zeros

Graph each of the polynomial functions defined as follows. See Examples 5 and 6.

33. $P(x) = x^3 - 7x - 6$

34. $P(x) = x^3 + x^2 - 4x - 4$

35. $P(x) = x^4 - 5x^2 + 6$

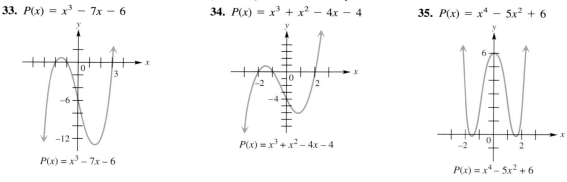

$P(x) = x^3 - 7x - 6$

$P(x) = x^3 + x^2 - 4x - 4$

$P(x) = x^4 - 5x^2 + 6$

36. $P(x) = x^3 - 3x^2 - x + 3$

37. $P(x) = 6x^3 + 11x^2 - x - 6$

38. $P(x) = x^4 - 2x^2 - 8$

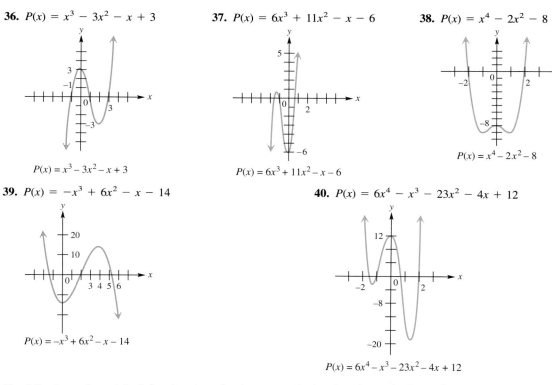

$P(x) = x^3 - 3x^2 - x + 3$

$P(x) = 6x^3 + 11x^2 - x - 6$

$P(x) = x^4 - 2x^2 - 8$

39. $P(x) = -x^3 + 6x^2 - x - 14$

$P(x) = -x^3 + 6x^2 - x - 14$

40. $P(x) = 6x^4 - x^3 - 23x^2 - 4x + 12$

$P(x) = 6x^4 - x^3 - 23x^2 - 4x + 12$

The following polynomials define functions that have zeros in the given intervals. Approximate these zeros to the nearest hundredth.

41. $P(x) = x^4 + x^3 - 6x^2 - 20x - 16$;   [3.2, 3.3] and [−1.4, −1.1]   3.24, −1.24

42. $P(x) = x^4 - 2x^3 - 2x^2 - 18x + 5$;   [.2, .4] and [3.7, 3.8]   .27, 3.73

43. $P(x) = x^4 - 4x^3 - 20x^2 + 32x + 12$;   [−4, −3], [−1, 0], [1, 2], and [6, 7]   −3.65, −.32, 1.65, 6.32

44. $P(x) = x^4 - 4x^3 - 44x^2 + 160x - 80$;   [−7, −6], [0, 1], [2, 3], and [7, 8]   −6.61, .61, 2.76, 7.24

45. A technique for measuring cardiac output depends on the concentration of a dye in the bloodstream after a known amount is injected into a vein near the heart. For a normal heart, the concentration of dye in the bloodstream at time x (in sec) is given by the function defined as follows.

$$g(x) = -.006x^4 + .140x^3 - .053x^2 + 1.79x$$

(a) Find $g(20)$.   175 (rounded)

(b) Graph g.

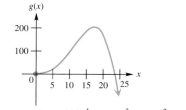

$g(x) = -.006x^4 + .140x^3 - .053x^2 + 1.79x$

46. The polynomial function defined by

$$A(x) = -.015x^3 + 1.058x$$

gives the approximate alcohol concentration (in tenths of a percent) in an average person's bloodstream x hours after drinking about 8 oz of 100-proof whiskey. The function is approximately valid for x in the interval [0, 8].

(a) Graph A.

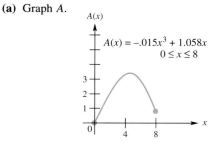

$A(x) = -.015x^3 + 1.058x$
$0 \le x \le 8$

(b) Using the graph you drew for part (a), estimate the time of maximum alcohol concentration.

about 4.8

(c) In one state, a person is legally drunk if the blood alcohol concentration exceeds .08 percent. Use the graph from part (a) to estimate the period in which the average person is legally drunk.

from before 1 hour to about 8 hours

47. Give an example of a polynomial function that is never negative and has -3 and 2 as zeros.

for example, $P(x) = (x + 3)^2 (x - 2)^2 = x^4 + 2x^3 - 11x^2 - 12x + 36$

48. Give an example of a polynomial function that has -3 and 2 as zeros and is positive only between -3 and 2.

for example, $P(x) = (x + 3)(-x + 2) = -x^2 - x + 6$

49. Explain why a seventh-degree polynomial with some terms missing, that is, with one or more zero coefficients, cannot have seven positive zeros.

50. Show that a polynomial of the form $x^4 \ldots + 1$ must have an even number of positive zeros and that a polynomial of the form $x^4 \ldots -1$ must have an odd number of positive zeros.

6.6   RESOURCES

Instructor's Test Manual
Test Items, p. 349–350
Instructor's Solution Manual
pp. 241–248, 660–667
Study Guide
pp. 229–236
Student's Solution Manual
pp. 241–248

Test Generator
IBM Disks
5.25″ Disk B
3.5″ Disk A
Mac Disk A
Tutorials
IBM Disks
5.25″ Disk F
3.5″ Disk C
Mac Disk C
Apple II Module 3, Lessons 1–5

HarperCollins Videotape   7
Transparency   T-31

6.6 EXERCISES   (TEXT PAGE 341) ■ ──────────

Use reflections and translations to graph the rational functions defined as follows. See Examples 1 and 2.

1. $f(x) = \dfrac{2}{x}$

2. $f(x) = -\dfrac{3}{x}$

3. $f(x) = \dfrac{1}{x + 2}$

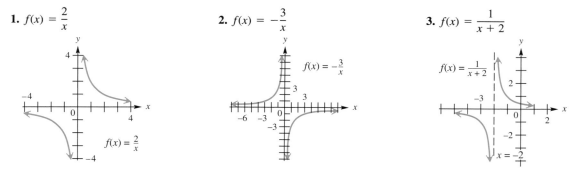

4. $f(x) = \dfrac{1}{x-3}$

5. $f(x) = \dfrac{1}{x} + 1$

6. $f(x) = \dfrac{1}{x} - 2$

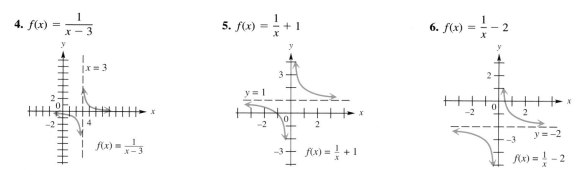

7. Sketch the following graphs and compare them with the graph of $f(x) = 1/x^2$.

(a) $f(x) = \dfrac{1}{(x-3)^2}$

(b) $f(x) = -\dfrac{2}{x^2}$

(c) $f(x) = \dfrac{-2}{(x-3)^2}$

8. Describe in your own words what is meant by an *asymptote* of the graph of a rational function.

Give the equations of any vertical, horizontal, or oblique asymptotes for the graphs of the rational functions in Exercises 9–22. See Example 3.

9. $f(x) = \dfrac{2}{x-5}$ vertical asymptote: $x = 5$;   horizontal asymptote: $y = 0$

10. $f(x) = \dfrac{-1}{x+2}$ vertical asymptote: $x = -2$;   horizontal asymptote: $y = 0$

11. $f(x) = \dfrac{-8}{3x-7}$ vertical asymptote: $x = 7/3$;   horizontal asymptote: $y = 0$

12. $f(x) = \dfrac{5}{4x-9}$ vertical asymptote: $x = 9/4$;   horizontal asymptote: $y = 0$

13. $f(x) = \dfrac{2-x}{x+2}$ vertical asymptote: $x = -2$;   horizontal asymptote: $y = -1$

14. $f(x) = \dfrac{x-4}{5-x}$ vertical asymptote: $x = 5$;   horizontal asymptote: $y = -1$

15. $f(x) = \dfrac{3x-5}{2x+9}$ vertical asymptote: $x = -9/2$;   horizontal asymptote: $y = 3/2$

16. $f(x) = \dfrac{4x+3}{3x-7}$ vertical asymptote: $x = 7/3$;   horizontal asymptote: $y = 4/3$

17. $f(x) = \dfrac{2}{x^2 - 4x + 3}$ vertical asymptotes: $x = 3, x = 1$;   horizontal asymptote: $y = 0$

18. $f(x) = \dfrac{-5}{x^2 - 3x - 10}$

vertical asymptotes: $x = 5, x = -2$;   horizontal asymptote: $y = 0$

19. $f(x) = \dfrac{x^2 - 1}{x + 3}$ vertical asymptote: $x = -3$;   oblique asymptote: $y = x - 3$

20. $f(x) = \dfrac{x^2 + 4}{x - 1}$ vertical asymptote: $x = 1$;   oblique asymptote: $y = x + 1$

21. $f(x) = \dfrac{(x - 3)(x + 1)}{(x + 2)(2x - 5)}$

vertical asymptotes: $x = -2, x = 5/2$;   horizontal asymptote: $y = 1/2$

22. $f(x) = \dfrac{3(x + 2)(x - 4)}{(5x - 1)(x - 5)}$

vertical asymptotes: $x = 1/5, x = 5$;   horizontal asymptote: $y = 3/5$

◎ **23.** Which one of the following has a graph that does not have a vertical asymptote?

(a) $f(x) = \dfrac{1}{x^2 + 2}$ **(b)** $f(x) = \dfrac{1}{x^2 - 2}$ **(c)** $f(x) = \dfrac{3}{x^2}$ **(d)** $f(x) = \dfrac{2x + 1}{x - 8}$ (a)

◎ **24.** Which one of the following has a graph that does not have a horizontal asymptote?

(a) $f(x) = \dfrac{2x - 7}{x + 3}$ **(b)** $f(x) = \dfrac{3x}{x^2 - 9}$ **(c)** $f(x) = \dfrac{x^2 - 9}{x + 3}$ **(d)** $f(x) = \dfrac{x + 5}{(x + 2)(x - 3)}$ (c)

Graph each of the following. See Examples 4–8.

25. $f(x) = \dfrac{4}{5 + 3x}$ **26.** $f(x) = \dfrac{1}{(x - 2)(x + 4)}$ **27.** $f(x) = \dfrac{3}{(x + 4)^2}$ **28.** $f(x) = \dfrac{3x}{(x + 1)(x - 2)}$

29. $f(x) = \dfrac{2x + 1}{(x + 2)(x + 4)}$ **30.** $f(x) = \dfrac{5x}{x^2 - 1}$ **31.** $f(x) = \dfrac{-x}{x^2 - 4}$ **32.** $f(x) = \dfrac{3x}{x - 1}$

◎ **CONCEPTUAL**   ✎ **WRITING**   ▲ **CHALLENGING**

33. $f(x) = \dfrac{4x}{1 - 3x}$

34. $f(x) = \dfrac{x + 1}{x - 4}$

35. $f(x) = \dfrac{x - 5}{x + 3}$

36. $f(x) = \dfrac{x}{x^2 - 9}$

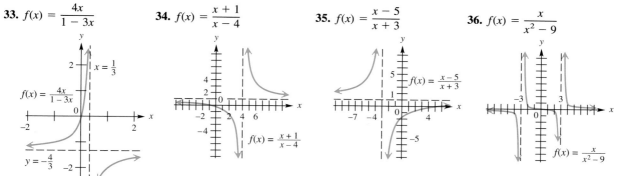

37. $f(x) = \dfrac{3x}{x^2 - 16}$

38. $f(x) = \dfrac{2x^2 + 3}{x - 4}$

39. $f(x) = \dfrac{x^2 + 1}{x + 3}$

40. $f(x) = \dfrac{x^2 + 2x}{2x - 1}$

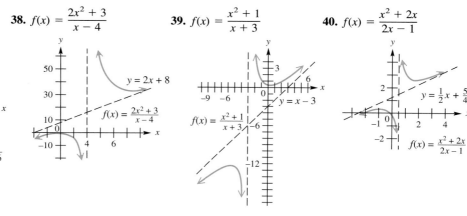

41. $f(x) = \dfrac{x^2 - x}{x + 2}$

42. $f(x) = \dfrac{(x - 3)(x + 1)}{(x - 1)^2}$

43. $f(x) = \dfrac{x(x - 2)}{(x + 3)^2}$

44. $f(x) = \dfrac{(x - 5)(x - 2)}{x^2 + 9}$

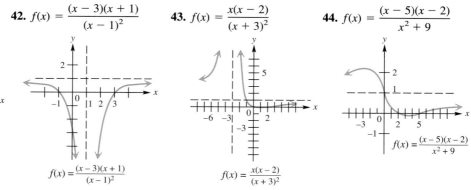

45. $f(x) = \dfrac{1}{x^2 + 1}$

46. $f(x) = \dfrac{x^2 - 9}{x + 3}$

47. $f(x) = \dfrac{x^2 - 16}{x + 4}$

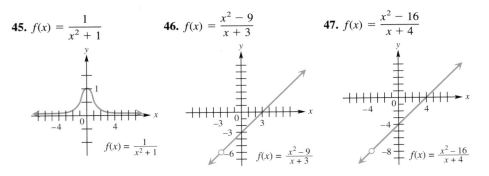

$f(x) = \dfrac{1}{x^2 + 1}$

$f(x) = \dfrac{x^2 - 9}{x + 3}$

$f(x) = \dfrac{x^2 - 16}{x + 4}$

◉ **48.** The figures below show the four ways that the graph of a rational function can approach the vertical line $x = 2$ as an asymptote. Identify the graph of each of the following rational functions defined as follows.

(a) $f(x) = \dfrac{1}{(x - 2)^2}$ **C**   **(b)** $f(x) = \dfrac{1}{x - 2}$ **A**   **(c)** $f(x) = \dfrac{-1}{x - 2}$ **B**   **(d)** $f(x) = \dfrac{-1}{(x - 2)^2}$ **D**

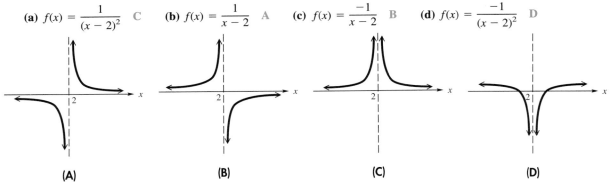

(A)   (B)   (C)   (D)

▦ *Work the following problems.*

49. Antique-car owners often enter their cars in a *concours d'elegance* in which a maximum of 100 points can be awarded to a particular car. Points are awarded for the general attractiveness of the car. The function defined by

$$C(x) = \dfrac{10x}{49(101 - x)}$$

expresses the cost, in thousands of dollars, of restoring a car so that it will win x points. Graph the function.

$C(x) = \dfrac{10x}{49(101 - x)}$

50. Suppose the average cost per unit in thousands of dollars, $C(x)$, to produce x units of margarine is given by

$$C(x) = \dfrac{500}{x + 30}.$$

(a) Find $C(10)$, $C(20)$, $C(50)$, $C(75)$, and $C(100)$.

12.5, 10, 6.25, 4.76, 3.85

(b) Graph the function.

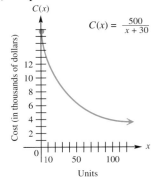

$C(x) = \dfrac{500}{x + 30}$

◉ CONCEPTUAL   ▨ WRITING   ▲ CHALLENGING

51. In situations involving environmental pollution, a cost-benefit model expresses cost as a function of the percentage of pollutant removed from the environment. Suppose a cost-benefit model is expressed as

$$C(x) = \frac{6.7x}{100 - x},$$

where y is the cost in thousands of dollars of removing x percent of a certain pollutant.

(a) Graph the function.

$$C(x) = \frac{6.7x}{100 - x}$$

(b) Is it possible, according to this function, to remove all of the pollutant?   **no**

52. In a recent year, the cost per ton, y, to build an oil tanker of x thousand deadweight tons was approximated by

$$C(x) = \frac{110,000}{x + 225}.$$

(a) Find y for $x = 25$, $x = 50$, $x = 100$, $x = 200$, $x = 300$, and $x = 400$.

440, 400, 338, 259, 210, 176

(b) Graph the function.

$$C(x) = \frac{110,000}{x + 225}$$

*In recent years the economist Arthur Laffer has been a center of controversy because of his **Laffer curve,** an idealized version of which is shown here.*

Tax rate percent

According to this curve, increasing a tax rate, say from x_1 percent to x_2 percent on the graph above, can actually lead to a decrease in government revenue. All economists agree on the endpoints, 0 revenue at tax rates of both 0% and 100%, but there is much disagreement on the location of the rate x_1 that produces maximum revenue.

53. Suppose an economist studying the Laffer curve produces the rational function defined by

$$R(x) = \frac{80x - 8000}{x - 110},$$

with y giving government revenue in tens of millions of dollars for a tax rate of x percent, with the function valid for $55 \le x \le 100$. Find the revenue for the following tax rates.

(a) 55%   **$65.5 tens of millions, or $655,000,000**
(b) 60%   **$64 tens of millions, or $640,000,000**
(c) 70%   **$60 tens of millions, or $600,000,000**
(d) 90%   **$40 tens of millions, or $400,000,000**
(e) 100%   **$0**
(f) Graph the function.

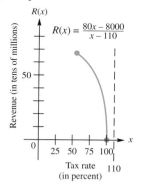

$$R(x) = \frac{80x - 8000}{x - 110}$$

54. Suppose an economist studies a different tax, this time producing

$$R(x) = \frac{60x - 6000}{x - 120}$$

where y is government revenue in millions of dollars from a tax rate of x percent, with $y = R(x)$ valid for $50 \le x \le 100$. Find the revenue from the following tax rates.

(a) 50% $42.9 million   **(b)** 60% $40 million
(c) 80% $30 million   **(d)** 100% $0
(e) Graph the function.

55. Let $f(x) = p(x)/q(x)$ define a rational function where the expression is reduced to lowest terms. Suppose that the degree of $p(x)$ is m and the degree of $q(x)$ is n. Write an explanation of how you would determine the nonvertical asymptote in each of the following situations.
(a) $m < n$ **(b)** $m = n$ **(c)** $m > n$

56. Suppose that a friend tells you that the graph of

$$f(x) = \frac{x^2 - 25}{x + 5}$$

has a vertical asymptote with equation $x = -5$. Is this correct? If not, describe the behavior of the graph at $x = -5$.

CHAPTER 6 REVIEW EXERCISES   (TEXT PAGE 345) ■

Sketch the graph of each polynomial function P as defined in Exercises 1–6.

1. $P(x) = x^3 + 5$

$P(x) = x^3 + 5$

2. $P(x) = 1 - x^4$

$P(x) = 1 - x^4$

3. $P(x) = x^2(2x + 1)(x - 2)$

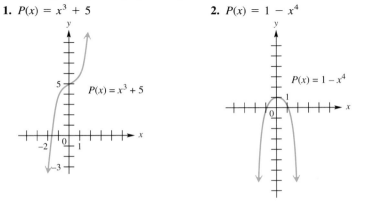

$P(x) = x^2(2x + 1)(x - 2)$

◎ CONCEPTUAL   ✎ WRITING   ▲ CHALLENGING

4. $P(x) = (4x - 3)(3x + 2)(x - 1)$ **5.** $P(x) = 2x^3 + 13x^2 + 15x$ **6.** $P(x) = x(x - 1)(x + 2)(x - 3)$

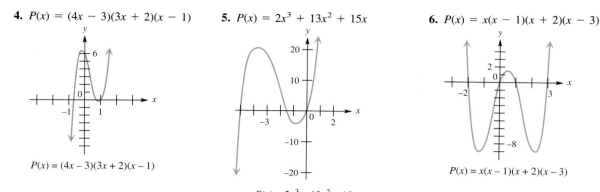

$P(x) = (4x - 3)(3x + 2)(x - 1)$

$P(x) = 2x^3 + 13x^2 + 15x$

$P(x) = x(x - 1)(x + 2)(x - 3)$

7. For the polynomial function defined by $P(x) = x^3 - 3x^2 - 7x + 12$, $P(4) = 0$. Therefore, we can say that 4 is a(n) _____ of the function, 4 is a(n) _____ of the equation $x^3 - 3x^2 - 7x + 12 = 0$, and that 4 is a(n) _____ of the graph of the function.

zero;   solution;   x-intercept

8. Which one of the following is not a polynomial function?
 (a) $f(x) = x^2$ **(b)** $f(x) = 2x + 5$ **(c)** $f(x) = 1/x$ **(d)** $f(x) = x^{100}$ (c)

Perform each division.

9. $\dfrac{15m^4n^5 - 10m^3n^7 + 20m^7n^3}{10mn^4}$ $\dfrac{3m^3n}{2} - m^2n^3 + \dfrac{2m^6}{n}$

10. $\dfrac{10r^6m^3 + 8r^4m - 3r^{10}m^2}{12r^3m^4}$ $\dfrac{5r^3}{6m} + \dfrac{2r}{3m^3} - \dfrac{r^7}{4m^2}$

11. $\dfrac{6p^2 - 5p - 56}{2p - 7}$ $3p + 8$

12. $\dfrac{72r^2 + 59r + 12}{8r + 3}$ $9r + 4$

13. $\dfrac{5m^3 - 7m^2 - 10m + 14}{m^2 - 2}$ $5m - 7$

14. $\dfrac{3b^3 - 8b^2 + 12b - 30}{b^2 + 4}$ $3b - 8 + \dfrac{2}{b^2 + 4}$

Use synthetic division to perform each division.

15. $(2x^3 - x^2 - x - 6) \div (x + 1)$ $2x^2 - 3x + 2 + \dfrac{-8}{x + 1}$

16. $(3t^3 + 10t^2 - 9t - 4) \div (t + 4)$ $3t^2 - 2t - 1$

17. $(4m^3 + m^2 - 12) \div (m - 2)$ $4m^2 + 9m + 18 + \dfrac{24}{m - 2}$

18. $(5y^3 + 24y^2 + 25y - 6) \div (y + 3)$ $5y^2 + 9y - 2$

Use synthetic division to find P(2) for each of the following.

19. $P(x) = x^3 + 3x^2 - 5x + 1$ 11   **20.** $P(x) = 2x^3 - 4x^2 + 3x - 10$ -4

21. $P(x) = 5x^4 - 12x^2 + 2x - 8$ 28   **22.** $P(x) = x^5 - 3x^2 + 2x - 4$ 20

Find a polynomial function P of lowest degree having the given zeros. **In Exercises 23–26, we give only one such polynomial. There are others.**

23. $-1, 4, 7$ $P(x) = x^3 - 10x^2 + 17x + 28$ **24.** $8, 2, 3$ $P(x) = x^3 - 13x^2 + 46x - 48$

25. $-\sqrt{7}, \sqrt{7}, 2, -1$ $P(x) = x^4 - x^3 - 9x^2 + 7x + 14$

26. $1 + \sqrt{5}, 1 - \sqrt{5}, -4, 1$ $P(x) = x^4 + x^3 - 14x^2 - 4x + 16$

27. Is -1 a zero of P, if $P(x) = 2x^4 + x^3 - 4x^2 + 3x + 1$?   no

28. Is -2 a zero of P, if $P(x) = 2x^4 + x^3 - 4x^2 + 3x + 1$?   no

29. Is $x + 1$ a factor of $P(x) = x^3 + 2x^2 + 3x - 1$?   no

30. Is $x + 1$ a factor of $P(x) = 2x^3 - x^2 + x + 4$?   yes

31. Find a function defined by a polynomial $P(x)$ of degree 3 with real coefficients, having -2, 1, and 4 as zeros, and $P(2) = 16$. $P(x) = -2x^3 + 6x^2 + 12x - 16$

32. Find a function defined by a polynomial $P(x)$ of degree 4 with real coefficients having 1, -1, and $3i$ as zeros, and $P(2) = 39$. $P(x) = x^4 + 8x^2 - 9$

33. Find a lowest-degree polynomial $P(x)$ with real coefficients defining a function having zeros 2, -2, and $-i$. $P(x) = x^4 - 3x^2 - 4$ (others are possible)

34. Find a lowest-degree polynomial $P(x)$ with real coefficients defining a function having zeros 2, -3, and $5i$. $P(x) = x^4 + x^3 + 19x^2 + 25x - 150$ (others are possible)

35. Find a polynomial $P(x)$ of lowest degree with real coefficients defining a function having -3 and $1 - i$ as zeros. $P(x) = x^3 + x^2 - 4x + 6$ (others are possible)

36. Find all zeros of P, if $P(x) = x^4 - 3x^3 - 8x^2 + 22x - 24$, given that $1 - i$ is a zero, and factor $P(x)$. $1 - i$, $1 + i$, 4, -3; $P(x) = (x - 1 + i)(x - 1 - i)(x - 4)(x + 3)$

37. Find all zeros of P, if $P(x) = x^4 - 6x^3 + 14x^2 - 24x + 40$, given that $3 + i$ is a zero, and factor $P(x)$. $3 + i$, $3 - i$, $2i$, $-2i$; $P(x) = (x - 3 - i)(x - 3 + i)(x - 2i)(x + 2i)$

38. Find all zeros of P, if $P(x) = x^4 + x^3 - x^2 + x - 2$, given that 1 and -2 are zeros, and factor $P(x)$.   1, -2, i, $-i$; $P(x) = (x - 1)(x + 2)(x - i)(x + i)$

Find all rational zeros of each polynomial function P, if $P(x)$ is defined as follows.

39. $P(x) = 2x^3 - 9x^2 - 6x + 5$ $1/2$, -1, 5

40. $P(x) = 3x^3 - 10x^2 - 27x + 10$ $1/3$, -2, 5

41. $P(x) = x^3 - \dfrac{17}{6}x^2 - \dfrac{13}{3}x - \dfrac{4}{3}$ 4, $-1/2$, $-2/3$

42. $P(x) = 8x^4 - 14x^3 - 29x^2 - 4x + 3$ 3, -1, $1/4$, $-1/2$

For the polynomial functions defined in Exercises 43 and 44, find all rational zeros and factor the polynomial. Then graph the function.

43. $P(x) = x^3 - 5x^2 - 13x - 7$

   -1 (multiplicity 2), 7; $P(x) = (x + 1)^2 (x - 7)$

44. $P(x) = 3x^3 - 2x^2 - 19x - 6$

   $-1/3$, 3, -2; $P(x) = (3x + 1)(x - 3)(x + 2)$

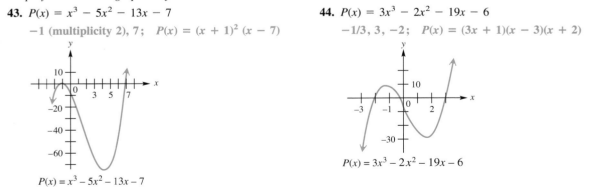

45. Use a polynomial to show that $\sqrt{11}$ is irrational.

46. Show that the function defined by $P(x) = x^3 - 9x^2 + 2x - 5$ has no rational zeros.

Show that each polynomial P defined as follows has real zeros satisfying the given conditions.

47. $P(x) = 3x^3 - 8x^2 + x + 2$, zero in $[-1, 0]$ and $[2, 3]$

48. $P(x) = 4x^3 - 37x^2 + 50x + 60$, zero in $[2, 3]$ and $[7, 8]$

49. $P(x) = x^3 + 2x^2 - 22x - 8$, zero in $[-1, 0]$ and $[-6, -5]$

50. $P(x) = 2x^4 - x^3 - 21x^2 + 51x - 36$, has no real zero greater than 4

51. $P(x) = 6x^4 + 13x^3 - 11x^2 - 3x + 5$, has no zero greater than 1 or less than -3

In Exercises 52 and 53, approximate the real zeros of each polynomial function P as defined. Give answers as decimals to the nearest tenth.

52. $P(x) = 2x^3 - 11x^2 - 2x + 2$ $-.5, .4, 5.6$ **53.** $P(x) = x^4 - 4x^3 - 5x^2 + 14x - 15$ $-2.3, 4.6$

54. (a) Find the possible numbers of positive and negative zeros of the polynomial function P defined by $P(x) = x^3 + 3x^2 - 4x - 2$.   **positive: 1;   negative: 2 or 0**

 (b) Show that the polynomial function P has a zero between -4 and -3. Approximate this zero to the nearest tenth. -3.9

Graph each polynomial function defined as follows.

55. $P(x) = 2x^3 - 11x^2 - 2x + 2$ (See Exercise 52.)

$P(x) = 2x^3 - 11x^2 - 2x + 2$

56. $P(x) = x^4 - 4x^3 - 5x^2 + 14x - 15$ (See Exercise 53.)

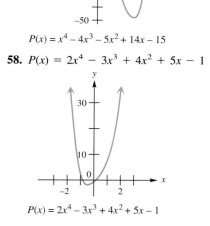

$P(x) = x^4 - 4x^3 - 5x^2 + 14x - 15$

57. $P(x) = x^3 + 3x^2 - 4x - 2$ (See Exercise 54.)

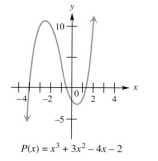

$P(x) = x^3 + 3x^2 - 4x - 2$

58. $P(x) = 2x^4 - 3x^3 + 4x^2 + 5x - 1$

$P(x) = 2x^4 - 3x^3 + 4x^2 + 5x - 1$

Graph each rational function defined as follows.

59. $f(x) = \dfrac{8}{x}$

60. $f(x) = \dfrac{2}{3x - 1}$

61. $f(x) = \dfrac{4x - 2}{3x + 1}$

62. $f(x) = \dfrac{6x}{(x - 1)(x + 2)}$

63. $f(x) = \dfrac{2x}{x^2 - 1}$

64. $f(x) = \dfrac{x^2 + 4}{x + 2}$

65. $f(x) = \dfrac{x^2 - 1}{x}$

66. $f(x) = \dfrac{-2}{x^2 + 1}$

67. $f(x) = \dfrac{4x^2 - 9}{2x + 3}$

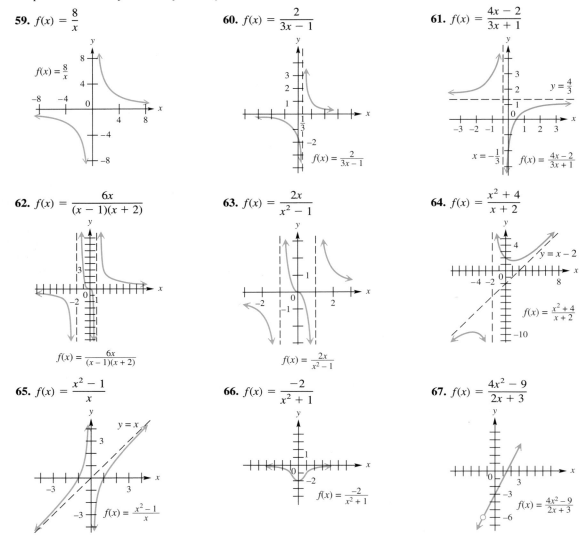

68. Under what conditions will the graph of a rational function defined by an expression reduced to lowest terms have an oblique asymptote?

CHAPTER 6 TEST   (TEXT PAGE 347) ■

Graph each polynomial function P as defined in Exercises 1 and 2.

1. $P(x) = (1 - x)^4$

2. $P(x) = (x + 1)(x - 2)x$

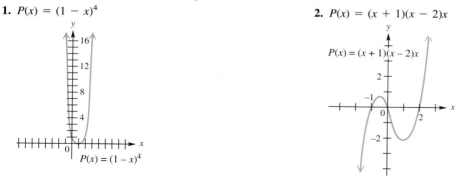

Use synthetic division to perform each division.

3. $(3x^3 + 4x^2 - 9x + 6) \div (x + 2)$ $3x^2 - 2x - 5 + \dfrac{16}{x + 2}$

4. $(2x^3 - 11x^2 + 28) \div (x - 5)$ $2x^2 - x - 5 + \dfrac{3}{x - 5}$

5. Use synthetic division to find $P(3)$ for $P(x) = 2x^3 - 9x^2 + 4x + 8$. -7

6. Find a polynomial function P of lowest degree with -1, -2, and 4 as zeros.
$P(x) = x^3 - x^2 - 10x - 8$ (others are possible)

7. Is 3 a zero of the function P defined by $P(x) = 6x^4 - 11x^3 - 35x^2 + 34x + 24$? Why or why not?   Yes, because when $P(x)$ is divided by $x - 3$, the remainder is 0.

8. Is $x + 2$ a factor of $P(x) = 6x^4 - 11x^3 - 35x^2 + 34x + 24$? Why or why not?
Yes, because $P(-2) = 0$.

9. Find a function P defined by a polynomial of degree 4 with real coefficients having 2, -1, and $-i$ as zeros, and with $P(3) = 80$. $P(x) = 2x^4 - 2x^3 - 2x^2 - 2x - 4$

10. Factor $P(x) = 2x^3 - x^2 - 13x - 6$, given that -2 is a zero of P.
$P(x) = (x + 2)(2x + 1)(x - 3)$

11. For the polynomial function defined by $P(x) = 6x^3 - 25x^2 + 12x + 7$,
(a) list all rational numbers that can possibly be zeros of P;
± 1, $\pm 1/2$, $\pm 1/3$, $\pm 1/6$, ± 7, $\pm 7/2$, $\pm 7/3$, $\pm 7/6$

(b) find all rational zeros of P. $-1/3$, 1, $7/2$

12. Explain why the function P defined by the polynomial $P(x) = x^4 + x^2 + 1$ cannot have any negative zeros. (*Hint:* Look at the exponents.)

13. Show that the polynomial function P defined by $P(x) = 2x^4 - 3x^3 + 4x^2 - 5x - 1$ has no real zeros greater than 2 or less than -1.

14. Find the possible numbers of positive and negative real zeros of the polynomial function P defined by $P(x) = x^4 + 3x^3 - x^2 - 2x + 1$.   positive: 2 or 0;   negative: 2 or 0

15. Graph the polynomial function
defined by $P(x) = x^3 - 5x^2 + 8x - 4$.

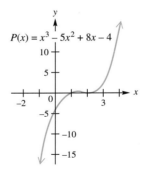

Graph the rational functions defined as follows.

16. $f(x) = \dfrac{-2}{x + 3}$ **17.** $f(x) = \dfrac{3x - 1}{x - 2}$ **18.** $f(x) = \dfrac{x^2 - 1}{x^2 - 9}$

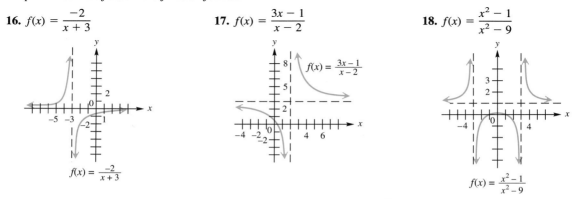

19. What is the equation of the oblique asymptote of the graph of $f(x) = \dfrac{2x^2 + x - 6}{x - 1}$?

$y = 2x + 3$

◎ **20.** Which one of the functions defined below has a graph with no x-intercepts?

     (a) $f(x) = (x - 2)(x + 3)^2$ **(b)** $f(x) = \dfrac{x + 7}{x - 2}$ **(c)** $f(x) = x^3 - x$ **(d)** $f(x) = \dfrac{1}{x^2 + 4}$ **(d)**

■ **CHAPTER 7 SYSTEMS OF EQUATIONS AND INEQUALITIES**

7.1 RESOURCES

Instructor's Test Manual
Test Items, p. 351–354
Instructor's Solution Manual
pp. 261–271, 675–682
Study Guide
pp. 244–250
Student's Solution Manual
pp. 261–271

Test Generator
IBM Disks
5.25″ Disk B
3.5″ Disk A
Mac Disk A
Tutorials
IBM Disks
5.25″ Disk G
3.5″ Disk D
Mac Disk D

HarperCollins Videotape 7
Transparency T-32

◎ CONCEPTUAL   ✎ WRITING   ▲ CHALLENGING

7.1 EXERCISES (TEXT PAGE 357) ■ ───────────────────

◉ *Decide whether the systems in Exercises 1–8 have a single solution, no solution, or infinitely many solutions.*

1. One line has positive slope and one line has negative slope.   single solution

2. One line has slope 0 and one line has undefined slope.   single solution

3. Both lines have slope 0 and have the same *y*-intercept.   infinitely many solutions

4. Both lines have undefined slope and have the same *x*-intercept.   infinitely many solutions

5. $y = 3x + 6$
$y = 3x + 9$ no solution

6. $y = 4x + 8$
$y = 4x + 12$ no solution

7. $x + y = 4$
$kx + ky = 4k$ $(k \neq 0)$ infinitely many solutions

8. $x + y = 10$
$kx + ky = 9k$ $(k \neq 0)$ no solution

Use the substitution method to solve each system. Check your answers. (Hint: In Exercises 13 and 14, clear fractions first by multiplying through by the least common denominator.) See Example 1.

9. $y = 2x + 3$
$3x + 4y = 78$ $\{(6, 15)\}$

10. $y = 4x - 6$
$2x + 5y = -8$ $\{(1, -2)\}$

11. $3x - 2y = 12$
$5x = 4 - 2y$ $\{(2, -3)\}$

12. $8x + 3y = 2$
$5x = 17 + 6y$ $\{(1, -2)\}$

13. $\dfrac{x}{2} = \dfrac{7}{6} - \dfrac{y}{3}$
$\dfrac{2x}{3} = \dfrac{3y}{2} - \dfrac{7}{3}$ $\{(1, 2)\}$

14. $\dfrac{3x}{4} = 4 + \dfrac{y}{2}$
$\dfrac{x}{3} + \dfrac{5y}{4} = -\dfrac{7}{6}$ $\{(4, -2)\}$

◉ **15.** Refer to Example 2 in this section. If we began solving the system by eliminating *y*, by what numbers might we have multiplied equations (4) and (5)?

Multiply equation (4) by 3 and equation (5) by 4.

📝 ◉ **16.** Explain how one can determine whether a system is inconsistent or has dependent equations when using the substitution or elimination method.

Use the addition method to solve each system. Check your answers. (Hint for Exercises 29–32: Let 1/x = t, 1/y = u.) See Examples 2–4.

17. $5x + 3y = 7$
$7x - 3y = -19$ $\{(-1, 4)\}$

18. $2x + 7y = -8$
$-2x + 3y = -12$ $\{(3, -2)\}$

19. $3x + 2y = 5$
$6x + 4y = 8$ $\emptyset$

20. $9x - 5y = 1$
$-18x + 10y = 1$ $\emptyset$

21. $2x - 3y = -7$
$5x + 4y = 17$ $\{(1, 3)\}$

22. $4x + 3y = -1$
$2x + 5y = 3$ $\{(-1, 1)\}$

23. $5x + 7y = 6$
$10x - 3y = 46$ $\{(4, -2)\}$

24. $12x - 5y = 9$
$3x - 8y = -18$ $\{(2, 3)\}$

25. $4x - y = 9$
$-8x + 2y = -18$ $\left\{\left(\dfrac{y + 9}{4}, y\right)\right\}$

26. $3x + 5y + 2 = 0$
$9x + 15y + 6 = 0$ $\left\{\left(\dfrac{-5y - 2}{3}, y\right)\right\}$

27. $\dfrac{x}{2} + \dfrac{y}{3} = 8$
$\dfrac{2x}{3} + \dfrac{3y}{2} = 17$ $\{(12, 6)\}$

28. $\dfrac{x}{5} + 3y = 31$
$2x - \dfrac{y}{5} = 8$ $\{(5, 10)\}$

29. $\dfrac{2}{x} + \dfrac{1}{y} = \dfrac{3}{2}$
$\dfrac{3}{x} - \dfrac{1}{y} = 1$ $\{(2, 2)\}$

30. $\dfrac{1}{x} + \dfrac{3}{y} = \dfrac{16}{5}$
$\dfrac{5}{x} + \dfrac{4}{y} = 5$ $\{(5, 1)\}$

31. $\dfrac{2}{x} + \dfrac{1}{y} = 11$
$\dfrac{3}{x} - \dfrac{5}{y} = 10$ $\{(1/5, 1)\}$

32. $\dfrac{2}{x} + \dfrac{3}{y} = 18$
$\dfrac{4}{x} - \dfrac{5}{y} = -8$ $\{(1/3, 1/4)\}$

▲ **33.** Find values of a and b so that the line with equation $ax + by = 5$ passes through the points $(-2, 1)$ and $(-1, -2)$. $a = -3;\quad b = -1$

▲ **34.** Find m and b so that the line $y = mx + b$ passes through $(4, 6)$ and $(-5, -3)$.

$m = 1;\quad b = 2$

▲ **35.** Find m and b so that the line $y = mx + b$ passes through $(2, 5)$ and $(-1, 4)$.

$m = 1/3;\quad b = 13/3$

▲ **36.** Find values of a and b so that the line with equation $ax + by = 23$ passes through $(-3, 2)$ and $(-4, -5)$. $a = -7;\quad b = 1$

Write each problem as a system of equations and then solve. See Example 5.

37. To start a new business, Shannon borrowed money from two financial institutions. One loan was at 7% interest and the other was for one-third as much money at 8% interest. How much did she borrow at each rate if the total amount of annual interest was $1160? **$12,000 at 7%; $4000 at 8%**

38. Mark LeBeau has invested a total of $20,000 in two ways. Part of the money is in certificates of deposit paying 3.75% interest, while the rest is in municipal bonds that pay 8.2% interest. How much is there in each account if the total annual interest is $1417.50?

$5000 in certificates of deposit; $15,000 in bonds

39. Alexis is a botanist who has patented a successful type of plant food that contains two chemicals, X and Y. Eight hundred kilograms of these chemicals will be used to make a batch of the food, and the ratio of X to Y must be 3 to 2. How much of each chemical should be used? **480 kg of X;   320 kg of Y**

40. A manufacturer of portable compact disc players shipped 200 of the players to its two Quebec warehouses. It costs $3 per unit to ship to Warehouse A, and $2.50 per unit to ship to Warehouse B. If the total shipping cost was $537.50, how many were shipped to each warehouse? **75 to A;   125 to B**

41. Octane ratings show the percent of isooctane in gasoline. An octane rating of 98, for example, indicates a gasoline that is 98% isooctane. How many gallons of 98-octane gasoline should be mixed with 92-octane gasoline to produce 40 gallons of 94-octane gasoline?

40/3 gal of 98 octane;   80/3 gal of 92 octane

42. A chemist needs 10 liters of a 24% alcohol solution. She has on hand a 30% alcohol solution and an 18% alcohol solution. How many liters of each should be mixed to get the required solution?

5 liters of the 30% solution;   5 liters of the 18% solution

43. A supplier of poultry sells 20 turkeys and 8 chickens for $74. He also sells 15 turkeys and 24 chickens for $87. Find the cost of a turkey and the cost of a chicken. **$3 for a turkey; $1.75 for a chicken**

44. Jose Ortega is a building contractor. If he hires 8 brick layers and 2 roofers, his daily payroll is $960, while 10 brick layers and 5 roofers require a daily payroll of $1500. What is the daily wage of a brick layer and the daily wage of a roofer?

$90 for a brick layer; $120 for a roofer

45. During summer vacation Hector and Ann earned a total of $6496. Hector worked 8 days less than Ann and earned $4 per day less. Find the number of days he worked and the daily wage made, if the total number of days worked by both was 72.

32 days at $88 per day

46. A bank teller has ten-dollar bills and twenty-dollar bills. He has 25 more twenties than tens. The value of the bills is $2900. How many of each kind does he have? **80 tens;   105 twenties**

47. Mike Karelius plans to invest $30,000 he won in a lottery. With part of the money he buys a mutual fund, paying 4.5% a year. The rest he invests in utility bonds paying 5% per year. The first year his investments bring a return of $1410. How much is invested at each rate? **$18,000 at 4.5%; $12,000 at 5%**

48. How much milk that is 3% butterfat should be mixed with milk that is 18% butterfat to get 25 gal of milk that is 4.8% butterfat?

22 gal of 3%;   3 gal of 18%

◎ **CONCEPTUAL**   ✐ **WRITING**   ▲ **CHALLENGING**

The break-even point for a company is the point where its costs equal its revenues. If both cost and revenue are expressed as linear equations, the break-even point is the solution of a linear system. In each of the following exercises, C represents the cost to produce x items, and R represents the revenue from the sale of x items. Use the substitution method to find the break-even point in each case, that is, the point where C = R.

49. $C = 1.5x + 252$
   $R = 5.5x$ $x = 63$; $R = C = 346.5$

50. $C = 2.5x + 90$
   $R = 3x$ $x = 180$; $R = C = 540$

51. $C = 20x + 10,000$
   $R = 30x - 11,000$ $x = 2100$; $R = C = 52,000$

52. $C = 4x + 125$
   $R = 9x - 200$ $x = 65$; $R = C = 385$

In each of the following exercises, p is the price of an item, while q represents the supply in one equation and the demand in the other. Find the equilibrium price and the equilibrium supply/demand. See Example 6.

53. $p = 80 - \dfrac{3}{5}q$

   $p = \dfrac{2}{5}q$ 32;   80

54. $p = 630 - \dfrac{3}{4}q$

   $p = \dfrac{3}{4}q$ 315;   420

55. $3p = 84 - 2q$
   $3p - q = 0$ 28/3;   28

56. $4p + q = 80$
   $3p - 2q = 5$ 15;   20

57. Suppose that the demand and price for a certain model of electric can opener are related by

$$p = 16 - \frac{5}{4}q,$$

where p is price and q is demand, in appropriate units. Find the price when the demand is at the following levels.
   (a) 0 units   16
   (b) 4 units   11
   (c) 8 units   6
   Find the demand for the electric can opener at the following prices.
   (d) \$6   8
   (e) \$11   4
   (f) \$16   0
   (g) Graph $p = 16 - (5/4)q$.

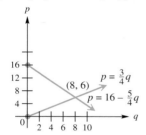

Suppose that the price and supply of the item above are related by

$$p = \frac{3}{4}q,$$

where q represents the supply and p the price. Find the supply at the following prices.
   (h) \$0   0   **(i)** \$10   40/3   **(j)** \$20   80/3
   (k) Graph $p = (3/4)q$ on the same axes used for part (g).
   (l) Find the equilibrium supply.   8
   (m) Find the equilibrium price.   6

58. Let the supply and demand equations for banana smoothies be

   supply: $p = \dfrac{3}{2}q$ and   demand: $p = 81 - \dfrac{3}{4}q$.

   (a) Graph these on the same axes.

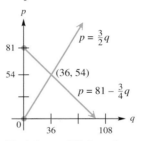

   (b) Find the equilibrium demand.   36
   (c) Find the equilibrium price.   54

59. Let the supply and demand equations for chocolate frozen yogurt be given by

supply: $p = \frac{2}{5}q$ and demand: $p = 100 - \frac{2}{5}q$.

(a) Graph these on the same axes.

(b) Find the equilibrium demand. 125
(c) Find the equilibrium price. 50

60. Let the supply and demand equations for onions be given by

supply: $p = 1.4q - .6$ and demand: $p = -2q + 3.2$.

(a) Graph these on the same axes.

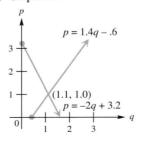

(b) Find the equilibrium demand. about 1.1
(c) Find the equilibrium price. about 1.0

7.2 RESOURCES

Instructor's Test Manual
Test Items, p. 354–356
Instructor's Solution Manual
pp. 271–280, 682–691
Study Guide
pp. 250–254
Student's Solution Manual
pp. 271–280

Test Generator
IBM Disks
5.25″ Disk B
3.5″ Disk A
Mac Disk A
Tutorials
IBM Disks
5.25″ Disk G
3.5″ Disk D
Mac Disk D

HarperCollins Videotape 7
Transparency T-33

7.2 EXERCISES (TEXT PAGE 365) ■

Use the addition method to solve each of the following systems of three equations in three un-knowns. Check your answers. See Example 1. (Hint: In Exercises 15–18, let $t = 1/x$, $u = 1/y$, and $v = 1/z$. Solve for t, u, and v, and then for x, y, and z.)

1. $x + y + z = 2$
 $2x + y - z = 5$
 $x - y + z = -2$ $\{(1, 2, -1)\}$

2. $2x + y + z = 9$
 $-x - y + z = 1$
 $3x - y + z = 9$ $\{(2, 1, 4)\}$

3. $x + 3y + 4z = 14$
 $2x - 3y + 2z = 10$
 $3x - y + z = 9$ $\{(2, 0, 3)\}$

4. $4x - y + 3z = -2$
 $3x + 5y - z = 15$
 $-2x + y + 4z = 14$ $\{(-1, 4, 2)\}$

5. $x + 2y + 3z = 8$
 $3x - y + 2z = 5$
 $-2x - 4y - 6z = 5$ $\emptyset$

6. $3x - 2y - 8z = 1$
 $9x - 6y - 24z = -2$
 $x - y + z = 1$ $\emptyset$

◎ CONCEPTUAL   ✐ WRITING   ▲ CHALLENGING

7. $x + 4y - z = 6$
$2x - y + z = 3$
$3x + 2y + 3z = 16$ {(1, 2, 3)}

8. $4x - 3y + z = 9$
$3x + 2y - 2z = 4$
$x - y + 3z = 5$ {(2, 0, 1)}

9. $5x + y - 3z = -6$
$2x + 3y + z = 5$
$-3x - 2y + 4z = 3$ {(-1, 2, 1)}

10. $2x - 5y + 4z = -35$
$5x + 3y - z = 1$
$x + y + z = 1$ {(-3, 5, -1)}

11. $x - 3y - 2z = -3$
$3x + 2y - z = 12$
$-x - y + 4z = 3$ {(4, 1, 2)}

12. $x + y + z = 3$
$3x - 3y - 4z = -1$
$x + y + 3z = 11$ {(2, -3, 4)}

13. $2x + 6y - z = 6$
$4x - 3y + 5z = -5$
$6x + 9y - 2z = 11$ {(1/2, 2/3, -1)}

14. $8x - 3y + 6z = -2$
$4x + 9y + 4z = 18$
$12x - 3y + 8z = -2$ {(-3/4, 5/3, 3/2)}

▲ **15.** $\dfrac{1}{x} + \dfrac{1}{y} - \dfrac{1}{z} = \dfrac{1}{4}$

$\dfrac{2}{x} - \dfrac{1}{y} + \dfrac{3}{z} = \dfrac{9}{4}$

$-\dfrac{1}{x} - \dfrac{2}{y} + \dfrac{4}{z} = 1$ {(2, 4, 2)}

▲ **16.** $\dfrac{3}{x} + \dfrac{2}{y} - \dfrac{1}{z} = \dfrac{11}{6}$

$\dfrac{1}{x} - \dfrac{1}{y} + \dfrac{3}{z} = -\dfrac{11}{12}$

$\dfrac{2}{x} + \dfrac{1}{y} + \dfrac{1}{z} = \dfrac{7}{12}$ {(3, 4, -3)}

▲ **17.** $\dfrac{2}{x} - \dfrac{2}{y} + \dfrac{1}{z} = -1$

$\dfrac{4}{x} + \dfrac{1}{y} - \dfrac{2}{z} = -9$

$\dfrac{1}{x} + \dfrac{1}{y} - \dfrac{3}{z} = -9$ {(-1, 1, 1/3)}

▲ **18.** $\dfrac{5}{x} - \dfrac{1}{y} - \dfrac{2}{z} = -6$

$-\dfrac{1}{x} + \dfrac{3}{y} - \dfrac{3}{z} = -12$

$\dfrac{2}{x} - \dfrac{1}{y} - \dfrac{1}{z} = 6$ {(-1/6, -1/12, -1/6)}

◉ **19.** Consider the linear equation in three variables $x + y + z = 4$. Find a pair of linear equations that, when considered together with the given equation, will form a system having the following.

(a) Exactly one solution

for example: $x + 2y + z = 5$, $2x - y + 3z = 4$ (There are others.)

(b) No solution   for example: $x + y + z = 5$, $2x - y + 3z = 4$ (There are others.)

(c) Infinitely many solutions

for example: $2x + 2y + 2z = 8$, $2x - y + 3z = 4$ (There are others.)

◉ **20.** Refer to Example 2 in this section. Write the solution set with x arbitrary.

{(x, 1 - x, x + 2)}

◉ **21.** Give an example using your immediate surroundings of three planes that intersect in a single point.

For example, the ceiling and two perpendicular walls of a standard room meet in a point.

◉ **22.** Give an example using your immediate surroundings of three planes that intersect in a line.

For example, three pages of this book intersect in a line (represented by the spine of the book).

Solve each of the following systems in terms of the arbitrary variable x. (Hint: Begin by elimi-nating either y or z.) See Example 2.

23. $x - 2y + 3z = 6$
$2x - y + 2z = 5$ $\{(x, -4x + 3, -3x + 4)\}$

24. $3x + 4y - z = 13$
$x + y + 2z = 15$ $\left\{\left(x, \dfrac{41 - 7x}{9}, \dfrac{47 - x}{9}\right)\right\}$

25. $5x - 4y + z = 9$
$x + y = 15$ $\{(x, -x + 15, -9x + 69)\}$

26. $x - y + z = -6$
$4x + y + z = 7$ $\left\{\left(x, \dfrac{-3x + 13}{2}, \dfrac{-5x + 1}{2}\right)\right\}$

27. $3x - 5y - 4z = -7$
$y - z = -13$ $\left\{\left(x, \dfrac{-15 + x}{3}, \dfrac{24 + x}{3}\right)\right\}$

28. $3x - 2y + z = 15$
$x + 4y - z = 11$ $\{(x, -2x + 13, -7x + 41)\}$

Solve the following problems. See Examples 3 and 4.

29. Find *a*, *b*, and *c* so that the graph of the equation $y = ax^2 + bx + c$ passes through the points $(2, 3)$, $(-1, 0)$, and $(-2, 2)$. $a = 3/4$; $b = 1/4$; $c = -1/2$

30. Find *a*, *b*, and *c* so that $(2, 14)$, $(0, 0)$, and $(-1, -1)$ lie on the graph of $y = ax^2 + bx + c$.
$a = 2, b = 3, c = 0$

▲ **31.** Find the equation of the circle $x^2 + y^2 + ax + by + c = 0$ that passes through the points $(2, 1)$, $(-1, 0)$, and $(3, 3)$. $x^2 + y^2 + x - 7y = 0$

▲ **32.** Find the equation of the circle $x^2 + y^2 + ax + by + c = 0$ that passes through the points $(4, 2)$, $(-5, -2)$, and $(0, 3)$. $x^2 + y^2 - \dfrac{7}{5}x + \dfrac{27}{5}y - \dfrac{126}{5} = 0$

33. A coin collection contains a total of 29 coins, made up of cents, nickels, and quarters. The number of quarters is 8 less than the number of cents. The total face value of the coins is $1.77. How many of each denomination are there?   12 cents; 13 nickels; 4 quarters

34. A sparkling water distributor wants to make up 300 gallons of sparkling water to sell for $6.00 per gallon. She wishes to mix three grades of water selling for $9.00, $3.00, and $4.50 per gallon, respectively. She must use twice as much of the $4.50 water as the $3.00 water. How many gallons of each should she use?

120 gal of $9.00;   60 gal of $3.00;   120 gal of $4.50

35. A glue company needs to make some glue that it can sell for $120 per barrel. It wants to use 150 barrels of glue worth $100 per barrel, along with some glue worth $150 per barrel, and glue worth $190 per barrel. It must use the same number of barrels of $150 and $190 glue. How much of the $150 and $190 glue will be needed? How many barrels of $120 glue will be produced?

30 barrels each of $150 and $190 glue;   210 barrels of $120 glue

36. Billy Dixon and the Topics sell three kinds of concert tickets, "up close," "middle," and "farther back." "Up close" tickets cost $6 more than "middle" tickets, while "middle" tickets cost $3 more than "farther back" tickets. Twice the cost of an "up close" ticket is $3 more than 3 times the cost of a "farther back" seat. Find the price of each kind of ticket.

"up close": $24;   "middle": $18;   "farther back": $15

37. The perimeter of a triangle is 59 inches. The longest side is 11 inches longer than the medium side, and the medium side is 3 inches more than the shortest side. Find the length of each side of the triangle.

28 in;   17 in;   14 in

38. The sum of the measures of the angles of any triangle is 180°. In a certain triangle, the largest angle measures 55° less than twice the medium angle, and the smallest measures 25° less than the medium angle. Find the measures of each of the three angles.

75°;   65°;   40°

39. Sam Abo-zahrah wins $100,000 in the Louisiana state lottery. He invests part of the money in real estate with an annual return of 5% and another part in a money market account at 4.5% interest. He invests the rest, which amounts to $20,000 less than the sum of the other two parts, in certificates of deposit that pay 3.75%. If the total annual interest on the money is $4450, how much was invested at each rate?

$50,000 at 5%; $10,000 at 4.5%; $40,000 at 3.75%

40. Ellen Keith invests $10,000 received in an inheritance in three parts. With one part she buys mutual funds which offer a return of 4% per year. The second part, which amounts to twice the first, is used to buy government bonds paying 4.5% per year. She puts the rest into a savings account that pays 2.5% annual interest. During the first year, the total interest is $415. How much did she invest at each rate?

$3000 at 4%; $6000 at 4.5%; $1000 at 2.5%

41. Write an application similar to the one in Exercise 33. (Do this by first starting with an answer, and writing the exercise to fit the answer.) Then solve your own problem using a system of equations.

42. Explain why your solution to the problem in Exercise 41 cannot have fractional answers. Can it have negative answers? Explain.

7.3 RESOURCES

Instructor's Test Manual
Test Items, p. 356–357
Instructor's Solution Manual
pp. 280–288, 691–698
Study Guide
pp. 254–260
Student's Solution Manual
pp. 280–288

Test Generator
IBM Disks
5.25" Disk B
3.5" Disk A
Mac Disk A
Tutorials
IBM Disks
5.25" Disk G
3.5" Disk D
Mac Disk D

HarperCollins Videotape   7
Transparency   T-34(a)–(d)

7.3 EXERCISES   (TEXT PAGE 374) ∎

Use the third row transformation to change each of the following matrices as indicated. See Example 1.

1. $\begin{bmatrix} 2 & 4 \\ 4 & 7 \end{bmatrix}$; -2 times row 1 added to row 2 $\begin{bmatrix} 2 & 4 \\ 0 & -1 \end{bmatrix}$

2. $\begin{bmatrix} -1 & 4 \\ 7 & 0 \end{bmatrix}$;   7 times row 1 added to row 2 $\begin{bmatrix} -1 & 4 \\ 0 & 28 \end{bmatrix}$

3. $\begin{bmatrix} 1 & 5 & 6 \\ -2 & 3 & -1 \\ 4 & 7 & 0 \end{bmatrix}$;   2 times row 1 added to row 2 $\begin{bmatrix} 1 & 5 & 6 \\ 0 & 13 & 11 \\ 4 & 7 & 0 \end{bmatrix}$

4. $\begin{bmatrix} 2 & 5 & 6 \\ 4 & -1 & 2 \\ 3 & 7 & 1 \end{bmatrix}$; -6 times row 3 added to row 1 $\begin{bmatrix} -16 & -37 & 0 \\ 4 & -1 & 2 \\ 3 & 7 & 1 \end{bmatrix}$

5. $\begin{bmatrix} -3 & 1 & -4 \\ 2 & 1 & 3 \\ -7 & 5 & 2 \end{bmatrix}$; -5 times row 2 added to row 3 $\begin{bmatrix} -3 & 1 & -4 \\ 2 & 1 & 3 \\ -17 & 0 & -13 \end{bmatrix}$

6. $\begin{bmatrix} 4 & 10 & -8 \\ 7 & 4 & 3 \\ -1 & 1 & 0 \end{bmatrix}$; -4 times row 3 added to row 2 $\begin{bmatrix} 4 & 10 & -8 \\ 11 & 0 & 3 \\ -1 & 1 & 0 \end{bmatrix}$

Write the augmented matrix for each of the following systems. Do not solve the system.

7. $2x + 3y = 11$ $\begin{bmatrix} 2 & 3 & | & 11 \\ 1 & 2 & | & 8 \end{bmatrix}$
$x + 2y = 8$

8. $3x + 5y = -13$ $\begin{bmatrix} 3 & 5 & | & -13 \\ 2 & 3 & | & -9 \end{bmatrix}$
$2x + 3y = -9$

9. $x + 5y = 6$ $\begin{bmatrix} 1 & 5 & | & 6 \\ 1 & 2 & | & 8 \end{bmatrix}$
$x + 2y = 8$

10. $2x + 7y = 1$ $\begin{bmatrix} 2 & 7 & | & 1 \\ 5 & 0 & | & -15 \end{bmatrix}$
$5x = -15$

11. $2x + y + z = 3$ $\begin{bmatrix} 2 & 1 & 1 & | & 3 \\ 3 & -4 & 2 & | & -7 \\ 1 & 1 & 1 & | & 2 \end{bmatrix}$
$3x - 4y + 2z = -7$
$x + y + z = 2$

12. $4x - 2y + 3z = 4$ $\begin{bmatrix} 4 & -2 & 3 & | & 4 \\ 3 & 5 & 1 & | & 7 \\ 5 & -1 & 4 & | & 7 \end{bmatrix}$
$3x + 5y + z = 7$
$5x - y + 4z = 7$

13. $x + y = 2$ $\begin{bmatrix} 1 & 1 & 0 & | & 2 \\ 0 & 2 & 1 & | & -4 \\ 0 & 0 & 1 & | & 2 \end{bmatrix}$
$2y + z = -4$
$z = 2$

14. $x = 6$ $\begin{bmatrix} 1 & 0 & 0 & | & 6 \\ 0 & 1 & 2 & | & 2 \\ 1 & 0 & -3 & | & 6 \end{bmatrix}$
$y + 2z = 2$
$x - 3z = 6$

Write the system of equations associated with each of the following augmented matrices. Do not try to solve.

15. $\begin{bmatrix} 2 & 1 & | & 1 \\ 3 & -2 & | & -9 \end{bmatrix}$ $2x + y = 1;\quad 3x - 2y = -9$

16. $\begin{bmatrix} 1 & -5 & | & -18 \\ 6 & 2 & | & 20 \end{bmatrix}$ $x - 5y = -18;\quad 6x + 2y = 20$

17. $\begin{bmatrix} 1 & 0 & 0 & | & 2 \\ 0 & 1 & 0 & | & 3 \\ 0 & 0 & 1 & | & -2 \end{bmatrix}$ $x = 2;\quad y = 3;\quad z = -2$

18. $\begin{bmatrix} 1 & 0 & 1 & | & 4 \\ 0 & 1 & 0 & | & 2 \\ 0 & 0 & 1 & | & 3 \end{bmatrix}$ $x + z = 4;\quad y = 2;\quad z = 3$

19. $\begin{bmatrix} 3 & 2 & 1 & | & 1 \\ 0 & 2 & 4 & | & 22 \\ -1 & -2 & 3 & | & 15 \end{bmatrix}$ $3x + 2y + z = 1;\quad 2y + 4z = 22;\quad -x - 2y + 3z = 15$

20. $\begin{bmatrix} 2 & 1 & 3 & | & 12 \\ 4 & -3 & 0 & | & 10 \\ 5 & 0 & -4 & | & -11 \end{bmatrix}$ $2x + y + 3z = 12;\quad 4x - 3y = 10;\quad 5x - 4z = -11$

Use the Gauss-Jordan method to solve each of the following systems of equations. See Examples 2–5.

21. $x + y = 5$
$x - y = -1$ $\{(2, 3)\}$

22. $x + 2y = 5$
$2x + y = -2$ $\{(-3, 4)\}$

23. $x + y = -3$
$2x - 5y = -6$ $\{(-3, 0)\}$

24. $3x - 2y = 4$
$3x + y = -2$ $\{(0, -2)\}$

25. $2x - 3y = 10$
$2x + 2y = 5$ $\{(7/2, -1)\}$

26. $4x + y = 5$
$2x + y = 3$ $\{(1, 1)\}$

27. $2x - 3y = 2$
$4x - 6y = 1$ $\emptyset$

28. $x + 2y = 1$
$2x + 4y = 3$ $\emptyset$

29. $6x - 3y = 1$
$-12x + 6y = -2$ $\left\{ \left(\dfrac{3y + 1}{6}, y \right) \right\}$

30. $x - y = 1$
$-x + y = -1$ $\{(y + 1, y)\}$

31. $x + y = -1$
$y + z = 4$
$x + z = 1$ $\{(-2, 1, 3)\}$

32. $x - z = -3$
$y + z = 9$
$x + z = 7$ $\{(2, 4, 5)\}$

33. $x + y - z = 6$
$2x - y + z = -9$
$x - 2y + 3z = 1$ $\{(-1, 23, 16)\}$

34. $x + 3y - 6z = 7$
$2x - y + 2z = 0$
$x + y + 2z = -1$ $\{(1, 0, -1)\}$

35. $-x + y = -1$
$y - z = 6$
$x + z = -1$ $\{(3, 2, -4)\}$

36. $x + y = 1$
$2x - z = 0$
$y + 2z = -2$ $\{(-1, 2, -2)\}$

37. $2x - y + 3z = 0$
$\quad x + 2y - z = 5$
$\quad\quad 2y + z = 1 \quad \{(2, 1, -1)\}$

38. $4x + 2y - 3z = 6$
$\quad x - 4y + z = -4$
$\quad\quad -x + 2z = 2 \quad \{(2, 2, 2)\}$

39. Compare the use of an augmented matrix as a shorthand way of writing a system of linear equations and the use of synthetic division as a shorthand way to divide polynomials.

40. Compare the use of the third row transformation on a matrix and the elimination method of solving a system of linear equations.

Use the Gauss-Jordan method to solve each of the following systems.

41. $\quad 3x + 2y - w = 0$
$\quad\quad 2x + z + 2w = 5$
$\quad\quad\quad x + 2y - z = -2$
$\quad 2x - y + z + w = 2 \quad \{(-1, 2, 5, 1)\}$

42. $\quad x + 3y - 2z - w = 9$
$\quad 4x + y + z + 2w = 2$
$\quad -3x - y + z - w = -5$
$\quad x - y - 3z - 2w = 2 \quad \{(0, 2, -2, 1)\}$

Solve each of the following problems by first setting up a system of equations. Use the Gauss-Jordan method.

43. George Esquibel deposits some money in a bank account paying 3% per year. He uses some additional money, amounting to 1/3 the amount placed in the bank, to buy bonds paying 4% per year. With the balance of his funds he buys a 4.5% certificate of deposit. The first year his investments bring a return of $400. If the total of the investments is $10,000, how much is invested at each rate?

$3000 at 3%; $1000 at 4%; $6000 at 4.5%

44. To get necessary funds for a planned expansion, a small company took out three loans totaling $25,000. The company was able to borrow some of the money at 8%. It borrowed $2000 more than 1/2 the amount of the 8% loan at 10%, and the rest at 9%. The total annual interest was $2220. How much did the company borrow at each rate?

$10,000 at 8%; $7000 at 10%; $8000 at 9%

45. In one day, a service station sold 400 gallons of premium gasoline, 150 gallons of regular gasoline, and 130 gallons of super gasoline for a total of $909. The next day 380 gallons of premium, 170 gallons of regular, and 150 gallons of super were sold for $931. The difference in price per gallon between super and regular is one-half the difference in price per gallon between premium and regular. How much does this station charge for each type of gasoline?

$1.40 for premium; $1.20 for regular; $1.30 for super

46. A biologist has three salt solutions: some 5% solution, some 15% solution, and some 25% solution. She needs to mix some of each to get 50 liters of 20% solution. She wants to use twice as much of the 5% solution as the 15% solution. How much of each solution should she use?

10 liters of 5%; 5 liters of 15%; 35 liters of 25%

Solve each of the systems in Exercises 47–52 by the Gauss-Jordan method. Let z be the arbitrary variable if necessary. See Example 5.

47. $x - 3y + 2z = 10$
$\quad 2x - y - z = 8 \quad \left\{\left(\dfrac{5z + 14}{5}, \dfrac{5z - 12}{5}, z\right)\right\}$

48. $3x + y - z = 12$
$\quad x + 2y + z = 10 \quad \left\{\left(\dfrac{3z + 14}{5}, \dfrac{-4z + 18}{5}, z\right)\right\}$

49. $\quad x + 2y - z = 0$
$\quad 3x - y + z = 6 \quad \left\{\left(\dfrac{12 - z}{7}, \dfrac{4z - 6}{7}, z\right)\right\}$
$\quad -2x - 4y + 2z = 0$

50. $\quad 3x + 5y - z = 0$
$\quad 4x - y + 2z = 1 \quad \left\{\left(\dfrac{5 - 9z}{23}, \dfrac{10z - 3}{23}, z\right)\right\}$
$\quad -6x - 10y + 2z = 0$

51. $\quad x - 2y + z = 5$
$\quad -2x + 4y - 2z = 2$
$\quad 2x + y - z = 2 \quad \emptyset$

52. $3x + 6y - 3z = 12$
$\quad -x - 2y + z = 16$
$\quad x + y - 2z = 20 \quad \emptyset$

▲ **53.** At rush hours, substantial traffic is encountered at the traffic intersections shown in the figure. The city wishes to improve the signals at these corners so as to speed the flow of traffic. The traffic engineers first gather data. As the figure shows, 700 cars per hour come down *M* Street to intersection *A*; 300 cars per hour come to intersection *A* on 10th Street. A total of x_1 of these cars leave *A* on *M* Street, while x_4 cars leave *A* on 10th Street. The number of cars entering *A* must equal the number leaving, so that

$$x_1 + x_4 = 700 + 300$$

or $\qquad x_1 + x_4 = 1000. \qquad$ **(1)**

For intersection *B*, x_1 cars enter *B* on *M* Street, and x_2 cars enter *B* on 11th Street. The figure shows that 900 cars leave *B* on 11th while 200 leave on *M*. We have

$$x_1 + x_2 = 900 + 200$$

or $\qquad x_1 + x_2 = 1100. \qquad$ **(2)**

At intersection *C*, 400 cars enter on *N* Street, 300 on 11th Street, while x_2 leave on 11th Street and x_3 leave on *N* Street. This gives

$$x_2 + x_3 = 400 + 300$$

or $\qquad x_2 + x_3 = 700. \qquad$ **(3)**

Finally, intersection *D* has x_3 cars entering on *N* and x_4 entering on 10th. There are 400 leaving *D* on 10th and 200 leaving on *N*.

(a) Set up an equation for intersection *D*.

$x_3 + x_4 = 600$

(b) Use equations (1), (2), (3), and your answer to part (a) to set up a system of equations. Solve the system by the Gauss-Jordan method.

The system has no unique solution; equations $x_2 + x_3 = 700$ and $x_3 + x_4 = 600$ are dependent.

(c) Since you got a row of zeros, the system of equations is dependent and does not have a unique solution. Solve equation (1) for x_1 and substitute the result into equation (2). $x_2 - x_4 = 100$

(d) Solve equation (1) for x_4. What is the largest possible value of x_1 so that x_4 is not negative?

$x_4 = 1000 - x_1;$ 1000

(e) Your answer to part (c) should be $x_2 - x_4 = 100$. Solve the equation for x_4 and then find the smallest possible value of x_2 so that x_4 is not negative.

$x_4 = x_2 - 100;$ 100

(f) Your answer to part (a) should be $x_3 + x_4 = 600$. Solve the equation for x_4 and then find the largest possible values of x_3 and x_4 so that neither is negative.

$x_4 = 600 - x_3;$ $600, 600$

(g) From your answers to parts (d)–(f), what is the maximum value of x_4 so that all the equations are satisfied and all variables are nonnegative? Of x_3? Of x_2? Of x_1? $600;$ $600;$ $700;$ 1000

7.4   RESOURCES

Instructor's Test Manual
Test Items, p. 358–359
Instructor's Solution Manual
pp. 288–296, 699–708
Study Guide
pp. 260–264
Student's Solution Manual
pp. 288–296

Test Generator
IBM Disks
5.25″ Disk B
3.5″ Disk A
Mac Disk A
Tutorials
IBM Disks
5.25″ Disk G
3.5″ Disk D
Mac Disk D

HarperCollins Videotape   8
Transparency   T-35–T-36

© CONCEPTUAL   ✎ WRITING   ▲ CHALLENGING

7.4 EXERCISES (TEXT PAGE 382) ∎

For each of Exercises 1–8, draw a sketch of the two graphs described, with the indicated number of points of intersection. (There may be more than one way to do this.) See Figures 7.4 and 7.5.

1. a line and an ellipse; two points

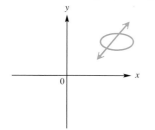

2. a line and a hyperbola; one point

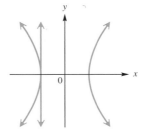

3. a circle and an ellipse; one point

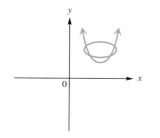

4. a circle and a hyperbola; two points

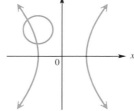

5. a line and a hyperbola; two points

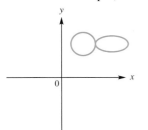

6. a circle and an ellipse; four points

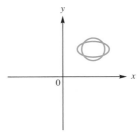

7. a parabola and an ellipse; four points

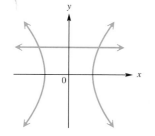

8. a parabola and a hyperbola; four points

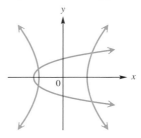

✐ ◉ **9.** In Example 3, we used a combination of methods to get the equation

$$x^4 - 10x^2 + 16 = 0$$

in order to find the solutions of the system. Explain why this equation cannot possibly have more than four solutions.

◉ **10.** In Example 5, there are four solutions to the system, but there are no points of intersection of the graphs. When will the number of solutions of a system be less than the number of points of intersection of the graphs of the equations of the system?   *never*

Give all solutions of the following nonlinear systems of equations, including those with imaginary values. See Examples 1–5.

11. $y = x^2$
  $x + y = 2$ $\{(1, 1), (-2, 4)\}$

12. $y = -x^2 + 2$
  $x - y = 0$ $\{(1, 1), (-2, -2)\}$

13. $y = (x - 1)^2$
  $x - 3y = -1$ $\{(2, 1), (1/3, 4/9)\}$

14. $y = (x + 3)^2$
  $x + 2y = -2$ $\{(-4, 1), (-5/2, 1/4)\}$

15. $y = x^2 + 4x$
  $2x - y = -8$ $\{(2, 12), (-4, 0)\}$

16. $y = 6x + x^2$
  $3x - 2y = 10$ $\{(-2, -8), (-5/2, -35/4)\}$

17. $3x^2 + 2y^2 = 5$
  $x - y = -2$ $\{(-3/5, 7/5), (-1, 1)\}$

18. $x^2 + y^2 = 5$
  $-3x + 4y = 2$ $\{(-2, -1), (38/25, 41/25)\}$

19. $x^2 + y^2 = 8$
  $x^2 - y^2 = 0$ $\{(2, 2), (2, -2), (-2, 2), (-2, -2)\}$

20. $x^2 + y^2 = 10$
  $2x^2 - y^2 = 17$ $\{(3, 1), (3, -1), (-3, 1), (-3, -1)\}$

21. $5x^2 - y^2 = 0$
  $3x^2 + 4y^2 = 0$ $\{(0, 0)\}$

22. $x^2 + y^2 = 4$
  $2x^2 - 3y^2 = -12$ $\{(0, 2), (0, -2)\}$

23. $3x^2 + y^2 = 3$
  $4x^2 + 5y^2 = 26$ $\{(i, \sqrt{6}), (-i, \sqrt{6}), (i, -\sqrt{6}), (-i, -\sqrt{6})\}$

24. $x^2 + 2y^2 = 9$
  $3x^2 - 4y^2 = 27$ $\{(-3, 0), (3, 0)\}$

25. $2x^2 + 3y^2 = 5$
  $3x^2 - 4y^2 = -1$ $\{(1, -1), (-1, 1), (1, 1), (-1, -1)\}$

26. $3x^2 + 5y^2 = 17$
  $2x^2 - 3y^2 = 5$ $\{(2, 1), (-2, 1), (2, -1), (-2, -1)\}$

27. $2x^2 + 2y^2 = 20$
  $3x^2 + 3y^2 = 30$ $\{(x, \pm\sqrt{10 - x^2})\}$

28. $x^2 + y^2 = 4$
  $5x^2 + 5y^2 = 28$ ∅

29. $9x^2 + 4y^2 = 1$
  $x^2 + y^2 = 1$ $\{(i\sqrt{15}/5, 2\sqrt{10}/5), (i\sqrt{15}/5, -2\sqrt{10}/5), (-i\sqrt{15}/5, 2\sqrt{10}/5), (-i\sqrt{15}/5, -2\sqrt{10}/5)\}$

30. $2x^2 - 3y^2 = 8$
  $6x^2 + 5y^2 = 24$ $\{(2, 0), (-2, 0)\}$

31. $xy = -15$
  $4x + 3y = 3$ $\{(-3, 5), (15/4, -4)\}$

32. $xy = 8$
  $3x + 2y = -16$ $\{(-4, -2), (-4/3, -6)\}$

33. $2xy + 1 = 0$
  $x + 16y = 2$ $\{(4, -1/8), (-2, 1/4)\}$

34. $-5xy + 2 = 0$
  $x - 15y = 5$ $\{(6, 1/15), (-1, -2/5)\}$

35. $x^2 + 4y^2 = 25$
  $xy = 6$ $\{(3, 2), (-3, -2), (4, 3/2), (-4, -3/2)\}$

36. $5x^2 - 2y^2 = 6$
  $xy = 2$ $\{(2i\sqrt{5}/5, -i\sqrt{5}), (-2i\sqrt{5}/5, i\sqrt{5}), (\sqrt{2}, \sqrt{2}), (-\sqrt{2}, -\sqrt{2})\}$

◉ CONCEPTUAL   ✐ WRITING   ▲ CHALLENGING

37. $x^2 + 2xy - y^2 = 14$
$x^2 - y^2 = -16$ $\{(3, 5), (-3, -5), (5i, -3i), (-5i, 3i)\}$

38. $3x^2 + xy + 3y^2 = 7$
$x^2 + y^2 = 2$ $\{(1, 1), (-1, -1)\}$

39. $x^2 - xy + y^2 = 5$
$2x^2 + xy - y^2 = 10$ $\{(\sqrt{5}, 0), (-\sqrt{5}, 0), (\sqrt{5}, \sqrt{5}), (-\sqrt{5}, -\sqrt{5})\}$

40. $3x^2 + 2xy - y^2 = 9$
$x^2 - xy + y^2 = 9$ $\{(\sqrt{3}, 2\sqrt{3}), (-\sqrt{3}, -2\sqrt{3}), (6\sqrt{7}/7, -3\sqrt{7}/7), (-6\sqrt{7}/7, 3\sqrt{7}/7)\}$

41. $x = |y|$
$x^2 + y^2 = 18$ $\{(3, -3), (3, 3)\}$

42. $2x + |y| = 4$
$x^2 + y^2 = 5$ $\{(1, 2), (1, -2)\}$

43. $y = |x - 1|$
$y = x^2 - 4$ $\left\{\left(\dfrac{1 + \sqrt{13}}{2}, \dfrac{-1 + \sqrt{13}}{2}\right), \left(\dfrac{-1 - \sqrt{21}}{2}, \dfrac{3 + \sqrt{21}}{2}\right)\right\}$

44. $2x^2 - y^2 = 4$
$|x| = |y|$ $\{(2, 2), (-2, -2), (2, -2), (-2, 2)\}$

Solve each problem using a system of equations in two variables.

45. Find two numbers whose ratio is 9 to 2 and whose product is 162.   27 and 6 or -27 and -6

46. Find two numbers whose ratio is 4 to 3 such that the sum of their squares is 100.   6 and 8 or -6 and -8

47. Does the straight line $3x - 2y = 9$ intersect the circle $x^2 + y^2 = 25$? (To find out, solve the system made up of these two equations.)   yes

48. Do the parabola $y = x^2 + 4$ and the ellipse $2x^2 + y^2 - 4x - 4y = 0$ have any points in common?   yes

▲ 49. Find the equation of the straight line through $(2, 4)$ that touches the parabola $y = x^2$ at only one point. (*Note:* Recall that a quadratic equation has a unique solution when the discriminant is 0.) $y = 4x - 4$

▲ 50. For what value of b will the line $x + 2y = b$ touch the circle $x^2 + y^2 = 9$ in only one point? $b = \pm 3\sqrt{5}$

▲ 51. For what nonzero values of a do the graphs of $x^2 + y^2 = 25$ and $x^2/a^2 + y^2/25 = 1$ have exactly two points in common? $a \neq 5, a \neq -5$

▲ 52. Find the equation of the line passing through the points of intersection of the graphs of $y = x^2$ and $x^2 + y^2 = 90$. $y = 9$

�'◎ 53. Suppose that you are given the equations of two circles that are known to intersect in exactly two points. Explain how you would find the equation of the only chord common to these circles.

▦ 54. In electronics, circuit gain is given by

$$G = \frac{Bt}{R + R_t}$$

where R is the value of a resistor, t is temperature, and B is a constant. The sensitivity of the circuit to temperature is given by

$$S = \frac{BR}{(R + R_t)^2}.$$

If $B = 3.7$ and t is 90K (Kelvin), find the values of R and R_t that will make $G = .4$ and $S = .001$. $R = 187$; $R_t = 645$ (rounded)

7.5 RESOURCES

Instructor's Test Manual
    Test Items, p. 359–361
Instructor's Solution Manual
    pp. 296–303, 708–715
Study Guide
    pp. 264–270
Student's Solution Manual
    pp. 296–303

Test Generator
    IBM Disks
        5.25" Disk B
        3.5" Disk A
        Mac Disk A
Tutorials
    IBM Disks
        5.25" Disk G
        3.5" Disk D
        Mac Disk D

HarperCollins Videotape 8
**"College Algebra: In Simplest
Terms" Program** 21
Transparency T-37(a)–(b)

7.5 EXERCISES (TEXT PAGE 389) ■

Graph the solution set of each system of inequalities. See Examples 1–3.

1. $x + y \le 4$
    $x - 2y \ge 6$

2. $2x + y > 2$
    $x - 3y < 6$

3. $4x + 3y < 12$
    $y + 4x > -4$

4. $3x + 5y \le 15$
    $x - 3y \ge 9$

5. $x + y \le 6$
    $2x + 2y \ge 12$

6. $3x + 4y < 15$
    $6x + 8y > 30$

7. $x + 2y \le 4$
    $y \ge x^2 - 1$

8. $4x - 3y \le 12$
    $y \le x^2$

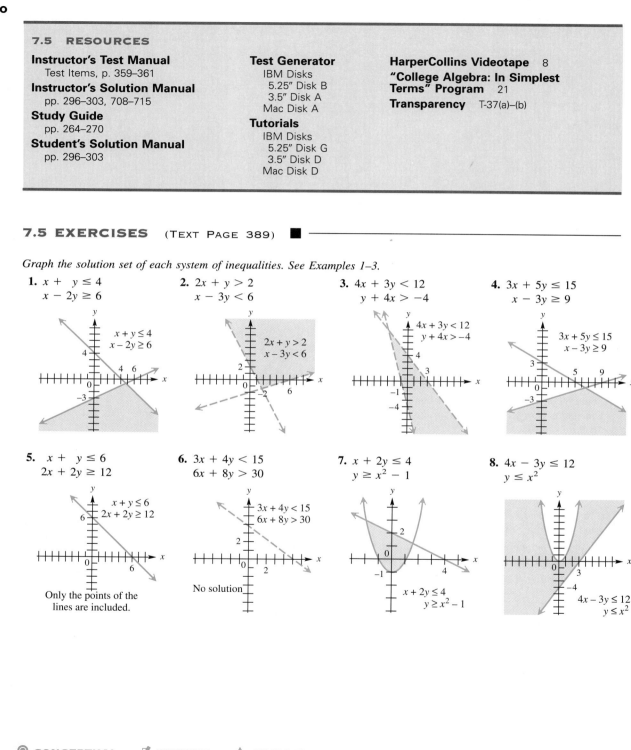

5. Only the points of the lines are included.

6. No solution

9. $y \leq -x^2$
$y \geq x^2 - 6$

10. $x^2 + y^2 \leq 9$
$x \leq -y^2$

11. $x^2 - y^2 < 1$
$-1 < y < 1$

12. $x^2 + y^2 \leq 36$
$-4 \leq x \leq 4$

13. $2x^2 - y^2 > 4$
$2y^2 - x^2 > 4$

14. $y \geq x^2 + 4x + 4$
$y < -x^2$

15. $\dfrac{x^2}{16} + \dfrac{y^2}{9} \leq 1$
$\dfrac{x^2}{4} - \dfrac{y^2}{16} \geq 1$

16. $\dfrac{x^2}{36} - \dfrac{y^2}{9} \geq 1$
$\dfrac{x^2}{81} + y^2 \leq 1$

17. $x + y \leq 4$
$x - y \leq 5$
$4x + y \leq -4$

18. $3x - 2y \geq 6$
$x + y \leq -5$
$y \leq 4$

19. $2y + x \leq -5$
$y \geq 3 + x$
$x \geq 0$
$y \geq 0$

20. $2x + 3y \leq 12$
$2x + 3y > -6$
$3x + y < 4$
$x \geq 0$
$y \geq 0$

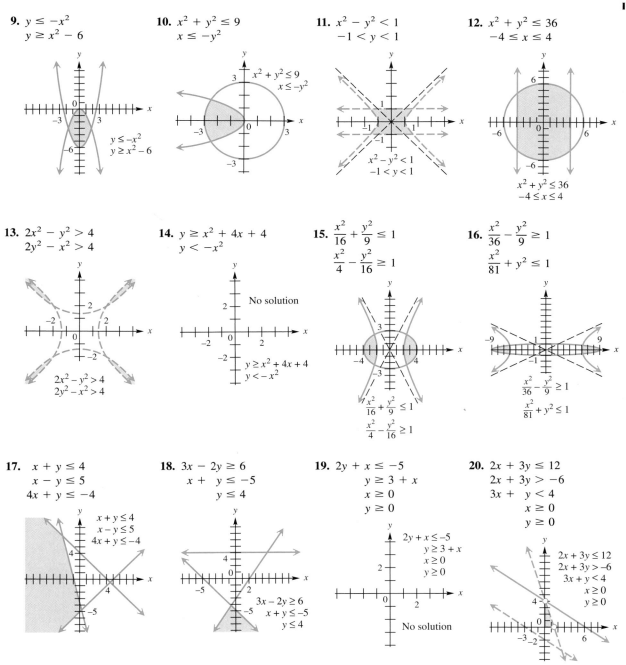

21. $\dfrac{x^2}{4} + \dfrac{y^2}{9} > 1$

$x^2 - y^2 \geq 1$

$-4 \leq x \leq 4$

22. $2x - 3y < 6$

$4x^2 + 9y^2 < 36$

$x \geq -1$

23. $y \geq 3^x$

$y \geq 2$

24. $y \leq \left(\dfrac{1}{2}\right)^x$

$y \geq 4$

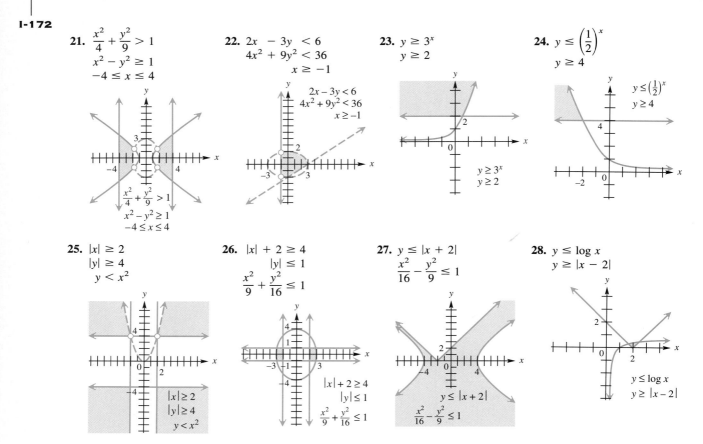

25. $|x| \geq 2$

$|y| \geq 4$

$y < x^2$

26. $|x| + 2 \geq 4$

$|y| \leq 1$

$\dfrac{x^2}{9} + \dfrac{y^2}{16} \leq 1$

27. $y \leq |x + 2|$

$\dfrac{x^2}{16} - \dfrac{y^2}{9} \leq 1$

28. $y \leq \log x$

$y \geq |x - 2|$

29. Which one of the following is a description of the solution set of the system below?

$$x^2 + 4y^2 < 36$$

$$y < x$$

(a) all points outside the ellipse $x^2 + 4y^2 = 36$ and above the line $y = x$

(b) all points outside the ellipse $x^2 + 4y^2 = 36$ and below the line $y = x$

(c) all points inside the ellipse $x^2 + 4y^2 = 36$ and above the line $y = x$

(d) all points inside the ellipse $x^2 + 4y^2 = 36$ and below the line $y = x$ **(d)**

30. Fill in the blanks with the appropriate responses. The graph of the system

$$y > x^2 + 2$$

$$x^2 + y^2 < 16$$

$$y < 7$$

consists of all points _____ the parabola $y = x^2 + 2$, _____

(above/below)   (inside/outside)

the circle $x^2 + y^2 = 16$, and _____ the line $y = 7$.   above, inside, below

(above/below)

In Exercises 31–34, use graphical methods to find values of x and y satisfying the given conditions. Find the value of the maximum or minimum.

31. Find $x \geq 0$ and $y \geq 0$ such that

$$2x + 3y \leq 6$$
$$4x + y \leq 6$$

and $5x + 2y$ is maximized. $(6/5, 6/5);$ $42/5$

32. Find $x \geq 0$ and $y \geq 0$ such that

$$x + y \leq 10$$
$$5x + 2y \geq 20$$
$$2y \geq x$$

and $x + 3y$ is minimized. $(10/3, 5/3);$ $25/3$

33. Find $x \geq 2$ and $y \geq 5$ such that

$$3x - y \geq 12$$
$$x + y \leq 15$$

and $2x + y$ is minimized. $(17/3, 5);$ $49/3$

34. Find $x \geq 10$ and $y \geq 20$ such that

$$2x + 3y \leq 100$$
$$5x + 4y \leq 200$$

and $x + 3y$ is maximized. $(10, 80/3);$ 90

Solve each of the following linear programming problems. See Example 4.

35. Farmer Jones raises only pigs and geese. She wants to raise no more than 16 animals with no more than 12 geese. She spends $50 to raise a pig and $20 to raise a goose. She has $500 available for this purpose. Find the maximum profit she can make if she makes a profit of $80 per goose and $40 per pig.

$1120, with 4 pigs, 12 geese

36. A wholesaler of party goods wishes to display her products at a convention of social secretaries in such a way that she gets the maximum number of inquiries about her whistles and hats. Her booth at the convention has 12 sq m of floor space to be used for display purposes. A display unit for hats requires 2 sq m, and for whistles, 4 sq m. Experience tells the wholesaler that she should never have more than a total of 5 units of whistles and hats on display at one time. If she receives three inquiries for each unit of hats and two inquiries for each unit of whistles on display, how many of each should she display in order to get the maximum number of inquiries?

5 units of hats and no whistles, for a maximum of 15 inquiries

37. An office manager wants to buy some filing cabinets. She knows that cabinet #1 costs $10 each, requires 6 sq ft of floor space, and holds 8 cu ft of files. Cabinet #2 costs $20 each, requires 8 sq ft of floor space, and holds 12 cu ft. She can spend no more than $140 due to budget limitations, while her office has room for no more than 72 sq ft of cabinets. She wants the maximum storage capacity within the limits imposed by funds and space. How many of each type of cabinet should she buy?

8 of #1 and 3 of #2, for 100 cu ft of storage

38. In a small town in South Carolina, zoning rules require that the window space (in square feet) in a house be at least one-sixth of the space used up by solid walls. The cost to heat the house is 20¢ for each square foot of solid walls and 80¢ for each square foot of windows. Find the maximum total area (windows plus walls) if $160 is available to pay for heat.

80 sq ft of window and 480 sq ft of wall, for a maximum area of 560 sq ft

39. The manufacturing process requires that oil refineries manufacture at least 2 gal of gasoline for each gallon of fuel oil. To meet the winter demand for fuel oil, at least 3 million gal a day must be produced. The demand for gasoline is no more than 6.4 million gal per day. If the price of gasoline is $1.90 and the price of fuel oil is $1.50 per gal, how much of each should be produced to maximize revenue?

6.4 million gallons of gasoline and 3.2 million gallons of fuel oil, for maximum revenue of $16,960,000

40. Seall Manufacturing Company makes color television sets. It produces a bargain set that sells for $100 profit and a deluxe set that sells for $150 profit. On the assembly line the bargain set requires 3 hr, while the deluxe set takes 5 hr. The cabinet shop spends 1 hr on the cabinet for the bargain set and 3 hr on the cabinet for the deluxe set. Both sets require 2 hr of time for testing and packing. On a particular production run the Seall Company has available 3900 work hours on the assembly line, 2100 work hours in the cabinet shop, and 2200 work hours in the testing and packing department. How many sets of each type should it produce to make maximum profit? What is the maximum profit?

800 bargain sets and 300 deluxe sets, for a maximum profit of $125,000

RESOURCES FOR REVIEW

Instructor's Test Manual
    Chapter Test Forms, pp. 151–176
Instructor's Solution Manual
    pp. 303–318, 715–722

Study Guide
    pp. 271–274
Student's Solution Manual
    pp. 303–318

Test Generator
    IBM Disks
    5.25" Disk B
    3.5" Disk A
    Mac Disk A

CHAPTER 7 REVIEW EXERCISES   (TEXT PAGE 392) ■

Use the substitution method to solve each of the following linear systems. Identify any systems with dependent equations or any inconsistent systems.

1. $4x - 3y = -1$
    $3x + 5y = 50$ $\{(5, 7)\}$

2. $10x + 3y = 8$
    $5x - 4y = 26$ $\{(2, -4)\}$

3. $7x - 10y = 11$
    $\dfrac{3x}{2} - 5y = 8$ $\{(-5/4, -79/40)\}$

4. $\dfrac{x}{2} + \dfrac{2y}{3} = -8$
    $\dfrac{3x}{4} + \dfrac{y}{3} = 0$ $\{(8, -18)\}$

5. $\dfrac{x}{2} - \dfrac{y}{5} = \dfrac{11}{10}$
    $2x - \dfrac{4y}{5} = \dfrac{22}{5}$ $\left\{ \left(\dfrac{2y + 11}{5}, y \right) \right\}$;   dependent

6. $4x + 5y = 5$
    $3x + 7y = -6$ $\{(5, -3)\}$

Use the addition method to solve each of the following linear systems. Identify any systems with dependent equations or any inconsistent systems.

7. $3x - 5y = -18$
    $2x + 7y = 19$ $\{(-1, 3)\}$

8. $6x + 5y = 53$
    $4x - 3y = 29$ $\{(8, 1)\}$

9. $\dfrac{2}{3}x - \dfrac{3}{4}y = 13$
    $\dfrac{1}{2}x + \dfrac{2}{3}y = -5$ $\{(6, -12)\}$

10. $3x + 7y = 10$
     $18x + 42y = 50$ inconsistent; $\emptyset$

11. $\dfrac{1}{x} + \dfrac{1}{y} = \dfrac{7}{10}$
     $\dfrac{3}{x} - \dfrac{5}{y} = \dfrac{1}{2}$ $\{(2, 5)\}$

 12. A student solves the system
$$x + y = 3$$
$$2x + 2y = 6$$
and obtains the result $0 = 0$. Then the student gives the solution set as $\{(0, 0)\}$. Is this correct? If not, explain.

Solve each of the following problems by writing a system of equations and then solving it.

13. A student bought some candy bars, paying 25¢ each for some and 50¢ each for others. The student bought a total of 22 bars, paying a total of $8.50. How many of each kind of bar did he buy?

    10 at 25¢;   12 at 50¢

14. Ink worth $25 a bottle is to be mixed with ink worth $18 per bottle to get 12 bottles of ink worth $20 each. How much of each type of ink should be used?

    3 3/7 bottles at $25;   8 4/7 bottles at $18

15. Mai Ling wins $50,000 in a lottery. She invests part of the money at 3%, twice as much at 3.5%, with $10,000 more than the amount invested at 3% invested at 4.5%. The total annual interest is $1900. How much is invested at each rate?

$10,000 at 3%; $20,000 at 3.5%; $20,000 at 4.5%

16. The sum of three numbers is 23. The second number is 3 more than the first. The sum of the first and twice the third is 4. Find the three numbers.   12, 15, −4

17. A gold merchant has some 12-carat gold (12/24 pure gold), and some 22-carat gold (22/24 pure). How

many grams of each should be mixed to get 25 g of 15-carat gold?

17.5 g of 12-carat;   7.5 g of 22-carat

18. A chemist has some 40% acid solution and some 60% solution. How many liters of each should be used to get 40 liters of a 45% solution?

30 liters of 40%;   10 liters of 60%

Use the addition method to solve each of the following linear systems.

19. $2x - 3y + z = -5$
$x + 4y + 2z = 13$
$5x + 5y + 3z = 14$ $\{(-1, 2, 3)\}$

20. $x - 3y = 12$
$2y + 5z = 1$
$4x + z = 25$ $\{(6, -2, 1)\}$

21. $x + y - z = 5$
$2x + y + 3z = 2$
$4x - y + 2z = -1$ $\{(1, 3, -1)\}$

22. $5x - 3y + 2z = -5$
$2x + 2y - z = 4$
$4x - y + z = -1$ $\{(0, 3, 2)\}$

Find solutions for the following systems in terms of the arbitrary variable z.

23. $3x - 4y + z = 2$
$2x + y - 4z = 1$ $\left\{\left(\dfrac{15z + 6}{11}, \dfrac{14z - 1}{11}, z\right)\right\}$

24. $2x + 3y - z = 5$
$-x + 2y + 4z = 8$ $\{(2z - 2, -z + 3, z)\}$

Solve each of the following problems by writing a system of equations and then solving it.

25. Three kinds of tea worth $4.60, $5.75, and $6.50 per pound are to be mixed to get 20 lb of tea worth $5.25 per pound. The amount of $4.60 tea used is to be equal to the total amount of the other two kinds together. How many pounds of each tea should be used?

10 lb of $4.60 tea;   8 lb of $5.75 tea;   2 lb of $6.50 tea

26. A 5% solution of a drug is to be mixed with some 15% solution and some 10% solution to get 20 ml of 8% solution. The amount of 5% solution used must be

2 ml more than the sum of the other two solutions. How many milliliters of each solution should be used?

11 ml of 5%;   3 ml of 15%;   6 ml of 10%

27. The cashier at an amusement park has a total of $2480, made up of fives, tens, and twenties. The total number of bills is 290, and the value of the tens is $60 more than the value of the twenties. How many of each type of bill does the cashier have?

164 fives;   86 tens;   40 twenties

28. Can a system consisting of two equations in three variables have a unique solution? Explain.

Use the Gauss-Jordan method to solve each of the following systems.

29. $2x + 3y = 10$
$-3x + y = 18$ $\{(-4, 6)\}$

30. $5x + 2y = -10$
$3x - 5y = -6$ $\{(-2, 0)\}$

31. $3x + y = -7$
$x - y = -5$ $\{(-3, 2)\}$

32. $x - z = -3$
$y + z = 6$
$2x - 3z = -9$ $\{(0, 3, 3)\}$

33. $2x - y + 4z = -1$
$-3x + 5y - z = 5$
$2x + 3y + 2z = 3$ $\{(0, 1, 0)\}$

34. $5x - 8y + z = 1$
$3x - 2y + 4z = 3$
$10x - 16y + 2z = 3$ $\emptyset$

Solve each of the following nonlinear systems of equations.

35. $y = x^2 - 1$
$x + y = 1$ $\{(-2, 3), (1, 0)\}$

36. $x^2 + y^2 = 2$
$3x + y = 4$ $\{(1, 1), (7/5, -1/5)\}$

37. $x^2 + 2y^2 = 22$
$2x^2 - y^2 = -1$ $\{(2, 3), (-2, 3), (2, -3), (-2, -3)\}$

38. $x^2 - 4y^2 = 19$
$x^2 + y^2 = 29$ $\{(3\sqrt{3}, \sqrt{2}), (-3\sqrt{3}, \sqrt{2}), (3\sqrt{3}, -\sqrt{2}), (-3\sqrt{3}, -\sqrt{2})\}$

39. $xy = 4$
$x - 6y = 2$ $\{(6, 2/3), (-4, -1)\}$

40. $x^2 + 2xy + y^2 = 4$
$x = 3y - 2$ $\{(-2, 0), (1, 1)\}$

41. Do the circle $x^2 + y^2 = 144$ and the line $x + 2y = 8$ have any points in common? If so, what are they?   yes; $\left\{ \left(\dfrac{8 - 8\sqrt{41}}{5}, \dfrac{16 + 4\sqrt{41}}{5} \right), \left(\dfrac{8 + 8\sqrt{41}}{5}, \dfrac{16 - 4\sqrt{41}}{5} \right) \right\}$

▲ **42.** Find a value of b so that the straight line $3x - y = b$ intersects the circle $x^2 + y^2 = 25$ in only one point. $b = \pm 5\sqrt{10}$

◉ **43.** Sketch a line and a hyperbola on the same set of axes so that their intersection contains each of the following numbers of points.

(a) 0   **(b)** 1   **(c)** 2

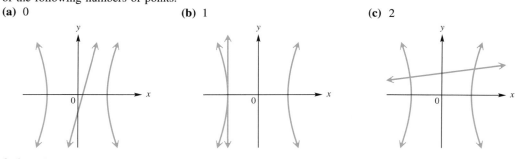

Graph the solution of each of the following systems of inequalities.

44. $x + y \le 6$
$2x - y \ge 3$

45. $x - 3y \ge 6$
$y^2 \le 16 - x^2$

46. $9x^2 + 16y^2 \ge 144$
$x^2 - y^2 \ge 16$

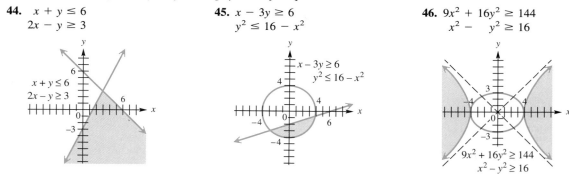

47. Find $x \ge 0$ and $y \ge 0$ such that
$$3x + 2y \le 12$$
$$5x + y \ge 5$$
and $2x + 4y$ is maximized.

maximum of 24 at $(0, 6)$

48. Find $x \ge 0$ and $y \ge 0$ such that
$$x + y \le 50$$
$$2x + y \ge 20$$
$$x + 2y \ge 30$$
and $4x + 2y$ is minimized.

minimum of 40 at $(0, 20)$ or $(10/3, 40/3)$ or at any point between them

◉ CONCEPTUAL   ✐ WRITING   ▲ CHALLENGING

Solve the following linear programming problems.

49. A bakery makes both cakes and cookies. Each batch of cakes requires 2 hr in the oven and 3 hr in the decorating room. Each batch of cookies needs 1 1/2 hr in the oven and 2/3 hr in the decorating room. The oven is available no more than 16 hr a day, while the decorating room can be used no more than 12 hr per day. A batch of cookies produces a profit of $20; the profit on a batch of cakes is $30. Find the number of batches of each item that will maximize profit.

44/19 ≈ 2 batches of cakes and 144/19 ≈ 8 batches of cookies, for a maximum profit of approximately $220

50. A candy company has 100 kg of chocolate-covered nuts and 125 kg of chocolate-covered raisins to be sold as two different mixtures. One mix will contain 1/2 nuts and 1/2 raisins, while the other mix will contain 1/3 nuts and 2/3 raisins. How much of each mixture should be made to maximize revenue if the first mix sells for $6.00 per kilogram and the second mix sells for $4.80 per kilogram?

150 kg of the half and half and 75 kg of the other, for a maximum revenue of $1260

CHAPTER 7 TEST (TEXT PAGE 394) ■ ——————————

Use the substitution method to solve each of the following systems. Identify any systems with dependent equations or any inconsistent systems. If a system has dependent equations, express the solution set with y arbitrary.

1. $x - 3y = -5$
$2x + y = 4$ {(1, 2)}

2. $3m - n = 9$
$m + 2n = 10$ {(4, 3)}

3. $6x + 9y = -21$
$4x + 6y = -14$ $\left\{ \left(\dfrac{-3y - 7}{2}, y \right) \right\}$; dependent

Use the addition method to solve each of the following systems. Identify any systems with dependent equations or any inconsistent systems. If a system has dependent equations, express the solution set with y arbitrary.

4. $4a + 6b = 31$
$3a + 4b = 22$ {(4, 5/2)}

5. $\dfrac{1}{4}x - \dfrac{1}{3}y = -\dfrac{5}{12}$
$\dfrac{1}{10}x + \dfrac{1}{5}y = \dfrac{1}{2}$ {(1, 2)}

6. $x - 2y = 4$
$-2x + 4y = 6$ ∅; inconsistent

7. Can a system of two linear equations in two variables have exactly two solutions? Explain.

Solve the following problem using a system of equations.

8. Christopher Michael has $1000 more invested at 5% than he has invested at 4%. If his annual income from the two investments together is $698, how much does Christopher have invested at each rate? $7200 at 4%; $8200 at 5%

Use the addition method to solve the following linear systems.

9. $3a - 4b + 2c = 15$
$2a - b + c = 13$
$a + 2b - c = 5$ {(5, 3, 6)}

10. $2x + y + z = 3$
$x + 2y - z = 3$
$3x - y + z = 5$ {(2, 0, -1)}

11. Find a solution for the following system in terms of the arbitrary variable *z*.

$$x - 2y + 3z = 2$$
$$4x + y - z = 1 \qquad \left\{ \left(\dfrac{4 - z}{9}, \dfrac{-7 + 13z}{9}, z \right) \right\}$$

12. The sum of three numbers is 2. The first number is equal to the sum of the other two, and the third number is the result of subtracting the first from the second. Find the numbers by solving a system of equations.   1, 1, 0

Use the Gauss-Jordan method to solve each of the following systems.

13. $3a - 2b = 13$
$4a - b = 19$ {(5, 1)}

14. $2x + 3y - 6z = 1$
$x - y + 2z = 3$
$4x + y - 2z = 7$ {(2, 2z − 1, z)}

Solve each of the following nonlinear systems of equations.

15. $2x^2 + y^2 = 6$
$x^2 - 4y^2 = -15$ {(1, 2), (−1, 2), (1, −2), (−1, −2)}

16. $x^2 + y^2 = 25$
$x + y = 7$ {(3, 4), (4, 3)}

⊚ **17.** If a system of two nonlinear equations contains one equation whose graph is a circle and another equation whose graph is a line, can the system have exactly one solution? If so, draw a sketch to indicate this situation.   yes

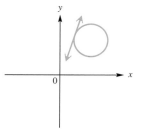

18. Find two numbers such that their sum is −1 and the sum of their squares is 61.   5 and −6

19. Graph the solution of the following system of inequalities.

$$9x^2 + 4y^2 \geq 36$$
$$y < x^2$$

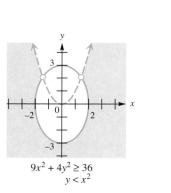

$9x^2 + 4y^2 \geq 36$
$y < x^2$

20. Gwen, who is dieting, requires two food supplements, I and II. She can get these supplements from two different products, A and B. Product A provides 3 grams per serving of supplement I and 2 grams per serving of supplement II. Product B provides 2 grams per serving of supplement I and 4 grams per serving of supplement II. Her dietician, Dr. Shoemake, has recommended that she include at least 15 grams of each supplement in her daily diet. If product A costs 25¢ per serving and product B costs 40¢ per serving, how can she satisfy her requirements most economically?

3 3/4 servings of A and 1 7/8 servings of B, for a minimum cost of $1.69

⊚ CONCEPTUAL   ✎ WRITING   ▲ CHALLENGING

■ **CHAPTER 8 MATRICES**

8.1 RESOURCES

Instructor's Test Manual
Test Items, p. 362–363
Instructor's Solution Manual
pp. 319–321, 723–725
Study Guide
pp. 279–283
Student's Solution Manual
pp. 319–321

Test Generator
IBM Disks
  5.25″ Disk B
  3.5″ Disk A
  Mac Disk A
Tutorials
IBM Disks
  5.25″ Disk H
  3.5″ Disk D
  Mac Disk D
  Apple II Module 5, Lessons 1–5

HarperCollins Videotape 8

8.1 EXERCISES (TEXT PAGE 402) ■ ────────────────────

◎ **1.** A 3×8 matrix has _____ columns and _____ rows. eight; three

Find the size of each of the following matrices. Identify any square, column, or row matrices. See Example 1.

2. $\begin{bmatrix} -4 & 8 \\ 2 & 3 \end{bmatrix}$ 2×2; square

3. $\begin{bmatrix} -9 & 6 & 2 \\ 4 & 1 & 8 \end{bmatrix}$ 2×3

4. $\begin{bmatrix} -6 & 8 & 0 & 0 \\ 4 & 1 & 9 & 2 \\ 3 & -5 & 7 & 1 \end{bmatrix}$ 3×4

5. $[8 \quad -2 \quad 4 \quad 6 \quad 3]$ 1×5; row

6. $\begin{bmatrix} 2 \\ 4 \end{bmatrix}$ 2×1; column

7. $[-9]$ 1×1; square; row; column

8. $\begin{bmatrix} -4 & 2 & 3 \\ -8 & 2 & 1 \\ 4 & 6 & 8 \end{bmatrix}$ 3×3; square

9. $\begin{bmatrix} -4 & 2 \\ 3 & 5 \end{bmatrix}$ 2×2; square

Find the values of the variables in each of the following matrices. See Examples 2 and 3.

10. $\begin{bmatrix} 2 & 1 \\ 4 & 8 \end{bmatrix} = \begin{bmatrix} x & 1 \\ y & z \end{bmatrix}$ $x = 2$; $y = 4$; $z = 8$

11. $\begin{bmatrix} -5 \\ y \end{bmatrix} = \begin{bmatrix} -5 \\ 8 \end{bmatrix}$ $y = 8$

12. $\begin{bmatrix} x + 6 & y + 2 \\ 8 & 3 \end{bmatrix} = \begin{bmatrix} -9 & 7 \\ 8 & k \end{bmatrix}$ $x = -15$; $y = 5$; $k = 3$

13. $\begin{bmatrix} 9 & 7 \\ r & 0 \end{bmatrix} = \begin{bmatrix} m - 3 & n + 5 \\ 8 & 0 \end{bmatrix}$ $r = 8$; $m = 12$; $n = 2$

14. $\begin{bmatrix} 3 & 5 \\ 8 & 9 \end{bmatrix} + \begin{bmatrix} m & 3 \\ 5 & n \end{bmatrix} = \begin{bmatrix} 9 & 8 \\ 13 & 0 \end{bmatrix}$ $m = 6$; $n = -9$

15. $[8 \quad p + 9 \quad q + 5] + [9 \quad -3 \quad 12] = [k - 2 \quad 12 \quad 2q]$ $k = 19$; $p = 6$; $q = 17$

16. $\begin{bmatrix} -7 + z & 4r & 8s \\ 6p & 2 & 5 \end{bmatrix} + \begin{bmatrix} -9 & 8r & 3 \\ 2 & 5 & 4 \end{bmatrix} = \begin{bmatrix} 2 & 36 & 27 \\ 20 & 7 & 12a \end{bmatrix}$ $z = 18$; $r = 3$; $s = 3$; $p = 3$; $a = 3/4$

17. $\begin{bmatrix} a+2 & 3z+1 & 5m \\ 4k & 0 & 3 \end{bmatrix} + \begin{bmatrix} 3a & 2z & 5m \\ 2k & 5 & 6 \end{bmatrix} = \begin{bmatrix} 10 & -14 & 80 \\ 10 & 5 & 9 \end{bmatrix}$ $a = 2;\; z = -3;\; m = 8;\; k = 5/3$

🖉 **18.** Your friend missed the lecture on adding matrices. In your own words, explain to him how to add two matrices.

🖉 **19.** Explain to a friend in your own words how to subtract two matrices.

◎ **20.** What is a scalar?   a real number

Perform each of the following operations, whenever possible. See Examples 4 and 5.

21. $3\begin{bmatrix} 6 & -1 & 4 \\ 2 & 8 & -3 \\ -4 & 5 & 6 \end{bmatrix} + 5\begin{bmatrix} -2 & -8 & -6 \\ 4 & 1 & 3 \\ 2 & -1 & 5 \end{bmatrix}$ $\begin{bmatrix} 8 & -43 & -18 \\ 26 & 29 & 6 \\ -2 & 10 & 43 \end{bmatrix}$

22. $4\begin{bmatrix} 1 & -4 \\ 2 & -3 \\ -8 & 4 \end{bmatrix} - 3\begin{bmatrix} -6 & 9 \\ -2 & 5 \\ -7 & -12 \end{bmatrix}$ $\begin{bmatrix} 22 & -43 \\ 14 & -27 \\ -11 & 52 \end{bmatrix}$

23. $\begin{bmatrix} -8 & 4 & 0 \\ 2 & 5 & 0 \end{bmatrix} + \begin{bmatrix} 6 & 3 \\ 8 & 9 \end{bmatrix}$ not possible

24. $\begin{bmatrix} 2 \\ 3 \end{bmatrix} - \begin{bmatrix} 8 & 1 \\ 9 & 4 \end{bmatrix}$ not possible

25. $\begin{bmatrix} 9 & 4 & 1 & -2 \\ 5 & -6 & 3 & 4 \\ 2 & -5 & 1 & 2 \end{bmatrix} - \begin{bmatrix} -2 & 5 & 1 & 3 \\ 0 & 1 & 0 & 2 \\ -8 & 3 & 2 & 1 \end{bmatrix} + \begin{bmatrix} 2 & 4 & 0 & 3 \\ 4 & -5 & 1 & 6 \\ 2 & -3 & 0 & 8 \end{bmatrix}$ $\begin{bmatrix} 13 & 3 & 0 & -2 \\ 9 & -12 & 4 & 8 \\ 12 & -11 & -1 & 9 \end{bmatrix}$

26. $\begin{bmatrix} 6 & -2 & 4 \\ -2 & 5 & 8 \\ 1 & 0 & 2 \end{bmatrix} + \begin{bmatrix} 3 & 0 & 8 \\ 1 & -2 & 4 \\ 6 & 9 & -2 \end{bmatrix} - \begin{bmatrix} -4 & 2 & 1 \\ 0 & 3 & -2 \\ 4 & 2 & 0 \end{bmatrix}$ $\begin{bmatrix} 13 & -4 & 11 \\ -1 & 0 & 14 \\ 3 & 7 & 0 \end{bmatrix}$

27. $\begin{bmatrix} -4x+2y & -3x+y \\ 6x-3y & 2x-5y \end{bmatrix} + \begin{bmatrix} -8x+6y & 2x \\ 3y-5x & 6x+4y \end{bmatrix}$ $\begin{bmatrix} -12x+8y & -x+y \\ x & 8x-y \end{bmatrix}$

28. $\begin{bmatrix} 4k-8y \\ 6z-3x \\ 2k+5a \\ -4m+2n \end{bmatrix} - \begin{bmatrix} 5k+6y \\ 2z+5x \\ 4k+6a \\ 4m-2n \end{bmatrix}$ $\begin{bmatrix} -k-14y \\ 4z-8x \\ -2k-a \\ -8m+4n \end{bmatrix}$

Let $A = \begin{bmatrix} -2 & 4 \\ 0 & 3 \end{bmatrix}$ *and* $B = \begin{bmatrix} -6 & 2 \\ 4 & 0 \end{bmatrix}$. *Find each of the following matrices. See Example 6.*

29. $2A$ $\begin{bmatrix} -4 & 8 \\ 0 & 6 \end{bmatrix}$

30. $-3B$ $\begin{bmatrix} 18 & -6 \\ -12 & 0 \end{bmatrix}$

31. $-4A$ $\begin{bmatrix} 8 & -16 \\ 0 & -12 \end{bmatrix}$

32. $5B$ $\begin{bmatrix} -30 & 10 \\ 20 & 0 \end{bmatrix}$

33. $2A - B$ $\begin{bmatrix} 2 & 6 \\ -4 & 6 \end{bmatrix}$

34. $-4A + 5B$ $\begin{bmatrix} -22 & -6 \\ 20 & -12 \end{bmatrix}$

35. $3A - 11B$ $\begin{bmatrix} 60 & -10 \\ -44 & 9 \end{bmatrix}$

36. $-2A + 4B$ $\begin{bmatrix} -20 & 0 \\ 16 & -6 \end{bmatrix}$

Work each of the following problems.

37. Richard Maciasz bought 7 shares of Sears stock, 9 shares of IBM stock, and 8 shares of Chrysler stock. The following month, he bought 2 shares of Sears stock, no IBM, and 6 shares of Chrysler. Write this in-   formation first as a 3×2 matrix and then as a 2×3 matrix.

$\begin{bmatrix} 7 & 2 \\ 9 & 0 \\ 8 & 6 \end{bmatrix}$; $\begin{bmatrix} 7 & 9 & 8 \\ 2 & 0 & 6 \end{bmatrix}$

38. Margie Bezzone works in a computer store. The first week she sold 5 computers, 3 printers, 4 disk drives, and 6 monitors. The next week she sold 4 computers, 2 printers, 6 disk drives, and 5 monitors. Write this information first as a 2 × 4 matrix and then as a 4 × 2 matrix.

$$\begin{bmatrix} 5 & 3 & 4 & 6 \\ 4 & 2 & 6 & 5 \end{bmatrix}; \quad \begin{bmatrix} 5 & 4 \\ 3 & 2 \\ 4 & 6 \\ 6 & 5 \end{bmatrix}$$

39. A recent study revealed that the average number of miles driven in the United States has been increasing steadily, with miles driven by women increasing much faster than miles driven by men. In 1969, 5411 thousand miles were driven by women and 11,352 thousand miles were driven by men. In 1990, 9371 thou-

sand miles were driven by women and 15,956 thousand miles by men. Write this information as a 2 × 2 matrix in two ways.

$$\begin{bmatrix} 5411 & 11,352 \\ 9371 & 15,956 \end{bmatrix}; \quad \begin{bmatrix} 5411 & 9371 \\ 11,352 & 15,956 \end{bmatrix}$$

40. The proportion of the population of China living in cities was slowed for a time by government-imposed birth-control policies. The urban population proportion has increased again in recent years. In 1952, the proportion was 12.5%; in 1960, 19.7%; in 1975, 12.1%; and in 1985, 19.7%. Write this information as a row matrix and as a column matrix.

$$\begin{bmatrix} 12.5 & 19.7 & 12.1 & 19.7 \end{bmatrix}; \quad \begin{bmatrix} 12.5 \\ 19.7 \\ 12.1 \\ 19.7 \end{bmatrix}$$

Addition of the real numbers satisfies certain properties, such as the commutative, associative, identity, and inverse properties. In the following exercises, you can check to see which of these properties are satisfied by addition of matrices.

▲ *Let* $\quad A = \begin{bmatrix} a & b \\ c & d \end{bmatrix}, \quad B = \begin{bmatrix} e & f \\ g & h \end{bmatrix}, \quad$ *and* $\quad C = \begin{bmatrix} j & m \\ k & n \end{bmatrix}.$ *Decide which of the follow-*
ing statements are true for these square matrices. Then decide if a similar property holds for any matrices of the same size.

41. $A + B = B + A$ (commutative property)   yes, always true

42. $A + (B + C) = (A + B) + C$ (associative property)   yes, always true

43. There exists a matrix O such that $A + O = A$ and $O + A = A$.   (identity property)
yes, always true

44. There exists a matrix $-A$ such that $A + (-A) = O$ and $-A + A = O$.   (inverse property)
yes, always true

45. Are these properties valid for matrices that are not square?   yes

▲ ✎ **46.** Does $kM = Mk$ for any scalar k and matrix M? Explain.

8.2   RESOURCES

Instructor's Test Manual
Test Items, p. 364–365
Instructor's Solution Manual
pp. 321–325, 725–730
Study Guide
pp. 283–287
Student's Solution Manual
pp. 321–325

Test Generator
IBM Disks
5.25" Disk B
3.5" Disk A
Mac Disk A
Tutorials
IBM Disks
5.25" Disk H
3.5" Disk D
Mac Disk D
Apple II Module 5, Lessons 1–5

HarperCollins Videotape   8
Transparency   T-38

8.2 EXERCISES (TEXT PAGE 410) ■ ────────────────

In each of the following exercises, the sizes of the two matrices A and B are given. Find the size
of the product AB and the product BA, whenever these products exist. See Example 3.

1. *A* is 2×2, *B* is 2×2 2×2; 2×2

2. *A* is 3×3, *B* is 3×3 3×3; 3×3

3. *A* is 4×2, *B* is 2×4 4×4; 2×2

4. *A* is 3×1, *B* is 1×3 3×3; 1×1

5. *A* is 3×5, *B* is 5×2 3×2; *BA* cannot be found

6. *A* is 4×3, *B* is 3×6 4×6; *BA* cannot be found

7. *A* is 4×2, *B* is 3×4 *AB* cannot be found; 3×2

8. *A* is 7×3, *B* is 2×7 *AB* cannot be found; 2×3

9. *A* is 4×3, *B* is 2×5 *AB* cannot be found; *BA* cannot be found

10. *A* is 1×6, *B* is 2×4 *AB* cannot be found; *BA* cannot be found

◎ **11.** The product *MN* of two matrices can be found only if the number of _____ of *M* equals
the number of _____ of *N*.   columns; rows

◎ **12.** True or false: For matrices *A* and *B,* if *AB* can be found, then *BA* can always be found, too.
false

◎ **13.** To find the product *AB* of matrices *A* and *B,* the first row, second column entry is found
by multiplying the _____ elements in *A* and the _____ elements in *B* and then
_____ these products.   first row; second column; adding

Find each of the following matrix products, whenever possible. See Examples 1 and 2.

14. $\begin{bmatrix} 1 & 2 \\ 3 & 4 \end{bmatrix}\begin{bmatrix} -1 \\ 7 \end{bmatrix}$ $\begin{bmatrix} 13 \\ 25 \end{bmatrix}$

15. $\begin{bmatrix} -1 & 5 \\ 7 & 0 \end{bmatrix}\begin{bmatrix} 6 \\ 2 \end{bmatrix}$ $\begin{bmatrix} 4 \\ 42 \end{bmatrix}$

16. $\begin{bmatrix} 3 & -4 & 1 \\ 5 & 0 & 2 \end{bmatrix}\begin{bmatrix} -1 \\ 4 \\ 2 \end{bmatrix}$ $\begin{bmatrix} -17 \\ -1 \end{bmatrix}$

17. $\begin{bmatrix} -6 & 3 & 5 \\ 2 & 9 & 1 \end{bmatrix}\begin{bmatrix} -2 \\ 0 \\ 3 \end{bmatrix}$ $\begin{bmatrix} 27 \\ -1 \end{bmatrix}$

18. $\begin{bmatrix} 5 & 2 \\ -1 & 4 \end{bmatrix}\begin{bmatrix} 3 & -2 \\ 1 & 0 \end{bmatrix}$ $\begin{bmatrix} 17 & -10 \\ 1 & 2 \end{bmatrix}$

19. $\begin{bmatrix} -4 & 0 \\ 1 & 3 \end{bmatrix}\begin{bmatrix} -2 & 4 \\ 0 & 1 \end{bmatrix}$ $\begin{bmatrix} 8 & -16 \\ -2 & 7 \end{bmatrix}$

20. $\begin{bmatrix} 2 & 2 & -1 \\ 3 & 0 & 1 \end{bmatrix}\begin{bmatrix} 0 & 2 \\ -1 & 4 \\ 0 & 2 \end{bmatrix}$ $\begin{bmatrix} -2 & 10 \\ 0 & 8 \end{bmatrix}$

21. $\begin{bmatrix} -9 & 2 & 1 \\ 3 & 0 & 0 \end{bmatrix}\begin{bmatrix} 2 \\ -1 \\ 4 \end{bmatrix}$ $\begin{bmatrix} -16 \\ 6 \end{bmatrix}$

22. $\begin{bmatrix} -1 & 2 & 0 \\ 0 & 3 & 2 \\ 0 & 1 & 4 \end{bmatrix}\begin{bmatrix} 2 & -1 & 2 \\ 0 & 2 & 1 \\ 3 & 0 & -1 \end{bmatrix}$ $\begin{bmatrix} -2 & 5 & 0 \\ 6 & 6 & 1 \\ 12 & 2 & -3 \end{bmatrix}$

23. $\begin{bmatrix} -2 & -3 & -4 \\ 2 & -1 & 0 \\ 4 & -2 & 3 \end{bmatrix}\begin{bmatrix} 0 & 1 & 4 \\ 1 & 2 & -1 \\ 3 & 2 & -2 \end{bmatrix}$ $\begin{bmatrix} -15 & -16 & 3 \\ -1 & 0 & 9 \\ 7 & 6 & 12 \end{bmatrix}$

24. $[-2 \quad 4 \quad 1]\begin{bmatrix} 3 & -2 & 4 \\ 2 & 1 & 0 \\ 0 & -1 & 4 \end{bmatrix}$ $[2 \quad 7 \quad -4]$

25. $[0 \quad 3 \quad -4]\begin{bmatrix} -2 & 6 & 3 \\ 0 & 4 & 2 \\ -1 & 1 & 4 \end{bmatrix}$ $[4 \quad 8 \quad -10]$

26. $\begin{bmatrix} -2 & 1 & 4 \\ 0 & 1 & 2 \end{bmatrix}\begin{bmatrix} -2 & 1 & 0 \\ 0 & -2 & 0 \\ 4 & 1 & 2 \end{bmatrix}$ $\begin{bmatrix} 20 & 0 & 8 \\ 8 & 0 & 4 \end{bmatrix}$

27. $\begin{bmatrix} -1 & 0 & 0 \\ 2 & 1 & 4 \end{bmatrix}\begin{bmatrix} 4 & -2 & 5 \\ 0 & 1 & 4 \\ 2 & -9 & 0 \end{bmatrix}$ $\begin{bmatrix} -4 & 2 & -5 \\ 16 & -39 & 14 \end{bmatrix}$

28. $\begin{bmatrix} -3 & 0 & 2 & 1 \\ 4 & 0 & 2 & 6 \end{bmatrix}\begin{bmatrix} -4 & 2 \\ 0 & 1 \end{bmatrix}$ not possible

29. $\begin{bmatrix} -1 & 2 & 4 & 1 \\ 0 & 2 & -3 & 5 \end{bmatrix}\begin{bmatrix} 1 & 2 & 4 \\ -2 & 5 & 1 \end{bmatrix}$ not possible

◎ CONCEPTUAL   ☑ WRITING   ▲ CHALLENGING

30. $[-2 \quad 4 \quad 6]\begin{bmatrix} 3 \\ -2 \\ 1 \end{bmatrix}$ [−8]

31. $[4 \quad 0 \quad 2]\begin{bmatrix} -5 \\ 1 \\ 6 \end{bmatrix}$ [−8]

32. $\begin{bmatrix} 3 \\ -2 \\ 1 \end{bmatrix}[-2 \quad 4 \quad 6]$ $\begin{bmatrix} -6 & 12 & 18 \\ 4 & -8 & -12 \\ -2 & 4 & 6 \end{bmatrix}$

33. $\begin{bmatrix} -5 \\ 1 \\ 6 \end{bmatrix}[4 \quad 0 \quad 2]$ $\begin{bmatrix} -20 & 0 & -10 \\ 4 & 0 & 2 \\ 24 & 0 & 12 \end{bmatrix}$

Let $A = \begin{bmatrix} -2 & 4 \\ 1 & 3 \end{bmatrix}$, $B = \begin{bmatrix} -2 & 1 \\ 3 & 6 \end{bmatrix}$, *and* $C = \begin{bmatrix} 5 & -2 & 1 \\ 0 & 3 & 7 \end{bmatrix}$. *Find each of the following products. See Examples 4 and 5.*

34. AB $\begin{bmatrix} 16 & 22 \\ 7 & 19 \end{bmatrix}$

35. BA $\begin{bmatrix} 5 & -5 \\ 0 & 30 \end{bmatrix}$

36. AC $\begin{bmatrix} -10 & 16 & 26 \\ 5 & 7 & 22 \end{bmatrix}$

37. CA not possible

◉ **38.** Did you get the same answer in Exercises 34 and 35? What about Exercises 36 and 37? Do you think that matrix multiplication is commutative?

no; the answers are different (CA is not possible); no

◉ **39.** For any matrices P and Q, what must be true for both PQ and QP to exist?

Both products will exist if P is $m \times n$ and Q is $n \times m$.

▲ *For the following exercises, let* $A = \begin{bmatrix} a & b \\ c & d \end{bmatrix}$, $B = \begin{bmatrix} e & f \\ g & h \end{bmatrix}$, *and* $C = \begin{bmatrix} j & m \\ k & n \end{bmatrix}$.

Decide which of the following statements are true for these three matrices. Then decide if a similar property holds for any square matrices of the same size.

40. $(AB)C = A(BC)$ (associative property) always true

41. $A(B + C) = AB + AC$ (distributive property) always true

42. $k(A + B) = kA + kB$ for any real number k always true

43. $(k + p)A = kA + pA$ for any real numbers k and p always true

◉ **44.** If matrix A represents costs/models and matrix B represents materials/costs, in what order should A and B be multiplied to get a meaningful product? What would that product represent?

as BA; materials/models

45. Yummy Yogurt sells three types of yogurt: nonfat, regular, and super creamy at three locations. Location I sells 50 gallons of nonfat, 100 gallons of regular, and 30 gallons of super creamy each day. Location II sells 10 gallons of nonfat and Location III sells 60 gallons of nonfat each day. Daily sales of regular yogurt are 90 gallons at Location II and 120 gallons at Location III. At Location II, 50 gallons of super creamy are sold each day, and 40 gallons of super creamy are sold each day at Location III.

(a) Write a 3 × 3 matrix that shows the sales figures for the three locations.

$\begin{bmatrix} 50 & 100 & 30 \\ 10 & 90 & 50 \\ 60 & 120 & 40 \end{bmatrix}$

(b) The income per gallon for nonfat, regular, and super creamy is $12, $10, and $15, respectively. Write a 1 × 3 or 3 × 1 matrix displaying the income.

$\begin{bmatrix} 12 \\ 10 \\ 15 \end{bmatrix}$ (If the rows and columns are interchanged in part (a), this should be a 1 × 3 matrix.)

(c) Find a matrix product that gives the daily income at each of the three locations.

$\begin{bmatrix} 2050 \\ 1770 \\ 2520 \end{bmatrix}$ (This may be a 1 × 3 matrix.)

(d) What is Yummy Yogurt's total daily income from the three locations? $6340

46. The Bread Box, a small neighborhood bakery, sells four main items: sweet rolls, bread, cakes, and pies. The amount of each ingredient (in cups, except for eggs) required for these items is given by matrix A.

$$\begin{array}{c} \\ \text{Rolls} \\ \text{(doz)} \\ \text{Bread} \\ \text{(loaves)} \\ \text{Cakes} \\ \\ \text{Pies} \\ \text{(crust)} \end{array} \begin{array}{ccccc} \text{Eggs} & \text{Flour} & \text{Sugar} & \text{Shortening} & \text{Milk} \\ \left[\begin{array}{ccccc} 1 & 4 & 1/4 & 1/4 & 1 \\ 0 & 3 & 0 & 1/4 & 0 \\ 4 & 3 & 2 & 1 & 1 \\ 0 & 1 & 0 & 1/3 & 0 \end{array}\right] \end{array} = A$$

The cost (in cents) for each ingredient when purchased in large lots or small lots is given in matrix B.

$$\begin{array}{c} \\ \text{Eggs} \\ \text{Flour} \\ \text{Sugar} \\ \text{Shortening} \\ \text{Milk} \end{array} \begin{array}{cc} \multicolumn{2}{c}{\text{Cost}} \\ \text{Large lot} & \text{Small lot} \\ \left[\begin{array}{cc} 5 & 5 \\ 8 & 10 \\ 10 & 12 \\ 12 & 15 \\ 5 & 6 \end{array}\right] \end{array} = B$$

(a) Use matrix multiplication to find a matrix giving the comparative cost per item for the two purchase options.

$$\begin{bmatrix} 47.5 & 57.75 \\ 27 & 33.75 \\ 81 & 95 \\ 12 & 15 \end{bmatrix}$$

Suppose a day's orders consist of 20 dozen sweet rolls, 200 loaves of bread, 50 cakes, and 60 pies.
(b) Write the orders as a 1×4 matrix and, using matrix multiplication, write as a matrix the amount of each ingredient needed to fill the day's orders.

[20 200 50 60]; [220 890 105 125 70]

(c) Use matrix multiplication to find a matrix giving the costs under the two purchase options to fill the day's orders. [11,120 13,555]

8.3 EXERCISES (TEXT PAGE 421) ■

◉ **1.** What is the product when an $n \times n$ matrix A is multiplied by the $n \times n$ identity matrix?
matrix A

◉ **2.** When an $n \times n$ matrix A is multiplied by the inverse matrix A^{-1}, the product is what special matrix? the $n \times n$ identity matrix

◉ CONCEPTUAL   ✍ WRITING   ▲ CHALLENGING

Decide whether the given matrices are inverses of each other.

3. $\begin{bmatrix} 2 & 3 \\ 1 & 1 \end{bmatrix}, \begin{bmatrix} -1 & 3 \\ 1 & -2 \end{bmatrix}$ yes

4. $\begin{bmatrix} 5 & 7 \\ 2 & 3 \end{bmatrix}, \begin{bmatrix} 3 & -7 \\ -2 & 5 \end{bmatrix}$ yes

5. $\begin{bmatrix} 2 & 1 \\ 3 & 2 \end{bmatrix}, \begin{bmatrix} 2 & 1 \\ -3 & 2 \end{bmatrix}$ no

6. $\begin{bmatrix} -1 & 2 \\ 3 & -5 \end{bmatrix}, \begin{bmatrix} -5 & -2 \\ -3 & -1 \end{bmatrix}$ no

7. $\begin{bmatrix} 1 & -2 & -3 \\ 2 & -2 & -5 \\ -1 & 1 & 4 \end{bmatrix}, \begin{bmatrix} -1 & 5/3 & 4/3 \\ -1 & 1/3 & -1/3 \\ 0 & 1/3 & 2/3 \end{bmatrix}$ yes

8. $\begin{bmatrix} 1 & 2 & -1 \\ 2 & -1 & 3 \\ 3 & -2 & 3 \end{bmatrix}, \begin{bmatrix} 3/10 & -2/5 & 1/2 \\ 3/10 & 3/5 & -1/2 \\ -1/10 & 4/5 & -1/2 \end{bmatrix}$ yes

9. $\begin{bmatrix} 1 & 2 & -1 \\ 0 & 1 & 3 \\ 2 & 1 & -2 \end{bmatrix}, \begin{bmatrix} 1 & 1 & 2 \\ 1 & 1 & 1 \\ 2 & 3 & 4 \end{bmatrix}$ no

10. $\begin{bmatrix} 2 & -1 & 4 \\ 0 & 5 & 0 \\ 3 & 2 & -1 \end{bmatrix}, \begin{bmatrix} 1 & 0 & 1 \\ 6 & 4 & 2 \\ 1 & 1 & 0 \end{bmatrix}$ no

Use row transformations to find any inverses that exist for the following matrices. See Examples 3–5.

11. $\begin{bmatrix} 1 & -1 \\ 2 & 0 \end{bmatrix}$ $\begin{bmatrix} 0 & 1/2 \\ -1 & 1/2 \end{bmatrix}$

12. $\begin{bmatrix} 3 & -1 \\ -5 & 2 \end{bmatrix}$ $\begin{bmatrix} 2 & 1 \\ 5 & 3 \end{bmatrix}$

13. $\begin{bmatrix} -6 & 4 \\ -3 & 2 \end{bmatrix}$ does not exist

14. $\begin{bmatrix} -1 & 2 \\ -2 & -1 \end{bmatrix}$ $\begin{bmatrix} -1/5 & -2/5 \\ 2/5 & -1/5 \end{bmatrix}$

15. $\begin{bmatrix} -1 & -2 \\ 3 & 4 \end{bmatrix}$ $\begin{bmatrix} 2 & 1 \\ -3/2 & -1/2 \end{bmatrix}$

16. $\begin{bmatrix} 5 & 10 \\ -3 & -6 \end{bmatrix}$ does not exist

17. $\begin{bmatrix} .6 & .2 \\ .5 & .1 \end{bmatrix}$ $\begin{bmatrix} -2.5 & 5 \\ 12.5 & -15 \end{bmatrix}$

18. $\begin{bmatrix} .8 & -.3 \\ .5 & -.2 \end{bmatrix}$ $\begin{bmatrix} 20 & -30 \\ 50 & -80 \end{bmatrix}$

19. $\begin{bmatrix} 1 & 0 & 0 \\ 0 & -1 & 0 \\ 1 & 0 & 1 \end{bmatrix}$ $\begin{bmatrix} 1 & 0 & 0 \\ 0 & -1 & 0 \\ -1 & 0 & 1 \end{bmatrix}$

20. $\begin{bmatrix} 1 & 3 & 3 \\ -1 & 0 & 0 \\ -4 & -4 & -3 \end{bmatrix}$ $\begin{bmatrix} 0 & -1 & 0 \\ -1 & 3 & -1 \\ 4/3 & -8/3 & 1 \end{bmatrix}$

21. $\begin{bmatrix} 3 & 6 & 3 \\ 6 & 4 & -2 \\ 0 & 1 & -1 \end{bmatrix}$ $\begin{bmatrix} -1/24 & 3/16 & -1/2 \\ 1/8 & -1/16 & 1/2 \\ 1/8 & -1/16 & -1/2 \end{bmatrix}$

22. $\begin{bmatrix} 2 & 6 & 0 \\ 1 & 5 & 3 \\ 0 & 0 & 1 \end{bmatrix}$ $\begin{bmatrix} 5/4 & -3/2 & 9/2 \\ -1/4 & 1/2 & -3/2 \\ 0 & 0 & 1 \end{bmatrix}$

23. $\begin{bmatrix} -1 & -1 & -1 \\ 4 & 5 & 0 \\ 0 & 1 & -3 \end{bmatrix}$ $\begin{bmatrix} 15 & 4 & -5 \\ -12 & -3 & 4 \\ -4 & -1 & 1 \end{bmatrix}$

24. $\begin{bmatrix} 2 & 0 & 4 \\ 3 & 1 & 5 \\ -1 & 1 & -2 \end{bmatrix}$ $\begin{bmatrix} -7/2 & 2 & -2 \\ 1/2 & 0 & 1 \\ 2 & -1 & 1 \end{bmatrix}$

25. $\begin{bmatrix} -.4 & .1 & .2 \\ 0 & .6 & .8 \\ .3 & 0 & -.2 \end{bmatrix}$ $\begin{bmatrix} -10/3 & 5/9 & -10/9 \\ 20/3 & 5/9 & 80/9 \\ -5 & 5/6 & -20/3 \end{bmatrix}$

26. $\begin{bmatrix} .8 & .2 & .1 \\ -.2 & 0 & .3 \\ 0 & 0 & .5 \end{bmatrix}$ $\begin{bmatrix} 0 & -5 & 3 \\ 5 & 20 & -13 \\ 0 & 0 & 2 \end{bmatrix}$

Write the matrix of coefficients, A, the matrix of variables, X, and the matrix of constants, B, for each of the following systems. Do not solve.

27. $x + y = 8$
$2x - y = 4$
$A = \begin{bmatrix} 1 & 1 \\ 2 & -1 \end{bmatrix}; \quad X = \begin{bmatrix} x \\ y \end{bmatrix}; \quad B = \begin{bmatrix} 8 \\ 4 \end{bmatrix}$

28. $2x + y = 9$
$x + 3y = 17$
$A = \begin{bmatrix} 2 & 1 \\ 1 & 3 \end{bmatrix}; \quad X = \begin{bmatrix} x \\ y \end{bmatrix}; \quad B = \begin{bmatrix} 9 \\ 17 \end{bmatrix}$

29. $4x + 5y = 7$
$2x + 3y = 5$
$A = \begin{bmatrix} 4 & 5 \\ 2 & 3 \end{bmatrix}; \quad X = \begin{bmatrix} x \\ y \end{bmatrix}; \quad B = \begin{bmatrix} 7 \\ 5 \end{bmatrix}$

30. $2x + 3y = -2$
$5x + 4y = -12$
$A = \begin{bmatrix} 2 & 3 \\ 5 & 4 \end{bmatrix}; \quad X = \begin{bmatrix} x \\ y \end{bmatrix}; \quad B = \begin{bmatrix} -2 \\ -12 \end{bmatrix}$

31. $\begin{aligned} x + y + z &= 9 \\ 2x + y + 3z &= 17 \\ 5x - 2y + 2z &= 16 \end{aligned}$ $A = \begin{bmatrix} 1 & 1 & 1 \\ 2 & 1 & 3 \\ 5 & -2 & 2 \end{bmatrix}$; $X = \begin{bmatrix} x \\ y \\ z \end{bmatrix}$; $B = \begin{bmatrix} 9 \\ 17 \\ 16 \end{bmatrix}$

32. $\begin{aligned} 3x + y - 5z &= 8 \\ 2x + 6y - 3z &= 6 \\ x + 5y - 9z &= -12 \end{aligned}$ $A = \begin{bmatrix} 3 & 1 & -5 \\ 2 & 6 & -3 \\ 1 & 5 & -9 \end{bmatrix}$; $X = \begin{bmatrix} x \\ y \\ z \end{bmatrix}$; $B = \begin{bmatrix} 8 \\ 6 \\ -12 \end{bmatrix}$

 33. A system of linear equations can be solved by using the matrix equation $X = A^{-1}B$. How are the matrices X, A^{-1}, and B related to the linear system?

X is a matrix of the variables of the system; A^{-1} is the inverse of matrix A, which is a matrix of the coefficients of the variables of the system; B is a matrix of the constants in the system.

Solve each of the following systems by using the inverse of the coefficient matrix. See Example 7.

34. $\begin{aligned} -x + y &= 1 \\ 2x - y &= 1 \end{aligned}$ $\{(2, 3)\}$

35. $\begin{aligned} x + y &= 5 \\ x - y &= -1 \end{aligned}$ $\{(2, 3)\}$

36. $\begin{aligned} 2x - y &= -8 \\ 3x + y &= -2 \end{aligned}$ $\{(-2, 4)\}$

37. $\begin{aligned} x + 3y &= -12 \\ 2x - y &= 11 \end{aligned}$ $\{(3, -5)\}$

38. $\begin{aligned} 2x + 3y &= -10 \\ 3x + 4y &= -12 \end{aligned}$ $\{(4, -6)\}$

39. $\begin{aligned} 2x - 3y &= 10 \\ 2x + 2y &= 5 \end{aligned}$ $\{(7/2, -1)\}$

40. $\begin{aligned} 2x - 5y &= 10 \\ 4x - 5y &= 15 \end{aligned}$ $\{(5/2, -1)\}$

41. $\begin{aligned} 2x - 3y &= 2 \\ 4x - 6y &= 1 \end{aligned}$ ∅

Use the inverse of the coefficient matrix to solve each of the following systems of equations. The inverses of the coefficient matrices for the first four problems are found in the text and exercises of this section.

42. $\begin{aligned} x + 3y + 3z &= 11 \\ -x &= -2 \\ -4x - 4y - 3z &= -10 \end{aligned}$ $\{(2, -7, 10)\}$

43. $\begin{aligned} x + z &= 3 \\ 2x - 2y - z &= -2 \\ 3x &= 3 \end{aligned}$ $\{(1, 1, 2)\}$

44. $\begin{aligned} 2x + 4z &= 14 \\ 3x + y + 5z &= 19 \\ -x + y - 2z &= -7 \end{aligned}$ $\{(3, 0, 2)\}$

45. $\begin{aligned} 3x + 6y + 3z &= 12 \\ 6x + 4y - 2z &= -4 \\ y - z &= -3 \end{aligned}$ $\{(1/4, 1/4, 13/4)\}$

46. $\begin{aligned} x + 3y + z &= 2 \\ x - 2y + 3z &= -3 \\ 2x - 3y - z &= 34 \end{aligned}$ $\{(12, -15/11, -65/11)\}$

47. $\begin{aligned} x + y - z &= 6 \\ 2x - y + z &= -9 \\ x - 2y + 3z &= 1 \end{aligned}$ $\{(-1, 23, 16)\}$

In Exercises 48 and 49, write a system of equations and solve it by using the inverse of the coefficient matrix.

48. Midtown Manufacturing Company makes two products, plastic plates and plastic cups. Both require time on two machines: plates require 1 hr on machine *A* and 2 hr on machine *B*; cups require 3 hr on machine *A* and 1 hr on machine *B*. Both machines operate 15 hr a day. How many of each product can be produced in a day under these conditions?   6 plates;   3 cups

49. A company produces two models of bicycles, model 201 and model 301. Model 201 requires 2 hr of assembly time and model 301 requires 3 hr of assembly time. The parts of model 201 cost $25 per bike and the parts for model 301 cost $30 per bike. If the company has a total of 34 hr of assembly time and $365 available per day for these two models, how many of each can be made in a day?   5 model 201;   8 model 301

50. List the methods presented so far in this text to solve a linear system. Which one do you prefer? Explain why.

Let $A = \begin{bmatrix} a & b \\ c & d \end{bmatrix}$. Show that the following are true.

51. $AA^{-1} = I$ and $A^{-1}A = I$

52. $IA = A$

53. $AI = A$

54. $A \cdot O = O$

◎ CONCEPTUAL   ✍ WRITING   ▲ CHALLENGING

8.4 RESOURCES

Instructor's Test Manual
Test Items, p. 367–368
Instructor's Solution Manual
pp. 333–337, 737–742
Study Guide
pp. 295–299
Student's Solution Manual
pp. 333–337

Test Generator
IBM Disks
5.25" Disk B
3.5" Disk A
Mac Disk A
Tutorials
IBM Disks
5.25" Disk H
3.5" Disk D
Mac Disk D
Apple II Module 5, Lessons 1–5

HarperCollins Videotape   8
Transparency   T-39–T-40(a)–(c)

8.4 EXERCISES   (Text Page 427) ■

Evaluate each of the following determinants. See Example 1.

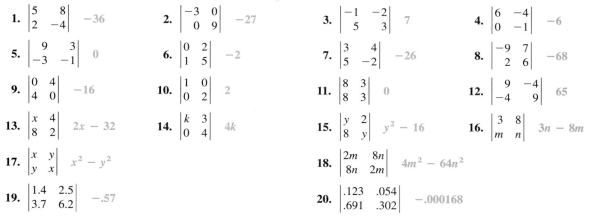

1. $\begin{vmatrix} 5 & 8 \\ 2 & -4 \end{vmatrix}$ -36

2. $\begin{vmatrix} -3 & 0 \\ 0 & 9 \end{vmatrix}$ -27

3. $\begin{vmatrix} -1 & -2 \\ 5 & 3 \end{vmatrix}$ 7

4. $\begin{vmatrix} 6 & -4 \\ 0 & -1 \end{vmatrix}$ -6

5. $\begin{vmatrix} 9 & 3 \\ -3 & -1 \end{vmatrix}$ 0

6. $\begin{vmatrix} 0 & 2 \\ 1 & 5 \end{vmatrix}$ -2

7. $\begin{vmatrix} 3 & 4 \\ 5 & -2 \end{vmatrix}$ -26

8. $\begin{vmatrix} -9 & 7 \\ 2 & 6 \end{vmatrix}$ -68

9. $\begin{vmatrix} 0 & 4 \\ 4 & 0 \end{vmatrix}$ -16

10. $\begin{vmatrix} 1 & 0 \\ 0 & 2 \end{vmatrix}$ 2

11. $\begin{vmatrix} 8 & 3 \\ 8 & 3 \end{vmatrix}$ 0

12. $\begin{vmatrix} 9 & -4 \\ -4 & 9 \end{vmatrix}$ 65

13. $\begin{vmatrix} x & 4 \\ 8 & 2 \end{vmatrix}$ $2x - 32$

14. $\begin{vmatrix} k & 3 \\ 0 & 4 \end{vmatrix}$ $4k$

15. $\begin{vmatrix} y & 2 \\ 8 & y \end{vmatrix}$ $y^2 - 16$

16. $\begin{vmatrix} 3 & 8 \\ m & n \end{vmatrix}$ $3n - 8m$

17. $\begin{vmatrix} x & y \\ y & x \end{vmatrix}$ $x^2 - y^2$

18. $\begin{vmatrix} 2m & 8n \\ 8n & 2m \end{vmatrix}$ $4m^2 - 64n^2$

19. $\begin{vmatrix} 1.4 & 2.5 \\ 3.7 & 6.2 \end{vmatrix}$ $-.57$

20. $\begin{vmatrix} .123 & .054 \\ .691 & .302 \end{vmatrix}$ $-.000168$

🖉 21. Explain how to evaluate a 2 × 2 determinant.

🖉 22. In your own words, define the minor of an element of a matrix. Then define the cofactor of the element.

For the following determinants, find the cofactor of each element in the second row of the corresponding matrix. See Example 2.

23. $\begin{vmatrix} -2 & 0 & 1 \\ 3 & 2 & -1 \\ 1 & 0 & 2 \end{vmatrix}$ $0, -5, 0$

24. $\begin{vmatrix} 0 & -1 & 2 \\ 1 & 0 & 2 \\ 0 & -3 & 1 \end{vmatrix}$ $-5, 0, 0$

25. $\begin{vmatrix} 1 & 2 & -1 \\ 2 & 3 & -2 \\ -1 & 4 & 1 \end{vmatrix}$ $-6, 0, -6$

26. $\begin{vmatrix} 2 & -1 & 4 \\ 3 & 0 & 1 \\ -2 & 1 & 4 \end{vmatrix}$ $8, 16, 0$

Evaluate each of the following determinants. See Examples 2 and 3.

27. $\begin{vmatrix} 1 & 0 & 0 \\ 0 & -1 & 0 \\ 1 & 0 & 1 \end{vmatrix}$ -1

28. $\begin{vmatrix} -2 & 0 & 1 \\ 0 & 1 & 0 \\ 0 & 0 & -1 \end{vmatrix}$ 2

29. $\begin{vmatrix} -2 & 0 & 0 \\ 4 & 0 & 1 \\ 3 & 4 & 2 \end{vmatrix}$ 8

30. $\begin{vmatrix} 3 & -2 & 0 \\ 0 & -1 & 1 \\ 4 & 0 & 2 \end{vmatrix}$ -14

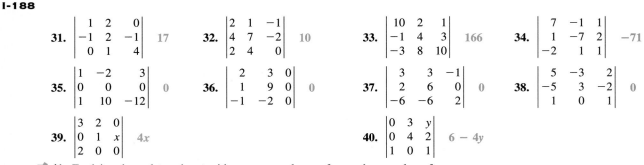

31. $\begin{vmatrix} 1 & 2 & 0 \\ -1 & 2 & -1 \\ 0 & 1 & 4 \end{vmatrix}$ 17   **32.** $\begin{vmatrix} 2 & 1 & -1 \\ 4 & 7 & -2 \\ 2 & 4 & 0 \end{vmatrix}$ 10   **33.** $\begin{vmatrix} 10 & 2 & 1 \\ -1 & 4 & 3 \\ -3 & 8 & 10 \end{vmatrix}$ 166   **34.** $\begin{vmatrix} 7 & -1 & 1 \\ 1 & -7 & 2 \\ -2 & 1 & 1 \end{vmatrix}$ −71

35. $\begin{vmatrix} 1 & -2 & 3 \\ 0 & 0 & 0 \\ 1 & 10 & -12 \end{vmatrix}$ 0   **36.** $\begin{vmatrix} 2 & 3 & 0 \\ 1 & 9 & 0 \\ -1 & -2 & 0 \end{vmatrix}$ 0   **37.** $\begin{vmatrix} 3 & 3 & -1 \\ 2 & 6 & 0 \\ -6 & -6 & 2 \end{vmatrix}$ 0   **38.** $\begin{vmatrix} 5 & -3 & 2 \\ -5 & 3 & -2 \\ 1 & 0 & 1 \end{vmatrix}$ 0

39. $\begin{vmatrix} 3 & 2 & 0 \\ 0 & 1 & x \\ 2 & 0 & 0 \end{vmatrix}$ 4x   **40.** $\begin{vmatrix} 0 & 3 & y \\ 0 & 4 & 2 \\ 1 & 0 & 1 \end{vmatrix}$ 6 − 4y

41. Explain why a determinant with a row or column of zeros has a value of zero.

42. Does a determinant with two identical rows have a value of zero?   **yes**

Solve each of the following equations for x.

43. $\begin{vmatrix} -2 & 0 & 1 \\ -1 & 3 & x \\ 5 & -2 & 0 \end{vmatrix} = 3$ {−4}   **44.** $\begin{vmatrix} 4 & 3 & 0 \\ 2 & 0 & 1 \\ -3 & x & -1 \end{vmatrix} = 5$ {−2}

45. $\begin{vmatrix} 5 & 3x & -3 \\ 0 & 2 & -1 \\ 4 & -1 & x \end{vmatrix} = -7$ {13}   **46.** $\begin{vmatrix} 2x & 1 & -1 \\ 0 & 4 & x \\ 3 & 0 & 2 \end{vmatrix} = x$ {−2/3}

Find the value of each of the following determinants. See Example 4.

47. $\begin{vmatrix} 4 & 0 & 0 & 2 \\ -1 & 0 & 3 & 0 \\ 2 & 4 & 0 & 1 \\ 0 & 0 & 1 & 2 \end{vmatrix}$ −88   **48.** $\begin{vmatrix} -2 & 0 & 4 & 2 \\ 3 & 6 & 0 & 4 \\ 0 & 0 & 0 & 3 \\ 9 & 0 & 2 & -1 \end{vmatrix}$ 720

49. $\begin{vmatrix} 1 & 1 & 0 & 1 \\ 2 & 1 & 0 & 2 \\ 0 & 1 & -1 & 1 \\ 1 & -1 & 1 & 1 \end{vmatrix}$ 1   **50.** $\begin{vmatrix} 2 & 7 & 0 & -1 \\ 1 & 0 & 1 & 3 \\ 2 & 4 & -1 & -1 \\ -1 & 1 & 0 & 8 \end{vmatrix}$ 129

Let $A = \begin{bmatrix} a_{11} & a_{12} & a_{13} \\ a_{21} & a_{22} & a_{23} \\ a_{31} & a_{32} & a_{33} \end{bmatrix}$ *for Exercises 51–52.*

51. Find $|A|$ by expansion about row 3 of the matrix. Show that your result is really equal to $|A|$ as given in the definition of the determinant of a 3 × 3 matrix.

52. Repeat Exercise 51 for column 3.

Determinants can be used to find the equation of a line passing through two given points. The next two exercises show how to do this.

53. Expand the determinant below and show that the result is the equation of the line through (2, 3), and (−1, 4).

$$\begin{vmatrix} x & y & 1 \\ 2 & 3 & 1 \\ -1 & 4 & 1 \end{vmatrix} = 0 \qquad x + 3y - 11 = 0$$

54. (a) Write the equation of the line through the points (x_1, y_1) and (x_2, y_2) using the point-slope formula. $y - y_1 = \dfrac{y_2 - y_1}{x_2 - x_1}(x - x_1)$

○ CONCEPTUAL   ✎ WRITING   ▲ CHALLENGING

(b) Expanding the determinant in the equation below, show that the equation is equivalent to the equation in part (a).

$$\begin{vmatrix} x & y & 1 \\ x_1 & y_1 & 1 \\ x_2 & y_2 & 1 \end{vmatrix} = 0$$

▲ **55.** The inverse of matrix $A = \begin{bmatrix} a & b \\ c & d \end{bmatrix}$ is

$$A^{-1} = \begin{bmatrix} \dfrac{d}{ad-bc} & \dfrac{-b}{ad-bc} \\ \dfrac{-c}{ad-bc} & \dfrac{a}{ad-bc} \end{bmatrix}.$$

Show that A^{-1} does not exist if $|A| = 0$. What does this tell you about the solution of a linear system with A as the coefficient matrix?

The system has either no solution or an infinite number of solutions if $|A| = 0$.

▲ **56.** Show that if $A = \begin{bmatrix} a & b \\ c & d \end{bmatrix}$ and $|A| \neq 0$, then

$$A^{-1} = \begin{bmatrix} \dfrac{d}{|A|} & \dfrac{-b}{|A|} \\ \dfrac{-c}{|A|} & \dfrac{a}{|A|} \end{bmatrix}.$$

This is an alternative method for finding the inverse of a 2×2 matrix A, where $|A| \neq 0$. Unfortunately, the method does not generalize to any $n \times n$ matrix.

8.5 RESOURCES

Instructor's Test Manual
  Test Items, p. 368–369
Instructor's Solution Manual
  pp. 337–341, 742–745
Study Guide
  pp. 299–303
Student's Solution Manual
  pp. 337–341

Test Generator
  IBM Disks
  5.25″ Disk B
  3.5″ Disk A
  Mac Disk A

Tutorials
  IBM Disks
  5.25″ Disk H
  3.5″ Disk D
  Mac Disk D
  Apple II Module 5, Lessons 1–5

HarperCollins Videotape 9

8.5 EXERCISES (TEXT PAGE 436) ■

Give the property that tells why each of the following determinants equals 0. (More than one property may be used.)

1. $\begin{vmatrix} 2 & 3 \\ 2 & 3 \end{vmatrix}$ Property 5 **2.** $\begin{vmatrix} -5 & -5 \\ 6 & 6 \end{vmatrix}$ Property 5 **3.** $\begin{vmatrix} 2 & 0 \\ 3 & 0 \end{vmatrix}$ Property 1 **4.** $\begin{vmatrix} -8 & 0 \\ -6 & 0 \end{vmatrix}$ Property 1

5. $\begin{vmatrix} -1 & 2 & 4 \\ 4 & -8 & -16 \\ 3 & 0 & 5 \end{vmatrix}$

By Property 4, multiply each element in the second row by $-1/4$. The determinant of the resulting matrix is 0, by Property 5.

6. $\begin{vmatrix} 2 & -8 & 3 \\ 0 & 2 & -1 \\ -6 & 24 & -9 \end{vmatrix}$

By Property 4, multiply each element in row 3 by $-1/3$. The resulting determinant is 0 by Property 5.

7. $\begin{vmatrix} 3 & 6 & 6 \\ 2 & 0 & 4 \\ 1 & 4 & 2 \end{vmatrix}$ Property 4 and Property 5

8. $\begin{vmatrix} 1 & 0 & 0 \\ 1 & 0 & 1 \\ 3 & 0 & 0 \end{vmatrix}$ Property 1

9. $\begin{vmatrix} m & 2 & 2m \\ 3n & 1 & 6n \\ 5p & 6 & 10p \end{vmatrix}$ Property 4 and Property 5

10. $\begin{vmatrix} 7z & 8x & 2y \\ z & x & y \\ 7z & 7x & 7y \end{vmatrix}$ Property 4 and Property 5

11. Explain why the determinant of a matrix with two identical rows (or columns) equals 0.

Use the appropriate properties from this section to tell why each of the following is true. Do not evaluate the determinants. See Examples 1–6.

12. $\begin{vmatrix} 2 & 1 & 6 \\ 3 & 0 & 2 \\ 4 & 1 & 8 \end{vmatrix} = \begin{vmatrix} 2 & 3 & 4 \\ 1 & 0 & 1 \\ 6 & 2 & 8 \end{vmatrix}$ Property 2

13. $\begin{vmatrix} 4 & -2 \\ 3 & 8 \end{vmatrix} = \begin{vmatrix} 4 & 3 \\ -2 & 8 \end{vmatrix}$ Property 2

14. $\begin{vmatrix} 2 & 6 \\ 3 & 5 \end{vmatrix} = -\begin{vmatrix} 3 & 5 \\ 2 & 6 \end{vmatrix}$ Property 3

15. $\begin{vmatrix} -1 & 8 & 9 \\ 0 & 2 & 1 \\ 3 & 2 & 0 \end{vmatrix} = -\begin{vmatrix} 8 & -1 & 9 \\ 2 & 0 & 1 \\ 2 & 3 & 0 \end{vmatrix}$ Property 3

16. $3\begin{vmatrix} 6 & 0 & 2 \\ 4 & 1 & 3 \\ 2 & 8 & 6 \end{vmatrix} = \begin{vmatrix} 6 & 0 & 2 \\ 4 & 3 & 3 \\ 2 & 24 & 6 \end{vmatrix}$ Property 4

17. $-\dfrac{1}{2}\begin{vmatrix} 5 & -8 & 2 \\ 3 & -6 & 9 \\ 2 & 4 & 4 \end{vmatrix} = \begin{vmatrix} 5 & 4 & 2 \\ 3 & 3 & 9 \\ 2 & -2 & 4 \end{vmatrix}$ Property 4

18. $\begin{vmatrix} 3 & -4 \\ 2 & 5 \end{vmatrix} = \begin{vmatrix} 3 & -4 \\ 5 & 1 \end{vmatrix}$

Property 6 (first row added to second row)

19. $\begin{vmatrix} -1 & 6 \\ 3 & -5 \end{vmatrix} = \begin{vmatrix} -1 & 6 \\ 2 & 1 \end{vmatrix}$

Property 6 (first row added to second row)

20. $\begin{vmatrix} 5 & 8 \\ 2 & -1 \end{vmatrix} = \begin{vmatrix} 5 & -2 \\ 2 & -5 \end{vmatrix}$

Property 6 (-2 times the elements of the first column added to the second column)

21. $\begin{vmatrix} 13 & 5 \\ 6 & 1 \end{vmatrix} = \begin{vmatrix} -2 & 5 \\ 3 & 1 \end{vmatrix}$

Property 6 (-3 times the elements of the second column added to the elements of the first column)

22. $\begin{vmatrix} 2 & 5 & 8 \\ 1 & 0 & 2 \\ 4 & 3 & 5 \end{vmatrix} = \begin{vmatrix} 2 & 5 & 8 \\ 1 & 0 & 2 \\ 7 & 3 & 11 \end{vmatrix}$

Property 6 (3 times the elements of the second row added to the third row)

23. $2\begin{vmatrix} 4 & 2 & -1 \\ m & 2n & 3p \\ 5 & 1 & 0 \end{vmatrix} = \begin{vmatrix} 4 & 2 & -1 \\ 2m & 4n & 6p \\ 5 & 1 & 0 \end{vmatrix}$ Property 4

24. $\begin{vmatrix} 3 & 5 & 0 \\ 2 & 1 & 3 \\ -5 & 1 & 6 \end{vmatrix} = \begin{vmatrix} 3 & 5 & 0 \\ 2+3k & 1+5k & 3+0k \\ -5 & 1 & 6 \end{vmatrix}$ Property 6

25. $\begin{vmatrix} -4 & 2 & 1 \\ 3 & 0 & 5 \\ -1 & 4 & -2 \end{vmatrix} = \begin{vmatrix} -4 & 2 & 1+(-4)k \\ 3 & 0 & 5+3k \\ -1 & 4 & -2+(-1)k \end{vmatrix}$ Property 6

Use Property 6 to evaluate each of the following determinants. See Examples 7 and 8.

26. $\begin{vmatrix} 2 & 4 \\ 3 & 6 \end{vmatrix}$ 0

27. $\begin{vmatrix} -5 & 10 \\ 6 & -12 \end{vmatrix}$ 0

28. $\begin{vmatrix} 4 & 8 & 0 \\ -1 & -2 & 1 \\ 2 & 4 & 3 \end{vmatrix}$ 0

29. $\begin{vmatrix} 6 & 8 & -12 \\ -1 & 16 & 2 \\ 4 & 0 & -8 \end{vmatrix}$ 0

30. $\begin{vmatrix} 3 & 1 & 2 \\ 2 & 3 & 1 \\ 1 & 0 & -2 \end{vmatrix}$ −19

31. $\begin{vmatrix} -2 & 2 & 3 \\ 0 & 2 & 1 \\ -1 & 4 & 0 \end{vmatrix}$ 12

32. $\begin{vmatrix} -4 & 2 & 3 \\ 2 & 0 & 1 \\ 0 & 4 & 2 \end{vmatrix}$ 32

33. $\begin{vmatrix} 6 & 3 & 2 \\ 1 & 0 & 2 \\ -1 & 4 & 1 \end{vmatrix}$ −49

34. $\begin{vmatrix} 1 & 0 & 2 & 2 \\ 2 & 4 & 1 & -1 \\ 1 & -3 & 1 & 0 \\ 1 & 1 & 0 & 1 \end{vmatrix}$ −32

35. $\begin{vmatrix} 2 & -1 & 1 & 0 \\ 1 & 1 & 0 & 1 \\ 0 & -1 & 1 & 1 \\ 1 & 2 & 1 & 2 \end{vmatrix}$ −6

36. The note following Example 8 advises working with sums of rows to get a column with only one nonzero number. Why is this good advice? What happens if you work with columns to get a column with just one nonzero number?

▲ *Use the determinant*

$$\begin{vmatrix} a & b & c \\ d & e & f \\ g & h & j \end{vmatrix}$$

to prove the following properties of determinants by finding the values of the original determinant and the new determinant and comparing them.

37. Property 2

38. Property 3

39. Property 4

8.6   RESOURCES

Instructor's Test Manual
Test Items, p. 369–370
Instructor's Solution Manual
pp. 341–351, 745–750
Study Guide
pp. 304–307
Student's Solution Manual
pp. 341–351

Test Generator
IBM Disks
5.25″ Disk B
3.5″ Disk A
Mac Disk A
Tutorials
IBM Disks
5.25″ Disk H
3.5″ Disk D
Mac Disk D
Apple II Module 5, Lessons 1–5

HarperCollins Videotape   9

8.6 EXERCISES   (TEXT PAGE 441)   ■ —————————

Use Cramer's rule to solve each of the following systems of equations. If D = 0, use another method to determine the solution. See Example 1.

1. $x + y = 4$
  $2x - y = 2$ $\{(2, 2)\}$

2. $3x + 2y = -4$
  $2x - y = -5$ $\{(-2, 1)\}$

3. $4x + 3y = -7$
  $2x + 3y = -11$ $\{(2, -5)\}$

4. $4x - y = 0$
  $2x + 3y = 14$ $\{(1, 4)\}$

5. $5x + 4y = 10$
  $3x - 7y = 6$ $\{(2, 0)\}$

6. $3x + 2y = -4$
  $5x - y = 2$ $\{(0, -2)\}$

7. $2x - 3y = -5$
   $x + 5y = 17$ $\{(2, 3)\}$

8. $x + 9y = -15$
   $3x + 2y = 5$ $\{(3, -2)\}$

9. $3x + 2y = 4$
   $6x + 4y = 8$
   can't use Cramer's rule, $D = 0$; $\left\{\left(\dfrac{4 - 2y}{3}, y\right)\right\}$

10. $1.5x + 3y = 5$
   $2x + 4y = 3$
   can't use Cramer's rule, $D = 0$; $\emptyset$

11. $12x + 8y = 3$
   $15x + 10y = 9$
   can't use Cramer's rule, $D = 0$; $\emptyset$

12. $15x - 10y = 5$
   $9x + 6y = 3$ $\{(1/3, 0)\}$

Use Cramer's rule to solve each of the following systems of equations. If $D = 0$, use another method to complete the solution. See Examples 2 and 3.

13. $4x - y + 3z = -3$
   $3x + y + z = 0$
   $2x - y + 4z = 0$ $\{(-1, 2, 1)\}$

14. $5x + 2y + z = 15$
   $2x - y + z = 9$
   $4x + 3y + 2z = 13$ $\{(3, -1, 2)\}$

15. $2x - y + 4z = -2$
   $3x + 2y - z = -3$
   $x + 4y + 2z = 17$ $\{(-3, 4, 2)\}$

16. $x + y + z = 4$
   $2x - y + 3z = 4$
   $4x + 2y - z = -15$ $\{(-4, 3, 5)\}$

17. $4x - 3y + z = -1$
   $5x + 7y + 2z = -2$
   $3x - 5y - z = 1$ $\{(0, 0, -1)\}$

18. $2x - 3y + z = 8$
   $-x - 5y + z = -4$
   $3x - 5y + 2z = 12$ $\{(4, 0, 0)\}$

19. $x + 2y + 3z = 4$
   $4x + 3y + 2z = 1$
   $-x - 2y - 3z = 0$
   can't use Cramer's rule, $D = 0$; $\emptyset$

20. $2x - y + 3z = 1$
   $-2x + y - 3z = 2$
   $5x - y + z = 2$
   can't use Cramer's rule, $D = 0$; $\emptyset$

21. $-2x - 2y + 3z = 4$
   $5x + 7y - z = 2$
   $2x + 2y - 3z = -4$
   can't use Cramer's rule, $D = 0$;

$$\left\{\left(\dfrac{-32 + 19z}{4}, \dfrac{24 - 13z}{4}, z\right)\right\}$$

22. $-3x + 2y - 2z = 4$
   $4x + y + z = 5$
   $3x - 2y + 2z = 1$
   can't use Cramer's rule, $D = 0$; $\emptyset$

23. $2x + 3y = 13$
   $2y - z = 5$
   $x + 2z = 4$ $\{(2, 3, 1)\}$

24. $3x - z = -10$
   $y + 4z = 8$
   $x + 2z = -1$ $\{(-3, 4, 1)\}$

25. $5x - y = -4$
   $3x + 2z = 4$
   $4y + 3z = 22$ $\{(0, 4, 2)\}$

26. $3x + 5y = -7$
   $2x + 7z = 2$
   $4y + 3z = -8$ $\{(1, -2, 0)\}$

27. $x + 2y = 10$
   $3x + 4z = 7$
   $-y - z = 1$ $\{(31/5, 19/10, -29/10)\}$

28. $5x - 2y = 3$
   $4y + z = 8$
   $x + 2z = 4$ $\{(24/19, 63/38, 26/19)\}$

29. In your own words, explain what it means in applying Cramer's rule if $D = 0$.

30. Describe D_x, D_y, and D_z in terms of the coefficients and constants in the given system of equations.

Use Cramer's rule to solve each of the following systems.

31.
$$x + 3y - 2z - w = 9$$
$$4x + y + z + 2w = 2$$
$$-3x - y + z - w = -5$$
$$x - y - 3z - 2w = 2 \quad \{(0, 2, -2, 1)\}$$

32.
$$3x + 2y - w = 0$$
$$2x + z + 2w = 5$$
$$x + 2y - z = -2$$
$$2x - y + z + w = 2 \quad \{(-1, 2, 5, 1)\}$$

33.
$$x + y - z + w = 2$$
$$x - y + z + w = 4$$
$$-2x + y + 2z - w = -5$$
$$x + 3z + 2w = 5$$
can't use Cramer's rule, $D = 0$

34. $x + 2y - z + w = 8$
$$2x - y + 2w = 8$$
$$y + 3z = 5$$
$$x - z = 4 \quad \{(5, 2, 1, 0)\}$$

▲ *Solve each system for x and y using Cramer's rule. Assume a and b are nonzero constants.*

35. $bx + y = a^2$
$ax + y = b^2 \quad \{(-a - b, a^2 + ab + b^2)\}$

36. $ax + by = \dfrac{b}{a}$

$x + y = \dfrac{1}{b} \quad \left\{ \left(\dfrac{-1}{a}, \dfrac{b + a}{ab} \right) \right\}$

37. $b^2x + a^2y = b^2$
$ax + by = a \quad \{(1, 0)\}$

38. $x + \dfrac{1}{b}y = b$

$\dfrac{1}{a}x + y = a \quad \left\{ \left(\dfrac{a(b^2 - a)}{ab - 1}, \dfrac{b(a^2 - b)}{ab - 1} \right) \right\}$

▲ *For the following two exercises, use the system of equations*
$$a_1x + b_1y = c_1$$
$$a_2x + b_2y = c_2.$$

39. Assume $D_x = 0$ and $D_y = 0$. Show that if $c_1c_2 \neq 0$, then $D = 0$, and the equations are dependent.

40. Assume $D = 0$, $D_x = 0$, and $b_1b_2 \neq 0$. Show that $D_y = 0$.

RESOURCES FOR REVIEW

Instructor's Test Manual
Chapter Test Forms, pp. 177–198
Instructor's Solution Manual
pp. 351–362, 750–757

Study Guide
pp. 308–311
Student's Solution Manual
pp. 351–362

Test Generator
IBM Disks
5.25″ Disk B
3.5″ Disk A
Mac Disk A

CHAPTER 8 REVIEW EXERCISES   (TEXT PAGE 446) ∎

Find the values of all variables in the following statements.

1. $\begin{bmatrix} 2 & z & 1 \\ m & 9 & -7 \end{bmatrix} = \begin{bmatrix} x & 5 & 1 \\ -8 & y & p \end{bmatrix}$ $m = -8;\ p = -7;\ x = 2;\ y = 9;\ z = 5$

2. $\begin{bmatrix} 5 & -10 \\ a & 0 \end{bmatrix} = \begin{bmatrix} b - 3 & c + 2 \\ 4 & 0 \end{bmatrix}$ $a = 4;\ b = 8;\ c = -12$

3. $\begin{bmatrix} 5 & x+2 \\ -6y & z \end{bmatrix} = \begin{bmatrix} a & 3x-1 \\ 5y & 9 \end{bmatrix}$ $a = 5;\ \ x = 3/2;\ \ y = 0;\ \ z = 9$

4. $\begin{bmatrix} 6+k & 2 & a+3 \\ -2+m & 3p & 2r \end{bmatrix} + \begin{bmatrix} 3-2k & 5 & 7 \\ 5 & 8p & 5r \end{bmatrix} = \begin{bmatrix} 5 & y & 6a \\ 2m & 11 & -35 \end{bmatrix}$

   $k = 4;\ \ y = 7;\ \ a = 2;\ \ m = 3;\ \ p = 1;\ \ r = -5$

Perform each of the following operations whenever possible.

5. $\begin{bmatrix} 3 & -4 & 2 \\ 5 & -1 & 6 \end{bmatrix} + \begin{bmatrix} -3 & 2 & 5 \\ 1 & 0 & 4 \end{bmatrix}$ $\begin{bmatrix} 0 & -2 & 7 \\ 6 & -1 & 10 \end{bmatrix}$

6. $\begin{bmatrix} 3 \\ 2 \\ 5 \end{bmatrix} - \begin{bmatrix} 8 \\ -4 \\ 6 \end{bmatrix} + \begin{bmatrix} 1 \\ 0 \\ 2 \end{bmatrix}$ $\begin{bmatrix} -4 \\ 6 \\ 1 \end{bmatrix}$

7. $\begin{bmatrix} 2 & 5 & 8 \\ 1 & 9 & 2 \end{bmatrix} - \begin{bmatrix} 3 & 4 \\ 7 & 1 \end{bmatrix}$ **not possible**

8. $3\begin{bmatrix} 2 & 4 \\ -1 & 4 \end{bmatrix} - 2\begin{bmatrix} 5 & 8 \\ 2 & -2 \end{bmatrix}$ $\begin{bmatrix} -4 & -4 \\ -7 & 16 \end{bmatrix}$

9. $-1\begin{bmatrix} 3 & -5 & 2 \\ 1 & 7 & -4 \end{bmatrix} + 5\begin{bmatrix} 0 & 2 \\ -1 & 3 \end{bmatrix}$ **not possible**

10. $10\begin{bmatrix} 2x+3y & 4x+y \\ x-5y & 6x+2y \end{bmatrix} + 2\begin{bmatrix} -3x-y & x+6y \\ 4x+2y & 5x-y \end{bmatrix}$ $\begin{bmatrix} 14x+28y & 42x+22y \\ 18x-46y & 70x+18y \end{bmatrix}$

🖉 **11.** Complete the following sentence. The sum of two $m \times n$ matrices A and B is found _____.

   by adding corresponding elements

12. The speed limits in Italy, Britain, and the U.S. are 87 mph, 70 mph, and 55 mph, respectively. The corresponding fatalities per 100 million miles driven in a recent year were 6.4, 4.0, and 3.3. Write this information as a 3×2 matrix and then as a 2×3 matrix.

   $\begin{bmatrix} 87 & 6.4 \\ 70 & 4.0 \\ 55 & 3.3 \end{bmatrix}$; $\begin{bmatrix} 87 & 70 & 55 \\ 6.4 & 4.0 & 3.3 \end{bmatrix}$

13. In a recent year, Dan Marino, quarterback for the Miami Dolphins, attempted 606 passes and completed 354 for 4434 yd and 28 touchdowns. Joe Montana, San Francisco 49'ers quarterback, attempted 397 passes and completed 238 for 2981 yd and 18 touchdowns. Phil Simms, quarterback for the New York Giants, attempted 479 passes and completed 263 for 3359 yd and 21 touchdowns. Write this information as a matrix in two ways.

   $\begin{bmatrix} 606 & 354 & 4434 & 28 \\ 397 & 238 & 2981 & 18 \\ 479 & 263 & 3359 & 21 \end{bmatrix}$; $\begin{bmatrix} 606 & 397 & 479 \\ 354 & 238 & 263 \\ 4434 & 2981 & 3359 \\ 28 & 18 & 21 \end{bmatrix}$

Find each of the following matrix products, whenever possible.

14. $\begin{bmatrix} -3 & 4 \\ 2 & 8 \end{bmatrix}\begin{bmatrix} -1 & 0 \\ 2 & 5 \end{bmatrix}$ $\begin{bmatrix} 11 & 20 \\ 14 & 40 \end{bmatrix}$

15. $\begin{bmatrix} 3 & 2 & -1 \\ 4 & 0 & 6 \end{bmatrix}\begin{bmatrix} -2 & 0 \\ 0 & 2 \\ 3 & 1 \end{bmatrix}$ $\begin{bmatrix} -9 & 3 \\ 10 & 6 \end{bmatrix}$

16. $\begin{bmatrix} 1 & -2 & 4 & 2 \\ 0 & 1 & -1 & 8 \end{bmatrix}\begin{bmatrix} -1 \\ 2 \\ 0 \\ 1 \end{bmatrix}$ $\begin{bmatrix} -3 \\ 10 \end{bmatrix}$

17. $\begin{bmatrix} 1 & 2 & 5 \\ -3 & 4 & 7 \\ 0 & 2 & -1 \end{bmatrix}\begin{bmatrix} 4 & 2 & 3 \\ 10 & -5 & 6 \end{bmatrix}$ **not possible**

18. $\begin{bmatrix} 4 & 2 & 3 \\ 10 & -5 & 6 \end{bmatrix}\begin{bmatrix} 1 & 2 & 5 \\ -3 & 4 & 7 \\ 0 & 2 & -1 \end{bmatrix}$ $\begin{bmatrix} -2 & 22 & 31 \\ 25 & 12 & 9 \end{bmatrix}$

19. $[3 \ \ -1 \ \ 0]\begin{bmatrix} 1 & 3 & 2 \\ 2 & -4 & 0 \\ 5 & 7 & 3 \end{bmatrix}$ $[1 \ \ 13 \ \ 6]$

◉ **20.** What must be true of two matrices to find their product?   **The number of columns of the first matrix must equal the number of rows of the second matrix.**

◎ CONCEPTUAL   🖉 WRITING   ▲ CHALLENGING

Decide whether or not each of the following pairs of matrices are inverses.

21. $\begin{bmatrix} 2 & -3 \\ 1 & -2 \end{bmatrix}, \begin{bmatrix} 2 & -3 \\ 1 & -2 \end{bmatrix}$ yes

22. $\begin{bmatrix} 1 & 0 \\ 2 & -3 \end{bmatrix}, \begin{bmatrix} 1 & 0 \\ 2/3 & -1/3 \end{bmatrix}$ yes

23. $\begin{bmatrix} 2 & 0 & 6 \\ 0 & 1 & 0 \\ 1 & 0 & 1 \end{bmatrix}, \begin{bmatrix} -1 & 0 & 3/2 \\ 0 & 1 & 0 \\ 1/4 & 0 & -1 \end{bmatrix}$ no

24. $\begin{bmatrix} 1 & 0 & 2 \\ 0 & 2 & 4 \\ 0 & 0 & 1 \end{bmatrix}, \begin{bmatrix} 1 & 0 & -2 \\ 0 & 1/2 & -2 \\ 0 & 0 & 1 \end{bmatrix}$ yes

Find the inverse, if it exists, for each of the following matrices.

25. $\begin{bmatrix} 2 & 1 \\ 5 & 3 \end{bmatrix}$ $\begin{bmatrix} 3 & -1 \\ -5 & 2 \end{bmatrix}$

26. $\begin{bmatrix} -4 & 2 \\ 0 & 3 \end{bmatrix}$ $\begin{bmatrix} -1/4 & 1/6 \\ 0 & 1/3 \end{bmatrix}$

27. $\begin{bmatrix} 2 & 0 \\ -1 & 5 \end{bmatrix}$ $\begin{bmatrix} 1/2 & 0 \\ 1/10 & 1/5 \end{bmatrix}$

28. $\begin{bmatrix} 2 & 0 & 4 \\ 1 & -1 & 0 \\ 0 & 1 & -2 \end{bmatrix}$ $\begin{bmatrix} 1/4 & 1/2 & 1/2 \\ 1/4 & -1/2 & 1/2 \\ 1/8 & -1/4 & -1/4 \end{bmatrix}$

29. $\begin{bmatrix} 2 & -1 & 0 \\ 1 & 0 & 1 \\ 1 & -2 & 0 \end{bmatrix}$ $\begin{bmatrix} 2/3 & 0 & -1/3 \\ 1/3 & 0 & -2/3 \\ -2/3 & 1 & 1/3 \end{bmatrix}$

30. $\begin{bmatrix} 2 & 3 & 5 \\ -2 & -3 & -5 \\ 1 & 4 & 2 \end{bmatrix}$ does not exist

Use matrix inverses to solve each of the following. Identify any inconsistent systems or systems with dependent equations.

31. $x + y = 4$
$2x + 3y = 10$ $\{(2, 2)\}$

32. $5x - 3y = -2$
$2x + 7y = -9$ $\{(-1, -1)\}$

33. $2x + y = 5$
$3x - 2y = 4$ $\{(2, 1)\}$

34. $x - 2y = 7$
$3x + y = 7$ $\{(3, -2)\}$

35. $3x - 2y + 4z = 1$
$4x + y - 5z = 2$
$-6x + 4y - 8z = -2$ not possible;   dependent equations

36. $x + 2y = -1$
$3y - z = -5$
$x + 2y - z = -3$ $\{(1, -1, 2)\}$

37. $x + y + z = 1$
$2x - y = -2$
$3y + z = 2$ $\{(-1, 0, 2)\}$

38. $x = -3$
$y + z = 6$
$2x - 3z = -9$ $\{(-3, 5, 1)\}$

Evaluate each of the following determinants.

39. $\begin{vmatrix} -1 & 8 \\ 2 & 9 \end{vmatrix}$ -25

40. $\begin{vmatrix} -2 & 4 \\ 0 & 3 \end{vmatrix}$ -6

41. $\begin{vmatrix} -2 & 4 & 1 \\ 3 & 0 & 2 \\ -1 & 0 & 3 \end{vmatrix}$ -44

42. $\begin{vmatrix} -1 & 2 & 3 \\ 4 & 0 & 3 \\ 5 & -1 & 2 \end{vmatrix}$ -1

Solve each of the following determinant equations for x.

43. $\begin{vmatrix} -3 & 2 \\ 1 & x \end{vmatrix} = 5$ $\{-7/3\}$

44. $\begin{vmatrix} 3x & 7 \\ -x & 4 \end{vmatrix} = 8$ $\{8/19\}$

45. $\begin{vmatrix} 2 & 5 & 0 \\ 1 & 3x & -1 \\ 0 & 2 & 0 \end{vmatrix} = 4$ $\{$all real numbers$\}$

46. $\begin{vmatrix} 6x & 2 & 0 \\ 1 & 5 & 3 \\ x & 2 & -1 \end{vmatrix} = 2x$ $\{1/31\}$

47. What are the determinant properties used for?

Give the property (or properties) that justifies each of the following statements.

48. $\begin{vmatrix} 8 & 9 & 2 \\ 0 & 0 & 0 \\ 3 & 1 & 4 \end{vmatrix} = 0$ **Property 1 (One row is all zeros, so the determinant is zero.)**

49. $\begin{vmatrix} 4 & 6 \\ 3 & 5 \end{vmatrix} = \begin{vmatrix} 4 & 3 \\ 6 & 5 \end{vmatrix}$ **Property 2 (Rows and columns are interchanged, so the determinants are equal.)**

50. $\begin{vmatrix} 8 & 2 \\ 4 & 3 \end{vmatrix} = 2\begin{vmatrix} 4 & 1 \\ 4 & 3 \end{vmatrix}$ **Property 4 (Multiplying the elements of row 1 of the second matrix by 2 gives the first matrix.)**

51. $\begin{vmatrix} 4 & 6 & 2 \\ -3 & 8 & -5 \\ 4 & 6 & 2 \end{vmatrix} = 0$ **Property 5 (Two identical rows indicate the determinant equals zero.)**

52. $\begin{vmatrix} 5 & -1 & 2 \\ 3 & -2 & 0 \\ -4 & 1 & 2 \end{vmatrix} = \begin{vmatrix} 5 & -1 & 2 \\ 8 & -3 & 2 \\ -4 & 1 & 2 \end{vmatrix}$ **Property 6 (In the first matrix, adding to the elements of row 2 the results of multiplying each element in row 1 by 1 does not change the value of the determinant.)**

53. $\begin{vmatrix} 8 & 2 & -5 \\ -3 & 1 & 4 \\ 2 & 0 & 5 \end{vmatrix} = -\begin{vmatrix} 8 & -5 & 2 \\ -3 & 4 & 1 \\ 2 & 5 & 0 \end{vmatrix}$ **Property 3 (Exchanging columns 2 and 3 produces a matrix with determinant equal to −1 times the value of the first determinant.)**

Use Determinant Property 6 to evaluate the following.

54. $\begin{vmatrix} 5 & -1 & 2 \\ 3 & -2 & 0 \\ -4 & 1 & 2 \end{vmatrix}$ **−24**

55. $\begin{vmatrix} 8 & 2 & -5 \\ -3 & 1 & 4 \\ 2 & 0 & 5 \end{vmatrix}$ **96**

56. Cramer's rule has the condition that $D \neq 0$. Why is this necessary? What is true of the system when $D = 0$?

Solve each of the following systems by Cramer's rule. Identify any systems with dependent equations or any inconsistent systems.

57. $3x + y = -1$
$5x + 4y = 10$ **{(−2, 5)}**

58. $3x + 7y = 2$
$5x - y = -22$ **{(−4, 2)}**

59. $2x - 5y = 8$
$3x + 4y = 10$ **{(82/23, −4/23)}**

60. $3x + 2y + z = 2$
$4x - y + 3z = -16$
$x + 3y - z = 12$
{(−4, 6, 2)}

61. $5x - 2y - z = 8$
$-5x + 2y + z = -8$
$x - 4y - 2z = 0$
can't use Cramer's rule, $D = 0$; dependent equations

62. $-x + 3y - 4z = 2$
$2x + 4y + z = 3$
$3x - z = 9$
{(172/67, −14/67, −87/67)}

CHAPTER 8 TEST   (TEXT PAGE 448) ■ ——————————————————

1. Find the values of all variables in the following equation.

$$\begin{bmatrix} 5 & x+6 \\ 0 & 4 \end{bmatrix} = \begin{bmatrix} y-2 & 4-x \\ 0 & w+7 \end{bmatrix}$$ $w = -3;\ x = -1;\ y = 7$

Perform each of the following operations whenever possible.

2. $3\begin{bmatrix} 2 & 3 \\ 1 & -4 \\ 5 & 9 \end{bmatrix} - \begin{bmatrix} -2 & 6 \\ 3 & -1 \\ 0 & 8 \end{bmatrix}$ $\begin{bmatrix} 8 & 3 \\ 0 & -11 \\ 15 & 19 \end{bmatrix}$

3. $\begin{bmatrix} 1 \\ 2 \end{bmatrix} + \begin{bmatrix} 4 \\ -6 \end{bmatrix} + \begin{bmatrix} 2 & 8 \\ -7 & 5 \end{bmatrix}$ **not possible**

4. At a small college, the freshman class consists of 108 women and 142 men. The sophomore class consists of 112 women and 98 men. The junior class consists of 123 women and 117 men. The senior class consists of 110 women and 130 men. Write this information as a 2×4 matrix.

$$\begin{bmatrix} 108 & 112 & 123 & 110 \\ 142 & 98 & 117 & 130 \end{bmatrix}$$

Find each of the following matrix products, whenever possible.

5. $\begin{bmatrix} 2 & 1 & -3 \\ 4 & 0 & 5 \end{bmatrix} \begin{bmatrix} 1 & 3 \\ 2 & 4 \\ 3 & -2 \end{bmatrix}$ $\begin{bmatrix} -5 & 16 \\ 19 & 2 \end{bmatrix}$

6. $\begin{bmatrix} 2 & -4 \\ 3 & 5 \end{bmatrix} \begin{bmatrix} 4 \\ 2 \\ 7 \end{bmatrix}$ not possible

7. Which of the following properties does not apply to multiplication of matrices?
 (a) commutative   **(b)** associative   **(c)** distributive   **(d)** identity   (a)

8. Decide whether the following matrices are inverses.

$$\begin{bmatrix} 1 & 0 & 1 \\ -1 & 0 & 1 \\ 0 & 1 & 0 \end{bmatrix}, \quad \begin{bmatrix} 1/2 & -1/2 & 0 \\ 0 & 0 & 1 \\ 1/2 & 1/2 & 0 \end{bmatrix}$$ yes

Find the inverse, if it exists, for each of the following matrices.

9. $\begin{bmatrix} -8 & 5 \\ 3 & -2 \end{bmatrix}$ $\begin{bmatrix} -2 & -5 \\ -3 & -8 \end{bmatrix}$

10. $\begin{bmatrix} 4 & 12 \\ 2 & 6 \end{bmatrix}$ does not exist

11. $\begin{bmatrix} 1 & 3 & 4 \\ 2 & 7 & 8 \\ -2 & -5 & -7 \end{bmatrix}$ $\begin{bmatrix} -9 & 1 & -4 \\ -2 & 1 & 0 \\ 4 & -1 & 1 \end{bmatrix}$

Use matrix inverses to solve each of the following.

12. $2x + y = -6$
$3x - y = -29$ $\{(-7, 8)\}$

13. $x + y = 5$
$y - 2z = 23$
$x + 3z = -27$ $\{(0, 5, -9)\}$

Evaluate each of the following determinants.

14. $\begin{vmatrix} 6 & 8 \\ 2 & -7 \end{vmatrix}$ -58

15. $\begin{vmatrix} 2 & 0 & 8 \\ -1 & 7 & 9 \\ 12 & 5 & -3 \end{vmatrix}$ -844

16. Solve the following determinant equation for x.

$$\begin{vmatrix} 1 & 3x & 3 \\ -2 & 0 & 7 \\ 4 & x & 5 \end{vmatrix} = -202 \quad \{-2\}$$

Give the property (or properties) that justifies the following statements.

17. $\begin{vmatrix} 6 & 7 \\ -5 & 2 \end{vmatrix} = -\begin{vmatrix} -5 & 2 \\ 6 & 7 \end{vmatrix}$ Property 3 (The rows are interchanged.)

18. $\begin{vmatrix} 7 & 2 & -1 \\ 5 & -4 & 3 \\ 6 & -2 & 1 \end{vmatrix} = \begin{vmatrix} 7 & 2 & -1 \\ -7 & 0 & 1 \\ 13 & 0 & 0 \end{vmatrix}$ Property 6 (-2 times row 3 was added to row 2 and row 1 was added to row 3.)

Solve each of the following systems by Cramer's rule. Identify any systems with dependent equations or any inconsistent systems.

19. $2x - 3y = -33$
$4x + 5y = 11$ $\{(-6, 7)\}$

20. $x + y - z = -4$
$2x - 3y - z = 5$
$x + 2y + 2z = 3$ $\{(1, -2, 3)\}$

■ CHAPTER 9 SEQUENCES AND SERIES; PROBABILITY

9.1 RESOURCES

Instructor's Test Manual
Test Items, p. 371–372

Instructor's Solution Manual
pp. 363–368, 758–763

Study Guide
pp. 316–321

Student's Solution Manual
pp. 363–368

Test Generator
IBM Disks
 5.25" Disk B
 3.5" Disk A
 Mac Disk A

Tutorials
IBM Disks
 5.25" Disk I
 3.5" Disk E
 Mac Disk E
 Apple II Module 6, Lessons 1–5

HarperCollins Videotape 9
"College Algebra: In Simplest Terms" Program 22

9.1 EXERCISES (TEXT PAGE 456) ■ ─────────────────

Write the first five terms of each of the following sequences. See Example 1.

1. $a_n = 6n + 4$ $10, 16, 22, 28, 34$

2. $a_n = 3n - 2$ $1, 4, 7, 10, 13$

3. $a_n = 2^n$ $2, 4, 8, 16, 32$

4. $a_n = 3^n$ $3, 9, 27, 81, 243$

5. $a_n = (-1)^{n+1}$ $1, -1, 1, -1, 1$

6. $a_n = (-1)^n(n + 2)$ $-3, 4, -5, 6, -7$

7. $a_n = \dfrac{2n}{n + 3}$ $1/2, 4/5, 1, 8/7, 5/4$

8. $a_n = \dfrac{-4}{n + 5}$ $-2/3, -4/7, -1/2, -4/9, -2/5$

9. $a_n = \dfrac{8n - 4}{2n + 1}$ $4/3, 12/5, 20/7, 28/9, 36/11$

10. $a_n = \dfrac{-3n + 6}{n + 1}$ $3/2, 0, -3/4, -6/5, -3/2$

11. $a_n = (-2)^n(n)$ $-2, 8, -24, 64, -160$

12. $a_n = (-1/2)^n(n^{-1})$ $-1/2, 1/8, -1/24, 1/64, -1/160$

13. $a_n = x^n$ x, x^2, x^3, x^4, x^5

14. $a_n = n \cdot x^{-n}$ $1/x, 2/x^2, 3/x^3, 4/x^4, 5/x^5$

Find the first ten terms for the sequences defined as follows. See Example 2.

15. $a_1 = 4$; for $n \geq 2$, $a_n = a_{n-1} + 5$ $4, 9, 14, 19, 24, 29, 34, 39, 44, 49$

16. $a_1 = -8$; for $n \geq 2$, $a_n = a_{n-1} - 7$

  $-8, -15, -22, -29, -36, -43, -50, -57, -64, -71$

17. $a_1 = 2$; for $n \geq 2$, $a_n = 2 \cdot a_{n-1}$ $2, 4, 8, 16, 32, 64, 128, 256, 512, 1024$

18. $a_1 = -3$; for $n \geq 2$, $a_n = 2 \cdot a_{n-1}$

  $-3, -6, -12, -24, -48, -96, -192, -384, -768, -1536$

19. $a_1 = 1, a_2 = 1$; for $n \geq 3$, $a_n = a_{n-1} + a_{n-2}$* $1, 1, 2, 3, 5, 8, 13, 21, 34, 55$

20. $a_1 = 3, a_2 = 2$; for $n \geq 3$, $a_n = a_{n-1} - a_{n-2}$ $3, 2, -1, -3, -2, 1, 3, 2, -1, -3$

---

*This is the Fibonacci sequence, a well-known sequence that occurs widely in nature.

◎ CONCEPTUAL ☑ WRITING ▲ CHALLENGING

For each of the following arithmetic sequences, write the indicated number of terms. See Example 4.

21. $a_1 = 4, d = 2, n = 5$ 4, 6, 8, 10, 12

22. $a_1 = 6, d = 8, n = 4$ 6, 14, 22, 30

23. $a_2 = 9, d = -2, n = 4$ 11, 9, 7, 5

24. $a_3 = 7, d = -4, n = 4$ 15, 11, 7, 3

25. $a_3 = -2, d = -4, n = 4$ 6, 2, −2, −6

26. $a_2 = -12, d = -6, n = 5$ −6, −12, −18, −24, −30

27. $a_3 = 6, a_4 = 12, n = 6$ −6, 0, 6, 12, 18, 24

28. $a_5 = 8, a_6 = 5, n = 6$ 20, 17, 14, 11, 8, 5

For each of the following sequences that are arithmetic, find d and a_n. See Examples 3 and 5.

29. 12, 17, 22, 27, 32, 37, . . . $d = 5, a_n = 7 + 5n$

30. 8, 17, 26, 35, 44, 53, . . . $d = 9, a_n = 9n - 1$

31. −19, −12, −5, 2, 9, . . . $d = 7, a_n = -26 + 7n$

32. −30, −20, −12, −6, −2, . . .   **not arithmetic**

33. $x, x + m, x + 2m, x + 3m, x + 4m, . . .$ $d = m, a_n = x + nm - m$

34. $k + p, k + 2p, k + 3p, k + 4p, . . .$ $d = p, a_n = k + np$

35. $2z + m, 2z, 2z - m, 2z - 2m, 2z - 3m, . . .$ $d = -m, a_n = 2z + 2m - mn$

36. $3r - 4z, 3r - 3z, 3r - 2z, 3r - z, 3r, 3r + z, . . .$ $d = z, a_n = 3r - 5z + zn$

Find a_8 and a_n for each of the following arithmetic sequences. See Examples 6 and 7.

37. $a_1 = 5, d = 2$ $a_8 = 19; \quad a_n = 3 + 2n$

38. $a_1 = -3, d = -4$ $a_8 = -31; \quad a_n = 1 - 4n$

39. $a_3 = 2, d = 1$ $a_8 = 7; \quad a_n = n - 1$

40. $a_4 = 5, d = -2$ $a_8 = -3; \quad a_n = 13 - 2n$

41. $a_1 = 8, a_2 = 6$ $a_8 = -6; \quad a_n = 10 - 2n$

42. $a_1 = 6, a_2 = 3$ $a_8 = -15; \quad a_n = 9 - 3n$

43. $a_{10} = 6, a_{12} = 15$ $a_8 = -3; \quad a_n = -39 + 9n/2$

44. $a_{15} = 8, a_{17} = 2$ $a_8 = 29; \quad a_n = 53 - 3n$

45. $a_1 = x, a_2 = x + 3$ $a_8 = x + 21; \quad a_n = x + 3n - 3$

46. $a_2 = y + 1, d = -3$ $a_8 = y - 17; \quad a_n = y - 3n + 7$

47. $a_6 = 2m, a_7 = 3m$ $a_8 = 4m; \quad a_n = mn - 4m$

48. $a_5 = 4p + 1, a_7 = 6p + 7$ $a_8 = 7p + 10; \quad a_n = -11 + (n - 1)(p + 3)$

◉ **49.** Give an example of an arithmetic sequence with $a_4 = 12$.

   One example is the arithmetic sequence with $a_1 = 6$ and $d = 2$.

✎ **50.** Explain in your own words what is meant by an *arithmetic sequence*.

◉ **51.** Which one of the following is not an arithmetic sequence?   **(a)** 4, 6, 8, 10, . . .
   (b) −2, 6, 14, 22, . . .   **(c)** 1/2, 1, 3/2, 2, . . .   **(d)** 5, 10, 20, 40, . . .   **(d)**

✎ **52.** Refer to the sequence in Exercise 51 that is not arithmetic. Explain in your own words how each term after the first is determined by using the previous term of the sequence.

Find a_1 for each of the following arithmetic sequences. See Example 7.

53. $a_9 = 47, a_{15} = 77$ $a_1 = 7$

54. $a_{10} = 50, a_{20} = 110$ $a_1 = -4$

55. $a_{15} = 168, a_{16} = 180$ $a_1 = 0$

56. $a_{10} = -54, a_{17} = -89$ $a_1 = -9$

✎ ◉ *Explain why each of the following sequences is arithmetic.*

57. log 2, log 4, log 8, log 16, log 32, . . .

58. log 12, log 36, log 108, log 324, . . .

9.2 RESOURCES

Instructor's Test Manual
  Test Items, p. 372
Instructor's Solution Manual
  pp. 368–372, 763–766
Study Guide
  pp. 321–323
Student's Solution Manual
  pp. 368–372

Test Generator
  IBM Disks
  5.25" Disk B
  3.5" Disk A
  Mac Disk A
Tutorials
  IBM Disks
  5.25" Disk I
  3.5" Disk E
  Mac Disk E
  Apple II Module 6, Lessons 1–5

HarperCollins Videotape 9

9.2 EXERCISES (TEXT PAGE 460) ■

For each of the following, write the first n terms of the geometric sequence that satisfies the given conditions.

1. $a_1 = 2$, $r = 3$, $n = 4$ $2, 6, 18, 54$

2. $a_1 = 4$, $r = 2$, $n = 5$ $4, 8, 16, 32, 64$

3. $a_1 = 1/2$, $r = 4$, $n = 4$ $1/2, 2, 8, 32$

4. $a_1 = 2/3$, $r = 6$, $n = 3$ $2/3, 4, 24$

5. $a_1 = -2$, $r = -3$, $n = 4$ $-2, 6, -18, 54$

6. $a_1 = -4$, $r = 2$, $n = 5$ $-4, -8, -16, -32, -64$

7. $a_1 = 3125$, $r = 1/5$, $n = 7$ $3125, 625, 125, 25, 5, 1, 1/5$

8. $a_1 = 729$, $r = 2/3$, $n = 5$ $729, 486, 324, 216, 144$

9. $a_1 = -1$, $r = -1$, $n = 6$ $-1, 1, -1, 1, -1, 1$

10. $a_1 = -1$, $r = 1$, $n = 5$ $-1, -1, -1, -1, -1$

11. $a_3 = 6$, $a_4 = 12$, $n = 5$ $3/2, 3, 6, 12, 24$

12. $a_2 = 9$, $a_3 = 3$, $n = 4$ $27, 9, 3, 1$

Find a_5 and a_n for each of the following geometric sequences. See Examples 2 and 3.

13. $a_1 = 4$, $r = 3$ $a_5 = 324$; $a_n = 4 \cdot 3^{n-1}$

14. $a_1 = 8$, $r = 4$ $a_5 = 2048$; $a_n = 8 \cdot 4^{n-1}$

15. $a_1 = -2$, $r = 3$ $a_5 = -162$; $a_n = -2 \cdot 3^{n-1}$

16. $a_1 = -5$, $r = 4$ $a_5 = -1280$; $a_n = -5 \cdot 4^{n-1}$

17. $a_1 = -3$, $r = -5$ $a_5 = -1875$; $a_n = -3(-5)^{n-1}$

18. $a_1 = -4$, $r = -2$ $a_5 = -64$; $a_n = -4(-2)^{n-1}$

19. $a_2 = 3$, $r = 2$ $a_5 = 24$; $a_n = (3/2)2^{n-1}$

20. $a_3 = 6$, $r = 3$ $a_5 = 54$; $a_n = (2/3)3^{n-1}$

21. $a_4 = 64$, $r = -4$ $a_5 = -256$; $a_n = -1(-4)^{n-1}$

22. $a_4 = 81$, $r = -3$ $a_5 = -243$; $a_n = (-3)^n$

For each of the following sequences that are geometric, find a_n. See Examples 1 and 3.

23. $6, 12, 24, 48, \ldots$ $a_n = 6 \cdot 2^{n-1}$

24. $4, 16, 64, 256, \ldots$ $a_n = 4 \cdot 4^{n-1}$ or 4^n

25. $3/4, 3/2, 3, 6, 12, \ldots$ $a_n = (3/4)2^{n-1}$

26. $5/6, 5/3, 10/3, 20/3, 40/3, \ldots$ $a_n = (5/6)2^{n-1}$

27. $-4, 2, -1, 1/2, \ldots$ $a_n = -4(-1/2)^{n-1}$

28. $49, -7, 1, -1/7, \ldots$ $a_n = 49(-1/7)^{n-1}$

29. $18, 20, 24, 32, 48, \ldots$ not geometric

30. $-7, -5, -3, -1, 1, 3, \ldots$ not geometric

31. $a_3 = 9$, $r = 2$ $a_n = (9/4)2^{n-1}$

32. $a_5 = 6$, $r = 1/2$ $a_n = 96(1/2)^{n-1}$

33. $a_3 = -2$, $r = 3$ $a_n = (-2/9)3^{n-1}$

34. $a_2 = 5$, $r = 1/2$ $a_n = 10(1/2)^{n-1}$

Work the following problems.

35. John Vasquez contracts to work for thirty days, receiving $.01 on the first day of work, $.02 on the second, $.04 on the third, $.08 on the fourth, and so on, with each day's pay double that of the previous day. What would Vasquez earn on the twentieth day? $5242.88

36. In Exercise 35, what would Vasquez earn on the thirtieth day? $5,368,709.12

© CONCEPTUAL   ✐ WRITING   ▲ CHALLENGING

◉ **37.** Give an example of a geometric sequence with $a_3 = 15$.

One example is the geometric sequence with $a_1 = 5/3$ and $r = 3$.

✎ **38.** Explain in your own words what is meant by a *geometric sequence*.

◉ **39.** Which one of the following is not a geometric sequence?   **(a)** $1, -2, 4, -8, \ldots$
(b) $1, 10, 100, 1000, \ldots$ **(c)** $3, 3/2, 3/4, 3/8, \ldots$ **(d)** $1, 1, 2, 3, 5, 8, \ldots$ (d)

✎ **40.** Refer to the sequence in Exercise 39 that is not geometric. Explain in your own words how each term after the first two terms is determined. (This is a famous sequence known as the Fibonacci sequence.)

Find a_1 and r for each of the following geometric sequences. See Example 4.

41. $a_2 = 6$, $a_6 = 486$ $a_1 = 2, r = 3$ or $a_1 = -2, r = -3$

42. $a_3 = -12$, $a_6 = 96$ $a_1 = -3;$ $r = -2$

43. $a_2 = 64$, $a_8 = 1$ $a_1 = 128, r = 1/2$ or $a_1 = -128, r = -1/2$

44. $a_2 = 100$, $a_5 = 1/10$ $a_1 = 1000;$ $r = 1/10$

✎ ◉ *Explain why the following sequences are geometric.*

45. $\log 6, \log 6^2, \log 6^4, \log 6^8, \ldots$ **46.** $\log 2, \log 4, \log 16, \log 256, \ldots$

9.3   RESOURCES

Instructor's Test Manual
Test Items, p. 373–374

Instructor's Solution Manual
pp. 372–379, 767–773

Study Guide
pp. 324–330

Student's Solution Manual
pp. 372–379

Test Generator
IBM Disks
  5.25″ Disk B
  3.5″ Disk A
  Mac Disk A

Tutorials
IBM Disks
  5.25″ Disk I
  3.5″ Disk E
  Mac Disk E
  Apple II Module 6, Lessons 1–5

9.3 EXERCISES   (TEXT PAGE 469) ∎

Evaluate each of the following sums. See Example 1.

1. $\displaystyle\sum_{i=1}^{5} (2i + 1)$ 35   **2.** $\displaystyle\sum_{i=1}^{6} (3i - 2)$ 51   **3.** $\displaystyle\sum_{j=1}^{4} \frac{1}{j}$ 25/12   **4.** $\displaystyle\sum_{i=1}^{5} (i + 1)^{-1}$ 29/20

5. $\displaystyle\sum_{i=1}^{4} i^{i}$ 288   **6.** $\displaystyle\sum_{k=1}^{4} (k + 1)^2$ 54   **7.** $\displaystyle\sum_{k=1}^{6} (-1)^k \cdot k$ 3   **8.** $\displaystyle\sum_{i=1}^{7} (-1)^{i+1} \cdot i^2$ 28

Find the sum of the first ten terms for each of the following arithmetic sequences. See Examples 2 and 3.

9. $a_1 = 8, d = 3$ 215   **10.** $a_1 = -9, d = 4$ 90   **11.** $a_3 = 5, a_4 = 8$ 125   **12.** $a_2 = 9, a_4 = 13$ 160

13. $5, 9, 13, \ldots$ 230   **14.** $8, 6, 4, \ldots$ -10

15. $a_1 = 10, a_{10} = 5 \ 1/2$ 155/2   **16.** $a_1 = -8, a_{10} = -5/4$ $-185/4$

Find a_1 and d for each of the following arithmetic sequences. See Example 4.

17. $S_{20} = 1090$, $a_{20} = 102$ $a_1 = 7, d = 5$

18. $S_{31} = 5580$, $a_{31} = 360$ $a_1 = 0, d = 12$

19. $S_{12} = -108$, $a_{12} = -19$ $a_1 = 1, d = -20/11$

20. $S_{25} = 650$, $a_{25} = 62$ $a_1 = -10, d = 3$

Evaluate each of the following sums. See Example 5.

21. $\sum_{i=1}^{3} (i + 4)$ 18

22. $\sum_{i=1}^{5} (i - 8)$ -25

23. $\sum_{j=1}^{10} (2j + 3)$ 140

24. $\sum_{j=1}^{15} (5j - 9)$ 465

25. $\sum_{i=1}^{12} (-5 - 8i)$ -684

26. $\sum_{k=1}^{19} (-3 - 4k)$ -817

27. $\sum_{i=1}^{1000} i$ 500,500

28. $\sum_{k=1}^{2000} k$ 2,001,000

Find the sum of the first five terms for each of the following geometric sequences. See Example 6.

29. 3, 6, 12, 24, . . .   93

30. 5, 20, 80, 320, . . .   1705

31. 12, -6, 3, $-3/2$, . . .   33/4

32. 18, -3, 1/2, $-1/12$, . . .   1111/72

33. $a_1 = 4$, $r = 2$ 124

34. $a_1 = 3$, $r = 3$ 363

35. $a_2 = 1/3$, $r = 3$ 121/9

36. $a_2 = -1$, $r = 2$ $-31/2$

Find each of the following sums. (Hint: In Exercises 43–44 note the beginning value of i.) See Example 7.

37. $\sum_{i=1}^{4} 2^i$ 30

38. $\sum_{j=1}^{6} 3^j$ 1092

39. $\sum_{i=1}^{4} (-3)^i$ 60

40. $\sum_{i=1}^{4} (-3^i)$ -120

41. $\sum_{i=1}^{6} 81(2/3)^i$ 1330/9

42. $\sum_{k=1}^{5} 9(5/3)^k$ 7205/27

43. $\sum_{i=3}^{6} 2^i$ 120

44. $\sum_{i=4}^{7} 3^i$ 3240

45. $\sum_{i=1}^{4} 5(3/5)^{i-1}$ 272/25

46. $\sum_{i=1}^{5} 256(-3/4)^{i-1}$ 181

Solve each problem. See Examples 3 and 8.

47. Find the sum of all the integers from 51 to 71.

1281

48. Find the sum of all the integers from -8 to 30.

429

49. If a clock strikes the proper number of chimes each hour on the hour, how many times will it chime in a month of 30 days?   4680

50. A stack of telephone poles has 30 in the bottom row, 29 in the next, and so on, with one pole in the top row. How many poles are in the stack?   465

51. A sky diver falls 10 m during the first second, 20 m during the second, 30 m during the third, and so on. How many meters will the diver fall during the tenth second? During the first ten seconds?

100 m;   550 m

52. Deepwell Drilling Company charges a flat $100 set-up charge, plus $5 for the first foot, $6 for the second, $7 for the third, and so on. Find the total charge for a 70-foot well. $2865

53. An object falling under the force of gravity falls about 16 ft the first second, 48 ft during the next second, 80 ft during the third second, and so on. How far would the object fall during the eighth second? What is the total distance the object would fall in 8 seconds?

240 ft;   1024 ft

54. The population of a city was 49,000 five years ago. Each year the zoning commission permits an increase of 580 in the population. What will the maximum population be five years from now?   54,800

55. A super slide of uniform slope is to be built on a level piece of land. There are to be twenty equally spaced supports, with the longest support 15 m long and the shortest 2 m long. Find the total length of all the supports.   170 m

56. How much material would be needed for the rungs of a ladder of 31 rungs, if the rungs taper uniformly from 18 in to 28 in?   713 in

◎ CONCEPTUAL   ✍ WRITING   ▲ CHALLENGING

Solve each problem. See Example 9.

57. Suppose you could save $1 on January 1, $2 on January 2, $4 on January 3, and so on. What amount would you save on January 31?

2^{30} or $1,073,741,824

58. See Exercise 57. What would be the total amount of your savings during January? $2,147,483,647

59. Richland Oil has a well that produced $4,000,000 of income its first year. Each year thereafter, the well produced half as much as it did the previous year. What total amount of income would be produced by the well in 6 years? $7,875,000

60. Fruit and vegetable dealer Olive Greene paid 10¢ per lb for 10,000 lb of onions. Each week the price she charges increases by .1¢ per lb, while her onions lose 5% of their weight. If she sells all the onions after 6 weeks, does she make or lose money? How much?

At the end of six weeks, she had 7351 pounds of onions, worth 10.6¢ per pound, for a total value of $779.21. She lost $220.79.

61. The final step in processing a black and white photographic print is to immerse the print in a chemical called "fixer." The print is then washed in running water. Under certain conditions, 98% of the fixer in a print will be removed with 15 min of washing. How much of the original fixer would then be left after 1 hour?   .00002%

62. A scientist has a vat containing 100 liters of a pure chemical. Twenty liters are drained and replaced with water. After complete mixing, 20 liters of the mixture are drained and replaced with water. What will be the strength of the mixture after nine such drainings?

13.42%

63. The half-life of a radioactive substance is the time it takes for half the substance to decay. Suppose that the half-life of a substance is 3 yr and that 10^{15} molecules of the substance are present initially. How many molecules will be present after 15 yr?

$(1/32) \times 10^{15}$ molecules

64. Each year a machine loses 20% of the value it had at the beginning of the year. Find the value of a machine at the end of 6 yr if it cost $100,000 new.

$26,214.40

Find the future value of each of the following annuities. See Example 10.

65. Payments of $1000 at the end of each year for 9 years at 8% interest compounded annually

$12,487.56

66. Payments of $800 at the end of each year for 12 years at 10% interest compounded annually

$17,107.43

67. Payments of $2430 at the end of each year for 10 years at 11% interest compounded annually

$40,634.48

68. Payments of $1500 at the end of each year for 6 years at 12% interest compounded annually

$12,172.78

9.4 RESOURCES

Instructor's Test Manual
Test Items, p. 375–376

Instructor's Solution Manual
pp. 379–382, 773–777

Study Guide
pp. 330–331

Student's Solution Manual
pp. 379–382

Test Generator
IBM Disks
5.25" Disk B
3.5" Disk A
Mac Disk A

Tutorials
IBM Disks
5.25" Disk I
3.5" Disk E
Mac Disk E
Apple II Module 6, Lessons 1–5

9.4 EXERCISES (TEXT PAGE 474) ■ ─────────────────────────

Find r for each of the following infinite geometric sequences. Identify any whose sums would exist. See Example 2.

1. 9, 18, 36, 72, 144, . . . 2   **2.** 3, 9, 27, 81, . . . 3   **3.** 10, 100, 1000, 10,000, . . . 10

4. −8, −4, −2, −1, −1/2, . . . 1/2; sum would exist

5. 12, 6, 3, 3/2, . . . **1/2**; sum would exist   **6.** −8, −16, −32, −64, . . . 2

7. 1, 1.1, 1.21, 1.331, . . . 1.1   **8.** −1, 1.2, −1.44, 1.728, . . . −1.2

◉ **9.** Under what conditions will an infinite geometric sequence have a sum? $-1 < r < 1$ may be written as .9999 . . . . The sum does exist, since $r = .1$. Use the formula to find this sum, and then complete this statement: The value of .9999 . . .

◉ **10.** The sum of the terms of

.9 + .09 + .009 + .0009 + . . .

is ────. 1

Find the sum of each geometric sequence by using the formula of this section where it applies. See Example 2.

11. 16 + 4 + 1 + . . . 64/3   **12.** 81 + 27 + 9 + 3 + 1 + . . . 243/2

13. 100 + 10 + 1 + . . . 1000/9   **14.** 128 + 64 + 32 + . . . 256

15. 90 + 30 + 10 + . . . 135   **16.** 25 + 5 + 1 + . . . 125/4

17. 256 − 128 + 64 − 32 + 16 − . . . 512/3   **18.** 120 − 60 + 30 − 15 + . . . 80

19. 108 − 36 + 12 − 4 + . . . 81   **20.** 10,000 − 1000 + 100 − 10 + 1 − . . . 100,000/11

21. $\frac{3}{4} + \frac{3}{8} + \frac{3}{16} + \ldots$ 3/2   **22.** $\frac{4}{5} + \frac{2}{5} + \frac{1}{5} + \ldots$ 8/5   **23.** $3 - \frac{3}{2} + \frac{3}{4} - \ldots$ 2

24. 9 − 3 + 1 − . . . 27/4   **25.** $\frac{1}{3} - \frac{2}{9} + \frac{4}{27} - \frac{8}{81} + \ldots$ 1/5   **26.** $1 + \frac{1}{1.01} + \frac{1}{(1.01)^2} + \ldots$ 101

27. $\frac{1}{36} + \frac{1}{30} + \frac{1}{25} + \ldots$ The sum does not exist.   **28.** $1 + \frac{1}{2^2} + \frac{1}{2^4} + \ldots$ 4/3

29. $\sum_{i=1}^{\infty} (1/4)^i$ 1/3   **30.** $\sum_{i=1}^{\infty} (9/10)^i$ 9

31. $\sum_{i=1}^{\infty} (1.2)^i$ The sum does not exist.   **32.** $\sum_{i=1}^{\infty} (1.001)^i$ The sum does not exist.

33. $\sum_{i=1}^{\infty} (-1/4)^i$ −1/5   **34.** $\sum_{i=1}^{\infty} (.3)^i$ 3/7   **35.** $\sum_{i=1}^{\infty} 1/5^i$ 1/4

36. $\sum_{i=1}^{\infty} (-1/2)^i$ −1/3   **37.** $\sum_{i=1}^{\infty} 10^{-i}$ 1/9   **38.** $\sum_{i=1}^{\infty} 4^{-i}$ 1/3

39. $\sum_{i=1}^{\infty} (1/2)^{-i}$ The sum does not exist.   **40.** $\sum_{i=1}^{\infty} (3/4)^{-i}$ The sum does not exist.

◉ **41.** Which one of the following does not have a sum?

(a) $\sum_{i=1}^{10} (1/2)^i$ (b) $\sum_{i=1}^{\infty} (3/2)^i$ (c) $\sum_{i=1}^{\infty} (1/2)^i$ (d) $\sum_{i=1}^{1000} 3 \cdot 4^i$ (b)

✐ **42.** Refer to Exercise 10. The same result can be obtained a different way. Use the fact that 1/3 = .3333 . . . to develop an argument that leads to the same conclusion.

◉ CONCEPTUAL   ✐ WRITING   ▲ CHALLENGING

Express each repeating decimal in the form p/q, where p and q are integers. See Example 3.

43. .55555 . . .   5/9   **44.** .77777 . . .   7/9   **45.** .121212 . . .   4/33   **46.** .858585 . . .   85/99

47. .313131 . . .   31/99   **48.** .909090 . . .   10/11

49. 1.508508508 . . .   1507/999   **50.** 2.613613613 . . .   2611/999

Use the formula of this section to solve each problem.

51. Mitzi drops a ball from a height of 10 m and notices that on each bounce the ball returns to about 3/4 of its previous height. About how far will the ball travel before it comes to rest? (*Hint:* Consider the sum of two sequences.)   70 m

52. A sugar factory receives an order for 1000 units of sugar. The production manager thus orders production of 1000 units of sugar. He forgets, however, that the production of sugar requires some sugar (to prime the machines, for example), and so he ends up with only 900 units of sugar. He then orders an additional 100 units, and receives only 90 units. A further order for 10 units produces 9 units. Finally seeing his mistake, the manager decides to try mathematics. He views the production process as an infinite geometric progression with $a_1 = 1000$ and $r = .1$. Using this, find the number of units of sugar that he should have ordered originally.   10,000/9 units

53. After a person pedaling a bicycle removes his or her feet from the pedals, the wheel rotates 400 times the first minute. As it continues to slow down, each minute it rotates only 3/4 as many times as in the previous minute. How many times will the wheel rotate before coming to a complete stop?   1600

54. A pendulum bob swings through an arc 40 cm long on its first swing. Each swing thereafter, it swings only 80% as far as on the previous swing. How far will it swing altogether before coming to a complete stop?   200 cm

55. A sequence of equilateral triangles is constructed. The first triangle has sides 2 m in length. To get the second triangle, midpoints of the sides of the original triangle are connected. If this process could be continued indefinitely, what would be the total perimeter of all the triangles?   12 m

56. What would be the total area of all the triangles in Exercise 55, disregarding the overlapping? $4\sqrt{3}/3$ m^2

9.5 RESOURCES

Instructor's Test Manual
Test Items, p. 376

Instructor's Solution Manual
pp. 382–389, 777–783

Study Guide
pp. 332–333

Student's Solution Manual
pp. 382–389

Test Generator
IBM Disks
    5.25″ Disk B
    3.5″ Disk A
Mac Disk A

Tutorials
IBM Disks
    5.25″ Disk I
    3.5″ Disk E
Mac Disk E
Apple II Module 6, Lessons 1–5

9.5 EXERCISES   (TEXT PAGE 480) ■

Use the method of mathematical induction to prove the following statements. Assume that n is a positive integer. See Examples 1–3. Except for Exercise 22, Exercises 1–25 are proofs, so no answers are supplied.

1. $2 + 4 + 6 + \cdots + 2n = n(n + 1)$

2. $1 + 3 + 5 + \cdots + (2n - 1) = n^2$

3. $3 + 6 + 9 + \cdots + 3n = \dfrac{3n(n + 1)}{2}$

4. $5 + 10 + 15 + \cdots + 5n = \dfrac{5n(n + 1)}{2}$

5. $2 + 4 + 8 + \cdots + 2^n = 2^{n+1} - 2$

6. $3 + 3^2 + 3^3 + \cdots + 3^n = \dfrac{3(3^n - 1)}{2}$

7. $1^2 + 2^2 + 3^2 + \cdots + n^2 = \dfrac{n(n + 1)(2n + 1)}{6}$

8. $1^3 + 2^3 + 3^3 + \cdots + n^3 = \dfrac{n^2(n + 1)^2}{4}$

9. $5 \cdot 6 + 5 \cdot 6^2 + 5 \cdot 6^3 + \cdots + 5 \cdot 6^n = 6(6^n - 1)$

10. $7 \cdot 8 + 7 \cdot 8^2 + 7 \cdot 8^3 + \cdots + 7 \cdot 8^n = 8(8^n - 1)$

11. $\dfrac{1}{1 \cdot 2} + \dfrac{1}{2 \cdot 3} + \dfrac{1}{3 \cdot 4} + \cdots + \dfrac{1}{n(n + 1)} = \dfrac{n}{n + 1}$

12. $\dfrac{1}{1 \cdot 4} + \dfrac{1}{4 \cdot 7} + \dfrac{1}{7 \cdot 10} + \cdots + \dfrac{1}{(3n - 2)(3n + 1)} = \dfrac{n}{3n + 1}$

13. $\dfrac{1}{2} + \dfrac{1}{2^2} + \dfrac{1}{2^3} + \cdots + \dfrac{1}{2^n} = 1 - \dfrac{1}{2^n}$

14. $\dfrac{4}{5} + \dfrac{4}{5^2} + \dfrac{4}{5^3} + \cdots + \dfrac{4}{5^n} = 1 - \dfrac{1}{5^n}$

15. $x^{2n} + x^{2n-1}y + \cdots + xy^{2n-1} + y^{2n} = \dfrac{x^{2n+1} - y^{2n+1}}{x - y}$

16. $x^{2n-1} + x^{2n-2}y + \cdots + xy^{2n-2} + y^{2n-1} = \dfrac{x^{2n} - y^{2n}}{x - y}$

17. $(a^m)^n = a^{mn}$ (Assume that a and m are constant.)

18. $(ab)^n = a^n b^n$ (Assume that a and b are constant.)

19. If $a > 1$, then $a^n > 1$.

20. If $a > 1$, then $a^n > a^{n-1}$.

21. If $0 < a < 1$, then $a^n < a^{n-1}$.

 22. A proof by mathematical induction allows us to prove that a statement is true for all _____.

    positive integers

23. Suppose that Step 2 in a proof by mathematical induction can be satisfied, but Step 1 cannot. May we conclude that the proof is complete? Explain.

24. What is wrong with the following proof by mathematical induction?

Prove: Any natural number equals the next natural number; that is, $n = n + 1$.

Proof. To begin, we assume the statement true for some natural number $n = k$:

$$k = k + 1.$$

We must now show that the statement is true for $n = k + 1$. If we add 1 to both sides, we have

$$k + 1 = k + 1 + 1$$
$$k + 1 = k + 2.$$

Hence, if the statement is true for $n = k$, it is also true for $n = k + 1$. Thus, the theorem is proved.

25. Suppose that n straight lines (with $n \geq 2$) are drawn in a plane, where no two lines are parallel and no three

lines pass through the same point. Show that the number of points of intersection of the lines is $(n^2 - n)/2$.

26. The series of sketches below starts with an equilateral triangle having sides of length 1. In the following steps, equilateral triangles are constructed on each side of the preceding figure. The lengths of the sides of these new triangles are 1/3 the length of the sides of the preceding triangles. Develop a formula for the number of sides of the nth figure. Use mathematical induction to prove your answer. $3 \cdot 4^{n-1}$

27. Find the perimeter of the nth figure in Exercise 26.

    $\dfrac{4^{n-1}}{3^{n-2}}$ or $3\left(\dfrac{4}{3}\right)^{n-1}$

28. Show that the area of the nth figure in Exercise 26 is

$$\sqrt{3}\left[\dfrac{2}{5} - \dfrac{3}{20}\left(\dfrac{4}{9}\right)^{n-1}\right].$$

9.6 RESOURCES

Instructor's Test Manual
Test Items, p. 377–378
Instructor's Solution Manual
pp. 389–391, 783–785
Study Guide
pp. 334–337
Student's Solution Manual
pp. 389–391

Test Generator
IBM Disks
5.25″ Disk B
3.5″ Disk A
Mac Disk A
Tutorials
IBM Disks
5.25″ Disk I
3.5″ Disk E
Mac Disk E
Apple II Module 6, Lessons 1–5

"College Algebra: In Simplest Terms" Program 25

9.6 EXERCISES (TEXT PAGE 485) ■

Evaluate each of the following. See Examples 4 and 5.

1. $P(7, 7)$ 5040

2. $P(5, 3)$ 60

3. $P(6, 5)$ 720

4. $P(4, 2)$ 12

5. $P(10, 2)$ 90

6. $P(8, 2)$ 56

7. $P(8, 3)$ 336

8. $P(11, 4)$ 7920

9. $P(7, 1)$ 7

10. $P(18, 0)$ 1

11. $P(9, 0)$ 1

12. $P(14, 1)$ 14

13. Explain in your own words what is meant by a *permutation*.

14. Explain why the restriction $r \leq n$ is needed in the formula for $P(n, r)$.

Use the multiplication principle to solve the following problems. (See Examples 1–3, 5 and 6.)

15. How many different types of homes are available if a builder offers a choice of 5 basic plans, 3 roof styles, and 2 exterior finishes? 30

16. An auto manufacturer produces 7 models, each available in 6 different colors, with 4 different upholstery fabrics, and 5 interior colors. How many varieties of the auto are available? 840

17. How many different 4-letter radio station call letters can be made
(a) if the first letter must be K or W and no letter may be repeated; 27,600
(b) if repeats are allowed (but the first letter is K or W)? 35,152
(c) How many of the 4-letter call letters (starting with K or W) with no repeats end in R? 1104

18. A menu offers a choice of 3 salads, 8 main dishes, and 5 desserts. How many different 3-course meals (salad, main dish, dessert) are possible? 120

19. A couple has narrowed down the choice of a name for their new baby to 3 first names and 5 middle names. How many different first- and middle-name arrangements are possible? 15

20. A concert to raise money for an economics prize is to consist of 5 works: 2 overtures, 2 sonatas, and a piano concerto. In how many ways can a program with these 5 works be arranged? 120

21. For many years, the state of California used three letters followed by three digits on its automobile license plates.
(a) How many different license plates are possible with this arrangement? 17,576,000
(b) When the state ran out of new plates, the order was reversed to three digits followed by three letters. How many additional plates were then possible?
17,576,000
(c) Several years ago, the plates described in (b) were also used up. The state then issued plates with one letter followed by three digits and then three letters. How many plates does this scheme provide? 456,976,000

22. How many 7-digit telephone numbers are possible if the first digit cannot be zero and
(a) only odd digits may be used; 78,125
(b) the telephone number must be a multiple of 10 (that is, it must end in zero); 900,000
(c) the telephone number must be a multiple of 100;
90,000

(d) the first 3 digits are 481; 10,000
(e) no repetitions are allowed? 544,320

Use permutations to solve each of the following problems. (See Examples 4–5.)

23. In an experiment on social interaction, 6 people will sit in 6 seats in a row. In how many ways can this be done? 720

24. In how many ways can 7 of 10 monkeys be arranged in a row for a genetics experiment? 604,800

25. A business school offers courses in typing, shorthand, transcription, business English, technical writing, and accounting. In how many ways can a student arrange a schedule if 3 courses are taken? 120

26. If your college offers 400 courses, 20 of which are in mathematics, and your counselor arranges your schedule of 4 courses by random selection, how many

schedules are possible that do not include a math course? 20,523,714,120 or 2.052371412 × 10^{10}

27. In a club with 15 members, how many ways can a slate of 3 officers consisting of president, vice-president, and secretary/treasurer be chosen? 2730

28. A baseball team has 20 players. How many 9-player batting orders are possible?
60,949,324,800 or 6.09493248 × 10^{10}

29. In how many ways can 5 players be assigned to the 5 positions on a basketball team, assuming that any player can play any position? In how many ways can 10 players be assigned to the 5 positions?
120; 30,240

▲ 30. Show that $P(n, n - 1) = P(n, n)$.

9.7 RESOURCES

Instructor's Test Manual
Test Items, p. 379–380
Instructor's Solution Manual
pp. 391–395, 785–789
Study Guide
pp. 338–341
Student's Solution Manual
pp. 391–395

Test Generator
IBM Disks
5.25″ Disk B
3.5″ Disk A
Mac Disk A
Tutorials
IBM Disks
5.25″ Disk I
3.5″ Disk E
Mac Disk E
Apple II Module 6, Lessons 1–5

Transparency   T-41

9.7 EXERCISES   (TEXT PAGE 489) ■

Evaluate each of the following. See Examples 1 and 2.

1. $\binom{6}{5}$ 6   2. $\binom{4}{2}$ 6   3. $\binom{8}{5}$ 56   4. $\binom{10}{2}$ 45   5. $\binom{15}{4}$ 1365   6. $\binom{9}{3}$ 84

7. $\binom{10}{7}$ 120   8. $\binom{10}{3}$ 120   9. $\binom{14}{1}$ 14   10. $\binom{20}{2}$ 190   11. $\binom{18}{0}$ 1   12. $\binom{13}{0}$ 1

✎ 13. Explain the difference between a permutation and a combination.

✎ 14. Is a telephone number an example of a permutation of digits or a combination of digits? Explain.

✎ 15. Padlocks with digit dials are often referred to as "combination locks." According to the mathematical definition of combination, is this an accurate description? Explain.

◉ 16. Describe each of the following as either a permutation or a combination.
(a) Your social security number   permutation
(b) A particular five-card hand of playing cards in the game of poker   combination
(c) A committee of school board members
combination
(d) An automobile license plate number
permutation

◉ CONCEPTUAL   ✎ WRITING   ▲ CHALLENGING

Work each problem using combinations. See Examples 1 and 2.

17. A club has 30 members. If a committee of 4 is selected at random, how many committees are possible?
27,405

18. How many different samples of 3 apples can be drawn from a crate of 25 apples? 2300

19. Hal's Hamburger Heaven sells hamburgers with cheese, relish, lettuce, tomato, mustard, or ketchup. How many different hamburgers can be made using any 3 of the extras? 20

20. Three students are to be selected from a group of 12 students to participate in a special class. In how many ways can this be done? In how many ways can the group that will not participate be selected?

220; 220

21. Five cards are marked with the numbers 1, 2, 3, 4, and 5, shuffled, and 2 cards are then drawn. How many different 2-card combinations are possible? 10

22. If a bag contains 15 marbles, how many samples of 2 marbles can be drawn from it? How many samples of 4 marbles? 105; 1365

23. In Exercise 22, if the bag contains 3 yellow, 4 white, and 8 blue marbles, how many samples of 2 can be drawn in which both marbles are blue? 28

24. In Exercise 18, if it is known that there are 5 rotten apples in the crate:
(a) How many samples of 3 could be drawn in which all 3 are rotten? 10
(b) How many samples of 3 could be drawn in which there are 1 rotten apple and 2 good apples? 950

25. A city council is composed of 5 liberals and 4 conservatives. Three members are to be selected randomly as delegates to a convention.
(a) How many delegations are possible? 84
(b) How many delegations could have all liberals?
10

(c) How many delegations could have 2 liberals and 1 conservative? 40
(d) If 1 member of the council serves as mayor, how many delegations are possible which include the mayor? 28

26. Seven workers decide to send a delegation of 2 to their supervisor to discuss their grievances.
(a) How many different delegations are possible?
21

(b) If it is decided that a certain employee must be in the delegation, how many different delegations are possible? 6
(c) If there are 2 women and 5 men in the group, how many delegations would include at least one woman?
11

Solve the following problems by using either combinations or permutations. See Examples 3 and 4.

27. How many ways can the letters of the word TOUGH be arranged? 120

28. If Matthew has 8 courses to choose from, how many ways can he arrange his schedule if he must pick 4 of them? 1680

29. How many samples of 3 pineapples can be drawn from a crate of 12? 220

30. Velma specializes in making different vegetable soups with carrots, celery, beans, peas, mushrooms, and potatoes. How many different soups can she make using any 4 ingredients? 15

31. From a pool of 7 secretaries, 3 are selected to be assigned to 3 managers, 1 secretary to each manager. In how many ways can this be done? 210

32. In a game of musical chairs, 12 children will sit in 11 chairs (1 will be left out). How many seatings are possible? 479,001,600

33. In an experiment on plant hardiness, a researcher gathers 6 wheat plants, 3 barley plants, and 2 rye plants. She wishes to select 4 plants at random.
(a) In how many ways can this be done? 330
(b) In how many ways can this be done if exactly 2 wheat plants must be included? 150

34. In a club with 8 men and 11 women members, how many 5-member committees can be chosen that have the following?
(a) all men 56
(b) all women 462
(c) 3 men and 2 women 3080
(d) no more than 3 women 8526

35. From 10 names on a ballot, 4 will be elected to a political party committee. In how many ways can the committee of 4 be formed if each person will have a different responsibility? 5040

36. In how many ways can 5 out of 9 plants be arranged in a row on a window sill? 15,120

9.8 RESOURCES

Instructor's Test Manual
Test Items, p. 380–383

Instructor's Solution Manual
pp. 395–398, 789–792

Study Guide
pp. 342–345

Student's Solution Manual
pp. 395–398

Test Generator
IBM Disks
  5.25" Disk B
  3.5" Disk A
Mac Disk A

Tutorials
IBM Disks
  5.25" Disk I
  3.5" Disk E
Mac Disk E
Apple II Module 6, Lessons 1–5

9.8 EXERCISES (TEXT PAGE 496) ■ ——————————————————

Write a sample space with equally likely outcomes for each of the following experiments.

1. A two-headed coin is tossed once. Let h = heads, t = tails. $S = \{h\}$

2. Two ordinary coins are tossed. $S = \{hh, ht, th, tt\}$

3. Three ordinary coins are tossed. $S = \{hhh, hht, hth, thh, htt, tht, tth, ttt\}$

4. Slips of paper marked with the numbers 1, 2, 3, 4, and 5 are placed in a box. After mixing well, two slips are drawn.
$S = \{(1, 2), (1, 3), (1, 4), (1, 5), (2, 3), (2, 4), (2, 5), (3, 4), (3, 5), (4, 5)\}$

5. An unprepared student takes a three-question true/false quiz in which he guesses the answer to all three questions.
Let c = correct, w = wrong. $S = \{ccc, ccw, cwc, wcc, wwc, wcw, cww, www\}$

6. A die is rolled and then a coin is tossed. $S = \{1h, 2h, 3h, 4h, 5h, 6h, 1t, 2t, 3t, 4t, 5t, 6t\}$

Write the following events in set notation and give the probability of each event. See Example 1.

7. In Exercise 1:
(a) the result of the toss is heads $\{h\}$; 1
(b) the result of the toss is tails. $\emptyset$; 0

8. In the experiment described in Exercise 2:
(a) both coins show the same face; $\{hh, tt\}$; 1/2
(b) at least one coin turns up heads.
$\{hh, ht, th\}$; 3/4

9. In Exercise 5:
(a) the student gets all three answers correct;
$\{ccc\}$; 1/8
(b) he gets all three answers wrong; $\{www\}$; 1/8
(c) he gets exactly two answers correct;
$\{ccw, cwc, wcc\}$; 3/8
(d) he gets at least one answer correct.
$\{cww, wcw, wwc, ccw, cwc, wcc, ccc\}$; 7/8

10. In Exercise 4:
(a) both slips are marked with even numbers;
$\{(2, 4)\}$; 1/10
(b) both slips are marked with odd numbers;
$\{(1, 3), (1, 5), (3, 5)\}$; 3/10
(c) both slips are marked with the same number;
$\emptyset$; 0
(d) one slip is marked with an odd number, the other with an even number.
$\{(1, 2), (1, 4), (2, 3), (2, 5), (3, 4), (4, 5)\}$; 3/5

11. A student gives the answer to a probability problem as 6/5. Explain why this answer must be incorrect.

12. If the probability of an event is .857, what is the probability that the event will not occur? .143

Work the following problems. See Examples 1–5.

13. A marble is drawn at random from a box containing 3 yellow, 4 white, and 8 blue marbles. Find the probabilities in (a)–(c).

(a) A yellow marble is drawn.   1/5

(b) A black marble is drawn.   0

(c) The marble is yellow or white.   7/15

(d) What are the odds in favor of drawing a yellow marble?   1 to 4

(e) What are the odds against drawing a blue marble?   7 to 8

14. A baseball player with a batting average of .300 comes to bat. What are the odds in favor of his getting a hit?   3 to 7

15. In Exercise 4, what are the odds that the sum of the numbers on the two slips of paper is 5?   1 to 4

16. If the odds that it will rain are 4 to 5, what is the probability of rain?   4/9

17. If the odds that a candidate will win an election are 3 to 2, what is the probability that the candidate will lose?   2/5

18. A card is drawn from a well shuffled deck of 52 cards. Find the probability that the card is

(a) a 9;   1/13   (b) black;   1/2

(c) a black 9;   1/26   (d) a heart;   1/4

(e) a face card (K, Q, J of any suit);   3/13

(f) red or a 3;   7/13

(g) less than a 4 (consider aces as 1s).   3/13

19. Mrs. Elliott invites 10 relatives to a party: her mother, two uncles, three brothers, and four cousins. If the chances of any one guest arriving first are equally likely, find the following probabilities.

(a) The first guest is an uncle or a brother.   1/2

(b) The first guest is a brother or cousin.   7/10

(c) The first guest is a brother or her mother.   2/5

20. Two dice are rolled. Find the probability of the following events.

(a) The sum of the points is at least 10.   1/6

(b) The sum of the points is either 7 or at least 10.   1/3

(c) The sum of the points is 2 or the dice both show the same number.   1/6

21. The table shows the probability that a customer of a department store will make a purchase in the indicated price range.

Cost	Probability
Below $5	.25
$5–$19.99	.37
$20–$39.99	.11
$40–$69.99	.09
$70–$99.99	.07
$100–$149.99	.08
$150 or more	.03

Find the probability that a customer makes a purchase that is

(a) less than $20;   .62   (b) $40 or more;   .27

(c) more than $99.99;   .11

(d) less than $100.   .89

22. One game in a state lottery requires you to pick one heart, one club, one diamond, and one spade, in that order, from the thirteen cards in each suit. What is the probability of getting all four picks correct and winning $5000?   1/28,561 ≈ .000035

23. If three of the four selections in Exercise 22 are correct, the player wins $200. Find the probability of this outcome.   48/28,561 ≈ .001681

24. The law firm of Alam, Bartolini, Chinn, Dickinson, and Ellsberg has two senior partners, Alam and Bartolini. Two of the attorneys are to be selected to attend a conference. Assuming that all are equally likely to be selected, find the following probabilities.

(a) Chinn is selected.   1/5

(b) Alam and Dickinson are selected.   1/10

▲ (c) At least one senior partner is selected.   7/10

25. The management of a firm wishes to survey the opinions of its workers, classified as follows for the purpose of an interview:

30% have worked for the company more than 5 years,

28% are female,

65% contribute to a voluntary retirement plan, and

1/2 of the female workers contribute to the retirement plan.

Find the following probabilities.

(a) A male worker is selected.   .72

(b) A worker is selected who has worked for the company less than 5 years.   .70

▲ (c) A worker is selected who contributes to the retirement plan or is female.   .79

RESOURCES FOR REVIEW

Instructor's Test Manual
Chapter Test Forms, pp. 199–226
Instructor's Solution Manual
pp. 398–411, 792–799

Study Guide
pp. 346–348
Student's Solution Manual
pp. 398–411

Test Generator
IBM Disks
5.25" Disk B
3.5" Disk A
Mac Disk A

CHAPTER 9 REVIEW EXERCISES (TEXT PAGE 500) ■

Write the first five terms for each of the following sequences.

1. $a_n = \dfrac{n}{n+1}$ 1/2, 2/3, 3/4, 4/5, 5/6

2. $a_n = (-2)^n$ $-2, 4, -8, 16, -32$

3. $a_n = 2(n+3)$ 8, 10, 12, 14, 16

4. $a_n = n(n+1)$ 2, 6, 12, 20, 30

5. $a_1 = 5$; for $n \geq 2$, $a_n = a_{n-1} - 3$ $5, 2, -1, -4, -7$

6. $a_1 = -2$; for $n \geq 2$, $a_n = 3a_{n-1}$ $-2, -6, -18, -54, -162$

7. $a_1 = 5$, $a_2 = 3$; for $n \geq 3$, $a_n = a_{n-1} - a_{n-2}$ $5, 3, -2, -5, -3$

8. $b_1 = -2$, $b_2 = 2$, $b_3 = -4$; for $n \geq 4$, $b_n = -2 \cdot b_{n-2}$ if n is even, and $b_n = 2 \cdot b_{n-2}$ if n
is odd. $-2, 2, -4, -4, -8$

9. Arithmetic, $a_1 = 6$, $d = -4$ $6, 2, -2, -6, -10$

10. Arithmetic, $a_3 = 9$, $a_4 = 7$ $13, 11, 9, 7, 5$

11. Arithmetic, $a_1 = 3 - \sqrt{5}$, $a_2 = 4$ $3 - \sqrt{5}, 4, 5 + \sqrt{5}, 6 + 2\sqrt{5}, 7 + 3\sqrt{5}$

12. Arithmetic, $a_3 = \pi$, $a_4 = 0$ $3\pi, 2\pi, \pi, 0, -\pi$

13. A certain arithmetic sequence has $a_6 = -4$ and $a_{17} = 51$. Find a_1 and a_{20}.

   $a_1 = -29$; $a_{20} = 66$

Find a_8 for each of the following arithmetic sequences.

14. $a_1 = 6$, $d = 2$ 20

15. $a_1 = -4$, $d = 3$ 17

16. $a_1 = 6x - 9$, $a_2 = 5x + 1$ $-x + 61$

17. $a_3 = 11m$, $a_5 = 7m - 4$ $m - 10$

Write the first five terms for the geometric sequences in Exercises 18–21.

18. $a_1 = 4$, $r = 2$ 4, 8, 16, 32, 64

19. $a_4 = 8$, $r = 1/2$ 64, 32, 16, 8, 4

20. $a_1 = -3$, $a_2 = 4$ $-3, 4, -16/3, 64/9, -256/27$

21. $a_3 = 8$, $a_5 = 72$ $8/9, \pm 8/3, 8, \pm 24, 72$

22. For a given geometric sequence, $a_1 = 4$ and $a_5 = 324$. Find a_6.   972   or -972

Find a_5 for each of the following geometric sequences.

23. $a_1 = 3$, $r = 2$ 48

24. $a_2 = 3125$, $r = 1/5$ 25

25. $a_1 = 5x$, $a_2 = x^2$ $x^5/125$

26. $a_2 = \sqrt{6}$, $a_4 = 6\sqrt{6}$ ± 36

🖉 **27.** Explain the difference between an arithmetic sequence and a geometric sequence.

Find S_{12} for each of the following arithmetic sequences.

28. $a_1 = 2$, $d = 3$ 222

29. $a_2 = 6$, $d = 10$ 612

30. $a_1 = -4k$, $d = 2k$ $84k$

Find S_4 for each of the following geometric sequences.

31. $a_1 = 1$, $r = 2$ 15

32. $a_1 = 3$, $r = 3$ 120

33. $a_1 = 2k$, $a_2 = -4k$ $-10k$

◎ CONCEPTUAL   🖉 WRITING   ▲ CHALLENGING

Evaluate each of the following sums.

34. $\sum_{i=1}^{4} \dfrac{2}{i}$ 25/6

35. $\sum_{i=1}^{7} (-1)^{i+1} \cdot 6$ 6

36. $\sum_{i=4}^{8} 3i(2i - 5)$ 690

37. $\sum_{i=1}^{6} i(i + 2)$ 133

38. $\sum_{i=1}^{6} 4 \cdot 3^i$ 4368

39. $\sum_{i=1}^{4} 8 \cdot 2^i$ 240

Solve each problem.

40. A stack of canned goods in a market display requires 15 cans on the bottom, 13 in the next layer, 11 in the next layer, and so on. How many cans are needed for the display? **64 cans**

41. Gale Stockdale borrows $6000 at simple interest of 12% per year. He will repay the loan and interest in monthly payments of $260, $258, $256, and so on. If he makes 30 payments, what is the total amount required to pay off the loan plus the interest? **$6930**

42. On a certain production line, new employees during their first week turn out 5/4 as many items each day as on the previous day. If a new employee produces 48 items the first day, how many will she produce on the fifth day of work? **about 117 items**

43. The half-life of a radioactive substance is 20 years. If 600 g are present at the start, how much will be left after 100 years? **18.75 g**

Evaluate each of the following sums that exists.

44. $18 + 9 + 9/2 + 9/4 + \ldots$ **36**

45. $20 + 15 + 45/4 + 135/16 + \ldots$ **80**

46. $-5/6 + 5/9 - 10/27 + \ldots$ **−1/2**

47. $.8 + .08 + .008 + .0008 + \ldots$ **8/9**

48. $\sum_{i=1}^{\infty} \left(\dfrac{5}{8}\right)^i$ 5/3

49. $\sum_{i=1}^{\infty} -10\left(\dfrac{5}{2}\right)^i$ **Sum does not exist.**

Convert each of the following repeating decimals to a quotient of integers.

50. $.6666 \ldots$ 2/3

51. $.512512512 \ldots$ 512/999

52. Explain the difference between a sequence and a series.

Use mathematical induction to prove that each of the following is true for every positive integer n.

53. $2 + 6 + 10 + 14 + \cdots + (4n - 2) = 2n^2$

54. $2^2 + 4^2 + 6^2 + \cdots + (2n)^2 = \dfrac{2n(n + 1)(2n + 1)}{3}$

55. $2 + 2^2 + 2^3 + \cdots + 2^n = 2(2^n - 1)$

56. $1^3 + 3^3 + 5^3 + \cdots + (2n - 1)^3 = n^2(2n^2 - 1)$

Evaluate each of the following.

57. $P(5, 5)$ **120**   **58.** $P(9, 2)$ **72**   **59.** $P(6, 0)$ **1**

60. $\dbinom{8}{3}$ **56**   **61.** $\dbinom{10}{5}$ **252**   **62.** $\dbinom{6}{0}$ **1**

Solve each problem.

63. Two people are planning their wedding. They can select from 2 different chapels, 4 soloists, 3 organists, and 2 ministers. How many different wedding arrangements are possible? **48**

64. John Jacobs, who is furnishing his apartment, wants to buy a new couch. He can select from 5 different styles, each available in 3 different fabrics, with 6 color choices. How many different couches are available? **90**

65. Four students are to be assigned to 4 different summer jobs. Each student is qualified for all 4 jobs. In how many ways can the jobs be assigned? **24**

66. A student body council consists of a president, vice-president, secretary-treasurer, and 3 representatives at large. Three members are to be selected to attend a conference.

(a) How many different such delegations are possible? **20**

(b) How many are possible if the president must attend? **10**

67. Nine football teams are competing for first-, second-, and third-place titles in a statewide tournament. In how many ways can the winners be determined? **504**

68. How many different license plates can be formed with a letter followed by 3 digits and then 3 letters? How many such license plates have no repeats?

456,976,000; 258,336,000

Solve each problem.

69. A marble is drawn at random from a box containing 4 green, 5 black, and 6 white marbles. Find the following probabilities.

(a) A green marble is drawn. 4/15
(b) A marble that is not black is drawn. 2/3
(c) A blue marble is drawn. 0

70. Refer to Exercise 69 and answer each question.
(a) What are the odds in favor of drawing a green marble? 4 to 11
(b) What are the odds against drawing a white marble? 3 to 2

A card is drawn from a standard deck of 52 cards. Find the probability that each of the following is drawn.

71. A black king 1/26

72. A face card or an ace 4/13

73. An ace or a diamond 4/13

74. A card that is not a diamond 3/4

75. A card that is not a diamond or not black 1

A sample shipment of 5 swimming pool filters is chosen. The probability of exactly 0, 1, 2, 3, 4, or 5 filters being defective is given in the following table.

Number defective	0	1	2	3	4	5
Probability	.31	.25	.18	.12	.08	.06

Find the probability that the following numbers of filters are defective.

76. No more than 3 .86

77. At least 2 .44

78. More than 5 0

CHAPTER 9 TEST (TEXT PAGE 502) ■ ───────────────────

Write the first five terms for each of the following sequences.

1. $a_n = (-1)^n(n^2 + 1)$ $-2, 5, -10, 17, -26$

2. $a_1 = 5;$ for $n \geq 2, a_n = n + a_{n-1}$ $5, 7, 10, 14, 19$

3. $a_n = n + 1$ if n is odd and $a_n = a_{n-1} + 2$ if n is even. $2, 4, 4, 6, 6$

4. Arithmetic, $a_2 = 1, a_4 = 25$ $-11, 1, 13, 25, 37$

5. Geometric, $a_1 = 81, r = 2/3$ $81, 54, 36, 24, 16$

6. A certain arithmetic sequence has $a_7 = -6$ and $a_{15} = -2$. Find a_{31}. 6

7. Find S_{10} for the arithmetic sequence with $a_1 = 37$ and $d = 13$. 955

8. For a given geometric sequence, $a_1 = 12$ and $a_6 = -3/8$. Find a_3. 3

9. Find a_7 for the geometric sequence with $a_1 = 2x^3$ and $a_3 = 18x^7$. $1458x^{15}$

10. Find S_4 for the geometric sequence with $a_1 = 4$ and $r = 1/2$. 15/2

Evaluate each of the following sums that exists.

11. $\sum_{i=1}^{30} (5i + 2)$ 2385

12. $\sum_{i=1}^{5} (-3 \cdot 2^i)$ -186

13. $75 + 30 + 12 + \dfrac{24}{5} + \ldots$ 125

14. $\sum_{i=1}^{\infty} 54(2/9)^i$ 108/7

Work each problem.

15. Fred Meyers deposited $50 in a new savings account on February 1. On the first of each month thereafter he deposited $5 more than the previous month's deposit. Find the total amount of money he deposited after twenty $5 payments. $2100

16. The number of bacteria in a certain culture doubles every 30 minutes. If 50 are present at noon, how many will be present at 4:30 P.M.? 25,600

Find each of the following.

18. $P(11, 3)$ 990

19. $P(7, 7)$ 5040

17. Use mathematical induction to prove that
$$8 + 14 + 20 + 26 + \cdots + (6n + 2) = 3n^2 + 5n$$
for every positive integer n.

20. $\binom{10}{2}$ 45

21. $\binom{12}{0}$ 1

22. A clothing manufacturer makes women's coats in four different styles. Each coat can be made from one of three fabrics. Each fabric comes in five different colors. How many different coats can be made? 60

23. A club with 30 members is to elect a president, secretary, and treasurer from its membership. If a member can hold at most one position, in how many ways can the offices be filled? 24,360

24. If there are 14 women and 16 men in the club in Exercise 23, in how many ways can 2 women and 2 men be chosen to attend a conference? 10,920

25. Discuss the similarities and differences between permutations and combinations.

A card is drawn from a standard deck of 52 cards. Find the probability that each of the following is drawn.

26. A red three 1/26

27. A card that is not a face card 10/13

28. A king or a spade 4/13

29. In the card-drawing experiment above, what are the odds in favor of drawing a face card? 3 to 10

30. A sample of 4 transistors is chosen. The probability of exactly 0, 1, 2, 3, or 4 transistors being defective is given in the following table.

Number defective	0	1	2	3	4
Probability	.19	.43	.30	.07	.01

Find the probability that at most two are defective.
.92

∎ APPENDIX A SETS

APPENDIX A RESOURCES

Instructor's Solution Manual
pp. 412–419, 800–807
Study Guide
pp. 349–352
Student's Solution Manual
pp. 412–419

APPENDIX A EXERCISES (TEXT PAGE 507) ∎ ─────────

List all the elements of each of the following sets.

1. {12, 13, 14, . . . , 20} 12, 13, 14, 15, 16, 17, 18, 19, 20

2. {8, 9, 10, . . . , 17} 8, 9, 10, 11, 12, 13, 14, 15, 16, 17

3. {1, 1/2, 1/4, . . . , 1/32} 1, 1/2, 1/4, 1/8, 1/16, 1/32

4. $\{3, 9, 27, \ldots, 729\}$ $3, 9, 27, 81, 243, 729$

5. $\{17, 22, 27, \ldots, 47\}$ $17, 22, 27, 32, 37, 42, 47$

6. $\{74, 68, 62, \ldots, 38\}$ $74, 68, 62, 56, 50, 44, 38$

7. {all natural numbers greater than 7 and less than 15} $8, 9, 10, 11, 12, 13, 14$

8. {all natural numbers not greater than 4} $1, 2, 3, 4$

Identify the sets in Exercises 9–16 as finite *or* infinite.

9. $\{4, 5, 6, \ldots, 15\}$ finite

10. $\{4, 5, 6, \ldots\}$ infinite

11. $\{1, 1/2, 1/4, 1/8, \ldots\}$ infinite

12. $\{0, 1, 2, 3, 4, 5, \ldots, 75\}$ finite

13. $\{x|x$ is a natural number larger than 5$\}$ infinite

14. $\{x|x$ is a person alive on the earth now$\}$ finite

15. $\{x|x$ is a fraction between 0 and 1$\}$ infinite

16. $\{x|x$ is an even natural number$\}$ infinite

Complete the blanks with either $\in$ *or* $\notin$ *so that the resulting statement is true.*

17. 6 _____ $\{3, 4, 5, 6\}$ $\in$

18. 9 _____ $\{3, 2, 5, 9, 8\}$ $\in$

19. -4 _____ $\{4, 6, 8, 10\}$ $\notin$

20. -12 _____ $\{3, 5, 12, 14\}$ $\notin$

21. 0 _____ $\{2, 0, 3, 4\}$ $\in$

22. 0 _____ $\{5, 6, 7, 8, 10\}$ $\notin$

23. $\{3\}$ _____ $\{2, 3, 4, 5\}$ $\notin$

24. $\{5\}$ _____ $\{3, 4, 5, 6, 7\}$ $\notin$

25. $\{0\}$ _____ $\{0, 1, 2, 5\}$ $\notin$

26. $\{2\}$ _____ $\{2, 4, 6, 8\}$ $\notin$

27. 0 _____ $\emptyset$ $\notin$

28. $\emptyset$ _____ $\emptyset$ $\notin$

Tell whether each statement is true *or* false.

29. $3 \in \{2, 5, 6, 8\}$ false

30. $6 \in \{-2, 5, 8, 9\}$ false

31. $1 \in \{3, 4, 5, 11, 1\}$ true

32. $12 \in \{18, 17, 15, 13, 12\}$ true

33. $9 \notin \{2, 1, 5, 8\}$ true

34. $3 \notin \{7, 6, 5, 4\}$ true

35. $\{2, 5, 8, 9\} = \{2, 5, 9, 8\}$ true

36. $\{3, 0, 9, 6, 2\} = \{2, 9, 0, 3, 6\}$ true

37. $\{5, 8, 9\} = \{5, 8, 9, 0\}$ false

38. $\{3, 7, 12, 14\} = \{3, 7, 12, 14, 0\}$ false

39. $\{x|x$ is a natural number less than 3$\} = \{1, 2\}$ true

40. $\{x|x$ is a natural number greater than 10$\} = \{11, 12, 13, \ldots\}$ true

41. $\{5, 7, 9, 19\} \cap \{7, 9, 11, 15\} = \{7, 9\}$ true

42. $\{8, 11, 15\} \cap \{8, 11, 19, 20\} = \{8, 11\}$ true

43. $\{2, 1, 7\} \cup \{1, 5, 9\} = \{1\}$ false

44. $\{6, 12, 14, 16\} \cup \{6, 14, 19\} = \{6, 14\}$ false

45. $\{3, 2, 5, 9\} \cap \{2, 7, 8, 10\} = \{2\}$ true

46. $\{8, 9, 6\} \cup \{9, 8, 6\} = \{8, 9\}$ false

47. $\{3, 5, 9, 10\} \cap \emptyset = \{3, 5, 9, 10\}$ false

48. $\{3, 5, 9, 10\} \cup \emptyset = \{3, 5, 9, 10\}$ true

49. $\{1, 2, 4\} \cup \{1, 2, 4\} = \{1, 2, 4\}$ true

50. $\{1, 2, 4\} \cap \{1, 2, 4\} = \emptyset$ false

51. $\emptyset \cup \emptyset = \emptyset$ true

52. $\emptyset \cap \emptyset = \emptyset$ true

Let $A = \{2, 4, 6, 8, 10, 12\},$ $B = \{2, 4, 8, 10\},$ $C = \{4, 10, 12\},$
$D = \{2, 10\},$ *and* $U = \{2, 4, 6, 8, 10, 12, 14\}.$

Tell whether each statement is true *or* false.

53. $A \subseteq U$ true

54. $C \subseteq U$ true

55. $D \subseteq B$ true

56. $D \subseteq A$ true

57. $A \subseteq B$ false

58. $B \subseteq C$ false

59. $\emptyset \subseteq A$ true

60. $\emptyset \subseteq \emptyset$ true

61. $\{4, 8, 10\} \subseteq B$ true

62. $\{0, 2\} \subseteq D$ false

63. $B \subseteq D$ false

64. $A \not\subseteq C$ true

Insert $\subseteq$ *or* $\not\subseteq$ *in each blank to make the resulting statement true.*

65. $\{2, 4, 6\}$ _____ $\{3, 2, 5, 4, 6\}$ $\subseteq$

66. $\{1, 5\}$ _____ $\{0, -1, 2, 3, 1, 5\}$ $\subseteq$

67. $\{0, 1, 2\}$ _____ $\{1, 2, 3, 4, 5\}$ $\not\subseteq$

68. $\{5, 6, 7, 8\}$ _____ $\{1, 2, 3, 4, 5, 6, 7\}$ $\not\subseteq$

69. $\emptyset$ _____ $\{1, 4, 6, 8\}$ $\subseteq$

70. $\emptyset$ _____ $\emptyset$ $\subseteq$

◎ CONCEPTUAL   ✍ WRITING   ▲ CHALLENGING

Let $U = \{0, 1, 2, 3, 4, 5, 6, 7, 8, 9, 10, 11, 12, 13\}$, $M = \{0, 2, 4, 6, 8\}$,
$N = \{1, 3, 5, 7, 9, 11, 13\}$, $Q = \{0, 2, 4, 6, 8, 10, 12\}$,   and $R = \{0, 1, 2, 3, 4\}$.

Use these sets to find each of the following. Identify any disjoint sets. See Examples 2–4.

71. $M \cap R$ $\{0, 2, 4\}$

72. $M \cup R$ $\{0, 1, 2, 3, 4, 6, 8\}$

73. $M \cup N$ $\{0, 1, 2, 3, 4, 5, 6, 7, 8, 9, 11, 13\}$

74. $M \cap N$ $\emptyset$

75. $M \cap U$ M or $\{0, 2, 4, 6, 8\}$

76. $M \cup Q$ Q or $\{0, 2, 4, 6, 8, 10, 12\}$

77. $N \cup R$ $\{0, 1, 2, 3, 4, 5, 7, 9, 11, 13\}$

78. $U \cap N$ N or $\{1, 3, 5, 7, 9, 11, 13\}$

79. N' Q or $\{0, 2, 4, 6, 8, 10, 12\}$

80. Q' N or $\{1, 3, 5, 7, 9, 11, 13\}$

81. $M' \cap Q$ $\{10, 12\}$

82. $Q \cap R'$ $\{6, 8, 10, 12\}$

83. $\emptyset \cap R$ $\emptyset$; $\emptyset$ and R are disjoint

84. $\emptyset \cap Q$ $\emptyset$; $\emptyset$ and Q are disjoint

85. $N \cup \emptyset$ N or $\{1, 3, 5, 7, 9, 11, 13\}$

86. $R \cup \emptyset$ R or $\{0, 1, 2, 3, 4\}$

87. $(M \cap N) \cup R$ R or $\{0, 1, 2, 3, 4\}$

88. $(N \cup R) \cap M$ $\{0, 2, 4\}$

89. $(Q \cap M) \cup R$ $\{0, 1, 2, 3, 4, 6, 8\}$

90. $(R \cup N) \cap M'$ N or $\{1, 3, 5, 7, 9, 11, 13\}$

91. $(M' \cup Q) \cap R$ R or $\{0, 1, 2, 3, 4\}$

92. $Q \cap (M \cup N)$ M or $\{0, 2, 4, 6, 8\}$

93. $Q' \cap (N' \cap U)$ $\emptyset$; Q' and $(N' \cap U)$ are disjoint

94. $(U \cap \emptyset') \cup R$ U or $\{0, 1, 2, 3, 4, 5, 6, 7, 8, 9, 10, 11, 12, 13\}$

Let $U = \{$all students in this school$\}$, $M = \{$all students taking this course$\}$,
$N = \{$all students taking calculus$\}$,   and $P = \{$all students taking history$\}$.

Describe each of the following sets in words.

95. M' all students in this school who are not taking this course

96. $M \cup N$ all students in this school who are taking this course or calculus or both

97. $N \cap P$ all students in this school who are taking calculus and history

98. $N' \cap P'$

all students in this school who are not taking calculus and are not taking history

99. $M \cup P$ all students in this school who are taking this course or history or both

100. $P' \cup M'$

all students in this school who are not taking history or not taking this course or taking neither course

In the study of probability, the set of all possible outcomes for an experiment is called the sample space for the experiment. For example, the sample space for tossing an honest coin is the set $\{h, t\}$. Find the sample space for each of these exercises.

101. Rolling an honest die $\{1, 2, 3, 4, 5, 6\}$

102. Tossing a coin twice $\{hh, ht, th, tt\}$

103. Tossing a coin three times $\{hhh, hht, hth, thh, htt, tht, tth, ttt\}$

104. Having three children $\{bbb, bgb, bbg, gbb, ggb, gbg, bgg, ggg\}$

▲ ◉ *State the conditions on sets X and Y for which the following are true.*

105. $X \cap Y = \emptyset$ X and Y are disjoint.

106. $X \cup Y = \emptyset$ $X = \emptyset$ and $Y = \emptyset$

107. $X \cap Y = X$ $X \subseteq Y$

108. $X \cup Y = X$ $Y \subseteq X$

109. $X \cap \emptyset = X$ $X = \emptyset$

110. $X \cup \emptyset = X$ no restriction on X

3.3–3.6 Summary of the Conic Sections

Equation	Graph	Description	Identification
$y = a(x - h)^2 + k$		Opens upward if $a > 0$, downward if $a < 0$. Vertex is at (h, k)	x^2 term y is not squared.
$x = a(y - k)^2 + h$		Opens to right if $a > 0$, to left if $a < 0$. Vertex is at (h, k).	y^2 term x is not squared.
$(x - h)^2 + (y - k)^2 = r^2$		Center is at (h, k), and radius is r.	x^2 and y^2 terms have the same positive coefficient.
$\dfrac{x^2}{a^2} + \dfrac{y^2}{b^2} = 1$		x-intercepts are a and $-a$. y-intercepts are b and $-b$.	x^2 and y^2 terms have different positive coefficients.
$\dfrac{x^2}{a^2} - \dfrac{y^2}{b^2} = 1$		x-intercepts are a and $-a$. Asymptotes found from $(a, b), (a, -b), (-a, -b)$, and $(-a, b)$.	x^2 has a positive coefficient. y^2 has a negative coefficient.
$\dfrac{y^2}{b^2} - \dfrac{x^2}{a^2} = 1$		y-intercepts are b and $-b$. Asymptotes found from $(a, b), (a, -b), (-a, -b)$, and $(-a, b)$.	y^2 has a positive coefficient. x^2 has a negative coefficient.

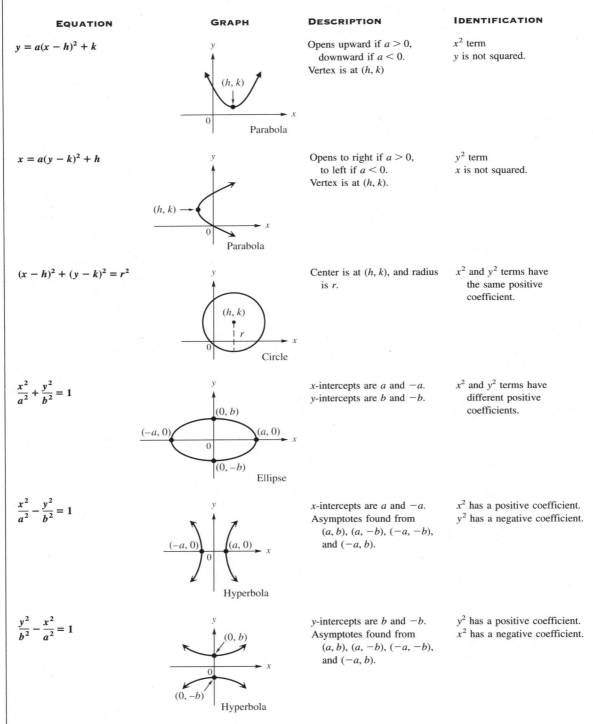